STORMWATER COLLECTION SYSTEMS DESIGN HANDBOOK

STORMWATER COLLECTION SYSTEMS DESIGN HANDBOOK

Larry W. Mays Editor in Chief

Department of Civil and Environmental Engineering
Arizona State University
Tempe, Arizona

McGRAW-HILL

New York   Chicago   San Francisco   Lisbon   London   Madrid
Mexico City   Milan   New Delhi   San Juan   Seoul
Singapore   Sydney   Toronto

Library of Congress Cataloging-in-Publication Data

Stormwater collection systems design handbook / Larry W. Mays, editor in chief.
p.   cm.
Includes index.
ISBN 0-07-135471-9
1. Storm sewers—Design and construction—Handbooks, manuals, etc.   2. Sewer design—Handbooks, manuals, etc.   I. Mays, Larry W.

TD665 .S757   2001
628′.212—dc21

00-069553

McGraw-Hill 

*A Division of The **McGraw·Hill** Companies*

ISBN 0-07-135471-9

The sponsoring editor for this book was Larry S. Hager and the production supervisor was Sherri Souffrance. It was set in Times Roman by Pro-Image Corporation.

Printed and bound by Quebecor/Martinsburg.

This book is printed on acid-free paper.

McGraw-Hill books are available at special quantity discounts to use as premiums and sales promotions, or for use in corporate training programs. For more information, please write to the Director of Special Sales, McGraw-Hill, Two Penn Plaza, New York, NY 10121-2298. Or contact your local bookstore.

CONTENTS

Chapter 8. Design of Off-line Detention Systems   *James C. Y. Guo*   **8.1**

Chapter 9. Design of Infiltration Basins for Stormwater   *James C. Y. Guo*   **9.1**

Chapter 10. Stormwater Treatment Wetland Design   *Roland D. Wass*   **10.1**

Chapter 11. Design of Distributed Stormwater Control and Re-use Systems   *Leonard T. Wright and Jim P. Heaney*   **11.1**

Chapter 16. Design of Flood Control and Drainage Channels   *George K. Cotton*   **16.1**

Chapter 17. Design for Erosion and Sediment Control   *Carlos Carriaga and I. Kaan Tuncok*   **17.1**

Chapter 18. Storm and Combined Sewer Overflow: Flow Regulators and Control   *Larry W. Mays*   **18.1**

Chapter 19. The Removal of Urban Litter from Stormwater Drainage Systems   *Neil Armitage*   **19.1**

Chapter 20. Design of Flow Transitions and Energy Dissipators for Culverts and Channels   *Larry W. Mays*

Chapter 21. Computer Models for Stormwater Systems Design   *Bing Zhao*   **21.1**

Chapter 22. Risk/Reliability Models for Design   *Yeou-Koung Tung, Larry W. Mays, and Ben C. Yen*

Chapter 23. Storm Drain Design and Analysis Using HYDRA   *Stuart Stein and Ken Young*

CONTRIBUTORS

A. Osman Akan   *Department of Civil and Environmental Engineering, Old Dominion University, Norfolk, Va.* (CHAP. 7)

Neil Armitage   *Department of Civil Engineering, University of Cape Town, Rondebosch, South Africa* (CHAP. 19)

Carlos Carriaga   *Stantec Consulting Inc., Phoenix, Ariz.* (CHAP. 17)

George K. Cotton   *Simons & Associates, Inc., Fort Collins, Colo.* (CHAP. 16)

Jacques W. Delleur   *Professor Emeritus, School of Civil Engineering, Purdue University, West Lafayette, Ind.* (CHAP. 14)

Roy D. Dodson   *Dodson & Associates Inc., Houston, Tex.* (CHAP. 2)

Richard H. French   *Desert Research Institute, University and Community College System of Nevada, Reno, Nev.* (CHAP. 3)

James C. Y. Guo   *Department of Civil Engineering, University of Colorado at Denver* (CHAPS. 8, 9)

James P. Heaney   *Department of Civil, Environmental and Architectural Engineering, University of Colorado, Boulder* (CHAP. 11)

Kevin Lansey   *Department of Civil Engineering and Engineering Mechanics, University of Arizona, Tucson, Arizona* (CHAP. 12)

Larry W. Mays   *Department of Civil and Environmental Engineering, Arizona State University, Tempe, Ariz.* (CHAPS. 1, 15, 18, 20, 22)

John W. Nicklow   *Southern Illinois University Carbondale, Carbondale, Ill.* (CHAP. 5)

John N. Paine   *BTG, Incorporated, Newport News, Va.* (CHAP. 7)

Brian Roberts   *National Corrugated Steel Pipe Association, Washington, DC* (CHAP. 13)

Walid El-Shorbagy   *Department of Civil Engineering, University of United Arab Emirates, Al Ain, United Arab Emirates* (CHAP. 12)

Stuart Stein   *GKY & Associates, Inc., Springfield, Va.* (CHAP. 23)

I. Kaan Tuncok   *Daniel, Mann, Johnson, Mendenhall (DMJM), Phoenix, Ariz.* (CHAPS. 15, 17)

Yeou-Koung Tung   *Department of Civil Engineering, Hong Kong University of Science & Technology, Clearwater Bay, Kowloon, Hong Kong* (CHAP. 22)

Roland D. Wass   *PBSJ, Phoenix, Ariz.* (CHAP. 10)

Jerome Westphal   *Cottonwood, Ariz.* (CHAP. 4)

Leonard T. Wright   *Department of Civil, Environmental and Architectural Engineering, University of Colorado, Boulder* (CHAP. 11)

Ben C. Yen   *Department of Civil and Environmental Engineering, University of Illinois at Urbana–Champaign, Urbana, Ill.* (CHAPS. 6, 22)

Ken Young   *GKY & Associates, Inc., Springfield, Va.* (CHAP. 23)

Bing Zhao   *Flood Control District of Maricopa County, Phoenix, Ariz.* (CHAP. 21)

PREFACE

At the beginning of this new century, it is an exciting time to be involved in writing about the design of facilities for the conveyance, storage, and disposal of stormwater. During the past three decades there has been a great effort at publishing on the subject of stormwater management and the use of computer models. However, even with this extensive amount of literature on the subject of stormwater management, particularly considering the quality aspects, there has not been one single effort that captures our present-day knowledge of the design of various components that make up what we call stormwater collection systems.

The various components of stormwater collections systems considered in this book include: stormwater inlets, storm sewers, combined sewers, detention, infiltration basins, stormwater wetlands, flow regulators, culverts, drainage channels, flow transitions and energy dissipators for channels, erosion and sediment control, and stormwater pump facilities. This handbook presents the newest technology for these components and also covers new concepts and methodologies for removal of urban litter, sediment movement in drainage systems, distributed stormwater control and reuse systems. Discussions are presented on many computer models that have been developed for stormwater collection systems design.

The *Stormwater Collections Systems Design Handbook*, referred to herein as the Handbook, has been an extensive effort to develop a comprehensive reference book on design. Over the years the need for such a comprehensive reference book has become more and more obvious to me, particularly for practicing engineers, who need a central reference to start to accumulate knowledge for the design process of various components of stormwater collections systems. They need a reference to find state-of-the-art design procedures, worked-out example problems, discussions of what can go wrong in the design process, up-to-date references for further reading, and limited discussions of the theoretical aspects as they relate to the design process. Also references and discussions are needed for many different computer codes that are used in design practice. This Handbook is an attempt to fulfill these needs of the practicing engineer.

Also I consider this Handbook to be a valuable reference, if not a text, for use in teaching design at both the undergraduate and graduate levels. Basically there are no published books (or textbooks) that comprehensively cover the design of stormwater collections systems. Hopefully the Handbook can be used effectively for such courses because the balance of design concepts and procedures, with worked-out examples and just enough theoretical coverage, should make this a valuable book for the design classroom.

A large amount of knowledge has accumulated over the years on the design of the various components of stormwater collections systems, mainly in various government publications and reports, in journals, and a limited amount, in textbooks. Unfortunately this knowledge has not been accumulated into one central location. Over the years many new methods have been developed and many computer programs have been written. Engineers need an authoritative source of guidance as to which methods and computer programs are appropriate for various given tasks. This Handbook is intended to provide this information in a concise and accessible form. This Handbook is not, however, an encyclopedia of every known fact.

As I reflect upon my own experiences in the field of water resources, I am continually reminded of how handbooks have always been a part of my learning. These handbooks have included the *Handbook of Hydraulics* by C. V. Davis and K. E. Sorenson, the *Handbook of Hydraulics* by E. F. Brater and H. W. King, the *Handbook of Applied Hydrology* by V. T. Chow, and the *Handbook of Fluid Dynamics* by V. L. Streeter. These have all been valuable resources to me throughout my engineering education, consulting experiences, and teaching and research. More recently, David Maidment developed the *Handbook of Hydrology* and I developed the *Water Resources Handbook,* the *Hydraulic Design Handbook,* and the *Water Distribution Systems Handbook.* This new effort, the *Stormwater Collection Systems Design Handbook,* hopefully will complement the previous handbooks to fill a gap with information that may not have been provided by these earlier handbooks.

ACKNOWLEDGMENTS

I must first acknowledge the authors who made the *Stormwater Collection Systems Design Handbook* possible. It has been a sincere privilege to have worked with such an excellent group of dedicated people. They are all experienced professionals who are among the leading experts in stormwater systems design. References to material in the Handbook should be attributed to the respective chapter authors.

During the past twenty-five years of my academic career as a professor I have received help and encouragement from so many people that it is not possible to name them all. These people represent a wide range of universities, research institutions, government agencies, and professions. To all of you I express my deepest thanks.

I would like to acknowledge Arizona State University, especially for the time afforded me to pursue this Handbook.

I would like to thank Larry Hager of McGraw-Hill for having faith in me through his willingness to publish this Handbook. I sincerely appreciated his advice and encouragement throughout this project. Larry has always been a great guy to work with on the four handbooks that we have done together. He is always a joy to talk to as he is one of the few willing to listen to my fly-fishing and snow-skiing experiences.

This Handbook has been a part of a personal journey that began years ago when I was a young boy with a love of water. Each book that I have worked on has been a part of my life long journey in water resources. The Handbook certainly is no exception. I have gained more from this experience than can ever be measured in words. From my early days, as a boy growing up in Illinois near Mark Twain Country between the Mississippi and Illinois Rivers, to now, I have always had this loving appreciation for water and a desire to continually learn more. Over the years even my spare time has been focused on water in one form or another, fly fishing, scuba diving, and snow skiing.

Books are companions along the journey of learning. I hope that you will be able to use this Handbook in your own journey of learning about water. Have a happy and wonderful journey.

I dedicate this Handbook to humanity and human welfare.

Larry W. Mays
Scottsdale, Arizona

CHAPTER 1
INTRODUCTION

Larry W. Mays
Department of Civil and Environmental Engineering
Arizona State University
Tempe, Arizona

1.1 HISTORICAL PERSPECTIVES OF STORM DRAINAGE

1.1.1 Ancient Urban Stormwater Drainage Systems

Since the first successful efforts to control the flow of water by the Egyptians and the Mesopotanians thousands of years ago, a very rich history of hydraulics has evolved. The first successful efforts to control the flow of water were probably made in Mesopotamia and Egypt, where the remains of the prehistoric irrigation works still exist.

The use of drainage systems by humans has a long history dating back to the early third millennium B. C. during the Indus civilization. Not far behind were the Mesopotamians (Adams, 1981). The Minoan civilization on Crete, in the second millenium B.C. also had extensive drainage systems. Knossos, approximately 5 kilometers from Herakleion, the modern capital of Crete, was one of the most ancient and most unique cities of the Aegean and of Europe.

The drainage systems at Knossos were most interesting, consisting of two separate systems, one to collect the sewage and the other to collect rain water (see Figs. 1.1a-c). After the collapse of the Minoan civilization and before the Greek influence, which was roughly from 1100 to 700 B.C., there was disarray in the Aegean society. The use of drains were fairly extensive in Minoan palaces and later their use was rediscovered by the Greeks, as they started living in settlements.

Community drainage systems were a relatively late development of the Greeks (Crouch, 1993). Drainage in Greek cities included sewers under the streets in residential areas and drainage channels in public areas. Components of the drainage systems included eavetroughs for individual buildings, drain pipes through walls or foundation of individual houses, collector channels in neighborhoods, and drains in public areas.

After the Greeks, many of the cities and towns were eventually taken over by the Romans. Many Roman cities did not have any type of drainage system, especially those in the outer parts of the Roman Empire. In the more developed communities, stone drains were provided. In the old established cities that were originally built without storm drains, it was difficult to install them during later times. This is why cities such as Pompeii did not have a full network of storm drains. The older parts of the cities had a somewhat random layout because of no urban planning, whereas the newer parts of the cities were built on a square grid street pattern. The downtown core of Pompeii, around the forum, does have the random layout;

(*a*)

(*b*)

FIGURE 1.1   Drainage system at Knossos. (Photos by L. W. Mays)

whereas rectangular city blocks were used in the later expansion of the city. Ironically, the older part of Pompeii was the only part that did have storm drains (Hodge, 1992).

The drains in Pompeii were located approximately 1 meter under the sidewalks or under the roadway. They were typically around 50 cm wide with a gabled roof and had manholes for purposes of inspection. These manholes were covered by round or square stone lids with a bronze ring for lifting. Street drainage flowed into the drain by either openings in the vertical face of the sidewalk curbstones or by openings cut in the paving slab set flat in the

(*c*)

FIGURE 1.1   (*Continued*)

roadway. Also flow from various other sources entered the drainage system from drains connected to private houses for toilets and kitchens and from drains connected directly to the system from public bathes and public toilets.

In Pompeii, the streets had stepping stones placed across them as shown in Fig. 1.2*a* to keep from getting wet. Keep in mind that the streets flowed with water not only from runoff as a result of storms but also had flow from the bath houses and overflow from the many fountains found in Pompeii. Figure 1.2*b* shows a small open channel for drainage along a street and Figure 1.2*c* shows a drainage pipe along a street. Figures 1.3*a* and *b* show tile drains that brought runoff from the roofs of houses and other buildings and flowed into a cistern (shown at the bottom of the tile in Fig. 1.3*b*). Figure 1.3*c* shows a tile gutter type pipe for drainage.

The overall drainage system sloped downhill at somewhat arbitrary gradients, usually discharging into a large central collector sewer. One of the best examples of a large collector sewer is the Cloaca Maxima in Rome illustrated in Fig. 1.4. The last stage of the drainage system was typically the discharge into a river. The Cloaca Maxima emptied into the Tiber River.

Another interesting Roman city was Ephesus, in present day Turkey, which was founded during the 10th Century B.C. as an Ionian city. During the 6th Century B.C., Ephesus was re-established at its present site where it further developed during the Roman period. Figures

(a)

(b)

FIGURE 1.2 Drainage system at Pompeii. (a) Stepping stones across street. (b) Drainage along street. (Photos by L. W. Mays)

(*c*)

FIGURE 1.2 (*Continued*) (*c*) Drainage pipe along street.

(*a*)

FIGURE 1.3 Tile for roof runoff that connects to cistern. (*a*) Tile. (Photos by L. W. Mays)

(b)

FIGURE 1.3 (*Continued*) (*b*) Tile draining to cistern.

1.5*a* and *b* show the types of clay pipes used at Ephesus. Figure 1.6*a* shows a drainage opening along a street in Ephesus. Figures 1.6*b* and *c* show stone covers for inspection manholes of the under street drainage. The Great Theatre at Ephesus is the largest and most impressive building and had a capacity to seat 24,000 spectators. Of notable interest from a water-resources viewpoint is the drainage system for this theatre. Figure 1.7 shows a drainage channel at the floor of the theatre.

The fall of the Roman Empire extended over a 1000-year transition period called the Dark Ages. During this period, the concepts of science related to water resources probably retrogressed. After the fall of the Roman Empire, water and sanitation—indeed, public health—declined in Europe. Historical accounts tell of incredibly unsanitary conditions: polluted water, human and animal wastes in the streets, and water thrown out of windows onto passers-by. Various epidemics ravaged Europe. During the same period, Islamic cultures, on the periphery of Europe, had religiously mandated high levels of personal hygiene, along with highly developed water supplies and adequate sanitation systems.

1.1.2 Early Methods for Computation

1.1.2.1 Rational and Lloyd-Davies Methods. The earliest known method for storm drainage design is the *rational method* attributed to Mulvaney (1850), Kuichling (1889) and Lloyd-Davies (1906). Some have distinguished between the Lloyd-Davies method and var-

(*c*)

FIGURE 1.3 (*Continued*) (*c*) Gutter for roof runoff.

ious formulations for the rational method. The differences are basically local adaptations of the same fundamental principles. The *rational equation* is usually expressed in the form:

$$Q = CiA \tag{1.1}$$

where Q = peak runoff rate (ft^3/s)
$\quad i$ = design rainfall intensity (in/hr)
$\quad A$ = area of catchment (acres)
$\quad C$ = runoff coefficient

This represents steady rainfall of intensity i falling on an area A giving rise to a catchment peak outflow rate Q. Since 1 acre inch per hour equals 1.008 ft^3/s, the conversion factor between the units is assumed to be unity.

If C is equated to the proportion of the catchment area consisting of *directly connected impermeable surface* (A_p) i.e. $C = A_p/A$, then we have the *Lloyd-Davies formula*:

$$Q = iA_p \tag{1.2}$$

The main variation between the Lloyd-Davies and the American rational method is that the latter permitted greater flexibility in the selection of the runoff coefficient according to the

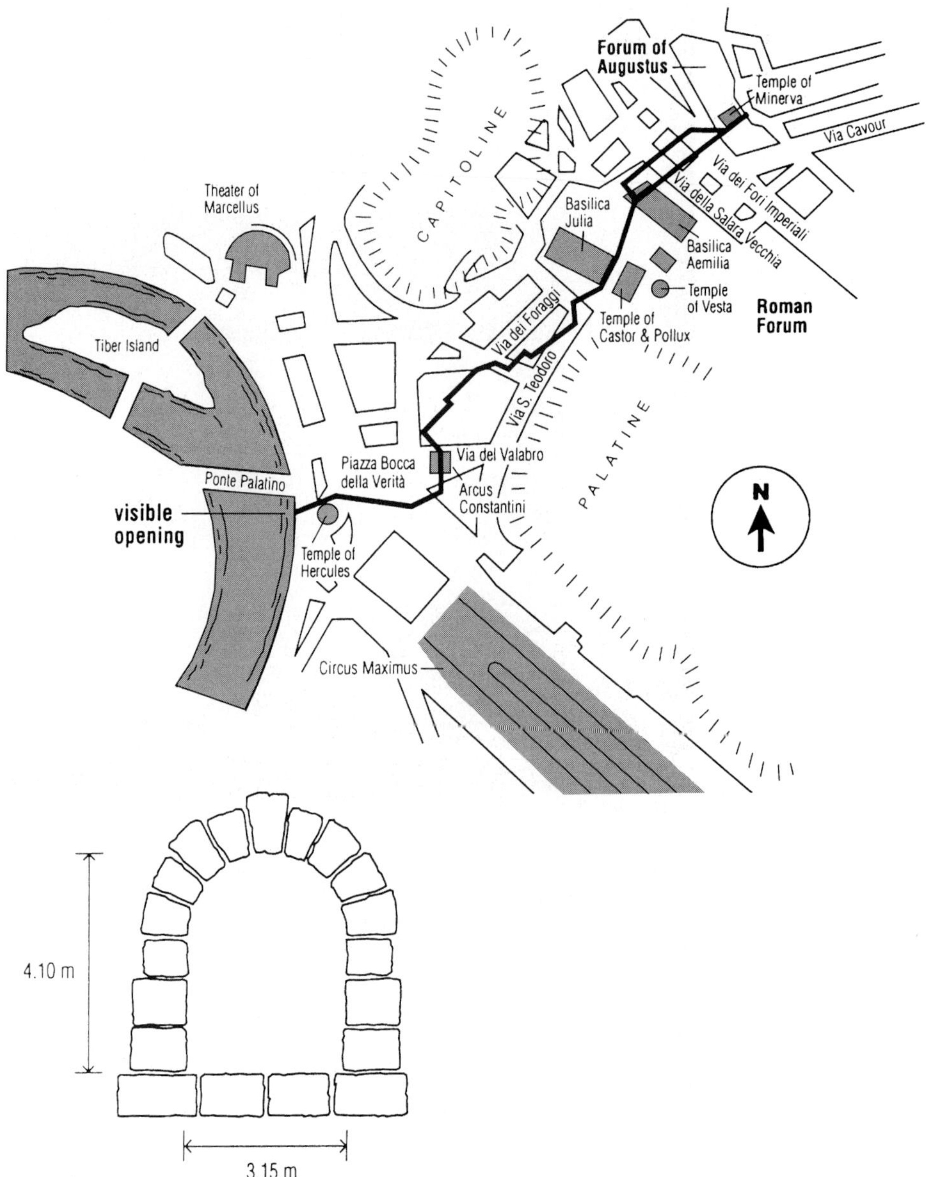

FIGURE 1.4   Ancient Rome's main sewer line, the Cloaca maxima. Still used for street run-off, the sewer's terminus, ancient stone intact, can be seen from Ponte Palatino (after Lamprecht as presented in Aicher, 1995).

(a)

(b)

FIGURE 1.5 Pipes used at Ephesus, Turkey. (Photos by L. W. Mays)

rainfall duration, intensity or frequency. The Lloyd-Davies method only considered 100% runoff from impermeable areas.

The rational and Lloyd-Davies methods are based upon the following assumptions (Colyer and Pethwick, 1976):

(a) The rainfall intensity is steady.

(b) The frequency of peak runoff equals the frequency of the rainfall causing it.

(a)

(b)

FIGURE 1.6 Drainage components at Ephesus, Turkey. (a) Drainage opening along street. (b) Cover to underground drainage. (Photos by L. W. Mays)

(*c*)

FIGURE 1.6   (*Continued*) (*c*) Cover to underground drainage.

FIGURE 1.7   Drainage channel on floor of the Greek Theater at Ephesus. (Photo by L. W. Mays)

(c) The routing velocity is equal to the full flow velocity.

(d) In the Lloyd-Davies method, the degree of permeability is constant and independent of rainfall intensity, duration and frequency.

The Lloyd-Davies method is applied to each pipe as follows (Colyer and Pethwick, 1976):

1. Determine the area (A_p) in acres of the directly connected impermeable surface which contributes runoff to the sewer section being considered.

2. Assume a suitable pipe size.

3. Calculate the *full bore velocity* (v_{fi}) using the Manning or Colebrook-White flow formula.

4. Calculate the time of concentration for flow from the most remote part of the catchment to the current pipe length. The time of concentration is assumed to be the sum of a time of entry (t_e) representing the time taken for flow over the surface into the sewer flow time is determined assuming full bore velocity. Hence time of flow to the downstream end of the jth pipe is expressed as:

$$t_{fj} = \sum_{i=1}^{N} \frac{L_i}{v_{fi}} \tag{1.3}$$

where L_i = length of ith pipe
v_{fi} = full bore velocity of ith pipe
N = number of pipes defining the longest flow path from any part of the catchment to the point being considered

Note that t_{fi} includes the time of flow down the current pipe.

5. Select a return period representing the average frequency at which the storm sewer is to be allowed to surcharge. This will depend on the economics of flooding of the area being sewered.

6. For the selected frequency, find the average rainfall intensity (i) corresponding to a duration equal to the time of concentration to the design point.

7. Calculate the resulting flow $Q = iA_p$:

8. Calculate the *full bore capacity* of the current pipe. If it is inadequate assume a larger pipe size and repeat the procedure from Step 3. Step 3 required the calculation of the so-called full-bore velocity using the Manning or Colebrook-White formula.

1.1.2.2 Manning's Formula.   Manning (1851, 1852, 1880, 1889, and 1895) also made a significant impact upon stormwater drainage design in that as pointed out above the pipe velocity is required in the rational and Lloyed-Davies methods. The first formula fit to the mean velocity-hydraulic radius relationship was (Manning 1889, p. 172)

$$V = 32[RS(1 + R^{1/3})]^{1/2} \tag{1.4}$$

Manning described this equation as "entirely empirical." To quote Manning (1889):

The second was found on the assumption that the exponent of S was constant and equal to the square root of that function. If then it was possible to represent the velocity by a monomial equation such as Chezy's, it should take the form

$$V = CS^{1/2}R^X \tag{1.5}$$

This was found to be the case, and for the mean value of the velocities in the table, the equation was found approximately to be

$$V = 46S^{1/2}R^{4/7} \qquad (1.6)$$

Manning commented that "numerous empirical equations closely approximating to that curve may be found" (Manning, 1889 p. 175). He must have been satisfied that the form of equation sufficiently approximated hydraulic principles, that it was simple enough to calibrate and use, and was sufficiently close to the mean result of experimenters to warrant a closer analysis (Dooge, 1991).

Manning described the method of calculation as:

> "The method adopted in these calculations was to take the first observation of each series as unity, and to reduce all the others to it, so that the exponent of R might be easily found; of course a similar operation should be performed successively on each of the other experiments, and a mean of all the results taken, this was not done, it being considered sufficiently accurate to take the value of the exponent at 0.666 or 2/3, and so the formula

$$V = CS^{1/2}R^{2/3} \qquad (1.7)$$

> was established and was applied to 170 experiments."

Willcocks and Holt (1899) in their "Elementary Hydraulics" appear to have been the first to publish a version of the Manning formula which explicitly included the Kutter roughness factor n (Dooge, 1991), in metric units

$$V = R^{2/3}S^{1/2}/n \qquad (1.8)$$

Bovey (1901) seems to have been the first to convert this to feet units (Dooge, 1991).

$$V = (1.486/n)R^{2/3}S^{1/2} \qquad (1.9)$$

Buckley (1911) cited Willcocks and Holt (1899) and used, in feet units

$$V = (1.4858/n)R^{2/3}S^{1/2} \qquad (1.10)$$

Parker (1913) presented the Manning formula as

$$V = (1.49)R^{0.67}S^{0.5} \qquad (1.11)$$

and this conversion factor (1.49) was also used by Dougherty (1916). King (1918) reverted to the value of 1.486 used by Bovey (1901). The reader is referred to Dooge (1987 a, 1987 b, and 1991) for more detail on the historical aspects.

1.1.2.3   The Beginning of Modern Hydrology.   Chow, et al (1988) pointed out that quantitative hydrology was still very immature at the beginning of the 20th century. The previous two sub-sections, 1.1.2.1 and 1.1.2.2, respectively discuss the history of the rational formula and Manning's formula, which were certainly two major advances for the beginning of modern hydrology. Empirical approaches were employed to solve practical hydrological problems. Gradually hydrologists replaced empiricism with rational analysis of observed data. Green and Ampt (1911) developed a physically based model for infiltration; Hazen (1914) introduced frequency analysis of flood peaks and water storage requirements; Richards (1931) derived the governing equation for unsaturated flow; Sherman devised the unit hydrograph method to transform effective rainfall to direct runoff (1932); Horton developed infiltration theory (1933) and a description of drainage basin form (1945); Gumbel proposed the extreme value law for hydrologic studies (1941); and Hurst (1951) demonstrated that hydrologic observations may exhibit sequences of low or high values that persist over many years.

1.1.3   Evolution of Today's Urban Stormwater Management Models

Today expenditures for urban drainage works and pollution control facilities are among the largest items in the budgets of most municipalities, and represent a significant percentage of federal funding of public works. Design and planning procedures firmly based on the fundamental processes governing the quantity and quality of urban runoff flows result in the most effective solutions to the problems facing planners and decision makers. Widespread access to computers and the instigation of sampling programs have led to the development of urban runoff models that have been calibrated and validated by comparisons with field data.

The need for comprehensive approaches for the simulation of flow quantities and the limitations of the rational method have been recognized since these methods only began in the later 1950's, even though hydrograph methods had been introduced much earlier. The principal reasons for the time lag appear to be the lack of rainfall and flow measurements and the fact that expenditures for the installation of storm sewers and culverts were, in the past, less significant than in other areas of water resources. The first uses of hydrologic models for urban flow simulation followed the development of the RRL Model (Road Research Laboratory) in the U.K., and the Chicago Model in the U.S. (Watkins, 1962, Kiefer, 1970). Many models have been developed in the U.S., such as the EPA's SWMM, the WRE model, the University of Cincinnati model, ILLUDAS, MIT, HYDROCOMP etc. and are described in Brandstetter (1977). Torno (1974) presented the characteristics of various models available during that time period. Linsley (1971) and James Maclaren (1975) made comparative studies of urban runoff models available in the early 1970's.

During the last 25 years, there has been a proliferation of computer models that can be used for various aspects of the design of stormwater collection, storage and conveyance structures. Computer modeling became an integral part of hydrologic and hydraulic design and analysis in the early to mid 1970's when several federal agencies began the development of software. Some of the more notable accomplishments of software development during that time were:

USACE Hydrologic Engineering Center

    HEC-1 (Flood hydrograph package)(U.S.A.C.E., 1973)
    HEC-2 (Water surface profiles)(U.S.A.C.E., 1976)
    STORM (*S*torage, *T*reatment, and *O*verflow *R*unoff) Model (U.S.A.C.E., 1977)

U.S. Soil Conservation Service

    TR-20 (Project formulation hydrology)(U.S.S.C.S, 1965)
    WSP2 (Water surface profile computations)(U.S.S.C.S., 1976)

U.S. Environmental Protection Agency

    SWMM (Stormwater Management) Model (Metcalf and Eddy, 1971)

Also some state agencies began the development of software that could be used for design, including the Illinois State Water Survey who developed the ILLUDAS (Illinois Urban Drainage Area Simulator) Model (Terstriep and Stall, 1979).

Brandstetter (1977) assessed the various models that could be used for storm and combined sewer management. This is an excellent source to read about the state-of-the-art of modeling during that time frame. Since that early development, there have been many new versions of several of these models. The newest versions of the HEC-1 and HEC-2 are, respectively, the HEC-HMS and HEC-RAS models. Over this time period, numerous versions of the SWMM model have been developed. There has been a proliferation of proprietary models, many of which have been variations of the original government models. Chapter 21 presents a summary of the many non-proprietary models available.

1.2 HYDROLOGIC PROCESSES AND THEIR ROLE

1.2.1 Hydrologic Cycle

The U.S. National Research Council (1991) presented the following definition of hydrology:

> "Hydrology is the science that treats the waters of the Earth, their occurrence, circulation, and distribution, their chemical and physical properties, and their reaction with the environment, including the relation to living things. The domain of hydrology embraces the full life history of water on Earth."

For purposes of this book, we are interested in the engineering aspects of hydrology or what we might call engineering hydrology. From this point of view, we are mainly concerned with quantifying amounts of water at various locations (spatially) as a function of time (temporally) for surface water applications. In other words, we are concerned with solving engineering problems using hydrologic principles. This chapter is not concerned with the chemical properties of water and their relation to living things.

The central focus of hydrology is the *hydrologic cycle* consisting of the continuous processes shown in Fig. 1.8. Water *evaporates* from the oceans and land surfaces to become water vapor that is carried over the earth by atmospheric circulation. The *water vapor* condenses and *precipitates* on the land and oceans. The precipitated water may be *intercepted* by vegetation, become overland flow over the ground surface, *infiltrate* into the ground, flow through the soil as *subsurface flow,* and discharge as *surface runoff.* Evaporation from the land surface comprises evaporation directly from soil and vegetation surfaces, and *transpiration* through plant leaves. Collectively these processes are called *evapotranspiration.* In-

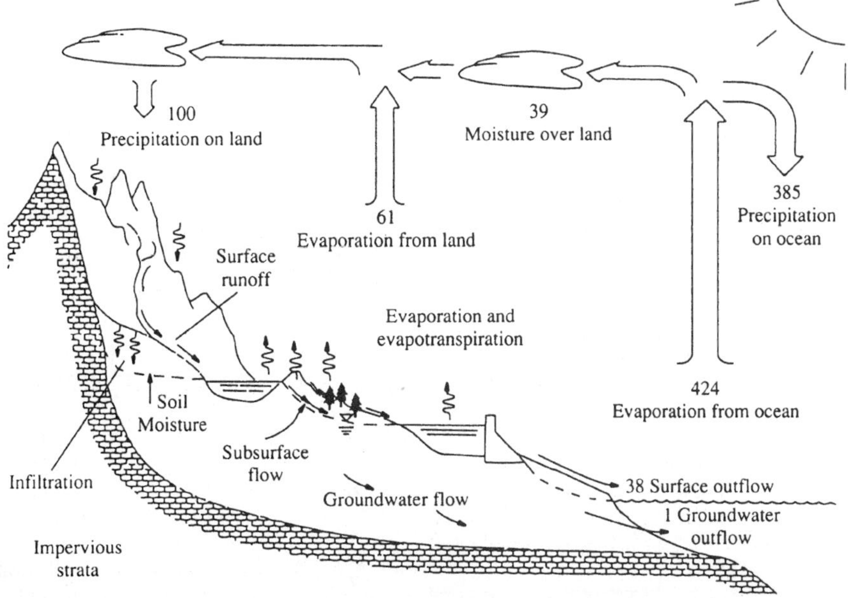

FIGURE 1.8 Hydrologic cycle with global annual average water balance given in units relative to a value of 100 for the rate of precipitation on land. (From Chow, Maidment, and Mays, 1988)

filtrated water may percolate deeper to recharge groundwater and later become *springflow* or *seepage* into streams to also become streamflow.

1.2.2 Hydrologic Systems

According to Chow, Maidment and Mays (1988), a *hydrologic system* is defined as a structure or volume in space, surrounded by a boundary, that accepts water and other inputs, operates on them internally, and produces them as outputs. The *structure* (for surface or subsurface flow) or volume in space (for atmospheric moisture flow) is the totality of the flow paths through which the water may pass as throughput from the point it enters the system to the point it leaves. The *boundary* is a continuous surface defined in three dimensions enclosing the volume or structure. A *working medium* enters the system as input, interacts with the structure and other media, and leaves as output. Physical, chemical, and biological processes operate on the working media within the system; the most common working media involved in hydrologic analysis are water, air, and heat energy.

The *global hydrologic cycle* can be represented as a system containing three subsystems: the *atmospheric water system,* the *surface water system,* and the *subsurface water system.* Another example is the *storm-rainfall-runoff process* on a watershed which can be represented as a hydrologic system. The input is rainfall distributed in time and space over the watershed and the output is streamflow at the watershed outlet. The boundary is defined by the *watershed divide* and extends vertically upward and downward to horizontal planes.

Drainage basins, catchments, and *watersheds* are three synonymous terms that refer to the topographic area that collects and discharges surface streamflow through one outlet or mouth. Catchments are typically referred to as small drainage basins but no specific area limits have been established. The *drainage basin divide, watershed divide,* or *catchment*

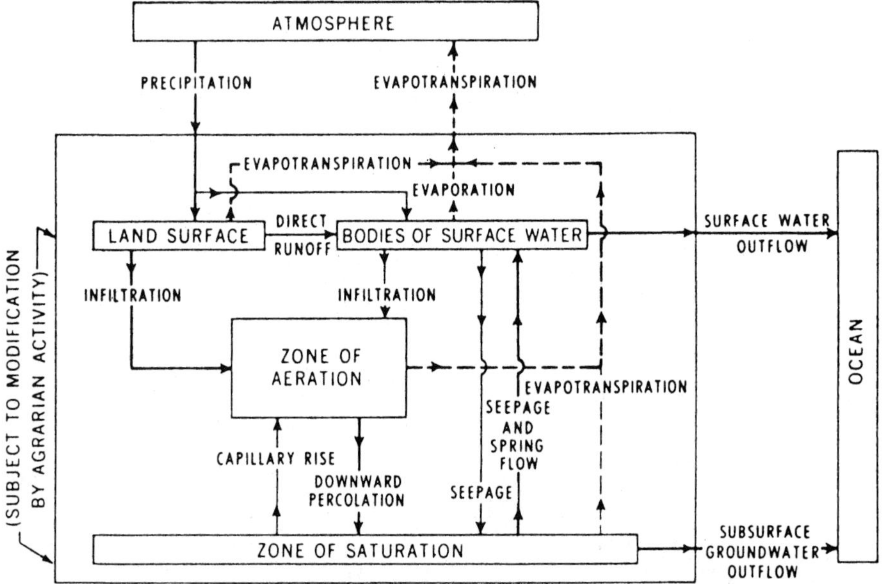

FIGURE 1.9   Pre-urban hydrologic system from *Cohen et al.* [1968] showing major flow paths (heavy lines), minor flow paths (thin lines), flow of liquid water (solid lines), and flow of water vapor (dashed lines).

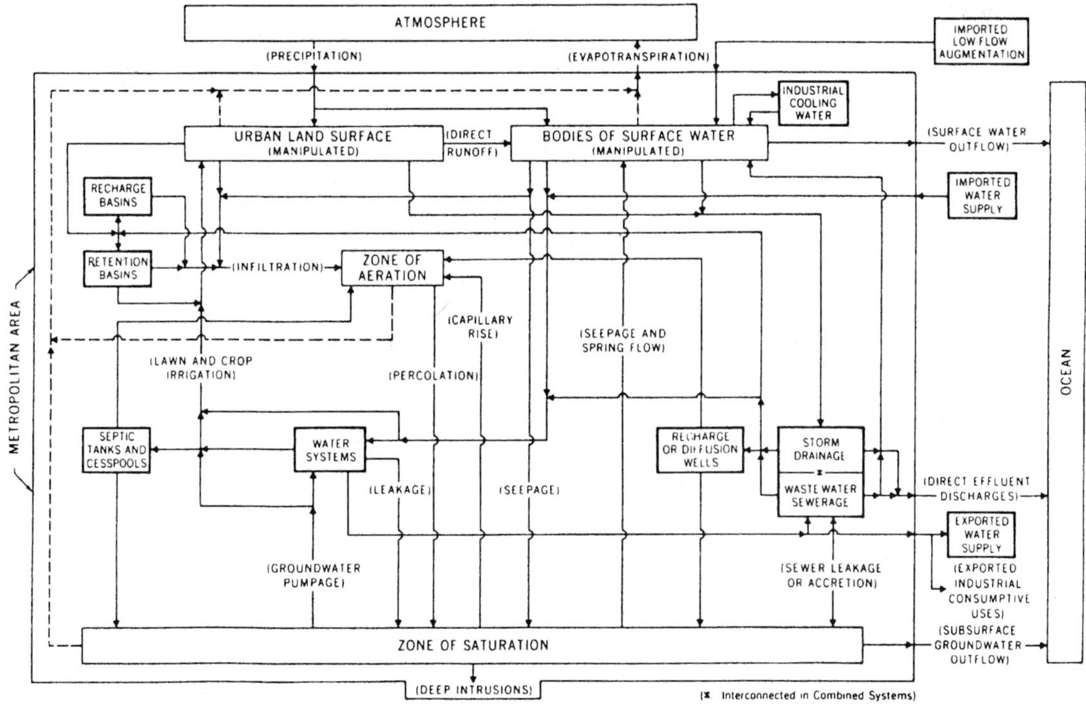

FIGURE 1.10   Urban hydrologic system. (From Franke and McClymonds, 1972)

divide are the lines dividing land whose drainage flows toward the given stream from land whose drainage flows away from that stream. Think of drainage basin sizes ranging from the Mississippi River drainage basin to small urban drainage basin in your local community or some small valley in the countryside near you.

Drainage basins can be pictured in a pyramidal fashion as the runoff from smaller basins (subsystems) combine to form larger basins (subsystem in system) and the runoff from these basins in turn combine to form even larger basins and so on (see Mays, 2001). Marsh (1987) refers to this mode of organization as a *hierarchy* or *nested hierarchy,* as each set of smaller basins is set inside the next layer. Along the same thinking is that streams which drain small basins combine to form larger streams and so on.

1.2.3   Hydrologic Effects of Urbanization

The hydrologic system of a drainage area prior to urbanization (*pre-urban hydrologic system*) is represented schematically in Fig. 1.9. This is a simplified version of the many processes of the pre-urban hydrologic system; however, a comparison with Fig. 1.10 showing a schematic of the *urban hydrologic system* illustrates the hydrologic effects of changes in land and water use associated with the progressive stager of urbanization. Both the pre-urban and urban hydrologic schematics ignore the water quality aspects, which is of particular importance in the urban hydrologic system. Hopefully these two figures illustrate the great increase in complexity of the hydrologic system that results from urbanization. Fig. 1.11 illustrates the *urban stormwater disposal subsystem.*

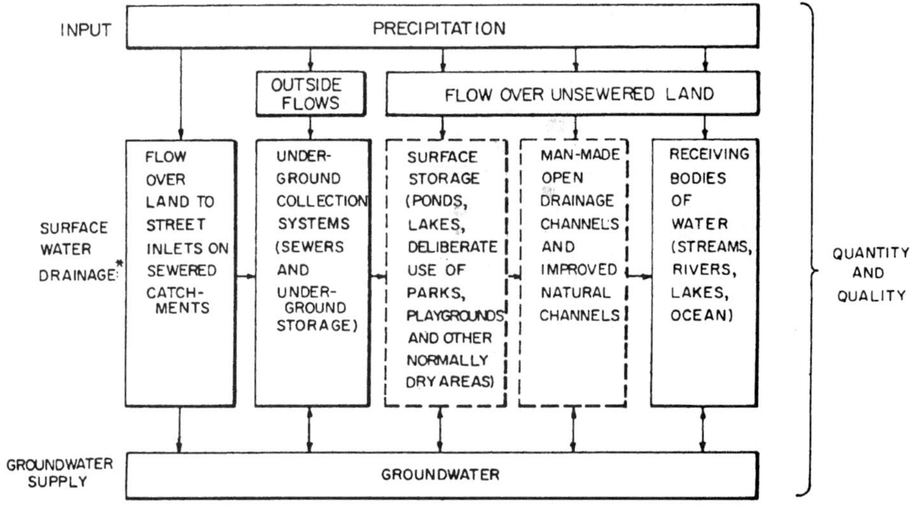

FIGURE 1.11   Urban stormwater disposal physical subsystem. (From American Society of Civil Engineers, 1969)

1.3   MODERN STORM DRAINAGE SYSTEMS

1.3.1   What are Storm Drainage Systems

Figure 1.12 illustrates the configuration of a storm-water system as consisting of two separate systems: (1) a *minor system* for storm drainage; and (2) a *major system* for emergency flows. Grigg (1996) refers to the minor drainage system as the "initial" system or the "convenience" system. Minor systems include gutters, small ditches, culverts, and storm drains, detention ponds, and small channels. Major systems include the streets and the urban streams, floodways, and flood fringe areas. Grigg (1996) superimposes the water quality subsystem on top of the minor and major systems, as problems arising from the wash-off of surface pollutants, from combined sewer overflows, or from the erosion of pollutants from the inside of sewers.

1.3.2   System Components

Figure 1.13 illustrates the principal hydraulic elements in *urban stormwater drainage systems* and Fig. 1.14 illustrates a typical urban *combined sewer system*. Figure 1.15 also shows schematically a simplified description of the major components of the urban stormwater disposal physical subsystem. The urban stormwater drainage system consists of basic sub-catchments in which the excess rainfall is transformed into *overland flow*. The overland flow hydrographs at the drainage manholes form the *inlet hydrographs* for the transport. Each sub-catchment is conceptualized as a flow plane over which overland flow occurs.

1.3.2.1   Why Detention: Effects of Urbanization.   *Urban stormwater management systems* typically include detention and retention facilities to help mitigate the negative impacts of urbanization on storm-water drainage. The effects of urbanization on storm-water runoff

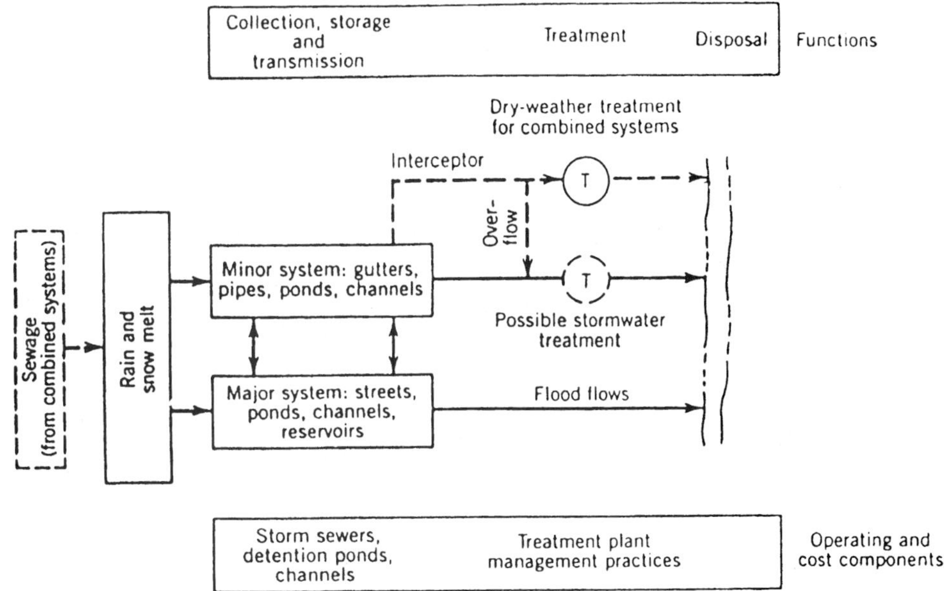

FIGURE 1.12   Urban stormwater management system. (From Grigg, 1996)

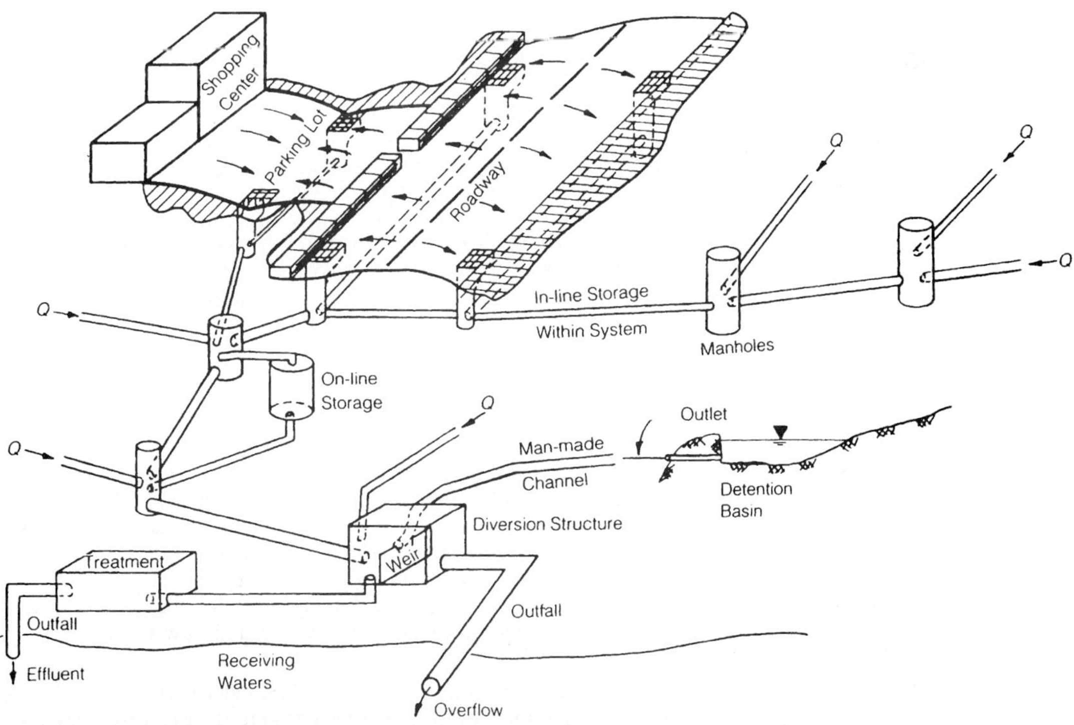

FIGURE 1.13   Principal hydraulic elements in urban storm drainage system. (From ASCE, 1992)

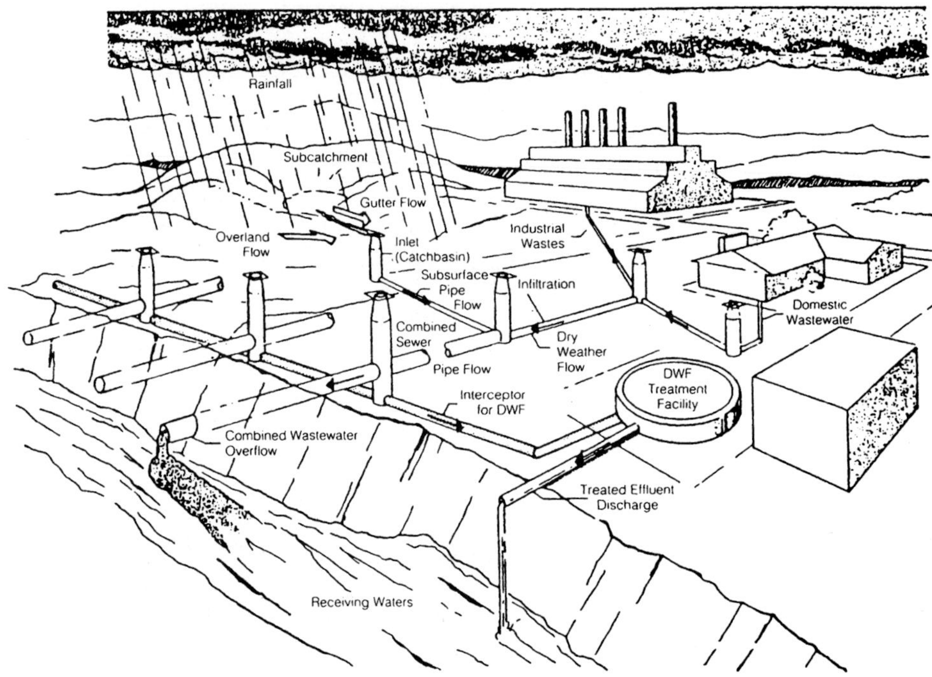

FIGURE 1.14 Typical urban combined sewer system. (From Kibler, 1982)

includes increased total volumes of runoff and peak flow rates, as depicted in Fig. 1.16. In general major changes in flowrates in urban watershed are the result of (Chow, et al., 1988):

1. The increase in the volume of water available for runoff because of the increased impervious cover provided by parking lots, streets, and roofs, which reduce the amount of infiltration.

2. Changes in hydraulic efficiency associated with artificial channels, curbing, gutters, and storm drainage collection systems increase the velocity of flow and the magnitude of flood peaks.

1.3.2.2 Major Types of Detention. The ASCE Manual 77/, (WEF Manual of Practice FD-20), "Design and Construction of Urban Stormwater Management System" (ASCE, 1992) defines the major types of stormwater detention as:

(a) *Detention*—The temporary storage of flood water which is usually released by a measured but uncontrolled outlet. Detention facilities typically flatten and spread the inflow hydrograph, lowering the peak. Structures that release storage over a period of 12 to 36 or (more) hours may also serve water quality purposes (State of New Jersey 1986).

(b) *Retention*—Storage provided in a facility without a positive outlet, or with a specially regulated outlet, where all or a portion of the inflow is stored for a prolonged period. Infiltration basins are a common type of retention facility. Ponds that maintain water permanently, with freeboard provided for flood storage, are probably the most common type retention facility.

The ASCE (WEF) manual further subdivides detention and retention facilities into:

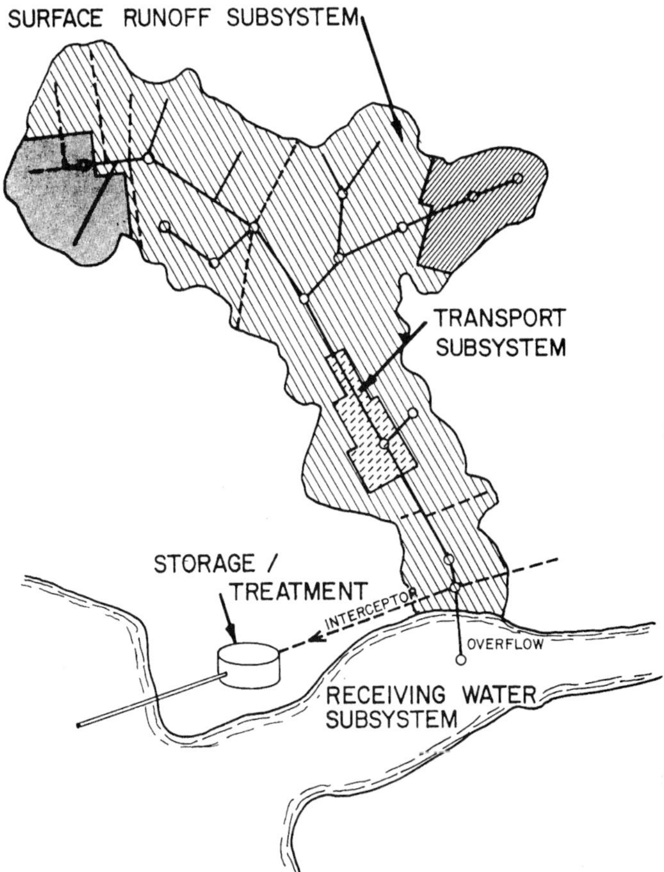

FIGURE 1.15 The urban drainage system. (From Proctor and Redfern, 1976)

(a) *On-Stream Storage*—A facility that intercepts the streamflow directly. On-stream storage occasionally is provided as an on-site facility, though it is more often an integral part of a watershed or a regional stormwater plan.

(b) *On-site Facilities*—Special attention must be given to the design of outlet structures for controlling runoff from rooftops, parking lots, and swales. Because runoff volumes from such areas are small, the required outlets are also small, and this increases the potential for plugging by debris. Also, the outlet must release temporarily-stored water in a reasonable amount of time. As an example, parking lots must drain relatively quickly in order not to be a nuisance. Roof top storage must be designed so as to provide safety of the structure if outlets are plugged.

(c) *Off-stream Storage*—Diversion of flow out of the stream into a separate storage facility. A typical example is a side channel spillway that diverts storm flows from the stream into a storage impoundment (or a structure that can divert and store the "first flush" of particularly contaminated runoff).

(d) *Conveyance Storage*—Conveyance storage is an often neglected form of storage, because it is dynamic and requires channel storage routing analysis to identify. Slower-

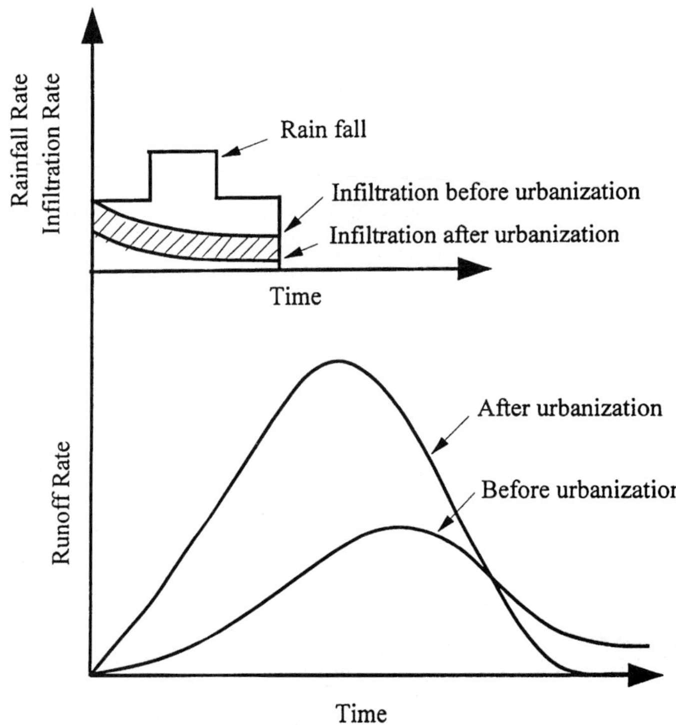

FIGURE 1.16   Effect of urbanization on stormwater runoff.

flowing conveyance caused by flatter slopes or rougher channels can markedly retard the buildup of flood peaks and alter the time response of the tributaries in a watershed. ·

(e) *Wet Basins and Infiltration Basins*—Wet basins are detention basins designed to maintain a permanent pool of water. In most aspects, their design is similar to other detention basins (dry basins), except for the permanent pool. Wet basins are used for aesthetic or water quality enhancement, or for the maintenance of fish or wildlife. All outlets are above the normal level of the pool. Infiltration basins resemble other detention basins in most respects, though they may be built without outlets. They may retain flood flows for a prolonged period of time, for the purpose of encouraging infiltration into the groundwater.

Stahre and Urbonas (1990) present the classification of storage facilities shown in Fig. 1.17. The major classification is *source control* or *downstream control*. Source control involves the use of smaller facilities that are located near the source allowing better use of the downstream conveyance system. Downstream control uses storage facilities which are larger and consequently at fewer locations, such as at watershed outlets. Referring to Fig. 1.14, source control consists of local disposal, inlet control, and on site detention with the various types for each listed in the figure. *Local disposal* refers to the use of infiltration or percolation. *Inlet control* refers to detaining stormwater where the precipitation occurs (such as roof tops and packing lots). *On-site detention* typically refers to detaining stormwater from larger areas than the previous two and includes swales, ditches, dry ponds, wet ponds, and concrete basins that are typically underground, and underground piping. *Wet ponds* have a permanent water pool as opposed to dry ponds.

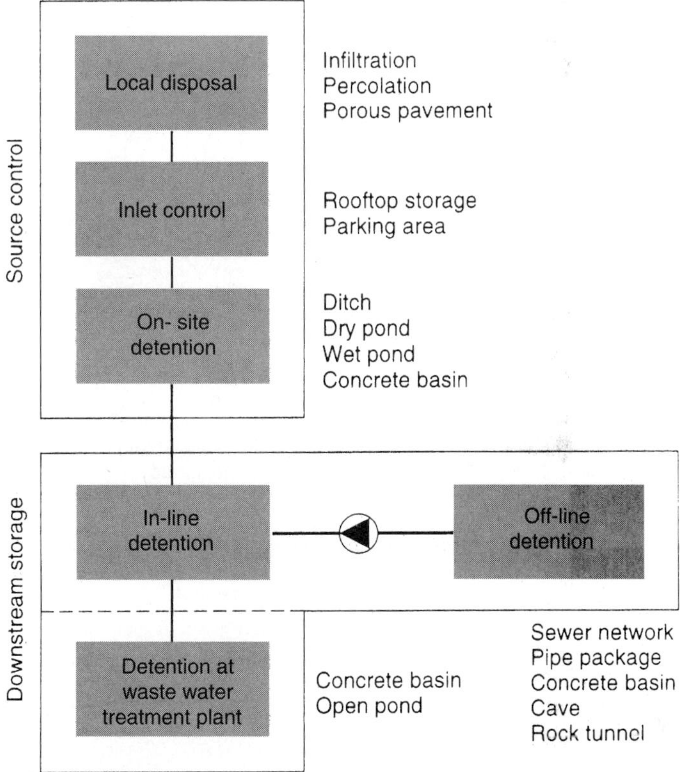

FIGURE 1.17 Classification of storage facilities. (From Stahre and Urbonas, 1990)

Downstream storage includes *in-line detention, off-line detention,* and *detention at waste water treatment plants.* In-line detention refers to detention storage in sewer lines, tunnels, storage vaults, pipes, surface ponds, or other facilities that are connected in-line with a stormwater conveyance network. Off-line storage facilities are not in line with the stormwater conveyance system.

Two other classifications of detention are *underground* or *sub-surface systems* and *surface systems.*

1.3.3 Types of Surface Detention

Surface detention, for purposes of this discussion refers to *extended detention basins* (or *dry detention basins*) and *retention ponds* (or *wet detention ponds*). Dry detention ponds empty after a storm whereas retention ponds retain the water much longer above a permanent pool of water. Dry detention is the most widely used in the U.S. and many other countries. Figure 1.18 illustrates an extended detention basin. Water enters the basin and is impounded behind the embankment and is slowly discharged through a perforated riser outlet. The coarse aggregate around the perforated riser minimizes clogging by debris. Typically once a required water quality volume is filled, the remaining inflow is diverted around the basin or the pond

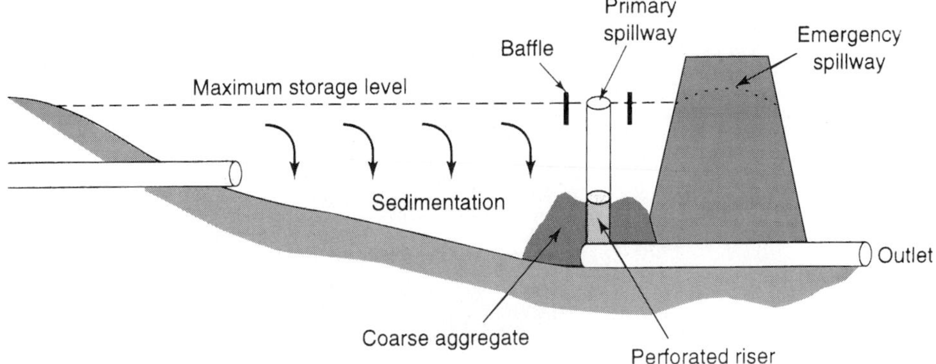

- Efficiency: Poor for detention times under 12 hours
        Good for detention times greater than 24 hours
- Function: Settle pollutants out; soluble pollutants pass through
- Maintenance is moderate if properly designed
- Improper design can make facilities an eyesore and a mosquito-breeding mudhole
- Newer designs are incorporating a shallow marsh around outlet
        Result: Better removal efficiency and no mosquito nuisance
- Regional detention facilities serving 100–200 acres can be aesthetically developed
        Result: Lower maintenance costs

FIGURE 1.18   Design of an extended detention basin. (From Urbonas and Roesner, 1993)

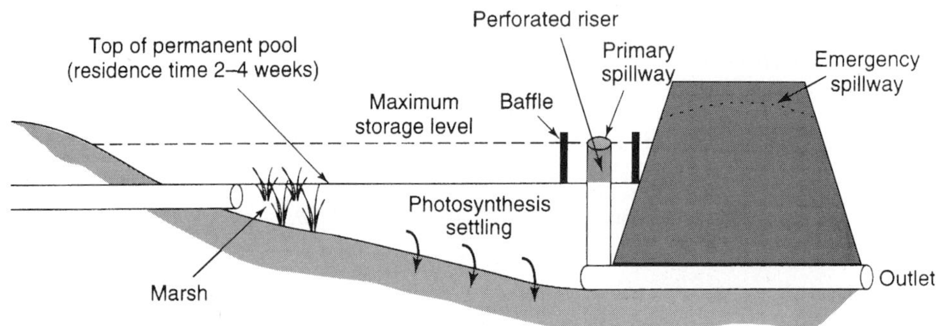

- Efficiency: Excellent if properly designed.
        Can be poor if bottom goes anoxic.
- Function: Removes pollutants by settling dissolved pollutants biochemically.
- Maintenance: Relatively free after first year except for major cleanout at
        about 10 years.
- Aesthetic design can make pond an asset to community.
    Excellent as a regional facility.

FIGURE 1.19   Design of a retention pond. (From Urbonas and Roesner, 1993)

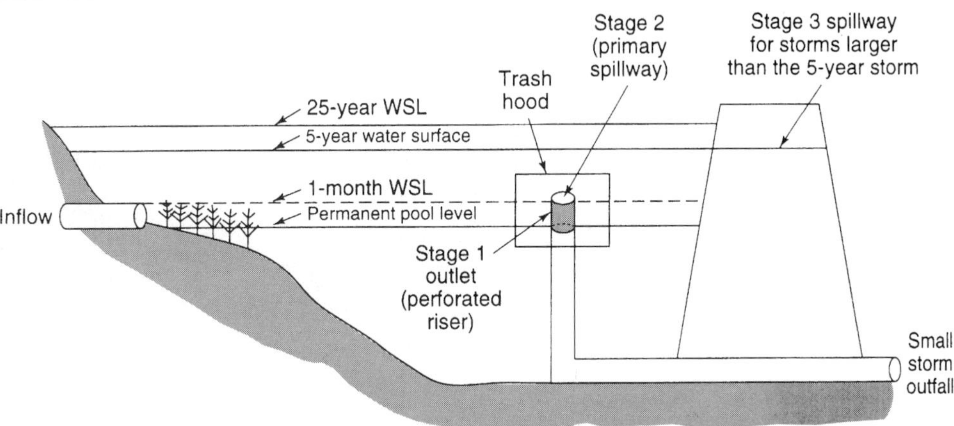

FIGURE 1.20   Conceptual design of a multipurpose pond. (From Urbonas and Roesner, 1993)

overflows through a primary spillway. A large part of the sediment from the stormwater settles in the basin.

Figure 1.19 illustrates a *retention pond,* which is basically a lake that can be designed to remove pollutants. The figure illustrates the basic treatment processes that occur in the retention pond. Pollutants are removed by settling. Nutrients are removed by photoplankton growth in the water column and by shallow marsh plants around the pond perimeter.

A *multipurpose detention basin* for quantity and quality is illustrated in Fig. 1.20. The outlet works are staged so that the water-quality design volume is released very slowly. The other stages provide storage and outlet peak discharges for erosion and flood control. Refer to Fig. 1.20 and note the various stages.

Figures 1.21 to 1.24 illustrate detention basins in Scottsdale, Arizona. Figure 1.25 illustrates retention ponds hear Pearland, Texas. Figure 1.26 shows detention ponds built to store highway drainage.

1.3.4   Urban Drainage: Design and Prediction

Urban drainage problems, from the hydraulic engineering viewpoint, can be classified into two types: (1) design, and (2) prediction for forecasting or operation. Some of the types of design and prediction problems are listed in Table 1.1.

1.4   STORMWATER MANAGEMENT

Stormwater management is knowledge used to understand, control, and utilize waters in their different forms within the hydrologic cycle (Wanielista and Yousef, 1993). The goal of this chapter is to provide an introduction to the various concepts and design procedures involved in stormwater management. The overall key component of stormwater management is the drainage system. Urbonas and Roesner (1993) point out the following vital functions of a drainage system:

1. It removes stormwater from the streets and permits the transportation arteries to function during bad weather. When this is done efficiently, the life expectancy of street pavement is extended.

(a)

(b)

FIGURE 1.21 Small detention basin in residential area in Scottsdale, Arizona. (Photos by L. W. Mays)

(*a*)

(*b*)

FIGURE 1.22 Inlet and outlet structures for a detention basin. (*a*) Inlet structure. (*b*) Outlet structure. (Photos by L. W. Mays)

(*a*)

(*b*)

FIGURE 1.23 Detention basin in Scottsdale, Arizona used as a park. (*a*) Detention basin. (*b*) Close-up of outlet structure. (Photos by L. W. Mays)

(*a*)

(*b*)

FIGURE 1.24   Detention basin used in commercial/office areas in Scottsdale, AZ. (*a*) Parking lot used for detention with outlet shown. Outlet drains to detention area in Figure 1.24(*b*). (*b*) Detention area draining parking lot detention area in 1.24(*a*).

(c)

(d)

FIGURE 1.24 (*Continued*) (*c*) Detention basin for commercial development, inflows from parking lot. (*d*) Detention basin for commercial development, inlet from parking area shown on left.

(*e*)

(*f*)

FIGURE 1.24   (*Continued*) (*e*) Detention basin in commercial development. (*f*) Detention basin for office building and connecting parking lot.

(a)

(b)

FIGURE 1.25 Retention ponds near Pearland, Texas. (a) Retention pond in residential area. (b) Retention pond in golf course. (Photos by L. W. Mays)

FIGURE 1.26 Detention pond at highway intersection of Beltway 8 and State Highway 288 South of Houston, Texas. (Photo by L. W. Mays)

2. The drainage system controls the rate and velocity of runoff along gutters and other surfaces in a manner that reduces the hazard to local residents and the potential for damage to pavement.

3. The drainage system conveys runoff to natural or manmade major driveways.

4. The system can be designed to control the mass of pollutants arriving at receiving waters.

5. Major open drainage ways and detention facilities offer opportunities for multiple use such as recreation, parks, and wildlife preserves.

Storm drainage criteria are the foundation for developing stormwater control. These criteria should set limits on development; provide guidance and methods of design; provide details of key components of drainage and flood control systems; and ensure longevity, safety, aesthetics, and maintainability of the system served (Urbonas and Roesner, 1993).

1.5 FLOODPLAIN MANAGEMENT

Floods are natural events that have always been an integral part of the geologic history of earth. Flooding occurs along rivers, streams and lakes, in coastal areas, on alluvial fans, in ground failure areas such as subsidence, in areas influenced by structural measures, and in areas that flood due to surface runoff and locally inadequate drainage. Human settlements and activities have always tended to use floodplains. Their use has frequently interfered with the natural floodplain processes causing inconvenience and catastrophe to humans. This section focuses on the management of water excess (floods).

1.5.1 Floodplain Definition

A *floodplain* is the normally dry land area adjoining rivers, streams, lakes, bays, or oceans that is inundated during flood events. The most common causes of flooding are the overflow

TABLE 1.1   Types of Urban Drainage Problems (a) Design Problems

Type	Design purpose	Hydro information sought	Required hydraulic level
Sewers	Pipe size (and slope) determination	Peak discharge, Q_p for design return period	Low
Drainage channels	Channel dimensions	Peak discharge, Q_p for design return period	Low to moderate
Detention/retention storage ponds	Geometric dimensions (and outlet design)	Design hydrograph, $Q(t)$	Low to moderate
Manholes and junctions	Geometric dimensions	Design hydrograph, $Q(t)$	Low to moderate
Roadside gutters	Geometric dimensions	Design peak discharge, Q_p	Low to moderate
Inlet catch basins	Geometric dimensions	Design peak discharge, Q_p	Low
Pumps	Capacity	Design hydrograph	Moderate to high
Control gates or valves	Capacity	Design hydrograph	Moderate to high

Types of Urban Drainage Problems (b) Prediction Problems

Type	Purpose	Hydro input	Hydro information sought	Required hydraulic level
Real-time operation	Real-time regulation of flow	Predicted and/or just measured rainfall, network data	Hydrographs, $Q(t, x_i)$	High
Performance evaluation	Simulation for evaluation of a system	Specific storm event, network data	Hydrographs, $Q(t, x_i)$	High
Storm event simulation	Determination of runoff at specific locations for particular past or specified events	Given past storm event or specified input hydrographs, network data	Hydrographs, $Q(t, x_i)$	Moderate-high
Flood level determination	Determination of the extent of flooding	Specific storm hyetographs, network data	Hydrographs and stages	High
Storm runoff quality control	Reduce and control of water pollution due to runoff from rainstorms	Event or continuous rain and pollutant data, network data	Hydrographs $Q(t, x_i)$ Pollutographs, $c(t, x_i)$	Moderate to high
Storm runoff master planning	Long-term, usually large spatial scale planning for stormwater management	Long-term data	Runoff volume Pollutant volume	Low

Source: Yen & Akan (1999).

of streams and rivers, and abnormally high tides resulting from severe storms. The floodplain can include the full width of narrow stream valleys, or broad areas along streams in wide, flat valleys. As shown in Fig. 1.27, the channel and floodplain are both integral parts of the natural conveyance of a stream. The floodplain carries flow in excess of the channel capacity and the greater the discharge, the further the extent of flow over the floodplain. Floodplains may be defined as either natural geologic features or from a regulatory perspective. The *100-year floodplain* is the standard (most commonly) used in the U.S. for management and regulatory purposes. Flooding concerns are not limited to riverine and coastal flooding. Also of concern are floods associated with alluvial fans, unstable channels, ice jams, mudflows, and subsidence.

Alluvial fans are characterized by a cone or fan-shaped deposit of boulders, gravel and fine sediments that have been eroded from mountain slopes and transported by flood flows,

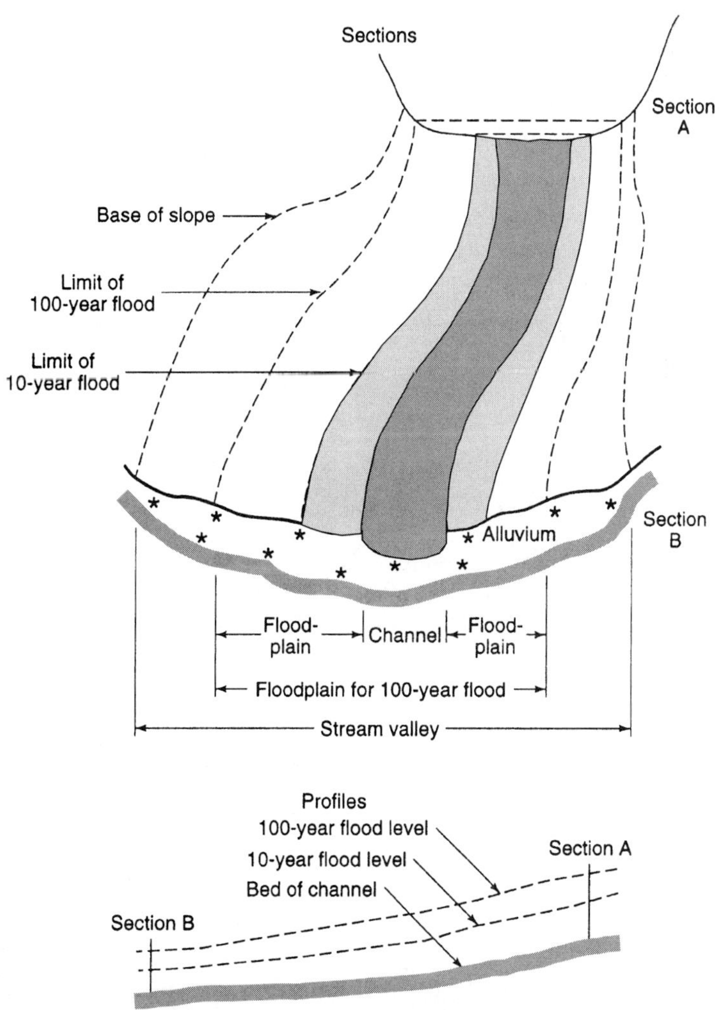

FIGURE 1.27 Typical sections and profiles in an unobstructed reach of stream valley. (From Waananen et al., 1977)

debris flows, erosion, sediment movement and deposition, and channel migration. Alluvial fans are common throughout many parts of the world. In the U.S., they are common in Arizona, California, Idaho, Montana, Nevada, New Mexico, Utah, Washington and Wyoming. Illustrated in Fig. 1.28 is the flood insurance rate zone defined for alluvial fan systems.

1.5.2 Hydrologic and Hydraulic Analysis of Floods

The hydrologic and hydraulic analysis of floods is required for the planning, design, and management of many types of facilities including hydrosystems within a floodplain or watershed. These analyses are needed for determining potential flood elevations and depths, areas of inundation, sizing of channels, levee heights, right of way limits, design of highway crossings and culverts, and many others. The typical requirements include (Hoggan, 1997):

1. *Floodplain information studies.* Development of information on specific flood events such as the 10-, 100-, and 500-year frequency events.

2. *Evaluations of future land-use alternatives.* Analysis of a range of flood events (different frequencies) for existing and future land uses to determine flood-hazard potential, flood damage, and environmental impact.

3. *Evaluation of flood-loss reduction measures.* Analysis of a range of flood events (different frequencies) to determine flood damage reduction associated with specific design flows.

4. *Design studies.* Analysis of specific flood events for sizing facilities to ensure their safety against failure.

5. *Operation studies.* Evaluation of a system to determine if the demands placed upon it by specific flood events can be met.

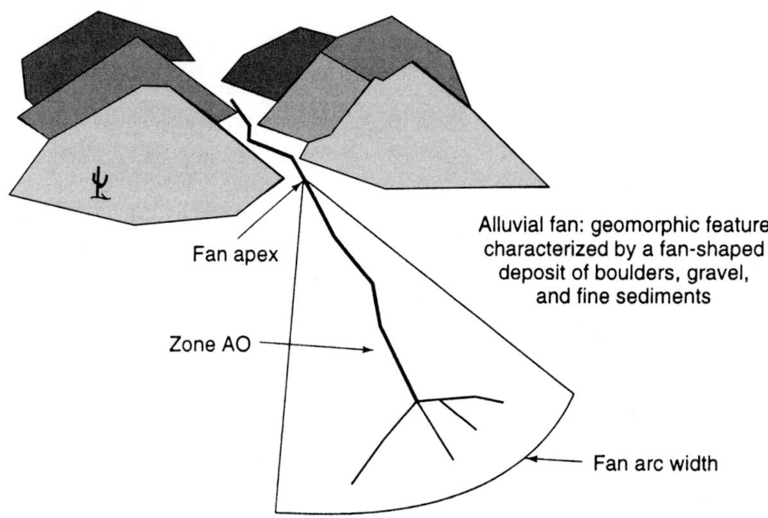

Zone AO: flood insurance rate zone that corresponds to the areas of 100-year shallow flooding where average depths are between 1 ft and 3 ft.

FIGURE 1.28   Alluvial fan system. (From Mays, 2001)

The methods used in hydrologic and hydraulic analysis are determined by the purpose and scope of the project and the data availability. Figure 1.29 is a schematic of the components of a hydrologic and hydraulic analysis for floodplain studies. The types of hydrologic analysis for flood plains are to perform either a rainfall-runoff analysis or a flood-flow frequency analysis. If an adequate number of historical annual instantaneous peak discharges (*annual maximum series*) are available, the flood-flow frequency analysis can be performed to determine peak discharges for various return periods. Otherwise, a rainfall-runoff analysis must be performed using a historical storm or design storm for a particular return period to develop a storm-runoff hydrograph.

Determination of water-surface elevations can be performed using a steady-state water-surface profile analysis if only peak discharges are known, or one can select the peak discharges from generated storm-runoff hydrographs. For a more detailed and comprehensive analysis, an unsteady-flow analysis based upon a hydraulic-routing model and requiring the storm-runoff hydrograph can be used to more accurately define maximum water-surface elevations. The unsteady-flow analysis also provides more detailed information such as the routed-discharge hydrographs at various locations throughout a river reach.

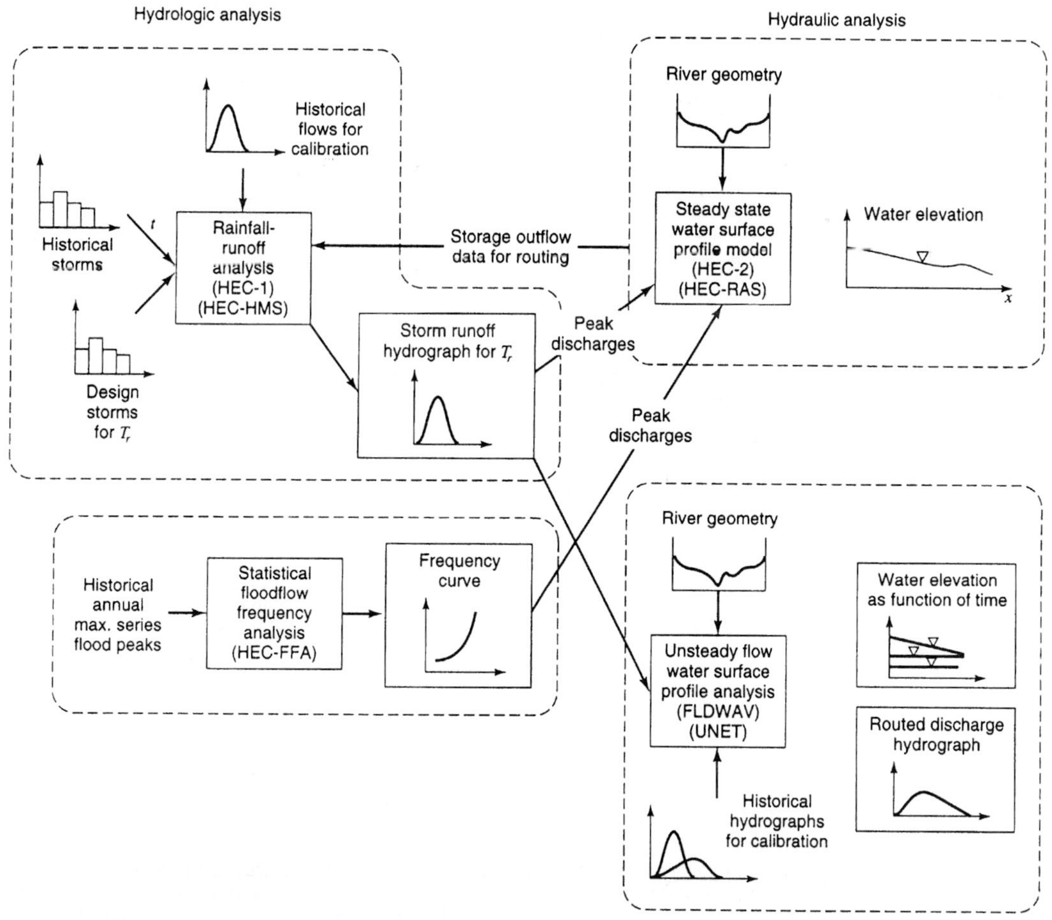

FIGURE 1.29 Components of a hydrologic-hydraulic floodplain analysis. (From Mays, 2001)

1.5.3 Floodways and Floodway Fringes

Encroachment on floodplains, such as by artificial fill material, reduces the flood-carrying capacity, increases the flood heights of streams, and increases flood hazards in areas beyond the encroachment. One aspect of floodplain management involves balancing the economic gain from floodplain development against the resulting increase in flood hazard. For purposes of Federal Emergency Management Agency (FEMA) studies, the 100-year flood area is divided into a floodway and a floodway fringe, as shown in Fig. 1.30. The *floodway* is the channel of a stream plus any adjacent floodplain areas that must be kept free of encroachment in order for the 100-year flood to be carried without substantial increases in flood heights. FEMA's minimum standards allow an increase in flood height of 1.0 foot, provided that hazardous velocities are not produced. The *floodway fringe* is the portion of the floodplain that could be completely obstructed without increasing the water surface elevation of the 100-year flood by more than 1.0 foot at any point.

Two types of floodplain inundation maps—flood-prone area and flood hazard maps—have been used. *Flood-prone area maps* show areas likely to be flooded by virtue of their proximity to a river, stream, bay, ocean, or other watercourse as determined from readily available information. *Flood hazards maps* show the extent of inundation as determined from a study of flooding at the given location. Flood hazard maps are commonly used in floodplain information reports and require updating when changes have occurred in the channels, on the floodplains, and in upstream areas.

1.5.4 Floodplain Management and Floodplain Regulations

According to the National Flood Insurance Program (NFIP) regulations administered by FEMA, *floodplain management* is "the operation of an overall program of corrective and preventive measures for reducing flood damage, including but not limited to emergency preparedness plans, flood control works, and floodplain management regulations." Floodplain management regulations are the most effective method for preventing future flood damage in developing communities with known flood hazards.

Floodplain management investigates problems which have arisen in developed areas and potential problems that can be forecasted due to future developments. The basic approaches to floodplain management are: actions to reduce susceptibility to floods; actions that modify

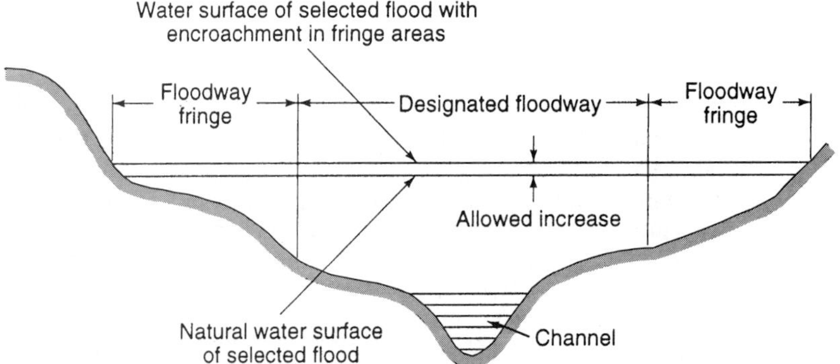

FIGURE 1.30 Definition of floodway and floodway fringe. The floodway fringe is the area between the designated floodway limit and the limit of the selected flood. The floodway limit is defined so that encroachment limited to the floodway fringe will not significantly increase flood elevation. The 100-year flood is commonly used and a 1-ft allowable increase is standard in the United States. (From Mays, 2001)

the flood; and actions that assist individuals and communities in responding to floods. *Floodplain regulation* is the centerpiece of any floodplain management program, and is particularly effective in underdeveloped areas, where the ability exists to control future development.

A key component in floodplain regulation is the definition of the *flood hazard area* (usually defined as the 100-year floodplain) and the floodway. The floodway includes the channel of the stream and the adjacent land areas that must be reserved in order to discharge the design flood without cumulatively increasing the water surface by more than a given amount. For example, the maximum rise allowed by the National Flood Insurance Program (NFIP) is one foot; but in many situations a lesser amount may be more appropriate. An adequate floodplain management plan that eliminates development from the flood hazard areas may be a major step in the right direction. Not allowing other obstructions such as fill and detention basins to be placed in the flood hazard area is another step forward. A floodplain provides both a conveyance mechanism and a temporary storage area for excess water. Allowing obstructions to be placed in the floodplain eliminates the temporary storage areas and in turn increases the hydraulic heads to increase flood levels both downstream and upstream of the floodplain developments.

1.5.5 National Flood Insurance Program

In 1968, the U.S. Congress created the National Flood Insurance Program (NFIP) through the passage of the National Flood Insurance Act. The Flood Disaster Protection Act of 1973 and the National Flood Insurance Reform Act of 1994 further defined the NFIP. The purposes of the NFIP is to minimize future flood loss and to allow the floodplain occupants to be responsible for flood damage costs instead of the taxpayer. The NFIP, administered by the Federal Emergency Management Agency (FEMA), provides federally backed flood insurance that encourages communities to enact and enforce floodplain regulations. If a state or community does not participate in the NFIP, the following consequences occur:

1. The community will not be eligible for flood disaster relief in the event of a federally-declared flood disaster.
2. Federal or federally-related financial assistance for acquisition or construction purposes for structures in flood-prone areas will not be available.
3. Flood insurance will not be available.

For a state or community to be eligible for participation in the NFIP, the state or the community must agree to adopt floodplain management regulations that meet minimum standards as defined by FEMA. These minimum standards include, but are not limited to:

1. Requiring permits for all proposed development within a flood hazard area.
2. Assuring that all necessary governmental permits have been obtained.
3. Ensuring that proper materials and methods are used in new construction to protect new buildings from future floods (for example: elevate the lowest finished floor of residential structures of floodproof non-residential structures above the base flood elevation).
4. Ensuring that all proposed development within a flood hazard area is consistent with the need to minimize flood damage within the flood-prone area.
5. Notifying adjacent communities and the state prior to the alterations or relocation of a watercourse.
6. Ensuring that the flood carrying capacity within the altered or relocated portion of any watercourse is maintained.
7. Prohibiting encroachments, including fill, new construction, substantial improvements, and other development within the adopted regulatory floodway unless it has been dem-

onstrated through hydrologic and hydraulic analysis performed in accordance with standard engineering practice that the proposed encroachment would not result in any increase in flood levels within the community during the occurrence of the base flood flow.

A state is considered a "community" and state agencies are required to comply with minimum standards just as local communities do. A state may comply with the floodplain regulations of the local community in which state land is located or the state may establish and enforce their own floodplain regulations for state agencies.

1.5.6 Stormwater Management and Floodplain Management

Stormwater management plans are most successful when they are implemented at the start of development in an area and should be administered as part of a land-use planning process. The implementation of a stormwater management plan, in a remedial mode, to correct stream deterioration resulting from previous uncontrolled development is a much more difficult task. Stormwater detention programs may have little effect because the flood peak caused by detention diminishes as the flood passes downsteam, while the increase in total runoff caused by the development swells the total mass of the flood wave. The cumulative effect downstream of any number of detention basins would mainly be to delay the arrival of the flood crest by a few hours, having little or no effect on reducing the peak discharge. A partial solution to this is to provide retention over a long time period. The increases in peak flood discharges can be controlled but only through coordinated, extensive planning prior to development. Zoning to preserve undeveloped areas, particularly those in the floodplain, can be a very effective measure.

Stormwater management and floodplain management are generally separate and different programs; however there are interfaces such as detention basins built in floodplains are unavoidable issues. Detention basins have been placed in the floodplains of many areas of the U.S. Detention basins generally should be placed out of the floodplain, particularly for small streams of relatively small drainage areas. In such cases, the same storms affect the development site and the floodplain simultaneously. In other words, the time during which the detention basin is needed to store stormwater from the development is basically the same time that the floodplain is flooded and the detention basin location would already be filled by floodwater. To the extent that the flood at the development coincides with the flooding in the floodplain, detention storage in the floodplain is ineffective. An additional factor that has decreased the effectiveness of detention basins in many areas is the large amount of fill incidental with the development. The placement of effective regional detention and retention, along with improved drainage and conveyance structures and other hydraulic structures may be required to alleviate drainage problems.

1.5.7 Flood Control Alternatives

Flooding results from conditions of hydrology and topography in floodplains such that the flows are large enough that the channel banks overflow, resulting in overbank flow that can extend over the floodplain. For large floods, the floodplain acts both as a conveyance and as a temporary storage for flood flows. The main channel is usually a defined channel that can meander through the floodplain carrying low flows. The overbank flow is usually shallow as compared to the channel flow and also flows at a much slower velocity than the channel flow.

The *objective of flood control* is to reduce or to alleviate the negative consequences of flooding. Alternative measures that modify the flood runoff are usually referred to as flood-control facilities and consist of engineering structures or modifications. Construction of flood-control facilities, referred to as *structural measures,* are usually designed to consider the flood characteristics including reservoirs, diversions, levees or dikes, and channel modifications. Flood-control measures that modify the damage susceptibility of floodplains are usually referred to as *nonstructural measures* and may require minor engineering works. Nonstructural measures are designed to modify the damage potential of permanent facilities and provide for reducing potential damage during a flood event. Nonstructural measures include flood proofing, flood warning, and land use controls. Structural measures generally require large sums of capital investment. *Floodplain management* considers the integrated view of all engineering, nonstructural, and administrative measures for managing (minimizing) losses due to flooding on a comprehensive scale.

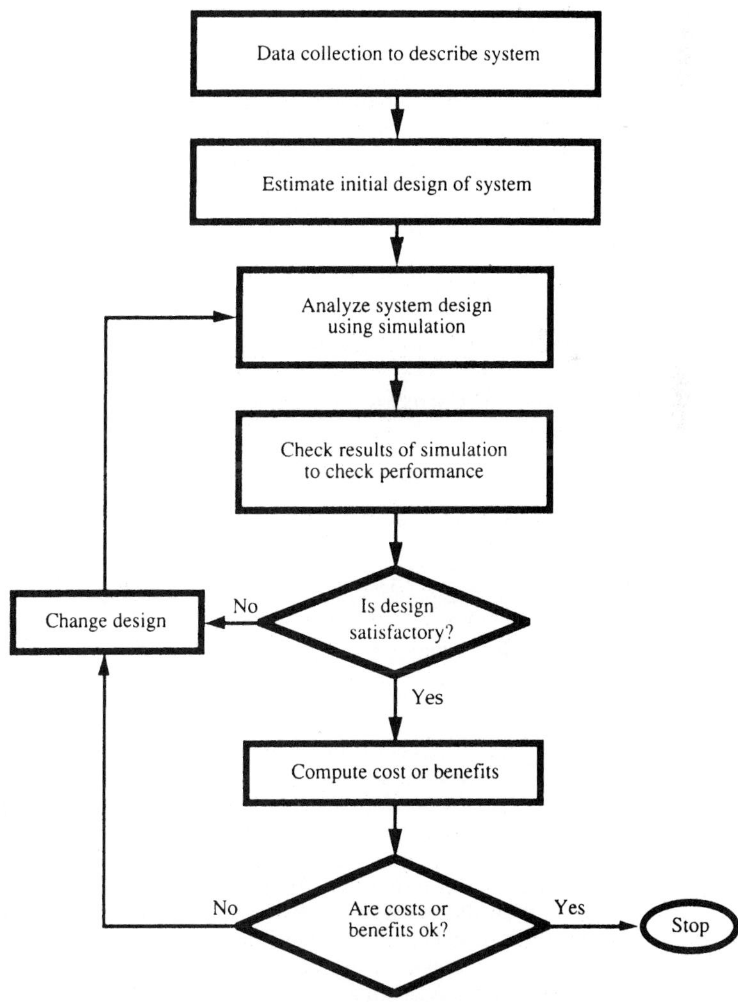

FIGURE 1.31 Conventional design and analysis process. (From Mays and Tung, 1992)

1.6   THE CONVENTIONAL DESIGN PROCESS

Conventional procedures for design are basically iterative trial-and-error procedures. The effectiveness of conventional procedures are dependent upon an engineer's intuition, experience, skill, and knowledge of the hydraulic system. Conventional procedures therefore are closely related to the human element, a factor which could lead to inefficient results for the design and analysis of complex systems. Conventional procedures are typically based upon using simulation models in a trial-and-error process. A procedure may be to iteratively use a simulation model to attempt to arrive at an optimal solution. Figure 1.31 presents a depiction of the conventional design and analysis procedure. As an example, to determine a least-cost pumping scheme for an aquifer dewatering problem would require a selection of pump sizes and location for the aquifer to be dewatered. Using a trial set of pump sizes and locations, a groundwater simulation model is solved to determine if the water levels are lowered below the desirable elevation. If the pumping scheme (pump size and location) does not satisfy the water levels, then a new pumping scheme is selected and simulated. This iterative process is continued, each time determining the cost of the pumping scheme.

Optimization eliminates the trial-and-error process of changing a design and re-simulating with each new design change. Instead, an optimization model automatically changes the design parameters. An optimization procedure has mathematical expressions that describe the system and its response to the system inputs for various design parameters. These mathematical expressions are constraints in the optimization model. In addition, constraints are used to define the limits of the design variables and the performance is evaluated through an objective function, which could be to minimize cost.

An advantage of the conventional process is that the engineer's experience and intuition are used in making conceptual changes in the system or to change or make additional specifications. The conventional procedure can lead to non-optimal or uneconomical designs and operation policies. Also the conventional procedure can be very time consuming. An optimization procedure requires the engineer to explicitly identify the design variables, the objective function of measure of performance to be optimized, and the constraints for the system. In contrast to the decision-making process in the conventional procedure, the optimization procedure is more organized using a mathematical approach to select the decisions.

1.7   THE ROLE OF ECONOMICS IN DESIGN

1.7.1   Engineering Economic Analysis

Engineering economic analysis is an evaluation process that can be used for comparing various hydraulic design alternatives and then applying a discounting technique to select the best alternative. In order to perform this analysis, several basic concepts such as equivalence of kind, equivalence of time, and discounting factors must be understood.

One of the first steps in economic analysis is to find a common value unit such as monetary units. Through the use of common value units, alternatives of rather diverse kinds can be evaluated. The monetary evaluation of alternatives generally occurs over a number of years. Each monetary value must be identified by amount and time. The time value of money results from the willingness of people to pay interest for the use of money. Consequently, money at different times cannot be directly combined or compared, but must first be made equivalent through the use of *discount factors*. Discount factors convert a monetary value at one date to an equivalent value at another date.

Discount factors are described using the notation: i is the annual interest rate; n is the number of years; P is the present amount of money; F is the future amount of money; and A is the annual amount of money. Consider an amount of money P that is to be invested for

n years at i-percent interest rate. The future sum F at the end of n years is determined from the following progression:

	Amount at beginning of year	+	Interest	=	Amount at end of year
First year	P	+	iP	=	$(1 + i)P$
Second year	$(1 + i)P$	+	$iP(1 + i)$	=	$(1 + i)^2P$
Third year	$(1 + i)^2P$	+	$iP(1 + i)^2$	=	$(1 + i)^3P$
	$\vdots$		$\vdots$	$\vdots$	
nth year	$(1 + i)^{n-1}P$	+	$iP(1 + i)^{n-1}$	=	$(1 + i)^nP$

The future sum is then

$$F = P(1 + i)^n \tag{1.12}$$

and the *single-payment compound amount factor* is

$$\frac{F}{P} = (1 + i)^n = \left(\frac{F}{P}, i\%, n\right) \tag{1.13}$$

This factor defines the number of dollars which accumulate after n years for each dollar initially invested at an interest rate of i percent. The *single-payment present worth* factor $(P/F, i\%, n)$ is simply the reciprocal of the single-payment compound amount factor. Table 1.2 summarizes the various discount factors.

   Uniform annual series factors are used for equivalence between present (P) and annual (A) monetary amounts, or between future (F) and annual (A) monetary amounts. Consider the amount of money A that must be invested annually (at the end of each year) to accumulate F at the end of n years. The last value of A in the nth year is withdrawn immediately upon deposit so it accumulates no interest. The future value F is

$$F = A + (1 + i)A + (1 + i)^2A + \cdots + (1 + i)^{n-1}A \tag{1.14}$$

Equation (1.14) is multiplied by $(1 + i)$, and subtract Eq. (1.14) from the result to obtain the *uniform annual series sinking fund factor,*

$$\frac{A}{F} = \frac{i}{(1 + i)^n - 1} = \left(\frac{A}{F}, i\%, n\right) \tag{1.15}$$

The sinking fund factor is the number of dollars A that must be invested at the end of each of n years at i percent interest to accumulate \$1. The *series compound amount factor* (F/A) is simply the reciprocal of the sinking fund factor (Table 1.2), which is the number of accumulated dollars if \$1 is invested at the end of each year. The *capital-recovery factor* can be determined by simply multiplying the sinking fund factor (A/F) by the single-payment compound amount factor (Table 1.2)

$$\left(\frac{A}{P}, i\%, n\right) = \frac{A}{F}\frac{F}{P} \tag{1.16}$$

This factor is the number of dollars that can be withdrawn at the end of each of n years if \$1 is initially invested. The reciprocal of the capital-recovery factor is *the series present*

TABLE 1.2 Summary of Discounting Factors

Type of discount factor	Symbol	Given*	Find	Factor
Single-Payment Factors				
Compound-amount factor	$\left(\dfrac{F}{P}, i\%, n\right)$	P	F	$(1 + i)^n$
Present-worth factor	$\left(\dfrac{P}{F}, i\%, n\right)$	F	P	$\dfrac{1}{(1 + i)^n}$
Uniform Annual Series Factors				
Sinking-fund factor	$\left(\dfrac{A}{F}, i\%, n\right)$	F	A	$\dfrac{i}{(1 + i)^n - 1}$
Capital-recovery factor	$\left(\dfrac{A}{P}, i\%, n\right)$	P	A	$\dfrac{i(1 + i)^n}{(1 + i)^n - 1}$
Series compound-amount factor	$\left(\dfrac{F}{A}, i\%, n\right)$	A	F	$\dfrac{(1 + i)^n - 1}{i}$
Series present-worth factor	$\left(\dfrac{P}{A}, i\%, n\right)$	A	P	$\dfrac{(1 + i)^n - 1}{i(1 + i)^n}$
Uniform Gradient Series Factors				
Uniform gradient series present-worth factor	$\left(\dfrac{P}{G}, i\%, n\right)$	G	P	$\dfrac{(1 + i)^{n+1} - (1 + ni + i)}{i^2(1 + i)^n}$

*The discount factors represent the amount of dollars for the given amounts of one dollar for P, F, A and G.
Source: Mays & Tung, 1992.

worth factor (P/A), which is the number of dollars initially invested to withdraw $1 at the end of each year.

A *uniform gradient series factor* is the number of dollars initially invested in order to withdraw $1 at the end of the first year, $2 at the end of the second year, and $3 at the end of the third year, etc.

1.7.2 Benefit-Cost Analysis

Water projects extend over time, incur costs ($c_0, \ldots, c_n$) throughout the duration of the project, and yield benefits ($b_0, \ldots, b_n$). Typically, costs (c_0) are large during the initial construction an startup period, followed by only operation and maintenance costs. Benefits typically build up to a maximum over time as depicted in Fig. 1.32. The present value of benefits (PVB) and costs (PVC) are, respectively,

$$PVB = b_0 + \frac{b_1}{(1 + i)} + \frac{b_2}{(1 + i)^2} + \cdots + \frac{b_n}{(1 + i)^n} \tag{1.17}$$

and

$$PVC = c_0 + \frac{c_1}{(1 + i)} + \frac{c_2}{(1 + i)^2} + \cdots + \frac{c_n}{(1 + i)^n} \tag{1.18}$$

The present value of net benefits is

$$PVNB = PVB - PVC \tag{1.19}$$

$$= (b_0 - c_0) + \frac{(b_1 - c_1)}{(1 + i)} + \frac{(b_2 - c_2)}{(1 + i)^2} + \cdots + \frac{(b_n - c_n)}{(1 + i)^n}$$

In order to carry out *benefit-cost analysis,* rules for economic optimization of the project design and procedures for ranking projects are needed. The most important point in project planning is to consider the broadest range of alternatives. The range of alternatives selected are typically restricted by the responsibility of the water resource agency and/or the planners. The nature of the problem to be solved may also condition the range of alternatives. Preliminary investigation of alternatives can help to rule out projects because of technical infeasibility or on the basis of costs.

Consider the selection of an optimal, single-purpose project design such as the construction of a flood-control system or a water-supply project. The optimum size can be determined by selecting the alternative such that the marginal or incremental present value of costs, ΔPVC, is equal to the marginal or incremental present value of the benefits, ΔPVB,

$$\Delta PVB = \Delta PVC$$

The *marginal* or *incremental value* of benefits and costs are for a given increase in the size of a project,

$$\Delta PVB = \frac{\Delta b_1}{(1 + i)} + \frac{\Delta b_2}{(1 + i)^2} + \cdots + \frac{\Delta b_n}{(1 + i)^n} \tag{1.20}$$

and

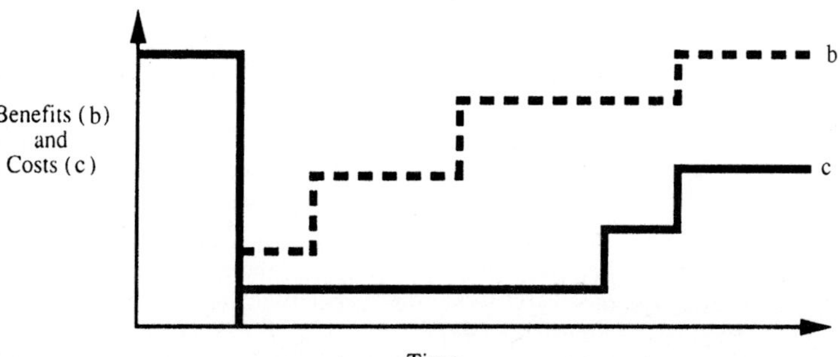

FIGURE 1.32   Benefits and costs over time. (From Mays & Tung, 1992)

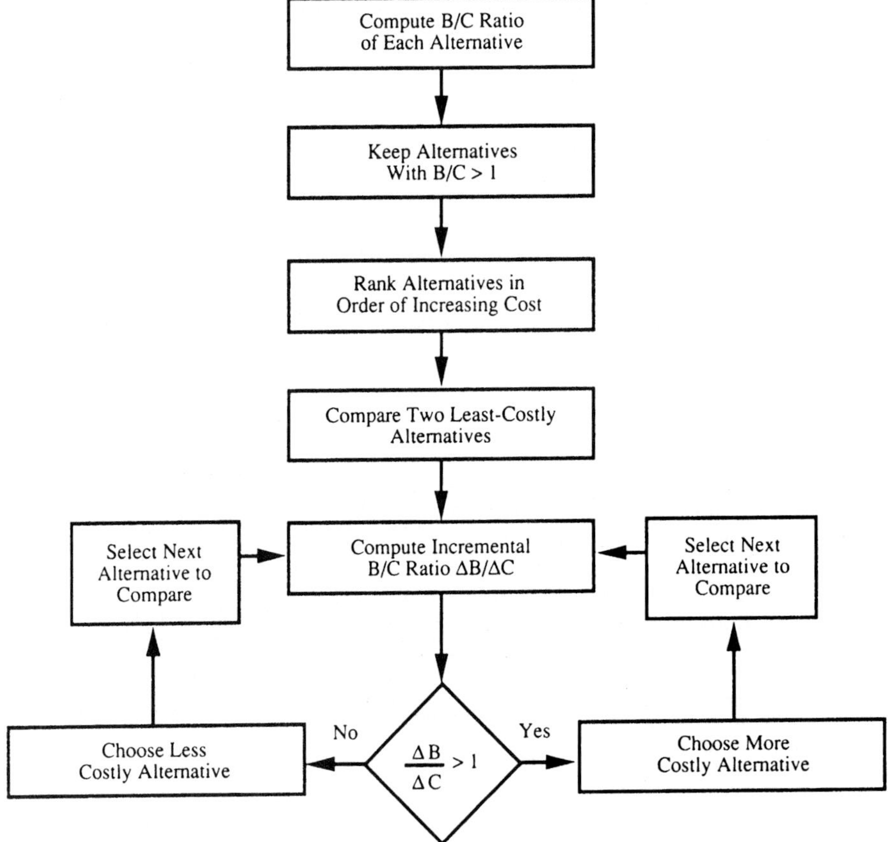

FIGURE 1.33 Flowchart for benefit-cost analysis. (From Mays & Tung, 1992)

$$\Delta PVC = \frac{\Delta c_1}{(1 + i)} + \frac{\Delta c_2}{(1 + i)^2} + \cdots + \frac{\Delta c_n}{(1 + i)^n} \tag{1.21}$$

When selecting a set of projects, one rule for optimal selection is to maximize the present value of net benefits. Another ranking criterion is to use the *benefit-cost ratio* (B/C), PVB/PVC.

$$\frac{B}{C} = \frac{PVB}{PVC} \tag{1.22}$$

This method has the option of subtracting recurrent costs from the annual benefits or including all costs in the present value of cost. Each of these options will result in a different B/C, with higher B/Cs when netting out annual costs, if the B/C is greater than one. The B/C is frequently used to screen infeasible alternatives whose $B/C < 1$ from further consideration.

Selection of the optimum alternative is based upon the incremental benefit-cost ratios, $\Delta B/\Delta C$, whereas the B/C ratio is used for ranking alternatives. The *incremental benefit-cost ratio* is

TABLE 1.3 Lives (in Years) for Elements of Hydraulic Projects

Barges	12	Penstocks	50
Booms, log	15	Pipes	
Canals and ditches	75	Cast iron	
Coagulating basins	50	2–4 in.	50
		4–6 in.	65
Construction equipment	5	8–10 in.	75
Dams		12 in. and over	100
Crib	25	Concrete	20–30
Earthen, concrete, or masonry	150	PVC	40
Loose rock	60	Steel	
Steel	40	Under 4 in.	30
Filters	50	Over 4 in.	40
Flumes		Wood stave	
Concrete or masonry	75	14 in. and larger	33
Steel	50	3–12 in.	20
Wood	25	Pumps	18–25
Fossil-fuel power plants	28	Reservoirs	75
Generators*		Standpipes	50
Above 3000 kva	28	Tanks	
1000–3000 kva	25	Concrete	50
50 hp–1000 kva	17–25	Steel	40
Below 50 hp	14–17	Wood	20
Hydrants	50	Tunnels	100
Marine construction equipment	12	Turbines, hydraulic	35
Meters, water	30	Wells	40–50
Nuclear power plants	20		

  *Alternating-current generators are rated in kilovolt-amperes (kva).
  Source: Linsley et al. (1992).

$$\frac{\Delta B}{\Delta C} = \frac{PVB(A_j) - PVB(A_k)}{PVC(A_j) - PVC(A_k)} \tag{1.23}$$

where $PVB(A_j)$ is the present value of benefits for alternative A_j. Figure 1.33 is a flowchart illustrating the *benefit-cost method*.

1.7.3 Estimated Lives of Hydraulic Structures

The Internal Revenue Service gives estimated average lives for many thousands of different types of industrial assets. The lives (in years) given for certain elements of hydraulic projects are listed in Table 1.3. Such estimates of average lives may be helpful even though they are not necessarily the most appropriate figures to use in any given instance.

1.8   *THE ROLE OF OPTIMIZATION IN DESIGN*

An optimization problem in water resources may be formulated in a general framework in terms of the decision variables (**x**) with an objective function to

$$\text{Optimize } f(\mathbf{x}) \tag{1.24}$$

subject to constraints

$$\mathbf{g(x)} = \mathbf{0} \tag{1.25}$$

and bound constraints on the decision variables

$$\underline{\mathbf{x}} < \mathbf{x} < \bar{\mathbf{x}} \tag{1.26}$$

where $\mathbf{x}$ is a vector of n decision variables (x_1, x_2, ..., x_n), $\mathbf{g(x)}$ is a vector of m equations called *constraints* and $\underline{x}$ and $\bar{x}$ represent the lower and upper bounds, respectively, on the decision variables.

Every optimization problem has two essential parts: the objective function and the set of constraints. The *objective function* describes the performance criteria of the system. *Constraints* describe the system or process that is being designed or analyzed and can be of two forms: equality constraints and inequality constraints. A *feasible solution* of the optimization problem is a set of values of the decision variables that simultaneously satisfy the constraints. The *feasible region* is the region of feasible solutions defined by the constraints. An *optimal solution* is a set of values of the decision variables that satisfies the constraints and provides an optimal value of the objective function.

Depending upon the nature of the objective function and the constraints, an optimization problem can be classified as: linear vs. nonlinear; deterministic vs. probabilistic; static vs. dynamic; continuous vs. discrete; and lumped parameter vs. distributed parameter. *Linear programming* problems consist of both a linear objective function and all constraints are linear. *Nonlinear programming* problems are represented by nonlinear equations, that is, part or all of the constraints and/or the objective function are nonlinear. *Deterministic* problems consist of coefficients and parameters that can be assigned fixed values, whereas *probabilistic* problems consist of uncertain parameters that are considered as random variables. *Static* problems do not explicitly consider the variable time aspect, whereas *dynamic* problems do consider the variable time. Static problems are referred to as mathematical programming problems and dynamic problems are often referred to as optimal control problems, which involve difference or differential equations. *Continuous* problems have variables that can take on continuous values, whereas with *discrete* problems the variables must take on discrete values. Typically discrete problems are posed as *integer programming* problems in which the variables must be integer values. A *lumped problem* considers the parameters and variables to be homogeneous throughout the system, whereas *distributed problems* must take into account detailed variations in the behavior of the system from one location to another. The method of optimization used depends upon: the type of objective function; the type of constraints; and the number of decision variables. Optimization is not covered in this handbook, but is discussed in detail in Mays and Tung (1992).

1.9   THE ROLE OF RISK ANALYSIS IN DESIGN

1.9.1   Existence of Uncertainties

Uncertainties and the consequent related risks in hydraulic design are unavoidable. Hydraulic structures are always subject to a probability of failure in achieving their intended purposes. As an example, a flood control project may not protect an area from extreme floods. A water supply project may not deliver demanded water. This failure may be due to failure of the delivery system or may be due to lack of supply. A water distribution system may not deliver water meeting quality standards even though the source quality does. The rational in the selection of the design and operation parameters and the design and operation standards are continually questioned. Water resource engineering design and operation procedures do not involve any required assessment and quantification of uncertainties and the resultant evaluation of a risk.

Risk is defined as the probability of failure. *Failure* is defined as the event in which system fails to function with respect to its desired objectives. *Reliability* is defined as the complement of risk, i.e. the probability of non-failure. Failure can be grouped into either structural failure or performance failure. A good example of this is for water distribution systems. A *structural failure* such as pipe breakage or pump failure can cause demands not to be met. Also operational aspects of a water distribution system such as the inability to meet demands at required pressure heads is a failure without any structural failure of any component in the system. *Uncertainty* can be defined as the occurrence of events that are beyond our control. The uncertainty of a hydraulic structure is an indeterministic characteristic and is beyond our rigid controls. In the design and operation of these systems, decisions must be made under various kinds of uncertainty.

The sources of uncertainties are many-fold. We will first discuss the ideas of *natural uncertainties, model structure uncertainties, model parameter uncertainties, data uncertainties,* and *operational uncertainties.* Natural uncertainties are associated with the random temporal and spatial fluctuations inherent in natural processes. Model structural uncertainties reflect the inability of a simulation model or design procedure to represent precisely the system's true physical behavior or process. Model parameter uncertainties reflect the variability in the determination of the parameters to be used in the model or design. Data uncertainties include measurement inaccuracy and errors, inadequacy of the data gauging network, data handling and transcription errors. Operational uncertainties are associated with construction, manufacture, deterioration, maintenance, and other human factors that are not accounted for in the modeling or design procedure.

The four major categories of uncertainties are: *hydrologic uncertainty; hydraulic uncertainty; structural uncertainty;* and *economic uncertainty.* Each of these uncertainties has various component uncertainties. Hydrologic uncertainty can be classified into three types: inherent, parameter, and model uncertainties. The occurrence of various hydrological events such as streamflow or rainfall events are considered as stochastic processes because of the observable natural, or inherent, randomness. Because of the lack of perfect hydrological information about these processes or events, there exist informational uncertainties about the processes. These uncertainties are referred to as the parameter uncertainties and the model uncertainties. The model uncertainty in many cases results from the lack of data and knowledge adequate to select the appropriate probability model or through the use of an over-simplified model such as the rational method for storm sewer design.

Hydraulic uncertainty is the uncertainty in the design of hydraulic structures and in the analysis of the performance of hydraulic structures. It mainly arises from three basic types: model, construction and material, and operational conditions of flow. The model uncertainty results from the use of a simplified or an idealized hydraulic model to describe flow conditions, which contribute to the uncertainty in determining the design capacity of hydraulic structures. Simplified relationships such as Manning's equation are typically used to model complex flow processes that cannot be adequately described, resulting in model errors.

Structural uncertainty refers to the failure from structural weakness. Physical failures of hydraulic structures can be caused by water saturation and loss of soil stability, erosion or hydraulic soil failures, wave action, hydraulic overloading, structural collapse, material failure, etc. An example is the structural failure of a levee system either in the levee or in the adjacent soil. The structure failure could be caused by water saturation and loss of soil stability. A flood wave can cause increased saturation of the levee through slumping. Levees can also fail because of hydraulic soil failures and wave action.

Economic uncertainty can arise from uncertainties in construction costs, damage costs, projected revenue, operation and maintenance costs, inflation, project life, and other intangible cost and benefit items. Construction, damage, and operation/maintenance costs are all subject to uncertainties because of the fluctuation in the rate of increase of construction materials, labor costs, transportation costs, economic losses, regional differences, and many others. There are also many other economic and social uncertainties that are related to inconvenience losses. An example of this is the failure of a highway crossing caused by flooding resulting in traffic related losses.

The objective in the analysis of uncertainties is to systematically incorporate the uncertainties into the evaluation of the loading and resistance. The most commonly used method is the first order analysis of uncertainties. These methods are used to determine the statistics of the random variables loading and resistance which are typically defined through the use of deterministic models but have uncertain parameter inputs.

1.9.2   Risk-Reliability Evaluation

1.9.2.1   Load-Resistance.   The *load* for a system can be defined as an external stress to the system and the *resistance* can be defined as the capacity of the system to overcome the external load. Load and resistance are terms that have been used in structural engineering but definitely have a place in the types of risk analysis that need to be performed for water resources engineering projects.

If we use the variable R for resistance and the variable L for load, then we can define a failure as when the load exceeds the resistance and the consequent risk as the probability of the loading exceeding the resistance, $P(L > R)$. A simple example of this would be the failure of a dam due to overtopping. The risk would be the probability that the water surface elevation in a reservoir exceeds the elevation of the top of the dam. In this case, the resistance is the elevation of the top of the dam and the loading is the maximum elevation of the water surface of a flood wave entering the reservoir.

Because many uncertain variables define both the resistance and loading, they are both considered as random variables. A simple example would be to use the rational equation, $Q = CiA$, to define the design discharge (loading) for a storm sewer. The loading, $L = Q$, is a function of three uncertain variables: the runoff coefficient C, the rainfall intensity i, and the drainage area A. Because none of these three variables can be determined with complete certainty, they are considered as random variables. So in this case the loading is a random variable consisting of three random variables. If the resistance is defined through the use of Manning's equation, then the resistance is a function of Manning's roughness factor, the pipe diameter, and the slope (friction slope). The two main contributors to uncertainty in this equation would be the friction slope and the roughness factor so that they are considered as random variables. The resistance is then also a random variable which is a function of the two random variables.

It is interesting to note that in the storm sewer example both the loading and the resistance are defined by deterministic equations, the rational equation and Manning's equation. Both are considered to have uncertain design parameters that result in the resistance and loading being uncertain, and consequently are considered as random variables. In the storm sewer example as in many types of hydraulic structures, the loading uncertainty is actually the hydrologic uncertainty and the resistance uncertainty is the hydraulic uncertainty.

1.9.2.2   Composite Risk.   The above discussion about the hydrologic and hydraulic uncertainties being the resistance and loading uncertainties leads to the idea of a *composite risk*. The probability of failure defined previously as the risk, $P(L > R)$, is actually a composite risk. If only the hydrologic uncertainty, in particular the inherent hydrologic uncertainty, were considered, then this would not be a composite risk. In the conventional design processes of water resources engineering projects, only the inherent hydrologic uncertainties have been considered. Essentially a large return period is selected and artificially considered as the safety factor without any regard to systematically accounting for the various uncertainties that actually exist.

1.9.2.3   Safety Factor.   The *safety factor* is defined as the ratio of the resistance to loading, R/L. Because the safety factor, $SF = R/L$, is the ratio of two random variables, it is also a random variable. The risk can be written as $P(SF < 1)$ and the reliability can be written as $P(SF > 1)$. Using the storm sewer example above, both the resistance and the loading were considered as random variables because they are both functions of random variables. Consequently because the resistance and the loading for the storm sewer design are random variables, the safety factor for storm sewer design would also be a random variable.

1.9.2.4   Risk Assessment.   *Risk assessment* requires several phases or steps which can vary for different types of water resources engineering projects. These steps include:

Step 1: Determine the risk of hazard identification.

Step 2: Assess load and resistance.

Step 3: Perform analysis of uncertainties.

Step 4: Quantify the composite risk.

Step 5: Develop the composite risk-safety factor relationships.

1.9.2.5   A Model for Risk-Based Design.   The *risk-based design* of hydraulic structures promises to be, potentially, the most significant application of uncertainty and risk analysis. The risk-based design of hydraulic structures integrates the procedures of economics, uncertainty analysis, and risk analysis in design practice. Such procedures can consider the tradeoffs among risk, economics, and other performance measures in hydraulic structure design. When risk-based design is embedded into an optimization framework, the combined procedure is called *optimal risk-based design.* The optimal risk-based design approach is the ultimate model for design, analysis and operation of hydraulic structures and water resources projects that we need to strive for in the future.

REFERENCES

Adams, R. M, *Heartland of Cities, Surveys of Ancient Settlements and Land Use on the Central Floodplain of the Euphrates,* University of Chicago Press, 1965.

Aicher, P. J., *Guide to the Aqueducts of Ancient Rome,* Bolchazy-Carducci Publishers, Inc., Wauconda, IL, 1995.

American Society of Civil Engineers, "Basic information needs in urban hydrology," Report to the U.S. Geological Survey, New York, N.Y., April 1969.

American Society of Civil Engineers (ASCE) and Water Environment Federation (WEF), *Design and Construction of Urban Stormwater Management Systems,* Reston, VA, 1992.

Bovey, H. T., *Treatise on Hydraulics,* 2nd ed., Wiley and Sons, New York, Chapman and Hall, London, 1901.

Brandstetter, A. B., *Assessment of Mathematical Models for Storm and Combined Sewer Management,* EPA-600/2-76-175a, (NTIS PB-259597), Environmental Protection Agency, Cincinnati, OH, 1977.

Buckley, R. B., *Irrigation Pocket Book,* 1st ed., Spon., New York, 1911, 2nd ed., 1913.

Chow, V. T., D. R. Maidment, and L. W. Mays, *Applied Hydrology,* McGraw-Hill, Inc,. New York, 1988.

Cohen, P., O. L. Franke, and B. L. Foxworthy, *An Atlas of Long Island's Water Resources,* Bull. 62, p. 57, N. Y. Water Resources Committee, Albany, 1968.

Colyer, P. J., and R. W. Pethick, "Storm Drainage Design Methods," Report INT 154, Hydraulics Research Station, Wallingford, England, March 1976.

Crouch, D. P., *Water Management in Ancient Greek Cities,* Oxford University Press, New York, 1993.

Dooge, J. C. I., "Manning and Mulvaney, River Improvement in 19th Century Ireland," *Hydraulics and Hydraulic Research, A Historical Review,* G. Garbrecht, (ed.), Balkema, Rotterdam and Boston, 1987a, pp. 173–183.

Dooge, J. C. I., "Historical Development of Concepts in Open Channel Flow," *Hydraulics and Hydraulic Research, A Historical Review,* G. Garbrecht, (ed.), Balkema, Rotterdam and Boston, 1987b, pp. 205–230.

Dooge, J. C. I., "The Manning Formula in Context," *Channel Flow Resistance: Centennial of Manning's Formula,* B. C. Yen (ed.), Water Resources Publications, Littleton, CO, 1991, pp. 136–185.

Dougherty, R. L., *Hydraulics,* 1st ed., McGraw-Hill, New York, 1916.

Franke, O. L., and N. E. McClymonds, "Summary of the Hydrological Situation in Long Island, N. Y., as a Guide to Water Management Alternatives," U.S. Geological Survey Professional Paper 622-F, 1972.

Green, W. H., and G. A. Ampt, "Studies on Soil Physics," 4:1 *J. Agric. Sci.* 1–24, 1911.

Gumbel, E. J., "The Return Period of Flood Flows," 12:2 *Ann. Math. Stat,* 163–190, 1941.

Grigg, N. S., *Water Resources Management,* McGraw-Hill, New York, 1996.

Hazen, A., "Storage to be Provided in Impounding Reservoirs for Municipal Water Supply," *Trans. Am. Soc. Civ. Eng.* 1539–1640, 1914.

Hodge, A. T., *Roman Aqueducts and Water Supply,* Gerald Duckworth & Co., Ltd., London, 1992.

Hoggan, D. H., *Computer-Assisted Floodplain Hydrology and Hydraulics,* 2nd ed., McGraw-Hill, Inc., New York, 1997.

Horton, R. E., "The Role of Infiltration in the Hydrologic Cycle," 14 *Trans. Am. Geophys. Union,* 446–460, 1933.

Horton, R. E., "Erosional Development of Streams and Their Drainage Basins: Hydrophysical Approach to Quantitative Morphology," 56 *Bull. Geol. Soc. Am.,* 275–370, 1945.

Hurst, H. E., "Long-term storage capacity of reservoirs," 116(2447) *Trans. Am. Soc. Civ. Eng.,* 770–799, 1951.

James F. Maclaren, Ltd., *Review of Canadian Storm Sewer Design Practice and Comparison of Urban Hydrologic Models, for Canada—U.S. Agreement Task Project 4,* October 1975, Canada Ontario Agreement Project No. 74-8-31, Research Report No. 26, Ottawa, Ontario, 1975.

Kibler, D. F., ed., Urban Stormwater Hydrology, *Water Resources Monograph* 7, American Geophysical Union, Washington, D. C., 1982.

Kiefer, C. J., J. P. Harrison, and T. Hixson, Unpublished Preliminary Report, Chicago Hydrograph Method, July 1970.

King, H. W., *Handbook of Hydraulics,* 1st ed., McGraw-Hill, New York, 1918.

Kuichling, E., "The Relation Between the Rainfall and the Discharge of Sewers in Populous Districts," 20 *Transactions, ASCE,* 1–56, 1889.

Linsley, R. K., *A Critical Review of Currently Available Hydrologic Models for Analysis of Urban Stormwater Runoff, Office of Water Resources Research,* Washington, D. C., 1971.

Linsley, R. K., J. B. Franzini, D. L. Freyberg, and G. Tchobanoglous, *Water Resources Engineering,* McGraw-Hill, Inc., New York, 1992.

Lloyd-Davies, D. E., "The Elimination of Stormwater from Sewerage," 164(2) *Proceedings ICE* 41–67, 1906.

Manning, R., "Observations on Subjects Connected with Arterial Drainage," *Trans. Inst. Civ. Eng. Ireland,* Part II, IV at 1851.

Manning R., *Tables for Calculating the Discharge, Inclination and Dimensions of River Courses and Drains,* Thom, Dublin, 51, 1852.

Manning, R., "Presidental Address to the Institution of Civil Engineers of Ireland, 6th February, 1878," *Trans. Inst. Civ. Eng. Ireland,* vol. 12 at 68–85, 1880.

Manning, R., "On the Flow of Water in Open Channels and Pipes," *Trans. Inst. Civ. Eng. Ireland,* vol. 20, at 161–195, 1889.

Manning, R., "On the Flow of Water in Open Channels and Pipes," *Trans. Inst. Civ. Eng. Ireland,* Supplement to 1889 paper, vol. 24 at 179–207, 1895.

Marsh, W. M., *Earthscape: A Physical Geography,* John Wiley and Sons, Inc., New York, 1987.

Mays, L. W., *Water Resources Engineering,* John Wiley and Sons, Inc., New York, 2001.

Mays, L. W., and Y. K. Tung, *Hydrosystems Engineering and Management,* McGraw-Hill, Inc., New York, 1992.

Metcalf and Eddy, Inc., "Stormwater Management Model," vol. 1, Final Report for the U. S. EPA, Contract No. 14-12-501, Water Pollution Control Research Series 11024DOC07/71, July 1971.

Mulvaney, T. J., "On the Use of Self-registering Rain and Flood Gauges in Making Observations of the Relations of Rainfall and of Flood Discharges in a Given Catchment," *Proc. Inst. Civ. Eng. Ireland,* vol. 4 at 18–31, 1850.

McPherson, M. B., and W. J. Schneider, "Problems in Modeling Urban Watersheds," *Water Resources Research,* vol. 1, no. 3 at 434–440, June 1974.

Parker, P., *The Control of Water,* Van Nostrand, New York, 1913.

Proctor and Redfern Ltd., and James F. MacLaren Ltd., "Stormwater Management Model Study," Vol. I, Final Report, Research Report No. 47, Ministry of the Environment, Environment Canada, Ontario, Canada, September 1976.

Richards, L. A., "Capillary Conduction of Liquids Through Porous Mediums," *Physics, A Journal of General and Applied Physics,* American Physical Society, Minneapolis, Minn., vol. 1 at 318–333, July–Dec. 1931.

Sherman, L. K., "Streamflow from Rainfall by the Unitgraph Method," *Engineering News Record,* vol. 108 at 501–505, 1932.

Stahre, P., and B. Urbonas, *Storm-water Detention for Drainage, Water Quality, and CSO Management,* Prentice-Hall, Englewoods Cliffs, New Jersey, 1990.

Terstriep, M. L., and J. B. Stall, "The Illinois Urban Drainage Area Simulator," *ILLUDAS,* Bull. 58, Illinois State Water Survey, Urbana, IL, 1974.

Torno, H. C., *Stormwater Management Models,* Urban Runoff Quantity and Quality, American Society of Civil Engineers, Proceedings of a Research Conference, Franklin Pierce College, Rindge, New Hampshire, August 1974.

Urbonas, B. R., and L. A. Roesner, Hydrologic Design for Urban Drainage and Flood Control, chap. 28, *Handbook of Hydrology,* D. R. Maidment (ed.), McGraw-Hill, Inc., New York, 1993.

U.S. Army Corps of Engineers (USACE), "HEC-1, Flood Hydrograph Package," Hydrologic Engineering Center, Davis, CA, 1973.

U.S. Army Corps of Engineers (USACE), "HEC-2, Water Surface Profiles," Generalized Computer Programs, Hydrologic Engineering Center, Davis, CA, 1976.

U.S. Army Corps of Engineers (USACE), "Storage, Treatment, Overflow, Runoff Model, STORM, User's Manual," Generalized Computer Program 723-S8-L7520, Hydrologic Engineering Center, Davis, CA, 1977.

U.S. Soil Conversation Service (USSCS), Computer Programs for Project Formulation Hydrology, Technical Release 20, Washington, D.C., 1965.

U.S. Soil Conversation Service (USSCS), WSP2 Computer Program, Technical Release 61, Washington, D.C., 1976.

U.S. National Research Council, Committee on Opportunities in the Hydrologic Sciences, Water Science and Technology Board, *Opportunities in the Hydrologic Sciences,* National Academy Press, Washington, D.C., 1991.

Waananen, A. O., J. T. Limerinos, W. J. Kockelman, W. E. Spangle, and M. L. Blair, "Floodprone Areas and Land-use Planning-selected Examples from the San Francisco Bay Region," U.S. Geological Survey Professional Paper 942, California 1977.

Wanielista, M. P., and Y. A. Yousef, *Stormwater Management,* John Wiley and Sons, Inc., New York, 1993.

Watkins, L. H., "The Design of Urban Sewer Systems," Road Research Technical Paper No. 55, Department of Scientific and Industrial Res., London, H. M. Stationary Office, 1962.

Willcocks, W., and R. Holt, *Elementary Hydraulics,* National Printing Office, Cairo, 1899.

Yen, B. C., and A. O. Akan, "Hydraulic Design of Urban Drainage Systems," chap. 14 in *Hydraulic Design Handbook,* L. W. Mays (ed.), McGraw-Hill, 1999.

CHAPTER 2

Regulation of Stormwater Collection Systems in the United States

Roy D. Dodson
Craig T. Maske
Dodson & Associates Inc.
Houston, Texas

2.1 REGULATION OF STORMWATER COLLECTION SYSTEMS IN THE UNITED STATES

Municipal stormwater collection systems discharge many substances in addition to water. As stormwater collects and flows over pavements, lawns, driveways, and other urban surfaces, it often picks up considerable quantities of pollutants, such as oil and grease, fertilizers, pesticides, and metals. Sediment from active construction sites often discharges into storm drainage systems.

Unaware of the consequences of their actions, private citizens may contribute to stormwater pollution by improper disposal of lawn clippings, used oil or household chemicals. Industrial and commercial facilities may discharge pollutants into stormwater collection systems through cross-connections of storm drains and sanitary sewers. Floor drains may be connected directly into the storm drainage system.

Because of the pollution resulting from these sources, the federal government has created the National Storm Water Program for regulating stormwater discharges throughout the United States. This chapter describes this program and its effects on state and local government agencies and those involved in industrial and construction activities.

2.2 FOCUS OF THIS CHAPTER

This chapter focuses on those practical considerations that should lead to cost-effective compliance with federal, state, and local stormwater pollution control requirements. It is important to note that compliance is a legal term[1] to indicate that a person's actions are not violating any law or regulation. However, most people would like to do more than simply

---

[1] Nothing in this chapter should be construed as legal advice, which should be obtained only from a qualified attorney.

avoid fines and stay out of jail. Therefore, this chapter provides information that should be useful in achieving real improvements in stormwater quality at reasonable costs.

2.3 CURRENT EXTENT OF THE NATIONAL STORM WATER PROGRAM

The *National Storm Water Program* is a federal government initiative, directed by the U.S. Environmental Protection Agency (EPA), with the voluntary cooperation of authorized states and mandatory participation of many local government agencies. The program is being implemented in two major phases, with effective dates of October 1992 and March 2003, respectively. The first phase includes discharges associated with industrial activity (including construction activity) and discharges from all public stormwater collection systems serving urban populations of 100,000 or more. The second phase includes all other public stormwater collection systems within urbanized areas[2] (with certain minor exceptions), plus other small public stormwater collection systems meeting EPA or state criteria for designation.

Outside urbanized areas, all stormwater collection systems serving a population center of at least 10,000 people with a population density of at least 1,000 people per square mile are included in the National Storm Water Program after the full implementation of phase 2. Within urbanized areas, almost all stormwater collection systems, as well as those serving a population of fewer than 10,000 people, are included in the program.

The second phase of the National Storm Water Program also extends the size of construction projects requiring permit coverage. Whereas the minimum amount of soil disturbance that would trigger a permit requirement under phase 1 was 5 acres (2 hectares), this minimum has been reduced to 1 acre (0.4 hectares) in phase 2, for most cases.

The EPA has the authority to require stormwater discharge permits from any discharger that contributes to a violation of a water quality standard or that contributes a substantial load of pollutants to the waters of the United States.[3] The EPA could exercise this authority to extend the National Storm Water Program as needed in the future.

Return flows from irrigated agriculture, agricultural stormwater runoff, and discharges from non-point silvicultural activities are exempt from *National Pollutant Discharge Elimination Systems* (NPDES) permit requirements (40 Code of Federal Regulation (CFR) 122.2; 40 CFR 122.3(e) and (f)).

---

[2] An urbanized area is defined as having a population of at least 50,000 with a minimum population density of 1,000 people per square mile.

[3] The term "waters of the United States" means:

All waters which are currently used, or were used in the past, or may be susceptible to use in interstate or foreign commerce, including all waters which are subject to the ebb and flow of the tide;
All interstate waters including interstate wetlands;
All other waters such as intrastate lakes, rivers, streams (including intermittent streams), mudflats, sandflats, wetlands, sloughs, prairie potholes, wet meadows, playa lakes, or natural ponds, the use, degradation or destruction of which could affect interstate or foreign commerce including any such waters:
Which are or could be used by interstate or foreign travelers for recreational or other purposes; or
From which fish or shellfish are or could be taken and sold in interstate or foreign commerce; or
Which are used or could be used for industrial purposes by industries in interstate commerce;
All impoundments of waters otherwise defined as waters of the United States under this definition;
Tributaries of waters identified in paragraphs (s)(1) through (4) of this section;
The territorial sea;
Wetlands adjacent to waters (other than waters that are themselves wetlands) identified in paragraphs (s)(1) through (6) of this section; waste treatment systems, including treatment ponds or lagoons designed to meet the requirements of CWA (other than cooling ponds as defined in 40 CFR 423.11(m) which also meet the criteria of this definition) are not waters of the United States.

2.4   LEGAL BASIS FOR STORMWATER REGULATIONS

The EPA developed the National Storm Water Program in response to legislation passed by Congress. The most important item of legislation was the *Federal Clean Water Act of 1972* (CWA) (Public Law 92-500), which established the NPDES. The CWA has been amended several times. One important set of amendments was the *Water Quality Act of 1987* (Public Law 100-4) that established the phased approach for stormwater discharge regulation in the United States.

The CWA has been setting the direction of water pollution control in the United States since 1972. The CWA is built on the premise that no one has a right to pollute the waters of the United States. Anyone wishing to discharge pollutants must obtain a permit to do so, and the permit must limit the composition of the discharge and the concentrations of the pollutants in it. Some permit conditions require specified levels of control based on a consideration of technology and cost, regardless of the receiving water's ability to purify itself naturally. However, tighter limits may be imposed, if necessary, to preserve or restore the quality of the water body that receives the discharge.

2.4.1   The Role of State Governments in Stormwater Permitting

The CWA allows states to request EPA authorization to administer the NPDES program within their borders. The EPA must approve a state's request to operate the permit program once the EPA determines that the state has adequate legal authorities, procedures, and the ability to administer the program. The EPA is also obligated to adopt standard requirements for state NPDES programs, including guidelines on monitoring, reporting, enforcement, personnel, and funding, and to develop uniform national forms for use by both EPA and approved states. At all times following authorization, state NPDES programs must be consistent with minimum federal requirements, although the programs may always be more stringent. (However, this does not mean that state permits must be as strict as the corresponding EPA permits.)

At present, more than 40 states have chosen to assume at least some stormwater permitting authority. Within these authorized states, all permit submissions are made to the state agency that administers and enforces the stormwater program. In nonauthorized states, the appropriate EPA regional office is responsible for permitting and permit enforcement.

All states are required to develop water quality standards for waters of the United States within their boundaries. States are required to review their water quality standards at least once every 3 years and, if appropriate, revise or adopt new standards. The minimum elements that must be included in a state's water quality standards include the use designations for all water bodies in the state, water quality criteria sufficient to protect those use designations, and an anti-degradation policy consistent with EPA's water quality standards (40 CFR 131.6).

2.4.2   Role of Local Governments in Stormwater Regulation

The role of local governments in the National Storm Water Program has become very significant. As stated previously, the first phase of the program involved only municipal entities[4] serving urban populations of 100,000 or more. The total number of phase 1 municipal

---

[4] Municipal entities include not only incorporated cities, but counties, special districts, townships, and other state or local government agencies. Any reference to municipal stormwater collection systems in this chapter should be understood to include all municipal entities, not just incorporated cities.

permits is fewer than 300. However, phase 2 requires several thousand additional municipal permits to be issued. Many of the phase 2 municipal dischargers are small local government agencies with limited technical resources.

Municipal dischargers have a very broad set of requirements under the National Storm Water Program. First, they are responsible for obtaining permit coverage for the discharges from their own stormwater collection system, and in meeting various requirements regarding the operation and overseeing that system. Second, they are responsible for obtaining permit coverage for any industrial facilities or construction sites that they own. Finally, they are also responsible for recordkeeping, inspection, and enforcement of stormwater permit requirements for construction activities and certain types of industrial operations within their jurisdiction.

The two primary types of public stormwater collection systems in the United States are separate systems and combined systems. Most cities use separate systems which are designed to carry only stormwater runoff and other wet weather flows. However, over 30 of the oldest cities in the United States rely on *Combined Sewer Systems* (CSS). A CSS carries sanitary sewage under dry weather conditions, but is surcharged with runoff under storm conditions. All of the requirements discussed in this chapter pertain to separate storm sewer systems only; CSS are regulated under another program.

2.5 TECHNOLOGY-BASED REQUIREMENTS OF STORMWATER DISCHARGE PERMITS

Stormwater permits are intended to achieve improvements in water quality by reducing or eliminating pollutant loadings from stormwater sources. The exact requirements for attaining this goal depend upon the type of permit, the type of discharge, and the characteristics of the body of water that receives the discharge.

Technology-based requirements represent the minimum level of control that must be imposed by an NPDES permit. The *best conventional technology* (BCT) standard applies to the control of conventional pollutants, while the *best available technology* (BAT) standard applies to the control of all toxic pollutants and all pollutants that are neither toxic nor conventional pollutants. BCT and BAT standards are generally applied to stormwater discharges associated with industrial or construction activity. These requirements are met by the development and implementation of *Best Management Practices* (BMPs) and pollution prevention measures as a part of a stormwater pollution prevention plan for the industrial facility or construction site.

Two technology-based standards have been established for discharges from public stormwater collection systems. The first standard provides that municipal permits must contain a requirement to effectively prohibit illicit non-stormwater discharges into the system. The other standard requires that permits for discharges from public stormwater collection systems reduce the discharge of pollutants to the *maximum extent practicable* (MEP), including management practices, control techniques, and system, design, and engineering methods.

2.5.1 Stormwater Discharge Permits and State Water Quality Standards

In addition to technology-based controls, *NPDES permits* must include any conditions more stringent than technology-based controls necessary to meet state water quality standards. *Water quality standards* establish the "goals" for a water body. The CWA states the national goal of achieving "water quality which provides for the protection and propagation of fish, shellfish, and wildlife and . . . recreation in and on the water," wherever attainable. These national goals are commonly referred to as the fishable/swimmable goals of the CWA. The

EPA requires that water quality standards provide for fishable/swimmable uses, unless those uses have been shown to be unattainable.

Scientific studies are performed to establish the *Total Maximum Daily Load* (TMDL) of a particular pollutant that is allowable without violation of the water quality standard. If TMDL studies indicate that too much of a particular pollutant is entering the stream system, then any discharge permit within that stream system may be subject to revision in order to lower the pollutant levels to the TMDL value.

2.5.2 Municipal Stormwater Discharge Permit Requirements

The EPA has identified six minimum control measures that are always necessary for municipal stormwater dischargers to comply with the statutory requirements of eliminating illicit non-stormwater discharges and reducing pollutant loading to the maximum extent possible:

1. Public education and outreach on stormwater impacts
2. Public involvement and participation
3. Illicit discharge detection and elimination
4. Construction site stormwater runoff control
5. Post-construction stormwater management in areas of new development and redevelopment
6. Pollution prevention and good housekeeping for municipal operations

Additional requirements are often imposed for larger systems. These include spill prevention and response, monitoring of wet weather flows and dry weather flows, and special inspection and enforcement requirements for high-risk industrial dischargers that contribute flows to the public drainage system.

Judgment is required in order to determine the best combination of control measures for a particular municipal stormwater collection system. The selection of control measures should consider such factors as the conditions of receiving waters, specific local concerns, and other aspects included in a comprehensive watershed plan. Various municipal entities may choose to cooperate in the development and implementation of the minimum control measures.

2.5.2.1 Types of Permits Required for Municipal Dischargers

Separate types of stormwater discharge permits are used for municipal stormwater discharges, for *construction stormwater discharges,* and for *industrial stormwater discharges.* Therefore, industrial facilities or construction sites that discharge into a permitted municipal separate storm sewer system are still required to obtain EPA or state permit coverage for the facility's discharge. This is true even if the industrial facility or construction site is operated by the same agency that operates the public drainage system. Therefore, many local and state government agencies should obtain two or three different types of stormwater discharge permits: one for the public drainage system as a whole, and separate permits for each industrial facility and construction site operated by the agency.

2.5.3 Public Education about Municipal Stormwater Discharges

Public education is essential to the success of a municipal stormwater pollution control effort because the actions of the public are important in determining the level of pollutants present in stormwater discharges. The public can reduce pollutant discharges by properly maintaining septic systems to reduce overflows and other discharges, by using lawn and garden chemicals sparingly and carefully, and by properly disposing of used motor oil or household hazardous wastes. However, the public generally will not be sufficiently informed or motivated to carry out these actions without public education and outreach efforts.

It is estimated that 15% to 20% of household hazardous wastes end up in storm drains or runoff (King County Solid Waste Division, 1990). In addition, EPA estimates that 40% (80 million gal/yr) of "do-it-yourself oil changers" pour oil onto roads, driveways, or yards or into storm sewers (EPA, 1993). Significant amounts of fertilizers and pesticides enter the water from lawn maintenance and landscaping activities.

Public education programs should have the following goals:

1. Public reporting of stormwater problems

2. Proper disposal of used oil and household hazardous wastes

3. Proper use of broadcast chemicals

In addition to educating the public about modifications it can make in its own actions to reduce pollutant discharges, municipalities can reach out to members of the public to participate in community activities that further reduce pollution. These activities might include such services as roadside litter pickup and storm drain stenciling.

Some public education and outreach efforts should be directed toward targeted groups of commercial, industrial, and institutional entities likely to have significant stormwater impacts. The operators of these types of entities (such as restaurants or automotive repair shops) should be given the opportunity to learn about and understand the impacts of their discharges on stormwater pollution.

2.6 ALLOWABLE NON-STORMWATER DISCHARGES

Municipal stormwater management plans must consider non-stormwater discharges. There are two basic categories of non-stormwater discharges: allowable and illicit. Allowable non-stormwater discharges may be mixed with stormwater and discharged without creating a permit violation.

Some types of non-stormwater discharges are relatively innocuous and could be allowable unless they contribute substantial quantities of pollutants. The following non-stormwater discharges are classified as allowable under the EPA's final rules for phase 2 of the National Storm Water Program:

- Water line flushing
- Landscape irrigation
- Diverted stream flows
- Rising groundwater
- Uncontaminated groundwater infiltration (as defined at 40 CFR 35.2005(20)) to separate storm sewers
- Uncontaminated pumped groundwater
- Discharges from potable water sources
- Foundation drains
- Air-conditioning condensation
- Irrigation water
- Springs
- Water from crawl space pumps
- Footing drains

- Lawn watering
- Individual residential car washing
- Flows from riparian habitats and wetlands
- Dechlorinated swimming pool discharges
- Street wash water
- Discharges or flows from fire fighting

The fact that these discharges are "allowable" does not imply that any pollutants present in these discharges may be disregarded. As noted previously, every discharger is obligated to reduce pollutant discharges to the maximum extent practicable. Therefore, all practicable means should be employed to reduce the pollutant levels in allowable stormwater discharges.

2.7 ILLICIT NON-STORMWATER DISCHARGES

Illicit discharges include all discharges of potential pollutants to waters of the United States that are not covered by a currently effective NPDES permit. Illicit discharges are not necessarily harmful; the term describes the legal status of the discharge, not the quality or quantity.

Illicit discharges enter the system through either direct connections (e.g., wastewater piping either mistakenly or deliberately connected to the storm drains) or indirect connections (e.g., infiltration into the storm drain system or spills collected by drain inlets).

One of the statutory requirements of the federal *Water Quality Act of 1987* is the elimination of illicit discharges from municipal separate storm sewer systems. Three major steps are generally required to deal successfully with illicit non-stormwater discharges:

1. Make sure that there is adequate knowledge of the public drainage system. This knowledge should be recorded in the form of system maps and related documents showing the location of major pipes, outfalls, and topography and the areas of concentrated activities that are likely to be a source of stormwater pollution.
2. The legal prohibition of such discharges, to the full extent of available government authority.
3. Informing key individuals about their responsibilities to properly dispose of wastes. These key individuals may include public employees, business owners and managers, and the general public.

2.8 MUNICIPAL OPERATIONS AND STORMWATER QUALITY

Municipal operations must be included in a municipal stormwater management program. These operations should include an effective operation and maintenance program, and adequate training for municipal employees and contractors, to prevent or reduce pollutant runoff from municipal operations. The plan should include at least the following elements:

- Maintenance activities, maintenance schedules, and long-term inspection procedures for structural and other stormwater controls
- Controls for reducing or eliminating the discharge of pollutants from streets, roads, highways, municipal parking lots, maintenance and storage yards, and waste transfer stations, including programs that promote recycling
- Programs to promote the minimal use of pesticides

- Procedures for the proper disposal of waste removed from the separate storm sewer systems and related areas, including dredge spoil, accumulated sediments, floatables, and other debris

- Methods of ensuring that new flood management projects assess the impacts on water quality and examine existing projects for incorporation of additional water quality protection devices or practices

The program should include local government employee training addressing these prevention measures in government operations, such as park, golf course, and open-space maintenance; fleet maintenance; and planning, building oversight, and stormwater system maintenance.

In areas where salt is used, reduced application or alternative agents, consistent with the need for safety, will reduce pollution of area water bodies. Storage facilities can be constructed or modified to prevent salt exposure to rainfall. Additional street sweeping should be scheduled after periods of freezing weather in order to remove road salt and sand from bridges and roadways as soon as feasible.

2.8.1   Municipal Permit Requirements for Stormwater Management in Areas of New Development and Redevelopment

Most municipal permits require that some form of structural BMPs be applied to new development or redevelopment. These structural BMPs may include storage practices (wet ponds and extended-detention outlet structures), filtration practices (grassed swales, sand filters, and filter strips), and infiltration practices (infiltration basins, infiltration trenches, and porous pavement). Because of constant improvements in these types of control measures, the municipal entity should allow flexibility in the requirements for implementation of these BMPs.

Not all BMPs for control of pollutant discharges from new construction and redevelopment are necessarily structural BMPs. Nonstructural BMPs may also be incorporated. These may include requirements to limit growth to identified areas, protect sensitive areas such as wetlands and riparian areas, minimize imperviousness, maintain open space, and minimize disturbance of soils and vegetation.

One of the major obstacles to the implementation of long-term stormwater pollution control measures in developed areas has been the lack of an effective method for ensuring the adequate long-term operation and maintenance of BMPs. The EPA has announced its intention to develop and publish guidelines in this area (63 FR 01735 (January 9, 1998)).

The primary guideline for the selection of stormwater pollution control measures for new development and redevelopment is that the water quality effects of the development should not be significantly different from the water quality effects of the same site before development.

2.8.2   Municipal Permit Requirements for Construction Site Stormwater Runoff Control

Construction sites can pose special problems for municipal stormwater collection systems. In a short time, stormwater discharges from construction site activity can contribute more pollutants, including sediment, to a receiving stream than had been deposited over several decades. Stormwater runoff from construction sites can include pollutants other than sediment, such as phosphorus and nitrogen from fertilizer, pesticides, petroleum derivatives, construction chemicals, and solid wastes that may become mobilized when land surfaces are disturbed.

Municipal stormwater discharge permits generally require that the municipal entity exercise some controls over stormwater discharges from construction sites within their jurisdiction. Usually, these construction sites will be subject to the requirements of their own stormwater discharge permit. Therefore, the responsibility of the municipal entity is often nothing more than ensuring that the construction operator is abiding by the terms of the stormwater discharge permit issued by the state or the EPA. However, because of the relatively large number of construction projects in the United States, many municipalities may find it difficult to manage the recordkeeping, inspection, and enforcement activities required in connection with construction sites. The municipal entity must implement a stormwater management plan that includes, at a minimum:

- Requirements for construction site owners or operators to implement appropriate BMPs, such as silt fences, temporary detention ponds, and hay bales
- Provisions for preconstruction review of site management plans
- Procedures for receipt and consideration of information provided by the public
- Regular inspections during construction
- Penalties to ensure compliance

More details on construction permit requirements are presented later in this chapter.

2.8.3 Special Requirements for Larger Municipal Dischargers

Certain control measures are generally applied to only the larger municipal entities. These include special requirements for inspections and monitoring of industrial discharges into the municipal drainage system, stormwater sampling and monitoring, and spill prevention and response.

Larger municipal permits often contain a requirement that the system operator identify and control pollutants that enter the public stormwater collection system from industrial and high-risk facilities such as municipal landfills; other treatment, storage, or disposal facilities for municipal waste (e.g., transfer stations, incinerators); hazardous waste treatment, storage, disposal, and recovery facilities and facilities that are subject to *Superfund Reauthorization Act* (SARA) Title III, section 313; and any other industrial or commercial discharges the permittee determines are contributing a substantial pollutant loading to the public stormwater collection system.

Large municipal permits also generally include a requirement to perform monitoring to estimate the quantity of pollutants discharged from the entire stormwater collection system. The system operator is required to monitor in wet weather a specified number of times per year and to screen the system in dry weather to locate illicit discharges. The purpose of the monitoring requirement is to provide data necessary to assess the effectiveness and adequacy of stormwater management plan (SWMP) control measures; estimate annual cumulative pollutant loadings from the MS4; estimate event mean concentrations and seasonal pollutants in discharges from major outfalls; identify and prioritize portions of the MS4 requiring additional controls; and identify water quality improvements or degradation (40 CFR 122.26(d)((2)(iii)(C) and (D)).

2.9 CONSTRUCTION STORMWATER DISCHARGES

Construction activities produce many different kinds of pollutants that may cause stormwater contamination problems. Grading activities remove grass, rocks, pavement, and other protective ground covers, resulting in the exposure of underlying soil to the elements. Because the soil surface is unprotected, soil and sand particles are easily picked up by wind and/or washed away by rain or snowmelt. The water carrying these particles eventually reaches a stream, river, or a lake where the water slows down, allowing the particles to fall onto the bottom of the streambed or lake. Gradually, layers of these clays and silt build up in the stream beds, choking the river and stream channels and covering the areas where fish spawn and plants grow. These particles also cloud waters, causing aquatic respiration problems, and can kill fish and plants growing in the river stream.

Sediment runoff rates from construction sites are typically 10 to 20 times those of agricultural lands and 1,000 to 2,000 times those of forest lands. Even a small amount of construction may have a significant negative impact on water quality in localized areas. Over a short time, construction sites can contribute more sediment to streams than was deposited previously over several decades.

In addition, the construction of buildings and roads may require the use of toxic or hazardous materials, such as petroleum products, pesticides, fertilizers, and herbicides, and building materials, such as asphalt, sealants, and concrete, which may pollute stormwater running off the construction site. Demolition of existing structures may cause the exposure of other toxic substances. These materials can be toxic to aquatic organisms and can degrade water for drinking and water-contact recreational purposes.

2.9.1 Construction Activities Requiring Stormwater Permit Coverage

Under phase 1 of the National Storm Water Program (beginning in 1992), the EPA regulates stormwater discharges from construction sites, including clearing, grading, and excavation activities, if the disturbed land area is 5 acres (2 hectares) or more. Construction sites with disturbed areas of fewer than 5 acres (2 hectares) may also be regulated, if they are part of a "larger common plan of development or sale."

Under phase 2 of the National Storm Water Program (beginning in 2003), the EPA will regulate stormwater discharges from construction sites, if the disturbed land area is 1 acre (0.4 hectares) or more.

Regulated construction activities may include road building; construction of residential houses, office buildings, or industrial buildings; demolition; and other activities. The key factor is the surface area of disturbed soil. At a demolition site, disturbed areas might include where building materials, demolition equipment, or disturbed soil is situated, which may alter the surface of the land (EPA, 1993).

2.9.2 Avoiding Construction Stormwater Permit Requirements

Stormwater discharge permits are not required for construction sites under 5 acres if there is no significant risk of erosion during the period of construction. Research by the US Department of Agriculture has established a relationship between rainfall characteristics and erosion potential. The *"rainfall erosivity factor,"* or R-value, has been developed. The R-value reflects the total amount of kinetic energy available in rainfall. Therefore, the R-value increases with the average annual rainfall. However, it also increases for areas in which the average raindrop size and speed increase. In many locations, rainfall erosivity is highly seasonal. Refer to chapter 17 for more detailed discussion.

Under the EPA final rules for phase 2 of the National Storm Water Program, any project that is smaller than 5 acres and expected to receive fewer than 5 units of total rainfall erosivity is not required to obtain stormwater discharge permit coverage for construction activities. In determining whether a particular project is eligible for this exemption, the following steps are required:

1. Determine the acreage of soil disturbance due to the project. If this acreage exceeds 5 acres (2 hectares), permit coverage is required regardless of the rainfall erosivity estimates.
2. For projects less than 5 acres (2 hectares), the total project duration and project start date should be estimated.
3. The total rainfall erosivity should be estimated using USDA methods, considering the project location, start date, and duration. If the rainfall erosivity exceeds 5 units, a stormwater discharge permit is required.

There are certain locations in the desert southwest in which the annual average rainfall erosivity is only about 5 units. In these locations, a stormwater discharge permit for construction activities would be required for small projects only if the project duration exceeded one year.

2.9.3 Industrial and Construction Permit Coverage

If a construction activity is undertaken at an industrial facility that already holds a permit for industrial stormwater discharges, a separate permit must be obtained for the construction activity (unless the construction activity falls below the minimum acreage requirement). Maintenance activities for flood control channels or roadside ditches (such as removal of vegetation) require stormwater discharge permit coverage if they involve grading, clearing, or excavation activities that exceed the minimum acreage cutoff, either individually or as part of a long-term maintenance plan (EPA, 1993).

2.9.4 Types of Construction Permits Available

In most states, two types of NPDES permits are available for stormwater discharges from construction sites: an individual permit and a general permit. An individual permit is specifically developed for only one discharger at one location. A general permit is developed for an entire class of dischargers at many locations. Most traditional NPDES permits have been individual permits. The EPA has used general permits as a tool to accommodate the large number of dischargers included in the National Storm Water Program.

Most general permits require the submission of a Notice of Intent (NOI), which states the permittee's desire to discharge according to the terms and provisions of the general permit. Compliance with the provisions of the general permit involves the preparation and maintenance of a *stormwater pollution prevention plan* (SWPPP).

Almost all construction activities that require NPDES permit coverage for stormwater discharges can be covered under a general permit. However, the EPA or state government administering the NPDES permit program has the right to require that a construction project submit an application for an individual permit under certain conditions. General permit coverage is often unavailable for projects that have adverse effects on endangered species or on historic properties, for example. These conditions are most likely to be encountered in large, complex, and controversial construction projects.

2.10 IDENTIFYING THE OPERATOR OF A CONSTRUCTION FACILITY

The "operator" of a discharge of stormwater associated with construction activity is required to obtain coverage under an NPDES permit. Therefore, it is important to understand that

EPA defines *operator* to mean any party associated with a construction project that meets either of the following two criteria:

1. Anyone who has operational control over construction plans and specifications, including the ability to make modifications to those plans and specifications, is an operator. Since the owner of the site generally has the ability to change the construction specifications, this definition tends to include the owner. This definition does not include government agencies or other organizations that develop or enforce building codes or other common design standards. In general, the owner is considered to be the party that owns the structure being built, but not necessarily the land on which the construction is occurring.

2. Anyone who has day-to-day operational control of those activities at a project which are necessary to ensure compliance with a stormwater pollution prevention plan for the site or other permit conditions. Since the general contractor usually maintains day-to-day control of the site activities, this definition tends to include the general contractor. Subcontractors and utility service companies and their subcontractors are not generally required to obtain stormwater permit coverage because they do not generally have enough operational control of site activities.

Most construction projects are covered by a general permit. The operators of the permit (for a particular construction site) are the project owner and the general contractor, and these operators are "co-permittees" who share permit obligations, at least in the view of the EPA. Several states have different requirements concerning the operator. Some states name the owner as the operator while others name the general contractor; some follow the EPA example and require them to become co-permittees.

In general, an owner can avoid permit requirements only by negotiating a "turnkey design-build" contract in which the general contractor develops and implements all design features and specifications. In such a project, the owner must effectively give up all operational control over site plans and specifications, including the ability to modify the plans and specifications during construction. However, most owners are involved in the construction process, at least to a certain extent. The involvement may be minimal, consisting of occasional site visits and review of submissions from the general contractor.

2.11 THE "COMMON PLAN OF DEVELOPMENT OR SALE"

For sites disturbing fewer than 5 acres (2 hectares), the first two steps in deciding whether a permit is needed for stormwater discharges associated with construction activity are to determine the following:

1. Is there a "common plan of development or sale" tying individual sites together? For example, if the lots are part of a subdivision plat filed with the local land-use planning authority, then this would be a "common plan."

2. Will the total area disturbed by all the individual sites (including the cumulative total disturbance necessary to completely build out the subdivision) come to 5 acres (2 hectares) or more?

If the answer to both questions is no, a stormwater discharge permit is not needed unless the EPA determines that discharges contribute to a violation of water quality standards or are a significant contributor of pollutants to waters of the United States and specifically requests a permit application.

The plan in a common plan of development or sale is broadly defined as any announcement or piece of documentation (including a sign, public notice or hearing, sales pitch, advertisement, drawing, permit application, zoning request, computer design, etc.) or physical demarcation (including boundary signs, lot stakes, surveyor markings, etc.) indicating that construction activities may occur on a specific plot.

2.12   *FUTURE CONSTRUCTION*

Once a residence or commercial building has been completed and occupied by the owner (or tenant), future activities by the owner on the individual's property are not considered part of the original common plan of development. For example, after a home is occupied by the homeowner or tenant, future construction activity on that particular lot is considered a new and distinct project and is compared to applicable disturbed area limits for permit applicability.

In many cases, a common plan of development or sale consists of many small construction projects that collectively come to 5 acres (2 hectares) of total disturbed land. After this initial plan is completed for a particular parcel, any subsequent development or redevelopment of that parcel will be regarded as a new plan of development, and will then be subject to the 5-acre (2-hectare) cutoff for stormwater permitting purposes.

Infill development occurs after most of a subdivision or similar development is mostly completed, but when isolated residential or commercial lots remain vacant. If the total area of these remaining lots that will be disturbed is less than 5 acres (2 hectares), then no permit coverage is required for stormwater discharges occurring during construction on these lots. However, if the total area of expected disturbance on the remaining lots is greater than 5 acres (2 hectares), then construction on any of the remaining lots must be covered by a stormwater discharge permit.

2.12.1   Support Activities at Off-site Areas

Off-site areas are commonly used for storage of fill material or soil excavated from the construction site, borrow areas to obtain fill material, storage of building materials, concrete or asphalt batch plants, or storage of construction equipment. Where activities at off-site locations would not exist without the construction project, discharges of pollutants in stormwater from these areas must be controlled. The pollution prevention plan for the construction project must include controls for all temporary off-site areas directly supporting the construction project.

2.13   *NOTICE OF INTENT (NOI) SUBMISSION REQUIREMENTS*

The EPA requires that the *notice of intent* (NOI) be postmarked at least 2 days before permit coverage is required. Some states have different deadlines and requirements. Only one NOI is required to cover all the activities of one operator on any one common plan of development or sale.

Before the NOI form can be submitted, the SWPPP must be completed to ensure that appropriate controls to meet *Endangered Species Act* (ESA) and *National Historic Preservation Act* (NHPA) certification requirements, if needed, to avoid or mitigate adverse effects to listed endangered or threatened species, critical habitat, or historic properties.

2.13.1   Authorized Signatories of Certification Forms

The NOI and most other forms used in the National Storm Water Program have specific requirements concerning the requirements for the person signing the form.

In general, the person signing the NOI must have a sufficient level of authority and responsibility within the organization to help ensure compliance with the terms and conditions of the permit. For a sole proprietorship, the proprietor must sign the NOI. For a partnership, a general partner must sign.

For a corporation, the person signing the NOI must be a responsible corporate officer, which includes the president, secretary, treasurer, or vice-president of the corporation in charge of a principal business function or any other person who performs similar policy- or decision-making functions for the corporation.

A corporate plant or facility manager may sign the NOI only under certain conditions. The plant or facility must employ more than 250 persons or have gross annual sales or expenditures exceeding $25 million (in second-quarter 1980 dollars). In addition, the authority to sign documents must have been assigned or delegated to the plant or facility manager in accordance with corporate procedures.

For a municipality, state, federal, or other public agency, the person signing the NOI must be a principal executive officer or ranking elected official.

2.14   ENDANGERED SPECIES ACT (ESA) REQUIREMENTS

The EPA is required to comply with the *Endangered Species Act* (ESA) of 1966. Part of the EPA's obligations under the ESA is to consider the effects of any permits that are issued by the EPA on endangered or threatened species (known as listed species) and their critical habitat. This type of consideration must be done on a site-specific basis. Therefore, anyone who applies for coverage of a stormwater discharge under the EPA construction general permit must supply the EPA with site-specific information that will provide some assurance that the possible effects of the discharge on listed species and their critical habitat have been adequately considered. Impacts to listed species and critical habitat can occur from development and construction even on fully developed sites. Often, the impacts occur at the point of discharge into surface waters.

In cooperation with the Fish and Wildlife Service (FWS) and the National Marine Fisheries Service (NMFS), the EPA has developed a six-step procedure for addressing the possible effects of a construction project on listed species and their critical habitat:

1. Is the construction site within a critical habitat area?
2. Are there listed species in the project county or counties?
3. Are listed species present in the project area?
4. Are listed species or critical habitat likely to be adversely affected?
5. Can adverse effects be avoided?
6. Can the proposed project meet minimum eligibility requirements?

The assistance of a qualified biologist is often essential in order to properly deal with these issues.

There is the possibility that several operators may apply for and receive permit coverage for stormwater discharges from the same construction project. In this case, the first operator (often the owner, developer, or general contractor) may have performed all the work necessary to allow ESA certification of all the stormwater discharges from the entire project, even those stormwater discharges that are under the authority of a different operator later in the construction process. If this is the case, then the later operator may achieve ESA certification by simply agreeing to comply with any measures or controls upon which the initial operator's certification was based. However, the initial operator's certification must apply to the later operator's project area and must address the effects from the stormwater discharges and stormwater discharge-related activities on listed species and critical habitat.

Later applicants or permittees may be liable for inadequacies or falsehoods in that certification. Thus, it important for those applicants who choose to rely on another operator's certification to carefully review that certification and its SWPPP for accuracy and completeness.

2.15 *CONSTRUCTION STORMWATER POLLUTION PREVENTION PLAN*

General permits for construction stormwater discharges require that a *Stormwater Pollution Prevention Plan* (SWPPP) be completed for each construction project for which an NOI has been submitted. The SWPPP must be completed and ready to implement at the time the project begins construction. The plan must be signed. A notice should be posted near the entrance of the construction site, listing the NPDES permit number, the name and telephone number of a local contact, a brief description of the project, and the location of the SPPP, if not on-site.

2.15.1 Site Description

The SPPP should include a site description covering the following items, if applicable:

- Description of the construction activity
- Sequence of major soil disturbing events
- Total area and disturbed area acreage
- Runoff coefficient pre/post construction
- Location of industrial discharges
- Name of receiving water(s)
- Copy of the permit
- Endangered or threatened species information
- Historical places information

2.15.2 Location and Site Map

A location map and site map must be included in the SWPPP indicating drainage patterns, approximate slopes, areas of soil disturbance, areas of no soil disturbance, locations of major controls, structural practices shown, stabilization practices, off-site materials storage, waste disposal areas, borrow areas, equipment storage areas, surface waters (including wetlands), and discharges into surface waters.

2.15.3 Erosion and Sediment Controls

Erosion and sediment controls must include interim and permanent stabilization practices including temporary vegetation, permanent vegetation, mulching, geotextiles, sod stabilization, vegetative buffer strips, protection of trees, and preservation of mature vegetation. In each case, the selected stabilization practices should be described along with the reasons for its selection.

Structural practices are used to divert flows from exposed soils and to recapture a portion of the sediment carried by runoff. These practices include silt fences, earth dikes, drainage, swales, sediment traps, check dams, subsurface drains, pipe slope drains, level spreaders, storm drain inlet protection, rock outlet protection, reinforced soil retaining systems, and temporary or permanent sediment basins.

2.15.4 Permanent Stormwater Management

The SPPP must include a description of measures to be installed to control pollutants in stormwater discharges that will occur after construction has ended. These may include stormwater detention structures (including wet ponds); stormwater retention structures; flow attenuation by use of open vegetated swales and natural depressions; infiltration of runoff on-site, and sequential systems (more than one combined).

2.15.5 Other Controls

The SWPPP must also deal with the potential pollution from construction and waste materials stored on site. The SWPPP should include a spill prevention and response plan or a reference to one that is readily available.

It is important to ensure consistency with state, tribal, or local requirements. The SWPPP must be updated as necessary when state, tribal, or local official approve changes and give a written notice.

BMPs must be maintained in effective operating condition; any repairs must be made before the next rain event or as soon as practicable.

2.15.6 Inspections

Inspections are crucial to stormwater compliance, and it is very important to perform all inspections on schedule and maintain full, written records of the inspection and all follow-up actions. Inspections are required every 14 calendar days and within 24 hours of .5 in or greater rainfall. Inspections should include all the following areas:

- Disturbed areas and storage areas exposed to precipitation
- Sediment and erosion controls
- Discharge points (looking for visible signs of erosion and impact to receiving waters)
- Entrances and exits (looking for evidence of off-site sediment tracking)

After each inspection, the SWPPP should be modified as necessary, including the following information on the inspection form:

- Date of inspection and major observations
- Location of any discharge off-site
- Location of BMPs that need maintenance
- Location of BMPs that failed to work
- Location where new BMPs are needed
- Certification of compliance or non-compliance
- Inspector's signature

2.16 CONSTRUCTION PERMIT NOTICE OF TERMINATION

In most states, a permittee must submit a *Notice of Termination* (NOT) form after project completion. The NOT certifies that specific activities in the SWPPP have ended and either (1) final stabilization is complete, and temporary erosion and sediment controls have been removed from this operator's portion of the site, or (2) the operator has changed, and the new operator is responsible for compliance in all portions of the project area that had been the responsibility of the earlier operator. The new operator is responsible for submitting an NOI if activities continue.

Final stabilization occurs when all soil-disturbing activities at the site have been completed and a uniform, for example, evenly distributed, without large bare areas, perennial vegetative cover with a density of 70 percent of the native background vegetative cover for the area has been established on all unpaved areas and areas not covered by permanent structures. As an alternative, equivalent permanent stabilization measures (such as the use of riprap, gabions, or geotextiles) may be employed.

2.17 STORMWATER ASSOCIATED WITH INDUSTRIAL ACTIVITY

For *industrial stormwater discharge permits,* only stormwater discharges associated with industrial activity are required to have permits. The definition of this term is lengthy and detailed. It includes "the discharge from any conveyance which is used for collecting and conveying stormwater and which is directly related to manufacturing, processing or raw materials storage areas at an industrial plant. The term does not include discharges from facilities or activities excluded from the NPDES program" (40 CFR 122.26(b)(14)).

The EPA definition also provides several examples of industrial activity:

the term includes, but is not limited to, stormwater discharges from industrial plant yards; immediate access roads and rail lines used or traveled by carriers of raw materials, manufactured products, waste material, or by-products used or created by the facility; material handling sites; refuse sites; sites used for the application or disposal of process waste waters; sites used for the storage and maintenance of material handling equipment; sites used for residual treatment, storage, or disposal; shipping and receiving areas; manufacturing buildings; storage areas (including tank farms) for raw materials, and intermediate and finished products; and areas where industrial activity has taken place in the past and significant materials remain and are exposed to stormwater (40 CFR 122.26(b)(14)).

One important aspect of the EPA definition is the exclusion of stormwater discharges from those portions of the industrial facility that are not actively involved in industrial activities; the term excludes areas located on plant lands separate from the plant's industrial activities, such as office buildings and accompanying parking lots, as long as the drainage from the excluded areas is not mixed with stormwater drained from the above-described areas (40 CFR 122.26(b)(14)). In addition, off-site stockpiles of final product from an industrial facility do not require permit coverage, because they are not located at the site of the industrial facility (EPA, 1992).

Areas associated with industrial activity do not include commercial or retail facilities. This is an important distinction; in some cases (such as construction), the EPA has chosen to regulate only those activities that are significant enough to be inherently industrial in nature.

2.17.1 Industries Required to Obtain NPDES Stormwater Discharge Permit Coverage

The EPA requires stormwater discharge permits only for specific types of industrial activities. The activities requiring permits are defined in two ways: by a narrative description or by a Standard Industrial Classification (SIC) code. SIC codes are standard numeric codes assigned to each type of industrial process in the United States by the President's Office of Management and Budget (Office of Management and Budget, 1987).

There is an important distinction between these two types of categories: industrial sites identified by a SIC code are required to obtain permit coverage only if the primary site activity is within the SIC codes listed. If the listed activity is not the primary site activity, it is considered an auxiliary activity, which does not require permit coverage. For categories defined by a narrative description, however, a permit is required if any of the described activity occurs on-site. Therefore, the narrative categories are more inclusive.

Seven categories of industrial activity are defined by narrative description:

1. Subchapter N industries
2. Hazardous waste treatment, storage, or disposal facilities
3. Landfills
4. Power generation facilities
5. Sewage treatment plants
6. Construction activities
7. Water quality violators or significant polluters

There are five main categories of industrial activity defined by SIC codes:

1. Heavy manufacturing
2. Light manufacturing
3. Mining
4. Recyclers
5. Industrial transportation

2.18 SUBCHAPTER N INDUSTRIES

Subchapter N of Title 40 of the CFR includes all the effluent guidelines and standards for various types of industrial facilities. Subchapter N contains 40 CFR sections 401 through 471. Facilities subject to any of the following types of limitations or guidelines under 40 CFR, subchapter N (except facilities that are exempt under the light industry exclusion) must obtain NPDES permits to discharge stormwater:

• Stormwater effluent limitation guidelines
• New source performance standards (NSPSs)
• Toxic pollutant effluent standards

According to the EPA, the industries in these categories have generally been identified by EPA as the most significant dischargers of process wastewaters in the country. As such, these facilities are likely to have stormwater discharges associated with industrial activity for which permit applications should be required (55 FR 47989 (November 16, 1990)).

2.18.1 Hazardous Waste Treatment, Storage, or Disposal Facilities

Hazardous waste treatment, storage, or disposal facilities, including those that are operating under interim status or a permit under subtitle C of the *Resource Conservation and Recovery Act* (RCRA), must obtain NPDES stormwater discharge permits. A facility that stores hazardous waste for less than 90 days is not considered to be a treatment, storage, or disposal facility and therefore is not required to submit a stormwater discharge permit application (EPA, 1992). Because these industries are described as a narrative category, a permit is required if any of the described activity occurs on-site.

Land disposal units and incinerators, as well as *boilers and industrial furnaces* (BIFs) that burn hazardous waste, may receive a diverse range of industrial wastes. Waste receiving, handling, storage, and processing, in addition to actual waste disposal, can be a significant source of pollutants at waste disposal facilities. The EPA has summarized case studies documenting surface water impacts and groundwater contamination of land disposal units. Evaluation of 163 case studies revealed surface water impacts at 73 facilities. Elevated levels of organic chemicals, including pesticides, and metals have been found in groundwater and/or surface water at many sites (55 FR 47989 (November 16, 1990)).

Landfills, land application sites, and open dumps that receive industrial wastes must obtain NPDES stormwater discharge permits. *Industrial waste* is waste received from the manufacturing portions of facilities under any of the other industrial categories under this program, and does include construction debris.

Landfills that are capped and closed must be judged on a case-by-case basis. A permit application should be filed for such facilities.

2.18.2 Power Generation Facilities

Steam electric power generating facilities, including coal handling sites, must obtain NPDES stormwater discharge permits. This would include single-user facilities, such as a steam electric power generating facility for a university campus. However, steam production for heating and cooling is not covered by permit requirements. Co-generation facilities are regulated if they are based on the use of dual fuels. However, co-generation facilities based on heat capture only are not regulated.

2.18.3 Sewage Treatment Plants

Sewage treatment plants have been required to obtain NPDES permits to discharge treated sewage effluent since the passage of the CWA. The *Water Quality Act of 1987,* however, now requires permit coverage for stormwater discharges from such facilities.

Sewage facilities with a design flow of 1.0 million gallons per day (MGD) or more, or which are required to have an approved pre-treatment program under 40 CFR part 403, are included. Farm lands, domestic gardens, and lands used for sludge management where sludge is beneficially reused and that are not physically located within the confines of the sewage treatment facility, or areas that are in compliance with section 405 of the CWA, are not included. (Section 405 of the CWA regulates the disposal of sewage sludge.) If the facility collects all stormwater from the plant site and treats it as part of the normal inflow that is processed through the treatment plant, no stormwater discharge permit is required.

2.18.4 Heavy Industries

The SIC code categories include the following:

SIC code 24: lumber and wood products (except 2434: wood kitchen cabinets). These facilities are engaged in operating sawmills, planing mills, and other mills engaged in producing lumber and wood basic materials.

SIC code 26: paper and allied products (except 265: paperboard containers and boxes, and 267: converted paper and paperboard products).

SIC code 28: chemicals and allied products (except 283: drugs, and 285: paints, varnishes, lacquers, enamels, etc.)

SIC code 29: petroleum refining and related activities.

SIC code 311: leather tanning and finishing. Such processes use chemicals such as sulfuric acid and sodium dichromate; detergents; and a variety of raw and intermediate materials.

SIC code 32: stone, clay, glass, and concrete products (except 323: glass products made of purchased glass). These facilities manufacture glass, clay, stone, and concrete products from raw materials in the form of quarried and mined stone, clay, and sand.

SIC code 33: primary metal industries, including facilities that smelt and refine ferrous and nonferrous metals from ore, pig, or scrap, and manufacturing related products

SIC code 3441: structural metal fabricating

SIC code 373: ship and boat building and repair

2.18.5 Light Industries

SIC code 20: food and kindred products, including process foods such as meats, dairy food, fruit, and flour

SIC code 21: tobacco products, including cigarettes, cigars, chewing tobacco, and related products

SIC code 22: textile mill products, producing yarn, and so on, and/or dye and finish fabrics

SIC code 23: apparel and other textile products, which produce clothing by cutting and sewing purchased woven or knitted textile products

SIC code 2434: wood kitchen cabinets

SIC code 25: furniture and fixtures

SIC code 265: paperboard containers and boxes

SIC code 267: converted paper and paperboard products (except containers and boxes)

SIC code 27: printing and publishing, including bookbinding and plate making

SIC code 283: drugs (pharmaceuticals)

SIC code 285: paints, varnishes, lacquers, enamels, and allied products

SIC code 30: rubber and miscellaneous plastic products

SIC code 31: leather and leather products (except 311: leather tanning and finishing)

SIC code 323: glass products made of purchased glass

SIC code 34: fabricated-metal products (except 3441: structural metal fabricating)

SIC code 35: industrial and commercial machinery and computer equipment

SIC code 36: electronic and other electric equipment and components

SIC code 37: transportation equipment (except 373: ship and boat building and repair)

SIC code 38: instruments and related products, including measuring, analyzing, and controlling instruments; photographic, medical, and optical goods; and watches and clocks

SIC code 39: miscellaneous manufacturing industries, including jewelry, silverware, plated ware, musical instruments, dolls, toys, games, sporting and athletic goods, pens, pencils, artists' materials, novelties, buttons, notions, brooms, brushes, signs, burial caskets, and hard surface floor coverings

SIC code 4221: farm products warehousing and storage

SIC code 4222: refrigerated warehousing and storage

SIC code 4225: general warehousing and storage

2.18.6 Mining Industries

SIC code 10: metal mining

SIC code 11: anthracite mining

SIC code 12: coal mining

SIC code 13: oil and gas extraction

SIC code 14: nonmetallic minerals, except fuels

2.18.7 Recycling Industries

SIC code 5015: motor vehicle parts, used—wholesale or retail

SIC code 5093: scrap and waste materials, including the following wholesale businesses—automotive wrecking for scrap, bag reclaiming, waste bottles, waste boxes, fur cuttings and scraps, iron and steel scrap, general junk and scrap, metal and waste scrap, nonferrous metals scrap, waste oil, plastics scrap, rags, rubber scrap, scavenging, scrap and waste materials, textile waste, wastepaper (including paper recycling), and wiping rags (including washing and reconditioning)

2.18.8 Transportation Industries

SIC code 40: railroad transportation

SIC code 41: local and interurban highway passenger transit

SIC code 42: trucking and warehousing (except 4221: farm products warehousing and storage, 4222: refrigerated warehousing trucking and warehousing and storage, and 4225: general warehousing and storage)

SIC code 43: US Postal Service

SIC code 44: water transportation

SIC code 45: transportation by air

SIC code 5171: petroleum bulk stations and terminals

2.18.9 Industrial Facilities that Do Not Discharge Contaminated Stormwater

Under current EPA regulations, industrial facilities can operate without obtaining a stormwater discharge permit, if no material handling equipment or activities, raw materials, intermediate products, final products, waste materials, byproducts, or industrial machinery is exposed to stormwater (40 CFR 122.26(b)(14)).

Some industrial facilities handle oil drums or other contained materials that are exposed during loading and unloading operations. If there is a reasonable potential for leaks or spills from these containers which could be exposed to stormwater, discharges from the exposed area would be subject to stormwater permitting requirements (EPA, July 1993).

If material handling equipment or activities, raw materials, intermediate products, final products, waste materials, byproducts, or industrial machinery is stored outside in a structure with a roof but with no sides, and if wind-blown rain, snow, or runoff comes into contact with the equipment, material, or activities, then discharges from the area will be subject to stormwater permitting requirements (EPA, July 1993).

Gas stations and commercial automotive repair facilities are not included in the definition of industrial transportation because these facilities are commercial or retail in nature.

2.18.9.1   Stormwater Discharge Permit Requirements for Airports.   Airports or airline companies must apply for a stormwater discharge permit for locations where de-icing chemicals are applied. This includes, but is not limited to, runways, taxiways, ramps, and areas used for the de-icing of airplanes. The operator of the airport should apply for the stormwater discharge permit, while the individual airline companies should be included as co-applicants. The EPA has the discretion to issue individual permits to each discharger or to issue an individual permit to the airport operator and have other dischargers to the same system act as co-permittees.

2.19   PENALTIES FOR STORMWATER VIOLATIONS

The *Clean Water Act* provides severe penalties for those who fail to obtain permit coverage for discharges and for those who do not comply with the terms and conditions of an NPDES permit. Section 309 of the *Clean Water Act* gives the EPA broad enforcement authority. In a state with an approved program (one with NPDES permitting and enforcement authority), the EPA notifies the state whenever a violation comes to the EPA's attention. The approved state must have the same or similar enforcement authority as the EPA has in states without approved programs. EPA retains its authority and exercises oversight of the approved state program. Operating agreements with approved states generally allow that if the state fails to take timely and appropriate action, the EPA can commence enforcement action (Stimson et al., 1993). In states with unapproved programs, the EPA regional offices are the primary enforcement agencies.

Violations may include actions that are inconsistent with provisions of the law itself, (for example, discharges without the required permit), or actions that are inconsistent with conditions of permits issued by the EPA or states under the Act. Violations may be cited where the discharge or potential to discharge is to waters of the United States or in some situations to sewer systems. Violations of recordkeeping, reporting, and inspection requirements may also be cited.

2.19.1   EPA Stormwater Enforcement Authority

The CWA gives to the EPA three broad categories of enforcement authority: civil administrative authority, civil judicial authority, and criminal enforcement authority.

Civil administrative authority is used to issue compliance orders and levy civil penalties for alleged violators of any of the law or permit requirements. Class I penalties are limited under section 309(g)(2)(A) of the *Clean Water Act* to $11,000 per violation or $27,500 total. To propose a class I penalty, the EPA issues an administrative complaint to the violator, giving notice of the violations alleged and the amount of the penalty proposed. The violator has 30 days to request a hearing and to directly contest any allegation the violator intends to raise as an issue in a hearing.

Class II penalties may be up to $11,000 per day with a $137,500 maximum under section 309(g)(2)(B) of the Act. As with class I actions, the EPA issues an administrative complaint proposing to assess the penalty. Alleged violators are entitled to a hearing within 30 days and must specifically contest the allegations.

The EPA also has civil enforcement authority that allows the agency to file civil cases in the district court to enforce the law and assess penalties to violators. Parties that violate the law or permit conditions are subject to court-imposed civil penalties of up to $27,500 per day for each violation. These penalties can accrue until each violation is corrected, with the penalty limited only by the number of violations and the number of days each violations continues, up to the five-year statute of limitations. Each day that a violation occurs is considered to be a separate violation for the purpose of computing maximum penalties.

2.19.2 Criminal Prosecution for Stormwater Violations

The most serious violations of the CWA involve criminal penalties, all of which would be imposed by federal courts. Criminal cases are prosecuted by the Department of Justice, usually after referral from the EPA, although prosecution may result from other sources of information. Criminal penalties may be invoked for negligent violations, knowing violations, knowing endangerment, making false statements, or tampering with monitoring equipment.

These criminal penalties can include jail sentences as well as fines. If the violator is a corporation, the corporation may be fined while the persons holding responsible positions within the corporation may face fines and/or imprisonment. The persons prosecuted may include not only those with hands-on compliance responsibilities but also those who manage or supervise such activities.

The CWA creates a separate class of criminal penalties for parties that file false statements or tamper with monitoring equipment. Violations are punishable by fines of up to $10,000 and jail sentences up to 2 years, with penalties doubled for repeat offences.

It is not advisable at any time to falsify or withhold information from the EPA or state regulators. The possible criminal penalties resulting from falsifying or withholding information are likely to be much more severe than the civil penalties resulting from honestly reporting the true facts of a violation, however unfavorable those facts may be.

2.19.2.1 Avoiding Prosecution for Stormwater Violations. The most important way to reduce or even avoid large penalties for stormwater discharges is to voluntarily contact the EPA or the appropriate state agency and notify it of the violations. By doing so, one can take full advantage of the EPA's audit/self-policing policy (60 FR 66706 (December 22, 1995)). The EPA will not seek heavy penalties and will not recommend criminal prosecution for companies that meet all the requirements of the audit/self-policing policy, including the following:

- The violation must have been discovered through the discharger's own systematic procedures or practices, such as a regularly scheduled environmental audit.
- The violation must have been identified voluntarily—not as a result of any legally prescribed process.
- The violation must have been disclosed to the EPA promptly (within 10 days).
- The violation and disclosure must be independent of any third-party action (i.e., citizens' suits, local government regulations, etc.).
- The violation must be corrected with 60 days, and any environmental harm must be remediated.
- The violation will be prevented from recurring by taking appropriate corrective and preventive actions.
- The discharger will cooperate fully with EPA in any investigations.

- The policy cannot be used for repeated violations, violations that result in serious actual harm, or violations that may present an imminent and substantial endangerment.

- Corporations will remain criminally liable for violations resulting from conscious disregard of their legal duties, and individuals remain liable for criminal wrongdoing.

REFERENCES

Birch, P. B., and H. E. Pressley eds., *Stormwater Management Manual for the Puget Sound Basin,* Review Draft Dept. of Ecology Publication No. 90–73, 1992.

Fields, S., "Regulations and Policies Relating to the Use of Wetlands for Nonpoint Source Pollution Control," in R. K. Olson, ed., *Created and Natural Wetlands for Controlling Nonpoint Source Pollution,* C. K. Smoley, CRC Press, Boca Raton, FL, pp. 151–158, 1993.

Hammer, D. A., "Guidelines for Design, Construction and Operation of Constructed Wetlands for Livestock Wastewater Treatment," in P. J. DuBowy and R. P. Reaves, eds., *Constructed Wetlands for Animal Waste Management: Proceedings of Workshop,* Department of Forestry and Natural Resources, Purdue University, West Lafayette, IN, pp. 155–181, 1994.

King County Solid Waste Division, "Local Hazardous Waste Management Plan for Seattle-King County: Final Plan and Environmental Impact Statement for the Management of Small Quantities of Hazardous Waste in the Seattle-King County Region, 1990."

Kovalic, J. M., *The Clean Water Act of 1987,* Water Environment Federation (formerly Water Pollution Control Federation), Alexandria, VA, 1987.

Office of Management and Budget, *Standard Industrial Classification Manual,* Executive Office of the President, Washington, DC, 1987.

Phillips, N., *Decisionmaker's Stormwater Handbook: A Primer,* The Terrene Institute, Washington, DC, 1992.

Schueler, T. R., *Controlling Urban Runoff: A Practical Manual for Planning and Designing Urban BMPs,* Publication no. 87703, Metropolitan Washington Council of Governments, Washington, DC, 1987.

Schueler, T. R., and J. Lugbill, *Performance of Current Sediment Control Measures at Maryland Construction Sites,* Metropolitan Washington Council of Governments, Washington, DC, 1990.

Shaver, E., "Sand Filter Design for Water Quality Treatment," in *Stormwater Management: Urban Runoff Management Workshop,* book 2, Environmental Protection Agency, Washington, DC, 1992.

U.S. Agricultural Research Service (ARS), *Predicting Soil Erosion by Water—A Guide to Conservation Planning with the Revised Universal Soil Loss Equation (RUSLE),* USDA/HB-703, January, 1997.

U.S. Environmental Protection Agency (EPA), "Region 6: Fact Sheet for Draft National Pollutant Discharge Elimination System (NPDES) Permit No. TXS000601," for the City of Corpus Christi Municipal Separate Storm Sewer System, Dallas, TX, date unknown.

U.S. Environmental Protection Agency (EPA), "Development Document for Proposed Effluent Limitations Guidelines and Standards for the Shipbuilding and Repair Point Source Category," EPA 440/1/-79/076-b, December 1979a.

U.S. Environmental Protection Agency (EPA), "Development Document for Effluent Limitations Guidelines and Standards for Pretreatment Standards for the Petroleum Refineries Point Source Category," EPA 440/1/-79/014b, 1979b.

U.S. Environmental Protection Agency (EPA), "Final Development Document for Effluent Limitations Guidelines and Standards and Pretreatment Standards for the Steam Electric Point Source Category," EPA-440 1-82 029, November 1982.

U.S. Environmental Protection Agency (EPA), *Urban Targeting and BMP Selection, Information and Guidance Manual for State Nonpoint Source Program Staff Engineers and Managers,* The Terrene Institute, EPA 68-C8-0034, 1990.

U.S. Environmental Protection Agency (EPA), NPDES Storm Water Program Question and Answer Document," vol. 1, EPA 833-F-93-002, March 16, 1992.

U.S. Environmental Protection Agency (EPA), *Storm Water Management for Industrial Activities: Developing Pollution Prevention Plans and Best Management Practices,* EPA 832-R-92-006, Office of Water, Washington, DC, 1992.

U.S. Environmental Protection Agency (EPA), "How to Set Up a Local Program to Recycle Used Oil," EPA 530-SW-89-039A, 1993.

U.S. Environmental Protection Agency (EPA), *Investigation of Inappropriate Pollutant Entries into Storm Drainage Systems—A User's Guide,* Office of Research and Development, EPA 600/R-92/238, Washington, DC, 1993.

U.S. Environmental Protection Agency (EPA), *Proposed Guidance Specifying Management Measures for Sources of Nonpoint Pollution in Coastal Waters,* EPA 340-B-92-002, Office of Water, January 1993.

U.S. Environmental Protection Agency (EPA), "NPDES Storm Water Program Question and Answer Document," vol. 2, EPA 833-F-93-002B, July 1993b.

U.S. Environmental Protection Agency (EPA), "Tulsa Municipal Separate Storm Sewer System Permit Fact Sheet," Dallas, TX, August 3, 1994.

U.S. Environmental Protection Agency (EPA), "Fact Sheet for Draft National Pollutant Discharge Elimination System (NPDES) Permit No. OKS000101," for the City of Oklahoma City Municipal Separate Storm Sewer System, Dallas, TX, December 22, 1994.

U.S. Environmental Protection Agency (EPA), Guidance Specifying Management Measures for Sources of Nonpoint Pollution in Coastal Waters, EPA 840-B-92-002, January 1993.

U.S. Environmental Protection Agency (EPA), "NPDES Storm Water Program Question and Answer Document," vol. 2, EPA 833-F-93-002B, September 1983.

U.S. Environmental Protection Agency (EPA), *Watershed Protection—NPDES Watershed Strategy,* Washington, DC, March 1994.

U.S. Government Printing Office (GPO), *Code of Federal Regulations,* Office of the Federal Register, National Archives and Records Administration, Washington, DC, July 1, 1992.

U.S. Government Printing Office (GPO), *Federal Register,* Office of the Federal Register, National Archives and Records Administration, Washington, DC, December 8, 1999.

Wetzel, R. G., "Constructed Wetlands: Scientific Foundations Are Critical," in G. A. Moshiri, ed., *Constructed Wetlands for Water Quality Improvement,* CRC Press, Boca Raton, FL, pp. 3–8, 1993.

CHAPTER 3
HYDRAULICS OF OPEN CHANNEL FLOW

Richard H. French
Desert Research Institute
University and Community College System of Nevada
Reno, Nevada

3.1 INTRODUCTION

By definition, an *open channel* is a flow conduit having a free surface: that is, a boundary exposed to the atmosphere. The free surface is essentially an interface between two fluids of different density. Open-channel flows are almost always turbulent, unaffected by surface tension, and the pressure distribution within the fluid is hydrostatic. Open channels include flows ranging from rivulets flowing across a field to gutters along residential streets and highways to partially filled closed conduits conveying waste water to irrigation and water supply canals to vital rivers.

In this chapter, the basic principles of open channel hydraulics are presented as an introduction to subsequent chapters dealing with design. By necessity, the material presented in this chapter is abbreviated—an abstract of the fundamental concepts and approaches—for a more detailed treatment, the reader is referred to any standard references or texts dealing with the subject: for example, Chow (1959), French (1985), Henderson (1966), or Chaudhry (1993)

As with any other endeavor, it is important that a common vocabulary be established and used:

Critical slope (S_c): A longitudinal slope such that uniform flow occurs in a critical state.

Flow area (*A*): The flow area is the cross-sectional area of the flow taken normal to the direction of flow (Table 3.1).

Froude number (*Fr*): The Froude number is the dimensionless ratio of the inertial and gravitational forces or

$$\text{Fr} = \frac{V}{\sqrt{gD}} \tag{3.1}$$

where V = average velocity of flow
g = gravitational acceleration
D = hydraulic depth

TABLE 3.1 Channel Section Geometric Properties

Channel definition (1)	Area A (2)	Wetted perimeter P (3)	Hydraulic radius R (4)	Top width T (5)	Hydraulic depth D (6)
Rectangle	by	$b + 2y$	$\dfrac{by}{b + 2y}$	b	y
Trapezoid with equal side slopes	$(b + zy)y$	$b + 2y\sqrt{1 + z^2}$	$\dfrac{(b + zy)y}{b + 2y\sqrt{1 + z^2}}$	$b + 2zy$	$\dfrac{(b + zy)y}{b + 2zy}$
Trapezoid with unequal side slopes	$by + 0.5y^2(z_1 + z_2)$	$b + y(\sqrt{1 + z_1^2} + \sqrt{1 + z_2^2})$	$\dfrac{by + 0.5y^2(z_1 + z_2)}{b + y(\sqrt{1 + z_1^2} + \sqrt{1 + z_2^2})}$	$b + y(z_1 + z_2)$	$\dfrac{by + 0.5y^2(z_1 + z_2)}{b + y(z_1 + z_2)}$
Triangle with equal side slopes	zy^2	$2y\sqrt{1 + z^2}$	$\dfrac{zy}{2\sqrt{1 + z^2}}$	$2zy$	$0.5y$
Triangle with unequal side slopes	$0.5y^2(z_1 + z_2)$	$y(\sqrt{1 + z_1^2} + \sqrt{1 + z_2^2})$	$\dfrac{0.5y^2(z_1 + z_2)}{y(\sqrt{1 + z_1^2} + \sqrt{1 + z_2^2})}$	$y(z_1 + z_2)$	$0.5y$
Circular	$\dfrac{1}{8}(\theta - \sin\theta)\,d_o^2$	$0.5\theta d_o$	$0.25\left(1 - \dfrac{\sin\theta}{\theta}\right)d_o$	$2\sqrt{y(d_o - y)}$	$\dfrac{1}{8}\left[\dfrac{\theta - \sin\theta}{\sin(0.5\theta)}\right]$

When Fr $= 1$, the flow is in a *critical* state with the inertial and gravitational forces in equilibrium; when Fr < 1, the flow is in a *subcritical* state and the gravitational forces are dominant; and when Fr > 1, the flow is in a *supercritical* state and the inertial forces are dominant. From a practical perspective, sub – and supercritical flow can be differentiated simply by throwing a rock or other object into the flow. If ripples from the rock progress upstream of the point of impact, the flow is subcritical; however, if ripples from the rock do not progress upstream but are swept downstream, the flow is supercritical.

Hydraulic depth (*D*). The *hydraulic depth* is the ratio of the flow area (*A*) to the top width (*T*) or $D = A/T$ (Table 3.1).

Hydraulic radius (*R*). The hydraulic radius is the ratio of the flow area (*A*) to the wetted perimeter (*P*) or $R = A/P$ (Table 3.1).

Kinetic energy correction factor (α). Since no real open-channel flow is one-dimensional, the true kinetic energy at a cross section is not necessarily equal to the spatially averaged energy. To account for this, the kinetic energy correction factor is introduced, or

$$\alpha \left(\gamma \frac{V^3}{2g} \right) A = \iint \gamma \frac{v^3}{2g} \, dA$$

and solving for α,

$$\alpha = \frac{\iint v^3 dA}{V^3 A} \tag{3.2}$$

When the flow is uniform, $\alpha = 1$ and values α of for various situations are summarized in Table 3.2.

Momentum correction coefficient (β): Analogous to the kinetic energy correction factor, the momentum correction factor is given by

$$\beta \rho Q V = \iint \rho v^2 dA$$

$$\beta = \frac{\iint v^2 dA}{V^2 A} \tag{3.3}$$

When the flow is uniform, $\beta = 1$ and values of β for various situations are summarized in Table 3.2.

Prismatic channel. A prismatic channel has both a constant cross-sectional shape and bottom slope (S_o). Channels not meeting these criteria are termed *nonprismatic*.

Specific energy (*E*). The specific energy of an open-channel flow is

TABLE 3.2 Typical Values of α and β for Various Situations

Situation	Value of α			Value of β		
	Min.	Avg.	Max.	Min.	Avg.	Max.
Regular channels, flumes, spillways	1.10	1.15	1.20	1.03	1.05	1.07
Natural streams and torrents	1.15	1.30	1.50	1.05	1.10	1.17
Rivers under ice cover	1.20	1.50	2.00	1.07	1.17	1.33
River valleys, overflooded	1.50	1.75	2.00	1.17	1.25	1.33

Source: After Chow (1959).

$$E = y + \alpha \frac{V^2}{2g} \tag{3.4}$$

where y = depth of flow and the units of specific energy are length in meters or feet.

Specific momentum (*M*). By definition, the specific momentum of an open-channel flow is

$$M = \frac{Q^2}{gA} + \bar{z}A \tag{3.5}$$

Stage: The stage of a flow is the elevation of the water surface relative to a datum. If the lowest point of a channel section is taken as the datum, then the stage and depth of flow (y) are equal if the longitudinal slope (S_o) is not steep or cos (θ) $\approx$ 1, where θ is the longitudinal slope angle. If $\theta \leq 10°$ or $S_o \leq 0.18$ where S_o is the longitudinal slope of the channel, then the slope of the channel can be assumed to be small.

Steady. The depth (y) and velocity of flow (v) at a location do not vary with time; that is, ($\partial y / \partial t = 0$) and ($\partial v / \partial t = 0$). In *unsteady* flow, the depth and velocity of flow at a location vary with time: that is, ($\partial y / \partial t \neq 0$) and ($\partial v / \partial t \neq 0$).

Top width (*T*). The top width of a channel is the width of the channel section at the water surface (Table 3.1).

Uniform flow. The depth (y), flow area (A), and velocity (V) at every cross section are constant, and the energy grade line (S_f), water surface, and channel bottom slopes (S_o) are all parallel.

Superelevation (Δy). The rise in the elevation of the water surface at the outer channel boundary above the mean depth of flow in an equivalent straight channel, because of centrifugal force in a curving channel.

Wetted perimeter (*P*). The wetted perimeter is the length of the line that is the interface between the fluid and the channel boundary (Table 3.1).

3.2   ENERGY PRINCIPLE

3.2.1   Definition of Specific Energy

Central to any treatment of open-channel flow is that of conservation of energy. The total energy of a particle of water traveling on a streamline is given by the *Bernoulli equation* or

$$H = z + \frac{p}{\gamma} + \alpha \frac{V^2}{2g}$$

where H = total energy, z = elevation of the streamline above a datum, p = pressure, γ = fluid specific weight, (p/γ) = pressure head, $V^2/2g$ = velocity head, and g = acceleration of gravity. H defines the elevation of the *energy grade line,* and the sum [$z + (p/\gamma)$] defines the elevation of the *hydraulic grade line.* In most uniform and gradually varied flows, the pressure distribution is hydrostatic (divergence and curvature of the streamlines is negligible) and the sum [$z + (p/\gamma)$] is constant and equal to the depth of flow y if the datum is taken at the bottom of the channel. The *specific energy* of an open-channel flow relative to the channel bottom is

$$E = y + \alpha \frac{V^2}{2g} = y + \alpha \frac{Q^2}{2gA^2} \tag{3.6}$$

where the average velocity of flow is given by

$$V = \frac{Q}{A} \qquad (3.7)$$

where Q = flow rate and A = flow area.

The assumption inherent in Eq. (3.6) is that the slope of the channel is small, or $\cos(\theta)$ α 1. If $\theta \leq 10°$ or $S_o \leq 0.18$, where S_o is the longitudinal slope of the channel, Eq. (3.6) is valid. If θ is not small, then the pressure distribution is not hydrostatic since the vertical depth of flow is different from the depth measured perpendicular to the bed of the channel.

3.2.2 Critical Depth

If y in Eq. (3.6) is plotted as a function of E for a specified flow rate Q, a curve with two branches results. One branch represents negative values of both E and y and has no physical meaning; but the other branch has meaning (Fig. 3.1). With regard to Fig. 3.1, the following observations are pertinent: 1) the portion designated AB approaches the line $y = E$ asymptotically, 2) the portion AC approaches the E axis asymptotically, 3) the curve has a minimum at point A, and 4) there are two possible depths of flow—the *alternate depths*—for all points on the E axis to the right of point A. The location of point A, the minimum depth of flow for a specified flow rate, can be found by taking the first derivative of Eq. (3.6) and setting the result equal to zero, or

$$\frac{dE}{dy} = 1 - \frac{Q^2}{gA^3}\frac{dA}{dy} = 0 \qquad (3.8)$$

It can be shown that $dA = (T = dy)$ or $(dA/dy = T)$ (French, 1985). Substituting this result, using the definition of hydraulic depth and rearranging, Eq. (3.8) becomes

$$1 - \frac{Q^2}{gA^3}\frac{dA}{dy} = 1 - \frac{Q^2}{gA^2}\frac{T}{A} = 1 - \frac{V^2}{gD} = 0$$

or

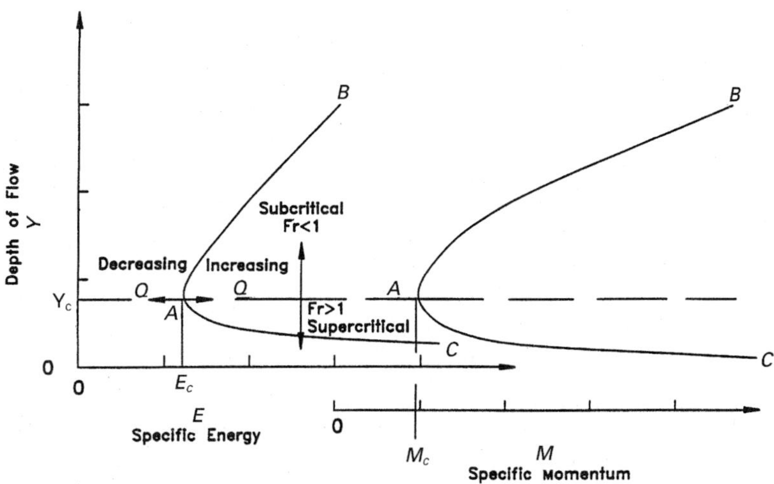

FIGURE 3.1 Specific energy and momentum as a function of depth when the channel geometry and flow rate are specified.

$$\frac{V^2}{2g} = \frac{D}{2}$$ (3.9)

and

$$\frac{V}{\sqrt{gD}} = Fr = 1$$ (3.10)

which is the definition of critical flow. Therefore, minimum specific energy occurs at the critical hydraulic depth and is the minimum energy required to pass the flow Q. With this information, the portion of the curve AC in Fig. 3.1 is interpreted as representing supercritical flows, where as AB represents subcritical flows.

With regard to Fig. 3.1 and Eq. (3.6), the following observations are pertinent. First, for channels with a steep slope and $\alpha \neq 1$, it can be shown that

$$Fr = \frac{V}{\sqrt{\dfrac{gD \, \cos(\theta)}{\alpha}}}$$ (3.11)

Second, $E - y$ curves for flow rates greater than Q lie to the right of the plotted curve, and curves for flow rates less than Q lie to the left of the plotted curve. Third, in a rectangular channel of width b, $y = D$ and the flow per unit width is given by

$$q = \frac{Q}{b}$$ (3.12)

and

$$V = \frac{q}{y}$$ (3.13)

Then, where the subscript c indicates variable values at the critical point,

$$y_c = \left(\frac{q^2}{g}\right)^{1/3}$$ (3.14)

$$\frac{V_c^2}{2g} = \frac{y_c}{2}$$ (3.15)

and

$$y_c = \frac{2}{3} E_c$$ (3.16)

In nonrectangular channels when the dimensions of the channel and flow rate are specified, critical depth is calculated either by the trial and error solution of Eqs. (3.8), (3.9), and (3.10) or by use of the semiempirical equations in Table 3.3.

3.2.3 Variation of Depth with Distance

At any cross section, the total energy is

$$H = \frac{V^2}{2g} = y + z$$ (3.17)

where y = depth of flow, z = elevation of the channel bottom above a datum, and it is

TABLE 3.3 Semiempirical Equations for the Estimation of y_c

Channel definition (1)	Equation for y_c in terms of $\Psi = \alpha Q^2/g$ (2)
Rectangle	$\left(\dfrac{\Psi}{b^2}\right)^{0.33}$
Trapezoid	$0.81\left(\dfrac{\Psi}{z^{0.75}b^{1.25}}\right)^{0.27} - \dfrac{b}{30z}$
Triangle	$\left(\dfrac{2\Psi}{z^2}\right)^{0.20}$
Circle	$\left(\dfrac{1.01}{d_o^{0.26}}\right)\Psi^{0.25}$

Source: From Straub (1982).

assumed that α and $\cos(\theta)$ are both equal to 1. Differentiating Eq. (3.17) with respect to longitudinal distance,

$$\frac{dH}{dx} = \frac{d\left(\dfrac{V^2}{2g}\right)}{dx} + \frac{dy}{dx} + \frac{dz}{dx} \tag{3.18}$$

where dH/dx = the change of energy with longitudinal distance (S_f), dz/dx = the channel bottom slope (S_o), and, for a specified flow rate,

$$\frac{d\left(\dfrac{V^2}{2g}\right)}{dx} = \frac{Q^2}{gA^3}\frac{dA}{dy}\frac{dy}{dx} = -\frac{Q^2T}{gA^3}\frac{dy}{dx} = -(\text{Fr})^2\frac{dy}{dx}$$

Substituting these results in Eq. (3.18) and rearranging,

$$\frac{dy}{dx} = \frac{S_o - S_f}{1 - \text{Fr}^2} \tag{3.19}$$

which describes the variation of the depth of flow with longitudinal distance in a channel of arbitrary shape.

3.2.4 Compound Section Channels

In channels of compound section (Fig. 3.2), the specific energy correction factor α is not equal to 1 and can be estimated by

$$\alpha = \frac{\displaystyle\sum_{i=1}^{N} \left(\frac{K^3_{\,i}}{A^2_{\,i}}\right)}{\dfrac{K^3}{A^2}} \tag{3.20}$$

where K_i and A_i as follows the conveyance and area of the ith channel subsection, respectively, K and A are conveyance are as follows:

$$K = \sum_{i=1}^{N} K_i$$

and

$$A = \sum_{i=1}^{N} A_i$$

N = number of subsections, and conveyance (K) is defined by Eq. (3.48) in Sec. 3.4. Equation (3.20) is based on two assumptions: (1) the channel can be divided into subsections by appropriately placed vertical lines (Fig. 3.2) that are lines of zero shear and do not contribute to the wetted perimeter of the subsection, and (2) the contribution of the nonuniformity of the velocity within each subsection is negligible in comparison with the variation in the average velocity among the subsections.

3.3 MOMENTUM

3.3.1 Definition of Specific Momentum

The one-dimensional momentum equation in an open channel of arbitrary shape and a control volume located between Sections 1 and 2 is

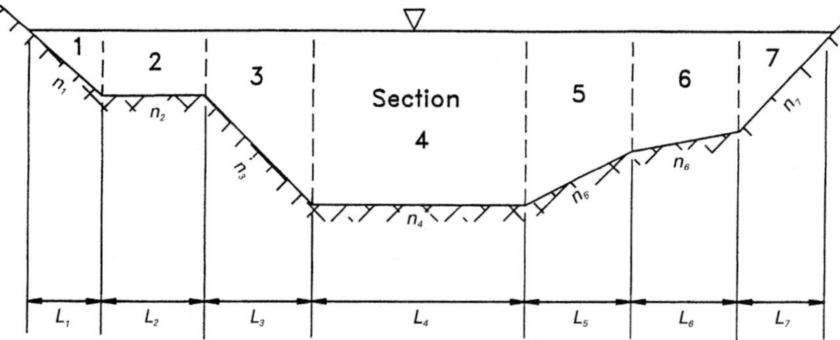

FIGURE 3.2   Channel with a compound section.

$$\gamma \bar{z}_1 A_1 - \gamma \bar{z}_2 A_2 - P_f = \frac{\gamma}{g} Q(V_2 - V_1) \tag{3.21}$$

where γ = specific weight of water, A_i = flow area at sections 1 and 2; V_i = average velocity of flow at sections 1 and 2, P_f = horizontal component of unknown force acting between Sections 1 and 2 and $\bar{z}_i$ = distances to the centroids of the flow areas 1 and 2 from the free surface. Substitution of the flow rate divided by the area for the velocities and rearrangement of Eq. (3.21) yields

$$\frac{P_f}{\gamma} = \left(\frac{Q^2}{gA_1} + \bar{z}_1 A_1 \right) - \left(\frac{Q^2}{gA_2} + \bar{z}_2 A_2 \right)$$

or

$$\frac{P_f}{\gamma} = M_1 - M_2 \tag{3.22}$$

where

$$M_i = \frac{Q^2}{gA_i} + \bar{z}_i A_i \tag{3.23}$$

and M is known as the *specific momentum or force function*. In Fig. 3.1, specific momentum is plotted with specific energy for a specified flow rate and channel section as a function of the depth of flow. Note that the point of minimum specific momentum corresponds to the critical depth of the flow.

The classic application of Eq. (3.22) occurs when $P_f = 0$ and the application of the resulting equation to the estimation of the sequent depths of a hydraulic jump. *Hydraulic jumps* result when there is a conflict between the upstream and downstream controls that influence the same reach of channel. For example, if the upstream control causes supercritical flow while the downstream control dictates subcritical flow, there is a contradiction that can be resolved only if there is some means to pass the flow from one flow regime to the other. When hydraulic structures, such as weirs, chute blocks, dentated or solid sills, baffle piers, and the like, are used to force or control a hydraulic jump, P_f in Eq. (3.22) is not equal to zero. Finally, the hydraulic jump occurs at the point where Eq. (3.22) is satisfied (French, 1985).

3.3.2 Hydraulic Jumps in Rectangular Channels

In the case of a rectangular channel of width b and $P_f = 0$, it can be shown (French, 1985) that

$$\frac{y_2}{y_1} = \sqrt{0.5[1 + 8(\mathrm{Fr}_1)^2 - 1]} \tag{3.24}$$

or

$$\frac{y_1}{y_2} = 0.5[\sqrt{1 + 8(\mathrm{Fr}_2)^2} - 1] \tag{3.25}$$

$$\frac{y_1}{y_2} = 2(\mathrm{Fr}_2)^2 - 4(Fr_2)^4 + 16(Fr_2)^6 - \dots$$

Equations (3.24) and (3.25) each contain three independent variables, and two must be known before the third can be found. It must be emphasized that the downstream depth of flow (y_2) is not the result of upstream conditions but is the result of a downstream control—that is, if the downstream control produces the depth y_2 then a hydraulic jump will form. The second form of Eq. (3.25) should be used when $(Fr_2)^2 \leq 0.05$ (French, 1985).

3.3.3 Hydraulic Jumps in Nonrectangular Channels

In analyzing the occurrence of hydraulic jumps in nonrectangular but prismatic channels, we see that no equations are analogous to Eqs. (3.24) and (3.25). In such cases, Eq. (3.22) could be solved by trial and error or by use of semiempirical equations. For example, in circular sections, Straub (1978) noted that the upstream Froude number (Fr_1) can be approximated by

$$\mathrm{Fr}_1 = \left(\frac{y_c}{y_1}\right)^{1.93} \tag{3.26}$$

and the sequent depth can be approximated by

$$\mathrm{Fr}_1 < 1.7 y_2 = \frac{y_c^2}{y_1} \tag{3.27}$$

$$\mathrm{Fr}_1 > 1.7 y_2 = \frac{y^{1.8}}{y_1^{0.73}} \tag{3.28}$$

For horizontal triangular and parabolic prismatic channel sections, Silvester (1964, 1965) presented the following equations.

For triangular channels:

$$\left(\frac{y_2}{y_1}\right)^{2.5} - 1 = 1.5(Fr_1)^2 \left[1 - \left(\frac{y_1}{y_2}\right)^2\right] \tag{3.29}$$

For parabolic channels with the perimeter defined by $y = aT^2/2$, where a is a coefficient:

$$\left(\frac{y_2}{y_1}\right)^{2.5} - 1 = 1.67(Fr_1)^2 \left[1 - \left(\frac{y_1}{y_2}\right)^{1.5}\right] \tag{3.30}$$

In the case of trapezoidal channels, Silvester (1964) presented a method for graphical solution in terms of the parameter

$$k = \frac{b}{zy_1} \tag{3.31}$$

In Fig. 3.3, the ratio of (y_2/y_1) is plotted as a function of Fr_1 and k.

3.4 UNIFORM FLOW

3.4.1 Manning and Chezy Equations

For computational purposes, the average velocity of a uniform flow can be estimated by any one of a number of semiempirical equations that have the general form

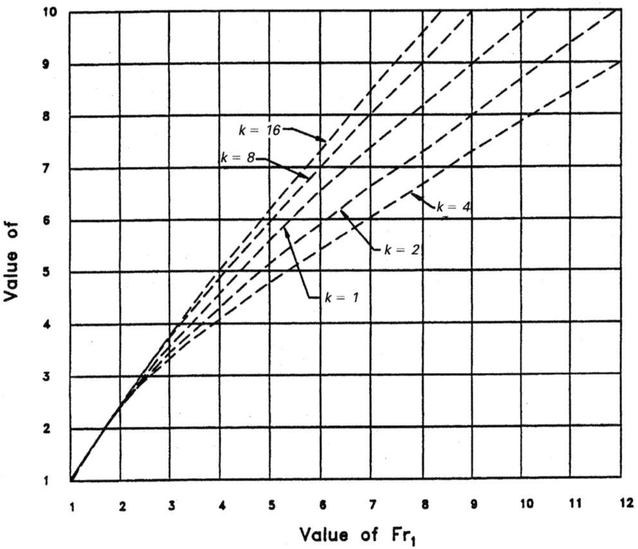

FIGURE 3.3 Analytic curves for estimating sequent depths in a trapezoidal channel. (From Silvester, 1964)

$$V = CR^x S^y \tag{3.32}$$

where C = a resistance coefficient, R = hydraulic radius, S = channel longitudinal slope, and x and y are exponents. At some point in the period 1768–1775 (Levi, 1995), Antoine Chezy, designing an improvement for the water system in Paris, France, derived an equation relating the uniform velocity of flow to the hydraulic radius and the longitudinal slope of the channel, or

$$V = C\sqrt{RS} \tag{3.33}$$

where C is the Chezy resistance coefficient. It can be easily shown that Eq (3.33) is similar in form to the Darcy pipe flow equation. In 1889, Robert Manning, a professor at the Royal College of Dublin (Levi, 1995) proposed what has become known as Manning's equation, or

$$V = \frac{\phi}{n} R^{2/3}\sqrt{S} \tag{3.34}$$

where n is Manning's resistance coefficient and $\phi = 1$ if SI units are used and $\phi = 1.49$ if English units are used. The relationship among C, n, and the Darcy-Weisbach friction factor (f) is

$$C = \frac{\phi}{n} R^{1/6} = \sqrt{\frac{8g}{f}} \tag{3.35}$$

At this point, it is pertinent to observe that n is a function of not only boundary roughness and the Reynolds number but also the hydraulic radius, an observation that was made by Professor Manning (Levi, 1995).

3.4.2 Estimation of Manning's Resistance Coefficient

Of the two equations for estimating the velocity of a uniform flow, Manning's equation is the more popular one. A number of approaches to estimating the value of n for a channel are discussed in French (1985) and in other standard references, such as Barnes (1967), Urquhart (1975), and Arcement and Schneider (1989). Appendix 3.A lists typical values of n for many types of common channel linings.

In an unvegetated alluvial channel, the total roughness consists of two parts: grain or skin roughness resulting from the size of the sediment particles and form roughness because of the existence of bed forms. The total coefficient n can be expressed as

$$n = n' + n'' \tag{3.36}$$

where n' = portion of Manning's coefficient caused by grain roughness and n'' = portion of Manning's coefficient caused by form roughness. The value of n' is proportional to the diameter of the sediment particles to the sixth power. For example, Lane and Carlson (1953) from field experiments in canals paved with cobbles with d_{75} in inches, developed

$$n' = 0.026 d_{75}^{1/6} \tag{3.37}$$

and Meyer-Peter and Muller (1948) for mixtures of bed material with a significant proportion of coarse-grained sizes with d_{90} in meters developed

$$n' = 0.038 d_{90}^{1/6} \tag{3.38}$$

In both equations, d_{xx} is the sediment size such that xx percent of the material is smaller by weight.

Although there is no reliable method of estimating n'', an example of the variation of f for the 0.19 mm sand data collected by Guy et al. (1966) is shown in Fig. 3.4. The n values commonly found for different bed forms are summarized in Table 3.4. The inability to

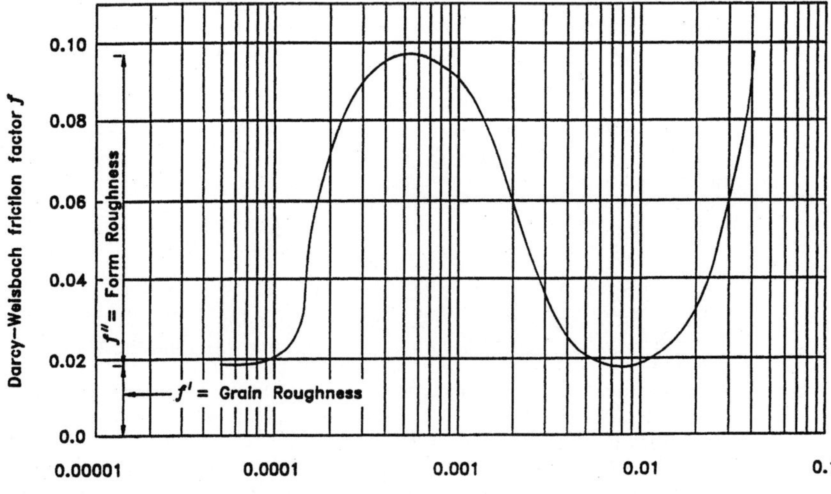

FIGURE 3.4 Variation of the Darcy-Weisbach friction factor as a function of unit stream power.

TABLE 3.4 Equivalent Roughness Values of Various Bed Materials

Material	k (ft)	k (m)
(1)	(2)	(3)
Brass, copper, lead, glass	0.0001–0.0030	0.00003048–0.0009
Wrought iron, steel	0.0002–0.0080	0.0001–0.0024
Asphalted cast iron	0.0004–0.0070	0.0001–0.0021
Galvanized iron	0.0005–0.0150	0.0002–0.0046
Cast iron	0.0008–0.0180	0.0002–0.0055
Wood stave	0.0006–0.0030	0.0002–0.0009
Cement	0.0013–0.0040	0.0004–0.0012
Concrete	0.0015–0.0100	0.0005–0.0030
Untreated gunite	0.01–0.033	0.0030–0.0101
Drain tile	0.0020–0.0100	0.0006–0.0030
Riveted steel	0.0030–0.0300	0.0009–0.0091
Rubble masonry	0.02	0.0061
Straight, uniform earth channels	0.01	0.0030
Natural streambed	0.1000–3.0000	0.0305–0.9144

Sources: From Ackers C (1958), Chow (1959), and Zegzhda (1938).

estimate or determine the variation of form roughness poses a major problem in the study of alluvial hydraulics (Yang, 1996).

Use of Manning's equation to estimate the velocity of flow in channels where the primary component of resistance is from drag rather than bed roughness has been questioned (Fischcnich, 1996). However, the use of Manning's equation has persisted among engineers because of its familiarity and the lack of a practical alternative. Jarrett (1984) recognized that guidelines for estimating resistance coefficients for high-gradient streams with stable beds composed of large cobbles and boulders and minimally vegetated banks ($S_o > 0.002$) were based on limited data. Jarrett (1984) examined 21 high-gradient streams in the Rocky Mountains and developed the following empirical equation relating n to S_o and R (in feet):

$$n = \frac{0.39 S_O^{0.38}}{R^{0.16}} \tag{3.39}$$

Jarrett (1984) stated the following limitations on the use of Eq. (3.39): First, the equation is applicable to natural main channels with stable bed and bank materials (gravels, cobbles, boulders) with no backwater. Second, the equation can be used for $0.002 \leq S_o \leq 0.04$ and $0.15 \leq R \leq 2.1$ m ($0.5 \leq R \leq 7.0$ ft). Results of the regression analysis indicated that for $R \geq 2.1$ m (7.0 ft), n did vary significantly with depth; therefore, as long as the bed and bank material remain stable, extrapolation to larger flows should not result in significant error. Third, the hydraulic radius does not include the wetted perimeter of the bed particles. Fourth, the streams used in the analysis had relatively small amounts of suspended sediment.

Vegetated channels present unique challenges from the viewpoint of estimating roughness. In grass-lined channels, the traditional approach assumed that n was a function of vegetal retardance and VR (Coyle, 1975). However, there are approaches more firmly based on the principles of fluid mechanics and the mechanics of materials (Kouwen, 1988; Kouwen and Li, 1980.) Data also exist that suggest that in such channels flow duration is not a factor as long as the vegetal elements are not destroyed or removed. Further, inundation times, and/ or hydraulic stresses, or both that are sufficient to damage vegetation have been found, as might be expected, to reduce the resistance to flow (Temple, 1991).

Petryk and Bosmajian (1975) presented a relation for Manning's n in vegetated channels based on a balance of the drag and gravitational forces, or

$$n = \phi R^{2/3} \left[\frac{C_d (Veg)_d}{2g} \right]^{1/2}$$

(3.40)

where C_d a coefficient accounting for the drag characteristics of the vegetation and $(Veg)_d$ the vegetation density. Flippin-Dudley (1997) has developed a rapid and objective procedure using a horizontal point frame to measure $(Veg)_d$. Equation (3.40) is limited because there is limited information regarding C_d for vegetation (Flippin-Dudley et al., 1997).

3.4.3 Equivalent Roughness Parameter k

In some cases, an equivalent roughness parameter k is used to estimate n. Equivalent roughness, sometimes called "roughness height," is a measure of the linear dimension of roughness elements but is not necessarily equal to the actual or even the average height of these elements. The advantage of using k instead of Manning's n is that k accounts for changes in the friction factor due to stage, whereas the Manning's n does not. The relationship between n and k for hydraulically rough channels is

$$n = \frac{\phi R^{1/6}}{\Gamma log_{10} \left(12.22 \, \dfrac{R}{k} \right)}$$

(3.41)

where $\Gamma = 32.6$ for English units and 18.0 for SI units.

With regard to Eq. (3.41), it is pertinent to observe that as R increases (equivalent to an increase in the depth of flow), n increases. Approximate values of k for selected materials are summarized in Table 3.4. For sand-bed channels, the following sediment sizes have been suggested by various investigators for estimating the value of k: $k = d_{65}$ (Einstein, 1950), $k = d_{90}$ (Meyer-Peter and Muller, 1948), and $k = d_{85}$ (Simons and Richardson, 1966).

3.4.4 Resistance in Compound Channels

In many designed channels and most natural channels, roughness varies along the perimeter of the channel, and it is necessary to estimate an equivalent value of n for the entire perimeter. In such cases, the channel is divided into N parts, each with an associated wetted perimeter (P_i), hydraulic radius (R_i), and roughness coefficient (n_i), and the equivalent roughness coefficient (n_e) is estimated by one of the following methods. Note that the wetted perimeter does not include the imaginary boundaries between the subsections.

1. Horton (1933) and Einstein and Banks (1950) developed methods of estimating n_e assuming that the average velocity in each of the subdivisions is the same as the average velocity of the total section. Then

$$n_e = \left[\frac{\sum\limits_{i=1}^{N} (P_i n_i^{3/2})}{P} \right]^{2/3}$$

(3.42)

2. Assuming that the total force resisting motion is equal to the sum of the subsection resisting forces,

$$n_e = \left[\frac{\sum\limits_{i=1}^{N} (P_i n_i^2)}{P}\right]^{1/2}$$

(3.43)

3. Assuming that the total discharge of the section is equal to the sum of the subsection discharges,

$$n_e = \frac{PR^{5/3}}{\sum\limits_{i=1}^{N} \dfrac{P_i R_i^{5/3}}{n_i}}$$

(3.44)

4. Weighting of resistance by area (Cox, 1973),

$$n_e = \frac{\sum\limits_{i=1}^{N} n_i A_i}{A}$$

(3.45)

5. The Colebatch method (Cox, 1973).

$$n_e = \left(\frac{\sum\limits_{i=1}^{N} A_i n_i^{3/2}}{A}\right)^{2/3}$$

(3.46)

3.4.5 Solution of Manning's Equation

The uniform flow rate is the product of the velocity of flow and the flow area, or

$$Q = VA = \frac{\phi}{n} AR^{2/3}\sqrt{S}$$

(3.47)

In Eq. (3.47), $AR^{2/3}$ is termed the *section factor* and, by definition, the conveyance of the channel is

$$K = \frac{\phi}{n} AR^{2/3}$$

(3.48)

Before the advent of computers, the solution of Eq. (3.34) or Eq. (3.47) to estimate the depth of flow for specified values of V (or Q), n, and S was accomplished in one of two ways: by trial and error or by the use of a graph of $AR^{2/3}$ versus y. In the age of the desktop computer, software is used to solve the equations of uniform flow. Trial and error and graphical approaches to the solution of the uniform flow equations can be found in any standard reference or text (e.g., French, 1985).

3.4.6 Special Cases of Uniform Flow

3.4.6.1 Normal and Critical Slopes. If Q, n, and y_N (normal depth of flow) and the channel section are defined, then Eq. (3.47) can be solved for the slope that allows the flow to occur as specified; by definition, this is a normal slope. If the slope is varied while the discharge and roughness are held constant, then a value of the slope such that normal flow occurs in a critical state can be found: that is, a slope such that normal flow occurs with $\text{Fr} = 1$. The slope obtained is the critical slope, but it also is a normal slope. The smallest critical slope, for a specified channel shape, roughness, and discharge is termed the *limiting critical slope*. The critical slope for a given normal depth is

$$S_c = \frac{gn^2 D_N}{\phi^2 R_N^{4/3}} \tag{3.49}$$

where the subscript N indicates the normal depth value of a variable and, for a wide channel,

$$S_c = \frac{gn^2}{\phi^2 y_c^{1/3}} \tag{3.50}$$

3.4.6.2 Sheetflow.   A special but noteworthy uniform flow condition is that of sheetflow. From the viewpoint of hydraulic engineering, a necessary condition for sheetflow is that the flow width must be sufficiently wide so that the hydraulic radius approaches the depth of flow. With this stipulation, the Manning's equation, Eq. (3.48), for a rectangular channel becomes

$$Q = \frac{\phi}{n} \, T y_N^{5/3} \sqrt{S} \tag{3.51}$$

where T = sheetflow width and y_N = normal depth of flow. Then, for a specified flow rate and sheetflow width, Eq. (3.51) can be solved for the depth of flow, or

$$y_N = \left(\frac{nQ}{\phi T \sqrt{S}} \right)^{3/5} \tag{3.52}$$

The condition that the value of the hydraulic radius approaches the depth of flow is not a sufficient condition. That is, this condition specifies no limit on the depth of flow, and there is general agreement that sheetflow has a shallow depth of flow. Appendix 3.A summarizes Manning's n values for overland and sheetflow.

3.4.6.3 Superelevation.   When a body of water moves along a curved path at constant velocity, it is acted for a force directed toward the center of the curvature of the path. When the radius of the curve is much larger than the top width of the water surface, it can be shown that the rise in the water surface at the outer channel boundary above the mean depth of flow in a straight channel (or *superelevation*) is

$$\Delta y = \frac{V^2 T}{2gr} \tag{3.53}$$

where r = the radius of the curve (Linsley and Franzini, 1979). It is pertinent to note that if the effects of the velocity distribution and variations in curvature across the channel are considered, the superelevation may be as much as 20 percent more than that estimated by Eq. (3.53) (Linsley and Franzini, 1979). Additional information regarding superelevation is available in Nagami et al., (1982) and U.S. Army Corps of Engineers (USACE, 1970).

3.5 GRADUALLY AND SPATIALLY VARIED FLOW

3.5.1 Introduction

The gradual variation in the depth of flow with longitudinal distance in an open channel is given by Eq. (3.19), or

$$\frac{dy}{dx} = \frac{S_o - S_f}{1 - Fr^2}$$

and two cases warrant discussion. In the first case, because the distance over which the change in depth is short it is appropriate to assume that boundary friction losses are small, or $S_f = 0$. When this is the case, important design questions involve abrupt steps in the bottom of the channel (Fig. 3.5) and rapid expansions or contractions of the channel (Fig. 3.6). The second case occurs when $S_f \neq 0$.

3.5.2 Gradually Varied Flow with $S_f = 0$

When $S_f = 0$ and the channel is rectangular in shape and has a constant width, Eq. (3.19) reduces to

$$(1 - Fr^2)\frac{dy}{dx} + \frac{dz}{dx} = 0 \tag{3.54}$$

and the following observations are pertinent (the observations also apply to channels of arbitrary shape):

1. If $dz/dx > 0$ (upward step) and $Fr < 1$, then dy/dx must be less than zero—depth of flow decreases as x increases.
2. If $dz/dx > 0$ (upward step) and $Fr > 1$, then dy/dx must be greater than zero—depth of flow increases as x increases.

Profile

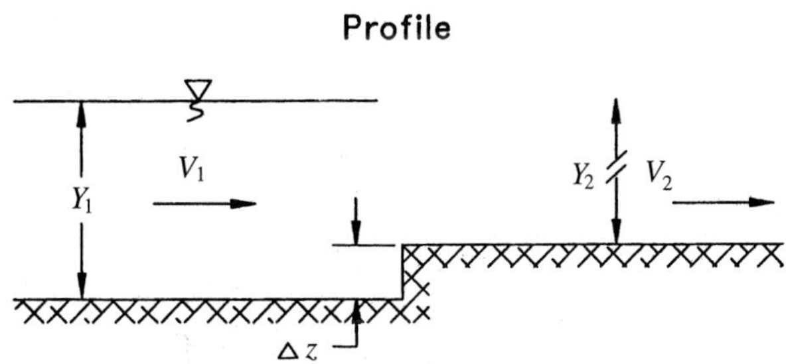

Profile

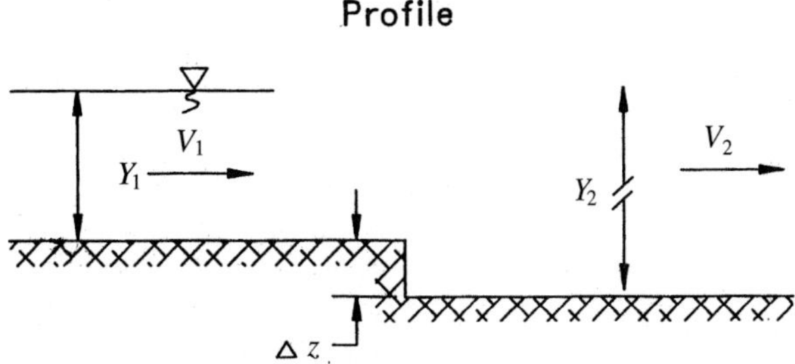

FIGURE 3.5 Definition of variables for gradually varied flow over positive and negative steps.

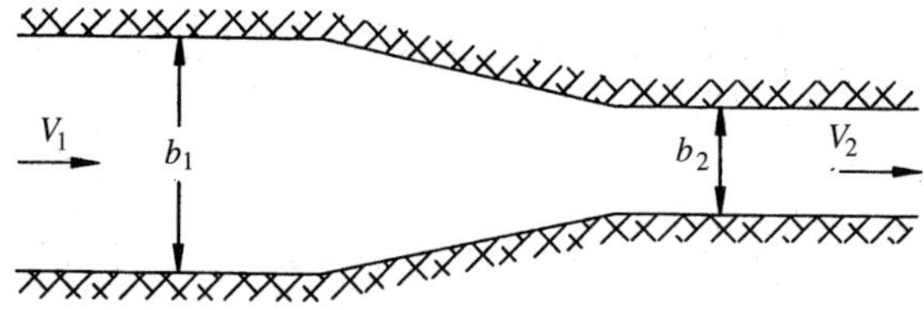

Plan

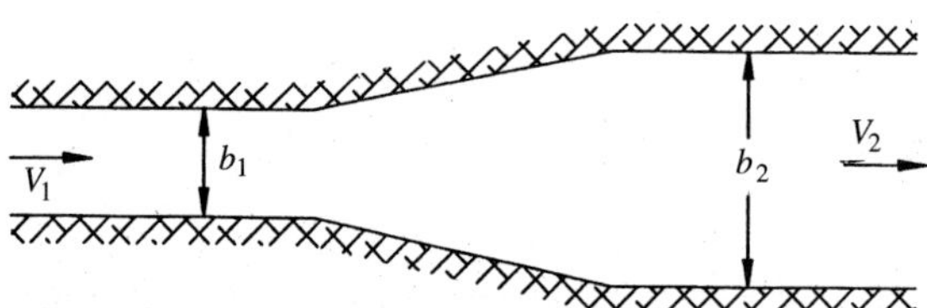

FIGURE 3.6 Definition of variables for gradually varied flow through contracting and expanding channel sections.

3. If $dz/dx < 0$ (downward step) and Fr < 1, then dy/dx must be greater than zero—depth of flow increases as x increases.

4. If $dz/dx < 0$ (downward step) and Fr > 1, then dy/dx must be less than zero—depth of flow decreases as x increases.

In the case of a channel of constant width with a positive or negative step, the relation between the specific energy upstream of the step and the specific energy downstream of the step is

$$E_1 = E_2 + \Delta z \tag{3.55}$$

In the case $dz/dx = 0$, if the channel is rectangular in shape but the width of the channel changes, it can be shown (French, 1985) that the governing equation is

$$(1 - \mathrm{Fr}^2)\frac{dy}{dx} - \mathrm{Fr}^2 \frac{y}{b}\frac{dT}{dx} = 0 \tag{3.56}$$

The following observations also apply to channels of arbitrary shape:

1. If $db/dx > 0$ (width increases) and Fr < 1, then dy/dx must be greater than zero–depth of flow increases as x increases.

2. If $db/dx > 0$ (width increases) and Fr > 1, then dy/dx must be less than zero—depth of flow decreases as x increases.

3. If $db/dx < 0$ (width decreases) and Fr < 1, then dy/dx must be less than zero—depth of flow decreases as x increases.

4. If $db/dx < 0$ (width decreases) and Fr > 1, then dy/dx must be greater than zero—depth of flow increases as x increases.

In this case, the relation between the specific energy upstream of the contraction (expansion) and the specific energy downstream of the step contraction (expansion) is

$$E_1 = E_2 \tag{3.57}$$

It is pertinent to note that in the case of supercritical flow, channel expansions and contractions may result in the formation of waves.

Additional information regarding steps, expansions, and contractions can be found in any standard reference or text on open-channel hydraulics (e.g., French, 1985).

3.5.3 Gradually Varied Flow with $S_f \neq 0$

In the case where S_f cannot be neglected, the water surface profile must estimated. For a channel of arbitrary shape, Eq. (3.19) becomes

$$\frac{dy}{dx} = \frac{S_o - S_f}{1 - \dfrac{Q^2 T}{gA^3}} = S_o - \frac{n^2 Q^2 P^{4/3}}{1 - \dfrac{Q^2 T}{gA^3}} \tag{3.60}$$

For a specified value of Q, Fr and S_f are functions of the depth of flow y. For illustrative purposes, assume a wide channel; in such a channel, Fr and S_f will vary in much the same way with y since $P \simeq T$ and both S_f and Fr have a strong inverse dependence on the flow area. In addition, as y increases, both S_f and Fr decrease. By definition, $S_f = S_o$ when $y = y_N$. Given the foregoing, the following set of inequalities must apply:

$$S_f > S_o \qquad \text{for} \qquad y < y_N$$

$$\text{Fr} > 1 \qquad \text{for} \qquad y < y_c$$

$$S_f < S_o \qquad \text{for} \qquad y > y_N$$

and

$$\text{Fr} < 1 \qquad \text{for} \qquad y > y_c$$

These inequalities divide the channel into three zones in the vertical dimension. By convention, these zones are labeled 1 to 3 starting at the top. Gradually varied flow profiles are labeled according to the scheme defined in Table 3.5.

For a channel of arbitrary shape, the *standard step methodology* of calculating the gradually varied flow profile is commonly used: for example, HEC-2 (USACE, 1990) or HEC-RAS (USACE, 1997). The use of this methodology is subject to the following assumptions: (1) steady flow, (2) gradually varied flow, (3) one-dimensional flow with correction for the horizontal velocity distribution, (4) small channel slope, (5) friction slope (averaged) constant between two adjacent cross sections, and (6) rigid boundary conditions.

The application of the energy equation between the two stations shown in Fig. 3.7 yields

$$z_1 + \alpha_1 \frac{V_1^2}{2g} = z_1 + \alpha_2 \frac{V_2^2}{2g} + h_f + h_e \tag{3.61}$$

where z_1 and z_2 = elevation of the water surface above a datum at Stations 1 and 2
$\quad\quad\quad h_e$ = eddy and other losses incurred in the reach
$\quad\quad\quad h_f$ = reach friction loss

The friction loss can be obtained by multiplying a representative friction slope, S_f, by the

TABLE 3.5   Classifications of Gradually Varied Flow Profiles

		Profile designation				
Channel slope (1)	Zone 1 (2)	Zone 2 (3)	Zone 3 (4)	Relation of y to y_N and y_c (5)	Type of curve (6)	Type of flow (7)
Mild $0 < S_o < S_c$	M1			$y > y_N > y_c$	Backwater $(dy/dx > 0)$	Subcritical
		M2		$y_N > y > y_c$	Drawdown $(dy/dx < 0)$	Subcritical
			M3	$y_N > y_c > y$	Backwater $(dy/dx > 0)$	Supercritical
Critical $S_o = S_c > 0$	C1			$y > y_c = y_N$	Backwater $(dy/dx > 0)$	Subcritical
		C2		$y = y_N = y_c$	Parallel to channel bottom $(dy/dx = 0)$	Uniform critical
			C3	$y_c = y_N > y$	Backwater $(dy/dx > 0)$	Supercritical
Steep $S_o > S_c > 0$	S1			$y > y_c > y_N$	Backwater $(dy/dx > 0)$	Subcritical
		S2		$y_c > y > y_N$	Drawdown $(dy/dx < 0)$	Supercritical
			S3	$y_c > y_N > y$	Backwater $(dy/dx > 0)$	Supercritical
Horizontal $S_o = 0$	None					
		H2		$y_N > y > y_c$	Drawdown $(dy/dx < 0)$	Subcritical
			H3	$y_N > y_c > y$	Backwater $(dy/dx > 0)$	Supercritical
Adverse $S_o < 0$	None					
		A2		$y_N > y > y_c$	Drawdown $(dy/dx < 0)$	Subcritical
			A3	$y_N > y_c > y$	Backwater $(dy/dx > 0)$	Supercritical

length of the reach, L. Four equations can be used to approximate the friction loss between two cross sections:

$$\overline{S}_f = \left(\frac{Q_1 + Q_2}{K_1 + K_2}\right)^2 \text{ (average conveyance)} \tag{3.62}$$

$$\overline{S}_f = \frac{S_{f1} + S_{f2}}{2} \text{ (average friction slope)} \tag{3.63}$$

$$\overline{S}_f = \frac{2S_{f1}S_{f2}}{S_{f1} + S_{f2}} \text{ (harmonic mean friction slope)} \tag{3.64}$$

and

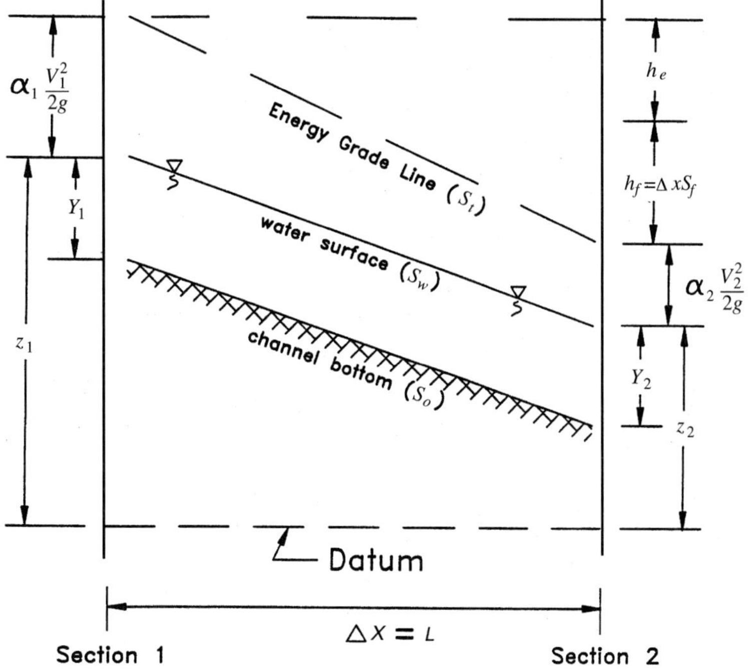

FIGURE 3.7 Energy relationship between two channel sections.

$$\overline{S}_f = \sqrt{S_{f1}S_{f2}} \text{ (geometric mean friction slope)} \tag{3.65}$$

The selection of a method to estimate the friction slope in a reach is an important decision and has been discussed in the literature. Laurenson (1986) suggested that the "true" friction slope for an irregular cross section can be approximated by a third-degree polynomial. He concluded that the average friction slope method produces the smallest maximum error, but not always the smallest error, and recommended its general use along with the systematic location of cross sections. Another investigation based on the analysis of 98 sets of natural channel data showed that there could be significant differences in the results when different methods of estimating the friction slope were used (USACE, 1986). This study also showed that spacing cross sections 150m (500 ft) a part eliminated the differences.

The eddy loss takes into account cross section contractions and expansions by multiplying the absolute difference in velocity heads between the two sections by a contraction or expansion coefficient, or

$$h_e = C_x \left| \alpha_1 \frac{V_2^2}{2g} - \alpha_2 \frac{V_2^2}{2g} \right| \tag{3.66}$$

There is little generalized information regarding the value of the expansion (C_e) or the contraction coefficient (C_c). When the change in the channel cross section is small, the coefficients C_e and C_c are typically on the order of 0.3 and 0.1, respectively (USACE, 1990). However, when the change in the channel cross section is abrupt, such as at bridges, C_e and C_c may be as high as 1.0 and 0.6, respectively (USACE, 1990).

With these comments in mind,

$$H_1 = z_1 + \alpha_1 \frac{V_1^2}{2g} \tag{3.67}$$

and

$$H_2 = z_2 + \alpha_2 \frac{V_2^2}{2g} \tag{3.68}$$

With these definitions, Eq. (3.61) becomes

$$H_1 = H_2 + h_f + h_e \tag{3.69}$$

Eq. (3.69) is solved by trial and error: that is, assuming H_2 is known and given a longitudinal distance, a water surface elevation at Station 1 is assumed, which allows the computation of H_1 by Eq. (3.67). Then, h_f and h_e are computed and H_1 is estimated by Eq. (3.67). If the two values of H_1 agree, then the assumed water surface elevation at Station 1 is correct.

Gradually varied water surface profiles are often used in conjunction with the peak flood flows to delineate areas of inundation. The underlying assumption of using a steady flow approach in an unsteady situation is that flood waves rise and fall gradually. This assumption is of course not valid in areas subject to flash flooding such as the arid and semiarid Southwestern United States (French, 1987).

In summary, the following principles regarding gradually varied flow profiles can be stated:

1. The sign of dy/dx can be determined from Table 3.5.
2. When the water surface profile approaches normal depth, it does so asymptotically.
3. When the water surface profile approaches critical depth, it crosses this depth at a large but finite angle.
4. If the flow is subcritical upstream but passes through critical depth, then the feature that produces critical depth determines and locates the complete water surface profile. If the upstream flow is supercritical, then the control cannot come from the downstream.
5. Every gradually varied flow profile exemplifies the principle that subcritical flows are controlled from the downstream while supercritical flows are controlled from upstream. Gradually varied flow profiles would not exist if it were not for the upstream and downstream controls.
6. In channels with horizontal and adverse slopes, the term "normal depth of flow" has no meaning because the normal depth of flow is either negative or imaginary. However, in these cases, the numerator of Eq. (3.60) is negative and the shape of the profile can be deduced.

Any method of solving a gradually varied flow situation requires that cross sections be defined. Hoggan (1989) provided the following guidelines regarding the location of cross sections:

1. They are needed where there is a significant change in flow area, roughness, or longitudinal slope.
2. They should be located normal to the flow.
3. They should be located in detail—upstream, within the structure, and downstream-at structures such as bridges and culverts. They are needed at all control structures.
4. They are needed at the beginning and end of reaches with levees.
5. They should be located immediately below a confluence on a main stem and immediately above the confluence on a tributary.

6. More cross sections are needed to define energy losses in urban areas, channels with steep slopes, and small streams than needed in other situations.

7. In the case of HEC-2, reach lengths should be limited to a maximum distance of 0.5 mi for wide floodplains and for slopes less than 38,550 m (1800 ft) for slopes equal to or less than 0.00057, and 370 m (1200 ft) for slopes greater than 0.00057 (Beaseley, 1973).

3.6 GRADUALLY AND RAPIDLY VARIED UNSTEADY FLOW

3.6.1 Gradually Varied Unsteady Flow

Many important open-channel flow phenomena involve flows that are unsteady. Although a limited number of gradually varied unsteady flow problems can be solved analytically, most problems in this category require a numerical solution of the governing equations. Examples of gradually varied unsteady flows include flood waves, tidal flows, and waves generated by the slow operation of control structures, such as sluice gates and navigational locks.

The mathematical models available to treat gradually varied unsteady flow problems are generally divided into two categories: models that solve the complete Saint Venant equations and models that solve various approximations of the Saint Venant equations. Among the simplified models of unsteady flow are the kinematic wave, and the diffusion analogy. The complete solution of the Saint Venant equations requires that the equations be solved by either finite difference or finite element approximations.

The one dimensional Saint Venant equations consist of the equation of continuity

$$\frac{\partial y}{\partial t} + y\frac{\partial v}{\partial x} + u\frac{\partial y}{\partial x} = 0 \tag{3.70a}$$

and the conservation of momentum equation

$$\frac{\partial v}{\partial t} + v\frac{\partial v}{\partial x} + g\frac{\partial y}{\partial x} - g(S_o - S_f) = 0 \tag{3.71a}$$

An alternate form of the continuity and momentum equations is

$$T\frac{\partial y}{\partial t} + \frac{\partial(Au)}{\partial x} = 0 \tag{3.70b}$$

and

$$\frac{1}{g}\frac{\partial v}{\partial t} + \frac{v}{g}\frac{\partial v}{\partial x} + \frac{\partial y}{\partial x} + S_f - S_o = 0 \tag{3.71b}$$

By rearranging terms, Eq. (3.71b) can be written to indicate the significance of each term for a particular type of flow, or

$$S_f = S_o \Big\rfloor_{\text{steady}} - \frac{\partial y}{\partial x} - \left(\frac{v}{g}\right)\frac{\partial v}{\partial x}\Big\rfloor_{\text{steady,nonuniform}} - \left(\frac{1}{g}\right)\frac{\partial v}{\partial t}\Big\rfloor_{\text{unsteady,nonuniform}} \tag{3.72}$$

Equations (3.70) and (3.71) compose a group of gradually varied unsteady flow models that are termed *complete dynamic models*. Being complete, this group of models can provide accurate results; however, in many applications, simplifying assumptions regarding the relative importance of various terms in the conservation of momentum equation (Eq. 3.71) leads to other equations, such the kinematic and diffusive wave models (Ponce, 1989).

The governing equation for the *kinematic wave model* is

$$\frac{\partial Q}{\partial t} + (\epsilon V)\,\frac{\partial Q}{\partial x} = 0 \tag{3.73}$$

where ϵ = a coefficient whose value depends on the frictional resistance equation used ($\epsilon = 5/3$ when Manning's equation is used). The kinematic wave model is based on the equation of continuity and results in a wave being translated downstream. The kinematic wave approximation is valid when

$$\frac{t_R S_o V}{y} \geq 85 \tag{3.74}$$

where t_R 5 time of rise of the inflow hydrograph (Ponce, 1989).

The governing equation for the *diffusive wave model* is

$$\frac{\partial Q}{\partial t} + (\epsilon V)\,\frac{\partial Q}{\partial x} = \left(\frac{Q}{2TS_o}\right)\frac{\partial^2 Q}{\partial x^2} \tag{3.75}$$

where the left side of the equation is the kinematic wave model and the right side accounts for the physical diffusion in a natural channel. The diffusion wave approximation is valid when (Ponce, 1989),

$$t_R S_o \left(\frac{g}{y}\right)^{0.5} \geq 15 \tag{3.76}$$

If the foregoing dimensionless inequalities (Eq. 3.74 and 3.76) are not satisfied, then the complete dynamic wave model must be used. A number of numerical methods can be used to solve these equations (Chaudhry, 1987; French; 1985, Henderson, 1966; Ponce, 1989).

3.6.2 Rapidly Varied Unsteady Flow

The terminology "rapidly varied unsteady flow" refers to flows in which the curvature of the wave profile is large, the change of the depth of flow with time is rapid, the vertical acceleration of the water particles is significant relative to the total acceleration, and the effect of boundary friction can be ignored. Examples of rapidly varied unsteady flow include the catastrophic failure of dams, tidal bores, and surges that result from the quick operation of control structures such as sluice gates. A surge producing an increase in depth is termed a positive surge, and one that causes a decrease in depth is termed a negative surge. Furthermore, surges can go either upstream or downstream, thus giving rise to four basic types (Fig. 3.8). Positive surges generally have steep fronts, often with rollers, and are stable. In contrast, negative surges are unstable, and their form changes with the advance of the wave.

Consider the case of a positive surge (or wave) traveling at a constant velocity (wave celerity) c up a horizontal channel of arbitrary shape (Fig. 3.8b). Such a situation can result from the rapid closure of a downstream sluice gate. This unsteady situation is converted to a steady situation by applying a velocity c to all sections; that is, the coordinate system is moving at the velocity of the wave. Applying the continuity equation between Sections 1 and 2

$$(V_1 + c)A_1 = (V_2 + c)A_2 \tag{3.77}$$

Since there are unknown losses associated with the wave, the momentum equation rather than the energy equation is applied between Sections 1 and 2 or

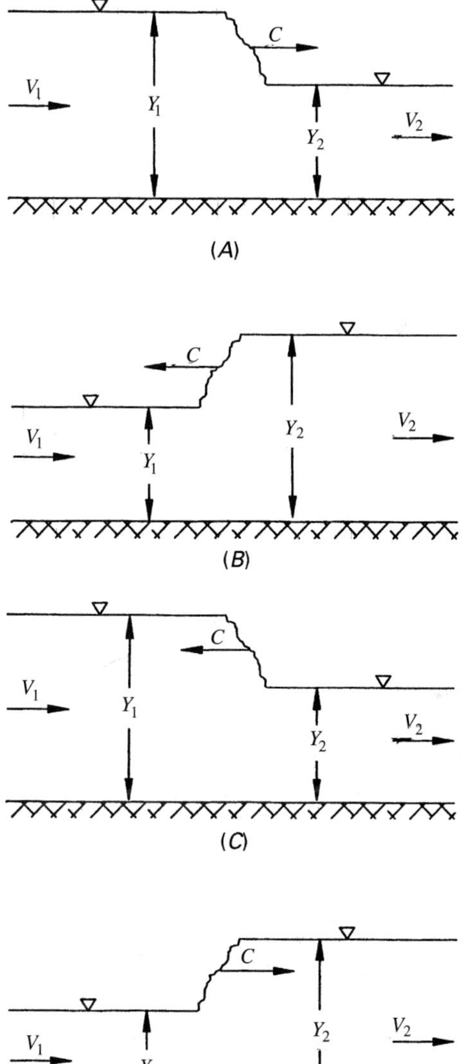

FIGURE 3.8 Definition of variables for simple surges moving in an open channel.

$$\gamma A_1 \bar{z}_1 - \gamma A_2 \bar{z}_2 = \frac{\gamma}{g} y_1 (V_1 + c)(V_2 + c - V_1 - c) \tag{3.78}$$

where boundary friction has been ignored. Eliminating V_2 in Eq. (3.78) by manipulation of Eq. (3.77) yields

$$V_1 + c = \left[\frac{g \left(\dfrac{A_2}{A_1} \right) (A_1 \bar{z}_1 - A_2 \bar{z}_2)}{A_1 - A_2} \right]^{0.5} \tag{3.79}$$

In the case of a rectangular channel, Eq. (3.79) reduces to

$$V_1 + c = \sqrt{gy_1} \left[\frac{y_2}{2y_1} \left(1 + \frac{y_2}{y_1} \right) \right]^{0.5} \tag{3.80}$$

When the slope of a channel becomes very steep, the resulting supercritical flow at normal depth may develop into a series of shallow water waves known as *roll waves*. As these waves progress downstream, they eventually break and form hydraulic bores or shock waves. When this type of flow occurs, the increased depth of flow requires increased freeboard, and the concentrated mass of the wavefronts may require additional structural factors of safety.

Escoffier (1950) and Escoffier and Boyd (1962) considered the theoretical conditions under which a uniform flow must be considered unstable. Whether roll waves form or not is a function of the *Vedernikov number* (Ve), the *Montuori number* (Mo), and the concentration of sediment in the flow. When the Manning equation is used, the Ve is

$$\text{Ve} = \frac{2}{3} \Gamma \text{Fr} \tag{3.81a}$$

and if the Chezy equation is used

TABLE 3.6   Shape Factor for Common Channel Sections

Channel definition (1)		Γ (2)
Rectangle		$\dfrac{b}{b + 2y}$
Trapezoid with unequal side slopes		$1 - \dfrac{R(\sqrt{1 + z_1^2} + \sqrt{1 + z_2^2})}{T}$
Circle		$1 - \dfrac{\theta - \sin(\theta)}{\theta[1 - \cos(\theta)]}$

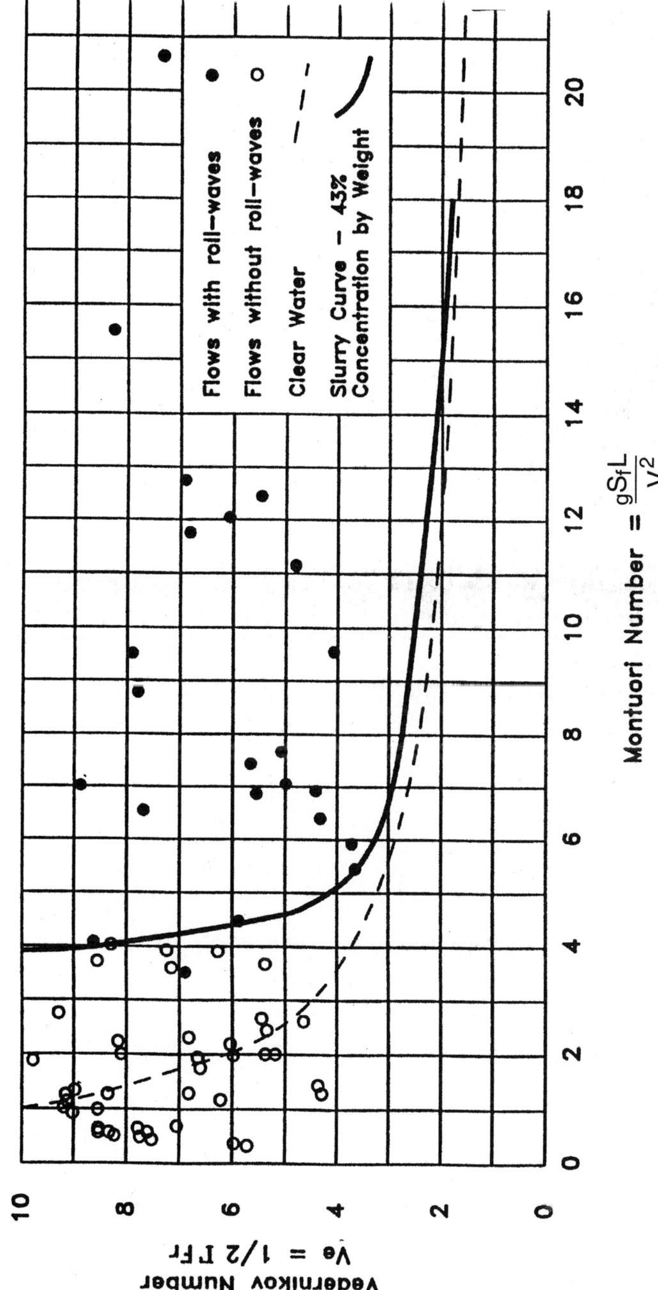

FIGURE 3.9 Flow stability as a function of the Vedernikov and Montuori numbers for clear water and slurry flow. (Based on data from Montuori, 1963; Niepelt and Locher, 1989)

$$Ve = \frac{1}{2}\Gamma Fr \tag{3.81b}$$

Fr should be computed using Eq. (3.11) and Γ = a channel shape factor (Table 3.6) or

$$\Gamma = 1 - R\frac{dP}{dA} \tag{3.82}$$

When Mo $>>$ 1, flow instabilities should expected. The Montuori number is given by

$$Mo = \frac{gS_fL}{V^2} \tag{3.83}$$

It is appropriate to note that in some publications (e.g., Aisenbrey et al., 1978) Mo is the inverse of Eq. (3.83). Figure 3.9 provides a basis for deciding whether roll waves will form in a given situation. In the figure, data from Niepelt and Locher (1989) for a slurry flow are also plotted. The Niepelt and Locher data suggest that flow stability also is a function of the concentration of sediment.

3.7 CONCLUSION

The foregoing sections provide the basic principles on which the following chapters on design are based. Two observations are pertinent. First, open-channel hydraulics is incrementally progressing. That is, over the past several decades, there have been incremental advances that primarily have added details, often important details, but no major new advances. Second, open-channel hydraulics remains a one-dimensional analytic approach. However, the assumption of a one-dimensional approach may not be valid in many situations: for example, nonprismatic channels, flow downstream of a partially breached dam, or lateral flow over a spillway. In some of these cases, the one-dimensional approach may provide an approximation that is suitable for design. In other cases, however, a two– or three– dimensional approach should be used. Additional information regarding two– and three– dimensional approaches can be found in Chaudhry (1993).

APPENDIX 3.A: VALUES OF THE ROUGHNESS COEFFICIENT n*

	Values of the roughness coefficient n		
Type of channel	Minimum	Normal	Maximum
A. *Closed Conduits flowing partly full*			
A–1 Metal			
a. Brass, smooth	0.009	**0.010**	0.013
b. Steel			
1. Lockbar and welded	0.010	0.012	0.014
2. Riveted and spiral	0.013	0.016	0.017
c. Cast iron			
1. Coated	0.010	0.013	0.014
2. Uncoated	0.011	0.014	0.016
d. Wrought iron			

Type of channel	Values of the roughness coefficient n		
	Minimum	Normal	Maximum
1. Black	0.012	0.014	0.015
2. Galvanized	0.013	0.016	0.017
e. Corrugated metal			
1. Subdrain	0.017	0.019	0.030
2. Storm drain	0.021	**0.024**	0.030
A–2 Non-metal			
a. Lucite	0.008	0.009	0.010
b. Glass	0.009	**0.010**	0.013
c. Cement			
1. Neat, surface	0.010	0.011	0.013
2. Mortar	0.011	0.013	0.015
d. Concrete			
1. Culvert, straight and free of debris	0.010	0.011	0.013
2. Culvert, with bends, connections, and some debris	0.011	**0.013**	0.014
3. Finished	0.011	0.012	0.014
4. Sewer and manholes, inlet, etc, straight	0.013	0.015	0.017
5. Unfinished, steel form	0.012	0.013	0.014
6. Unfinished, smooth wood form	0.012	**0.014**	0.016
7. Unfinished, rough wood form	0.015	0.017	0.020
e. Wood			
1. Stave	0.010	0.012	0.014
2. Laminated, treated	0.015	0.017	0.020
f. Clay			
1. Common drainage tile	0.011	**0.013**	0.017
2. Vitrified sewer	0.011	0.014	0.017
3. Vitrified sewer with manholes, inlet, etc.	0.013	0.015	0.017
4. Vitrified subdrain with open joint	0.014	0.016	0.018
g. Brickwork			
1. Glazed	0.011	0.013	0.015
2. Lined with cement mortar	0.012	0.015	0.017
h. Sanitary sewers coated with sewage slimes with bends and connections	0.012	0.013	0.016
i. Paved invert, sewer, smooth bottom	0.016	0.019	0.020
j. Rubble masonry, cemented	0.018	0.025	0.030
k. Polyethylene pipe	0.009	—	—
l. Polyvinyl chloride	0.010	—	—
B. *Lined or Built–up Channels*			
B–1 Metal			
a. Smooth steel surface			
1. Unpainted	0.011	**0.012**	0.014
2. Painted	0.012	0.013	0.017
b. Corrugated	0.021	0.025	0.030
B–2 Nonmetal			
a. Cement			

Type of channel	Values of the roughness coefficient n		
	Minimum	Normal	Maximum
1. Neat, surface	0.010	0.011	0.013
2. Mortar	0.011	0.013	0.015
b. Wood			
1. Planed, untreated	0.010	0.012	0.014
2. Planed, creosoted	0.011	0.012	0.014
3. Unplaned	0.011	0.013	0.015
4. Plank with battens	0.012	0.015	0.018
5. Lined with roofing paper	0.010	0.014	0.017
c. Concrete			
1. Trowel finish	0.011	**0.013**	0.015
2. Float finish	0.013	0.015	0.016
3. Finished, with gravel on bottom	0.015	0.017	0.020
4. Unfinished	0.014	0.017	0.020
5. Gunite, good section	0.016	0.019	0.023
6. Gunite, wavy section	0.018	0.022	0.025
7. On good excavated rock	0.017	0.020	—
8. On irregular excavated rock	0.022	0.027	—
d. Concrete bottom float with sides of			
1. Dressed stone in mortar	0.015	0.017	0.020
2. Random stone in mortar	0.017	0.020	0.024
3. Cement, rubble masonry, plastered	0.016	0.020	0.024
4. Cement rubble masonry	0.020	0.025	0.030
5. Dry rubble or riprap	0.020	0.030	0.035
e. Gravel bottom with sides of			
1. Formed concrete	0.017	0.020	0.025
2. Random stone in mortar	0.020	0.023	0.026
3. Dry rubble or riprap	0.023	0.033	0.036
f. Brick			
1. Glazed	0.011	**0.013**	0.015
2. In cement mortar	0.012	**0.015**	0.018
g. Masonry			
1. Cemented rubble	0.017	0.025	0.030
2. Dry rubble	0.023	0.032	0.035
h. Dressed ashlar	0.013	0.015	0.017
i. Asphalt			
1. Smooth	0.013	0.013	—
2. Rough	0.016	0.016	—
j. Vegetal cover	0.030	—	0.500
C. *Excavated or Dredged*			
C–1 General			
a. Earth, straight and uniform			
1. Clean and recently completed	0.016	0.018	0.020
2. Clean, after weathering	0.018	**0.022**	0.025
3. Gravel, uniform section, clean	0.022	0.025	0.030
4. With short grass, few weeds	0.022	0.027	0.033
b. Earth, winding and sluggish			
1. No vegetation	0.023	0.025	0.030
2. Grass, some weeds	0.025	0.030	0.033

| | Values of the roughness coefficient n | | |
Type of channel	Minimum	Normal	Maximum
3. Dense weeds or aquatic plants in deep channels	0.030	0.035	0.040
4. Earth bottom and rubble sides	0.028	0.030	0.035
5. Stony bottom and weedy banks	0.025	0.035	0.040
6. Cobble bottom and clean sides	0.030	0.040	0.050
c. Dragline-excavated or dredged			
1. No vegetation	0.025	0.028	0.033
2. Light brush on banks	0.035	0.050	0.060
d. Rock cuts			
1. Smooth and uniform	0.025	0.035	0.040
2. Jagged and irregular	0.035	0.040	0.050
e. Channels not maintained, weeds and brush uncut			
1. Dense weeds, high as flow depth	0.050	0.080	0.120
2. Clean bottom, brush on sides	0.040	0.050	0.080
3. Same, highest stage of flow	0.045	0.070	0.110
4. Dense brush, high stage	0.080	0.100	0.14
C–2 Channels with maintained vegetation and velocities of 2 and 6 ft/s			
a. Depth of flow up to 0.7 ft			
1. Bermuda grass, Kentucky bluegrass, buffalo grass			
Mowed to 2 in	0.07		0.045
Length 4 to 6 in	0.09		0.05
2. Good stand, any grass			
Length approx. 12 in	0.18		0.09
Length approx. 24 in	0.30		0.15
3. Fair stand, any grass			
Length approx. 12 in	0.014		0.08
Length approx. 24 in	0.25		0.13
b. Depth of flow up to 0.7–1.5 ft			
1. Bermuda grass, Kentucky bluegrass, buffalo grass			
Mowed to 2 in	0.05		0.035
Length 4–6 in	0.06		0.04
2. Good stand, any grass			
Length approx. 12 in	0.12		0.07
3. Length approx. 24 in	0.20		0.10
Fair stand, any grass			
Length approx. 12 in	0.10		0.16
Length approx. 24 in	0.17		0.09
D. *Natural streams*			
D–1 Minor streams (top width at flood stage < 100 ft)			
a. Streams on plain			
1. Clean, straight, full stage no rifts or deep pools	0.025	**0.030**	0.033
2. Same as above, but with more stones and weeds	0.030	0.035	0.040
3. Clean, winding, some pools and shoals	0.033	0.040	0.045

	Values of the roughness coefficient n		
Type of channel	Minimum	Normal	Maximum
4. Same as above, but with some weeds and stones	0.035	0.045	0.050
5. Same as above, lower stages more ineffective slopes and sections	0.040	0.048	0.055
6. Same as no. 4, more stones	0.045	0.050	0.060
7. Sluggish reaches, weedy, deep pools	0.050	0.070	0.080
8. Very weedy, reaches, deep pools or floodways with heavy stand of timber and underbrush	0.075	0.100	0.150
b. Mountain streams, no vegetation in channel, banks usually steep, trees and brush along banks submerged at high stages			
1. Bottom: gravels, cobbles and few boulders	0.030	0.040	0.050
2. Bottom: cobbles with large boulders	0.040	0.050	0.070
D–2 Floodplains			
a. Pasture, no brush			
1. Short grass	0.025	0.030	0.035
2. High grass	0.030	0.035	0.050
b. Cultivated areas			
1. No crop	0.020	0.030	0.040
2. Mature row crops	0.025	0.035	0.045
3. Mature field crops	0.030	0.040	0.050
c. Brush			
1. Scattered brush, heavy weeds	0.035	0.050	0.070
2. Light brush and trees in winter	0.035	0.050	0.060
3. Light brush and trees in summer	0.040	0.070	0.110
4. Medium to dense brush in winter	0.045	0.070	0.110
5. Medium to dense brush in summer	0.070	0.100	0.160
d. Trees			
1. Dense willows, summer, straight	0.110	0.150	0.200
2. Cleared land with tree stumps, no sprouts	0.030	0.040	0.050
3. Same as above but with a heavy growth of sprouts	0.050	0.060	0.080
4. Heavy stand of timber, a few down trees, little undergrowth, flood stage below branches	0.080	0.100	0.120
5. Same as above, but with flood stage reaching branches	0.100	0.120	0.160
D–3 Major streams (top width at flood stage > 100 ft); the n value is less that for minor streams of similar description because banks offer less effective resistance			

| | Values of the roughness coefficient n | | |
Type of channel	Minimum	Normal	Maximum
a. Regular section with no boulders or brush	0.025	—	0.060
b. Irregular and rough section	0.035	—	0.100
D–4 Alluvial sandbed channels (no vegetation and data is limited to sand channels with $D50 < 1.0$ mm			
a. Tranquil flow, Fr < 1			
1. Plane bed	0.014	—	0.020
2. Ripples	0.018	—	0.030
3. Dunes	0.020	—	0.040
4. Washed out dunes or transition	0.014	—	0.025
b. Rapid flow, Fr > 1			
1. Standing waves	0.010	—	0.015
2. Antidunes	0.012	—	0.020
E. *Overland Flow* (*Sheetflow*)			
E–1 Vegetated areas			
a. Dense turf	0.17	—	0.80
b. Bermuda and dense grass	0.17	—	0.48
c. Average grass cover	0.20	—	0.40
d. Poor grass cover on rough surface	0.20	—	0.30
e. Short prairie grass	0.10	—	0.20
f. Shrubs and forest litter, pasture	0.30	—	0.40
g. Sparse vegetation	0.05	—	0.13
h. Sparse rangeland with debris			
1. 0% cover	0.09	—	0.34
2. 20% cover	0.05	—	0.25
E–2 Plowed or tilled fields			
a. Fallow—no residue	0.008	—	0.012
b. Conventional tillage	0.06	—	0.22
c. Chisel plow	0.06	—	0.16
d. Fall disking	0.30	—	0.50
e. No till—no residue	0.04	—	0.10
f. No till (20–40% residue cover)	0.07	—	0.17
g. No till (100% residue cover)	0.17	—	0.47
E–3 Other surfaces			
a. Open ground with debris	0.10	—	0.20
b. Shallow flow on asphalt or concrete	0.10	—	0.15
c. Fallow fields	0.08	—	0.12
d. Open ground, no debris	0.04	—	0.10
f. Asphalt or concrete	0.02	—	0.05

Source: From Chow (1959), Richardson et al. (1987), Simons, Li, & Associates (SLA), 1982, and others.
*The values in bold are recommended for design.

REFERENCES

Ackers, P., "Resistance to Fluids flowing in Channels and Pipes," Hydraulic Research Paper No. 1, Her Majety's Stationery Office, London, 1958.

Aisenbrey, A. J., Jr., R. B., Hayes, H. J., Warren, D. L., Winsett, and R. B. Young, *Design of Small Canal Structures,* U.S. Department of Interior, Bureau of Reclamation, Washington, DC 1978.

Arcement, G. J., and V. R. Schneider, "Guide for Selecting Manning's Roughness Coefficients for Natural Channels and Flood Plains," Water Supply Paper 2339, U.S. Geological Survey, Washington, DC, 1989.

Barnes, H. H., "Roughness Characteristics of Natural Channels," U.S. Geological Survey Water Supply Paper No. 1849, U.S. Geological Survey, Washington, DC. 1967.

Beasley, J. G., *An Investigation of the Data Requirements of Ohio for the HEC-2 Water Surface Profiles Model,* Master's thesis, Ohio State University, Columbus, 1973.

Chaudhry, M. H., *Open-Channel Flow,* Prentice-Hall, New York 1993.

Chaudhry, M. H., *Applied Hydraulic Transients,* Van Nostrand Reinhold, New York, 1987.

Chow, V. T., *Open-Channel Hydraulics,* McGraw-Hill, New York, 1959.

Cox, R. G., "Effective Hydraulic Roughness for Channels Having Bed Roughness Different from Bank Roughness," Miscellaneous Paper H-73-2, U.S. Army Engineers Waterways Experiment Station, Vicksburg, MS, 1973.

Coyle, J. J. "Grassed Waterways and Outlets," *Engineering Field Manual,* U.S. Soil Conservation Service, Washington, DC, April, 1975, pp. 7-1–7-43.

Einstein, H. A., "The Bed Load Function for Sediment Transport in Open Channel Flows. Technical Bulletin No. 1026, U.S. Department of Agriculture, Washington, DC, 1950.

Einstein, H. A., and R. B. Banks, "Fluid Resistance of Composite Roughness," *Transactions of the American Geophysical Union,* 31(4): 603–610, 1950.

Escoffier, F. F., "A Graphical Method for Investigating the Stability of Flow in Open Channels or in Closed Conduits Flowing Full," *Transactions of the American Geophysical Union,* 31(4), 1950.

Escoffier, F. F., and M. B. Boyd, "Stability Aspects of Flow in Open Channels," *Journal of the Hydraulics Division,* American Society of Civil Engineers, 88(HY6): 145–166, 1962.

Fischenich, J. C., "Hydraulic Impacts of Riparian Vegetation: Computation of Resistance," EIRP Technical Report EL-96-XX, U.S. Army Engineers Waterways Experiment Station, Vicksburg, MS, August 1996.

Flippin-Dudley, S. J., "Vegetation Measurements for Estimating Flow Resistance," Doctoral dissertation, Colorado State University, Fort Collins, 1997.

Flippin-Dudley, S. J., S. R. Abt, C. D. Bonham, C. C. Watson, and J. C. Fischenich, "A Point Quadrant Method of Vegetation Measurement for Estimating Flow Resistance," Technical Report No. EL-97-XX, U.S. Army Engineers Waterways Experiment Station, Vicksburg, MS, 1997.

French, R. H., *Hydraulic Processes on Alluvial Fans.* Elsevier, Amsterdam, 1987.

French, R. H., *Open-Channel Hydraulics,* McGraw-Hill, New York, 1985.

Guy, H. P., D. B. Simons, and E. V. Richardson, "Summary of Alluvial Channel Data from Flume Experiments, 1956-61," Professional Paper No. 462-1, U.S. Geological Survey, Washington, DC, 1966.

Henderson, F. M., *Open Channel Flow,* Macmillan, New York, 1966.

Hoggan, D. H., *Computer-Assisted Floodplain Hydrology & Hydraulics,* McGraw-Hill, New York, 1989.

Horton, R. E., "Separate Roughness Coefficients for Channel Bottom and Sides," *Engineering News Record,* 3(22): 652–653, 1933.

Jarrett, R. D., "Hydraulics of High-Gradient Streams," *Journal of Hydraulic Engineering,* American Society of Civil Engineers, 110(11): 1519–1539, 1984.

Kouwen, N., "Field Estimation of the Biomechanical Properties of Grass," *Journal of Hydraulic Research,* International Association and Hydraulic Research, 26(5): 559–568, 1988.

Kouwen, N., and R. Li, "Biomechanics of Vegetative Channel Linings," *Journal of the Hydraulics Division,* American Society of Civil Engineers, 106(HY6): 1085–1103, 1980.

Lane, E. W., and E. J. Carlson, "Some Factors Affecting the Stability of Canals Constructed in Coarse Granular Materials," *Proceedings of the Minnesota International Hydraulics Convention,* September 1953.

Laurenson, E. M., "Friction Slope Averaging in Backwater Calculations," *Journal of Hydraulic Engineering,* American Society of Civil Engineers 112(12),1151–1163 1986.

Levi, E., *The Science of Water: The Foundation of Modern Hydraulics,* Translated from the Spanish by D. E. Medina, ASCE Press, New York, 1995.

Linsley, R. K. and J. B. Franzini, *Water Resources Engineering,* 3rd ed., Mc-Graw-Hill, New York, 1979.

Meyer-Peter, P. E., and R. Muller, "Formulas for Bed Load Transport," *Proceedings of the 3rd International Association for Hydraulic Research,* Stockholm, 1948, pp. 39–64.

Montuori, C., Discussion of "Stability Aspects of Flow in Open Channels," *Journal of the Hydraulics Division,* American Society of Civil Engineers 89(HY4): 264–273, 1963.

Nagami, M., R. Scavarda, G. Pederson, G. Drogin, D. Chenoweth, C. Chow, and M. Villa, *Design Manual: Hydraulic, Design Division,* Los Angeles County Flood Control District, Los Angeles, CA, 1982.

Niepelt, W. A., and F. A. Locher, "Instability in High Velocity Slurry Flows, *Mining Engineering,* Society for Mining, Metallurgy and Exploration, 1989, pp. 1204–1209.

Petryk, S., and G. Bosmajian, "Analysis of Flow Through Vegetation," *Journal of the Hydraulics Division,* American Society of Civil Engineers, 101(HY7): 871–884, 1975.

Ponce, V. M., *Engineering Hydrology: Principles and Practices,* Prentice–Hall, Englewood Cliffs, NJ, 1989.

Richardson, E. V., D. B. Simons, and P. Y. Julien, *Highways in the River Environment,* U.S. Department of Transportation, Federal Highway Administration, Washington, DC, 1987.

Silvester, R., "Theory and Experiment on the Hydraulic Jump," *Proceedings of the 2nd Australasian Conference on Hydraulics and Fluid Mechanics,* 1965, pp. A25–A39.

Silvester, R., "Hydraulic Jump in All Shapes of Horizontal Channels," *Journal of the Hydraulics Division,* American Society of Civil Engineers, 90(HY1): 23–55, 1964.

Simons, D. B., and E. V. Richardson, "Resistance to Flow in Alluvial Channels," Professional Paper 422-J, U.S. Geological Survey, Washington, DC, 1966.

Simons, Li & Associates, SLA Engineering Analysis of Fluvial Systems, Fort Collins, CO, 1982.

Straub, W. O. "A Quick and Easy Way to Calculate Critical and Conjugate Depths in Circular Open Channels," *Civil Engineering,* 70–71, December 1978.

Straub, W. O., Personal Communication, Civil Engineering Associate, Department of Water and Power, City of Los Angeles, January 13, 1982.

Temple, D. M., "Changes in Vegetal Flow Resistance During Long-Duration Flows," *Transactions of the ASAE,* 34: 1769–1774, 1991.

Urquhart, W. J. "Hydraulics," in *Engineering Field Manual,* U.S. Department of Agriculture, Soil Conservation Service, Washington, DC, 1975.

U.S. Army Corps Engineers, *HEC-RAS River Analysis System, User's Manual,* U.S. Army Corps of Engineers Hydrologic Engineering Center, Davis, CA, 1997.

U.S. Army Corps Engineers, "HEC-2, Water Surface Profiles, User's Manual," U.S. Army Corps of Engineers Hydrologic Engineering Center, Davis, CA, 1990.

U.S. Army Corps Engineers, "Accuracy of Computed Water Surface Profiles," U.S. Army Corps of Engineers Hydrologic Engineering Center, Davis, CA, 1986.

U.S. Army Corps Engineers, "Hydraulic Design of Flood Control Channels," EM 1110-2-1601. U.S. Army Corps of Engineers, Washington, DC, 1970.

Yang, C. T., *Sediment Transport: Theory and Practice,* McGraw-Hill, NewYork, 1996.

Zegzhda, A. P., *Theroiia Podobija Metodika Rascheta Gidrotekhnickeskikh Modele (Theory of Similarity and Methods of Design of Models for Hydraulic Engineering),* Gosstroiizdat, Leningrad, 1938.

CHAPTER 4
HYDROLOGY FOR DRAINAGE SYSTEM DESIGN AND ANALYSIS

Jerome A. Westphal
Cottonwood, Arizona

4.1 PEAK RUNOFF ESTIMATES

4.1.1 The Rational Formula

The hydraulic sizing of drainage and conveyance structures in urban settings always requires estimation of peak flow rates. Historically, the venerable *"Rational method"* has been the tool of choice for most practicing engineers around the world. Although the method definitely has its place in hydrologic design, it is routinely misapplied and overextended.

The roots of this methodology date as far back as 1851 (Mulvaney, 1851), and certainly as far back as 1889 (Kuichling, 1889). See discussion on chapter 1. The concept is attractive and easy to understand. If rainfall occurs over a basin at a constant intensity for a period of time that is sufficient to produce steady state runoff at the outlet or design point, then the peak outflow rate will be proportional to the product of rainfall intensity and basin area. In the United States, the method is commonly expressed by the equation known as the *"Rational formula"*:

$$Q = C \cdot I \cdot A \qquad (4.1)$$

where Q = peak runoff rate (cfs)
$\quad C$ = dimensionless runoff coefficient used to adjust for abstractions from rainfall
$\quad I$ = rainfall intensity for a duration that equals time of concentration of the basin (in/hr)
$\quad A$ = basin area (ac)

In English units, it turns out that the dimensions of the product $I \cdot A$ are ac·in/hr, and 1.0 ac·in/hr is very nearly equivalent to 1.0 cfs. In SI units, the equation must be made dimensionally homogeneous (e.g. if A is hectares and I is cm/hr, then the product $C \cdot I \cdot A$ must be multiplied by 0.00278 to make the dimensions on Q equal to cms).

Since its inception, the Rational formula has been discussed extensively in the published literature and in theses. Most of its limitations and shortcomings are well documented, but these constraints are largely ignored by most practicing engineers. For credible engineering design, engineers must observe the constraints which limit the applicability of the Rational formula. An exhaustive discussion of the rational method was done by Rossmiller (1982). The reader is urged to read Rossmiller for a cogent discussion of limitations that should be considered in its application.

There are several assumptions inherent to the Rational method:

- The rainfall intensity is constant over a period that equals the time of concentration of the basin.
- The rainfall intensity is constant throughout the basin.
- The frequency distribution of the event rainfall and the peak runoff rate are identical (this assumption is true for all event-based computations).
- The time of concentration of a basin is constant and is easily determined (this assumption is also shared by other event based methods).
- Despite the natural temporal and spacial variability of abstractions from rainfall, the percentage of event rainfall that is converted to runoff (the runoff coefficient, C) can be estimated reliably.
- The runoff coefficient is invariant, regardless of season of the year or depth or intensity of rainfall.

There are a number of practical limitations to the application of the Rational formula. Records of mass rainfall curves for heavy storms invariably show that peak periods of constant rainfall intensity are usually of comparatively short duration (a few minutes or, at most, a few tens of minutes). This indicates that the assumption of constant rainfall intensity is more likely to be realized in basins that have a short time of concentration.

Isohyetal maps for storms on areas of less than 50 mi^2 usually have elliptical isohyetal patterns. They show that the area of greatest rainfall depth is near the storm center and that depth decreases away from the storm center. Thus, for basin-centered storms, smaller basins are more likely to have a reasonably uniform spacial distribution of rainfall than are larger basins.

Like everything else in our physical world, rainfall-runoff relations in watersheds obey the Law of Conservation of Mass (and all other physical laws). In narrative form, this may be stated for any watershed (or stream reach) as: mass inflow rate minus mass outflow rate equals time rate of change of mass in storage. For instance, if we take a lined channel and introduce a hydrograph at the upstream end, we observe that the peak outflow rate at the downstream end is smaller than the peak inflow rate and it occurs later in time. The attenuation of the peak flow rate and the displacement in time is due to the channel storage in the intervening reach. It is a comparatively simple numerical task to demonstrate similar behavior on an idealized, impervious watershed that receives rainfall at a constant intensity. In the absence of storage, the peak outflow rate equals the steady-state rate of supply to the basin and it is coincident in time with the time of concentration of the basin. When storage is introduced into the computation, the peak outflow rate is attenuated and it occurs later in time than the time of concentration of the basin. Except in the case of zero storage in the system, the peak outflow rate from a basin must always be less than the peak rate of supply (e.g. the product of rainfall intensity and basin area). The storage in a basin (disregarding impoundments) will decrease as the area decreases and the percentage of the impervious area increases.

From the foregoing, it is clear that the application of the Rational formula should be limited to "small" watersheds. Although the term "small" is subjective, a reasonable upper limit under most circumstances would be about 200 acres (800 km^2, 80 ha).

Time of concentration is defined as the length of time it takes for water to travel from the hydraulically most remote point in a basin to the outlet. Although this definition is attractive, its simplicity is deceiving. There is no practical way of measuring the time of concentration. From elementary considerations of free-surface flow (e.g. velocity of flow increases with increasing depth of flow), we know that for any given storm duration, greater rainfall depths will induce greater depths of flow in the drainage network, and travel times through the basin than will be less than those that will occur during smaller, more frequent rainfall events. In the Rational formula, design rainfall intensity is a direct function of the time of concentration. Generally, the disparity between estimates of time of concentration by the various methods decreases as basin area decreases. In the case of runoff estimates for residential areas and other urbanized areas where the percentage of pervious surface is comparatively high compared to the area of impervious surfaces, the time of concentration should be based on travel time through the connected impervious part of the basin, and basin area should be taken as the area of the connected pervious surface.

The notion that the runoff coefficient, C, is a constant for any given watershed is flawed. As a percentage of storm rain, runoff increases with increasing rainfall depth and rainfall intensity. Published tables of suggested values of C are seldom explicit about the justifications for their guidance, but the general consensus in the literature seems to be that the Rational method be limited to events with return periods of 10-years or less (exceedence probabilities of 10% or more). For less frequent (more rare) events, there are published multipliers for C, but in no case should a value of C ever exceed 1.0.

Application of the Rational method requires that the duration of the design rainfall event be equal to the time of concentration of the basin. Many equations, nomographs and charts have been developed for estimation of the time of concentration of a basin. McCuen, et al. (1984) reviewed a number of equations that are in use for the determination of the time of concentration. No single approach has been demonstrated (or even claimed) to be superior to any of the others.

Time of concentration may be conceptualized as the sum of three components: sheet (overland) flow, concentrated flow (swales, natural channels), and flow in lined canals or closed conduits. Usually, these components occur sequentially beginning with sheet flow and ending with flow in a constructed conduit. However, sometimes only one or two of these components exist.

Starting in the vicinity of the hydraulically most remote point, runoff initially moves at shallow depths as overland or sheet flow until it reaches a swale or channel. Often, this component is estimated using the Kerby/Hathaway equation. The Kerby/Hathaway equation is an empirical relation developed by Kerby (1959) on the basis of published research on airport drainage done by Hathaway (1945). The FSS unit form of this equation is

$$t_o = \left[\frac{0.67 \cdot n \cdot L_o}{\sqrt{S_o}} \right]^{0.467} \tag{4.2}$$

where t_o = time of overland flow, minutes
$\quad n$ = overland flow resistance factor, dimensionless
$\quad L_o$ = length of the overland flow segment, feet
$\quad \sqrt{S_o}$ = slope of overland flow segment, feet vertical/feet horizontal

The distance L_o should always be less than 300 feet. The variable, n, is analogous to Manning's coefficient of friction. However, unlike channelized flow, the percentage of flow depth represented by roughness elements is much greater for sheet flow than for channelized flow, so the resistance imparted by these elements is considerably greater for sheet flow. The usual values of Manning's n that are considered for various surfaces and channel linings for channelized flow should not be used for overland flow computations. The following table (Table 4.1) was excerpted from Table 3.5, HEC-1 Flood Hydrograph Package, Users Manual (Sept 1990).

One of the more common methods for estimating the travel time of concentrated flow is the so-called Kirpich (1945) equation. Results of Kirpich's work are based on published research done by Ramser (1927) on a small, partially wooded, agricultural watershed in

TABLE 4.1 Resistance Factors for Overland Flow

Resistance Factor for Overland Flow	
Surface	n value
Asphalt/concrete	0.05–0.15
Bare packed soil, free of stone	0.10
Poor grass cover on moderately rough ground	0.30
Light turf	0.20
Average grass cover	0.40
Dense turf	0.17–0.80
Dense grass	0.17–0.30
Bermuda grass	0.30–0.48

Tennessee. From nearly three years of record, Kirpich chose the five shortest times to rise on measured hydrographs emanating from the principal outlet of the study watershed and from the outlets of the four sub basins and plotted them versus the commensurate quotient of channel length divided by the square root of stream slope. The points plotted as a straight line on logarithmic coordinates, so the Kirpich equation is a power function that fits the plotted data (5 data pairs). In FSS units, the *Kirpich equation* is

$$t_{ch} = 0.0078 \cdot L^{0.77} \cdot S^{-0.385} \tag{4.3}$$

where t_{ch} = time of channelized flow, minutes
$\quad\quad L$ = length of the channelized flow reach, feet
$\quad\quad S$ = slope of the channelized flow reach, dimensionless

and S is computed as

$$S = \frac{\Delta H}{L} \tag{4.4}$$

where ΔH is the difference in elevation (feet) between the upper and lower ends of the channelized reach of length, L.

Time of flow in lined channels, curbside gutters, and closed conduits can be computed by dividing the conduit length by velocity as computed for full-flow conditions using the Manning equation

$$t_{lc} = \frac{L_{lc}}{\dfrac{89.4}{n} \cdot R^{0.67} \cdot S_{lc}^{0.5}} \tag{4.5}$$

where t_{lc} = time of travel in the lined conduit, minutes
$\quad\quad n$ = Manning's roughness coefficient, dimensionless
$\quad\quad R$ = hydraulic radius of the lined conduit, feet. R can be approximated by depth of flow at bank full in gutters and in channels, and as $0.25 \cdot$ diameter for closed conduit flow
$\quad\quad S_{lc}$ = longitudinal slope of the lined conduit, dimensionless

Manning's equation is very sensitive to the roughness coefficient, and under some conditions will give unrealistically high values for channel velocities. Computed velocities should be checked independently for reasonableness.

TABLE 4.2   Time of Concentration for Small,
Impervious Areas

Location	Standard t_c (minutes)
Roof and property drainage	5
Road inlet pits	5
Small areas <1 acre	10

Rather than attempting more detailed analysis, for areas up to one acre, values in Table 4.2 (taken from Urban Storm Water Management Manual for Malaysia, 2000) can be used.

To obtain an estimate of the time of concentration for the basin, the component values should be added. However, the addition must be done only for those components of flow in series along the principal path of flow from the hydraulically most remote point to the design point. Do not add travel time in intermediate tributaries. In the case of residential areas, consider only the interconnected impervious area.

In the selection of a time of concentration for any individual component, the engineer should probably use two or three of the methods with which he or she is familiar and for which the necessary independent variables and parameters exist or can be economically determined. The procedure that gives a reasonable value and is conservative (shorter times will result in higher peak flow rates) with respect to the design objective should be the value that is selected.

Remember, the equations are based on average velocities, not point velocities. Therefore, in many, if not most cases, you can walk faster than water flows. For overland flow and for concentrated flow in swales or natural channels, velocities will normally be less than 5 ft/sec or 6 ft/sec, and in larger, lined channels and closed conduits, they will usually be less than 10 ft/sec to 12 ft/sec.

Typically, rainfall intensities are determined from *Intensity-Duration-Frequency curves* (*IDF curves*) or *Depth-Duration-Frequency curves* (*DDF curves*). These are plots of rainfall intensity (or depth) versus duration of event rainfall. Usually, there are several curves on a single graph, one for each of several different rainfall frequencies (e.g. the 10-year event, the 25-year event, the 50-year event and the 100-year event). Figure 4.1 is a set of IDF curves for Rolla, MO.

Often, one or both of these graphs (or a tabular form thereof) are supplied by the local governmental authority under whose auspices the work is being done. First, the time of concentration is determined. Then, in the case of the IDF curve, the graph is entered with time of concentration and the rainfall intensity or depth is read at the intersection of the time coordinate and the corresponding design frequency. In the case of DDF data, the design depth is divided by the time of concentration so as to obtain the desired rainfall intensity, *I*.

In the absence of preexisting IDF or DDF curves, the engineer must develop the information that is required. In the best of all worlds, an engineer should develop IDF or DDF curves from local or nearby rainfall data. However, for all practical purposes, the necessary rainfall data are seldom available, and if they are available, the cost of developing the rainfall frequency relations is seldom justifiable in terms of any design project where the Rational method is applicable (e.g. drainage design for a subdivision, an industrial park, a shopping center, etc.). In this case, IDF or DDF curves can be developed from rainfall data contained in published rainfall frequency atlases (e.g. Hershfield, 1961; Miller et al., 1973; Frederick et al., 1977; Huff and Angel, 1992). For a given frequency, rainfall depths are interpolated from isohyets in the atlas at the published durations. Each rainfall depth is divided by the corresponding duration to obtain an average rainfall intensity for that particular duration. The computed intensities for the given frequency are then plotted on either logarithmic or on semilogarithmic coordinates (time on the logarithmic scale in both cases). Use of loga-

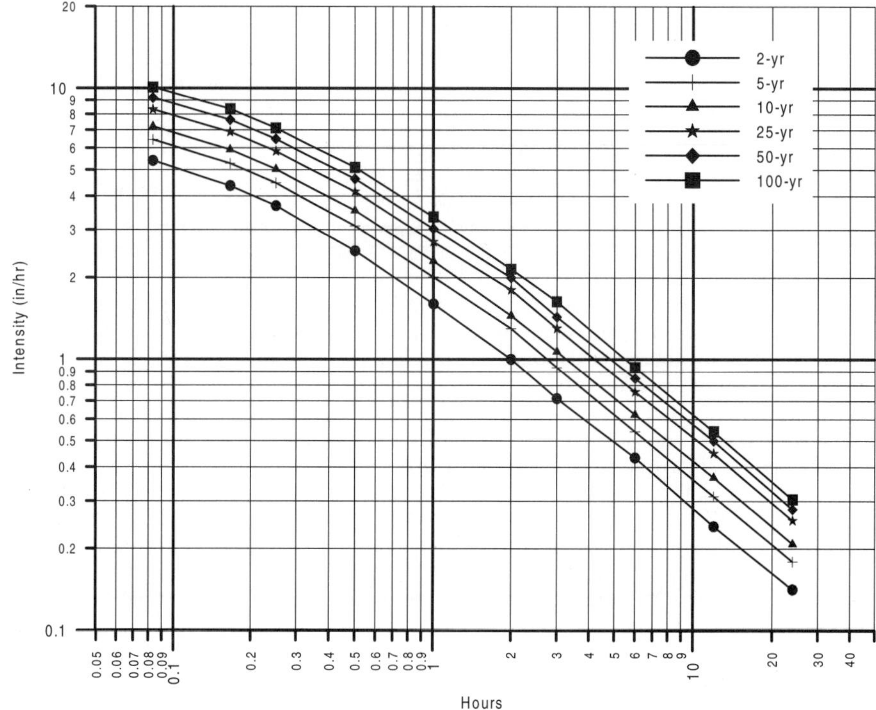

FIGURE 4.1   IDF curves for Rolla, MO.

rithmic coordinates is purely for plotting convenience. There is no reason to suspect that the data will fit either an exponential or a power function form. Therefore, a least-squares fit to the data is not only unwarranted, but it may be misleading. Consequently, a smooth curve is fitted manually to the plotted data, one curve for each frequency.

The runoff coefficient, C, is selected from an appropriate table. It represents the decimal fraction of event rainfall that becomes runoff. The justifications for values in these tables seem to have been lost to posterity, so selection of a particular table and a particular value from a table is a highly subjective matter. Table 4.3 is taken from the California Department of Transportation Highway Design Manual, Chapter 810 (July 1, 1995).

If it is necessary to use the Rational equation for a return period that exceeds 10 years, the runoff coefficient, C, from the preceding table should be multiplied by a frequency adjustment factor, C_f. Table 4.4 shows suggested values of C_f.

4.1.2 Regression Estimates

The U.S. Geological Survey (USGS) has developed equations for estimation of peak flow rates for various return periods for urban and non-urban watersheds in each of the 50 states and for Puerto Rico. The equations were derived from a generalized least squares regression procedure that uses independent variables that characterize physical and physiographic features of watersheds. Where it was appropriate to do so, each state was divided into hydrologic regions, a hydrologic region being an area which, by virtue of physiography and precipitation-runoff characteristics, can be considered to be hydrologically homogeneous. Usually, in

TABLE 4.3 Runoff Coefficients for Urban Areas

Type of drainage area	Runoff coefficient
Business:	
Downtown areas	0.70–0.95
Neighborhood areas	0.50–0.70
Residential:	
Single-family areas	0.30–0.50
Multi-units, detached	0.40–0.60
Multi-units, attached	0.60–0.75
Suburban	0.25–0.40
Apartment dwelling areas	0.50–0.70
Industrial:	
Light areas	0.50–0.80
Heavy areas	0.60–0.90
Parks, cemeteries:	0.10–0.25
Playgrounds:	0.20–0.40
Railroad yard areas:	0.20–0.40
Unimproved areas:	0.10–0.30
Lawns:	
Sandy soil, flat, 2%	0.05–0.10
Sandy soil, average, 2–7%	0.10–0.15
Sandy soil, steep, 7%	0.15–0.20
Heavy soil, flat, 2%	0.13–0.17
Heavy soil, average, 2–7%	0.18–0.25
Heavy soil, steep, 7%	0.25–0.35
Streets:	
Asphaltic	0.70–0.95
Concrete	0.80–0.95
Brick	0.70–0.85
Drives and walks	0.75–0.85
Roofs:	0.75–0.95

TABLE 4.4 Rational Coefficient Frequency Adjustment Factor

Frequency (yrs)	$C(f)$
25	1.1
50	1.2
100	1.25

each region, an individual equation was developed for the 2-year, 10-year, 25-year, 50-year, 100-year, 200-year, and in some cases, the 500-year event. The result was roughly 1500 equations nationwide.

Individual equations vary with return period and with region. The equations have the form of a power function

$$Q_p = a \cdot W^b \cdot X^c \cdot Y^d \cdots Z^n \tag{4.6}$$

where

$$
\begin{aligned}
Q_p &= \text{peak flow rate for some specified return period, f (cfs)} \\
a &= \text{ordinate intercept of the regression line} \\
W, X, Y, \ldots, Z &= \text{variables that characterize basin features} \\
b, c, d, \ldots, n &= \text{exponents on independent variables}
\end{aligned}
$$

The independent variables used to characterize basin features almost always include drainage area and usually stream slope. Variables such as basin shape, elevation of the basin centroid, impervious area, and others are sometimes included if, during analysis, they resulted in a significant reduction in the variance of the residuals about the equation. In any event, the variables used for the equations can be determined in a reasonably straightforward manner from 7-1/2 minute topographic mapping or other readily available mapping.

Unless there is a theoretical basis for the equation form (linear, exponential, power function), regression equations are unreliable when they are used to extrapolate beyond the range of data that were used for their creation. The bases for the USGS equations are entirely empirical, so the reader is urged to read the supporting documentation prior to using the equations, and to make certain that each of the individual independent variables falls within the range of the commensurate variables used for the equation. Failure to observe this limitation may result in estimated peak flow rates that are unacceptably poor extrapolations from the equation. Unfortunately, unless the result is a negative flow rate or a flow rate that is inordinately high or low for the basin being considered, an unreliable result may be difficult to recognize.

Because of the empirical nature of the USGS equations, it is imperative that the engineer adhere closely to both the units and the definitions of the independent variables. For instance, there are several ways to define "slope" and at least three ways to locate the centroid of a basin.

The USGS publishes a measure of the relative power of the regression equations called the "standard error of prediction" (SEP). This measure is sometimes cited as a percentage, and sometimes as a decimal fraction. For practical purposes, the SEP is similar to the more formally defined (in statistics texts) "standard error of the estimate" (SEE) and, in the context of the USGS peak flow prediction equations, can be interpreted similarly. The SEE is the standard deviation of the residuals from the regression where the "residual" is the difference between the regression estimate of peak flow rate and the observed peak flow rate. Thus, one may infer that about 67% of all regression estimates will be within ± 1.0 SEE.

Because the USGS equations are a power functions, they were developed by doing multiple linear regression on the logarithms of peak flow rates and of the associated independent variables. The decimal form of the SEP relates to the difference between the logarithms of the estimated and observed peak flow rates. Thus, to obtain a physically meaningful measure of the uncertainty in the equation: 1) operate the equation to find peak flow rate (e.g., say the 100-yr peak flow rate is 100,000 cfs); 2) take the common logarithm of the peak flow rate (e.g. log 100000 = 5.0); 3) add and subtract the decimal form of the SEP to the logarithm of the peak flow rate (e.g. SEP = 0.2 5.0-0.2 = 4.8, 5.0 + 0.2 = 5.2); and 4) take the antilogs of both values (e.g. $10^{4.8} = 63100$; $10^{5.2} = 158500$). Although the regression equation gives the best estimate of the peak flow rate, the SEP gives an idea of the scatter about the regression line. In this case, in very general terms, one might expect that the true value of the peak flow would be between 63100 cfs and 158500 cfs. Obviously, as the SEP increases, the certainty of the regression estimate decreases.

The choice of reporting the SEP as a percentage has the advantage of informing the reader about the approximate scatter about the regression in a direct way. However, where it is reported as a single percentage, it may be somewhat misleading. For instance, in the previous example, the difference between the regression estimate and the lower bound is greater than the difference for the upper bound. The average difference is 47.7% [0.5 · (39100 + 58500)/100000]. Thus, reportage of a singe SEP percentage implies a symmetry about the regression estimate that doesn't exist, but it is still true that as the SEP increases, the reliability of the regression estimate decreases.

In any event, the SEP should be used as a qualitative indicator of the relative reliability of any particular peak flow equation. An informal scan of equations for a few states indicates that the SEP may vary from roughly 20% to 150%, with most being in the range of 30% to 50%. Although there seems to be no published formal guidance, alternative methods of estimating peak flow rates should be considered whenever the SEP is greater than 50%.

4.2 HYDROGRAPH METHODS

When watersheds are large, that is, when they are comprised of two or more smaller watersheds whose streamflow at the confluence with common collector channel can be expected to be displaced in time, where storage influences the time distribution of flow in a stream, or where storage is a part of the design problem, peak flow methods are inappropriate for hydrologic design. In these instances, it is necessary to estimate the entire flow hydrograph. A number of computer programs (models) are available to do the requisite hydrologic and hydraulic computations. Most of these programs have a number of options for each element of the process that begins with rainfall and ends with a hydrograph at some point in the system. Conceptually, the process starts with rainfall over a sub watershed(s) at the periphery of a larger system. The rainfall is transformed into a hydrograph of direct runoff at the outlet of the sub watershed. The hydrograph is then combined with a hydrograph from an adjacent basin and/or is routed through a channel to the next downstream point of interest.

There are two types of computer programs (models) for doing hydrologic and hydraulic computations for a system: continuous simulation models and event-based models. Event-based models are used for nearly all design problems. Discussion in this chapter will be restricted to methods embedded in event-based models. Furthermore, it will be restricted to elements related to hydrograph computations. Flood routing is covered in chapters 7 and 8.

4.2.1 Rainfall Events for Design—Design Hyetographs

The process of computing a hydrograph begins with selection of a design storm, the first step of which is to select a design frequency. In an event-based design using methods of synthetic hydrology, the frequency of the storm event is assumed to equal the frequency of the resulting computed peak flow rate on the hydrograph. This is probably not true for individual events, but it is hoped that it approaches reality over the long term. In any event, there is currently no acceptable alternative.

Often, the local approving authority (city, county, drainage district, etc.) will specify the level of design to be used for any particular type of structure. In the absence of statutory or regulatory specifications, the Table 4.5 (excerpted from Table 13.1.1 in Chow et al., 1988) shows recurrence intervals that are commonly used in the practice.

Next, duration of the rainfall event should be selected so as to be at least as long as the time of concentration of the entire system that is under analysis. Time of concentration has been discussed Section 4.1.1. In published rainfall atlases, depth of rain is directly proportional to duration while average intensity is inversely proportional to duration. Everything

TABLE 4.5   Common Design Frequencies for Hydraulic Structures

Type of structure	Return period (years)
Highway culverts:	
Low traffic	5–10
Intermediate traffic	10–25
High traffic	50–100
Highway bridges	
Secondary system	10–50
Primary system	50–100
Urban drainage	
Storm sewers in small cities	2–25
Storm sewers in large cities	25–50
Airfields	
Low traffic	5–10
Intermediate traffic	10–25
High traffic	50–100

being equal, higher rainfall intensities result in higher runoff rates, while greater rainfall depths result in greater volumes of runoff. It is seldom possible to know in advance whether design of a hydraulic structure will be more sensitive to peak runoff rates or to runoff volumes. Therefore, it is good practice to select several rainfall durations and compute the runoff for each.

Often, regulatory authorities will supply IDF or DDF data. However, peak flows computed by hydrograph methods do not require that rainfall durations equal time of concentration. Therefore, it is usually more efficient to choose durations that equal or exceed the time of concentration, and that are divisible by some convenient fraction of an hour (e.g. 15 minutes, 20 minutes, 30 minutes). As mentioned earlier, it is always best, but seldom practicable, to use rainfall frequency relations that are derived from local rainfall records, provided they have a sufficient period of record and suitable quality. Existing rainfall atlases (e.g. Hershfield, 1961; Miller, et al., 1973; Frederick, et al., 1977; Huff and Angel, 1992) show rainfall for a number of frequencies for a commensurate number of durations such that design values can be taken directly from an appropriate atlas.

The design rainfall must be distributed in time to approximate (in a gross sense) a naturally occurring event comprised of a series of short duration segments whose intensity varies from segment to segment. A histogram (or table) that depicts rainfall intensity versus time is called a *hyetograph*. In design, the sequential increments of rainfall must be of equal duration. A good rule of thumb is to select the time increment to be

$$\frac{t_c}{5} \le \Delta t \le \frac{t_c}{3} \tag{4.7}$$

where t_c = time of concentration
    Δt = the duration of each time segment of the hyetograph (period of constant intensity)

This guideline ensures that steady state runoff cannot occur during any individual segment of constant intensity (as is the case in nature). At the same time, it gives reasonable detail to the mass arrival characteristics of the rainfall. For convenience of computation, Δt should be an integer number of minutes and the total event duration should be an integer multiple of Δt.

A number of procedures have been developed for synthesizing hyetographs. Hereafter, these will be called hyetograph methods. Some (Kiefer and Chu, 1957; Huff, 1967; Pilgrim and Cordery, 1975; Yen and Chow, 1980; Soil Conservation Service, 1986) have developed procedures that derive from an analysis of temporal distributions of naturally occurring rainfall. Pilgrim and Cordery (1975) take a quasi-probabilistic approach that tends to preserve the position in time of the periods of highest intensity. Their procedure usually results in a multimodal distribution, whereas the other methods derived from analysis of naturally occurring rainfall result in unimodal distributions. Other arbitrary methods such as the alternating block method (Chow, Maidment, Mays; 1988) and a similar unnamed approach for creating a Probable Maximum Precipitation hyetograph (U.S. Bureau of Reclamation, 1974) rearrange rainfall segments so that the greatest depth of rainfall occurs prior to the period of peak intensity, and peak intensity is centered in the storm.

Application of the Pilgrim and Cordery method (1975) requires analysis of local or regional rainfall. Because this is a time consuming process, and because applicable rainfall data are not always present, this method has not been widely applied in the U.S. However, it has been adopted as a standard method of hydrologic design in Australia (The Institution of Engineers Australia, 1987) and has been recommended by Greene County, Missouri (Green County Storm-Water Design Standards, 1999).

Kiefer and Chu's procedure (1957) is generally known as the Chicago method. It presupposes an IDF relation of the form

$$ i = \frac{a}{t_D^b + c} \tag{4.8} $$

where i = rainfall intensity, in/hr
t_D = total duration of rainfall, hr
a,b,c = shape and location parameters, dimensionless

An equation taken from Modern Sewer Design (1980) proposes an IDF relation (which they attribute to Kiefer and Chu (1957) of the form

$$ i = \frac{a}{(t_D + c)^b} \tag{4.9} $$

where the variables and parameters are as defined above. Hyetographs derived from the Modern Sewer Design formulation are very similar to those that are derived from Kiefer and Chu. The peak intensity is slightly smaller and intensities preceding and following the period of peak intensity are slightly larger than those that derive from the Kiefer and Chu procedure. However, the differences are small, and in the application they result in no practical differences in either the computed peak rate or volume of runoff. Furthermore, the Kiefer and Chu Method requires trial and error fitting of periods of peak intensity so as to approximate the continuous curve of intensity versus time that derives from the method. The Modern Sewer Design formulation results in equations that can be integrated to find the proper intensities directly. It also has the advantage that the dimensionless parameters can be determined directly from IDF data.

Figure 4.2 shows a hyetograph for a 25-year, 2-hour rainfall in Rolla, MO as derived from the Modern Sewer Design approach. Rather than a continuously changing rainfall intensity, practical applications demand a discrete representation. By shifting the origin to the time of occurrence of peak intensity, the following equations can be used to find the depth of rainfall under the curve. The following equation can be used to determine the rainfall depth between the peak intensity and any time prior to the peak,

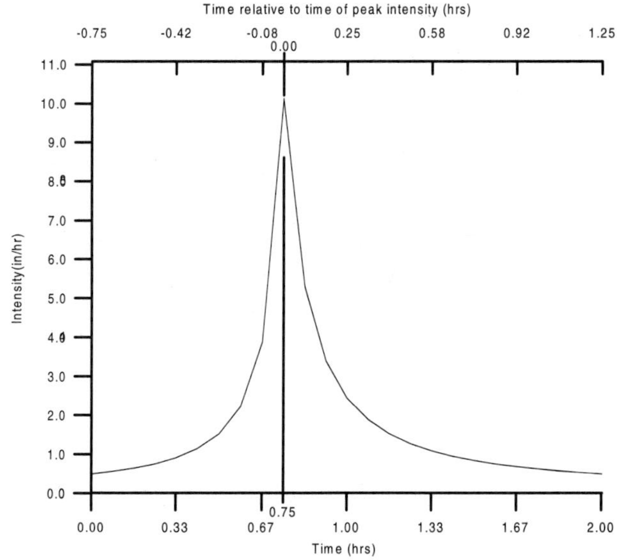

FIGURE 4.2 2-hr, 25-yr hyetograph for Rolla, MO. (Modified Chicago Method)

$$P_B = \frac{-a \cdot t_B}{\left(\dfrac{t_B}{r} + c\right)^b}$$

(4.10)

where t_B = length of time measured forward from the time of peak intensity, hours
P_B = the depth of rain in the t_B hours immediately prior to the peak intensity, inches
r = time from start of rainfall to the peak intensity, hours
a, b, c = shape and location parameters derived from the IDF curve

Ideally, the variable, r, should be based on an analysis of local rainfall data. However, Kiefer and Chu (1957) and Yen and Chow (1975), in the investigations leading to their published studies, indicate that the peak rainfall intensity tends to fall in the second quarter of the event rainfall of duration, t_D, with the average near $0.375 \cdot t_D$. In the absence of detailed local analysis, it is acceptable to set $r = 0.375 \cdot t_D$.

The equation for depth of rain following the peak is similar to the one for computing depth prior to the peak intensity.

$$P_A = \frac{a \cdot t_A}{\left(\dfrac{t_A}{1 - r} + c\right)^b}$$

(4.11)

where the subscript A refers to that part of the hyetograph following the peak intensity.

Example 4.1
Use the Modified Chicago Method to construct a 2-hour, 25-year hyetograph for a watershed in Rolla, MO.

Parameters a, b, and c are derived from the IDF curve for the design frequency. On logarithmic coordinates, a plot of IDF data is concave downward as shown below in Fig. 4.3.

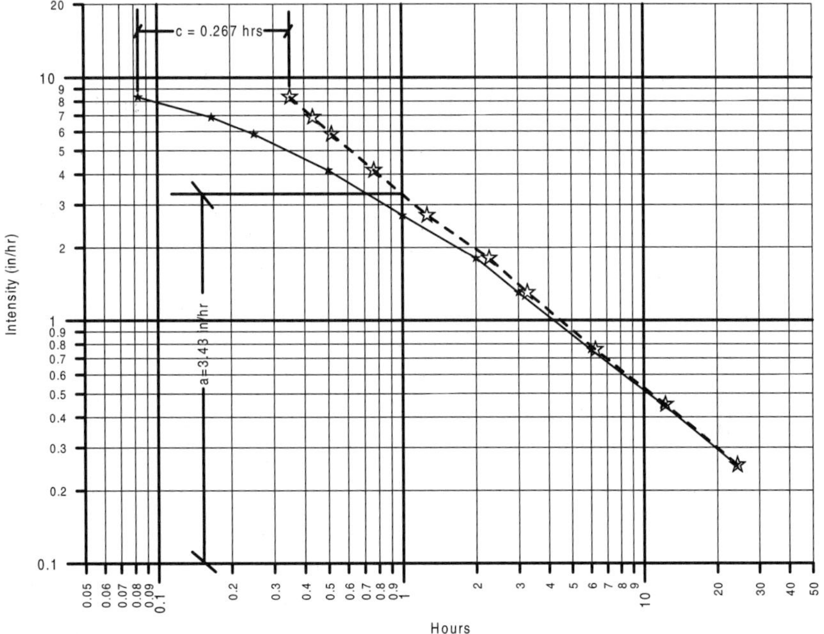

FIGURE 4.3   2-hr, 25-yr IDF curve for Rolla, MO.

This curve can be linearized by adding a time constant to the duration data. In Fig. 4.3, this was done by adding 0.2667 hours (16 minutes) to each of the time ordinates. The time shift, 0.2667 hours, is the parameter c in the equations above. The open stars show the shifted time positions of each intensity and the dashed line is the adjusted IDF curve. This line can be fitted manually, or alternatively, it can easily be computed by linear regression with $\log(t + c)$ versus $\log(i)$. The parameter c is adjusted until the coefficient of determination for the regression is maximized. In any event, parameter a is the ordinate intercept (the value of i when $t + c$ is 1.0 on the logarithmic scale), and b is the slope of the line.

After the distribution parameters have been determined, computation of the hyetograph ordinates is straightforward. For the Rolla data, $a = 3.43$ in/hr, $b = 0.816$, $c = 0.267$ hr, and $r = 0.375$ hr. For an event duration of 2 hours, the peak intensity will be at 0.75 hours ($2.0 \times 0.375 = 0.75$).

Assume that $\Delta t = 0.33$ hours. Therefore, the intensity will change at intervals of 0.33 hours.

Use Eq. 4.10 to compute cumulative precipitation with respect to cumulative time prior to the peak intensity,

$$P_B = \frac{-3.43 \cdot t_B}{\left(\dfrac{t_B}{0.375} + 0.2667\right)^{0.816}}$$

and then repeat the process using Eq. 4.11 for the receding part of the hyetograph.

$$P_A = \frac{3.43 \cdot t_A}{\left(\dfrac{t_A}{0.625} + 0.2667\right)^{0.816}}$$

Note that on Fig. 4.2, the peak intensity occurs between 0.67 hours and 1.0 hours. Thus,

TABLE 4.6   Design Computations for 2-hr, 25-yr Hyetograph at Rolla, MO

Time from start of rain (hrs)	Time from peak rain intensity (hrs)	Cumulative rain (inches)	Incremental rain (inches)	Rainfall intensity (in/hr)
0.000	−0.750	1.32		
0.333	−0.417	1.10	0.22	0.66
0.667	−0.083	0.51	0.59	1.77
0.750	0.000	0.00	0.51	
1.000	0.250	1.19	1.19	5.10
1.333	0.583	1.72	0.53	1.59
1.667	0.917	2.01	0.29	0.87
2.000	1.250	2.20	0.19	0.57

part of the rain in the interval from 0.67 hours to 1.0 hours falls prior to the peak, and part falls after the peak. Once the cumulative rain is known at the end of each time increment, the incremental rain can be obtained by differencing and the incremental intensity can be computed by dividing incremental rain by Δt. Table 4.6 summarizes the computation.

From Table 4.6, total event rain is 3.52 inches. From the Rolla IDF curve, the 2-hr, 25-yr rain has an average intensity of about 1.8 in/hr (3.6 inches of rain in 2 hours). Therefore, the computation checks reasonably well. Figure 4.4 is a plot of the design hyetograph superimposed on Fig. 4.2.

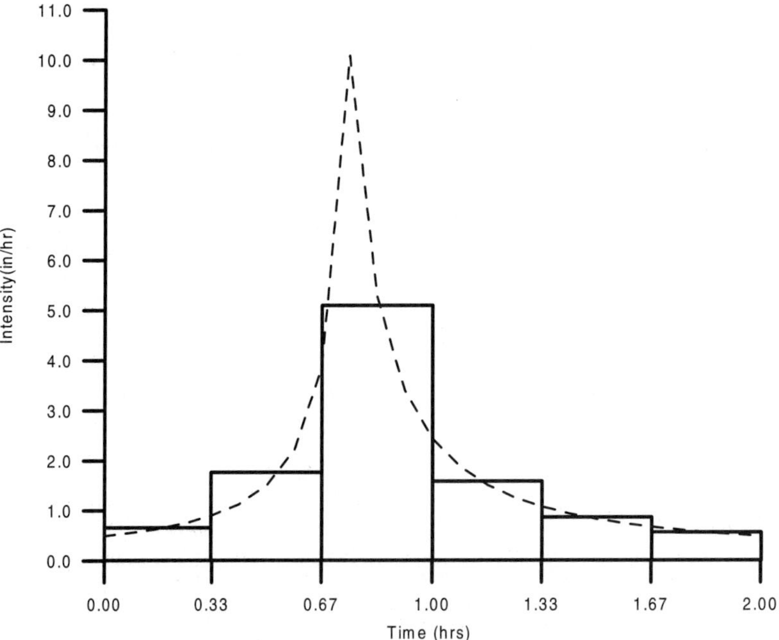

FIGURE 4.4   Design hyetograph for the 2-hr, 25-yr event at Rolla, MO.

The Yen and Chow method (1980) for creating design hyetographs is attractive in its simplicity. Because it is based on analysis of nearly 10000 storms at four locations: Boston, MA; Elizabeth City, NJ; Urbana, IL; and San Luis Obispo, CA, the breadth of its geographic applicability is reinforced. The basic hyetograph is represented by a triangular shape. The peak intensity is computed as

$$i_p = \frac{2 \cdot P}{t_D} \qquad (4.12)$$

where i_p = peak intensity, in/hr
$\quad P$ = total depth of rainfall, in
$\quad t_D$ = rainfall duration, hrs

Yen and Chow define the location of the peak intensity, r, as the storm advancement coefficient. In their study, they showed that r always lay in the second quarter of the storm time. The average r was approximately $0.375 \cdot t_D$.

Example 4.2
Use 2-hour, 25-year event at Rolla, MO to create a design hyetograph using the Yen and Chow method.
From the Rolla IDF data, t_D = 2 hours and P_{25} = 3.6 inches. Equation 3.2.6 is used to compute the peak intensity.

$$i_P = \frac{2 \cdot 3.6}{2} = 3.6 \text{ in/hr}$$

Locate the peak intensity at 0.75 hrs ($0.375 \cdot t_D$). Figure 4.5 shows the resulting triangular shaped hyetograph.
The runoff computation requires that the hyetograph be represented in successive discrete time increments with constant intensity. Table 4.7 shows the computations for Δt = 2 hours. For each tabulated time, intensity on the rising limb is computed as

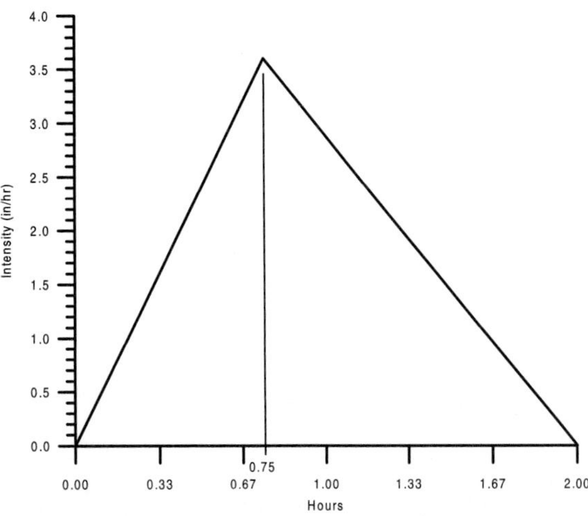

FIGURE 4.5   2-hr, 25-yr triangular hyetograph for Rolla, MO.

TABLE 4.7   Computation for 2-hr, 25-yr
Hyetograph Rolla, MO (Yen and Chow Method)

Time (hrs)	Intensity (in/hr)	Average incremental intensity (in/hr)
0.000	0.00	
0.333	1.60	0.80
0.667	3.20	2.40
0.750	3.60	
1.000	2.88	3.25
1.333	1.92	2.40
1.667	0.96	1.40
2.000	0.00	0.70

$$i = \frac{i_p}{r \cdot t_D} \cdot t \tag{4.13}$$

where i is the intensity at time t. For the falling limb,

$$i = \frac{i_p}{(1 - r) \cdot t_D} \cdot (t - r \cdot t_D) \tag{4.14}$$

The last column is the average intensity over the period, Δt, calculated as the incremental area under the triangular hyetograph divided by 0.333 (Δt).

Figure 4.6 is a plot of the design hyetograph superimposed on the triangular hyetograph.

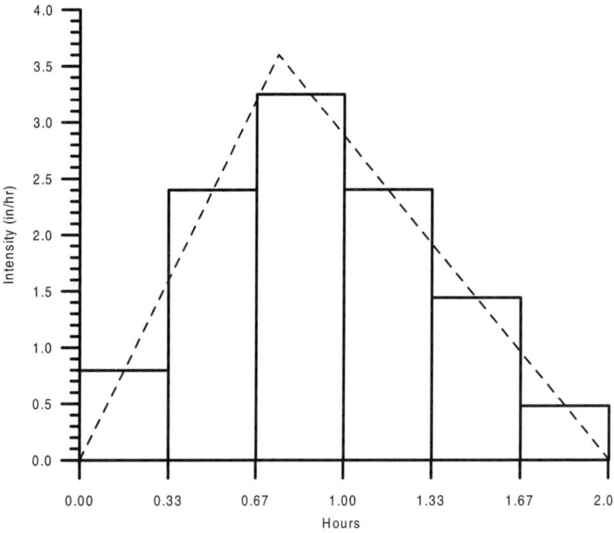

FIGURE 4.6   2-hr, 25-yr hyetograph for Rolla, MO. (Yen and Chow Method)

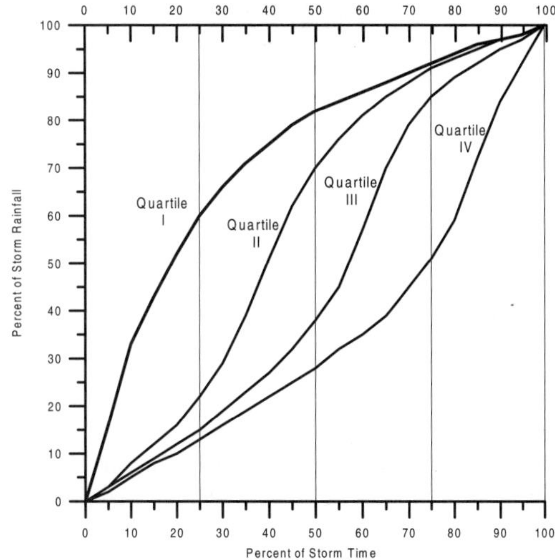

FIGURE 4.7   Huff's dimensionless mass curves for areas <50 mi².

TABLE 4.8   Huff's Dimensionless Mass Curves for Areas Less than 50 mi²

	% Cumulative storm rain			
% Storm time	QI	QII	QIII	QIV
0	0	0	0	0
5	16	3	3	2
10	33	8	6	5
15	43	12	9	8
20	52	16	12	10
25	60	22	15	13
30	66	29	19	16
35	71	39	23	19
40	75	51	27	22
45	79	62	32	25
50	82	70	38	28
55	84	76	45	32
60	86	81	57	35
65	88	85	70	39
70	90	88	79	45
75	92	91	85	51
80	94	93	89	59
85	96	95	92	72
90	97	97	95	84
95	98	98	97	92
100	100	100	100	100

TABLE 4.9 Relation Between Quartile Distribution and Storm Duration

Storm duration	Quartile receiving most rain
$t_D < 12$ hours	I
$t_D < 12$ hours	II
12 hours $< t_D < 24$ hours	III
$t_D > 24$ hours	IV

Huff (1967) published results of a study of heavy storms of 3 hours duration and greater that occurred over a 400 mi^2 area in north-central Illinois. By comparing dimensionless mass curves, he showed that storms could be classified according to the quarter (quartile) of the storm during which most of the rainfall occurred. Figure 4.7 shows plots of the median dimensionless mass curves for each quartile for areas less than 50 mi^2. The mass curves were slightly different for storms that covered areas of 50 mi^2 to 400 mi^2. Table 4.8 is a tabulation of the median curves shown in Fig. 4.7.

Huff found that storm duration tended to be associated with the quartile in which most of the rain fell. Table 4.9 shows the relationship between duration and quartile. More than 50% of the storms in each quartile met the duration criteria shown in Table 4.9. Quartile I and II type storms occurred with roughly the same frequency. For a given storm depth, Quartile I will produce a slightly higher peak intensity than Quartile II, but depending on the method of computing abstractions, it may produce a smaller depth of excess rain.

Example 4.3

Use Huff's method to create a hyetograph for a 3-hour, 25-year rainfall in Rolla, MO.

The 3-hour, 25-year rainfall in Rolla, MO is 3.9 inches. Because it is a 3-hour rainfall, choose either a Quartile I or Quartile II distribution. For this example, choose a Quartile I distribution. To demonstrate the method, use a Δt of 0.5 hours. One half hour is 16.7% of 3 hours, so tabulations will be at intervals of 16.7%. The following computations are shown in Table 4.10.

1. Interpolate values of Cumulative % storm rain from the column labeled QI in Table 4.10.

2. Multiply Cumulative % storm rain by 3.9 inches to get Cumulative storm rain.

TABLE 4.10 Computations for a 3-hr, 25-yr hyetograph from a Quartile I Distribution

% Storm time	Cumulative % storm rain	Cumulative time (hours)	Cumulative storm rain (inches)	Incremental rain (inches)	Rainfall intensity (in/hr)
0.0	0.0	0.0	0.00	0.00	0.00
16.7	46.0	0.5	1.79	1.79	3.58
33.3	69.0	1.0	2.69	0.90	1.80
50.0	82.0	1.5	3.20	0.51	1.02
66.7	88.7	2.0	3.46	0.26	0.52
83.3	95.3	2.5	3.72	0.26	0.52
100.0	100.0	3.0	3.90	0.18	0.36

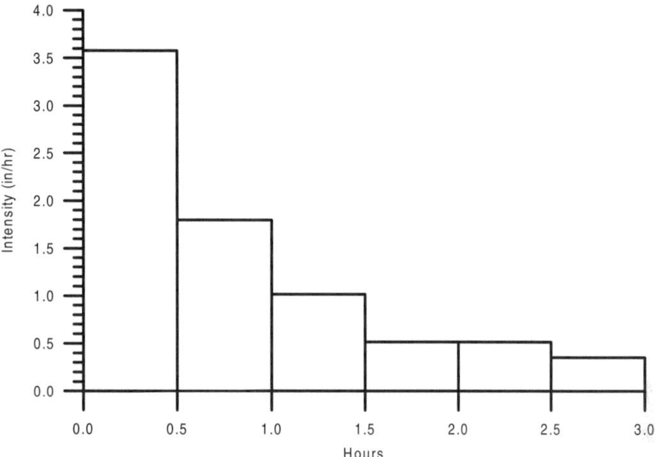

FIGURE 4.8   Hyetograph for the 3-hr, 25-yr rain at Rolla, MO. (Huff's Method)

3. Incremental rain is the difference between cumulative storm rain at the end of a time increment and cumulative rain at the beginning of the time increment.

4. Average rainfall intensity over each segment of time is incremental rain divided by 0.5 hours, the time increment. In Table 4.10, the average intensity is tabulated at the end of the time increment over which the average applies.

Figure 4.8 shows the resulting 3-hour, 25-year design hyetograph for Rolla, MO.

The Soil Conservation Service (1986) rainfall distributions are commonly used to derive hyetographs for hydrologic design in urban settings. These distributions are based on data taken from Hershfield (1961), from Miller, Frederick and Tracey (1973), and from additional storm data. There are four quasi-dimensionless mass curves for application with 24-hour rainfall events, and one for application with 6-hour events. Selection of a particular 24-hour mass curve is dependent on geographic location. Figure 4.9 is a map showing the regions of application for each of the 24-hour mass curves. The paired time and cumulative percent of storm rainfall data for each of the 24-hour distributions and for the 6-hour distribution are shown in Table 4.11.

The procedure for deriving a design hyetograph from the SCS rainfall mass curves is almost identical to the procedure used in conjunction with Huff's (1967) dimensionless mass curves. The differences are that cumulative storm time is presented directly instead of as a percent of the total, and the choices for storm duration are 24 hours and 6 hours.

4.2.2 Abstractions from Rainfall

For practical design considerations, water that makes it overland directly to the surface drainage network during a rainfall event is termed *direct runoff* (synonymously, *excess rainfall, effective rainfall*). The difference between the rainfall supply and the direct runoff is comprised of interception storage, depression storage, and infiltration. During periods of active rainfall, losses due to evaporation and evapotranspiration are negligible.

For event-based computations, abstractions from rainfall are considered to be comprised of two components, initial abstractions and infiltration. Interception storage, depression stor-

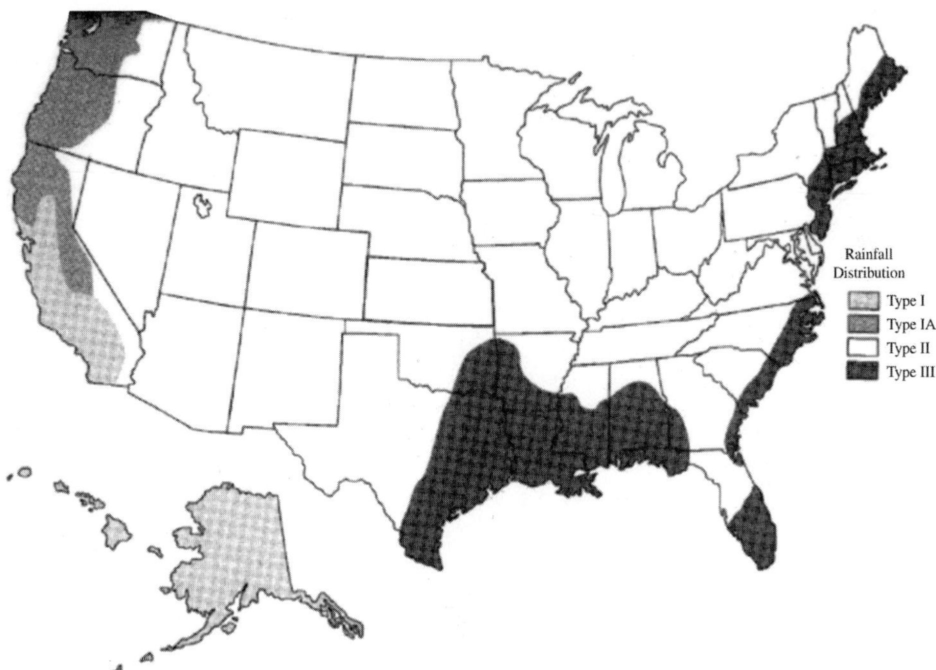

FIGURE 4.9   Regions of applicability of SCS 24-hr rainfall distributions.

age, and infiltration that occurs prior to the onset of direct runoff comprise the initial abstractions. Conceptually, initial abstractions must be satisfied before direct runoff can occur. Thereafter, direct runoff is the difference between rainfall rate and infiltration capacity. Infiltration capacity decreases exponentially with time. For rainfall events of the magnitude considered for hydrologic design, infiltration losses are generally much larger than initial abstractions.

With the possible exception of the Soil Conservation Service (SCS) Curve Number Method (CN Method;1985), there is no methodology for estimating initial abstractions that has widespread use in hydrologic design. For small depths of rainfall (e.g. for design storms with recurrence intervals less than 10 years) on watersheds with dense foliage or a high density of surface storage features, it might be appropriate to subtract a few tenths of an inch of rainfall from the design storm to account for initial abstractions. However, the usual practice (except for the CN Method) is to ignore initial abstractions. Doing so adds a small, but usually acceptable, measure of conservatism to the design.

Infiltration capacity is influenced by a number of factors such as soil texture, antecedent moisture content, type and density of vegetation, land use, surface slope, and rainfall intensity. A number of empirical methods have been developed to account for infiltration losses. The Richards equations (Richards, 1931; Swartzendruber, 1969) are physically-based equations that have been solved for both single phase (e.g. water) and two-phase (e.g. water and air), one dimensional infiltration. However, because of their computational complexity and the paucity of soil moisture/pore-water pressure/hydraulic conductivity data, they are not used for practical engineering design. The Green and Ampt (1911) method is a simplified physically-based method that has been incorporated into some computerized runoff computation models. A number of empirical equations (e.g. Kostiakov, 1932; Horton, 1939; Phillip, 1957; Holtan, 1961; Smith, 1972) have been developed to describe the infiltration

TABLE 4.11   SCS Cumulative Rainfall Distributions

| | | 24-hour storm | | | | 6-hour storm | | |
| | | P_t/P_{24} | | | | | | |
Hour t	$t/24$	Type I	Type IA	Type II	Type III	Hour t	$t/6$	P_t/P_6
0	0	0	0	0	0	0	0	0
2.0	0.083	0.035	0.050	0.022	0.020	0.60	0.10	0.04
4.0	0.167	0.076	0.116	0.048	0.043	1.20	0.20	0.10
6.0	0.250	0.125	0.206	0.080	0.072	1.50	0.25	0.14
7.0	0.292	0.156	0.268	0.098	0.089	1.80	0.30	0.19
8.0	0.333	0.194	0.425	0.120	0.115	2.10	0.35	0.31
8.5	0.354	0.219	0.480	0.133	0.130	2.28	0.38	0.44
9.0	0.375	0.254	0.520	0.147	0.148	2.40	0.40	0.53
9.5	0.396	0.303	0.550	0.163	0.167	2.52	0.42	0.60
9.75	0.406	0.362	0.564	0.172	0.178	2.64	0.44	0.63
10.0	0.417	0.515	0.577	0.181	0.189	2.76	0.46	0.66
10.5	0.438	0.583	0.601	0.204	0.216	3.00	0.50	0.70
11.0	0.459	0.624	0.624	0.235	0.250	3.30	0.55	0.75
11.5	0.479	0.654	0.645	0.283	0.298	3.60	0.60	0.79
11.75	0.489	0.669	0.655	0.357	0.339	3.90	0.65	0.83
12.0	0.500	0.682	0.664	0.663	0.500	4.20	0.70	0.86
12.5	0.521	0.706	0.683	0.735	0.702	4.50	0.75	0.89
13.0	0.542	0.727	0.701	0.772	0.751	4.80	0.80	0.91
13.5	0.563	0.748	0.719	0.799	0.785	5.40	0.90	0.96
14.0	0.583	0.767	0.736	0.820	0.811	6.00	1.00	1.00
16.0	0.667	0.830	0.800	0.880	0.886			
20.0	0.833	0.926	0.906	0.952	0.957			
24.0	1.000	1.000	1.000	1.000	1.000			

Source: U.S. Dept. of Agriculture, Soil Conservation Service, 1973, 1986.

capacity decay process. The two principal index methods in use are the Φ-index method and the SCS CN method (1985).

As a practical matter for watersheds that are larger than a few acres, the factors that control or otherwise influence infiltration vary with position in the watershed and with time. Also, the Hortonian concept upon which event-based runoff computations are founded rarely occurs in nature, and then probably only on relatively small, impervious watersheds. Thus, in terms of accuracy or reliability, there is no reason to believe that one of the existing methods is generally superior to any other.

The Φ-*index* is a commonly used method for accounting for losses from rainfall. The Φ-index is a constant loss rate. Where there are contemporaneous rainfall and runoff records in a watershed, it can sometimes be computed for individual events. The volume of direct runoff is computed on the basis of gauged flow at the basin outlet. Direct runoff volume is then converted to a depth of excess rain by dividing by basin area. By plotting (or tabulating) the storm hyetograph, a constant loss rate can be found above which the area of the hyetograph equals the depth of effective rain. This rate is known as the Φ-index, and the period for which rainfall intensity exceeds the Φ-index is called the *effective duration*.

Derivation of the Φ-index has several practical drawbacks. First, the analytical procedure generally requires a unimodal hydrograph. These usually only occur during frequent, comparatively short duration rainfall events. Thus, computed values of the Φ-index are likely to be considerably higher than would be the case for larger, longer duration storms. Second, in perennial streams, computation of excess rain is a subjective process that is subject to the prejudices of the analyst and different assumptions pertaining to base flow can lead to substantially different values for effective rain. Third, the computed Φ-index is intended to reflect only the losses due to infiltration. Depending on the location of the rain gauge in relation to the stream flow gauging station, on the temporal distribution of rain, and on the track of the storm in the basin, it may not be possible to evaluate initial abstractions. Last, the Hortonian assumption of uniform generation of direct runoff across the watershed is undoubtably incorrect.

For practical design purposes, the rainfall-runoff records that are requisite for determination of a Φ-index are rarely present. Therefore, it is the usual practice to assume that the Φ-index is equal to the estimated final infiltration capacity. Assuming that a reasonable estimate of the final infiltration capacity can be made, this approach will lead to an overestimate of the volume of excess rain, and ultimately of the peak runoff rate. Table 4.12 (after Musgrave, 1955) provides general guidance for selection of final infiltration capacity.

In recent years, the SCS Curve Number Method (1985; hereafter the CN Method) has become one of the most pervasive methods in the practice for computing abstractions from rainfall. This may be due in part to the imprimatur of the SCS (now National Resource Conservation Service, NRCS) and in part due to the standardized procedures that are required to achieve a result. Despite the elegant appearance of the equations developed in the SCS National Engineering Handbook, Section 4 (1985) and in various hydrology textbooks, the theoretical underpinning of the method is tenuous, and in fact, it is classifiable as an index method. However, it seems to produce results consistently that are acceptable for engineering design.

The formulation used for the direct runoff computation is

$$\sum Q = \frac{\left(\sum P - I_a\right)^2}{\sum P - I_a + S} = \frac{\left(\sum P - 0.2 \cdot S\right)^2}{\sum P + 0.8 \cdot S} \tag{4.15}$$

where $\sum Q$ = cumulative direct runoff, inches
$\quad\quad \sum P$ = cumulative event rainfall, inches
$\quad\quad\ I_a$ = initial abstractions (interception, depression storage, and infiltration prior to onset of direct runoff, inches
$\quad\quad\ S$ = maximum potential retention, inches

TABLE 4.12   Criteria for Selection of a Hydrologic Soil Group

SCS hydrologic soil group	Soil description	Final infilt. cap. (in/hr)
A	Deep, well to excessively drained sands and gravels. Have high infiltration rates even when thoroughly wetted.	0.45–0.30
B	Moderately deep to deep, well drained soils with moderately fine to moderately coarse textures	0.30–0.15
C	Soils with a subsurface layer that impedes downward movement of water, or soils with moderately fine to fine texture	0.15–0.05
D	Soils consisting chiefly of clays with a high swelling potential, or soils with a permanent high water table, or soils with a near-surface clay pan or clay layer, or shallow soils over a nearly impervious material	0.05–0.00

The writer has departed from SCS symbols (ΣQ instead of Q; ΣP instead of P) so as to enhance the understanding of the application when computing excess rain from a design hyetograph. According to the SCS, $I_a = 0.2 \cdot S$. However, some writers have suggested that for urban areas with a substantial percentage of impervious area, that I_a be reduced to $0.1 \cdot S$ so as to reflect a decrease in interception and depression storage that might occur in the change from natural to urban land use.

TABLE 4.13   Runoff Curve Numbers for Urban Areas[1,6]

Cover description		Curve numbers for hydrologic soil group			
Cover type and hydrologic condition	Average percent impervious area[2]	A	B	C	D
Fully developed urban areas (vegetation established)					
Open space (lawns, parks, golf courses, cemeteries, etc.)[3]:					
Poor condition (grass cover <50%)		68	79	86	89
Fair condition (grass cover 50% to 75%)		49	69	79	84
Good condition (grass cover >75%)		39	61	74	80
Impervious areas:					
Paved parking lots, roofs, driveways, etc. (excluding right-of-way)		98	98	98	98
Streets and roads:					
Paved; curbs and storm sewers (excluding right-of-way)		98	98	98	98
Paved; open ditches (including right-of-way)		83	89	92	93
Gravel (including right-of-way)		76	85	89	91
Dirt (including right-of-way)		72	82	87	89
Western desert urban areas:					
Natural desert landscaping (pervious areas only)[4]		63	77	85	88
Artificial desert landscaping (impervious weed barrier, desert shrub with I- to 2-inch sand or gravel mulch and basin borders)		96	96	96	96
Urban districts:					
Commercial and business	85	89	92	94	95
Industrial	72	81	88	91	93
Residential districts by average lot size:					
1/8 acre or less (town houses)	65	77	85	90	92
1/4 acre	38	61	75	83	87
1/3 acre	30	57	72	81	86
1/2 acre	25	54	70	80	85
1 acre	20	51	68	79	84
2 acres	12	46	65	77	82
Developing urban areas					
Newly graded areas (pervious areas only, no vegetation)[5]		77	86	91	94

[1] Average runoff condition and $I_z = 0.2S$

[2] The average percent impervious area shown was used to develoop the composite CN's. Ohter assumptions are as follows: impervious areas are directly connected to the drainage system, impervious areas have a CN = 98, and pervious areas are considered equivalent to open space in good hydrologic condition.

[3] CN's shown are equivalent o those of pasture. Composite CN's may be computed for other combinations of open space cover type.

[4] Composite CN's for natural desert landscaping should be computed based on a CN = 98 and the pervious area CN. The pervious area CN's are assumed to be equivalent to desert shrub in poor hydrologic condition.

[5] Composite CN's should be computed based on impervious area percentage and CN's for newly graded pervious areas

[6] Table taken from SCS Urban Hydrology for Small Watersheds, 2nd ed., (TR-55), June 1986

Maximum potential retention, S, is defined as the maximum depth of water that can infiltrate (following the onset of direct runoff) under existing conditions if the storm continues without limit. The variable, S, is an implicit function of soil infiltration characteristics and of land use. The following equation is used to compute S:

$$S = \frac{1000}{CN} - 10 \qquad (4.16)$$

where CN is a *curve number.* The CN is tabulated according to *hydrologic soil group, HSG,* and land use. Table 4.13, taken from SCS TR-55 manual (1986) shows suggested values of the CN for urban areas.

The first step in the application is to determine the hydrologic soil groups represented by soils in the watershed. The SCS has mapped the soils for much of the US. These maps are published by county in each of the states. Where detailed mapping is unavailable, there are often generalized soil maps, and occasionally hydrologic atlases (usually done jointly by State departments of natural resources and the U.S. Geological Survey) that depict generalized soils maps over a region or a state. The maps outline areas of different soils. The SCS (NRCS) has determined the HSG for many of the mapped soils in the U.S., and has published an alphabetical list of these soils and the HSG to which each belongs in NEH-4. Often, similar lists are also contained in the engineering manuals of State agencies such as departments of conservation or natural resources. In the absence of specific guidance, Table 4.12 (after Musgrave, 1955) can be used to estimate the HSG for specific soils or soil associations.

After the HSG has been determined for each soil in the basin, a CN is determined from Table 4.13 for each land use. The area that comprises each CN is usually estimated visually. The inherent uncertainty of the soil mapping and the idea that a soil must fall within one of only four hydrologic soil groups make more detailed analysis unwarranted. An area-weighted CN is then computed for the basin, the value of S is determined, and Eq. (4.15) is operated in conjunction with the design hyetograph to compute the hyetograph of excess rainfall.

Example 4.4
Determine the hyetograph of effective rainfall for a small urban watershed in Rolla, MO. Assume that the appropriate area-weighted CN is 85. To find S:

$$S = \frac{1000}{85} - 10 = 1.76 \text{ in}$$

The intial abstraction, I_a, is 0.35 inches (0.2 × 1.76). Therefore, runoff cannot occur until a depth of at least 0.35 inches has been supplied by the rainfall event. The runoff computation equation becomes

$$\sum Q = \frac{\left(\sum P - 0.35\right)^2}{\sum P + 1.41}$$

using the hyetograph derived in Example 4.4 for the 2-hour, 25-year rain in Rolla, MO, the direct runoff computation is Table 4.14. The last column in Table 4.14 is the hyetograph of excess precipitation.

There are a number of empirical equations that have been devised to describe the infiltration process. As used herein, "empirical" means that infiltrometer tests have been conducted in the field and a curve has been fitted to a plot of infiltration rate (or cumulative infiltration) versus time. Although a rough equivalence between empirical equations and physically-based equations has been shown for some restrictive conditions (e.g. Eagleson, 1970, and Raudkivi, 1979), have shown that Horton's equation can be derived from a

TABLE 4.14   2-hr, 25-yr Hyetograph of Excess Rain by the CN Method

Time from start of rain (hrs)	Rainfall intensity (in/hr)	ΣP (inches)	ΣQ (inches)	ΔQ (inches)	i_{eff} (in/hr)
0.000		0.00	0.00	0.00	
0.333	0.66	0.22	0.00*	0.00	
0.667	1.77	0.81	0.10	0.10	0.30
1.000	5.10	2.51	1.19	1.09	3.27
1.333	1.59	3.04	1.63	0.44	1.32
1.667	0.87	3.33	1.87	0.24	0.72
2.000	0.57	3.52	2.04	0.17	0.51

*$\Sigma P < I_a$ (in this case, 0.22 in $<$ 0.35 in so no runoff is generated)

Richard's equation), these equations must be considered the result of curve fitting for specific soils and land uses.

Horton's (1939) equation is

$$f = f_c + (f_o - f_c) \cdot e^{-kt} \tag{4.17}$$

where f = infiltration capacity (in/hr) at any time, t (hr)
$\quad f_o$ = initial infiltration capacity, in/hr
$\quad f_c$ = final infiltration capacity, in/hr
$\quad k$ = infiltration decay constant, hr^{-1}
$\quad e$ = base of the natural logarithms

All of the parameters in the equation are dependent on the various factors that control infiltration in general. In combination, these factors exert an infinite number of influences on infiltration rate such that it is probably impossible to find typical values of the equation parameters that are meaningful. McCuen (1989) reports that f_c can vary from 0.01 in/hr to 2 in/hr, that f_o can be 3 to 5 times greater than f_c, and that values of k can range from 1 hr^{-1} to more than 20 hr^{-1}. Because they are similarly derived, similar variations can be expected for the parameters in other empirical equations. In the absence of reliable data from field tests that are conducted on the watershed of interest, utilization of empirical equations to compute infiltration should be used only with great circumspection.

The Green and Ampt Method is a quasi-physically based method. It derives from the application of Darcy's Law and the Law of Conservation of Mass under the assumption of plug flow (e.g. the soil above the wetting front is fully saturated and the soil below is at the initial residual moisture content). Solution of the Green and Ampt equations is tractable even with a calculator. The equations to be solved are the infiltration capacity equation

$$f = K_s \cdot \left(\frac{S_f \cdot (\theta_s - \theta_i)}{F} + 1 \right) \tag{4.18}$$

and the equation for mass infiltration

$$F = K_s \cdot t + S_f \cdot (\theta_s - \theta_i) \cdot \ln \left(1 + \frac{F}{S_f \cdot (\theta_s - \theta_i)} \right) \tag{4.19}$$

where f = infiltration capacity, in/hr
$\quad F$ = mass infiltration, in
$\quad K_s$ = saturated hydraulic conductivity, in/hr
$\quad S_f$ = pore-water pressure at the wetting front (suction head), in

TABLE 4.15   Green and Ampt Infiltration Parameters for Different Soil Classes

Soil class	Porosity η	Effective porosity θ_e	Pore-water pressure at wetting front S_f (in)	Saturated hydraulic conductivity K_s (in/hr)
Sand	0.437 (0.374–0.500)	0.417 (0.354–0.480)	1.95 (0.38–9.98)	4.64
Loamy sand	0.437 (0.363–0.506)	0.401 (0.329–0.473)	2.41 (0.53–11.00)	1.18
Sandy loam	0.453 (0.351–0.555)	0.412 (0.283–0.541)	4.33 (1.05–17.90)	0.43
Loam	0.463 (0.375–0.551)	0.434 (0.334–0.534)	3.50 (0.52–23.38)	0.13
Silt loam	0.501 (0.420–0.582)	0.486 (0.394–0.578)	6.57 (1.15–37.56)	0.26
Sandy clay loam	0.398 (0.322–0.464)	0.330 (0.235–0.425)	8.60 (1.74–42.52)	0.06
Clay loam	0.464 (0.409–0.519)	0.309 (0.279–0.501)	8.22 (1.89–35.87)	0.04
Silty clay loam	0.471 (0.418–0.524)	0.432 (0.347–0.517)	10.75 (2.23–51.77)	0.04
Sandy clay	0.430 (0.370–0.490)	0.321 (0.207–0.435)	9.41 (1.61–55.20)	0.02
Silty clay	0.479 (0.425–0.533)	0.423 (0.334–0.512)	11.50 (2.41–54.88)	0.02
Clay	0.475 (0.427–0.523)	0.385 (0.269–0.501)	12.45 (2.52–61.61)	0.01

Note: Terms in parentheses are 1 standard deviation from the parameter value shown.

θ_s = saturated volumetric water content, decimal fraction
θ_i = initial volumetric water content, decimal fraction
t = time, hours

These equations were derived under the assumption that the depth of ponding on the soil surface is negligible.

Rawls, Brakensiek, and Miller (1983) analyzed roughly 5000 soils across the United States and published values for the Green and Ampt parameters η, θ_e, S_f, and K_s as shown in Table 4.15 (note, table entries were originally in SI units and the parameter symbols have been changed to conform with text in this chapter).

When a saturated soil is allowed to drain thoroughly, there will always be residual moisture content. In the table, the effective porosity, θ_e, is the difference beween the porosity, η, and the residual moisture content, $\theta_r (\theta_e = \eta - \theta_r)$. In other words, it represents the maximum volumetric fraction of water that a well drained soil can hold. For runoff computations that are done in conjunction with design, it is acceptable in most cases to assume that $\theta_i = \theta_r$. Thus, $\theta_s - \theta_i = \theta_e$.

Example 4.5
Use the Green and Ampt method to derive a 2-hour, 25-year hyetograph of effective rain for a clay loam near Rolla, MO.

From Table 4.15, select $\theta_e = 0.309$, $S_f = 8.22$ inches, and $K_s = 0.04$ in/hr. Use the hyetograph derived in Example 4.4 for the 2-hour, 25-year rain in Rolla, MO.

At the end of 0.33 hours, the design storm has produced 0.22 inches of rain.

At this point, we do not know whether 0.22 inches has all infiltrated, or whether some runoff has been produced. As a first trial, assume it has all infiltrated. That is $F = 0.22$ inches. Now solve Eq. (4.18) for f.

$$f = (0.04) \cdot \left(\frac{(8.22) \cdot (0.309)}{0.22} + 1.0 \right) = 0.50 \text{ in/hr}$$

Because the computed f is less than i for the design storm, it is clear that runoff commences somewhere in the interval 0 hrs $\leq t \leq$ 0.33 hrs. Equation (4.18) assumes that the rainfall supply rate is always greater than the infiltration capacity. Therefore, it is not yet clear whether all of the rain infiltrated or if runoff commences part way through the first rain period. To check, use Eq. (4.19) to determine how much infiltration must occur in order to drive the infiltration capacity down to 0.66 in/hr (equal to the rainfall intensity).

$$F = \frac{K_s \cdot S_f \cdot \theta_e}{f - K_s} = \frac{(0.04) \cdot (8.22) \cdot (0.309)}{0.66 - 0.04} = 0.16 \text{ in}$$

Next, determine how long it would take to infiltrate 0.16 inches of rain if the rainfall intensity were always greater than the infiltration capacity. Use Eq. (4.19), but denote the time variable as t' to differentiate from the time scale of the design hyetograph.

$$t' = \frac{1}{0.04} \cdot \left(0.16 - (8.22) \cdot (0.309) \cdot \ln \left(\frac{(0.16)}{(8.22) \cdot (0.309)} + 1 \right) \right) = 0.12 \text{ hrs}$$

Thus, if rainfall intensity exceeded infiltration capacity for all $t > 0$ hours, it would take 0.12 hours to infiltrate 0.16 inches and for the infiltration capacity to drop to 0.66 inches per hour. Obviously, for this to be the case, rainfall intensity would have to be greater than the rainfall intensity during the first rain period, 0.66 in/hr.

The next step is to determine how long it takes the storm to deliver 0.16 inches of rain.

$$t = \frac{0.16}{0.66} = 0.24 \text{ hrs}$$

Therefore, for the first 0.24 hours, all of the rain that falls is infiltrated. At $t = 0.24$ hours, runoff commences. Beyond 0.24 hours, the infiltration capacity drops below 0.66 in/hr.

The next step is to determine what the total infiltration will be in the first 0.33 hours of rainfall. There are at least two acceptable ways to approach this problem. One way is to adjust the time scale of the problem and operate Eq. (4.19). If supply rate always exceeded infiltration capacity, then it would take 0.12 hours for the infiltration capacity to drop to 0.66 in/hr (the point where runoff begins). The duration between the onset of runoff and the end of the first rain period is 0.09 hours (0.33 hrs − 0.24 hrs). Thus, under capacity conditions, it would take only 0.21 hours to infiltrate the same amount of water as would infiltrate during 0.33 hours when the supply rate is 0.66 in/hr. Solve Eq. 4.19 for the depth of infiltration over 0.21 hours.

$$F = (0.04) \cdot (0.21) + (8.22) \cdot (0.309) \cdot \ln \left(\frac{F}{(8.22) \cdot (0.309)} + 1 \right)$$

This equation must be solved using a root solver or by trial and error. In this case, $F = 0.21$ inches. Thus, after 0.33 hours of rain at an intensity of 0.66 in/hr, a total of 0.21 inches will infiltrate. The direct runoff will be 0.01 inches (0.22 − 0.21).

Next, use Eq. (4.18) to compute the infiltration capacity at 0.33 hours.

$$f = (0.04) \cdot \left(\frac{(8.22) \cdot (0.309)}{0.21} + 1.0 \right) = 0.52 \text{ in/hr}$$

The infiltration capacity at 0.33 hours is 0.52 in/hr. Commencing at 0.33 hours, the rainfall intensity for the next period of rain increases to 1.77 in/hr. Because $i > 0.52$ in/hr, runoff will be generated throughout the entire period from 0.33 hours to 0.67 hours. The next step in the computation is to compute the total infiltration for the period 0.0 hrs $\leq t \leq$ 0.67 hrs. In terms of continuous infiltration at capacity rates, the time to be used in Eq. (4.19) should be 0.54 hours (0.21 + 0.33).

$$f = (0.04) \cdot (0.54) + (8.22) \cdot (0.309) \cdot \ln \left(\frac{F}{(8.22) \cdot (0.309)} + 1 \right)$$

The solution is $F = 0.35$ inches. The incremental infiltration for the second rainfall period is 0.14 inches (0.35 − 0.21). The excess rain during this period is 0.45 inches (0.59 − 0.14). Compute the infiltration capacity at 0.67 hours (Eq. 4.18) as

$$f = (0.04) \cdot \left(\frac{(8.22) \cdot (0.309)}{0.35} + 1 \right) = 0.33 \text{ in/hr}$$

Beginning at 0.67 hours, the rainfall intensity increases to 5.10 in/hr, so the entire next rainfall segment will produce direct runoff. Increment the time by 0.33 hours to 0.88 hours (0.54 + 0.33), and solve Eq. (4.19) for total infiltration. The result will be $F = 0.44$ inches. Thus, the net infiltration during the third rainfall segment will be 0.09 inches (0.44 − 0.35; F is the total cumulative rain to any time, t, during the storm). Depth of excess rain during the third rainfall period will be 1.61 inches. Solving Eq. (4.18) for f at 1.00 hours yields 0.27 in/hr.

The procedure continues, time segment by time segment, until the storm ends or until the computed value of f is less than the rainfall intensity for the next rainfall segment. The effective rain is calculated as the difference between the depth of rainfall and the depth of infiltration in each storm segment.

4.2.3 Unit Hydrographs

The *unit hydrograph* is defined as the hydrograph of direct runoff that results from one inch of excess rainfall that is generated uniformly over a watershed at a constant rate during a specified time. The unit hydrograph is the transfer function that is used to convert a hyetograph of excess rainfall into a design hydrograph. According to Brater (1940), unit hydrographs can be used for areas as small as four acres.

The concept of the unit hydrograph is attributed to Sherman (1932). The idea is almost intuitive. If one assumes that an idealized (simple geometry) watershed behaves similarly to a linear reservoir, it is a comparatively simple matter to demonstrate that effective rain from storms of a constant intensity and the same duration will produce hydrographs of direct runoff that have identical times to peak and identical durations of runoff. The flow rate ordinates are in proportion to one another. If the ordinates are known for a hydrograph that results from one inch of excess rain, then the hydrograph ordinates that result from a storm with any other depth of excess can be computed, provided all of the storms have the same effective duration.

Rather than thinking of a design storm as being an event with variable intensity, a design hyetograph of excess rainfall is visualized as a sequential series of back-to-back storms of a fixed duration. If there is a unit hydrograph that is based on the same duration storm as the sequential storms in the design hyetograph, then a hydrograph of excess is computed for each segment of excess rain and the hydrograph ordinates are added to obtain the hydrograph of direct runoff for the entire design storm.

Because there is a severe paucity of basins with the paired stream gauges and rainfall gauges that are requisite for deriving unit hydrographs from actual rain-runoff events, hydrologic design depends on methods of synthetic hydrograph formulation. Basically, this means that someone in a region or in a state analyzes existing paired rainfall-runoff records and develops relationships that relate unit hydrograph ordinates to various basin parameters that can be measured from or estimated using information from topographic mapping. Typically, these parameters include basin area, and indirectly, slope of the stream, but other features may be included as well. Thus, *synthetic unit hydrographs* are extrapolations from other people's analysis of actual data.

There are a number of different procedures for development of synthetic unit hydrographs. No single method has proven to be generally superior or inferior to other methods. No matter how a synthetic unit hydrograph is derived, it's application is identical to every other unit hydrograph. Some Methods that are widely used in the practice, or are incorporated into computerized rainfall-runoff programs are Snyder's Method (1938), SCS dimensionless unit hydrograph (1985), SCS triangular unit hydrograph (1985, Clark's unit hydrograph (1945), Espey, Altman and Graves Method (1977) and Gray's unit hydrograph (1961). The analyst/engineer should use the method that seems to give realistic results for the area or region where the hydrograph is being computed.

Snyder's method (1938) entails computation of selected ordinates of the unit hydrograph, followed by a trial and error adjustment of the remaining ordinates until a volume equivalent to a depth of one inch on the watershed is achieved. In the writer's experience, the method tended to result in peak flow rates that were too small and time bases that were too long for small watersheds in Missouri. However, the Corps of Engineers has successfully adapted the method for their purposes (e.g. for runoff simulation at Wapapello Lake, MO).

Gray's method (1961) was developed from watershed data taken from Missouri, Illinois, Nebraska, Ohio and Wisconsin. The method fits a two-parameter gamma distribution to streamflow data from small watersheds in the above named states. In the writer's experience, unit hydrographs developed by Gray's Method compare favorably with those derived from the SCS dimensionless unit hydrograph in terms of peak flow rate, time to peak, and time-base of runoff. However, the exponents required for Gray's unit hydrograph were developed only for the above named states.

The *SCS dimensionless unit hydrograph* (1985) is frequently used in the practice. The SCS (1985) credits Victor Mockus with development of the dimensionless unit hydrograph. According to the SCS (1985), "it was derived from a large number of natural unit hydrographs from watersheds varying widely in size an geographical locations." Table 4.16 below shows the ordinates of the dimensionless unit hydrograph.

The dimensionless time and flow rate ordinates are dimensionalized by multiplying the ratios in the columns by t_p (in either hours or minutes) and Q_p (cfs), respectively. The result is a t_r-hour (or minute) unit hydrograph where t_r is the effective duration, calculated as

$$t_r = 0.133 \cdot t_c \qquad (4.20)$$

and t_c is the time of concentration for the watershed under consideration. Time to peak, t_p, is calculated as

$$t_p = \frac{t_r}{2} + 0.6 \cdot t_c \qquad (4.21)$$

The peak flow rate, Q_p (cfs), is calculated as

$$Q_p = \frac{484 \cdot A}{t_p} \qquad (4.22)$$

where A = watershed area, mi^2
  t_p = time to peak flow rate, hours

TABLE 4.16 SCS Dimensionless Unit Hydrograph

t/t_p	Q/Q_p
0.0	0.000
0.2	0.100
0.4	0.310
0.6	0.660
0.8	0.930
1.0	1.000
1.2	0.930
1.4	0.780
1.6	0.560
1.8	0.390
2.0	0.280
2.2	0.207
2.4	0.147
2.6	0.107
2.8	0.077
3.0	0.055
3.2	0.040
3.4	0.029
3.6	0.021
3.8	0.015
4.0	0.011
4.2	0.008
4.4	0.006
4.6	0.004
4.8	0.002
5.0	0.000

t_p = time to peak
Q_p = peak flow rate

It is important to use the correct units in Eq. (4.22). The coefficient, 484, embodies both a units conversion and a multiplier to ensure that area under the unit hydrograph corresponds to one inch of direct runoff. It is also important to note that, unlike the corresponding table in the source document (NEH-4, 1985), the ordinates in Table 4.16 have been tabulated at time-ratio intervals of $0.2 \cdot t_p$. If $t_r/0.133$ is substituted for t_c in Eq. (4.21), and Eq. (4.22) is then solved for t_r, it can be shown that the effective duration of the unit hydrograph will always be equal to $0.2 \cdot t_p$. The convolution process (conversion of a hyetograph to a hydrograph) must always be done at even intervals of t_r. Thus, for purposes of hydrograph application, tabulation of dimensionless hydrograph ordinates at intervals other than integer multiples of $0.2 \cdot t_p$ are superfluous and may lead to confusion.

The SCS (NEH-4, 1985) has also promulgated a dimensionless triangular hydrograph that is an approximation to the dimensionless curvilinear unit hydrograph discussed in the preceding paragraph. The effective duration, time to peak, and peak flow rate are all calculated using the same equations as for the curvilinear unit hydrograph. However, unlike the curvilinear unit hydrograph which has a time base (T_b, duration of runoff) of 5 t/t_p, the time base of the triangular unit hydrograph is 3.67 t/t_p.

An advantage of the triangular dimensionless unit hydrograph is that the entire unit hydrograph is defined by three points, Q_p, t_p, and T_b. This may be an attractive feature when doing hand- or spreadsheet calculations for short duration rainfall events. However, for longer, more complex design storms, the direct runoff hydrograph computed from the triangular hydrograph may have smaller peak flows and will have a shorter time base than the direct runoff hydrograph that is derived from the curvilinear unit hydrograph.

Clark (1945) devised a unit hydrograph procedure that translates (that is, moves without storage) direct runoff from one inch of rainfall excess through the watershed to the design point. The translation process is dependent on being able to define sequential areas that are equidistant in time from the design point. The translated hydrograph is then routed through a linear reservoir that has storage properties that are similar to those of the basin. The result is a unit hydrograph. The procedure has been coded into the U.S. Army Corps of Engineers' runoff computation program, HEC-1 (1990) and into the Illinois State Water Survey's Illinois Urban Drainage Simulator (ILLUDAS, 1974).

Determination of the areas that are equidistant in time from the design point is a nontrivial, time-consuming process. The result is called a *time-area diagram.* Singh (1992) gives a clear account of the procedure. In the absence of a derived time-area diagram, HEC-1 (1992) uses a dimensionless relationship for the purpose:

$$AI = 1.414 \cdot T^{1.5} \qquad\qquad 0 \le T \le 0.5 \qquad\qquad (4.23)$$

$$1 - AI = 1.414 \cdot (1 - T)^{1.5} \qquad 0.5 \le T \le 1.0 \qquad (4.24)$$

where AI = the cumulative fraction of the total basin area $(0 \le AI \le 1)$
T = the cumulative fraction of the basin time of concentration $(0 \le T \le 1)$

AI is calculated at fixed intervals of ΔT. Next, the values of AI for each increment of ΔT are calculated. The incremental values of AI (e.g. ΔAI) can be non-dimensionalized by multiplying by basin area to obtain incremental area ΔA in acres. The incremental values of dimensionless time, ΔT, can be converted to incremental time, Δt by multiplying by basin time of concentration, in hours. If the quotient, $\Delta A / \Delta t$ is plotted as a histogram with $\Delta A / \Delta t$ on the ordinate, this represents the shape of a hydrograph that would result if one inch of rain were deposited instantaneously over the entire watershed and was then allowed to drain in the absence of storage effects (a process of pure translation). When conceived in this manner, the ordinate is actually $(\Delta A \times 1)/\Delta t$ with units of acre $\cdot$ inch/hr.

To account for storage effects, the translated hydrograph is routed through an imaginary linear reservoir using a mass balance equation. If basin outflow rate is assumed to be a linear function of basin storage, the mass balance equation can be manipulated to compute outflow from the basin.

$$Q_{O,t+\Delta t} = C_0 \cdot \bar{Q}_I + C_1 \cdot Q_{O,t} \qquad\qquad (4.25)$$

where t, Δt = time and incremental time, respectively, hrs
Q_O = outflow from the basin, cfs
$\bar{Q}_I$ = flow rate from translation hydrograph for time increment Δt, cfs
C_o, C_1 = routing coefficients, dimensionless

$$C_0 = \frac{\Delta t}{\left(R + \dfrac{\Delta t}{2}\right)} \qquad\qquad (4.26)$$

and

$$C_1 = 1 - C_0 \qquad\qquad (4.27)$$

where R is the basin storage factor in hours. Usually, basin lag is used as an approximation to R.

The translation hydrograph is routed using Eq. (4.25). The result is an *instantaneous unit hydrograph,* so called because it represents the hydrograph that would result from a instantaneous, uniformly distributed deposit of one inch of excess rain over the watershed. The Clark unit hydrograph ordinates are found by averaging adjacent flow rate ordinates. Because of the routing procedure used in this method, the recession limb on the unit hydrograph returns to zero at infinity. Therefore, it is customary to truncate the recession limb after 0.995 inches of excess rain have accrued as outflow.

Example 4.6

Use the Clark method to develop a unit hydrograph for a 5 mi^2 with a time of concentration, t_c, of 1.0 hours. Assume a 0.1 hour unit hydrograph is needed.

First, calculate the AI data using Eqs. (4.25) and (4.26), then determine the ΔAI. Finally, convert T and ΔAI to dimensioned variables. Table 4.17 shows the results of the computations.

Let $R = 0.6$ hrs (lag $\approx 0.6 \cdot t_c$) and $\Delta t = 0.1$ hours. Use Eqs. (4.26) and (4.27) to calculate the routing coefficients, C_0 and C_1.

$$C_0 = \frac{0.1}{0.6 + \dfrac{0.1}{2}} = 0.154$$

$$C_1 = 1.000 - 0.154 = 0.846$$

The routing equation, Eq. (4.25) becomes

$$Q_{O,t+\Delta t} = 0.154 \cdot \bar{Q}_I + 0.846 \cdot Q_{O,t}$$

Table 4.18 shows the results of routing the first several ordinates of the translation hydrograph, $\bar{Q}_I$, through the imaginary reservoir. The final column in Table 4.26 is calculated as the average of each sequential pair of Q_O (e.g. $0.5 \times (0 + 222) = 111$; $0.5 \times (587 + 222) = 405$).

TABLE 4.17 Computed Translation Hydrograph (Clark's Method)

T ($\Sigma\%t_c$)	AI ($\Sigma\%A$)	ΔAI (%A)	t (hrs)	$\bar{Q}$ (ac $\cdot$ in $\cdot$ hr^{-1})
0.0	0.000		0.0	0.0
0.1	0.045	0.045	0.1	1440
0.2	0.126	0.081	0.2	2592
0.3	0.232	0.106	0.3	3392
0.4	0.358	0.126	0.4	4032
0.5	0.500	0.142	0.5	4544
0.6	0.642	0.142	0.6	4544
0.7	0.768	0.126	0.7	4032
0.8	0.874	0.106	0.8	3392
0.9	0.955	0.081	0.9	2592
1.0	1.000	0.045	1.0	1440

TABLE 4.18 Routing of Translation Hydrograph, and Computation
of First 15 Ordinates of 0.1-hr Unit Hydrograph (Clark's Method)

t (hrs)	$\bar{Q}$ (1 ac $\cdot$ in $\cdot$ hr^{-1} = 1 cfs)	Q_o (cfs)	0.1-hr unit hydrograph (cfs)
0.0	0	0	0
0.1	1440	222	111
0.2	2592	587	405
0.3	3392	1019	1251
0.4	4032	1483	1251
0.5	4544	1954	1719
0.6	4544	2352	2153
0.7	4032	2611	2482
0.8	3392	2731	2671
0.9	2592	2710	2721
1.0	1440	2514	2612
1.2	0	2127	2321
1.3	0	1799	1963
1.4	0	1522	1661
1.5	0	1288	1405

Espey, Altman and Graves (1977) and Espey and Altman (1978) analyzed rainfall-runoff relations from 41 watersheds in 8 different states with areas ranging from 9 acres to 15 mi^2 and with impervious percentages ranging from 2% to 100%. Their analysis resulted in a set of regression equations that provide guidance to be used sketching, either visually or by using curve-fitting meaures, 10-minute unit hydrographs. The equations for estimating the unit hydrograph parameters are:

$$T_P = 3.1 \cdot L^{0.23} \cdot S^{-0.25} \cdot I^{-0.18} \cdot \phi^{1.57} \tag{4.28}$$

where T_P = time to peak flow rate, minutes
 L = total distance along the principal flow path from the watershed boundary to the design point, feet
 S = slope of the main channel, ft/ft
 I = percent impervious area within the watershed, %
 ϕ = watershed conveyance factor, dimensionless

$$S = \frac{H}{0.8 \cdot L} \tag{4.29}$$

where H is the elevation difference between the channel invert at the design point and a point along the principal flow path that is $0.2 \cdot L$ downstream from the channel boundary.

$$Q_P = 31.62 \times 10^3 \cdot A^{0.96} \cdot T_P^{-1.07} \tag{4.30}$$

where Q_P = peak flow rate, cfs
 A = watershed area, mi^2

$$T_B = 125.89 \times 10^3 \cdot A \cdot Q_P^{-.95} \tag{4.31}$$

where T_B = time base of the 10-minute unit hydrograph, min

$$W_{50} = 16.22 \times 10^3 \cdot A^{0.93} \cdot Q_P^{-0.92} \tag{4.32}$$

$$W_{75} = 3.24 \times 10^3 \cdot A^{0.79} \cdot Q_P^{-0.78} \tag{4.33}$$

where W = width of the 10-minute unit hydrograph

$_{50, 75}$ = subscripts to denote the percentage of peak flow rate at which the widths are to be applied

The shape parameters, W_{50} and W_{75}, are to be drawn parallel to the abscissa and are usually located such that one third of the width is located to the left of the time to peak. The watershed conveyance factor, ϕ, is a function of percent impervious area and the length-weighted Manning n for the principal path of flow. It is taken from Fig. 4.10.

Application of the method entails determination of the variables L, S, A, I and a length-weighted n value for the mainstem drainage channel. The first four variables can usually be determined from existing topographic maps and aerial photographs. Manning's n must be based on field inspections, a task that is not always easy to accomplish. After values have been determined for the required variables, the equations are solved to find the seven flow rate-time ordinates that govern the general shape of the 10-minute unit hydrograph. A smooth curve is then fit through these points and adjusted until the volume of direct runoff is equivalent to one inch over the watershed.

Of the methods covered in Section 4.2.3, the Espey-Altman-Graves Method is the only one that was developed specifically for urban applications. However, because of the curve-fitting procedure that is required for its derivation, it must usually be entered into computerized runoff computation programs as a "user defined" unit hydrograph.

4.2.4 Using Unit Hydrographs to Compute Hydrographs of Direct Runoff

As mentioned earlier in Section 4.2.3, the unit hydrograph is the transfer function that is used to transform a hyetograph of excess rain into a hydrograph of direct runoff at the design

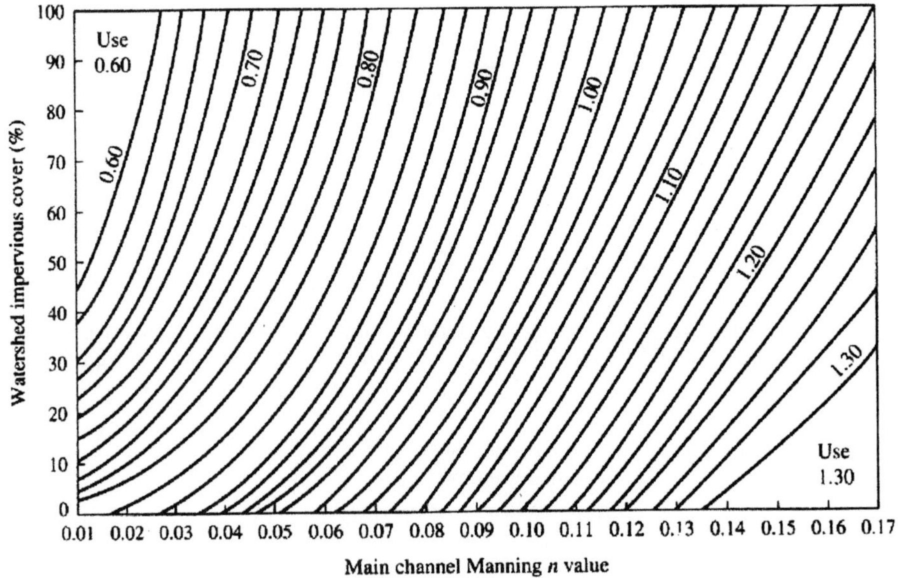

FIGURE 4.10   Watershed conveyance factors as a function of Manning's n and percentage of impervious area.

point. Although the computational process is straightforward and intuitive, terminology (e.g. convolution, deconvolution, impulse-response function) is sometimes distracting to the practitioner. Fortunately, mastery of the terminology is not required for the understanding of the computations or what the computations represent.

Recall that a unit hydrograph for a given watershed is a hydrograph of direct runoff that results from a storm that produces 1.0 inch of excess rain in a ***known period of time*** which is known as an effective duration. On a given watershed, all events with a given effective duration produce direct runoff hydrographs that have the same time to peak and same time base. If the effective duration of one storm is different from that of another on the same basin, the resulting direct runoff hydrographs will have different times to peak and different time bases. That is why unit hydrographs are identified by the effective duration of the storm which produces them.

In runoff computations, the hyetograph of effective duration is viewed as a sequence of back-to-back events with a constant effective duration, but differing depths (and therefore intensities). Each event produces its own hydrograph. In theory, once water is set in motion toward the outlet of the basin, its behavior is independent of any runoff that preceded it or follows it. Therefore, to obtain a hydrograph from a complex storm, hydrographs are computed for each segment and their ordinates, displaced according to time of origin of excess rain on the watershed, are added. The following schematic (Fig. 4.11) demonstrates the concept.

The hyetograph of excess rain is shown inverted over the hydrographs. The depth of excess rain in each segment of the hyetograph is shown. The dashed triangular hydrographs are the 0.5-hour unit hydrograph, lagged in time so that the start of the rising limb corresponds in time to the beginning of a new rainfall segment. The solid lines are the hydrographs of excess rain that result, in sequence, from each of the segments of excess rain. Ordinates of the solid hydrographs are 1.2, 1.4, and 1.1 times larger than the ordinates of the unit hydrograph.

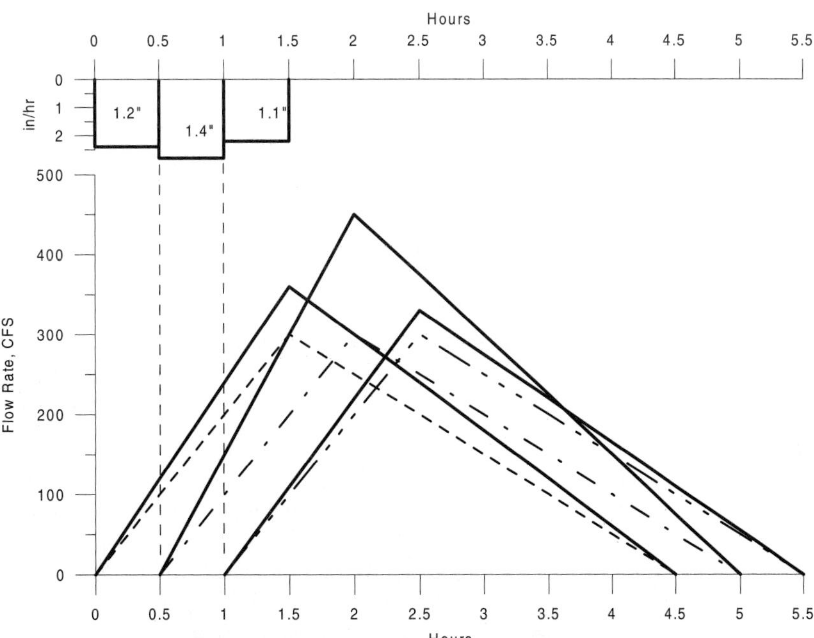

FIGURE 4.11   Schematic showing relation of incremental storm segments to segmental components of the storm hydrograph.

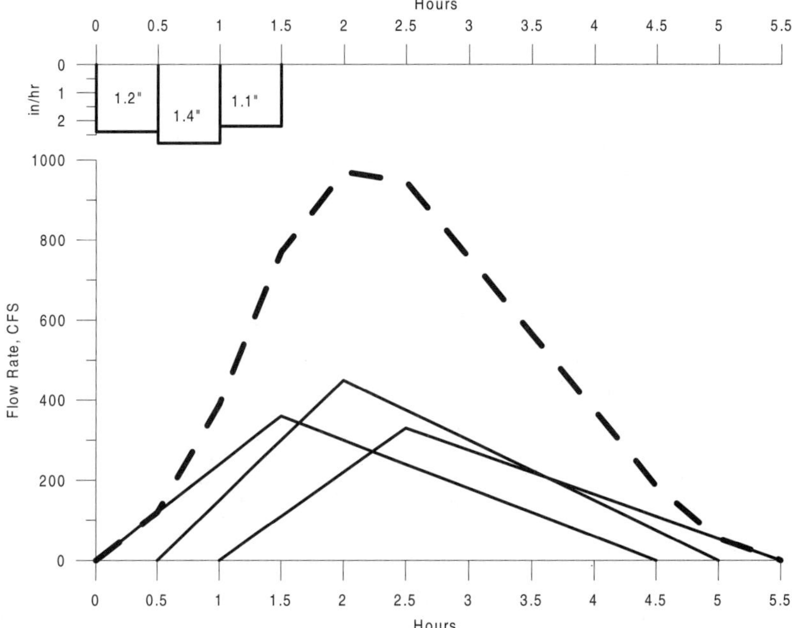

FIGURE 4.12 Storm hydrograph as the sum of hydrographs from incremental storm segments.

Figure 4.12, shown below, is the same as the previous figure, but the underlying unit hydrographs have been eliminated, and the ordinates of the individual component hydrographs have been added to obtain the design hydrograph.

Table 4.19 shows results of the computation. The unit hydrograph has been tabulated only once for reference. Component hydrographs are lagged so that the rising limb begins with the corresponding segment of effective rain and the columns are labeled according to the depth of excess rain in each segment. The final column shows the ordinates of the hydrograph that results from the 3-segment storm.

4.2.5 Converting a *t*-hr Unit Hydrograph to a (*t* + Δ*t*)-hr Unit Hydrograph

By most methods, the effective duration for synthetic unit hydrographs is fixed by physiographic features of the watershed for which the unit hydrograph is derived. Depending on the design problem, this effective duration is not always convenient or desirable to work with. For example, it is always necessary to divide the design storm into an integer number of time segments. The effective duration that derives directly from basin parameters will not often divide into the design storm duration an integer number of times. In other instances, the derived duration may be of such a magnitude that the total number of computations can be significantly reduced if a larger effective duration is used.

For any basin, a unit hydrograph of a given effective duration can be converted into a unit hydrograph of any other effective duration through use of the *S-curve* (sometimes called *S-hydrograph*) method. The *S-curve* is the entire rising limb of the hydrograph that would result from effective rainfall that is uniformly deposited over the watershed at a constant intensity for a duration that is sufficient to produce steady-state (equilibrium) flow at the

TABLE 4.19 Unit Hydrograph Transformation of Design Hyetograph to Storm Hydrograph

Time (hr)	.5-h unit hydrograph (cfs)	Q from 1.2″ excess rain (cfs)	Q from 1.4″ excess rain (cfs)	Q from 1.1″ excess rain (cfs)	Storm hydrograph (cfs)
0.0	0	0			0
0.5	100	120	0		120
1.0	200	240	150	0	390
1.5	300	360	300	110	780
2.0	250	300	450	220	970
2.5	200	240	375	330	945
3.0	150	180	300	275	755
3.5	100	120	225	220	565
4.0	50	60	150	165	375
4.5	0	0	75	110	185
5.0			0	55	55
5.5				0	0

design point. The S-curve is derived by lagging an infinite series of t_r-hr unit hydrographs and adding the flow ordinates just as would be done by computing runoff for any other storm. The only difference here is that the effective rainfall intensity, $1/t_r$ in/hr never changes. Because the effective intensity is constant, the usual lag and add procedure that is done for effective rain with variable intensity can be accomplished by running a cumulative total on the t_r-unit hydrograph **at intervals of t_r**. Use of any other interval of accumulation will invalidate the procedure.

By superimposing an S-curve on itself, but displaced in time by some interval, $t_r' \neq t_r$, and subtracting flow rate ordinates at intervals of t_r', a new hydrograph is generated that has a volume equivalent to a depth of $t_r' \cdot (1/t_r)$ inches over the watershed, and t_r' is the effective duration of the storm that produced the hydrograph. The t_r'-hr unit hydrograph is then calculated by dividing the ordinates t_r'-hr hydrograph by the depth, t_r'/t_r. The reader is referred to a contemporary hydrology textbook (e.g. Chow, Maidment, and Mays, 1988; Singh, 1992; Viessman and Lewis, 1996) for a complete justification of the S-curve procedure.

For an example, change a 15-minute unit hydrograph to a 30-minute unit hydrograph using the S-curve procedure. The 15-minute unit hydrograph is shown in Fig. 4.13. It is important to note that the 15-minute unit hydrograph derives from a 15-minute period of excess rain at an intensity of 4 in/hr. Therefore, the S-curve will be the rising limb of a hydrograph that results from a continuous rainfall intensity of 4 in/hr.

The ordinates of the 15-minute unit hydrograph are tabulated in Table 4.20. The S-curve ordinates, a cumulative summation of the unit hydrograph ordinates **at a fixed interval of 15 minutes** ($t_r = 15$ minutes), is also tabulated in Table 4.20. Note, that once equilibrium on the S-curve is reached, it continues at that flow rate infinitely because as one segment of excess rain drains completely from the watershed, another is being added. Figure 4.14 shows a plot of the S-curve as derived from the 15-minute unit hydrograph and a portion of the hyetograph of excess rain from which it results.

Table 4.21 shows a tabulation of the computations of the 30-minute unit hydrograph. The objective is a 30-minute unit hydrograph, so only those ordinates that are at 30-minute intervals are needed. In instances where the S-curve ordinates doesn't contain ordinates at the desired intervals (t_r' is not an integer multiple of t_r), the S-curve is plotted and ordinates are taken from the plot. The same is true if it is necessary to obtain a unit hydrograph where $t_r' < t_r$.

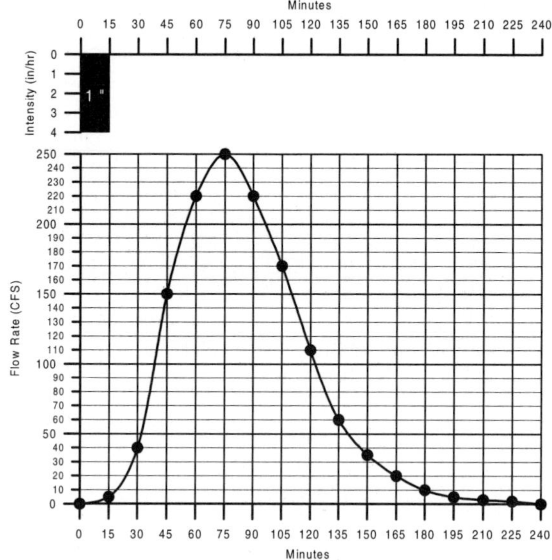

FIGURE 4.13   15-min unit hydrograph.

TABLE 4.20   *S*-curve from a 15-min unit Hydrograph

Time (min)	15-m U.G. (cfs)	*S*-curve (cfs)
0	0	0
15	5	5
30	40	45
45	150	195
60	220	415
75	250	665
90	220	885
105	170	1055
120	110	1165
135	60	1225
150	35	1260
165	20	1280
180	10	1290
195	5	1295
210	3	1298
225	2	1300
240	0	1300
255	0	1300
270	0	1300

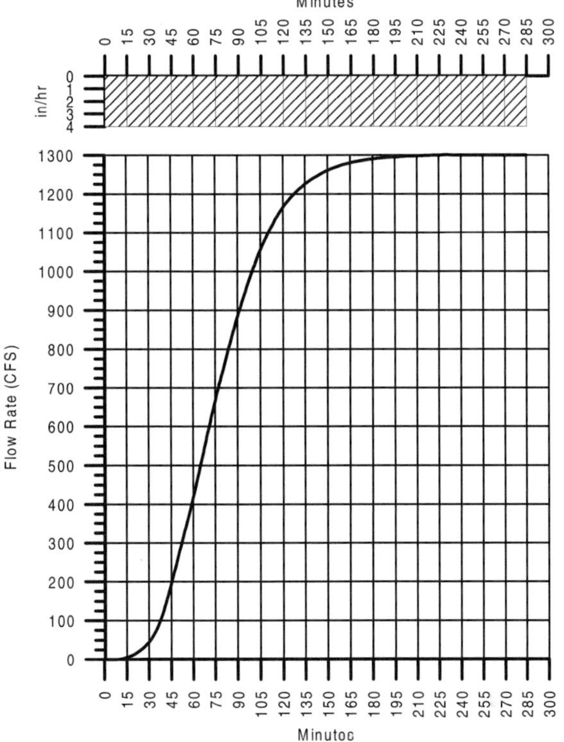

FIGURE 4.14 S-curve resulting from a 15-min unit hydrograph.

TABLE 4.21 Computation of a 30-min Unit Hydrograph from *S*-Curves Base Derived from a 15-min Unit Hydrograph

Time (min)	*S*-curve (cfs)	Lagged *S*-curve (cfs)	Hydrograph from 30-min rain (cfs)	30-min unit hydrograph (cfs)
0	0		0	0
30	45	0	45	22
60	415	45	370	185
90	885	415	470	235
120	1165	885	280	140
150	1260	1165	95	47
180	1290	1260	30	15
210	1298	1290	8	4
240	1300	1298	2	1
270	1300	1300	0	0

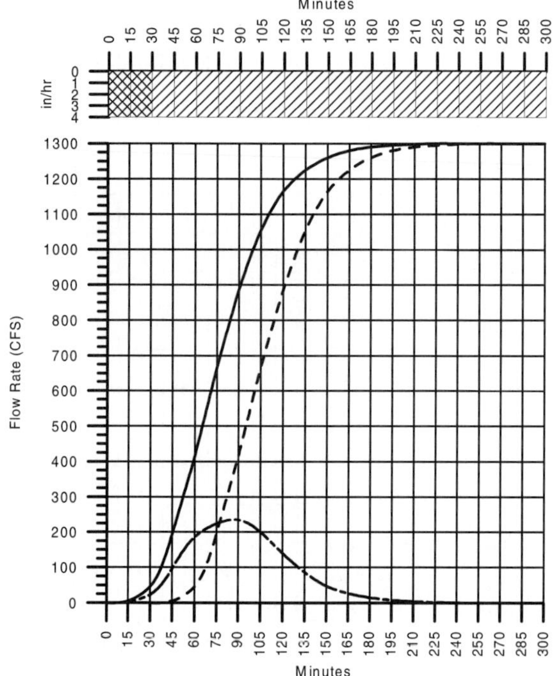

FIGURE 4.15 Lagged S-curves and hydrograph of 2 inches of excess rain.

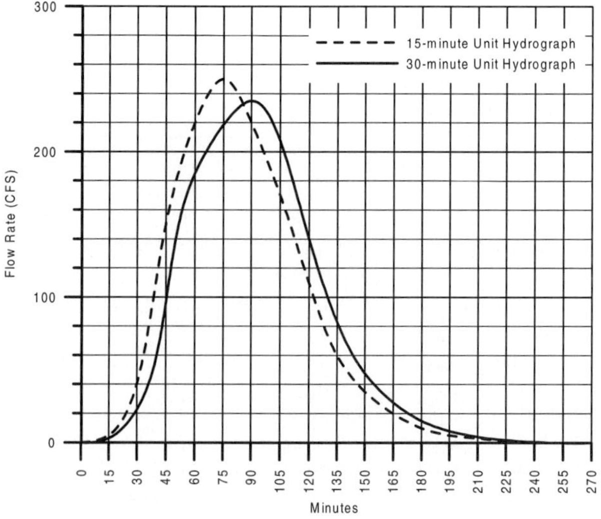

FIGURE 4.16 Comparison of 15-min and 30-min unit hydrographs.

Entries in the column labeled Hydrograph from 30-minute rain are the differences between the S-curve and lagged S-curve columns. They represent the ordinates of a hydrograph that results from 2 inches of excess rain produced during an effective duration of 30 minutes $(t'_r \cdot 1/t_r; 30 \cdot 1/15)$. Because hydrograph ordinates for storms of a given effective duration are all proportional, one to another, the 30-minute unit hydrograph is computed by dividing the ordinates of the 30-minute hydrograph (column 4) by 2 inches, the depth of excess rain.

Figure 4.15 shows a plot of the S-curve and the lagged S-curve. The 30-minute hydrograph is a plot of the of the vertical differences between the two curves. The 30-minute duration, double cross hatchured part of the hyetograph is the increment of effective rain (2 inches) associated with the 30-minute hydrograph.

Figure 4.16 is a plot showing both the original 15-minute unit hydrograph and the 30-minute unit hydrograph. This figure demonstrates why it is necessary for a unit hydrograph to be representative of a storm with the same effective duration as the duration of the incremental rainfall segments in the design hyetograph. Different effective durations result in hydrographs with different shapes (peak flow rate, time to peak, and time base) are unique to a specific effective duration.

4.3   *KINEMATIC WAVE METHODS*

Unit hydrograph methods are empirical approaches to runoff computation that circumvent the necessity of physical representation of the laws of conservation of mass and of momentum that govern the actual movement of water over the surface of the land. The practical reasons for this approach are well justified, namely: 1) the various components of runoff generation and movement of runoff are not well understood; and 2) the complexity of the processes is so enormous that it would be impractical (and probably impossible), in terms of journeyman design, to collect the requisite data and code them into a computerized computational scheme. However, in some cases (e.g. very small areas, small areas with a high percentage of impervious surface), kinematic wave methods may be preferable to unit hydrograph methods for computing hydrographs of direct runoff.

The kinematic wave approach derives from the one-dimensional continuity and momentum equations (St. Venant equations) that describe the physical processes of the movement of water. The kinematic wave equation results when the inertial terms and the pressure terms have minimal influence on the transmission of water. In terms of computing runoff, kinematic wave methods are amenable to solution for situations where the input to the collector system is accrued at a uniform rate along the path of flow. Constant intensity rainfall on a strip of unit width is an example.

Manual solutions of kinematic wave formulations in relation to hydrologic design are impractical. However, they have been incorporated into the Penn State Runoff Model (1987) as the method for routing overland flow, and into the U.S. Army Corps of Engineers Model, HEC-1 (1992), as an option for routing overland flow. The form of the equations to be solved are different for the two programs. However, both programs require that the overland flow runoff elements be visualized as elements that are "typical" of those that actually exist in the watershed for which the analysis is being performed. As a rule, these "typical" elements never exist in reality. Therefore, despite the approach toward conformance with physical laws that the kinematic wave approach implies, results from kinematic wave routing of overland flow may not be any more realistic than those derived from unit hydrograph approaches. As with most intelligent engineering design, the reasonableness of the results of the computation is dependent on the experience level that goes into the characterization of the overland flow elements, and into the appraisal of the model output.

4.4 *EVENT-BASED RUNOFF MODELS*

There are a number of computer programs that are available to do runoff computations. These are termed "models", and not only have the capability of generating hydrographs from design storms, but most have flood routing routines that make them useful for computing hydrographs at various locations in systems that are comprised of several subasins.

The U.S. Geological Survey's (USGS) Distributed Routing Rainfall-Runoff Model (DR3M) was developed principally for simulation of runoff from urban basins. The model can be used to compute runoff from a system of watersheds. Each component is comprised of an overland flow element and a collector channel. Reservoirs or detention basins can be incorporated into the model. Both overland flow routing and channel routing are done with kinematic wave procedures. DR3M can be obtained directly from the USGS in Reston, VA. Program documentation is published as a USGS Open File Report 82-344 (Alley and Smith, 1982).

HEC-1 Flood Hydrograph Package (1992) was developed by the U.S. Army Corps of Engineers at their Hydrologic Engineering Center(HEC). The model has an impressive array of options for development of design hyetographs and for computing effective rainfall. It has a number of unit hydrograph procedures built in, including a transparent *S*-curve procedure, and a kinematic wave runoff routine. It can accept user-generated rainfall and user generated hydrographs. In terms of hydraulics, it has several options for flood routing in both channels and in reservoirs. In the event that the analyst has rainfall or hydrograph records, it contains parameter optimization routines. The Corps of Engineers has a Windows-based program under development, HEC-HMS, which will probably ultimately replace HEC-1. HEC-HMS can be downloaded from the Hydrologic Engineering Center website. HEC-1 must be purchased from private vendors, some of whom add their own enhancements. The HEC maintains a list of vendors.

The Illinois Urban Drainage Simulator (ILLUDAS) was developed by the Illinois State Water Survey (Tierstrup and Stall, 1974). The program is based on the British Road Research Laboratory Model (Watkins, 1962). It requires partitioning of the watershed into pervious and connected impervious areas. Separate hydrographs are computed for each and then combined to form a hydrograph at the basin outlet. Like other models, a watershed system that is comprised of a number of subasins can be modeled. Unlike HEC-1, ILLUDAS requires that a design hyetograph be supplied as input. For residents of Illinois, the program can be obtained from the Illinois State Water Survey. Other users must purchase the software from CE Software in Champaign, IL.

The Penn State Runoff Model (PSRM; Aron, 1987) was originally developed as the result of class projects at The Pennsylvania State University. It was subsequently improved under joint funding from the Office of Water Research and Technology (U.S. Dept. of Interior), The Pennsylvania State University and the City of Philadelphia. Rainfall data must be supplied for input, either as a design hyetograph or as station rainfall. PSRM has the capability of simulating runoff from areally distributed rainfall. Abstractions from rainfall are computed by the SCS CN Method. Overland flow hydrographs are generated by a kinematic wave procedure. The user specifies travel time through channels and pipes, but the modified Puls Method is used to route hydrographs through impoundments. A current version of PSRM, now called KU-Penn State Runoff Model, HY-5, is available for purchase from the Kansas University Transportation Center, Lawrence, KS.

The SCS program, TR-20 (1973), is a single event model that generates runoff hydrographs and routes them through a drainage system that may be comprised of several subasins. The methodology employed by the model utilizes SCS procedures for all computations. TR-55 (1986) is a simplified version that utilizes procedures of TR-20 for a single watershed. TR-55 procedures are amenable to hand calculations. The TR-20 program can be obtained from the NRCS (formerly SCS), and documentation of TR-20 and TR-55 are available through NTIS.

REFERENCES

Alley, W. M., and P. E. Smith, "Distributed Routing Rainfall-runoff Model-Version II, User's Manual," U.S. Geological Survey Open-File Report 82–344, 1982.

American Iron and Steel Institute (AISI), *Modern Sewer Design,* Washington, D.C., 1980.

Aron, G., "Penn State Runoff Model for IBM PC," Users Manual, Dept. of Civil Engineering and Institute for Research on Land and Water Resources, Pennsylvania State University, 1987.

Calif. Department of Transportation (Cal Trans), *Highway Design Manual,* chap. 810, July 1995.

Chow, V. T., D. R. Maidment and L. W. Mays, *Applied Hydrology,* McGraw-Hill, New York, 1988.

Clark, C. O., "Storage and the Unit Hydrograph," Transactions of the Amer. Soc. Civ. Eng., vol. 110, 1945.

Eagleson, P. S., *Dynamic Hydrology,* McGraw-Hill, New York, 1970.

Espey, W. H., Jr., D. G. Altman, and C. B. Graves, "Nomographs for 10-minute Unit Hydrographs for Small Urban Watersheds," *Tech. Memorandum No. 32,* Urban Water Resources Research Program, Amer. Soc. Civ. Eng., 1977.

Espey, W. H. and D. G. Altman, "Nomographs for 10-minute Unit Hydrographs for Small Urban Watersheds," U.S. Environmental Protection Agency, Report EPA-600/9-78-035, 1978.

Frederick, R. H., V. A. Meyers, and E. P. Auciello, "Five to 60-minute Precipitation Frequency for the Eastern and Central United States," NOAA Technical Memorandum NWS HYDRO-35, Office of Hydrology, June 1977.

Gray, D. M., "Synthetic Unit Hydrographs for Small Watersheds," Proc. Amer. Soc. Civ. Eng., *Journal of the Hydraulics Division,* 88(HY4), 1961.

Green County, MO, Green County Storm-water Design Standards, 1999.

Green, W. H., and G. Ampt, "Studies of Soil Physics, Part 1—The Flow of Air and Water Through Soils," *J. Agricultural Science,* 1911.

Hathaway, G. A., "Design of Drainage Facilities," *Transactions of the Amer. Soc. Civ. Eng.,* vol. 110, 1945.

Hershfield, D. M., "Rainfall Frequency Atlas of the United States for Durations from 30 Minutes to 24 Hours and Return Periods from 1 to 100 Years," Tech. Paper 40, U.S. Dept. of Commerce, Weather Bureau, Washington, D.C. May 1961.

Holtan, H. N., "A Concept for Infiltration Estimates in Watershed Engineering," U.S. Dept. of Agriculture, *Agricultural Research Service Bulletin,* 1961.

Horton, R. E., "The Role of Infiltration in the Hydrologic Cycle," *Transactions of the American Geophysical Union,* vol. 14, 1933.

Horton, R. E., "Analysis of Runoff Plat Experiments with Varying Infiltration Capacity," *Transactions of the American Geophysical Union,* vol. 20, 1939.

Huff, F. A., "Time Distribution of Rainfall in Heavy Storms," 3(4) *Water Resources Research,* 1967.

Huff, F. A.. and J. R. Angel, "Rainfall Frequency Atlas for the Midwest," Bulletin 71, *Midwestern Climate Center Report* 92-03, Midwestern Climate Center and Illinois State Water Survey, 1992.

Institution of Engineers Australia, *Australian Rainfall and Runoff,* vol. 1 D. H. Pilgrim (ed.), vol. 2 R. P. Canterford (ed.), Canberra, Australia, 1987.

Kerby, W. S, "Time of Concentration for Overland Flow," 29(3) *Civil Engineering,* March 1959..

Kiefer, C. J., and H. H. Chu, "Synthetic Storm Pattern for Drainage Design," *Journal of the Hydraulics Division,* Amer. Soc. Civ. Eng., 84(HY1), 1958.

Kirpich, Z. P., "Time of Concentration of Small Agricultural Watersheds," 10(6) *Civil Engineering,* June 1940.

Kostiakov, A. N., "On the Dynamics of the Coefficient of Water-percolation in Soils and on the Necessity for Studying It from a Dynamic Point of View for Purposes of Amelioration," Trans. 6th Comm. Intern. Soil Sci. Soc., Russian Part A, 1932.

Kuichling, E., "The Relation between Rainfall and the Discharge of Sewers in Populous Districts," Trans. Amer. Soc. Civ. Eng., vol. 20, 1889.

Malaysia Department of Irrigation and Drainage, *Urban Storm Water Management Manual for Malaysia, 2000,* River Engineering Division.

McCuen, R. H., S. L. Wong, and W. J. Rawls, "Estimating Urban Time of Concentration,"110 (HY7) *J. Hydraulic Engineering,* Amer. Soc. Civ. Eng., 1984.

McCuen, R. H., *Hydrologic Analysis and Design,* 2nd ed., Prentice-Hall, Englewood Cliffs, NJ, 1988.

Miller, J. F., R. H. Frederick, and R. J. Tracey, "Precipitation Frequency Analysis of the Coterminous Western United States (By States)," NOAA Atlas 2, 11 volumes, National Weather Service, 1973.

Mulvaney, T. J., "On the Use of Self-registering Flood and Rain Gauges on Making Observations of the Relations of Rainfall and of Flood Discharges in a Given Catchment," *Trans. Inst. Civ. Eng., Ireland,* 4(2) 1851.

Musgrave, G. W., "How Much of the Rain Enters the Soil?", U.S. Dept. of Agriculture, *Yearbook of Agriculture, Water,* 1955.

Phillip, J. R., "The Theory of Infiltration: 1. The Infiltration Equation and Its Solution," 83 *Soil Science,* 1957.

Phillip, J. R., "The Theory of Infiltration: 4. Sorptivity and Algebraic Infiltration Equations," 84 *Soil Science,* 1957.

Phillip, J. R., "The Theory of Infiltration: 5. The Influence of Initial Water Content," 84 *Soil Science,* 1957.

Pilgrim, D. H., and I. Cordery, "Rainfall Temporal Patterns for Design Floods," *Journal of the Hydraulics Division,* Amer. Soc. Civ. Eng., 101(HY1), 1975.

Ramser, C. E., "Runoff from Small Agricultural Areas," 34(9) *Journal of Agricultural Research,* 1927.

Raudkivi, A. J., *Hydrology,* Pergamon Press, Oxford, 1979.

Rawls, W. J., D. L. Brakensiek, and N. Miller, "Green-Ampt Infiltration Parameters from Soils Data," *Journal of the Hydraulics Division,* Amer. Soc. Civ. Eng., 109(HY1) 1983.

Richards, L. A., "Capillary Conduction Through Porous Mediums," *Physics,* 1931.

Rossmiller, R. L., "Rational Formula Revisited," Proceedings of Conference on Stormwater Detention Facilities: Planning, Design, Operation, and Maintenance, Engineering Foundation Urban Water Resources Research Council, ASCE, New England College, Henniker, N.H., Aug. 2–6, pp. 146–162, 1982.

Sherman, L. K., "Stream Flow from Rainfall by the Unit-graph Record," 108 *Eng. News Record,* April 7, 1932.

Smith, R. E., "The Infiltration Envelope: Results from a Theoretical Infiltrometer," 17 *Journal of Hydrology,* 1972.

Snyder, F. F., "Synthetic Unit-graphs, *Transactions of the American Geophysical Union,* vol. 19, 1938.

Swartzendruber, D., "The Flow of Water in Unsaturated Soils," in R. M. Dewiest (ed.), *Flow Through Porous Media,* Academic Press, 1969, New York, N.Y.

Terstriep, M. L., and J. B. Stall, "The Illinois Urban Drainage Area Simulator," *ILLUDAS,* Bull. 58, Illinois State Water Survey, Urbana, IL, 1974.

U.S. Army Corps of Engineers, "HEC-1, Flood Hydrograph Package," Hydrologic Engineering Center, Sept 1990.

U.S. Dept. of Agriculture, Soil Conservation Service, "Computer Program for Project Formulation: Hydrology," Technical Release no. 20 (TR-20), 1983.

U.S. Dept. of Agriculture, Soil Conservation Service, "National Engineering Handbook," Section 4— Hydrology, March 1985.

U.S. Dept. of Agriculture, Soil Conservation Service, "Urban Hydrology for Small Watersheds," Technical Release no. 55, June 1986.

U.S. Dept. of the Interior, Bureau of Reclamation, "Design of Small Dams," GPO, 1974.

Viessman, W. Jr., and G. L. Lewis, *Introduction to Hydrology,* 4th ed., Harper-Collins, 1996, New York, N.Y.

Singh, V. P., *Elementary Hydrology,* Prentice Hall, 1992.

Watkins, L. H., *The Design of Urban Sewer Systems,* Road Research Paper no. 55, Dept. of Scientific and Industrial Research London: Her Majesty's Stationery Office, 1962.

Yen, B. C., and V. T. Chow, "Design Hydrographs for Small Drainage Structures," *Journal of the Hydraulics Division,* Amer. Soc. Civ. Eng., 106(HY6) 1980.

CHAPTER 5
DESIGN OF STORMWATER INLETS

John W. Nicklow
Southern Illinois University Carbondale
Carbondale, Illinois

5.1 INTRODUCTION

When rain falls on a sloped pavement surface, it forms a thin layer of water that increases in thickness as it flows to the sides of the roadway. This accumulation of water on travel lanes can defeat the purpose of a highway by disrupting traffic flow, reducing vehicle skid resistance, increasing potential for motorist hydroplaning and visibility problems, and accelerating roadway deterioration. The water may also freeze, making vehicle maneuvering even more difficult. The objective in highway drainage design is to minimize such problems by collecting runoff in gutters and intercepting it using stormwater inlets that subsequently direct flow to subsurface conveyance systems, culverts, or ditches. Proper design of these facilities is thus essential to maintaining safe vehicular and pedestrian travel conditions and ensures that highway service levels will avoid disruption.

This chapter provides guidance on the selection of runoff frequency and spread for design computations, design flow estimation, flow in gutters, inlet design, and bridge deck drainage. For additional information, the reader is referred to Federal Highway Administration guidance documents, including Hydraulic Engineering Circular (HEC) No. 12 (Johnson and Chang, 1984), HEC-21 (Young et al., 1993), and HEC No. 22 (Brown et al., 1996). In addition, Anderson et al. (1995) provides further information on the mechanics of surface drainage.

5.2 FACTORS AFFECTING INLET DESIGN

The *spread* of water onto pavement, or the top width of gutter flow measured perpendicular to the edge of the roadway, is of primary concern to the hydraulic engineer. As spread increases, the risks associated with accidents and delays are increased. Stormwater inlets should be designed and located at intervals along the pavement to reduce spread to safe levels as shown in Fig. 5.1. The following is a list of key factors that affect spread, and thus require special consideration in specification of inlets:

- Frequency of the design runoff event and a corresponding rainfall intensity
- Physical characteristics of the drainage surface, including size, pavement grade or longitudinal slope, lateral cross slope, drainage length, and roughness

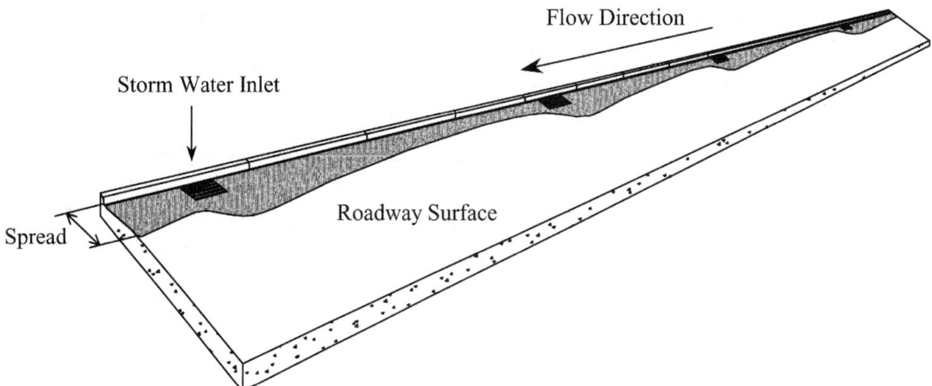

FIGURE 5.1   Spatial variation of spread.

- Physical characteristics of the inlet, including the type of inlet, and its dimensions, capacity, and efficiency
- Location or spacing of inlets along the drainage surface

Without an understanding of each of these factors, a system is likely to be either over- or under-designed. The former is undesirable from an economic viewpoint, while the latter sacrifices highway functionality and may result in consequent damages. Surface drainage problems are, therefore, best solved with accurate knowledge of the relative merits of each of these factors.

5.3   DESIGN FREQUENCY AND ALLOWABLE SPREAD

Two of the major design variables to be considered in sizing and locating highway drainage facilities are the frequency of the design runoff event and the maximum portion of a traffic lane that can be covered by water, or *allowable spread*. These two criteria are integrally related, since the implications of one allowable spread can be considerably different for storms of differing recurrence intervals (Brown et al., 1996). The challenge for the designer is to collect runoff in a way that provides for safe passage of traffic during the design runoff event and at a reasonable cost.

5.3.1   Risk versus Cost

The runoff frequency and allowable spread used for design computations implies an acceptable level of tradeoff between capital and maintenance costs and the risk of traffic accidents and disruption. Selection of design criteria, therefore, requires a full assessment of budgetary limitations and relative risks. The specific elements to consider in selecting design frequency and spread include highway classification, design speed, projected traffic volumes, rainfall intensity, and capital costs (Brown et al., 1996). The influence of these factors is summarized as follows:

- Functional classification of the highway serves as an indicator to public expectations and acceptability regarding allowable spread. Ponding on high-speed, high-volume roadways is contrary to public expectations and greatly increases the risk of traffic disruption. Associated risks are reduced, however, on highways having a lower classification.

- Larger allowable spreads will be less tolerable as design speeds increase. At speeds greater that 70 km/h (44 mph), however, water on pavement can cause hydroplaning and rainfall intensity becomes more important than spread.
- Economic importance of maintaining functionality of the highway can be evaluated using projected traffic volumes since the costs of delays and accidents rise with increasing traffic volumes.
- Rainfall intensity may significantly impact the selection process. The associated risks of water on pavements may be less in arid regions subject to high intensity storms events than in regions subjected to frequent but less intense storms.
- Capital costs may make it necessary to compromise between desirable and practical criteria. In some cases, it may be feasible to significantly upgrade drainage facilities and reduce risks at a moderate cost. In other instances, such as where extensive outfalls or pumping stations are required, costs may be particularly sensitive to design criteria.

Another factor to consider in selection of design frequency and spread is the potential inconvenience and hazards to pedestrian traffic, especially in urbanized regions. Additionally, the relative elevation of the highway with respect to the surrounding area may be important in cases where water can only be drained through a collection system. For example, ponding to hazardous depths significantly increases risks in underpasses and depressed highway regions.

5.3.2 Recommended Design Criteria

The hydraulic engineer should select a design frequency and spread that meets the needs of a particular project. Table 5.1 provides suggested minimum design criteria according to roadway classification and design speed. Additionally, the associated depth at the curb may limit design spread. The recommended frequency for sag locations and depressed areas is a 50-year runoff event.

5.3.3 Relationship to Design Flow

The most common method of determining runoff for a predetermined storm frequency is the *rational method*. The method is based on the principle that the maximum rate of runoff from a drainage basin occurs when all parts of the watershed contribute to flow and that rainfall is distributed uniformly over the catchment area. The *rational formula* is expressed as

TABLE 5.1 Suggested Minimum Design Frequency and Spread

Road classification		Design frequency	Design spread
High volume or divided or bi-directional	<70 km/hr (45 mph)	10-year	Shoulder + 1 m (3 ft)
	>70 km/hr (45 mph)	10-year	Shoulder
	Sag point	50-year	Shoulder + 1 m (3 ft)
Collector	<70 km/hr (45 mph)	10-year	½ Driving lane
	>70 km/hr (45 mph)	10-year	Shoulder
	Sag point	10-year	½ Driving lane
Local streets	Low ADT	5-year	½ Driving lane
	High ADT	10-year	½ Driving lane
	Sag point	10-year	½ Driving lane

Source: Brown et al., 1996.

$$Q = \frac{CiA}{K} \tag{5.1}$$

where Q is peak runoff in m³/s (cfs); C is a dimensionless runoff coefficient, listed in Table 5.2 as a function of land use; i is the average rainfall intensity in mm/hr (in/hr) for a duration equal to the *time of concentration,* or time required for water to travel from the most remote portion of the basin to the inlet or other point of concern; A is the drainage area in hectares (acres); and K is a conversion constant equal to 360 (1.0 in English units). The intensity for Eq. (5.1) is obtained from regional intensity-duration-frequency curves (Fig. 5.2) for the design runoff frequency. These curves are often available from state highway agencies, the National Oceanic and Atmospheric Administration, or the Federal Highway Administration's HYDRAIN computer model (see Chapters 21 and 23).

For nonhomogeneous drainage areas that contain varying land use, a composite runoff coefficient, C_w, should be used in Eq. (5.1) and is obtained by (McCuen et al., 1996)

TABLE 5.2 Typical Runoff Coefficients for Rational Method (2 to 10 Year Return Periods)

Description of drainage area	Runoff coefficient, C
Business	
Downtown areas	0.70–0.95
Neighborhood areas	0.50–0.70
Residential	
Single-family areas	0.30–0.50
Multi-unit, detached	0.40–0.60
Multi-unit, attached	0.60–0.75
Suburban	0.25–0.40
Apartment dwelling areas	0.50–0.70
Industrial	
Light areas	0.50–0.80
Heavy areas	0.60–0.90
Parks and cemeteries	0.10–0.25
Railroad yards	0.20–0.35
Unimproved areas	0.10–0.30
Pavement	
Asphalt and concrete	0.70–0.95
Brick	0.75–0.85
Roofs	0.75–0.95
Lawns	
Sandy soil, flat (2%)	0.05–0.10
Sandy soil, average (2 to 7%)	0.10–0.15
Sandy soil, steep (7% or more)	0.15–0.20
Heavy soil, flat (2%)	0.13–0.17
Heavy soil, average (2 to 7%)	0.18–0.22
Heavy soil, steep (7% or more)	0.25–0.35

Source: ASCE, 1992, *Design and Construction of Urban Stormwater Management Systems,* pp. 91–92. Reprinted by permission of ASCE.

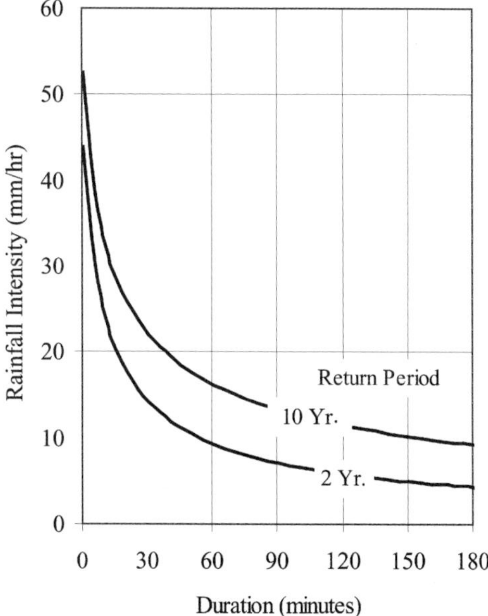

FIGURE 5.2 Examples of intensity-duration-frequency curves.

$$C_w = \frac{\sum\limits_{j=1}^{n} C_j A_j}{\sum\limits_{j=1}^{n} A_j} \tag{5.2}$$

where A_j = area in hectare (acre) for land use j
$\quad\quad C_j$ = runoff coefficient for area j
$\quad\quad n$ = total number of land covers
$\quad\quad C_w$ = composite runoff coefficient

Application of the rational method is generally valid for drainage areas less than 80 hectares (200 acres), which typically have times of concentration of less than 20 minutes (ASCE, 1992; Wanielista et al., 1997).

5.3.4 Check Events

For cases in which significant ponding can occur, the drainage design should be verified using a flow associated with an event of lesser frequency, such as a 100-year storm. Such a recurrence interval is referred to as a *check storm* or *check event*. Using a check event is particularly important for assessing hazards at critical locations, such as sag vertical curves where highway grade is continually reduced and ponding to hazardous depths can occur. The frequency of the check event should be based on the same factors used to select the design storm. Criteria used for allowable spread during the check event are (1) one lane open to traffic, or (2) one lane free of water during the event (Brown et al., 1996). Although these criteria differ significantly, each sets a standard by which the design can be evaluated.

5.4 FLOW IN GUTTERS

A *gutter* is a section of pavement adjacent to the curb that is designed to convey water to drainage inlets during a runoff event. In this respect, the gutter may include all or only a portion of a traffic lane. Typical gutter sections are classified as either *conventional* or *shallow swale* type.

5.4.1 Conventional Gutters

Lateral cross slopes for conventional gutters may be uniform, composite or parabolic, as shown in Fig. 5.3. Uniform gutters have a cross slope equal to that of the shoulder or adjacent travel lane, whereas composite sections are depressed in relation to the adjacent pavement slope. Parabolic sections are less common, but are frequently found along older city streets and roadways having curved pavement sections.

5.4.1.1 Uniform Sections. Uniform gutters have a shallow, triangular cross section, with a curb forming the near-vertical leg of the triangle, and extend 0.3 to 1 m (1 to 3.3 ft) toward the centerline of the roadway. In addition to confining runoff, the curb prevents erosion of fill slopes and serves to delineate adjacent property from the highway. Izzard (1946) showed that for this case, flow could be derived by integrating Manning's equation for an increment of cross-sectional width. Assuming resistance due to the curb face is negligible, a reasonable assumption for uniform cross slopes less than 10 percent, the integration yields

$$Q = \frac{K_c}{n} S_x^{5/3} S_L^{1/2} T^{8/3} \tag{5.3}$$

where Q = gutter flow rate in m³/s (cfs)
K_c = empirical constant equal to 0.376 (0.56 in English units)
n = Manning's roughness coefficient from Table 5.3
S_x = gutter cross slope in m/m (ft/ft)
S_L = longitudinal slope, or grade, of the highway in m/m (ft/ft)
T = spread of water onto the pavement in m (ft), or top width of flow

Equation (5.3) includes an adjustment factor since the hydraulic radius is incapable of

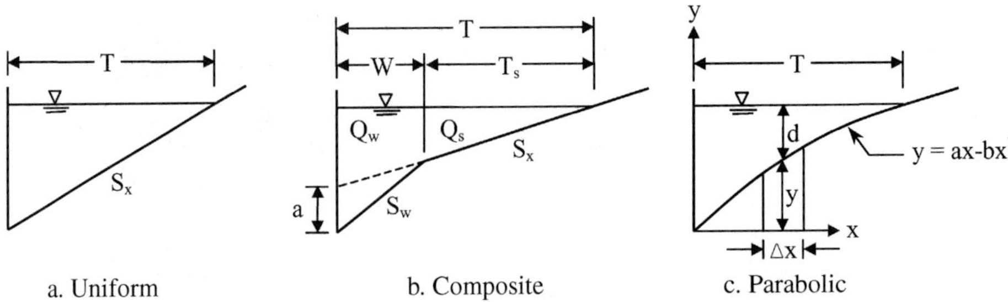

FIGURE 5.3   Conventional gutter sections. (Adapted from Brown et al., 1996)

TABLE 5.3   Manning's Roughness Coefficient for Gutters

Gutter or pavement type	n
Concrete gutter, trowel finish	0.012
Asphalt pavement:	
Smooth	0.013
Rough	0.016
Concrete gutter-asphalt pavement:	
Smooth	0.013
Rough	0.015
Concrete pavement:	
Float finish	0.014
Broom finish	0.016
For gutters with small slope, where sediment may accumulate, increase the above values of n by	0.02

Source: FHWA, 1977a.

fully describing a shallow, gutter cross section, particularly when the spread can exceed 40 times the flow depth (ASCE, 1992). Subsequently, spread can be related to flow depth at the curb, d, by

$$d = TS_x \tag{5.4}$$

Evidenced by the dependent terms in Eq. (5.3), the effects of cross slope on gutter capacity can be relatively large. Consider, for example, that at a cross slope of 4 percent, a gutter has ten times the capacity of a gutter having a one-percent cross slope. The designer should, however, balance the need for steeper cross slopes for effective drainage and the need for flatter cross slopes for driver comfort and safety. Table 5.4 provides an acceptable range of pavement cross slopes for drainage practice. Highway grade can similarly affect gutter capacity, but the hydraulic engineer generally has little latitude in varying roadway grade.

TABLE 5.4   Recommended Pavement Cross Slopes

Surface type	Range of cross slope
High-type surface	
2-Lanes	0.015–0.020
3 or more lanes, each direction	0.015 minimum; increase 0.005 to 0.010 per lane; 0.040 maximum
Intermediate surface	0.015–0.03
Low-type surface	0.020–0.060
Shoulders	
Bituminous or concrete with curbs	0.020–0.060 $\geq$0.040

Adapted from AASHTO, 1990.

AASHTO (1990) provides recommended minimum values of longitudinal slope for pavement drainage. From the exponents in Eq. (5), however, the effect of spread on gutter capacity is much greater than that of either cross slope or grade. Consider that gutter capacity with a 3-m (9.8-ft) spread is nearly nineteen times greater than with a 1-m (3.3-ft) spread, and three times greater than that having a 2-m (6.6-ft) spread.

Example 5.1
Evaluate the spread and depth at the curb for a triangular gutter section carrying a design discharge of 0.09 m³/s and having a uniform cross slope of 0.022 m/m, Manning's roughness of 0.015, and longitudinal slope of 0.014 m/m.

Solution
Step 1. Compute the spread, T, from Eq. (5.3).

$$T = \left(\frac{Qn}{K_c S_x^{5/3} S_L^{1/2}}\right)^{3/8} = \left[\frac{(0.09)(0.015)}{(0.376)(0.022)^{5/3}(0.014)^{1/2}}\right]^{3/8} = 2.9 \text{ m}$$

Step 2. Determine the depth at the curb, d, from Eq. (5.4).

$$d = TS_x = (2.9)(0.022) = 0.064 \text{ m}$$

5.4.1.2 Composite Sections.   Design computations for composite gutter sections require additional consideration of flow in the depressed section. The depression serves to retain more water above inlet entrances and thus increases gutter capacity and inlet efficiency. The relationship between total discharge and the depressed gutter flow is given by

$$Q = Q_w + Q_s \tag{5.5}$$

where Q = total gutter flow rate in m³/s (cfs)
$\quad Q_w$ = flow rate in the depressed section in m³/s (cfs)
$\quad Q_s$ = discharge in the gutter above the depressed section in m³/s (cfs)

Q_s can be evaluated using Eq. (5.3) if T is taken as only the spread over the undepressed portion of the gutter, T_s. Eq. (5.5) can be used in conjunction with the following expressions for computing flow in a composite cross section (Brown et al., 1996):

$$E_o = \left[1 + \frac{\left(\dfrac{S_w}{S_x}\right)}{\left\{1 + \dfrac{(S_w/S_x)}{(T/W) - 1}\right\}^{8/3} - 1}\right]^{-1} \tag{5.6}$$

and

$$Q = \frac{Q_s}{(1 - E_o)} \tag{5.7}$$

where E_o = is the ratio of flow in the depressed section to total gutter flow (i.e., Q_w/Q)
$\quad W$ = width of the depressed section in m (ft)
$\quad S_w$ = cross slope of the depressed portion of the gutter in m/m (ft/ft)

S_w is expressed as

$$S_w = S_x + \frac{a}{W} \tag{5.8}$$

where a = depth of gutter depression in m (ft)

The procedure for evaluating flow in composite gutter sections is demonstrated by Example (5.2).

Example 5.2
Compute the discharge in a composite gutter section having a cross slope of 0.022 m/m, Manning's roughness of 0.015, and longitudinal slope of 0.014 m/m. The spread is 2.9 m and the gutter depression is 50-mm deep and 0.60-m wide.

Solution
Step 1. Determine the cross slope of the depressed gutter section, S_w, from Eq. (5.8).

$$S_w = S_x + \frac{a}{W} = 0.022 + \frac{\left(\dfrac{50}{1000}\right)}{0.60} = 0.11 \text{ m/m}$$

Step 2. Compute the flow, Q_s, in the undepressed portion of the gutter using Eq. (5.3).

$$T_s = T - W = 2.9 - 0.60 = 2.3 \text{ m}$$

$$Q_s = \frac{K_c}{n} S_x^{5/3} S_L^{1/2} T_s^{8/3} = \frac{(0.376)}{(0.015)}(0.022)^{5/3}(0.014)^{1/2}(2.3)^{8/3} = 0.047 \text{ m}^3/\text{s}$$

Step 3. Calculate the ratio of depressed flow to total gutter discharge, E_o, by Eq. (5.6).

$$E_o = \left[1 + \frac{\left(\dfrac{S_w}{S_x}\right)}{\left\{1 + \dfrac{(S_2/S_x)}{(T/W) - 1}\right\}^{8/3} - 1}\right]^{-1}$$

$$= \left[1 + \frac{\left(\dfrac{0.11}{0.022}\right)}{\left\{1 + \dfrac{(0.11/0.022)}{(2.9/0.60) - 1}\right\}^{8/3} - 1}\right]^{-1} = 0.62$$

Step 4. Evaluate the total gutter discharge, Q, from Eq. (5.7).

$$Q = \frac{Q_s}{(1 - E_o)} = \frac{0.047}{(1 - 0.62)} = 0.12 \text{ m}^3/\text{s}$$

If the design flowrate is initially known, spread in a composite gutter can be computed using an iterative approach, whereby Q_s is assumed and an estimate for spread is found from Eqs. (5.6) and (5.7). The spread is used to compute a new value of Q_s, which can be compared to that assumed. The method is repeated until the assumed and computed values are sufficiently matched.

5.4.1.3 Parabolic Sections.   Where the pavement cross section is curved, gutter flow rate varies with the configuration of the roadway. For this reason, discharge-spread relationships developed for one parabolic configuration cannot be universally applied to other sections unless approximations are made. A parabolic cross section can be described by the expression

$$y = ax - bx^2 \qquad (5.9)$$

where $a = 2H/B$
$\qquad b = H/B^2$
$\qquad H$ = roadway crown height in m (ft)
$\qquad B$ = gutter width from the curb to the crown in m (ft)

To compute the gutter flow, the parabolic cross section is divided into segments of equal width, Δx, and the discharge for each segment is computed using Manning's equation. The total gutter discharge will be the sum of all flows for all segments, as illustrated by Example 5.3.

Example 5.3
Compute the discharge in a parabolic gutter section having a spread of 1.2 m, longitudinal slope of 0.014 m/m, and Manning's roughness of 0.015. The gutter has a vertical rise of 0.20 m over a pavement width of 9.75 m.

Solution
Step 1. Choose Δx and determine the coefficients, a and b, in Eq. (5.9).
$\qquad$ Assume $\Delta x = 0.60$ m

$$a = \frac{2H}{B} = \frac{2(0.20)}{9.75} = 0.041 \quad \text{and} \quad b = \frac{H}{B^2} = \frac{0.20}{(9.75)^2} = 0.0021$$

Step 2. Compute the depth at the curb using Eq. (5.9).

$$y = (0.041)x - (0.0021)x^2 = (0.041)(1.2) - (0.0021)(1.2)^2 = d = 0.046 \text{ m}$$

Step 3. Evaluate the mean depth for Δx_1.
$\qquad$ The total rise over Δx_1 is computed by

$$y = (0.041)x - (0.0021)x^2 = (0.041)(0.6) - (0.0021)(0.6)^2 = 0.024 \text{ m}$$

and the average rise of pavement over Δx_1 is (0.024)/2 or 0.012 m. The mean depth is then the depth at the curb minus the average rise over Δx_1, or 0.034 m.

Step 4. Compute the discharge in Δx_1 using Manning's equation.

$$Q_1 = \frac{1.0}{n} AR^{2/3}S_L^{1/2} = \frac{1.0}{n}(\Delta x)d_1^{5/3}S_L^{1/2} = \frac{1.0}{0.015}(0.60)(0.034)^{5/3}(0.014)^{1/2} = 0.017 \text{ m}^3/\text{s}$$

Step 5. Repeat steps 3 and 4 for subsequent computational segments.
$\qquad$ The total rise over Δx_2, measured from the curb, is computed by

$$y = (0.041)x - (0.0021)x^2 = (0.041)(1.2) - (0.0021)(1.2)^2 = 0.046 \text{ m}$$

and the average rise over Δx_2 is (0.024 + 0.046)/2 or 0.035 m. The mean depth is then the depth at the curb minus the average rise over Δx_2, or 0.011 m.

$$Q_2 = \frac{1.0}{0.015}\,(0.60)(0.011)^{5/3}(0.014)^{1/2} = 0.0026 \text{ m}^3/\text{s}$$

Step 6. Evaluate the total gutter flow by summing the discharges from each segment.

$$Q = \sum Q_i = 0.017 + 0.0026 = 0.020 \text{ m}^3/\text{s}$$

5.4.2 Shallow Swale Gutters

In cases where curbs are not warranted, a small, V-shaped or circular swale section can be used to convey runoff along the shoulder (Fig. 5.4).

5.4.2.1 V-Shaped Sections. Equation (5.3) can be used to evaluate flow in V-shaped sections if the cross slope is modified as

$$S_x = \frac{S_{x1}S_{x2}}{S_{x1} + S_{x2}} \tag{5.10}$$

where S_{x1} and S_{x2} = cross slopes of each side the of gutter section, respectively, in m/m (ft/ft)

Example 5.4
Determine the spread in a V-shaped swale carrying 0.090 m³/s and having cross slopes of 0.33 m/m and 0.022 m/m, Manning's roughness of 0.015, and a longitudinal slope of 0.014 m/m.

Solution
Step 1. Compute the composite cross slope using Eq. (5.10).

$$S_x = \frac{S_{x1}S_{x2}}{S_{x1} + S_{x2}} = \frac{(0.33)(0.022)}{(0.33) + (0.022)} = 0.021$$

Step 2. Determine the spread, T, from Eq. (5.3)

$$T = \left(\frac{Qn}{K_c S_x^{5/3} S_L^{1/2}}\right)^{3/8} = \left[\frac{(0.09)(0.015)}{(0.376)(0.021)^{5/3}(0.014)^{1/2}}\right]^{3/8} = 3.0 \text{ m}$$

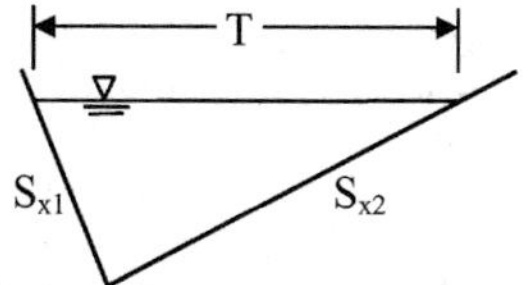

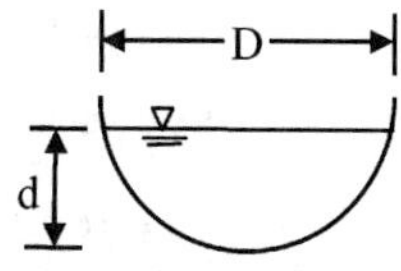

a. V-Shaped   b. Circular

FIGURE 5.4 Shallow swale gutter sections. (Adapted from Brown et al., 1996)

5.4.2.2 Circular Sections.   Shallow, circular gutters can be evaluated using the expression

$$\frac{d}{D} = K_c \left[\frac{Qn}{D^{8/3}S_L^{1/2}} \right]^{0.488} \tag{5.11}$$

where d = depth of flow at its deepest point in m (ft)
$\quad D$ = diameter of the circular gutter in m (ft)
$\quad K_c$ = empirical constant equal to 1.179 (0.972 in English units)

The spread of flow, or top width, for the section can then be computed by

$$T = 2 \left[\left(\frac{D}{2} \right)^2 - \left(\frac{D}{2} - d \right)^2 \right]^{1/2} \tag{5.12}$$

Example 5.5
Compute the discharge in a 1.0-m diameter circular swale if the spread is 0.85 m. The longitudinal slope is 0.014 m/m, and Manning's roughness is 0.015.

Solution
Step 1. Compute the depth of flow, d, by rearranging Eq. (5.12).

$$d = \frac{D}{2} - \sqrt{\left(\frac{D}{2} \right)^2 - \left(\frac{T}{2} \right)^2} = \frac{1.0}{2} - \sqrt{\left(\frac{1.0}{2} \right)^2 - \left(\frac{0.85}{2} \right)^2} = 0.24 \text{ m}$$

Step 2. Determine the flowrate using Eq. (5.11).

$$Q = \frac{D^{8/3}S_L^{1/2}}{n} \left(\frac{d}{DK_c} \right)^{1/0.488} = \frac{(1.0)^{8/3}(0.014)^{1/2}}{0.015} \left(\frac{0.24}{(1.0)(1.179)} \right)^{1/0.488} = 0.30 \text{ m}^3/\text{s}$$

5.4.3 Flow Travel Time

Travel time for flow in gutters is an important aspect of time of concentration used for designing drainage inlets. Assuming flow varies spatially from Q_1 at the beginning of a gutter section to Q_2 at the drainage inlet (Fig. 5.5), the gutter component of time of concentration, t_g, is found by dividing the average flow velocity into the length of gutter section.

The overland flow portion of time of concentration can be computed using the *kinematic wave equation*, which was based on work by Ragan (1971) and can be expressed as

$$t_s = \frac{K_c}{i^{0.4}} \left(\frac{nL_s}{\sqrt{S}} \right)^{0.6} \tag{5.13}$$

where t_s = overland flow component of time of concentration in minutes
$\quad K_c$ = empirical coefficient equal to 6.943 (0.933 in English units)
$\quad i$ = rainfall intensity in mm/hr (in/hr) for a duration equal to the time of concentration for overland flow
$\quad n$ = Manning's roughness coefficient, from Table 5.5
$\quad L_s$ = overland flow length in m (ft)
$\quad S$ = surface slope in m/m (ft/ft)

Note that since rainfall intensity is dependent on t_s, which is initially unknown, solution to this relationship is an iterative process. Using an assumed estimate of t_s, the intensity is

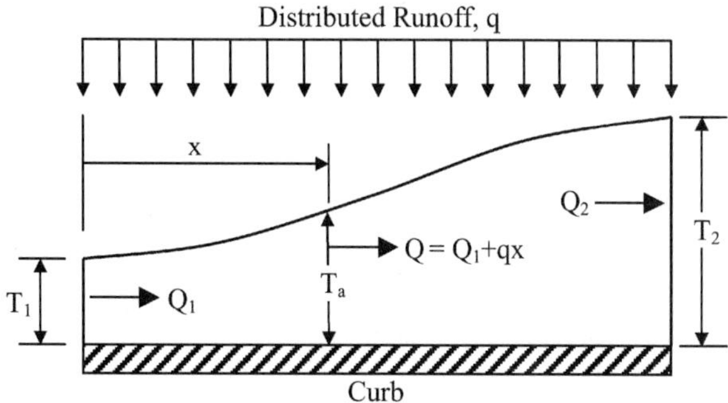

FIGURE 5.5   Spatially varied gutter flow. (Adapted from Brown et al., 1996)

TABLE 5.5   Manning's Roughness Coefficient for Overland Flow

Surface	n
Smooth asphalt	0.011
Smooth concrete	0.012
Ordinary concrete lining	0.013
Good wood	0.014
Brick with cement mortar	0.014
Vitrified clay	0.015
Cast iron	0.015
Corrugated metal pipes	0.024
Cement rubble surface	0.024
Fallow (no residue)	0.05
Cultivated soils	
Residue cover ≤20%	0.06
Residue cover >20%	0.17
Range (natural)	0.13
Grass	
Short, prairie	0.15
Dense	0.24
Bermudagrass	0.41
Woods	
Light underbrush	0.40
Dense underbrush	0.80

Source: McCuen et al., 1996.

obtained from regional intensity-duration-frequency data. The value computed from Eq. (5.13) is then compared with the assumed value. If they are not equal, the process is repeated until the successive estimates of t_s are the same (McCuen et al., 1996).

The total time of concentration is found by summing the overland flow travel time and gutter flow component, or

$$t_c = t_s + t_g = \frac{K_c}{i^{0.4}} \left(\frac{nL_s}{\sqrt{S}} \right)^{0.6} + \left(\frac{L_g}{60\, V_a} \right) \tag{5.14}$$

where t_c = time of concentration in minutes
$\quad V_a$ = average gutter flow velocity in m/s (ft/s)
$\quad L_g$ = length of the gutter section in m (ft)

and other terms are as previously defined.

Additional terms could be incorporated in Eq. (5.14) when appropriate. For example, if channelized flow occurs prior to entering the gutter, the channel flow time should also be considered and can be computed through use of Manning's equation. For pavement drainage design, a minimum time of concentration of 5 minutes is recommended.

The average gutter velocity in Eq. (5.14) is obtained by integrating Manning's equation with respect to time and distance (Brown et al., 1996). For a triangular, curbed gutter section, the resultant can be expressed as

$$V_a = \frac{K_m}{n} S_x^{2/3} S_L^{1/2} T_a^{2/3} \tag{5.15}$$

where V_a = average velocity in m/s (fps)
$\quad K_m$ = empirical constant equal to 0.752 (1.12 in English units)
$\quad T_a$ = spread in m (ft) at the average velocity, which can be evaluated by (Brown et al., 1996)

$$T_a = (0.65)(T_2) \left[\frac{1 - \left(\dfrac{T_1}{T_2} \right)^{8/3}}{1 - \left(\dfrac{T_1}{T_2} \right)^2} \right]^{3/2} \tag{5.16}$$

where T_1 and T_2 are the spread at the upstream and downstream ends of the gutter section being evaluated, respectively, in m (ft).

Example 5.6
Using the partial intensity-duration-frequency data below, determine the time of concentration for a stormwater inlet draining an area of short grass prairie ($n = 0.15$) flowing to a 150-m long triangular gutter section. The overland flow length and slope for the grassland are 200 m and 0.036 m/m, respectively. The gutter section has a cross slope of 0.025 m/m, Manning's roughness of 0.016, and a longitudinal slope of 0.020 m/m. Assume that the spread at the upstream end of the gutter section is 0.80 m, as a result of upstream bypass flows, and the design spread at the downstream inlet is 3.0 m.

Duration (min)	Rainfall intensity (mm/hr)
10	147
20	112
30	88
40	72
50	60

Solution
Step 1. Determine the overland flow portion of time of concentration.

a. Assume $t_s = 10$ minutes

b. From the IDF data, rainfall intensity is 147 minutes at a duration of 10 minutes

c. Compute t_s from Eq. (5.13)

$$t_s = \frac{K_c}{i^{0.4}} \left(\frac{nL_s}{\sqrt{S}}\right)^{0.6} = \frac{6.943}{(147)^{0.4}} \left(\frac{(0.15)(200)}{\sqrt{0.036}}\right)^{0.6} = 19.7 \text{ mins}$$

d. Since the assumed and computed values are not equal, repeat steps a through c with an assumed $t_s = 19.7$ minutes. The following table summarizes the convergence upon the actual value of 22.4 minutes.

Assumed t_s	Rainfall intensity (mm/hr)	Computed t_s
10	147	19.7
19.7	113	21.9
21.9	107	22.3
22.3	106	22.4 (OK)

Step 2. Compute the gutter flow portion of time of concentration.

a. Evaluate the average spread using Eq. (5.16).

$$T_a = (0.65)(T_2) \left[\frac{1 - \left(\frac{T_1}{T_2}\right)^{8/3}}{1 - \left(\frac{T_1}{T_2}\right)^{2}}\right]^{3/2} = (0.65)(3.0) \left[\frac{1 - \left(\frac{0.80}{3.0}\right)^{8/3}}{1 - \left(\frac{0.80}{3.0}\right)^{2}}\right]^{3/2} = 2.08 \text{ m}$$

b. Determine average velocity in the gutter by Eq. (5.15).

$$V_a = \frac{K_m}{n} S_x^{2/3} S_L^{1/2} T_a^{2/3} = \frac{0.752}{0.016} (0.025)^{2/3}(0.02)^{1/2}(2.08)^{2/3} = 0.93 \text{ m/s}$$

c. Calculate the gutter travel time

$$t_g = \frac{L_g}{60\, V_g} = \frac{150}{60(0.93)} = 2.69 \text{ mins}$$

Step 3. Compute the total time of concentration by summing the overland flow and gutter flow components, or

$$22.4 + 2.69 = 25.1 \text{ mins}$$

5.5 *HIGHWAY DRAINAGE INLETS*

As flow accumulates in gutters and spread encroaches upon prespecified design values, a storm drain inlet is provided to intercept all or a portion of the flow. The common types of inlets used in practice are illustrated in Fig. 5.6 and include (1) *grate,* (2) *curb-opening,* (3) *combination,* and (4) *slotted drain inlets.* The design characteristics of inlets ultimately control the rate at which runoff is removed from the highway and enters a subsurface drainage system. Subsequently, inadequate inlet capacity or poorly located inlets can cause hazardous roadway flooding. The responsibility of the designer is to determine the type, size, and spacing of inlets to intercept a sufficient portion of the design gutter flow, while preserving attention to costs. In addition, the designer should ensure that inlets do not project significantly above a pavement surface or pose as an obstacle to oncoming traffic.

5.5.1 Interception Capacity and Efficiency

Inlet capacity, Q_i, is the amount of gutter flow intercepted by an inlet. Any flow that is not intercepted at an inlet is termed *bypass,* or *carryover flow,* and is expressed as

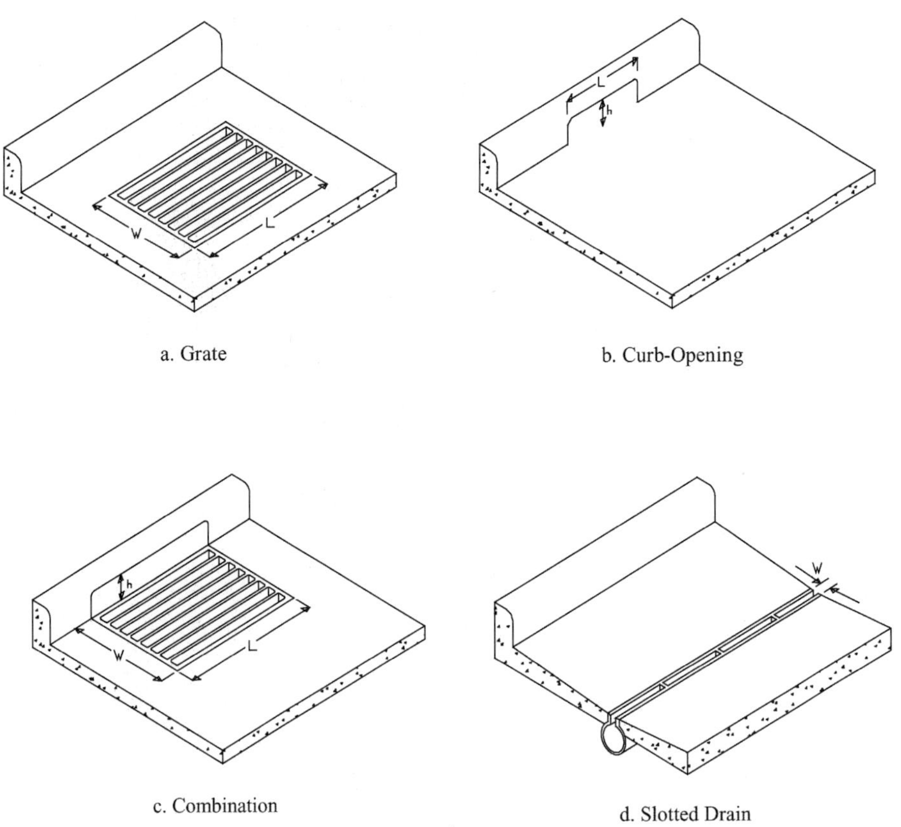

FIGURE 5.6 Types of storm drain inlets. (Adapted from Brown et al., 1996)

$$Q_b = Q - Q_i \qquad (5.17)$$

where Q_b = bypass flow in m³/s (cfs)
  Q = total gutter flow in m³/s (cfs)
  Q_i = interception capacity in m³/s (cfs)

The *interception efficiency* of the inlet, E, is the percent of gutter flow that the inlet will intercept under a given set of conditions, and is expressed as

$$E = \frac{Q_i}{Q} \qquad (5.18)$$

Interception capacity is a function of gutter cross slope, and to a smaller extent, pavement roughness, longitudinal gutter slope, total gutter flow, and inlet geometry. Efficiency of an inlet is generally dependent on the same factors, in addition to its capacity. Whereas interception capacity of all inlets increases with increasing gutter flowrates, efficiency generally decreases with increasing gutter flows (Brown et al., 1996).

5.5.2 Grate Inlets

Grate inlets consist of an opening in the gutter covered by one or more, flush-mounted grates placed parallel to the flow (Fig. 5.6.a). These inlets perform satisfactorily over a wide range of gutter grades, but their flow interception capacity generally degreases as gutter grade steepens. Additional factors that can affect their interception capacity include the depth of water next to the curb, amount of runoff flowing over the grate itself, the geometric configuration of the grate, and velocity of flow in the gutter.

A number of grate configurations have been hydraulically tested (Burgi et al., 1977; Burgi, 1978a; Burgi, 1978b; Pugh, 1980; FHWA 1977b). From these, Brown et al. (1996) lists the following grates for which design procedures have been developed.

- P − 50   A parallel bar grate with 48 mm (1-7/8 in) bar spacing on center (Fig. 5.7)
- P − 50 × 100   A parallel bar grate with 48 mm (1-7/8 in) bar spacing on center and 10 mm (3/8 in) diameter lateral rods spaced at 102 mm (4 in) on center (Fig. 5.7)
- P − 30   A parallel bar grate with 29 mm (1-1/8 in) bar spacing on center (Fig. 5.8)
- Curved Vane   Curved vane grate with 83 mm (3-1/4 in) longitudinal bar and 108 mm (4-1/4 in) transverse bar spacing on center (Fig. 5.9)
- 45°–60 Tilt Bar   45° tilt-bar grate with 57 mm (2-1/4 in) longitudinal bar and 102 mm (4 in) transverse bar spacing on center (Fig. 5.10)
- 45°–85 Tilt Bar   45° tilt-bar grate with 83 mm (3-1/4 in) longitudinal bar and 102 mm (4 in) transverse bar spacing on center (Fig. 5.10)
- 30°–85 Tilt Bar   30° tilt-bar grate with 83 mm (3-1/4 in) longitudinal bar and 102 mm (4 in) transverse bar spacing on center (Fig. 5.11)
- Reticuline   Honeycomb pattern of lateral bars and longitudinal bearing bars (Fig. 5.12)

The primary advantages of grate inlets are that they are installed in the direct path of conveyed runoff and can function well in sump locations. They are, however, highly susceptible to clogging by floating trash or debris. For those cases where clogging may be a chronic problem, grates should be considered only partially effective. Table 5.6 can be used for a relative comparison of grate susceptibility to clogging. Consideration must also be

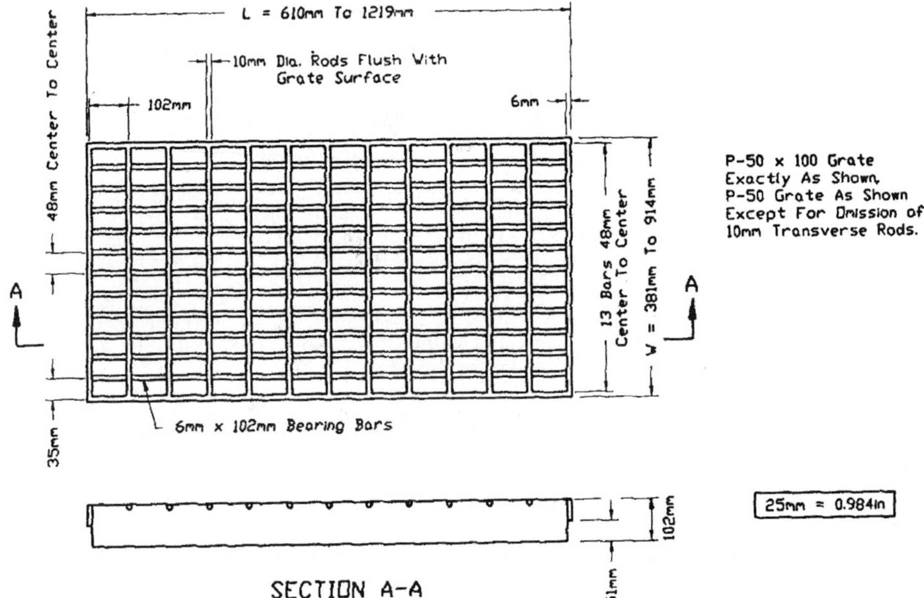

FIGURE 5.7 P−50 and P−50 × 100 grate. (Source: Brown et al., 1996)

given to the potential threat of a grate inlet to bicycle or pedestrian traffic. Table 5.7 ranks seven of the grate types according to bicycle and pedestrian safety. Note that while a P−50 grate is hydraulically efficient relative to several other types, it is not bicycle safe and is, therefore, not listed in the table. Finally, structural strength must be considered during the selection process. Grates in traffic zones must be able to withstand traffic loads, whereas grates draining yard areas do not generally need to be as structurally rigid.

In designing a grate inlet, it is important to distinguish between *frontal, side,* and *splash over flows.* The frontal flow is that portion of total gutter flow that directly passes over the upstream side of the grate, whereas side flow is that portion traveling around the perimeter of the grate when spread exceeds the grate width. Part of the side flow will be intercepted as it travels around the grate, depending on the cross slope, flow velocity, and grate length. When flow velocity is high or the grate length is short, only a portion of the frontal flow is intercepted. Splash over is that fraction of frontal flow that splashes over the grate and is not intercepted (Brown et al., 1996).

The ratio of frontal flow to total gutter flow, E_o, can be found using Eq. 5.6 for composite gutters. For a uniform cross slope, this ratio can be expressed as

$$E_o = \frac{Q_w}{Q} = 1 - \left(1 - \frac{W}{T}\right)^{8/3} \tag{5.19}$$

where Q = total gutter flow in m³/s (cfs)
Q_w = flow in m³/s (cfs) over the width of the grate (W)
T = spread in m (ft)

Similarly, the ratio of side flow to gutter flow is

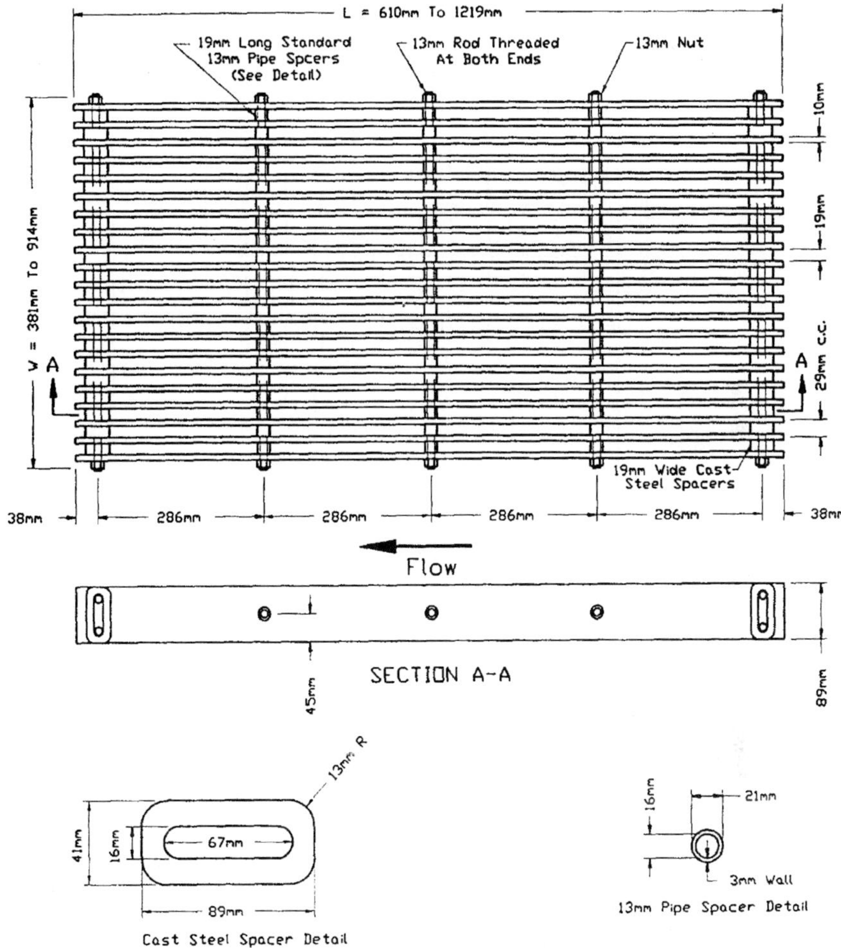

FIGURE 5.8 P–30 grate. (Source: Brown et al., 1996)

$$\frac{Q_s}{Q} = 1 - \left(\frac{Q_w}{Q}\right) = 1 - E_o \tag{5.20}$$

where Q_s = side flow in m³/s (cfs)

The ratio of intercepted frontal flow to total frontal flow, or *frontal flow efficiency, R_f*, is expressed as

$$R_f = 1 - K_f(V - V_o) \tag{5.21}$$

where K_f = empirical constant equal to 0.295 (0.09 in English units)
V = gutter flow velocity in m/s (ft/s)
V_o = critical gutter velocity in m/s (ft/s) at which splash over first occurs, also termed *splash-over velocity*

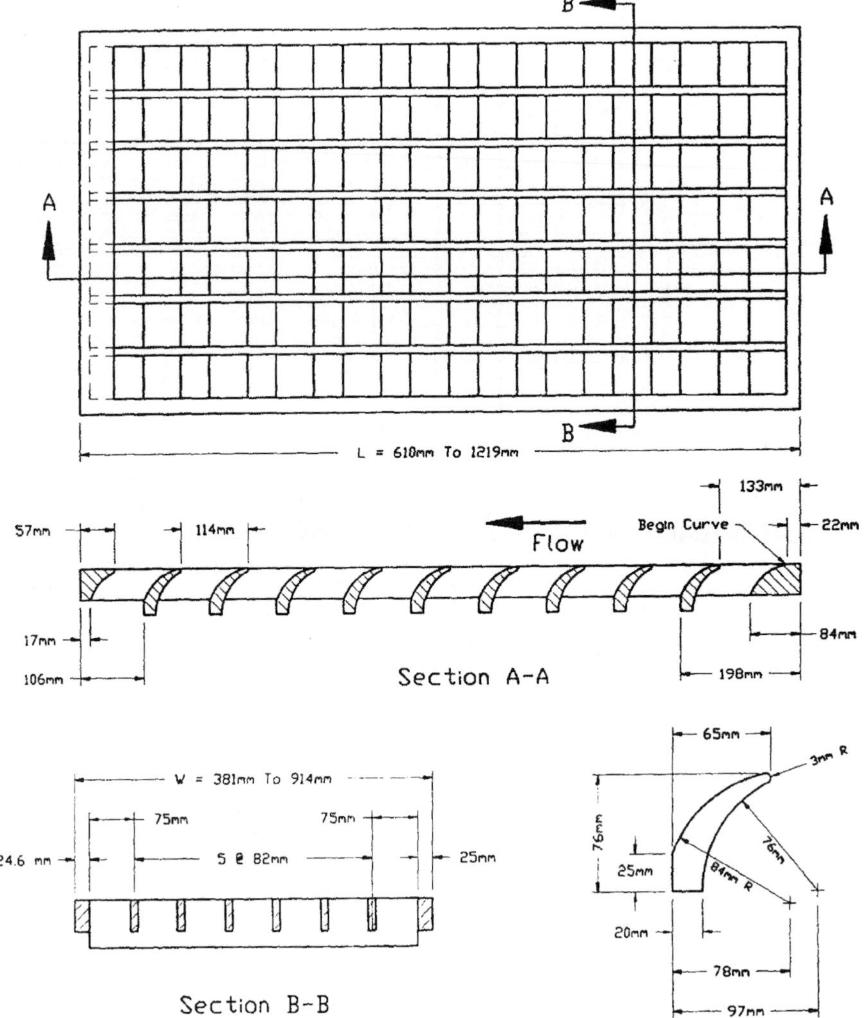

FIGURE 5.9 Curved vane grate. (Source: Brown et al., 1996)

The splash-over velocity and frontal flow efficiency can be determined graphically using the curves in Fig. 5.13, which are formulated according to bar configuration, grate length, and gutter flow velocity.

The ratio of intercepted side flow to total side flow, also termed the *side flow efficiency,* R_s, is given as

$$R_s = \frac{1}{\left(1 + \dfrac{K_s V^{1.8}}{S_x L^{2.3}}\right)} \tag{5.22}$$

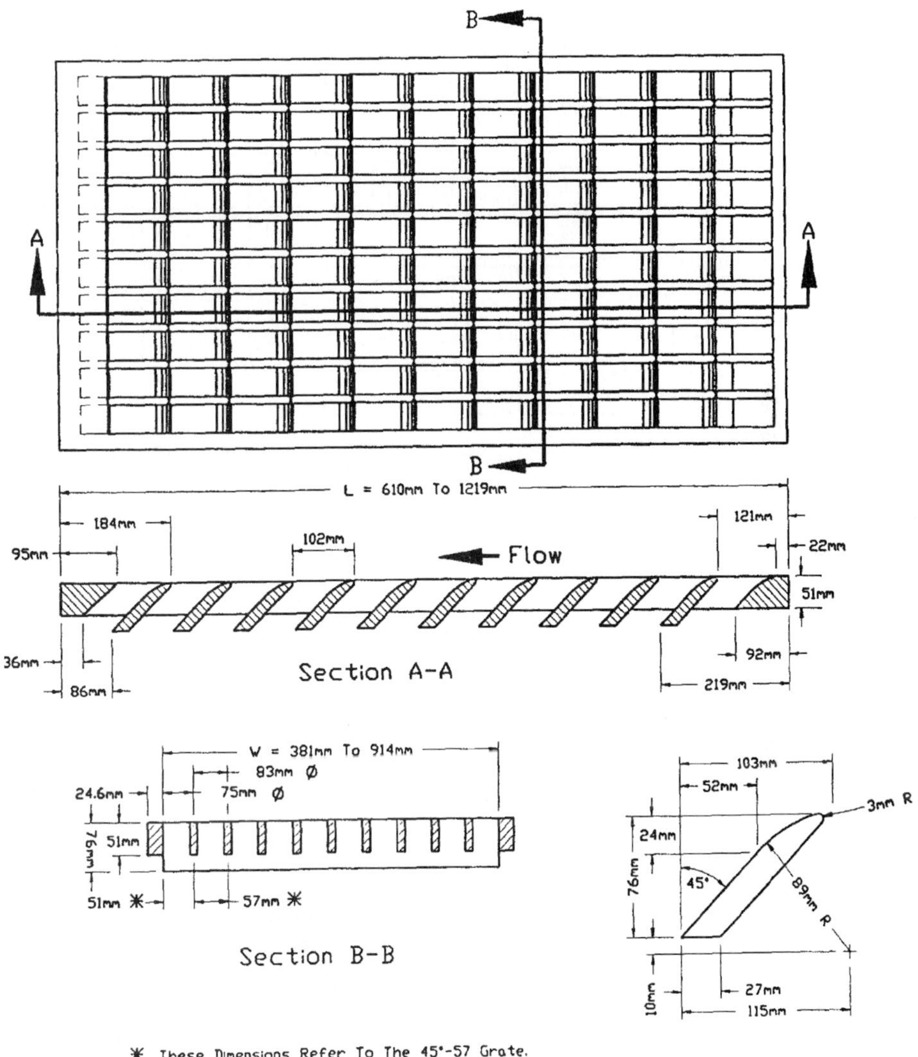

FIGURE 5.10   45°–60 and 45°–85 tilt bar grate. (Source: Brown et al., 1996)

where K_s = empirical constant equal to 0.0828 (0.15 in English units)
L = length of gutter section in m (ft)

The overall efficiency, E, of the grate can be evaluated as a function of the frontal and side flow efficiencies by using

$$E = R_f E_o + R_s(1 - E_o)$$

(5.23)

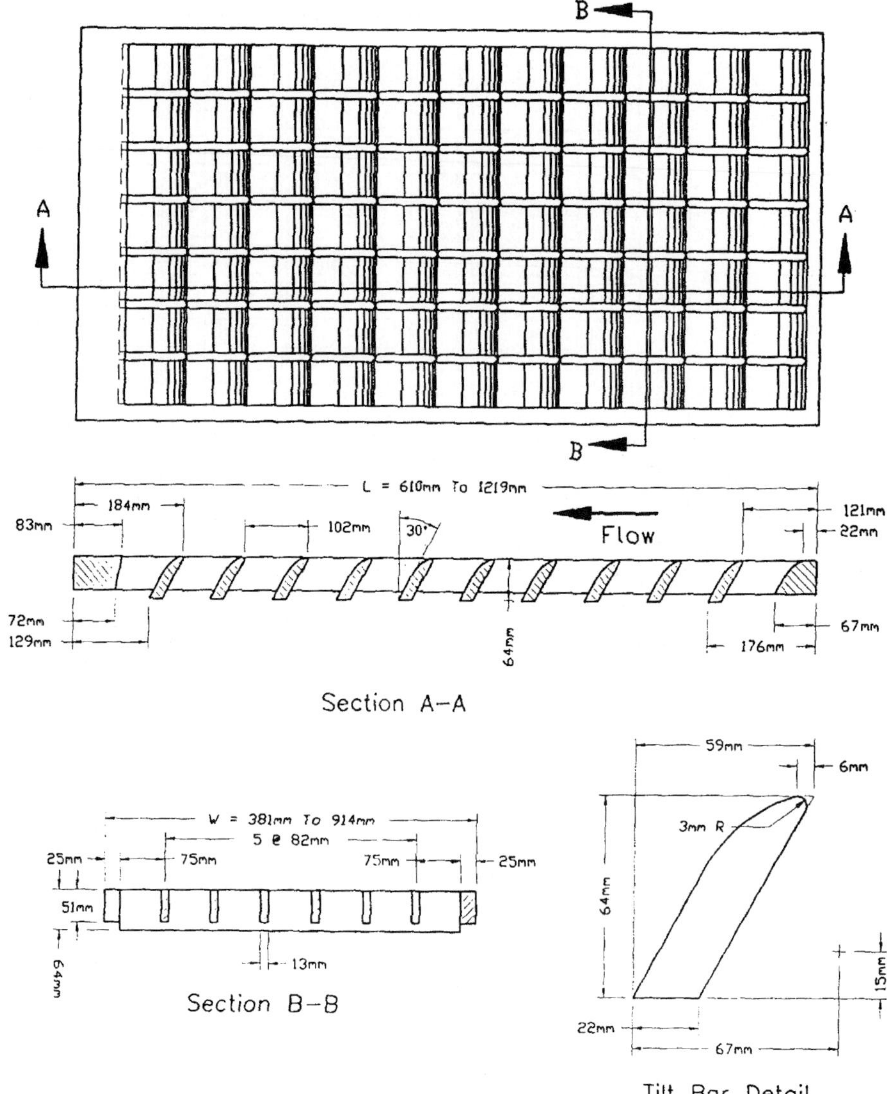

FIGURE 5.11   30°–85 tilt bar grate. (Source: Brown et al., 1996)

The first term on the right-hand side of Eq. (5.23) represents a ratio of intercepted frontal flow to total gutter discharge. The second term is a ratio of intercepted side flow to total side flow, which becomes insignificant at high velocities and longitudinally short grates. From Eq. (5.18), the *interception capacity of a grate inlet on grade, Q_i,* can be obtained by multiplying Eq. (5.23) by the total gutter flow, or

$$Q_i = EQ = (R_f E_o + R_s(1 - E_o))Q \tag{5.24}$$

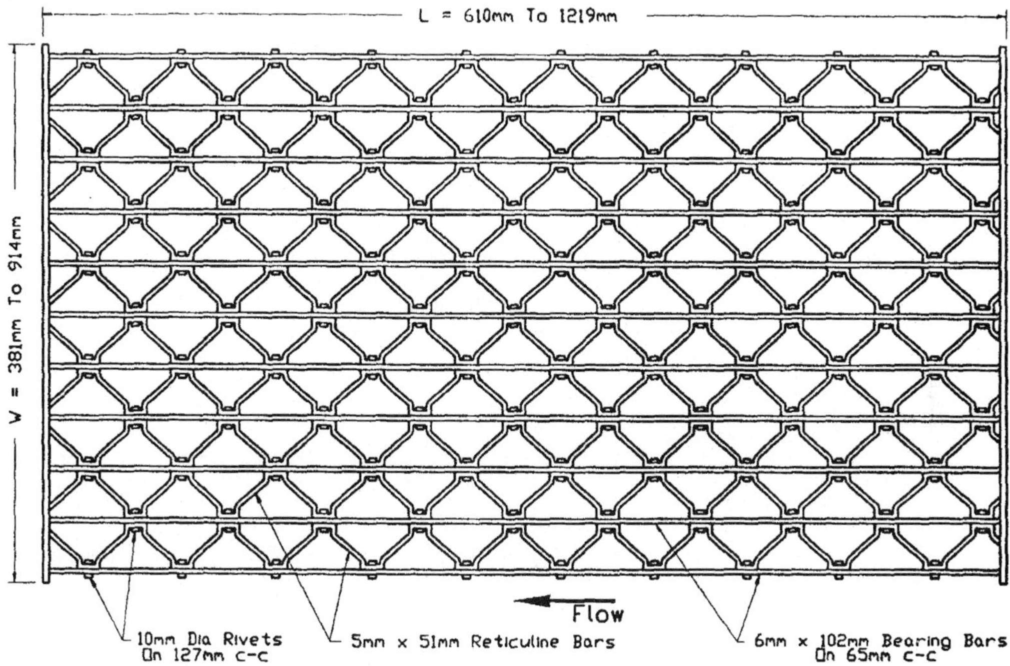

FIGURE 5.12   Reticuline grate. (Source: Brown et al., 1996)

TABLE 5.6   Average Debris Handling Efficiencies of Eight Grate Configurations

		Longitudinal slope	
Rank	Grate type	0.005	0.040
1	Curved vane	46	61
2	30°–85 tilt bar	44	55
3	45°–85 tilt bar	43	48
4	P–50	32	32
5	P–50 × 100	18	28
6	45°–60 tilt bar	16	23
7	Reticuline	12	16
8	P–30	9	20

Source: Brown et al., 1996.

TABLE 5.7   Ranking of Grate Types
According to Bicycle and Pedestrian Safety

Rank	Grate type
1	P-50×100
2	Reticuline
3	P-30
4	45°-85 tilt bar
5	45°-60 tilt bar
6	Curved vane
7	30°-85 tilt bar

Source: Burgi, 1978a.

Example 5.7
Determine the interception capacity and bypass flow for a 0.6 m $\times$ 0.6 m curved vane grate inlet in a composite gutter section having a cross slope of 0.022 m/m, Manning's roughness of 0.015, and longitudinal slope of 0.014 m/m. The gutter depression is 50.0-mm deep and 0.60-m wide, and the gutter discharge and spread at the inlet are 0.12 m³/s and 2.9 m, respectively.

Solution
Step 1. Calculate the flow velocity from gutter discharge, Q, and effective area, A.

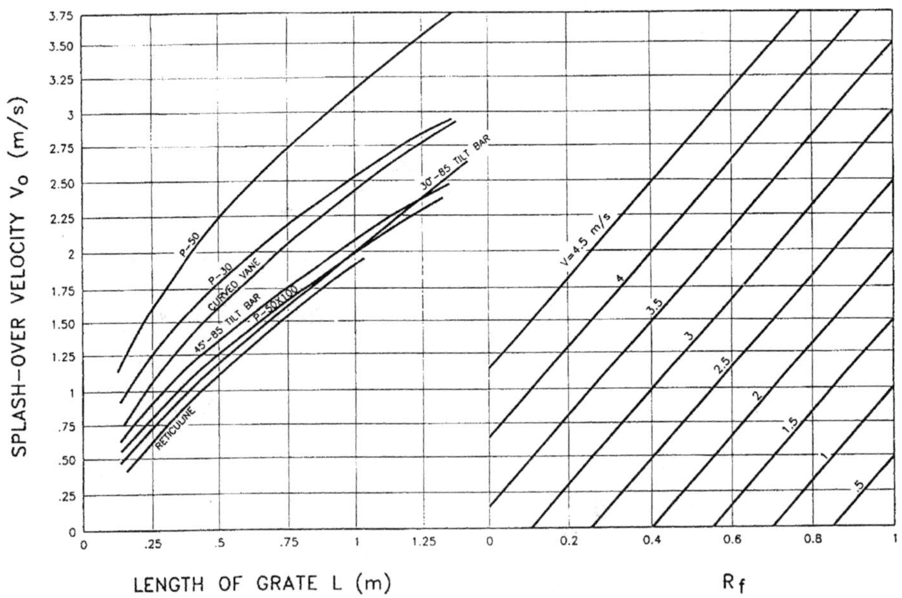

FIGURE 5.13   Grate inlet frontal flow interception efficiency. (Source: Brown et al., 1996)

$$A = \frac{T^2 S_x}{2} + \frac{aW}{2} = \frac{(2.9)^2(0.022)}{2} + \frac{\left(\dfrac{50}{100}\right)(0.60)}{2} = 0.11 \text{ m}^2$$

$$V = \frac{Q}{A} = \frac{0.12}{0.11} = 1.1 \text{ m/s}$$

Step 2. Determine the frontal flow efficiency, R_f, and side flow efficiency, R_s. From Fig. 5.12 with $V = 1.1$ m/s, $R_f = 1.0$

$$R_s = \frac{1}{\left(1 + \dfrac{K_s V^{1.8}}{S_x L^{2.3}}\right)} = \frac{1}{\left(1 + \dfrac{(0.0828)(1.1)^{1.8}}{(0.022)(0.6)^{2.3}}\right)} = 0.065$$

Step 3. Determine the cross slope of the depressed gutter, S_w, from Eq. (5.8).

$$S_w = S_x + \frac{a}{W} = 0.022 + \frac{\left(\dfrac{50}{1000}\right)}{0.60} = 0.11 \text{ m/m}$$

Step 4. Determine the ratio of depressed flow to gutter discharge, E_o, using Eq. (5.6).

$$E_o = \left[1 + \frac{\left(\dfrac{S_w}{S_x}\right)}{\left\{1 + \dfrac{(S_w/S_x)}{(T/W) - 1}\right\}^{8/3} - 1}\right]^{-1}$$

$$= \left[1 + \frac{\left(\dfrac{0.11}{0.022}\right)}{\left\{1 + \dfrac{(0.11/0.022)}{(2.9/0.60) - 1}\right\}^{8/3} - 1}\right]^{-1} = 0.62$$

Step 5. Evaluate the interception efficiency, E, from Eq. (5.23).

$$E = R_f E_o + R_s(1 - E_o) = (1.0)(0.62) + (0.65)(1 - 0.62) = 0.65$$

Step 6. Determine the inlet capacity, Q_i, by Eq. (5.24).

$$Q_i = EQ = (0.65)(0.12) = 0.078 \text{ m}^3/\text{s}$$

Step 7. Evaluate bypass flow, Q_b, from Eq. (5.17).

$$Q_b = Q - Q_i = 0.12 - 0.078 = 0.042 \text{ m}^3/\text{s}$$

5.5.3 Curb-opening Inlets

Curb-opening inlets, or vertical openings in the curb (Fig. 5.6.b), are most effective on flat slopes (less than 3%) and are typically longer than grate inlets. While they are less susceptible to clogging by debris and do not interfere with vehicular, bicycle, or pedestrian traffic, curb openings are less hydraulically efficient than grate inlets. In addition, they typically lose interception capacity as gutter grade increases, although at a slower rate than grate inlets. The primary factors affecting curb-opening capacity and inlet efficiency are the depth of water next to the curb and length of curb opening. Design procedures for curb-opening inlets are based on work conducted at Colorado State University for the Federal Highway Administration, as reported in Izzard (1946) and Bauer et al. (1964).

Although curb-opening heights vary in dimension, typical maximum heights range from 100 to 150 mm (4 to 6 in). For uniform cross slopes, the length of a curb-opening inlet on grade required to intercept 100 percent of gutter discharge can be expressed as

$$L_T = K_o Q^{0.42} S_L^{0.3} (nS_x)^{-0.6} \tag{5.25}$$

where L_T is the curb opening length in m (ft) required to intercept all of the gutter flow; K_o is an empirical constant equal to 0.817 (0.6 in English units); and other terms are as previously defined. The efficiency of curb-opening inlets shorter in length than L_T can be computed by

$$E = 1 - \left(1 - \frac{L}{L_T}\right)^{1.8} \tag{5.26}$$

where L is the curb-opening length in m (ft).

As shown by the terms in Eq. 5.25, increasing the cross slope can reduce the required length of curb opening for total interception. The cross slope can be increased through the use of locally or continuously depressed gutter sections, as shown in Fig. 5.14. For this case, the length of inlet required for 100 percent interception can be found by Eq. (5.25) if the equivalent cross slope, S_e, is substituted in place of S_x. The term S_e can be determined by

$$S_e = S_x + S_w' E_o \tag{5.27}$$

where E_o is the ratio of flow in the depressed section to total gutter flow, defined in Eq. (5.6); and S_w' is the cross slope of the depressed section measured from the cross slope, which can be expressed as

$$S_w' = \frac{a}{W} \tag{5.28}$$

where a is the *depth of gutter depression* in m (ft); and W is the width of the depressed section in m (ft).

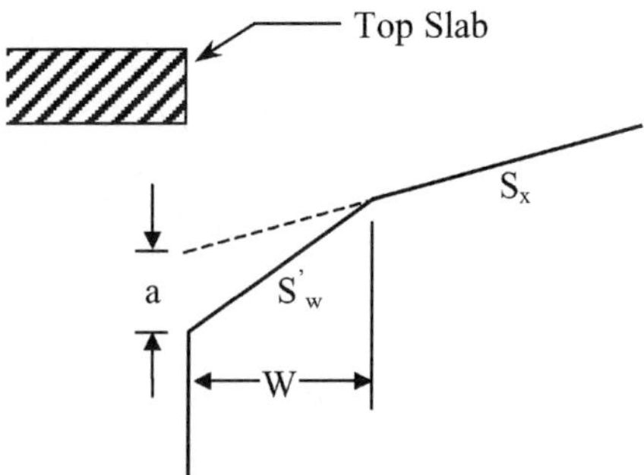

FIGURE 5.14 Depressed curb-opening inlet. (Adapted from Brown et al., 1996)

For curb-opening inlets with less than 100 percent interception, depressed sections can significantly increase the interception capacity and efficiency for a given length of curb opening. Efficiency can be evaluated by using Eq. (5.26), which is applicable to both uniform and composite cross slopes. Top slab supports placed flush with the curb line, however, can reduce interception capacity by as much as 50 percent and should, therefore, be avoided or be recessed from the curb line and rounded in shape.

Example 5.8
Compute the interception capacity of a 3.5-m long curb-opening inlet in a triangular gutter section having a cross slope of 0.025 m/m, longitudinal slope of 0.03 m/m, and Manning's roughness of 0.015. The gutter carries a design discharge of 0.08 m³/s.

Solution
Step 1. Compute the curb opening length, L_T, required for complete interception from Eq. (5.25).

$$L_T = K_o Q^{0.42} S_L^{0.3} (nS_x)^{-0.6} = (0.817)(0.08)^{0.42}(0.03)^{0.3}[(0.016)(0.025)]^{-0.6} = 10.8 \text{ m}$$

Step 2. Determine the inlet efficiency, E, using Eq. (5.26).

$$E = 1 - \left(1 - \frac{L}{L_T}\right)^{1.8} = 1 - \left(1 - \frac{3.5}{10.8}\right)^{1.8} = 0.506 \text{ or } 50.6\%$$

Step 3. Determine the inlet interception capacity using Eq. (5.18).

$$Q_i = EQ = (0.506)(0.08) = 0.041 \text{ m}^3/\text{s}$$

5.5.4 Combination Inlets

Combination inlets consist of both a vertical curb opening and an adjacent grate inlet (Fig. 5.6.c). They are frequently used in sag locations or in locations where chronic clogging of the grate may be a problem. For an *equal-length inlet,* in which both the grate and curb opening are of the same length, the interception capacity and efficiency of the combination inlet does not differ greatly from that of the grate alone. Capacity and efficiency can, therefore, be computed by neglecting the curb opening and using Eqs. 5.19 to 5.24 provided for grate inlets.

When the curb opening precedes the grate in a *sweeper* configuration, the inlet has a higher capacity equal to that of the curb-opening length upstream of the grate plus that of the grate. It is important to note, however, that the spread and frontal flow, and thus capacity, of the grate are reduced by interception of flow by the sweeper inlet. The sweeper inlet also serves to intercept floating debris that may otherwise clog the grate during the initial phases of a storm.

Example 5.9
Determine the interception capacity of a combination curb opening grate inlet in a triangular gutter section that carries 0.15 m³/s and has a cross slope of 0.03 m/m, longitudinal slope of 0.035 m/m, and Manning's roughness of 0.016. The grate inlet is a 0.6 m × 0.6 m P−50 × 100 grate and the curb-opening length is 3.5 m with 2.9 m of the opening located upstream of the grate in a sweeper configuration.

Solution

Step 1. Compute the length of curb opening, L_T, required for 100 percent interception using Eq. (5.25).

$$L_T = K_o Q^{0.42} S_L^{0.3} (nS_x)^{-0.6} = (0.817)(0.15)^{0.42}(0.035)^{0.3}[(0.016)(0.03)]^{-0.6} = 13.2 \text{ m}$$

Step 2. Evaluate the efficiency, E, of the curb opening inlet using Eq. (5.26).

$$E = 1 - \left(1 - \frac{L}{L_T}\right)^{1.8} = 1 - \left(1 - \frac{3.5}{13.2}\right)^{1.8} = 0.426 \text{ or } 42.6\%$$

Step 3. Compute the interception capacity of the curb opening, Q_{ic}, using Eq. (5.18).

$$Q_{ic} = EQ = (0.426)(0.15) = 0.064 \text{ m}^3/\text{s}$$

Step 4. Determine the spread, T, at the grate inlet by Eq. (5.3). Flow at the grate equals total gutter discharge minus that intercepted by the curb opening, or $0.15 - 0.064 = 0.086$ m^3/s

$$T = \left(\frac{Qn}{K_c S_x^{5/3} S_L^{1/2}}\right)^{3/8} = \left[\frac{(0.086)(0.016)}{(0.376)(0.03)^{5/3}(0.035)^{1/2}}\right]^{3/8} = 2.05 \text{ m}$$

Step 5. Compute the flow velocity from gutter discharge, Q, and effective area, A.

$$A = \frac{T^2 S_x}{2} + \frac{aW}{2} = \frac{(2.05)^2(0.03)}{2} = 0.063 \text{ m}^2$$

$$V = \frac{Q}{A} = \frac{0.086}{0.063} = 1.37 \text{ m/s}$$

Step 6. Evaluate the frontal flow efficiency, R_f, and side flow efficiency, R_s. From Fig. 5.13 with $V = 1.37$ m/s, $R_f = 1.0$

$$R_s = \frac{1}{\left(1 + \dfrac{K_s V^{1.8}}{S_x L^{2.3}}\right)} = \frac{1}{\left(1 + \dfrac{(0.0828)(1.37)^{1.8}}{(0.03)(0.6)^{2.3}}\right)} = 0.060$$

Step 7. Calculate the frontal flow ratio, E_o, using Eq. (5.19).

$$E_o = 1 - \left(1 - \frac{W}{T}\right)^{8/3} = 1 - \left(1 - \frac{0.6}{2.05}\right)^{8/3} = 0.60$$

Step 8. Compute the interception capacity of the grate inlet, Q_{ig}, using Eq. (5.24).

$$Q_{ig} = (R_f E_o + R_s(1 - E_o))Q = ((1.0)(0.60) + (0.06)(1 - 0.60))0.086 = 0.054 \text{ m}^3/\text{s}$$

Step 9. Evaluate the total interception capacity, Q_i.

$$Q_i = Q_{ic} + Q_{ig} = 0.064 + 0.054 = 0.12 \text{ m}^3/\text{s}$$

5.5.5 Slotted Drain Inlets

Slotted drains consist of a pipe that is cut longitudinally along the crown with bars placed perpendicular to maintain a slotted opening (Fig. 5.6.d). They are particularly useful for intercepting wide sections of sheet flow before it flows onto a roadway. Slotted drains offer the benefits of being used on either curbed or uncurbed gutter sections and interfering little with normal traffic operations. While they are highly susceptible to clogging from sediment deposition and floating debris, their configuration makes them accessible for cleaning with a high-pressure water jet.

If debris is not a factor, slotted drain inlets have approximately the same hydraulic characteristics as curb openings (Brown et al., 1996). Thus, inlet capacity is a function of flow depth and inlet length, and the drain acts as a side weir in which the flow is subjected to lateral acceleration due to the cross slope. Analysis of data from Federal Highway tests of slotted drains with slot widths greater than 45 mm (1.75 in) indicates that inlet length required for 100 percent interception can be computed using Eq. (5.25) for curb openings and that the efficiency of shorter lengths can be computed using Eq. (5.26). It is important to note, however, that it is far less costly to add length to an existing slotted inlet in order to raise interception capacity than it is to add length to an existing curb-opening inlet.

5.5.6 Inlets in Sag Locations

Inlets in vertical sag curves operate as weirs for shallow ponding depths and as orifices at larger depths. At intermediate depths, flow is in a transitional stage whereby conditions are ill defined and flow may fluctuate between weir and orifice control. The depth at which orifice flow begins is a function of the grate size, curb opening dimensions, or the slot width of an inlet (Brown et al., 1996). For example, larger grates will operate as weirs to greater depths than smaller grates or grates with less clear opening area.

The debris-passing ability of sag inlets is critical since the inlet must intercept all runoff entering a vertical sag curve. Total or partial clogging of the effective area of a sag inlet can lead to hazardous ponding conditions. Consequently, combination or curb-opening inlets should be used in sag locations, and the use of grate inlets alone is not recommended.

5.5.6.1 Grate Inlets. The capacity of grate inlets operating as weirs is given by

$$Q_i = C_w P d^{3/2} \tag{5.29}$$

where Q_i = interception capacity in m³/s (cfs)
P = perimeter of the grate in m (ft), not including the side adjacent to the curb, if present
C_w = weir discharge coefficient for grates, equal to 1.66 (3.0 in English units)
d = flow depth in m (ft) at the curb

When flow transitions to orifice control, capacity can be expressed as

$$Q_i = C_o A_g \sqrt{2gd} \tag{5.30}$$

where C_o = orifice discharge coefficient, equal to 0.67
A_g = clear opening, or effective, area of the grate in m² (ft²)
g = gravitational constant

Tests conducted for the Federal Highway Administration (Burgi, 1978a) indicate that for flat bar grates, the effective area in Eq. (5.30) is equal to the total area of the grate minus the area occupied by bars. Curved vane grates were shown to perform about 10% better than a grate with an effective area equal to the total area minus the area of the bars projected on a horizontal plane. Tilt-bar grates were not tested and are not recommended for sag locations where orifice flow control may occur.

Example 5.10
Determine the interception capacity of a 0.9 m × 1.2 m grate inlet in a sump location for a design spread of 2.0 m. The curbed gutter has a cross slope of 0.05 m/m and Manning's roughness of 0.016. Assume that 50% clogging occurs along the grate length.

Solution
Step 1. Compute the depth at the curb, d, from Eq. (5.4).

$$d = TS_x = (2.0)(0.05) = 0.1 \text{ m}$$

Assume weir flow controls for $d = 0.1$ m
Step 2. Compute the perimeter, P, of the grate.

$$P = (2)(0.9)(0.5) + 1.2 = 2.1 \text{ m}$$

Step 3. Evaluate the inlet capacity, Q_i, from Eq. (5.29).

$$Q_i = C_w P d^{3/2} = (1.66)(2.1)(0.1)^{3/2} = 0.11 \text{ m}^3/\text{s}$$

5.5.6.2 *Curb-opening Inlets.* Curb-opening inlets operate as weirs for ponding depth at the curb less than or equal to the height of the curb opening (Brown et al., 1996). In this range, interception capacity is given by

$$Q_i = C_w L d^{3/2} \tag{5.31}$$

where C_w = weir discharge coefficient for curb-openings, taken as 1.60 (3.0 in English units)
L = length of the curb opening in m (ft)

If the curb opening is depressed, the capacity is computed by

$$Q_i = C_w(L + 1.8W)d^{3/2} \tag{5.32}$$

where W = lateral width of depression in m (ft)

The weir coefficient for depressed inlets is reduced to 1.25 (2.3 in English units) and depth at the curb, d, is measured from the normal cross slope. Since Eq. (5.32) is for weir control, its application is limited to depths at the curb less than or equal to the height of the opening plus the depth of depression. Additionally, at curb-opening lengths greater than 3.6 m (12 ft), Eq. (5.31) for non-depressed inlets yields intercepted flows that exceed those values for depressed inlets. Since depressed inlets will perform at least as well as non-depressed inlets of the same length, Eq. (5.31) should be used for all curb-opening inlets having lengths greater than 3.6 m (12 ft).

At depths of approximately 1.4 times the opening height, curb-opening inlets operate as orifices (Brown et al., 1996). For this case, capacity is computed by

$$Q_i = C_o A_g \left[2g \left(d_i - \frac{h}{2} \right) \right]^{1/2} \tag{5.33}$$

where C_o = orifice discharge coefficient, equal to 0.67
A_g = effective area of the curb opening in m^2 (ft^2)
g = gravitational constant
d_i = depth in m (ft) at the lip of the curb opening, including any gutter depression
h = height of curb opening

Equation (5.33) assumes a horizontal orifice throat, illustrated in Fig. 5.15.a. For other throat configurations (Figs. 5.15.b and 5.15.c), this expression is generalized as

$$Q_i = C_o h L (2g d_o)^{1/2} \tag{5.34}$$

where h = defined as the throat width in m (ft)
d_o = effective head in m (ft) on the center of the orifice

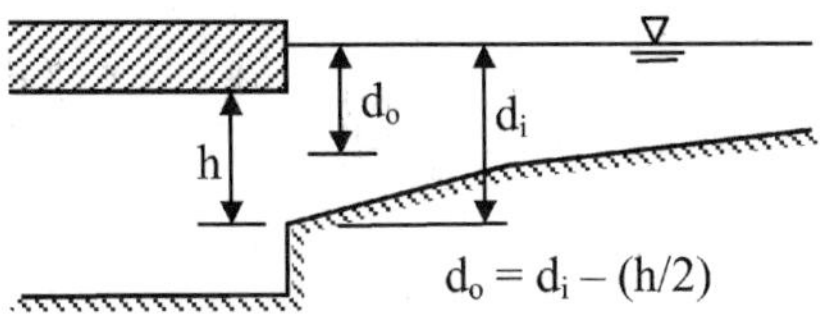

a. Horizontal Throat

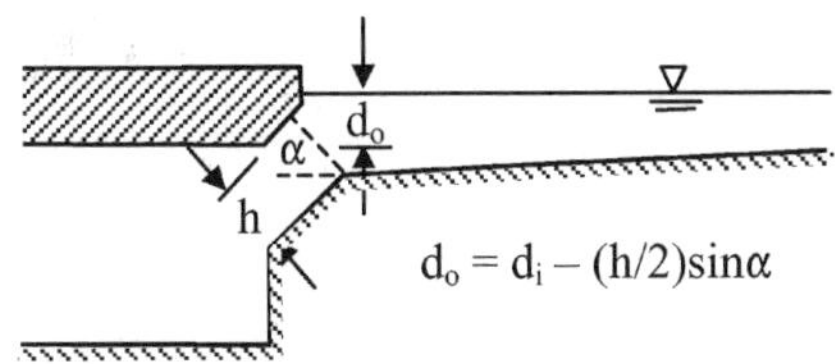

b. Inclined Throat

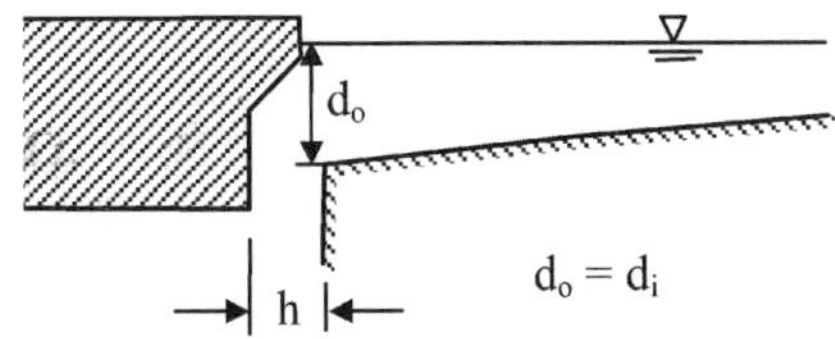

c. Vertical Throat

FIGURE 5.15 Curb-opening inlets. (Source: Brown et al., 1996)

Example 5.11
Evaluate the interception capacity of a 3-m long, 0.15-m high, depressed curb-opening inlet located in a sag vertical curve. The depression is 50.0-mm deep and 0.6-m wide, and the design spread and cross slope are 2.5 m and 0.03 m/m, respectively.

Solution
Step 1. Determine the depth, d, at the curb using Eq. (5.4).

$$d = TS_x = (2.5)(0.03) = 0.075 \text{ m}$$

$[d = 0.075 \text{ m}] < [h + a = 0.15 + 50/1000 = 0.2 \text{ m}]$, therefore assume weir control

Step 2. Compute the inlet capacity, Q_i, from Eq. (5.32).

$$Q_i = C_w(L + 1.8W)d^{3/2} = 1.25(3 + 1.8(0.6))(0.075)^{3/2} = 0.10 \text{ m}^3/\text{s}$$

5.5.6.3 Combination Inlets.   In sag locations where hazardous ponding can occur, the use of combination grate and curb-opening inlets is highly recommended. A sweeper configuration is generally more efficient than an equal-length arrangement and allows for improved debris handling. When used in vertical sags, the sweeper inlet can include a curb opening at both ends of the grate.

The interception capacity of an equal-length combination inlet is approximately equal to that of the grate alone operating under weir conditions, given by Eq. (5.29). Under orifice control, the capacity increases to that of the grate, from Eq. (5.30), plus the capacity of the curb opening, from Eq. (5.33) or (5.34). Additionally, combination inlets in sag locations are frequently designed assuming complete clogging of the grate.

5.5.6.4 Slotted Drain Inlets.   Due to susceptibility to clogging from debris, slotted drain inlets are not recommended for sag locations. Nevertheless, they generally operate as weirs when depth at the slot is below approximately 60 mm (0.2 ft), depending on slot width and length, and orifices for depths greater than 120 mm (0.4 ft) (Brown et al., 1996). Interception capacity of a slotted drain operating under weir conditions can be evaluated by

$$Q_i = C_w L d^{3/2} \qquad (5.35)$$

where C_w = weir discharge coefficient that varies with flow depth and slot length, having a typical value of 1.4 (2.48 in English units)
L = slot length in m (ft)
d = flow depth in m (ft)

Under orifice conditions, capacity is expressed as

$$Q_i = 0.8LW(2gd)^{1/2} \qquad (5.36)$$

where W is the slot width in m (ft); and g is the gravitational constant.

Example 5.12
Determine the capacity of a 2.0-m long slotted drain in a sump location. The slot width is 50 mm, design spread is 3.5 m, and the gutter cross slope is 0.04 m/m. Assume no tendency for clogging.

Solution
Step 1. Evaluate the depth, d, at the curb using Eq. (5.4).

$$d = TS_x = (3.5)(0.04) = 0.14 \text{ m}$$

Assume orifice control at $d = 140$ mm

Step 2. Compute interception capacity, Q_i, using Eq. (5.36).

$$Q_i = 0.8LW(2gd)^{1/2} = (0.8)(2.0)\left(\frac{50}{1000}\right)((2)(32.2)(0.14))^{1/2} = 0.13 \text{ m}^3/\text{s}$$

5.5.7 Inlet Structures

After surface water has been intercepted by stormwater inlets, inlet drop structures and catch basins allow runoff to enter a subsurface drainage system. These structures also serve a secondary purpose as access locations for cleaning and inspection. Figure 5.16 illustrates several typical structures used for grate, combination and curb-opening inlets. Inlet structures are most often constructed of cast-in-place concrete or pre-cast concrete. The height of the structure is likely to be dictated by the storm drain profile and surface topography.

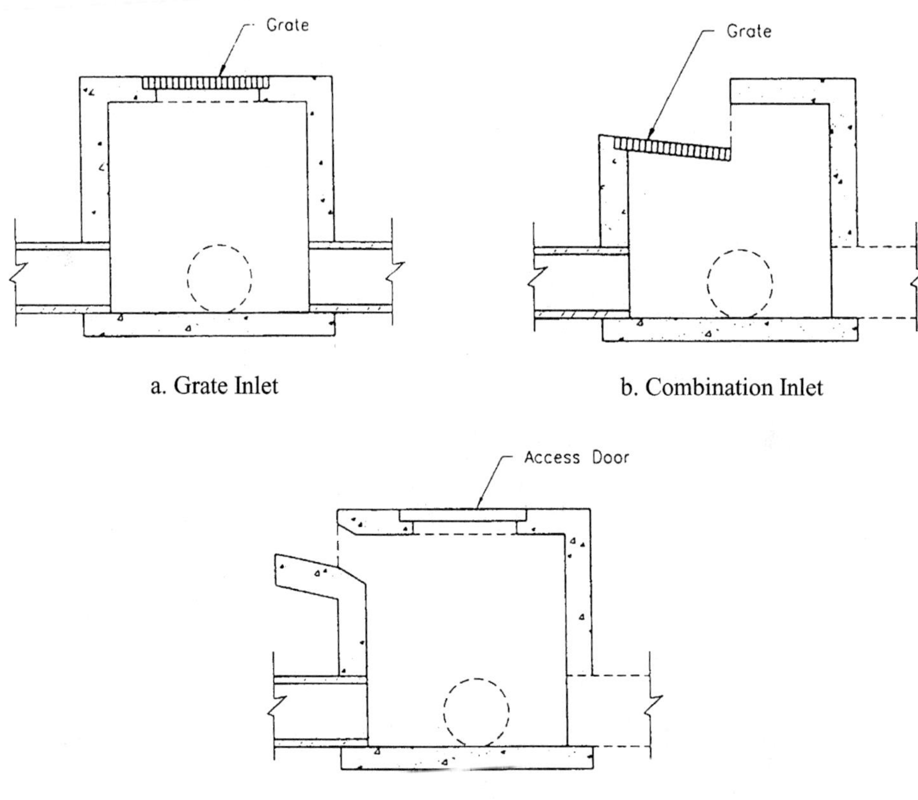

FIGURE 5.16   Inlet structures. (Source: Brown et al., 1996)

5.6 DESIGN OF INLET LOCATIONS

The spacing of inlets is determined by highway geometry and the design spread of water on the pavement. The following is a list of information that is needed for locating inlets:

- Plan and profile sheets indicating existing and proposed highway facilities
- Typical roadway cross sections
- Grading cross sections
- Superelevation diagrams
- Topographic maps of the drainage regions
- Intensity-duration-frequency data for the design storm
- Applicable local drainage criteria and design standards

5.6.1 Geometric Considerations

A number of locations require inlets solely based on geometry of the roadway system, regardless of contributing drainage area (Brown et al., 1996). These locations should be marked on the plans prior to computations regarding runoff, spread, and inlet capacity, and generally include:

- Sag locations on the gutter grade
- Upstream of median breaks, entrance/exit ramps, street intersections, and pedestrian cross-walks
- Upstream and downstream of bridges
- Upstream of cross slope reversals
- End of any channel in cut sections
- Behind curbs, shoulders or sidewalks to drain low areas

In addition to these areas, stormwater from cut slopes and adjacent regions draining toward the highway should be intercepted before it reaches the pavement using roadside channels. Curbed pavement sections and highway drainage inlets are generally a less efficient means for handling this extraneous runoff. Utilizing roadside channels minimizes deposition of sediment and other debris on the roadway, as well as the amount of water that must be conveyed in a gutter section.

5.6.2 Spacing on Continuous Grade

Design spread is the criterion used for locating drainage inlets between those required for geometric controls. The method for evaluating inlet spacing is outlined as follows;

1. Select a trial inlet location and evaluate its contributing drainage area.
2. Compute the peak runoff from the selected area using the rational equation (Eq. 5.1)).
3. The gutter flowrate is set equal to the peak runoff plus additional bypass flow from upgrade inlets, and Eqs. (5.3) and (5.4) are solved for the resulting spread and depth at the curb.
4. If the depth at the curb is greater than the actual curb height or computed spread is greater than that allowable, return to step 1 with a reduced drainage area, and thus spacing. Likewise, if the computed spread is significantly less than the allowable, return to step 1 with an increased spacing. Otherwise, compute the interception capacity and bypass flow for the inlet.
5. Repeat steps 1 to 4 for subsequent inlets.

Uniform spacing can be established for continuous grades and in cases where the drainage area consists of pavement only or has uniform runoff properties and is rectangular in shape. In this case, the inherent assumption is that time of concentration is the same for all inlets. The location of the first inlet from the crest can be found directly by solving the rational equation for the length of pavement that will generate the design runoff, expressed as

$$L_1 = \frac{QK'}{CiW_p} \tag{5.37}$$

where L_1 = length from the crest to the first inlet in m (ft)
$\quad Q$ = gutter discharge in m³/s (cfs) computed from Eq. (5.3) using the design spread
$\quad K'$ = conversion constant equal to 3.6×10^6 (43,560 in English units)
$\quad W_p$ = lateral distance in m (ft) from the pavement crown to the curb
$\quad C$ = dimensionless runoff coefficient
$\quad i$ = rainfall intensity in mm/hr (in/hr)

Interception capacity and subsequent evaluation of bypass flow of this upstream inlet then determines the spread at the inlet. Downstream inlets should be spaced according to where the design spread is reached, or

$$L_i = \frac{QK'}{CiW_p} E \tag{5.38}$$

where L_i = spacing between subsequent inlets in m (ft)
E = interception efficiency of the upstream inlet

Note that the last inlet, which will likely be placed at the low point in the roadway grade, should be designed for complete interception. In addition, more frequent spacing of smaller inlets, and consequently the allowance of larger bypass flows, can reduce capital costs in some cases. Furthermore, individual transportation agencies frequently have spacing limitations due to maintenance constraints. The following example demonstrates the design procedure for locating inlets on continuous grades.

Example 5.13
Determine the spacing required for a series of 0.6 m × 0.6 m reticuline grate inlets that drain a 7.5-m width of pavement at a design spread of 2.0 m. The gutter has a uniform cross slope of 0.02 m/m, longitudinal slope of 0.018 m/m, and Manning's roughness of 0.015. The design rainfall intensity and runoff coefficient are estimated at 150 mm/hr and 0.95, respectively.

Solution
Step 1. Compute the gutter discharge, Q, using Eq. (5.3).

$$Q = \frac{K_c}{n} S_x^{5/3} S_L^{1/2} T^{8/3} = \frac{0.376}{0.015} (0.02)^{5/3}(0.018)^{1/2}(2.0)^{8/3} = 0.032 \ \text{m}^3/\text{s}$$

Step 2. Determine the location of the first inlet, L_1, from Eq. (5.37).

$$L_1 = \frac{QK'}{Ciw_p} = \frac{(0.032)(3.6 \times 10^6)}{(0.95)(150)(7.5)} = 108 \ \text{m}$$

Step 3. Evaluate the efficiency of a 0.6 m × 0.6 m reticuline grate inlet

a. Calculate the flow velocity from gutter discharge, Q, and effective flow area, A.

$$A = \frac{T^2 S_x}{2} = \frac{(2.0)^2(0.02)}{2} = 0.040 \ \text{m}^2$$

$$V = \frac{Q}{A} = \frac{0.032}{0.040} = 0.80 \ \text{m/s}$$

b. Determine the frontal flow efficiency, R_f, and side flow efficiency, R_s. From Fig. 5.13 with $V = 0.80$ m/s, $R_f = 1.0$

$$R_s = \frac{1}{\left(1 + \dfrac{K_s V^{1.8}}{S_x L^{2.3}}\right)} = \frac{1}{\left(1 + \dfrac{(0.0828)(0.80)^{1.8}}{(0.02)(0.6)^{2.3}}\right)} = 0.10$$

c. Calculate the frontal flow ratio, E_o, using Eq. (5.19).

$$E_o = 1 - \left(1 - \frac{W}{T}\right)^{8/3} = 1 - \left(1 - \frac{0.6}{2.0}\right)^{8.3} = 0.61$$

d. Evaluate the interception efficiency, E, from Eq. (5.23).

$$E = R_f E_o + R_s(1 - E_o) = (1.0)(0.61) + (0.10)(1 - 0.61) = 0.65$$

Step 4. Determine the spacing of subsequent inlets, L_i, from Eq. (5.38).

$$L_i = \frac{QK'}{CiW_p} E = \frac{(0.032)(3.6 \times 10^6)}{(0.95)(150)(7.5)}(0.65) = 70 \text{ m}$$

The first inlet should be placed 108 m from the crest and subsequent inlets should be spaced at 70-m increments.

Where the highway grade changes, computations proceed in a similar manner, but spacing will vary according to longitudinal slope. As the grade becomes flatter, both capacity and spacing will be reduced and additional inlets will be required. Conversely, as slope increases, fewer inlets will be needed due to the increased capacity of gutter sections.

5.6.3 Flanking Inlets

In addition to locating inlets at sag locations, *flanking inlets* should be placed on each side of a sag curve as illustrated in Fig. 5.17. These supplementary inlets should be located as to limit the spread in the low gradient approaches to the sump and relieve the sag inlet if it should become clogged. The use of flanking inlets will also serve to reduce the deposition of sediment on the roadway.

Locations for flanking inlets can be determined by Johnson and Chang (1984)

$$x = \sqrt{200\, d_f K_f} \tag{5.39}$$

where x = distance from sag point to flanking inlet in m
$\quad d_f$ = depth at the curb, not including the sump depth
$\quad K_f$ = rate of vertical curvature in meters, defined by

$$K_f = \frac{L_c}{S_2 - S_1} \tag{5.40}$$

where $\quad L_c$ = length of the vertical curve in m
$\quad S_2$ and S_1 = approach grades, in percent, on opposite sides of the sag

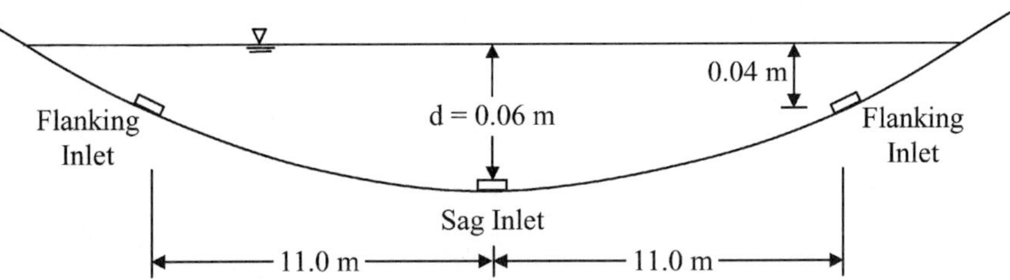

FIGURE 5.17   Example of flanking inlets. (Adapted from Brown et al., 1996)

If the flanking inlets have the same dimensions as the sag inlet, each will intercept one-half of the design flow when they are located so that the depth of ponding at the flanking inlets is 63 percent of the depth of ponding at the sag point. If the flanking inlets are of a different size, the capacity of each inlet as a function of depth should be determined using the weir equation with assumed depths (AASHTO, 1991).

Example 5.14
Determine the location of flanking inlets for a 150-m long sag vertical curve having approach grades of −2.5% and +2.5% on either side of the curve. The design spread is 3.0 m, and the gutter cross slope is 0.02 m. Assume the flanking inlets are designed to function as a relief when the inlet at the sag point is clogged. (Adapted from Brown et al., 1996.)

Solution
Step 1. Determine the rate of vertical curvature, K_f, from Eq. (5.40).

$$K_f = \frac{L_c}{S_2 - S_1} = \frac{150}{(2.5) - (-2.5)} = 30 \text{ m}$$

Step 2. Determine the depth at the curb, not including the sump depth, assuming that each flanking inlet will be positioned to intercept one-half of the design flow.

$$d_f = TS_x - (0.63)TS_x = TS_x(1 - 0.63) = (3.0)(0.02)(1 - 0.63) = 0.02 \text{ m}$$

Step 3. Compute the distance, x, from the sag point to the flanking inlet using Eq. (5.39).

$$x = \sqrt{200\ d_f K_f} = \sqrt{200(0.02)(30)} = 11.0 \text{ m}$$

The flanking inlets should be spaced 11.0 m from the sag point (Fig. 5.17).

5.7 MEDIAN INLETS

Median and *roadside channels* are designed to convey runoff from uncurbed highways and cut slopes. Drop inlets, such as the type shown in Fig. 5.18, are frequently used in medians to reduce the erosive nature of channel flow. These are similar to grate inlets that are used for pavement drainage. Drop inlets should be placed flush with the channel bottom and be

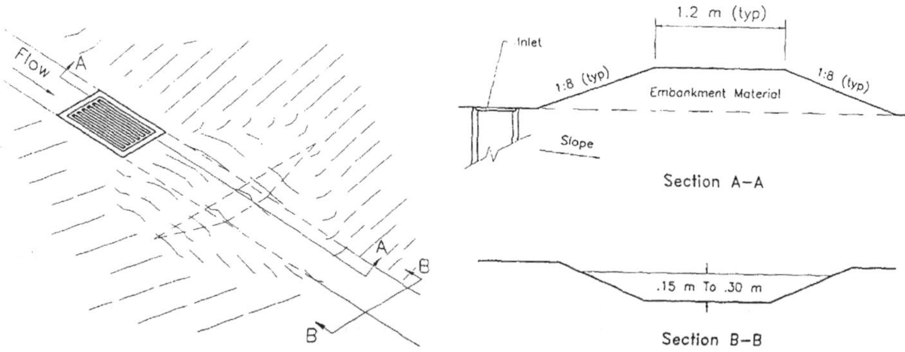

FIGURE 5.18   Median inlet. (Source: Brown et al., 1996)

constructed of traffic-safe bar grates. In addition, paving around the inlet perimeter can help prevent erosion and may slightly increase interception capacity of the inlet by accelerating the flow.

The total flow in a trapezoidal median channel is determined by solving Manning's equation, given as

$$Q = \frac{K_m}{n}(Bd + zd^2)\left(\frac{Bd + zd^2}{B + 2d\sqrt{z^2 + 1}}\right)^{2/3} S_L^{1/2} \tag{5.41}$$

where Q = channel discharge in m³/s (cfs)
$\quad K_m$ = empirical constant equal to 1.0 (1.486 in English units)
$\quad n$ = Manning's roughness coefficient
$\quad B$ = channel bottom width in m (ft)
$\quad d$ = flow depth in m (ft)
$\quad S_L$ = channel slope in m/m (ft/ft)
$\quad z$ = horizontal component of side slope in m (ft) for a rise of 1 m (ft) vertical distance

In addition, the ratio of frontal flow to total flow in the channel is expressed as

$$E_o = \frac{W}{B + dz} \tag{5.42}$$

where W = grate width in m (ft)

The preceding equations can be used with those outlined previously for grate inlets on grade to evaluate interception capacity of a drop inlet in a continuous-grade median.

Example 5.15
Evaluate the interception capacity and bypass flow for a median drop inlet that utilizes a 0.6 m × 0.6 m P − 30 grate. The trapezoidal median ditch drains a design flow of 0.40 m³/s and has a bottom width of 0.6 m, side slopes of 1V:5H, a cross slope of 0.012, a longitudinal slope of 0.025, Manning's roughness of 0.035.

Solution
Step 1. Determine the flow depth by solving Eq. (5.41) for d.

$$Q = \frac{K_m}{n}(Bd + zd^2)\left(\frac{Bd + zd^2}{B + 2d\sqrt{z^2 + 1}}\right)^{2/3} S_L^{1/2}$$

$$= \frac{1.0}{0.035}((0.6)d + (5)d^2)\left(\frac{(0.6)d + (5)d^2}{(0.6) + 2d\sqrt{(5)^2 + 1}}\right)^{2/3}(0.025)^{1/2} = 0.40$$

Solving for depth yields $d = 0.21$ m.

Step 2. Compute the flow velocity, V, using discharge and effective flow area.

$$V = \frac{Q}{Bd + zd^2} = \frac{0.4}{(0.6)(0.21) + (5)(0.21)^2} = 1.15 \text{ m/s}$$

Step 3. Determine the frontal flow ratio, E_o, from Eq. (5.42).

$$E_o = \frac{W}{B + dz} = \frac{0.6}{(0.6) + (0.21)(5)} = 0.36$$

Step 4. Determine the frontal flow efficiency, R_f, and side flow efficiency, R_s. From Fig. 5.13 with $V = 1.15$ m/s, $R_f = 1.0$

$$R_s = \cfrac{1}{\left(1 + \cfrac{K_s V^{1.8}}{S_x L^{2.3}}\right)} = \cfrac{1}{\left(1 + \cfrac{(0.0828)(1.15)^{1.8}}{(0.012)(0.6)^{2.3}}\right)} = 0.034$$

Step 5. Evaluate the interception efficiency, E, from Eq. (5.23).

$$E = R_f E_o + R_s(1 - E_o) = (1.0)(0.36) + (0.034)(1 - 0.36) = 0.38$$

Step 6. Compute the interception capacity, Q_i, by Eq. (5.24).

$$Q_i = EQ = (0.38)(0.40) = 0.15 \text{ m}^3/\text{s}$$

Step 7. Evaluate the bypass flow, Q_b, from Eq. (5.17).

$$Q_b = Q - Q_i - 0.40 - 0.15 = 0.25 \text{ m}^3/\text{s}$$

Small dikes are often placed downstream of drop inlets to impede bypass flow. The height of dike required for 100 percent interception is not large and can be determined by using methods outlined previously for grate inlets in sags. Note that since no curb is present in an open channel with a dike, the full grate perimeter should be used in Eqs. (5.29) and (5.30).

5.8 EMBANKMENT INLETS

Where adequate vegetation can be established on embankments, runoff should be discharged down slopes with as little flow concentration as possible. Due to the erosion potential of fill slopes, however, it is sometimes necessary to collect stormwater through embankment inlets, illustrated in Fig. 5.19, and discharge flow to the toe of fill slopes through chutes, swales, or downdrains. *Embankment inlets* are also frequently used for intercepting stormwater upgrade or downgrade of bridges. These inlets differ from other pavement drainage structures in several aspects. Overall efficiencies that can be attained for pavement drainage systems are often not possible for embankment systems because multiple inlets are not used. Additionally, embankment drainage frequently demands 100 percent interception in order to limit bypass flow from discharging onto a bridge deck. Finally, a closed subsurface storm drain system is sometimes not available to dispose of collected runoff, and the means for disposal must be provided as part of an individual inlet.

Design procedures for embankment drainage are similar to those outlined previously for continuous grade inlets, with the added criteria of complete interception of flow. Some of the most common inlet designs include combination inlets with a sweeper configuration, wide grates, or slotted drains of extended length (Brown et al., 1996). Pipe downdrains are recommended to convey runoff to the toe of fill slopes since flow is confined and erosion is limited. Although downdrains can be either open or closed, open chutes are often damaged by erosion from water spilling over the sides of the chute at bends and as a result of flow oscillations. Riprap or other energy dissipation devices should be considered for preventing erosion and undercutting at the outlet of the downdrain.

5.9 BRIDGE-DECK DRAINAGE

Effective drainage of bridges is important for minimizing the corrosion of their structural components, controlling the formation of ice on their decks, and controlling hydroplaning on their surfaces. Although designs are similar to that of curbed roadway sections in most aspects, bridge inlets are often less hydraulically efficient because cross slopes are flatter

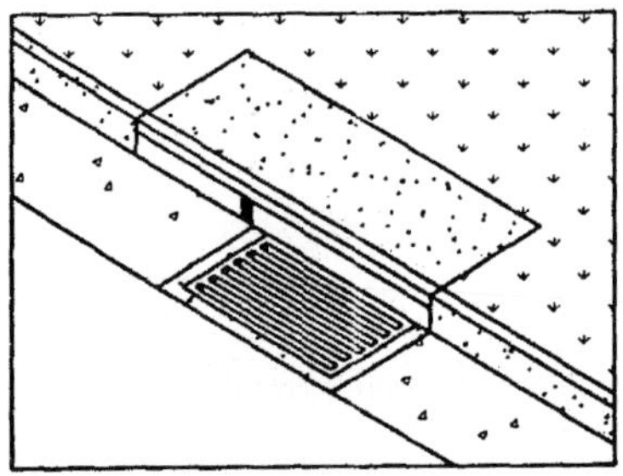

a. Perspective

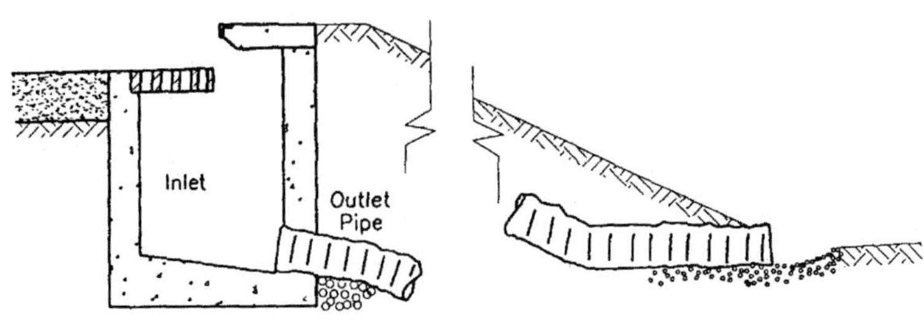

b. Section

FIGURE 5.19   Embankment inlet and downdrain. (Source: Brown et al., 1996)

and tend to be more easily clogged by debris (Young et al., 1993). As a result, the gutter flow should be intercepted upgrade of the bridge, and near zero gradients and sag vertical curves should be avoided whenever possible. Additionally, runoff should be intercepted immediately downgrade of the bridge, where larger grates and inlet structures can be used.

For long spanned bridges, increasing spread on the bridge deck may necessitate placement of small drainage inlets, also called *scuppers,* illustrated in Fig. 5.20. Design procedures for sizing and spacing bridge deck inlets are essentially the same as those used for pavement sections. From a hydraulic standpoint, however, an additional factor of safety should be applied to inlet sizing for maximizing deck drainage and minimizing problems associated with debris handling. A practical upper size limit on inlets placed within deck slabs is approximately 1 m (3 ft) on a side. To further promote drainage, most states specify uniform cross slopes at 1 to 2 percent minimum. Steep slopes greater than 4 percent, however, can make finishing the bridge deck difficult and can be troublesome for slow moving vehicles when the deck becomes icy. In addition, inlets should be designed for 100 percent interception upgrade of expansion joints and located so as to provide room for maintenance crews to work safely. Finally, in designing bridge deck inlets, the hydraulic engineer must simultaneously be aware of additional, and potentially conflicting, geometric and structural constraints. For example, more frequent and smaller inlets associated with reinforcing bar schedules or post-tensioned cable spacing may be required so as to not interfere with structural integrity of the bridge.

Runoff collected at bridge inlets either falls directly onto underlying surfaces or is collected and conveyed to downspouts fixed on bridge support columns. Collector pipes should have a minimum number of T-connections and should be sloped at least 2 percent to provide

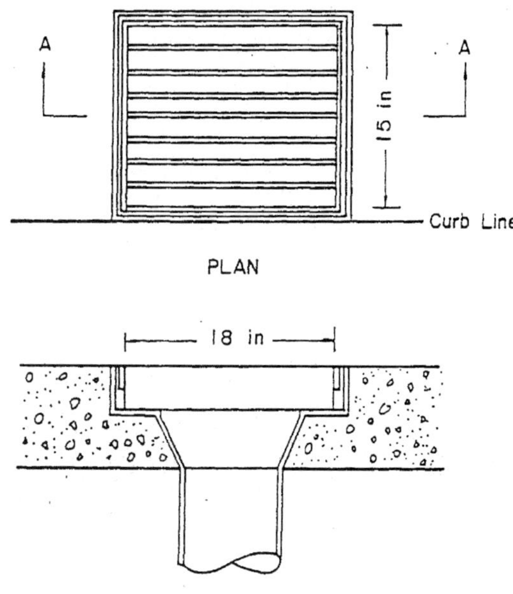

FIGURE 5.20   Bridge deck inlet. (Source: Johnson and Chang, 1984)

for self-cleansing velocities and prevent clogging. In the event that clogging does occur, provisions should be made for cleanouts in all pipes and downspouts.

REFERENCES

American Association of State Highway and Transportation Officials, *A Policy on Geometric Design of Highways and Streets,* Washington, DC, 1990.

American Association of State Highway and Transportation Officials, *Model Drainage Manual,* "Chapter 13: Storm Drainage Systems," Washington, DC, 1991.

Anderson, D. A., J. R. Reed, R. S. Huebner, J. J. Henry, W. P. Kilareski, and J. C. Warner, *Improved Surface Drainage of Pavements,* NCHRP Project I-29, The Pennsylvania Transportation Institute, The Pennsylvania State University, Federal Highway Administration, Washington, DC, 1995.

American Society of Civil Engineers, *Design and Construction of Urban Stormwater Management Systems,* Manuals and Reports on Engineering Practice No. 77, WEF Manual of Practice RD-20, New York, NY, 1992.

Bauer, W. J., and D. C. Woo, *Hydraulic Design of Depressed Curb-Opening Inlets,* Highway Research Record No. 58, Highway Research Board, Washington, DC, 1964.

Brown, S. A., S. M. Stein, and J. C. Warner, *Urban Drainage Design Manual,* Hydraulic Engineering Circular No. 22, FHWA-SA-96-078, Federal Highway Administration, U.S. Department of Transportation, Washington, DC, 1996.

Burgi, P. H., and D. E. Gober, *Bicycle-Safe Grate Inlets Study; Volume 1–Hydraulic and Safety Characteristics of Three Selected Grate Inlets on Continuous Grades,* FHWA-RD-77-24, Federal Highway Administration, U.S. Department of Transportation, Washington, DC, 1977.

Burgi, P. H., *Bicycle-Safe Grate Inlets Study; Volume 2–Hydraulic Characteristics of Three Selected Grate Inlets on Continuous Grades,* FHWA-RD-78-4, Federal Highway Administration, U.S. Department of Transportation, Washington, DC, 1978a.

Burgi, P. H., *Bicycle-Safe Grate Inlets Study; Volume 3–Hydraulic Characteristics of Three Selected Grate Inlets in a Sump Condition,* FHWA-RD-78-70, Federal Highway Administration, U.S. Department of Transportation, Washington, DC, 1978b.

Federal Highway Administration, *Design Charts for Open-Channel Flow,* Hydraulic Design Series No. 3, U.S. Department of Transportation, Washington, DC, 1977a (reprint).

Federal Highway Administration, *Bicycle-Safe Grate Inlets Study; Volumes 1 and 2–Hydraulic and Safety Characteristics of Selected Grate Inlets on Continuous Grades,* FHWA-RD-77-4, U.S. Department of Transportation, Washington, DC, 1977b.

Izzard, C. F., *Hydraulics of Runoff from Developed Surfaces,* Proc. Highway Research Board, Vol. 26, pp. 129–150, Highway Research Board, Washington, DC, 1946.

Johnson, F. L., and F. M. Chang, *Drainage of Highway Pavements,* Hydraulic Engineering Circular No. 12, FHWA-TS-84-202, Federal Highway Administration, U.S. Department of Transportation, Washington, DC, 1984.

McCuen, R. H., P. A. Johnson, and R. M. Ragan, *Hydrology,* Hydraulic Design Series No. 2, FHWA–SA-96-067, Federal Highway Administration, U.S. Department of Transportation, Washington, DC, 1996.

Pugh, C. A., *Bicycle-Safe Grate Inlets Study; Volume 4–Hydraulic Characteristics of Slotted Drain Inlets,* FHWA-RD-79-106, Federal Highway Administration, U.S. Department of Transportation, Washington, DC, 1980.

Ragan, R. M., "A Monograph Based on Kinematic Wave Theory for Determining Time of Concentration of Overland Flow," Report No. 44, Department of Civil Engineering, University of Maryland at College Park, MD, 1971.

Wanielista, M., R. Kersten, and R. Eaglin, *Hydrology: Water Quantity and Control,* 2nd ed., Wiley, New York, NY, 1997.

Young, G. K., S. E. Walker, and F. Chang, *Design of Bridge Deck Drainage,* Hydraulic Engineering Circular No. 21, FHWA-SA-92-010, Federal Highway Administration, U.S. Department of Transportation, Washington, DC, 1993.

CHAPTER 6
HYDRAULICS OF SEWER SYSTEMS

Ben Chie Yen
Department of Civil & Environmental Engineering
University of Illinois at Urbana-Champaign
Urbana, Illinois 61801

6.1 INTRODUCTION

The desired level of hydraulics to solve sewer system flow problems varies, depending on the type of problem and accuracy required. The purposes of analysis of sewer hydraulic problems can be classified into two types: (1) design and (2) prediction for forecasting or operation (Table 6.1). The required hydraulic level of the latter is often higher than the former.

In the design type, the drainage facility, including all the sewers in a network, is to be sized for construction and is supposed to serve for all future rainstorm events that do not exceed the specified design hydrologic level. Implicitly, the size of the apparatus is so determined that all rainstorms equal to and smaller than the design storm are presumably considered and accounted for. Each sewer in a drainage network has its own time of concentration and hence its own design storm. Thus, to capture the most critical design flow condition for all of the different sewers in the design of sewers in a network, all these different rainstorms should be considered. On the other hand, in runoff prediction, the drainage apparatus has already been built or pre-determined, its dimensions are known, and simulation of the flow from a particular single rainstorm event is made for the purpose of real-time forecasting, to be used for operation and runoff control, or sometimes for the determination of the flow of a past event, as in the case of legal events (Table 6.1b). The hydrologic requirements for these two different types of problems are different. For the latter, the flow prediction type, a particular rainstorm with its specific temporal and spatial distributions is considered. For the former, the design type, hypothetical rainstorms with an assigned design return period or acceptable risk level and assumed temporal and spatial distributions of the rainfall are used.

In the case of designing sanitary sewers, the problem becomes the estimation of the critical runoffs in both quantity and quality, from domestic, commercial, and industrial sources over the service period in the future. For real-time control problems, it involves simulation and prediction of the sanitary runoff in conjunction with the control measures.

Selected hydraulic-based computer models for sewer system flow simulation are listed according to their hydraulic levels in Table 6.2 and will be discussed further in the last section of this chapter. Selected computed-based models for design of storm sewers in a network are listed in Table 6.3.

TABLE 6.1   Major Types of Sewer Drainage Problems
(*a*) Typical Design Problems

Type of design	Design parameter	Known parameters	Hydro information needed	Required hydraulic level
Sewers	Pipe size	Network layout and sewer slope	Peak discharge, Q_p for design return period	Low
Sewers	Pipe size and pipe slope	Network layout	Peak discharge, Q_p for design return period	Low
Sewers	Network layout; pipe size and pipe slope	Locations of manholes and junctions	Peak discharge, Q_p for design return period	Low
Junctions and manholes	Geometric dimensions	Joining pipe inverts	Preferably design hydrographs $Q_i(t)$	Low to moderate
Sewer side weirs	Weir height and length	Sewer pipe size	Peak discharge Q_p	Moderate to high

TABLE 6.1   (*b*) Flow Prediction Problems

Type	Purpose	Hydro input	Hydro information sought	Required hydraulic level
Real-time operation	Real-time regulation of flow	Predicted and/or just measured rainfall, network data	Hydrographs, $Q(t, x_i)$	High
Performance evaluation	Simultion for evaluation of a system	Specific storm event, network data	Hydrographs, $Q(t, x_i)$	High
Storm event simulation	Determination of runoff at specific locations for particular past or specified events	Given past storm event or specified input hyetographs, network data	Hydrographs, $Q(t, x_i)$	Moderate-high
Flood level determination	Determination of the extent of flooding	Specific storm hyetographs	Hydrographs and stages	High
Storm runoff quality control	Reduce and control of water pollution due to runoff from rainstorms	Event or continuous rain and pollutant data, network data	Hydrographs, $Q(t, x_i)$ Pollutographs, $c(t, x_i)$	Moderate-high
Storm runoff master planning	Long-term, usually large spatial scale planning for stormwater	Long term data	Runoff volume Pollutant volume	Low

Source: From Yen and Akan (1999).

TABLE 6.2 Summary of Hydraulic Properties of Selected Simulation Models for Flow in Sewer Network

(*a*) Dynamic Wave Models

| | Open-channel flow | | | | Interior junction | | | Surcharge flow | | | |
Model	Numerical scheme	Parameters	S_f	Sewer downstream condition	Solution scheme	Detention storage	Equations	Transition condition	Surcharge hydraulics	Numerical solution scheme	References
SWMM-EXTRAN	Explicit	h, Q	Manning	Junction water surface or assumed condition	One sweep, pipe by pipe	Yes	$\Sigma Q = ds/dt$ and $h_i = h_o$	All pipes entering a junction are full or the highest entering pipe is submerged	Employ open-channel equations with junction head computed using assumed adjustment factors, excess water lost	Explicit, one sweep pipe by pipe	Roesner and Shubinski (1982); Roesner et al. (1988)
ISS	Characteristic	h, V	Darcy-Weisbach	Junction water surface or critical depth	Simultaneous on overlapping segments; pipe by pipe	Yes	$\Sigma Q = ds/dt$ and $h_o = H - (V^2/2g)$ or $\Sigma Q = O$ and $h_i = h_o$	NA	NA		Sevuk et al. (1973); Sevuk and Yen (1982)
CAREDAS	Four-point implicit	h, Q	Chezy or Manning	Junction water surface	Simultaneous (double sweep)	Yes	$\Sigma Q = ds/dt$ and $h_i = h_o$	$h/d > 0.91$	Preissmann slot		Chevereau et al. (1978); Cunge and Mazaudou (1984)
UNSTDY	Four-point implicit	h, Q	Manning	Junction water surface or sluice gate	Simultaneous (double sweep)	Yes	$\Sigma Q = ds/dt$ and $h_i = h_o$	Not given	Preissmann slot		Book et al. (1982); Chen and Chai (1991); Labadie et al. (1978)
MOUSE	Six-point implicit, $w = 0.5$	h, Q	Manning	Junction water surface	Simultaneous (double sweep)	Yes	$\Sigma Q = ds/dt$ and $h_o = H - (KV^2/2g)$ or $\Sigma Q = 0$ and $h_i = h_o$	Not given	Preissmann slot		Abbott et al. (1982); DHI (1994); Hoff-Clausen et al. (1982)
HYDRO-WORKS / SPIDA	Four-point implicit	h, Q	Colebrook-White or Manning	Junction water surface or critical depth	Simultaneous (double sweep)	Yes	$\Sigma Q = ds/dt$ and $h_o = H - (KV^2/2g)$		Preissmann slot		Wallingford Software (1991, 1997)

TABLE 6.2 (*b*) Noninertia Models

Model	Numerical scheme	Parameters	S_f	Sewer downstream condition	Solution scheme	Detention storage	Junction equation	Surcharge flow	References
	Open-channel flow					Interior junction			
HVM		h, Q	Colebrook-White	Unspecified or rating curve	Pipe by pipe	No	$\Sigma Q = 0$ and $h_i = h_o$	Standard pressurized pipe flow	Geiger and Dorsch (1980); Klym et al. (1972); Vogel and Klym (1973)
DAGVL-DIFF	Implicit, six-point continuity four-point momentum $w = 0.55$	h, Q	Manning	Junction water surface or critical depth	Simultaneous (double sweep)	Yes	$\Sigma Q = ds/dt$ and $h_o = H - (KV^2/2g)$ or $\Sigma Q = 0$ and $h_i = h_o$	Preissmann slot	Sjoberg (1982)
MOUSE	Six-point implicit $w = 0.5$	h, Q	Manning	Junction water surface	Simultaneous (double sweep)	Yes	$\Sigma Q = 0$ and $h_i = h_o$	Preissmann slot	DHI (1994)
NISN	Four-point implicit	h, Q	Manning	Junction water surface or critical depth	Simultaneous, over-lapping segment	Yes	$\Sigma Q = 0$ and $h_i = h_o$ or $\Sigma Q = ds/dt$ and $h_o = H - (KV^2/2g)$	Preissman slot	Pagliara and Yen (1997)

TABLE 6.2 (*c*) Nonlinear Kinematic Wave Models

Model	Open-channel flow							Surcharge flow				References
	Sewer hydraulics	Parameters	S_f	Numerical scheme	Solution scheme	Interior junction	Transition condition	Surcharge hydraulics	Interior junction	Solution scheme		
USGS	Nonlinear kinematic wave	A, Q	Manning	Explicit	Cascade	$\Sigma Q = ds/dt$	$Q > Q_f$	$Q = Q_f$	Store excess water, release later	Pipe by pipe		Dawdy et al. (1978)
ILSD-B2	Nonlinear kinematic wave	h, Q	Manning	Four-point implicit	Cascade	$\Sigma Q = 0$	NA	NA	NA	NA		Yen and Sevuk (1975); Yen et al. (1976)
ILSD-B3	Muskingum-Cunge	h, Q	Manning	Four-point implicit	Cascade	$\Sigma Q = 0$	NA	NA	NA	NA		Yen et al. (1976)
WASSP-SIM	Muskingum-Cunge	h, Q	Darcy-Weisbach and Colebrook-White	Quasi explicit	One sweep, pipe by pipe	$\Sigma Q = ds/dt$	$Q > Q_f$ or assumed submergence conditions	Unsteady dynamic equation	H calculated, headlosses considered	Implicit simultaneous relaxation		Price (1982a,b)
SWMM-TRANS-PORT	Improved nonlienar kinematic wave	A, Q	Manning	Four-point implicit, $w = 0.55$	One sweep, pipe by pipe	$\Sigma Q = 0$ unless Storage Block is used	$Q > Q_f$	$Q = Q_f$	Store excess water, release later	Pipe by pipe		Huber and Heaney (1982); Huber and Dickinson (1988); Metcalf & Eddy Inc. et al (1971)

TABLE 6.3 Select Sewer System Design Models
(*a*) Hydraulic/Hydrologic Design Models

Model	Design	Catchment hydrology	Sewer hydraulics	Input	Reference
Rational method	Sewer diam and elevations	Rational formula	Manning's formula	Sewer layout, ground elevations, min soil cover, acceptable max and min velocities, catchment area and runoff coefficient, design rainfall	Various
ISSD	Sewer diam and elevations	Design rainfall, Horton's infiltration, time-area method hydrographs	Manning's formula and hydrograph time-lag	Sewer layout, ground elevations, min soil cover, acceptable max and min velocities, catcment area and soil type, percentage impervious and pervious areas, design rainfall	Yen et al. (1984)

TABLE 6.3 (*b*) Least-Cost Sewer System Design Models

Model	Design	Optimization scheme	Catchment hydrology	Sewer hydraulics	Considering risk	Input	Reference
ILSD-1	Sewer diam, crown elevations, manhole depths, detention volume	DDDP*	Design rainfall, Horton's infiltration, time-area method hydrographs	Manning's formula and hydrograph time-lag	No	Sewer layout, ground elevations, min soil cover, acceptable max and min velocities, cost functions, time and space increments for routing computations, catchment area, percentage of pervious and impervious surfaces, design rainfall, optimization parameters	Yen et al. (1984)
ILSD-2	Sewer diam, crown elevations, manhole depths, detention volume	DDDP	Design rainfall, Horton's infiltration, time-area method hydrographs	Manning's formula and hydrograph time-lag	Yes	Same as above, in addition, design service life, risk-safety factor relationship	Yen et al. (1984)

*Discrete differential dynamic programming.

6.2 FLOW IN A SEWER

6.2.1 Types of Sewer Flow

The flow in a sewer is usually turbulent and changes with respect to time and location in the sewer pipe—hydraulically referred to as *unsteady* and *nonuniform flow*. Subcritical flows occur more often than supercritical flows. For flows varying slowly with time such that the flow travel time through the entire length of the sewer is much smaller than the rising time of the flow hydrograph, the flow can often be treated approximately as stepwise steady without significant error.

The flow in a sewer can be divided into three regions: (1) the entrance, (2) the pipe flow, and (3) the exit. Figure 6.1 shows a classification of 10 different cases of nonuniform pipe flow, based on whether the flow at a given instant is subcritical, supercritical, or surcharge. The four cases of entrance condition are shown in Fig. 6.2 and Table 6.4. Case I is associated with downstream control of the pipe flow. Case II is associated with upstream control. In case III, the pipe flow under the air pocket may be subcritical, supercritical, or transitional. In case IV, the sewer flow is often controlled by both the upstream and downstream conditions.

The exit condition also can be grouped into four cases as shown in Fig. 6.3 and Table 6.5. In case A, the pipe flow is under exit control. In case B, the flow is under upstream

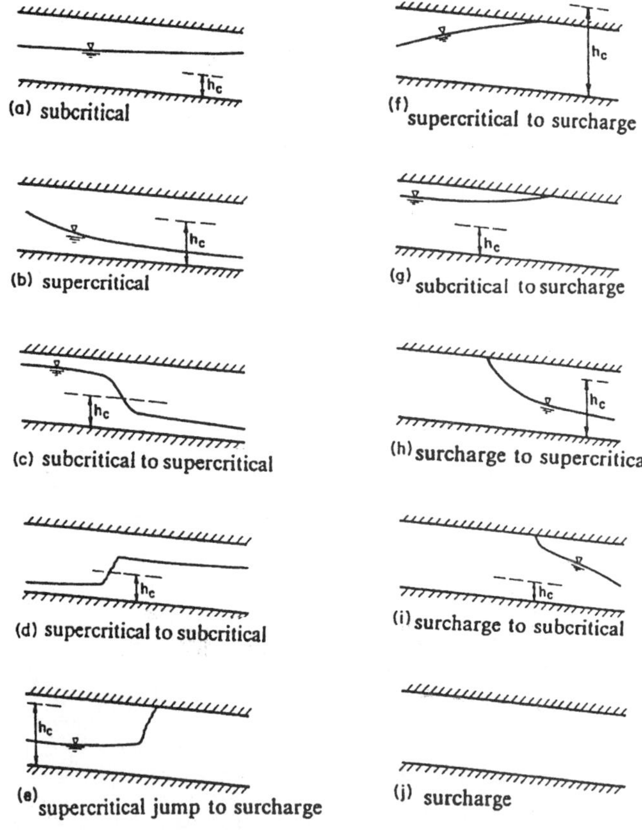

FIGURE 6.1   Classification of flow in a sewer pipe. (After Yen, 1986)

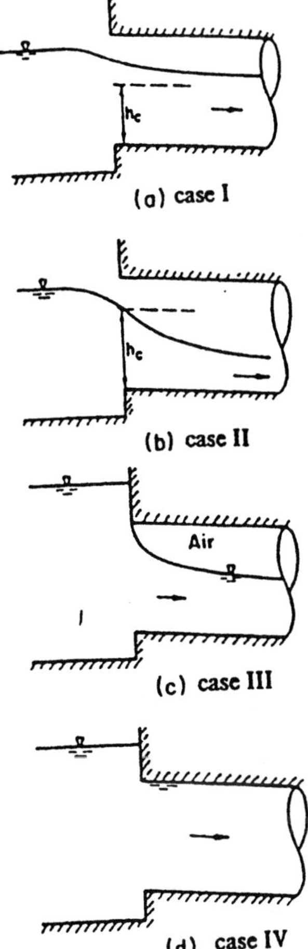

FIGURE 6.2 Types of sewer entrance flow. (After Yen, 1986)

TABLE 6.4 Pipe Entrance Conditions

Case	Hydraulic condition
I	Nonsubmerged entrance, subcritical flow
II	Nonsubmerged entrance, supercritical flow
III	Submerged entrance, air pocket
IV	Submerged entrance, water pocket

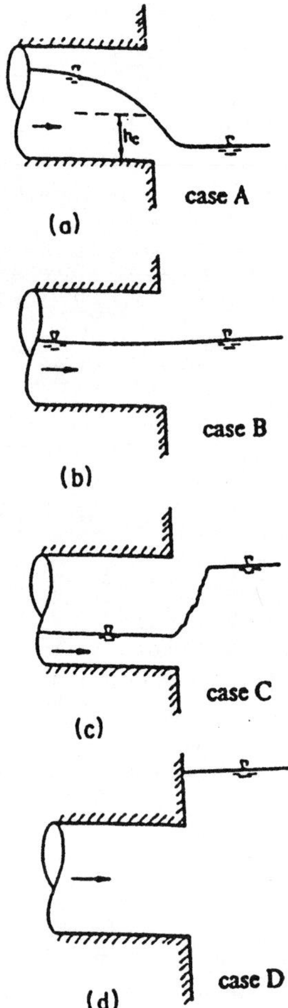

FIGURE 6.3 Types of sewer exit flow. (After Yen, 1986)

TABLE 6.5 Pipe Exit Conditions

Case	Hydraulic condition
A	Nonsubmerged, free fall
B	Nonsubmerged, continuous
C	Nonsubmerged, hydraulic jump
D	Submerged

control if it is supercritical, and downstream control, if subcritical. In case C, the pipe flow is under upstream control while the junction water surface is under downstream control. In case D, the pipe flow is often under downstream control, but can also be under both upstream and downstream control.

The possible combinations of the 10 cases of pipe flow with the entrance and exit conditions are shown in Table 6.6 for nonuniform sewer flow. Some of these 27 possible combinations are rare for unsteady flow and nonexistent for steady flow, for example, cases 3 and 6. The types of water surface profiles in the sewer corresponding to the different steady flow backwater curves of the mild-slope M and steep-slope S types (Fig. 6.4) are also listed in Table 6.6, of which 6 types were reported by Bodhaine (1968). Detailed discussion on the steady backwater curves and their computation methods can be found in standard open-channel hydraulics books, such as that of Chow (1959).

The nonuniform pipe flows shown in Fig. 6.1 are classified without considering the different modes of air entrainment. The types of the water surface profile, equivalent to the adverse or horizontal slope types of backwater curves for steady flow shown in Fig. 6.4, are also not taken into account. Adverse sewer slope exists due to flow reversal. Additional subcases of the 10 pipe flow cases can also be classified according to rising, falling, or stationary water surface profiles. For the cases with a hydraulic jump or drop, subcases can be grouped according to the movement of the jump or drop, be it moving upstream or downstream or stationary.

Partly due to the large number of possible cases of flow, in the past the flow in a sewer has not been analyzed hydraulically as it should have been. With the recent, improved understanding of sewer hydraulics and advance in computational capability a more realistic and accurate analysis of sewer flow is now possible and feasible.

6.2.2 Time Variation of Storm Runoff in a Sewer

During a runoff event, the change in magnitude of the flow in a sewer can range from only a few times of dry weather low flow in a sanitary sewer to as much as manyfold for a heavy rainstorm runoff in a storm sewer. The time variation of storm sewer flow is usually much

TABLE 6.6 Pipe Flow Conditions

Case	Pipe flows	Possible entrance conditions	Possible exit conditions	Steady backwater surface profiles
1	Subcritical	I, III	A	M2
			B	M1, S1
2	Supercritical	II, III	B, C	M3, S3, S2
3	Subcritical → hydraulic drop → supercritical	I, III	B, C	
4	Supercritical → hydraulic jump → subcritical	II, III	A	M3 → M2
			B	M3 → M1, or S2 or S3 → S1
5	Supercritical → hydraulic jump → surcharge	II, III	D	M3, S3, S2 → J
6	Supercritical → surcharge	II, III	D	
7	Subcritical → surcharge	I, III	D	M1, S1
8	Surcharge → supercritical	IV	B, C	S2
9	Surcharge → subcritical	IV	A, B	M2
10	Surcharge	IV	D	

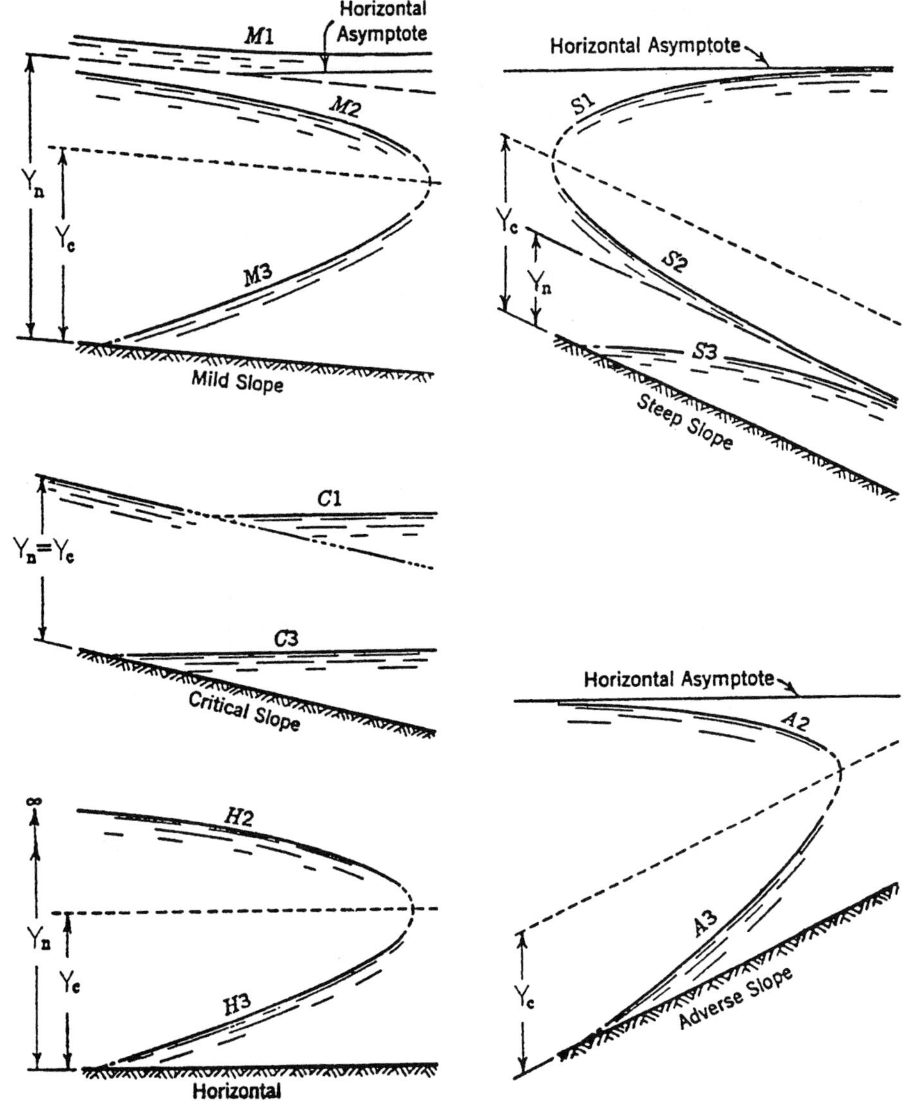

FIGURE 6.4   Types of backwater surface profiles.

more rapid than that of sanitary sewers. Therefore, the approximation of assuming steady flow is more acceptable for sanitary sewers than for storm and combined sewers.

In the case of a heavy storm runoff entering an initially dry sewer, as the flow enters the sewer, both the depth and discharge start to increase as illustrated in Fig. 6.5 at times t_1, t_2, and t_3 for the open-channel phase. If the flow continues to rise, the sewer pipe becomes completely filled and surcharges, as shown at t_4 and t_5 in Fig. 6.5. *Surcharge flow* occurs when the sewer is under-designed, when the flood exceeds that of the design return period, when the sewer is not properly maintained, or when storage or pumping occur.

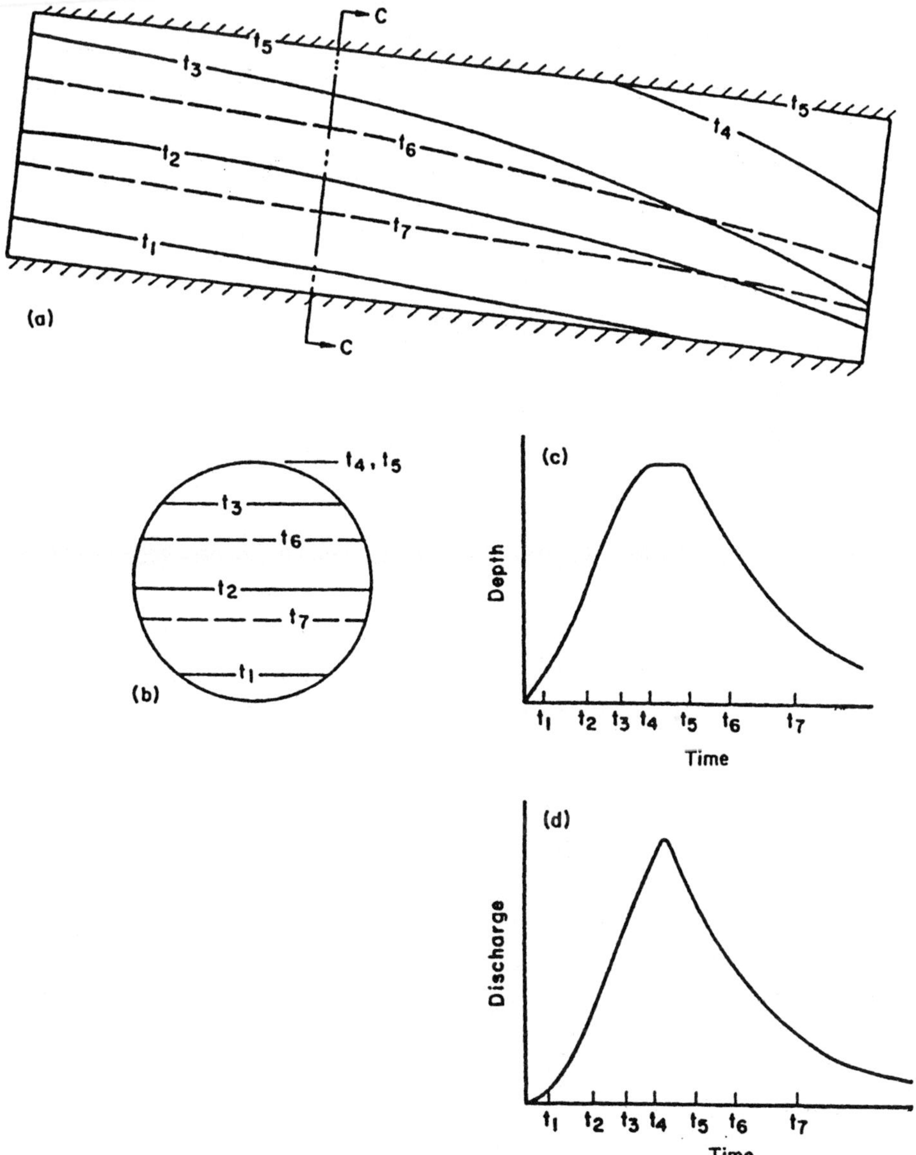

FIGURE 6.5 Time variation of flow in a sewer: (a) water surface profile along a sewer pipe; (b) time variation of water surface at section C–C; (c) depth hydrograph; (d) discharge hydrograph. (After Yen, 1986)

Under surcharge conditions, the flow cross-sectional area and depth can no longer increase because of the sewer pipe boundary. However, as the flood inflow continues to increase, the discharge in the sewer will also increase, due to the increasing difference in head between the upstream and downstream ends of the sewer, as sketched in the discharge hydrograph in Fig. 6.5. Even under surcharge conditions, while the sewer diameter remains constant, the flow is usually nonuniform. This is due to the effects of the entrance and exit on the flow inside the sewer, and, hence, the streamlines will not be parallel.

As the flood starts to recede, the aforementioned flow process is reversed. The sewer will return from surcharged pipe flow to open-channel flow, shown at t_6 and t_7 in the figure. Since the recession is usually, but not always, more gradual than the rising of the flood, the water surface profile in the sewer is usually more gradual during flow recession than during rising.

The differences in the gradient of the water surface profiles during the rising and recession of the flood bear importance in the self-cleaning and pollutant-transport abilities of the sewer. During the rising period, with relatively steep gradient, the flow can carry not only the sediment it brings into the sewer but erodes the deposit at the sewer bottom from previous storms. For a given discharge and gradient, the amount of erosion increases with the antecedent duration of wetting and softening of the deposit. During the recession, with a flatter water surface gradient and deceleration of the flow, the sediment being carried into the sewer by the flow tends to settle on the sewer bottom.

If the storm is not heavy and the flood is not severe, the rising flow will not reach a surcharge state. The flood may rise, for example, to the stage at t_3 shown in Fig. 6.5 and then start to recede. The sewer will remain under open-channel flow throughout the storm runoff. For such frequent, small storms, the flow in the sewer will be so small that it will be unable to transport out the sediment it carries into the sewer, resulting in deposition to be cleaned up by later heavy storms or through artificial means.

For a single-peak flood entering a long circular sewer having a diameter D and Nikuradse pipe surface roughness k, Yen (1973a, pp. 38–39) reported that for open-channel flow, the attenuation of the flood peak, Q_{px}, at a distance x downstream from the pipe entrance ($x = 0$) and the corresponding occurrence time of this peak, t_{px}, can be described dimensionlessly as

$$\frac{Q_{px}}{Q_{p0}} = \exp\left\{-0.0771\left(\frac{x}{D}\right)\left(\frac{k}{D}\right)^{0.17}\left(\frac{R_b}{D}\right)^{-0.42}\left(\frac{Q_{p0}}{D^{2.5}\sqrt{g}}\right)^{-0.16}\right.$$
$$\left.\times\left[t_{p0}^{1.64}(t_g - t_{p0})\frac{g^{1.32}}{D^{1.32}}\right]^{-4}\right\} \tag{6.1}$$

$$(t_{px} - t_{p0})\frac{V_w}{D} = \left\{6.03\log_{10}\left[\left(\frac{Q_{px}}{Q_{p0}}\right) - 0.18\right] - 0.520\right\}\left(\frac{k}{D}\right)^{-0.11}\left(\frac{R_b}{D}\right)^{0.66}$$
$$\times\left[\frac{Q_{p0}^{4.4}}{Q_b^4 g^{0.2}D}\right]^{0.1}\left[t_{p0}^{0.68}(t_g - t_{p0})\frac{g^{0.82}}{D^{0.82}}\right]^{0.5} \tag{6.2}$$

where Q_{p0} and t_{p0} = peak discharge and its time of occurrence at $x = 0$

Q_b = steady base flow rate

R_b = hydraulic radius of the base flow

t_g = time to the centroid of the inflow hydrograph at $x = 0$ above the base flow

g = gravitational acceleration

$V_w = (Q_b/A_b) + (gA_b/B_b)^{1/2}$ is the wave celerity of the base flow, where A_b is the base flow cross-sectional area and B_b the corresponding water-surface width.

In both equations, the second nondimensional parameter in the right side k/D is a pipe property parameter; the third parameter R_b/D is a base flow parameter; the fourth nondimensional parameter represents the influence of the flood discharge; whereas the fifth, the last nondimensional parameter, reflects the shape of the inflow hydrograph.

The single-peak hydrograph shown in Fig. 6.5 is an ideal case for the purposes of illustration. In reality, because of the phase shift of the peak flows in upstream sewers and the time-varying nature of rainfall and inflow, the real hydrographs are usually multipeak.

Because the flow is nonuniform and unsteady, the depth-discharge relationship, also known as the rating curve in hydrology, is nonunique. Even if one is willing to consider the flow to be steady and uniform as an approximation, the depth-discharge relation is nonlinear, and within a certain range, nonunique, as shown nondimensionally and ideally in Fig. 6.6 for a circular pipe. The nonunique depth-discharge relationship for nonuniform flow, aided by the poor quality of the water and restricted access to the sewer, makes it difficult to measure reliably the time-varying flow in sewers. Among the many simple and sophisticated

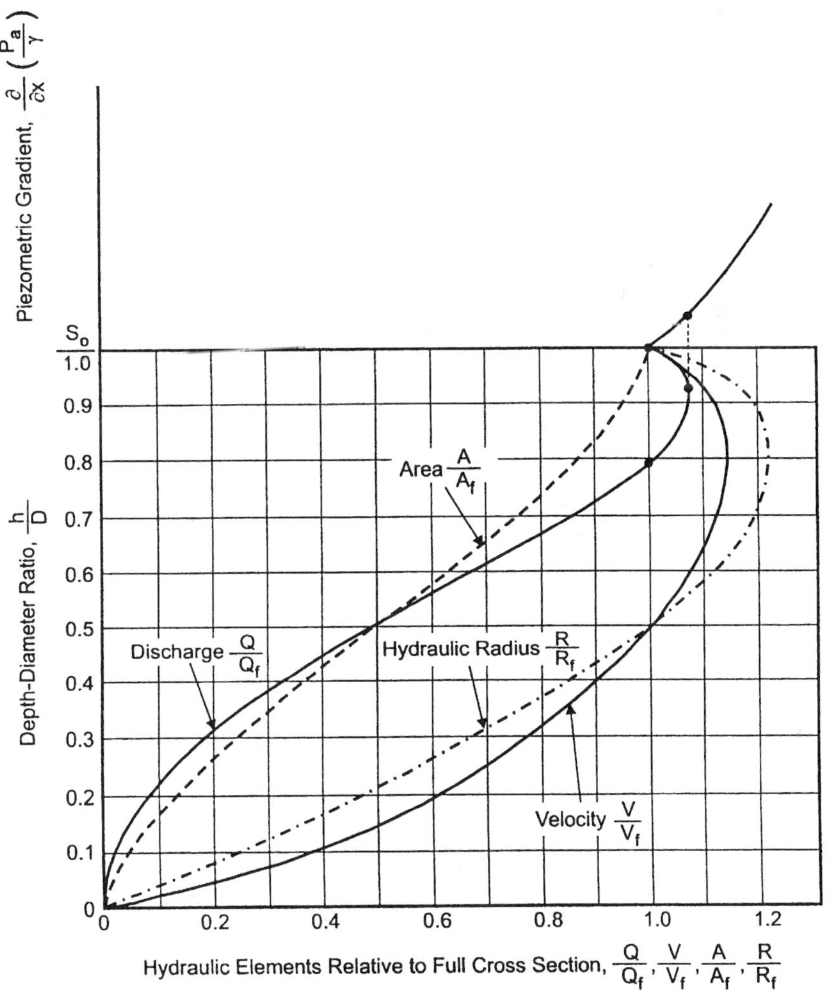

FIGURE 6.6 Rating curves for steady uniform flow in circular pipe.

mechanical or electronic measurement devices that have been attempted on sewers and reported in the literature, the simple, mechanical Venturi-type meter, which has side constriction instead of bottom constriction to minimize the effect of sediment clogging, still appears to be the most practical measurement means, that is, if it is properly designed, constructed, and calibrated, and if it is located at a sufficient distance from the entrance and exit of the sewer. On the other hand, the hydraulic performance graph described in sec. 6.1 can be used to establish the rating curve for a steady nonuniform flow.

6.3 SEWER HYDRAULIC EQUATIONS

Flows in sewers usually are open-channel flows with a free water surface. However, sewer pipes, culverts, and similar conduits under high flow could become surcharged and pressurized conduit flows do occur. Strictly speaking, the flow is always unsteady, that is, changing with time. Nevertheless, in a number of situations, such as in most cases of flow in sanitary sewers and for some rainstorm runoffs, change of flow with time is slow enough that the flow can be regarded as time-stepwise approximately steady.

6.3.1 Unsteady Flow Equations

Unsteady sewer flow with a free surface, as in Fig. 6.7, can be described by a momentum equation which is given below in both velocity (nonconservative) and discharge (conservative) forms:

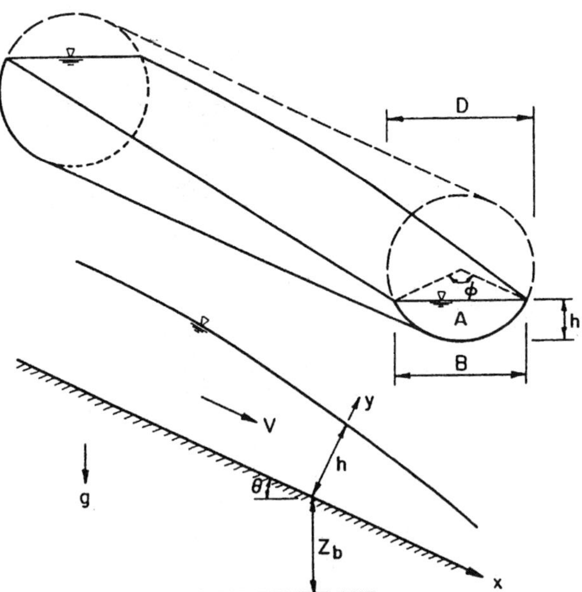

FIGURE 6.7 Sewer flow with a free surface.

$$\frac{1}{g}\frac{\partial V}{\partial t} + (2\beta - 1)\frac{V}{g}\frac{\partial V}{\partial x} + (\beta - 1)\frac{V^2}{gA}\frac{\partial A}{\partial x} + \cos\theta\frac{\partial h}{\partial x} - S_0 + S_f = 0 \qquad (6.3)$$

$$\qquad\;(1)\qquad\qquad\quad(2)\qquad\qquad\quad(2)\qquad\quad(3)\qquad\qquad(4)\quad(5)$$

$$\frac{1}{gA}\frac{\partial Q}{\partial t} + \frac{1}{gA}\frac{\partial}{\partial x}\left(\frac{\beta Q^2}{A}\right) + \cos\theta\frac{\partial h}{\partial x} - S_0 + S_f = 0 \qquad (6.4)$$

$$\qquad(1)\qquad\qquad(2)\qquad\qquad\quad(3)\qquad\qquad(4)\quad(5)$$

where x = longitudinal direction of sewer
    A = flow cross-sectional area normal to x
    y = coordinate direction normal to x on a vertical plane
    h = depth of flow of the cross section, measure along y direction
    Q = discharge through A
    V = Q/A, cross-sectional average velocity along x direction
    S_o = channel slope, equal to $\sin\theta$
    θ = angle between sewer bottom and horizontal plane
    S_f = friction slope
    g = gravitational acceleration
    t = time
    β = Boussinesq momentum flux correction coefficient for velocity distribution

$$\beta = \frac{1}{V^2 A}\int_A \bar{u}^2\, dA = \frac{A}{Q^2}\int_A \bar{u}^2\, dA \qquad (6.5)$$

where $\bar{u}$ = x-component of local (point) velocity averaged over turbulence

The continuity equation is

$$\frac{\partial A}{\partial t} + \frac{\partial Q}{\partial x} = 0 \qquad (6.6)$$

Since the sewer channel is prismatic, Eq. (6.6) can be written as

$$\frac{\partial h}{\partial t} + \frac{A}{B}\frac{\partial V}{\partial x} + V\frac{\partial h}{\partial x} = 0 \qquad (6.7)$$

where B = the water surface width of the cross-section

For circular pipes the parameters B, h, and A of the flow cross-section can be computed from the geometric equations in terms of the free surface central angle ϕ as shown in Fig. 6.8.

The physical meaning of the terms in the momentum Eqs. (6.3) and (6.4) follows. The first term (1) is the inertial term due to *local acceleration.* Term (2) is the inertial term from *convective acceleration.* Term (3) is the pressure term due to water surface gradient. Term (4) represents the body force. Term (5) denotes the resistance force. For steady uniform flow, only the last two terms are retained, that is, $S_o = S_f$, and the other three terms will disappear.

The pair of momentum and continuity equations (Eqs. (6.3) and (6.7) or Eqs. (6.4) and (6.6)) with $\beta = 1$ is often referred to as the *Saint-Venant equations* or *full dynamic wave*

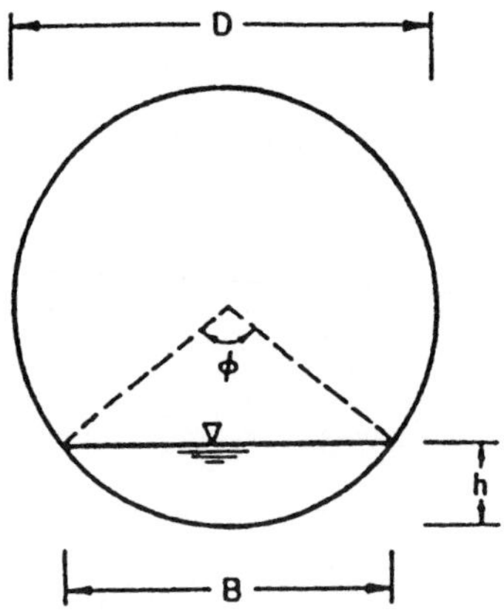

FIGURE 6.8   Part-full pipe flow geometry. (From Yen, 1986)

equations. Actually they are not an exact representation of the unsteady flow because they involve at least the following assumptions:

1. Hydrostatic pressure distribution over A
2. Uniform velocity distribution over A (hence $\beta = 1$)
3. Negligible spatial gradient of the force due to internal stresses

Those interested in the more exact form of the unsteady flow equations should refer to Yen (1973b, 1975, 1996).

6.3.2 Flow Resistance

The friction slope S_f is usually estimated by using a semiempirical formula such as *Manning's formula:*

$$S_f = \frac{n^2 V|V|}{K_n^2} R^{-4/3} = \frac{n^2 Q|Q|}{K_n^2 A^2} R^{-4/3} \tag{6.8}$$

or with the *Darcy-Weisbach formula*

$$S_f = \frac{f}{8gR} V|V| = \frac{f}{8gR} \frac{Q|Q|}{A^2} \tag{6.9}$$

where n = Manning's roughness factor
K_n = 1.486 for English units and 1.0 for SI units
f = Weisbach resistance coefficient
R = A/wetted perimeter is the hydraulic radius

The absolute sign is used to account for the occurrence of flow reversal.

Theoretically, the values of n and f for unsteady nonuniform open-channel and pressurized conduit flows have not been established. They depend on the pipe surface roughness and bed form if sediment is transported, Reynolds number, Froude number, and unsteadiness and nonuniformity of the flow (Rouse, 1965; Yen, 1991). One should be careful since, for unsteady nonuniform flow, the friction slope is different from the pipe slope, the dissipated energy gradient, the total-head gradient, or the hydraulic gradient. Only for steady uniform flow without lateral flow are these different gradients equal to one another.

At present, the steady uniform flow values of n and f given in the literature are usually used as approximations. The value of f for steady uniform flow can be found from the *Moody diagram* in Fig. 6.9 or the *Colebrook-White formula* (Colebrook, 1939):

$$\frac{1}{\sqrt{f}} = -2.0 \log \left[\frac{k_s}{14.83R} + \frac{2.52}{4Re\sqrt{f}} \right] \tag{6.10}$$

which is applicable for the Reynolds number $Re = VR/\nu > 25{,}000$, where ν is kinematic viscosity of the fluid, and k_s is the Nikuradse pipe roughness. There are three significant problems in using Eq. (6.10) for sewer flows. First, Eq. (6.10) is for full pipes, while sewer pipes are usually only partially full. Second, Eq. (6.10) is implicit in terms of f, requiring

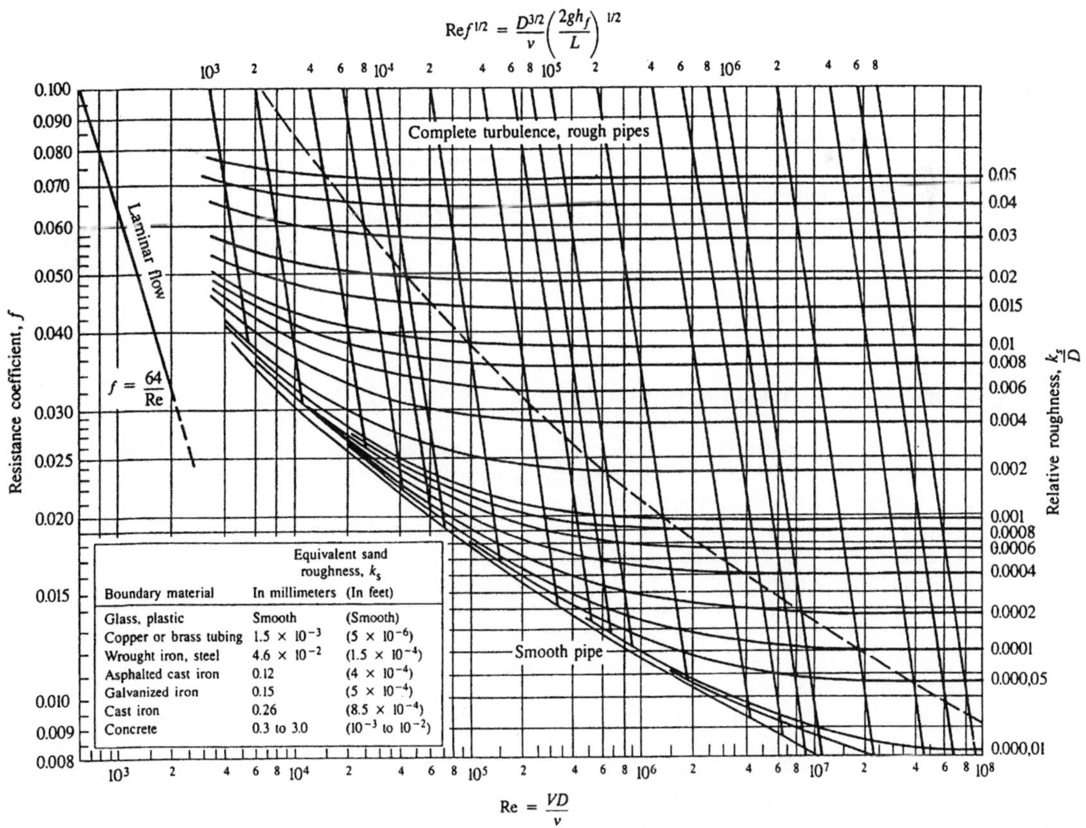

FIGURE 6.9 Moody diagram for steady flow in full pipe (permission from ASME).

iteration to determine its value. Third, for a given pipe and surface roughness, the value of f varies not merely with the Reynolds number but with the flow depth and geometry of the flow cross-section (Rouse, 1965; Yen, 1991). In other words, as the flow depth in the sewer changes during a storm runoff, f must be recomputed repeatedly. To alleviate the first two problems, Yen (1991) has suggested the following equation for wide open-channel flow with $Re > 30,000$ and $k_s/Re < 0.05$:

$$f = \frac{1}{4}\left[-\log\left(\frac{k_s}{12R} + \frac{1.95}{Re^{0.9}}\right)\right]^{-2} \tag{6.11}$$

The value of Manning's coefficient n is usually obtained from tables, among which the most comprehensive is the one of Chow (1959). The part corresponding to pipes is reproduced in Table 6.7.

Manning's n was originally derived empirically. Its major disadvantage is its troublesome dimension of length to $\frac{1}{6}$ power that is often misunderstood. Its main advantage is that for flows with sufficiently high Reynolds number over a rigid boundary with a given surface roughness in a prismatic channel, the value of n is nearly constant over a wide range of depth (Yen, 1991). Values of n can be found in Chow (1959) or chapter 3.

Other resistance coefficients and formulas, such as Chezy's or Hazen-Williams's, have also been used. They possess neither the direct fluid mechanics justification as f nor independence of depth as n for rigid pipes. Therefore, they are not recommended here. In fact, the *Hazen-Williams formula,*

$$V = 1.318C_w R^{0.63} S^{0.54} \tag{6.12}$$

which is popular with water supply distribution full-flow pipe analysis, may be considered as a special situation of the Darcy-Weisbach formula (Christensen, 2000). Discussion on the preference of the resistance coefficients can be found in Yen (1991).

Equations (6.8) and (6.9) are applicable to both surcharged and open-channel flows. For the open-channel case, the pipe is flowing partially filled and the geometric parameters of the flow cross-section are computed from the geometry equations given in Fig. 6.8.

6.3.3 Approximations to Unsteady Flow Equations

The dynamic wave equation, Eq. (6.3) or (6.4), is often referred to as "complete" because it contains all the terms describing the dynamic effects of an unsteady open-channel flow. However, it is not exact because of the assumptions stated in the last paragraph of Sec. 6.3.1, resulting in setting the value of the coefficients of the terms to unity and neglecting the effects of spatial change of internal stresses (Yen, 1973b, 1975). Even in the simplified "complete" form, the dynamic wave equation, together with the continuity equation and appropriately specified initial and boundary conditions, is rather tedious and computationally costly to solve, especially for large networks. Therefore, both improved solution methods and acceptable simplifications of the equation itself have been sought.

Different levels of approximation of the dynamic wave equation can be obtained by dropping terms in the equation as shown in Table 6.8. Referring to Eqs. (6.3) and (6.4), if the local acceleration term $\partial Q/\partial t$ is dropped, the approximation is a *quasi-steady dynamic wave.* The *noninertia* approximation is formed by dropping both the local and convective acceleration terms. If the pressure term $(\partial h/\partial x)$ is dropped in addition to dropping both inertia terms, the approximation is known as *kinematic wave.* Rewriting Eq. (6.3) in a different form the approximations can be expressed as follows:

TABLE 6.7 Values of Manning's Coefficient n for Conduits Flowing Partly Full

Type of channel	Minimum	Normal	Maximum
A. Metal			
a. Brass, smooth	0.009	0.010	0.013
b. Steel			
1. Lockbar and welded	0.010	0.012	0.014
2. Riveted and spiral	0.013	0.016	0.017
c. Cast iron			
1. Coated	0.010	0.013	0.014
2. Uncoated	0.011	0.014	0.016
d. Wrought iron			
1. Black	0.012	0.014	0.015
2. Galvanized	0.013	0.016	0.017
e. Corrugated metal			
1. Subdrain	0.017	0.019	0.021
2. Storm drain	0.021	0.024	0.030
B. Nonmetal			
a. Lucite	0.008	0.009	0.010
b. Glass	0.009	0.010	0.013
c. Cement			
1. Neat, surface	0.010	0.011	0.013
2. Mortar	0.011	0.013	0.015
d. Concrete			
1. Culvert, straight and free of debris	0.010	0.011	0.013
2. Culvert with bends, connections, and some debris	0.011	0.013	0.014
3. Finished	0.011	0.012	0.014
4. Sewer with manholes, inlet, etc., straight	0.013	0.015	0.017
5. Unfinished, steel form	0.012	0.013	0.014
6. Unifnshed, smooth wood form	0.012	0.014	0.016
7. Unfinished, rough wood form	0.015	0.017	0.020
e. Wood			
1. Stave	0.010	0.012	0.014
2. Laminated, treated	0.015	0.017	0.020
f. Clay			
1. Common drainage tile	0.011	0.013	0.017
2. Vitrified sewer	0.011	0.014	0.017
3. Vitrified sewer with manholes, inlet, etc.	0.013	0.015	0.017
4. Vitrified subdrain with open joint	0.014	0.016	0.018
g. Brickwork			
1. Glazed	0.011	0.013	0.015
2. Lined with cement mortar	0.012	0.015	0.017
h. Sanitary sewers coated with sewage slimes, with bends and conncections	0.012	0.013	0.016
i. Paved invert, sewers, smooth bottom	0.016	0.019	0.020
j. Rubble masonry, cemented	0.018	0.025	0.030

Source: From Chow (1959).

TABLE 6.8 Approximations to Dynamic Wave Equation

Approximation	Kinematic wave	Noninertia	Quasi-steady dynamic wave	Dynamic wave
Terms retained in momentum equation (6.3 or 6.4)	$4 + 5$	$3 + 4 + 5$	$2 + 3 + 4 + 5$	All terms
Momentum equation*	$S_o = S_f$	$h \cos \theta \dfrac{\partial h}{\partial x} = S_o - S_f$	$\dfrac{1}{gA} \dfrac{\partial}{\partial x}\left(\dfrac{\beta}{A} Q^2\right) + h \cos \theta \dfrac{\partial h}{\partial x} = S_o - S_f$	$\dfrac{1}{gA} \dfrac{\partial Q}{\partial x} + \dfrac{1}{gA} \dfrac{\partial}{\partial x}\left(\dfrac{\beta}{A} Q^2\right) + h \cos \theta \dfrac{\partial h}{\partial x} = S_o - S_f$
Number of wave characteristics	$1\ (c^+)$	$1\ (c^+)$	$1(c^+)$	$2\ (c^+ \text{ and } c^-)$
Boundary conditions required	1	2	2	2
Account for downstream backwater effect and flow reversal	No	Yes	Yes	Yes
Damping of flood peak	No	Yes	Yes	Yes
Account for flow acceleration	No	No	Only convective acceleration	Yes

*For all approximations, continuity equation is $\dfrac{\partial A}{\partial t} + \dfrac{\partial Q}{\partial x} = 0$

$$Q = CA \sqrt{S_o - \cos\theta\,\frac{\partial h}{\partial x} - (2\beta - 1)\frac{V}{g}\frac{\partial V}{\partial x} - \frac{1}{g}\frac{\partial V}{\partial t} + \frac{\beta - 1}{g}\frac{VB}{A}\frac{\partial h}{\partial t}} \qquad (6.13)$$

$\longleftarrow$ kinematic wave $|$

$\longleftarrow$ noninertia $|$

$\longleftarrow$ quasi-steady dynamic wave $|$

$\longleftarrow$ dynamic wave $\longrightarrow |$

in which the coefficient C depends on the formula used to represent the friction slope S_f

$$C = (K_n/n)R^{2/3} \qquad (6.14a)$$

for Manning's formula, and

$$C = (8gR/f)^{1/2} \qquad (6.14b)$$

for the Darcy-Weisbach formula.

6.3.3.1 Quasi-Steady Dynamic Wave Approximation.   In Eq. (6.3) or (6.4), if only the local acceleration term, $\partial V/\partial t$ or $\partial Q/\partial t$, is dropped, the approximation is a quasi-steady dynamic wave. The solution to this approximation requires two boundary conditions as for the full dynamic wave.

If the flow is steady, $\partial A/\partial t = 0$ in Eq. (6.6) and hence $Q = AV = $ constant, the quasi-steady dynamic wave equation can be transformed into the well-known steady backwater curve equation appearing in standard open-channel references (e.g., Chow, 1959; Yen, 1996),

$$\frac{dh}{dx} = \frac{S_o - S_f}{\cos\theta - (V^2 B/gA)} \qquad (6.15)$$

which can be used to compute the steady flow backwater in sewers as suggested in the manual of the American Society of Civil Engineers (ASCE, 1969).

6.3.3.2 Noninertia Approximation.   The *noninertia* (misnomer diffusion wave, Yen and Tsai, 2001) approximation is perhaps the most useful among the approximations of the dynamic wave equation, because it offers a balance between accuracy and simplicity to a large number of field situations. It ignores both inertia terms (local and convective accerlerations) in Eq. (6.3) or (6.4) but retains the pressure, body force, and resistance terms.

Inclusion of the pressure term requires two boundary conditions. Hence, it accounts for downstream backwater effects and permits peak attenuation, distortion, and translation of the flood hydrograph, just as for the full and quasi-dynamic wave models. Because the inertia terms are ignored, it requires the flow to vary gradually with both space and time. It does not have computational problems when the flow passes through $F = 1$, from supercritical to subcritical or vice versa, a difficulty that is often encountered in the full or quasi-dynamic wave models. Readers may refer to Akan and Yen (1977), Strelkoff and Katopodes (1977), and Hromadka and Yen (1987) for versions of the application of the noninertia approximation to open-channel flows.

6.3.3.3 Kinematic Wave Approximation.   Referring to Eq. (6.3) or (6.4), the *kinematic wave model* is the simplest approximation of the Saint-Venant momentum equation. It retains only the last two slope terms and ignores other terms representing the effect of inertia and pressure. In other words, for any Δt, within the computational space interval Δx, the water surface is assumed to be parallel to the channel bed, that is, the flow is uniform. This, of course, is only an approximation of the reality.

The set of nonlinear kinematic wave continuity and momentum equations requires only one boundary condition in addition to the initial condition to be properly posed for unique solution. The lone boundary condition is specified at the upstream end of the channel, usually the inflow hydrograph, and hence it permits the solution to proceed from upstream toward downstream irrespective of the downstream conditions. On the other hand, it is unreliable for subcritical flow when the downstream backwater effect is important, as in the case of short channels in a network.

Neglecting the inertial and pressure terms in the momentum equation also eliminates the mechanism for flood wave peak attenuation. Theoretically, the continuity equations permits translation and some distortion of the flow hydrograph as the flood propagates downstream (Lighthill and Whitham, 1955). However, in seeking a solution through a finite difference numerical procedure, inevitably numerical damping will be introduced. This numerical attenuation usually acts advantageously in the same direction as the actual flood peak attenuation, and it is often misunderstood by those who, mistakenly thinking that small computational steps will give more accurate results, reduce the sizes of computational steps in hopes of getting more attenuation. If fact, they are incurring more computational expense to obtain a solution that is closer to what the kinematic wave equations represent, namely, no attenuation, and usually farther away from the actual physical condition. In other words, the numerical solution, if converged, gives a value that depends on the grid size Δx and Δt used. Attempts have been made to match the numerical attenuation with the real hydraulic attenuation through properly selecting the sizes of Δx and Δt by relating them to the flood and channel properties, for example, see Koussis (1980), Ponce and Theurer (1982). However, such matching, even carefully executed, is only approximate due to (among other factors) the fact that the flow changes with time and space, whereas, in a computation, usually a fixed set of Δx and Δt is used.

Despite its deficiency of not accounting for the downstream backwater effect, the kinematic wave approximation, because of its relative simplicity, is the most extensively studied among the dynamic wave equation and its various approximations. Readers should refer to Bettess and Price (1976), Weinmann and Laurenson (1979), Koussis (1980), Smith (1980), and others for information on the various kinematic wave methods. Among them the most noticeable is the modification suggested by Cunge (1969), who showed that the hydrologic routing Muskingum method can be regarded as a special case of the kinematic wave approximation. The Muskingum coefficients can be computed from the channel width, reach length, discharge, kinematic wave celerity, and slope of the flow. Solution procedures of the Muskingum-Cunge method can be found in Fread (1993) and Ponce (1989).

6.3.3.4 *General Comments on the Approximations.*

Table 6.8 summarizes comparatively some differences and similarities between the dynamic wave model and its three hydraulic approximations. The dynamic wave or Saint-Venant equations model represents fully the dynamic behaviors of the unsteady flow. Hydrodynamically it possesses two characteristic waves; for subcritical flows one travels downstream while the other travels upstream. Together with the pressure term it fully accounts for the backwater effect from the downstream boundary, and two boundary conditions are required for the solution of the equations.

By eliminating the local acceleration term the quasi-steady dynamic wave approximation and noninertia approximation each carries only one characteristic wave traveling downstream and no upstream traveling characteristic wave for subcritical flows. But both approximations retain the pressure term through the depth gradient and hence the pressure part of the downstream boundary condition is reflected instantly upstream. Thus, two boundary conditions are required for a solution. In fact, the traditional steady flow backwater profile differential equation is a special case of the quasi-steady dynamic wave approximation with a constant Q, as given in Eq. (6.16).

The kinematic wave approximation has only one characteristic wave traveling downstream. By eliminating the inertia and pressure terms it has no mechanism to account for the downstream backwater effect. In fact, within the computational space interval Δx the water surface is considered parallel to the sewer invert during the computational time interval Δt. It requires only one boundary condition at the upstream for solution. Hydrodynamically it possesses no mechanism for flow peak attenuation. Conversely, the noninertia, quasi-dynamic wave, and full dynamic wave approximations allow different degrees of flow peak attenuation. All these above approximations have mechanism for flow hydrograph distortion and translation.

Hydraulically the dynamic wave is the most accurate model among the approximations, but it is also the most tedious for solutions. For gradually varied flow in channels without longitudinal boundary discontinuity the values of the local and convective acceleration terms carry opposite signs. For prismatic channels, including sewers, the magnitudes of these two terms are of the same order and hence, for the approximations, neglecting one term gives worse results than neglecting both terms. In other words, for sewers in terms of accuracy and solution simplicity the noninertia approximation is usually a better compromise.

Akan and Yen (1981), among others, compared the application of the dynamic wave, noninertia, and kinematic wave equations for flow routing in networks and found the non-inertia approximation to generally agree well with the dynamic wave solutions, whereas the solution of the kinematic wave approximate is clearly different from the dynamic wave solution, especially when the downstream backwater effect is important.

6.3.4 Surcharge Flow

Sewers, culverts, and other drain pipes sometimes flow full with water under pressure; this is often known as surcharge flow (Fig. 6.10). Such pressurized conduct flow occurs under extreme heavy rainstorms or under designed pipes. There are two ways to simulate unsteady surcharge flow in urban drainage: (1) the standard transient pipe flow approach and (2) the hypothetical piezometric open-slot approach.

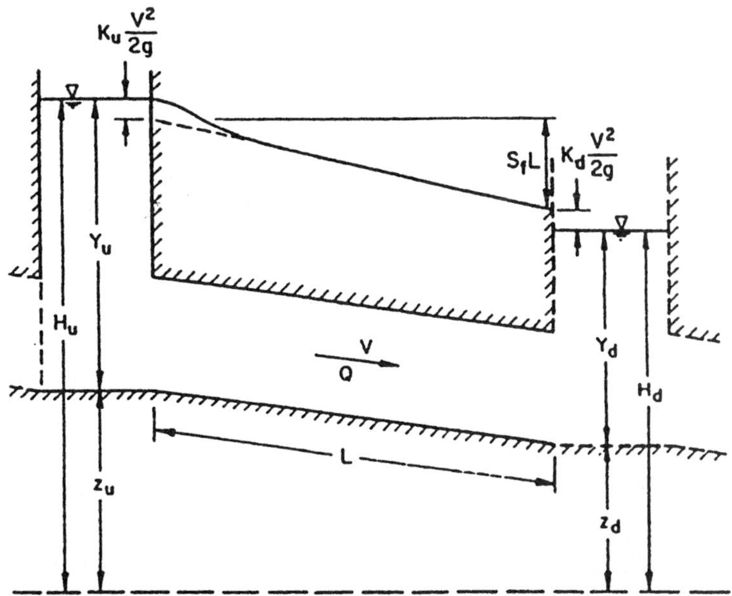

FIGURE 6.10   Surcharge flow in a sewer.

6.3.4.1 *Standard Transient Pipe Flow Approach.* In this approach, the flow is considered as it is physically, that is, pressurized transient pipe flow. For a uniform size pipe, the flow cross-sectional area is constant, being equal to the full pipe area A_f; hence, $\partial A/\partial t = 0$ and $\partial A/\partial x = 0$. The continuity and momentum equations, Eqs. (6.6) and (6.3) can be rewritten as

$$Q = A_f V \tag{6.16}$$

$$\frac{1}{g}\frac{\partial V}{\partial t} + \frac{\partial}{\partial x}\left(\frac{\beta V^2}{g} + \frac{P_a}{\gamma}\right) = -S_f \tag{6.17}$$

where P_a = piezometric pressure of the flow
γ = specific weight of the fluid

If the pipe has a constant cross-section and flowing full with an incompressible fluid throughout its length, then $\partial V/\partial x = 0$. By further neglecting the spatial variation of β, integration of Eq. (6.17) over the entire length, L, of the sewer pipe yields

$$-\frac{PA}{\gamma}\bigg|_{\text{entrance}}^{\text{exit}} = H_u - K_u\frac{V^2}{2g} - H_d - K_d\frac{V^2}{2g} = L\left(S_f + \frac{1}{g}\frac{\partial V}{\partial t}\right) \tag{6.18}$$

or

$$\frac{L}{gA}\frac{\partial Q}{\partial t} = H_u - H_d - (K_u + K_d)\frac{Q^2}{2gA_f^2} - S_f L \tag{6.19}$$

where H_u = total head at the entrance of the pipe
H_d = water surface outside the pipe exit
K_u and K_d = entrance and exit loss coefficients, respectively (Fig. 6.10)

Equations (6.16) and (6.17) can also be derived as a special case of the commonly used, general and basic closed conduit transient flow continuity equation for waterhammer and pressure surge analysis (see, e.g., Wylie and Streeter (1983); Chaudhry (1979); Stephenson (1984); Wood (1980)),

$$\frac{1}{A}\frac{dA}{dt} + \frac{1}{\rho}\frac{d\rho}{dt} + \frac{\partial V}{\partial x} = 0 \tag{6.20}$$

and the momentum equation

$$\frac{1}{V}\frac{\partial H}{\partial t} + \frac{\partial H}{\partial x} + \frac{c^2}{gV}\frac{\partial V}{\partial x} + \sin\theta = 0 \tag{6.21}$$

or

$$\frac{1}{g}\frac{\partial V}{\partial t} + \frac{V}{g}\frac{\partial V}{\partial x} + \frac{\partial H}{\partial x} + S_f = 0 \tag{6.22}$$

where ρ = bulk density of the fluid
$H = P_a/\gamma$ is the piezometric head above the reference datum
c = celerity of pressure surge

The fact that Eqs. (6.16) and (6.17) can be derived from Eqs. (6.6) and (6.3) is the theoretical basis of the Preissmann hypothetical slot concept, which will be discussed in the following.

6.3.4.2 Hypothetical Slot Approach. This approach introduces hypothetically a continuous, narrow, piezometric slot attached to the pipe crown and over the entire length of the pipe as shown in Fig. 6.11. The idea is to transform the pressurized conduit flow situation into a conceptual open-channel flow situation by introducing a virtual free surface to the flow. This idea was suggested by Preissmann (Cunge and Wegner, 1964). The hypothetical open-top slot should be narrow so that it will not introduce appreciable error in the volume of water. Conversely, the slot cannot be too narrow in order to avoid the numerical problem associated with a rapidly moving pressure surge.

A theoretical basis for the determination of the width of the slot is to size the width such that the wave celerity in the slotted sewer is the same as the surge celerity of the compressible water in the actual elastic pipe. The celerity c_1 of the slot pipe is

$$c_1 = \sqrt{gA/b} \tag{6.23}$$

where b = slot width
$\quad\quad A$ = flow cross-sectional area

Neglecting the area contribution of the slot and hence $A = A_f = \pi D^2/4$ for a circular pipe, and equating c_1 to the pressure wave speed c in the elastic pipe without the hypothetical slot, the theoretical slot width results in

$$b = \pi g D^2/4c^2 \tag{6.24}$$

The surge speed in a pipe usually ranges from a few hundred ft/second to a few thousand ft/second. For an elastic pipe with a wall thickness e and Young's modulus of elasticity E_p, assuming no pressure force from the soil acting on the pipe, the surge speed c is (Wylie and Streeter, 1983, pp. 2–11, 23):

$$c^2 = (E_f/\rho_f)/[1 + (\eta E_f D/E_p e)] \tag{6.25}$$

where E_f is the bulk modulus of elasticity and ρ_f is the bulk density, respectively, of the

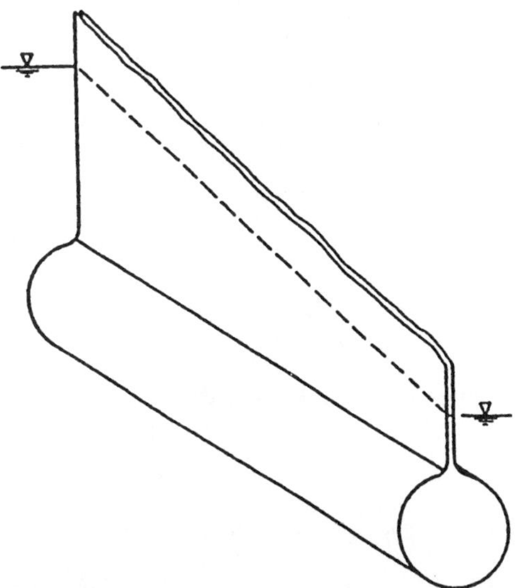

FIGURE 6.11 Preissmann hypothetical piezometric open slot.

flowing water. Special conditions of pipe anchoring against longitudinal expansion or contraction and elasticity relevant to the surge speed c are given in Table 6.9 in which ω is Poisson's ratio for the pipe wall material, that is, ω is the ratio of the lateral unit strain to axial unit strain, and α is a constant to account for the rigidity with respect to axial expansion of the pipe.

For small pipes, Eq. (6.24) may give too narrow a slot width that would cause numerical problems. Cunge et al. (1980) recommend a width of 1 cm or more.

Use of the hypothetical Preissmann slot has the following potential advantages:

1. It uses only the Saint-Venant equations and avoids switching between the surcharge equation and open-channel flow equations and avoids the associated separate treatment of the boundary conditions.

2. There is no need to define surcharge criteria.

3. It is not necessary to keep an inventory of the pipes that are surcharged at different times.

4. It permits the flow transition to progress computationally reach by reach in a sewer, as in the open-channel case, and, hence, it can account for the situation when only part of the length of the pipe is full.

5. It requires fewer additional assumptions than the standard approach to achieve numerical stability.

6. It is simpler to program.

However, it also has the following potential disadvantages:

1. It introduces a potential accuracy problem in the mass and momentum balance of the flow if the slot is too wide, and stability problems if it is too narrow.

2. It still requires computation of two equations (continuity and momentum) for each of the reaches within a sewer when the sewer is fully surcharged, whereas in the standard surcharge computation, only one equation is applied to the entire length of the sewer.

3. It is hypothetical rather than real.

The transition between part-full pipe flow and slot flow is by no means computationally smooth and easy, and assumptions are necessary (Cunge and Mazadou, 1984). One approach is to assume a gradual width transition from the pipe to the slot. Sjöberg (1982) suggested two alternatives for the slot width, based on two different values of the wave speed c in Eq.

TABLE 6.9   Special Conditions of Surge Speed c in Full Pipe, Eq. (6.25)

Factor	Condition			
Pipe Anchor	$\eta = \dfrac{2e}{D}(1 + \omega) + \alpha \dfrac{D}{D + e}$			
	Freedom of Pipe Longitudinal Expansion	Entirely free (expansion joints at both ends)	Only one end anchored	Entire length anchored
	Value of Axial Expansion Factor α	1	$1 - 0.5\,\omega$	$1 - \omega^2$
Elasticity E	Rigid pipe	Air entrainment	No air	
	$E_p = \infty$ $c^2 = E_f/\rho_f$	$\rho_f = \rho_w V_w + \rho_a V_a$ $E_f = \dfrac{E_w}{1 + V_a[(E_w/E_a) - 1)]}$	$\rho_f = \rho_w$ $E_f = E_w$	

V = volume; subscript w denotes water (liquid); subscript a denotes air, and subscript f denotes liquid-air mixture.

(6.24). For the alternative applicable to $h/D \geq 0.9999$, his suggested slot width b can be expressed as

$$b/D = 10^{-6} + 0.05423 \, \exp[-(h/D)^{24}] \tag{6.26}$$

He further proposed to compute the flow area A and hydraulic radius R when the depth h is greater than the pipe diameter D as

$$A = (\pi D^2/4) + (h - D)b \tag{6.27}$$

$$R = D/4 \tag{6.28}$$

A slight improvement to Sjöberg's suggestion to provide a smoother computational transition is to use

$$A = A_{9999} + b(h - 0.9999D) \tag{6.29}$$

for $h/D > 0.9999$ and assume the transition starts at $h/D = 0.91$. Between $h/D = 0.91$ and 0.9999, real pipe area A and surface width B are used. However, for $h/D > 0.91$, the hydraulic radius R is computed from Manning's formula using pipe slope and a discharge, Q_{91} equal to the steady uniform flow at $h = 0.91$D; thus, for $h/D > 0.91$

$$R = (A_{91}/A)R_{91} \tag{6.30}$$

Because of lack of reliable data, neither the standard surcharge sewer solution method nor the Preissmann hypothetical open-slot approach has been verified for a single pipe or a network of pipes. Past experiences with waterhammer and pressure surge problems in closed conduits may provide some indirect verification of the applicability of the basic flow equations to unsteady sewer flows. Nevertheless, direct verification is highly desirable.

Jun and Yen (1985) performed a numerical testing and found no clear superiority of one approach over the other. Nevertheless, specific comparison between them is given in Table 6.10. They suggested that if the sewers in a network are each divided into many computational reaches and a significant part of the flow duration is under surcharge, the standard approach will save computer time. Conversely, if the transition between open-channel and pressurized conduit flows occurs frequently and the transitional stability problem is important, the slot model would be preferable.

6.3.5 Flow Instability in a Sewer

The flow in sewers is perhaps one of the most complicated hydraulic phenomena. Even for a single sewer, there are a number of transitional flow instability problems. Yen (1978a) described five types of hydraulic instabilities in sewer systems. One of them is the surge instability of the flow in pipes of a network, which will be discussed later. The other four types of instabilities that could occur in a single sewer pipe are the following:

1. The instability at the transition between open-channel flow and full conduit flow

2. In the open-channel phase, the transitional instability between supercritical flow and subcritical flow

3. The water-surface roll-wave instability of supercritical open-channel flow

4. A near dry-bed flow instability

6.3.5.1 Transition Between Open-Channel Flow and Surcharge Flow. This hydraulic instability of transition between open-channel flow and pressurized full conduit flow is most relevant to those who want to model sewer flow along with surcharge. A number of factors contribute to this instability. The major factors are (1) a nonunique discharge—depth rela-

TABLE 6.10 Comparison Between Standard Surcharge Approach and Slot Approach

Item	Standard surcharge approach	Hypothetical slot approach
Concept	Direct physical	Conceptual
Flow equations	Two different sets, one equation for surcharge flow, two equations for open-channel flow	Same set of two equations (continuity and momentum) for surcharge and open channel flows
Discretization for solution	Whole pipe length for surcharge flow	Divide into Δx's
Water volume within pipe	Constant	Varies slightly with slot size, inaccurate if slot is too wide, stability problems if slot is too narrow
Discharge in pipe at given time	Same	Varies slightly with Δx, thus allows transition to progress within pipe
Transition between open channel flow and surcharge flow	Specific criteria	Slot width transition to avoid numerical instability
Part full over pipe length	Assume entire pipe length full or free	Assume full or free Δx by Δx
Time accounting for transition	Yes, specific inventory of surcharged pipes at different times	No, implicit
Programming efforts	More complicated because of two sets of equations and time accounting and computer storage for transition	Relatively simple because of one equation set and no specific accounting and storage for transition between open-channel and full-pipe flows
Computational effort	Depending mainly on accounting for transition times	Depending mainly on space discretization Δx

tionship when the pipe is nearly full and (2) an air supply insufficient to maintain an air pocket at the pipe entrance and in the pipe. Other factors include (3) the geometry of the pipe; (4) the flow and geometry conditions at the sewer entrance and exit, particularly the degree of submergence and the pipe crown condition that would affect the surface tension effect when the pipe is nearly full; (5) the water surface waves; and (6) the occurrence of hydraulic jump to surcharge or hydraulic drop from surcharge. These factors may act individually or in combination to cause the instability problem.

Concerning the first factor of the nonunique discharge–depth relationship, one may consider, as a simplified example, the case of steady uniform flow in a circular pipe without the complication of insufficient air supply or water surface waves. For the rating curve shown in Fig. 6.6, Q_f is the discharge when the pipe is just full, D the pipe diameter, Q the discharge corresponding to a given depth h, or when surcharge corresponding to a given longitudinal, x, gradient of P_a/γ, where P_a is the piezometric pressure and γ the specific weight of the water. In the open-channel flow regime, the maximum discharge does not occur when the pipe is full. It occurs at approximately $h/d = 0.94$, varying slightly with the Reynolds number of the flow. This decrease in discharge when the pipe is approaching full is due to the rapid increase in the wetted perimeter as h approaches D, and the consequent increase in the pipe boundary resistance to the flow. As shown in Fig. 6.6, the relationship between the discharge and depth of piezometric gradient is unique above the highest dot point or below the lowest dot point. Between these two points a given discharge can have different depths or piezometric gradient. For an unsteady flow, this nonunique discharge relationship is even more complicated.

The second factor of transitional instability is due to insufficient air supply and it can be classified into two aspects according to the effect of air on the flow. One is the entrainment of air into the water, reducing the density of the water, making the water surface easy to fluctuate, resulting in a very wavy surface, and even breaking up into droplets and causing interference with air movement. The other aspect is the movement of the air pocket or air mass in the sewer that is usually time varying and interacting with the water flow. This is essentially a two-phase flow phenomenon. The reader may refer to Wallis (1969), Rhodes and Scott (1968), Vermeuleu and Ryan (1971), Dukler (1972), and Taitel et al. (1978) for the hydrodynamics of this type of flow.

To further illustrate the effect of insufficient air supply on pipe flow instability, consider the case of a sewer with a submerged entrance. Initially at time t_0, the pipe is surcharged at its downstream part and has an air pocket in its upstream part. This is shown as the water surface profile marked t_0 in Fig. 6.12. This profile is classified as type III-5-D with a small exit submergence in Table 6.6. At this instance, the discharge into the sewer is Q. For clarity and simplicity in illustration, assume that the water surfaces at the upstream entrance and outside the exit of the pipe are constant, independent of the flow into and out from the sewer. Since the sewer is not ventilated and its downstream part is sealed by the hydraulic jump in the pipe, entrainment of the trapped air into the flowing water creates a low pressure in the air pocket (a situation similar to under-ventilated weir or sluice gate). Subsequently, the discharge into the sewer will increase ($> Q$). Suppose that the sewer pipe is not very long and the upstream entrance water surface is within an appropriate range; then the increased discharge will push the hydraulic jump outside the pipe. This is shown as surface profile t_1 in Fig. 6.12 and is case III-2-C in Table 6.6. The opening of the downstream end allows air to enter into the pipe, resulting in atmospheric pressure for the air in the sewer. The increase of air pressure in the sewer reduces the discharge into the pipe. The reduced discharge ($< Q$) is unable to keep the hydraulic jump beyond the exit. Consequently, the jump is pushed back into the pipe, the exit is resealed, as indicated by the surface profile t_2 in Fig. 6.12, and the cycle repeats itself. In an actual sewer, the water surfaces in both the upstream and downstream junctions vary, and the instability process becomes more complicated.

The effect of air supply in the pipe is more important for supercritical flow than for subcritical flow. In the extreme case of the instability of the moving transition of hydraulic jump or drop, which will be described later, the sequent depth of the supercritical flow may exceed the diameter of the pipe, and hence the pipe will become partially surcharged. Kalinske and Robertson (1943) described an experimental and theoretical study on the air entrainment and the transition from free surface flow to surcharge through a hydraulic jump.

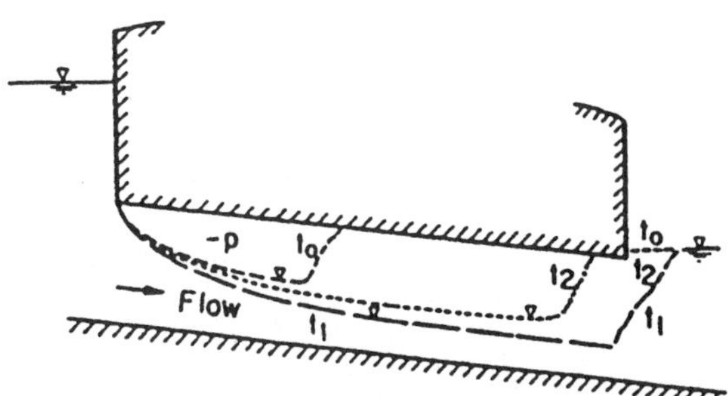

FIGURE 6.12   Flow instability due to insufficient air supply. (After Yen, 1986)

Haindl (1957) investigated the transition of steady flow from an open channel to a full pipe through a hydraulic jump. He found that the transition depends on the pre-jump Froude number and air supply in the pipe, and that the energy loss of the restrained hydraulic jump is less than the free jump. De Somer (1984) conducted a laboratory study of a slowly varying transition between open-channel flow and surcharge in a circular pipe. Haman and McCorquodale (1982) performed a laboratory investigation on the surcharge–open-channel flow transition with carefully observed behavior of the air in the pipe. Wiggert (1972) reported the transient surge in a large conduit. Information on air entrainment demand for flowing water in pipes has also been reported by Mura et al. (1959), Sharma (1976), and Volkart (1982). Observed problems on air entrainment and surge in pipes have been reported by Burton and Nelson (1971) and Wisner et al. (1975).

Concerning the other factors contributing to the transitional instability between open-channel and surcharge flows, the effect of pipe geometry is obvious. The effect of water surface waves on the transitional instability occurs when the wave peak touches the crown of the sewer pipe. This wave action is random in both space and time, and occasional touching of the wave peak to the pipe crown may not provide sufficient disturbance to cause a flow transition. However, when sufficient wave action is present, the air in the pipe may be isolated into pockets, causing a change in water flow. More often, when the depth of flow is nearly full, the waves may cause flow transition instability because of the nonunique depth–discharge relationship.

6.3.5.2 *Subcritical–Supercritical Transitional Instability.*

The second type of hydraulic instability occurs during the transition between supercritical flow and subcritical flow. It can further be divided into the transition from supercritical flow to subcritical, commonly known as a hydraulic jump, and the transition from subcritical to supercritical, a hydraulic drop. The hydraulic drop is also called a surge or bore. One can refer to standard open-channel flow reference books (e.g., Chow, 1959) to find the characteristics of steady-flow hydraulic jumps and constant speed surges. However, most of the information pertains to flow over a wide channel and not to circular channels.

Since sewer flows are usually unsteady, the subcritical–supercritical transition water surface discontinuity, or front, in a sewer is not stationary. It may move from upstream toward downstream or vice versa, depending on the slope of the sewer, the discharge, and the conditions at the upstream and downstream junctions. If the front is moving at a constant speed, it can be analyzed by superimposing a constant velocity equal but opposed to the celerity of the transition front. However, for sewer flow, usually the celerity varies with both time and space. No theoretical solution has been obtained for such a transition flow with a front of variable speed. Figure 6.13 illustrates the movement of the hydraulic jump or drop initiated from downstream (a and b), as well as from upstream (c and d). Starting from the initial time t_0 with no surface discontinuity, the surface profiles are shown with a jump or drop at subsequent times t_1, t_2, t_3, and finally at the time t_4 when the surface discontinuity is swept away. In Fig. 6.13a, the water surface discontinuity is moving upstream. The upstream moving hydraulic jump can be initiated, for example, from the high water level in the downstream junction. The upstream-moving hydraulic drop may occur because of a rapid withdrawal of water at the downstream end, for example, as a result of the opening of a flow control gate or valve, or the pumping at the downstream junction. In Fig. 6.13b, the transition front is moving downstream. The downstream-moving hydraulic drop may occur due to a surge of inflow into the sewer, for example, resulting from rapid filling of the upstream junction or the sudden opening of an upstream gate or valve. The downstream-moving hydraulic jump is associated with a rapid flow sweeping down from upstream; although this is a rare case, sometimes it occurs during a rising flow.

6.3.5.3 *Surface Waves and Roll Waves.*

There are basically two types of surface waves in a partially filled pipe. One type is the *air–water interfacial waves* generated by the interfacial shear between the flowing air and water. This occurs for both subcritical and su-

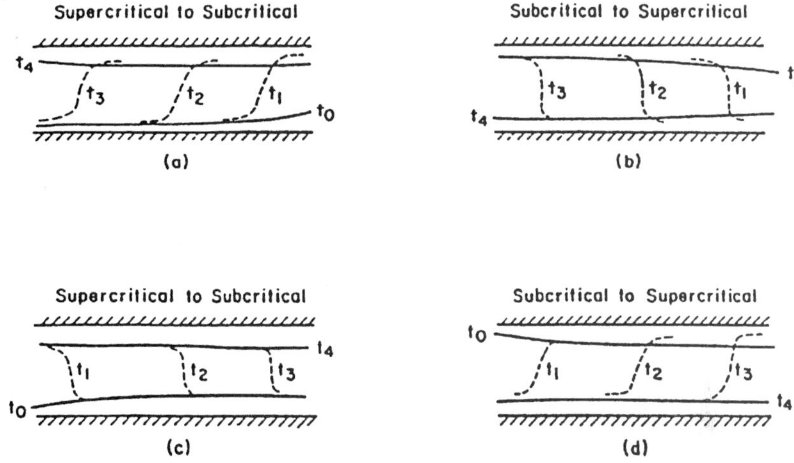

FIGURE 6.13 Moving water surface discontinuity in a sewer. (After Yen, 1986)

percritical open-channel flows. The wave amplitude is usually small and its frequency high (Killen and Anderson, 1969). Hence, its effect in causing a flow transition into surcharge occurs only when the water surface is near the pipe crown and for sewer pipes usually with a flow depth greater than 90% of the pipe diameter.

The other type of wave in supercritical flow is a *roll wave,* also called a *slug wave.* Roll waves usually occur in flow with Froude number greater than 2 for circular conduits. Contrary to the surface wave, which is an interfacial shear phenomenon, the roll wave is related to the friction in the channel bed, an instability caused mainly by the water moving considerably faster near the free surface than near the bed. Hence, it can be characterized by the velocity, depth, and flow resistance or, more appropriately, by the Froude number and, to some degree, the Reynolds number of the flow. A sketch of a roll wave train is shown in Fig. 6.14. Compared to the interfacial shear waves, roll waves are more regular and periodic and have larger amplitude. With roll waves in a sewer, the transitional instability may occur with the pipe far below 90% full.

For a unidirectional flow down a wide plane, roll waves start to occur at Froude number approximately equal to 1.6. For circular, rectangular, and trapezoidal channels, this stability critical value is higher because of the sidewall effect. Mayer (1961) and Brock (1969) described the generation and propagation of different types of roll waves. On the basis of momentum consideration, Iwasa (1954) proposed a stability criterion for a constant discharge

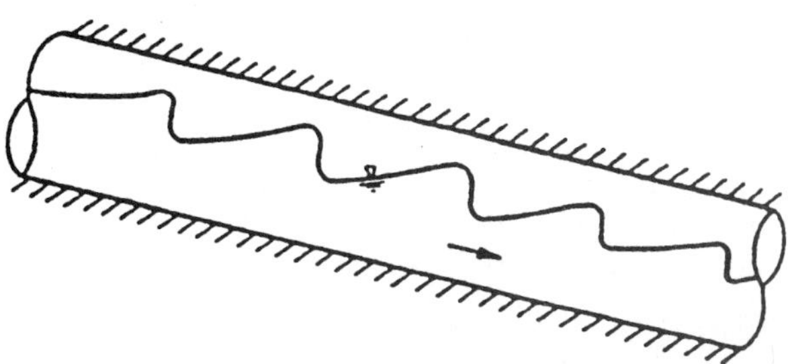

FIGURE 6.14 Roll waves in a sewer.

(quasi-steady) flow in a smooth rectangular channel. Koloseus and Davidian (1966) evaluated this criterion experimentally. Berlamont (1976) showed that for a given Reynolds number of flow, roll waves occur only within a range of the Froude number.

Roll wave problems are also of great interest to chemical and mechanical engineers. In fact, studies on roll waves in circular pipes have been reported by them but not by civil engineers. Most of these studies consider only laminar flow through solving the Navier-Stokes equations. The gravity effect is expressed in terms of the channel angle instead of the Froude number. Representative literature includes publications by Dukler (1972), Javdani and Goren (1972), Yih (1965), and Lin (1974).

6.3.5.4 Dry-bed Instability.   *Dry-bed instability* occurs when the channel is nearly dry and the surface tension effect becomes important. It may occur at the beginning or the end of the runoff, for overland flow as well as for sewers. As an illustration, consider a clean, smooth surface with a gentle slope. When a small amount of water is spread on it, the water will not flow downslope as a thin, uniform film. Instead, isolated small pools of water will be formed due to surface tension between the water and the surface, holding the water from flowing down the slope. As the inflow continues, the size of the pools increases. Hence, the gravity force increases while the surface tension decreases relatively. Soon, the water starts to move, pools join together, a thin-film flow occurs, the surface tension effect diminishes, and the pools disappear.

The thin-film instability is important to many chemical and mechanical engineering problems. Hence, research work on this subject has been done mostly by investigators in these two areas. For instance, one may refer to Wallis (1969), Javdani and Goren (1972), and Goren (1962) for information on film flow instabilities. Obviously, the Weber number is the most important parameter, although the Reynolds and Froude numbers of the flow also have their roles. Discussion of the effect of surface tension of film flow stability can also be found in Yih (1965, pp. 180–194). However, in sewers this stability problem is further complicated by the quality of the water and the roughened and curved boundary of the sewer pipe. No results exist that can be adopted directly for sewer flow dry-bed instabilities. Fortunately, this instability is of little practical significance for sewer flow, partly because of its occurrence only at the commencement and end of the runoff when the discharge is insignificant and partly because of the rough sewer pipe surface after a period of service, making the surface tension effect relatively small and negligible.

6.4 STEADY SEWER FLOW

Strictly speaking, as discussed in Sec. 6.2.2 storm sewer flows are generally unsteady and nonuniform. However, often the duration of the hydrograph of single rainstorm events is hours, the time of water flowing through individual sewers is minutes, whereas unsteady sewer flow computational time interval is often seconds. Therefore, within the time scale of the flow passing through the sewer or the computational time interval, the flow can be treated as steady without losing accuracy, that is, stepwise steady or quasi-steady. Extensive knowledge exists in the literature on the hydraulics of steady uniform and nonuniform free surface flow, for example, Chow (1959). To describe the water surface profiles of steady flows, two reference depths are often used. One is the normal depth and the corresponding normal flow, which will be described in the following subsection. The other is the *critical depth* at which the Froude number is equal to unity. The critical depth, which separates the supercritical flow from subcritical, is determined from

$$\frac{A_c^3}{B_c} = \frac{\beta Q^2}{g} \tag{6.31}$$

where the critical flow cross-sectional area A_c and critical flow water surface width B_c are each a function of either the critical depth h_c or the central angle ϕ (in radians) subscribed by the free surface (Fig. 6.8). For a given discharge Q the right side of Eq. (6.31) is a constant, provided β is constant. For circular sewer pipes Eq. (6.31) can be written as

$$\frac{(\phi_c - \sin \phi_c)^3}{512 \sin\frac{\phi_c}{2}} D^5 = \frac{\beta Q^2}{g} \tag{6.32a}$$

Straub suggested an approximation of Eq. (6.32a) for h_c over the range $0.02 = h_c/D \le 0.85$ (French, 1986),

$$h_c = \frac{1.01}{D^{0.26}} \left(\frac{\beta}{g}\right)^{1/2} Q \tag{6.32b}$$

Corresponding formulas of h_c for sewer conduits of noncircular cross-section can be found in open-channel references, for example, Yen (1996).

6.4.1 Steady Uniform Flow in Sewer Pipe

In open-channel hydraulics a *normal flow* is often used as a reference to describe the water surface profiles. For sewers, the normal flow can be defined for a specified discharge as the steady uniform flow for which the streamlines are parallel and the dynamic relationship (Eqs. (6.3) or (6.4)) can be simplified as $S_o - S_f = 0$, whereas the continuity equation (Eq. 6.6) is reduced to $Q_n = Q = AV = $ constant. The friction slope S_f can be represented by Manning's formula, Eq. (6.8) or the Darcy-Weisbach formula, Eq. (6.9). The normal depth, h_n, can be determined using Manning's formula with $S_f = S_o$ as

$$AR^{2/3} = F_1(h_n) = \frac{nQ}{K_n S_o^{1/2}} \tag{6.33a}$$

where F denotes a function. For a given discharge in a given pipe, the right side of Eq. (6.33a) is a constant. For circular sewer pipes the geometry relationships of the flow cross-section are given in Fig. 6.8 in terms of the central angle ϕ subscripted by the free surface and

$$AR^{2/3} = \frac{1}{16}\left[\frac{(\phi - \sin \phi)^5 D^8}{2\phi^2}\right]^{1/3} \tag{6.33b}$$

$$h = \frac{D}{2}\left(1 - \cos \frac{\phi}{2}\right) \tag{6.34}$$

Likewise, if the Darcy-Weisbach formula is used

$$AR^{1/2} = F_2(h_n) = Q\sqrt{f/8gS_o} \tag{6.35a}$$

$$AR^{1/2} = \frac{1}{16}\left[\frac{(\phi - \sin \phi)^3}{\phi} D^5\right]^{1/2} \tag{6.35b}$$

Solutions of Eqs. (6.33) and (6.35) for the central angle ϕ are plotted nondimensionally in Fig. 6.15 for graphic determination of ϕ knowing Q, or computing Q with known ϕ or h.

Equations (6.33) and (6.35) can also be used to compute the required pipe diameter to carry a discharge Q if the maximum allowable relative depth h_n/D is known. For the special case of a just-full pipe, $\phi = 2\pi$ and the required pipe diameter, d_r, to carry the specified Q is

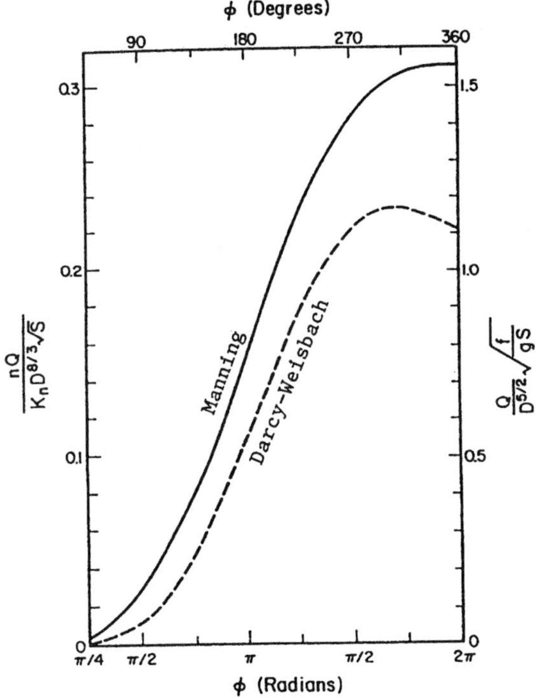

FIGURE 6.15   Central angle ϕ of water surface in circular pipe.

$$d_r = \left[\frac{8}{\pi} 2^{1/3} \frac{n}{K_n} \frac{Q}{\sqrt{S_o}} \right]^{3/8} \tag{6.36}$$

$$d_r = \left[\frac{8}{\pi^2} \frac{f}{gS_o} Q^2 \right]^{1/5} \tag{6.37}$$

6.4.2 Steady Nonuniform Flow

Nonuniform flows can be classified into (1) spatially gradually varying nonuniform flow for which there is a continuous water surface profile and the pressure distribution is essentially hydrostatic, and (2) spatially rapidly varying flow, which includes hydraulic jump and hydraulic drop.

As illustrated in Fig. 6.1 and described in Sec. 6.2, nonuniform varied flows are common in sewers. It is always helpful to know the flow conditions, including the water surface profile in the sewer, in order to check that the design or operational conditions are satisfied hydraulically. For gradually varied steady nonuniform flows, the water surface profiles can be classified according to the flow depth relative to the critical depth, h_c (Eq. (6.31)) and normal flow depth, h_n (Eq. (6.33) or (6.35)) and the type of sewer slope. A channel is steep if $h_n < h_c$ and is mild if $h_n > h_c$. The types of backwater surface profiles are shown in Fig. 6.4. The differential equation that describes such profiles is Eq. (6.15). However, the most often used computation method for backwater profiles is the *standard step method*, which is described as follows.

Consider a short reach of the sewer of length Δx between sections 1 and 2 as shown in Fig. 6.16. Following the *Bernoulli principle*, equating the total heads at the two sections

$$\frac{\alpha_2 V_2^2}{2g} + (Y_2 + y_{b2}) = \frac{\alpha_1 V_1^2}{2g} + (Y_1 + y_{b1}) + h_c \tag{6.38}$$

in which the subscripts 1 and 2 denote sections 1 and 2, respectively; Y is flow depth measured vertically; $Y + y_b$ is the stage of the water surface where the channel bed stage at section 1 is y_{b1} and that at section 2 is $y_{b2} = y_{b1} + S_o\Delta x$; the energy loss $h_e = S_e\Delta x$ where S_e is the slope of the energy line. If there is any other energy loss h_1, it should also be added to the right side of the equation.

The standard step method compares the total head at cross-section 2 computed from the left side of Eq. (6.38), H_2, with the total head computed from the right side of the equation, H_{12}, through trial and error until $H_2 = H_{12}$. In hand calculation, this is best done in a tabular form. An example is given in Table 6.11. In this table, H_2 is shown in column 6, which is computed as the sum of column 2 and column 6, and H_{12} in column 14 is computed as columns 12 and 13 added to column 14 of the previous station. The *energy-loss slope S_e* in column 9 is computed according to Manning's formula,

$$S_e = \frac{n^2 V^2}{K_n^2 R^{4/3}} \tag{6.39}$$

In the trial process, a likely flow stage is assumed, as shown in the second line of numbers in the table. The geometry parameters A, P, and R are obtained from the parameter-stage relationships predetermined for the channel at specified sections. Other values in the table are computed accordingly. Line 2 of computational numbers in Table 6.11 is crossed out because in the first trial $H_2 < H_{12}$, and hence a smaller river stage is assumed for the second trial at this station as shown in line 3 of the table.

For an experienced person, the trial required for the standard step method is rather minimal. It may be noted that in the trial-and-error process the true value of the total head H is much closer to H_{12} in column 14 than H_2 in column 6. Nevertheless, the computation is preferably done by computer. A computational algorithm is shown in Fig. 6.17 to aid pro-

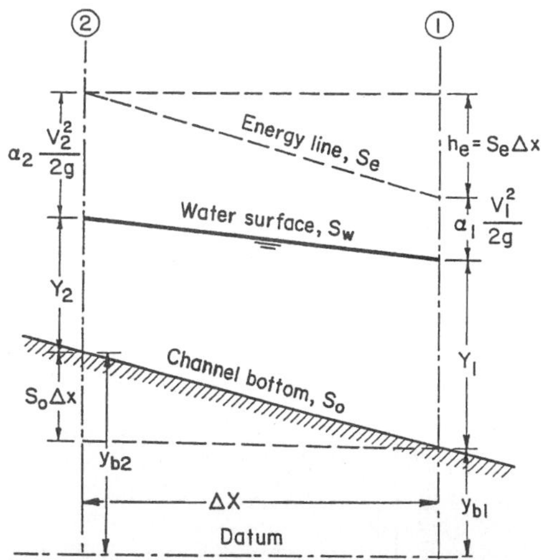

FIGURE 6.16   Channel reach for computational step method.

TABLE 6.11 Standard Step Method for Backwater Profile Computation

Station x ft (1)	Water surface $(Y + z_b)$ ft (2)	Depth Y ft (3)	Cross sectional area A ft^2 (4)	Velocity $V = Q/A$ fps (5)	$\dfrac{\alpha V^2}{2g}$ ft (6)	Total head $H = (2) + (6)$ ft (7)	Hydraulic radius R ft (8)	Energy slope S_e in 10^{-3} (9)	Average energy slope over reach $\bar{S}_e$ in 10^{-3} (10)	Length of reach Δx ft (11)	Energy loss in reach h_f ft (12)	Other losses h_1 ft (13)	Total head $H = (14)_1 + (12)_2 + (13)_2$ ft (14)
0	1.0	1.0	1.571	1.831	0.0573	1.0573	0.500	0.643					1.0573
10	1.007	0.997	1.565	1.838	0.0577	1.0647	0.499	0.650	0.646	10	0.00646	0	1.0637
10	1.0058	0.9958	1.563	1.841	0.0579	1.0637	0.499	0.652	0.648	10	0.00648	0	1.0637
20	1.012	0.992	1.555	1.850	0.0584	1.0704	0.497	0.661	0.656	20	0.00656	0	1.0703
—	—	—	—	—	—	—	—	—	—	—	—	—	—
—	—	—	—	—	—	—	—	—	—	—	—	—	—
—	—	—	—	—	—	—	—	—	—	—	—	—	—

Circular channel, diam = 10 ft, S = 0.0001, α = 1.1, n = 0.013, Q = 2.876 cfs, Y = 0.88 ft, Y = 0.59 ft.
Initial depth at downstream end x = 0 is y = 1.00 ft. M1 type profile.

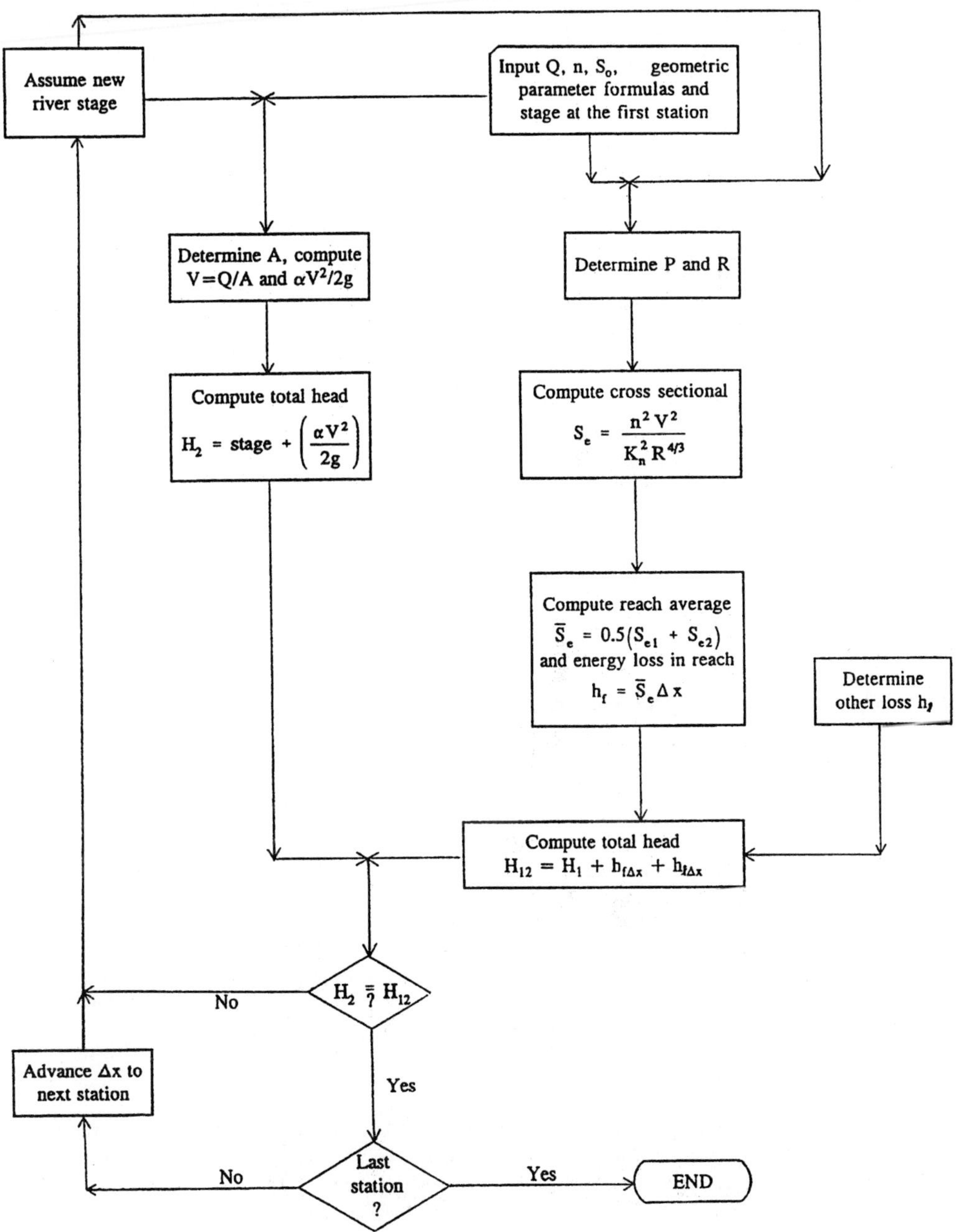

FIGURE 6.17 Flow chart for standard step method of backwater computation.

gramming. Since the method is applicable to nonprismatic channels, many existing computer programs, such as HEC-RAS (HEC-2) and WSPRO, use this method.

For sewers that are prismatic channels, the backwater surface profile can also be computed by using the direct step method or direct integration method instead of the standard step method. Those interested in these methods can refer to Chow (1959) for details.

Backwater profile computation usually starts from a control section along the channel. For subcritical flows the control is located at the downstream end, for example, the water stage at the large receiving water body. Hence, computation for subcritical steady flows usually proceeds from downstream toward upstream despite the flow being from upstream downward. Conversely, computation for supercritical flows proceeds from upstream downward. However, as shown in cases (d), (e), and (c) of Fig. 6.1, a hydraulic jump or drop may occur within a sewer reach such that it serves as the control for both upstream and downstream flows. The location of such internal control is determined through matching the *sequent depths* before and after the hydraulic jump, Y_1 and Y_2 respectively, with the backwater profile depths immediately before and after the jump. The sequent depths is computed with the following formula (Yen, 1996).

$$\frac{Y_2}{Y_1} = \frac{1}{2}\left[\left(1 + \frac{8Q^2 b_1}{gA_1^3 \cos^2 \theta / \beta}\right)^{1/2} - 1\right]$$

(6.40)

The length of the hydraulic jump, L_j, can be estimated as

$$L_j = 10(Y_2 - Y_1)\left(\frac{Q^2 b_1}{gA_1^3 / \beta}\right)^{-0.08}$$

(6.41)

6.5 STORM SEWER DESIGN WITH RATIONAL METHOD

6.5.1 Basics for Sewer Design

The most important components of an urban storm drainage system are the storm sewers. A number of methods exist for designing the size of such sewers. Some are hydraulically highly sophisticated, using the Saint-Venant equations, whereas others are relatively simple. Occasionally, a sophistaticated method is used to design one or a few sewer pipes. Each pipe has its own design storm of specified duration. Hence, the design computation is repeated for different pipes in a network, making the method inefficient for large networks with many sewers. Conversely, the simple methods, particularly the rational method, which does not require re-computation of the flow in upstream sewers, offer a practical option that is used often. In contrast, using the models for storm runoff prediction/simulation hydraulically sophisticated storm sewer design methods do not necessarily provide a better design than the simpler methods, mainly because of the discrete sizes of commercially available sewer pipes.

If the peak design discharge, Q_p, for a sewer is known, the required sewer dimensions can be computed by using Manning's formula, Eq. (6.36) or the Darcy-Weisbach formula, Eq. (6.37), which are obtained from Eq. (6.8) or (6.9) by assuming that the friction slope, S_f, is equal to the sewer slope, S_o. For a circular sewer pipe, the minimum required diameter, d_r, is

$$d_r = \left[3.208 \frac{n}{K_n} \frac{Q_p}{\sqrt{S_o}}\right]^{3/8}$$

(6.42)

in which $k_n = 1$ for SI units and 1.486 for English units. If the Darcy-Wesibach formula, Eq. (6.10) is used,

$$d_r = \left[0.811 \frac{f}{gS_o} Q_p^2 \right]^{1/5} \tag{6.43}$$

These two equations are plotted in Fig. 6.18 for design applications. The assumption $S_o = S_f$ essentially implies that around the time of peak discharge, the flow can well be regarded approximately as steady uniform flow for the design, despite the fact that the actual spatial and temporal variations of the flow are far more complicated as described in Sec. 6.2.2.

In sewer designs, a number of constraints and assumptions are commonly used in engineering practice. Those pertinent to sewer hydraulic design follow:

1. Free surface flow exists for the design discharge, that is, the sewer is under "gravity flow" or open-channel flow. The design discharge used is the peak discharge of the total inflow hydrograph of the sewer.

2. The sewers are commercially available circular sizes no smaller than, say, 8 in or 200 mm in diameter. In the United States, usually the commercial sizes in inches are 8, 10,

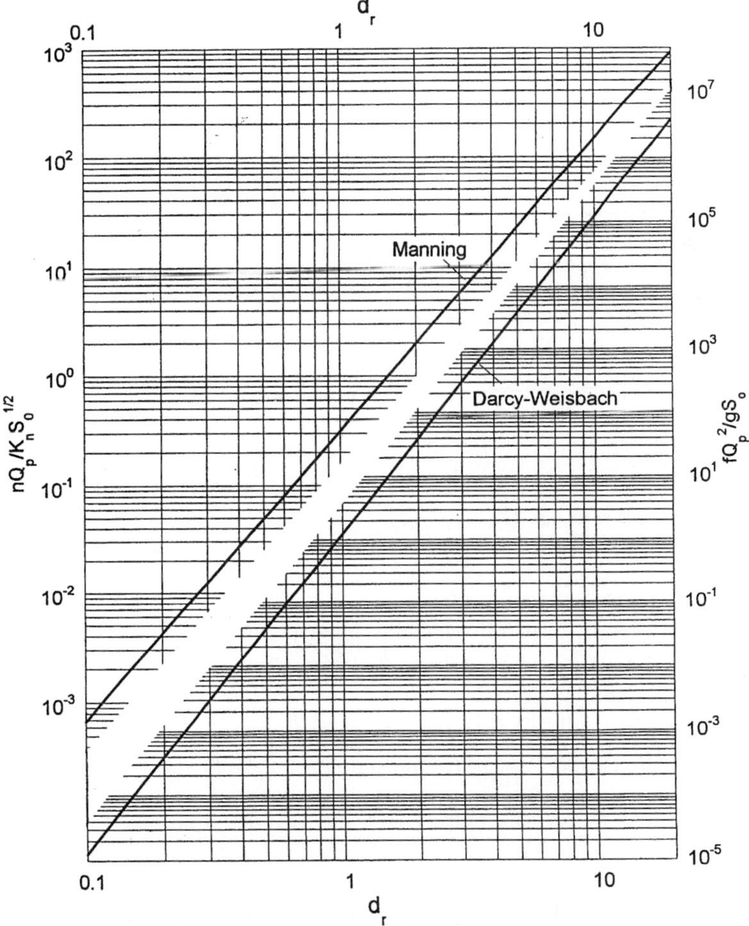

FIGURE 6.18   Required sewer diameter (m or ft).

12, from 15 to 30 with a 3-in increment, and from 36 to 120 with an increment of 6 in. In SI units, commercial sizes, depending on location, include most if not all of the following: (in mm) 150, 175, 200, 250, 300, 400, 500, 600, 750, 1000, 1250, 1500, 1750, 2000, 2500, and 3000.

3. The design diameter is the smallest commercially available pipe that has a flow capacity equal to or greater than the design discharge and satisfies all the appropriate constraints.

4. To prevent or reduce permanent deposition in the sewers, a nominal minimum permissible flow velocity at design discharge or at nearly full-pipe gravity flow is specified. A minimum full-pipe flow velocity of 2 ft/second or 0.5 m/second at the design discharge is usually recommended or required.

5. To prevent the occurrence of scour and other undesirable effects of high velocity flow, a maximum permissible flow velocity is also specified. The most commonly used value is 10 ft/second or 3 m/second. However, recent studies indicate that due to the improved quality of modern concrete and other sewer pipe materials, the acceptable velocity can be considerably higher.

6. Storm sewers must be placed at a depth that will allow sufficient cushioning to prevent breakage due to ground surface loading and will not be susceptible to frost. Therefore, a minimum cover depth must be specified.

7. The sewer system is a tree-type network, converging toward downstream.

8. The sewers are joined at junctions or manholes with specified alignment, for example, the crowns aligned, the inverts aligned, or the centerlines aligned.

9. At any junction or manhole, the downstream sewer cannot be smaller than any of the upstream sewers at that junction, unless the junction has significantly large detention storage capacity or pumping. There also is evidence that this constraint is unnecessary for very large sewers.

Various hydrologic and hydraulic methods exist for the determination of the design discharge Q_p. Among them, the rational method is perhaps the simplest and most widely used method for storm sewer design. With this method, each sewer is designed individually and independently, except that the upstream sewer flow time may be used to estimate the time of concentration. The design peak discharge for a sewer is computed by using the rational formula

$$Q_p = i \sum C_j a_j \tag{6.44}$$

where i = intensity of the design rainfall
   C = runoff coefficient (see Table 6.12 for its values)
   a = drained land surface area

The subscript "j" represents the j-th subarea upstream to be drained. Note that Σa_j includes all the subareas upstream of the sewer being designed. Each sewer has its own design rain intensity i because each sewer has its own flow time of concentration and design storm. In the design of a particular sewer in a network the only information needed from the upstream sewers is the upstream flow time for the determination of the time of concentration to the current sewer. The procedure of the rational method is illustrated as the flow chart shown in Fig. 6.19.

The rational formula is dimensionally homogenous and is applicable to any consistent measurement units. The runoff coefficient C is dimensionless. It is a peak discharge coefficient but not a runoff volume fraction coefficient. However, in English units usually the formula is used with the area a_j in acres and rain intensity i in in./hr. For Q_p in cfs the conversion factor 1.0083 is approximated as unity.

TABLE 6.12 Values of Runoff Coefficient C for Rational Formula

Land use	C
Business:	
Downtown areas	0.70–0.95
Neighborhood areas	0.50–0.70
Residential:	
Single family areas	0.30–0.50
Multi-units, detached	0.40–0.60
Multi-units, attached	0.60–0.75
Suburban	0.25–0.40
Apartment dwelling areas	0.50–0.70
Industrial:	
Light areas	0.50–0.80
Heavy areas	0.50–0.90
Park, cemeteries	0.10–0.25
Playgrounds	0.20–0.35
Railroad yard areas	0.10–0.30
Unimproved areas	0.10–0.30
Streets:	
Asphaltic	0.70–0.95
Concrete	0.80–0.95
Brick	0.70–0.85
Drives and walks	0.75–0.85
Roofs	0.75–0.95
Lawns:	
Sandy soil, flat, 2%	0.05–0.10
Sandy soil, average, 2–7%	0.10–0.15
Sandy soil, steep, 7%	0.15–0.20
Heavy soil, flat, 2%	0.13–0.17
Heavy soil, average, 2–7%	0.18–0.22
Heavy soil, steep, 7%	0.25–0.35
Agricultural land:	
Bare packed soil	
Smooth	0.30–0.60
Rough	0.20–0.50
Cultivated rows	
Heavy soil, no crop	0.30–0.60
Heavy soil, with crop	0.20–0.50
Sandy soil, no crop	0.20–0.40
Sandy soil, with crop	0.10–0.25
Pasture	
Heavy soil	0.15–0.45
Sandy soil	0.05–0.25
Woodlands	0.05–0.25

Notes: Use lower values for large areas; use higher values for steep slopes.

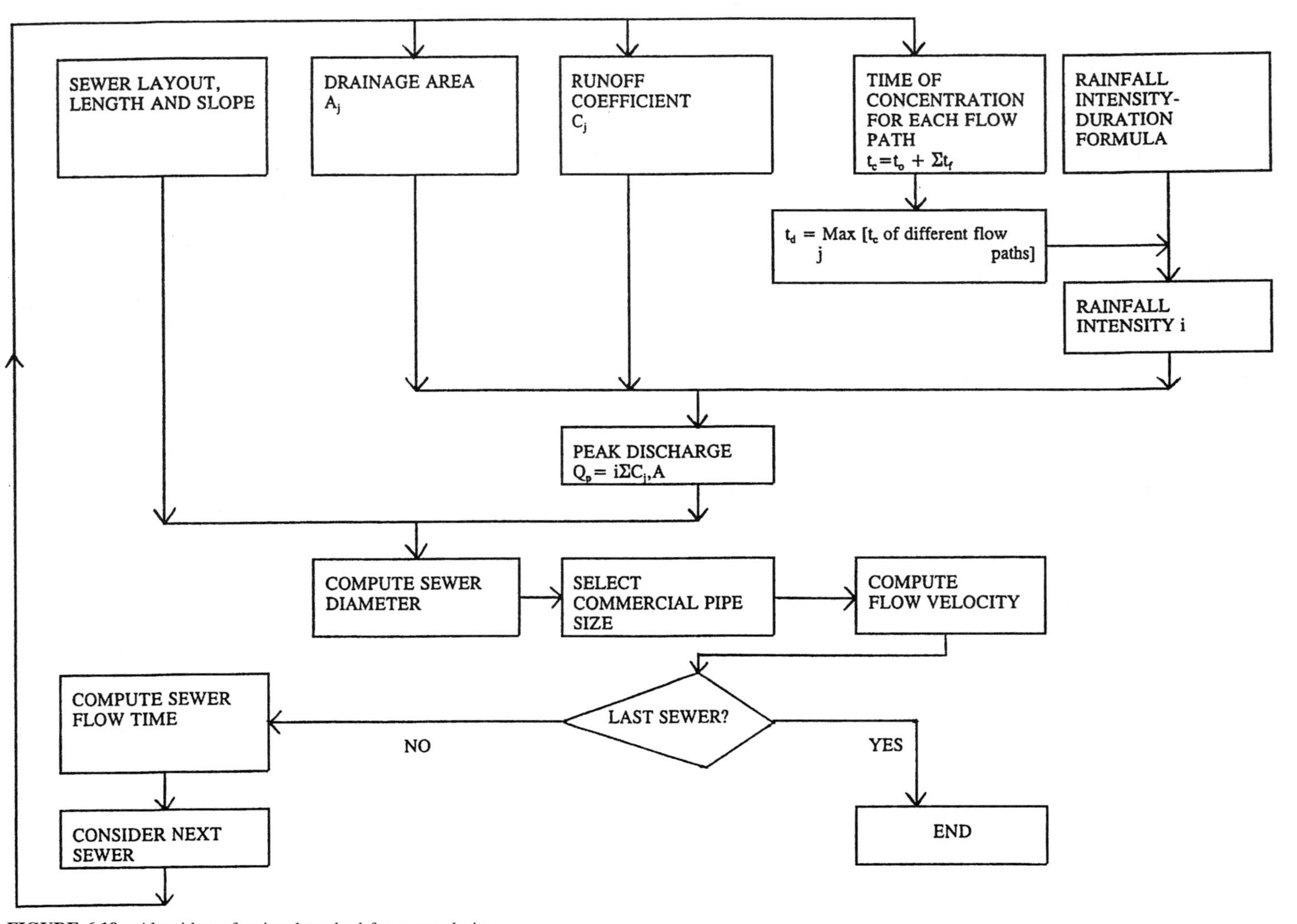

FIGURE 6.19 Algorithm of rational method for sewer design.

6.5.2 Rational Method Design Example

An example of using the rational method for sewer design is given in the following in English units for the design of the sewers of the simple drainage basin shown schematically in Fig. 6.20. The catchment properties are given in Table 6.13. For each catchment, the length L_o and slope S_o of the longest flow path—or better, the largest $L_o/\sqrt{S_o}$—should first be identified. A number of formulas are available to estimate the overland inlet time or time of concentration of the catchment to the inlet, t_o. They will be discussed in Sec. 6.5.3. In this example, the Yen and Chow (1983) formula is used with $K = 0.7$ for English units and heavy rain, that is, $t_o = 0.7(NL_o)/\sqrt{S_o})^{0.6}$. The catchment overland surface texture factor N is determined from Table 6.18.

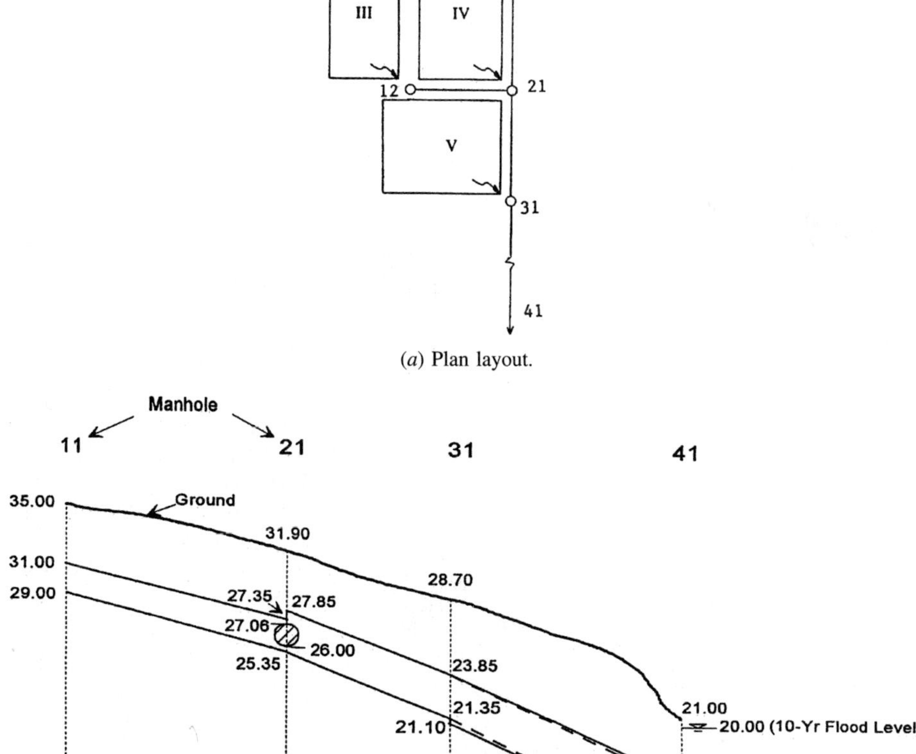

FIGURE 6.20 Example drainage basin for sewer design.

TABLE 6.13   Characteristics of Catchments of Example Drainage Basin

| | | Longest overland path | | | | Runoff coefficient |
Catchment	Area (acres)	Length L_o (ft)	Slope	Surface texture N	Inlet time t_o (min)	C
I	2	250	0.010	0.015	6.2	0.8
II	3	420	0.0081	0.016	9.3	0.7
III	3	400	0.012	0.030	11.7	0.4
IV	5	640	0.010	0.020	12.9	0.6
V	5	660	0.010	0.021	13.1	0.6

The design rainfall intensity is computed from the intensity–duration–frequency relation for this location:

$$i(\text{in./hr}) = \frac{100 T_r^{0.2}}{t_d + 25} \tag{6.45}$$

where t_d = rain duration in minutes, which is assumed equal to the time of concentration, t_c, of the area described

T_r = design return period in years

For this example, T_r = 10 years. Determination of i for the sewers is shown in Table 6.14a. The entries in this table are explained as follows:

Column 1:   Sewer number identified by the inlet numbers at its two ends

Column 2:   Sewer number immediately upstream or the number of the catchment that drains directly through manhole or junction into the sewer being considered

Column 3:   Size of the directly drained catchment

Column 4:   Value of the runoff coefficient for each catchment

Column 5:   Product of C and the corresponding catchment area

Column 6:   Summation of $C_j a_j$ for all the areas drained by the sewer—equal to the sum of contributing values in column 5

Column 7:   Values of inlet time to the sewer for the catchments drained, that is, the overland flow inlet time for directly drained catchments, or the time of concentration for the immediate upstream connecting sewers

Column 8:   Sewer flow time of the immediate upstream connecting sewer as given in column 9 in Table 6.14b

Column 9:   Time of concentration t_c for each of the possible critical flow paths, t_c = inlet time (column 7) + sewer flow time (column 10) for each flow path

Column 10:   Design rainfall duration t_d, assumed equal to the longest of the different times of concentration of different flow paths to arrive at the entrance of the sewer being considered, for example, for sewer 31, t_d is equal to 13.9 min from sewer 21, which is longer than that from directly contributing catchment V (13.1 min)

Column 11:   Rainfall intensity i for the duration given in column 10 is obtained from the intensity–duration relation for the given location, in this case, Eq. (6.45) for the 10-year design return period.

TABLE 6.14 Rational Method Design of Sewers of Example Drainage Basin
(a) Design Rain Intensity

Sewer (1)	Directly drained catchment or contributing upstream sewer (2)	Area a_j (acres) (3)	Runoff coefficient C_j (4)	$C_j a_j$ (5)	$\Sigma C_j a_j$ (6)	Inlet time (min) (7)	Upstream sewer flow time (min) (8)	Time of concentration t_c (min) (9)	Design rain duration t_d (min) (10)	Design rain intensity i (in./h) (11)
11–21	I	2	0.8	1.6		6.2	—	6.2		
	II	3	0.7	2.1		9.3	—	9.3		
					3.7				9.3	4.62
12–21	III	3	0.4	1.2	1.2	11.7	—	11.7	11.7	4.32
21–31	IV	5	0.6	3.0		12.9	—	12.9		
	11–21			3.7		9.3	1.4	10.7		
	12–21			1.2		11.7	0.9	12.6		
					7.9				12.9	4.18
31–41	V	5	0.6	3.0		13.1	—	13.1		
	21–31			7.9		12.9	1.0	13.9		
					10.9				13.9	4.07

TABLE 6.14 (*b*) Sewer Design

Sewer (1)	Upstream manhole ground elev. (ft) (2)	Length L (ft) (3)	Slope S (4)	Design discharge Q_p (ft³/s) (5)	Required diam. d_r (ft) (6)	Diameter used d_n (ft) (7)	Flow velocity V (ft/s) (8)	Sewer flow time (min) (9)	SL (ft) (10)	Upstream crown elev. (ft) (11)	Upstream invert elev. (12)	Downstream crown elev. (ft) (13)	Downstream invert elev. (ft) (14)
11	35.00	450	0.0081	17.1	1.98	2.00	5.4	1.4	3.65	31.00	29.00	27.35	25.35
12	41.50	360	0.0290	5.2	0.99	1.00	6.6	0.9	10.44	37.50	36.50	27.06	26.06
21	31.90	400	0.0100	33.0	2.43	2.50	6.7	1.0	4.00	27.85	25.35	23.85	21.35
31	28.70	500	0.0144	44.4	2.53	2.75			7.20	23.85	21.10	16.65	13.90
31	28.70	500	0.0156	44.4	2.50	2.50			7.80	23.85	21.35	16.05	13.55

Table 6.14*b* shows the design of the sewers for which the Manning $n = 0.015$, minimum soil cover is 4.00 ft, and minimum nominal design velocity is 2.5 ft/second. The exit sewer of the system (sewer 31) flows into a creek for which the bottom elevation is 11.90 ft, the ground elevation of its bank is 21.00 ft, and its 10-yr flood water level is 20.00 ft.

Column 1:	Sewer number identified by its upstream inlet (manhole) number
Column 2:	Ground elevation at the upstream manhole of the sewer
Column 3:	Length of the sewer
Column 4:	Slope of the sewer, usually follows approximately the average ground slope along the sewer
Column 5:	Design discharge Q_p computed according to Eq. (6.44); thus, the product of columns 6 and 11 in Table 6.14*a*
Column 6:	Required sewer diameter, as computed by using Eq. (6.42) or (6.43) or Fig. 6.18; for Manning's formula with $n = 0.015$ and d_r in ft, Eq. (6.42) yields

$$d_r = \left(0.0324 \, \frac{Q_p}{\sqrt{S_o}} \right)^{3/8}$$

in which Q_p, in cfs, is given in column 5 and S_o is in Column 4

Column 7:	The nearest commercial nominal pipe size that is not smaller than the required size is adopted.
Column 8:	Flow velocity computed as $V = Q/A_f$; that is, it is calculated as column 5 multiplied by $4/\pi$ and divided by the square of column 7. As discussed in Yen (1978a), there are several ways to estimate the average velocity of the flow through the length of the sewer. Since the flow is actually unsteady and nonuniform, usually the one used here, using full pipe cross section, is a good approximation.
Column 9:	Sewer flow time is computed as equal to L/V, that is, column 3 divided by column 8 and converted into minutes.
Column 10:	Product of columns 3 and 4; this is the elevation difference between the two ends of the sewer.
Column 11:	The upstream pipe crown elevation of sewer 11 is computed from the ground elevation minus the minimum soil cover, 4.00 ft, to save soil excavation cost. In this example, sewers are assumed invert aligned except the last one (sewer 31), which is crown aligned at its upstream (23.85 ft for upstream of sewer 31 and downstream of sewer 21) to reduce backwater influence from the water level at sewer exit.
Column 12:	Pipe invert elevation at the upstream end of the sewer, equal to column 11 minus column 7
Column 13:	Pipe crown elevation at the downstream end of the sewer, equal to column 11 minus column 10
Column 14:	Pipe invert elevation at the downstream end of the sewer, equal to column 13 minus column 7. For the last sewer, the downstream invert elevation should be higher than the creek bottom elevation, 11.90 ft

The above example has demonstrated that, in the rational method, each sewer is designed individually and independently, except the computation of sewer flow time for the purpose of rainfall duration determination for the next sewer, that is, the values of t_f in column 8 of Table 6.14*a* are taken from those in column 9 of Table 6.14*b*.

The profile of the designed sewers are shown as the solid lines in Fig. 6.20*b*. If the water level of the creek downstream of sewer 31 is ignored, theoretically a cheaper design can be achieved by putting the exit sewer 31 on a slightly steeper slope, from 0.0144 to 0.0156 to reduce the pipe diameter from 2.75 ft to 2.50 ft. The new slope is completed from

$$S_o = [Q^2 n^2 K_n^{-2} \pi^{-2} 4^{10/3}] \, d^{-16/3} \tag{6.46}$$

This alternative is shown with the parentheses in Table 16.11*b* and as dashed lines in Fig. 6.20*b*. However, one should be aware that the water level of a 10-yr flood in the creek is 20.00 ft and, hence, the last sewer is actually surcharged and its exit is submerged. The sewer will not achieve the design discharge unless its upstream manhole is surcharged by almost 4 ft (20.00 − 16.05). Therefore, the original design of 2.75 ft diameter is a safer and preferred option in view of the backwater effect from the tailwater level in the creek. In fact, sewer 21–31 may also be surcharged due to the downstream backwater effect.

Sometimes backwater profile analysis can be performed on the sewer network to assess the degree of surcharge in the sewers and manholes. In such analyses, energy losses in the pipes and manholes junctions should be realistically accounted for.

However, the intensity–duration–frequency based design rainfall used in the rational method design is an idealistic, conceptual, and simplistic rainfall whose future occurrence probability is nil. The actual performance of the sewer system varies with different rainstorms having different temporal and spatial distributions of the rain. But it is impossible to know in advance these distributions for future rainstorms, whereas the design rainstorms are used as a consistent protection level measure. Although designing sewers by using the rational method is a relatively simple and straightforward matter, checking the performance of the sewer system is a far more complex task that requires thorough understanding of the hydrology and hydraulics of watershed runoff. For instance, checking the network performance by using an unsteady flow simulation model would require simulation of the unsteady flow in various locations in the network, accounting for losses in sewer pipes as well as in manholes and junctions; the latter will be discussed in Sec. 6.7.

Moreover, for a given sewer network layout, by using different sewer slopes, alternative designs of the network sewers can be obtained. A cost analysis should be conducted to select the most economically feasible design. This can be done with a system optimization model such as ILSD listed in Table 6.3*b* (Yen et al., 1984).

6.5.3 Time of Concentration

Time of concentration, t_c, can be regarded loosely as the time required for water to flow from the furthermost point of the drainage basin to its outlet. Precise determination of this flow time requires complicated hydraulic analysis using the Saint-Venant equations. However, a number of empirical or semiempirical formulas have been proposed to estimate the time of concentration for a drainage basin. Some of these formulas require a division of the basin into regions of overland flow and channel flow, whereas others treat the entire basin as a unit. For the former type, the time of concentration is

$$t_c = t_o + t_f \tag{6.47}$$

where t_o = overland flow travel time
t_f = channel flow travel time

In a given watershed, different possible flow paths have different flow times. The longest time among different possible paths from the boundary to the outlet of the basin is the time t_c used in the rational formula. The implicit reason is that under this situation of longest flow path time, the entire basin contributes.

The *channel flow time* is computed by

$$t_f = \frac{L_c}{V} \tag{6.48}$$

where L_c = channel length
V = average velocity

If there is more than one channel reach along the same path, then

$$t_f = \sum \frac{L_{cj}}{V_j} \tag{6.49}$$

The *overland flow time* t_o can be estimated by using one of the empirical or semiempirical formulas given in Table 6.15. When applied to actual catchments, among these formulas, the kinematic wave and Morgali-Linsley (1965) formulas usually yield a t_o value smaller than found empirically. A number of reasons possibly cause this discrepancy, including the following:

1. The catchment surface is usually not homogeneous as is assumed in the derivation of these two formulas; the surface undulation is far more than the sand-equivalent roughness assumed in the derivation.
2. For shallow depth Manning's n is not a constant (Yen, 1991) and raindrop impact increases n.
3. The sensitivity to rain input, $i^{-0.4}$ is far greater than reality. In the derivation, the input i is assumed as evenly distributed over the surface and without momentum—a pattern different from real rainfall.

TABLE 6.15 Selected Time of Concentration Formulas for Overland Flow

Name	Formula, t_o in min	Remarks
Izzard (1946)	$= 41 \left(0.0007 i^{1/3} + \dfrac{k}{i^{2/3}}\right)\left(\dfrac{L}{S}\right)^{1/3}$	Experimental, for $iL < 500$, value of k see Table 6.16
Kerby (1959)	$= 0.83 \left(\dfrac{N_k L}{\sqrt{S}}\right)^{0.467}$	Empirical, for $L \le 1200$ ft., see Table 6.17 for values of N_k
Airport	$= 0.39(1.1 - C)\dfrac{\sqrt{L}}{S^{1/3}}$	Applicable to areas similar to airport conditions, see Table 6.12 for runoff coefficient C
Schaake et al. (1967)	$= \dfrac{0.503 L^{0.24}}{S^{0.16}\alpha^{0.26}}$	α = percent imperviousness of area
Morgali and Linsley (1965)	$= 0.99 \dfrac{n^{0.605} L^{0.593}}{i^{0.388} S^{0.38}}$	Works best for turbulent flow on homogeneous surface
Kinematic wave	$= \dfrac{0.93}{i^{0.4}}\left(\dfrac{nL}{\sqrt{S}}\right)^{0.6}$	Same as above
Yen and Chow (1983)	$= K_y \left(\dfrac{NL}{\sqrt{S}}\right)^{0.6}$	Modified from kinematic wave formula, see Table 6.19 for values of K_y and Table 6.18 for values of N

Note: Length L in feet; Slope S dimensionless (i.e., ft/ft); Rainfall intensity i in inches per hour.

TABLE 6.16 Values of Retardance Coefficient k for Izzard's Formula

Surface	k
Very smooth asphalt pavement	0.0070
Tar and sand pavement	0.0075
Crushed-slate roofing paper	0.0082
Concrete pavement	0.012
Tar and gravel pavement	0.017
Closely clipped sod	0.016
Dense bluegrass turf	0.060

Source: Izzard (1946).

TABLE 6.17 Values of N_k for Kerby's Formula

Surface type	N_k
Smooth impervious surface	0.02
Smooth bare-packed soil	0.10
Poor grass, cultivated row crops, or moderately rough bare surface	0.20
Pasture or average grass	0.40
Deciduous timberland	0.60
Conifer timberland, deciduous timberland with deep forest litter, or dense grass	0.60

Source: Kerby (1959).

TABLE 6.18 Overland Texture Factor N for Yen and Chow Formula

Overland surface	Low	Medium	High
Smooth asphalt pavement	0.010	0.012	0.015
Smooth impervious surface	0.011	0.013	0.015
Tar and sand pavement	0.012	0.014	0.016
Concrete pavement	0.014	0.017	0.020
Rough impervious surface	0.015	0.019	0.023
Smooth bare packed soil	0.017	0.021	0.025
Moderate bare packed soil	0.025	0.030	0.035
Rough bare packed soil	0.032	0.038	0.045
Gravel soil	0.025	0.032	0.045
Mowed poor grass	0.030	0.038	0.045
Average grass, closely clipped sod	0.040	0.050	0.060
Pasture	0.040	0.055	0.070
Timberland	0.060	0.090	0.120
Dense grass	0.060	0.090	0.120
Shrubs and bushes	0.080	0.120	0.180
Land use			
Business	0.014	0.022	0.035
Semibusiness	0.022	0.035	0.050
Industrial	0.020	0.035	0.050
Dense residential	0.025	0.040	0.060
Surburban residential	0.030	0.055	0.080
Parks and lawns	0.040	0.075	0.120

Source: Yen and Chow (1983).

TABLE 6.19 Values of K_y for Yen and Chow Formula

		Light rain	Moderate rain	Heavy rain
	(in./h)	<0.8	0.8–1.2	>1.2
Rain intensity	(mm/h)	<20	20–30	>30
For L_o in feet with	t_o in min	1.5	1.1	0.7
For L_o in meters with	t_o in min	3.0	2.2	1.4

4. These two formulas are based on the time reaching the peak flow considering the influence of the flood wave propagation, which is different from the travel time of the water particle along the longest (or largest $L/\sqrt{S_o}$) flow path.

5. The hydraulic time of peak flow is measured from the commencement of rainfall excess, whereas the hydrologic time of concentration is measured from the commencement of rainfall.

As a remedy to the last comment, for infiltrating overland flow planes Akan (1989) obtained a numerical solution to the kinematic overland flow and the Green and Ampt infiltration equations and fitted the following equation to the numerical results by regression:

$$t_o = \left(\frac{nL}{K_n\sqrt{S_o}}\right)^{0.6}(i - K)^{-0.4} + 3.10K^{1.33}P_f\phi(1 - S_i)i^{-2.33} \qquad (6.50)$$

where K = soil hydraulic conductivity
ϕ = porosity
P_f = characteristic suction head
S_i = initial degree of saturation of the soil

Note that this equation reduced to the kinematic wave formula for impervious surfaces $K = 0$.

The kinematic wave, Morgali-Linsley, Izzard as well as Akan's Eq. (6.50), all contain the rain intensity i. In practical applications, often i is unknown *a priori*. Hence, t_o in these equations are computed iteratively with the aid of a rainfall intensity relationship.

6.6 CAPACITY AND BOTTLENECK DETERMINATION

One of the most useful pieces of information in solving urban drainage problems is knowing the flow carrying capacity of the channel or sewer. Knowing the sewer capacities allows a new approach in solving flood drainage problems by separating them into two parts: (1) the demand part of how much water needs to be drained, which is essentially a hydrology problem and (2) the supply part of how much can the sewer handle, that is, the capacity, which is a hydraulic problem. The flood drainage problem can be analyzed by comparing the two parts and then searching for a solution.

There are actually various kinds of sewer discharge capacities. There is the capacity for a single sewer. There is the capacity of the sewer network as a system, which usually is different from the capacity of individual sewers. For an open channel, often the maximum steady uniform flow that the channel can carry without spilling over bank is quoted as its capacity. For a sewer, the just-about-full gravity (open-channel) steady uniform flow is usually quoted.

In fact, for a subcritical free-surface flow, the discharge that the channel can carry depends on the downstream water level. For a sewer with a range of possible exit water levels, this requires repeated backwater profile computations. For a sewer network that has a number of connected channels, the number of backwater computations can easily become very large, making it nearly impossible, if not impractical, for a network capacity determination. Yen and González (1994) developed a method to summarize the backwater information of a channel into a hydraulic performance graph from which the network capacity can be determined.

6.6.1 Hydraulic Performance Graph

A *hydraulic performance graph* (HPG) is a plot of a set of curves, showing the relationship between water depths, y_u and y_d, at the upstream and downstream ends of the channel reach for different specified discharges, that is, $y_u = F(y_d, Q)$. Depicted in Figs. 6.21 and 6.22 are the HPGs for the mild-slope channel and steep-slope channel, respectively.

For a channel with mild slope (i.e., $y_n > y_c$ where y_n is the normal flow depth and y_c is the critical depth for the given Q), the HPG has the following main characteristics:

1. The hydraulic performance curves, each for a given discharge, never intercept each other. The curves with higher discharges are located above those with lower discharges.

2. The left bound of the curves indicated as "*C*-curve" in Fig. 6.21 represents the locus of critical flow condition at the downstream end of the channel reach. The downstream critical depth $y_d = y_c(Q)$ is obtained through computing the water surface angle ϕ_c for the given Q using Eq. (6.32a).

3. The hydraulic performance curves are bounded at the right by the 45° straight line $y_u = y_d - S_oL$, designated as Z-line in Fig. 6.21, which is the line representing a horizontal water surface and no flow. The performance curves approach asymptotically to the Z-line for very large values of y_d or y_u, where the flow has very small convective acceleration.

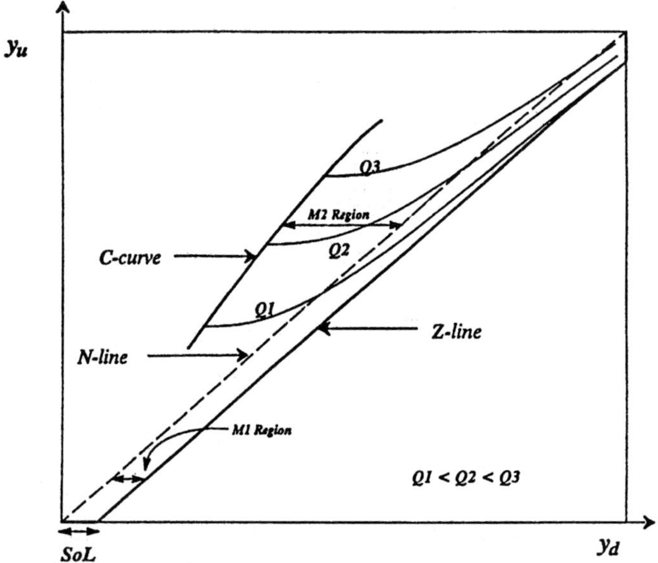

FIGURE 6.21   Hydraulic performance graph for mild-slope channel. (From Yen and Gonzalez, 1994)

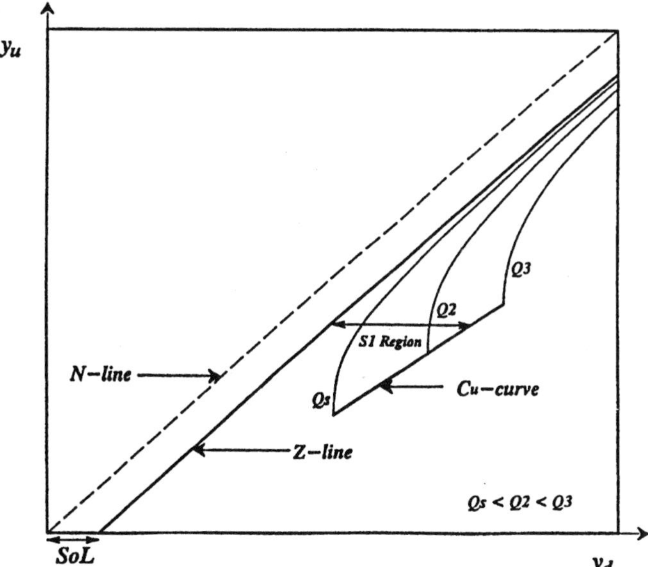

FIGURE 6.22   Hydraulic performance graph for steep-slope channel. (From Yen and Gonzalez, 1994)

4. The locus of normal flow depths for the possible discharges is named N-line. It can be expressed as $y_u = y_d$ and is located at a distance S_oL left to the Z-line.

5. The N-line divides the hydraulic performance curves into two regions. The region between the C-curve and the N-line contains all the possible pairs of upstream and downstream end water depths for which the backwater profiles are the $M2$-type (Fig. 6.4), whereas the water depth pairs within the region between the N-line and the Z-line correspond to all the possible $M1$-type profiles.

The HPGs for horizontal-slope and adverse-slope channels are similar to those in Fig. 14.21 without the N-line because for these two cases, the normal depth is infinity and imaginary, respectively.

The HPG of S1 profiles in a steep-slope channel (Fig. 6.22) has the following major characteristics:

1. The hydraulic performance curves of $S1$ profiles, each for a given discharge, never intercept each other. The curves with higher discharges are located below those with lower discharges.

2. The right bound of the curves indicated as C_u-curve in Fig. 6.22 represents the locus of critical flow condition at the upstream end of the channel reach. The upstream depth y_u is computed by using Eq. (6.32a) with $y_u = y_c(Q)$.

3. The region of $S1$-profile hydraulic performance curves is bounded on the left by the hydraulic performance curve that corresponds to the discharge Q_s, which starts at C_u curve and is asymptotic to the Z-line for large y_u and y_d. The threshold discharge Q_s is the discharge for which the channel slope is critical. For smaller discharges, the channel slope becomes mild instead of steep. The value of Q_s is determined by setting the critical depth equal to the normal depth, or equivalently the critical discharge equal to the normal flow discharge with $y_n = y_c$, which yields

$$\frac{R_s^{4/3} T_s}{A_s} = \frac{gn^2}{K_n^2 S_0} \tag{6.51}$$

For a given channel reach, the procedure to establish the hydraulic performance graph (HPG) is as follows:

1. Determine the ranges of depths or water surface elevations to be considered at the two ends of the channel reach.

2. Determine and plot the Z-line for which the water surface elevations are equal at the upstream and downstream ends, or $y_d = y_u + S_o L$, where S_o is the channel slope and L is the reach length.

3. Determine and plot the N-line, which is the 45°-line at a distance equal to $S_o L$ to the left of the Z-line, and on which $y_d = y_u$.

4. For a mild-slope channel with $M1$- or $M2$-type backwater profiles, choose a discharge Q:
   a. Compute the normal flow depth $y_n = y_u = y_d$ and mark this point on the N-line for this Q.
   b. Compute the critical depth, y_c, at the downstream end of the channel reach by using Eq. (6.32a).
   c. Perform the backwater computation for the given Q and y_c at the downstream end to determine the corresponding upstream depth y_u, the backwater computation can be done by using the standard step method, direct step method, or any other methods described, for example, in Chow (1959).
   d. Plot the result of this set of (y_u, $y_d = y_c$) for the specified Q as one point of the the the C-curve on the HPG; it is also the beginning point of the hydraulic performance curve of the chosen discharge.

   For a steep-slope channel with $S1$-type backwater profiles,
   i.   Determine the value of Q_s which corresponds to the condition at which the critical depth is equal to the normal depth, that is, $y_n = y_c$, or equivalently the critical discharge is equal to the normal flow discharge, that is, $Q_n = Q_c$. The value of $y_n = y_c$ can be obtained by using Eq. (6.51). This depth is the water depth that corresponds to the minimum discharge Q_s, for which the channel slope remains steep, and with this depth, Q_s can be computed by using Eq. (6.32a).
   ii. For a chosen $Q \geq Q_s$, use Eq. (6.32a) to compute the critical depth y_c at the upstream end of the channel reach.
   iii. Perform the backwater computation for the specified Q starting with the corresponding y_c at the upstream end of the reach to determine the downstream depth y_d.
   iv. Plot the result of this set of (y_d, $y_u = y_c$) for the specified Q as one point of the C_u-curve on the HPG, and it serves as the beginning point of the hydraulic performance curve of the chosen discharge.

5. For the Q chosen in step (4), select a feasible downstream depth y_d and perform a backwater computation to determine the upstream depth y_u. This pair of (y_d, y_u) constitutes a point of the hydraulic performance curve for this Q.

6. Repeat step (5) for a few selected y_d's to provide sufficient pairs of (y_d, y_u) values to plot the hydraulic performance curve for the chosen Q. The curve starts at the C-curve and approaches asymptotically to the Z-line for large y_u or y_d. For a mild-slope channel, the constant Q curve crosses the N-line between the downstream C_d-curve and the Z-line. For a steep-slope channel, the constant Q curve starts at the upstream C_u-curve.

7. Select different feasible discharges and repeat steps (4) to (6) to establish the hydraulic performance curves for different Q's.

8. Connect the critical-depth C points computed for different discharges as the C-curve on the HPG.

A typical M-type HPG is shown in Fig. 6.23 as an example.

6.6.2 Flow Capacities of a Channel Reach

The following representative flow capacities can be defined for an individual channel reach (Yen and González, 1994):

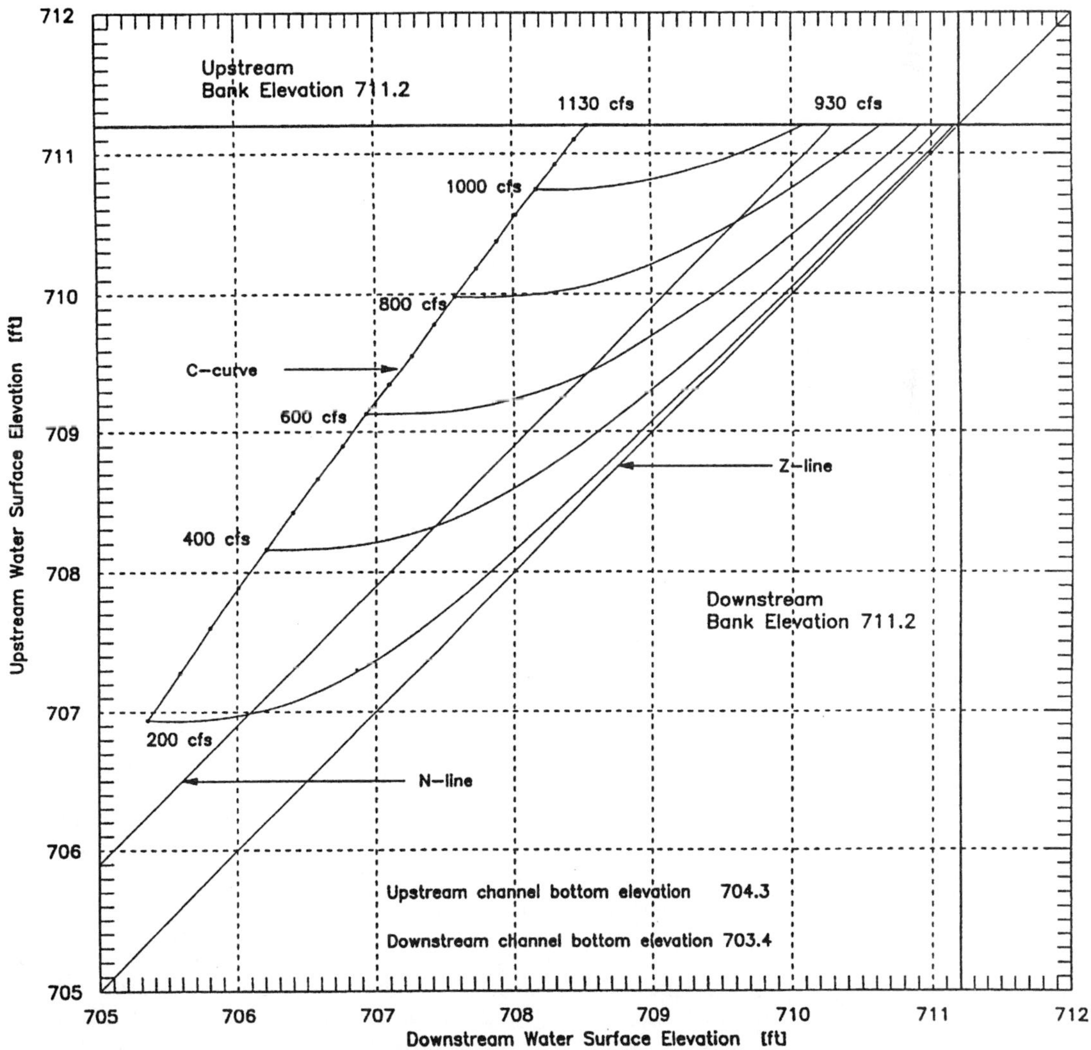

FIGURE 6.23 Example HPG for Reach 1 of Boneyard Creek.

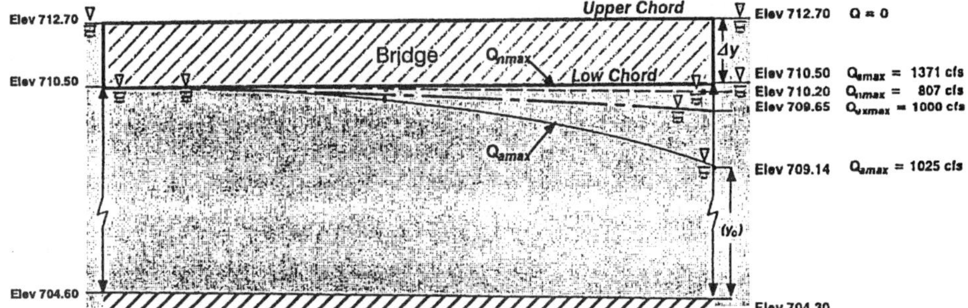

FIGURE 6.24 Water surface profiles for different hydraulic capacities in Reach 2.

1. Absolute maximum carrying capacity of a channel reach (Q_{amax})—the largest discharge the reach is able to convey when the water depth is at its exit cross-section—is critical while there is no bank overflow.

2. Maximum uniform flow capacity (Q_{nmax})—the maximum steady, uniform, flow discharge that the reach can convey either as the flow is just about to spill overbank, or as the flow reaches a surcharged condition, with the free surface parallel to the channel bottom.

3. Maximum flow capacity for a given exit water level (Q_{exmax})—for a given tailwater stage, the maximum steady gradually varied open-channel flow discharge that the reach can carry without spilling overbank. Obviously, this capacity varies with the tailwater lever, having Q_{amax} as its upper limit.

4. Maximum surcharged-flow capacity (Q_{smax})—for a channel reach with a top cover such that under high flow the open-channel flow changes to pressurized conduit flow; this capacity is the discharge that this closed-top reach can convey when the upstream water surface is at the bankfull stage and the downstream water elevation is at the crown level of the opening of the bridge, culvert, or sewer.

The water surface profiles corresponding to these four different capacities are shown in Fig. 6.24 for reach 2 in Fig. 6.25 as an example. The value of Q_{smax} is determined from the closed-conduit flow rating curve shown in Fig. 6.26, whereas the open-channel flow capac-

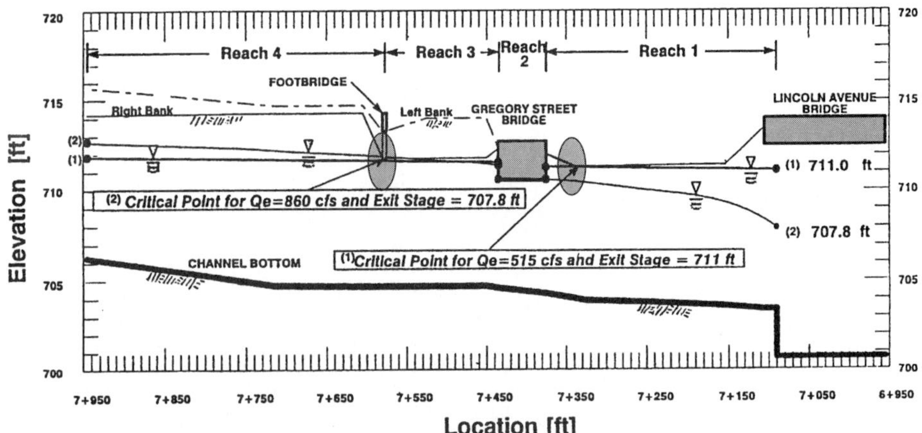

FIGURE 6.25 Example flow capacities and water surface profiles for two exit water levels.

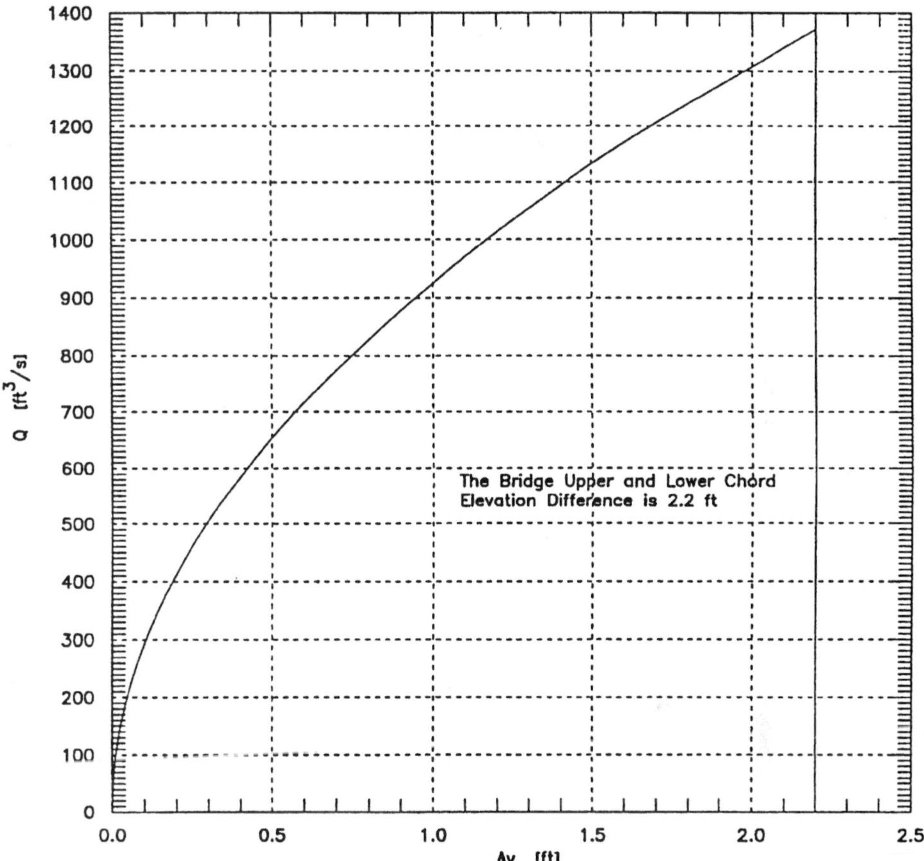

FIGURE 6.26   Rating curve for surcharge flow.

ities are determined from Fig. 6.27. The values of Q_{amax} and Q_{nmax} for the other three open-channel reaches individually are also listed in Table 6.20, which are read from their individual HPG shown in Figs. 6.23, 6.28, and 6.29.

The HPGs for channels of similar geometries can be nondimensionalized for more general uses. Shown in Fig. 6.30 is such a graph for open-channel flows in circular sewers with $S_oL/D = 0.05$ and $Q_f/\sqrt{gD^5} = 0.224$, where S_o, L, and D are the slope, length, and diameter of the sewer pipe; Q_f is the just-full steady uniform flow sewer capacity; y is the depth of flow, subscripts 1 and 2 refer to upstream (entrance) and downstream (exit) cross-section of the sewer, respectively; and the subscript n denotes normal (steady uniform) flow.

6.6.3   Bottleneck and Channel System Capacity Determination

The bottleneck of a network of drainage channels or sewers is the critical location within the network where the water is about to spill overbank or violate specified restriction. Therefore, for a given exit water level, the bottleneck location is identified in the process of determining the channel network system capacity. Different exit water levels may have different bottleneck locations. Because the flows in the channels of the network interact mutually, this interaction must be accounted for in the system capacity determination. The set

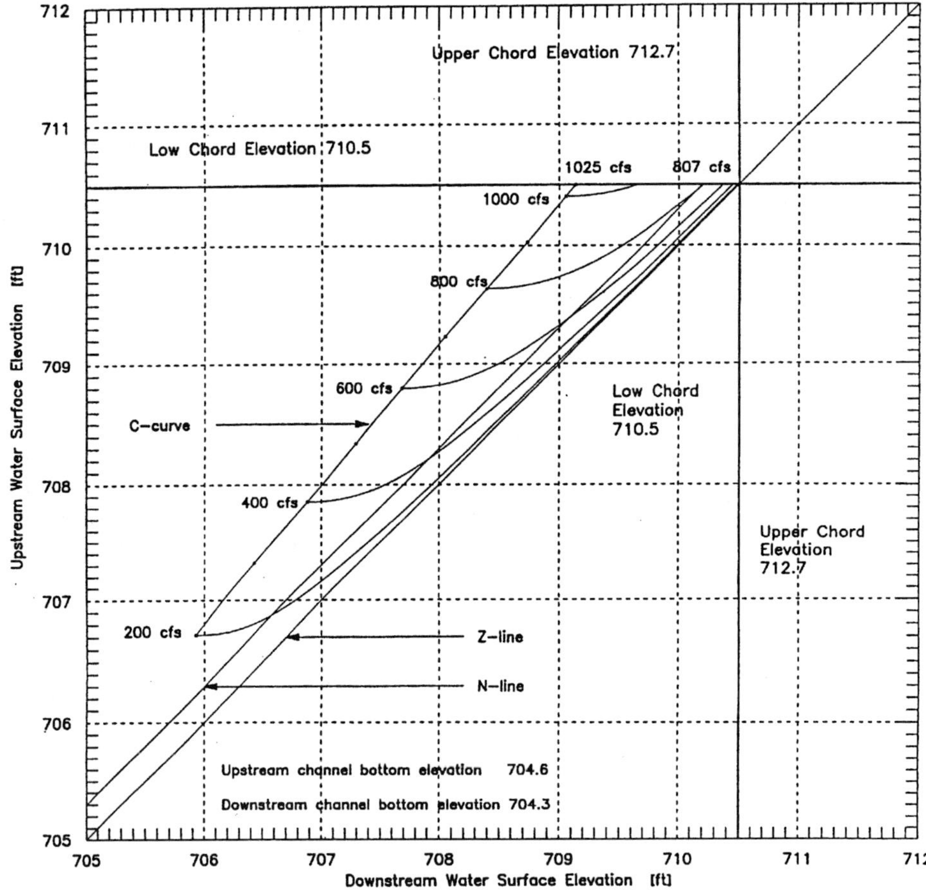

FIGURE 6.27   Example HPG for Reach 2.

TABLE 6.20   Maximum Capacities of Individual Reaches

Channel reach	Q_{amax} cfs	Q_{nmax} cfs	Q_{smax} cfs
Reach 1: Lincoln Ave.-Gregory St.	1,130	930	
Reach 2: Gregory St. Bridge	1,025	807	1,371
Reach 3: Gregory St.-Footbridge	1,100	800	
Reach 4: Footbridge-Loomis Lab (7 + 945)	1,370	1,070	

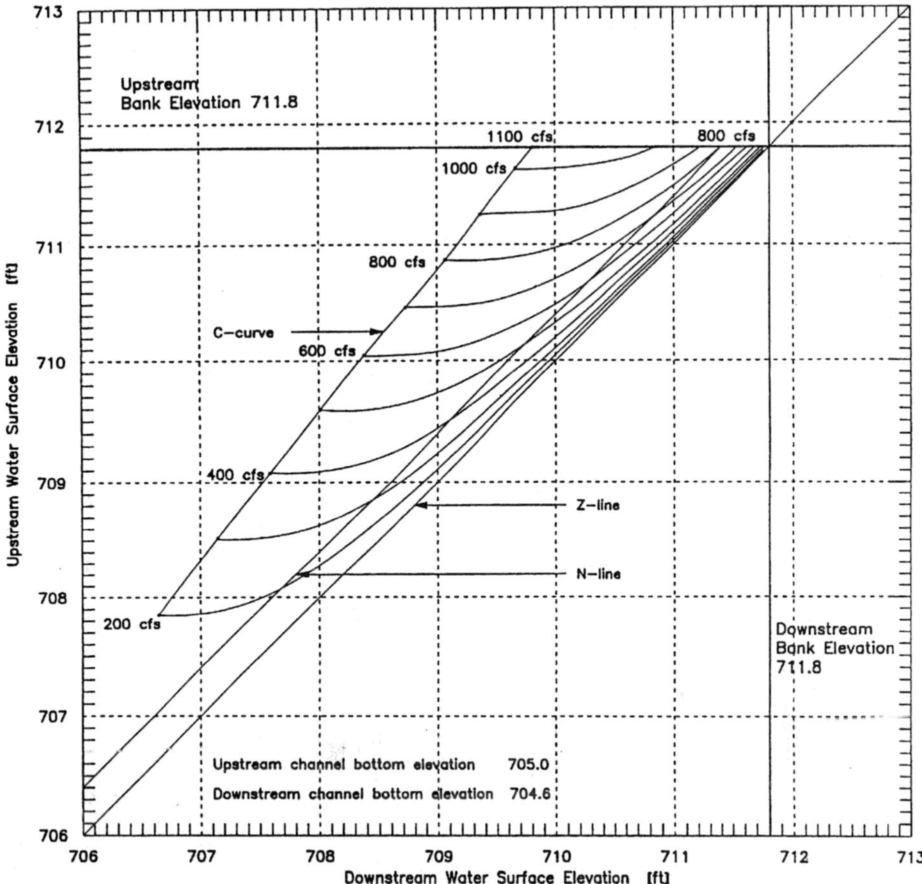

FIGURE 6.28   Example HPG for Reach 3.

of the HPGs and rating curves for the individual reaches of the system can be used together in sequence to determine the bottleneck and capacity of the channels as a system. To apply the HPG method of Yen and Gonzalez (1994) in determining the bottleneck and flow capacity of a drainage channel system, the channel system first is subdivided into reaches such that within each reach the geometry, alignment, and roughness are approximately the same and there is no significant lateral inflow within the reach. For subcritical flow, the system capacity is a function of the tailwater level at the exit of the most downstream reach. The maximum system capacity occurs when depth at the system exit is critical. While the flow capacity of each reach can be determined individually from its HPG, the flow capacity of the channel flowing as a part of the system should be determined by accounting for the backwater effect between the reaches, the losses at the junctions, if any, and the significant lateral flow joining the channel.

Concerning the lateral flow entering the channel system at the junctions between the reaches, a simple approximate method of Yen and Gonzalez (1994) can be used if no better method is available. As shown schematically in Fig. 6.31, the area drained by channel reach $j - 1$ is A_u and the corresponding discharge is $Q_u = C_u i A_u$, where C_u is the runoff coefficient for the area drained. The lateral flow joining the channel at the downstream end of reach

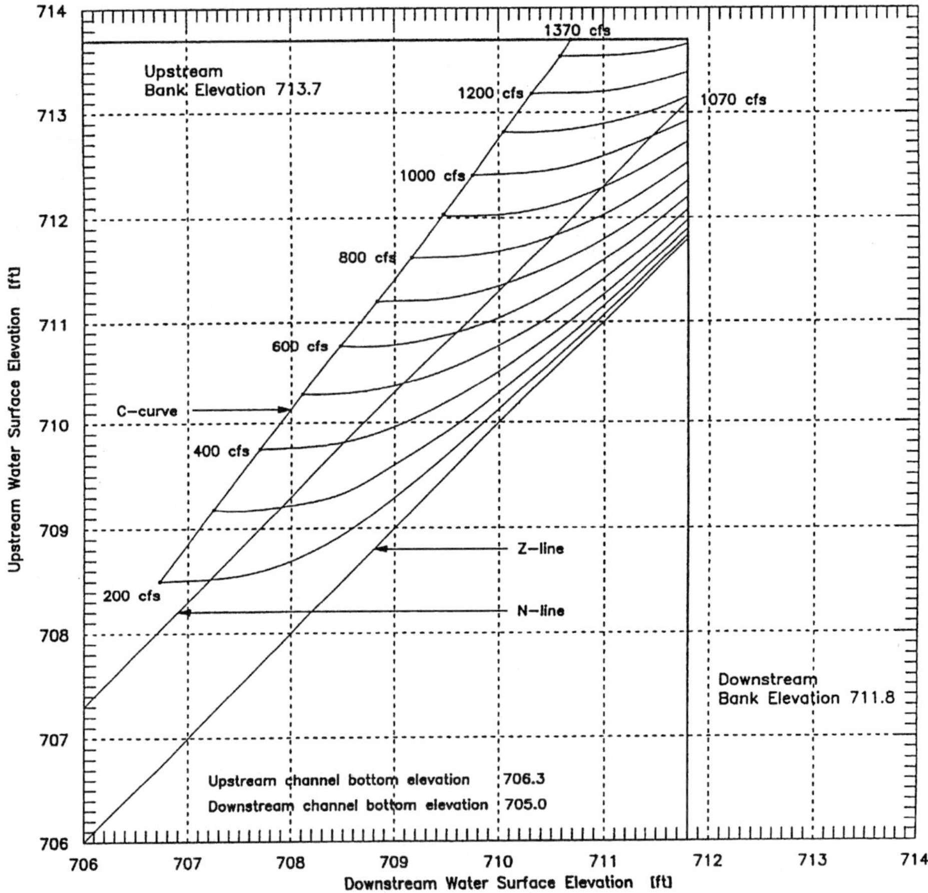

FIGURE 6.29   Example HPG for Reach 4.

$j - 1$ is $Q_{L_{(j-1)}}$. The sewer delivering the lateral flow peak discharge $Q_{L_{(j-1)}}$ drains an incremental local area $A_{L_{(j-1)}}$, having a $C_{L_{(j-1)}}$ runoff coefficient. Therefore, under the rainstorm with intensity i, $Q_{L_{(j-1)}} = C_{L_{(j-1)}} i A_{L_{(j-1)}}$. At the junction between reaches $j - 1$ and j, $Q_j = Q_u + Q_{L_{(j-1)}}$. Hence, the ratio between the lateral flow and the flow in reach $j - 1$ is

$$\frac{Q_{L_{(j-1)}}}{Q_u} = \frac{C_{L_{(j-1)}} A_{L_{(j-1)}}}{C_u A_u} \tag{6.52}$$

For a given tailwater level at the system exit, for each reach the upstream water level is determined from the HPG, knowing the downstream water level, progressing reach by reach toward upstream, with junction head losses included if they exist. If the reach is surcharged, the rating curve is used, instead of HPG. It may require a trial of several discharges to locate the bottleneck and identify the system capacity iteratively. Details and examples of this procedure can be found in Yen and Gonzalez (1994). By applying this procedure to the example 4-reach system shown in Fig. 6.25, with the rating curve of Fig. 6.26 and HPGs of Figs. 6.23, 6.28, and 6.29, the network capacity for exit tailwater level equal to 711.0 ft is determined as 515 cfs. This capacity is controlled by spilling near the upstream end of reach 1 as shown in Fig. 6.25. The water surface profile is also shown in this figure.

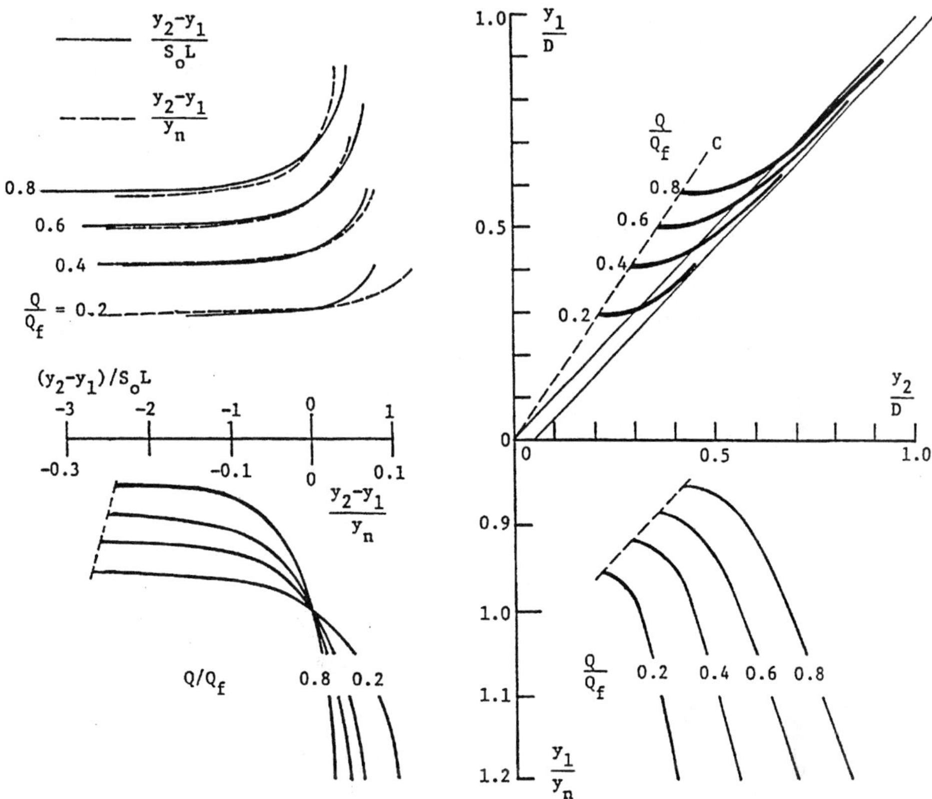

FIGURE 6.30   Nondimensional HPG for sewer pipes. (After Yen, 1987).

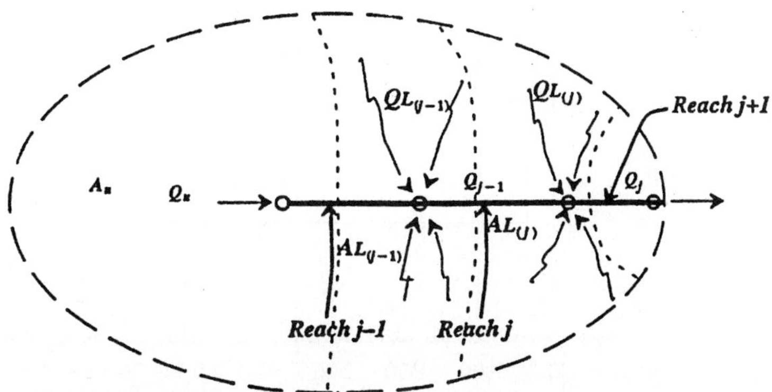

FIGURE 6.31   Schematic of lateral runoff contribution to a sewer. (After Yen and Gonzalez, 1994)

The channel system hydraulic capacity and the location of the bottleneck vary with the water surface elevation at the exit. When the exit water level is low, the bottlenecks tend to locate in the upstream parts of the system. As the exit tailwater level rises, the bottlenecks tend to move downstream and the system capacity decreases as demonstrated by the example shown in Fig. 6.32. Identification of the most likely range of exit tailwater levels and removal of the bottlenecks in this range may be a relatively simple and effective way for system capacity improvement. For the example system shown in Figs. 6.25 and 6.32, the removal (raising) of the Gregory Street Bridge (the first two obstacles) improves the system capacity as indicated by the dash curve in Fig. 6.32.

If the flood frequency (discharge–return period) relationship is known, the system capacity curve and the locations of bottlenecks shown in Fig. 6.32 can be converted into a system capacity curve in terms of return period versus exit water level as demonstrated in Fig. 6.33 for the system shown in Fig. 6.25. It can be seen that with the improvement of the Gregory Street Bridge removal, the system absolute maximum capacity is increased from 860 cfs to 970 cfs, or a return period improvement from 25 yrs to 40 yrs. At a likely exit water level of 710.75 ft, the system capacity is increased from 655 cfs to 730 cfs, or a return period improvement from 11 yrs to 15 yrs.

Thus, it is obvious that when the connecting reaches are considered to be interacting as a system, the overall channel capacity is different from any of the capacity values of the individual reaches. For an open-channel system, the backwater effects of connecting reaches usually prevent the exit depth of interior reaches from becoming critical. Therefore, the absolute maximum capacity, Q_{amax}, of a reach serves as the upper bound, provided that the

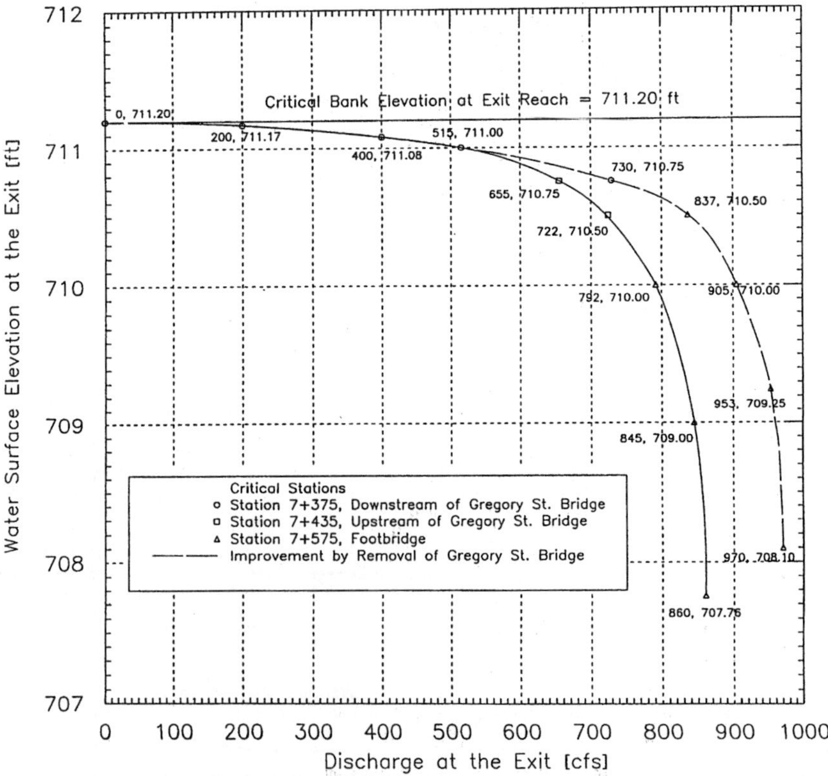

FIGURE 6.32 Example hydraulic capacity curve of channel system.

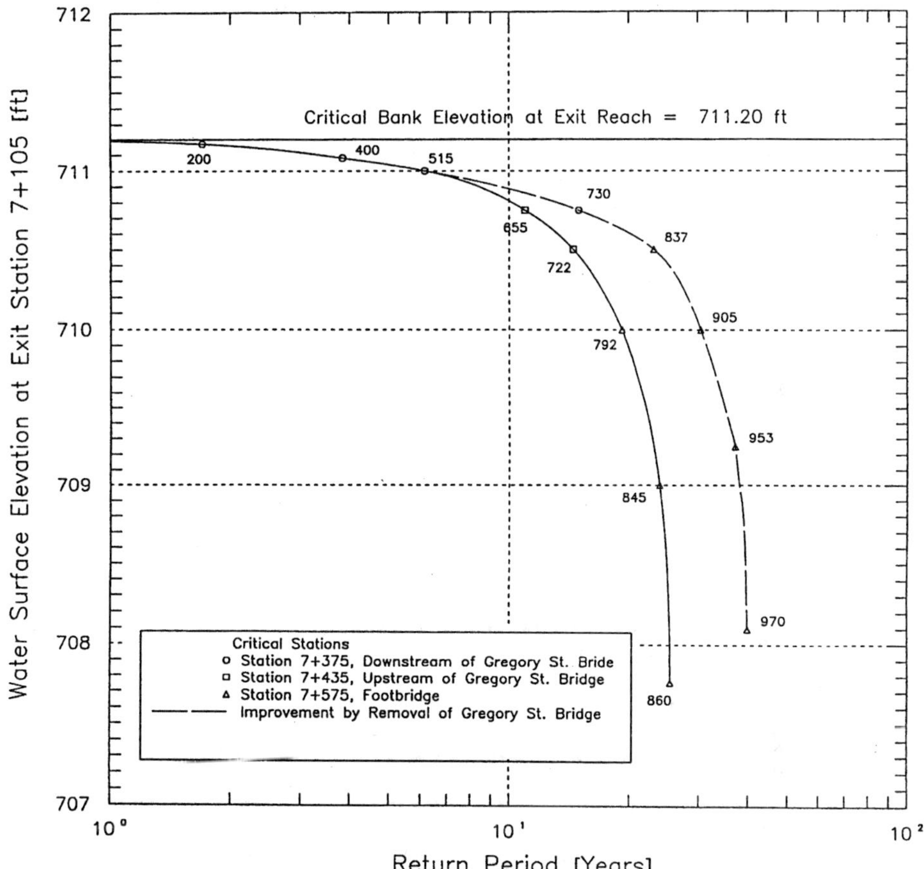

FIGURE 6.33 Example flood stage frequency of channel system.

open-channel flow prevails in the reach and also in adjacent reaches upstream and downstream. For a closed-top reach, the upper bound is the larger of Q_{amax} and Q_{smax}. For an open-channel reach connected to a closed-top reach at either its upstream, downstream, or both, the upper bound is the larger between Q_{amax} and the largest discharge allowed under submerged exit or entrance condition of the open-channel reach.

The capacity of the system of reaches as a whole cannot exceed the smallest of the upper bound of the individual reaches just mentioned, adjusted for lateral flow entering the interior reaches in the channel system. However, the location of the bottleneck, which determines the capacity of the channels as a system, may not and often is not in this reach of smallest Q_{amax} upper bound, and, in such a case, the system capacity is smaller than this smallest upper bound.

6.7 HYDRAULICS OF SEWER JUNCTIONS

Various auxiliary hydraulic structures, such as junctions, manholes, weirs, siphons, pumps, valves, gates, transition structures, outlet controls, and drop shafts can be found in sewer networks. Information relevant to design of most of these apparatuses are described in standard fluid mechanics textbooks and references, particularly in Hager (1999) and FHWA (1996). In this section, the most important auxiliary component in sewer flow modeling, namely, the sewer junctions, is discussed. For common sizes and lengths of sewers, the head loss for the flow through a sewer is usually two to five times the velocity head, while the junction loss is in the order of approaching velocity head. Thus, the head loss through a junction is comparable to the sewer pipe loss, not a minor loss.

6.7.1 Classification of Sewer Junctions

A sewer junction usually has three or four sewer pipes joined to it. Under usual flow condition, one downstream pipe receives the outflow from the junction, and other pipes flow into the junction. However, junctions with only two or more than four joining pipes are not uncommon. The most upstream junctions in a sewer network are usually one-way junctions, having only one sewer connected to the junction. The horizontal cross-section of the junction can be circular, square, or another shape. The diameter or horizontal dimension of a junction normally is not smaller than the largest diameter of the joining sewers. In order to allow the workman room for operation, usually junctions are not smaller than 3 ft (1 m) in diameter. For large sewers, the access to the junction can be smaller than the diameter of the largest joining sewer.

Sewers may join a junction with different vertical and horizontal alignments, and they may have different sizes and slopes. Vertically, the pipes may join at the junction with their centerlines, inverts or crowns aligned, or with any line of alignment in between. There is no clearly preferred alignment that could simultaneously satisfy the requirements of good hy-

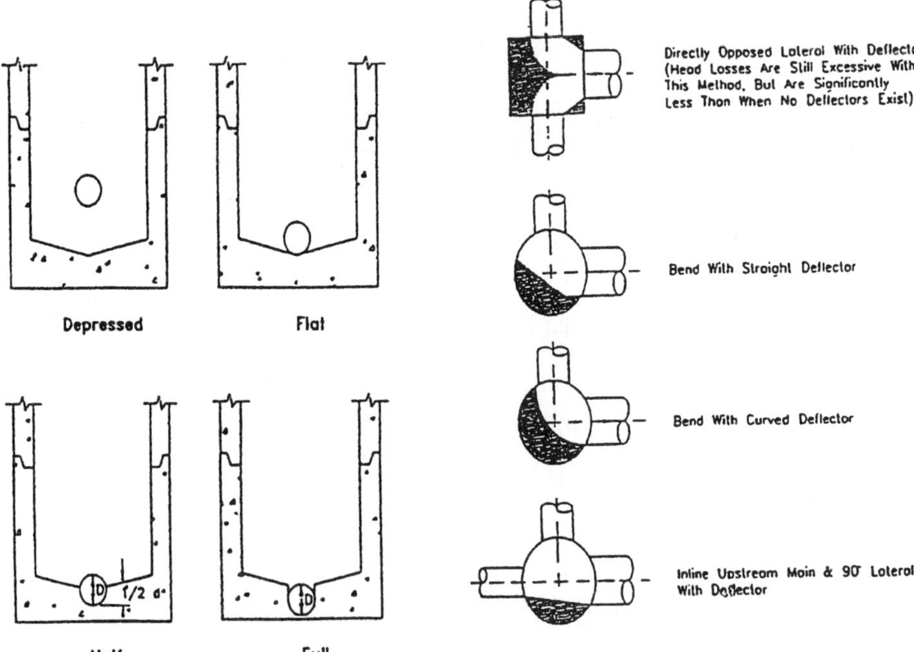

FIGURE 6.34 Benching and flow guide of sewer junction.

draulics at low and high flows without complicating either construction cost or design. The bottom of the junction is usually at or slightly lower than the lowest invert of the joining sewers.

In the horizontal alignment, often the outflow sewer is aligned with one (usually the major) inflow sewer in a straight line with other sewers joining at an angle. For cities with square blocks, right-angle junctions are most common. Typical sewer benching and flow guides in junctions are shown in Fig. 6.34.

With the alignment of the joining pipes and the shape and dimensions of junctions not standardized, the precise, quantitative hydraulic characteristics of the junctions vary considerably, resulting in many individual studies of specified junctions. A general comprehensive quantitative description of junctions has yet to be produced.

For the purpose of hydraulic analysis, junctions can be classified according to the following schemes (Yen, 1986):

1. According to the geometry:
    a. One-way junction
    b. Two-way junction
    c. Three-way junction
        i. Merging (two pipes flow into one pipe)
        ii. Dividing (one pipe flows into two pipes)
    d. Four- or more-way junction
        i. Merging
        ii. Dividing
        iii. Merging and dividing

2. According to the flow in the joining pipes:
    a. Open-channel junction (with open-channel flow in all joining pipes)
    b. Surcharge junction (with all joining pipes surcharged)
    c. Partially surcharged junction (with some, but not all, joining pipes surcharged)

3. According to the significance of the junction storage on the flow:
    a. Storage junction
    b. Point junction

Hydraulically, the most important feature of a junction is that it imposes backwater effects to the sewers connected to it. A junction provides, in addition to a volume—however small—of temporal storage, redistribution and dissipation of energy, and mixing and transfer of momentum of the flow and of the sediments and pollutants it carries. The precise, detailed hydraulic description of the flow in a sewer junction is complicated because of the high degree of mixing, separation, turbulence, and energy losses. However, correct representation of the junction hydraulics is important in realistic simulation and reliable computation of the flow in a sewer system (Sevuk and Yen, 1973).

6.7.2 Junction Hydraulic Equations

The continuity equation of the water in a junction is

$$\sum Q_i + Q_j = \frac{ds}{dt} \tag{6.53}$$

where Q_i = flow into or out from the junction by the i-th joining sewer, being positive for inflow and negative for outflow

Q_j = direct, temporally variable water inflow into (positive) or the pumpage or overflow or leakage out from (negative) the junction, if any

s = water storage in the junction
t = time

For a two-way junction, the index $i = 1, 2$; for a three-way junction, $i = 1, 2, 3$, etc. The energy equation in a one-dimensional analysis form is

$$\sum Q_i \left(\frac{V_i^2}{2g} + \frac{P_i}{\gamma} + Z_i \right) + Q_j H_j = s \frac{dY}{dt} + \sum Q_i K_i \frac{V_i^2}{2g} \qquad (6.54)$$

where Z_i, P_i, and V_i = pipe invert elevation above the reference datum, piezometric pressure above the pipe invert, and velocity of the flow at the end cross section of the i-th sewer where it meets the junction

H_j = net energy input per unit volume of the direct inflow expressed in water head

K_i = entrance or exit loss coefficient for the i-th sewer

Y = depth of water in the junction

g = gravitational acceleration

The first summation term in Eq. (6.54) is the sum of the energy input and output by the joining pipes. The second term at the left side of the equation is the net energy brought in by the direct inflow. The first term to the right of the equal sign is the energy stored in the junction as its water depth rises. The last term is the energy losses.

The momentum equations for the two horizontal orthogonal directions x and z are

$$\sum (Q_i V_{ix}) = \int g \frac{p_x}{\gamma} dA \qquad (6.55)$$

$$\sum (Q_i V_{iz}) = \int g \frac{p_z}{\gamma} dA \qquad (6.56)$$

where p_x and p_z = x and z components of the pressure acting on the junction boundary

A = solid and water boundary surface of the junction

The direct flow Q_j is assumed to enter the junction without horizontal velocity component. The right side term of Eqs. (6.55) and (6.56) is the x or z component force, where the integration is over the entire junction boundary surface A. The left side term is the sum of momentum of the inflow and outflow of the joining pipes. Note that for a three-way merging junction, two of the Q_i's are positive and one Q_i is negative, whereas, for a three-way dividing junction, two of the Q_i's are negative.

Taylor (1944), Kanda and Kitada (1977), Joliffe (1982), and others have suggested the momentum approach to deal with high velocity situations. To illustrate this, consider the three-way junction shown in Fig. 6.35. The control volume of water at the junction enclosed by the dash lines is regarded as a point, and there is no volume change associated with a change of depth within it. One of the two merging sewers is along the direction of the downstream sewer, whereas the branch sewer makes an angle φ with it. When one assumes

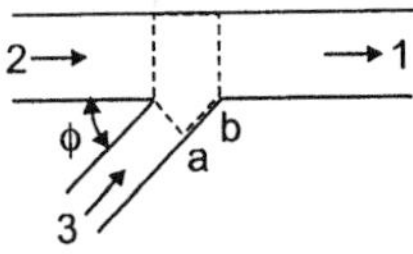

FIGURE 6.35   Control volume of junction for momentum analysis.

that the pressure distribution is hydrostatic and the flow is steady, the force–momentum relation can be written as

$$\gamma \bar{h}_2 A_2 + \gamma \bar{h}_3 A_3 \cos \varphi - \gamma \bar{h}_b A_b \sin \varphi - \gamma \bar{h}_1 A_1 + F$$
$$= \rho Q_1 V_1 - \rho Q_2 V_2 - \rho Q_3 V_3 \cos \varphi \tag{6.57}$$

where A = flow cross-sectional area in a sewer
  h = depth of the centroid of A
  γ = specific weight
  ρ = density of water
  Q = discharge
  $V = Q/A$, the cross-sectional mean velocity
  F = sum of other forces that are normally neglected

Some of these neglected forces are the component of the water weight in the control volume along the small bottom slope, the shear stresses on the walls and bottom, and the force due to geometry of the junction if the sewers are not invert aligned or the longitudinal sewers are of different dimensions. The subscripts 1, 2, and 3 identify the sewers as shown in Fig. 6.35, and "b" represents the exposed wall surface of the branch in the control volume shown as "ab" in the figure. For the special case of invert aligned sewers with the branch (pipe 3) joining at right angle, $\varphi = 90°$, Eq. (6.57) can be simplified as

$$A_2(g\bar{h}_2 + V_2^2) = A_1(g\bar{h}_1 + V_1^2) \tag{6.58}$$

or

$$\frac{Q_2}{Q_1} = \left[\left(\frac{A_2}{A_1} \right) \frac{(g\bar{h}_1/V_1^2) + 1}{(g\bar{h}_2/V_2^2) + 1} \right]^{1/2} \tag{6.59}$$

Based on experimental results of invert-aligned equal-size pipes merging with $\varphi = 90°$, Joliffe (1982) observed that the upstream depths $h_3 = h_2$ and proposed that

$$\frac{h_3}{h_{c1}} = \frac{h_2}{h_{c1}} = \xi F_3^{-b} \tag{6.60}$$

where h_{c1} = critical depth in the downstream sewer
  F_3 = Froude number of the flow in the branch sewer

and

$$\xi = 0.999 - 0.408 \left(\frac{Q_2}{Q_1} \right) - 0.381 \left(\frac{Q_2}{Q_1} \right)^2 \tag{6.61}$$

$$b = 0.514 - 0.067 \left(\frac{Q_2}{Q_1} \right) + 0.197 \left(\frac{Q_2}{Q_1} \right)^2 - 0.122 \left(\frac{Q_2}{Q_1} \right)^3 \tag{6.62}$$

The equation describing the load of sediment or pollutants, expressed in terms of concentration c, can be derived from the principle of conservation as

$$\frac{d}{dt} \int_s cds + \sum Q_i c_i + Q_j c_j = G \tag{6.63}$$

in which G denotes a source (positive) or sink (negative) of the sediment or pollutants in the junction.

TABLE 6.21 Experiments on Three-way Junction of Merging Surcharged Channels

Reference	Type of junction	Shape of channels	Channel slope	Pipe alignment at junction		Type of flow	Remarks
				Vertical	Longitudinal		
Sangster et al (1958, 1961)	Square, rectangular or round box	Circular, D = 3.0 in., 3.75 in., 4.75 in., or 5.72 in.	Horizontal	Flushed bottom	Straight through and one 90° merging channel	Steady	Also tests of opposed lateral pipes; tests with grate inflow into junction
Lindvall (1984, 1987)	Round box	Circular, D_{main} = 144 mm, D_{br}/D_{main} = 1.0 or 0.686 or 0.389	Horizontal	Center aligned	Straight through and one 90° merging channel	Steady	Loss coefficient dependent on junction diameter, lateral pipe diameter, and flow ratio
Johnston & Volker (1990)	Square box	Circular, $D_{main\ d}$ = 70 mm, D_{main} up/ $D_{main\ d}$ = 0.64, $D_{br}/D_{main\ d}$ = 0.91	Horizontal	Flushed bottom	Centerline aligned with slight deflector for lateral in junction	Steady	
Blaisdell & Mason (1967)	Enclosed pipe junction	Circular, D_{br}/D_{main} = 0.25 ~ 1.0		Center or top aligned	Straight through and one merging channel at 15° to 165° by 15° increments	Steady	Reynolds number effect insignificant
Serre et al. (1994)	Enclosed pipe junction	Circular, D_{main} = 444 mm, D_{br}/D_{main} = 0.14, 0.23, 0.34, or 0.46	(Horizontal)	Center aligned	Straight through and one 90° merging channel	Steady	
Ramamurthy & Zhu (1997)	Enclosed rectangular conduit junction	Rectangular, 4.14 mm high, main width 91.5 mm, branch width 20.4 mm, 70.5 mm, or 91.5 m	(Horizontal)	Same height	Straight through and one 90° emerging branch	Steady	

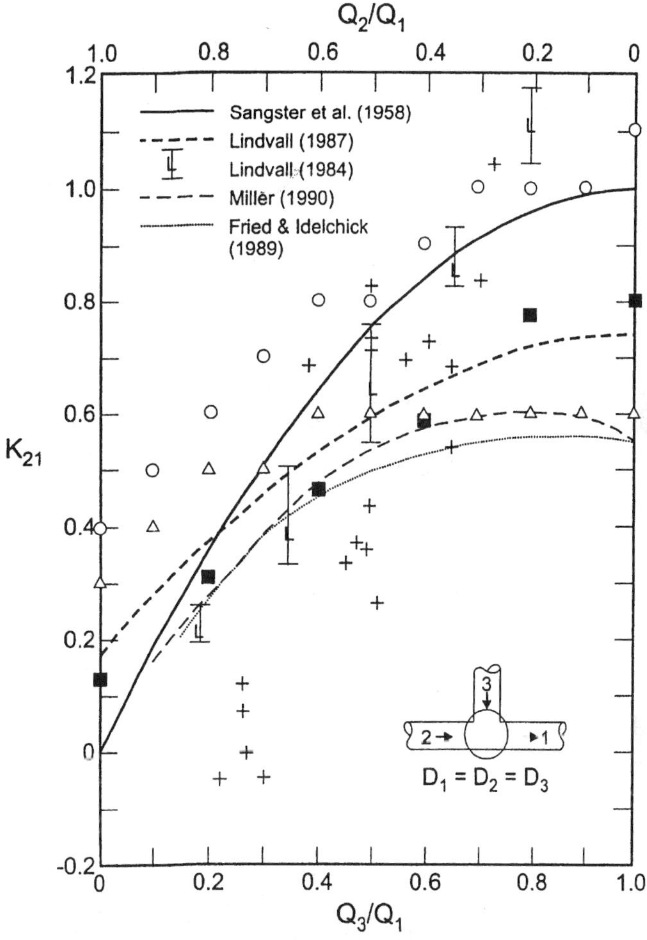

(*a*) Mainline loss coefficient K_{21}

FIGURE 6.36 Experimental headloss coefficients for surcharged 3-way sewer junction with identical joining pipe sizes and 90° merging lateral.

Equations (6.53) to (6.62) are the basic equations for sewer junctions. They are applicable to junctions under surcharge as well as open-channel flows in the joining pipes. However, more specific equations can be written for point-type and storage-type junctions. Selection of the equations for application to a particular junction depends on the junction geometry and flow condition.

Proper handling of flow in sewer networks requires information about the loss coefficients at the junctions. Unfortunately, practically no useful quantitative information exists about energy and momentum losses of unsteady flow passing through a junction. Therefore, steady flow information on sewer junction losses are commonly used as approximations.

6.7.3 Loss Coefficients for Three-Way Sewer Junctions

Table 6.21 summarizes the experimental conditions of top-open sewer junctions of three-way merging and surcharged pipes conducted by Sangster et al. (1958, 1961), Lindvall

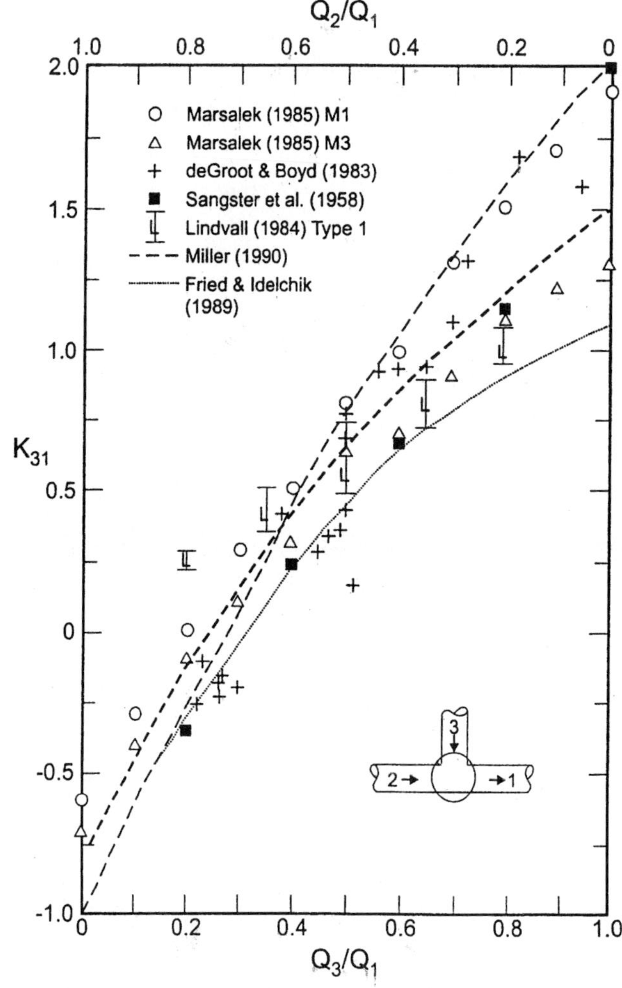

(*b*) Branch loss coefficient K_{31}

FIGURE 6.36   (*Continued*)

(1984), and Johnston and Volker (1990). Also listed in the table are experiments by Blaisdell and Mason (1967), Serre et al. (1994), and Ramamurthy and Zhu (1997), which were not conducted on open-top sewer junctions but on three-way merging closed-top junctions. They are listed as an example because these tests were conducted with different branch and main diameter ratios, and with different pipe alignments. Hence, they may provide helpful information for sewer junctions. Considerably more information about merging or dividing branched closed conduits exists than about sewer junctions. The reader may find information elsewhere (e.g., Fried and Idelchik, 1989; Miller, 1990) about centerline-aligned, three-way joining pipes as an approximation to sewer junctions.

The loss coefficients K_{21} and K_{31} for the merging flow are defined as

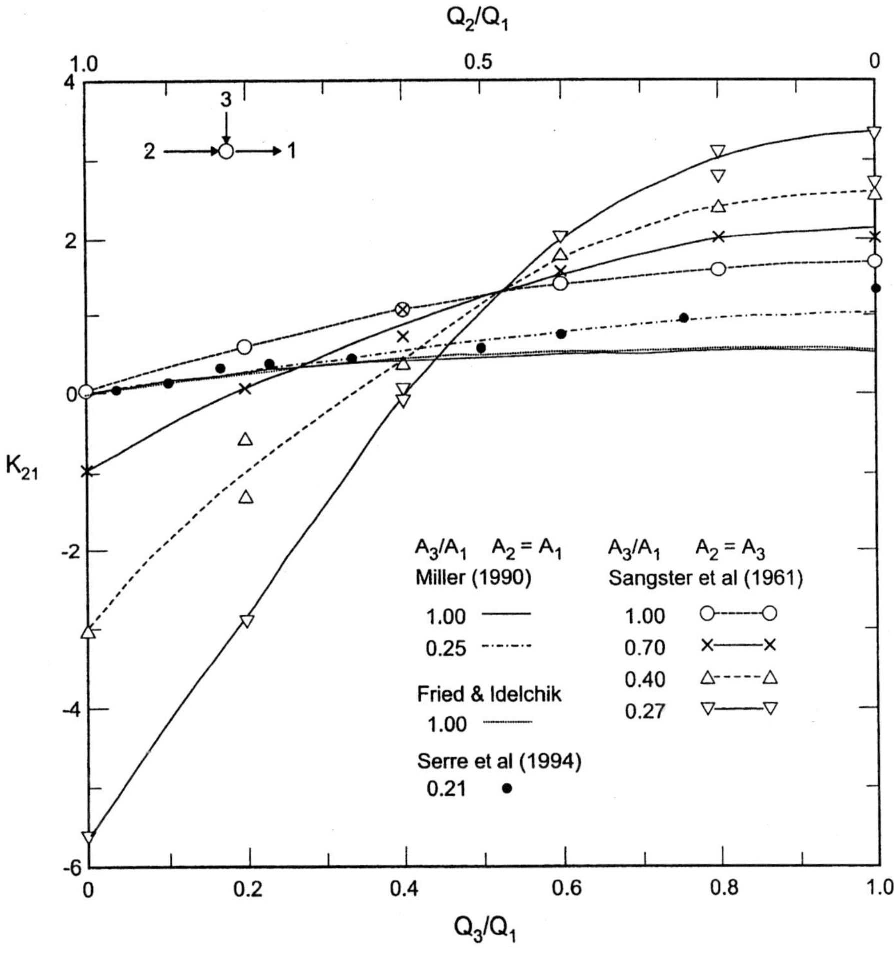

(*a*) Mainline loss coefficient K_{21}

FIGURE 6.37   Headloss coefficients for surcharged 3-way junction with 90° merging lateral of different sizes.

$$k_{ij} = \frac{\left[\left(\dfrac{V_i^2}{2g} + h_i + Z_i\right) - \left(\dfrac{V_j^2}{2g} + h_j + Z_j\right)\right]}{\left(\dfrac{V_1^2}{2g}\right)} \qquad (6.64)$$

Figure 6.36 shows the experimental results of Sangster et al. (1958) and Lindvall (1984, 1987) for the case of identical pipe size of the main and 90° merging lateral. The corresponding curves suggested by Miller (1990) and Fried and Idelchik (1989) for three-way identical closed pipe junctions are also shown as a reference. The values of the loss coefficients in a sewer junction that is open to air on its top are expected to be slightly higher

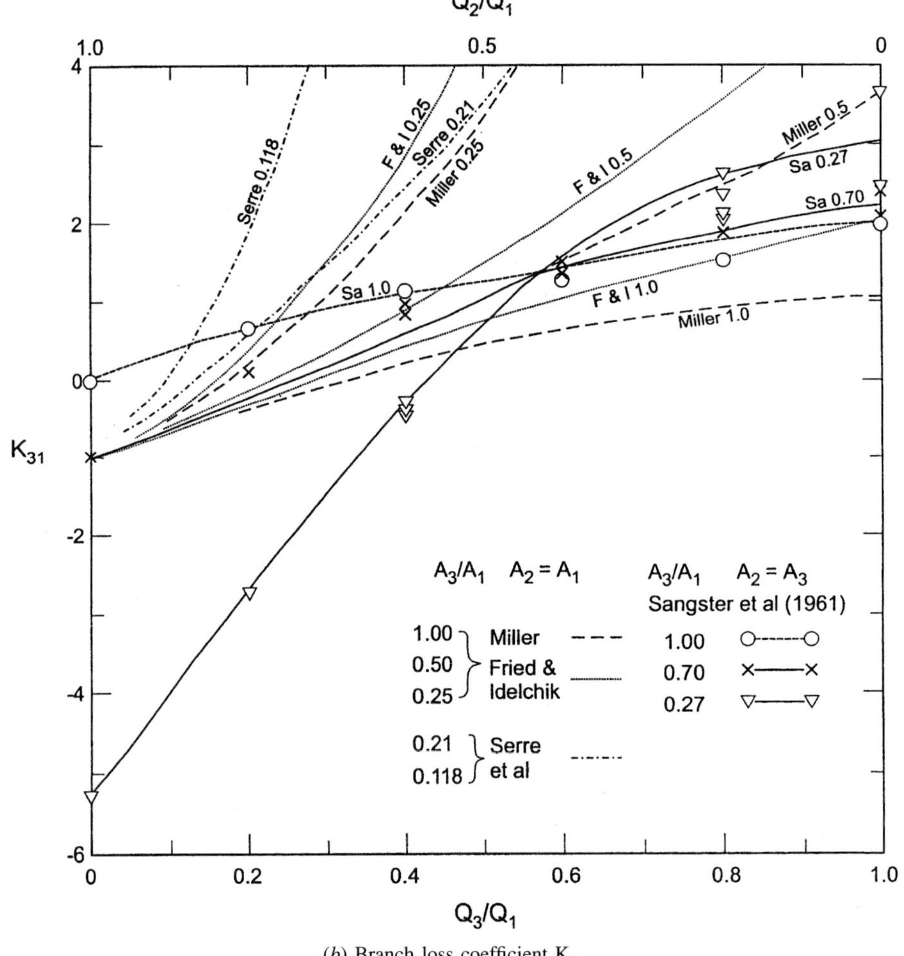

(*b*) Branch loss coefficient K_{31}

FIGURE 6.37   (*Continued*)

than the enclosed pipe junction cases given by Miller because of the water volume at the junction above the pipes.

The effect of the relative size of the joining branch pipe is shown in Fig. 6.37. The experimental data of Sangster et al. (1961) have identical upstream pipe sizes, $D_2 = D_3$ for four different values of lateral branch to downstream main pipe area ratio, A_3/A_1. The data of Johnston and Volker (1990) on surcharged, circular, open-top sewer junction are not plotted in Fig. 6.37 because of the mainline pipe area ratio $A_2/A_1 = 0.41$, instead of unity in the figure. Conversely, as a comparison, the smoothed curve of K_{21} for $A_3/A_1 = 0.21$ of the three-way pipe junction of Serre et al. (1994) with $A_1 = A_2$ is plotted in Fig. 6.37*a*, and their experimental curves of K_{31} for $A_3/A_1 = 0.21$ and 0.118 are plotted in Fig. 6.37*b*. Also shown in the figure, as reference, are the three-way pipe junction curves for different values of A_3/A_1 suggested by Miller (1990) and Fried and Idelchik (1989) for identical sizes of main pipes, $A_2 = A_1$. The experiments of Sangster et al. (1961) indicated that for a given

A_3/A_1, the effect of A_2/A_1 on the loss coefficients is minor. Therefore, their curves should be comparable with those of Fried and Idelchik (1989), Miller (1990), and Serre et al. (1994). However, Fig. 6.37 depicts considerable disagreement among the different sources, indicating the need for more reliable investigations.

The joining angle of the lateral branch is a significant factor in affecting the loss coefficients, particularly on K_{31}. The values of the loss coefficients decrease if the joining angle more or less aligns with the flow direction of the main, and increase if the lateral flow is directed against the main. The references of Miller (1990) and Fried and Idelchik (1989) provide some idea on the adjustment needed for the K values due to the joining angle.

Listed in Table 6.22 is a summary of experiments on steady flow in three-way merging open-channel junctions. Most of the studies were done with point-type junctions. The experimental subcritical flow results of storage-type junctions by Townsend and Prins (1978) and Marsalek (1985) are plotted in Fig. 6.38 for lateral joining 90° to same-size mainline pipes. Yevjevich and Barnes (1970) gave the combined main and lateral loss coefficient but not the separate coefficients, making the result difficult to use in routing simulation. The points in the figure scatter considerably, but they are generally in the same range of the loss coefficient values for a surcharged three-way 90° merging junction, except K_{31}, for Townsend and Prins's data. Interestingly, the most frequently encountered sewer junctions are three- and four-way box junctions with unsteady subcritical flow in the joining circular sewers. None of the open-channel experiments was conducted under this condition. All were tested with steady flow. Obviously, existing experimental evidence and theory do not yield reliable quantitative information on the loss coefficients of three-way sewer junctions. Before more reliable information is obtained, provincially for the design and simulation of three joining identical size sewers, for K_{21} a curve drawn between Lindvall's (1987) and that of Sangster et al. (1961) can be used as an approximation. For K_{31}, the curve of Lindvall can be used. For joining pipes of unequal sizes, the curves of Sangster et al. appear to be tentatively acceptable.

6.7.4   Loss Coefficient for Two-Way Sewer Junctions

Two-way junctions are used for changes of pipe slope, pipe alignment, or pipe size. Experimental studies on two-way, surcharged, and top-open sewer junctions are listed in Table 6.23. All the experiments were conducted with the same size upstream and downstream pipes joining the junction. Only Sangster et al. (1958, 1961) tested also the effect of different joining pipe sizes. These experimental results show that for a straight-through, two-way junction, the value of the loss coefficient is usually no higher than 0.2. Alignment of the joining pipes and benching in the junction are also important factors in determining the value of the loss coefficient.

Figure 6.39*a* shows the head loss of a surcharged, two-way open-top junction connecting two pipes of identical diameters aligned centrally given by the experiments of Archer et al. (1978), Lindvall (1984), Howarth and Saul (1984), and Johnston and Volker (1990). Noticeable are the swirl and instability phenomena when the junction submergence (junction-depth-to-pipe-diameter ratio) is close to 2 and the corresponding high head loss coefficient. The ranges of loss coefficient given by Sangster et al. (1958), Ackers (1959), and Marsalek (1984) are also indicated in Fig. 6.39*a*, but the data on the variation with the pipe-to-junction-size ratio were not given by these investigators. Sangster et al. (1958) also tested the effect of different sizes of joining pipes for a surcharged two-way junction. Some of their results are plotted in Fig. 6.39*b*. They did not indicate a clear influence of the effect of the size of the junction box. However, Bo Pedersen and Mark (1990) demonstrated that the loss coefficient of a two-way junction can be estimated as a combination of the exit head loss due to a submerged discharging jet and the entrance loss of flow contraction. They suggested that the loss coefficient K depends mainly on the size ratio between the junction and the joining

TABLE 6.22 Experiments on Three-Way Junction of Merging Open Channels

References	Type of junction	Shape of channels	Channel slope	Alignment at junction		Type of flow		Remarks
				Vertical	Longitudinal	Upstream channels	Downstream channel	
Taylor (1944)	Point	Rectangular, identical width, $B = 4$ in.	Horizontal	Flushed bottom	Straight through and one merging channel at 45° or 135°	Subcritical	Subcritical	Also theoretical analysis based on momentum, good agreement with 45° merging but not with 135° merging
Bowers (1950)	Point	Trapezoidal, identical width, $B = 7.2$ in.	0.0062, 0.012	Flushed bottom	Straight through and one merging channel at 51°	Supercritical	Supercritical	Structure P7, hydraulic jumps formed upstream of junction, other structures with lateral bottom up to 3 ft above main
Behlke & Pritchett (1966)	Point	Rectangular or trapezoidal (side slope 1:1)	Each channel slope varied independently	Flushed bottom	Straight through and one merging channel at 15°, 30°, or 45°	Supercritical	Supercritical	Use of tapered wall in the junction to diminish diagonal wave and pile-up problems
Webber and Greated (1966)	Point	Rectangular, $B = 5$ in.	Horizontal	Flushed bottom	Straight through and one merging channel at 30°, 60°, or 90°	Subcritical	Subcritical	Greater losses associated with increasing merging angles of branch channel
Yevjevich and Barnes (1970)	Square box	Circular, $D_{main} = 6.25$ in., $D_{br} = 1.87$ in.	0.0008 0.00054 0.00107	Flushed bottom or crown aligned	Straight through and one 90° merging pipe	Subcritical	Subcritical	Greater loss for the case of crown aligned lateral

Kanda and Kitada (1977)	Point	Rectangular, B_{main} = 100, 200, 400 mm, B_{br} = 100 mm; Circular, different sizes	Horizontal	Flushed bottom	Straight through and one merging channel at 30°, 60°, or 90°	Supercritical	Supercritical	Also theoretical analysis based on momentum
Radojkovic and Maksimovic (1977)	Point	Circular, different sizes	Horizontal	Flushed bottom	Straight through and one merging channel	Supercritical	Supercritical	Junction zone with expansion and without expansion
Townsend and Prins (1978)	Rectangular box	Circular, D_{main} = 160 mm, D_{br} = 102 mm	Less than 0.01	Invert drop across junction box	Straight through and one merging channel at 45°, or 90°	Subcritical	Subcritical	Simple junction box and special junction box with flow deflector
Lin and Soong (1979)	Point	Rectangular, B = 457 mm	Horizontal	Invert drop (15 or 18 mm)	Straight through and one 90° merging channel	Subcritical	Subcritical	Energy loss coef as a function of lateral to total flow rate ratio
Joliffe (1982)	Point	Circular, equal diameter, D = 69 mm	Horizontal, 0.0001, 0.0075, 0.005, or 0.01	Flushed bottom	Straight through and one 90° merging channel	Subcritical	Sub- or Supercritical	Upstream flow depth depends on critical depth in downstream pipe
Best and Reid (1984)	Point	Rectangular, identical width, B = 0.5 ft	Channel slope adjustable to achieve equilibrium water depth	Not available	Straight through and one merging channel at 15°, 45°, 70°, or 90°	Subcritical	Subcritical	Systems operated for $0.1 < F < 0.3$
Ramamurthy et al. (1988)	Point	Rectangular, B_{main} = 248 mm B_{br} = 245 mm	Horizontal	Flushed bottom	Straight through and one merging channel at 90°	Subcritical	Subcritical	
Hsu et al. (1998)	Point	Rectangular, identical width, B = 155 mm	Horizontal	Flushed bottom	Straight through and one merging channel at 30°, 45°, or 60°	Subcritical	Subcritical	
Marsalek (1985)	Square box or round box	Circular, identical diameter	Horizontal	Flushed bottom		Subcritical	Subcritical	

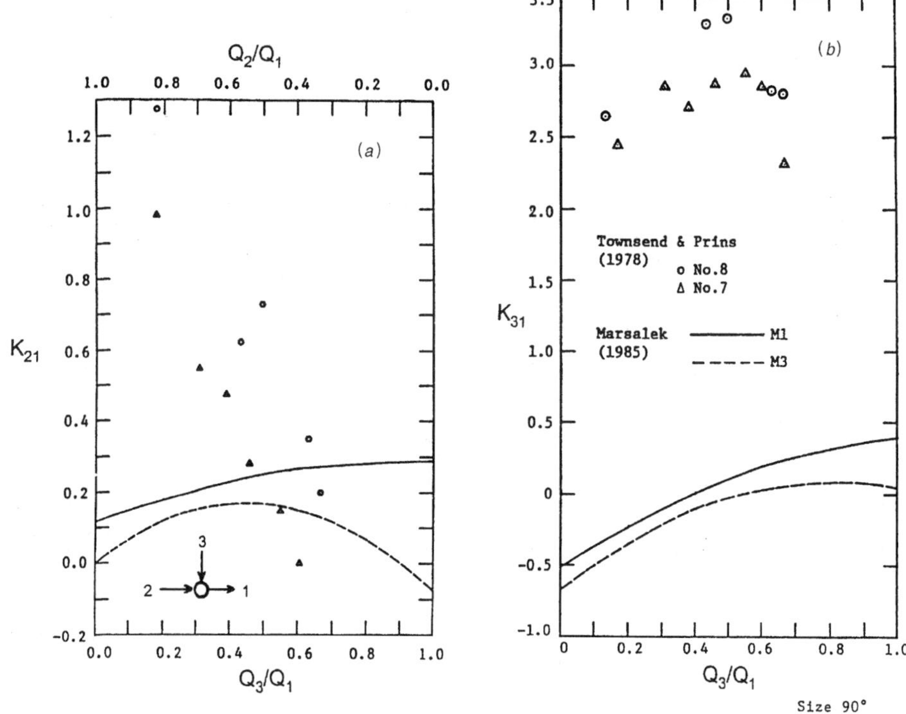

FIGURE 6.38 Headloss coefficients for 3-way open-channel sewer junction with identical joining pipe sizes and 90° merging lateral. (a) Mainline loss coefficient K_{21}. (b) Branch loss coefficient K_{31}.

pipes of identical size. For an infinitely large storage junction, the theoretical limit of K is 1.5. For the junction-diameter-to-pipe-diameter ratio, D_M/D less than four, they proposed to estimate the K values according to benching as shown in Fig. 6.40.

6.7.5 Storage Junctions

For a *storage-* (or *reservoir-*) *type junction,* the storage capacity of the junction is relatively large in comparison to the flow volume, and hydraulically it behaves like a reservoir. A water surface, and, hence, the depth in the junction can be defined without great difficulty. A significant portion of the energy carried in by the flows from upstream sewers is dissipated in the junction. If the horizontal cross-sectional area of the junction A_j remains constant, independent of the junction depth Y, the storage is $s = A_j Y$. Hence, $ds/dt = A_j(dY/dt) = A_j(dH/dt)$, where $H = Y + Z$ is the water surface elevation above the reference datum, and Z is the elevation of the junction bottom. Therefore, from Eq. (6.53)

$$\sum Q_i + Q_j = A_j \frac{dH}{dt} \tag{6.65}$$

Either the energy equation, Eq. (6.54), or the momentum equations, Eqs. (6.55) and (6.56) can be used as the dynamic equation of the junction. If the energy loss coefficient K_i in Eq. (6.54) can be determined, use of the energy equation is appropriate. On the other hand, if the pressure on the junction boundary can be determined, the momentum equation is also applicable. If the junction were truly a large reservoir, both the loss coefficients and the pressure could reasonably be estimated on the basis of information on steady flow entering or leaving a reservoir.

Customarily, for the convenience of computation, instead of Eq. (6.54), the junction energy relationship is divided for each joining sewer by relating the total head of the sewer flow to the total head in the junction. Assuming that the energy contribution from the direct lateral inflow Q_j is negligible, the component of Eq. (6.54) for each joining sewer i can be written as

$$H = (1 - K_i)(V_i^2/2g) + (P_i/\gamma) + Z_i \qquad (6.66)$$

For open-channel flow in the joining pipes, the piezometric term P_i/γ is

$$P_i/\gamma = h_i \qquad (6.67)$$

where h_i = open-channel flow depth of the i-th pipe at the junction

It should be cautioned that Eq. (6.66) is applicable only when there is no free surface discontinuity between the junction and the sewer. In other words, they are applicable to cases B and D in Fig. 6.3 and all four cases in Fig. 6.2. The flow equations for these pipe exit and entrance cases are given in Table 6.24.

6.7.6 Point Junctions

A *point-type junction* is one whose storage capacity is negligible, and the junction is treated as a single confluence point. Hence, Eq. (6.53) is reduced to

$$\sum Q_i + Q_j = 0 \qquad (6.68)$$

For subcritical flow in sewers that empty into the point junction, the flow can discharge freely into and without the influence of the junction only when a free fall exists over a nonsubmerged drop at the end of the pipe (case A in Fig. 6.3). Otherwise, the subcritical flow in the sewer is subject to backwater effect from the junction. Since the junction is treated as a point, the dynamic conditions of the junction is usually represented by a kinematic compatibility condition of common water surface at the junction for all the joining pipes without a free fall (Harris, 1968; Larson et al., 1971; Sevuk and Yen, 1973; Roesner et al., 1988; and Yen, 1986). Thus, by neglecting the junction storage and for subcritical sewer flow into the junction,

$$h_i = h_{ic} \qquad \text{if } Z_i + h_{ic} > Z_o + h_o \qquad \text{(case A in Fig. 6.3)}$$

$$h_i + Z_i = h_o + Z_o \qquad \text{otherwise} \qquad \text{(case B in Fig. 6.3)} \qquad (6.69)$$

in which Z_o and h_o are the invert elevation and depth of the flow at the entrance of the downstream sewer which takes the outflow from the junction.

For a supercritical flow in a sewer flowing into a point junction, case C in Fig. 6.3 would not occur. Only a subcase of case B in Fig. 6.3 with $h_i < h_{ic}$ exists where Eq. (6.69) applies. For case D of Fig. 6.3 with submerged exit,

$$Z_i + (P_i/\gamma) = (P_o/\gamma) + Z_o > Z_i + D_i \qquad (6.70)$$

The flow in the downstream sewer, which takes water from the junction, may be subcritical, supercritical, or submerged, depending on the geometry and flow conditions. The

TABLE 6.23 Experiments on Two-Way Open-Top Junction of Surcharged Channels

References	Type of junction	Shape of channels	Channel slope	Pipe alignment at junction		Type of flow	Remarks
				Vertical	Longitudinal		
Sangster et al. (1958, 1961)	Square, rectangular or round box	Circular, $D = 3.0$ in., 3.75 in., 4.75 in., or 5.72 in.	Horizontal	Flushed bottom	Straight through (or 90° bend)	Steady	Also test of grate inflow into junction
Ackers (1959)	Rectangular box	Circular, identical diameter, $D = 6$ in.	0.0094 to 0.0192	Flushed bottom	Straight through (or 45° bend in junction or 52° bend downstream of junction)	Steady	Also studied partfull supercritical flow
Archer et al. (1978)	Rectangular box or round box	Circular, identical diameter, $D = 102$ mm	0.002 and 0.010	Flushed bottom	Straight through, (or 30° or 60° bend in junction)	Steady	
Howarth & Saul (1984)	Rectangular box or round box	Circular, identical diameter, $D = 88$ mm	Horizontal	Flushed bottom	Straight through	Steady or unsteady	Loss coefficient increases as junction diameter increases
Lindvall (1984)	Round box	Circular, identical diameter, $D = 144$ mm	Horizontal	Center aligned	Straight through	Steady	
Marsalek (1984)	Square box or round box	Circular, identical diameter, $D = 6$ in.	Horizontal	Flushed bottom	Straight through	Steady	Manhole diameter $\leq 2.26D$, head loss coefficient constant for given junction geometry
Johnston & Volker (1990)	Square box	Circular, identical diameter, $D = 88$ mm	Horizontal	Center aligned	Straight through	Steady	Three types of benching

Bo Pedersen and Mark (1990)	Round box	Circular, identical diameter, $D = 90$ mm	Horizontal	Center aligned	Straight through	Steady	Four types of benching
Merlein & Valentin (1999)	Round box	Circular, identical diameter, $D = 242$ mm	Horizontal	Flushed bottom	Straight through (also, 45° bend)	Unsteady	Manhole diameter = 1 m and 1.2 m. Fully benched. Also studied effect of perforated cover plate
Sakakibara et al. (1996)	Rectangular box	Circular, identical diameter, $D = 200$ mm	Horizontal	Flushed bottom	Straight through	Steady	Short manhole longitudinal dimension affects loss coef.
Kusuda & Arao (1996)	Round box	Circular, identical diameter, $D_{up} = 50$ mm, $D_{down} =$ 50 mm, 60 mm	Horizontal	Flushed bottom (also different drop heights)	Straight through	Steady	Manhole diameter = 70 − 180 mm
Arao & Kusuda (1999)	Round box	Circular, identical diameter, $D = 50$ mm	Horizontal	Flushed bottom (also three drop heights)	Straight through (also, 45° or 90° bend in junction)	Steady	Manhole diameter = 180 mm

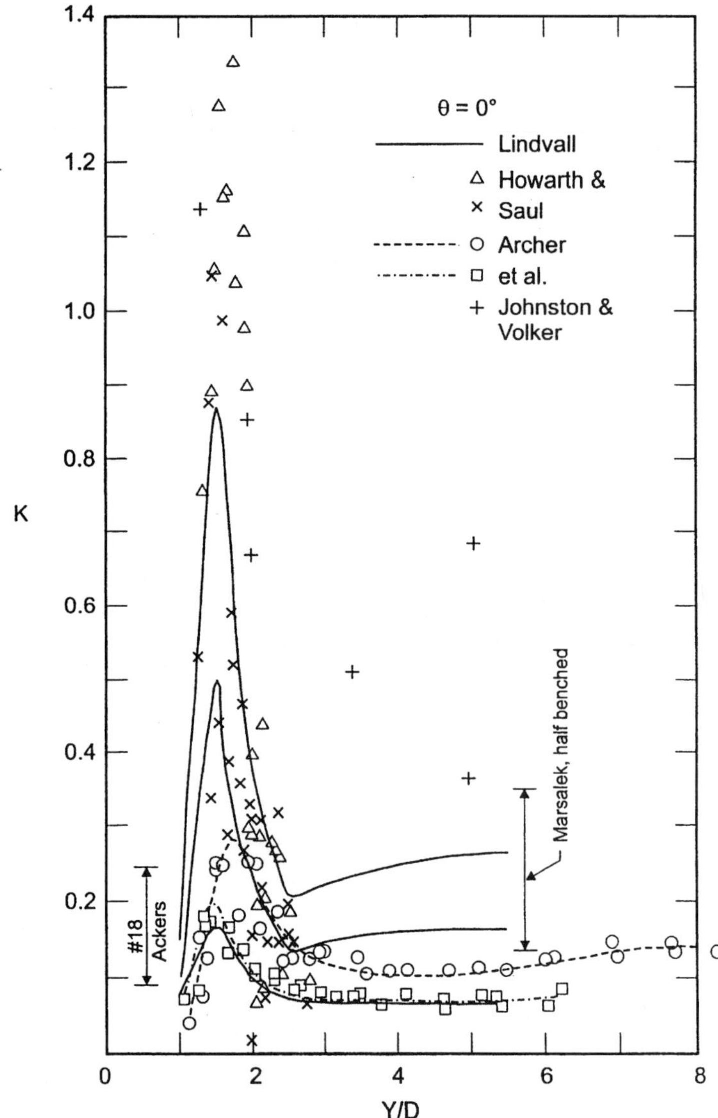

(*a*) Same size sewers upstream and downstream

FIGURE 6.39 Headloss coefficient for surcharged 2-way open-top straight-through sewer junction. (After Yen, 1987)

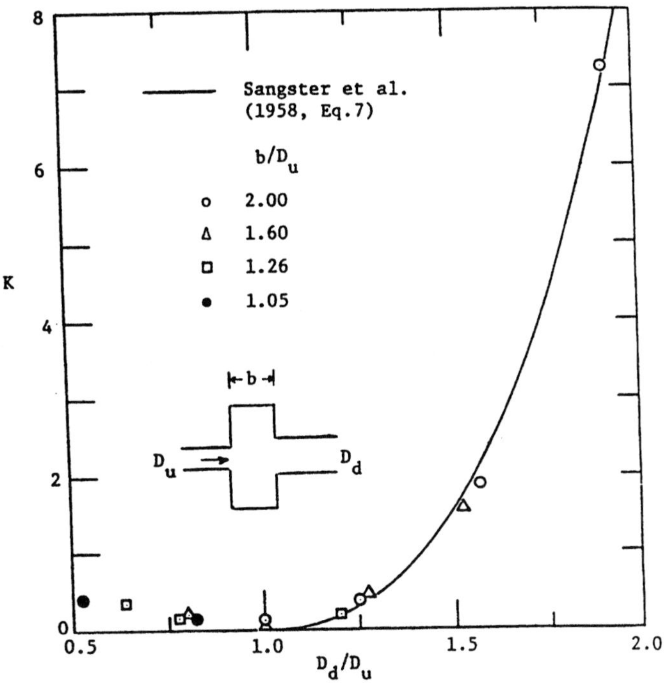

(*b*) Different joining sewer sizes

FIGURE 6.39   (*Continued*)

$$K = \zeta \left(\frac{D_M}{D} \right) \qquad \text{for } \frac{D_M}{D} \le 4$$

Shape				
ζ	0.24	0.12	0.07	0.025

FIGURE 6.40   Effect of benching on loss coefficient of surcharged 2-way sewer junction according to Bo Petersen and Mark (1990).

TABLE 6.24 Storage Junction Flow Equations

Case in Fig. 6.2 or 6.3	Condition		Equation	Remarks
		For sewer flowing into junction		
D	Submerged sewer exit	$H > Z + D$	$H = (1 - K)(V^2/2g) + (P/\gamma) + Z$	For reservoir junction $K = 1$
A	Free fall at sewer exit	$H < Z + h_c F < 1$	$h = h_c$	
B	Subcritical flow in sewer	$Z + D > H$ $H > Z + h_c F < 1$	$H = (1 - K)(V^2/2g) + h + Z$	For reservoir junction $K = 1$
B	Supercritical flow in sewer	$Z + D > H$ $Z + h_c > H$ $F > 1$	$H = h + Z$	Upstream control for sewer
C	Supercritical flow in sewer, hydraulic jump in junction	$H > Z + h_c$ but $F > 1\ h_c > h$		Upstream control for sewer; H determined by other sewers
		For outflow from junction in sewer		
I	Subcritical flow in sewer	$h > h_c$ $H < Z + D + \delta\,(V^2/2g)$ $H > Z + h_c + (V_c^2/g)$	$H = (1 - K)\,(V^2/2g) + h + Z$	$\delta = 1$
II	Supercritical flow in sewer	$H < Z + D + \delta\,(V_c^2/2g)$ $(H = h_c)$	$H = (1 - K)(V_c^2/2g) + h_c + Z$	$\delta = 1$
III	Submerged sewer entrance, open-channel flow in sewer	$H < Z + D + \delta\,(V_c^2/2g)$ $h < D$	$Q = = A_v V_v$ $V_v = C_v\,\sqrt{2g\,(H - Z)}$	$\delta = 1$
IV	Surcharged sewer	$H < Z + D + \delta\,(V_c^2/2g)$ $h < D$	$H = (1 - K)\,(V^2/2g) + (P/\gamma) + Z$	$\delta = 1$

Source: From Yen (1986).

flow equations are the same as the storage junction outflow cases I to IV given in Table 6.24.

6.8 NUMERICAL TECHNIQUES FOR UNSTEADY FLOW IN A SEWER

For drainage problems that require unsteady flow simulation, using the dynamic wave Eqs. (6.3) and (6.7) or Eqs. (6.2) and (6.6) or their lower hydraulic level approximations described in Sec. 6.3.3, numerical solutions are in order because no analytical solutions are known for these sets of partial differential equations except for a few simple cases with noninertia or kinematic wave approximations. Numerical solution of the unsteady flow equations involves discretization with space and time and specifying the initial and boundary conditions.

6.8.1 Discretization of Space–Time Domain of Sewer for Simulation

No analytical solutions are known for the Saint-Venant equations or the surcharged sewer flow equation. Therefore, these equations for sewer flows are solved numerically with appropriate initial and boundary conditions. The differential terms in the partial differential equations are approximated by finite differences of selected grid points on a space and time domain, a process known as *discretization*. Substitution of the finite differences into a partial differential equation transforms it into an algebraic equation. Thus, the original set of differential equations can be transformed into a set of finite difference algebraic equations for numerical solution.

Theoretically, the computational grid of space and time need not be rectangular. Neither do the space and time differences Δx and Δt need be kept constant. Nonetheless, it is usually easier for computer coding to keep Δx and Δt constant throughout a computation. For surcharge flow, Eq. (6.19) dictates the application of the equation to the entire length of the sewer, and the discretization applies only to the time domain. In an open-channel flow, it is normally advisable to subdivide the length of a sewer into two or three computational reaches of Δx, unless the sewer is unusually long or short. One computational reach tends to carry significant inaccuracy due to the entrance and exit of the sewer and is usually incapable of sufficiently reflecting the flow inside the sewer. Conversely, too many computational reaches would increase the computational complexity and costs without significant improvement in accuracy.

The selection of the time difference Δt is often an unhappy compromise of three criteria. The first criterion is the physically significant time required for the flow to pass through the computational reach. Consider a typical range of sewer length between 100 and 1,000 ft and divide it into two or three Δx, and a high flow velocity of 5 to 10 ft/second; a suitable computational time interval would be approximately 0.2 to 2 minutes. For a slowly varying unsteady flow, this criterion is not important and a larger computational Δt will suffice. For a rapidly varying, unsteady flow, this criterion should be taken into account to ensure the computation is physically meaningful.

The second criterion is a sufficiently small Δt to ensure numerical stability. An often used guide is the Courant criterion

$$\Delta x / \Delta t \geq V + \sqrt{gA/B} \tag{6.71}$$

In sewers that usually have small Δx compared to rivers and estuaries, this criterion often requires a Δt less than half a minute and sometimes 1 or 2 seconds.

The third criterion is the time interval of the available input data. It is rare to have rainfall or corresponding inflow hydrograph data with a time resolution as short as 2, 5, or even 10 minutes. Values for Δt that are smaller than the data time resolution can only be interpolated. This criterion becomes important if the in-between values cannot be reliably interpolated. In a realistic application, all three criteria should be considered. Unfortunately, in many computations, only the second numerical stability is considered.

Many numerical schemes can be adopted for the solution of the Saint-Venant equations or their approximate forms. They can be classified as explicit schemes, implicit schemes, and the method of characteristics. Many of these methods are described in Lai (1986), Abbott and Basco (1990), and Cunge et al. (1980).

6.8.2 Initial and Boundary Conditions

As discussed previously and indicated in Table 6.8, boundary conditions, in addition to the initial condition, must be specified in order to obtain a unique solution of the Saint-Venant equations or their approximate simplified equations.

The initial condition is, of course, the flow condition in the sewer pipe when computation starts, $t = 0$, that is, either the discharge $Q(x, 0)$, or the velocity $V(x, 0)$, paired with the depth $h(x, 0)$. For a combined sewer, this is usually the dry weather flow or base flow. For a storm sewer, theoretically, this initial condition is dry bed with zero depth, zero velocity, and zero discharge. However, this zero initial condition imposes a singularity in the numerical computation. To avoid this singularity problem, either a small depth or a small discharge is assumed so that the computation can start. This assumption is justifiable because physically there is dry-bed film flow instability, and the flow, in fact, does not start gradually and smoothly from dry bed. Based on dry-bed stability consideration, an initial depth on the order of 0.25 in., or less than 5 mm, appears reasonable.

However, in sewers, this small initial depth usually is unsatisfactory because negative depth is obtained at the end of the initial time step of the computation. The reason is that the continuity equation of the reach often requires a water volume much bigger than the amount of water in the sewer reach with a small depth. Hence, an initial discharge, or base flow, that permits the computation to start is assumed. For a storm sewer, the magnitude of the base flow depends on the characteristics of the inflow hydrograph, the sewer pipe, the numerical scheme, and the size of Δt and Δx used. It is not uncommon to see that in the first few time steps of the computation, the calculated depth and discharge decrease as the flood propagates, a result that contradicts the actual physical process rising depth and discharge. Nonetheless, if the base flow is reasonably selected and the numerical scheme is stable, this anomaly will soon disappear as the computation progresses. An alternative to this assumed base flow approach in avoiding the numerical problem is to use an invert Priessmann hypothetical slot throughout the pipe bottom and assigning a small initial depth, discharge, or velocity to start the computation. Currey (1998) reported satisfactory use of slot width between 0.001 ft and 0.01 ft.

As to boundary conditions, when the Saint-Venant equations are applied to an interior computational reach of a sewer not connected to its entrance or exit, the upstream condition is simply the depth and discharge (or velocity) at the downstream end of the preceding reach, which are identical with the depth and discharge at the upstream of the present reach. Likewise, the downstream condition of the reach is the shared values of depth and discharge (or velocity) with the following reach. Therefore, the boundary conditions for an interior reach need not be explicitly specified because they are implicitly accounted for in the flow equations of the adjacent reaches.

For the exterior computational reaches containing either the sewer entrance or the exit, the upstream boundary conditions required depend on whether the flow is subcritical or supercritical as indicated in Table 6.25.

TABLE 6.25 Typical Boundary Conditions for Simulation of Exterior Reaches of Sewers

Location	Upstream end of sewer entrance reach ($x = 0$)	Downstream end of sewer exit reach ($x = L$)
Subcritical flow	One of $Q(0, t)$ $h(0, t)$ $V(0, t)$ for all t to be simulated	One of $h(L, t)$; e.g., ocean tides, lakes $Q(L, t)$; release hydrograph $Q(h)$; rating curve $V(h)$; storage-velocity relation for all t to be simulated
Supercritical flow	Two of the above	None

Source: From Yen and Akan (1999).

6.8.3 Number of Equations for a Sewer Divided into Computational Reaches

For a sewer that is divided into M computational reaches and $M + 1$ stations or nodes, a continuity equation and a momentum equation is written in finite difference algebraic form for each reach. There are $2(M + 1)$ unknowns, namely, the depth and discharge (or velocity) at each station. The $2(M + 1)$ equations required to solve for the unknowns come from M continuity equations and M momentum equations for the M reaches, plus the two boundary conditions. If the flow is subcritical, one boundary condition is at the sewer entrance ($x = 0$) and the other is at the sewer exit ($x = L$). If the flow is supercritical, both boundary conditions are at the upstream end, the entrance; one of them often is a critical depth criterion.

If at one instant a hydraulic jump occurs in an interior reach inside the sewer, two upstream boundary conditions at the sewer entrance and one downstream boundary condition at the sewer exit should be specified. The jump becomes an interior boundary condition to be solved. If the flow is not chaning rapidly with time, the sequent depth relation for steady flow, Eqs. (6.40) and (6.41) can be used as approximation. If a hydraulic drop occurs inside the sewer, one boundary condition each at the entrance and exit of the sewer is needed; the drop is described with a critical depth relation as an interior boundary condition. Handling the moving surface discontinuity, shown schematically in Fig. 6.16, is not a simple matter. The moving front may travel from reach to reach slowly in different Δt, or it may move through the entire sewer in one Δt. If, for any reason, it is desired to compute the velocity of the moving front V_w between two computational stations i and $i + 1$ in a sewer, the following equation can be used as an approximation:

$$V_w = \frac{A_i V_i - A_{i+1} V_{i+1}}{A_{i+1} - A_i} \tag{6.72}$$

6.9 FLOW IN SEWER NETWORK

Hydraulically, flows in sewers in a network interact, and the mutual flow interaction must be accounted for to achieve realistic results. In designing the sewers of a network, the constraints and assumptions on sizing sewers as discussed in Sec. 6.5.1 should be noted.

The rational method is the most commonly and traditionally used method for the *design* of sewer sizes. As described in Sec. 6.5, in the rational method each sewer is designed independently without direct, explicit consideration of the flow in other sewers. This can be done because to design a sewer, only the peak discharge, not the entire hydrograph, of the design-storm runoff is required. As previously explained in Sec. 6.5, each sewer has its own design storm. The information needed from upstream sewers is for the alignment and bury depth of the sewer in addition to the upstream flow time needed to estimate the time of concentration for the determination of the rainfall intensity i for the sewer to be designed.

Contrarily, in simulation of flow in an existing or predetermined sewer network for urban stormwater control and management, often the hydrograph, not merely the peak discharge, is sought, and a higher level of hydraulic analysis that considers the interaction of the sewer flows in the network is required. This network system analysis involves combining the hydraulics of individual sewers as described in Secs. 6.2 and 6.8, together with the hydraulics of junctions described in Sec. 6.7.

A sewer network can be considered as a number of nodes joined together by a number of links. The nodes are the manholes, junctions, and network outlets. The links are the sewer pipes. Depending on their locations in the network, the nodes and links can be classified as exterior or interior. The exterior links are the most upstream sewers or the last sewer having the network exit at its downstream end. An exterior sewer has only one end connected to other sewers. Interior links are the sewers inside the network that have both ends connected to other sewers. Exterior nodes are the junctions or manholes connected to the upstream end of the most upstream sewers, or the exit node of the network. An exterior node has only one link connected to it. Interior nodes (junctions) inside the network have more than one link connected to each node.

A systematic numeric representation of the nodes or links is important for a computer simulation of a network. One approach is to number the links (sewers) according to the branches and the order of the pipes in the branch—similar to Horton's numbering of river systems. Another approach is to identify the links by the node numbers at the two ends of the link. Using the node number identification system, a numbering order technique similar to topographic contour lines called the *isonodal line method,* can be used. This method was proved effective for computer manipulation of sewer network simulation (Yen et al., 1976; Mays, 1978).

Special hydraulic features of the networks are often ignored to simulate flow in sewers. Usually, much attention is given to single sewers in the network but little attention to the junctions, and the mutual effects among the sewers are often neglected. If the pipes in a sewer network are all surcharged, obviously the network should be solved as a whole similar to water supply networks. Conversely, if all the pipes carry supercritical flow, one can simply solve the flow starting from the most upstream sewers and complete the solution for the upstream sewers before proceeding to the solution of the downstream sewers in sequence. Likewise, if each one of the sewers in the network has a drop of sufficient height at its exit to ensure a free fall at its downstream end, the downstream boundary condition will be specified when the flow is subcritical. Hence, the solution can still progress sewer by sewer, starting from the most upstream exterior sewers, completing the solution for the upstream sewers before proceeding to the next downstream sewer.

However, more often, flows in sewer networks do not fall into one of these three types just mentioned. For example, for a subcritical sewer flow that can be mathematically represented by the Saint-Venant equations or their simplified approximations, the downstream boundary condition at the sewer exit depends on the hydraulic condition of the downstream junction. Except for the network exit, this downstream junction condition by itself is unknown, and its solution is affected by the sewers joining to it. The *junction continuity* equation is obtained from Eq. (6.65) and is written in a finite difference form, using the average values between time levels n and $n + 1$,

$$\sum (Q_{i,n} + Q_{i,n+1}) + Q_{j,n} + Q_{j,n+1} - \frac{2A_j}{\Delta t}(H_{n+1} - H_n) = 0 \qquad (6.73)$$

in which the summation of i is over the number of the sewers connected to the junction. For a point-type junction, the last term vanishes. It should be cautioned that if surcharge flow is considered, point-type junctions should not be used. When the junction is completely filled and ground flooding occurs, A_j is so large that H_{n+1} practically equals H_n. The junction dynamic equation depends on the condition between the junction and the connecting pipes, as shown in Table 6.24 and discussed in Secs. 6.7.5 and 6.7.6.

Thus, for a simple network consisting of a single four-way junction with three upstream sewers and one downstream sewer, and dividing each sewer into two computational reaches, by using an implicit numerical scheme, the matrix of the equations to be solved for the 24 unknowns is shown in Fig. 6.41. There is one continuity equation and one momentum equation in its complete or simplified form for each reach. The two boundary conditions for each sewer are described by the junction equations for an interior junction, and by the network outlet or inflow conditions for an exterior junction. Therefore, in general, if there are N sewers in the network and each sewer is divided into M reaches for open-channel flow computation, there are $N(2M + 2)$ algebraic equations to solve for the $N(2M + 2)$ unknowns.

If all the sewers of the simple four-way junction network are surcharged and the Preissmann hypothetical slot is not used, there are nine algebraic equations to be solved for the nine unknowns, that is, one surcharge equation and one upstream junction continuity equation for each sewer, plus the network outlet boundary condition. The matrix for implicit solution of the simple network is shown in Fig. 6.42.

Likewise, if the simple four-way junction network has one sewer surcharged and the other three sewers under open-channel flow, for which each sewer is divided into two computational reaches, there are 20 equations for the 20 unknowns. The corresponding matrix is shown in Fig. 6.43. In general, if a network has N_o sewers under open-channel flow, each divided into M computational reaches, and N_s sewers under surcharge flow, the total number of equations is $N_o(2M + 2) + 2N_s$ if the last (exit) pipe of the network is not surcharged, or $N_o(2M + 2) + 2N_s + 1$ if the exit pipe of the network is surcharged.

Thus, it is obvious that the number of unknowns and equations to be handled can easily become large for a sewer network consisting of many sewers. Consequently, the method of solution becomes important to achieve an efficient solution of the entire network. The network solution techniques can be classified into four groups as follows:

1. *Cascade method.* In this method, the solution is sought sewer by sewer, starting from the most upstream sewers. Each sewer is solved for the entire duration of runoff before moving downstream to solve the immediately following sewer, that is, the entire inflow hydrograph is routed through the sewer before the immediate downstream sewer is considered. For the noninertia, quasi-steady dynamic wave and dynamic wave equations, this cascading routing is possible only for the following two conditions: (1) the flow in all the sewers is supercritical throughout, or (2) the exit flow condition of each sewer is specified, independent of the downstream junction condition, for example, a free fall over a drop of sufficient height at the downstream end of each sewer. However, in some sewer models, the downstream condition of the sewer is arbitrarily assumed, for example, using a forward or backward difference and approaching normal flow, so that the cascading routing computation can proceed, and the solution, of course, bears no relation to reality. Conversely, in the kinematic wave approximation, only one boundary condition is needed and it is usually the inflow discharge or depth hydrograph of the sewer. At the junction, only the downstream sewer dynamic equation and junction continuity equations are used. The junction dynamic equations for the upstream sewers are ignored. Thus, the computation can proceed downstream sewer by sewer in a cascading manner, completely ignoring the downstream backwater effects. This method of solving each sewer individually for the entire hydrograph before proceeding to solve for the next downstream sewer is relatively simple and inexpensive, but it is inaccurate if the downstream backwater effect is significant.

2. *One-sweep explicit solution method.* In this method, the flow equations of the sewers and junctions are formulated by using an explicit finite difference scheme such that the flow

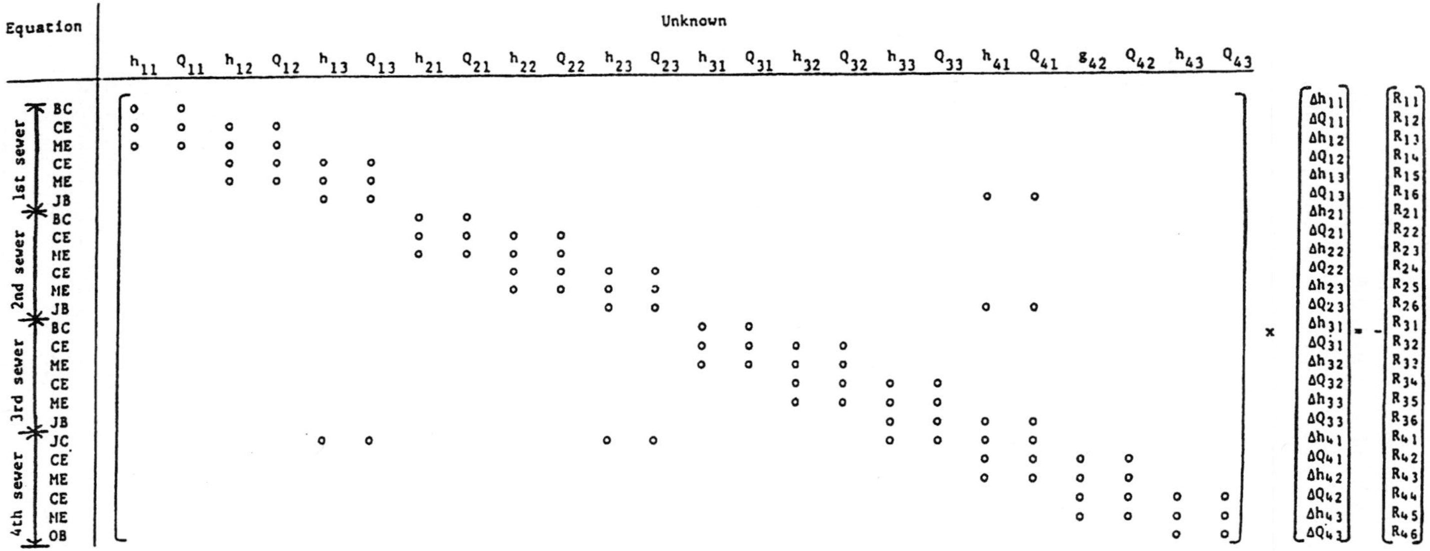

upstream inflow boundary condition; CE, continuity equation; ME, momentum equation; JB, interior junction boundary condition; JC, interior junction continuity equation; OB, outlet boundary condition; R, residual; o, open-channel flow.

FIGURE 6.41 Matrix configuration for 4-way junction with open-channel flow in all four joining sewers. (After Yen, 1986)

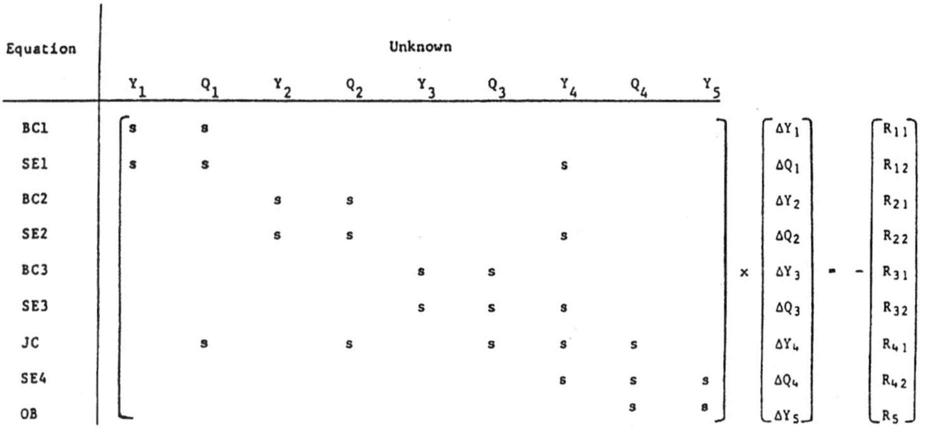

Equation	Y_1	Q_1	Y_2	Q_2	Y_3	Q_3	Y_4	Q_4	Y_5			
BC1	s	s									ΔY_1	R_{11}
SE1	s	s					s				ΔQ_1	R_{12}
BC2			s	s							ΔY_2	R_{21}
SE2			s	s			s				ΔQ_2	R_{22}
BC3					s	s				$\times$	ΔY_3	$= -\;R_{31}$
SE3					s	s	s				ΔQ_3	R_{32}
JC		s		s		s	s	s			ΔY_4	R_{41}
SE4							s	s	s		ΔQ_4	R_{42}
OB								s	s		ΔY_5	R_5

s, surcharge flow; BC, upstream boundary condition; SE, surcharge equation; JC, interior junction continuity equation; OB, outlet boundary condition; R, residual.

FIGURE 6.42 Matrix configuration for 4-way junction with all four joining sewers surcharged. (After Yen, 1986)

depth, discharge, or velocity at a given computation station $x = i\Delta x$ and the current time level $t = n\Delta t$ can be solved explicitly from the known information at the previous locations $x < i\Delta x$ at the same time level, as well as known information at the previous time level $t = (n - 1)\Delta t$. Thus, the solution is sought reach by reach, sewer by sewer, and junction by junction, individually from upstream to downstream, over a given time level for the entire network before progressing to the next time level for another sweep of

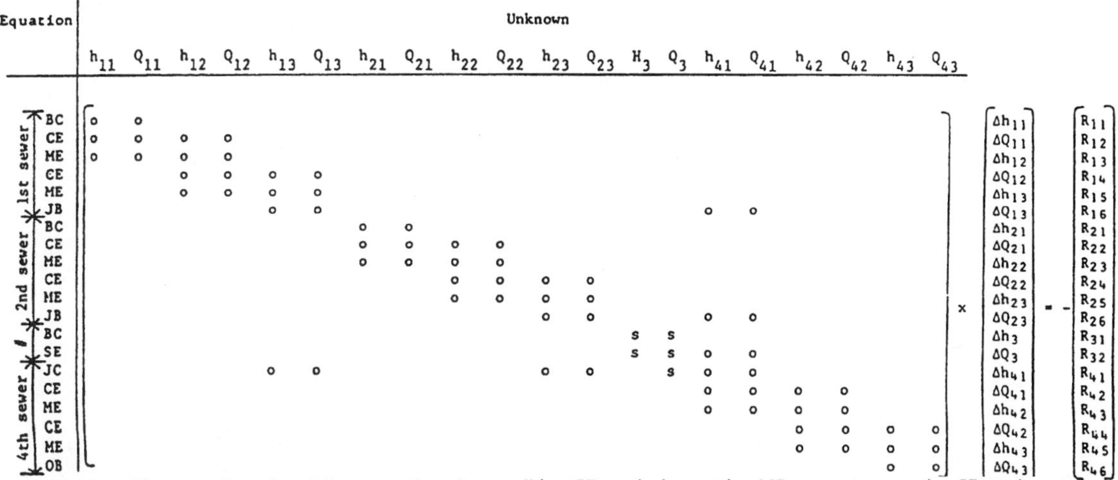

o, open-channel flow; s, surcharge flow; BC, upstream boundary condition; CE, continuity equation; ME, momentum equation; SE, surcharge equation; JB, interior junction boundary conditions; JC, interior junction continuity equation; OB, outlet boundary condition; R, residual; #, third sewer surcharge equations.

FIGURE 6.43 Matrix configuration for 4-way junction with open-channel flow in three joining sewers and one sewer surcharged. (After Yen, 1986).

individual solutions of the sewers and junctions for the entire network. In this method, only one or a few equations are solved for each station, avoiding the large matrix manipulation as in the implicit simultaneous solution of the entire network, and computer programming is relatively direct and simple compared to the simultaneous solution and overlapping segment methods to be discussed later. Nevertheless, this one-sweep method bears the drawback of computational stability and accuracy problems of explicit schemes that requires the use of small Δt. An example of this approach for sewer flow is the EXTRAN Block of SWMM (Roesner et al., 1988), in which each sewer is treated as a single computational reach. Variations of this one-sweep approach do exist. For example, each reach of a sewer can be solved explicitly, or the sewer is solved implicitly and then the junction flow condition is solved explicitly.

3. *Simultaneous solution method.* When the implicit difference formulations of the dynamic wave, quasi-steady dynamic wave, or noninertia equations and the corresponding junction equations are applied to the entire sewer network and solved for the unknowns of flow variables together, a simultaneous solution is ensured. The simultaneous solution usually involves solving a matrix similar to those shown in Figs. 6.41 to 6.43, but much bigger for most networks. There are numerically stable and efficient solution techniques for sparse matrices of implicit schemes. Nonetheless, since the matrix is not banded, solution of the sparse matrix may still require high computational cost and large computer capacity if the network is large.

4. *Overlapping segment method.* To avoid the costly implicit simultaneous solution for large networks and still preserve the advantages of stability and numerical efficiency of the implicit schemes, Sevuk (1973) and Yen (1973a) proposed a technique called the overlapping segment method. Similar to the well-known Hardy Cross method for solving flow in distribution networks, this method is a successive iteration technique. Unlike the Hardy Cross method, which applies only to looped networks, the overlapping segment method can be applied to dendritic as well as loop-type networks. It decomposes the network into a number of small, overlapped subnetworks or segments, and solutions are sought for the segments in succession. Thus, it is suitable in programming for models for simulating large sewer networks in different locations.

A simple, single-step overlapping segment example of a network consisting of three segments is shown in Fig. 6.44. Each segment is formed by a junction together with all the sewers joining to it. Thus, except for the most upstream and downstream (exterior) sewers, each interior sewer belongs to two segments—as the downstream sewer for one segment and then as an upstream sewer for the sequent segment, that is, "overlapped." Each segment is solved as a unit.

The flow equations are first applied to each of the branches of the most upstream segment for which the upstream boundary condition is known and solved numerically and simultaneously with appropriate junction equations for all the sewers in the segment.

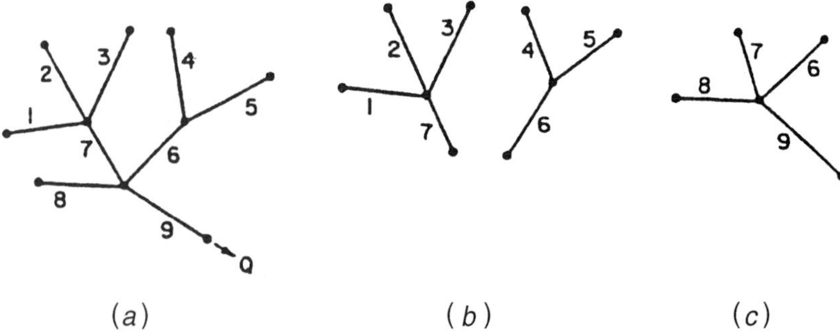

FIGURE 6.44 Overlapping segment scheme for network solution. (After Yen, 1986).

If the flow is subcritical and the boundary condition at the lower exit of the downstream sewer of the segment is unknown, the forward or backward differences, depending on the numerical scheme, are used as an approximate substitution. A simultaneous numerical solution is obtained for all the sewers and junctions of the segment for each time step, and is repeated until the entire flow duration is completed.

For example, for the network shown in Fig. 6.44a, solutions are first sought for each of the two segments shown in Fig. 6.44b. Since the downstream boundary condition of the segment is assumed, the solution for the downstream sewer is doubtful and discarded, whereas the solutions for the upstream sewers are retained. The computation then proceeds to the next immediate downstream segment (e.g., the segment shown in Fig. 6.44c. The upstream sewers of this new segment were the downstream sewers of each of the preceding segments for which solutions have already been obtained. The inflows into this new segment are given as the outflow from the junctions of the preceding segments. With the inflows known, the solution for this new segment can be obtained. This procedure is repeated successively, segment by segment, in the downstream direction, until the entire network is solved. For the last (most downstream) segment of the network, the prescribed boundary condition at the network exit is used.

The method of overlapping segments reduces the requirements on computer size and time when solving for large networks. It accounts for downstream backwater effect and simulates flow reversal, if it occurs. Its accuracy and practical usefulness have been demonstrated by Sevuk (1973) and Yen and Akan (1976).

Solution by the single-step overlapping segment method accounts for the downstream backwater effect of subcritical flow only for the adjacent upstream sewers of the junction, but it cannot reflect the backwater effect from the junction to sewers farther upstream if such case occurs. However, by considering the length-to-depth ratio of actual sewers, in most cases the effect of backwater beyond the immediate upstream branches is small, and, hence, this imposes no significant error in routing sewer flows. For the rare case of two closely located junctions, the overlapping segment method can be modified to perform double iteration by recomputing the upstream segment after the approximate junction condition is computed from the first iteration of the downstream segment. Alternatively and perhaps better, a two-step overlapping procedure can be employed by making a segment containing two neighboring junctions, forming the segment with three levels of sewers instead of two, and retaining only the solutions of the top level sewers for each iteration. Thus, interior sewers are iterated twice. The overlapping segment method can be modified to account for divided sewers or loop networks, in addition to tree-type networks.

The entrance and exit loss coefficients of the sewers at a junction vary with the submergence Y/D at the exit or entrance of the sewers. The values of these loss coefficients may be approximated from the information given in Sec. 6.7. In order to minimize the chance of computation instability due to discrete values of the loss coefficient K, Pansic (1980) assumed an S-curve-type smooth transition, starting from a minimum value K_1 at incipient submergence $Y/D = 1$. The loss coefficient will attain a maximum value K_m at some maximum submergence Y_m/D, for example, at the level of ground flooding. The variation in between is

$$K = K_1 + \alpha \left(\frac{Y}{D} - 1\right)^2 \qquad \text{for } 1 < \frac{Y}{D} \le \frac{Y_m + D}{2D} \qquad (6.74)$$

and

$$K = K_m - \alpha \left[\frac{Y}{D} - \frac{1}{2}\left(\frac{Y_m}{D} + 1\right)\right]^2 \qquad \text{for } \frac{Y_m + D}{2D} \le \frac{Y}{D} \le \frac{Y_m}{D} \qquad (6.75)$$

where

$$\alpha = 2\,\frac{K_m - K_1}{\left[\dfrac{Y_m}{D} - 1\right]^2} \tag{6.76}$$

6.10 SEWER HYDRAULIC SIMULATION MODELS

In designing the size of most drainage sewers, usually knowing the design peak discharge, Q_p, suffices. Conversely, for operation, planning, stormwater quality control and design involving runoff volume (such as detention storage), the stage or discharge hydrograph of the design rainstorm is needed. For the former, Q_p, traditionally simple hydrologic methods such as the rational method can be used. For the latter, the runoff hydrograph can be determined using a hydrologic or hydraulic simulation model. Hydraulic-based simulation models employ a momentum or energy equation (either Eq. (6.3) or (6.4), or any of their simplified approximations), together with the continuity equation, whereas hydrologic models do not consider momentum or energy equations.

In a sewer network the flows in the sewers and junctions interact; therefore, for reliable simulation results, they must be considered as a system. Modeling of flow in a sewer has been described in Sec. 6.8 and the hydraulics of junctions in Sec. 6.7. Linking the components of junctions and sewers for modeling of a network system has been presented in the preceding section. According to this approach, many sewer network models have been developed. Selected ones in terms of popular uses or representative of its kind are listed in Table 6.2 and will be described in the following.

With so many models available, the selection of the appropriate model that is suitable and effective for a particular sewer network flow simulation job is itself a challenging task. Listed below are some factors affecting the selection of models:

1. Objective of the simulation

2. Details and accuracy of the results required from the simulation

3. Input data required to run the simulation, including the data that are readily available and the data still to be collected. For a high level model the required data are often extensive and costly to collect. Also, existing data may not be compatible with the accuracy of data required for simulation.

4. Capability of the computer hardware and the software required by the model to perform simulations

5. Efficiency, including the running time and costs in running the model, to achieve the objective

6. Availability of modeler's support when difficulties arise in running the model

7. Budget and time available for simulation using the model

6.10.1 Classification of Sewer Network Models

A model is defined here as a method or simulation algorithm that has been coded into a computer program for computations and applications. Numerous models have been developed for sewer networks. These sewer models can be classified in different ways as follows:

1. According to the purpose of the model:
  a. Design models
    i. Hydraulic design
    ii. Optimal design
    iii. Risk-based design
  b. Evaluation/predictions models
  c. Planning models

2. According to the objective of the project:
  a. Flood control
  b. Pollution control

3. According to the extent of space consideration:
  a. Sewer system only
  b. Sewer system and overland surface

4. According to the nature of waste water:
  a. Sanitary sewer models
  b. Storm and combined sewer models

5. According to water-quality consideration:
  a. Quantity only
  b. Quality only (rare)
  c. Quantity and quality

6. According to time consideration of rainfall input:
  a. Single-event models
  b. Multiple-even continuous models

7. According to probability consideration:
  a. Deterministic
  b. Probabilistic
    i. Pure statistical
    ii. Stochastic

8. According to systems concept:
  a. Lumped system
  b. Distributed system

9. According to hydrologic principles considered:
  a. Hydrologic (principle of mass conservation)
  b. Hydraulic (principles of conservation of mass and momentum or energy)

In the first classification, the design models are for determining the size of the sewers, and perhaps also their slope, and layout of either a new sewer system or an extension or modification of an existing system. The evaluation/prediction models are those used to simulate the flow in an existing or predetermined sewer system for which the size, slope, and layout are already specified. The use of the model is to compute the flow in the sewers in order to check the adequacy of sewer capacity, system performance, operation, management of pollution abatement, flood mitigation, etc. Or the model may be incorporated as part of a real-time operation system. The planning models are those models used for strategy planning and decision-making for urban or regional stormwater and waste water management, usually applied to a larger time and spatial frame than the design or evaluation models.

The design models design the sewers in a network for a specific future event, which is represented by the design storms of specified return period or risk level. The evaluation/prediction models simulate the runoff produced by a rainstorm of the past, present, future, or the flow from other sources. The planning models usually consider a relatively long continuous period of time covering many rainstorms and dry periods in between. The planning models utilize the least hydraulic consideration of flow on overland areas and in sewers. Often, a simple water budget balance suffices. A typical example is the STORM model (Hydrologic Engineering Center, 1974). Supposedly, for the purpose of reliable flow simulation, the evaluation/prediction models require the highest level of hydraulic sophistication and accuracy. However, many lower level models exist. Due primarily to the discrete sizes of commercial pipes, usually a moderate level of hydraulics is adequate for the design models (Yen and Sevuk, 1975; Yen et al., 1976). Most of the existing sewer models are evaluation/prediction models. Aside from the design models derived from the rational method, there are actually very few true sewer design models; among them only two models, ILSD (Yen et al., 1976, 1984) and WASSP (Price, 1982b), have published user's guides and arrangements for the release of programs. Some of the evaluation/prediction models have the ability to compute the diameter required for gravity flow of a specific discharge. However, they are not true design models because different sewers should be designed for different rainstorms of different durations corresponding to the different time of concentration of the sewers. Hence, many computer runs are required to complete the design of a network using these models.

In the last classification, the hydraulic models can be classified further according to the level of hydraulics shown in Eq. (6.3) or (6.4) as follows: dynamic wave models, noninertia models, nonlinear kinematic wave models, and linear kinematic wave models. It is impossible to summarize and report the hydraulic properties of all the existing sewer models in this short chapter. Therefore, only selected models are made in this presentation and listed in Tables 6.2 and 6.3. Since this article deals with the hydraulics of sewers, in the following section, only the hydraulics of selected models are discussed. Models that allow more than one hydraulic level for flow routing are presented according to their respective highest hydraulic level. For information and comparison of the nonhydraulic aspects of the models, the reader should refer to such other references as Chow and Yen (1976), Brandstetter (1976), and Colyer and Pethick (1976), in addition to the original model developer's reports or papers. Models without hydraulic consideration of the sewer flow, such as the rational method models, have been excluded. When water quality transport simulation is sought, nearly all the models perform the flow routing first and allow another pollutant transport model—usually in concentration form—coupled with the routing result for simulation. Only SWMM has the quality included integrated in the model as a modular block.

6.10.2 Hydraulic Properties of Selected Dynamic Wave Sewer Models

Well-known models, in which the highest hydraulic level, the dynamic wave simulation, is employed are listed in Table 6.2a. In the table, the subscript "o" denotes the sewer receiving outflow from the junction. All these models were developed for flow simulation rather than for design of sewers in a network. CAREDAS, UNSTDY, HYDROWORKS, and MOUSE are proprietary models. Among the nonproprietary models, only two (ISS and SWMM-EXTRAN) have user's manuals published and available to the public. User's information for the other models has not been released to the public. For dynamic wave and noninertia models, the junction conditions and surcharge transition conditions—if surcharge is allowed—are important for reliable and realistic simulation of the flow. However, for most of the models listed in Table 6.2a, information about the details and assumption on the surcharge transition and on junctions is inadequately given. Also, except for ISS, which cannot handle flow with a Froude number greater than 1.6, it is not known whether the other models can handle supercritical flow with roll waves, and, if so, what assumptions are involved. In the following, the dynamic wave models listed in Table 6.2a are briefly discussed in three groups, namely, the explicit scheme model (SWMM-EXTRAN), the models that handle only open-channel sewer flows, and the models that handle both open-channel and surcharge sewer flows.

6.10.2.1 Explicit Scheme Model: SWMM-EXTRAN. The Storm Water Management Model (SWMM) developed under continuous support of the US Environmental Protection Agency (EPA) is one of the best known among all the sewer models. The Extended Transport block (EXTRAN) (Roesner and Shubinski, 1982; Roesner et al., 1988) was added to the SWMM to provide the model with dynamic wave simulation capability. The entire length of the sewer is considered as a single computational reach, and the dynamic wave equation is written in backward time difference between the time levels $n + 1$ and n for the sewer, and expressed explicitly as

$$Q_{n+1} = \left(1 + \frac{gn^2 \Delta t}{2.21 R_n^{4/3}} |V_n|\right)^{-1} Q_n + 2\overline{V}_n \Delta A + \overline{V}_n \frac{A_{u,n} - A_{d,n}}{L} \Delta t$$

$$- g\overline{A}_n \frac{h_{u,n} - h_{d,n}}{L} \Delta t \tag{6.77}$$

in which all the symbols have been defined previously, the subscript u denotes the upstream end of the sewer (i.e., entrance) and d denotes the downstream end of the sewer (i.e., exit), the bar indicates the average of values at the entrance and exit, and presumably $\Delta A = A_{n+1} - A_n$ is also the average of the values at the sewer ends. The junction condition used is the continuity equation, Eq. (6.65), expressed explicitly in terms of the depth and discharge values at the time $n\Delta t$ as

$$H_{n+1} = H_n + \frac{\Delta t}{A_j} (\Sigma Q_{i,n} + Q_{j,n}) \tag{6.78}$$

where subscript j = junction

The junction dynamic relation is simplified as a common water surface, Eq. (6.69). Equations (6.77) and (6.78) are solved explicitly by using a modified Euler method and half-step and full-step calculations. Courant's stability criterion is adopted to select the computational Δt.

In EXTRAN, when a junction is surcharged, on the basis of the junction continuity equation, an expression of the junction water head is derived through an improper application of the chain rule of differentiation, for which a Taylor expansion would be more appropriate. The unsatisfactory result had been apparently recognized, and remedies were attempted through the introduction of an adjustment factor and the assumption on the numerical iterations to either reach a maximum number set by the user or the algebraic sum of the inflows and outflows of a junction being less than a tolerance. In an earlier version of EXTRAN that was applied to a project in San Francisco, an attempt was made to artificially modify the geometry of the junction so that numerical solution could be obtained.

The SWMM-EXTRAN, with its explicit difference formulation, solves the flow sewer by sewer by using the one-sweep explicit solution method with no need for a simultaneous solution of the sewers of the network. Therefore, it is relatively easy to program. Nonetheless, because of the assumptions on the surcharge condition, and also the stability and convergence (accuracy) problems of the explicit scheme for the open-channel condition, on a theoretical basis EXTRAN is inferior to other dynamic wave models listed in Table 6.2*a*. The other models, of course, have their share of problems concerning the assumptions on the transitions between open-channel and surcharge flows, between supercritical and subcritical flows, and on roll waves.

6.10.2.2 Dynamic Wave Model Handling Only Open-Channel Flow: ISS. The Illinois Storm Sewer System (ISS) model (Sevuk et al., 1973) solves the dynamic wave equation using the first-order scheme of the method characteristics. The Saint-Venant equations, Eqs.

(6.3) and (6.4), or similar type partial differential equations are transformed mathematically into two sets of characteristic equations, each set consisting of a pair of ordinary differential equations. These equations are discretized, written in algebraic form, and solved numerically by using a semi-implicit scheme. The formulation can be found in Sevuk and Yen (1982).

The junction conditions used for a storage junction are Eq. (6.65), together with the equations in Table 6.24, for sewer exits, and for sewer entrance Eq. (6.66) with $K_i = 0$, that is,

$$H = (V^2/2g) + h + Z \qquad (6.79)$$

For a point junction, the equations are Eq. (6.68), together with Eq. (6.69) or (6.70). The ISS model program considers direct backwater effects for up to three sewers in a junction. For a junction with more than three joining sewers, the excess sewers (preferably those with small backwater effects from the junction) are treated as direct inflow, that is, Q_j in Eq. (6.65) or (6.68). The flow in the network is solved by using the overlapping segment method. The outlet of the network can be any one of the following: (1) a free fall, (2) a flow continuing to approach normal flow, (3) a stage hydrograph $h = f(t)$, (4) a rating curve $Q = f(h)$, (5) a velocity-depth relationship $v = f(h)$, and (6) a discharge-time relationship $Q = f(t)$.

When used to compute the required pipe diameter of a sewer, ISS is the only model among those listed in Table 6.2(a) that uses a maximum depth criterion to ensure gravity flow in the sewer for the design situation. Other models compute the required pipe diameter on the basis of the peak discharge that does not guarantee gravity flow because, for unsteady flow, maximum depth usually does not occur at the same time as the maximum discharge in a sewer. The ISS model can easily be modified to account also for surcharge flow by adding the Preissmann hypothetical slot.

6.10.2.3 Dynamic Wave Models Handling Both Open-Channel and Surcharge Flows.

Among the five models belonging to this group listed in Table 6.2a, three of them—CAREDAS, UNSTDY, and HYDROWORKS—are numerically similar, using a four-point implicit scheme and adopting the Preissmann fictitious open slot to simulate surcharge flow. Details of the four-point implicit scheme can be found in Liggett and Cunge (1975) and Lai (1986). SWMM-EXTRAN uses an explicit numerical scheme as described in Sec. 6.10.2.1.

HYDROWORKS, MOUSE, UNSTDY, and CAREDAS are proprietary models. They are briefly introduced in the following:

1. HYDROWORKS. The dynamic wave sewer flow routing option of HYDRO-WORKS is based on an earlier model SPIDA from the same company, Hydraulics Research, in England. HYDROWORKS also contains noninertia (WALLRUS) and nonlinear kinematic wave (WASSP-SIM) sewer routing options. The model can handle a looped-type network as well as dendritic type. For dynamic wave routing, the inertia terms are linearly phased out from Froude number equal to 0.8 to 1.1. Essentially, for supercritical flow, the non-inertia approximation is used. For pressurized flow, the hypothetical slot width is assumed to be $\frac{1}{20}$ of the maximum pipe diameter.

2. MOUSE. This model was upgraded from Danish Hydraulic Institute's (DHI) system 11-sewer (S11-S) model. It was first released in 1985 and subsequently upsted with PC technology advancements. It uses the Abbott-Ionescu six-point implicit scheme (Abbott and Basco, 1990), which is relatively stable and consistent but costly in computation. The model allows loop network. In addition to dynamic wave routing, it also has noninertia (identified in the model as diffusion wave) and kinematic wave routing options for sewers.

3. UNSTDY. The UNSTDY model uses four-point noncentral implicit schemes to solve the Saint-Venant equations for subcritical flow. Supercritical flow is simulated by using the kinematic wave approximation. The model can solve a looped–type network in the system.

4. CAREDAS. This is one of the earliest full dynamic wave sewer flow routing models developed by SOGREAH at Grenoble, France. This is the first model to incorporate the Preissmann slot to simulate surcharge flow. In applying CAREDAS, a sewer network is first checked for the sewers with sufficiently steep slope for which the kinematic wave equation can be applied as an adequate approximation. The dynamic wave model is applied to each group of the connected, gently sloped sewers.

6.10.3 Hydraulic Properties of Noninertia Sewer Models

The noninertia approximation of the unsteady flow momentum equation, Eq. (6.4), is probably the most efficient option among the dynamic-wave momentum equation options to solve unsteady open-channel sewer flow problems. It accounts for downstream backwater effect and allows reversal flow. Computationally, it is much simpler than the full dynamic wave option. It is only for rare highly unsteady cases that the noninertia option is inadequate and that the full dynamic wave or the exact momentum options are required. However, only a few noninertia sewer models have been developed; only four are reported in the literature and they are summarized in Table 6.2b.

The proprietary HVM-QQS model was developed by Dorsch Consult (Klym et al., 1972; Vogel and Klym, 1973) at Munich, Germany. It has been misquoted as a dynamic wave model (e.g., Brandstetter, 1976). Examination of the equations (Eqs. (3) and (4) in Vogel and Klym, 1973) reveals that, in fact, it is a noninertia model. It was stated that in order to avoid a simultaneous solution of all the sewers in the network, further assumptions were made. One assumption is to let $S_f = S_o(Q/Q_o)^2$, where Q_o is defined as a normal flow discharge corresponding to S_o, but it is not clear what depth is used in computing Q_o. Another issue is that the sewer downstream boundary condition at the exit is either unspecified, or a rating curve $h = h(Q)$, or using a known exit depth hydrograph $h(t)$. In fact, with unspecified downstream boundary condition, this model does not account for the backwater effect, and thus, it omits one of the important, advantageous properties of the noninertia model. No information is given on whether the flow equations are solved implicitly or explicitly.

The DAGVL-DIFF model was developed at the Chalmers University of Technology (Sjöberg, 1982) at Göteborg, Sweden. The equations in the model are solved in a manner similar to the dynamic wave model DAGVL-A and were found generally satisfactory. No further development or support of the DAGVL models has been provided since the development of S11-S/MOUSE.

The proprietary DHI (1994) model MOUSE contains noninertia and kinematic wave sewer routing options in addition to dynamic wave routing. The noninertia option simulates the flow the same way as the dynamic wave option; thus, it does not take full advantage of the simplicity and computational efficiency of the noninertia modeling.

The NISN model (Pagliara and Yen, 1997) utilizes the overlapping segment method to solve for the flow in a network. For each segment, the flow equations are solved simultaneously using the Preissmann four-point implicit scheme. Junction storage and head loss are allowed. There is no network size limit for this model.

6.10.4 Nonlinear Kinematic Wave Models

Unlike the dynamic wave and noninertia sewer models, many kinematic wave models exist. Only a few nonlinear kinematic wave models are listed in Table 6.2c for discussion. All of the models listed in this table, except the USGS model, have provision to compute the required diameter for a specified discharge using Manning's formula or the Darcy-Weisbach formula. All the nonlinear kinematic wave models listed in Table 6.2c consider the backwater effect from upstream (entrance) of the sewer within the realm of a single sewer and not beyond, and not the backwater effect from downstream (sewer exit) for subcritical flow.

The kinematic wave models, unable to compute reliably the sewer flow cross-section area A, depth h, and velocity V, for subcritical flow when the downstream backwater effect is significant, are of questionable usefulness in coupling with a water-quality equation for water-quality evaluation. Unless the downstream backwater effect is always insignificant, a water-quantity model having a hydraulic level of noninertia approximation or higher should be used.

Nonlinear kinematic wave models may be classified further according to the manner in which the flow equations are formulated for solution. The first group includes the models solving directly the nonlinear kinematic wave equations. The first two models in Table 6.2c, USGS (US Geological Survey's Distributed Routing Rainfall-Runoff Model (Dawdy et al., 1978)) and ILSD-B2 (Illinois Least-Cost Sewer System Design Model, option B2 (Yen et al., 1976)) belong to this group. The second group includes the models that solve an explicit linear algebraic equation of the Muskingum equation form. The ILSD-B3 (Yen et al., 1976) and the British Hydraulics Research's Wallingford Storm Sewer Design and Analysis Package (Price, 1982a, b) Simulation Method (WASSP-SIM) belong to this group. The third group consists of the models using other modified nonlinear kinematic wave equations for solution such as the TRANSPORT Block in the Storm Water Management Model (Metcalf and Eddy et al., 1971).

6.10.4.1 ILSD-B2 and USGS Models.

In the first group, the continuity equation is written as a finite difference algebraic equation of one variable (usually h or Q) or two variables (e.g., h, Q or A, Q) and solved iteratively with the aid of the simplified momentum equation, $S_o = S_f$, where S_f is approximated by Manning's or similar formulas to relate the depth or area to discharge. A formulation used in Yen and Sevuk (1975) and adopted in ILSD-B3 is given in the following as an example. Noting that $b(h) = \partial A/\partial h$ and $G(H) = \partial Q/\partial h$, Eq. (6.6) can be rewritten as

$$B(h)\,\frac{\partial h}{\partial t} + G(h)\,\frac{\partial h}{\partial x} = 0 \qquad (6.80)$$

For partially filled circular pipes (Fig. 6.8),

$$B(h) = D\,\sin(\phi/2) \qquad (6.81)$$

and using Manning's formula

$$G(h) = \frac{K_n}{n}\,S_0^{1/2}R^{2/3}\,\frac{B}{3}\left[5 + \frac{1}{\sin^2(\phi/2)}\left(\frac{\sin\phi}{\phi} - 1\right)\right]$$

$$= \frac{0.132 K_n}{n}\,S_0^{1/2}D^{5/3}\left(1 - \frac{\sin\phi}{\phi}\right)^{2/3}\left[5\,\sin(\phi/2) + \frac{1}{\sin(\phi/2)}\left(\frac{\sin\phi}{\phi} - 1\right)\right] \qquad (6.82)$$

in which the central angle ϕ in radians is (Fig. 6.8)

$$\phi = 2\,\cos^{-1}[1 - (2h/D)] \qquad (6.83)$$

Consider the four computational grid points boxed by the time levels n and $n + 1$ and space levels i and $i + 1$, Eq. (6.80) can be transformed into the following implicit four-point forward-difference equation:

$$\frac{1}{2\Delta t}\,[(B_{i,n+1} + B_{i+1,n+1})(h_{i,n+1} + h_{i+1,n+1} - h_{i,n} - h_{i+1,n})]$$

$$+ \frac{1}{\Delta x}\,[(G_{i,n+1} + G_{i+1,n+1})(h_{i+1,n+1} - h_{i,n+1})] = 0 \qquad (6.84)$$

This equation is nonlinear only with respect to the unknown flow depth $h_{i+1,n+1}$ since $B_{i+1,n+1}$

and $G_{i+1,n+1}$ are both expressed in terms of the depth (Eqs. (6.81) and (6.82)), and hence it can readily be solved by using Newton's iteration method. The solution proceeds sewer by sewer from upstream toward downstream. Within each sewer, the flows for all the reaches are determined for a given time before proceeding to the next time step.

In ILSD, there are actually several sewer flow routing schemes of different hydraulic levels, including options B2 and B3 listed in Table 6.2c and the option of hydrograph time lag adopted in ILSD-1 & 2. The objective of ILSD is to develop an efficient and practical optimization model for the least cost system design of sewer networks. Therefore, the sensitivity and significance of the sophistication of hydraulics on optimal design of sewer systems have been investigated. It was found that for the purpose of designing sewers, because of the discrete sizes of commercially available pipes, unsophisticated hydraulic schemes often suffice, and hence the hydrograph time lag method, instead of options B2 and B3, is adopted in ILSD-1 & 2. Yen and Sevuk (1975) also arrived at a similar conclusion that for design, a low hydraulic level routing method is often acceptable, whereas for evaluation and simulation of flow in sewers, a high hydraulic level routing is usually required.

In the USGS model, the finite difference equation is formulated from Eq. (6.65), similar to ILSD-B2. However, the nonlinear relation between Q/Q_f and A/A_f is approximated by a straight line, and the flow area A is expressed explicitly as

$$A_{i+1,n+1} = f(A_{i,n+1}, A_{i+1,n}) \qquad (6.85)$$

Hence, solution for all the reaches within a sewer must be obtained at each time for the time increments. However, for the sewers in a network, the solution technique can be either the cascade method or the one-sweep method. No information on which one is used in the model is given in the literature.

6.10.4.2 SWMM-TRANSPORT.

Only one model in the third group of modified nonlinear kinematic wave models is listed in Table 6.2c. The SWMM is a comprehensive urban stormwater runoff quality and quantity simulation model for evaluation and management. A good summary of the model is given in Metcalf & Eddy et al. (1971), Huber and Heaney (1982), Huber (1995), and Huber and Dickinson (1988). It has two sewer flow routing options, TRANSPORT and EXTRAN, not counting the crude gutter-type routing in the RUNOFF block. EXTRAN was discussed previously. TRANSPORT is the original sewer-routing submodel built in the progam (Metcalf and Eddy et al., 1971). In TRANSPORT, the continuity equation is first normalized using the just-full steady uniform flow discharge Q_f and area A_f, and then the equation is written in finite differences and expressed as a linear function of the normalized unknowns A/A_f and Q/Q_f at the grid point $x = (i + 1)\Delta x$ and $t = (n + 1)\Delta t$,

$$(Q/Q_f)_{i+1,n+1} + C_1(A/A_f)_{i+1,n+1} + C_2 = 0 \qquad (6.86)$$

in which C_1 and C_2 are functions of known quantities. From the simplified dynamic equation $S_f = S_o$ and Manning's formula, one has

$$Q/Q_f = AR^{2/3}/A_f R_f^{2/3} = f(A/A_f) \qquad (6.87)$$

Accordingly, curves of normalized discharge-area relationship Q/Q_f versus A/A_f for steady uniform flow in pipes of different cross-sectional geometries are established and solved together with Eq. (6.86) for Q/Q_f and A/A_f. In the kinematic wave method of solving Eqs. (6.86) and (6.87), in addition to the initial condition, only one boundary condition is needed, which is usually the inflow hydrograph at the sewer entrance. No downstream boundary condition is required, and, hence, no backwater effect from the downstream can be accounted for if the flow is subcritical. However, in TRANSPORT through a formulation of friction slope calculation using the previous time h values, the downstream backwater effect is partially accounted for at one time step behind. In routing the unsteady nonuniform flow by using Eqs. (6.86) and (6.87), the value of Q_f is not calculated as the steady uniform full-pipe discharge. Instead, it is adjusted by assuming that

$$S_f = S_0 - \frac{\partial h}{\partial x} - \frac{V}{g}\frac{\partial V}{\partial x} = S_0 - \frac{h_{i+1,n} - h_{i,n}}{\Delta x} - \frac{V_{i+1,n}^2 - V_{i,n}^2}{2g\Delta x} \tag{6.88}$$

To improve computational stability, it is further assumed in TRANSPORT that at any iteration k, Q_{fk} is taken as the average of previous and current values, that is,

$$Q_{fk} = \frac{1}{2}Q_{f(k-1)} + \frac{K_n}{2n\sqrt{\Delta x}}A_f R_f^{2/3}$$
$$\times \left[S_0\Delta x + h_{i(k-1)} - h_{i+1(k-1)} + \frac{V_{i(k-1)}^2}{2g} - \frac{V_{i+1(k-1)}^2}{2g} \right]^{1/2} \tag{6.89}$$

where all the values of h and V are those at the previous time $n\Delta t$ that are known if the one-sweep or implicit solution method is used to solve for the flow in individual sewers at incremental times. Incorporating Eq. (6.88) for S_f in Manning's formula yields a quasi-steady dynamic wave approximation instead of the kinematic wave. Thus, use of Eq. (6.89) to compute Q_f indirectly gives a partial consideration of the downstream backwater effect with a time lag.

This improvement of the kinematic wave approximation makes SWMM-TRANSPORT hydraulically more attractive than the standard nonlinear kinematic wave models. Presumably, the partial accounting of the downstream backwater effect is effective as long as the flow does not change rapidly with time, and no hydraulic jump or hydraulic drop is allowed. A comparison of EXTRAN to improvement and advantages over TRANSPORT has not been reported and such a hydraulic comparison would be interesting.

Nonetheless, since the downstream boundary condition is not truly accounted for, it is recommended in SWMM-TRANSPORT that for a sewer with a large downstream storage element from which the backwater effect is severe, the water surface is assumed as horizontal from the storage element going backward until it intercepts the sewer invert. Moreover, when the sewer slope is steep, presumably implying high-velocity supercritical flow, the flood may simply be translated through the sewer without routing, that is, shifting of the hydrograph without time lag. Also, if the backwater effect is expected to be small and the sewer is circular in cross-section, the gutter flow routing method in the RUNOFF Block may be applied to the sewer as an approximation.

In SWMM, large junctions with significant storage capacity and storage facilities are called storage elements, equivalent to the case of storage junction (i.e., $ds/dt \neq 0$), which was discussed earlier. Only the continuity equation, Eq. (6.53) is used in storage element routing. No dynamic equation is considered except for the cases with weir or orifice outlets. Small junctions are treated as point-type junctions with $ds/dt = 0$.

6.10.4.3   ILSD-B3 and WASSP-SIM.

In the second group of nonlinear kinematic wave models, both ILSD-B3 and WASSP-SIM adopt the *Muskingum-Cunge method*. The *Muskingum routing formula* can be written for discharge at $x = (i + 1)\Delta x$ and $t = (n + 1)\Delta t$ as

$$Q_{i+1,n+1} = C_1 Q_{i,n} + C_2 Q_{i,n+1} + C_3 Q_{i+1,n} \tag{6.90}$$

where

$$C_1 = \frac{KX + 0.5\Delta t}{K(1 - X) + 0.5\Delta t} \tag{6.91a}$$

$$C_2 = \frac{0.5\Delta t - KX}{K(1 - X) + 0.5\Delta t} \tag{6.91b}$$

$$C_3 = \frac{K(1 - K) - 0.5\Delta t}{K(1 - X) + 0.5\Delta t} \tag{6.91c}$$

where K is known as the storage constant having a dimension of time and X a factor expressing the relative importance of inflow. Cunge (1969) showed that by taking K and Δt as constants, Eq. (6.90) is an approximate solution of the nonlinear kinematic wave equation (Eqs. (6.6) and $S_0 = S_f$). He further demonstrated that

$$K = \Delta x / c \tag{6.92}$$

and

$$X = \frac{1}{2} - \left(\frac{\varepsilon}{c\Delta x}\right) \tag{6.93}$$

in which ε is the "diffusion" coefficient and c is the celerity of the flood peak that can be approximated as the length of the reach divided by the flood peak travel time through the reach. Assuming $K = \Delta t$ and denoting $\alpha = 1 - 2X$, Eq. (6.90) can be rewritten as

$$Q_{i+1,n+1} = \frac{2 - \alpha}{2 + \alpha} Q_{i,n} + \frac{\alpha}{2 + \alpha} Q_{i,n+1} + \frac{\alpha}{2 + \alpha} Q_{i+1,n} \tag{6.94}$$

In the traditional Muskingum method, X and, consequently, α are regarded as constant. In the Muskingum method as modified by Cunge, α is allowed to vary according to the channel geometry and is computed as

$$\alpha = KQ / S_0(\Delta x)^2 B \tag{6.95}$$

where B = surface width of the flow
S_o = sewer slope

The values of α are restricted to being between 0 and 1 so that C_1, C_2, and C_3 in Eq. (6.91) will not be negative. It is the variation of α, and hence C_1, C_2, and C_3, that classifies the Muskingum-Cunge method as a nonlinear kinematic wave approximation.

The Muskingum-Cunge method offers two advantages over the standard nonlinear kinematic wave methods. First, the solution is obtained through a linear algebraic equation (Eq. (6.90) or Eqs. (6.94) and (6.95)) instead of a partial differential equation, permitting the entire hydrograph to be obtained at successive cross-sections instead of solving for the flow over the entire length of the sewer pipe for each time step, as for the standard nonlinear kinematic wave method. Second, because of the use of Eq. (6.95), a limited degree of wave attenuation is included, permitting a more flexible choice of the time and space increments for the computations as compared to the standard nonlinear kinematic wave method.

In ILSD-B3, the coefficient α in Eq. (6.94) is computed at each grid point by using Eq. (6.95), while B and K both change with respect to time and space. The values of K are computed by using Eq. (6.92) with the celerity c evaluated by

$$c = \partial Q / \partial A \tag{6.96}$$

or for a partially filled pipe using Manning's formula

$$ c = \frac{0.132 K_n}{n} S_0^{1/2} D^{2/3} \left(1 - \frac{\sin \phi}{\phi} \right)^{2/3} \left[5 + \sin^{-2} \left(\frac{\phi}{2} \right) \left(\frac{\sin \phi}{\phi} - 1 \right) \right] \qquad (6.97) $$

The initial flow condition is the specified base flow as in ILSD-B2. The upstream boundary condition of the sewer is the given inflow hydrograph. The flow depth and other geometric parameters at the sewer entrance can be computed from the geometric equations given in Fig. 6.8. The junction condition used is the continuity relationship, Eq. (6.73). The solution is obtained over the entire time period at a flow cross-section before proceeding to the next cross-section. The solution then proceeds downstream section by section and then sewer by sewer in a cascading sequence. More details on the computational procedure of ILSD-B3 can be found in Yen et al. (1976).

The British model WASSP is a sewer design and analysis package consisting of four submodels (Price, 1982b): a modified rational method for design of sewers, a hydrograph method for design of sewers using the Muskingum-Cunge routing, an optimal design method, and a simulation method using the fixed parameter Muskingum-Cunge technique for open-channel routing in sewers and the unsteady dynamic equation for surcharge flow computations. Open-channel flow is routed using Eq. (6.90) with the coefficients C_1, C_2, and C_3 expressed as functions of c and $\mu = Q/2BS_{bo}$. In computation, c is taken as the full-pipe velocity and μ is evaluated at $h/D = 0.6$. Sewers under open-channel flow are solved pipe by pipe, using a directionally explicit algorithm to calculate the discharge at the sewer exit. The space increment Δx along the sewer is selected automatically in terms of Δt to enhance computational accuracy. Connected surcharged sewers are solved simultaneously. For surcharge flow, a time increment as small as a few seconds may be necessary if surges occur. The transition between open-channel flow and surcharge flow is assumed to occur when the discharge exceeds Q_f, when the sewer entrance and exit are submerged, or when the water depth in the junction is higher than the sewer flow depth plus the entrance or exit head loss (Bettess et al., 1978). At a junction, only the continuity equation is considered for open-channel flow. For surcharge flow, in addition to the continuity equation, junction head loss is considered and incorporated into the surcharge unsteady dynamic wave equation. The head loss coefficient is assumed to be 0.15 for a junction with straight pipes, 0.50 for 30° bend pipes, and 0.90 for 60° bend pipes. Some details of WASSP-SIM are reported in Price (1982b).

6.10.5 Verification and Calibration of Models

Models should never be used without being tested and verified. It has happened again and again that in the enthusiasm in model development, the models are used or suggested to be used without verification. All models have their own assumptions and simplifications. Moreover, most urban runoff models contain coefficients, exponents, or adjustment factors that require calibration with data to determine their values.

Besides verification and application for predictions, there are other operational modes of models such as calibration, parameter identification, and sensitivity analysis. In calibration, one tries to determine the most suitable values of the coefficients of the parameters variables knowing the input and output from observed data. In verification, one has the parameters and their coefficient values all determined for the model, and the data on both input and output. The input is run through the model to produce output, which is compared to the known output in the data set to verify the agreement between the computed and observed outputs. On the other hand, verification is different from validation. Validation is to ascertain if the correct equation or model is used to solve the problem. Verification is to find out if the equation or model is solved correctly.

No model can do everything. For example, a good flow simulation model may not produce a good design of the drainage system. Conversely, a good design model may not—and often

TABLE 6.26 Error Measures for Simulation or Measurements

	Definition	Remarks*
Magnitude errors		
Peak rate error	$\varepsilon_{QP} = (Q_p - Q_{pm})/Q_{pm}$	
Mean rate error	$\varepsilon_d = (\bar{Q} - \bar{Q}_m)/\bar{Q}_m$	
Cumulative volume error	$\varepsilon_V = (V - V_m)/V_m$	$V = \int_0^t Q\,dt \simeq \sum_i Q_i \Delta t$ $V_m = \int_0^t Q_m\,dt \simeq \sum_i Q_m \Delta t$
Absolute volume error	$\varepsilon_{va} = \dfrac{1}{V_m}\int_0^t |Q - Q_m|\,dt \simeq \dfrac{1}{V_m}\sum_i |Q_i - Q_{mi}|\Delta t$	
Rate moment error	$\varepsilon_{Qr} = (Q_r - Q_{rm})/Q_{rm}$	$Q_r = \dfrac{1}{V}\int_0^t \dfrac{1}{2}Q^2\,dt \simeq \dfrac{1}{2V}\sum_i Q_i^2 \Delta t$ $Q_{rm} = \dfrac{1}{V_m}\int_0^t \dfrac{1}{2}Q_m^2\,dt \simeq \dfrac{1}{2V_m}\sum_i Q_{mi}^2 \Delta t$
Root-mean-square	$\varepsilon_{RMS} = \dfrac{T}{V_m}\left[\dfrac{1}{T}\int_0^T (Q - Q_m)^2\,dt\right]^{1/2}$ $\simeq \dfrac{\sqrt{T}}{V_m}\left[\sum_i (Q_i - Q_{mi})^2 \Delta t\right]^{1/2}$	
Time errors		
Peak-rate time error	$\varepsilon_{tp} = (t_p - t_{pm})t_{pm}$	
Peak time first-moment error	$\varepsilon_{t1} = (\tau_1 - \tau_{1m})/\tau_{\_m}$	$\tau_1 = \dfrac{1}{V}\int_0^t tQ\,dt \simeq \dfrac{1}{V}\sum_i Q_i\left(t_i + \dfrac{\Delta t}{2}\right)\Delta t$
Graph dispersion error	$\varepsilon_g = (G - G_m)/G_m$	Second moments with respect to t_{pm}: $G = \dfrac{1}{V}\int_0^T Q(t - t_{pm})^2\,dt$ $G_m = \dfrac{1}{V_m}\int_0^T Q_m(t_m - t_{pm})^2\,dt$

*V or V_m becomes total volume if t = flow duration considered, T.
Subscript m = measured or reference base values.
Subscript i = summation index. Subscript p = peak magnitude of the time graph.
Source: From Yen (1982).

need not—be an accurate flow simulation model. Therefore, models should be verified and validated according to their objectives and their applications. In verifying a model, the verification criteria should be set up to confirm with the model objectives. For example, the verification can be made according to the peak discharge, time to peak, or to the fitting of the hydrograph as desired by the objective. Various verification fitting error measures have been suggested in the literature (e.g., Yen, 1982; ASCE Task Committee, 1993). Some measures are listed in Table 6.26. In the table, the magnitude parameter Q can be discharge, depth, velocity, or concentration as appropriate to the problem being investigated. The subscript p denotes the peak magnitude of the time graph. The subscript m represents the measured or true values used as the gauge for the curve fitting and verification of the model simulation.

The selection of the error measures to evaluate the merit of simulation models depends on the objective of the simulation. For example, if the accuracy of the peak rate and peaking time are of paramount importance, ε_{Qp} and ε_{tp} would be the most appropriate error measures. If the overall fitting of the curves is the main objective, ε_{RMS} would be the most important measure, while ε_{T1}, ε_{va}, ε_{Qp}, and ε_{tp} could be used as auxiliary measures.

In calibration, since the reliability of a single set of data is uncertain, the more sets of data used, the better. Different sets of data will produce different sets of coefficient values. Normally, a weighted average (e.g., through optimization) of the values is adopted for each of the coefficients.

It is not infrequent to see a model being misused or abused. Sometimes this is due to the lack of understanding about how the model works. Sometimes it is due to the lack of appreciation of the operational modes. For example, data used for calibration should not be used again for verification. Yet, this situation happens again and again. In such a case of using the same data for calibration and verification, the difference between the model output and the recorded data itself is simply a reflection of the numerical errors and the deviation of the particular data set from the weighted average situation.

Not all models require calibration. Presumably, some strictly physically based models have their coefficient values assigned based on available information and no calibration is needed. However, in rainfall-runoff modeling, some degree of spatial and temporal aggregation of the physical process is unavoidable. Therefore, calibration is desirable, if not necessary.

REFERENCES

Abbott, M. B., and D. R. Basco, *Computational Fluid Dynamics,* John Wiley & Sons, New York, 1990.

Abbott, M. B., K. Havnø, N. E. Hoff-Clausen, and A. Kej, "A Modelling System for the Design and Operation of Storm-Sewer Networks," in M. B. Abbott and J. A. Cunge, eds., *Engineering Applications of Computational Hydraulics, Vol. 1,* Pitman, London, pp. 11–36, 1982.

Ackers, P., "An Investigation of Head Losses at Sewer Manholes," *Civil Engineering Public Works Review,* 54:882–884 and 1033–1036, 1959.

Akan, A. O., "Time of Concentration Formula for Overland Flow," *Journal of Irrigation & Drainage Engineering,* ASCE, 115(4):733–735, 1989.

Akan, A. O., and B. C. Yen, "A Nonlinear Diffusion-Wave Model for Unsteady Open-Channel Flow," *Proceedings of the 17th Congress of the International Association for Hydraulic Research,* Baden Baden, Germany, 2:181–190, 1977.

Akan, A. O., and B. C. Yen, "Diffusion Wave Flood Routing in Channel Networks," *Journal of Hydraulics Division,* ASCE 107(HY6):719–732, 1981.

American Society of Civil Engineers (ASCE), "Design and Construction of Sanitary and Storm Sewers," *Manuals and Reports on Engineering Practice,* no. 37, Reston, VA, 1969.

American Society of Civil Engineers, Task Committee on Definition of Criteria for Evaluation of Watershed Models of the Watershed Management Committee, "Criteria for Evaluation of Watershed Models," *Journal of Irrigation Drainage Engineering,* ASCE 119(3):429–443, 1993.

Arao, S. and T. Kusuda, "Effects of Pipe Bending Angle on Energy Losses at Two-Way Circular Drop Manholes," *Proc. 8th Int. Conf. Urban Storm Drainage,* 4:2163–2168, Sydney, Australia, 1999.

Archer, B., F. Bettess, and P. J. Colyer, "Head Losses and Air Entrainment at Surcharged Manhole," Report IT185, Hydraulics Research Station, Wallingford, England, 1978.

Behlke, C. E., and H. D. Pritchett, "The Design of Supercritical Flow Channel Junctions," *Highw. Res. Rec. No.* 123:17–35, National Research Council Highway Research Board, Washington, DC, 1966.

Berlamont, J., "Roll-Waves in Inclined Rectangular Open Channels," *Unsteady Flow in Open Channels,* BHRA Cranfield, England, pp. A2.13–26, 1976.

Bermeuleu, L. R., and J. T. Ryan, "Two-Phase Slug Flow in Horizontal and Inclined Tubes," *Canadian Journal of Chemical Engineering,* 49:195–201, 1971.

Best, J. L., and I. Reid, "Separation Zone at Open-Channel Junctions," *Journal of Hydraulics Engineering,* ASCE 110(HY11):1588–1594, 1984.

Bettess, R., R. A. Pitfield, and R. K. Price, "A Surcharging Model for Storm Sewer Systems," in P. R. Helliwell, ed., *Urban Storm Drainage, Proceedings of the 1st International Conference,* Pentech Press, London and Wiley-Interscience, New York, pp. 306–316, 1978.

Bettess, R., and R. K. Price, "Comparison of Numerical Methods for Routing Flow Along a Pipe," Report no. IT162, Hydraulics Research Station, Wallingford, England, 1976.

Blaisdell, F. W., and P. W. Mason, "Energy Loss at Pipe Junction," *Journal of Irrigation & Drainage Division,* ASCE 93(IR3):59–78, 1967; *Discussions,* 94(IR2):280–282, 1968.

Bodhaine, G. L., "Measurement of Peak Discharge at Culvert by Indirect Methods," *Techniques of Water Resources Investigations,* book 3, chapter A3, US Geological Survey, 1968.

Book, D. E., J. W. Labadie, and D. M. Morrow, "Dynamic vs. Kinematic Routing in Modeling Urban Storm Drainage," in B. C. Yen, ed., *Urban Stormwater Hydraulics and Hydrology,* Water Resources Publications, Highlands Ranch, CO, pp. 154–163, 1982.

Bo Pedersen, F. B., and O. Mark, "Head Losses in Storm Sewer Manholes: Submerged Jet Theory," *Journal of Hydraulics Engineering,* ASCE 116(11):1317–1328, 1990.

Bowers, C. E., "Studies of Open-Channel Junctions," Technical Paper, no. 6, Series B, St. Anthony Falls Hydraulic Laboratory, University of Minnesota, Minneapolis, MN, 1950.

Brandstetter, A., "Assessment of Mathematical Models for Urban Storm and Combined Sewer Management," *Environ. Prot. Technol. Ser., EPA*-600/2-76-175a, Municipal Environ. Res. Lab., US EPA, Cincinnati, OH, 1976.

Brock, R. R., "Development of Roll-Wave Trains in Open Channels," *Journal of Hydraulics Engineering,* ASCE 95(HY4):1401–1427, 1969.

Burton, L. H., and D. F. Nelson, "Surge and Air Entrainment in Pipelines," in J. P. Tullis, ed., *Control of Flow in Closed Conduits,* Colorado State University, Fort Collins, CO, 1971.

Chaudhry, M. H., *Applied Hydraulic Transients,* Van Nostrand-Reinhold, Princeton, NJ, 1979.

Chen, Y. H., and S.-Y. Chai, *UNSTDY Combined Storm Sewer Model User's Manual,* Chen Engineering Technology, Fort Collins, CO, 1991.

Chevereau, G., F. Holly, and A. Preissmann, "Can Detailed Hydraulic Modeling be Worthwhile When Hydrologic Data is Incomplete?" in P. R. Helliwell, ed., *Urban Storm Drainage, Proceedings of the 1st International Conference,* Pentech, London and Wiley-Interscience, New York, pp. 317–326, 1978.

Chow, V. T., *Open-Channel Hydraulics,* McGraw-Hill Book Co., New York, 1959.

Chow, V. T., and B. C. Yen, "Urban Stormwater Runoff—Determination of Volumes and Flowrates," *Environ. Prot. Technol. Ser. EPA*–600/2-76-116, Municipal Environ. Res. Lab., US EPA, Cincinnati, OH, 1976.

Christensen, B. A., "Discussion of 'Limitations and Proper Use of the Hazen-Williams Equations' by C. P. Liou," *Journal of Hydraulic Engineering,* ASCE 126(2):167–168, 2000.

Colebrook, C. F., "Turbulent Flow in Pipes with Particular Reference to the Transition Region Between the Smooth and Rough Pipe Laws," *Journal,* Institution of Civil Engineers (London), 11:133–156, 1938–1939.

Colyer, P. J., and R. W. Pethick, "Storm Drainage Design Methods: A Literature Review," Report no. INT 154, Hydraulics Research Station, Wallingford, England, 1976.

Cunge, J. A., "On the Subject of a Flood Propagation Computation Method," *Journal Hydraul. Res.,* 7(2): 205–230, 1969.

Cunge, J. A., and B. Mazaudou, "Mathematical Modelling of Complex Surcharge Systems: Difficulties in Computation and Simulation of Physical Situations," in P. Balmer, P. A. Malmqvist, and A. Sjöberg, eds., *Proceedings of the 3rd International Conference on Urban Storm Drainage,* 1:363–373, Chalmers University of Technology, Göteborg, Sweden, 1984.

Cunge, J. A., and M. Wegner, "Intégration Numérique des Équations d'Écoulement de Barré de Saint-Venant par un Schéma Implicite de Différences Finies: Application au Cas d'Une Galerie Tantôt en Charge, Tantôt é Surface Libre," *La Houille Blanche,* 1:33–39, 1964.

Cunge, J. A., F. M. Holly, and A. Verwey, *Practical Aspects of Computational River Hydraulics,* Pitman Publishing, London, 1980.

Currey, D. L., "A Two-Dimensional Distributed Hydrologic Model for Infiltrating Watershed with Channel Networks," M.S. thesis, Dept. of Civil & Environ. Eng., Old Dominion University, Norfork, VA, 1998.

Dawdy, D. R., J. C. Schaake, Jr., and W. M. Alley, "User's Guide for Distributed Routing Rainfall-Runoff Model," *Water Resour. Invest.,* US Geological Survey, pp. 78–90, 1978.

DeGroot, C. F., and M. J. Boyd, "Experimental Determination of Head Losses in Stormwater Systems," *Proceedings of the 2nd National Conference on Local Government Engineering,* Brisbane, Australia, pp. 19–22, September 1983.

DeSomer, M., "Flow Instability Occurred During the Transition of Full Pipe Flow to Free Surface Flow and Vice Versa in a Closed Conduit," in P. Balmer, P. A. Malmqvist, and A. Sjöberg, eds., *Proceedings of the 3rd International Conference on Urban Storm Drainage,* 1:427–433, Chalmers University of Technology, Göteborg, Sweden, 1984.

DHI, "MOUSE: Reference Manual Version 3.2," Danish Hydraulic Institute, Copenhagen, Denmark, 1994.

Dukler, A. E., "Characterization, Effects and Modeling of the Wavy Gas-Liquid Interface," *Prog. Heat Mass Transfer,* 6:207–234, 1972.

Federal Highway Administration, *Urban Drainage Design Manual,* Hydraulic Engineering Circular no. 22, Washington, DC, 1996.

Fread, D. L., "Chapter 10: Flow Routing," in D. R. Maidment, ed., *Handbook of Hydrology,* McGraw-Hill, New York, pp. 10.1–10.36, 1993.

French, R. H., *Open-Channel Hydraulics,* McGraw-Hill, New York, 1986.

Fried, E., and I. E. Idelchik, *Flow Resistance,* Hemisphere Publishing Corp., New York, 1989.

Geiger, W. F., and H. R. Dorsch, "Quantity-Quality Simulation (QQS): A Detailed Continuous Planning Model for Urban Runoff Control," Report EPA-600/2-80-011, US EPA, 1980.

Goren, S. L., "The Instability of an Annular Thread of Fluid," *J. Fluid Mech.,* 12:309–319, 1962.

Hager, W. H., *Wastewater Hydraulics,* Springer-Verlag, Berlin, Germany, 1999.

Haindl, K., "Hydraulic Jump in Closed Conduits," *Proceedings of the 7th Congress of the International Association for Hydraulic Research,* Lisbon, Portugal, 2:D32.1–D32.12, 1957.

Haman, M. A., and J. A. McCorquodale, "Transient Conditions in the Transition from Gravity to Surcharged Sewer Flow," *Canadian Journal of Civil Engineering,* 9:189–196, 1982.

Harris, G. S., "Development of a Computer Program to Route Runoff in the Minneapolis-St. Paul Interceptor Sewers," *Memo M121,* St. Anthony Falls Hydraulic Laboratory, University of Minnesota, Minneapolis, MN, 1968.

Hoff-Clausen, N. E., K. Havnø, and A. Kej, "System 11 Sewer—A Storm Sewer Model," in B. C. Yen, ed., *Urban Stormwater Hydraulics and Hydrology,* Water Resources Publications, Highlands Ranch, CO, pp. 137–146, 1982.

Howarth, D. A., and A. J. Saul, "Energy Loss Coefficients at Manholes," in P. Balmer, P. A. Malmqvist, and A. Sjöberg, eds., *Proceedings of the 3rd International Conference on Urban Storm Drainage,* 1: 127–136, Chalmers University of Technology, Göteborg, Sweden, 1984.

Hromadka, T. V., and C. C. Yen, "A Diffusion Hydrodynamic Model," *Water Resources Investigation Report* 87-4137, US Geological Survey, 1987.

Hsu, C. C., W.-J. Lee, and C. H. Chang, "Subcritical Open-Channel Junction Flow," *Journal of Hydraulic Engineering,* ASCE, 124 (8):847–855, 1998.

Huber, W. C., "Chapter 22: EPA Storm Water Management Model—SWMM," in V. P. Singh, ed., *Computer Models of Watershed Hydrology,* Water Resources Publications, Highlands Ranch, CO, pp. 783–808, 1995.

Huber, W. C., and R. E. Dickinson, "Storm Water Management Model User's Manual: Version 4," *Environ. Prot. Technol. Ser.* EPA 600/3-88/001a (NTIS PB88-236641/AS), US EPA, Athens, GA, 1988.

Huber, W. C., and J. P. Heaney, "The USEPA Storm Water Management Model, SWMM: A Ten Year Perspective," in B. C. Yen, ed., *Urban Stormwater Hydraulics and Hydrology,* Water Resources Publications, Highlands Ranch, CO, pp. 247–256, 1982.

Hydrologic Engineering Center, "Urban Runoff: Storage, Treatment and Overflow Runoff Model—STORM," US Army Corps of Engineers Hydrologic Engineering Center Computer Program 723-58-L2520, Davis, CA, 1974.

Iwasa, Y., "The Criterion for Instability of Steady Uniform Flows in Open Channels," *Mem. Faculty Eng.,* Kyoto University, Kyoto, Japan, 16:264–275, 1954.

Izzard, C. F., "Hydraulics of Runoff from Developed Surfaces," *Proceedings Highway Research Board,* 26:129–146, 1946.

Javdani, K., and S. L. Goren, "Finite Amplitude Wavy Flow of Thin Films, *Prog. Heat Mass Transfer,* 6:253–262, 1972.

Johnston, A. J., and R. E. Volker, "Head Losses at Junction Boxes," *Journal of Hydraulic Engineering,* ASCE 116(3): 326–341, 1990.

Joliffe, I. B., "Accurate Pipe Junction Model for Steady and Unsteady Flows," in B. C. Yen, ed., *Urban Stormwater Hydraulics and Hydrology,* Water Resources Publications, Highlands Ranch, CO, pp. 92–100, 1982.

Joliffe, I. B., "Free Surface and Pressurized Pipe Flow Computations," in P. Balmer, P. A. Malmqvist, and A. Sjöberg, eds., *Proceedings of the 3rd International Conference on Urban Storm Drainage,* 1: 397–405, Chalmers University of Technology, Göteborg, Sweden, 1984a.

Joliffe, I. B., "Computation of Dynamic Waves in Channel Networks," *Journal of Hydraulic Engineering,* ASCE 110(10):1358–1370, 1984b.

Jun, B. H., and B. C. Yen, "Dynamic Wave Simulation of Unsteady Open Channel and Surcharge Flows in Sewer Network," *Civil Engineering Studies Hydraulic Engineering Series* no. 40, University of Illinois at Urbana-Champaign, Urbana, IL, 160 pp., 1985.

Kalinske, A. A., and J. M. Robertson, "Entrainment of Air in Flowing Water: Closed Conduit Flow," *Trans.,* ASCE 108:1435–1447; 1513–1516, 1943.

Kanda, T., and T. Kitada, "An Implicit Method for Unsteady Flows with Lateral Inflows in Urban Rivers," *Proceedings of the 17th Congress of the International Association for Hydraulic Research,* Baden-Baden, Germany, 2:213–220, 1977.

Kerby, J.H., "Time of Concentration for Overland Flow," *Civil Engineering,* 60:174, 1959.

Killen, J. M., and A. G. Anderson, "A Study of the Air-Water Interface in Air-Entrained Flow in Open-Channels," *Proceedings of the 13th Congress of the International Association for Hydraulic Research,* Tokyo, Japan, 2:339–348, 1969.

Klym, H., W. Königer, F. Mevius, and G. Vogel, "Urban Hydrological Processes," paper presented in the Seminar on Computer Methods in Hydraulics, Swiss Federal Institute of Technology, Zurich, Switzerland, 1972.

Koloseus, H. J., and J. Davidian, "Free-Surface Instability Correlations," Water-Supply Paper 1959-C, US Geological Survey, 1966.

Koussis, A. D., "Comparison of Muskingum Method Different Schemes," *J. Hydraul. Div.,* ASCE 108(HY5):925–929, 1980.

Kusuda, T., and S. Arao, "Energy Losses at Circular Drop Manhole," *Proceedings of the 7th International Conference on Urban Storm Drainage,* 1:85–90, Hannover, Germany, 1996.

Labadie, J. W., D. M. Morrow, and R. C. Lazaro, "Urban Stormwater Control Package for Automated Real-Time Systems," Project Report no. C6179, Department of Civil Engineering, Colorado State University, Fort Collins, CO, 1978.

Lai, C., "Numerical Modeling of Unsteady Open-Channel Flow," in B. C. Yen, *Advances in Hydroscience, vol. 14,* Academic Press, Orlando, FL, pp. 162–333, 1986.

Larson, C. L., T. C. Wei, and C. E. Bowers, "Numerical Routing of Flood Hydrographs Through Open Channel Junctions," *Water Resources Research Center Bulletin,* no. 40, University of Minnesota, Minneapolis, MN, 1971.

Liggett, J. A., and J. A. Cunge, "Numerical Methods of Solution of the Unsteady Flow Equation," in K. Mahmood and V. Yevjevich, eds., *Unsteady Flow in Open Channels,* vol. 1, Water Resources Publications, Highlands Ranch, CO, 1975.

Lighthill, M. J., and G. B. Whitham, "On Kinematic Waves: I. Flood Movements in Long Rivers," *Proc. Royal Soc.* London, 229A(1178):281–316, 1955.

Lin, J. D., and H. K. Soong, "Junction Losses in Open Channel Flows," *Water Resour. Res.,* 15:414–418, 1979.

Lin, S. P., "Finite Amplitude Side-Band Stability of a Viscous Film," *J. Fluid Mech,* 63:417–429, 1974.

Lindvall, G., "Head Losses at Surcharged Manholes with a Main Pipe and a 90° Lateral," in P. Balmer, P. A. Malmqvist, and A. Sjöberg, eds., *Proceedings of the 3rd International Conference on Urban Storm Drainage,* 1:137–146, Chalmers University of Technology, Göteborg, Sweden, 1984.

Lindvall, G., "Head Losses at Surcharged Manholes," in B. C. Yen, ed., *Urban Stormwater Hydraulics and Hydrology,* (joint *Proceedings of the 4th International Conference on Urban Storm Drainage* and *IAHR 22nd Congress,* Lausanne, Switzerland), Water Resources Publications, Highlands Ranch, CO, pp. 140–141, 1987.

Marsalek J., "Head Losses at Sewer Junction Manholes," *Journal of Hydraulic Engineering,* ASCE 110(8):1150–1154, 1984.

Marsalek, J., "Head Losses at Selected Sewer Manholes," Special Report no. 52, American Public Works Association, Chicago, IL, 1985.

Mayer, P. G., "Roll Waves and Slug Flows in Inclined Open Channels," *Trans.,* ASCE 126, Part I:505–564, 1961.

Mays, L. W., "Sewer Network Scheme For Digital Computations," *J. Environ. Eng. Div.,* ASCE 104(EE3):535–539, 1978.

Merlein, J., and F. Valentin, "Hydraulic Conditions and Energy Loss in Submerged Manholes," *Proceedings of the 8th International Conference on Urban Storm Drainage,* 4:2155–2162, Sydney, Australia, 1999.

Metcalf & Eddy, Inc., University of Florida, and Water Resources Engineers, Inc., "Storm Water Management Model," *Water Pollution Control Res. Ser.,* 11024 *DOC,* vol. 1–4, U.S. EPA, 1971.

Miller, D. S., *Internal Flow Systems,* 2d ed., Gulf Publishing Co., Book Division, Houston, TX, 1990.

Moody, L. F., "Friction Factors for Pipe Flow," *Transactions,* ASME, 66:671–684, 1944.

Morgali J., and R. K. Linsley, "Computer Analysis of Overland Flow," *Journal of Hydraulics Division,* ASCE 91(HY3):81–100, 1965.

Mura, Y., S. Ijuin, and H. Nakagawa, "Air Demand on Conduits Partially Filled with Flowing Water," *Proceedings of the 8th Congress International Association for Hydraulic Research,* Montreal, Canada, 2:12D.1–12D.4, 1959.

Pagliara, S., and B. C. Yen, "Sewer Network Hydraulic Model: NISN," *Civil Eng. Studies Hydraul. Series* 53, University of Illinois at Urbana-Champaign, Urbana, IL, 1997.

Pansic, N., "Dynamic-Wave Modeling of Storm Sewers with Surcharge," M.S. thesis, Department of Civil Engineering, University of Illinois at Urbana-Champaign, Urbana, IL, 1980.

Ponce, V. M., *Engineering Hydrology,* Prentice-Hall, Englewood Cliffs, NJ, 1989.

Ponce, V. M., and F. D. Theurer, "Accuracy Criteria in Diffusion Routing," *Journal of Hydraulics Division,* ASCE 108(HY6):747–757, 1982.

Price, R. K., "A Simulation Model for Storm Sewers," in B. C. Yen, ed., *Urban Stormwater Hydraulics and Hydrology,* Water Resources Publications, Highlands Ranch, CO, pp. 184–192, 1982a.

Price, R. K., "The Wallingford Storm Sewer Design and Analysis Package," in B. C. Yen, ed., *Urban Stormwater Hydraulics and Hydrology,* Water Resources Publications, Highlands Ranch, CO, pp. 213–220, 1982b.

Radojkovic, M., and C. Maksimovic, "Internal Boundary Conditions for Free Surface Unsteady Flow in Expansions and Junctions," *Proceedings of the 17th Congress of International Association for Hydraulic Research,* Baden-Baden, Germany, 2:367–372, 1977.

Ramamurthy, A. S., and W. Zhu, "Combining Flows in 90° Junctions of Rectangular Closed Conduits," *Journal of Hydraulic Engineering,* ASCE 123(11):1012–1019, 1997.

Ramamurthy, A. S., L. B. Carballada, and D. M. Tran, "Combining Open Channel Flow at Right Angled Junctions," *Journal of Hydraulic Engineering,* ASCE, 114(12):1449–1460, 1988.

Rhodes, E., and D. S. Scott, "Cocurrent Gas-Liquid Flow," *Proc. Int. Symp. Res. in Cocurrent Gas-Liquic Flow,* University of Waterloo, ON, Canada, pp. 1–17, 1968.

Roesner, L. A., and R. P. Shubinski, "Improved Dynamic Routing Model for Storm Drainage Systems," in B. C. Yen, ed., *Urban Stormwater Hydraulics and Hydrology,* Water Resources Publications, Highlands Ranch, CO, pp. 146–173, 1982.

Roesner, L. A., J. A. Aldrich, and R. E. Dickinson, "Stormwater Management Model User's Manual Version 4: Addendum I, EXTRAN," *Environ. Prot. Technol. Ser. EPA–600/3-88/001b* (NTIS PB88-236658/AS), US EPA, Athens, GA, 1988.

Rouse H., "Critical Analysis of Open-Channel Resistance," *Journal of the Hydraulics Division,* ASCE 91(HY4):1–25, 1965.

Sakakibara, T., S. Takana, and T. Imaida, "Energy Loss at Surcharged Manholes–Model Experiment," *Proceedings of the 7th International Conference on Urban Storm Drainage,* 1:79–84, Hannover, Germany, 1996.

Sangster, W. M., H. W. Wood, E. T. Smerdon, and H. G. Bossy, "Pressure Changes at Storm Drain Junctions," Bulletin no. 41, Engineering Experiment Station, University of Missouri, Columbia, MO, 1958.

Sangster, W. M., H. W. Wood, E. T. Smerdon, and H. G. Bossy, "Pressure Changes at Open Junctions in Conduit," *Trans.,* ASCE 126, Part I:364–396, 1961.

Schaake, J. C., J. C. Geyer, and J. W. Knapp, "Experimental Examination of the Rational Formula," *Journal of Hydraulic Division* ASCE, 93 (HY6):353–370, 1967.

Serre, M., A. J. Odgaard, and R. A. Elder, "Energy Loss at Combining Pipe Junction," *Journal of Hydraulic Engineering,* ASCE, 120(7):808–830, 1994.

Sevuk, A. S., "Unsteady Flow in Sewer Networks," Ph.D. thesis, Department of Civil Engineering, University of Illinois at Urbana-Champaign, Urbana, IL, 1973.

Sevuk, A. S., and B. C. Yen, "Sewer Network Routing by Dynamic Wave Characteristics," *Journal of Hydraulics Division,* ASCE 108(HY3):379–398, 1982.

Sevuk, A. S., B. C. Yen, and G. F. Peterson, II, "Illinois Storm Sewer System Simulation Model: User's Manual," Res. Report no. 73, Water Resources Center, University of Illinois at Urbana-Champaign, Urbana, IL, 1973.

Sharma, H. R., "Air-Entrainment in High Head Gated Conduits," *Journal of Hydraulics Division,* ASCE 102(HY11), 1629–1646, 1976.

Shen, H. W., and R. M. Li, "Rainfall Effects on Sheet Flow over Smooth Surface," *Journal of Hydraulics Division,* ASCE99(HY5):771–792, 1973.

Sjöberg, A., "Calculation of Unsteady Flows in Regulated Rivers and Storm Sewer Systems," Report, Division of Hydraulics, Chalmers University of Technology, Göteborg, Sweden, 1976.

Sjöberg, A., "Sewer Network Models DAGVL-A and DAGVL-DIFF," in B. C. Yen, ed., *Urban Stormwater Hydraulics and Hydrology,* Water Resources Publications, Highlands Ranch, CO, pp. 127–136, 1982.

Smith, A. A., "A Generalized Approach to Kinematic Flood Routing," *J. Hydrol.,* 45:71–87, 1980.

Stephenson, D., *Pipeflow Analysis,* Elsevier, Amsterdam, The Netherlands, 1984.

Strelkoff, T., and N. D. Katopodes, "Border Irrigation Hydraulics with Zero-Inertia," *Journal of Irrigation & Drainage Division,* ASCE 103(IR3):325–342, 1977.

Taitel, Y., N. Lee, and A. E. Dukler, "Transient Gas-Liquid Flow in Horizontal Pipes: Modeling the Flow Pattern Transitions," *J. AICHE,* 24:920–934, 1978.

Taylor, E. H., "Flow Characteristics at Rectangular Open Channel Junctions," *Trans.,* ASCE 109:893–902, 1944.

Townsend, R. D., and J. R. Prins, "Performance of Model Storm Sewer Junctions," *Journal of Hydraulics Division,* ASCE 104(HY1):99–104, 1978.

Vermeuleu, L. R., and J. T. Ryan, "Two-Phase Slug Flow in Horizontal and Inclined Tubes," *Canadian Journal of Chemical Engineering,* 49:195–201, 1971.

Vogel, G., and H. Klym, "Die Ganglinien-Volumen-Methode," paper presented at the Workshop on Methods of Sewer Network Calculation, Dortmund, Germany, 1973.

Volkart, P. U., "Self-Aerated Flow in Steep, Partially Filled Pipes," *Journal of Hydraulics Division,* ASCE 108(HY9):1029–1046, 1982.

Wallingford Software, "SPIDA User Manual—Version Alpha-3," Hydraulic Research Ltd., Wallingford, UK, 1991.

Wallingford Software, "HYDROWORKS User Manual," Hydraulic Research Ltd., Wallingford, UK, 1997.

Wallis, G. B., *One-Dimensional Two-Phase Flow,* McGraw-Hill Book Co., New York, 1969.

Webber, N. B., and C. A. Greated, "An Investigation of Flow Behaviour at the Junction of Rectangular Channels," *Proc. Inst. Civ. Eng.,* (London), 34:321–334, 1966.

Weinmann, P. E., and E. M. Laurenson, "Approximate Flood Routing Methods: A Review," *Journal of Hydraulics Division,* ASCE 105(HY12):1521–1536, 1979.

Wiggert, D. C., "Transient Flow in Free-Surface Pressurized Systems," *Journal of Hydraulics Division,* ASCE 98(HY1):11–27, 1972.

Wisner, P. E., F. N. Mohsen, and N. Kouwen, "Removal of Air from Water Lines by Hydraulic Means," *Journal of Hydraulics Division,* ASCE 101(HY2):243–257, 1975.

Wood, D. J., "The Analysis of Flow in Surcharged Storm Sewer Systems," *Proceedings of the International Symposium on Urban Storm Runoff,* University of Kentucky, Lexington, Kentucky, pp. 29–35, 1980.

Wylie, E. B., and V. L. Streeter, *Fluid Transients,* FEB Press, Ann Arbor, MI, 1983.

Yen, B. C., "Methodologies for Flow Prediction in Urban Storm Drainage Systems," Research Report no. 72, Water Resources Center, University of Illinois at Urbana-Champaign, Urbana, IL, 1973a.

Yen, B. C., "Open-Channel Flow Equations Revisited," *Journal of Engineering Mechanics Division,* ASCE 99(EM5):979–1009, 1973b.

Yen, B. C., "Further Study on Open-Channel Flow Equations," *Sonderforschungsbereich 80,* Report no. SFB80/T/49, University of Karlsruhe, Karlsruhe, Germany, 1975.

Yen, B. C., "Hydraulic Instabilities of Storm Sewer Flows," in P. R. Helliwell, ed., *Urban Storm Drainage, Proceedings of the 1st International Conference,* Pentech Press, London and Wiley-Interscience, New York, pp. 282–293, 1978a.

Yen, B. C., *Storm Sewer System Design,* Department of Civil Engineering, University of Illinois at Urbana–Champaign, Urbana, IL, 1978b.

Yen, B. C., "Some Measures for Evaluation and Comparison of Simulation Models," in B. C. Yen, ed., *Urban Stormwater Hydraulics and Hydrology,* Water Resources Publications, Highlands Ranch, CO, pp. 341–349, 1982.

Yen, B. C., "Hydraulics of Sewers," in B. C. Yen, ed., *Advances in Hydroscience, vol. 14,* Academic Press, Orlando, FL, pp. 1–122, 1986.

Yen, B. C., "Urban Drainage Hydraulics and Hydrology: From Art to Science" (Joint Keynote at 22nd IAHR Congress and 4th International Conference on Urban Storm Drainage, EPF-Lausanne, Switzerland), *Urban Drainage Hydraulics and Hydrology,* Water Resources Publications, Highlands Ranch, CO, pp. 1–24, 1987.

Yen, B. C., "Hydraulic Resistance in Open Channels," in B. C. Yen, ed., *Channel Flow Resistance: Centennial of Manning's Formula,* Water Resources Publications, Highlands Ranch, CO, pp. 1–135, 1991.

Yen, B. C., "Chapter 25: Hydraulics for Excess Water Management," in L. W. Mays, ed., *Handbook of Water Resources,* McGraw-Hill, New York, pp. 25-1–25-55, 1996.

Yen, B. C., and A. O. Akan, "Flood Routing Through River Junctions," *Rivers '76,* ASCE, 1:212–231, New York, 1976.

Yen, B. C., and A. O. Akan, "Chapter 14: Hydraulic Design of Urban Drainage Systems," in L.W. Mays, ed. *Hydraulic Design Handbook,* McGraw-Hill, New York, pp. 14.1–14.114, 1999.

Yen, B. C., and J. A. Gonzalez, "Determination of Boneyard Creek Flow Capacity by Hydraulic Performance Graph," *Research Report 219,* Water Resources Center, University of Illinois at Urbana-Champaign, Urbana, IL 1994.

Yen, B. C., and A. S. Sevuk, "Design of Storm Sewer Networks," *J. Environ. Eng. Div.,* ASCE, 101(EE4):535–553, 1975.

Yen, B. C. and C. W.-S. Tsai, "Noninertia Wave vs. Diffusion Wave in Flood Routing," *J. Hydrology,* 2001.

Yen, B. C., H. G. Wenzel, Jr., L. W. Mays, and W. H. Tang, "Advanced Methodologies for Design of Storm Sewer Systems," Res. Report no. 112, Water Resources Center, University of Illinois at Urbana-Champaign, Urbana, IL, 1976.

Yen, B. C., S. T. Cheng, B.-H. Jun, M. L. Voorhees, H. G. Wenzel, Jr., and L. W. Mays, "Illinois Least-Cost Sewer System Design Model: ILSD-1&2 User's Guide," Res. Report no. 188, Water Resources Center, University of Illinois at Urbana-Champaign, Urbana, IL, 1984.

Yevjevich, V., and A. H. Barnes, "Flood Routing Through Storm Drains, Parts I–IV," Hydrologic Papers 43–46, Colorado State University, Fort Collins, CO, 1970.

Yih, Y.S., *Dynamics of Nonhomogeneous Fluids,* Macmillan, New York, 1965.

CHAPTER 7
DESIGN OF DETENTION SYSTEMS

John N. Paine
BIG, Incorporated
Newport News, Virginia

A. Osman Akan
Department of Civil and Environmental Engineering
Old Dominion University
Norfolk, Virginia

Urban development results in increased runoff volumes and flowrates, which may cause frequent flooding and severe stream erosion downstream. Many communities use flow retardation structures to limit adverse downstream effects of urban storm runoff. These structures can also partially settle out the particulate pollution contained in the storm runoff.

Various types of flow retardation measures include detention basins, retention basins, roof top storage, infiltration basins, and dry wells. *Detention basins* are small impoundments of water with a capacity of normally 10 acre-ft or less. *Retention basins* are usually larger, and they release stored water at a slower rate mostly through controlled outlets. *Infiltration basins* are used to allow the stored water to percolate into the ground. *Dry wells* are small trenches excavated in porous soil and backfilled with rock. Although different flow retardation measures may be used under different circumstances, detention basins are probably most common. A detention pond can be created by damming a channel or by excavating a pond into the existing ground. Often, ponds are constructed by a combination of cut and fill. Figure 7.1 is a schematic diagram of a typical detention basin.

This chapter first presents a brief review of hydrology and hydraulics for detention systems. Then various methods are summarized for preliminary sizing of detention basins, extended detention basins, and retention basins. Finally, a detailed discussion is presented on the design of detention basins with a complete design example. Off-site detention basins are covered in another chapter.

7.1 HYDROLOGY AND HYDRAULICS FOR DETENTION SYSTEMS

7.1.1 Flood Routing

We can evaluate the effect of a detention basin on a given flood by routing the flood hydrograph through the basin. In a typical routing problem, the givens are the inflow hydrograph, the pond characteristics, and the initial condition. The outflow hydrograph is sought. The

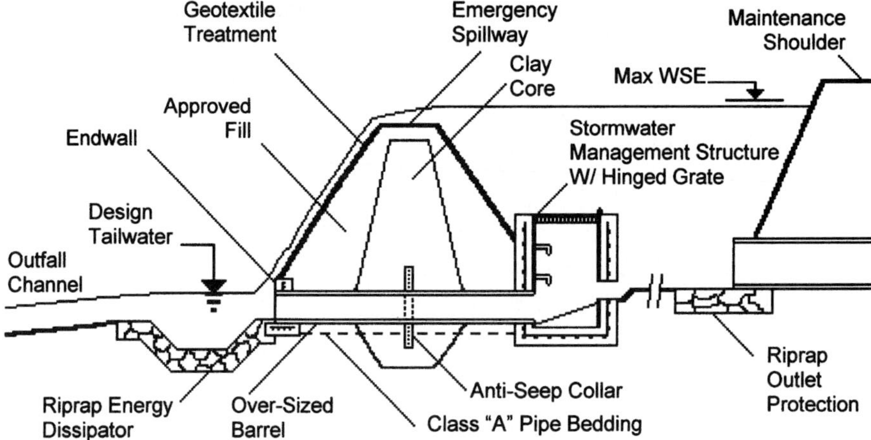

FIGURE 7.1   Schematic of a detention basin.

pond characteristics are normally prescribed as stage-storage and stage-discharge (outflow) relationships. These relationships can be in graphical, tabular, or equation form. The *stage* represents the elevation of water surface in the pond measured from a horizontal datum. The *storage* is the volume of water in the pond. Needless to say, the storage increases with increasing stage, and the relationship depends on the shape and the size of the detention pond. The *stage-discharge relationship* is governed by the hydraulics of the outlet structures. From the given stage-storage and stage-discharge relationships, we can develop a relationship between the storage and the outflow rate. The initial condition required is normally the water level in the pond at the time the incoming flood reaches the detention basin.

The change of volume of water in storage in the pond is described by the equation

$$I - Q = \frac{dS}{dt} \tag{7.1}$$

where I = inflow rate
$\quad Q$ = outflow rate
$\quad S$ = storage volume
$\quad t$ = time

For a finite time period, Δt, Eq. (7.1) can be written in finite difference form and rearranged as

$$(I_1 + I_2) + \left(\frac{2S_1}{\Delta t} - Q_1\right) = \left(\frac{2S_2}{\Delta t} + Q_2\right) \tag{7.2}$$

where I_1 = inflow rate at start of the time period
$\quad I_2$ = inflow rate at end of the time period
$\quad \Delta t$ = duration of the time period
$\quad S_1$ = storage at the beginning of the time period
$\quad S_2$ = storage at the end of the time period
$\quad Q_1$ = outflow rate at the beginning of the time period
$\quad Q_2$ = outflow rate at the end of the time period

The unknowns in Eq. (7.2) are Q_2 and S_2. Using the storage-discharge relationship of the pond along with Eq. (7.2), we can determine Q_2 and S_2. However, in many cases the storage-discharge relationship is not in equation form, and a semi-graphical procedure is needed. We can summarize a solution procedure as follows:

1. From the given stage-storage and stage-discharge relationship obtain a storage-discharge (S versus Q) relationship.

2. Select a time increment, Δt. Calculate the quantity $[(2S/\Delta t) + Q]$ as a function of Q and prepare a plot such as shown in Fig. 7.2.

3. For any time step computations calculate $(I_1 + I_2)$ from the inflow hydrograph, and $[(2S_1/\Delta t) - Q_1]$ from either the initial conditions or previous time step calculations.

4. Calculate $[(2S_2/\Delta t) + Q_2]$ from Eq. 7.2.

5. Obtain Q_2 from the graph developed in step 2. This will be the outflow rate at time t_2.

6. To proceed to the next time step, calculate $[(2S_2/\Delta t) - Q_2]$ by subtracting $2Q_2$ from $[(2S_2/\Delta t) + Q_2]$ and go back to step 3. Obviously, the value of $[(2S_2/\Delta t) - Q_2)]$ calculated at any time step will become $[(2S_1/\Delta t) - Q_1)]$ for the next time step.

7. Repeat the same procedure until the routing is completed.

Example 7.1

The stage-discharge-storage relationship for a reservoir is given in tabular form in columns 1, 2 and 3 of Table 7.1. We are to route the inflow hydrograph tabulated in columns 2 and 4 of Table 7.2 through this reservoir. Initially, that is at time zero; the reservoir water level and the outflow rate are both zero. We will use a time increment of $\Delta t = 10$ minutes $= 600$ seconds.

To solve this problem, first the reservoir storage outflow function $[(2S/\Delta t) + Q]$ is obtained and tabulated in column 4 of Table 7.1. These values are found by multiplying the entries in column 3 by 2.0, dividing by 600 seconds and adding them to the entries in column 2. Then, a plot of Q versus $[(2S/\Delta t) + Q]$ is prepared as shown in Fig. 7.2. The routing calculations are summarized in Table 7.2. For the first time step, $t_1 = 0$, $t_2 = 10$ minutes, $I_1 = 0$, $I_2 = 50$ cfs, $Q_1 = 0$ and $S_1 = 0$. The entry, 50 cfs, in column 8 is calculated from Eq. (7.2). The entry, 2.5 cfs, in column 9 is obtained from Fig. 7.2. Thus the outflow rate at 10 minutes is 2.5 cfs. The entry, 45.0 cfs, in column 10 is found by multiplying the entry in column 9 by 2.0 and subtracting it from the entry in column 8. The 45.0 cfs so calculated becomes the value of $[(2S_1/\Delta t) - Q_1]$ for the next time step calculations, and it is entered in the second row of column 7. The same procedure is repeated until the routing is completed.

The inflow and the calculated outflow hydrographs are plotted in Fig. 7.3. An inspection of the results show that the peak outflow rate is 247.5 cfs, which is less than the peak inflow rate. Also, the peak outflow rate occurs at $t = 80$ minutes; that is, 120 minutes after the

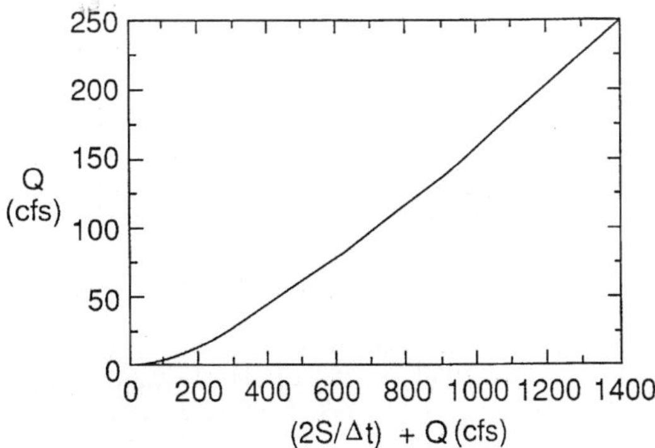

FIGURE 7.2 Example of storage-outflow function.

TABLE 7.1 Storage-outflow Relationship for Example 7.1

(1) Stage h (ft)	(2) Discharge Q (cfs)	(3) Storage S (10^3 ft^3)	(4) $(2S/\Delta t) + Q$ (cfs)
0.0	0.0	0	0
0.5	3.5	20	70.2
1.0	10.0	40	143.3
1.5	18.3	60	218.3
2.0	28.3	80	295.0
2.5	39.5	100	372.8
3.0	52.0	120	452.0
3.5	65.5	140	532.2
4.0	80.0	160	613.3
4.5	95.5	180	695.5
5.0	111.8	200	778.5
5.5	129.0	220	862.3
6.0	147.0	240	947.0
6.5	165.7	260	1032.6
7.0	185.2	280	1118.5
7.5	204.9	300	1204.9
8.0	223.6	320	1290.3
8.5	244.1	340	1377.4
9.0	263.0	360	1463.0

peak of the inflow hydrograph. The attenuation and the lagging of the peak discharge represent typical effects of reservoirs on flood waves. Large reservoirs and smaller outlet structures will result in more pronounced effects.

7.1.2 Mathematical Stage-Storage Expressions

For regular-shaped basins, the stage-storage relationship can be obtained from the geometry of the basin. For instance, for a trapezoidal detention basin that has a rectangular base of W by L and a side slope of z, the relationship between the volume (or storage), S, and the flow depth d is

$$S = LWd + (L + W)\, zd^2 + \tfrac{4}{3}z^2d^3 \tag{7.3}$$

For irregular shaped detention basins, first the surface area, A_s, versus elevation, h, relationship is obtained from the contour maps of the detention basin site. Then

$$S_2 = S_1 + (h_2 - h_1)\, \frac{A_{s1} + A_{s2}}{2} \tag{7.4}$$

in which S_1 and A_{s1} correspond to elevation h_1, and S_2 and A_{s2} correspond to h_2. Eq. (7.4) is applied to sequent elevations. A more accurate relationship is

TABLE 7.2 Example on Detention Pond Routing

(1) Time step	(2) t_1 (min)	(3) t_2 (min)	(4) I_1 cfs	(5) I_2 cfs	(6) $I_1 + I_2$ cfs	(7) $[(2S_1/\Delta t) - Q_1]$ cfs	(8) $[(2S_2/\Delta t) + Q_2]$ cfs	(9) Q_2 cfs	(10) $[(2S_2/\Delta t) - Q_2]$ cfs
1	0	10	0	50	50	0	50.	2.5	45.0
2	10	20	50	100	150	45.0	195.	15.0	165.0
3	20	30	100	150	250	165.	415.	45.0	325.0
4	30	40	150	200	350	325.	675.	90.	495.0
5	40	50	200	250	450	495.	945.	145.0	655.0
6	50	60	250	300	550	655.	1205.	204.	797.0
7	60	70	300	270	570	797.	1367.	240.5	886.0
8	70	80	270	240	510	886.	1396.	247.5	901.0
9	80	90	240	210	450	901.	1351.	237.5	876.0
10	90	100	210	180	390	876.	1266.	215.5	835.0
11	100	110	180	150	330	835.	1165.	194.3	776.4
12	110	120	150	120	270	776.4	1046.4	167.5	711.4
13	120	130	120	90	210	711.4	921.4	140.5	640.4
14	130	140	90	60	150	640.4	790.4	112.5	565.4
15	140	150	60	30	90	565.4	655.4	86.0	483.4
16	150	160	30	0	30	483.4	513.4	60.5	392.4
17	160	170	0	0	0	392.4	392.4	42.5	307.4
18	170	180	0	0	0	307.4	307.4	30.0	247.4
19	180	190	0	0	0	247.4	247.4	20.75	205.9
20	190	200	0	0	0	205.9	205.9	17.25	171.4
21	200	210	0	0	0	171.4	171.4	13.0	145.4

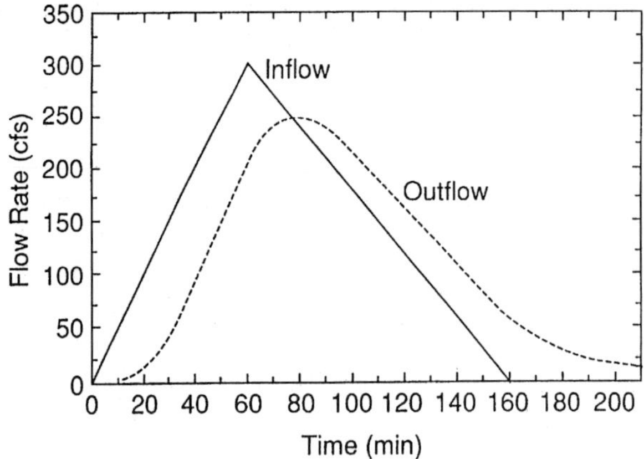

FIGURE 7.3   Inflow and outflow hydrographs for Example 7.1.

$$S_2 = S_1 + \frac{h_2 - h_1}{3}(A_{S1} + A_{S2} + \sqrt{A_{S1}A_{S2}}) \tag{7.5}$$

The stage-storage relationship for most man-made and natural basins can be approximated by

$$S = bh^c \tag{7.6}$$

where S = storage above the outlet
$\quad h$ = stage above the outlet
$\quad b, c$ = constant parameters

The constant c is dimensionless, and the constant b has the dimension of $[\text{length}]^{3-c}$. These constant parameters depend on the size and the shape of the pond. For instance, for a pond that has vertical sidewalls, $c = 1$ and b = horizontal area. Figure 7.4 displays approximate relationships between the parameter b and c, the base area, length to width ratio, and the side slope for trapezoidal ponds that has a rectangular base area and a constant side slope of z.

7.1.3 Mathematical Stage-Discharge Expressions

The outflow from a detention basin depends on the type and the size of the outlet structures. A relationship between the stage and the discharge can be obtained based on the hydraulics of these structures. Most common types of outlets can be categorized into three groups: orifice-type, weir-type, and riser-pipe structures.

7.1.3.1 Orifice-Type Outlets.   The discharge through an orifice outlet is determined as

$$Q = k_o a_o \sqrt{2gh} \tag{7.7}$$

where Q = discharge
$\quad k_o$ = dimensionless orifice discharge coefficient
$\quad a_o$ = orifice area
$\quad g$ = gravitational acceleration

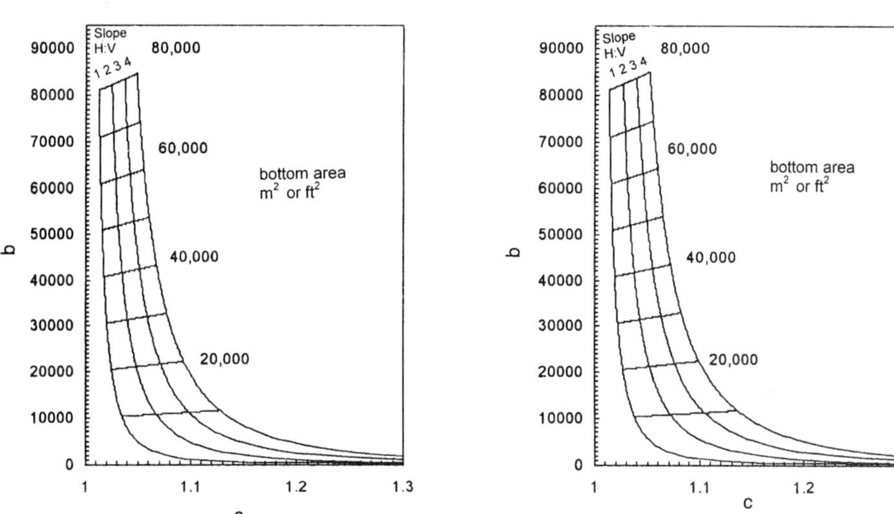

FIGURE 7.4   Values of b and c. (After Currey and Akan, 1998)

$$h = \text{depth of water above the orifice}$$

If the orifice is submerged by the tailwater as shown in Fig. 7.5, h is the difference between the headwater and tailwater elevations. If the orifice is not submerged by the tailwater, it is assumed that $Q = 0$ if the headwater is below the centroid of the orifice. Otherwise, h is set equal to the difference between the headwater elevation and the orifice centroid as shown in Fig. 7.5. This approximation is acceptable for small orifices. To account for partial flow in large orifices, the inlet control culvert flow formulation discussed elsewhere in this handbook can be used to determine the orifice flow rates setting the culvert length equal to zero. Typical values of k_o are 0.6 for square-edge uniform entrance conditions, and 0.4 for ragged edge orifices (FHWA, 1996). Short outflow pipes and culverts smaller than 0.3 m (1.0 ft) in diameter can also be treated as an orifice provided that h is greater than 1.5 times the diameter. For a circular orifice

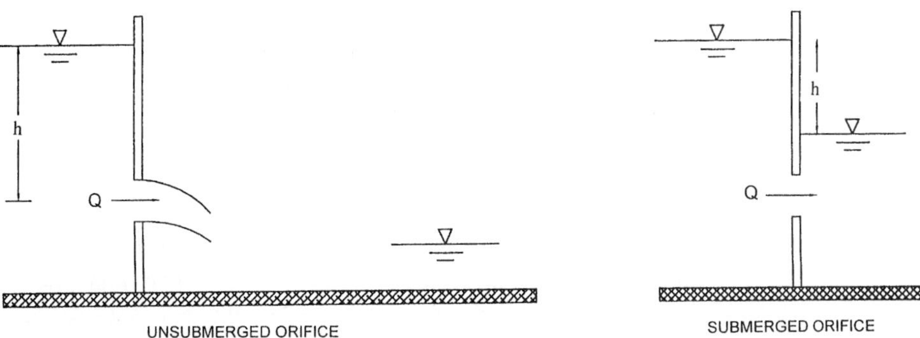

FIGURE 7.5   Orifice outlets.

$$a_o = \frac{\pi D^2}{4} \qquad (7.8)$$

where D = pipe diameter. For a rectangular orifice

$$a_o = bD \qquad (7.9)$$

where b and D represent the side lengths of the rectangular opening.

7.1.3.2 Weir-Type Outlets.   Rectangular broad-crested weirs and overflow spillways are included in this group. With reference to Fig. 7.6 the discharge over these structures is determined using

$$Q = k_w L \sqrt{2g} h^{3/2} \qquad (7.10)$$

where k_w = dimensionless weir discharge coefficient
     L = effective crest length
     h = water depth above the crest

Simon (1981) reported the discharge coefficients for many different types of broad-crested weirs. Table 7.3 may be used to guide the selection of the discharge coefficients, k_w, for Ogee spillways. Table 7.3 should not be used if $(P/h_D) < 1.0$ where P = height of the spillway, and h_d = the head for which the spillway crest is shaped.

7.1.3.3 Stand Pipes and Inlet Boxes.   Stand pipes and inlet boxes have intake openings that are parallel to the water surface as shown in Fig. 7.7. This type of structure is typically called a *stand pipe* if it has a circular cross section and an inlet box if it has a rectangular cross section. Both structures discharge into a barrel that should be large enough not to cause surcharge.

Stand pipes and inlet boxes operate as weirs when the head over the structure is low and we use Eq. (7.10) to represent this situation. The crest length, L, is calculated as

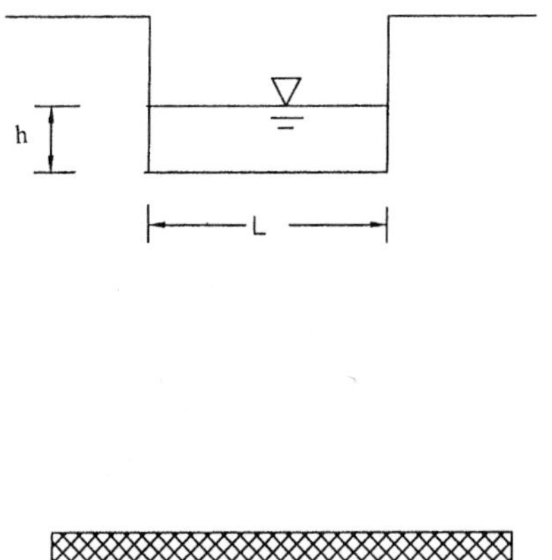

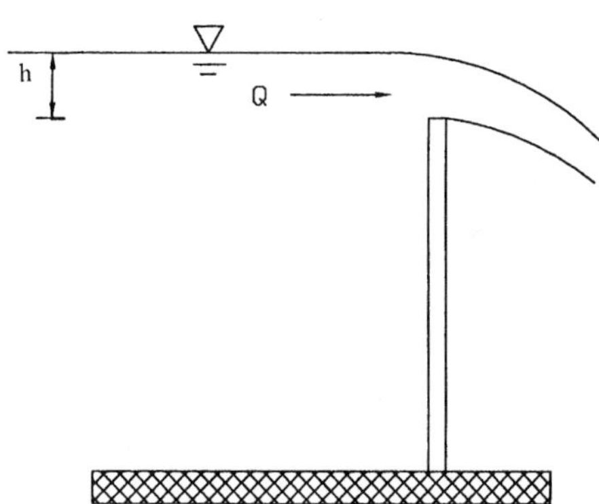

FIGURE 7.6 Weir outlets.

TABLE 7.3 Discharge Coefficient, k_w, for Ogee Spillways (Provisional)

h/h_D	k_w
0.2	0.41
0.4	0.44
0.6	0.46
0.8	0.48
1.0	0.49
1.2	0.59

$$L = \pi D \tag{7.11}$$

and

$$L = 2B + 2D \tag{7.12}$$

respectively for stand pipes and inlet boxes where D = diameter of stand pipe and B and D represent the side lengths of an inlet box. It should be noted that the k_w coefficient for these types of structures take different values from those of rectangular weirs, and they should be selected carefully.

At higher heads, stand pipes and inlet boxes will behave as an orifice, and Eqs. (7.8) and (7.9) will apply. The ranges over which the weir and orifice equations apply have not been well-established. Indeed the change from weir behavior to orifice behavior occurs gradually over a transition zone. However, for simplicity we can define a transition head, h_T, as

$$h_T = \frac{k_0 a_0}{k_w L} \tag{7.13}$$

and use the weir equation for $h < h_T$, and the orifice equation for $h > h_T$.

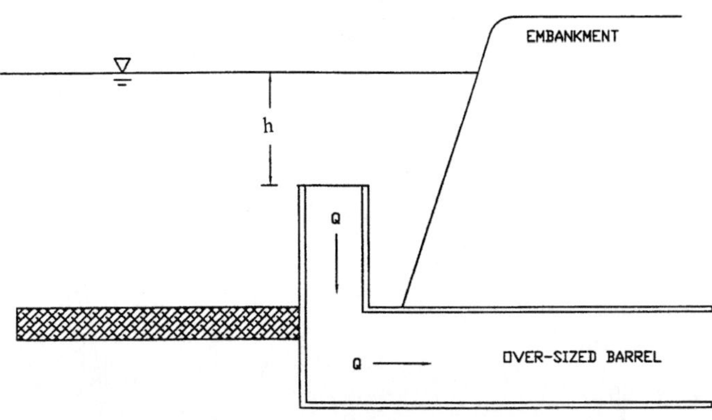

FIGURE 7.7 Stand pipes.

7.1.4 Flood Routing Charts for Detention Basins

Pre-determined solutions to the reservoir routing problem are available for quick estimates of the peak outflow rates from detention basins. These solutions are obtained and presented in terms of dimensionless parameters assuming a certain general shape for the inflow hydrographs. Those included herein are for inflow hydrographs that can be represented by a gamma function.

Figure 7.8 displays a set of pre-determined solutions in chart form for *orifice-type* outlets (Currey and Akan, 1998). In this figure, the parameters $Q*$, P, and F are dimensionless and defined, respectively, as

$$Q* = \frac{Q_p}{I_p} \tag{7.14}$$

$$P = \frac{t_p k_0 a_0 \sqrt{2g}}{b^x S_R^{1-x}} \tag{7.15}$$

and

$$F = \frac{I_p t_p}{S_R} \tag{7.16}$$

where Q_p = peak outflow rate from the pond
I_p = peak rate of the inflow hydrograph
k_o = dimensionless orifice discharge coefficient
a_o = orifice area
g = gravitational acceleration
t_p = time of occurrence of the peak inflow rate
b = coefficient in stage-storage relationship (Eq. (7.6))
c = exponent in stage-storage relationship (Eq. (7.6))
x = $0.5/c$ for an orifice outlet
S_R = total volume of runoff

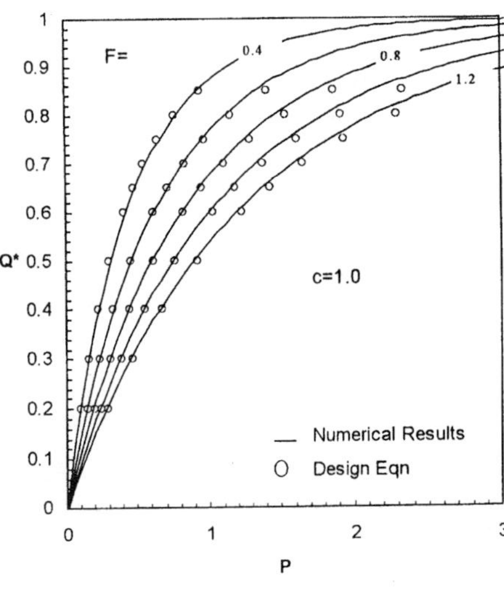

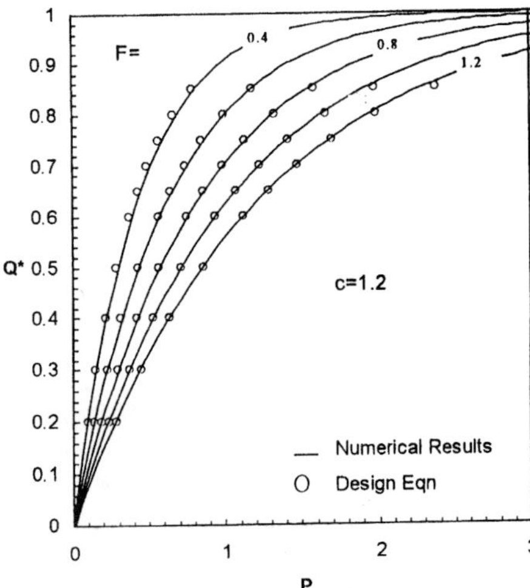

FIGURE 7.8 Pond charts for orifice outlet.

For detention basins with weir-type outlets, Fig. 7.9 presents a set of pre-determined solutions in chart form. Eqs. (7.14) and (7.16) are valid for this chart also. However, for weir-type outlets, the dimensionless parameter, P, is defined as

$$P = \frac{t_p k_w L \sqrt{2g}}{b^x} S_R^{x-1} \tag{7.17}$$

where k_w = dimensionless weir discharge coefficient
L = crest length
$x = 1.5/c$ for a weir outlet

Example 7.2
The stage-storage relationship at a pond site has the form of Eq. (7.6) with $c = 1.2$ and $b = 5000$ m$^{1.8}$. The outlet is a circular orifice with $D = 1.5$ m and $k_o = 0.6$. The watershed has a drainage area of 2.7 km^2 and the runoff for the event being considered is 3 cm. In other words, $S_R = (2.7 \times 10^6 \text{ m}^2)(0.03 \text{ m}) = 81{,}000$ m^3. For the inflow hydrograph, $I_p = 18$ m^3/sec and $t_p = 1$ hour $= 3600$ seconds. We are to find the peak outflow rate, the maximum water level in the pond, and the maximum storage volume.

First the orifice area is determined as $a_o = (3.14)(1.5)^2/4 = 1.77$ m^2 by using Eq. (7.8). Also, by using Eq. (7.16), $F = (18)(3600)/(81000) = 0.80$. Next from Eq. (7.15)

$$P = \frac{(3600)(0.60)(1.77)\sqrt{2(9.81)}}{(5000)^{0.5/1.2}(81000)^{(1.0-0.5/1.2)}} = 0.67$$

Then, with $F = 0.80$, $c = 1.2$ and $P = 0.67$, we obtain $Q^* = 0.55$ from Eq. (7.8). Next, from Eq. (7.14)

$$Q_p = (0.55)(18) = 9.9 \text{ m}^3/\text{sec.}$$

The maximum water level, h_p, above the outlet invert is found by solving Eq. (7.7) as

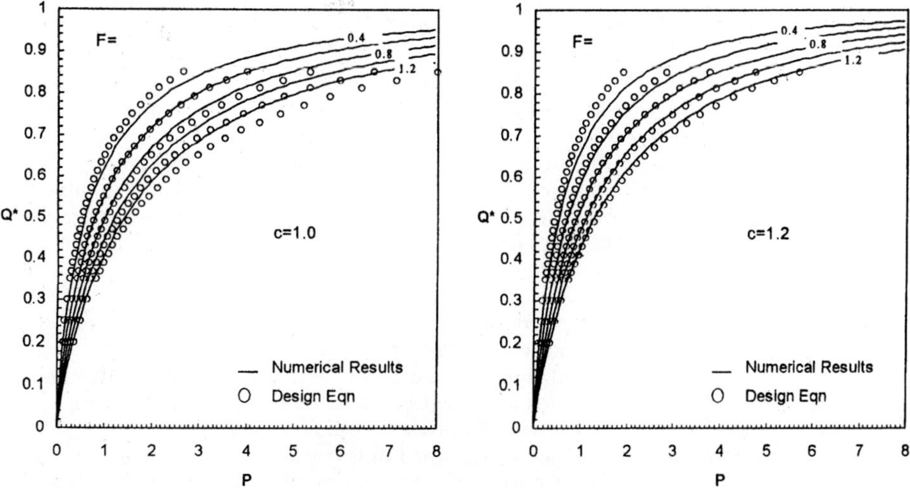

FIGURE 7.9 Pond charts for weir outlets.

$$h_p = [(9.9)/(0.60)(1.77)]^2/[2(9.81)] = 4.43 \text{ m}$$

Finally, from Eq. (7.6), the maximum storage volume, S_p, is obtained as

$$S_p = 5000(4.43)^{1.2} = 29830 \text{ m}^3$$

7.2 SIZING REQUIREMENTS FOR DETENTION BASINS

Specific design criteria for detention basins vary in different local ordinances. Some general guidelines are summarized herein. Similar guidelines can be found elsewhere in the literature (Yu and Kaighn, 1992; FHWA, 1996; Stahre and Urbonas, 1990; ASCE, 1998; Loganathan et al., 1996; Urbonas and Stahre, 1993).

The main objective of a detention basin is to control the peak runoff rates. The outfall structures should be designed to limit the peak outflow rates to allowable rates. A detention basin should also provide sufficient volume for temporary storage of runoff. The inlet, outlet, and side slopes should be stabilized where needed to prevent erosion. The side slopes should be $3H/1V$ or flatter. The channel bottom should be sloped no less than 2% towards the trickle ditch. Detention basin length to width ratio should be no less than 3.0. Outlets should have trash racks. Coarse gravel packing should be provided if a perforated riser outlet is used. An emergency spillway should be built to provide controlled overflow relief for large storms. A 100-year storm event can be used for the emergency spillway design.

A detailed discussion of the design process is given in Sec. 7.5 along with a complete example. The discussion in this section is limited to pre-sizing of detention basins. The conventional procedure for the hydraulic design of a detention basin is a trial-and-error procedure, and it consists of the following steps:

1. Calculate the detention basin inflow hydrograph(s) for the design return period(s) being considered. A rainfall-runoff model, such as HEC1, TR-20, or SCSHYDRO, can be used for this purpose. For urbanizing areas, the inflow hydrograph(s) are normally those calculated for post-development conditions.

2. Set the hydraulic design criteria. In most applications, the post-development peak(s) are required to be reduced to the magnitude(s) of the pre-development peak(s) for the design return period(s). If pre-development peak(s) are not available, a rainfall-runoff model can be used to calculate them. The hydraulic design criteria may also restrict the maximum water surface elevation in the detention basin.

3. Trial-design a detention basin. A trial design consists of the stage-storage relationship, and the types, sizes, and elevations of the outlet structures. Route the inflow hydrograph(s) through the trial-designed detention basin and check if the design criteria set are met. If not, go back to step 3. Also, if the criteria are met, but the outflow peak(s) are much smaller than the allowable value(s), then the trial basin is overdesigned. Again, go back to step 3. The level-pool routing procedure described in Sec. 7.1 is adequate for most detention basin design situations.

Obviously a good trial design is the key in this procedure. Designing a detention basin can become a tedious and lengthy task if the trial designed basins are not chosen carefully.

Various charts and equations are available in the literature that can be used as trial-design aids. Most of these aids are based on pre-determined solutions to Eq. (7.1) in dimensionless form (Akan, 1989; Akan, 1990; Kessler and Diskin, 1991; McEnroe, 1992; Currey and Akan, 1998). Others are based on assumed inflow and outflow shapes (Abt and Grigg, 1978; Aron and Kibler, 1990) or results of numerous routings for many detention basins (Wycoff and Singh, 1976; Soil Conservation Service, 1986). Table 7.4 presents various design aid equa-

TABLE 7.4 Design-Aid Equations

Equation	Number of outlets	Outlet types(s)	Remarks	Reference
$\dfrac{S_{\max}}{S_R} = \dfrac{1.291(1 - Q_p/I_p)^{0.753}}{(T_b/t_p)^{0.411}}$	Not specified	Not specified	Based on numerical simulation T_6 = time base of inflow hydrograph	Wycoff and Singh (1976)
$\dfrac{S_{\max}}{S_R} = \left(1 - \dfrac{Q_p}{I_p}\right)^2$	Not specified	Not specified	Triangular inflow hydrograph, trapezoid	Abt and Grigg (1978)
$\dfrac{S_{\max}}{S_R} = 1 - \dfrac{Q_p}{I_p}$	Not specified	Not specified	Triangular inflow and outflow hydrograph	Baker (1979)
$\dfrac{S_{\max}}{S_R} = 0.660 - 1.76\left(\dfrac{Q_p}{I_p}\right) + 1.96\left(\dfrac{Q_p}{I_p}\right)^2 - 0.730\left(\dfrac{Q_p}{I_p}\right)^3$	Not specified	Not specified	For SCS 24-Hour Types I and I	Soil Conservation Service (1986)
$\dfrac{S_{\max}}{S_R} = 0.682 - 1.43\left(\dfrac{Q_p}{I_p}\right) + 1.64\left(\dfrac{Q_p}{I_p}\right)^2 - 0.804\left(\dfrac{Q_p}{I_p}\right)^3$	Not specified	Not specified	For SCS 24-Hour Types II and I	
$S_{\max} = I_p t_d - Q_p\left(\dfrac{t_d + T_c}{2}\right)$	Not specified	Not specified	t_d = storm duration, T_c = time of concentration Trapezoidal inflow hydrograph, rising limb of outflow hydrograph is linear,	Aron and Kibler (1990)
$\dfrac{S_{\max}}{S_R} = 0.932 - 0.792\,\dfrac{Q_p}{I_p}$	Single	Weir type	$0.2 < \dfrac{Q_p}{I_p} < 0.9$ Constant reservoir surface area	Kessler and Diskin (1991)
$\dfrac{S_{\max}}{S_R} = 0.872 - 0.861\,\dfrac{Q_p}{I_p}$	Single	Orifice type	$0.2 < \dfrac{Q_p}{I_p} < 0.9$ Constant reservoir surface area	Kessler and Diskin (1991)
$\dfrac{S_{\max}}{S_R} = 0.98 - 1.17\,\dfrac{Q_p}{I_p} + 0.77\left(\dfrac{Q_p}{I_p}\right)^2 - 0.46\left(\dfrac{Q_p}{I_p}\right)^3$	Single	Weir type	Gamma function inflow hydrograph	McEnroe (1992)
$\dfrac{S_{\max}}{S_R} = 0.97 - 1.42\,\dfrac{Q_p}{I_p} + 0.82\left(\dfrac{Q_p}{I_p}\right)^2 - 0.46\left(\dfrac{Q_p}{I_p}\right)^3$	Single	Orifice type	Gamma function inflow hydrograph	McEnroe (1992)

TABLE 7.4 Design-Aid Equations (*Continued*)

Equation	Number of outlets	Outlet types(s)	Remarks	Reference
$\dfrac{S_{max}}{S_R} = 0.922 - 0.787\left(\dfrac{Q_p}{I_p}\right)$ $h_{max} = \left(\dfrac{0.922 S_R - 0.787\dfrac{Q_p}{I_p} S_R}{b}\right)^{1/c}$ $L_c = \left(\dfrac{b}{0.922 S_R - 0.787\dfrac{Q_p}{I_p} S_R}\right)^{1.5/c} \dfrac{Q_p}{k_w\sqrt{2g}}$	Single	Weir Type	Gamma function inflow hydrograph relationship: $S = bh^c$, h = stage, L_c = weir crest length k_w = weir discharge coefficient, g = gravitational acceleration	Currey and Akan (1998)
$a_o = \left(\dfrac{b}{0.847 S_R - 0.841\dfrac{Q_p}{I_p} S_R}\right)^{0.5/c} \dfrac{Q_p}{k_o\sqrt{2g}}$ $\dfrac{S_{max}}{S_R} = 0.847 - 0.841\left(\dfrac{Q_p}{I_p}\right)$ $h_{max} = \left(\dfrac{0.847 S_R - 0.841\dfrac{Q_p}{I_p} S_R}{b}\right)^{1/c}$	Single	Orifice type	Gamma function inflow hydrograph relationship: $S = bh^c$, h = stage, a_o = orifice area, k_o = orifice discharge coefficient, g = gravitational acceleration	Currey and Akan (1998)

tions in which I_p is the peak inflow rate (peak discharge of post-development hydrograph), and Q_p is the allowable peak outflow rate, S_{max} is the required storage volume, and S_R is the volume of runoff.

7.2.1 Pre-sizing of Single-outlet Detention Basins

In a typical detention basin design problem, the givens are the inflow hydrograph, the maximum allowable outflow rate, and the maximum allowable reservoir water level. The design engineer would need to determine if the stage-storage relationship for the detention basin and the type, size, and elevation of the outlet structure(s). As discussed in the preceding section, a final hydraulic design of a detention basin requires routing of the design inflow through the detention pond to ensure that the design criteria are satisfied. However, the equations presented in Table 7.4 can be used to trial-design a detention basin.

Example 7.3

Suppose the rainfall excess resulting from a design rainfall is 3.5 in over a 50-acre $=$ 2178000 ft² urban watershed, and the runoff hydrograph has a peak of $I_p = 212$ cfs occurring at $t_p = 30$ minutes $= 1800$ seconds. A detention basin is to be designed to reduce the peak flow rate to $Q_p = 120$ cfs. A weir-type outlet will be used that has $k_w = 0.40$. It is also required that the depth of water above the weir crest not to exceed 6.50 ft.

Let us try a trapezoidal detention basin with a length to width ratio of 4 and side slopes of $3H/1V$. Let the surface area of the detention basin at the weir crest elevation be 40,000 ft². Then from Fig. 7.4 we obtain $b = 42500$ and $c = 1.055$. By definition the volume of runoff is determined as $S_R = (3.5/12)(2178000) = 635250$ ft³. Then using the equations suggested by Currey and Akan (1998) from Table 7.4

$$S_{max} = (635250)\{(0.922 - (0.787)(120/212)\} = 302715 \text{ ft}^3$$

$$h_{max} = \left(\frac{(0.922)(635250) - (0.787)(120/212)(635250)}{42500}\right)^{1/1.055} = 6.43 \text{ ft}$$

$$L = \left(\frac{42500}{(0.922)(635250) - (0.787)(120/212)(635250)}\right)^{1.5/1.055} \frac{120}{(0.40)\sqrt{2(32.2)}} = 2.29 \text{ ft}$$

Note that $h_{max} < 6.50$ ft, and also h_{max} is close to 6.50 ft, so the suggested basin with a base area of 40,000 ft² should be acceptable. This is, however, just one of the many hydraulically feasible preliminary designs. Other alternatives should also be considered.

7.2.2 SCS Method for Pre-sizing Detention Basins

The U.S. Soil Conservation Service developed an approximate method for quick estimates of required storage volumes for detention basins (SCS, 1986). This method is based on the chart given in Fig. 7.10

where V_s = required storage volume
V_r = total volume of runoff
Q_p = peak rate of outflow from the detention basin
I_p = peak inflow rate

Note that this chart is also presented in equation form in Table 7.4. The types I, IA, II, and III refer to the SCS standard design rainfall distributions. The volume of runoff, V_r, and the

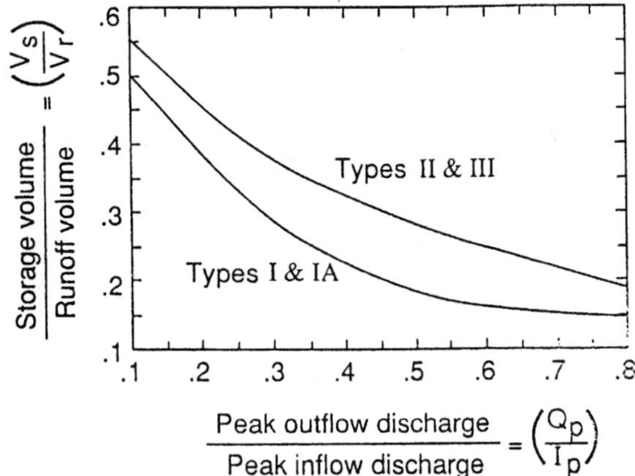

FIGURE 7.10   SCS chart for detention basin sizing. (After SCS, 1986)

peak inflow rate, I_p, can be found using a rainfall-runoff model based on SCS methodologies such as TR-20, TR-55, and SCSHYDRO.

Both single-outlet and double-outlet detention basins can be pre-sized using the SCS method. The following examples are adopted from the SCS Technical Release 55 (1986).

Example 7.4

A detention basin will be built at the outlet of a 75-acre watershed. The stage-storage relationship for the planned basin is shown in Fig. 7.11. A weir-type outlet that has a discharge coefficient of $k_w = 0.40$ will be used. The depth of runoff resulting from a 25-year type-II design storm is 3.4 inches, and the peak runoff rate is 360 cfs. This peak should be reduced to 180 cfs at the detention basin. We are to determine the required storage volume and size of the weir if the weir crest is at elevation 100 ft.

From the problem statement $I_p = 360$ cfs, $Q_p = 180$ cfs, $Q_p/I_p = 0.50$, and $V_r = (3.4)(75)/12 = 21.2$ acre-ft. Using Fig. 7.10, for a type-II storm, $V_s/V_r = 0.28$. The required storage volume becomes, $V_s = (0.28)(21.2) = 5.94$ acre-ft.

The maximum storage volume of 5.94 acre-ft represents a water stage of 105.7 ft as we can see from Fig. 7.11. Since the crest elevation is 100 ft, the maximum head above the crest becomes, $h_p = 105.7 - 100 = 5.7$ ft. Finally, from Eq. (7.10), we can find the corresponding crest length as

$$L = 180/[(0.40)\sqrt{2(32.2)}(5.7)^{1.5}] = 4.1 \text{ ft}$$

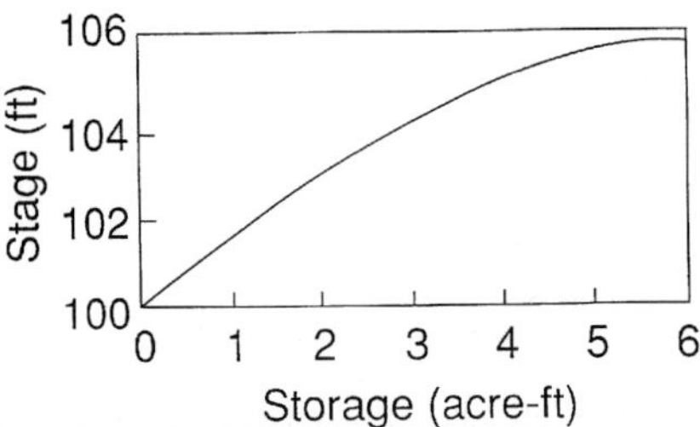

FIGURE 7.11   Stage-storage relationship for Examples 7.4 and 7.5.

Example 7.5

The watershed and the detention basin of Example 7.4 is considered again. In addition to the requirements of Example 7.4, it is decided to limit the 2-year outflow rate to 50 cfs. A 2-year type-II design storm produces a runoff depth of 1.5 in and a peak runoff rate of 91 cfs. The outlet structure will be a two-stage weir. The lower crest elevation is 100 ft. We are to determine the crest length for the lower and upper stages.

First, we consider the 2-year storm. The lower crest of the weir will be designed to control the runoff from this storm. From the problem statement, for the 2-year type-II storm, $I_p = 91$ cfs, $Q_p = 50$ cfs, $Q_p/I_p = 0.55$, and $V_r = (1.5)(75)/12 = 9.4$ ac-ft. From Fig. 7.10, $V_s/V_r = 0.26$, and therefore $V_s = (0.26)(9.4) = 2.4$ ac-ft. This represents a stage of 103.6 as obtained from Fig. 7.11. Thus, the maximum head over the lower crest is $h_p = 103.6 - 100 = 3.6$ ft, and from Eq. (7.10), the corresponding crest length will be

$$L = 50/[(0.40)\sqrt{(2)(32.2)}(3.6)^{1.5}] = 2.3 \text{ ft}$$

Now, we will consider the second stage. The upper weir crest is meant to operate for the 25-year storm but not for the 2-year storm. Therefore, the crest elevation for the second stage is chosen as $100 + 3.6 = 103.6$ ft. From Example 7.4, the maximum water stage resulting from a 25-year storm is 105.7 ft. The head over the lower crest at this stage is $105.7 - 100 = 5.7$ ft. From Eq. (7.10), the peak discharge over the lower crest, Q_{p1}, will be

$$Q_{p1} = 0.40\sqrt{2(32.2)}(2.3)(5.7)^{1.5} = 100 \text{ cfs}$$

Then, the peak discharge to be accommodated over the upper crest is $Q_{p2} = Q_p - Q_{p1} = 180 - 100 = 80$ ft. This discharge will occur under a head of $105.7 - 103.6 = 2.1$ ft. Therefore, from Eq. (7.13)

$$L = 80/[0.40\sqrt{2(32.2)}(2.1)^{1/5}] = 8.2 \text{ ft}$$

for the upper stage.

7.2.3 Modified Rational Method

The rational method was originally meant for the peak design discharge design only. Accordingly the runoff coefficients represent ratio of the peak discharge per unit area to average intensity of a storm that has the same return period. The runoff volume was not considered in developing the rational formula, and the method was not meant for detention basin design. However, a so-called *modified rational method*, which is an extension of the conventional rational method, has found widespread use in the engineering practice for preliminary sizing of detention basins in the recent years.

This method assumes that an urban stormwater runoff hydrograph (inflow hydrograph for the detention basin) under the design storm is trapezoidal in shape. The peak runoff rate is calculated using the rational formula

$$I_p = CiA \tag{7.18}$$

where I_p = peak discharge (peak inflow rate for detention basin)
    C = runoff coefficient
    i = design storm intensity
    A = area of the urban watershed

It is also assumed that the peak of the outflow hydrograph falls on the recession leg of the inflow hydrograph and the rising limb of the outflow hydrograph can be approximated by a straight line. With these assumptions, as reported by Aron and Kibler (1990)

TABLE 7.5 Example for Modified Rational Method

(1) t_d (min)	(2) i (in/hr)	(3) t_d (sec)	(4) i (ft/sec)	(5) I_p (cfs)	(6) S_d (ft³)
30	4.0	1,800	9.25×10^{-5}	56.41	65,538
40	3.5	2,400	8.10×10^{-5}	49.40	76,560
60	2.7	3,600	6.25×10^{-5}	38.11	83,196
90	2.0	5,400	4.62×10^{-5}	28.17	80,188

$$S_d = I_p t_d - Q_p \left(\frac{t_d + T_c}{2} \right) \tag{7.19}$$

where S_d = detention volume required
$\quad Q_p$ = allowable peak outflow rate
$\quad t_d$ = design-storm duration
$\quad T_c$ = time of concentration of the watershed

The *design storm duration* is the one that tends to maximize the detention storage volume, S_d, for a given return period. Aron and Kibler (1990) suggest that the storm duration be found by trial and error using the local intensity-duration-return period relationships.

The modified rational method should be used cautiously for detention basin sizing, and it should not be employed for watersheds greater than 20 to 30 acres.

Example 7.6
An urban watershed has a drainage area of $A = 20$ acres $= 871,200$ ft², runoff coefficient of $C = 0.70$, and time of concentration of $T_c = 30$ minutes $= 1,800$ seconds. A detention basin is needed to reduce the peak discharge for a design return period of 10 years to $Q_p = 20$ cfs. The rainfall duration and intensity for this design return period is tabulated in columns 1 and 2 of Table 7.5. Determine the size of the detention basin needed at this site.

The calculations for this example are presented in Table 7.5. First, the given storm duration and rainfall intensities are converted to seconds and feet per hour and tabulated in columns 3 and 4 to allow the use of consistent units in the problem. Then the peak runoff rate is calculated using Eq. (7.18) for each storm duration and intensity pair included in the table and entered in column 5. For example for $t_d = 1,800$ sec and $i = 9.25 \times 10^{-5}$ ft/sec, $I_p = (0.7)(9.25 \times 10^{-5})(871200) = 56.41$ cfs. Finally, the detention volume is calculated using Eq. (7.19) and tabulated in column 6. For instance, for $t_d = 1,800$ sec and $i = 9.25 \times 10^{-5}$ ft/second

$$S_d = (56.41)(1800) - (20) \left(\frac{1800 + 1800}{2} \right) = 65,538 \text{ ft}^3$$

The maximum detention volume included in column 6 is 83,196 ft³, and this value should be chosen as the size of the detention basin required. Also, it can be seen from the table that design storm for this detention basin has a duration of 60 minutes and an intensity of 2.7 in/hr.

7.3 EXTENDED DETENTION BASINS

Extended detention basins are effective means of removing particulate pollutants from urban storm water runoff. As shown in Fig. 7.12, an extended detention pond has two stages. The

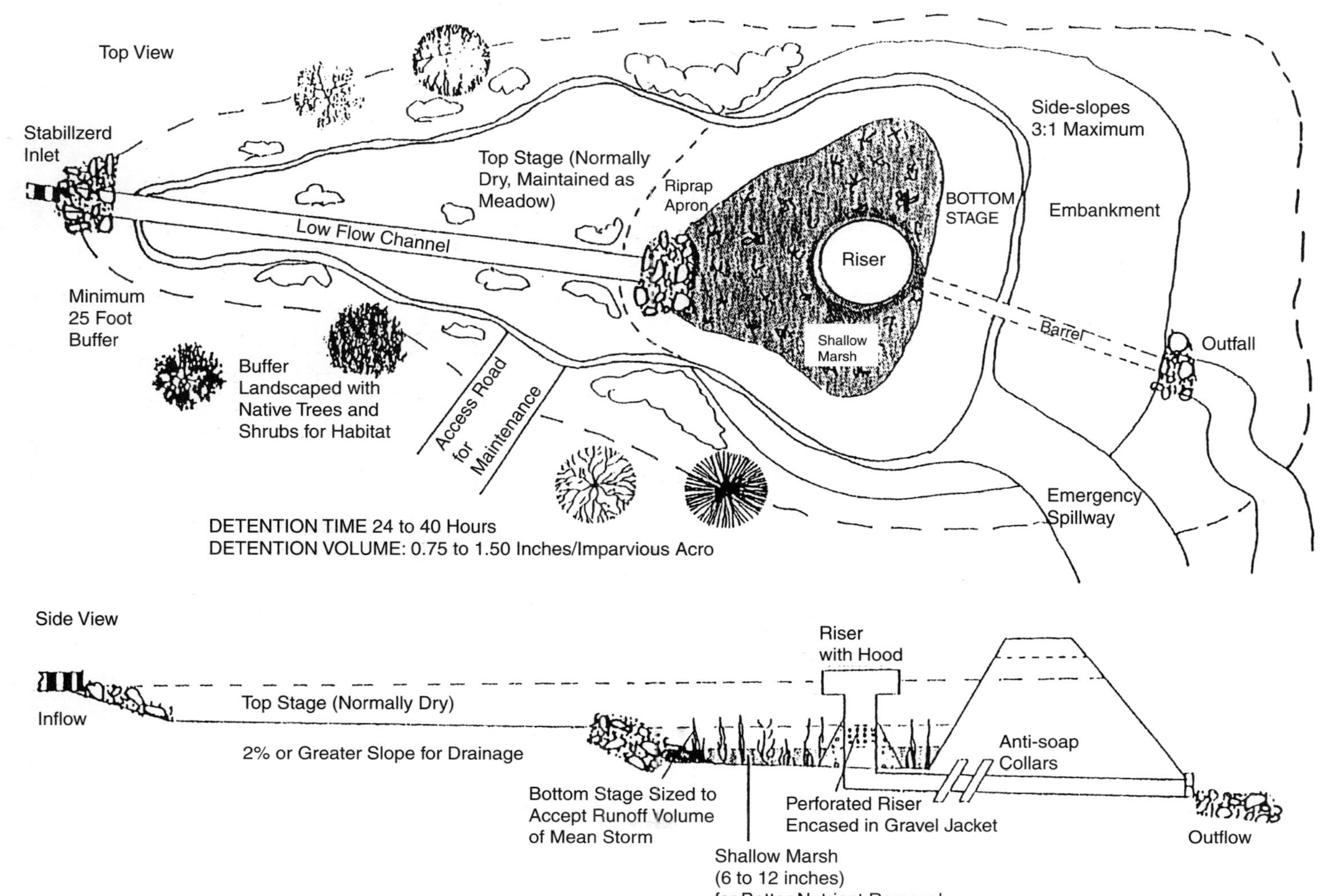

FIGURE 7.12 Schematic of an extended detention basin. (After Schueler, 1987)

bottom stage is expected to be inundated frequently. The top stage remains dry except during large storms. The design of the top stage is similar to that of regular detention basins.

7.3.1 Sizing Extended Detention Basins

The bottom stage of an extended detention basin is designed to detain a certain quantity of runoff, sometimes referred to as the *water quality volume,* for a certain period of time to achieve the targeted level of pollutant removal. The volume to be detained and the duration over which this volume to be released vary in different stormwater management policies. For example, Hampton Roads Planning District Commission (1991) requires that a quantity of runoff calculated as 0.5 in times the impervious watershed area be released over 30 hours in Southeastern Virginia. Prince George County Department of Environmental Resources (1984) requires the runoff volume generated from the one year, 24-hour storm be released over a minimum of 24 hours. The American Society of Civil Engineers (1998) outlines a procedure to size extended basins serving up to 1.0 km² (0.6 sq. mile) watersheds. In this procedure, the volume of water to be detained per unit watershed area, P_o, is estimated as

$$P_o = a_r(0.858I^3 - 0.78I^2 + 0.774I + 0.04)P_6 \tag{7.20}$$

where a_r = regression coefficient
I = watershed imperviousness expressed as a fraction
P_6 = mean storm precipitation depth that can be obtained from Fig. 7.13

The value of the regression coefficient a_r is 1.109, 1.299, and 1.545 for detention volume

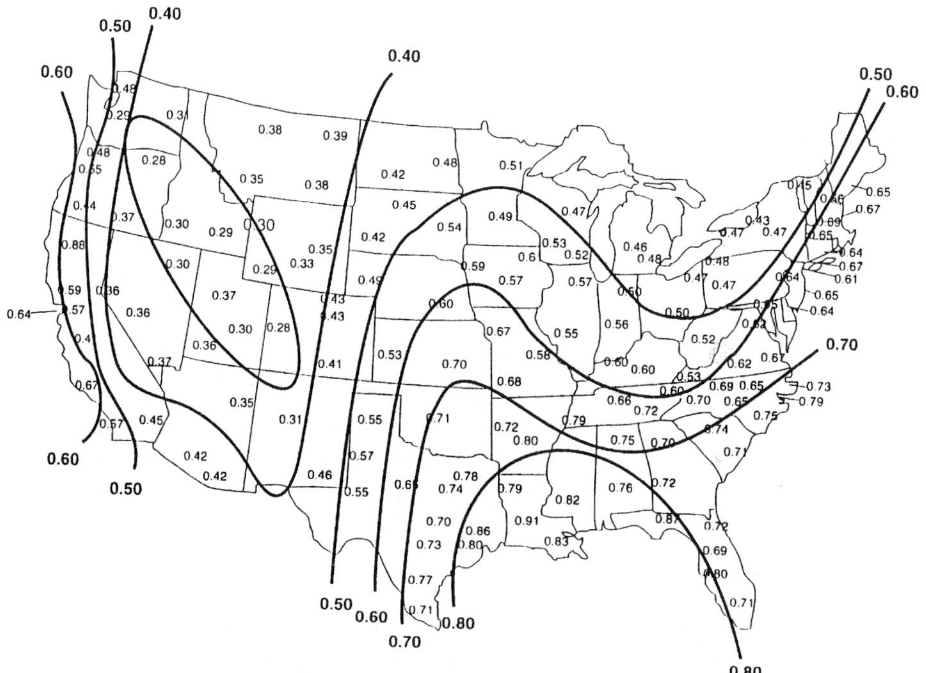

FIGURE 7.13   Mean-storm precipitation depth in inches. (After ASCE, 1998)

release times of 12, 24, and 48 hours, respectively. Interpolation of these values is allowed for durations between 24 and 48 hours.

The outlets for extended detention basins are designed to slowly release the captured runoff from the basin over the specified emptying time to allow settling of particulate pollutants. We sometimes refer to these outlets as water quality outlets. Low-flow orifices are often used as outlet structures.

A hydrograph routing approach is probably the best way to size an extended detention basin and the water quality outlet. However, this requires an inflow hydrograph. In practice, as discussed in the preceding section, only the volume of captured runoff is considered for pollutant removal. There are no broadly accepted procedures to convert this volume to an inflow hydrograph. Therefore, the water quality outlets are often sized by using approximate hydraulics. This can be illustrated by a simple example.

Example 7.7
An extended detention basin is to be designed for a 150-acre watershed in Norfolk, Virginia that is 40% impervious. Determine the required size if the detained runoff is to be released over 36 hours.

From Fig. 7.13, $P_6 = 0.67$ in. for Norfolk, Virginia. Because the watershed is 40% impervious, $I = 0.40$. Also, interpolating the a_r values between 24 and 48 hours, $a_r = 1.422$ for 36 hours. Then, from Eq. (7.20), we obtain $P_o = 0.27$ in. Therefore, the volume of runoff to be detained is $(150)(0.27/12) = 3.38$ acre ft $= 147,233$ ft^3. It is advisable to increase this volume by about 20% for sedimentation.

Example 7.8
An extended detention basin has a bottom length of 80 ft, a width of 20 ft, and side slopes of 3:1(H:V). The outlet is to be sized so that it will release a water quality volume of 10,200 ft^3 over a period of 40 hours.

To determine the depth of water corresponding to the water quality volume, Eq. (7.3) is written as

$$10200 = (80)(20)d + (80 + 20)3d^2 + (4/3)3^2\,d^3$$

By trial and error, $d = 3.6$ ft. Let the outlet structure be composed of ½-in circular ragged edge orifice holes cut around a riser pipe. Let the average elevation of the holes be 1 ft above the pond bottom. Therefore, the average head over the orifice holes is $(3.6 - 1.0)/2 = 1.3$ ft. To empty 10200 ft^3 over 40 hours $= 144,000$ seconds, the average release rate is $10200/144000 = 0.0708$ cfs. Noting that the orifice area of a ½-in hole is 0.00136 ft^2, and $k_o = 0.40$ for ragged edge orifices, we can write Eq. (7.7) as

$$0.0708 = N(0.40)(0.00136)\sqrt{2(32.2)}\sqrt{1.3}$$

where N = number of orifice holes

Solving for N we obtain $N = 14.22$. Therefore, we use 14 holes evenly distributed.

7.3.2 Design Considerations for Extended Detention Basins

Additional design considerations for extended detention basins can be found in various publications (Schueler, 1987; ASCE, 1998; Urbonas and Stahre, 1993; FHWA, 1996). Briefly, the basin should gradually expand from the inlet, toward the outlet. A length to width ratio of 2 or higher is recommended. Side slopes should not be steeper than 3:1 (H:V) and flatter than 20:1 (H:V). A riprap, concrete, or paved low flow channel is required to convey trickle flows. A two-stage design is recommended with a 1.5 to 3.0 ft deep bottom stage and a 2.0 to 6.0 ft deep upper stage. A wetland marsh created in the bottom stage will help remove soluble pollutants that cannot be removed by settling. The detention basin inlet should be protected to prevent erosion. If the outlet is not protected by a gravel pack, some form of trash rack should be used. A sediment forebay is recommended to encourage sediment deposition to occur near the point of inflow

7.4 RETENTION BASINS

Retention basins or *wet ponds* retain a permanent pool during dry weather as shown in Fig. 7.14. A high removal rate of sediment, BOD, organic nutrients, and trace metals can be achieved if stormwater is retained in the wet pond long enough. During wet weather, the incoming runoff displaces the old stormwater from the permanent pool from which significant amounts of pollutants have been removed. The new runoff is retained until it is displaced by subsequent storms. The permanent pool therefore will capture and treat the small and frequently occurring stormwater runoff that generally contain high levels of pollutant loading. The storage volume provided above the permanent pool is used to control the runoff peaks caused by the specified design storm events.

7.4.1 Permanent Pool Volume

Among all the factors influencing the pollutant removal efficiency of a retention basin, the size of the permanent pool is the most important. As pointed out by Schueler (1987), in general, "bigger is better." However, after a threshold size is reached, further removal by sedimentation is negligible.

The required size of the permanent pool in relation to the contributing watershed area varies in different stormwater management policies. For example, Yu and Kaign (1992) and FHWA (1996) recommend a permanent pool size three times the water quality volume defined for extended detention basins. Montgomery County Department of Environmental Protection (1984), Maryland, requires a volume greater than 0.5 inch times the total watershed area. ASCE (1998) recommends that Eq. 7.20, be used with a drain time of 12 hours to determine the permanent pool volume. It is also recommended that a surcharge extended detention volume, equal to the permanent pool volume, be provided above the permanent pool. U.S. Environmental Protection Agency (USEPA, 1986) provides geographically based design curves to determine the permanent pool surface area as percent of the contributing watershed area (see Fig. 7.15). Walker (1987) and Hartigan (1989) treat a retention basin as a small euthrophic lake and employ empirical models to size the retention pond. This procedure is outlined by ASCE (1998).

The U.S. Environmental Protection Agency (1986) presented a procedure to evaluate the long-term pollutant removal efficiency of retention basins depending on the basin size and the rainfall statistics of the project area. This procedure was developed by DiToro and Small (1979), and is outlined in various publications (Stahre and Urbonas, 1990; Akan, 1993; Urbonas and Stahre, 1993).

7.4.2 Design Considerations for Retention Basins

Wet ponds can be designed to control the peak runoff rates from rare and large storm events if additional storage volume is provided above the permanent pool. The size of the additional volume can be determined by using the procedures described for detention basins.

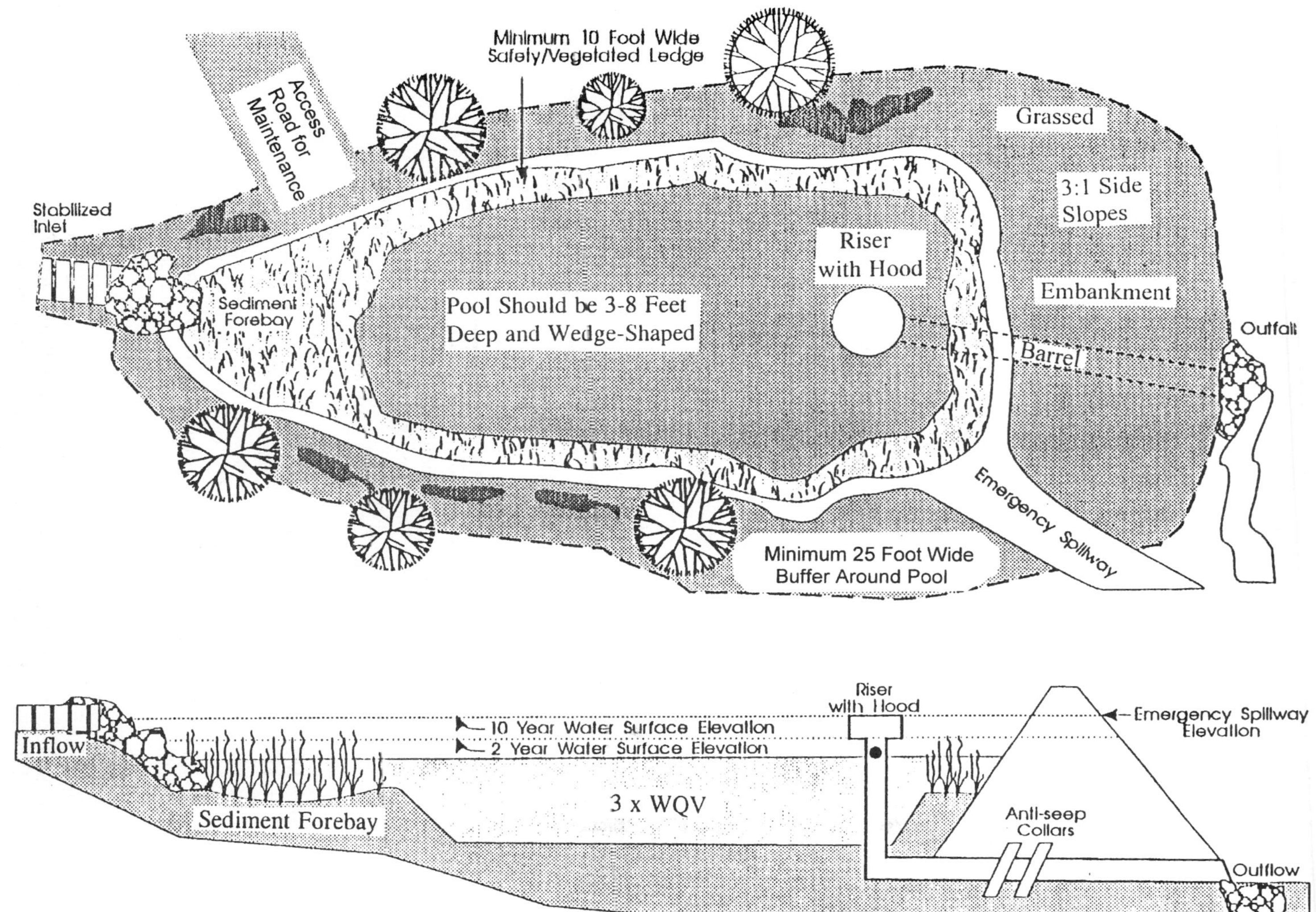

FIGURE 7.14 Schematic of a retention basin. (After Schueler, 1987)

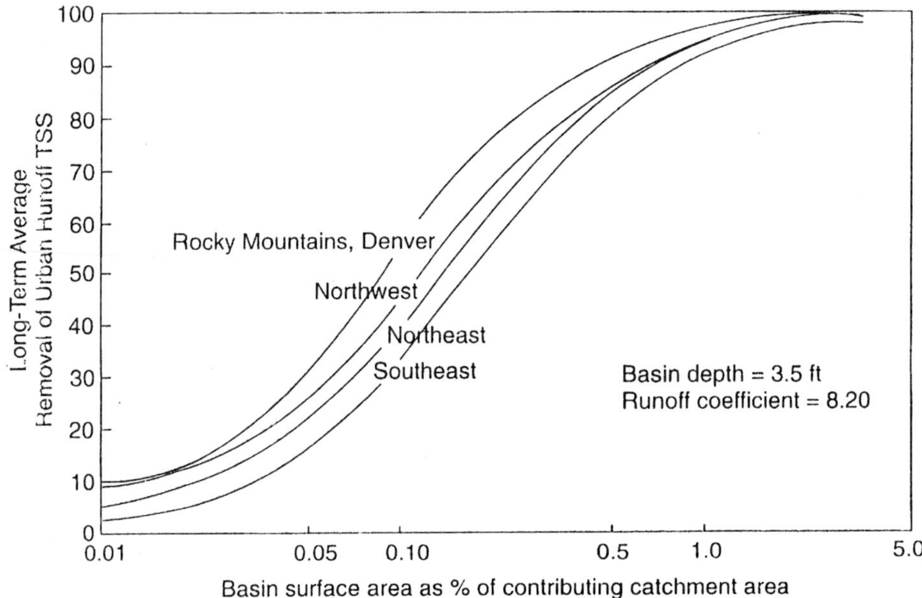

FIGURE 7.15 Retention basin solids settling for single family residential areas. (After USEPA, 1986)

The outlet structures for retention basins include a low-flow outlet to control the runoff from frequent storm events and overflow devices to control the runoff from larger storms.

Additional design considerations have been presented by Schueler (1987). In summary, the pond should be wedge-shaped, and narrowest at the inlet and widest at the outlet. A minimum length to width ratio of 3:1 should be used. The pond depth should average 3 to 6 ft., with a shallow underwater bench around the pond's perimeter. Side-slopes should be no steeper than 3:1 (H:V) and not flatter than 20:1 (H:V). If the soils at the pond site are highly permeable, the pond's bottom should be lined by impervious geotextile or a 6-in clay liner. The inlets and outfalls should be protected by riprap or other means to prevent erosion. Wet ponds should be surrounded by a 25-ft buffer strip planted with water tolerant grasses and shrubs. A sediment forebay should be constructed near the inlet of the pond with extra storage equal to the projected sediment trapping over a 20- to 40-year period.

7.5 DESIGN OF DETENTION BASINS

7.5.1 Preliminary Thinking

One of the most overlooked aspects of detention basin design is giving due consideration to whether or not the basin is truly needed. Unfortunately, much of the demand for detention basins is driven by regulatory fervor and rote interpretation of regulatory requirements. Before designing any detention basin, the first questions that should always be asked are "Is this basin really required?" and "Is there some way that we might avoid having to use a detention basin in this particular situation?" There are several valid practical reasons for asking these questions. Detention basins create drowning hazards. They kill forests. They destroy habitat for woodland species. They can cost more than $500,000 to construct. They require perpetual maintenance. When they fail, such as when an outlet structure becomes blocked, they can create more damage than they are designed to prevent. In some cases, they can actually contribute to increased flood peaks and prolonged erosion downstream.

For example, the hydraulic routing effect of a detention basin modifies the outflow hydrograph in two distinct ways. First, routing an inflow hydrograph through a basin produces an outflow hydrograph with a peak discharge that occurs later than the inflow hydrograph peak discharge. If this delayed peak is added to other downstream hydrographs, such as those at stream confluences, the delayed peak can actually increase the combined downstream total peak discharge. It all depends on the combined routing of all the hydrographs. This condition has the greatest likelihood of occurrence when sharp-peak design storm methodologies are being used and there is a significant disparity between the size of the contributing watershed area routed through the detention basin and the total watershed area at the downstream confluence. Second, in addition to producing a lagged peak, the hydraulic routing effect typically lengthens the time flows occur past a downstream outfall. If the discharge levels are sufficiently high after routing, and the receiving stream is subject to erosion at relatively low discharges, then the result can be increased erosion downstream. Both of these cases should always be checked, but due to the additional work required, frequently the computations are terminated just beyond the basin outfall.

Too often the design process simply moves forward and creates a worse situation than if no basin were constructed at all. In many instances, using a detention basin as an erosion control device is a flawed concept. Destroying woods to construct a large mud hole in the name of environmental protection can be a flawed concept. It is clearly understood that development in a watershed produces increased runoff, and that this runoff exacerbates erosion and flooding problems downstream. However, stream erosion is a natural process, and in many cases accepting some incremental increase in the erosion of a natural channel or addressing the issue of stream bank stabilization directly would be far better than stripping a site of mature woods. As basins get smaller and smaller, and serve individual sites rather than regional watersheds, ridiculous situations can occur as a direct result of following the regulations.

However, regulators have heard more than their fair share of absurd arguments against constructing detention facilities. The fact remains that detention basins are needed in many situations to protect against the adverse effects of increased flooding and erosion resulting from watershed development. With a little common sense and engineering judgment, it may be possible to reduce the size of the required detention basin, or even eliminate it altogether, and still follow the regulatory intent.

Can a regional basin be used to mitigate the effects downstream from several sites, rather than each individual site having its own small detention basin? Can an off-site basin be constructed, or modified, such that the post-development impacts at the true point of concern are adequately controlled? Can the proposed upstream hydrologic conditions be improved? Can the development be better planned so that impervious cover is minimized? Can multistory construction be used to increase pervious land cover? Can parking lot or rooftop detention be used safely? (Keep in mind with parking lot and rooftop detention that freezing can create very dangerous conditions.) Can over-sized underground pipes or storage devices be used in lieu of a basin? (Underground detention can be very expensive, but becomes more feasible as the contributing site becomes smaller.)

The effect of the watershed delineation itself can largely affect the outcome of the design. Defining a watershed is as much an art as it is a science, particularly on larger watersheds where the exact location for a detention basin has not yet been determined. In highway drainage design, for example, roadway widening is required due to increased traffic, which comes with urbanization. Typically, it is very difficult to even find an available plot of ground to use along the roadway, so the detention basins have to be sited farther down the outfalls. As we move farther down an outfall in the design process, the contributing watershed area increases, and the impacts of the roadway widening become less on a total percentage basis. In cases such as these, it can be easy to lose focus while concentrating on peak discharge mitigation, and overlook the channel erosion issue.

Under the best scenario, a regional stormwater management plan can be implemented that will provide adequate, reasonable, cost-efficient detention storage for all parties in the watershed. Unfortunately, upfront planning, funding, and political requirements have to be nearly ideal for regional cooperation. Communicating with the approval agency on a preliminary basis can help identify potential pratfalls, and may lead to cost-effective opportunities for regional cooperation—after all, they see everyone's plans and are in communication with other engineers working in the watershed. Without cooperation, the burden falls back to the developer, and individual site facilities proliferate the landscape.

It is very important for the design engineer to make sure that adequate preliminary thinking and communication with the developer and approval agencies has taken place, before launching into the design process.

7.5.2 Site Considerations

In the process of selecting a site for a detention basin, there are many factors to consider. The first is safety. Detention basins can be an attractive nuisance, and always create a drowning hazard. The potential for fatality increases as the basins are sited close to roadways where traffic accidents may result in submerged victims. Conditions have a propensity to combine adversely, such as a drunk driver on a stormy night, when the roads are slippery, and the basin is full of water. For this reason, many localities have minimum setback distances from roadways. Likewise, constructing a basin close to a residential area, particularly high density housing, increases the potential for fatality. Fencing can be used to deter small children, but many designers also specify that thick, prickly vegetation be copiously planted around the basin perimeter to prevent people and pets from wandering too close.

In addition to creating a safety concern, freezing can create dangerous conditions upstream as well, particularly if the basin outlet becomes blocked with ice, and additional inflows back up through the drainage system onto a parking lot or roadway. There are also more evident safety concerns such as damage from embankment failure and upstream flooding that could result from blocked outlet structures. From the designer's perspective, it is always prudent to advise the owner of safety considerations and to document the owner's response.

Maintenance access is a major factor in the siting of a detention basin, particularly in ensuring that adequate width and turnaround areas are provided. Sufficient width should be provided at least along one side of the major axis of the basin, so that a dump truck and excavator can easily access the site. Tight-turn radii should be avoided in the maintenance access lane, and frequently a turnaround area is required down near the outfall structures. A 25-ft minimum turn radius is sometimes specified, although a dump truck cannot make a sharp turn within a 25-ft radius. For this reason sharp turns should be avoided in the maintenance access lane. In drawing slope lines for basin grading, allowances should be made for shoulder rounding and slippery conditions. For example, if 15 ft is required for vehicular access to the site, the slope lines on the grading plan might show the lane to be 18 ft wide to allow for shoulder rounding, as shown in Fig. 7.16. Sometimes, maintenance can be performed from inside the basin, rather than on the shoulder, although soft, mucky bottom conditions do not lend themselves to dump truck operations.

Maintenance access lanes are sometimes protected with a lock and chain to prevent unauthorized vehicular access, and frequently a gravel apron will be specified up at the roadway entrance to help keep construction and maintenance vehicles from tracking excess soil onto the roadway. One of the greatest obstacles to drainage system maintenance is future overgrowth of vegetation at the site. If not properly and routinely cut back, eventually trees can effectively preclude maintenance access, sometimes requiring a much larger and drastic operation to restore access. This situation can become critical if access is required during an emergency.

Generally, layouts having longer flow paths through the basin are preferable to designs where short-circuiting of the flow may occur. Longer flow paths increase the retention time

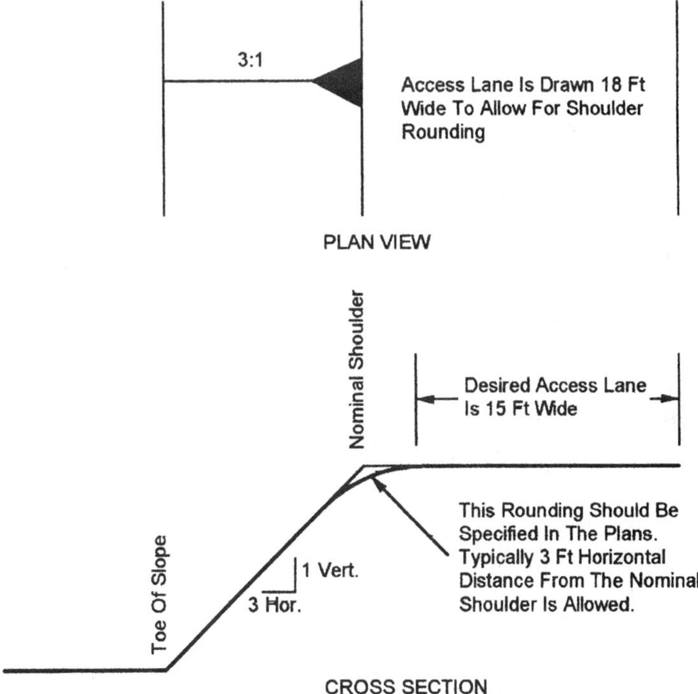

FIGURE 7.16 Grading lines allowing for shoulder rounding, showing maintenance access lane.

for lower flows, promoting more efficient settling of sediment and suspended pollutants. A longer length-to-width ratio is therefore preferable, as is a curvilinear or segmented alignment that follows the existing site contours more naturally than a purely rectangular layout. Fancy, over-complicated layouts should be avoided to keep construction and real estate costs to a minimum.

A number of competing environmental issues should be addressed in siting the basin. Wetlands regulations make it very difficult or expensive to construct a basin in a wetland area, or they simply forbid such construction altogether. In many jurisdictions, basins may not be constructed on live or perennial streams. In all cases, water flows to the lowest part of the watershed, and the most logical place to put a basin is at the downstream end of the watercourse. However, often the basin must be constructed offline, in purely upland areas. Ironically, a constant base flow from a live stream source is ideal for promoting healthy flushing and water quality conditions in a retention basin.

Some design literature suggests that it is a good idea to promote waterfowl habitat in the design of a retention basin. Unfortunately, waterfowl can overtake these facilities, causing a number of nuisances and safety hazards, such as nesting in airplane runway approaches, and creating traffic hazards. Frequently a population of waterfowl will cause fecal coliform levels in receiving streams to skyrocket. Careful consideration should be given in the design before doing anything to promote such conditions.

Outlet structures should be designed to operate in a hydraulically independent manner as much as possible. For example, a metal riser stand pipe that has a low-flow orifice cut in its bottom, and a slotted weir cut in its side is going to be computed using three sets of independent equations at high head conditions: vertical orifice, slotted weir, and weir-and-orifice flow for the top of the pipe. In reality, the flows from these three components will interfere with each other significantly. Yet there are virtually no laboratory tests to allow us to mathematically model such combined conditions with any degree of reliability. If the structures are more isolated, they become more redundant in the sense that failure or blockage of one structure does not keep the other structures from functioning. This layout concept can become critical if upstream flood damage potential is significant. One of the best ways to apply this concept is to construct the emergency spillway separate from the other structures, preferably in ground that has not been filled during construction. Constructing spillways through fill material increases the likelihood that they will fail due to scour failure.

Materials also play a role in siting a basin. In order to construct a retention basin, the "pond" must be capable of holding water. Geotechnical engineers should be consulted if there is any doubt as to the ability of the in-situ soils to retain water. A quick indicator test can be made in the field by wetting and rolling the layer of soil just below the topsoil in your hand—if the material sticks together in a clump, it will probably hold water. A clay layer may be specified as a liner if needed, or geotechnical materials may be used to retard leakage. Over time, as sediment collects in the basin, most bottom layers will become less leaky. Riprap is a great material to use around outfall structures to prevent scour, although it is difficult and costly to obtain in many geographic regions.

Flooding potential created by the proposed basin should also be carefully evaluated in selecting the site. If the embankment fails, what are the downstream consequences? Is there an opportunity to have a redundant outfall path? If the outlet becomes blocked, what are the upstream consequences? Ideally the site should be selected to carefully minimize these risks. If the facilities involved are sufficiently large, a dam break analysis may be warranted, or a set of urban flooding computations might be run to assess upstream flooding potential from a blocked outlet.

Outfall hydraulic boundary conditions are also very important. If the basin is located too far downstream, there may be high design tailwater elevations to overcome. In flat, low-lying coastal areas, during severe storm conditions such as a hurricane, there may be no storage available in the basin due to storm surge and tide conditions. We would like to avoid such conditions; however, the reality is that people live on the coast. Likewise, floodway hydraulics must be carefully considered in locating a basin. Flood regulations usually preclude construction of any kind of facility in a floodway.

And last, but by no means least, the costs to construct the basin should be considered carefully in siting the basin. How much is real estate going for under worst-case condemnation scenarios? Can we move the facility to a bottomland location where the real estate costs won't be as high? What kind of right-of-way takes will be required: permanent takes, temporary construction easements, drainage easements? Can we combine this basin with some other basin to achieve regulatory compliance at a point farther downstream when all flows from the project are evaluated together? Where will the excavated material be hauled, and can it be used as fill on another project? How will this site be maintained in perpetuity? What are the aesthetic costs? What are the political costs for the owner?

7.5.3 Design Methodologies

Design methodology is typically limited by the approving jurisdiction. There are certain guidelines to be followed, such as precluding the use of the modified rational method on watersheds over a few acres in size. Many engineers refuse to use the modified rational method at all, on the basis that its fundamental assumptions are flawed. Some states liberally allow the modified rational method, writing it into their stormwater regulations, seemingly preferring its use because on many projects it can be used to design much smaller facilities than would be required using other design methodologies. At some point, the integrity and judgment of the design engineer must come into play.

In some respects, the most theoretically accurate design methodology we might use is continuous simulation hydraulic modeling using local, historical rainfall records. Engineers must work in real-world competitive environments, and performing such studies for typical detention basin designs is very time consuming and costly—even with today's high-speed computers and powerful software programs. Typically detention basins and storm sewer systems are designed by using single-event design storm methodology. In some very real respects, the ultimate justification for this approach is that over time it seems to have worked. Such is reality. Selection of the design storm has a significant impact on the outcome of the basin design (Paine, 1989). In cases where continuous simulation hydrology is used to "design" a basin, it is often easier to design the basin with single-event hydrology, and then tweak the results after the continuous simulation in order to save significant design time and effort.

Design details are widely available from approval agencies, state highway departments, and product manufacturers. Standard design details should be favored over custom design items due to cost and construction issues. In general, simpler is better, and engineers will do better to learn and abide by the approved methods.

A significant amount of attention should be given to outfall scour computations. Good design practice and regulations require that the outfall channel be checked for scour and erosion potential, and that the final design prevent erosion to points well downstream. A number of computer programs are available to perform these computations, the best of which are based on tractive force methodology described in Chen and Cotton (1988) and Chow (1959).

In some locations water tables are sufficiently high and the topography sufficiently flat, such that detention basin hydraulics are impossible to isolate from basin to basin. In this type of environment, the basins are considered to be "inter-connected," and special software is used to analyze and design the basin in concert with design conditions on a wider regional basis. This approach is particularly relevant if the outfall channel hydraulic grade line at the basin outfall is significantly affected by the hydraulic response of some other basin to the same design storm.

In recent years, sophisticated software has become available to optimize the design of detention basins. This type of program seeks to minimize the surface expression and required excavation of the basin in order to minimize construction and real estate costs. However, minimizing the required surface area has additional benefits such as limiting the number of surrounding trees that need to be cleared. Unfortunately, in the past this type of analysis has required considerable trial-and-error design time, so there are a lot of over-designed basins in use today. Section 7.5.5 describes the optimization process in more detail.

7.5.4 Design Criteria and Considerations

7.5.4.1 *Tailwater Elevations.* *Design tailwater elevations* comprise hydraulic boundary conditions, and in many cases significantly affect the outcome of the basin design. Any excavation or pond volume that is lower than the design tailwater elevation is basically dead storage—unless some sort of tide gate is used on the outfall structure. There are several approaches to obtain a working tailwater elevation. First, we can simply assume there is no tailwater, although this assumption is usually not valid. In performing preliminary calculations however, we may choose to ignore tailwater effects until we have the basin laid out on the site. Tailwater effects should never be ignored in the final design.

Sometimes, a flood study will provide the design tailwater elevation, such as the 100-year floodplain elevation at the outfall. Caution should be exercised to avoid over-design due to combined probabilities, such as assuming that the 100-year downstream flood occurs at the same time as the 100-year rainfall event. In some cases it is easier to simply combine the two events without accounting for the combined probability effect, thereby providing an extra measure of conservatism. Separating such statistical events can require more engineering time than a project design budget can afford.

A third approach is to compute the normal depth elevation of each peak outflow from the pond, then add the normal depth to the channel invert to obtain a design tailwater elevation. Some trial-and-error is required to obtain the correct tailwater elevation, but the peak outflows and discharges can normally be balanced in four or five trials. This approach is somewhat conservative in that it will only be applicable during peak discharge conditions.

A fourth approach is to actually route the downstream hydrographs through the downstream channel using an open-channel flow computer program, and compute the variation in head over time at the outfall. This approach is more theoretically correct, but the amount of work involved usually precludes its use. Attempting to match a design hydrograph to a timed tailwater event, such as a high-tide elevation at a river outfall, is a risky undertaking, so most engineers simply use a constant design tailwater elevation and introduce some possible conservatism in the design. In other words, most of us would rather not be in the position of having to defend that the design storm will begin at 2:45 a.m.

7.5.4.2 Tide Gates.   *Tide gates* can be used to preserve useful flood storage in basins located in low-lying areas, by preventing water from the outfall from flooding back up into the basin. These devices are extremely useful when they work properly; however, there are reliability issues. Mechanical tide gates are basically flapped valves that allow flow in the downstream direction. They have the distinct disadvantage that they can easily become clogged with debris, preventing full closure of the valve. Once the valve closure is compromised, high tailwater conditions will flood the basin. A relatively new technology has been developed in recent years to solve several problems associated with mechanical-valve tide gates. A so-called "duckbill" tide gate, as shown in Fig. 7.17, offers several distinct advantages over mechanical valves. First and foremost, it can prevent backflow, even when partially clogged with debris. It also eliminates the nuisance of having neighbors report gunshots to the police that are actually the result of quick closure of a mechanical valve. The valve is passive in that forward hydraulic pressure opens the valve, and reverse hydraulic head closes the valve. The main disadvantage of duckbill tide gates is that if they become cut by debris, they lose their integrity.

7.5.4.3 Freeboard.   *Freeboard* is the vertical distance between the maximum computed water surface elevation in the pond and the embankment elevation at its lowest point. Freeboard requirements for basin design are generally less severe than those used for open-channel flow, because the water surface is essentially flat and calm during operation. Typically freeboards on the order of one foot are used. The freeboard requirement is often increased based on adjacent property flooding concerns and topographic constraints.

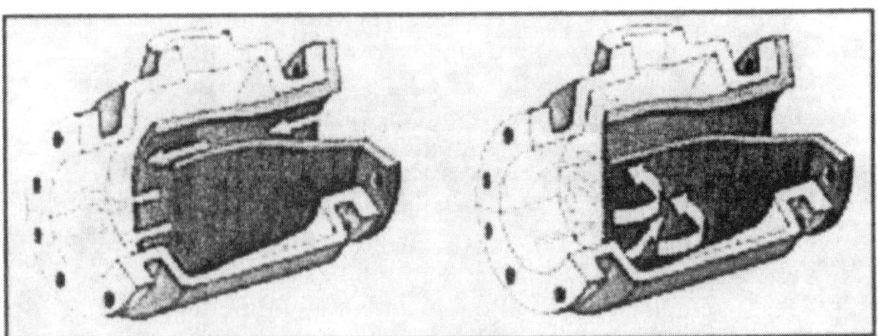

FIGURE 7.17   Duckbill tide gate. (After Red Valve Company, Inc., 2000)

7.5.4.4 *Embankment Slopes.*

Embankment slopes can be a point of strong contention in the design review process. Steeper embankment slopes are harder and more dangerous to maintain, and fail more easily than milder slopes. However, milder slopes require more real estate and material to construct than steeper slopes and can therefore be much more costly. A rule of thumb used in the design industry is that an embankment slope of 3 horizontal to 1 vertical provides a stable slope that can be maintained and mowed with relatively little difficulty. Milder slopes may be specified if there is a high likelihood that people will walk along the slopes, such as would be the case with a retention pond that is also intended to serve as a site amenity. In such a situation, an aquatic bench would be introduced just under the surface to mitigate the drowning potential. An aquatic bench is simply a graded ledge 6 in to 1 ft below the permanent pool elevation, extending out for a distance of 6 ft or so, so that if someone slips into the water, they end up on a ledge of shallow water, not in a deep pond with steep, slippery side slopes.

7.5.4.5 *Vegetation.*

Vegetation must be adequately considered and specified in the design. Generally, lawn grasses such as fescues will die if submerged for several hours. Basins are designed to have prolonged high water levels, which results in bare earth and slippery perimeter conditions if the vegetation is not adequately specified and planted. Careful selection of plant species, such as planting dense briars and thorn bushes can deter people and pets from approaching the basin. A competent botanist or nurseryman can help to ensure that the plantings will work in the required climate and environment. Generally, monotypic plantings and conditions, where one species predominates the layout, should be avoided. Without proper upfront planning, many plantings are simply doomed. Plant requirements, such as cutting and harvesting, should also be addressed in the long-term maintenance plan. If the basin is intended to have water quality benefits, pollutant uptake by the plants can significantly contribute to improvements in water quality; but to achieve the maximum removal efficiency, the plants must be cut and harvested, and not allowed to die and rot in place.

7.5.4.6 *Structure Materials and Outlet Structures.*

Pipe and structure materials used in the design should be compatible with local soils conditions and should be designed for long-term wear. Metal pipes in acidic soils can corrode quickly and are much more costly to replace than the cost to use concrete in the original construction. Likewise, metal risers have the same problem.

Outlet structures should be designed to operate as independently as possible, to maximize redundant backup and hydraulic independence. It is common to cut a low-flow orifice into the bottom of a stand pipe or inlet box, but as the orifice diameter increases, it will experience severely hampered flow efficiency as the head in the basin overtops the crest of the stand pipe or inlet box. Where structures are combined and flow through a shared outfall pipe, the barrel of the outfall pipe should be over-designed so that choking does not occur. If the flow becomes choked in the outlet pipe barrel, then the governing hydraulic equations used in the routing are moot, and all bets are off as to hydraulic response of the basin. If the barrel is sufficiently large, the hydraulic control of the outflow hydrograph remains at the structures, as intended. The outfall pipe should be laid on a hydraulically steep longitudinal pipe slope, say 0.005 or greater (vertical drop across the inverts divided by pipe length), to avoid outlet control culvert conditions.

A number of outlet structure design concepts use small diameter, reverse-gradient pipes at the outlet structure. These pipes are intended to prevent clogging, but are susceptible to breaking at the point of attachment to the inlet box. For this reason, many designers avoid reverse-gradient pipes.

All outlet structures have a propensity to clog, particularly grated structures and low-flow orifices. Clogging of the low-flow orifice is especially troublesome, because it tends to prolong saturation of the pond environment, which promotes very soggy bottom conditions, dead vegetation, slippery slopes, and other safety consequences. Design of debris racks helps to reduce clogging potential, but there is no substitute for regular maintenance inspections to ensure proper operation of the outlet structures. A general rule of thumb is that the smallest orifice should be no less than three inches in diameter; however, in reality, orifices this small are highly susceptible to clogging.

By their very nature, basin bottoms are wet, soggy, and soft. It should never be assumed that a concrete outlet structure can be adequately supported in this environment without piles or other geotechnical treatment. Without proper support, the outlet structure can settle and separate from the outfall pipe, creating a guaranteed failure scenario, or at least an expensive maintenance situation.

As the height of the inlet box or stand pipe increases, conditions become more chimney-like. In order to increase stability, the structure should be tucked back in the embankment, rather than located out in the open bottom of the basin. Generally, the structure should be located in the embankment at a point where the rim has a minimum 1-ft vertical clearance. In retention or wet basin design, the outlet structure should incorporate a valve to allow the basin top be drained for maintenance purposes, such as removal of sediment.

Rectangular and V-notch weirs are among the most studied hydraulic devices, and discharge coefficients and governing equations are readily available in any number of sources. However, in practice, weirs are often constructed using riprap, and they are often trapezoidal. Although we would like very much to specify a smooth, concrete rectangular weir, in reality such construction can be excessively expensive, and when joint settling occurs, concrete weirs can easily fail from scour. Riprap weirs can be specified, if the engineer is cautious in how the computations are handled, and the construction is carefully executed. Trapezoidal weirs can be broken into two components: a rectangular weir and a V-notch weir, with the combined flows added to obtain the composite flow. Riprap weir coefficients can be obtained by treating the riprap weir as a channel, developing a stage-discharge relationship using computer software, then back solving to obtain the weir coefficients. Some designers will specify that riprap weirs be covered and graded with a mixture of sand and topsoil to fill the voids, and prevent basin seepage through the riprap, while promoting small plant growth that can tighten the stone placement.

7.5.4.7 Grates.   Grates on outlet structures present several problems. They greatly reduce the effective flow area through the structure, and this reduction must be taken into full account in the computations. Second, they must be constructed of lightweight material to allow easy access for maintenance, and they must be secured so that they do not become displaced through constant operation. Mathematically, a grate as shown in Fig. 7.18 should be treated for orifice computations as an orifice having an effective flow area equivalent to the combined total slot opening area of the grate, and for weir computations as a weir having a length equivalent to the sum of the slot openings around the grate's perimeter.

7.5.4.8 Anti-Vortex Devices.   *Anti-vortex devices* may be specified on certain types of structures—typically stand pipes—in order to prevent vortex conditions from retarding flow through the structure. If a vortex forms, the basin will take longer to drain than intended and may overtop during extreme events. All the expense and trouble to put the basin in place is then largely wasted, and the overtopping can lead to embankment scour conditions and failure. There are several types of anti-vortex devices, and pre-cast concrete tops have become very popular. These tops should be carefully checked to ensure that they provide more than the computed inlet box or stand pipe cross-sectional flow area—otherwise, they become choking devices. A reliable and easy-to-maintain metal construction anti-vortex device and trash rack is depicted in Fig. 7.19. The corresponding dimensions are listed in Table 7.6.

7.5.4.9 Anti-Seep Collars.   *Anti-seep collars* are used to increase the effective flow-length along a pipe through saturated ground, resulting in slower, reduced seepage along the pipe. A typical approach is to attempt to increase the seepage length along the barrel by 10 percent. A simplified approach for sizing the collars proceeds as follows (after Edmonds, 1994):

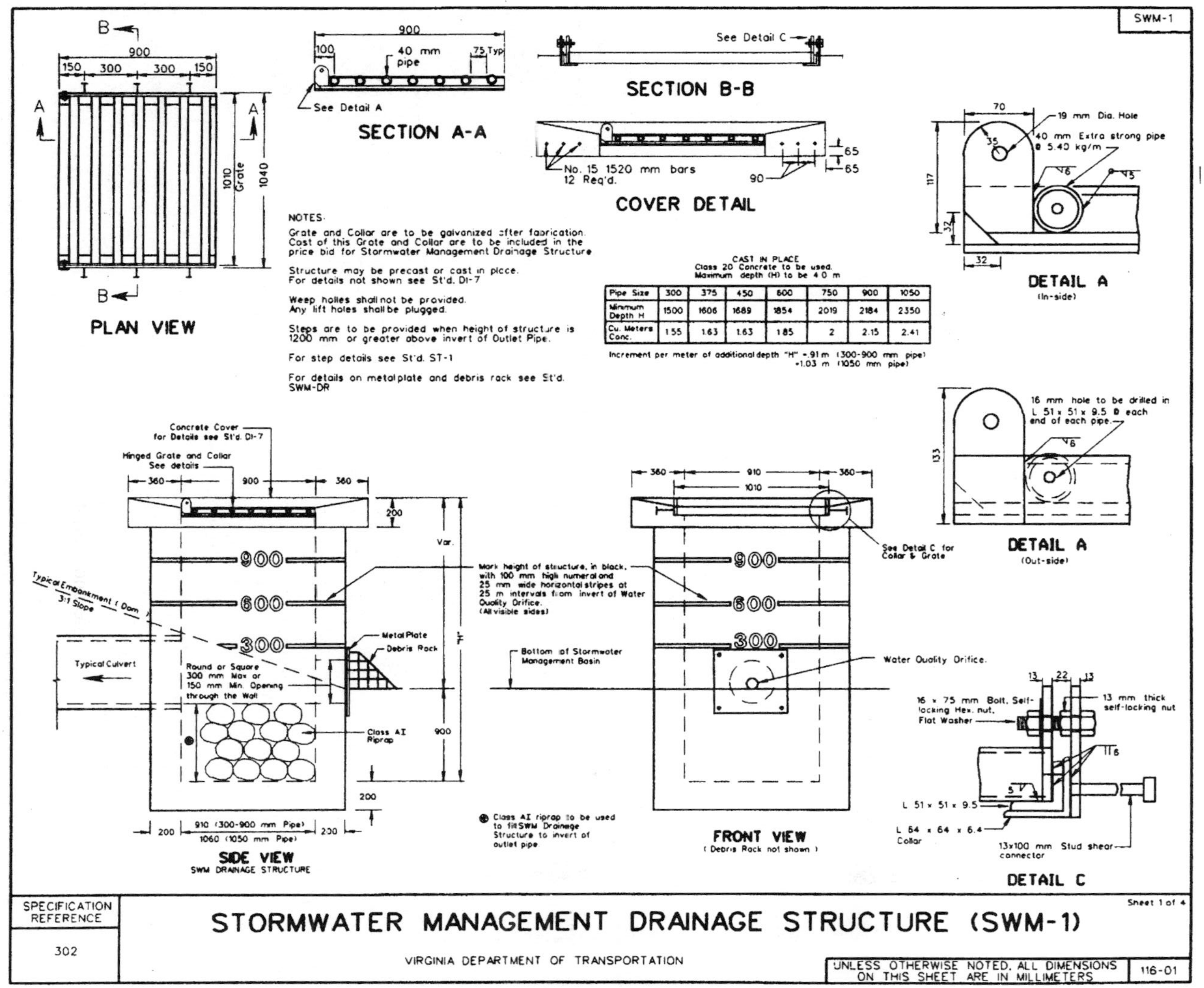

FIGURE 7.18 Standard storm water management inlet box. (After Virginia Dept. of Transportation, 1996)

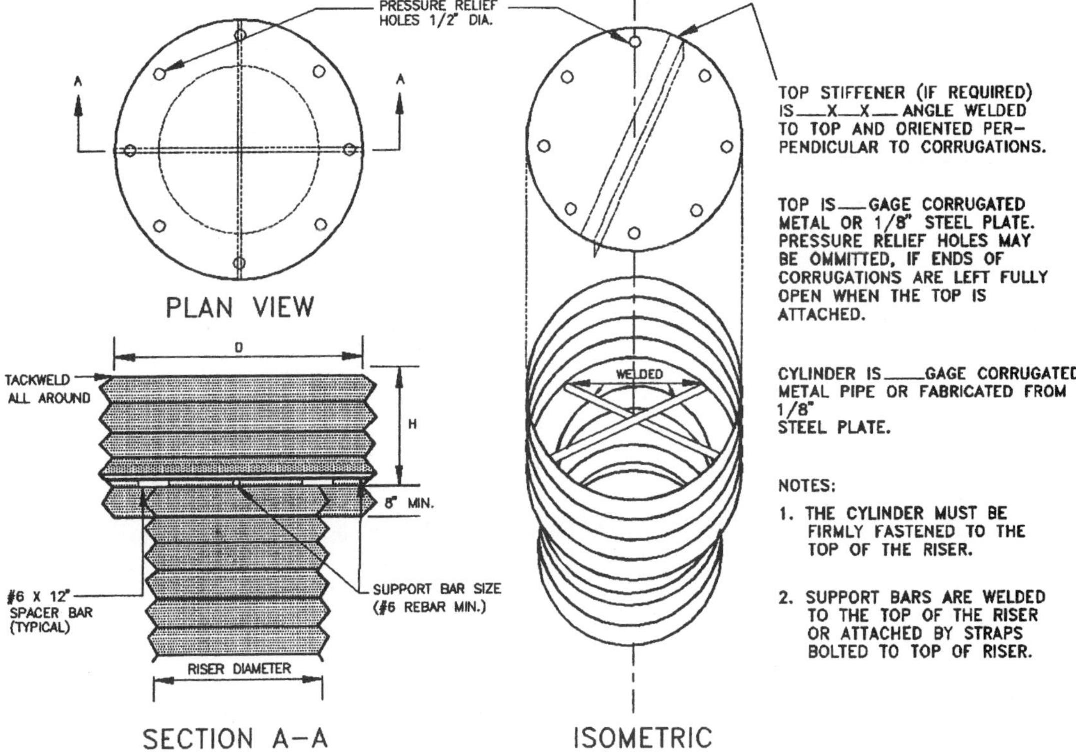

FIGURE 7.19 Anti-vortex device and trash rack. (After Soil Conservation Service, 1975 and Virginia Dept. of Conservation and Recreation, 1992)

1. Determine the length of the barrel in the saturated zone using

$$Ls = Y(Z + 4)(1 + S/(0.25 - S))   \tag{7.21}$$

where Ls = length of the barrel in the saturated zone, in ft
Y = depth of water at the principal spillway crest, in ft
Z = slope of the upstream embankment face, in Z ft horizontal to one ft vertical
S = longitudinal slope of the barrel in feet per ft

2. Let V = projection of the anti-seep collar, in ft.

3. Let n = the number of anti-seep collars.

4. The ratio of the line of seepage ($Ls + 2nV$) to Ls must be in the range from 1.10 to 1.30.

5. When more than one collar is used, spacing between collars should be $7V$ to $10V$ within the saturated zone, however the collars should not be spaced more than 25 ft apart.

6. Under no circumstances should V be less than 1.5 ft, and the collars should be located at least 2 ft away from the nearest pipe joint.

7. Check to make sure that all collars have sufficient cover (i.e., that they do not protrude above the ground).

TABLE 7.6 Concentric Trash Rack and Anti-Vortex Device Design Table (After Virginia Dept. of Conservation and Recreation, 1992)

Riser diameter (in)	Cylinder		Height (in)	Minimum size support bar	Minimum top	
	Diameter (in)	Thickness (gage)			Thickness	Stiffener
12	18	16	6	#6 rebar or 1½ × 3/16 angle	16 ga (F&C)	—
15	21	16	7	As above	As above	—
18	27	16	8	As above	As above	—
21	30	16	11	As above	16 ga (C) 14 ga (F)	—
24	36	16	13	As above	As above	—
27	42	16	15	As above	As above	—
36	54	14	17	#8 Rebar	14 ga (C) 12 ga (F)	—
42	60	16	19	As above	As above	—
48	72	16	21	1¼″ pipe or 1¼ × 1¼ × ¼ angle	14 ga (C) 10 ga (F)	—
54	78	16	25	As above	As above	—
60	90	14	29	1½″ pipe or 1½ × 1½ × ¼ angle	12 ga (C) 8 ga (F)	—
66	96	14	33	2″ pipe or 2 × 2 × 3/16 angle	12 ga (C) 8 ga (F) w/stiffener	2 × 2 × ¼ angle
72	102	14	36	As above	As above	2½ × 2½ × ¼ angle
78	114	14	39	2½″ pipe or 2 × 2 × ¼ angle	As above	As above
84	120	12	42	2½″ pipe or 2½ × 2½ × ¼ angle	As above	2½ × 2½ × 5/16 angle

Source: Adapted from SCS (1975) and Carl M. Henshaw Drainage Products Information.
Notes:
1. The criterion for sizing the cylinder is that the area between the inside of the cylinder and the outside of the riser is equal to or greater than the area inside the riser. Therefore, the above table is invalid for use with concrete pipe risers.
2. Corrugation for 12″–36″ pipe measures 2· ″ × ½″; for 42″–84″ the corrugation measures 5″ × 1 ″ or 8″ × 1″.
3. C = corrugated; F = flat.

Example 7.9
A 36-in reinforced concrete outfall pipe will be constructed through a proposed embankment on a longitudinal slope of 0.005. The upstream embankment face has a 3 horizontal to 1 vertical slope, and the peak depth in the pond under the 100-year storm event is 5.29 ft. Design the anti-seep collar(s). The embankment elevation over the outfall pipe is 88.7 ft, and the pipe invert under the center of the embankment is 81.91 ft, and the 36-in pipe has a wall thickness of 4 in.

Solution
From Eq. (7.21), $Ls = 5.29 (3 + 4)((1 + (0.005/(0.25 - 0.005)) = 37.79$ ft. We will try a minimum projection of $V = 1.5$ feet. Assuming that we will use only one anti-seep collar,

$n = 1$. The ratio of the line of seepage to the saturated length is $(37.79 + 2(1.5))/37.79 = 1.079$, which is not enough to satisfy the criteria. If we increase the collar projection to $V = 2$ ft, then the ratio becomes $(37.79 + 2(2.0))/37.79 = 1.106$, which is acceptable. Checking for adequate cover, the top elevation of the collar is $81.91 + 36/12 + 4/12 + 2 = 87.24$ feet, which gives us $88.7 - 87.24 = 1.46$ ft of cover.

7.5.4.10 Debris Racks. *Debris racks* are extremely important design items, particularly where there is no emergency spillway out of the basin. Debris will block most outfall structures at some point, and a well-designed debris rack can help to ensure that the outfall structure functions even under a heavy trash load. Debris racks can be as simple as welded wire and metal-hinged plates over low-flow orifices, as shown in Fig. 7.18, or as complicated as a large, hinged gate over a larger diameter outfall culvert. On larger diameter structures, the debris racks are locked to prevent children and animals from wandering into the culvert. A slanted horizontal alignment seems to work best to keep larger structures functioning during extreme events.

As watershed size increases, so does the size and volume of debris that will be deposited on the outfall structure. If it can float, and sometimes even if it can't, it can block an outfall. Trees, sheets of plywood from construction sites, large empty cable spools, discarded furniture, discarded large plastic toys, and items as seemingly harmless as shopping carts and cardboard can render the best design and construction of a detention basin completely useless, if the outfall becomes blocked.

The importance of the perpetual inspection and maintenance plan for the basin should be obvious in this respect, particularly where there is no emergency spillway.

7.5.4.11 Pipe Bedding and Cradle. Settling and separation of pipe joints is always a cause for concern, particularly when the pipe is part of an outfall structure. Pipe bedding must be designed and specified very carefully. In storm sewer design, most pipe bedding is actually a course of gravel that can be easily hand-graded as the installation crew sets each pipe joint. When the pipe passes through an embankment where seepage is a concern, using gravel is tantamount to constructing a French drain through the embankment. For this reason, outlet pipes should be constructed on *impervious* bedding materials such as concrete. Figure 7.20 presents a reasonable bedding design approach that can be adopted for designing impervious pipe cradles from outlet structures. "Class A" bedding works very well in this situation. The primary interest in this type of cradle is the imperviousness, not so much the load factor in small embankment applications.

7.5.4.12 Embankment Design. As development of the landscape has escalated, so has the number of detention basins in recent years. A considerable amount of attention is now being focused on small dam failures. Aside from the safety issues already noted, the engineer must keep several issues in mind in the design of a basin embankment. The core should be constructed of impervious material, typically some sort of clay. If clay is not available, grouting of in-situ materials or use of geomembranes, or even cement can be employed to make the core impervious. Although not an issue on most small basin embankments, in some jurisdictions dam safety regulations will come into play if the embankment height exceeds a certain level, or the volume of water stored in the facility exceeds a certain level. Generally these conditions should be avoided to reduce administrative burdens and to protect the safety of the public.

Most basin embankments will not be subject to traffic loads, although the entire design has to be approached very carefully if traffic is expected. In all cases, it is prudent to have a competent geotechnical engineer design or review the embankment details and specifications.

The impervious core should be keyed into undisturbed existing ground, to a sufficient depth to prevent excess seepage and sliding failure of the embankment. The top of the

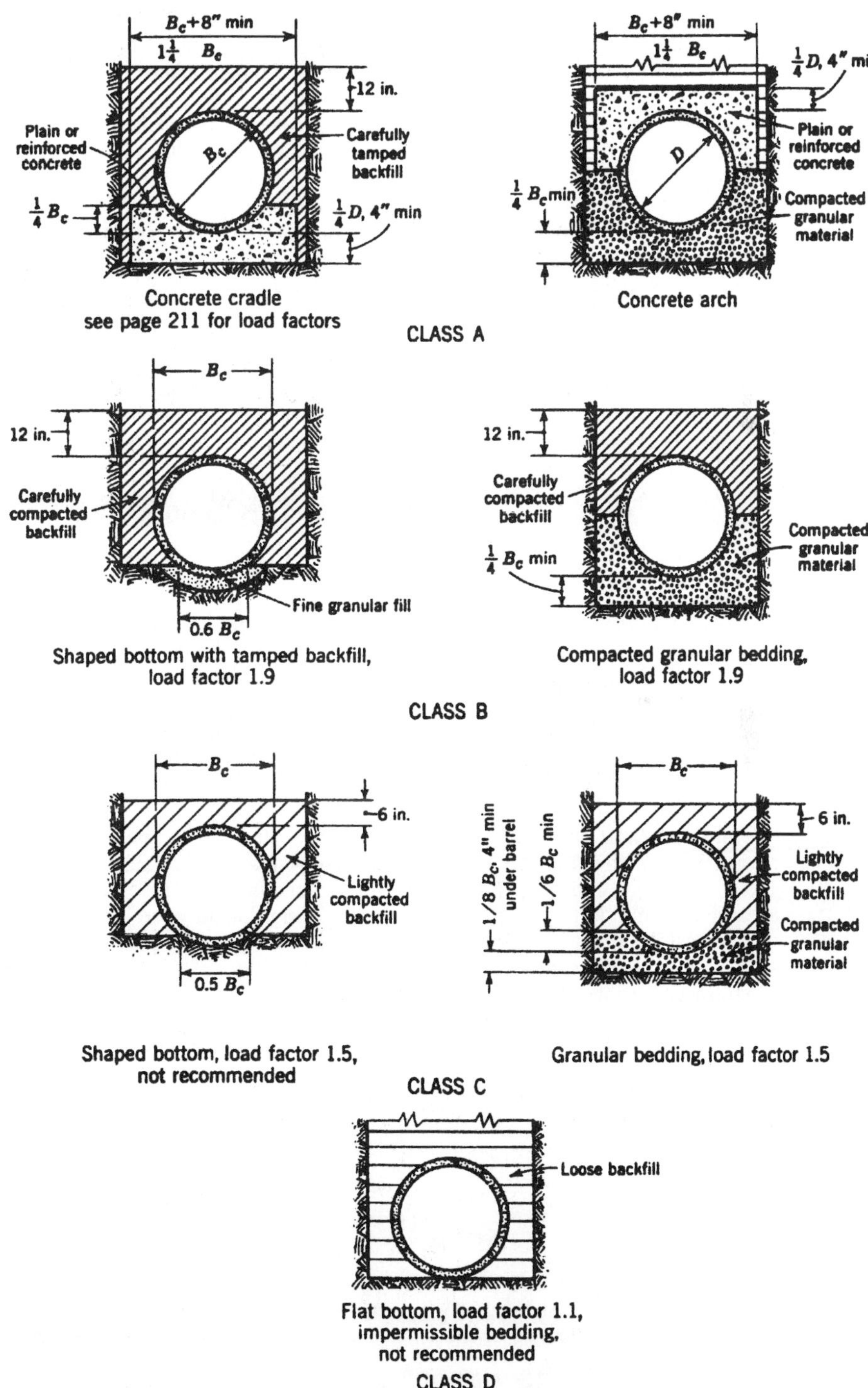

FIGURE 7.20 Pipe bedding and cradles. (After American Soc. of Civil Engineers and Water Pollution Control Federation, 1970)

embankment should be level to prevent a concentrated flow from occurring during overtopping; in other words, the best overtopping condition would be a widely spread weir flow. Careful consideration needs to be given to the embankment cover. Typically 6 in of topsoil and seed is sufficient to promote stabilizing grass growth, but in sensitive situations, geotextile treatment of the cover may be well-advised. A reasonable embankment design is presented in Fig. 7.21, although in no way should this design be considered a "template" or "go-by" for other situations.

7.5.4.13 Maintenance and Safety Issues.   Maintenance issues are relatively straightforward in the design process; however, the engineer needs to be leery of designing anything that might be unsafe. In particular, confined space entry regulations are in place in virtually all jurisdictions to prevent fatalities in sewer system maintenance work. Any design that departs from normal standards should directly consider the issue of confined space entry.

Adequate access must be maintained to the outlet structure, and it should be relatively easy for manual debris removal to take place. Constructing tall inlet box and stand pipe structures out away from the embankment means that a crew will have to bring a ladder to gain access to the structure. This situation can be avoided by tucking the structure into the embankment.

If possible, a line of sight should be maintained from the nearest roadway to allow constant visual checks from passing vehicles. Steep access lanes should be avoided because the basin in effect becomes inaccessible during rain and snow conditions. An adequate line of sight should also be provided to allow fully loaded maintenance vehicles to exit the site safely.

Grading the bottom of the basin to drain will help keep the bottom as stiff and workable as possible. Some designers call for paved swales to be used in the bottom of the basin, but on relatively flat grades they tend to get covered with silt. Paved low-flow channels are best utilized when there is a high propensity for scour during low-flow conditions.

Constructing retention basins near runways or along runway approaches can also cause airplane instrumentation interference, and should be avoided if possible. Waterfowl inhabiting a retention basin anywhere near a runway is cause for an enormous safety concern.

7.5.4.14 Inlet and Outlet Scour Protection.   It is good practice to protect pipe ends and structures with riprap scour protection, as depicted in Fig. 7.22. The riprap is self-healing to some degree, and will not fail due to settlement. If sufficiently large, standard riprap is specified, the outlet protection also provides significant energy dissipation benefits as well. The main outlet protection should be checked for proper riprap sizing and placement using methods described in Corry et al. (1983), using the peak discharge from the largest design storm event considered.

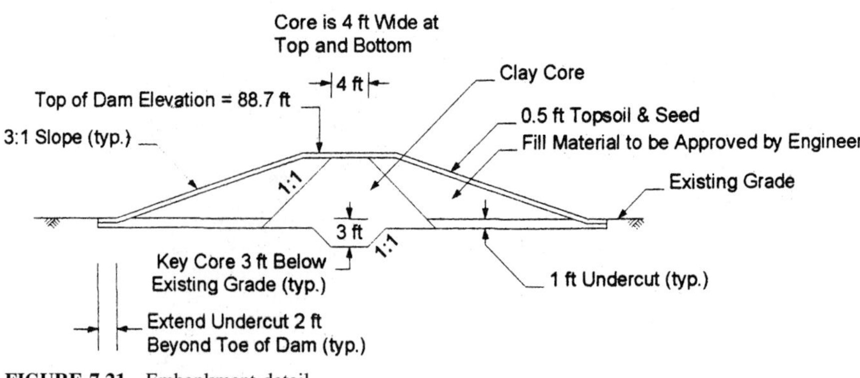

FIGURE 7.21   Embankment detail.

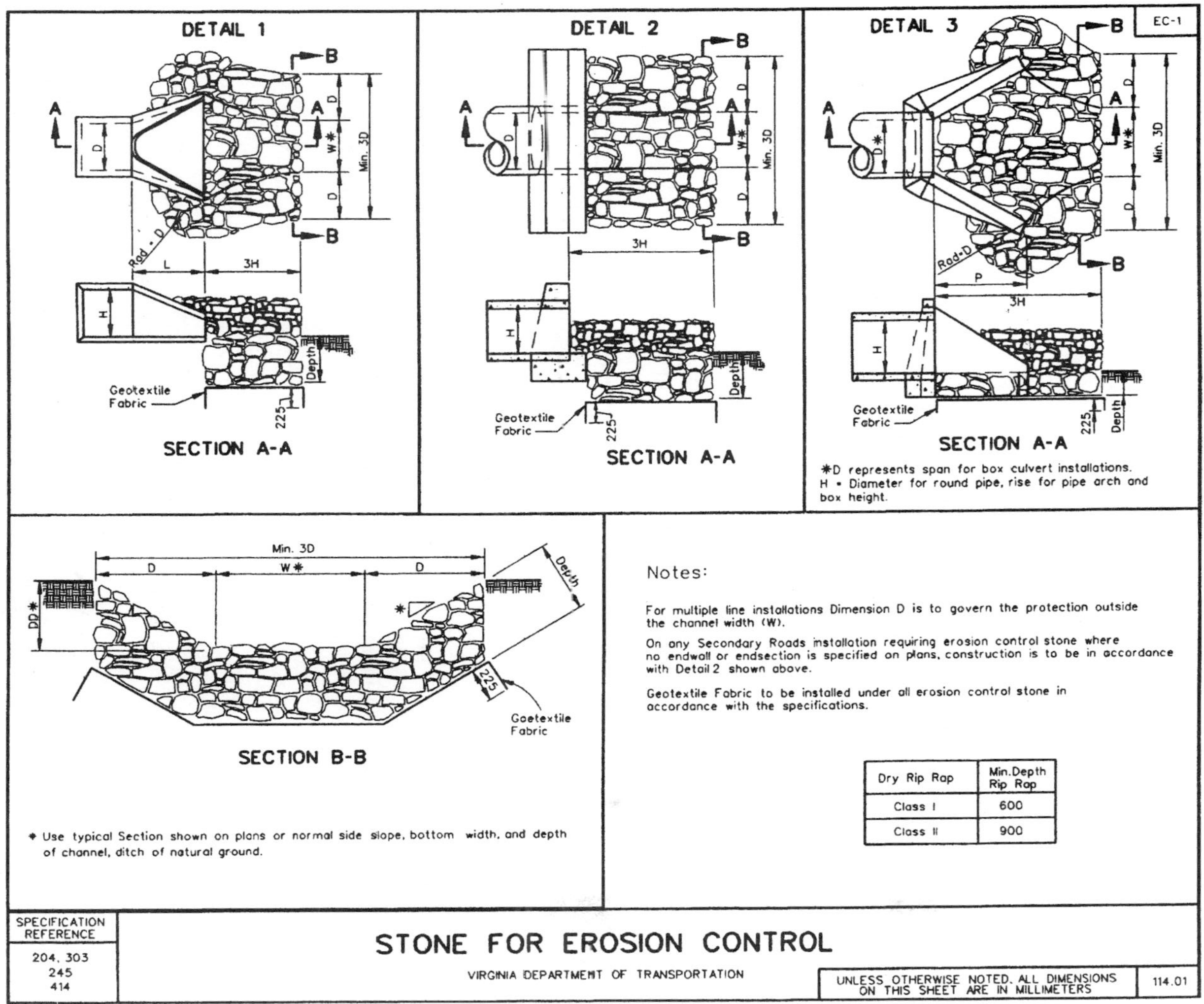

Dry Rip Rap	Min. Depth Rip Rap
Class I	600
Class II	900

FIGURE 7.22 Inlet and outlet scour protection. (After Virginia Dept. of Transportation, 1996)

7.5.5 Design Computations

7.5.5.1 Computational Approach.   With the help of computers, the computations involved in designing a stormwater basin are not particularly difficult; however, a tremendous amount of engineering effort and discipline goes into the design of each basin before the plans are finalized. The single most important thing to realize in performing this kind of work is that all of the calculations are sequential, and any mistake made is in all likelihood going to involve re-doing all of the subsequent work—not just computations, but design and drafting work as well. In the real world, the basin design process typically requires 24 to 48 hours of total labor. Every step should be checked carefully before continuing with the design effort. Imagine how much time is wasted if the engineer checking the work discovers an error in the pre-development hydrologic computations.

Boundary conditions and assumptions should also be carefully checked at the outset. Sometimes it is easy to miss the obvious. For example, we might draw flow arrows on a plan sheet to indicate which direction the flow will take. However, often the flow will follow the laws of physics rather than our flow arrows. This can be a source of great embarrassment as hydraulic conditions change during the modeling event, and some portion of the flow reverses direction due to energy grade line differential. Good designs will also work even if low-flow structures become blocked, typically by providing an overflow or emergency spillway.

Often the tailwater elevations for design purposes will be set by floodplain computations or storm surge predictions. Sometimes it is necessary to perform channel routing computations to compute flow profiles in the receiving stream. And sometimes normal depth assumptions are just as valid as anything else we might adopt. Once the design tailwater elevations are set, however, any change can cause all the subsequent work to be re-done.

7.5.5.2 Basin Design and Related Software.   Although there are numerous hydraulic routing and design software programs that can be used to perform the hydraulic routing computations required for basin design, most use some form of the reservoir routing equations described in Sec. 7.1. In addition to commercial software programs, there are a number of public domain and low-cost commercial programs available for download over the Internet for use in designing detention basins, storm sewers, and receiving channels. These include:

1. U.S. Army Corps of Engineers *HEC-RAS* program (for channel computations), available from *http://www.wrc-hec.usace.army.mil/*.

2. U.S. Army Corps of Engineers *HEC-HMS* and *HEC-1* programs (for hydrologic computations and development of inflow hydrographs), available from *http://www.wrc-hec.usace.army.mil/*.

3. Several implementations of the Storm Water Management Model "*SWMM*," which has its roots in the U.S. EPA. An excellent, low-cost derivation called "*PCSWMM*," may be obtained from the Computational Hydraulics International web site at *http://www.chi.on.ca/*.

4. *SCSHYDRO,* a hydrograph generation program based on SCS methodology, available from the Computer-Aided Hydrology & Hydraulics web site at *http://www.cahh.com/*.

5. *CHANNEL,* an open-channel design and analysis program, available from the Computer-Aided Hydrology & Hydraulics web site at *http://www.cahh.com/*.

6. *BASINOPT,* a detention basin design, optimization, and analysis program, available from the Computer-Aided Hydrology & Hydraulics web site at *http://www.cahh.com/*.

7. *Hydraflow Storm Sewers,* sewer design and analysis software, available from the Intelisolve web site at *http://www.intellisolve.com/*.

8. *Hydraflow Hydrographs,* watershed modeling and basin design software, available from the Intelisolve web site at *http://www.intelisolve.com/*.

If the underlying assumptions are understood and appreciated by the engineer, all of these programs can be used quite successfully, at relatively little or no licensing cost. One of the important caveats is that engineers cannot turn real-world systems into equations without some simplifying assumptions. Although models like SWMM can perform very sophisticated hydraulic computations, they cannot model every hydraulic nuance and wrinkle in the sewer system. Most sewer systems and detention basins are not designed using SWMM technology, because the design effort would be more laborious than using design-oriented tools like Hydraflow Storm Sewers and BASINOPT, and simplifying assumptions would still be required, because even SWMM, with all of its sophisticated algorithms, cannot model all of the simultaneous hydraulic conditions that occur in a sewer system, and it lacks a design mode.

The vast majority of storm sewer systems constructed today are designed using the rational method, as they have been for more than 110 years since the method was first proposed. Where rational-method-based storm sewer design programs are used to design the storm sewer system leading up to the detention basin, a measure of caution needs to be applied when switching to another program, such as HEC-HMS to develop the basin inflow hydrographs. Many of the hydrograph generation programs used today are based on empirical methods, and as such treat storm sewers in lumped or simplified fashions. After designing the storm sewer system based on the rational method, if a switch is made to an empirically based software program, a reality check must be made to ensure that the sewer system and watershed can deliver the computed discharge *into* the basin. For example, if the maximum flow that can enter the basin through a pipe culvert is 128 cfs, and the hydrology program computes 240 cfs entering the basin, just how is the extra 112 cfs going to be delivered to the basin?

And just to avoid any potential confusion, although it is good and standard practice to design storm sewer systems using the rational method, using the *modified rational method* to compute hydrographs does *not* produce a consistent basin design by any stretch of the imagination. The rational method and modified rational method are two totally different design procedures, and are not linked or related in any physical or computational sense.

7.5.5.3 *Checking the Computations.*

So how can we efficiently check the computations, and what sorts of things might we look for in the review process? What are some of the most common mistakes? Are there rule-of-thumb values that we can use?

The first rule of thumb is that there should be independent reviews throughout the design process. Although it may be hard to carve out time in the design process, it is better to have the checks occur along the way rather than at the end of the process, because of the total labor required and the sequential nature of the computations. The checks mentioned in the following text should be made throughout the design process, not all at the end.

The "Designed By" and "Checked By" blocks on the plan, computation, and report sheets should have different initials.

Can this basin be constructed? What about wetlands and permitting issues? What about the cost of the right-of-way and easements? Is it remotely feasible and likely to fly?

The first round of detailed checks should focus on the hydrologic computations. What type of hydrologic method is being used? Is the method applicable for the size watershed involved? What are the peak discharges entering the pond in terms of cfs per acre? In an urban setting, for 100-year storms, peak discharges entering detention basins are typically in the 2.0 to 4.5 cfs per acre range when using Soil Conservation Service design hyetographs such as the type-II or type-III storms.

Are there any safety issues? Is fencing required? What about a locked chain at the maintenance access lane entrance? Is there anything that we might change to improve safety? What about plantings and lines of sight? Generally the safety aspects are addressed upfront and shouldn't require any modification in the latter portion of the design process.

What happens if the outlets become blocked? Are we betting the farm that the outfall will be operational during severe storm events, or is there some emergency spillway scenario available? How likely are the outlets to become clogged? Have adequate debris racks been specified? Can the outlet structures be accessed easily? Can excavation equipment booms and buckets access the entire basin, and load dump trucks without double-handling the material?

Next we focus on the performance of the basin. Have all the design criteria for peak discharge, draw down and retention time, maximum allowable water surface elevation and freeboard, and volume been satisfied?

Do the final computations used to support this design represent the grading, structures and conditions shown on the final plans? This may seem like a silly question, but it happens far too frequently that some aspect of the design will be changed after the "final" computations have been performed, and somehow the computations remain the same. This is particularly true in putting the proposed contours on the plans and grading the site.

How were the tailwater elevations computed? Have probabilities been combined by assuming that the 100-year tailwater event, such as a hurricane storm surge, occurs simultaneously with the 100-year rainfall event? What was the initial stage used for performing the hydraulic routing? Have the pond discharges been computed for scour in the receiving channel? How far downstream were the scour computations carried? Were hydrograph timing conditions checked downstream to make sure that attenuated and lagged flows from the proposed pond do not worsen downstream conditions when combined with other downstream flows?

Are the discharge coefficients used in the computations correct, at least for peak discharge conditions? How were they determined? If the values were chosen from tabulated references, are the units consistent with those used in the program formulas? For example, a weir with one foot of head on it may have a discharge coefficient equal to 2.64, or it may be 0.33, depending on whether the equation used in the routing is in the traditional weir equation form, $Q = CLH^{3/2}$, or in the more modern form shown in Eq. (7.10).

Are assumptions used in the routing program valid? For example, suppose that the allowable 2-year outflow from the basin is such that the low-flow orifice would require a diameter of 3.5 ft. Most reservoir routing programs assume that there is no flow through the orifice until the water surface elevation reaches the centroid of the orifice. As the diameter increases, this assumption becomes less reliable. If we want to be more accurate, we should probably model orifices over 6 in in diameter as culverts to account for partially full flow that occurs for extended periods of time during most storms.

Finally, we can look at cost issues. Is there anything funky that will require special detailing and construction that might be avoided if we adapt a standard design item? Has the engineer been diligent in the trial-and-error portions of the design process, to arrive at a reasonable, cost-effective design? Might a little more work produce a more economical result? Is there anything that should be done to improve this design? Many of these questions are offset by tight design budgets and engineering fees, but with proper estimating, planning, execution, and checking, the answers should be favorable.

7.5.6 Operating Specifications

The design engineer's work is not complete when the plans are delivered to the owner. In cases where the owner is a public agency, long-term care and maintenance programs and procedures are typically in place, and the basin is simply added to the inventory of facilities maintained by the agency. However, the perpetual maintenance of private facilities is often overlooked and under-specified.

After construction is complete and the site is stabilized, regular maintenance inspections should occur at least on an annual basis. The outlet structures should be checked for free and clear operation, and repair items should be documented and communicated back to the owner in a written report. Debris should be removed during the annual inspections, and any time it is observed to be collecting on the outlet structures. Stand pipes and inlet boxes should be checked for differential settlement, which is much easier to correct if caught and corrected before it accelerates. Pipe connections and joints should be checked for true alignment and joint separation. Metal structures should be checked for corrosion and excessive rust, and hinges and locks should be lubricated.

The maintenance access lane should be checked for excessive erosion and rilling, and vegetation should be cut back to a mowable level if needed. Small saplings should be removed if they impede access to the site. Mowing should take place at least twice a year to avoid woody growth buildup that is much harder to remove. It is especially important to mow the embankments to prevent trees from growing. Trees growing and falling over on the embankment can easily lead to embankment failure.

Sediment build-up should be monitored and sediment should be removed as needed to ensure that the basin functions as intended. Typically, a basin serving a stable watershed can go 10 to 20 years between sediment cleanouts.

Pond vegetation should be harvested if the pond is designed to function as a water quality basin. Dry basins should be re-graded if necessary to promote as dry a bottom as possible. Wet basins should have the sediment forebay cleaned.

On-site personnel should be advised not to use fertilizers, pesticides, herbicides, and other chemicals in the proximity of the basin to avoid degrading water quality.

In some cases, emergency flood relief mechanisms are built into detention basins. If these emergency mechanisms involve nonpassive controls, such as valves and gates that require human intervention to operate, all operation and maintenance issues should be clearly communicated to the owner, and the engineer should be satisfied that the public safety and property will be adequately protected during construction of the basin, and perpetually thereafter.

7.5.7 Design Example of a Dry Detention Basin

Example 7.10 Dry Detention Basin

A new roadway widening project will increase discharges to the outfall shown in Fig. 7.23. The upstream contributing area is 28.8 acres. When analyzed with SCS type-III design-storm methodology, the peak discharges shown in Table 7.7 are generated.

The applicable stormwater management regulations require that the 2- and 10-year post-development flows do not exceed pre-development discharges. The 100-year post-development flow does not have to be attenuated, but the 100-year event must pass through the detention basin with at least one foot of freeboard. Additionally, the 100-year outflow from the basin cannot cause erosion downstream in the outfall channel.

Recent development has heightened homeowner concerns about increased runoff in the area. The highway department has had preliminary meetings with project stakeholders, and anticipates difficult right-of-way negotiations. It is expected that the required real estate cost for this basin will approach $300,000 per acre.

The likely site for the required detention facility has a lateral inflow joining the main channel through a 24-in concrete pipe from an adjacent parcel. The highway department would prefer to avoid a real estate take from this parcel owner if possible. Likewise, it would like to avoid backing any water up onto this parcel, even from the 100-year storm, for flood damage purposes. For this reason, the maximum allowable water surface elevation in the basin becomes 87.4 ft.

Maintenance access to detention facilities has become a significant concern in recent years with the proliferation of basins throughout the state. A working width of 15 ft must be provided for permanent dump truck and maintenance vehicle access to the site. A minimum inside turn radius of 25 ft must be used in the design. Access must be extended to the outfall structures, and a working ledge of 5 ft must be maintained on the opposite side of the basin

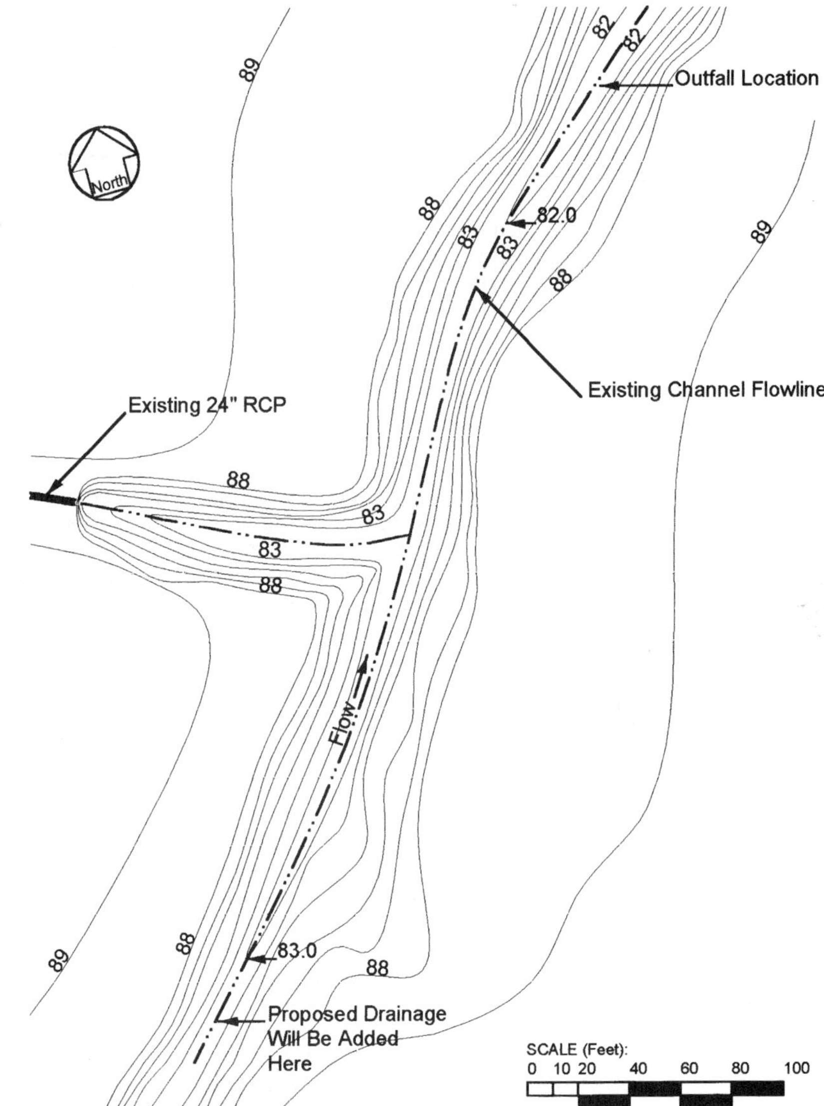

FIGURE 7.23   Existing outfall site.

TABLE 7.7   Example 7.10 Hydrologic Computations

Return period (yrs)	PRE-Development discharge (cfs)	POST-Development discharge (cfs)
2	7.6	16.7
10	12.1	35.3
100	39.3	64.7

from the 15-ft access lane. Highway department policy now requires that all new detention facilities in this region be constructed on 3:1 embankment slopes.

There is no design flood elevation in the downstream channel to use as a tailwater in the basin computations. It can be assumed that normal depths in the downstream channels can be used for tailwater elevations. Wetlands and other environmental concerns are negligible.

Design a dry detention basin that satisfies all of these criteria.

Solution

The design process requires a preliminary layout, which will be refined as the computations progress. The first step is to plot out a likely basin alignment. In general, simpler designs are better and less costly to construct and maintain than elaborate designs. The lateral inflow pipe and channel are sufficiently small that we can add a catch basin and extend the 24-in pipe, filling in the lateral channel to simplify the basin geometry. Otherwise, our design would require a dogleg layout that would complicate construction and maintenance, and require a real estate take from a difficult property owner. By extending the 24-in pipe and filling in the ditch, the highway department can actually create more usable space for the parcel owner, and avoid a real estate take.

By far, the two biggest cost items are real estate and excavation. By using the existing channel for detention storage, we can keep the excavation to a minimum. The main alignment will follow the existing outfall channel. Construction costs will be lower if the proposed design has a minimal number of bends, curves, and zigzags. Approximating the alignment of the existing channel with three tangents, we develop a proposed basin flow line and proposed "west edge" based on property line constraints, as shown in Fig. 7.24. Everything will be laid out based on this flow line, and trial-and-error computations will hold the west alignment as shown in the figure.

The bottom elevation of the proposed basin is set by the existing stream invert elevation at the outfall point, which in our case is approximately 82.0 ft. The top of the proposed embankment is determined by the existing topography to be 88.7 ft. Working within the surrounding property lines, after some sketching, we estimate that the basin will be on the order of three times longer than it will be wide. Having already performed the hydrologic analysis, as summarized in Table 7.7, we are ready to begin our preliminary computations.

The first step is to come up with a rough size for the basin. In the past, this process involved considerable trial and error; however, new computer software can help reduce this drudgery. Using the design mode of the BASINOPT program (Sec. 7.5.5.2), we can approximate the basin as a rectangle having a 3:1 length-to-width ratio. In this case, because the downstream channel is relatively wide, we will ignore tailwater elevations in the preliminary computations. BASINOPT will prompt us for structure types and sizes to try in the computations, and will build the basin volume to satisfy the maximum allowable discharges and stage criteria. We are careful to specify the structures such that standard sizes are used in the design computations. We would like to use a standard stormwater management inlet box structure, as shown in Fig. 7.18, if at all possible.

For the emergency spillway, we would like to use a trapezoidal riprap weir. Technically speaking, there is no such thing as a trapezoidal riprap weir. However, specification and construction of riprap trapezoidal weirs are commonplace. Riprap has several advantages over concrete in this type of application: it is self-healing and does not lead to structural failure when displaced (concrete does fail when displaced); it is usually cheaper to construct and maintain than concrete; and it has energy dissipation and anti-scour properties that are useful in the design. There is no available tabulation of coefficients for a riprap weir, but we can obtain these by modeling the weir as an open channel, set the computed stage-discharge relationship equal to the stage-discharge relationship for a regular weir, and back solve for the coefficients. Trapezoidal weir equations and coefficient values do not exist in the engineering literature. To solve this problem, BASINOPT breaks the trapezoidal weir into two components, a rectangular weir and a V-notch weir, and simply adds the discharges together to obtain the total discharge from the trapezoidal weir. As engineers, we do not like having to use such approximations, but we must design for real-world construction scenarios.

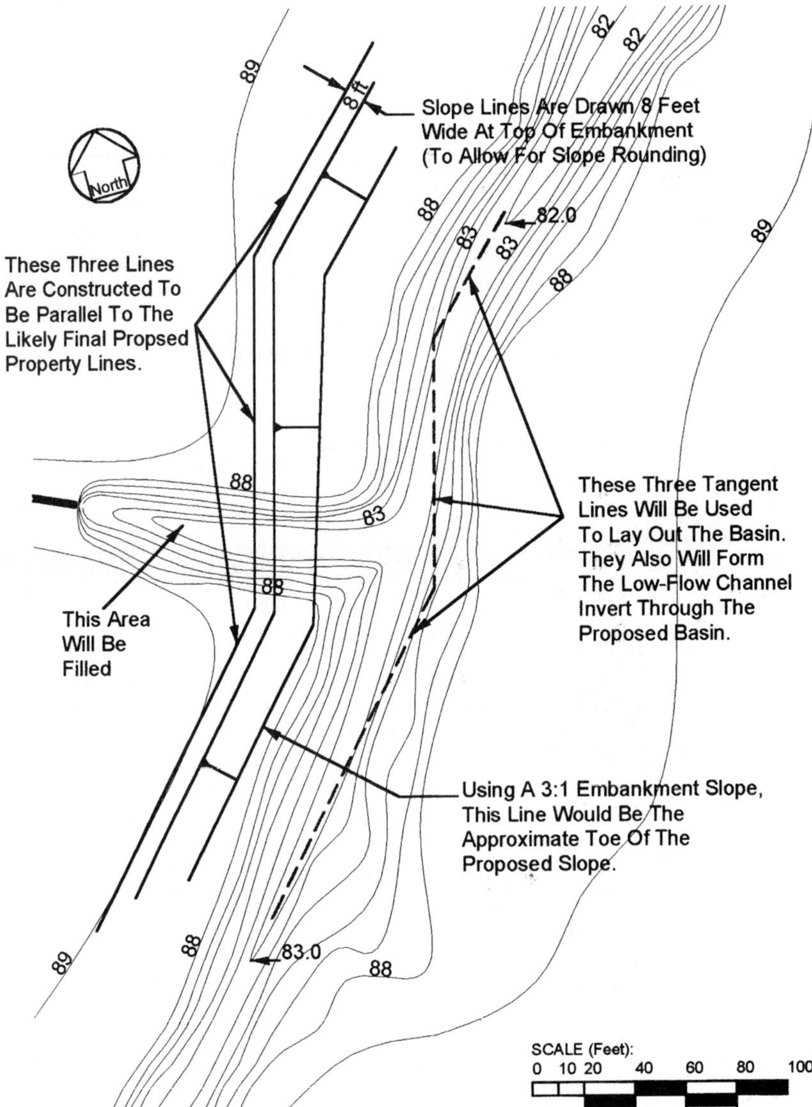

FIGURE 7.24 Preliminary basin construction layout.

Finally, a riprap weir must be constructed carefully to avoid excess variations in the specified crest elevation, and seepage through the riprap layer below the crest elevation.

In design mode, BASINOPT takes our inflow hydrographs and prompts us for input as shown in Table 7.8.

In design mode, BASINOPT sizes the structures and sets the invert and crest elevations, and designs the excavation in the form of a stage-area relationship, producing in this case two reasonable designs that meet our criteria. However our work is not yet complete—we must still site the basin and complete the computations and design details. BASINOPT has saved us a lot of trouble by eliminating the trial-and-error process up to this point. After selecting BASINOPT's suggested "Alternative 1" design, and reviewing the design mode results, shown in Fig. 7.25 (Fig. 7.25*b*), we know that our basin needs to cover approximately one acre. Suggested outlet structure design parameters for Alternative 1 are listed in Table 7.9.

At this point, all we have is a rough approximation of a working design. We have not yet accounted for tailwater effects at the outlets, have not fit the proposed basin contours on the site, and have not completed any design details.

Using the design boundary and alignment information shown in Fig. 7.24, we can now draw in the slopes, shoulders, embankments, and access geometry for our basin, as shown in Fig. 7.26. The maintenance access lane is drawn 18 ft wide, providing 15 ft of clear working lane while allowing for three feet of shoulder rounding in the grading. Likewise the maintenance shoulder is drawn 8 ft wide on the opposite (west) side.

After drawing the slope grading, we can draw in the proposed contours and measure their enclosed areas using CAD techniques, as also shown in Fig. 7.26. These stage-area measurements define the basin geometry for our final calculations and are perfectly consistent with our final site layout. The bottom of the basin is graded to promote low-flow drainage toward a center swale, to keep the entire bottom from remaining soft and soggy. This grading improves safety and provides a stiffer bottom for equipment access for routine maintenance. Many designers will specify a paved swale for the invert, but this is not necessary in very flat conditions—where silt will simply cover the paved swale. If we expected erosive low-flow conditions, we might specify a paved swale invert through the pond. In this case it is not needed.

After laying in the grading and proposed contours, the hydraulic routing computations must be re-run. This time, we are going to be very careful to ensure that nothing is overlooked, or inappropriately assumed, in the final computations. Some trial-and-error work is now required, and we must switch to simulation mode in BASINOPT to complete our work. Simulation mode allows us to completely specify the basin characteristics and outlet structures, without having the program take over and change the design. Several other hydraulic routing computer programs could also be used for this process.

One of the most common mistakes in designing storm water management basins is to skip the next few steps. After running a set of computations, and obtaining a successful design by whatever means, too often the engineer fails to make sure that the computations accurately reflect the final grading design. To check for this situation, look at the stage-area values used in the hydraulic routing program, and make sure they match the contours shown on the design plans.

With the proposed contours laid out on the site, we can focus on finalizing the design of the outlet structures and computing the final hydraulic routing results. After a few iterations, we develop the design as specified in Table 7.10.

The final routing results are presented in Fig. 7.28. The basin satisfies all of the design criteria, and the computations are completely tailored to the site.

There are several design details to complete. Fig. 7.29 shows the completed design. First, the existing and proposed inflow drainage pipes are extended to the pond limits, and terminated with flared end sections or endwalls, and riprap scour protection, as shown in Fig. 7.22. Specifications for these types of treatments can be found in state highway department drainage design standards. As we anticipated, the lateral inflow channel can be filled.

TABLE 7.8   Example 7.10 BASINPOPT Design Mode Inputs

BASINOPT input required	Values or data given to BASINOPT
2-, 10-, and 100-year inflow hydrographs, having peak inflows of 16.7, 35.3, and 64.7 cfs respectively	Computed from SCS Type III design storm analysis, prepared in machine-readable form prior to BASINOPT run. Summarized in output that follows.
Allowable outflows from the basin	$Q2_{allow}$ = 7.6 cfs $Q10_{allow}$ = 12.1 cfs $Q100_{allow}$ = 64.7 cfs (no attenuation required)
Design tailwater elevations	Use 82.0 ft for all events (i.e. we are ignoring tailwater effects in the preliminary sizing— we will account for them later).
Outlet structure types	Lowest (No. 1) = circular culvert (we are anticipating using a large orifice, but prefer to model it as a circular culvert to account for partially full flow). Middle (No. 2) = inlet box Highest (No. 3) = weir
Circular culvert design parameters	Min. Trial Diameter = .25 ft Max. Trial Diameter = 2.5 ft Diameter Increment = .25 ft K = 0.519 M = 0.64 C = 0.0289 Y = 0.9 Ks = −0.5 Ke = 0.5 Bottom Slope = 0.005 ft/ft Manning n = 0.013 Length = 0.75 ft TLIM1 = 3.5 TLIM2 = 4.0
Inlet box design parameters	Min. Trial Opening Length = 1.5 ft Max. Trial Opening Length = 6.0 ft Opening Length Increment = 1.5 ft Weir Discharge Coefficient = 0.3 Orifice Discharge Coefficient = 0.6
Weir design parameters	Min. Trial Weir Length = 2 ft Max Trial Weir Length = 20 ft Weir Length Increment = 2 ft Weir Discharge Coefficient = .32 Weir Side Slope = 1 (horizontal/vertical) Slope Discharge Coefficient = .26
Geometry	We will select a rectangular geometry for the preliminary site layout. Bottom Elevation = 82.0 ft Minimum Bottom Area = 0.5 acres (we are anticipating a fully excavated bottom on this site because real estate costs are much higher than excavation costs.) Maximum Allowable Stage = 87.4 ft Steepest Allowable Side Slope = 3.0 (hor/vert) Length/Width Ratio = 3.0 Lowest Structure Invert Elevation = 82.0 ft Embankment Top Elevation = 88.7 ft

ALTERNATIVE 1, 100-Yr MAXIMUM STAGE PLOT

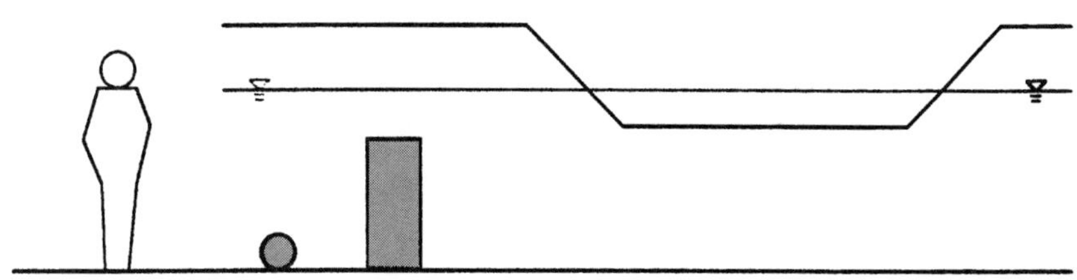

FIGURE 7.25a   BASINOPT design mode output.

STAGE-AREA RELATIONSHIP
ALTERNATIVE 1

STAGE
(ft)

This is a RECTANGULAR basin design.

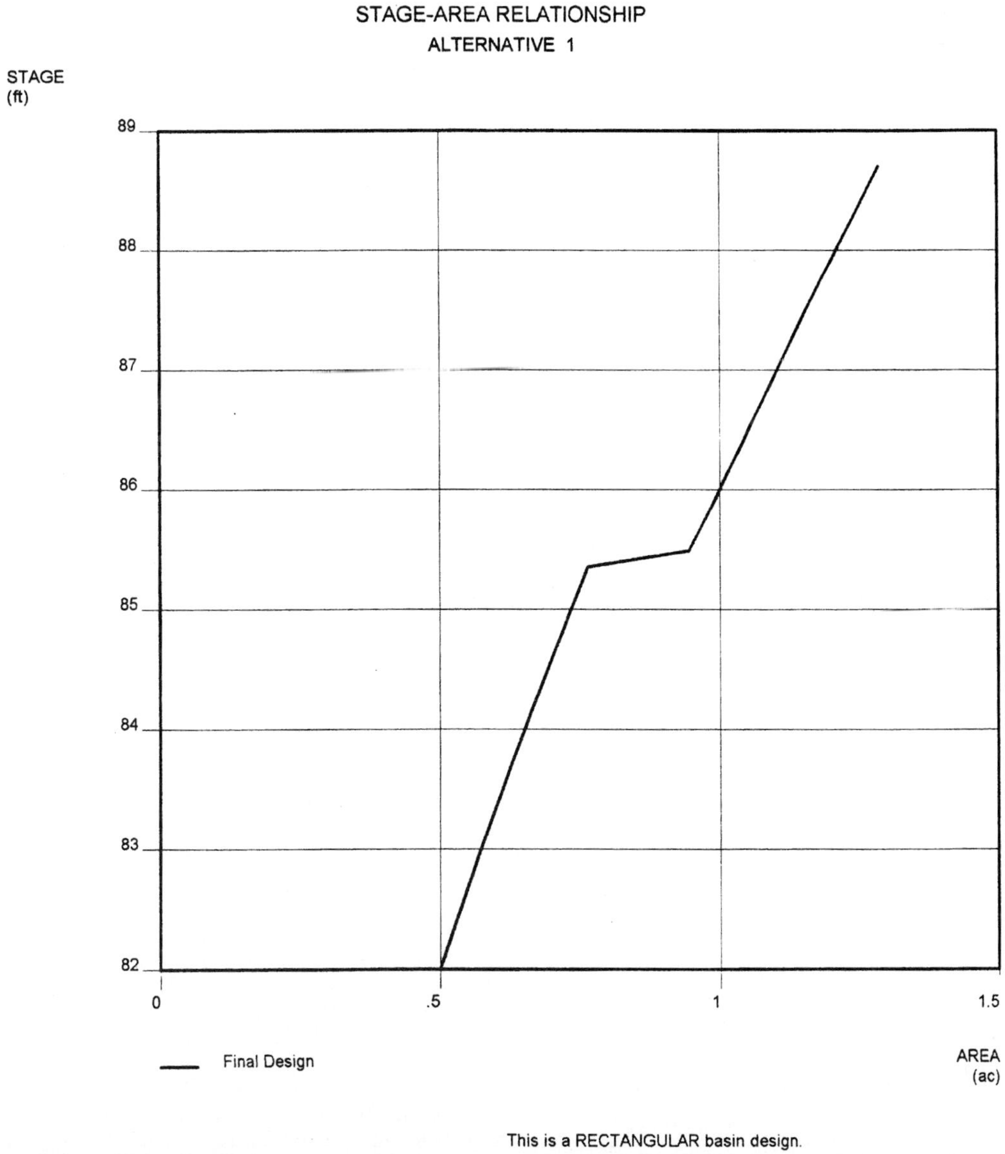

FIGURE 7.25b   BASINOPT design mode output.

SUMMARY OF HYDRAULIC RESULTS
ALTERNATIVE 1

Basin VOLUME: **5.797 acre-feet**

Basin BOTTOM AREA: **.5 acres**   This is a RECTANGULAR Basin.

Basin TOP AREA: **1.287 acres**

	STORM		
	2-yr	**10-yr**	**100-yr**
Peak INFLOW (cfs)	16.708	35.332	64.737
Peak OUTLOW (cfs)	4.852	10.000	43.358
Peak STORAGE VOLUME (ac-ft)	1.155	2.604	3.707
Peak STAGE (ft)	84.011	85.886	86.944
Time of Peak INFLOW (min)	756.	756.	750.
Time of Peak OUTFLOW (min)	816.	810.	774.
Time of Peak STAGE (min)	816.	810.	774.
Continuity Error (%)	-1.14E-02	-6.10E-03	-3.47E-03

FIGURE 7.25c   BASINOPT design mode output.

```
OPTIMIZED ALTERNATIVE IS NUMBER    2



===============    DESIGN RESULTS OF ALTERNATIVE NUMBER    1   ===============


        STRUCTURE NUMBER 1

             Type 1: CIRCULAR CULVERT
              DIAMETER  :      1.00 feet.

             INVERT ELEVATION :      82.00 feet.
             USER INPUT SLOPE :      .00500 feet horizontal per foot vertical.
               MODIFIED SLOPE :      .01690 feet horizontal per foot vertical.
               CULVERT LENGTH :        .75 feet.
                    MANNING n :     .01300
                            K :     .51900
                            M :     .64000
                            c :     .02890
                            Y :     .90000
             SLOPE COEF ks :       -.50000
           ENTRANCE LOSS Ke :       .50000
         WEIR LIMIT, TLIM1 :      3.50000
         ORIF LIMIT, TLIM2 :      4.00000


        STRUCTURE NUMBER 2

             Type 5: SQUARE-SHAPED INLET BOX
             SIDE LENGTH :     1.50 feet.

             CREST ELEVATION :        85.62 feet.
                   WEIR COEF :       .300
                 ORIFICE COEF :      .600


        IF THIS STRUCTURE WILL HAVE A GRATE:
            Transition head =     86.37 feet.
            Peak pond head =      86.94 feet.
            At peak pond elevation this structure probably operates
            as an ORIFICE.
```

FIGURE 7.25d BASINOPT design mode output.

```
(Continued)

              Select a grate that has a clear opening (i.e. effective
              orifice opening) of approximately   2.250 square feet.
                                                  --------

              The grate should also have an effective weir length of
              approximately   6.000 feet.

          OBVIOUSLY, IF A GRATE IS BEING USED, THEN THIS OUTLET
              STRUCTURE WILL HAVE TO BE LARGER THAN THE STRUCTURE
              DIMENSION(S) LISTED ABOVE: THIS STRUCTURE MUST ACCOMMODATE
              A GRATE HAVING THE CHARACTERISTIC DIMENSION UNDERLINED
              ABOVE.

     STRUCTURE NUMBER 3

        Type 3: WEIR
           LENGTH    :       8.00 feet.

         CREST ELEVATION :      85.95 feet.
          DISCHARGE COEF :        .320

        TRAPEZOIDAL WEIR
        Side Slope=         1.000000
        Slope Discharge Coeff=    2.600000E-01
```

FIGURE 7.25e BASINOPT design mode output.

The inlet box outfall structure can be set back in the dam embankment to a point where the top of the structure is 1 ft above the embankment slope. This setback helps to keep the structure stable and avoids having a chimney-like situation in soft, mucky ground. A trash rack and scour protection are added as well. The final structure is similar to that shown in Fig 7.18. The outfall weir is constructed in undisturbed (not fill) material. The riprap is extended down beyond the toe of both slopes to help stabilize the stones. The riprap at the end of the outfall pipe acts as an energy dissipator while providing outlet scour protection. In order to prevent the riprap from becoming a French drain, we will specify that a mix of sand and topsoil will be placed and graded into all riprap voids. Details of the outlet design are presented in Fig. 7.30.

To avoid choking in the outlet barrel, the outfall pipe is sized to have at least 25 percent more capacity than the portion of the 100-year design discharge passing through the outfall pipe. From the BASINOPT results, at peak 100-year discharge levels, the orifice and inlet

TABLE 7.9 Example 7.10 Suggested Outlet Structure Design Parameters from BASINOPT

Design parameter	Value suggested by BASINOPT
Culvert invert elevation	82.0 ft
Culvert diameter	1.0 ft
Inlet box rim elevation	85.62 ft
Inlet box effective side length	1.5 ft
Weir crest elevation	85.95 ft
Weir length	8.0 ft

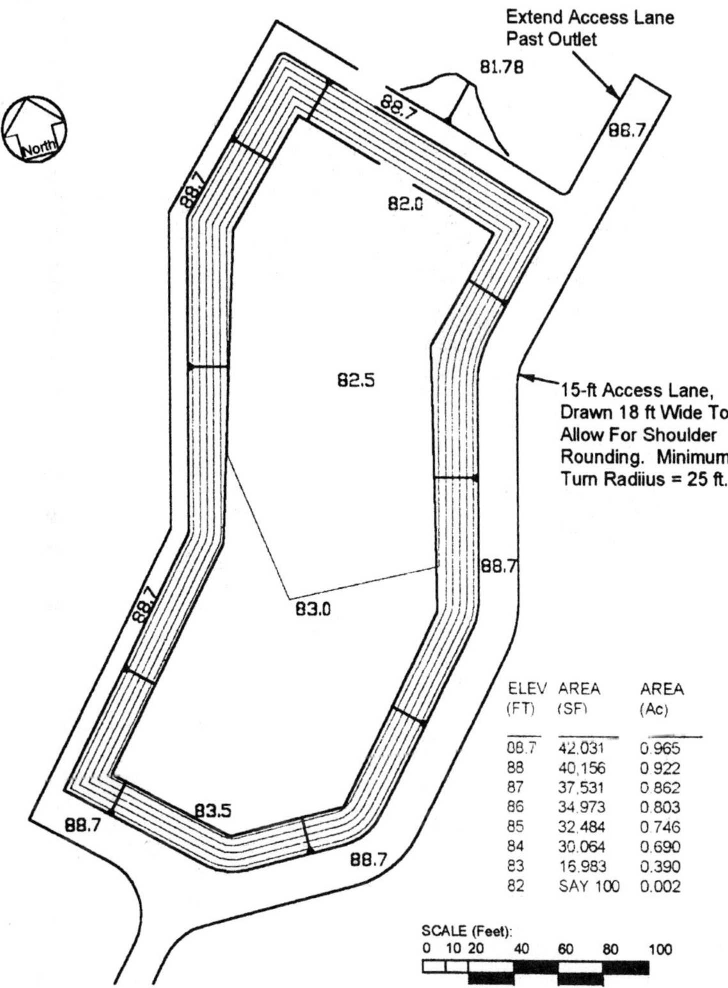

ELEV (FT)	AREA (SF)	AREA (Ac)
88.7	42,031	0.965
88	40,156	0.922
87	37,531	0.862
86	34,973	0.803
85	32,484	0.746
84	30,064	0.690
83	16,983	0.390
82	SAY 100	0.002

FIGURE 7.26 Proposed basin geometry.

TABLE 7.10 Example 7.10 BASINPOPT Simulation Mode Inputs (Finalizing the Design after Tweaking the Design Mode Results)

BASINOPT input required	Values or data given to BASINOPT
2-, 10-, and 100-year inflow hydrographs, having peak inflows of 16.7, 35.3, and 64.7 cfs respectively	Same inflow hydrographs as used in the design mode run

Initial basin stage and tailwater elevation for each storm event: These values are determined by trial and error. We start be guessing the resultant outflow from the basin for each storm event. Those discharges are then used to compute normal depths in the downstream channel—here using the CHANNEL program, and assuming that the channel is lined with retardance class C vegetation. The computed normal depths are added to the outfall channel invert elevation (81.78 ft) to determine the tailwater stages. The pond routing computations are then re-run and compared against our discharge guesses. Within three or four trials we have the following results.

Return period (yrs)	Outfall discharge (cfs)	Normal depth (ft)	Tailwater elevation (ft)
2	5.6	0.88	82.66
10	11.7	1.12	82.90
100	53.6	2.00	83.78

The 100-year CHANNEL Program output is shown in Fig. 7.27. For all three discharges, the outfall channel will not scour (because the computed velocity and shear stress are less than the permissible velocity and shear stress for retardance class C vegetation).

In this case, the initial basin stage for each event is set equal to the tailwater elevation. Therefore,

Initial basin stage$_2$ = Tailwater$_2$ = 82.66 ft
Initial basin stage$_{10}$ = Tailwater$_{10}$ = 82.90 ft
Initial basin stage$_{100}$ = Tailwater$_{100}$ = 83.78 ft

Circular culvert design parameters (this outlet is actually an orifice, but by computing it as a culvert, the routing computations will be more accurate):

Invert Elevation = 82.0 ft
Diameter = 1.0 ft
K = 0.519
M = 0.64
C = 0.0289
Y = 0.9
Ks = −0.5
Ke = 0.5
Bottom Slope = 0.005 ft/ft
Manning n = 0.013
Length = 0.75 ft
TLIM1 = 3.5
TLIM2 = 4.0

Inlet box design parameters:

Rim Elevation = 85.90 ft
Opening Length = 3.0 ft
Opening Width = 1.5 ft
Weir Discharge Coefficient = 0.3
Orifice Discharge Coefficient = 0.6
Note: just as we iterated to get the tailwater elevations, we also must iterate to get the weir discharge coefficients. The final results are checked to make sure that all our coefficients are applicable and correct.

Weir design parameters:

Crest Elevation = 86.50 ft
Weir Length = 10 ft
Weir Discharge Coefficient = .32
Weir Side Slope = 1 (horizontal/vertical)
Slope Discharge Coefficient = .26
Note: we are designing a riprap weir. The best way to do this is to use a channel routing program to compute the water surface profile through the riprap, then back solve for the coefficients. The final results are checked to make sure that all our coefficients are applicable and correct.

TABLE 7.10 Example 7.10 BASINPOPT Simulation Mode Inputs (Finalizing the Design after Tweaking the Design Mode Results) (*Continued*)

BASINOPT input required	Values or data given to BASINOPT	
Stage-area data	Stage (ft)	Surface area (ac)
	82	0.002
	83	0.390
	84	0.690
	85	0.746
	86	0.803
	87	0.862
	88	0.922
	88.7	0.965

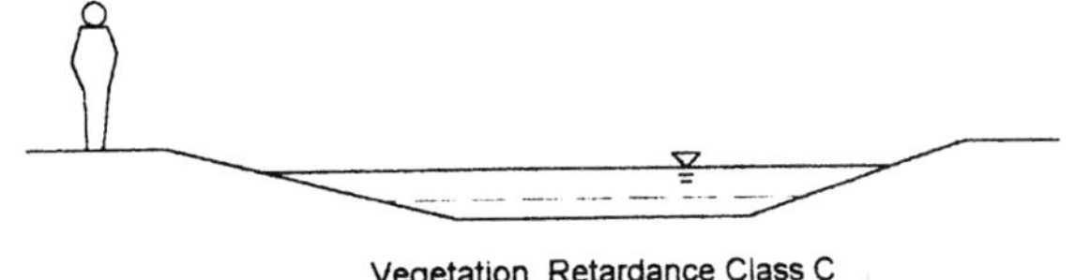

FLEXIBLE (HEC-15) ANALYSIS

INPUTS:

DISCHARGE	53.6 (cfs)
Bottom Slope (ft/ft)	.003
Left Side Slope (hor/vert)	4
Right Side Slope (hor/vert)	3
Bottom Width (ft)	12
Freeboard (ft)	1

LINING MATERIAL:

Vegetation, Retardance Class C

OUTPUTS:

MILD CHANNEL	NORMAL	CRITICAL
FLOW DEPTH	2.002 ft	.786 ft
Velocity (fps)	1.409	4.621
Flow Area (sf)	38.044	11.6
Hydraulic Radius (ft)	1.431	.654
Froude Number	0.2053	1.0000
Width at Depth (ft)	26.012	17.504
Width at Freeboard (ft)	33.012	
Computed Manning n	0.0734	

	COMPUTED	PERMISSIBLE
Velocity (fps)	1.409	2.733
Shear Stress (psf)	.375	1.0

FIGURE 7.27 100-year outfall channel computations.

SIMULATION MODE, 100-Yr MAXIMUM STAGE PLOT

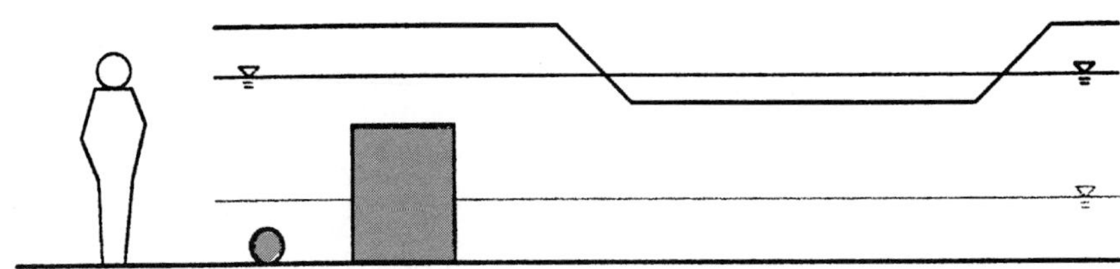

FIGURE 7.28a Final routing results.

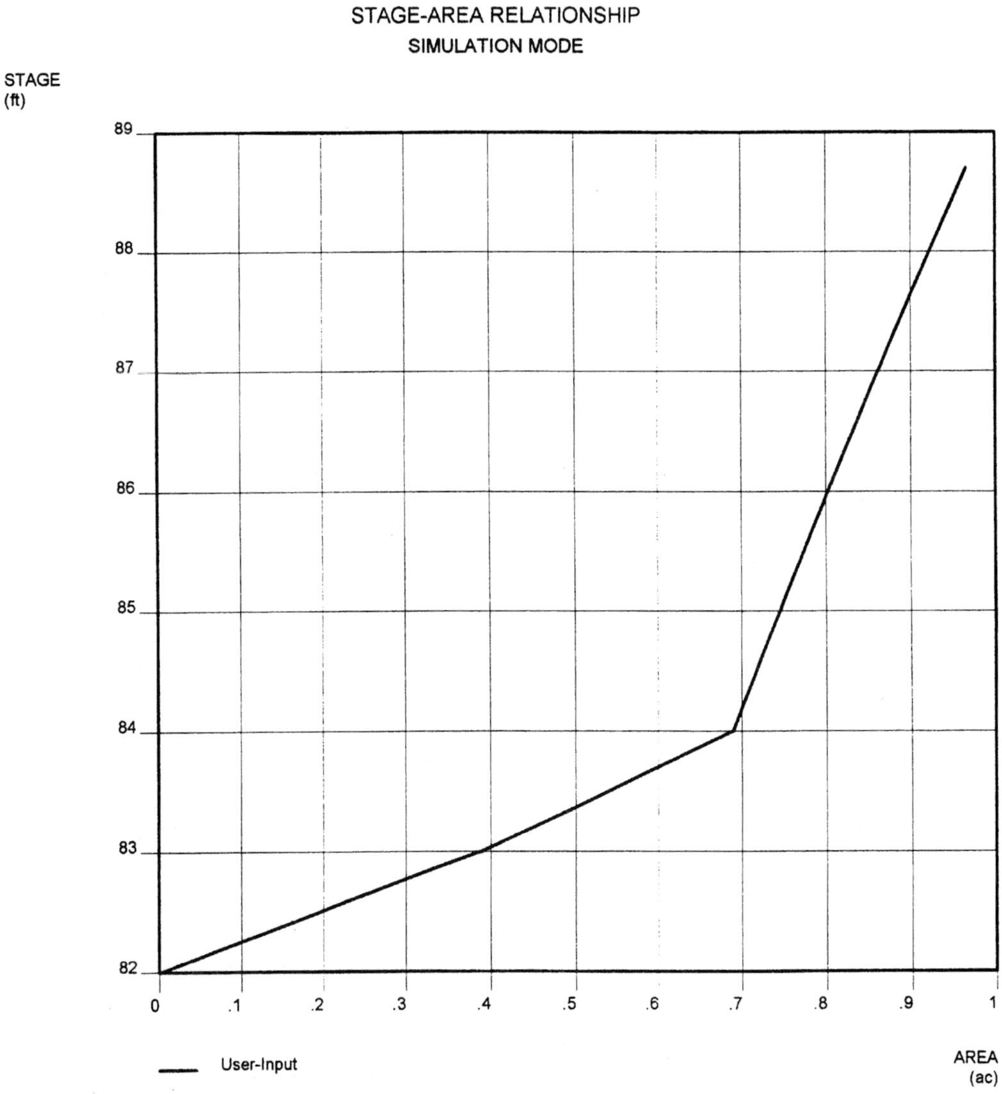

STAGE-AREA RELATIONSHIP
SIMULATION MODE
STAGE
(ft)
89
88
87
86
85
84
83
82
0
.1
.2
.3
.4
.5
.6
.7
.8
.9
1
User-Input
AREA
(ac)

FIGURE 7.28b Final routing results.

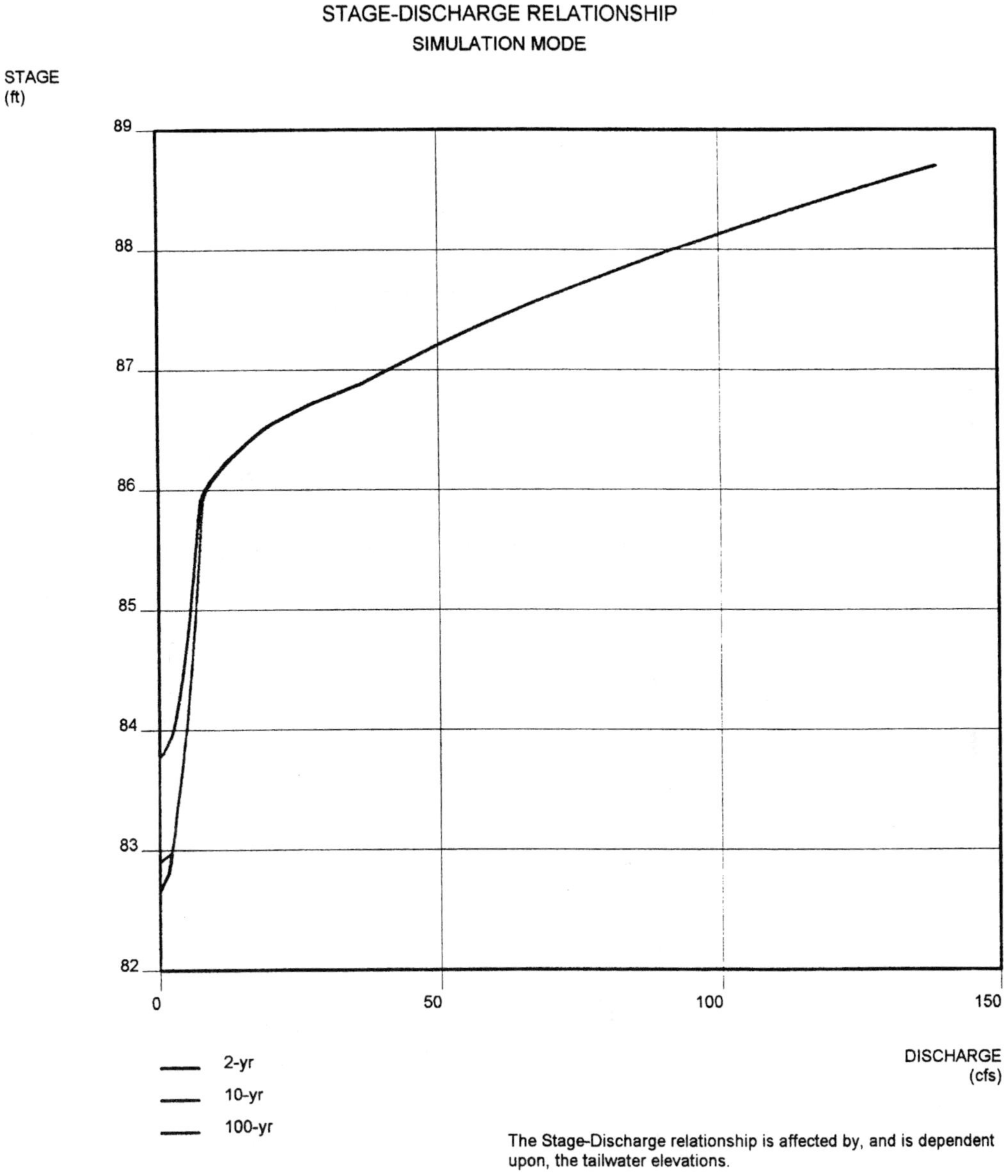

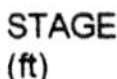

FIGURE 7.28c   Final routing results.

SUMMARY OF HYDRAULIC RESULTS
SIMULATION MODE

Basin VOLUME: **4.611 acre-feet**

Basin BOTTOM AREA: **.002 acres**

Basin TOP AREA: **.965 acres**

	STORM		
	2-yr	**10-yr**	**100-yr**
Peak INFLOW (cfs)	16.708	35.332	64.737
Peak OUTLOW (cfs)	5.578	11.670	53.622
Peak STORAGE VOLUME (ac-ft)	.987	2.372	3.308
Peak STAGE (ft)	84.361	86.180	87.285
Time of Peak INFLOW (min)	756.	756.	750.
Time of Peak OUTFLOW (min)	804.	804.	768.
Time of Peak STAGE (min)	804.	804.	768.
Continuity Error (%)	3.72E-05	5.27E-05	7.66E-05

FIGURE 7.28d   Final routing results.

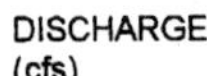

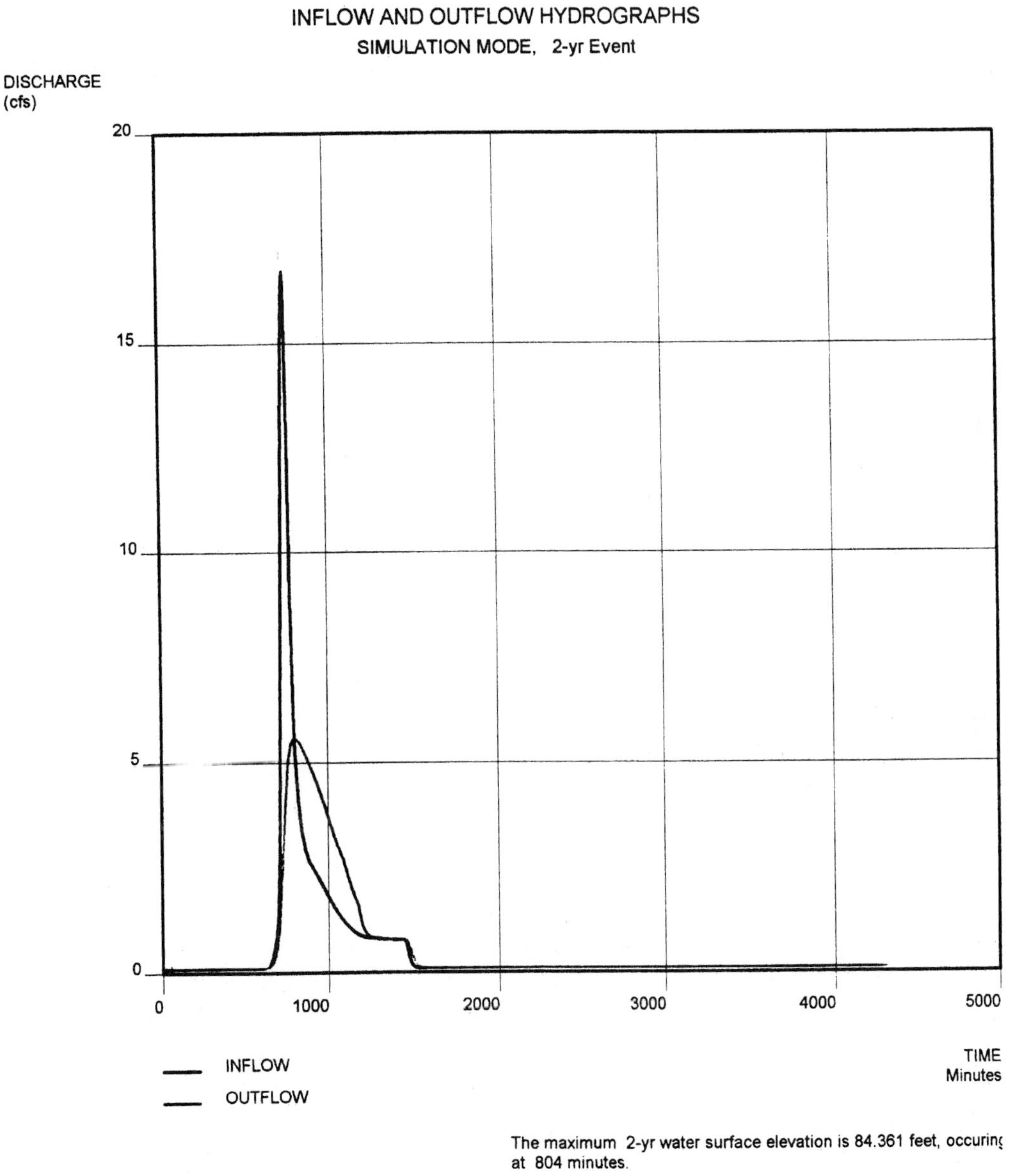

FIGURE 7.28e   Final routing results.

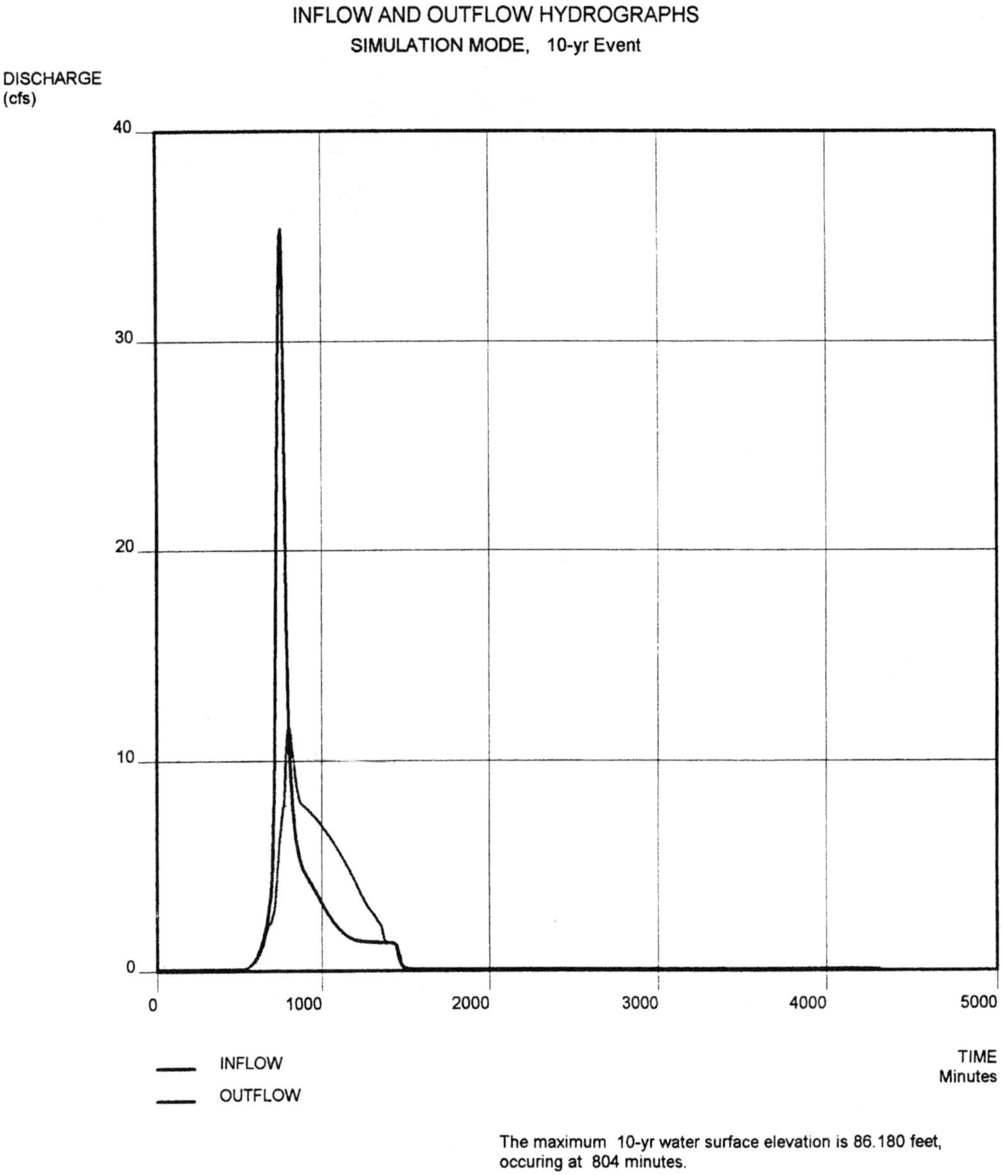

FIGURE 7.28f   Final routing results.

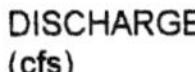

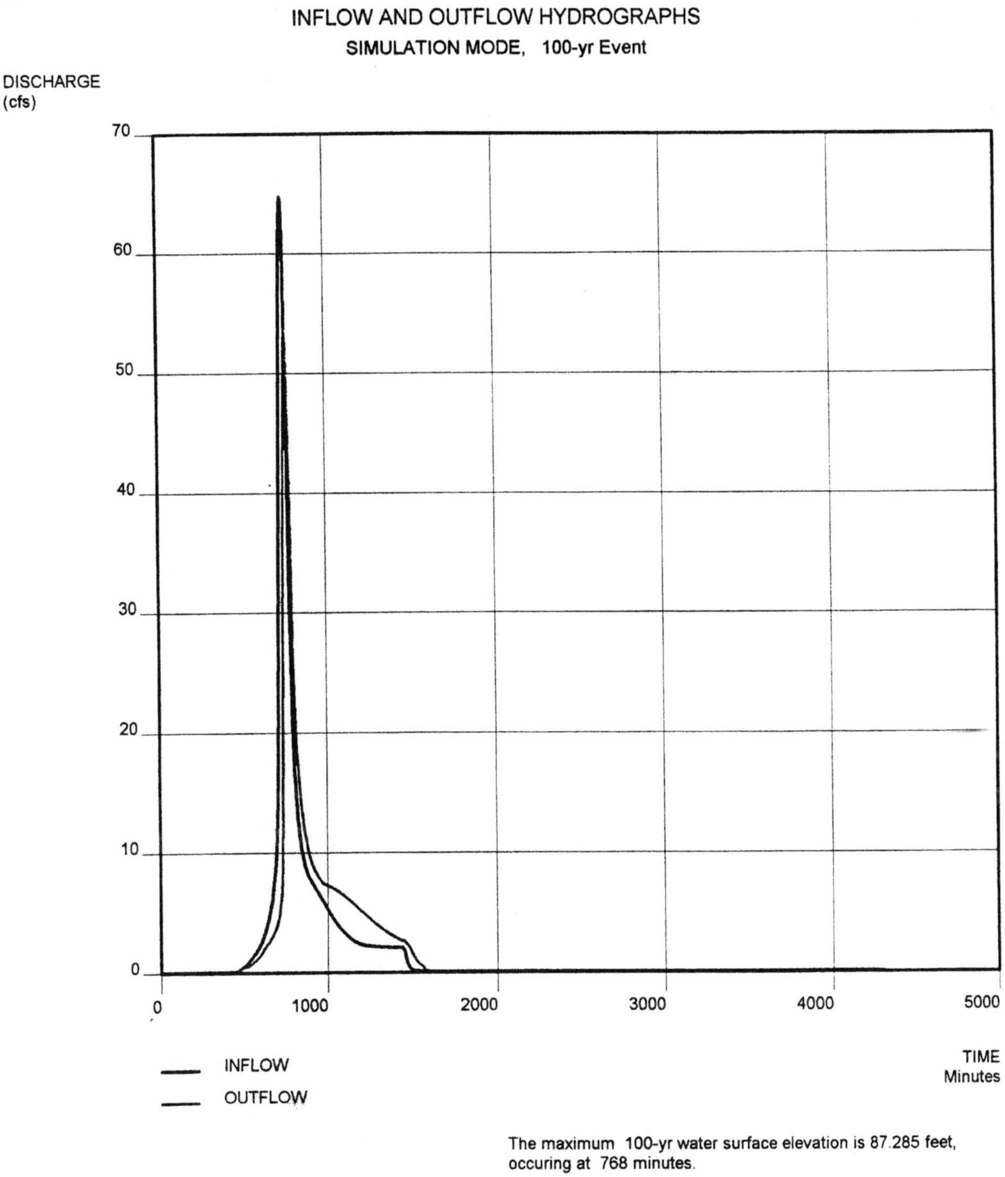

FIGURE 7.28g   Final routing results.

```
=========================    OUTLET STRUCTURE DATA    =============================


          STRUCTURE NUMBER    1  --------------------------------------------------


             Type 1: CIRCULAR CULVERT

                          DIAMETER :      1.00 feet.
                  INVERT ELEVATION :     82.00 feet.
                                K :      .51900
                                M :      .64000
                                c :      .02890
                                Y :      .90000
                   SLOPE COEF ks :     -.50000
               ENTRANCE LOSS Ke :      .50000
                     BARREL SLOPE :      .00500 feet horizontal per foot vertical.
                        MANNING n :      .01300
                  CULVERT LENGTH :        .75 feet.
              WEIR LIMIT, TLIM1 :    3.50000
              ORIF LIMIT, TLIM2 :    4.00000

          STRUCTURE NUMBER    2  --------------------------------------------------


             Type 5: INLET BOX

                           LENGTH :      3.00 feet.
                            WIDTH :      1.50 feet.
                  CREST ELEVATION :     85.90 feet.
                        WEIR COEF :       .300
                     ORIFICE COEF :       .600



             Note that the dimensions listed here are EFFECTIVE dimensions.
             If this structure will have a grate, make sure the grate has
             these effective dimensions.

             CAUTION: Verify that the outlet pipe from this structure
             is not submerged by tailwater.  The outlet pipe must flow
             partly full.

     (Continued)
          STRUCTURE NUMBER    3  --------------------------------------------------

             Type 3: WEIR

                           LENGTH :     10.00 feet.
                  CREST ELEVATION :     86.50 feet.
                   DISCHARGE COEF :       .320


             TRAPEZOIDAL WEIR
                       SIDE SLOPE :     1.000
              SLOPE DISCHARGE COEF :       .260
```

FIGURE 7.28h Final routing results.

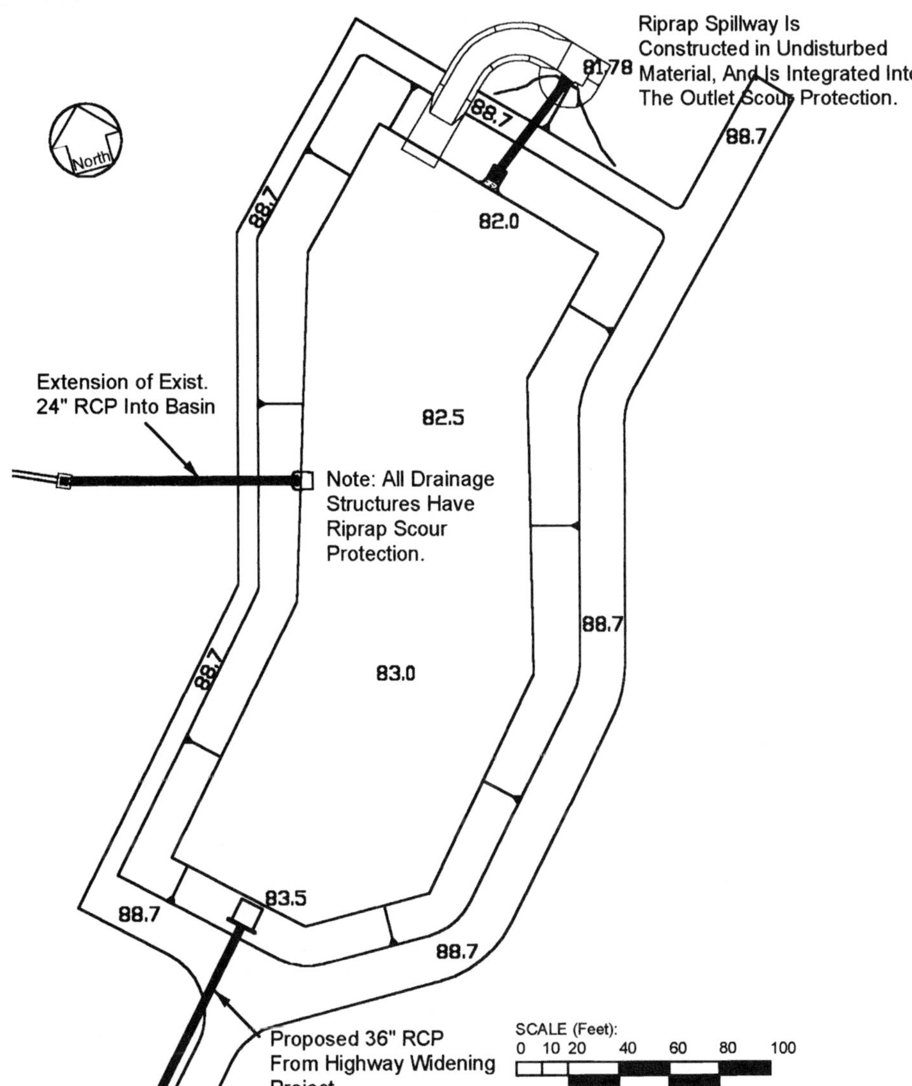

FIGURE 7.29   Completed final design.

box pass 9.56 and 25.49 cfs respectively. The combined 100-year discharge through the outlet barrel is therefore 35.05 cfs, which becomes 44 cfs after applying the 25 percent factor of safety. Using Manning's equation for just-full pipe flow, at 0.005 percent longitudinal slope, a 36-in pipe would have a just-full capacity of 47.16 cfs, with an exit velocity of 6.67 fps. In other words, a 36-in concrete pipe on a 0.005 ft/ft longitudinal slope will provide enough capacity that the outlet barrel will not choke the flow for the 100-year event. We design the outlet such that it takes whole lengths of pipe, to avoid any cutting—making the barrel 36 ft (six 6-ft sections) long in this case will result in a cleaner end condition, and is less bothersome to construct. The outlet barrel is also fitted with a flared end section. An

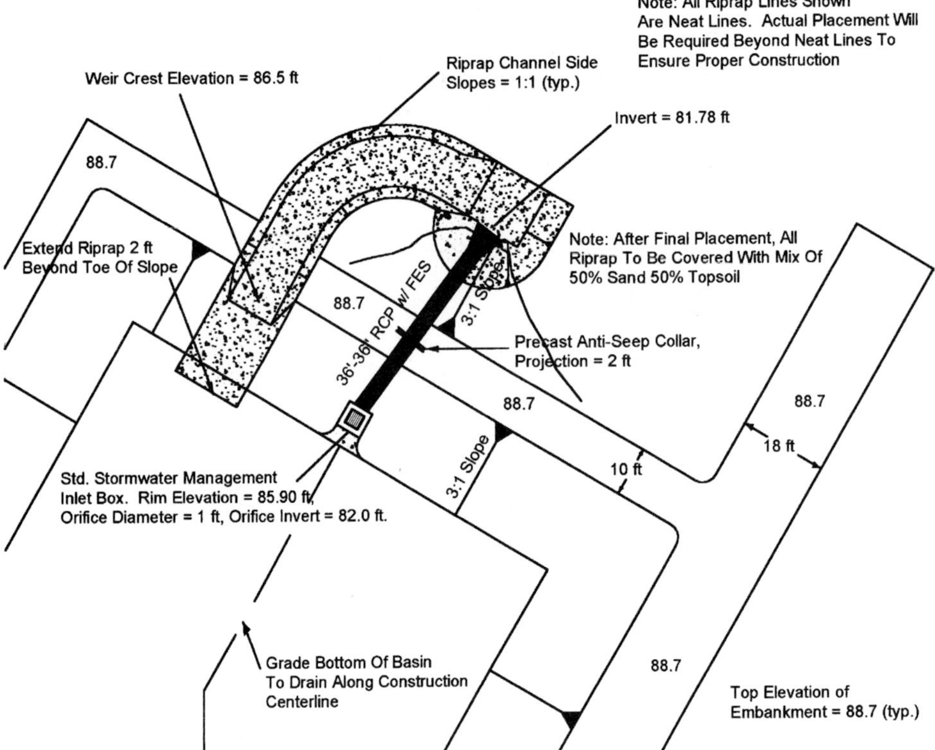

FIGURE 7.30   Outlet detail.

anti-seep collar can be designed using procedures described in Sec. 7.5.4.9 and Example 7.9. We will specify one pre-cast concrete anti-seep collar, placed under the center of the top of the embankment, having a 2-ft projection. The collar will have 1.46 ft of cover. We specify the pipe bedding as a "Class A Concrete Cradle," using the procedures in Sec. 7.5.4.11 and Fig. 7.20, 9 inches below the pipe. Why a concrete cradle? If gravel were used, we would be building a French drain right underneath the pipe, and this drain would promote a tremendous amount of free seepage—just what we are trying to prevent. A note on the drawing will specify that no gravel will be placed under the outlet barrel.

A geotechnical engineer is consulted to design the embankment. Her results are shown in Fig. 7.21. Likewise, she determines that the inlet box structure does not have to be constructed on piles. The highway department decides to fence the entire facility for safety reasons, given its proximity to an apartment complex. A vehicle entrance is designed at the main road, with a chain and lock to prevent access by unauthorized vehicles. An erosion control plan is also completed to minimize downstream water pollution during construction of the basin. Six inches of topsoil and seeding are specified for the dressed grading. Permanent right-of-way takes are added to the final site plan, along with construction easements. The final cost, including purchasing the required real estate, exceeds $490,000 for this facility.

So how good is our design? We have met all the design criteria, without over-designing the facility. If our computations were challenged, only two items would have any technical potential for discussion: the conservatism in our tailwater assumptions, and the assumption of constant discharge coefficients in the hydraulic routing. We have assumed that the tailwater elevation remains constantly fixed at the peak discharge normal elevation downstream, throughout the length of the simulation. This makes our computations somewhat conservative, but has very little appreciable effect on this design. Second, orifice discharge coefficients are relatively stable under a wide range of heads. The culvert coefficients we have used are specified as constants in reliable references, such as the U.S. Federal Highway Administration Culvert Design Manual. Our weir discharge coefficients do vary, but to check the effects of the variation, we would have to compute a stage-area-discharge relationship, using spreadsheet software and writing the structure discharge equations as a function of head, then perform the routing of the inflow hydrographs though the structures using equations presented earlier in this chapter. This procedure is tedious, and is not normally performed. The assumption to use discharge coefficients based on peak flow conditions is standard practice in this type of design.

And what about the cost to design this basin? A good design engineer and technician, having all field survey information on hand and having already completed the hydrologic computations could complete and check the design computations and prepare the design drawings for this facility in 28 to 40 labor hours.

REFERENCES

Abt, S. R., and N. S. Grigg, "An Approximate Method for Sizing Detention Reservoirs," *Water Resources Bulletin,* 14(4): 956–965, 1978.

Akan, A. O., "Detention Pond Sizing for Multiple Return Periods," *Journal of Hydraulic Engineering,* ASCE, 115 (5): 650–655, 1989.

Akan, A. O., "Single Outlet Pond Analysis and Design," *Journal of Irrigation and Drainage Engineering,* ASCE, 116, 1990.

American Society of Civil Engineers and Water Pollution Control Federation, *Design and Construction of Sanitary and Storm Sewers,* ASCE Manual on Engineering Practice No. 37 and WPCF Manual of Practice No. 9, published jointly by both societies, 1978.

American Society of Civil Engineers, *Urban Runoff Quality Management,* ASCE Manual and Report on Engineering Practice No. 87, also Water Environment Federation Manual of Practice No. 23, Alexandria, VA, 1998.

Aron, G., and D. F. Kibler, "Pond Sizing for Rational Formula Hydrographs," *Water Resources Bulletin,* 26(2): 255–258, 1990.

Baker, W. R., "Stormwater Detention Design for Small Drainage Areas," *Public Works,* 108(3): 75–79, 1979.

Chen, Y. H., and G. K. Cotton, *Design of Roadside Channels with Flexible Linings,* Hydraulic Engineering Circular No. 15, Federal Highway Administration Publication No. FHWA-IP-87-7, Federal Highway Administration, Office of Implementation, HRT-10, McLean, VA, 1988.

Chow, V. T., *Open-Channel Hydraulics.* McGraw-Hill, Inc., New York, NY, 1959.

Corry, M. L., P. L. Thompson, F. J. Watts, J. S. Jones, and D. L. Richards, *Hydraulic Design of Energy Dissipators for Culverts and Channels,* Hydraulic Engineering Circular No. 14, Federal Highway Administration, Office of Implementation, McLean, VA, 1983.

Currey, L., and A. O. Akan, "Single Outlet Detention Pond Design and Analysis Equation," *Water Resources Engineering 1998,* Vol. 1, pp. 796–801, ASCE, Reston, VA, 1998.

DiToro, D. M. and M. J. Small, "Stormwater Interception and Storage," *Journal of the Environmental Engineering Division,* ASCE, 105 (EE1), 1979.

Edmonds, J., Personal communication with the author, based on staff consultations at the Virginia Dept. of Conservation and Recreation, Division of Soil and Water Conservation, Richmond, VA, 1994.

Federal Highway Administration (FHWA), Urban Drainage Design Manual, Hydraulic Engineering Circular No. 22, FHWA-SA-96-078, Washington, DC, 1996.

Hampton Roads Planning District Commission, Best Management Practices Design Guidance Manual for Hampton Roads, 1991.

Hartigan, J. P., "Basis for Design of Wet Detention Basin BMP's," *Design of Urban Runoff Quality Controls,* ASCE, New York, NY, 1989.

Hydrologic Engineering Center, "HEC-1 Flood Hydrograph Package, User's Manual," US Army Corps of Engineers, Davis, CA, 1991.

Huber, W. C., and R. E. Dickinson, "Storm Water Management Model, Version 4: User's Manual," Department of Environmental Engineering Sciences, University of Florida, Gainesville, FLA, 1988.

Kessler, A., and M. H. Diskin, "The Efficiency Function of Detention Reservoirs in Urban Drainage Systems," *Water Resources Research,* 27(3): 253–258, 1991.

Loganathan, G. V., D. F. Kibler, and T. J. Grizzard, "Chapter 26—Urban Stormwater Management," in *Handbook of Water Resources Engineering,* L. W. Mays, ed., McGraw-Hill, New York, NY, 1996.

Montgomery County Department of Environmental Protection, Stormwater Management Pond Design Check List, Rockville, MD, 1984.

McEnroe, B. M., "Preliminary Sizing of Detention Resrvoirs to Reduce Peak Discharges," *Journal of Hydraulic Engineering,* ASCE, 118(11): 1540–1549, 1992.

Northern Virginia Planning District Commission, BMP Handbook for the Occoquan Watershed, Annandale, 1987.

Northern Virginia Planning District Commission, Evaluation of Regional BMP's in the Occoquan Watershed, Annandale, 1990.

Paine, J. N., "Effects of Hyetograph Shape on Detention Pond Sizing," *Proceedings of the International Conference on Channel Flow and Catchment Runoff: Centennial of Manning's Formula and Kuichling's Rational Formula,* Dept. of Civil Engineering, University of Virginia, Charlottesville, VA, pp. 36–44, 1989.

Prince George County Department of Environmental Resources, Stormwater Managememt Design Manual, Washington Suburban Sanitary Commission, 1984.

Red Valve Company, Inc., Tideflex© product literature, online at *www.redvalve.com,* 2000.

Schueler, T. B., Controlling Urban Runoff: A Practical Manual for Planning and Designing Urban BMP's, Washington Metropolitan Water Resources Planning Board, 1987.

Simon, A. L., *Practical Hydraulics,* John Wiley and Sons, New York, NY, 1981.

Soil Conservation Service, Standards and Specifications for Soil Erosion and Sediment Control in Developing Areas, U.S. Dept. of Agriculture, Soil Conservation Service, College Park, MD, 1975.

Soil Conservation Service, "Urban Hydrology for Small Watersheds," Technical Release 55, Washington, DC: U.S. Dept. Of Agriculture, 1986.

Stahre, P. and B. Urbonas, *Stormwater Detention for Drainage, Water Quality and CSO Management,* Prentice-Hall, Englewood Cliffs, NJ, 1990.

Urbonas, B. and P. Stahre, *Stormwater Best Management Practices Including Detention,* Prentice-Hall, Inc., Englewood Cliffs, NJ, 1993.

U.S. Environmental Protection Agency, Methodology for Analysis of Detention Basins for Control of Urban Runoff Quality, EPA-440/5-87-001, Washington, DC, 1986.

Virginia Dept. of Conservation and Recreation, *Virginia Erosion and Sediment Control Handbook,* 3rd ed., Virginia Dept. of Conservation and Recreation, Division of Soil and Water Conservation, 1992.

Virginia Dept. of Transportation, *Road and Bridge Standards,* Vol. I, Richmond, VA., pp. 116.01–166.03, 1996.

Walker, W. W., "Phosporus Removal by Urban Runoff Detention Basins," *Lake and Reservoir Management:* Vol. 3, North America Lake Management Society, Washington, DC, 1987.

Wycoff, R. L., and U. P. Singhm, "Preliminary Hydrologic Design of Small Flood Detention Reservoirs," Water Resources Bulletin, 12(2): 337–349, 1976.

Yu, S. L., and R. J. Kaighn Jr., "VDOT Manual of Practice for Planning Stormwater Management," Virginia Transportation Research Council, Charlottesville, VA, 1992.

CHAPTER 8
DESIGN OF OFF-LINE DETENTION SYSTEMS

James C. Y. Guo
Department of Civil Engineering
University of Colorado at Denver

The natural condition of a watershed is termed *undeveloped condition*. Natural streams, creeks, and waterways have been continuously shaped by storm runoff. Development of a watershed results in more paved areas which increase the frequency, magnitude, and volume of storm runoff. Man-made drainage facilities cause stormwater to move faster and to become more concentrated. Urbanization in a watershed is a process that leads to the *developed condition*. Development of a watershed takes several stages. At each stage, the watershed may have different flooding and stormwater drainage problems. To understand the current drainage problems in the watershed, the *existing condition* shall be studied. To mitigate a flooding potential or problem, various *scenarios* can be developed to evaluate the effectiveness and economics of possible remedial *alternatives*. The impact of a specific development or activity introduced to a watershed can be quantified by the comparison between the *pre-development* and *post-development* studies.

The increase of urban storm runoff is closely related to the areal ratio of imperviousness in a watershed (USEPA 1983). Table 8.1 presents a sensitivity study of the areal imperviousness on the 100-yr storm runoff. Let the undeveloped condition with a 5% areal imperviousness be the basis for comparison; it is found that the peak discharge increases 3.84 times and the runoff volume increases 1.61 times when the catchment imperviousness increases to 90%. This example indicates that urbanization process results in faster and more concentrated storm runoff flows.

8.1 CONCEPT OF FLOOD DETENTION

The concept of *stormwater detention* is to temporarily store the excess storm runoff for subsequent release at a rate not to exceed the capacities of the downstream existing drainage facilities. The required storage volume depends on the difference between the developed runoff rate and the allowable release rate. As shown in Fig. 8.1, the operation of a flood detention basin is divided into a *filling period* when the inflow is greater than the outflow, and a *depletion period* when the outflow rate is greater than the inflow rate. These two distinctive periods are separated by the peak outflow. The peak outflow occurs on the recession limb of the inflow hydrograph at the time when the outflow rate is equal to the inflow

TABLE 8.1   Impact of Watershed Imperviousness on Storm Runoff

Case number	Basin imperviousness percentage	Time to peak ratio to the undeveloped	Peak runoff ratio to the undeveloped	Runoff volume ratio to the undeveloped
1.00	5.00	1.00	1.00	1.00
2.00	10.00	0.91	1.05	1.03
3.00	30.00	0.79	1.52	1.19
4.00	50.00	0.74	2.37	1.32
5.00	70.00	0.70	3.19	1.48
6.00	90.00	0.66	3.84	1.61

rate. According to the continuity principle, the total stored (detention) volume during the filling period is equal to the total release volume during the depletion period.

A flood-control *detention basin* is designed to store the excess storm runoff associated with the increased watershed imperviousness. Between large storm events, a detention basin will remain dry. In order to provide a large storage volume, stormwater detention systems are often incorporated into floodplains, depressed areas, recreational parks, and sport fields. Although some detention facilities are merely designed and operated for the flood-control purpose, it is much more favorable, economically and socially, to provide multiple uses. A *regional detention basin* shall be placed at the best strategic site based on the regional well-being and benefits, and *local detention basins* are designed to control the local releases to collectively meet the regional flood control purposes. A stormwater storage facility placed close to the points of rainfall occurrence is termed *upstream detention*. Storage basins installed downstream of the watershed is called *downstream detention*. A wide floodplain in a flat reach of a waterway may provide *channel detention* that has a storage volume under a backwater wave form. To broaden the floodplain width and construct an embankment across a channel creates *on-line detention*. The storage of water in depressed open areas or closed conduits next to a stream is termed *off-line detention*. A *stormwater retention storage facility* is designed with an emphasis on stormwater quality control. A retention basin is designed to be a wet pond that has a permanent pool (Shueler and Helfrich, 1989). Through the extended release, the pollutant solids in stormwater may be reduced by the sedimentation process. Often a retention pool or wet land pool can be a part of the detention basin for stormwater quality enhancement.

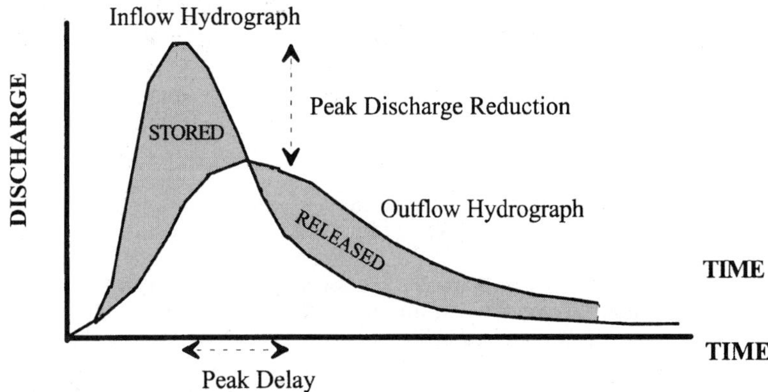

FIGURE 8.1   Concept of flood detention.

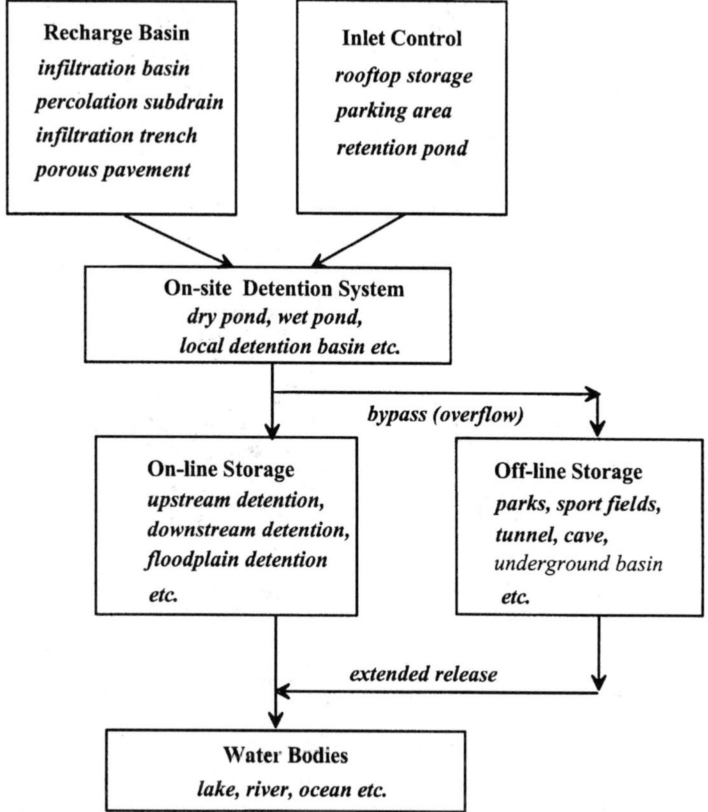

FIGURE 8.2   Detention basins in a waterway system.

Figure 8.2 illustrates how stormwater detention facilities are incorporated into a waterway system. Both recharge basins and inlet storage facilities shall be implemented at the source of runoff for water quality control purposes. An *on-site detention system* is often located at the outfall point of a major tributary watershed. *On-stream detention systems* are designed for the attenuation of peak flows in the waterway. When the flow in the waterway exceeds the maximum allowable flow rate, the excess stormwater shall be diverted into an off-stream storage facility. Figures 8.3 and 8.4 respectively show examples of channel detention on a floodplain and by backwater depth. Figure 8.5 shows detention system in a parking lot. Figure 8.6 shows infiltration and retention system.

8.2   ON-LINE AND OFF-LINE DETENTION BASINS

Utilizing the excess storage capacity available in a conveyance system is referred to as *on-line storage practices*. On the other hand, an off-line basin collects the diverted flow when the system is overloaded. For example, the concrete basin in Fig. 8.7 is connected to the combined sewer trunk line with an inlet having a greater hydraulic capacity than that of the outlet. The on-line basin in a sewer system may be designed to pass dry weather low flows without detention. But during a storm event, the volume difference between the inflow and

FIGURE 8.3 Example of channel detention on floodplain.

FIGURE 8.4 Example of channel detention by backwater depth.

FIGURE 8.5   Detention system in parking area.

FIGURE 8.6   Infiltration and retention system.

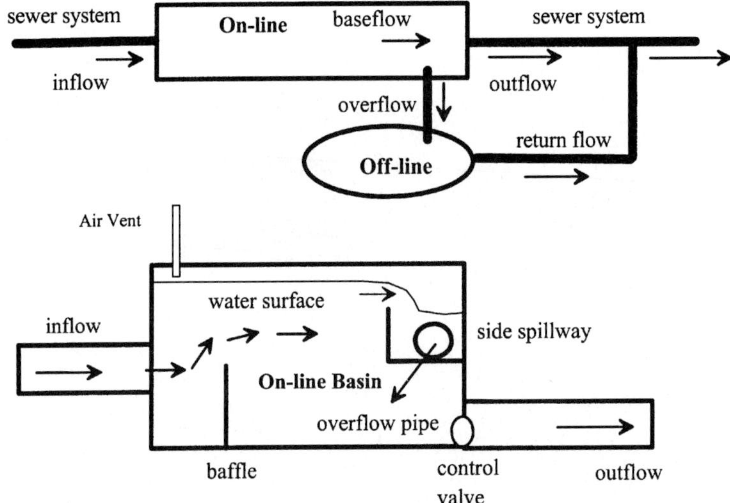

FIGURE 8.7   On-line and off-line storage basins in a sewer system.

the outflow will be detained in the basin. The inlet of an on-line basin shall be equipped with baffle systems to slow down the inflow jet and to spread sediment-laden water into the basin. The outlet of an on-line basin is controlled by a fixed orifice, adjusted sluice gate, pumps, or/and flow regulators. It is important to have adequate air ventilation for underground detention basins. The size of vent openings depends on the size of the basin and the exchange rate of air (Urbonas and Stahre, 1992).

Under an overflow condition, the on-line basin or the sewer trunk diverts the excess stormwater into off-line storage basins which provide temporary storage and releases the return flow back to the system over an extended period of time. The spillage arrangement for flow diversion can be controlled by electrical devices such as flow valves and pumps, or gravity-flow devices such as side spillway or overflow weir.

In an open system, an off-line basin is located near or adjacent to the waterway. Off-line basins can be designed to capture the first flush volume for stormwater quality enhancement. The runoff water captured is then diverted into vegetative beds, porous pavements, or infiltration/percolation basins for sediment treatment. Of course, off-line basins can also be designed as a flood control detention system which provides a large storage volume. Such off-line basins are usually placed in open spaces like parks or sport fields adjacent to the waterway. The flow diversion is designed to trigger water spillage when the flow in the

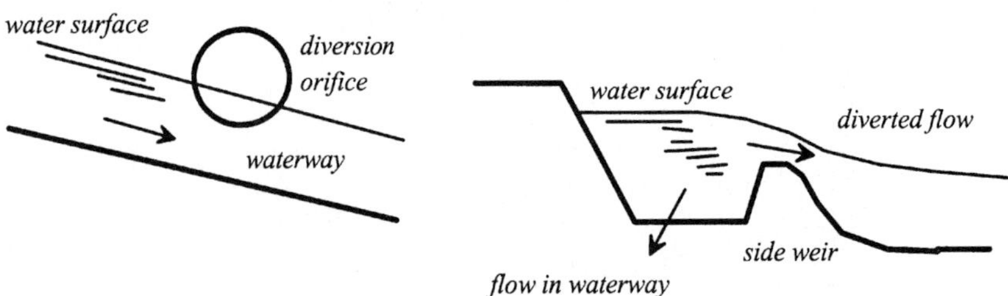

FIGURE 8.8   Examples of flow diversion in an open system.

FIGURE 8.9   Off-line detention basin with a side spillway for flow diversion.

FIGURE 8.10   Parking lot detention basin using side walk as overflow diversion.

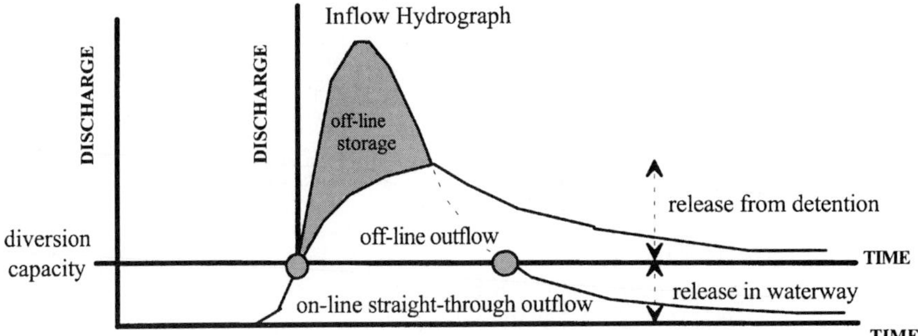

FIGURE 8.11 Inflow and outflow hydrographs for off-line detention.

waterway exceeds a pre-determined flow rate. There are many ways to design a flow diversion using orifice, weir, side spillway, sluice gate, or branch conduit (Metcalf and Eddy Inc., 1972; Bremen et al., 1989). Figure 8.8 presents examples of flow diversion by side weir and orifice and Fig. 8.9 shows an off-line detention basin with a side spillway for flow diversion. Figure 8.10 shows a parking lot detention basin using a sidewalk as the overflow diversion.

A diversion is designed to keep the allowable release through the waterway, and to spill the excess water into an off-line detention system. Figure 8.11 illustrates how the inflow hydrograph is divided into one straight-through flow in the waterway, and another diverted overflow through the off-line detention basin.

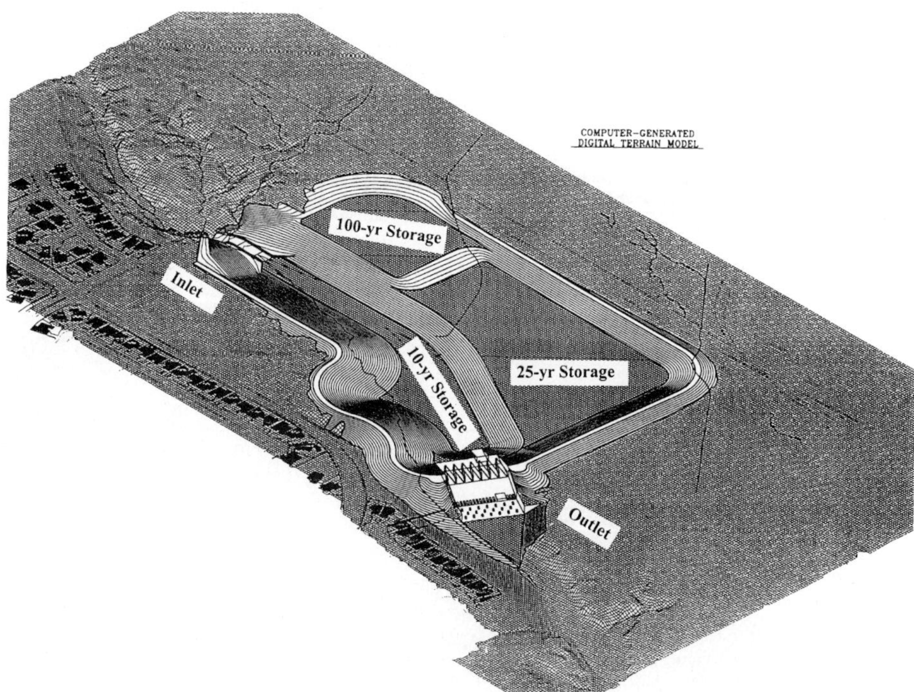

FIGURE 8.12 Lower detention basin in North Las Vegas, Nevada (courtesy of VTN-nevada Inc.).

Figure 8.12 is the Lower Detention Basin, which was constructed in North Las Vegas City, Nevada (VTN, 1995). The Lower Detention Basin is located on the Western Tributary Wash to the Las Vegas Wash. The storage capacity on the floodplain can only provide the 10-year detention volume. The 25-year and 100-year storage volumes are diverted into off-line storage basins.

In an urban area, the off-line stormwater detention practice is popular because of limited space and high cost land. This chapter focuses on the design of off-line stormwater detention basins in an open system.

8.3 *DESIGN CONSIDERATIONS*

The design of a detention system is an integration of functional integrity, land value, aesthetics, recreation, and safety into the urban environment. From the engineering aspect, the design of a stormwater detention system shall take the following factors into consideration.

8.3.1 Location

In an urban area, stormwater detention storage volumes can be provided by parking lots, parks, sport fields, road embankments, and depressed areas. The selection of a basin site depends on costs, public safety, and maintenance. The concept of multiple uses is an important practice to combine parks, ball fields, and green belts into the flood-control detention system. In a large detention basin, recreation activities may include jogging, walking, bicycling, playgrounds, fitness training, equestrian activities, skating, golfing, etc. Different specialists shall work together as a joint effort to develop desirable and acceptable criteria that fit the community recreational needs as well as the local flood-control purposes.

8.3.2 Basin Layout

An off-line detention basin has a *diversion structure* to control inflows, *energy dissipators* at the entrance for erosion control, a *trickle channel* to pass frequent nuisance flows, a *storage basin* for temporary stormwater detention, and an *outlet structure* to release the stored water. Because the waterway passes the base flow which tends to carry most sediment loads, the necessity of a sediment *forebay* at the entrance of an off-line detention basin depends on the amount of the diverted flow and sediment characteristics.

The basin length-to-width ratio must be greater than 2 so that the flood flows can sufficiently expand and diffuse into the water body to enhance the sedimentation process. At the entrance, proper energy dissipators shall be designed for erosion protection. A trickle channel or a low-flow channel shall be installed through the bottom of the basin to pass frequent nuisance flows. In general, the capacity of a trickle channel is 1.0 to 3.0% of the 100-year peak discharge, and the low-flow channel shall pass the 2-year event. Proper drop structures shall be placed along the trickle channel to control erosion. The trickle channel may drain into a permanent pool for stormwater quality control, or directly release low flows through the outlets (Pima County Drainage Criteria in 1989, Denver Stormwater Design Criteria Manual in 1985).

The geometry of the basin shall be designed for multiple events. As shown in Fig. 8.13, the lower storage volume in a basin is shaped by the frequent events such as the 5- or 10-year events. The 10-year water level to the weir crest shall provide an additional storage volume to accommodate the 100-year storage volume. The weir crest to the brimful of the basin is the height of the freeboard. Slopes on embankments have to maintain the bank slope

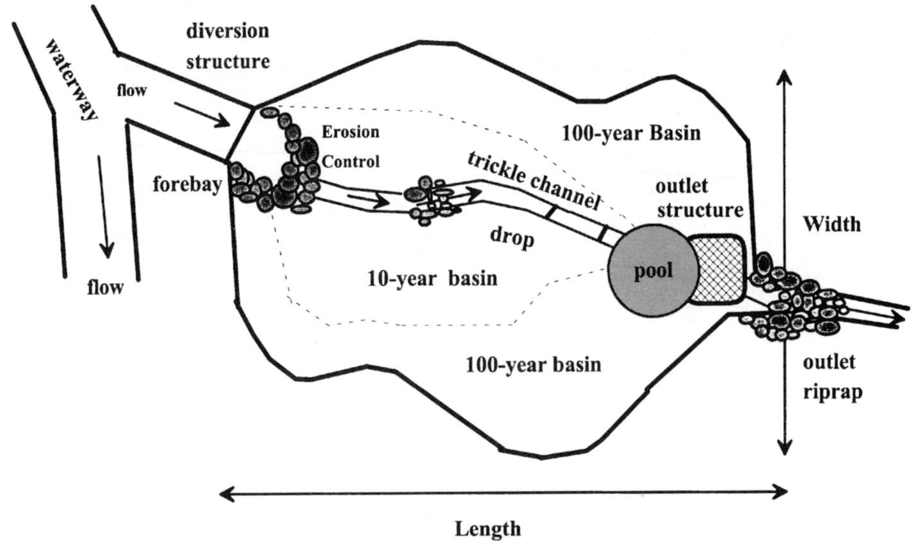

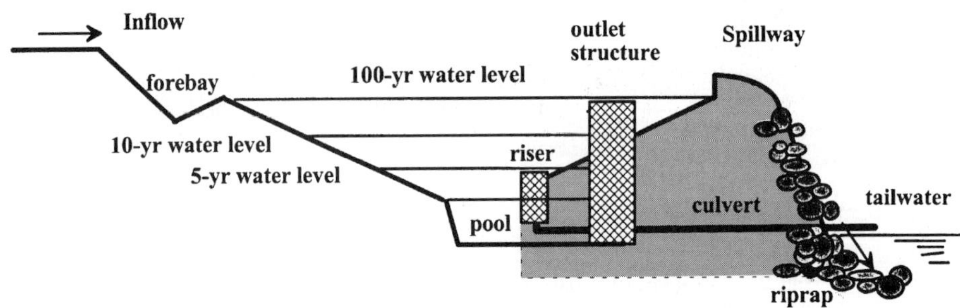

FIGURE 8.13 Geometry of detention basin shaped for multiple storm events.

stability. As a rule of thumb, slopes on earthen embankments shall not be steeper than $1V$:$4H$ and on riprap embankments shall not be steeper than $1V$:$3H$.

8.3.3 Allowable Release Rate

Nothing can be more important than public safety from flood hazards when determining the stormwater release from an urbanized catchment. As a rule of thumb, the stormwater release shall not worsen the flooding problems in the downstream properties. As a result, the following constraints may serve as guidelines to define the allowable release.

1. Pre-development peak discharge
2. Critical capacity of the downstream existing drainage facility

3. Local design criteria, such as Table 8.2 recommended for the Denver areas (Denver Stormwater Design Criteria Manual in 1985)

4. Regional empirical formulas such as those developed for the Denver areas (Urbonas and Glidden, 1983)

$$V_{100} = A(1.78 \, I - 0.002 \, I^2 - 3.56)/1000 \text{ for the 100-yr event} \tag{8.1}$$

$$V_{10} = A(0.95 \, I - 2.90)/1000 \qquad \text{for the 10-yr event} \tag{8.2}$$

where V = storage volume in acre-ft
A = drainage area in acres
I = basin imperviousness percentage

In practice, the recommended design criteria and the constraints in the existing drainage facilities must be considered, and the critical (smallest) one shall be adopted for design.

8.3.4 Groundwater Impacts

When the detention basin is to be operated like a dry pond such as a playground use, it is necessary to ensure that the average time interval between storms allows the basin to drain to a dry condition. On the contrary, if the detention basin is designed to be a wet pond, care must be taken in assessing infiltration to and exfiltration from the local ground water table. It is necessary to carefully evaluate the water budget among groundwater, surface water, and associated hydrologic losses. To design a detention basin without an outlet, soil infiltration tests must be carefully investigated. Although vertical drainage wells backfilled with aggregate gravel can be installed to increase the infiltration rate, the subsurface soil hydraulic conductivity must be carefully examined to make sure that the subsurface geometry sustains the infiltration rate on the land surface. Otherwise, the soil medium below the basin will become saturated and result in a backup to the flow system (Guo, 1998).

8.3.5 Inlet and Outlet Works

Inlets and outlets of a detention basin shall be protected from erosion and deposition of sediments. Design parameters shall be selected with the consideration of the effects of trash racks, inlet grates, back water surcharge, etc. Selections of orifice and weir coefficients must reflect the actual operations in the future. To design the outlet system for a basin, the effects of upstream and downstream submergence must be considered. The performance of the outfall culverts must be evaluated under various headwater depths at the entrance, and tailwater depths at the exit. Figure 8.14 illustrates the inlet and outlet of a small detention system.

TABLE 8.2 Allowable Release Rate Recommended for Denver Areas

Design event	Basin soil type A	Basin soil type B	Basin soil type C and D
10-yr	0.13 cfs/acre	0.23 cfs/acre	0.30 cfs/acre
100-yr	0.50 cfs/acre	0.85 cfs/acre	1.00 cfs/acre

FIGURE 8.14   Example of inlet and outlet for detention system.

8.3.6 Others

The successful implementation of a stormwater detention system also involves many institutional issues including the infrastructure needed to ensure proper planning, design, construction, operation, and maintenance. A monitoring or regulatory mechanism is required to ensure that the approved design is constructed, the operational integrity is implemented, and the maintenance is regularly provided. Other considerations also include public safety, access facilities, landscaping, and aesthetics.

8.4 DESIGN PROCEDURE

Stormwater detention is an effective means to reduce and delay flood peak flows. A detention basin shall be designed to accommodate multiple events with multiple objectives in land use and water re-use. Design of a detention basin begins with the watershed studies for the undeveloped, pre-development, and post-development conditions. The pre-development study is an attempt to identify the potential and existing flooding problems. It will serve as a basis for impact evaluations and alternative selections. Figure 8.15 outlines the design steps, beginning with the site selection for a detention basin. During the stage of preliminary design, the basin's stage-storage-outflow curve can be approximated by the inlet control capacity applied to weirs and orifices. During the final design, the preliminary design shall be refined with more information. It is an iterative procedure to take the downstream tailwater conditions and safety concerns into consideration. The final stage-storage-outflow curve must be revised until all design constraints are satisfied.

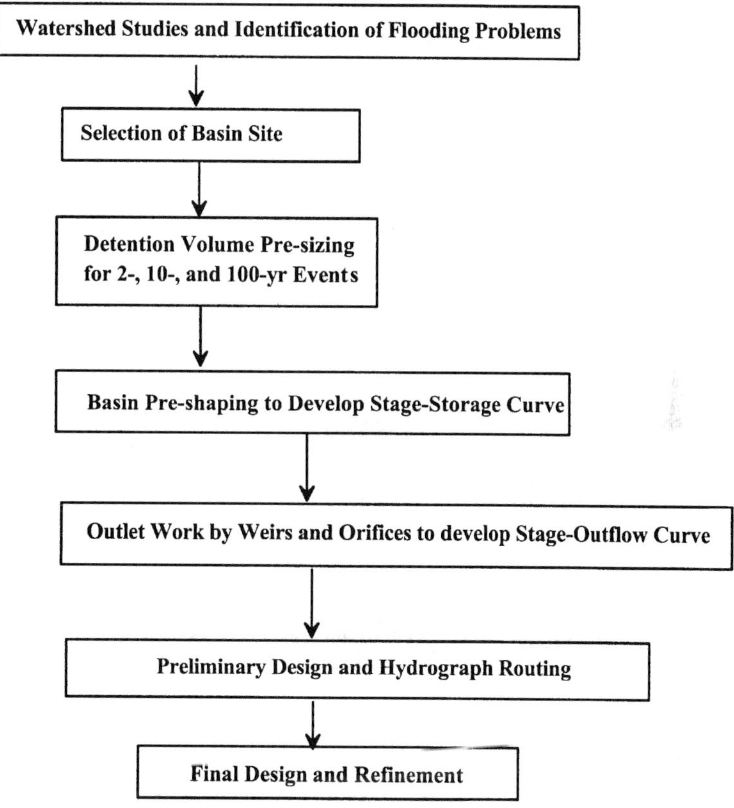

FIGURE 8.15   Procedure for detention basin design.

8.5   DETENTION VOLUME BY HYDROGRAPH METHOD

The inflow to an off-line detention basin is prescribed by the inflow hydrograph of the diverted flow as shown in Fig. 8.11. After knowing the peak inflow to the detention basin, the allowable release from the detention basin has to be determined based on the safety criteria. The detention storage volume is then determined by the volume difference between the inflow and outflow hydrographs from the beginning of the event to the time when the allowable release occurs. At the planning or preliminary design stage, sizing a small detention basin is more a hydrologic problem than a hydraulic problem. Without the prior knowledge of outlet hydraulics, the estimation of the storage volume can be achieved by assuming that the rising limb of the outflow hydrograph is linear (Malcom, 1982; Guo, 1999b). As shown in Fig. 8.16, the outflow rate, $O(t)$, at time t on the linear rising limb is estimated as

$$O(t) = \frac{O_a}{T_p} t \quad \text{for} \quad 0 \le t \le T_p \tag{8.3}$$

where $O(t)$ = outflow rate
$\quad O_a$ = allowable release
$\quad T_p$ = time to peak on outflow hydrograph
$\quad t$ = elapsed time

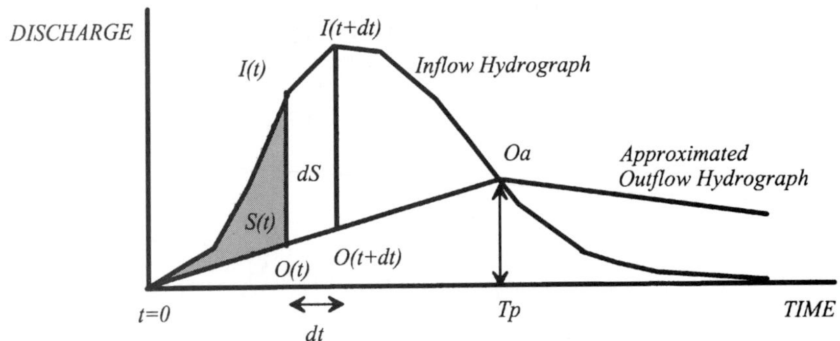

FIGURE 8.16   Detention volume by hydrograph method.

The accumulated storage volume, $V(t)$, is the volume difference between the inflow hydrograph and the rising limb of the outflow hydrograph as

$$V(t) = \sum_{t=0}^{t=Tp} (I(t) - O(t)\Delta t \quad \text{for} \quad 0 \leq t \leq T_p \tag{8.4}$$

where $I(t)$ = inflow rate at time t,
  Δt = time increment, such as 5 minutes

The required storage volume can then approximated as

$$V_m = V(T_p) \tag{8.5}$$

where V_m = detention storage volume

The hydraulic performance of a storage facility is described by its *storage-outflow curve*. For instance, HEC-1, SWMM, and TR-20 computer models define the performance of a detention basin by the prior knowledge of such a curve. During the alternative study, design information is not adequately available, the pairs (V, O) in Eqs. (8.3) and (8.4) can serve as the preliminary *storage-outflow curve* for the basin under design. Of course, such a preliminary relationship shall be refined by more detailed design information and confirmed by hydrograph routing techniques as well.

Example 8.1
As shown in Table 8.3, the diverted inflow hydrograph has a peak flow of 1200 cfs. The peak release rate from the detention basin is not to exceed 500 cfs. Under the assumption of a linear rising outflow hydrograph, Eq. (8.3) becomes

$$O(t) = \frac{500}{45} t \tag{8.6}$$

With a time increment of 5 minutes, the accumulative storage, $V(t)$ is computed by

$$V(t) = \sum_{t=0}^{t=45} [I(t) - \frac{500}{45} t] * (5 * 60)/43560 \tag{8.7}$$

As shown in Table 8.3, the accumulated storage volume $V(T_p = 45 = 22.52$ acre-ft. The pairs of $O(t)$ and $V(t)$ in Table 8.3 can be employed as the preliminary storage-outflow curve for the basin.

TABLE 8.3 Preliminary Storage-Outflow Curve by Hydrograph Method

Time minutes t	Inflow cfs $I(t)$	Outflow cfs $O(t)$ Eq 8.6	Acre-ft $V(t)$ Eq 8.7
0.00	0.00	0.00	0.00
5.00	120.00	55.56	0.44
10.00	250.00	111.11	1.40
15.00	400.00	166.67	3.01
20.00	750.00	222.22	6.64
25.00	1,200.00	277.78	12.99
30.00	1,000.00	333.33	17.58
35.00	850.00	388.89	20.76
40.00	700.00	444.44	22.52
45.00	500.00	500.00	22.52

8.6 DETENTION VOLUME BY VOLUME METHOD

To model a small watershed less than 150 acres, the assumption of uniform rainfall is acceptable for runoff volume predictions. Therefore, the required detention storage volume for a small watershed can be estimated directly by the volume difference between the inflow and outflow volumes through a detention basin. For a small urban watershed, the Rational method is applicable. For simplicity, a trapezoidal inflow hydrograph is considered.

As shown in Fig. 8.17, the inflow hydrograph has a linear rising limb over the time of concentration of the tributary watershed, and the peaking portion of the inflow hydrograph is a plateau from the time of concentration, T_c, to the end of the rainfall event. Runoff water is diverted into the off-line basin at a pre-set flow rate, Q_1. Let Q_2 be the allowable release

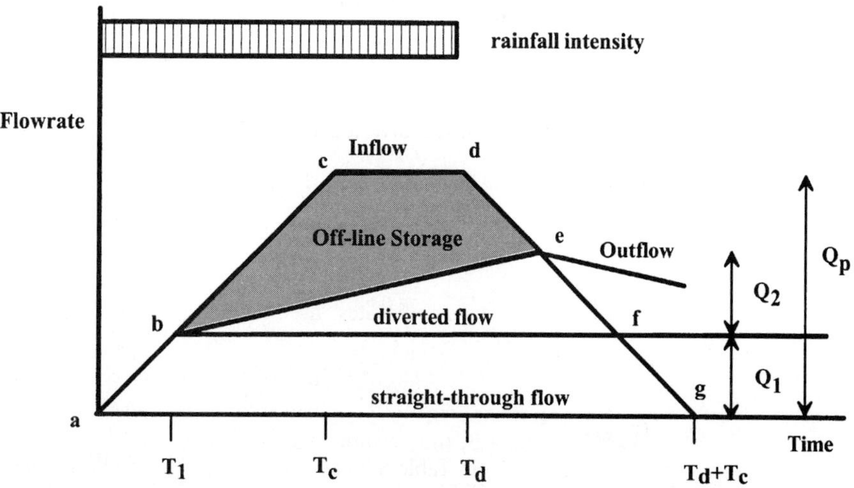

FIGURE 8.17 Detention volume by volume method.

from the detention basin. The outflow volume can be calculated by the area of **abeg** in Fig. 8.17 as

$$V_o = \frac{Q_2}{2}(T_d + T_c - 2T_1) + Q_1(T_d + T_c - T_1) \tag{8.8}$$

where V_o = outflow volume
$\quad Q_2$ = peak outflow from detention basin
$\quad Q_1$ = release rate from the tributary watershed
$\quad T_c$ = time of concentration of the tributary watershed
$\quad T_d$ = rainfall duration
$\quad T_1$ = diversion time.

As a linear approach, the diversion time can be approximated by

$$T_1 = \frac{Q_1}{Q_P} T_c \tag{8.9}$$

As recommended (FAA in 1970, U.S. Army and Air Force in 1977), the detention volume can be calculated using the rainfall duration as the base time. Therefore, the detention volume is

$$V_d = \alpha C I_d A T_d - \bar{Q} T_d \tag{8.10}$$

where C = runoff coefficient in Table 8.4
$\quad I_d$ = rainfall intensity
$\quad A$ = area of watershed
$\quad \alpha$ = unit conversion factor
$\quad \bar{Q}$ = average outflow rate

TABLE 8.4 Runoff Coefficients Recommended for Urban Land Uses

Land use	Impervious percentage	Runoff C 2-yr	Runoff C 5-yr	Runoff C 10-yr	Runoff C 100-yr
Business/Commercial	95.00	0.87	0.87	0.88	0.89
Business/Neighborhood	70.00	0.60	0.65	0.70	0.80
Single-familiy Residential	50.00	0.40	0.45	0.50	0.60
Multi-unit(detached)	50.00	0.45	0.50	0.60	0.70
Multi-unit(attached)	70.00	0.60	0.65	0.70	0.80
1/2 acre lot or larger	45.00	0.30	0.35	0.40	0.60
Apartments	70.00	0.65	0.70	0.70	0.80
Light Industrial	80.00	0.71	0.72	0.76	0.82
Heavy Industrial	90.00	0.80	0.80	0.85	0.90
Parks/Cemeteries	7.00	0.10	0.10	0.35	0.60
Playground	13.00	0.45	0.25	0.35	0.65
Schools	50.00	0.45	0.50	0.60	0.70
Railroad Yard Area	20.00	0.40	0.45	0.50	0.60
Paved Streets	100.00	0.87	0.88	0.90	0.93
Gravel Streets	40.00	0.15	0.25	0.35	0.65
Drive and Walks	96.00	0.87	0.87	0.88	0.89
Roofs	90.00	0.80	0.85	0.90	0.90
Lawns, sandy soil	2.00	0.00	0.01	0.05	0.20
Lawns, clay soil	2.00	0.05	0.10	0.20	0.40

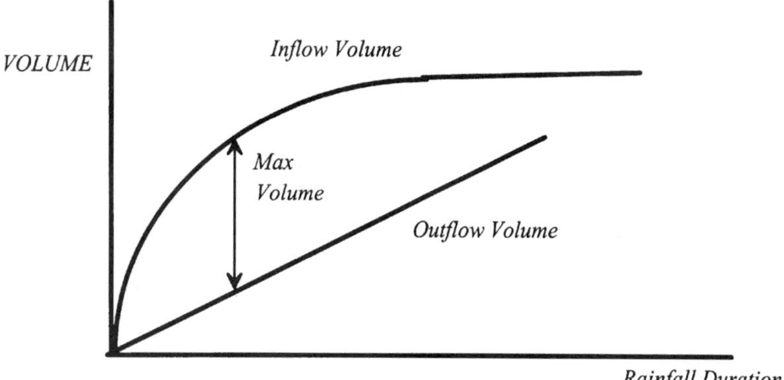

FIGURE 8.18   Maximization of detention storage by volume method.

Care must be taken on the consistency of units when using Eq. (8.10). The average release can be related to the allowable or peak outflow rate from the basin as

$$\bar{Q} = mQ_2 \tag{8.11}$$

Aided by Eq. (8.8), the value of m for an off-line basin is

$$m = \frac{1}{2}\left[1 + \frac{(T_c - 2T_1)}{T_d}\right] - \frac{Q_1}{Q_2}\left(1 + \frac{T_c - T_1}{T_d}\right) \quad \text{for } T_d \geq T_c \tag{8.12}$$

For an on-line detention basin, the pre-set diversion is zero, i.e., $Q_1 = 0$ and $T_1 = 0$. As a result, Eq. (8.12) is reduced to

$$m = \frac{1}{2}\left(1 + \frac{T_c}{T_d}\right) \quad \text{for } T_d \geq T_c \tag{8.13}$$

The basic concept used in the volume method is to find the maximum volume difference between the inflow and outflow volumes for a range of storm events in terms of rainfall duration. The design detention storage, V_m, is set to be the maximal volume difference as

$$V_m = \max(\alpha CI_d AT_d - mQ_2 T_d) \quad \text{for } T_d \geq T_c \tag{8.14}$$

As illustrated in Fig. 8.18, the method starts with the 5-minute event and then uses an increment of 5-minute for storm duration to compute the inflow and outflow volumes and their differences until the maximum storage volume is identified.

Example 8.2
A watershed is located upstream of Interstate Highway 25 in Denver, Colorado. The watershed has a drainage area of 100 acres and a runoff coefficient of 0.68. The time of concentration of the watershed is 25 minutes. The outfall system for the watershed includes two culverts under the highway. Culvert A can pass a peak flow of 40 cfs, and Culvert B can pass a peak flow of 30 cfs. Denver's IDF curve includes $a = 74.1$, $b = 10.0$, and $c = 0.786$ for the 100-year event. Determine the off-line detention storage volume.

1. Check the allowable Release Rate
   As recommended by Table 8.2 for the 100-yr event, the allowable release rate, Q_a, is 0.85 cfs/acre for type-B soils. As a result, the allowable release rate, Q_a, is:

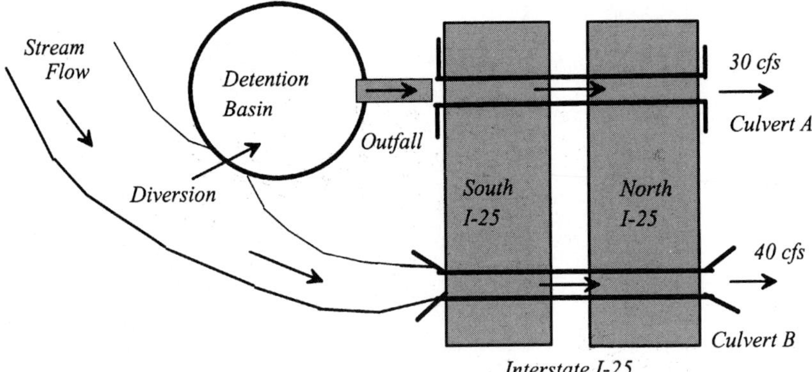

FIGURE 8.19   Outfall system for example watershed.

$$Q_a = 0.85 \ \text{cfs/acre} \ (100.0) \ \text{acre} = 85.0 \ \text{cfs}$$

For this example, both culverts will jointly release 70 cfs, which is less than the allowable release. Beginning with $T_d = T_c$, i.e., 25 minutes, the calculation steps are summarized as follows:

2. Inflow volume produced by the 100-yr 25-minute event

$$I_d = \frac{74.1}{(10 + T_d)^{0.786}} = \frac{74.1}{(10 + 25)^{0.786}} = 4.53 \quad \text{in/hr}$$

$$Q_p = CI_dA = 0.68(4.53)(100) = 308.1 \ \text{cfs}$$

$$V_i = I_dAT_d = \frac{60}{43560}(0.68)(4.53)(100)(25) = 10.70 \quad \text{acre ft}$$

3. Diversion time, T_1

$$T_1 = \frac{Q_1}{Q_p}T_c = \frac{40}{308.1}(25.0) = 3.25 \ \text{minutes}$$

4. Outflow volume

$$m = 0.5 * \left(1 + \frac{25 - (2.0)(3.25)}{25}\right) + \frac{40.0}{30.0}\left(1 + \frac{25.0 - 3.25}{25.0}\right) = 3.36$$

$$\bar{Q} = mQ_2 = 3.36(30.0) = 100.91 \ \textit{cfs}$$

$$V_o = m \, \bar{Q} \, T_d = \frac{3.36(100.91)(25.0)(60.0)}{43560.0} = 3.48 \ \text{acre-ft}$$

5. Stormwater storage volume, V_d, for the 25-minute rain storm

$$V_d = V_i - V_o = 10.70 - 3.48 = 7.22 \ \text{acre-ft}$$

Repeating this process for the range of rainfall duration as shown in Table 8.5 and Fig. 8.20, the maximum storage volume for this example is identified to be 9.04 acre-ft under the event with a duration of 70.0 minutes.

TABLE 8.5 Calculations of Detention Volumes for Example Watershed in Denver, CO.

Duration minutes	Rainfall intensity inch/hr	Inflow volume acre-ft	Peak runoff cfs	Diversion time T1 minutes	Average coeff m	Average outflow cfs	Outflow volume acre-ft	Storage volume acre-ft
25.00	4.53	10.70	308.10	3.25	3.36	100.91	3.48	7.22
30.00	4.08	11.56	277.40	3.61	3.08	92.42	3.82	7.74
35.00	3.72	12.29	252.87	3.96	2.88	86.38	4.16	8.13
40.00	3.42	12.93	232.78	4.30	2.73	81.86	4.51	8.42
45.00	3.18	13.50	215.97	4.63	2.61	78.35	4.86	8.64
50.00	2.97	14.01	201.70	4.96	2.52	75.56	5.20	8.80
55.00	2.79	14.47	189.40	5.28	2.44	73.28	5.55	8.92
60.00	2.63	14.89	178.68	5.60	2.38	71.39	5.90	8.99
65.00	2.49	15.28	169.25	5.91	2.33	69.79	6.25	9.03
70.00	2.37	15.64	160.88	6.22	2.28	68.43	6.60	**9.04**
75.00	2.26	15.98	153.39	6.52	2.24	67.25	6.95	9.03
80.00	2.16	16.30	146.65	6.82	2.21	66.22	7.30	9.00
85.00	2.07	16.59	140.55	7.12	2.18	65.32	7.65	8.95
90.00	1.99	16.88	135.00	7.41	2.15	64.52	8.00	8.88
95.00	1.91	17.14	129.92	7.70	2.13	63.80	8.35	8.79
100.00	1.84	17.40	125.25	7.98	2.11	63.16	8.70	8.70

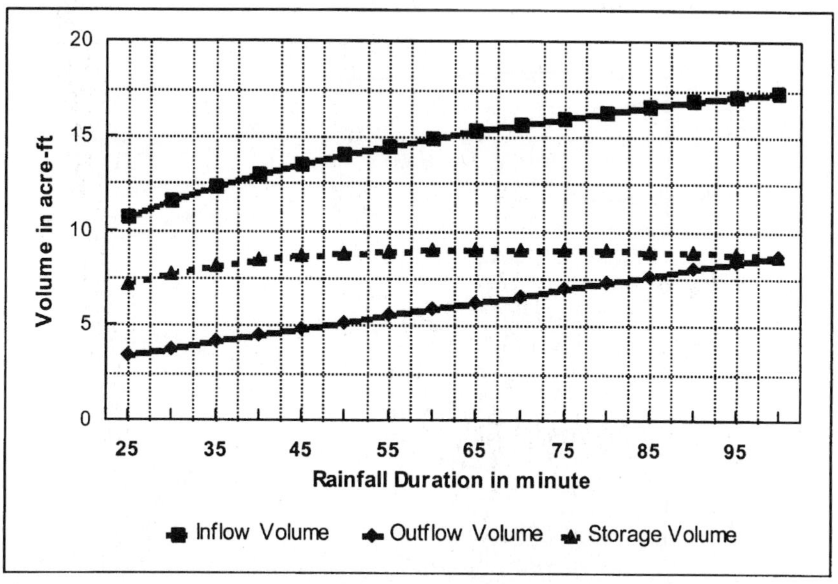

FIGURE 8.20 Maximization of detention volume between inflow and outflow volumes.

8.7 PRELIMINARY SHAPING

Under the concept of multiple events, Eq. (8.4) for a large watershed or Eq. (8.10) for a small watershed shall be repeatedly applied to 2-, 10-, and 100-year events. The detention basin is then shaped by multiple layers to accommodate the 2-, 10-, and 100-year storage volumes. During the feasibility study or preliminary design, neither the grading plan at the basin site nor the details of outlet structure is available. As a result, it is suggested that the preliminary shaping and storage sizing of a detention basin be approximated by a regular geometry such as ellipse or rectangle.

For instance, an inverted circular or elliptical cone, as shown in Fig. 8.21, truncated at both the top and the bottom can be adopted to estimate the shape and storage volume for the basin under design. The bottom area and side slope are required geometric parameters. Between layers, the volume is calculated as

$$L_2 = L_1 + 2zH \tag{8.15}$$

$$B_2 = B_1 + 2zH \tag{8.16}$$

$$A_2 = B_2L_2 \qquad \text{for a rectangular shape} \tag{8.17a}$$

$$A_2 = 0.5B_2L_2 \qquad \text{for a triangular shape} \tag{8.17b}$$

$$A_2 = \frac{\pi B_2 L_2}{4} \qquad \text{for an elliptical shape} \tag{8.17c}$$

$$V = \tfrac{1}{3}(A_1 + A_2 + \sqrt{A_1 A_2})H \tag{8.18}$$

$$\text{or} \quad V \cong 0.5(A_1 + A_2)H \tag{8.19}$$

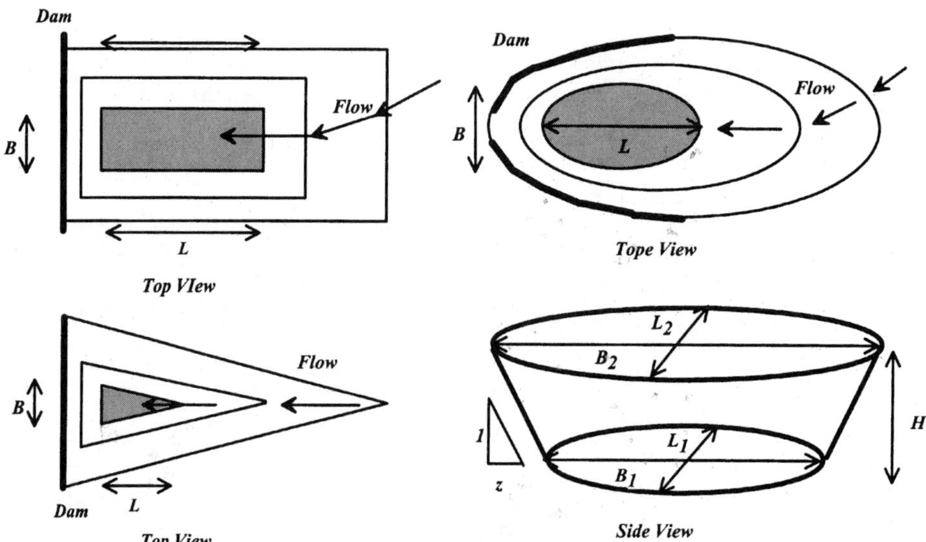

FIGURE 8.21 Pre-shaping scheme for detention basin in preliminary design.

TABLE 8.6 Preliminary Stage-Storage Curve for Example 8.3

Stage ft	Side slope ft/ft	Diameter ft	Area acre	Storage acre-ft
0.00	2.00	333.05	2.00	0.00
1.00	2.00	337.05	2.05	2.02
2.00	4.00	345.05	2.15	4.12
4.00	4.00	361.05	2.35	8.62
5.00	4.00	369.05	2.46	11.02
9.00	10.00	449.05	3.64	23.20
10.00	10.00	469.05	3.97	27.01

where D = width of cross-sectional area
$\quad L$ = length of cross sectional area
$\quad A$ = cross-sectional area
$\quad H$ = vertical distance between two cross sections
$\quad z$ = average side slope of the detention basin
$\quad V$ = volume

subscript 1 represents variables on the bottom area, and 2 represents the variables on the top area.

Applying Eqs. (8.14) through (8.19) to the 2-, 10-, and 100-year storage volumes, cross sections at various stages can be approximated with various side slopes as required. The *stage-storage curve* and *stage-contour area curve* can then be established. Upon completion of pre-shaping and pre-sizing a basin, the engineer can begin to work on the topography and contours at the basin site and further define the detailed shape of the detention basin and outlet works.

Example 8.3
The detention volume for the basin in Example 8.1 is determined to be 22.52 acre-ft. Distribute this volume of 22.52 acre-ft on a circular basin with a bottom area of two acres or 333.05 ft in diameter. The side slope varies from $1V{:}2H$ for water depths less than 2 ft, $1V{:}4H$ for depths between 2 and 5 ft, and $1V{:}10H$ for depths greater than 5 ft.

Table 8.6 are the calculations of the *stage-cross sectional area curve* and *stage-storage curve* for this example. The diameter of a circle increases according to Eq 8.15. The area of a circle is calculated by Eq. (8.17c) with diameter = $L = B$.

At the depth of 9-ft, this circular basin provides a storage volume of 23.20 acre-ft. With one-ft freeboard, the brimful storage volume for this basin, as shown in Table 8.6 is approximately 27.01 acre-ft.

8.8 OUTLET WORKS

Outlets used in a detention basin include risers, orifices, weirs, and culverts (ASCE, 1984). There is not a standardized procedure to find the optimal outlet combination for a detention basin. In fact, many different combinations among orifices, culverts, and weirs achieve multiple-event outlet structures. Figure 8.22 illustrates example outlet works for detention basins.

Figure 8.23 presents a more complicated outlet structure which consists of a riser for low flows and a concrete vault for high flows. A riser has a perforated vertical pipe and a cap on the top. The density of holes on the vertical pipe of a riser depends on the diameter of the vertical pipe (see the table in Fig. 8.23). A riser is designed to have an extended release on stormwater control volume such as 24-hr or longer drawdown time. All holes on the riser pipe under water shall function as an orifice. Under a low-flow condition, water is collected through the perforated holes at various elevations, and then drains into the concrete vault. During a high flow event, in addition to the riser's collection, the grate on the top opening of the concrete vault operates like a weir when the water surface is just above the top of the opening. Such an operation will switch to orifice hydraulics when the water surface is much above the top of the vault. The concrete vault has outfall pipes that are subject to downstream tailwater effects. The capacity of an outfall pipe is determined by culvert hydraulics. The accumulation of water inside the concrete vault reflects its culvert hydraulics under a condition of either inlet or outlet control. When the collection of water is faster than the release of water, the water depth in the vault increases; and it is an outlet control condition. Otherwise, it is an inlet control condition. In design, both collection and release capacities shall be calculated, whichever smaller dominates the operation of the outlet structure.

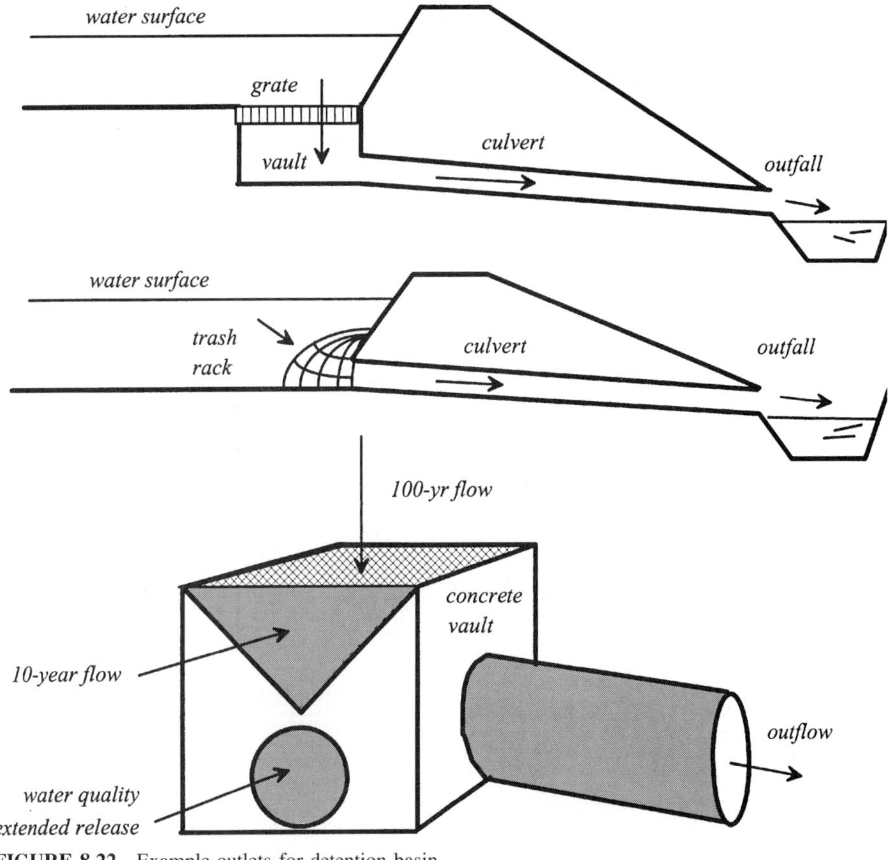

FIGURE 8.22 Example outlets for detention basin.

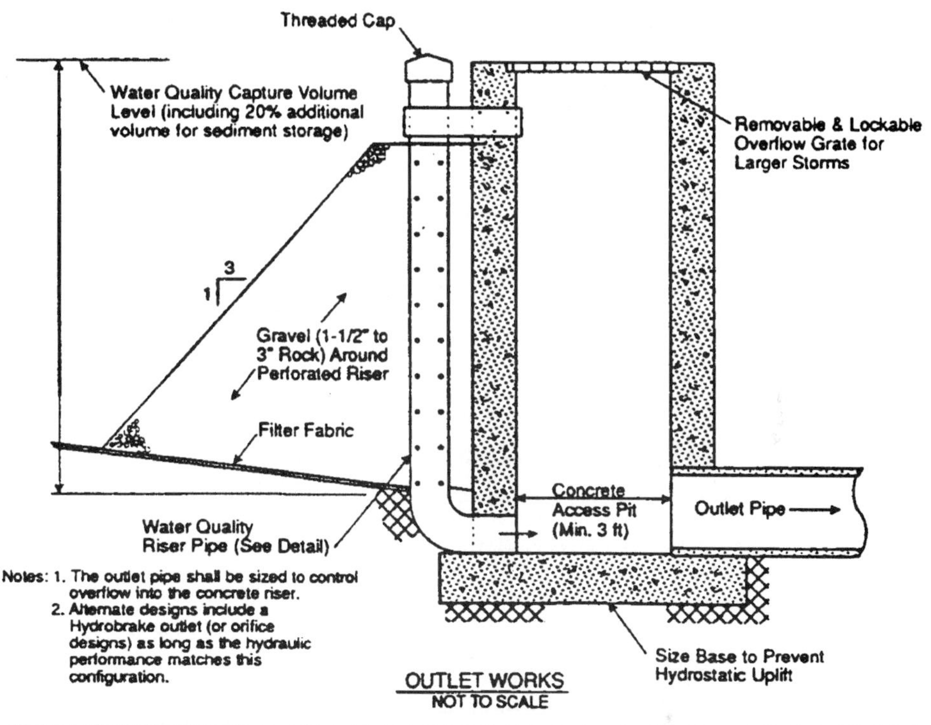

Notes: 1. Minimum number of holes = 8
2. Minimum hole diameter = 1/8" dia.

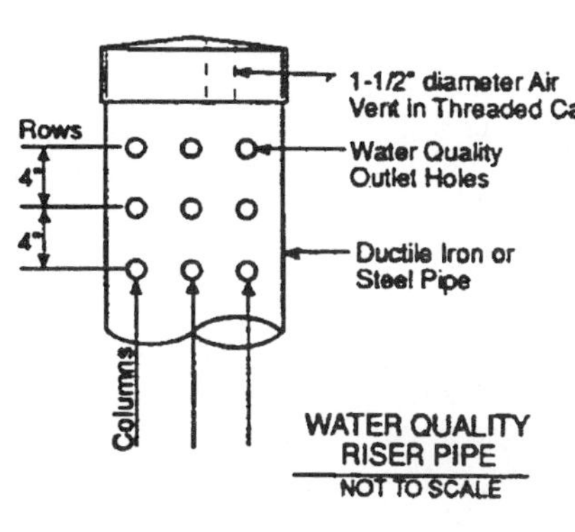

Maximum Number of Perforated Columns				
Riser Diameter (in.)	Hole Diameter, in.			
	1/4"	1/2"	3/4"	1"
4	8	8	--	--
6	12	12	9	--
8	16	16	12	8
10	20	20	14	10
12	24	24	18	12

Hole Diameter (in.)	Area of Hole (in.2)
1/8	0.013
1/4	0.049
3/8	0.110
1/2	0.196
5/8	0.307
3/4	0.442
7/8	0.601
1	0.785

FIGURE 8.23   Outlet system with riser, grate, and culvert (Urbonas and Stahre, 1992).

Trash racks shall be installed around a riser. A metal grate shall be installed on the top of the vault opening. When water is to directly drain into a culvert, a trash rack shall be installed at the entrance of the culvert. For a large outfall structure such as a 72-in pipe, the trash rack shall be a metal gate because of the high hydrostatic pressure.

8.8.1 Orifice Hydraulics

The capacity of an orifice is prescribed by the center elevation of the orifice and the net opening area of the orifice. The outflow rate through an orifice depends on the water surface elevation in the basin.

An orifice can be horizontally installed, either on the ground or as a projecting orifice on top of a vertical riser pipe. Or, the orifice may be vertically installed as the entrance of an outlet pipe as illustrated in Fig. 8.24. It is noted that both vertical and horizontal orifices behave like a weir when water is so shallow that the opening is not entirely submerged.

8.8.1.1 Unsubmerged Orifice   When the water depth above the center of a horizontal orifice is less than $0.5D$ in which D is the diameter of the orifice, the horizontal orifice operates like a weir with a crest length equal to the orifice circumference, i.e., a morning-glory-type outlet. The release rate of a horizontal orifice under a low head is described as

$$Q_o = C_w(\pi D)[H - E_o)]^{1.5} \qquad \text{provided that } (H - E_o) < 0.5\ D \qquad (8.20)$$

where Q_o = orifice outflow
C_w = weir coefficient between 2.7 to 3.0 (Table 8.6)
D = diameter of orifice
H = water surface stage
E_o = center elevation of orifice

8.8.1.2 Submerged Orifice.   As shown in Fig. 8.24, when the water depth is deep enough to submerge the entire orifice opening area, the release rate is described by the orifice equation as

$$Q_o = C_o A_o \sqrt{2g(H - E_o)} \qquad \text{provided that } (H - E_o) > 0.5\ D \qquad (8.21)$$

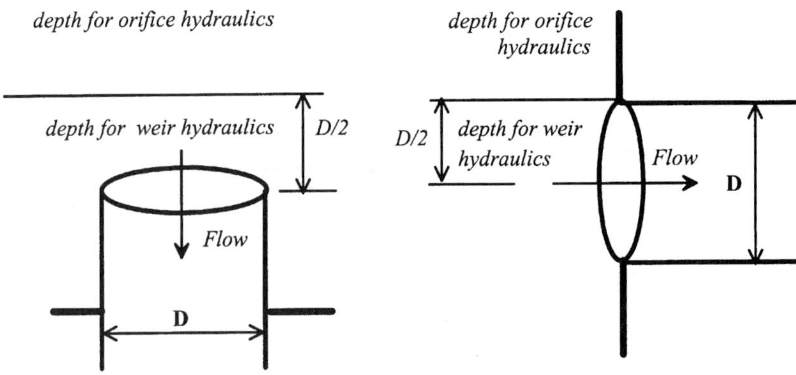

FIGURE 8.24   Weir and orifice hydraulics.

TABLE 8.7   Orifice Coefficients (Brater and King, 1976)

Shape of orifice	C
circular	0.614
square with vertical walls	0.616
rectangle, side ratio of 4:1, long side in vertical direction	0.626
rectangle, side ratio of 4:1, long side in horizontal direction	0.627
rectangle, side ratio of 10:1, long side in vertical direction	0.637
rectangle, side ratio of 10:1, long side in horizontal direction	0.637
triangle	0.615

Source: From Brater and King, 1976.

where Q_o = orifice outflow
C_o = orifice discharge coefficient
g = gravitational acceleration
A_o = net opening area.

The values of C_o are recommended on Table 8.7. Without knowing the orifice specifics, a value of 0.65 is suggested for C_o. In practice, both Eqs. (8.20) and (8.21) shall be applied to an orifice for a specified depth, and the smaller one dictates.

8.8.2   Weir Hydraulics

Weirs are classified by their cross-sectional shapes, such as rectangular, triangular, and trapezoidal weirs. The outflow rate of a *rectangular weir* is determined by its length while the release rate of a *triangular weir* is governed by its angle as illustrated in Fig. 8.25.

$$Q_w = C_w L_w [H - E_w]^{3/2} \tag{8.22}$$

$$Q_t = C_t [H - E_w]^{5/2} \tan\left(\frac{\theta}{2}\right) \tag{8.23}$$

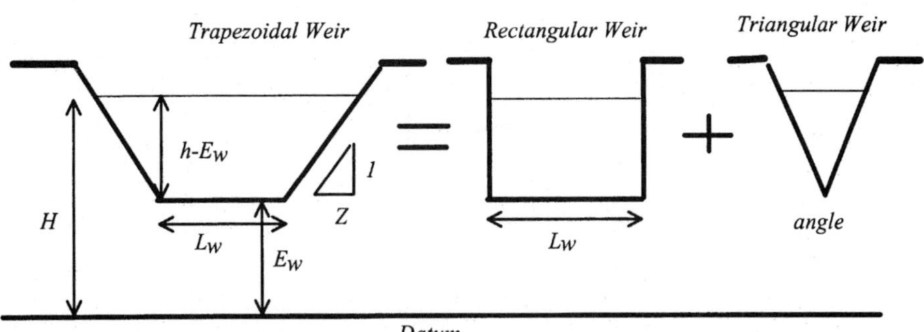

FIGURE 8.25   Weir hydraulics.

TABLE 8.8   Rectangular Weir Coefficients

Breath of weir crest feet	Headwater 1.0 ft	Headwater 2.0 ft	Headwater 3.0 ft	Headwater 4.0 ft	Headwater 5.0 ft
5.00	2.68	2.65	2.66	2.70	2.79
10.00	2.68	2.64	2.64	2.64	2.64
15.00	2.63	2.63	2.63	2.63	2.63

Source: From Brater and King, 1976.

where Q_w = flow rate in cubic ft/second
$\quad\quad C_w$ = weir coefficient for a rectangular weir between 3.0 to 2.60 in Table 8.8
$\quad\quad L_w$ = effective crest width in ft
$\quad\quad E_w$ = weir crest elevation in ft
$\quad\quad C_t$ = weir coefficient for a triangular weir
$\quad\quad \theta$ = center angle of triangular weir

Theoretically speaking, under the same hydraulic condition, a rectangular weir coefficient, C_t, is 20% less than a triangular weir coefficient, C_w (Table 8.9). As illustrated in Fig. 8.25, a *trapezoidal weir* is composed of a rectangular weir with a crest length equal to the trapezoidal bottom width and a triangular weir with the central angle equal to the trapezoidal side slope.

Typically, a weir discharge coefficient C_w for a broad- or sharp-crested weir ranges between 2.65 and 3.10. Without knowing the weir specifics and downstream tailwater conditions, the value of $C_w = 3.0$ is recommended for sharp weirs and 2.65 is recommended for broad-crested weirs. However, care must be taken in selecting the appropriate discharge coefficient after the downstream tailwater condition becomes well-understood.

8.8.3   Culvert Hydraulics

Water enters the concrete vault in Fig. 8.26 from the perforated riser and/or from the grate on the top. The outlet pipes are usually short enough, 200 to 300 ft, that they act like a culvert. The conveyance capacity of a culvert is dictated by its inlet and outlet conditions. Under inlet control, the capacity of a culvert is independent of the tailwater at the culvert outlet. For instance, a culvert laid on a steep slope can be operated under inlet control because its capacity is solely determined by the critical depth at the entrance. On the contrary, a pipe with a submerged outlet may become under outlet control. As a result, the capacity of such a pipe is dictated by the tailwater depth at the outlet. Therefore, the capacity of a culvert needs to be examined by both inlet and outlet controls. The condition requiring a higher headwater depth at the entrance dictates the culvert capacity.

TABLE 8.9   Weir Coefficients for Triangular Weirs With a Side Slope H:V

Headwater feet	H:V 1.0:1.0	H:V 2.0:1.0	H:V 3.0:1.0	H:V 5.0:1.0	H:V 10.0:1.0
0.50	3.85	3.49	3.22	3.05	2.84
1.0	3.85	3.50	3.40	3.13	2.91

Source: From Brater and King, 1976.

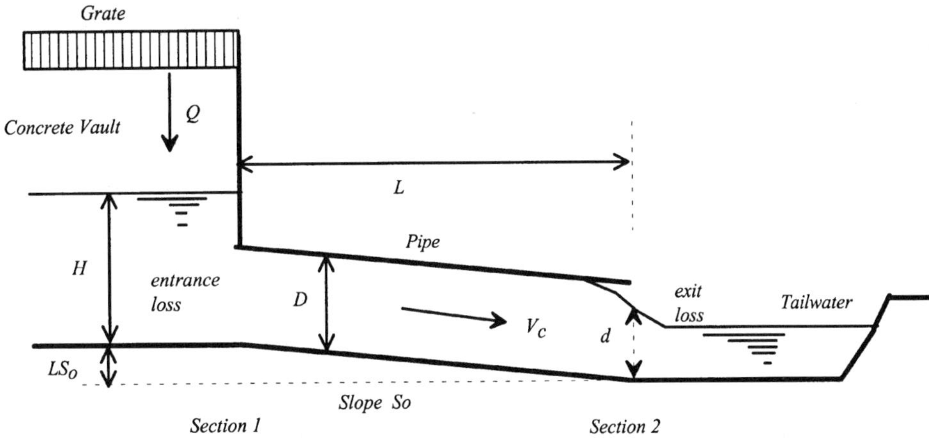

FIGURE 8.26   Culvert hydraulics.

During the preliminary design, the tailwater information may not be available. The outlet capacity of a detention system may temporarily be estimated by orifice and weir hydraulics without any consideration of tailwater effects. In other words, the outlet capacity is determined by the collection capacity of the riser and the grate only. At the final design, the outlet capacity must be revised by culvert hydraulics applied to the outfall pipes under tailwater effects.

When the culvert is under a full flow condition, the energy balance between Sec 1 and 2 is written as

$$H + LS_o = (K_e + K_x + K_b + K_n) \frac{V_c^2}{2g} + \frac{V_c^2}{2g} + d \tag{8.24}$$

$$K_n = 184.1 \frac{N^2 L}{D^{\frac{4}{3}}} \tag{8.25}$$

where H = headwater stage in the vault
L = length of the pipe in feet
S_o = pipe slope in ft/ft,
K_e = entrance loss coefficient (Tables 8.10 and 8.11)
K_x = exit loss coefficient between 0.5 to 1.0

TABLE 8.10   Entrance Loss Coefficients for Box Culverts

Structure of box culvert and entrance	Coefficient
Headwall parallel to embankment (no wingwall)	
a. square-edged on three edges	0.50
b. three edges rounded	0.20
Headwall with wingwalls at 15 to 45 degrees to barrel	
a. square-edge top corner	0.40
b. top corner rounded	0.20

Source: From U.S. Dept. of Commerce, 1965.

TABLE 8.11   Entrance Loss Coefficients for Circular Culverts

Structure of circular culvert and entrance	Coefficient
Concrete pipe projecting from fill (no headwall)	
a. socket end of pipe	0.20
b. square cut end of pipe	0.50
Concrete pipe with headwall or headwall and wingwalls	
a. socket end of pipe	0.10
b. square cut end of pipe	0.50
c. rounded entrance	0.10
Corrugated metal pipe	
a. projecting from fill (no headwall)	0.80
b. headwall or headwall and wingwalls	0.50

Source: From U.S. Dept. of Commerce, 1965.

K_b = bend loss coefficient as shown in Fig. 8.27,
K_n = friction coefficient
V_c = flow velocity
N = Manning's roughness coefficient, such as 0.015 for concrete pipe
d = tailwater depth

If the tailwater depth is not known, it is suggested that the average of the culvert diameter and critical depth, d_c, be the approximation as

$$d = \frac{D + d_c}{2} \tag{8.26}$$

Equation (8.24) is used during the final design when the tailwater and back-water effects can be incorporated into the outfall system. Let K be the sum of all the loss coefficients as

$$K = K_e + K_x + K_b + K_n \tag{8.27}$$

The flow velocity, V_c, through the pipe under tailwater effects is estimated as

$$V_c = \sqrt{\frac{1}{K + 1}} \; \sqrt{2g(H + LS_o - d)} \tag{8.28}$$

The discharging capacity of the pipe, Q_D, is

$$Q_D = \sqrt{\frac{1}{K + 1}} \; A_c \sqrt{2g(H + LS_o - d)} \tag{8.29}$$

in which A_c = pipe cross-sectional area.

Equation (8.29) is similar to the orifice equation except that the orifice coefficient is computed by the sum of all the loss coefficients.

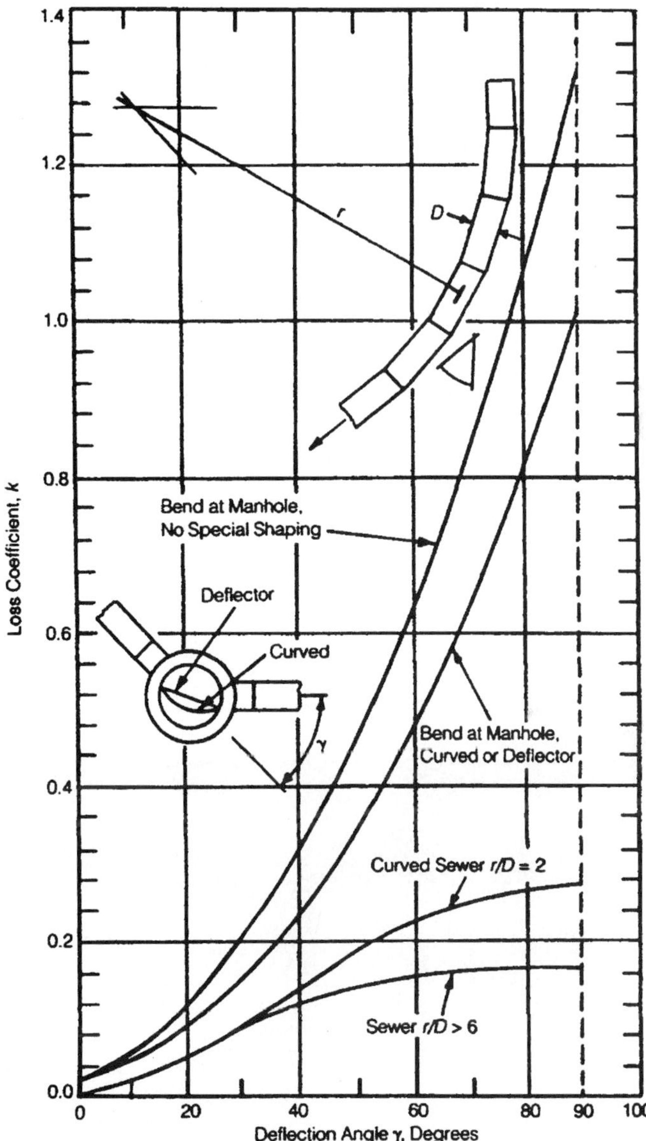

FIGURE 8.27   Bend loss coefficients. (From ASCE, 1992)

8.9   EMERGENCY SPILLWAY

Regardless of the design-storm for which the detention basin is designed, it is important that the engineer considers how to accommodate larger flood flows. A safe overflow passage can be accomplished by emergency outlets. Spillways are often used as emergency outlets. Water released from a spillway is modeled by weir hydraulics. As the name implies, an emergency spillway (Fig. 8.28) is designed to provide a safe overflow when situations arise (e.g., the blockage of the primary outlet structures or the occurrence of an event larger than the design capacity of the basin.) For an off-site detention basin that has a relatively high embankment, such as 25 ft or above, and a large storage volume, it is advisable that the emergency overflow capacity withstands up to a Probable Maximum Flood (Chow, 1964). The consequences of the embankment potential failure shall have been sufficiently studied, and various scenarios shall also have been developed for emergency plans. For a small on-site facility, either an emergency spillway is designed to pass $1/2$ *PMF*, or to make assure that the embankment withstands any overtopping flows.

The overtopping flow from a spillway drains into either a collector channel or a stilling basin through a concrete chute with baffle blocks. An emergency spillway must be accompanied by adequate erosion control and energy dissipating measures to ensure the stability of the embankment. Figure 8.29 is an example of a labyrinth weir, energy dissipator, and stilling basin. Design of a labyrinth weir can be found elsewhere (Hay and Taylor, 1970 and Tullis et al., 1995).

8.9.1 Side Spillway and Collector Channel

A collector channel in Fig. 8.30 is also termed side-spillway channel that receives the overtopping water from a spillway. The discharge in a collector channel increases downstream. As a spatially varied flow, the flow regime may change from a subcritical to a supercritical flow. For design, the length of a collector channel can be set as the length before the occurrence of critical flow (Li, 1983). In other words, the flow analysis in a collector channel begins with locating the critical flow section. Using the concept of singular point (Bremen

FIGURE 8.28 Spillway for a regional detention basin, Denver, Colorado.

FIGURE 8.29   Emergency spillway for a regional detention basin, City of Henderson, Nevada.

et al., 1989), the critical flow section in a collector channel can be located by solving the dynamic equation of spatially varied flow.

The flow condition in a collector channel is a typical spatially varied flow. The discharge increases linearly with respect to the distance as

$$Q_x = qx \tag{8.30}$$

Q_x = discharge at the location x in collector channel
 q = lateral inflow rate per unit length
 x = distance along with collector channel

The normalized governing equation for a spatially varied flow is derived as (Guo, 1999a)

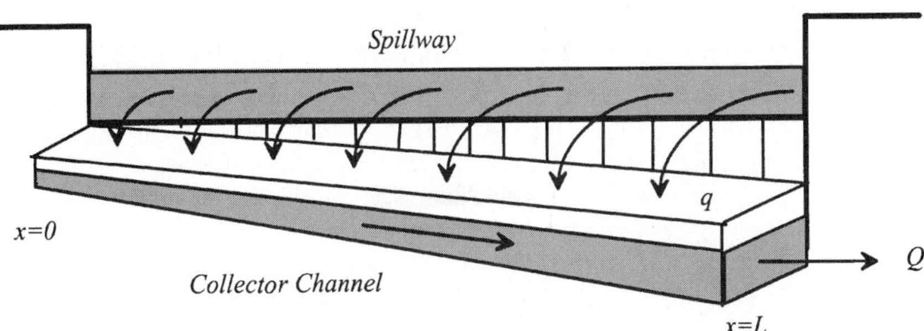

FIGURE 8.30   Illustration of side spillway and collector channel.

$$\frac{N}{S_o}\frac{P_*^{4/3}}{T_*A_*^{1/3}} + 2\frac{F_q}{S_o}\frac{1}{\sqrt{T_*A_*}} = 1 \tag{8.31}$$

$$X_* = \frac{\sqrt{A_*^3/T_*}}{F_q/S_o} \tag{8.32}$$

in which

$$N = \left[\frac{n\sqrt{g}}{1.49B^{1/6}}\right]^2 \tag{8.33}$$

$$F_q = \frac{q}{\sqrt{gB^3}} \tag{8.34}$$

$$T_* = 1 + 2zY_* \tag{8.35}$$

$$A_* = Y_* + zY_*^2 \tag{8.36}$$

$$P_* = 1 + 2\sqrt{1 + z^2}\,Y_* \tag{8.37}$$

$$Y_* = \frac{Y_c}{B} \tag{8.38}$$

$$X_* = \frac{S_o x_c}{B} \tag{8.39}$$

in which z = channel side slope and subscript , $*$, represents the normalized parameters at the critical flow section

For a rectangular channel, Eq. (8.31) can be further reduced to

$$\frac{N}{S_o}\frac{(1 + 2Y_*)^{4/3}}{Y_*^{1/3}} + 2\frac{F_q}{S_o}\frac{1}{\sqrt{Y_*}} = 1 \tag{8.40}$$

A higher ratio of N/S_o implies a rough channel on a mild slope, and a higher ratio of F_q/S_o means that the channel is subject to a high lateral inflow. As a part of a spillway system, the collector channel is usually designed to have a narrow cross section. Solutions of Eq. (8.31) are developed for trapezoidal channels with a side slope of $1V:1H$ or $2V:H$ plotted in Figs. 8.31 and 8.32. Solutions of Eq. (8.40) for rectangular channel sections are also plotted in Fig. 8.33. Examining Eq. (8.31), it is noticed that solutions do not exist if either one of the two dimensionless groups exceeds unity. For instance, a critical flow section can be developed in trapezoidal collector channels with a high value of N/S_o from 0.01 to 0.50 and F_q/S_o from 0.18 to 17.45, but not for rectangular collector channels whose N/S_o ranges from 0.005 to 0.05 and F_q/S_o ranges from 0.095 to 0.83.

Example 8.4
To explain how to apply Figs. 8.31 to 8.33 to the design of a collector channel, the example developed by Hinds in 1926, illustrated by Chow (1959), and discussed by French (1985) was used. Design information include: q = 40 cfs/ft, n = 0.015, S_o = 0.1505 ft/ft, B = 10.0 ft, and z = 0.5. The recommended procedures are

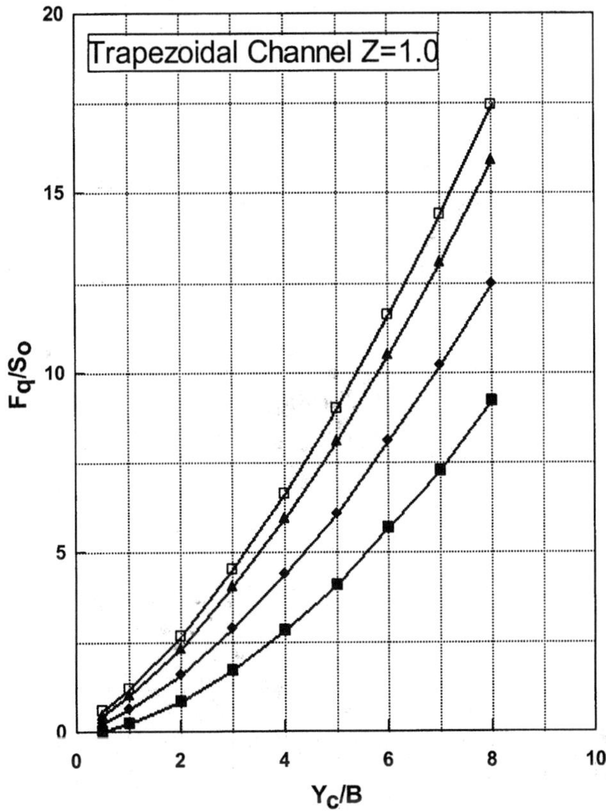

FIGURE 8.31   Critical flow condition in collector channels with side slope $Z = 1.0$.

1. Inflow Froude Number, F_q, to Channel Slope Ratio

$$F_q = \frac{q}{\sqrt{gB^3}} = \frac{40}{\sqrt{32.2 * 10^3}} = 0.223$$

$$\frac{F_q}{S_o} = \frac{0.223}{0.1505} = 1.481$$

2. Roughness Coefficient to Channel Slope Ratio

$$N = \left(\frac{n\sqrt{g}}{1.49B^{1/6}}\right)^2 = \left(\frac{0.015\sqrt{32.2}}{1.49 * 10.0^{1/6}}\right)^2 = 0.001515$$

$$\frac{N}{S_o} = \frac{0.001515}{0.1515} = 0.01$$

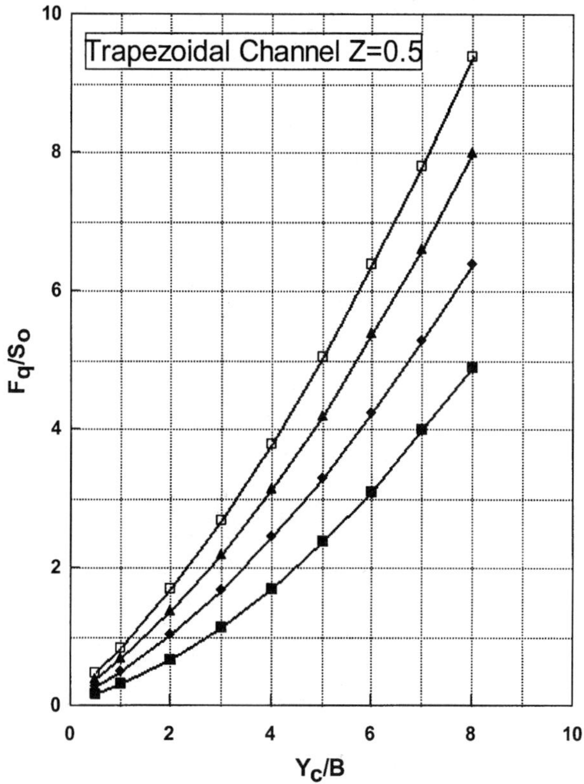

FIGURE 8.32   Critical flow condition in collector channels with side slope $Z = 0.5$.

3. Critical Flow Depth

Using Fig. 8.32 with $N/S_o = 0.01$, the ratio, $Y_* = Y_c/B$, is found to be 1.76 for $F_q/S_o = 1.481$, or the critical depth for this case is 17.6 ft.

4. Location of the Critical Flow Section using Eqs (8.36) through (8.39).

$$A_* = Y_* + zY_*^2 = 1.76 + 0.5(1.76)^2 = 3.31$$

$$T_* = 1 + 2zY_* = 1 + 2(0.5)(1.76) = 2.76$$

$$D_* = \frac{A_*}{T_*} = \frac{3.31}{2.76} = 1.20$$

$$X_* = \frac{\sqrt{D_* A_*^2}}{F/S_0} = \frac{\sqrt{1.20(3.31)^2}}{1.481} = 2.448$$

$$x_c = \frac{BX_*}{S_o} = 162.5 \; ft$$

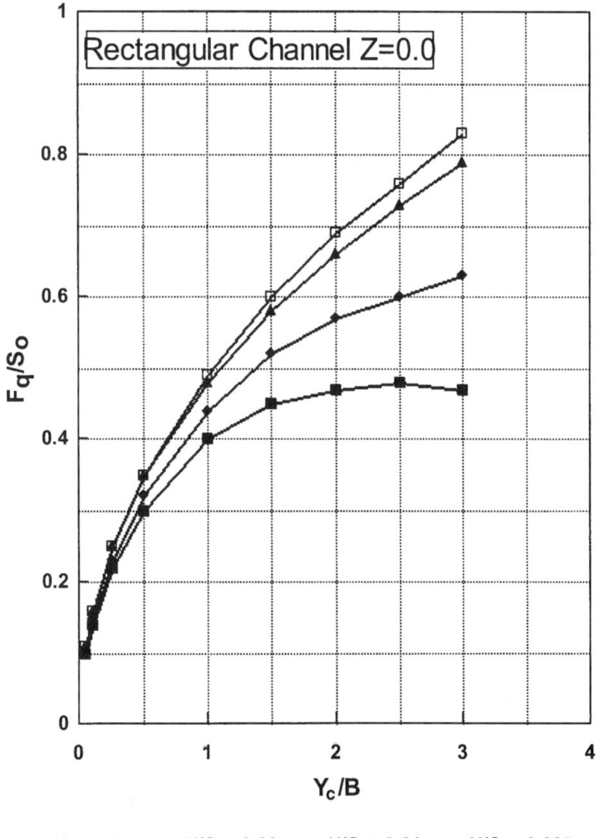

FIGURE 8.33 Critical flow condition in rectangular collector channels.

Solutions obtained from the design charts agree with Hind's graphical solutions. It is interesting to notice that this example has been used for decades to illustrate the water surface profile computations in a spatially varied flow. The channel slope of 0.15015 is not practical to construct in the field. Aided by Figs. 8.31 through 8.33, several alternative channel sections can also be developed for this case with $S_o = 0.10$, 0.15, or 0.20, and $B = 10.0$ or 15.0 ft. They are summarized in Table 8.12 for comparison of sensitivity among variables.

TABLE 8.12 Alternatives Developed for the Case Study of Collector Channel Design

B (ft)	S_o	F_q/S_o	N/S_o	Z	Y_c/B	S_oX_c/B	Y_c (ft)	X_c (ft)
10.00	0.15	1.485	0.0100	0.5	1.760	2.44	17.60	162.52
10.00	0.15	1.485	0.0100	1.0	1.200	1.57	12.00	104.36
15.00	0.10	1.213	0.0132	1.0	0.995	1.333	6.20	199.95
15.00	0.20	0.607	0.0066	1.0	0.500	0.757	10.00	56.78
15.00	0.15	0.808	0.0088	1.0	0.67	0.97	10.10	96.50
15.00	0.15	0.808	0.0088	0.5	0.94	1.44	14.10	144.20

8.10 BASIN CHARACTERISTIC CURVE: STAGE-OUTFLOW RELATIONSHIP

An *outfall structure* is a catch basin that has orifices and/or weirs to control the collection capacity at its inlet, and culverts to discharge the collected flow at its outlet. When the *collection capacity* is greater than the *discharging capacity,* the outfall structure is under *outlet control.* Otherwise, it is a condition of *inlet control.* It is essential to take both inlet and outlet controls into consideration when deriving the stage-outflow characteristic curve.

Example 8.5
The outfall structure is designed as described in Fig. 8.34. To maintain a permanent pool of 2-ft deep, a 12-in vertical riser is erected with perforated holes between 2 and 4 ft above the basin floor. The density of holes is 12 one-inch holes per layer and three layers per linear foot of the riser pipe length. The top opening of the concrete vault is covered by a horizontal rectangular grate of 3.5 by 10 ft. The grate is installed at 5 ft above the basin floor. After subtracting the surface area of steel bars, the net opening area ratio of the grate is 70.0%. The outfall system includes two 48-in concrete pipes laid on a slope of 3.0%. The length of these two pipes is 200 ft, and the roughness Manning's N is 0.020. These two pipes have a 45-degree bend between the entrance and the exit. The spillway for the basin is designed to have a weir length of 30 ft with a crest at 10.0 ft above the basin floor. The weir coefficient is 3.0 and the orifice coefficient is 0.60. Determine the stage-outflow curve.

Step 1. Determination of Collection Capacity Orifice Hydraulics for the Riser
   On the riser, each layer has 12 holes. The sum of the 12 opening areas, A_o, is

$$A_o = 12 \frac{\pi}{4} (1/12)^2 = 0.065 \text{ sq. ft}$$

Let $C_o = 0.60$. The water collection capacity, Q_{oi}, is calculated by the orifice formula. For the the layer of holes at E_{ei}, the collection capacity is:

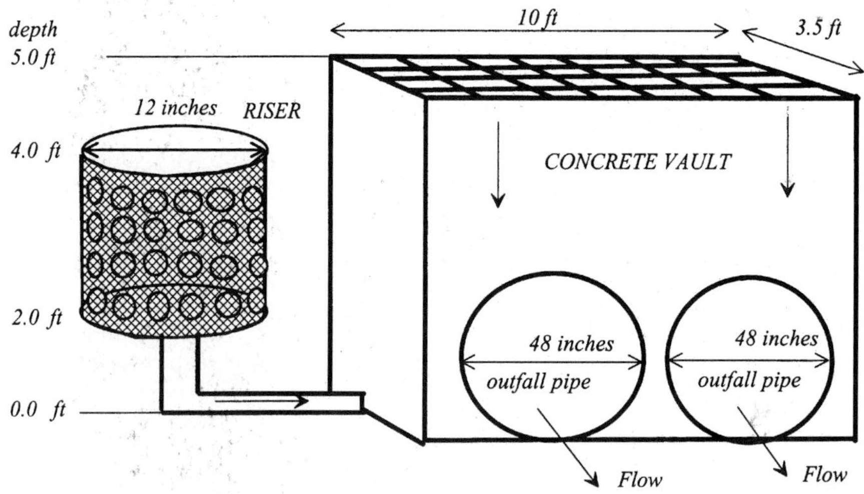

FIGURE 8.34   Design example of outlet structure.

$$Q_{oi} = (0.60)(0.065) \sqrt{64.4(H - E_{oi})} = 0.313 \sqrt{H - E_{oi}} \text{ cfs for } H > E_{oi}$$

Repeat the above procedure to every layer of holes at various elevations, E_{oi}. The collection of the riser is

$$Q_o = \sum_{i-1}^{i=n} 0.313 \sqrt{H - E_{oi}} \text{ cfs}$$

Weir or Orifice Flow through the Grate
   The grate has a wetted perimeter for the weir operation as

$$L_g = 2.0(3.5 + 10.0) = 27.0 \text{ ft}$$

The grate has a net opening area for the orifice operation as

$$A_g = (0.70)(3.5)(10.0) = 24.5 \text{ sq ft.}$$

Let $C_w = 3.0$. When $H > 5.0$ feet, the discharge, Q_{wg}, for the grate to operate like a weir is

$$Q_{wg} = C_w L_g (H - 5.0)^{1.5} = 81.0(H - 5.0)^{1.5} \text{ cfs} \qquad \text{for } H > 5.0 \text{ ft}$$

Let $C_o = 0.60$. When $H > 5.0$ feet, the discharge, Q_{og}, for the grate to operate like an orifice is

$$Q_{og} = C_o A_g \sqrt{2g(H - 5.0)} = 117.97 \sqrt{H - 5.0} \text{ cfs} \qquad \text{for } H > 5.0 \text{ ft}$$

The grate collection capacity, Q_g, at a specified water depth, H, is determined as:

$$Q_g = \min(Q_{wg}, Q_{og}) \qquad \text{for } H > 5.0 \text{ ft}$$

For a given H, the collection capacity, Q_U, for the outlet structure is calculated as:

$$Q_U = Q_o \qquad \text{for } H < 5.0 \text{ feet}$$

or

$$Q_U = Q_o + Q_g \qquad \text{for } H > 5.0 \text{ feet}$$

Table 8.13 presents the collection capacity at various depths.

Step 2. Determination of Discharging Capacity for Tailwater Depth of 4.0 ft. Culvert Hydraulics
   Knowing that $N = 0.02$, $K_e = K_x = 0.50$, and $K_b = 0.25$ for a 45-degree bend (Fig 8.27). The flow velocity in a single concrete pipe is determined as

$$K_n = 184.1 \frac{(0.02)^2(200.0)}{4.0^{4/3}} = 2.32$$

$$K = 0.5 + 0.5 + 0.45 + 2.32 = 3.77$$

$$LS_0 = (200.0)(3.0\%) = 6.0 \text{ ft}$$

$$V_c = 0.515 \sqrt{2g(H + 2)} \text{ fps}$$

The discharge, Q_D, for the dual 48-in pipes is

TABLE 8.13 Collection Capacity of Riser and Grate

Stage				Center of Holes (feet)						Riser capacity	Grate weir	Grate orifice	Collection capacity
ft	2.00	2.25	2.50	2.75	3.00	3.25	3.50	3.75	4.00	cfs	cfs	cfs	cfs
2.00										**0.00**			**0.00**
2.25	0.16									**0.16**			**0.16**
2.50	0.22	0.16								**0.38**			**0.38**
2.75	0.27	0.22	0.16							**0.65**			**0.65**
3.00	0.32	0.27	0.22	0.16						**0.97**			**0.97**
3.25	0.35	0.32	0.27	0.22	0.16	0.00				**1.32**			**1.32**
3.50	0.39	0.35	0.32	0.27	0.22	0.16	0.00			**1.71**			**1.71**
3.75	0.42	0.39	0.35	0.32	0.27	0.22	0.16	0.00		**2.12**			**2.12**
4.00	0.45	0.42	0.39	0.35	0.32	0.27	0.22	0.16	0.00	**2.57**			**2.57**
4.25	0.47	0.45	0.42	0.39	0.35	0.32	0.27	0.22	0.16	**3.04**			**3.04**
4.50	0.50	0.47	0.45	0.42	0.39	0.35	0.32	0.27	0.22	**3.38**			**3.38**
4.75	0.52	0.50	0.47	0.45	0.42	0.39	0.35	0.32	0.27	**3.68**			**3.68**
5.00	0.55	0.52	0.50	0.47	0.45	0.42	0.39	0.35	0.32	**3.96**			**3.96**
6.00	0.63	0.61	0.59	0.57	0.55	0.52	0.50	0.47	0.45	**4.88**	81.00	117.97	**85.88**
7.00	0.70	0.69	0.67	0.65	0.63	0.61	0.59	0.57	0.55	**5.65**	229.10	**166.83**	**172.48**
8.00	0.77	0.76	0.74	0.72	0.70	0.69	0.67	0.65	0.63	**6.33**	420.89	**204.32**	**210.65**
9.00	0.83	0.82	0.80	0.79	0.77	0.76	0.74	0.72	0.70	**6.94**	648.00	**235.93**	**242.87**
10.00	0.89	0.88	0.86	0.85	0.83	0.82	0.80	0.79	0.77	**7.50**	905.61	**263.78**	**271.28**
11.00	0.95	0.93	0.92	0.91	0.89	0.88	0.86	0.85	0.83	**8.02**	1,190.45	**288.96**	**296.97**
12.00	1.00	0.98	0.97	0.96	0.95	0.93	0.92	0.91	0.89	**8.50**	1,500.14	**312.11**	**320.61**

$$Q_D = 0.515(2A)\sqrt{2g(H + 2)} = 103.88 \sqrt{H + 2} \text{ cfs} \qquad \text{(discharging capacity)}$$

Table 8.14 presents the calculations of the discharging capacity of the dual pipe system. For a specified stage, H, both collection and discharging capacities shall be calculated, and the smaller one dictates the capacity, O, of the outlet structure.

$$O = \min(Q_u, Q_D) \qquad \text{for } H < 10.0 \text{ ft}$$

TABLE 8.14 Stage-outflow Relationship for Example Outlet Structure

Water depth (feet)	Discharging capacity (cfs)	Collection capacity (cfs)	Outflow capacity (cfs)	Spillway capacity (cfs)	Total release (cfs)
1.00		**0.00**			0.00
2.00		**0.00**			0.00
3.00		**0.97**			0.97
4.00		**2.57**			2.57
5.00	103.87	**3.96**	3.96	0.00	3.96
6.00	146.89	**85.88**	85.88	0.00	85.88
7.00	179.90	**172.48**	172.48	0.00	172.48
8.00	**207.73**	210.65	207.73	0.00	207.73
9.00	**232.25**	242.87	232.25	0.00	232.25
10.00	**254.42**	271.28	254.42	0.00	254.42
11.00	**274.81**	296.97	274.81	90.00	364.81
12.00	**293.78**	320.61	293.78	254.56	548.34

Spillway
The above procedure shall be repeated for the range of $H = 0.0$ at the basin floor to $H = 10.0$ ft at the crest of spillway. Let $C_w = 3.0$. When $H > 10.0$ foot, the discharge, Q_s, by the spillway is

$$Q_s = 3.0(30.0)(H - 10.0)^{1.5} = 90.0(H - 10.0)^{1.5} \qquad \text{for } H > 10 \text{ ft}$$

The stage-outflow curve for the outlet structure in Fig. 8.25 is computed as

$$O = Q_s + \min)(Q_U, Q_D) \qquad \text{for } H > 10.0 \text{ ft}$$

Table 8.14 is the summary of the stage-outflow relationship for the outlet structure.

8.11 PERFORMANCE EVALUATION BY HYDROGRAPH ROUTING

The preliminary design of a detention basin begins with approximations on the basin's performance. After knowing outflow works and basin shape, the performance of the detention basin needs to be examined by hydrograph routing. Hydrograph routing is a mathematical algorithm to calculate the outflow hydrograph from the inflow hydrograph under storage effects. A *hydraulic routing method* such as the dynamic wave approach takes both principles of continuity and momentum into consideration while a *hydrologic routing method* such as the reservoir routing technique only solves the continuity principle. The continuity principle is

$$I - O = \frac{dV}{dt} \tag{8.41}$$

where I = inflow
O = outflow
V = detention storage volume
t = elapsed time

Applying the finite difference scheme to Eq. (8.41) yields

$$\frac{I(t) + I(t + \Delta t)}{2} + \frac{O(t) + O(t + \Delta t)}{2} = \frac{V(t + \Delta t) - V(t)}{\Delta t} \tag{8.42}$$

where the subscript t = time at the beginning of the time interval
$t + \Delta t$ = time the end of the time interval
Δt = time interval

For convenience, Eq. (8.42) can be re-arranged as

$$\frac{2V(t + \Delta t)}{\Delta t} + O(t + \Delta t) = [I(t) + \Delta t)] + \left[\frac{2V(t)}{\Delta t} - O(t) \right] \tag{8.43}$$

There are two unknowns in Eq. (8.43): outflow rate, $O(t + \Delta t)$, and storage, $V(t + t)$, at time $t + \Delta t$. Both unknowns are related to stage, $h(t + \Delta t)$. There are many methods developed to solve Eq. (8.43), such as modified Puls method (Puls, 1928; Chow, 1964). They all require the *stage-storage-outflow curves* derived from the basin.

8.11.1 Storage Routing Method

The *storage routing* method is a mathematical algorithm to seek the solutions at $t + \Delta t$ based on flow volumes at time t (Guo, 2000). For convenience, Eq. (8.43) is re-arranged as:

$$[I(t) + I(t + \Delta t) - O(t)]\Delta t + 2V(t) = O(t + \Delta t)\,\Delta t + 2V(t + \Delta t) \qquad (8.44)$$

Let the *storage routing function* in volume unit be defined as

$$SO = (O)(\Delta t) + 2V \qquad (8.45)$$

The storage routing function can be established by the pairs (V, O). Aided by Eq. (8.45), Eq. (8.44) is converted to

$$SO(t + \Delta t) = [I(t) + I(t + \Delta t) - O(t)]\Delta t + 2V(t) \text{ (known flow volumes at time } t)$$

$$(8.46)$$

Figure 8.35 illustrates how solutions can be obtained from the storage-routing function and storage-outflow curve. The value of $SO(t + \Delta t)$ is prescribed by the known variables

$$I(t),\ I(t + \Delta t),\ O(t),\ \text{and}\ S(t).$$

Solutions for the two unknowns, $O(t + \Delta t)$ and $V(t + \Delta t)$, are the pair (V, O) that has a SO-function to satisfy:

$$SO(t + \Delta t) = O(t + \Delta t)\Delta t + 2V(t + \Delta t) \text{ (solutions at } t + \Delta t) \qquad (8.47)$$

Repeating Eqs. (8.46) and (8.47) for each time step, the outflow hydrograph can be generated.

8.11.2 Outflow Routing Method

The *outflow function method* is mathematical algorithm to provide solutions at $t + \Delta t$ based on flow rates at time t (Guo, 2000). Using flow rate as a base, Eq. (8.43) is re-arranged as:

$$\frac{2V(t) + \Delta t)}{\Delta t} + O(t + dt) = [I(t + \Delta t)] + \left[\frac{2V(t)}{\Delta t} + O(t)\right] - 2O(t) \qquad (8.48)$$

Let the *outflow routing function, OS-function,* be defined as:

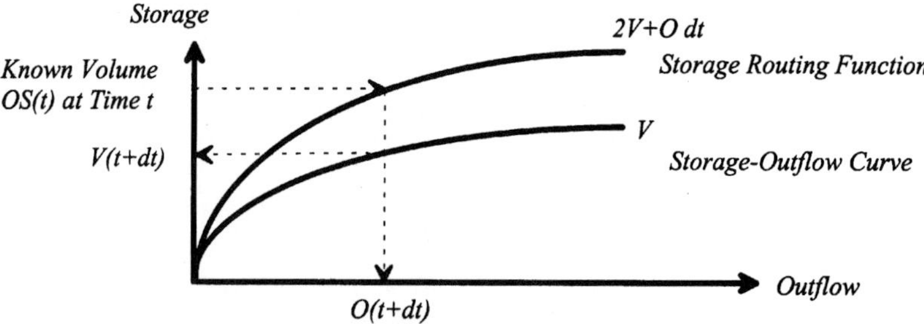

FIGURE 8.35   Graphic solution using storage routing function.

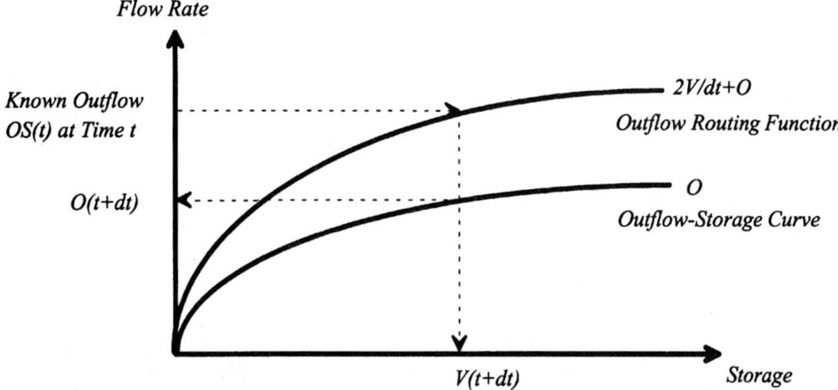

FIGURE 8.36 Graphic solution using outflow routing function.

$$OS = \frac{2V}{dt} + O \qquad (8.49)$$

It is noted that the outflow routing function has the same unit as flowrate. Aided by Eqs. (8.49), Eq. (8.48) can be separated into two parts as

$$OS(t + \Delta t) = I(t) + I(t + \Delta t) + OS(t) - 2O(t) \text{ ((known flow rates at time } t) \qquad (8.50)$$

$$\text{and} \quad OS(t + \Delta t) = \frac{2V(t + \Delta t)}{\Delta t} + O(t + \Delta t) \text{ (solutions at time } t + \Delta t) \qquad (8.51)$$

At a time step, t, we first compute the sum of known variables: $I(t)$, $I(t + \Delta t)$, $O(t)$ by Eq. (8.50), and $OS(t)$. The solution to Eq. (8.48) is the pair $(V(t + \Delta t), O(t + \Delta t))$ which has an OS function equal to Eq. (8.50). Repeat this process as illustrated in Fig. 8.36 to advance one Δt at a time until the inflow hydrograph is completely processed.

Example 8.6
Considering the time increment of 5 minutes, derive the *SO* routing function and *OS* routing function for the basin described in Table 8.15.

TABLE 8.15 Example for SO and OS Routing Functions

Stage h ft	Storage $V(h)$ acre-ft	Outflow $O(h)$ cfs	SO function $SO(h)$ acre-ft	OS-fucntion $OS(h)$ cfs
5,760.00	0.00	0.00	0.00	0.00
5,764.00	10.44	56.77	21.27	3,088.55
5,768.00	29.00	86.71	58.60	8,508.31
5,772.00	56.70	108.70	114.15	16,574.38
5,774.00	74.32	203.02	150.04	21,785.55
5,776.00	95.22	366.94	192.97	28,018.83

$$OS(h) = \frac{2V(h)(43560)}{5(60)} + O(h) \tag{17.14}$$

$$SO(h) = 2V(h) + O(h)(5)(60)/43560 \tag{17.15}$$

The *SO* and *OS* functions are developed in Table 8.15

Example 8.7
Table 8.16 is an example to route the inflow hydrograph through the basin in Example 8.6 using the *OS* routing function. The purpose of this exercise is to confirm that the peak outflow does not exceed the target value of 80 cfs. At time $t = 0$, the initial condition includes: at $t = 0$, $O(t) = 0$ and $S(t) = 0$.

At time $t + dt$, the value of *OS* function is calculated by Eq. (8.50) in column 3. The solution at time $t + dt$ is the pair (V, O) in columns 4 and 5. Solution finding is a trial-and-error procedure using the stage-storage-outflow relationship in Table 8.15 as a guidance. The pair (V, O) is acceptable if the *OS* function by Eq. (8.51) in column 6 closely agrees with the value in Column 3. For this case, the outflow reaches its peak value of 75 cfs at $t = 95$ minutes. The basin under design will control the outflow to be close to, but not to exceed, the allowable release of 80 cfs. The peak storage volume is 22.68 acre-ft at $t = 95$ minutes.

TABLE 8.16 Hydrograph Routing Using OS Function

(1) Time	(2) Inflow	(3) Calculated OS fct	(4) Storage	(5) Outflow	(6) Check OS fct
t $t + dt$ (minutes)	$I(t)$ $I(t + dt)$ (cfs)	$[I(t) + I(t + dt) + OS(t) - 2O(t)]$ (cfs)	$V(t)$ $V(t + dt)$ (acre-ft)	$O(t)$ $O(t + dt)$ (cfs)	$2V*43560/300 + O$ (cfs)
0.00	0.00	0.00	0.00	0.00	
5.00	3.00	3.00	0.01	0.00	3.00
10.00	16.00	22.00	0.08	0.00	22.00
15.00	48.00	86.00	0.29	2.00	86.00
20.00	136.00	266.00	0.91	3.00	266.00
25.00	296.00	692.00	2.37	4.00	692.00
30.00	424.00	1,404.00	4.82	5.00	1,404.00
35.00	452.00	2,270.00	7.77	13.00	2,270.00
40.00	427.00	3,123.00	10.67	25.00	3,123.00
45.00	383.00	3,883.00	13.23	41.00	3,883.00
50.00	337.00	4,521.00	15.38	55.00	4,521.00
55.00	298.00	5,046.00	17.17	60.00	5,046.00
60.00	265.00	5,489.00	18.68	64.00	5,489.00
65.00	228.00	5,854.00	19.93	66.00	5,854.00
70.00	193.00	6,143.00	20.92	69.00	6,143.00
75.00	161.00	6,359.00	21.65	71.00	6,359.00
80.00	132.00	6,510.00	22.17	72.00	6,510.00
85.00	108.00	6,606.00	22.50	73.00	6,606.00
90.00	85.00	6,653.00	22.65	74.00	6,653.00
95.00	*70.00*	*6,660.00*	*22.68*	*75.00*	*6,660.00*
100.00	60.00	6,640.00	22.61	75.00	6,640.00
105.00	53.00	6,603.00	22.48	75.00	6,603.00
110.00	48.00	6,554.00	22.31	75.00	6,554.00
115.00	43.00	6,495.00	22.11	75.00	6,495.00

8.12 CLOSING

Stormwater detention is an effective measure to mitigate the impacts of development on urban environment. A detention basin can reduce the runoff peak flows by storage effects, entrap pollutant solids during the sedimentation process, recharge groundwater through infiltration, and provide wildlife habitats by wetlands or retention pools. In an urban area, on-line detention storage volumes are often very limited. As a result, the practice of off-line detention basins using open space adjacent to the waterway is more practical. Design of an off-line detention basin shall take multiple events, and multiple land use into consideration. In addition to hydrologic and hydraulic objectives, public safety, aesthetics, restoration of wildlife habitat, and maintenance are also important aspects associated with stormwater detention practices.

In an urban area, stormwater must be managed as a natural resource. Re-use of stormwater is an important concept. Detention basins shall be planned and designed for much more than just a stormwater storage facility. It is necessary for a basin to also function as a visual amenity or focal point in the community. Open space in an urban area provides psychological relief and enjoyment of out-of-doors. It is essential to design a detention basin as open space that is landscaped and structured to accommodate passive and active recreational activities.

Although detention practices have many merits for stormwater quality and quantity control, it is undeniable that a stormwater detention basin also imposes geomorphic effects on streams. The impacts of a detention system on stream stability include the reduction of peak flows and sediment washload delivered to a stream while lengthening the duration of high flow for possible bank and streambed erosion. Therefore, stormwater detention shall be part of the overall strategy for the entire watershed management. Each and every detention basin shall collectively work together to achieve the goals set forth by the regional master drainage plan.

REFERENCE

American Society of Civil Engineers (ASCE), "Final Report of the Task Committee on Stormwater Detention Outlet Structures," New York, 1984.

American Society of Civil Engineers (ASCE), "Design and Construction of Urban Stormwater Management Systems," Water Environmental Federation Manual of Practice FD-20, and ASCE Manual and Reports No. 77, 1992.

Brater, E. F., and H. W. King, "Handbook of Hydraulics," McGraw-Hill Book Company, New York, 1976.

Bremen, R., and W. H. Hager, "Experiments in Side-Channel Spillways," *ASCE Journal of Hydraulic Engineering,* 115(5), May, 1989.

Chow, V. T., (1959) "Open-Channel Hydraulics," McGraw Hill Book Company, New York, 1959.

Chow, V. T., "Handbook of Applied Hydrology," McGraw Hill Book Company, New York, 1964.

French, R. H., *Open-Channel Hydraulics,* McGraw-Hill Book Company, New York, 1985.

Guo, J. C. Y., and R. Ben Urbonas, "Maximized Detention Volume Determined by Runoff Capture Rate," *Journal of Water Resources Planning and Management,* American Society of Civil Engineers, 122(1) January 1996.

Guo, J. C. Y., "Subsurface-surface Hydrologic Model for Infiltration Trenches," *ASCE Journal of Water Resources Planning and Management,* 24(5) Sept./Oct. 1998.

Guo, J. C. Y., "Critical Flow Section in a Collector Channel," *ASCE Journal of Hydraulic Engineering,* 124(6), Nov./Dec. 1999a.

Guo, J. C. Y., "Detention Storage Volume for Small Urban Catchment," *ASCE Journal of Water Resources Planning and Management,* 125(6), Nov./Dec. 1996.

Guo, J. C. Y., "Detention Basin Design and Sizing," Research Report, Department of Civil Engineering, University of Colorado at Denver, Denver, CO, 2000.

Hay, N. and G. Taylor, "Performance and Design of Labyrinth Weirs," *ASCE Journal of Hydraulic Division,* 11:2337–2356, Nov. 1970.

Henderson, F. M., *Open Channel Flow,* MacMillan Publishing Co., Inc., New York, 1966.

Li, W. S., *Fluid Mechanics in Water Resources Engineering,* Allyn and Bason, Inc. Boston, MA., 1983.

Malcom, H. R., "Some Detention Design Ideals," ASCE Proceedings of the Conference on Stormwater Detention Facilities, held in New England College, Hanniker, NH, William DeGroot, ed., 1982.

Metcalf and Eddy, Inc. "Wastewater Engineering, Collection, Treatment, Disposal," McGraw Hill Book Company, New York, 1972.

Pima County, Department of Transportation and Flood Control District, "Stormwater Detention/Retention Manual" distributed by City of Tuson, AZ, 1989.

Puls, L. G., "Construction of Flood Routing Curves," House Document 185, U.S. 70th Congress, 1st Session, 46–52 Washington, DC, 1928.

Shueler, T. R., "Controlling Urban Runoff," Department of Environmental Programs, Metropolitan Washington Council of Governments, 1987.

Shueler, T. R., and Mike Helfrich, "Design of Extended Detention Wet Pond System," Design of Urban Runoff Quality Controls, Larry A. Roesner, Ben Urbonas, and Michael B. Sonnen, eds. American Society of Civil Engineers, New York, NY, 1989.

Tullis, J. P., N. Amanian, and D. Waldron, "Design of Labyrinth Spillways," *ASCE Journal of Hydraulic Engineering,* 121(3):247–255, Dec. 1995.

Urbonas, B., and J. C. Y. Guo, "Hydrology for Optimal Sizing of Urban Runoff Treatment Control System," *Journal of Water Quality International* London, SW1H9BT, UK, Feb. 1996.

Urban Drainage and Flood Control District, "Urban Storm Drainage Design and Criteria," Vol. 1, 2, and 3, Denver, CO, 1985.

Urbonas, B. R., and M. W. Glidden, "Potential Effect of Detention Policies," Proceedings of the Second Southwest Regional Symposium on Urban Stormwater Management, Texas A&M University, Nov. 1983.

Urbonas, B., and P. Stahre "Stormwater Best Management Practices and Detention," Prentice Hall, Englewood Cliffs, NJ, 1992.

U.S. Department of the Army and Air Force, "Surface Drainage Facilities for Airfields and Heliports," Technical Manual No. 5-820-1, Washington DC, April 1977.

U.S. Department of Commerce, HEC 5, "Hydraulic Charts for the Section of Highway Culverts," Bureau of Roads, Washington DC, 1965.

U.S. Department of Transportation, Federal Aviation Administration, "Airport Drainage," Report 150/5320-5B, U.S. Government Printing Office, Washington DC, 1970.

U.S. Environmental Protection Agency (USEPA), *Results of the Nationwide Urban Runoff Program, Final Report,* U.S. Environmental Protection Agency, NTIS No. PB84-185545, Washington, DC, 1983.

U.S. Environmental Protection Agency (USEPA), *Methodology for Analysis of Detention Basins for Control of Urban Runoff Quality,* U.S Environmental Protection Agency, EPA440/5-87-001, Sept. 1986.

U.S. Environmental Protection Agency (USEPA), "Urbanization and Water Quality," prepared by Terrene Institute for U.S. Environmental Protection Agency, Washington, DC, 1994.

VTN-Nevada Report, "General Basin Site Study for the Lower Detention Basin," submitted to City of North Las Vegas, NV, Nov. 1995.

CHAPTER 9

DESIGN OF INFILTRATION BASINS FOR STORMWATER

James C. Y. Guo
Department of Civil Engineering
University of Colorado at Denver

The traditional approach of disposing stormwater is to collect surface runoff and to drain it to the nearest streams or lakes as rapidly as possible. Under this concept, an urban area is equipped with street gutters, swales, storm sewers, and roadside ditches. Under the 1987 Clean Water Act, many Best Management Practices (BMP's) have been developed to improve stormwater system designs. Many studies have revealed that the traditional approach in stormwater drainage has increased stream erosion and adversely impacted on downstream flooding problems (ASCE in 1984, Athayde in 1976). Increasing concerns have been developed regarding how the traditional approach affects the balance of a water body, creates adverse shock loads of pollutants to the receiving waters, or, in the case of combined sewers, contributes to treatment plant malfunctions. In response to various concerns, some communities have elected to encourage local disposal of stormwater at its source of runoff. This is done by having a portion of the stormwater infiltrate into the soil. The benefits of stormwater infiltration practices include groundwater recharge, low stream flow augmentation, water quality enhancement, and reduction in the total runoff volume. These multiple benefits make the infiltration practices suitable for meeting the intent and goals of maintaining the pre-development runoff characteristics.

To design infiltration basins or trenches for stormwater quality control, the catch basin requires a long and wide excavation and backfilled with gravel aggregate. Stored runoff water then infiltrates into the surrounding soil. The infiltration efficiency and amount depend on the distance to the local groundwater table and the soil properties. To design an infiltration basin, it is necessary to know the storage volume for the design rainfall event, soil infiltration rate on the land surface, seepage rate through the soil medium, assessment of water mounding effects on the local groundwater table, and overflow risk between two storm events. The challenge in the design of infiltration facilities is to assure that the facility will capture the target runoff amount, and the subsurface geometry can sustain the design infiltrating flow.

This chapter summarizes the required information and procedures for the design of infiltration facilities. It includes the design considerations for selection of basin site, estimation of detention volume, calculation of basin geometry, determination of infiltration rate on the land surface, and evaluation of subsurface seepage capacity. The analysis of infiltrating flow underneath a basin often includes water mounding effects. This chapter also summarizes the development of the potential flow model for predicting the growth and recession processes of the water mound under a basin. The potential flow model is applicable to small stormwater infiltration basins and can be a useful tool to investigate the impact of infiltration on the local groundwater table.

9.1 INFILTRATION DEVICES

Infiltration practices are essential to the stormwater management programs in many metropolitan cities in the United States. Infiltration practices have the ability to remove watershed born pollutants, to reduce the runoff volume, to provide groundwater recharge, to minimize thermal impacts on fisheries, and to augment low flow stream conditions (Harrington, 1988). The primary goal of infiltration practices in stormwater management is to maintain the post-development runoff characteristics, as nearly as possible to the pre-development. Stormwater infiltration devices include

1. *Infiltration beds.* Infiltration beds in Fig. 9.1 are often placed in the areas of vegetation and bushes around a building or business area. Runoff from roof areas or parking areas is spread onto vegetative areas. The ponding depth of infiltration beds shall not exceed 6 to 9 inches.

2. *Percolation subdrain.* Subdrains in Fig. 9.2 are used to filter the stormwater from roof areas or small paved areas. A roof gutter is extended into perforated pipes which deliver the stormwater into a rock backfilled subsurface catch basin.

3. *Sand filter and dry well.* Stormwater with greasy or fine pollutants go through filtering processes provided by sand filters or dry wells as shown in Fig. 9.2. For instance, precast concrete manhole segments can be stacked one on another to form a dry well which extends downward from the surface to permeable soils. Stormwater goes through the granular material and fiber fabric placed in and around the dry well or sand filter. When disposing a large amount of stormwater, clusters of dry wells can be used.

4. *Porous pavement.* Porous asphalt as well as parking pavements shown in Fig. 9.1 built of open-cell concrete blocks are commonly used to increase the infiltration amount. Designs of porous pavements take a conservative approach to allow for the less than satisfactory construction quality or the sporadic maintenance that often prevails.

5. *Infiltration trench.* Infiltration trenches in Fig. 9.1 are the devices frequently used for infiltration practices (Clement and Pensyl, 1987). An infiltration trench has an excavated width of 3 to 10 ft and then backfilled with stone aggregate to form a subsurface catch basin. Runoff is captured by depressing the surrounding ground surface. An infiltration trench is constructed to intercept the storm runoff from small paved areas less than 10 acres. To enhance the infiltration efficiency, a series of berms or check dams can be built across a trench. Each backwater pool acts like a small infiltration basin to increase the infiltration amount.

6. *Infiltration pond or basin.* An infiltration basin in Fig. 9.3 is used for stormwater quality or/and quantity controls. The storage basin is designed to have a large surface area under a shallow depth. Runoff water enters the basin through adequate energy dissipation. A rock-backfilled trench may be installed as a trickle channel. At the low area, a sand filter or drainage wells can be implemented to enhance the infiltration capacity. It is advisable to have a backup subdrain system in case of standing water problems. The emergency outlet can be an overflow spillway or a concrete vault.

An infiltration basin consists of four basic elements. They are: (a) inlet to receive and to spread storm runoff, (b) catch basin to store runoff volume, (c) filtering outlet to release water, and (d) emergency by-pass facility. Standards and specifications for urban stormwater

INFILTRATION BEDS

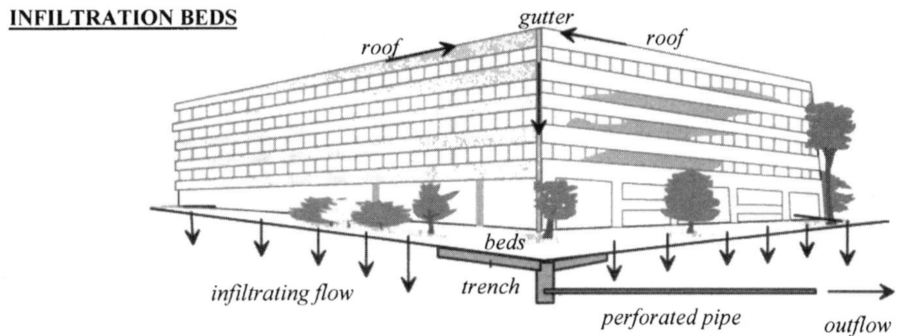

POROUS PAVEMENT

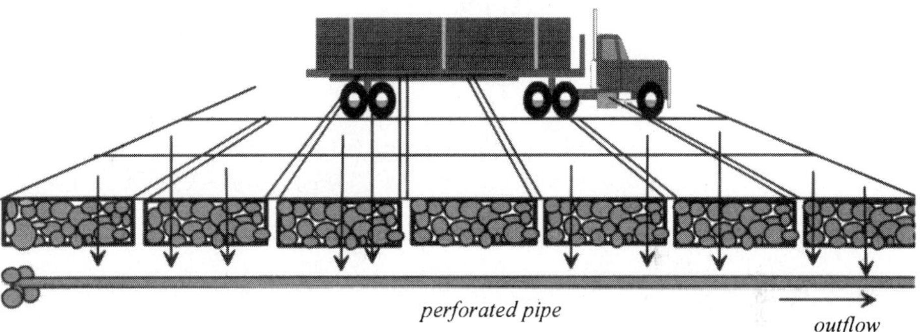

INFILTRATION TRENCH

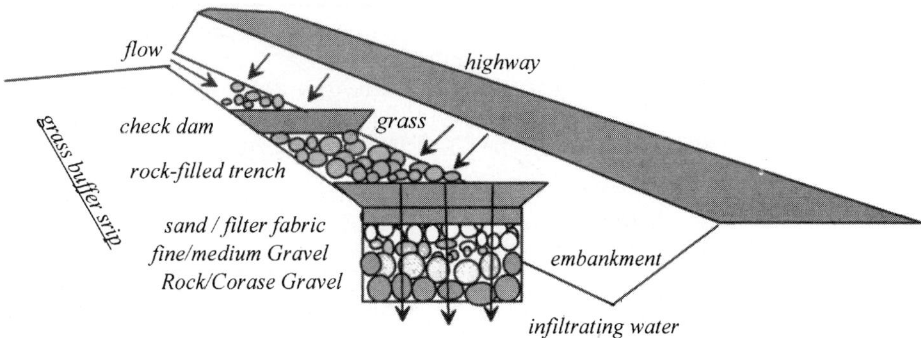

FIGURE 9.1   Examples of infiltration bed, porous pavement, and infiltration trench.

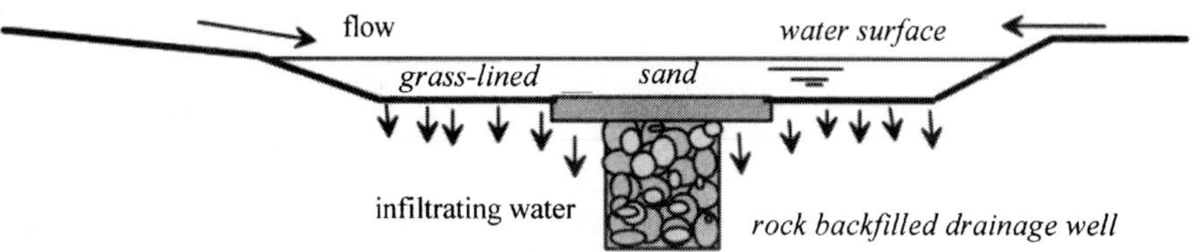

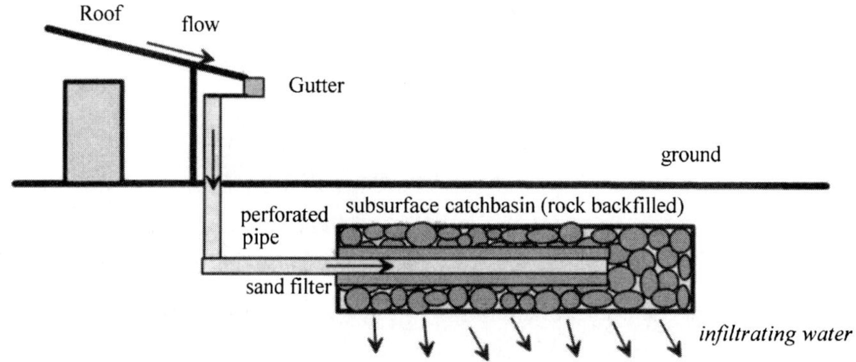

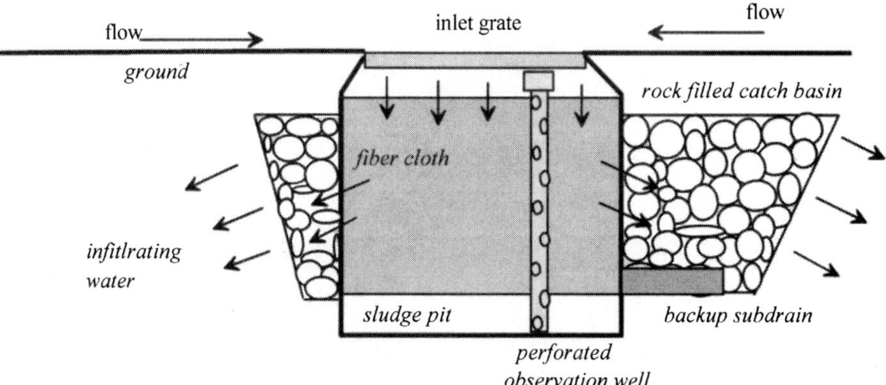

FIGURE 9.2 Examples of infiltration basin, percolated subdrain, and dry well.

infiltration devices are available in many stormwater drainage criteria manuals. For instance, *Underground Disposal of Storm Runoff Design Guideline* published by the Federal Highway Administration (Hannon, J. B, 1980)*, Stormwater Quality Design Criteria Manual* published by the Urban Drainage and Flood Control District, Denver, Colorado in 1999. The purpose of standards and specifications is to establish the recommended criteria for design, review,

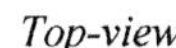

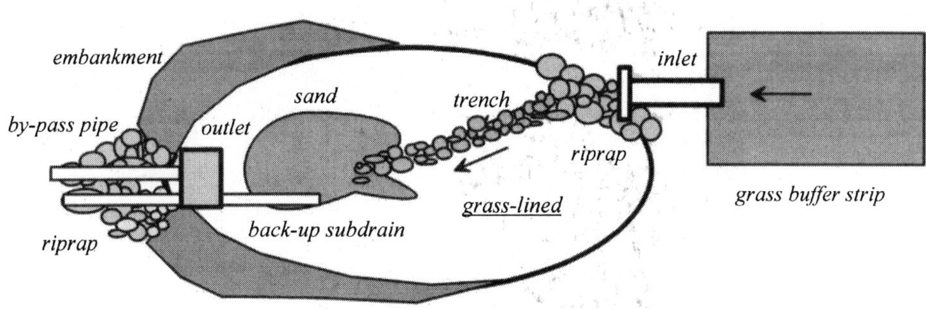

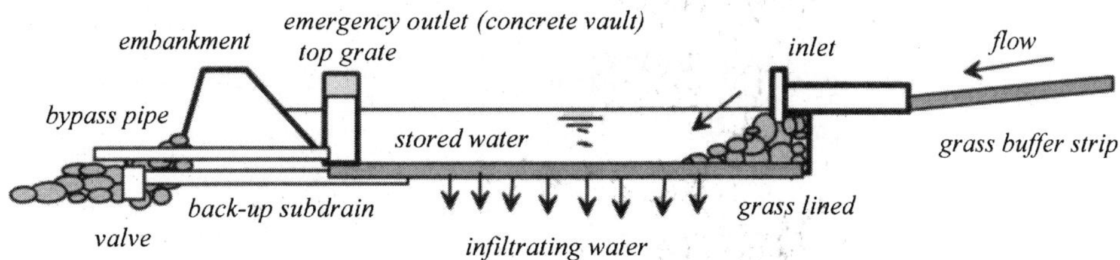

FIGURE 9.3   Illustration of infiltration basin.

approval, construction, and maintenance of infiltration practices. Therefore, the detailed layout and dimensions of various infiltration devices will not be repeated. Figures 9.4 through 9.8 illustrate infiltration trenches and infiltration basins.

9.2   GENERAL DESIGN CONSIDERATIONS

Infiltration practices can only control runoff from small drainage areas. The operational efficiency of an infiltration device is highly dependent on the underlying soil conditions. Design of an infiltration basin must take both surface and subsurface hydrologic systems into consideration. The surface hydrologic study estimates the stormwater storage volume, maximum water ponding depth, basin geometry, predominant pollutants, and sedimentation requirements. The subsurface hydrologic study is to evaluate whether the soil infiltration rate on the land surface can be sustained by the subsurface seepage capability. The basic design information for an infiltration facility includes:

1. Hydrologic parameters of the tributary watershed
2. Predominant pollutants in storm runoff and their residence times for sedimentation
3. Stormwater storage volume
4. Feasibility and safety factors for basin site

FIGURE 9.4 Infiltration trench built in business district.

FIGURE 9.5 Infiltration trench installed along highway.

FIGURE 9.6 Small infiltration basin to catch stormwater from parking area.

FIGURE 9.7 Large infiltration basin with 12- to 18-inch shallow depth.

FIGURE 9.8   Standing water in infiltration basin as a result of clogging.

5. Surface soil texture and infiltration capacity

6. Subsurface soil seepage capacity

7. Groundwater information

Design information of the tributary catchment to the infiltration facility includes the local rainfall pattern, catchment runoff coefficient, tributary area, and watershed-borne pollutants. For the purpose of stormwater peak flow attenuation, an infiltration basin may be designed for 2- to 10-year events. For water quality enhancement, an infiltration basin may be designed for the stormwater quality control capture volume. The type of pollutants dictates the selection of the infiltration device. For instance, oil and grits in stormwater can be more efficiently removed by a sand filter than vegetative beds. The sedimentation process for solid particles requires a sufficient residence time which dictates the operation of the basin.

At a site, it is necessary to examine the feasibility factors, including the topographic slope, the nature of the soil, the proximity of building foundations and water supply wells, and any documented groundwater pollution. The topographic slope for porous pavements and vegetated swales should be relatively flat, not exceeding 5.0%. It is advisable to restrict the use of infiltration practices on steep slopes because a steep slope increases the chance of downstream water seepage from the subgrade and reduces the amount of infiltrating runoff into the soil. The use of infiltration practices on fill material is not recommended due to the possibility of creating an unstable subgrade. Fill areas can be very susceptible to slope failure due to slippage along the interface of in-situ and fill materials. This condition might be aggravated if the fill material is on a slope and is allowed to become saturated by infiltrating water.

Building foundations should be at least 10 ft up the hydraulic gradient from infiltration practices to prevent possible flooding basements. Any infiltration basin receiving runoff from commercial or industrial parking areas should be located a minimum of 100 ft from any water supply well. Infiltration basins and trenches should be at least 10 ft down from any building foundations. The impact of an infiltration basin on the local groundwater table and groundwater quality must be evaluated.

An infiltration trench is installed adjacent to a paved area. Sheet flows of runoff water shall have gone through a grass buffer strip before entering the trench. The buffer strip shall have a slope less than 10 to 15% and maintain a flow depth less than 0.20 ft. An infiltration trench is backfilled with stone aggregate of 1 to 3 in in diameter. The aggregate void ratio can range from 30 to 40%, depending on the stone size. A filter fabric is required around the walls or a 6-in sand layer shall be installed on the bottom of the trench to prevent the migration of fine soil particles from entering into the gravel voids.

An infiltration basin is designed to remove pollutants from a watershed up to 10 acres. At the basin site, the minimum distance between the basin bottom to the groundwater elevation is 5 to 10 ft. The distance to the bed rock shall be greater than 5 ft. An infiltration basin must have the capability to drain off stored water between storm events. To enhance the infiltrating amount, a large and shallow basin is preferred to a small and deep basin. The basin floor shall be as flat as possible. Additional drainage wells or high infiltration medium like sand may be placed at the low point on the basin floor. An infiltration basin is very susceptible to sediment clogging. All construction activities which generate sediment from the areas upstream of the basin must be minimized. Protective measures to prevent sediment from reaching the basin should be implemented. For instance, grass butter strips shall be utilized where sheet flows enter the basin and a sediment forebay may be installed where concentrated flows enter the basin. A method to remove coarse sediment should be installed near the entrance of the basin. If a base flow exists, the infiltration basin shall be equipped like a wetland or retention pond.

The primary factors affecting soil infiltration are soil type, antecedent soil moisture (AMC), vegetative cover, and the soil surface texture such as crusted or frozen. Soils are hardly homogeneous. Adequate field samples shall be surveyed to estimate the infiltration capacities at a site. Table 9.1 presents the recommended design infiltration rates and water storage capacities in terms of soil pores for different types of soils. (Yousef et al., 1986)

In general, soil textures with infiltration rates less than 0.20 in per hour are not suitable for infiltration practices. Such unsuitable soils may consist of at least 25.0% clay. Soils with a poor drainage capacity are also susceptible to frost heaving and swelling expansion which may cause possible structural instability. Soil infiltration rates can be significantly reduced by clogging effects because of the accumulation of pollutants in soil pores. If an infiltration device is not properly designed, it has the tendency to clog rapidly due to sediment entry during and after construction. It is important to protect the basin site from clogging of soil

TABLE 9.1   Soil Infiltration Rates and Water Storage Capacities

Soil group	Infiltration rate (inch/hr)	Water storage capacity (inch of water)
Sand, open-structured	8.27	0.35
Loamy sand	2.41	0.31
Sandy loam	1.02	0.25
Loam	0.52	0.19
Silty loam	0.27	
Sandy clay loam	0.17	
Clay loam	0.09	
Silty clay loam	0.06	
Clay	0.02	

Source: Maryland Standards and Specification for Stormwater Management Infiltration Practices, 1984

TABLE 9.2 Permeability Coefficients (Hydraulic Conductivity) for Various Porous Media

Material	Hydraulic conductivity, K, (cm/sec)
Gravel	10^{-1} to 10^{-2}
Sand	10^{-5} to 1.0
Silt/loam	10^{-7} to 10^{-3}
Clay	10^{-9} to 10^{-5}

pores. A constant standing water pool implies possible clogging due to sediment yields from the catchment. It is necessary to check the soil erosion controls in the catchment. To warrant the functional integrity of an infiltration device requires adequate maintenance and backwash of the filtering layers.

The quantity of water that can infiltrate into the soil depends on the soil storage volume in terms of the soil porosity, and the conveyance capacity in terms of the subsurface hydraulic gradient and conductivity. *Darcy's Law* describes a steady laminar flow through saturated soil medium as:

$$q = KiA \tag{9.1}$$

where i = hydraulic gradient
  q = rate of flow through the cross-sectional area A
  i = hydraulic gradient
  K = hydraulic conductivity

The *permeability* is also called *hydraulic conductivity* which is a soil property to reflect how fast the water can flow through the soil. Permeability coefficients are listed in Table 9.2 for various porous media.

The flow pattern of infiltrating water through the bottom of an infiltration basin can be described by a flownet which consists of stream lines and equipotential lines. Stream lines are defined as the lines tangent to the velocity vectors throughout the flow field, and equipotential lines depict the equal potential in the flow field. The stream lines have to cross the equipotential lines at right angles. *Hydraulic gradient* is defined as the energy or head loss per unit length along the flow path. The hydraulic gradient between any two adjacent equipotential lines is computed by the difference of potential heads divided by the distance traversed as:

$$i = \frac{\Delta h}{\Delta x} \tag{9.2}$$

in which Δh = energy loss, and Δx = distance traveled. The impacts of the infiltrating water on the groundwater table can be studied by the flownet which requires the prior knowledge of soil infiltration and seepage rates.

Among the aforementioned infiltration devices, this chapter focuses on the design of infiltration basins.

9.3 DESIGN STORAGE VOLUME

An infiltration basin can be designed as a *flood detention basin* for peak flow attenuation purposes, or a *water quality control basin* for water quality enhancement purposes. A flood detention basin shall have the storage capacity determined by the minor or major event. Since soil infiltration is a slow process, a flood detention infiltration basin often requires a large surface area to increase the infiltration capacity. On the contrary, a stormwater quality control basin is designed to capture daily runoff. Its capacity shall be determined by the local continuous rainfall record rather than the extreme events. From the stormwater capture point of view, the larger the basin is, the lower the overflow risk will be. However, from the cost point of view, the smaller the better. Under such a tradeoff, an infiltration basin is often designed as a stormwater outfall device for a small and paved catchment such as parking lot, industrial park, or business area. The storage capacity of an infiltration basin depends on whether the basin is designed for stormwater quantity or quality control.

9.3.1 Detention Storage Volume

A flood detention infiltration basin is designed to mitigate the increased stormwater volumes due to development. Such a basin shall be sized to cope with the minor or major storm event. For a small urban catchment, the *Rational method* is applicable to the predictions of peak runoff rates and runoff volumes using the local *rainfall intensity-duration-frequency* (IDF) *curve* as:

$$Q_d = CI_d A \tag{9.3}$$

$$I_d = \frac{a}{(T_d + b)^n} \tag{9.4}$$

where C = runoff coefficient
A = watershed area
I_d = rainfall intensity
T_d = rainfall duration
Q_d = peak runoff rate
a, b, and n = constants on the IDF formula.

Soil infiltration can be described by the Horton's formula as:

$$f(t) = f_c + (f_o - f_c)e^{-kt} \tag{9.5}$$

where $f(t)$ = infiltration in in/hr (or mm/hr) at elapsed time t
f_o = initial infiltration rate
f_c = final infiltration rate
k = decay coefficient in 1/hr.

Integration of Eq. (9.5) yields

$$F(t) = f_c t + \frac{(f_o - f_c)}{k}(e^{-kt} - 1) \tag{9.6}$$

where $F(t)$ = accumulative infiltration depth in in (or mm) at elapsed time t

The design storage volume for an infiltration basin can be determined by the volume-based approach. As illustrated in Fig 9.9, the concept in a volume-based method is to find the maximum volume difference between the inflow and outflow volumes under a series of storm events with a different duration (Guo, 1999). For a specified storm duration, the runoff detention volume is equal to:

$$V_d = \alpha CI_d AT_d - \beta A_b F(T_d) \tag{9.7}$$

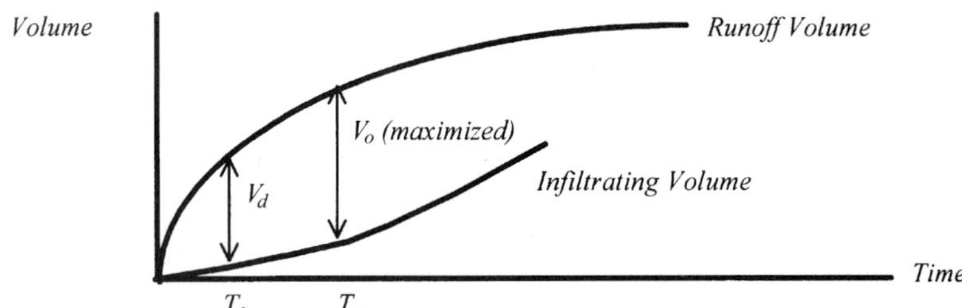

FIGURE 9.9   Maximization of storage volume for infiltration basin design.

where V_d = storage volume
  A_b = infiltrating area of the basin
 α and β = unit conversion factors

The maximal value of Eq. (9.7) is achieved by setting its first derivative with respect to T_d equal to zero, and it results:

$$T_m = \frac{1}{n}\left[(T_m + b) - (T_m + b)^{n+1}\frac{\beta A_b}{a\alpha CA}f(T_m)\right] \tag{9.8}$$

in which T_m = design rainfall duration described by Eq. (9.8). The maximum detention storage volume, V_o, is:

$$V_o = \alpha CI_m AT_m - \beta A_b F(T_m) \quad at \quad T_d = T_m \tag{9.9}$$

The *infiltration volumetric rate, Q,* is

$$Q = \eta f A_b \tag{9.10}$$

in which η = unit conversion factor.

When the rainfall IDF formula is not available, it is recommended that a tabulation approach be applied to a range of rainfall duration. For each rainfall duration, the detention volume for each storm duration can be directly calculated by Eq. (9.7). Repeat such a process until the maximal detention volume is achieved.

Example 9.1
A 2.1-acre residential subdivision is located in Denver, Colorado, and has a runoff coefficient of 0.65. The local 10-year IDF formula is described by Eq. (9.4) with $a = 45.92$, $b = 10.0$ and $n = 0.786$. The 10-year storm runoff from this watershed will drain into a 180-ft by 20-ft trench basin. The infiltration rates of the trench basin are: $f_o = 6.00$ in/hr, $f_c = 1.80$ in/hr, and $k = 6.50$ /hour. Based on the given information, the unit conversion factors and the bottom area of the basin are:

$$\alpha = 60, \ \beta = \frac{1}{12}, \ \eta = \frac{1}{12*3600}, \quad and \quad A_b = \frac{20.0*180.0}{43560.0} = 0.083 \text{ acre.}$$

Substituting these variables into Eqs. (9.5) and (9.8) yields

$$f(T_m) = 1.80 + (6.00 - 1.80)e^{-(T_m/60*6.5)}$$

$$T_m = \frac{1}{0.786}\left[(T_m + 10) - (T_m + 10)^{0.786+1}\frac{1/12*0.083*f(T_m)}{45.92*60.0*0.65*2.1}\right]$$

TABLE 9.3 Calculation of Detention Storage Volume for Various Events

Duration minutes	Rainfall intensity inch/hr	Inflow volume acre-ft	Infiltration rate inch/hr	Infiltration depth inches	Outflow volume acre-ft	Storage volume acre-ft
120	1.00	0.23	1.80	4.25	0.03	0.197
240	0.60	0.27	1.80	7.85	0.05	0.216
300	0.51	0.29	1.80	9.65	0.07	0.219
340	0.46	0.29	1.80	10.85	0.08	0.220
360	0.44	0.30	1.80	11.45	0.08	0.219
420	0.39	0.31	1.80	13.25	0.09	0.217
480	0.35	0.32	1.80	15.05	0.10	0.215

The design storm duration described by these two equations was found to be: $T_m = 340.0$ minutes or 5.67 hours. Using Eq. (9.9), the required detention volume, V_o, for this case is 0.220 acre-ft and the total infiltration depth, $F(T_m)$, calculated by Eq. (9.6) is 4.33 in. The same results can also be achieved by applying Eq. (9.7) to a range of rainfall duration as shown in Table 9.3.

The infiltration volume rate released from the basin is calculated by Eq. (9.10) as:

$$Q_o = \eta f_c A_b = \frac{1}{12 * 3600} * 1.80 * 0.083 = 0.151 \quad cfs$$

The drain time or ponding time, T_p, for the design volume of 0.219 acre-ft is

$$T_p = \frac{43560 A_b}{60 Q_0} = 17.55 \text{ hours}$$

If this release rate from the basin can be sustained by the subsurface conductivity, it will take 5.67 hours to load the basin, and another 17.5 hours to empty the basin, or a total of 23.22 hours to complete the cycle of operation.

9.3.2 Water Quality Capture Volume

The storage capacity of a stormwater quality control basin is termed *water quality control volume* (WQCV) expressed in watershed mm or watershed inch. The WQCV for a stormwater quality infiltration basin is to capture and to treat daily runoff volumes (micro events), not the extreme events. Analyses of the 40-year continuous rainfall records observed in Denver, Colorado reveal that less than 5% of rainfall events exceed the 2-year precipitation depth, and less than 1% of rainfall events exceed the 10-year rainfall depth. Of note is that 95% of a complete rainfall data series studied fall into the category of *micro or daily events* (Guo and Urbonas, 1996). As an alternative to the traditional extreme event approach, the *storm runoff capture curve* at the basin site shall be developed to describe the non-exceedance probability distribution of runoff volumes. Investigations on the rainfall event-depth distributions for seven metropolitan cities in various meteorological zones indicate that the density function of rainfall event-depths can be described by the exponential function as (Waniellista and Yousef, 1993, Guo and Hughes, 2001):

$$f(p) = \frac{1}{P_m} e^{-p/P_m} \tag{9.11}$$

where $f(p)$ = density function for rainfall event-depth
p = rainfall event-depth
P_m = average rainfall event-depth

The non-exceedance probability for rainfall event-depth is the integration of Eq. (9.11) as:

$$P_D(0 \le p \le P_o) = 1 - e^{-P_o/P_m} \tag{9.12}$$

in which P_o = precipitation depth for design. The runoff volume produced by a rainfall event is estimated by

$$d = C(p - P_i) \tag{9.13a}$$

where d = runoff volume produced by rainfall depth, p
C = runoff coefficient
P_i = incipient runoff depth which is the minimal required rainfall depth for producing runoff

The EPA reports (1983 and 1986) suggest that the incipient runoff depth be 0.1 in. Similarly, the relationship between the design rainfall depth, P_o, and runoff depth, D_o, is

$$D_o = C(P_o - P_i) \tag{9.13b}$$

Substituting Eqs. (9.13a) and (9.13b) into Eq. (9.12) yields

$$P_D(0 \le d \le D_o) = 1 - e^{-P_i/P_m} e^{-D_o/CP_m} \tag{9.14}$$

$$V_o = D_o A \tag{9.15}$$

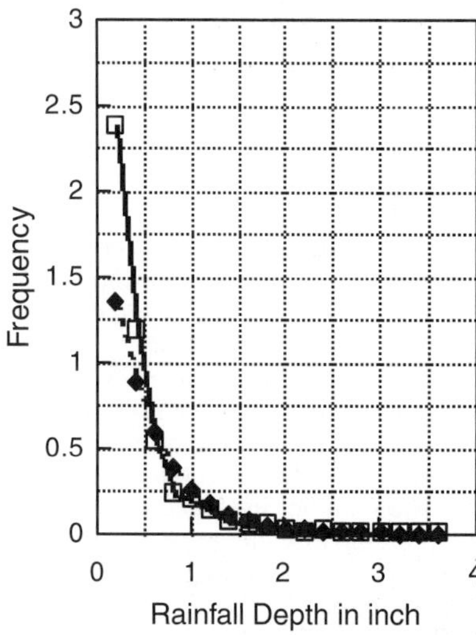

FIGURE 9.10 Comparison between observed and predicted rainfall depth distributions at Seattle, WA.

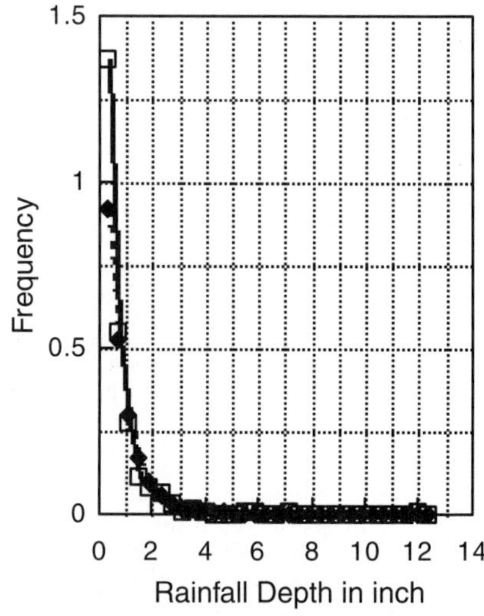

FIGURE 9.11 Comparison between observed and predicted rainfall depth distributions at Boston MA.

where D_o = WQCV or design runoff depth in watershed mm
      V_o = basin storage volume
      A = tributary catchment area

Equation (9.14) is termed the *runoff capture curve* which provides the non-exceedance probability distribution for runoff depths (Guo and Hughes, 2001). Figures 9.10 and 9.11 are the comparison of the predicted rainfall event-depth distributions by Eq. (9.11) and the observed records at Seattle, WA, and Boston, MA. Figures 9.12 and 9.13 are the runoff capture curves predicted by Eq. (9.14) and derived from the continuous rainfall records observed at Denver, CO, and Tampa, FLA. Similar to the traditional approach in which the magnitude of the design event depends on the selected overflow risk, the WQCV for an infiltration basin can

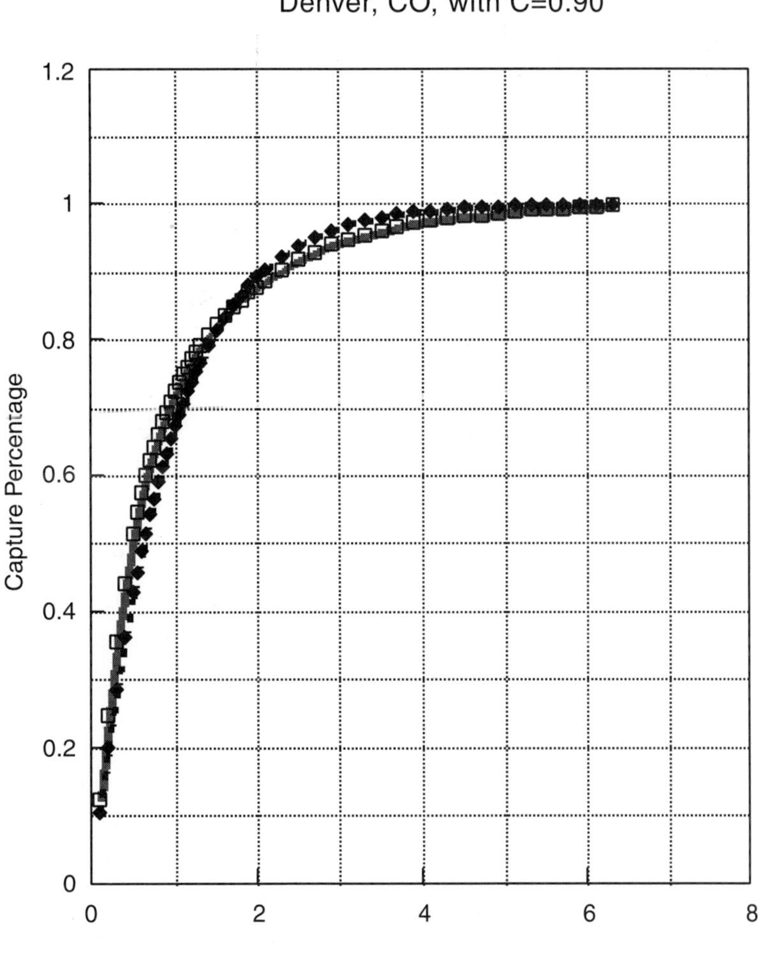

FIGURE 9.12   Comparison between observed and predicted runoff capture curves for Denver, CO.

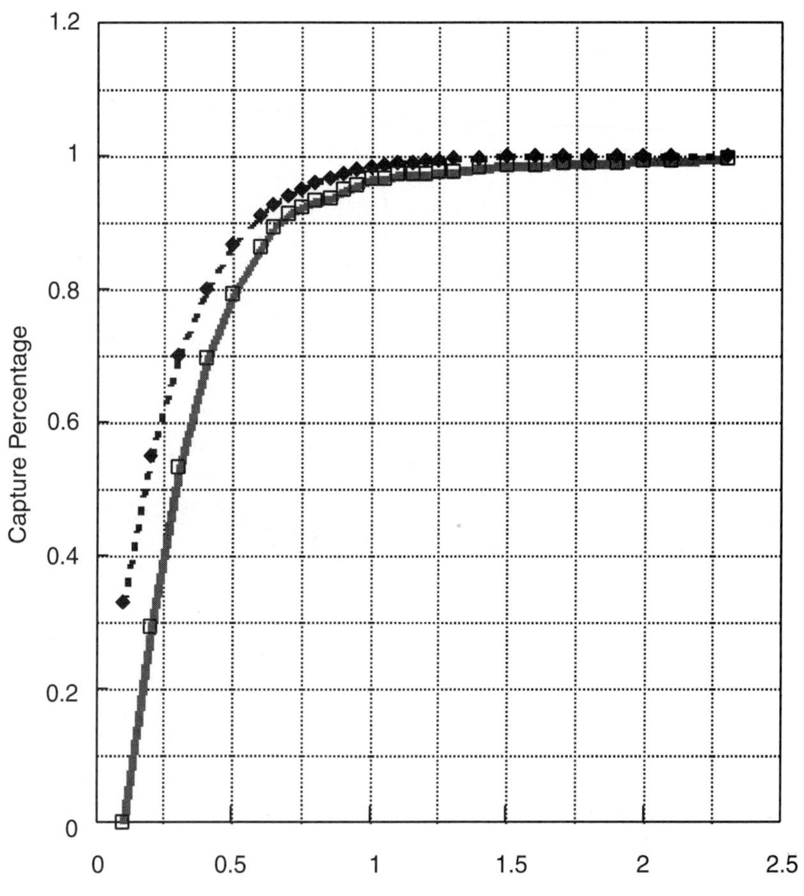

FIGURE 9.13   Comparison between observed and predicted runoff capture curves for Tampla, Florida.

be determined according to the target runoff capture ratio. To apply Eq. (9.14) to a basin site in the United States, the required average rainfall event-depth can be found in Fig. 9.14 (EPA Report, 1986).

Example 9.2

A stormwater quality control infiltration basin is located in Tampa, FLA and designed to capture 75% of runoff volumes from a 2.0-acre catchment. The runoff coefficient for the catchment is 0.65. Determine the WQCV for the basin.

From Fig. 9.14, the average rainfall event-depth at Tampa, FLA is 0.80 in. Substituting $P_i = 0.1$ in, $C = 0.65$ and $P_m = 0.8$ in into Eq. (9.14) yields:

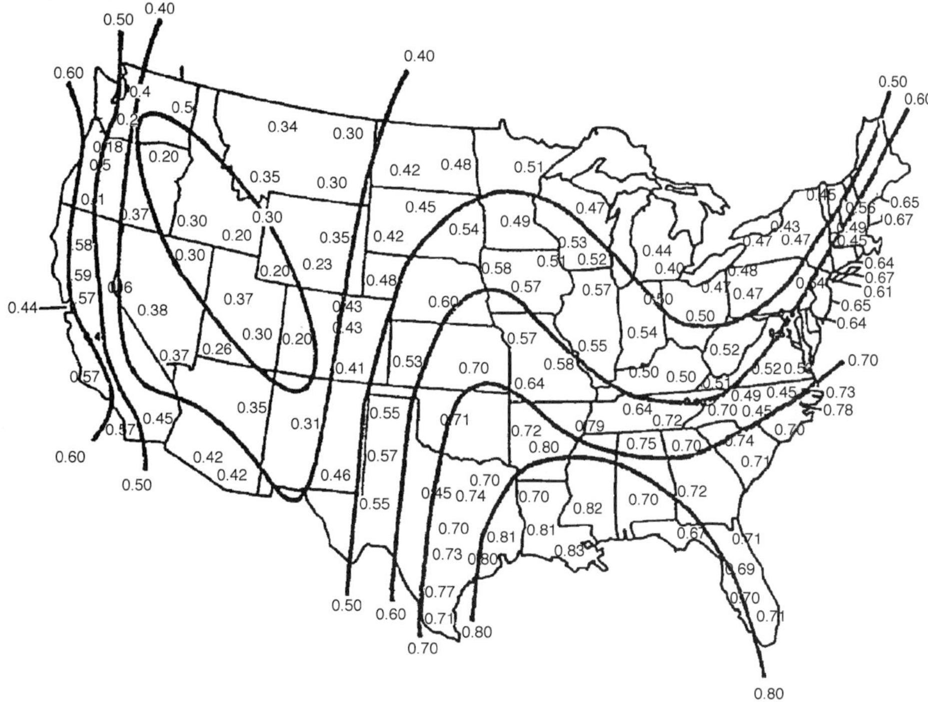

FIGURE 9.14   Average event-rainfall depths for the United States. (EPA, 1986)

$$P_D = 1 - e^{-0.1/0.8}\, e^{-D_o/0.65*0.80} = 1 - 0.88e^{-1.923D_o}$$

in which P_D = target runoff capture ratio, i.e. 0.75 for this example. The corresponding WQCV, D_o, is found to be 0.66 in per watershed, or the basin storage volume, V_o, by Eq. (9.15) is

$$V_o = 0.66/12 * 2.0 = 0.11 \text{ acre-ft}$$

9.4 *BASIN GEOMETRY*

The above procedure applies the surface hydrologic principles to size an infiltration basin. At the basin site, it is also necessary to make sure that the soil medium below the basin can provide an adequate subsurface reservoir by soil pores, i.e. the distance to the groundwater table. The infiltration process begins with filling up the soil pores beneath the basin. After the soil becomes saturated, the infiltrating water will then flow through the soil medium to recharge the groundwater table. An equilibrium condition is defined by the infiltration rate at the land surface equal to the seepage rate to the groundwater table. During this process, a water mound will be developed. (Hantush, 1968).

As illustrated in Fig. 9.15, the subsurface storage capacity in the soil below the basin can be estimated as

$$V_s = Z(\theta_s - \theta_o)A_o \tag{9.16}$$

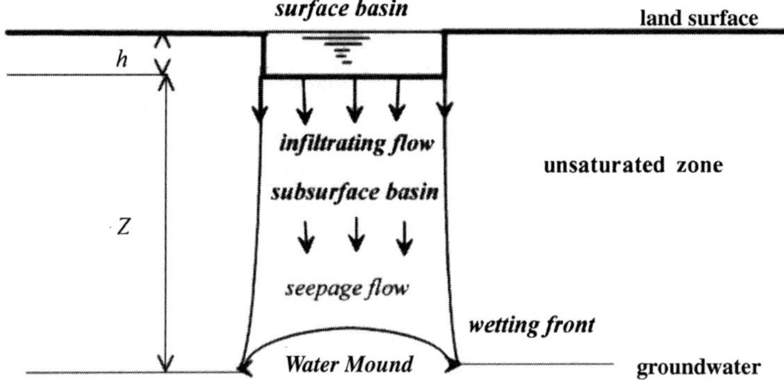

FIGURE 9.15   Illustration of soil saturation beneath a basin.

where V_s = subsurface storage capacity
A_o = basin bottom area
θ_s = soil porosity
θ_o = soil initial water content or specific retention
Z = distance between basin bottom and groundwater table

To be conservative, the required runoff detention volume determined by the surface hydrology shall be less than or equal to the water storage volume available in the soil pores below the basin. Therefore, the dimensions of the infiltration basin can be determined by

$$V_o \leq V_s \tag{9.17}$$

Substituting Eq. (9.16) into Eq. (9.17), the minimum basin bottom area, A_b, is determined as:

$$A_b \geq \frac{V_o}{Z(\theta_s - \theta_o)} \tag{9.18}$$

The maximum water ponding depth in the basin is determined as:

$$h \leq Z(\theta_s - \theta_o) \tag{9.19}$$

As a rule of thumb, the volume of 1-ft water is approximately equivalent to the soil pore volume in a 4-ft soil medium.

Example 9.3
At the basin site, the soil porosity is 0.35 and the soil specific retention is 0.10. The distance to the local groundwater table is 8.0 ft. Determine the basin bottom area and maximum water depth for the basin in Example 9.2.

   With V_o = 0.11 acre-ft, θ_s = 0.35, θ_o = 0.10, and Z = 8.0 ft, the basin bottom area is found

$$A_o \geq \frac{0.11}{8.0(0.35 - 0.10)} = 0.055 \text{ acre-ft} = 2395.8 \text{ sq. ft}$$

The basin bottom area is approximately equivalent to a 50-ft square. The maximum ponding depth for this infiltration basin is determined as:

$$h \leq 8.0(0.35 - 0.10) = 2.0 \text{ ft}$$

9.5 INFILTRATION RATE AND DRAIN TIME

The drain time of an infiltration basin may also serve as a design criterion. The draining process includes a *surface ponding time* to infiltrate water from the basin, and a *subsurface storage time* to fill up the subsurface aggregate reservoir. A *surface ponding time* is defined by the infiltration rate on the land surface as:

$$T_p = \frac{h}{f} \tag{9.20}$$

in which T_p = surface ponding time. A *subsurface storage time* is defined by the subsurface seepage rate to fill up the soil pores as:

$$T_s = \frac{Z(\theta_s - \theta_o)}{v} \tag{9.21}$$

where T_s = subsurface storage time
v = seepage rate in inch/hr or mm/hr

When $f \geq v$, the subsurface reservoir undergoes a filling process. The seepage rate gradually increases as the subsoils become saturated. It is desirable that the subsurface hydraulic gradients are sufficient to maintain $f \leq v$ when the subsurface reservoir becomes fully saturated. Otherwise, the system will be backed up and controlled by the outgoing seepage rate.

It is important that a proper ponding time is selected for the infiltration basin under design. However, between stormwater quality and quantity controls exists a tradeoff when selecting the ponding time for a basin. For instance, a longer ponding time is preferred for sediment removal, but a longer ponding time also increases the chance of overflow by the next rainfall event because the design storage volume is being occupied by the current event. Therefore, based on the local rainfall pattern, public safety, and pollutant characteristics, the maximum allowable ponding time can be established as a design criteria. For instance, the Department of Natural Resources, Maryland recommends that both the maximum allowable ponding and storage times be not to exceed a 72-hour period in which the stored water within the basin must be completely drained, with the exception of vegetative swales which have a maximum ponding period of 24 hours (Shaver, 1986). In practice, the maximum allowable ponding and storage times provide a guidance to design the soil infiltration rate.

Example 9.4
Using the basin in Example 9.3, determine the soil infiltration rate with a maximum allowable ponding and storage times of 72 hours.
Substituting T_p = 24 hours, and h = 2.0 ft or 24 in into Eq. (9.20) yields

$$24 \; hours = \frac{24 \; inches}{f}$$

The design soil infiltration rate is 1.0 in/hr for the case. According to Table 9.1, the basin bottom can be covered by sandy-loam soils. To sustain the continuity of flow between surface and subsurface systems, the seepage rate must be greater than or equal to 1.0 in/hr. Such a sustainability depends on the hydraulic gradients and soil permeability below the basin.

9.6 SUSTAINABILITY OF INFILTRATING FLOW

The *infiltrating capacity* of a basin is not only dependent on the soil texture on the land surface, but also the hydraulic conductivity through the soil medium. If the surface soil infiltration rate is higher than the subsurface seepage rate, the system is backed up and may even cause a failure in the operation. The soil hydraulic conductivity is generally related to its saturated condition by the soil-water characteristic curve (Stankovich and Lockingtom, 1995). To model the infiltrating water through soils typically requires such a functional representation of soil hydraulic properties (Brook and Corey, 1964). Infiltration of stormwater through soils is a transient flow process. When an infiltration basin is loaded with stormwater, the soil medium between the basin and the groundwater table will undergo a storage process in which the soil water content varies from an unsaturated to a saturated condition. When the infiltrating water reaches the groundwater table, a water mound will begin to build up. The shape and growth of a mound depend on infiltration rate, size of the basin, and hydraulic properties of soil medium (Ferguson, 1990). During a water mounding process, the soil medium undergoes a wetting and draining cycle. The soil hysteretic functional relationship between moisture content and pressure head makes the determination of the hydraulic conductivity even more challenging. In practice, either soil characteristic curve (Sumner et al., 1999) or a constant hydraulic conductivity (Bouwer, 1999; Rastogi and Pandy, 1998) was used to model a water mound.

The water mounding process has been investigated under different assumptions (Hantush in 1968; Ortiz, et al. in 1979; Morel-Seytous, et al. in 1988 and 1990). Although water mounding is a transient process, it may reach a quasi steady state when the inflow and outflow rates are balanced. For instance, the numerical simulations of the water mound below a 500 m by 100 m rectangular basin indicate that the water mound begins with a fast and significant build-up in the first 30 days, and then gradually reduces its height for the next 52.3 days until a steady state is achieved (Rastigo and Pandy, 1998). Such a cycle of increase, and decrease, in the mound height was caused by the change of water storage volume. During the initial period, the soil is unsaturated with a lower hydraulic conductivity, and the soil water storage capacity has a significant effect on the growth of water mound. As the time of recharge increases, the growth of a mound on a groundwater table results in higher hydraulic gradients from the top of the mound toward the initial water table. The mound growth reaches its steady state when the rate of infiltration is equalized by the rate of recharge to the groundwater table (Bouwer, 1999).

When dealing with a low-permeability environment, it is necessary to select a site which can provide sufficient hydraulic gradients to increase the seepage rate. If the seepage rate determined by the permeability and hydraulic gradient is higher than the surface infiltration rate, it is a case of direct recharge to the aquifer. Otherwise, the slow groundwater movement will cause saturation to the surroundings beneath the basin. As the result, the operation of the basin is rather controlled by the outgoing seepage rate to the groundwater table than the incoming infiltration rate from the land surface. Under this condition, the basin designed with a high infiltration rate becomes undersized. Therefore, when selecting a basin site, it is important to know if the subsurface geometry at the site will lead to an operation of surface infiltration control or subsurface soil seepage control.

9.7 POTENTIAL FLOW MODEL FOR A TRENCH

As illustrated in Fig. 9.16, the infiltrating water begins with a vertical downward velocity through the unsaturated zone underneath the basin. As the soil water content increases, the diffusive nature of the wetting front results in flow movements in both vertical and lateral directions.

Many studies on water mounds were performed with a potential function using the groundwater depth as the variable (Hantuch, 1968; Bouwer, 1999). On the other hand, the seepage flow through soils can also be depicted by a potential flow using a stream function. With the consideration of vertical and lateral movements, the stream function was developed for the infiltrating flow under a trench as (Guo, 1998):

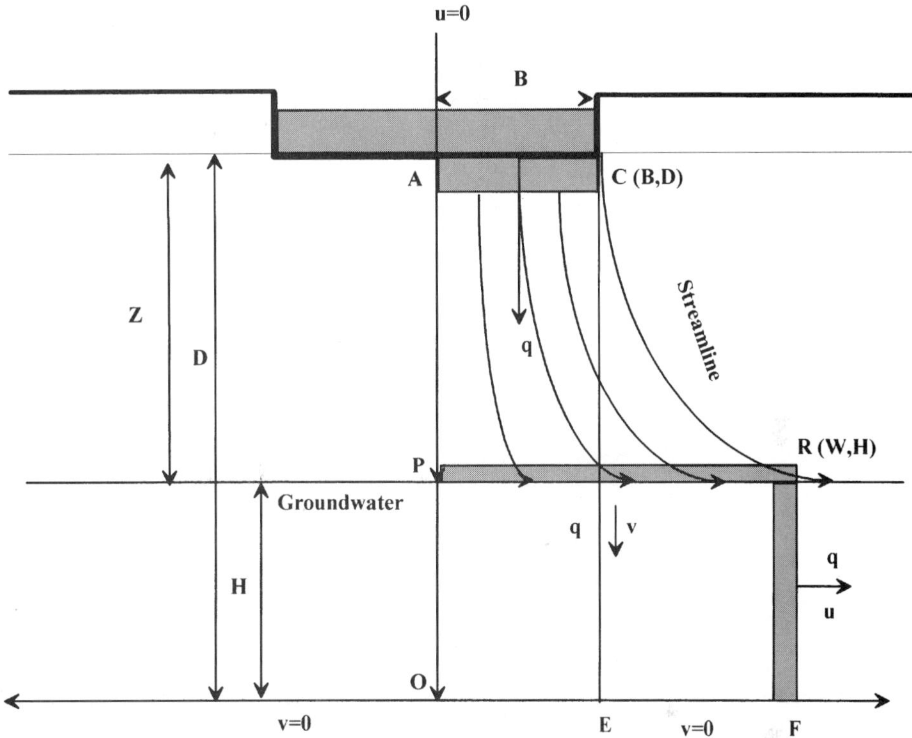

FIGURE 9.16   Potential flow pattern for infiltrating water under a trench.

$$\psi = \frac{f}{D}\,xy \tag{9.22}$$

where ψ = stream function
f = infiltration rate
D = saturated depth
x = horizontal from the central axis of the basin
y = vertical distance below the basin.

Equation (9.22) satisfies the two dimensional continuity and irrotationality equations. The stream function value, ψ_c, at point C (B, D) is

$$\psi_c = q = fB \tag{9.23}$$

where q = infiltration volume rate
B = half width of the trench

To maintain the continuity along the streamline, ψ_c, between the basin and any cross section at a distance y, Eq. (9.22) yields

$$w = \frac{BD}{y} \tag{9.24}$$

where w = width of the wetting front and y = vertical distance to the width w. Derivatives of Eq. (9.22) with respect to y and x represent the velocity vectors of the seepage flow as:

$$u = \frac{f}{D}\,x \tag{9.25}$$

$$v = -\frac{f}{D}\,y \tag{9.26}$$

Under the condition of a steady state, the flow rate remains as a constant at Sections AC, PR, and RF in Fig. 9.11. On Section PR, the hydraulic gradient, i, can be approximated by unity, $i = -1$ (downward) when the soil suction is ignored (Bouwer, 1999; Guo; 1998). According to Darcy's law, the flow of recharge at Section PR is:

$$v = \frac{q}{W} = \frac{K_i i W}{W} = -K_y \tag{9.27}$$

in which K_y = hydraulic conductivity in the vertical direction. Equation (9.27) must agree with Eq. (9.26) at $y = H$. As a result, it is concluded that

$$D = \lambda_y H \tag{9.28}$$

$$\lambda_y = \frac{f}{K_y} \tag{9.29}$$

Streamlines of recharge flow near the mound are concentrated in the upper or active thickness of the aquifer, with much less flow and almost stagnant water in the deeper portion of the aquifer (Bouwer, 1999). In this study, such an active flow depth, H, below the initial groundwater table can be estimated by the *Dupuit-Forchheimer equation* as:

$$q = \psi_c = \frac{K_x}{2}\frac{(D^2 - H^2)}{(W - B)} \tag{9.30}$$

in which K_x = hydraulic conductivity in the lateral direction. The difference, $(D^2 - H^2)$, is an important parameter and can in fact be incorporated into the governing equation to estimate the height of a water mound (Griffin and Warrington, 1988). In general, stormwater infiltration basins are small in size and operated with short loading periods. Unlike the long term recharging basin, the height of a water mound under a small infiltration basin decays toward the initial groundwater table. Under this condition, Eq. (9.30) can be an approximation to the edge of the area of recharge, i.e. Point R (W, H) in Fig. 9.11. Eq. (9.30) divided by the effective flow depth, H, yields a sectional velocity which must be equal to Eq. (9.25) at $x = W$. As a result, the effective thickness of an aquifer is derived as

$$\frac{H}{B} = \sqrt{\frac{2\lambda_x}{\lambda_y + 1}} \tag{9.31}$$

$$\lambda_x = \frac{f}{K_x} \tag{9.32}$$

In practice, the value of H is the smaller one between Eq. (9.31) and the available thickness, H_e. Aided by Eq. (9.28), the saturation depth is

$$\frac{D}{B} = \lambda_y \sqrt{\frac{\lambda_x}{\lambda_y + 1}} \tag{9.33}$$

For convenience, an anisotropic subsurface environment can be converted to an isotropic condition by an appropriate coordinate transformation, and the equivalent homogenous hydraulic conductivity, K_u, permeability is equal to the geometric mean (Morris and Wiggert, 1972) as:

$$K_u = \sqrt{K_x K_y} \tag{9.34}$$

Applications of the potential flow model to a water mound prediction requires the priori knowledge of K_u. During a water mounding process, the value of K_u varies with respect to soil moisture content. Care has to be taken when selecting the proper value for K_u.

Example 9.5

An infiltration trench has a width of 40 ft and a length of 180 ft. The soil infiltration rate at the land surface is 1.80 in/hr. Determine the required subsurface geometry under the trench in order to sustain the infiltration rate.

For this example, the half width, B, of the trench is 10 ft. The bottom area of the trench is:

$$A = 20(180)/43560 = 0.083 \text{ acres}$$

The infiltration flow rate is

$$Q_o = \eta f_c(A_b) = \frac{1}{12(3600)} (1.80)(0.083)(43560) = 0.151 \text{ cfs}$$

The unit flow rate released through the half width of the trench is:

$$q_o = \frac{1}{12(3600)} (1.80)10 = 0.00042 \text{ cfs/ft}$$

Assuming that the subsurface medium is isotropic, $K_x = K_y = K$ or $\lambda_x = \lambda_y$. Several cases in term of f_a/K ratio are evaluated by Eqs. (9.31) and (9.33) for the required depths D and H in order to sustain the design infiltrating flow rate of 0.151 cfs. Table 9.4 presents a comparison. When the infiltration rate is slightly greater than the coefficient of permeability such as $f_a/K = 1.50$, the required distance to the groundwater table, Z, is as shallow as 5.48 ft, but it sharply increases to 24.49 ft when the f_a/K ratio increases to 3.0. For instance, Table 9.4 indicates that it requires a saturation depth, Z, of 24.49 ft for this trench basin

TABLE 9.4 Saturation Depths for Steady Flow Condition under Various Coefficient of Permeability for $f_a = 1.80$ inch/hr, and $B = 10.0$ ft

f_a/K	K fpd	D/B	H/B	Z/B	D ft	H ft	$Z = D - H$ ft
1.1	3.27	1.13	1.02	0.1	11.26	10.24	1.02
1.5	2.4	1.64	1.1	0.55	16.43	10.95	5.48
2.0	1.8	2.31	1.15	1.15	23.09	11.55	11.55
2.5	1.44	2.99	1.2	1.79	29.88	11.95	17.93
	1.2	3.67	1.22	2.45	36.74	12.25	24.49

in order to produce sufficient hydraulic gradients when the infiltration to permeability ratio is 3.

9.8 POTENTIAL FLOW MODEL FOR A CIRCULAR BASIN

The flow pattern below a circular basin is described in Fig. 9.17. Such a three-dimensional axially symmetric flow can be described by the stream function as

$$\psi = \pi \frac{f}{D} r^2 y \tag{9.35}$$

in which r = radius at distance y. Equation (9.35) satisfies the continuity and irrotationality equations for a three-dimensional axi-symmetric flow. The streamlines below the basin are distributed as concentric circles with $\psi = 0$ along the y-axis and ψ = infiltration volume rate at the circumference of the basin bottom, i.e. Point $C\,(r, y) = (R_o, D)$ in Fig. 9.17. The infiltration volume rate released from the circular basin is:

$$q = f\,\pi R_o^2 \tag{9.36}$$

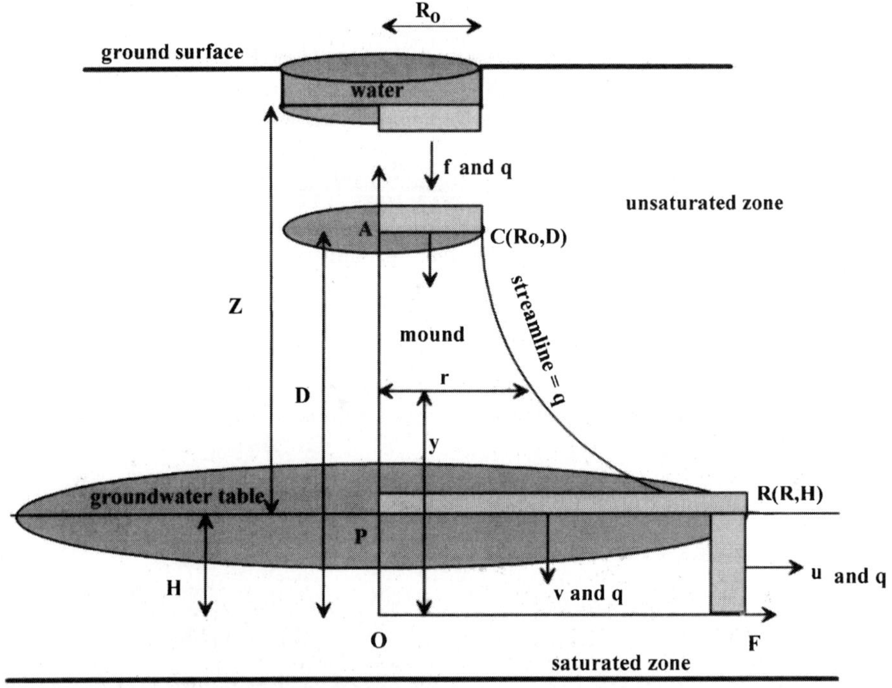

FIGURE 9.17   Flow pattern under circular basin.

where q = infiltration volume rate
f = soil infiltration rate
R_o = radius of circular basin

To maintain the continuity of flow, the radius, r, of a horizontal cross section at a specified y, can be obtained by Eq. (9.35) as:

$$r = \sqrt{\frac{D}{y}}\, R_o \tag{9.37}$$

For instance, at Section PR in Fig. 9.17, the radius of the area of recharge must pass Point R (R, H) and is estimated as:

$$R = \sqrt{\frac{D}{H}}\, R_o \tag{9.38}$$

To repeat the similar approach to an infiltration trench, the required subsurface geometry below a circular basin can be described as:

$$\frac{H}{R_o} = \sqrt{\frac{\lambda_r \ln \lambda_y}{2(\lambda_y^2 - 1)}} \tag{9.39}$$

$$\lambda_r = \frac{f}{K_r w} \tag{9.40}$$

in which K_r = hydraulic conductivity in the radial direction. Again, in practice, the value of H is the smaller one between Eq. (9.39) and the existing thickness of aquifer. Substituting Eq. (9.38) into Eq. (9.39) yields

$$\frac{D}{R_o} = \lambda_y \sqrt{\frac{\lambda_r \ln \lambda_y}{2(\lambda_y^2 - 1)}} \tag{9.41}$$

Similarly, the equivalent homogenous hydraulic conductivity, K_u, is equal to

$$K_u = \sqrt{K_r K_y} \tag{9.42}$$

Example 9.6
Applications of this model are illustrated by the design of a circular infiltration basin with a diameter of 68.0 ft. The basin has a loamy sand lining whose average infiltration rate, f_a, is 1.80 in/hr. With R_o = 34.0 ft, the bottom area of the basin is 0.083 acre. The design infiltration volume rate from the basin is estimated as:

$$Q = f_a \pi R_o^2 = \frac{1.80}{12.0(3600.0)}\,(3.1416)(34.0)^2 = 0.15 \text{ cfs}$$

Under the assumption of an isotropic environment, $K_r = K_y = K$ or $\lambda_r = \lambda_y$. Table 9.5 presents a comparison among cases with different f_a/K ratios. When f_a/K = 1.50, the required distance, Z, to the groundwater table is 8.38 ft, but it increases at an accelerating rate with respect to the soil infiltration to permeability ratio. As shown in Table 9.5 when f_a/K = 3.0, the required saturated depth, Z, is as high as 30.86 ft in order to produce sufficient hydraulic gradients.

TABLE 9.5 Saturation Depths for Various Conditions with $R_o = 34$ ft, $f_a = 1.80$ inch/hr

f_a/K	K fpd	H/R_o	D/R_o	Z/R_o	H ft	D ft	$Z = D - H$ ft
1.20	3.00	0.50	0.60	0.10	16.95	20.34	3.39
1.50	2.40	0.49	0.74	0.25	16.77	25.15	8.38
2.00	1.80	0.48	0.96	0.48	16.34	32.69	16.34
2.50	1.44	0.47	1.17	0.70	15.88	39.7	23.82
3.00	1.20	0.45	1.36	0.91	15.34	46.29	30.86

9.9 GROWTH FUNCTION OF WATER MOUND

Consider the flow underneath an infiltration trench. Before the steady state is developed, the storage effect of the soil as shown in Fig. 9.18 can be depicted as a water volume balance between the inflow at Section AC and the outflow at Section PR as:

$$\frac{dV}{dt} = q - q_o \tag{9.43}$$

where V = soil water storage volume
      t = time

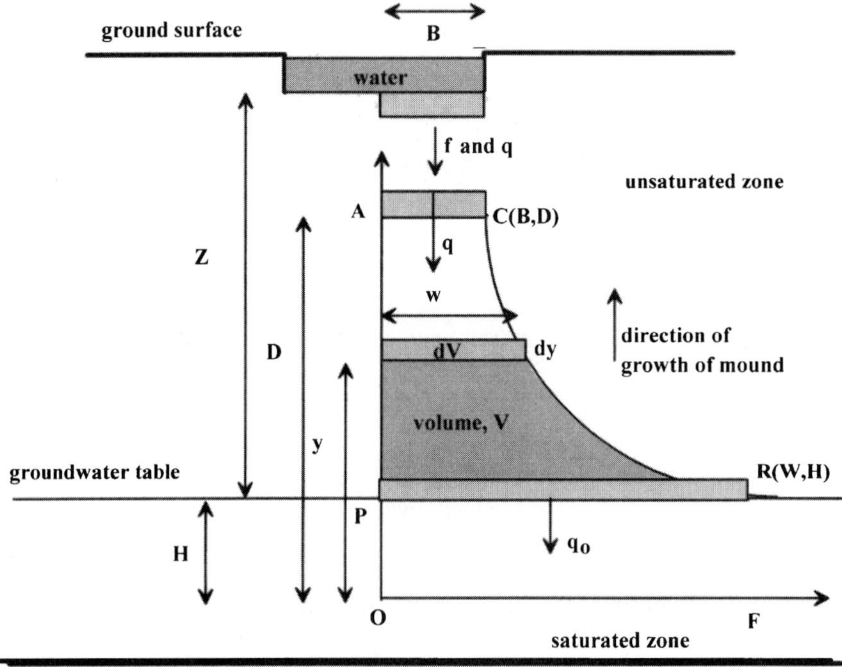

FIGURE 9.18 Growth of water mound underneath a trench.

q = infiltrating volumetric rate
q_o = outflow rate to recharge

Since the infiltrating flow must maintain the continuity as described by the stream function, the water flow during the filling process of a water mound shall be within the soil volume outlined by the stream lines. Aided by Eq. (9.24), the incremental volume, dV, at a depth y can be expressed as:

$$dV = S_y w\, dy = S_y \frac{D}{y} B dy \tag{9.44}$$

in which S_y = soil yield. Based on Darcy's law, the outflow rate at Section PR is described as:

$$q_o = S_y K_u W = S_y K_u \frac{D}{H} B \tag{9.45}$$

in which K_u = representative hydraulic conductivity for the growth of a water mound. Aided by Eqs. (9.24), and (9.43), Eq. (9.44) can be integrated from the thickness of aquifer, H, to a depth, y, and yields the growth function of water mound as

$$y = He^{-kt/D} \tag{9.46}$$

$$k = \left(\frac{f}{S_y} - K_u \frac{D}{H}\right) \tag{9.47}$$

in which k = pseudo hydraulic conductivity for the growth of a water mound. Generally, the exponent in Eq. (9.46) is so small that Eq. (9.46) reduces to

$$y = H\left(1 - \frac{k\,t}{D}\right) \tag{9.48}$$

Eq. (9.46) is an increasing function with a depth y not to exceed D in Eq. (9.33) or Eq. (9.41) and can be used to estimate the growth of a water mound when an elapsed time, t, is specified.

Similarly, the growth of a water mound under a circular basin can be derived. During the growing process, the incremental water storage volume in the soil medium is

$$dV = S_y \pi r^2\, dy = S_y \pi \frac{D}{y} R_o^2\, dy \tag{9.49}$$

The flow rate of recharge at Section PR in Fig. 9.17 is:

$$q_o = S_y K_u \pi R^2 = S_y K_u \frac{D}{H} R_o^2 \tag{9.50}$$

Aided by Eq. (9.37) and (9.50), Eq. (9.49) yields an identical growth function to Eq. (9.46). Therefore, it will not be repeated.

9.10 RECESSION FUNCTION OF WATER MOUND

The mound begins to recede by gradually recharging the groundwater table during the unloading period. As shown in Fig. 9.19, the volume of recharge at Section PR to the groundwater table must be equal to the volume released from the soil medium.

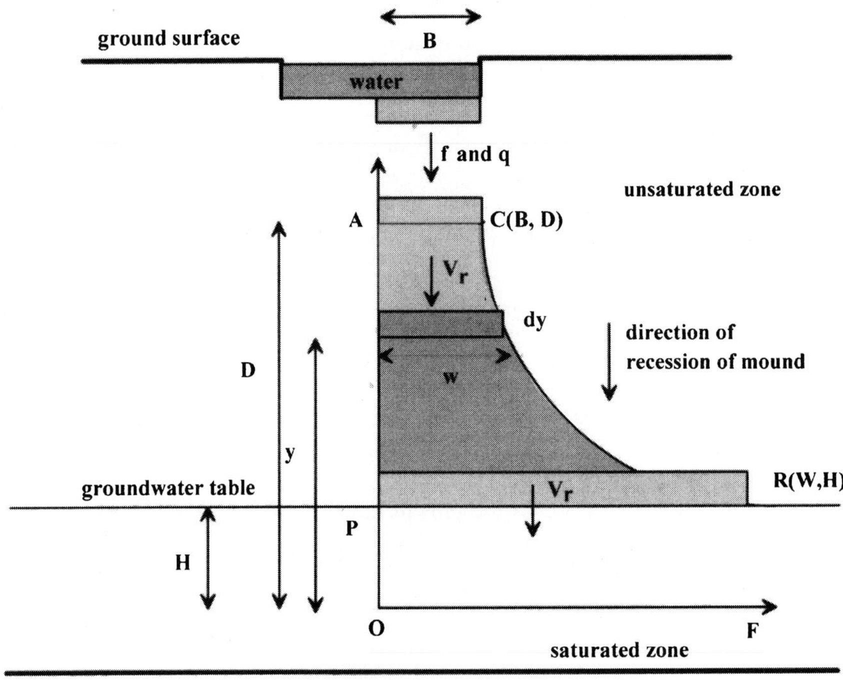

FIGURE 9.19   Recession of water mound.

The recession of a water mound begins from the bottom of the basin. Assuming that the receding water front maintains the shape as the saturated, the volume released from the soil can be estimated as:

$$V_r = S_y \int_{y=y}^{y=D} w \, dy = S_y \int_{y=y}^{y=D} \frac{D}{y} \, dy = S_y DB \ln\left(\frac{D}{y}\right) \tag{9.51}$$

With $i = -1$ for the vertical flow through Section PR, the volume of recharge to the groundwater table is

$$V_r = S_y K_u Wt = S_y K_u \frac{D}{H} Bt \tag{9.52}$$

Equating Eq. (9.51) to Eq. (9.52) yields the recession function of water mound as:

$$y = De^{-(K_u t / H)} \tag{9.53}$$

When the exponent in Eq. (9.53) is small enough, the recession function is reduced to a linear equation as:

$$y = D\left(1 - \frac{K_u t}{H}\right) \tag{9.54}$$

To repeat the similar exercise to a circular basin, an identical recession function to Eqs. (9.53) and (9.54) can also be derived for the water mound below a circular basin.

9.11 CALIBRATION OF GROWTH FUNCTION

To predict the growth of a water mound, the most sensitive parameter is the unsaturated hydraulic conductivity, K_u, which is not a constant, but varies with respect to soil moisture content. Table 9.6 presents nine cases for water mound studies, and their analyses by the potential flow model. For the cases involved a steady state, the saturated depth, D, of the water mound is predicted by Eq. (9.33) for a trench or Eq. (9.41) for a circular basin. The value of K_u for each case was determined based on the agreement with the observed height. For the cases with a known operation time, Eq. (9.46) is used with the selected K_u to match with the observed height at a specified elapsed time. For all cases, the effective thickness of aquifer, H, is the smaller one between the existing thickness and the required depth by Eq. (9.31) or (9.39). Using these nine cases as a data base, a dimensional analysis is performed as:

$$\frac{K_u}{K_s} = \phi(P_s) \tag{9.55}$$

$$P_s = \frac{f}{K_s}\frac{L}{H} \tag{9.56}$$

where K_s = saturated hydraulic conductivity
     P_s = subsurface geometric parameter
     $L = B$ for a trench basin or $L = R_o$ for a circular basin

Figure 9.20 presents the functional relation for Eq. (9.53) and leads to

$$\frac{K_u}{K_s} = 2.97 \frac{f}{K_s}\frac{L}{H} \ (r^2 = 0.91 \ \text{for water mound growth process}) \tag{9.57}$$

in which r^2 = correlation coefficient.

Equation (9.57) is empirical and derived for K_u to predict the growth of a water mound. Equation (9.57) implies that a higher K_u can be developed when an intense infiltration rate is applied to a larger basin with a shallower aquifer. This tendency can be explained by the water retention process in the soil.

9.12 EVALUATION OF RECESSION FUNCTION

During the recession period, the saturated depth, D, begins to diffuse into the groundwater through Section PR in Fig. 9.19. For a given time t, Eq. (9.53) predicts the location of the receding water front, and Eq. (9.52) estimates the water volume of recharge or the water volume released from the soil medium. Equation (9.53) essentially depicts the diffusion of the saturated depth, D, into the effective thickness, H, of the aquifer below the basin by a decay coefficient determined by the hydraulic conductivity. Four cases in Table 9.7 reported by Sumner, et al. in 1999 are used as examples. The basin under tests had a radius of 25 m and was operated under a loading of 0.3 or 1.05 m per day. The thickness of the aquifer under tests was set to be either 12 or 4 m.

Due to the soil hysteretic behavior, Eq. (9.57) developed for the growth function is not always suitable for the recession function. For instance, Cases 1 and 2 presented in Fig. 9.21 were operated with a 12-m aquifer which is close to the required depth of 12.37 m by Eq. (9.39). The loading of 0.3 m/d was applied to both Cases 1 and 2. But the best-fitted K_s for Cases 1 and 2 is found to be 2.10 m/d instead of 1.81 m/d by Eq. (9.57). However, Cases

TABLE 9.6 Cases of Water Mound Studies Analyzed by Potential Flow Model

First author of the case reported	Rastogi	Ortiz	Rastogi	Swamee	Sumner	Sumner	Sumner	Sumner
Year of the publication	1998	1979	1998	1997	1999	1999	1999	1999
Basin shape: (S)quare, (R)ectangle, (C)ircle	R	S	C	R	C	C	C	C
Units—day and meter	m/d	m/d	m/d	m/d	m/d	m/d	m/d	m/d
Radius R_o or Half Width, B	50.00	177.00	126.15	5.00	25.00	25.00	25.00	25.00
Saturated Conductivity, K_s	11.11	90.00	11.11	2.00	2.1	2.1	2.1	2.1
Infiltration Rate, f	0.20	3.00	0.80	0.20	0.30	0.30	0.30	1.05
Ratio f/K_s	0.02	0.033	0.07	0.10	0.14	0.14	0.14	0.50
Representative Hydraulic Conductivity, K_u	0.180	2.990	0.750	0.050	0.220	0.240	0.50	0.740
Ratio K_u/K_s	0.02	0.03	0.07	0.03	0.10	0.11	0.24	0.35
Ratio f/K_u	1.11	1.00	1.07	4.00	1.36	1.25	0.60	1.42
Subsurface Parameter, $f/K_s(B/H)$ or R_o/H	0.036	0.118	0.144	0.233	0.298	0.298	0.893	1.042
Soil Moisture	not available	not available	not available	not available	0.0300	0.0350	0.1200	0.1850
Soil-water Pressure	not available	not available	not available	not available	0.9500	0.9000	0.8500	0.8200
Saturation Depth, D	27.75	88.80	67.42	8.60	16.91	15.56	7.34	17.56
Radius or Half Width on Groundwater Table, R	52.70	235.88	130.45	10.00	29.68	28.47	33.87	30.24
Depth of Aquifer Used, H	24.98	50.00	63.05	2.15	12.00	12.00	4.00	12.00
Existing Thickness of Aquifer, H_e	90.00	50.00	90.00	5.33	12.00	12.00	4.00	12.00
Required Depth of Aquifer, H	24.98	88.50	63.05	2.15	12.40	12.45	12.23	12.37
Saturated Depth above the Table, $(H - D)$	2.78	38.80	4.37	6.45	4.91	3.56	3.34	5.56
Ratio D/H	1.11	1.78	1.07	4.00	1.41	1.30	1.84	1.46
Soil Yield, S_y	0.20	0.20	0.20	0.15	0.28	0.28	0.28	0.28
Storage Factor $= (f/S_y - K_u D/H)$	0.80	9.69	3.21	1.13	0.76	0.76	0.15	2.67
Operation Time in Days	steady	2.00	steady	6.00	14.00	7.00	7.00	2.00
Operation Time in Days	steady	2.00	steady	6.00	14.00	7.00	7.00	2.00
Predicted Mound Height at Operation Time	27.75	62.20	67.42	4.74	16.91	15.56	4.63	16.26
Observed Mound Height	28.00	66.90	67.25	8.80	17.00	15.70	3.50	17.00

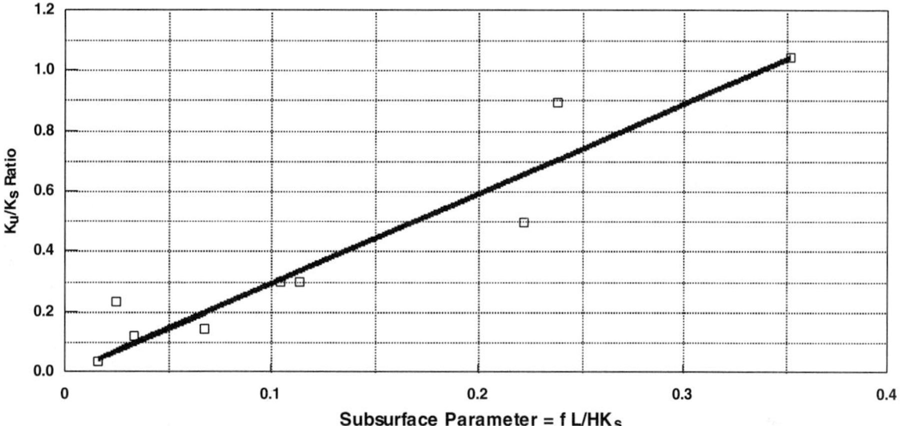

FIGURE 9.20   Hydraulic conductivity during growth of water mound.

TABLE 9.7   Mound Heights Predicted by Recession Function and VS2D Model

Variable	Case	One	Case	Two	Case	Three	Case	Four
Basin Radius, R_o	25.000 m		25.000 m		25.000 m		25.000 m	
Thickness of Aquifer, H	12.000 m		12.000 m		12.000 m		4.000 m	
Saturated Depth, D	15.000 m		16.400 m		16.000 lm		7.600 m	
Infiltration Rate, f	0.300 m/d		0.300 m/d		1.050 m/d		0.300 m/d	
Saturated Hy Conductivity	2.100 m/d		2.100 m/d		2.100 m/d		2.100 m/d	
K_u by Eq. (9.34)	0.088 m/d		0.088 m/d		0.309 m/d		0.265 m/d	
K_u by Best Fitted	0.178 m/d		0.245 m/d		0.275 m/d		0.245 m/d	

Time for recession	Recession function	VS2D model	Recession function	VS2D model	Recession function	VS2D model	Recession function	VS2D model
day	meter	meter	meter	meter	meter	meter	meter	meter
0	15.00	15.00	16.40	16.40	16.00	16.00	7.60	7.60
1	14.78	14.75	16.07	15.96	15.64	15.56	7.15	6.90
2	14.56	14.50	15.74	15.51	15.28	15.12	6.72	6.20
3	14.35	14.25	15.43	15.07	14.94	14.68	6.32	5.50
4	14.14	14.00	15.11	14.65	14.60	14.24	5.95	5.34
5	13.93	13.75	14.81	14.51	14.27	13.80	5.60	5.18
6	13.72	13.50	14.51	14.36	13.94	13.68	5.26	5.02
7	13.52	13.43	14.22	14.22	13.63	13.56	4.95	4.94
8	13.32	13.35	13.93	14.07	13.32	13.44	4.66	4.86
9	13.13	13.28	13.65	13.92	13.02	13.32	4.38	4.78
10	12.93	13.20	13.37	13.75	12.72	13.20	4.12	4.70
Sum of squared error		0.22		0.77		0.85		2.16

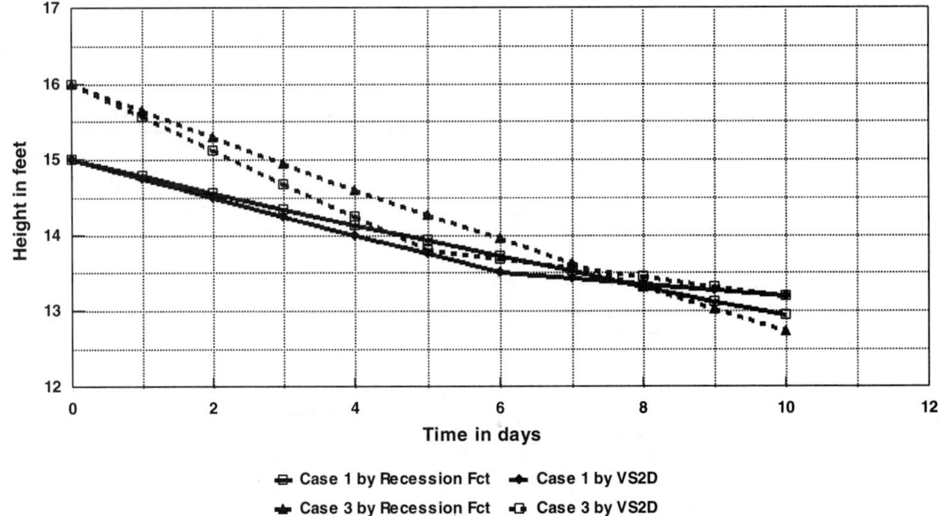

FIGURE 9.21   Comparison between recession function and observed recession of water mound.

3 and 4, observed under a high loading of 1.05 m/day and an aquifer as shallow as 4.0 m, can well be predicted using the same K_u for both the water mounding growth and recession processes. It implies that during an intense storm event, the operation of an basin is subject to a high infiltration rate. Equation (9.59) may serve as an estimator for K_u. A further field measurement is required to confirm the value of K_u.

9.13   CLOSING

Infiltration facilities are designed for a wide variety of land use conditions, but most frequently applied to small and highly paved areas such as commercial, residential, and highway environment. Figure 9.22 provides the recommended design steps.

At the selected basin site, the required flood detention volume or water quality capture volume shall be calculated by the local surface hydrology. The geometry of the infiltration basin or trench shall be designed to satisfy the equivalence between the surface and subsurface storage capacities. By following the design criteria on the maximum ponding time and maximum ponding depth, the design soil infiltration rate can be determined. Impacts on the local groundwater table depends on whether the subsurface geometry under the infiltration facility sustains the infiltration rate on the land surface. The ratio of the distance to groundwater table to the the radius of infiltration basin is an important factor to produce adequate hydraulic gradients. Table 9.8 is the summary of design criteria for infiltration basins.

The storage volume of an infiltration facility is relatively small, compared to the extreme storm events. Therefore, it is important to have an overflow by-pass outlet to copy with events larger than the design capacity. Since an infiltration facility is susceptible to sediment clogging, it is essential to maximize the removal of sediment particles by the grass buffer strip along the flow path upstream of the basin. An infiltration basin or trench must be constructed after the watershed is stabilized or with sediment laden runoff diverted from the construction area. After construction, it is necessary that an effective vegetated buffer strip is provided to minimize the risk of sediment contamination. With a careful design and construction, infiltration facilities can continue their operations as expected.

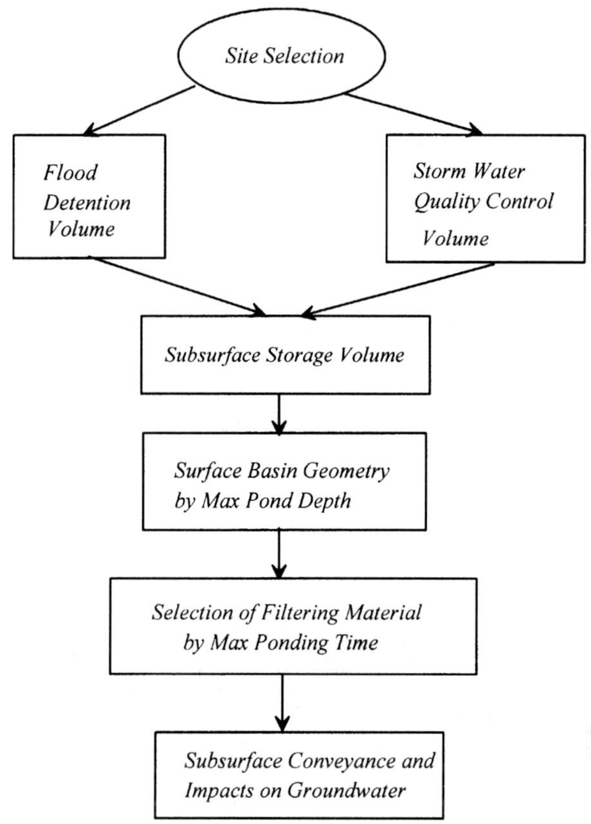

FIGURE 9.22   Design steps for infiltration basin.

TABLE 9.8   Summary of Design Guidelines for Infiltration Basins

Design parameter	Recommended guidelines
tributary watershed	up to 10 acres
storage volume	flood detention volume or capture volume
infiltration time	maximum 72 hours
infiltration rate	minimum 0.20 inch/hr
distance to groundwater	minimum 5 to 10 feet to seasonal high level
distance to bed rock	minimum 5 feet from the basin to bedrock
basin side slope	minimum 3H:1V
riprap	dissipator at entrance and outfall
cover	grass
bypass facility	spillway and back-up subdrain

The impact of an infiltration basin on the groundwater table can be assessed by the potential flow model which is developed for a short-term infiltration operation. The potential flow model does not require detailed site specifics, but does provide reasonable assessments on water mounding effects. It can be a useful tool at the planning stage when little design information is available, but it does not replace the necessity of more complicated modeling studies to refine the predictions.

REFERENCES

American Society of Civil Engineers, *Final Report of the Task Committee on Stormwater Detention Outlet Structures* 1984.

Athayde, D. N. (1976), "Best Management Practices (BMP)," *Proceedings: Urban Stormwater Seminars,* U.S. Environmental Protection Agency, Water Quality Management Guide, WPD 03-76-04, 1976.

Bouwer, H., J. T. Back, and J. M. Oliver, "Predicting Infiltration and Ground-water Mounds for Artificial Recharge." *Journal of Hydrologic Engineering, ASCE,* 4(4):350–357, October, 1999.

Brooks, R. H., and A. T. Corey, "Hydraulic Properties of Porous Media," Hydrologic Paper No. 3, Colorado State University, Ft. Collins, CO, 1964.

Clement, P., and K. Pensyl, *Results of the State of Maryland Infiltration Practice Survey,* Stormwater Division, Maryland Department of the Environment, Annapolis, MD, 1987.

Ferguson, B. K., "Role of the Long-term Water Balance in Management of Stormwater Infiltration," *Journal of Environmental Management,* Vol. 30, pp. 221–233.

Giffin, D. M. Jr., and R. O. Warrington, "Examination of 2-D Groundwater Recharge Solution," *ASCE Journal of Irrigation and Drainage Engineering,* 114(4):691–704, Nov. 1988.

Guo, J. C. Y., and B. Urbonas, "Maximized Detention Volume Determined by Runoff Capture Rate," *ASCE Journal of Water Resources Planning and Management,* 122(1), Jan. 1996.

Guo, J. C. Y. "Surface-Subsurface Model for Trench Infiltration Basins," *ASCE Journal of Water Resources Management and Planning,* 124(5):280–284, Sept./Oct. 1998.

Guo, J. C. Y., "Design of Detention Basin for Small Catchments," *ASCE Journal of Water Resources Management and Planning,* 125(6), Nov./Dec. 1999.

Guo, J. C. Y. "Sand Recovery for Highway Drainage Designs," *ASCE Journal of Drainage and Irrigation Engineering,* 125(6), Nov./Dec. 2000.

Guo, J. C. Y., "Design of Circular Infiltration Basin under Water Mound Effects," *ASCE Journal of Water Resources Management and Planning,* Jan. 2001.

Guo, J. C. Y., and W. Hughes, "Runoff Storage Volume for Infiltration Basin," submitted to *ASCE Journal of Irrigation and Drainage Engineering,* 2001.

Hannon, J. B. *Underground Disposal of Stormwater Runoff Design Guidelines Manual,* U.S. Department of Transportation, FHA, Feb. 1980.

Hantuch, M. S., "Growth and Decay of Groundwater-mounds in Response to Uniform Percolation," Water Resources Research, Vol. 3, pp. 227–234, 1967.

Harringron, B. W., *Design and Construction of Infiltration Trenches,* Proceedings of ASCE Engineering Foundation Conference on Design of Urban Runoff Quality Controls, Roesner, Urbonas, and Sonnen, (eds.) held at Potosi, MO, July, 1988.

Maryland Department of Natural Resources, *Maryland Standards and Specification for Stormwater Management Infiltration Practices,* Stormwater Management Division, 1984.

McDonald, M. G., and A. Harbaugh, *A Modular Three-dimensional Finite-difference Groundwater Flow Mode,* U.S. Department of the Interior, USGS, National Center, Reston, VA, 1984.

Morel-Seytoux, H. J., and C. Miracapillo, "Prediction of Infiltration, Mound Development, and Aquifer Recharge from a Spreading Basin or Intermittent Stream," *HYDROWAR Rep.,* No. 88.3, Hydrology Days Publication, Atherton, CA, 1988.

Morel-Seytoux, H. J., C. Miracapillo, and M. J. Abdulrazzak, "A Reductionist Physical Approach to Unsaturated Aquifer Recharge from a Circular Spreading Basin," *Water Resources Research.,* Vol. 26, pp. 771–777, 1990.

Morris, H. M., and J. M. Wiggert, *Applied Hydraulic Engineering,* The Ronald Press Company, New York, N.Y., 1972.

Lappala, E. G., R. W. Healy, and E. P. Weeks, "Documentation of Computer Program VS2D to Solve the Equations of Fluid Flow in Variably Saturated Porous Media," *USGS Water Resources Investigations Report,* No. 83-409, U.S. Geology Survey, Lakewood, CO, 1987.

Rastogi, A. K., and S. N. Pandy, "Modeling of Artificial Recharge Basins of Different Shapes and Effect on Underlying Aquifer System," *ASCE Journal of Hydrologic Engineering,* 123(3):62–68, Jan. 1998.

Ortiz, N. V., D. W. Zachmann, D. B. McWhorter, and D. K. Sunada, "Effects of In-Transit Water on Groundwater Mounds Beneath Circular and Rectangular Recharge Areas," *Water Resources Research,* 15(3):577–582, 1979.

Press, W. H., B. P. Flannery, S. A. Teukolsky, and W. T. Vetterling, *Numerical Recipes—the Art of Scientific Computing,* Cambridge University Press, Cambridge, U.K., 1989.

Shansai, A., and N. Sitar, "Method for Determination of Hydraulic Conductivity in Unsaturated Porous Media," *ASCE Journal of Irrigation and Drainage Engineering,* 117(1):64–78, Jan./Feb. 1991.

Shaver, H. E., *Infiltration as a Stormwater Management Component,* Proceeding of Engineering Foundation Conference on Impact and Quality Enhancement Technology, Henniker, NH, June 23–27, 1986.

Stankovich, J. M., and D. A. Lockington, "Brook-Corey and Van Genuchten Soil-water-retention Models," *ASCE Journal of Irrigation and Drainage Engineering,* 121(1):1–7, Jan./Feb., 1995.

Swanee, P. K. and C. S. P. Ojha, "Ground-water Mound Equation for Rectangular Recharge Area," *ASCE Journal of Irrigation and Drainage Engineering,* 123(3):215–217, May/June 1997.

Sumner, D. M., D. E. Rolston, and M. A. Marino, "Effects of Unsaturated Zone on Groundwater Mounding," *ASCE Journal of Hydrologic Engineering,* 4(1):65, 1999.

Stahre, P., and B. Urbonas, *Swedish Approach to Infiltration and Percolation Design,* Proceedings of ASCE Engineering Foundation Conference on Design of Urban Runoff Quality Controls, Roesner, Urbonas, and Sonnen (eds.), held at Potosi, MO, July, 1988.

Urbonas, B. R., "Design of a Sand Filter for Stormwater Quality Enhancement," *Journal of Water Environment Research,* 17(1)Jan./Feb., 1999

Wanielista, M. P., and Y. A. Yousef, *Stormwater Management,* John Wiley and Sons, Inc., New York, 1993.

Youself, Y. A., M. P. Wanielista, and H. H. Harper, *Design and Effectiveness of Urban Retention Basins,* Proceeding of Engineering Foundation Conference on Impact and Quality Enhancement Technology, Henniker, NH, June 23–27, 1986.

CHAPTER 10
STORMWATER TREATMENT WETLAND DESIGN

Roland D. Wass
PBSJ
Phoenix, Arizona

10.1 INTRODUCTION

The U.S. Environmental Protection Agency includes nonpoint source pollution as one of the remaining major stressors on our nations receiving water bodies. Stormwater runoff emanating from various land uses, for example, commercial, residential, industrial, and undeveloped, can transport a myriad of particulate and dissolved materials. These can take the form of oxygen-demanding substances, nutrients, and toxic materials that can negatively impact biota residing in or relying upon aquatic systems. In addition, stormwater runoff can cause physical degradation of aquatic systems, including erosion, sedimentation, and thermal stress.

In many cases, the pollutant concentration in stormwater runoff is dependent upon buildup during antecedent dry periods and subsequent exposure to runoff during precipitation events. Often, storm events produce differing amounts of runoff that, when coupled with multiple pollutant sources, result in varying degrees of contaminated runoff. The variability in rainfall amount, duration, and pollutant concentrations presents significant challenges to the engineer whose job is to treat these waters.

Although conventional treatment technologies exist that can remove the pollutants typically found in stormwater runoff, the variability in flow volume and pollutant strength would likely require conventional structures that would not be economically justified. Even though source control is the most preferred treatment technology, there are instances where treatment of runoff will be necessary. A potential treatment alternative is the use of engineered free water surface (FWS) constructed treatment wetlands.

10.1.1 Wetland Definitions

Wetlands can be perennial or intermittent, vegetated or non-vegetated ecosystems, and persist upon many different substrates in various geographic regions. They represent transitional areas between dry upland ecosystems and deep-water aquatic environments. In North America, wetlands are known by many names, including bogs, pocosins, marshes, swamps, seeps, fens, and/or bottomlands. Recognized today as providing numerous beneficial functions including water-quality improvement, flood attenuation, erosion control, recreation, environmental education, endangered species habitat, maintenance of stream flows, aesthetics, and commercial materials, in the past, wetlands have been feared by man. This is evidenced by such references as swamps, bogs or other sinister names and because, at times, these systems do present difficult places to access and possess strange sights and smells. Also, diseases such as malaria and yellow fever are associated with organisms that reside or utilize wetlands. Because of such stigmas, prior to the 1970s, it was common practice in the United States to drain and replace wetlands with agricultural fields, industrial complexes, commercial facilities, and even residential neighborhoods. Today wetland functions and benefits are more appreciated and a real interest exists in protecting and preserving natural wetlands, as well as in constructing systems.

Historically, wetland definitions were developed by individuals such as botanists, soil scientists, wildlife biologists, and hydrologists—each describing the systems from his own perspective. The insight provided by those efforts were studied and used to develop a wetlands classification scheme by Cowardin and others in 1979 that has since been adopted by the U.S. Fish and Wildlife Service (FWS) to map and inventory the nation's wetlands. At the most general classification level, wetlands are grouped into five ecological systems: palustrine, lacustrine, riverine, estuarine, and marine. The 1979, U.S. FWS comprehensive definition is:

> Wetlands are lands transitional between terrestrial and aquatic systems where the water table is usually at or near the land surface or the land is covered by shallow water. Wetlands must have one of the following three attributes: 1) at least periodically, the lands support predominantly hydrophytic vegetation, 2) the substrate is predominantly undrained hydric soil, and 3) the substrate is nonsoil and is saturated with water or covered by shallow water at some time during the growing season of each year.

Although this definition has been adopted by the scientific community and is used by engineers, planners, and land use managers for wetland inventories and mapping, it was not a usable definition for determining the jurisdictional boundary of a wetland for purposes of permitting dredge-and-fill activities as administered by the U.S. Army Corps of Engineers (COE), and as mandated by the Clean Water Act, section 404 requirements (Tammi, 1994). As such, in 1989 the FWS, COE, U.S. EPA, and the Soil Conservation Service for use in determining the jurisdictional status of wetland systems developed an interagency wetland definition. This legal definition states:

> The term "wetlands" means those areas that are inundated or saturated by surface or ground water at a frequency and duration sufficient to support, and that under normal circumstances do support, a prevalence of vegetation typically adapted for life in saturated soil conditions. Wetlands generally include swamps, marshes, bogs, and similar areas (EPA, 40 CFR 230.3 and CFR 328.3).

Constructed wetlands used for waste water treatment may or may not be considered as jurisdictional, for example, "Waters of the United States." If waste treatment systems, including treatment ponds or lagoons, are designed to meet the requirements of the CWA (other than cooling ponds as defined in 40 CFR 123.11(m), which also meet the criteria of this definition), they are currently not considered jurisdictional waters (33 CFR Part 328–Definition of "Waters of the United States"). Further guidance on this and other policy and permitting issues associated with constructed wetlands can be found in Guiding Principles for Constructed Treatment Wetlands: Providing for Water Quality and Wildlife Habitat, prepared by the Interagency Workgroup on Constructed Wetlands (this document is available on-line at *www.epa.gov/owow/wetland/constructed/guide.html*).

10.2   TREATMENT WETLAND HISTORY

The water "treatment" capabilities of wetlands were recognized very early by biologists and ecologists, but the engineering community did not consider them viable treatment systems until relatively recently. Some of the earliest research concerning pollutant removal in wetlands was conducted in Plon, Germany by K. Seidel and R. Kickuth (Seidel, 1976). They investigated the role aquatic that plants play in water purification and, in particular, the treatment of phenol and dairy wastewater using bulrush. Since that pioneering work, both natural and constructed wetlands have been examined for their pollutant removal/transformation capabilities using a variety of contaminated waters, including stormwater runoff. In 1972 R. H. Kadlec and associates investigated the use of a natural wetland (Houghton Lake, MI) to treat municipal waste water. In 1973, fish-processing waste was discharged to a fresh water marsh to investigate for nutrient removal (U.S. EPA ETI, 2000). In the late 1970s to the late 1980s treatment wetland form and vegetation was studied. B. C. Wolverton used gravel-based subsurface flow wetlands to treat municipal waste water for subsequent re-use and for removal of priority pollutants. In the 1980s Tom Debusk conducted pilot scale studies to look at vegetation issues, as did Ramesh Reddy and Smith (1987). In the mid-1990s, heavy-metal removal in wetland treatment systems was studied at the Phoenix, AZ and Sacramento, CA wetland demonstration projects.

10.3   STORMWATER WETLANDS

In all, almost 15 years of research directed at stormwater treatment has led the way to using natural and constructed wetland systems to improve runoff quality. Much of the wetland design was based upon rules of thumb with little or no formal design guidance. In the 1980s the U.S. EPA conducted the Nationwide Urban Runoff Project, which concluded that detention basins with a permanent pool are among the most effective systems for controlling sediments, organic compounds, nutrients, and heavy metals found in stormwater runoff. Further, the results of this project produced water quantity and quality data sets that indicated a correlation between land use and runoff quality. More recently information has been presented in conference proceedings and reports to local governments with stormwater regulatory roles that indicates that wetland treatment will improve stormwater runoff.

In general, *stormwater treatment wetlands* provide the potential to improve runoff quality for a number of commonly found pollutants. These systems can remove solid particles effectively with authors reporting total suspended solids (TSS) reductions on the order of 65 to 80% (Strecker, 1992; Bingham, 1994). Nutrients also appear to be removed or at least attenuated in these systems. Significant reductions in nitrogen species have been indicated in treatment wetlands with removals as high as 83% for nitrite + nitrate nitrogen and roughly 45% for ammonia (ASCE, 2000; Strecker, 1994). Phosphorous, both total and dissolved forms, are removed in wetlands with total phosphorous removals on the order of 40 to 50%. Concentration reduction of several heavy-metal species is also likely as a result of wetland treatment of runoff. Those heavy metals associated with particulate matter such Pb exhibit the greatest removals, ranging from 26 to 83%.

Constructed stormwater treatment wetlands have shown more consistent pollutant removal performance than have natural systems because designed systems considered the range of incoming flows to be treated, allowed for sediment control, and minimize preferential flow (Scheuler et al., 1992). Constructed wetland systems can be designed and operated such that they receive direct stormwater runoff inputs, or as off-line systems receiving runoff up to a given magnitude and then bypassing the greater flows. In any event, a stormwater treatment wetland will likely experience pulses in hydraulic and mass loadings. Such systems can be designed to discharge flows or retain stormwater until infiltration or evapotranspiration losses dewater the wetland, depending upon site conditions and project goals.

Two notable stormwater treatment wetlands include the DUST Marsh and Coyote Hills Regional Park, both of which are in Fremont, CA. Wetlands at the Coyote Hills Regional Park demonstrated problems and opportunities in funding and permitting of wetland treatment systems and were used to assess future stormwater wetland design parameters. At the DUST Marsh, a fresh/brackish wetland was constructed to treat urban runoff using three configurations representing: (1) lagoon pretreatment, (2) overland flow followed by a pond with vegetated underwater sills, and (3) a braided channel system. Overall, the DUST systems removed or attenuated 42–45% of the incoming solids (mass basis). Several heavy metals were also monitored and the following percent removals were reported: Pb 30–83%, Cr 40–53%, Ni 12–34%, Zn 6–51%, and Cu 5–32%. The author suggests that these removals are the result of settling of heavier particles, which in turn are functions of residence time and flow regime. This was substantiated in that the system exhibiting the greatest removal efficiency (3) also had the longest hydraulic retention time.

A review of the American Civil Engineer Society's (ASCE) stormwater best management practice (BMP) database currently provides background and performance data for at least eight wetlands treating stormwater runoff, with more sites being added (ASCE, 2000). These systems receive contaminated stormwater runoff from areas characterized by residential, commercial, industrial, and undeveloped/rangeland land uses. Those systems reporting watershed characteristics, hydrology, and in some cases water quality data, are listed in Table 10.1.

The overall performance of these systems appears parameter and site specific with both pollutant removal and enrichment reported. As more sites are added to the database and monitoring practices and reporting are coordinated, this will become an excellent source for stormwater wetland design insight.

10.3.1 Stormwater Runoff Quantity

Stormwater quantity is variable in duration, frequency, and location, which can represent more of a challenge to water treatment professionals than do other more predictable waste streams. For example, municipal waste water discharges vary in quantity and quality over a 24-hour period as a function of the collection system size, extent, and retention time within the treatment facilities, but on an annual or even a monthly basis they can be considered as

TABLE 10.1 Wetlands Treating Stormwater Runoff Currently Represented in the ASCE Stormwater BMP Database

Wetland name	State	City
Hidden River Wetland	Florida	Tampa
Queen Anne's Pond	Maryland	Centreville
Prince George's Pond	Maryland	Clinton
Mays Chapel Wetland Basin	Maryland	Mays Chapel
Carver Ravine Wetland	Minnesota	Woodbury
Franklin Wetland	Virginia	Chantilly
Silver Star Rd. Wetland	Florida	Orlando
Hidden Lake Wetland	Florida	Sanford

Source: ASCE, 2000.

reliable water sources. In fact, it could be argued that in arid areas, treated waste water effluents are the only truly drought resistant water supplies. Conversely, stormwater runoff is episodic in occurrence and variable in magnitude and may not occur for extended time periods (e.g., droughts). The quality of municipal waste water may also fluctuate but typically remains within well-documented limits (Metcalf & Eddy, 1991). These two factors have allowed flow and waste load relationships to be developed for a given waste stream permitting conventional wastewater treatment technologies, including treatment wetland systems, to be successfully designed and operated.

As previously mentioned, precipitation events—snow or rain—can be temporally and geographically specific. In some instances, most of the annual precipitation occurs in the winter as large regional events that may last several hours to several days or it may be spread evenly throughout the year. In the southwestern United States, two distinct rainfall periods occur, one in the summer and again in the winter months. Summer storm events in this area are characterized by high intensity, short duration thunderstorms, while in the winter, the events are generally more regional and may last up to a period of days. In any case, the total volume of runoff and time of concentration can be radically different depending upon the duration of the event, time of year, and geographic location. Such extremes must be addressed in the design of stormwater treatment wetlands and is usually accomplished by providing structure(s) upstream of the wetlands to attenuate and meter flows.

Stormwater runoff is the result of precipitation falling on the landscape and subsequently traveling over surfaces. Some of this water may be stored in depressions, travel as sheet flow over uniform surfaces, or collect into small channels that convey water to larger receiving water bodies. The amount of runoff reaching a receiving water body will often times depend upon the surface over which it travels. As a result, changing land use in a watershed can alter the amount and time of concentration of stormwater runoff. The most significant alteration with respect to runoff volume and velocity is the loss of pervious surfaces within the contributing watershed. Structures placed upon the landscape, for example, housing, commercial, and industrial facilities and the parking lots and roadways needed to access them, represent impervious surfaces where storage in depressions and subsequent infiltration cannot occur. This results in a larger total runoff volume and shorter time of concentration for a given storm event in a developed watershed (Figure 10.1). In turn, this can lead to bank erosion, channel down cutting, and increased sediment loads, all of which are deleterious to the system of receiving waters and the biota residing within.

10.3.1.1 Estimating Stormwater Runoff Quantity

Given the land use(s) in the contributing watershed, estimates of stormwater runoff peak flow and volume can be developed for small areas using the Rational method (see chapters 4 and 6). The Rational method produces only an estimate of the peak flow for a given storm event, and some deficiencies exist, particularly as the contributing watershed gets larger or less homogenous. The Maricopa County Flood Control District, (Arizona), for example, allows the use of the rational method on watersheds with contributing areas of 160 acres or less.

The Rational method assumes that if rainfall were applied evenly over an impervious surface, the amount of runoff would eventually equal the amount of rainfall applied. The time of concentration (T_c) is that time required for this equilibrium to be reached. Further, if the rainfall duration is at least equal to the T_c, then the peak of runoff will equal the rate of rainfall. Calculation of T_c is an iterative process the procedure of which is detailed elsewhere (Linsley and Franzini, 1972; FCDMC, 1992; see chapters 4 and 6).

$$Q_p = CiA_d$$

Theoretical Hydrographs for Pre and Post Development Conditions

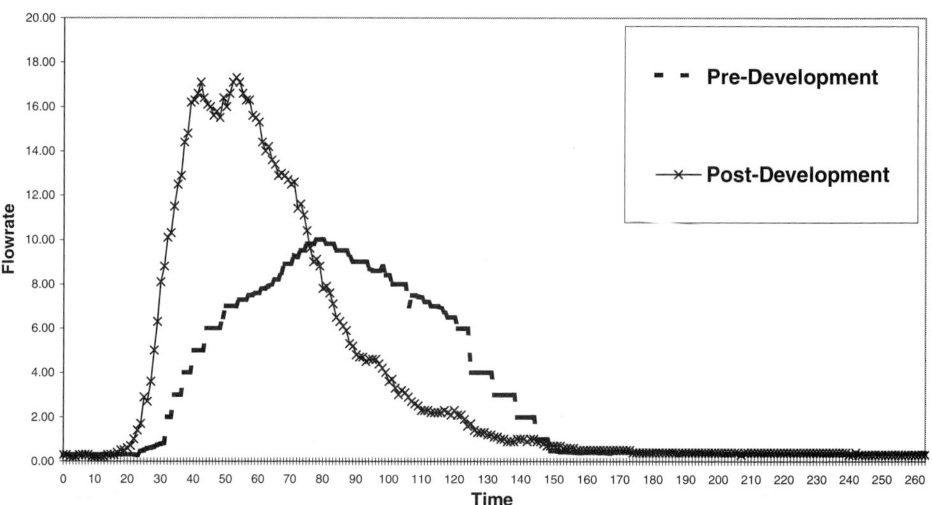

FIGURE 10.1   Pre- and post-development theoretical hydrographs depicting the impact of developments, e.g., increased impervious area results in a faster and greater runoff peak.

where Q_p = peak rate of runoff [ft^3/s]
  C = runoff coefficient [unitless]
  i = rainfall intensity for duration t_c [in/hr]
  A_d = watershed area [acres]
  T_c = time of concentration [hr]

Adjustments to both the T_c and C are typically made to account for the surfaces encountered as the runoff travels through the watershed and for different storm event return periods. Larger, less frequent storm events occurring on smooth, impervious surfaces result in faster T_c and more runoff (higher C) than do rough, pervious surfaces. Table 10.2 lists runoff coefficients based upon land use and return period for use in the Rational method in Maricopa County, AZ.

As the reader can see, the runoff coefficient depends not only on land use and storm event return frequency, but also on a degree of subjectivity supplied by the designer. Other methods exist that can be used to estimate stormwater runoff quantity that takes into account losses due to infiltration and depression storage. In areas where runoff losses to infiltration are important, theory-based models such as the Green-Ampt or Philips can be adapted for use (Maidment, 1992). Both models will impact hydrographs by decreasing runoff rates with respect to time as a function of infiltration.

If large watersheds are involved and more precise stormwater quantity estimates are required, the use of a computer model such as HEC 1 may be justified (see chapter 21). Developed by the U.S. COE at the Hydrologic Engineering Center (HEC), HEC-1 is a rainfall runoff and routing model. Hydrographs are generated from single historical or theoretical rainfall events for one or more locations within a drainage basin.

The Natural Resource Conservation Service, formerly the U.S. Soil Conservation Service (SCS), have also developed runoff equations to compute runoff hydrographs. The computer model TR-55 utilizes the SCS methods, often referred to as the CN or Curve Number method. When the model is run, surface runoff from one or more sub-basins can be computed from historical or theoretical storm data. Runoff losses are computed using the SCS curve

TABLE 10.2 Runoff Coefficient (C) for Use in the Rational Method in Maricopa County, AZ

Land use	Return period			
Streets & roads	2–10 year	25 year	50 year	100 year
Paved roads	0.75–0.85	0.83–0.94	0.9–0.95	0.94–0.95
Gravel roadways & shoulders	0.6–0.7	0.66–0.77	0.72–0.84	0.75–0.88
Industrial areas				
Heavy	0.7–0.8	0.77–0.88	0.84–0.95	0.88–0.95
Light	0.6–0.7	0.66–0.77	0.72–0.84	0.75–0.88
Business areas				
Downtown	0.75–0.85	0.83–0.94	0.9–0.95	0.94–0.95
Neighborhood	0.55–0.65	0.61–0.72	0.66–0.78	0.69–0.81
Residential areas				
Lawns–flat	0.1–0.25	0.11–0.28	0.12–0.3	0.13–0.31
Lawns–steep	0.25–0.4	0.28–0.44	0.3–0.48	0.31–0.50
Suburban	0.3–0.4	0.33–0.44	0.36–0.48	0.38–0.5
Single family	0.45–0.55	0.5–0.61	0.54–0.66	0.56–0.69
Multi-unit	0.5–0.6	0.55–0.66	0.6–0.72	0.63–0.75
Apartments	0.6–0.7	0.66–0.77	0.72–0.84	0.75–0.88
Parks/cemeteries	0.1–0.25	0.11–0.28	0.12–0.3	0.13–0.31
Playgrounds	0.4–0.5	0.44–0.55	0.48–0.6	0.5–0.63
Agricultural areas	0.1–0.2	0.11–0.22	0.12–0.24	0.13–0.25
Bare ground	0.2–0.3	0.22–0.33	0.24–0.36	0.25–0.38
Undeveloped desert	0.3–0.4	0.33–0.44	0.36–0.48	0.38–0.5
Mountain terrain (slopes > 10%)	0.6–0.8	0.66–0.88	0.72–0.95	0.75–0.95

Source: Adapted from the "Drainage Design Manual for Maricopa County, Arizona."

number method and the unit hydrograph used is the standard dimensionless unit SCS hydrograph (Maidment, 1992; see chapter 4).

10.3.2 Stormwater Runoff Quality

The quality of stormwater runoff is often a function of several mechanisms, of which any one or combination may be most important for a given storm event and location. During dry periods pollutants that will ultimately show up in stormwater runoff buildup on the various surfaces in the watershed. These materials may be generated locally or at great distance and carried aloft by winds only to ultimately settle onto surfaces within the local watershed. This phenomenon is often termed dry-fall or aerial deposition. In addition, during precipitation events, materials in the atmosphere are captured by raindrops as they fall to the land surface, thereby introducing more pollutants to the runoff. Even more material is added to the runoff as pollutants deposited during dry periods are washed from rooftops, parking lots, and other impervious surfaces. The type(s) and amount of pollutants found in such runoff are often associated with a given land use or activity. In any case, literally any pollutant is likely to show up in stormwater runoff if it was used in the watershed and is exposed to direct precipitation and/or runoff.

As is the case with the design of any waste water treatment system, the incoming quality should be known or estimated. As was previously mentioned, the pollutants found in stormwater runoff can be a function of buildup during antecedent dry periods, the land use comprising and the activities occurring within the contributing watershed. One of the first attempts at developing stormwater quality estimates came from the Nationwide Urban Runoff Project (NURP). In this program regional regression equations were developed that can be used to predict/estimate pollutant loads and concentrations as a function of drainage basin and storm characteristics (Driver and Trasker, 1990). Such predictions can be compared with measured values to evaluate the accuracy of the regional-regression equations. The regional-regression equations can also be adjusted using local data so that they may be applied to local unmonitored basins. Since the adjusted equations are based on local and regional data sets, the amount of local data needed to estimate pollutant loads and concentrations is reduced, but statistical strength is maintained. Procedures for adjusting the regional-regression equations are described by Hoos and Sisolak (1993).

Computer programs have been developed which combine stormwater quantity and quality issues. When validated by measured data, they can provide an efficient means of supplying hydraulic and pollutant mass loading for stormwater treatment wetland design. Two such models are available from the U.S. EPA, the Stormwater Management Model (SWMM) and the Hydrological Simulation Program-FORTRAN (HSPF). SWWM can be used for the analysis of urban stormwater quantity and quality and the modeler can simulate all aspects of urban hydrology, including storage and treatment. HSPF simulates watershed hydrology and water quality with simulations resulting in a timeline of runoff flowrate, sediment load, and nutrient and pesticide concentrations. The time history of water quantity and quality for any point in a watershed can also be provided. Both models operate in the DOS platform and can be obtained from the U.S. EPA Center for Exposure Assessment Modeling (CEAMS).

In addition to the computed and calculated pollutant concentrations in stormwater runoff, several authors have reported ranges of expected pollutant concentrations in stormwater runoff from a given land use category. The following Table (10.3) provides a compilation of select parameters from numerous sources.

10.4 *WETLAND PROCESSES THAT ALTER WATER QUALITY*

Free water surface constructed wetlands produce a range of effluent qualities, depending on the influent characteristics, constituent operational loading rates, climate, and aereal extent

TABLE 10.3 Selected Pollutant Median Event Mean Concentrations (mg/L) and Associated Coefficient of Variation (CV) in Stormwater Runoff as a Function of Land Use

Parameter:	BOD	COD	TSS	$NO_2 + NO_3$	Total P	Total Cu	Total Pb	Total Zn
Landuse:								
Residential	10.0	73	101	736	383	33	144	135
	(0.41)	(0.55)	(0.96)	(0.83)	(0.69)	(0.99)	(0.75)	(0.84)
Mixed	7.8	65	67	558	263	27	114	154
	(0.52)	(0.52)	(1.14)	(0.67)	(0.75)	(1.32)	(1.35)	(0.78)
Commercial	9.3	57	69	572	201	29	104	226
	(0.31)	(0.39)	(0.85)	(0.48)	(0.67)	(0.81)	(0.68)	(1.07)
Open/nonurban	—	40	70	534	121	—	30	195
	(—)	(0.78)	(2.92)	(0.91)	(1.66)	(—)	(1.52)	(0.66)

Source: USEPA, 1983.

of the system. When designed and operated properly, constructed treatment wetlands have performed within predictable ranges of effluent values and meet their permit limitations. Wetlands accomplish this through a combination of physical, chemical, and biological mechanisms.

Physical mechanisms that influence the water quality from a wetland include gas transfer, sedimentation, adsorption/desorption, filtration impaction, flocculation, photochemical reactions, and volatilization. Gas transfers involve the movement of gases such as O_2, N_2, CH_4, and sulfides across the air-water interface and to and from the bottom sediments. Sedimentation is an extremely important pollutant removal mechanism because many constituents, such as heavy metals, are often associated with the particulate phase. In treatment wetlands sedimentation is also a primary removal mechanism for BOD, TSS, and heavy metals, while it provides a secondary mechanism for the removal of nitrogen and phosphorous species. Adsorption and desorption can impact the parameters of BOD, TSS, bacteria and viruses, and heavy metals either by increasing (desorption) or decreasing (adsorption) their concentrations in the water column. Filtration/impaction refers to particulates being filtered mechanically as the water passes through substrates and plant materials, which contributes to the reduction of Biological Oxygen Demand (BOD), TSS, and heavy metals (Stowell et al., 1980). Flocculation precedes and can enhance sedimentation, thereby assisting in the removal of BOD, TSS, bacteria, and viruses. Photochemical reactions facilitate the degradation of organic compounds and contribute to bacteria and virus reduction. Volatilization is a physical mechanism that can contribute to the removal of ammonium and other pollutants with low partial pressures.

The previously mentioned physical mechanisms work in concert with the chemical processes in a wetland to further alter wetland water quality. The chemical mechanisms at work in a wetland include chemical adsorption, chelation, oxidation/reduction reactions, and chemical precipitation. Chemical adsorption onto surfaces within wetlands can result in water column reductions of phosphorous, bacteria and viruses, and heavy metals. Chelation reactions also impact phosphorous concentrations, but are a primary mechanism for heavy-metal removal. Chemical oxidation/reduction (redox) reactions are a primary removal mechanism for BOD and heavy metals, and can also have an impact on TSS. Certain heavy-metal species can be bound as metal sulfides under appropriate redox conditions. If reducing conditions are maintained these compounds can become buried and essentially immobile (USEPA ETI, 2000). Chemical precipitation reactions can be a primary removal mechanism for phosphorous.

The biological processes of wetlands also influence water quality. Important biological mechanisms include algal synthesis, assimilation into higher plants, bacterial metabolism, and predation. Algal synthesis or incorporation of nutrients into cell tissue can influence nutrient concentration in the water column. If these algal cells are allowed to exit the wetland system, they will export nutrients and show up analytically as TSS. Assimilation or the uptake and metabolism by plants can increase or decrease BOD, TSS, nutrients, dissolved oxygen, bacteria and viruses, and heavy metals depending on the life cycle stage of the vegetation. Bacterial metabolism, both aerobic and anaerobic, can have a profound effect on wetland water quality. Aerobic bacteria are responsible for nitrifying ammonia species to nitrite and nitrate, while anaerobic bacteria convert nitrate to dinitrogen (N_2) gas. Aerobic bacteria also reduce BOD and TSS concentrations. Aerobic bacteria may also depress dissolved oxygen concentration due to the use of O_2 in bacterially mediated reactions. Phosphorous concentrations are influenced by microorganisms as they uptake this nutrient for cell tissue growth and metabolic activities. Finally, zooplankton, aquatic insect larvae, and even small fish larva will feed upon suspended solids that can harbor bacteria and viruses.

These are all internal processes that potentially affect the quality of water exiting a wetland. It must be stressed that FWS wetlands are "open systems" and as such are subject to various perturbations. Even though properly designed wetland systems perform in a predicable range of effluent values, a limitation to using FWS-constructed wetlands as a stormwater runoff treatment system is not only the variability in hydraulic and mass loadings, but also the background concentration of constituents produced by the external loading and the internal wetland processes.

Background concentrations of BOD, COD, turbidity, total phosphorus, total nitrogen, and total and fecal coliform can control the effluent quality achievable in a FWS constructed wetland. The natural cycle of nutrients and the potential re-release of constituents incorporated in the wetland biomass must be considered in the effluent permit requirements for discharges from FWS constructed wetlands.

10.5   TREATMENT WETLAND CONFIGURATION

Treatment wetlands are complex systems offering a setting where physical components facilitate chemical reactions and biological processes to remove and/or transform pollutants. Typically treatment wetlands can be configured into three general classes: (1) *Free-water Surface (FWS)*, (2) *Subsurface Flow (SSF)*, and (3) *Floating Aquatic Plant (FAP)*. FAP systems rely upon the uptake of pollutants (generally nutrients, but perhaps dissolved constituents as well) by floating vegetation for treatment with true pollutant removal requiring the harvesting of the plants. SSF systems convey the design flow below ground while providing a suitable environment for wetland plants. Stormwater flows are often turbulent and can transport significant amounts of solid materials that may hydraulically overload FAP and/or clog SSF systems. Because of this, the required retention time for pollutant uptake by the vegetation may not be achievable or the size needed to meet those requirements may be cost prohibitive in FAP systems. SSF wetlands provide efficient treatment and should be considered only for treating runoff with low TSS or in systems with pre-treatment structures because of potential clogging issues. The remainder of this chapter will only consider FWS wetland design, if interested in SSF systems, the reader is encouraged to consult the municipal waste water treatment wetland literature for design insight (Kadlec and Knight, 1996; Reed, Crites, and Middlebrooks, 1995; Metcalf and Eddy, 1991; US EPA, 1993).

10.6   FREE WATER SURFACE (FWS) STORMWATER WETLAND DESIGN

A primary goal of stormwater treatment wetland design will be to accommodate design flows in a manner that dissipates energy to the extent that floating and suspended materials are allowed to settle in quiescent areas and/or can be mechanically filtered by wetland vegetation. This can be achieved through wetland inlet and forebay design or by providing upstream flow attenuation and/or pre-treatment structures separate from the wetlands. The internal wetland design should attempt to minimize or mitigate preferential flow paths and allow for maximum contact with the wetland surfaces while the water is retained within the system.

FWS treatment wetlands are comprised of several compartments, water, soil, and vegetation, each of which contributes to the alteration of water quality (Fig. 10.2). Unless the site topography permits, some type of containment will also be necessary to retain the stormwater and allow time for water quality transformations to occur. Within the saturated areas and where depth permits, wetland vegetation can be established which shades the water surface, dampen temperature fluctuations, and provide habitat. Emergent wetland vegetation is often established in dense stands that allow for mechanical removal of particulate matter and the pollutants associated with it. Dead and live stems provide for gas exchange between the sediments, water column, and atmosphere. The submerged portions of the plants and litter-fall serve as surfaces for the attachment of biological films. Upon senescence and subsequent decay, plant material can release nutrients that can sometimes be detected as

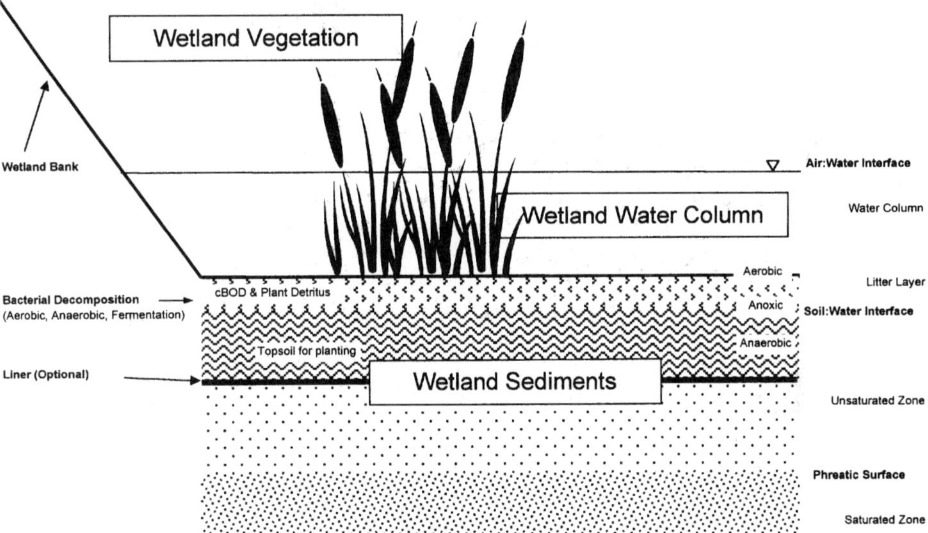

FIGURE 10.2   Treatment wetland compartments that work together to alter water quality.

seasonal pulses in the wetland discharge. Carbon is also released and can be used by microbial communities to satisfy metabolic needs and is important for microbial mediated nutrient transformations and removals. Carbon compounds resulting from the decay of wetland plants and microbial communities will also likely show up in the wetland effluent and can be measured as BOD or COD.

In FWS wetlands, depth can be used to restrict emergent vegetation and provide open or un-vegetated areas. As a rule of thumb, most emergent macrophytes are restricted to depths less than 1.5 m (Payne, 1998). Open-water features can provide mixing of flows, an air: water boundary for gas exchange between the water column and the atmosphere, quiescent conditions for removal of readily settleable materials, and refuge for wildlife during times when emergent areas are drained because of seasonal water level fluctuations or operations and maintenance (O&M) considerations. Some use has been made of floating aquatic plants in these open-water deep zones and these likely enhance water quality and the habitat value of the system. The floating plants tend to have dangling root structures that facilitate the uptake of dissolved constituents, permit gas exchange with the water column, and provide an environment suitable for zooplankton communities to establish within. In some cases, submerged aquatic plants such as *Potamogeton, Cerratophylum,* and/or *Myriophylum* species have been used in areas too deep to sustain emergent macrophytes. Submerged plants also alter water quality by exchanging gases with the water column, (O_2 during daylight hours, CO_2 at night), and can provide substantial surface area within the water column for the attachment of algae and bacteria.

The challenge to the designer is to arrange the compartments such that for a given range and character of pollutant loadings, improved effluent water quality is achieved. A balance must be struck between the incoming hydraulic and pollutant loads and the size/containment necessary to provide the residence time contact needed for the physical, chemical, and biological mechanisms to work. To begin to relate these parameters a couple of hydraulic and physical relationships can be used. First is the water balance that accounts for all inflows, outflows, and storage to, from, and within the wetland. In the case of a stormwater treatment wetland, inflows to the system will likely include the surface runoff (Q_{sw}), the precipitation falling directly upon the wetland water surface and contributing land surface (Q_{precip}), and in areas of shallow groundwater the potential exists for groundwater intrusion (Q_{gw}). If the treatment wetland is designed to discharge, the outflows that will likely need to be considered include the surface discharge (Q_{eff}), evapotranspiration (Q_{ET}), and if the system is not lined with an impermeable material, losses to groundwater (Q_{infil}).

Water Balance

$$Q_{in} = Q_{out} + S$$

where $Q_{in} = Q_{sw} + Q_{gw} + Q_{precip}$
$Q_{out} = Q_{eff} + Q_{ET} + Q_{infil}$
$S = $ storage

In a well-designed treatment wetland, surface inflow(s) can be measured at the inlet(s) to the system and likewise, the outlet(s) design should allow for the measurement of surface discharges. Evapotranspiration (ET) can be estimated in a number of ways, including adjusting pan evaporation data or the use of a reference evapotranspiration. Thus all inflows and outflows can be measured or estimated: if the basin volume is known this leaves the groundwater interactions as the only unknowns to be solved for. The utility of the water balance includes the ability to determine the wetland size and configuration necessary to achieve a desired system hydroperiod, which in turn, will dictate the vegetative makeup and ultimately the water quality performance of the system.

The second parameter that can be used to relate the *wetland size* (area and/or volume) to the potential treatment performance is the nominal or theoretical hydraulic residence time (HRT). Simply defined the theoretical HRT of a system is

$$\tau = V\Phi/Q$$

where $\tau = $ hydraulic retention time $[T]$
$V = $ wetland volume $[L^3]$
$\Phi = $ porosity
$Q = Q_{in}$, inlet-based, or $Q = (Q_{in} + Q_{out})/2$ accounts for water losses and gains $[L^3/T]$

Initially, the *wetland volume* (V) can be easily determined from construction drawings, but after filling and operation, changes in bottom elevations can occur, which may make this method less accurate as the system matures. Another means of estimating the wetland volume is to measure or estimate the average depth of the system and use a perimeter survey or aerial photography to determine the wetted surface area. Because wetlands are often large in aerial extent, the wetted surface area is a good approximation of the wetland bottom (Kadlec and Knight, 1996).

Φ, the *wetland porosity,* accounts for the volume occupied by the plant material. In practice, Φ can range from 0.65 to 0.95, with newly planted systems being associated with higher Φ values (Kadlec and Knight, 1996; Reed et al.,1995; Gearheart, 2000). Two choices are presented for use as the flowrate term; the first uses only the measured inlet flow. In systems where water losses and gains are negligible, this is probably a reasonable approximation. In most instances it is prudent, however, to incorporate the water losses and gains, which can be done by using the average of the flows, measured at the inlet(s) and outlet(s).

The theoretical retention time in treatment wetlands is often longer than the true or measured retention time because of the inefficient use of the wetland volume (dead zones) and preferential flow paths (short-circuiting). Tracer testing of several treatment wetlands has shown that the tracer retention time, the time necessary for the centroid of the tracer mass to exit the basin, can be 65 to 95% of the theoretical retention time (Kadlec and Knight, 1996; Wass, 1996).

10.6.1 Stormwater Wetland Sizing

Stormwater treatment wetland design is as much an art as it is a science. This is reflected in the literature by the variation in reported design guidelines and insight from practitioners. Part of this is due to site constraints and the uncertainties associated with rainfall events and the subsequent runoff water quantity and quality. Many of the systems currently treating stormwater were sized not for water quality enhancements, but rather to accommodate local or regional drainage and retention requirements. Much of the current design guidance relates FWS wetland surface area to the area of the contributing watershed. A *wetland to watershed area ratio* (*WWAR*) of 2% or greater is often recommended; if a detention basin precedes the wetland this ratio may be reduced to 1 to 2% (Schueler, 1992). The California Stormwater Best Management Practice Handbook for Municipalities (CDM, 1993) recommends a similar wetland to watershed area of 1% with 2% being ideal. Washington State recommends 1.5% while the largest WWAR recommendation came from Maryland, at 3%. In general, a WWAR of between 1% and 3% appears to be the rule, and for carefully designed and constructed wetlands the WWAR could be reduced to less than 2% (Strecker, 1992).

Other means of estimating the size of FWS stormwater treatment wetlands includes determining the basin volume needed to capture and treat a given fraction of runoff or to design for the volume of runoff produced from an event with a given return frequency. To provide approximately 80% TSS removal Shaver and Maxted (1994) report that stormwater wetlands should be sized to intercept and treat the first 1.0 in of runoff and release over a 24-hr period (Schueler, 1987). This is slightly more than the 0.75 in treatment requirement using conventional BMPs as was recently adopted by the Los Angeles Regional Water Quality Control Board (Lee, 2000). In terms of peak discharges, some authors advise containment of the 2- to 10-year return frequency events, although to satisfy local stormwater ordinances and regulatory concerns, higher peak discharges will likely need to be controlled (Shaver and Maxted, 1994; Anderson, 2000). Currently, stormwater treatment wetland designers in the Pacific Northwest size systems based upon controlling and treating runoff from a 2-year, 24-hour storm event with a 48-hour drawdown (Wilson, 2000).

10.6.2 Site Layout

If possible, the site layout and treatment wetlands should be designed with the overall goal(s) of the project in mind. Sometimes the treatment wetland goals will include more than merely managing runoff quantity and quality that at times may seem more important. Aspects such as aesthetics, habitat, passive recreation, and environmental education and land stewardship may all be important. In fact truly sustainable systems will probably combine several of the goals. Aesthetics, for example, can appear to be more important than the water quality functions if, for example, the system is located in a highly visible location or is surrounded by residential landuse. A wetland with stressed or rotting vegetation and floating trash, and that produces nuisance organisms will not persist long if a city councilman lives nearby or if such issues are brought to the attention of municipal leaders by distraught citizens. Such a scenario may also discourage the funding of similar projects in the future even if water quality objectives have been met.

To the extent practicable, the site layout should use existing topography and incorporate stormwater runoff controls to minimize earth-moving costs and so that treatment can begin as soon as possible. As an example of an on-site system, runoff from parking lots and roof drains could be routed through conveyances equipped with trash racks and/or oil/grease skimmers to remove floating pollutants and trash. Further, inlets into stormwater conveyance structures could be depressed, thereby allowing some solids removal. The conveyance system itself may also consist of grass or otherwise vegetated channels that follow existing contours and also tend to improve water quality. In some instances, the stormwater runoff conveyance system can also serve to complement landscaping and other aesthetic features of the complex prior to treatment in a wetland.

Systems that will be receiving stormwater runoff from larger watersheds or several sub-basins that produce high velocity flows may require attenuation of such flows prior to introduction into a treatment wetland. Reduced velocities will minimize the re-suspension of sediments and protect vegetation from scour and burial in the inlet zone. Provision of an upstream sedimentation basin will also reduce the amount of sediments entering the wetland so that sensitive benthic organisms are protected from burial and will also likely increase the length of time between maintenance dredging. Further, the hydraulic connection between the sedimentation basin and the wetland can be outfitted to allow for oil and grease removal, as well as control of floating materials and trash. The size of such a structure will likely be governed by local flood control or drainage regulations. Upstream retention structures also allow flow to be metered into the wetland basin, thus giving the operator the potential to regulate hydraulic and pollutant loadings.

10.6.3 Wetland Basin Shape

Free water surface treatment wetlands can be constructed using various shapes and still achieve water quality goals (Fig. 10.3). Recta-linear, oval, kidney-shaped, and amorphous shoreline configurations can be used successfully when attention is given to internal wetland features such as inlet outlet placement, vegetation, islands, and deep zones (USEPA ETI, 2000). Stormwater treatment wetland shapes that utilize existing topography and minimize the amount of aboveground containment and/or earth moving should result in cost effective designs (Figs. 10.4, 10.5, and 10.6).

10.6.4 Stormwater Wetland Inlet and Forebay Considerations

Stormwater runoff will enter the wetland via an inlet structure that should be stabilized to prevent erosion. Many stormwater treatment wetland designers use a simple open-channel inlet stabilized with freely placed or grouted riprap. Turbulent flowdown such an inlet structure can begin the treatment process by aerating the water and allowing compounds with low partial pressure to volatilize. In instances where flow velocity and channel slope are low and access for maintenance equipment is provided elsewhere, soil-cement or grass-lined inlets have been successfully used. Such inlets are easy to construct and need little maintenance.

Depending upon the scale of the stormwater wetland, the inlet width and stability should be constructed so that maintenance vehicles can use it to access the wetland forebay for sediment removal activities. The slope is maintained to minimize erosion down the slope of the inlet and to protect the inlet bottom from scour. Bank areas immediately adjacent to the inlet structure should be vegetated or otherwise stabilized. Flow measurement can be accomplished immediately upstream of the wetland inlet in an open channel by providing a Parshall flume or similar device (ISCO-BOOK Grant, 1992; Replogle, et al. 1999).

More complex inlet structures can be designed that allow flow to be metered into or diverted from a given wetland basin, thus adding operational flexibility to the system. Although many configurations can exist, a typical structure could consist of a stilling basin or concrete box where incoming flows are stilled and then metered/split out through control device(s) such as a series of stop-logs or movable weir gate(s). In any event the inlet structure should introduce runoff to the forebay with minimal disturbance to the wetland banks and bottom.

The *wetland inlet* introduces runoff into a deep-zone or forebay. The primary purpose of this structure is to protect the remainder of the wetland system from the erosive force of the incoming runoff, to promote sedimentation and to provide an area for easy solids removal. In addition, the forebay provides an area for initial mixing and aeration of waters prior to entering a vegetated zone. Forebay size can range from 5 to 20% of the total wetland area

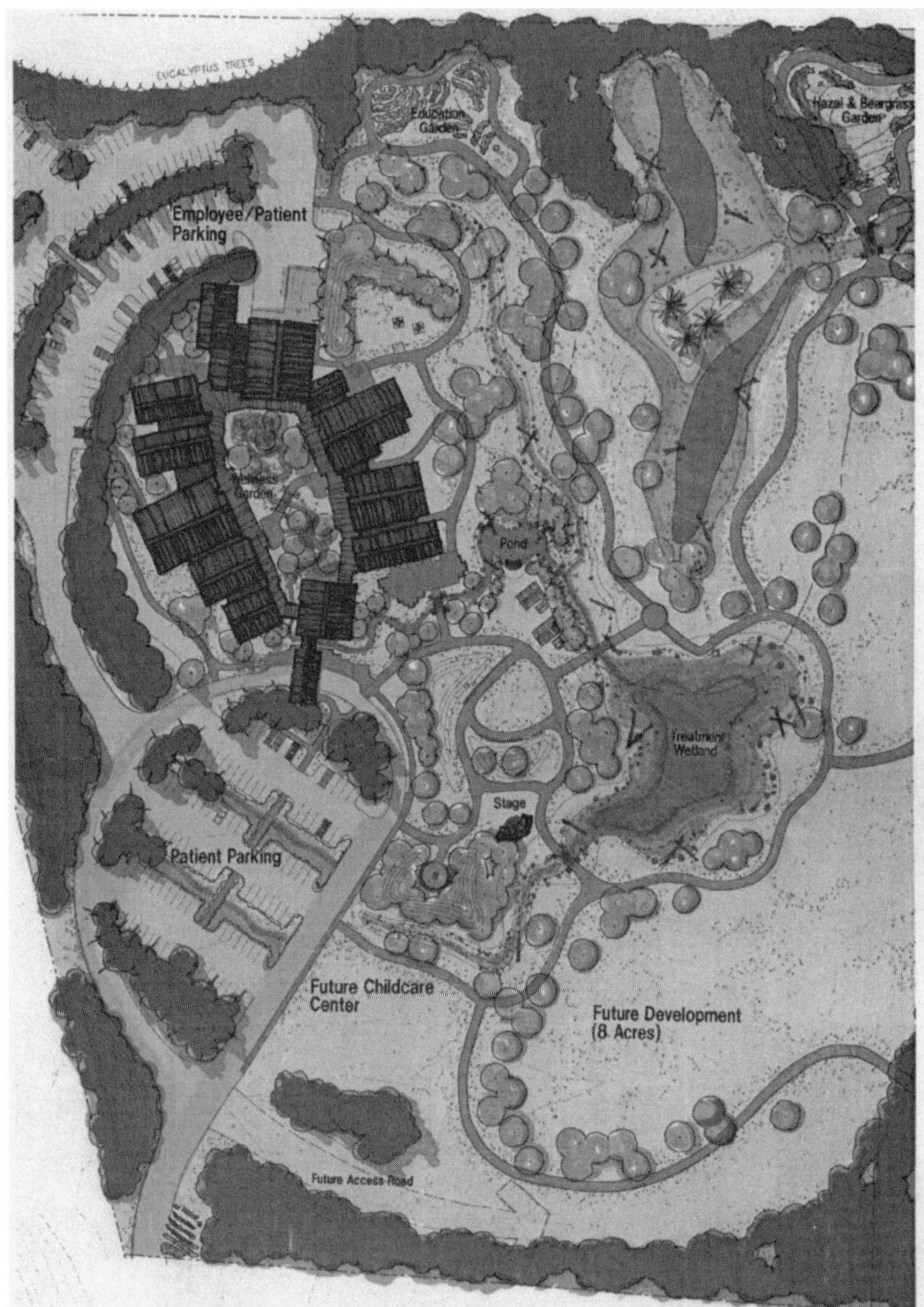

FIGURE 10.3   Arcata California conceptual site plan showing stormwater collected in parking areas and from impervious surfaces are conveyed via gravity in open, vegetated channels to a stormwater wetland for treatment. Discharge from the treatment wetland is used to support wetland restoration and interpretative areas. (Adapted with permission from HWR, Inc. Arcata, California)

FIGURE 10.4   This stormwater treatment wetland was designed within an existing swale. Appropriate wetland vegetation was established at varying elevations above the dry-weather pool to provide treatment, soil stabilization, and aesthetics to the overall treatment wetland complex. Note this stormwater wetland does not discharge until the water surface elevation reaches the outlet works that is located at the upper right side of the impoundment area shown in this photograph. (Photograph by HWR, Inc. Arcata, California)

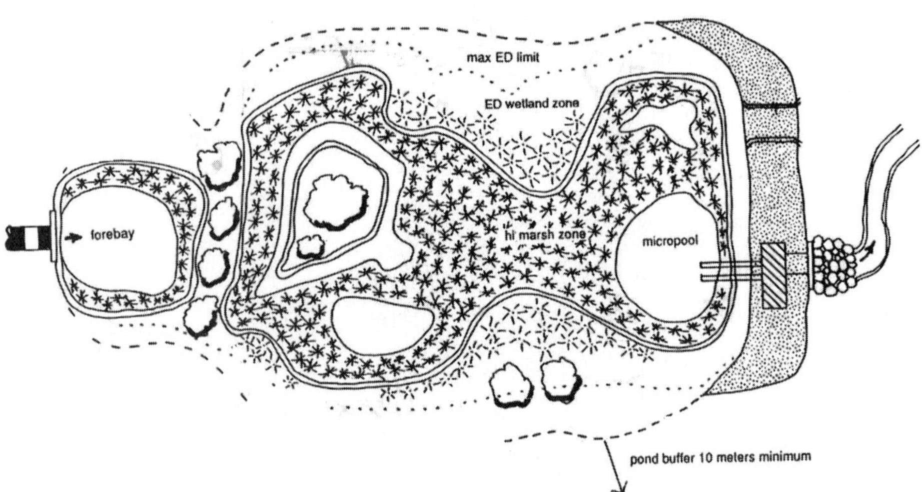

FIGURE 10.5   Example stormwater treatment wetland conceptual layout showing many design components, including extended detention (ED) zones that contribute to water quality improvements in such systems. Note the irregular wetland perimeter and general amorphous shape that could be tailored to use existing site topography. (Adapted with permission from Schueler, T.R., 1992)

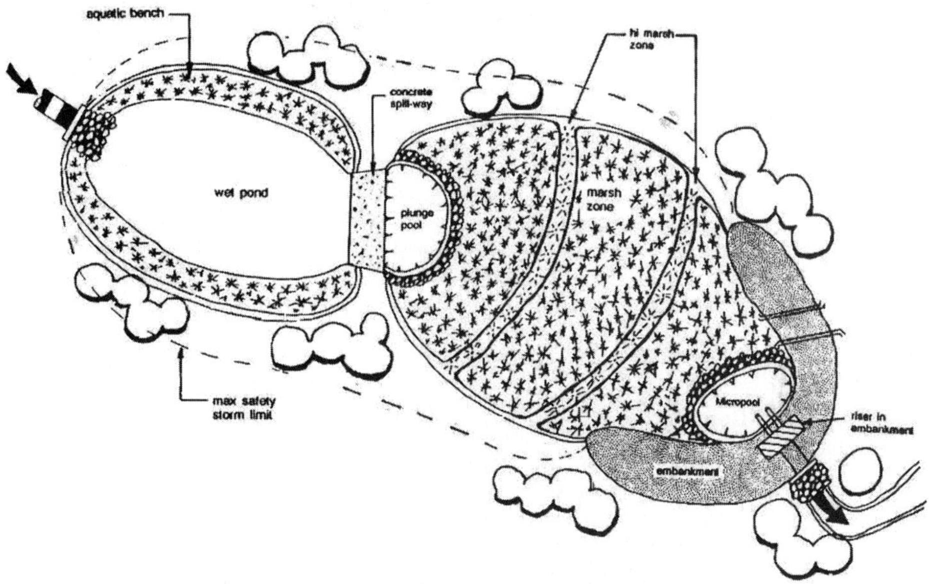

FIGURE 10.6 Example stormwater treatment wetland conceptual layout utilizing deepwater pond and emergent marsh configuration areas. Multiple marsh elevation zones allow for more diverse assemblage of aquatic macrophytes to be used, which in turn can provide additional habitat value. (Adapted with permission from Schueler, T.R., 1992)

depending upon the energy within incoming flows, while the depth should be 1.5 m or deeper to prevent encroachment by marsh vegetation and allow for sediment buildup. Side slopes should be maintained in the 10:1 to 4:1 range, with flatter slopes being more desirable. Since this structure will receive the highest energy flows, shoreline and bank stabilization is required and can be accomplished using a variety of techniques from semi-aquatic grasses and riparian trees and shrubs to geotextiles and or soil cement.

The *forebay* should be separated from the downstream wetlands and be capable of being drained while water still persists in the wetland areas. This separation can be accomplished in several ways including: (1) an underwater earthen sill planted with dense emergent macrophytes such as *Phragmites, Typha* or *Scirpus* sp. (300 to 800 stems/m²), (2) Concrete spillway structure, and (3) physical separation using either an open channel or piped connection to downstream wetlands.

The underwater earthen sill should be situated at the elevation equal to the downstream emergent marsh section. This will allow water depth to fluctuate on the sill between moist soils and the operating depth of the marsh zone(s), approximately 0.1 to 0.6 m. Additionally, this will allow the forebay to be isolated and dried out for maintenance while water remains in the downstream wetland areas. This structure must be planted with a dense stand of emergent macrophytes to provide structural integrity to the sill and provide for physical capture of floating materials and to allow impaction, mechanical straining, and settling of solids and particulate matter.

Another means of separating the forebay from downstream wetland areas is to construct a concrete spillway or a gabion wall structure on the downstream end of the forebay. Use of a concrete spillway may require provision of an armored plunge-pool or deep-zone immediately downstream of the structure to dissipate energy and minimize re-suspension of wetland sediments. If a concrete spillway is used it can serve as access for maintenance equipment for sediment removal from both the forebay and plunge-pool. Using a simple gabion wall for separation may require the least amount of space for the separation structures presented if the designer is challenged for space. If it is expected that high flows will overtop the structure, a downstream plunge-pool is again recommended.

The forebay can also be physically separated from the wetland system(s) and water subsequently conveyed to the wetland via an open channel or pipeline. In such a scenario, the forebay is designed with an inlet as described above, but the outlet is via a weir structure or piping. In the case of a weir, a skimming device should be positioned immediately upstream of the outlet to prevent trash and floating debris from clogging the conveyance structure to the wetland or entering the wetland itself. An adjustable piped outlet that draws from beneath the waters surface can also be used with pipe inlet screened to prevent debris from clogging. Discharging water from below the water surface will also allow the forebay to act as an oil skimmer. In any configuration, a means of draining the forebay should be provided to accommodate maintenance activities.

10.6.5 Wetland Containment

Very rarely does a site afford the treatment wetland designer a land surface suitable for wetland construction without requiring earth-moving activities. As previously mentioned, the most cost-effective designs will be those that balance cut and fill activities and minimizes earth moving associated with constructing the aboveground containment berms and excavating interior deep-zones. A typical exterior wetland berm should be designed with a minimum top-width of 1.5 m for foot traffic or 3.5 m for vehicular access (Fig. 10.7). The interior berm should be vegetated and the slope should be greater than or equal to 3:1 with flatter slopes more desirable. If constrained to using steeper slopes, consideration should be given to providing access points down the berm slope to the waters edge for data collection and O&M activities. This can be easily accomplished by placing a 20- to 30-cm deep course of clean angular rock with a mean diameter of 7.5 to 10.0 cm on the bank (personal communication, Wilson, 2000). The exterior slope of the wetland berm can be steeper, 2:1 to 3:1, and stabilized with shallow rooted vegetation such as grass or stabilized with rock. Total berm height should include consideration of wetland operating depths, typically 0.1 to 0.5 m, and freeboard of 1.0 m. If warranted by site conditions, the berms may be designed with an impervious plug to prevent seepage or a seepage collection trench can be constructed at the toe of slope on the exterior of the berm (potential borrow site for berm construction material). Berm penetrations for pipelines and conduits should be outfitted with seepage control collars. Internal berms for flow control are constructed so that the operational water

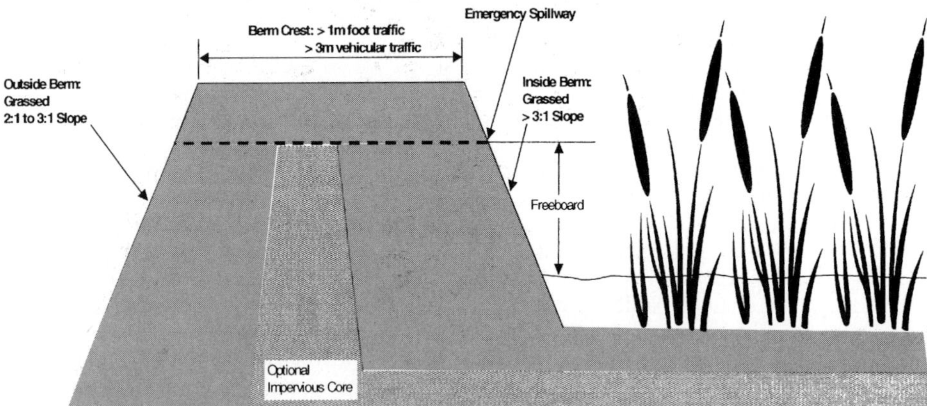

FIGURE 10.7   Wetland berm details. Adapted from Kadlec and Knight, 1996b.

surface does not cover the structure. The top width can be 0.5 to 1.0 m and should be grassed. Side slopes of interior berms should be in the 2:1 to 3:1 range to minimize sloughing and erosion of the berm slopes.

In some cases the wetland bottom and banks will need to be lined to prevent interaction and exchange with the underlying groundwater and/or to conserve water during times between storm events. Lining of the wetland bottom and side slopes can be accomplished using compacted native soil if the permeability is sufficiently low. If such soils do not exist at the site, a clay or Bentonite liner can be constructed to prevent groundwater exchange. Finally, a synthetic liner can be used such an HDPE. Consideration should be given to the strength of the liner and, since runoff also contains oil, grease, and petroleum products, also to the chemical compatibility. In any case, the liner should be covered with 31–61 cm of topsoil to serve as a rooting medium for the aquatic plants and to protect the liner material.

10.6.6 Inlet and Outlet Structures

Placement of wetland inlet and outlet structures with respect to each other is important for water quality improvements in wetlands. Ideally, the inlet and outlet should be separated to allow water to spend the most time and to come into contact with the most surfaces possible within the wetland. Further, they should be designed to minimize turbulence and energy transfer to the wetland. With the exception of the submerged earthen vegetated sill configuration, the wetland inlet should be an open-water zone with a depth of 1 to 1.5 m. This zone should extend across the full width of the wetland and be situated perpendicular to the main flow path, thus effectively providing additional area for settling, aeration, and mixing prior to introduction into shallow vegetated marsh zones.

If the stormwater treatment wetland is designed to discharge, the outlet should also be situated in a deep-water zone extending across the full width of the flow path. The top-width of the outlet zone can be 3.5 to 10 m or more depending upon the overall scale of the wetland, with the depth between 1.0 to 1.5 m deeper than the maximum operating emergent zone depth. Since open-water zones expose the water column to sunlight, the potential for algal development exists in such a structure. Floating aquatic vegetation (water hyacinth) or that which roots in a bank (*Hydrocotyle* sp.) and spreads across the water surface have been successfully used to minimize algal production in areas such as these (Fig. 10.8). The wetland outlet will be the chief means of controlling the water surface elevation within the treatment wetland. As such adjustable weirs or stop logs should be considered. As an alternative, the outlet can be configured such that water is drawn through adjustable perforated pipes located below the water surface. This will still allow depth manipulation and do away with the need of an outlet skimmer board or similar device to prevent floating debris from exiting the systems.

Finally, provision of an emergency spillway in addition to the primary outlet structure(s) is necessary. Even though freeboard has been designed into the wetland containment structures, it is normally not sufficient to contain all potential flood events. To protect the berms and prevent overtopping, an emergency spillway is usually provided so that flood flows are maintained 0.5 to 1 m below the top of berm elevation. Final design berm and emergency spillway elevations and flood storage requirements should be confirmed with local drainage and flood control authorities.

10.7 INTERNAL CONSTRUCTED TREATMENT WETLAND DESIGN FEATURES

10.7.1 Emergent Areas

Emergent areas or *wetland marsh zones* are typified by operating depths that range from essentially saturated soil conditions to a water column depth of 0.5 m, although most systems

FIGURE 10.8 Outlet deep-zone to permit even discharge of wetland waters planted with floating aquatic plants to reduce sunlight transmission as a means of controlling algal development in an open-water zone.

operate with less than 0.3 m (ASCE, 2000) water depth in the emergent zones. Such zones are densely vegetated with emergent aquatic macrophytes that set up conditions conducive for microbiologically mediated pollutant transformations and removals (Fig. 10.9). Depending upon the goals of the treatment wetland, emergent areas may occupy 80% or more of the wetland surface area with those systems that emphasize treatment having the larger emergent marsh zone percentages. As an example, the City of Arcata, CA has a stormwater wetland program whose goals emphasize both stormwater treatment and wildlife habitat. Currently the design guidance suggests 40% stormwater treatment and 60% wildlife habitat (HSU, 2000). These criteria would imply a wetland system with 40% of its surface area configured as emergent marsh, and 60% configured as open-water.

Emergent vegetation in these marsh zones is often quite dense (300 to 800 stems/m^2) and as they mature and undergo life cycle progression the aboveground vegetation will senesce, lodge, clump, or wildlife will otherwise create preferential flow paths within it. These short-circuits reduce the efficiency of the wetlands by preventing contact of the water with the wetland surfaces, for example, bottom sediments, detritus (litter-layer), and the submerged portions of the vegetation. Further, short-circuiting reduces the retention time within the wetland. The un-idealities of flow introduced by maturing vegetation or impacts from wildlife (e.g., muskrats or beaver) in the emergent zones can be mitigated to some extent by providing internal deep-zones within the wetland.

The decay of the wetland vegetation in the emergent zones and the oxygen demanding substances in the stormwater runoff depress dissolved oxygen levels in these areas. Again, this phenomenon can often be mitigated by provision of open-water deep-zones downstream.

10.7.2 Internal Deep-Zones

Internal deep-zones are treatment wetland structures that serve both water treatment and wildlife functions (Fig. 10.10). These features should be oriented perpendicular to the main

FIGURE 10.9 Emergent zones in treatment wetlands are densely vegetated features that can enhance mechanical straining of particulate matter, and provide an environment for bacteria and algae that can improve water quality.

FIGURE 10.10 Internal deep-zone provides an open water surface for gas exchange and waterfowl, while at the same time enhancing mixing and mitigating short-circuiting in upstream shallow emergent marsh zones.

wetland flow path and extend across the full width of the basin. Side slopes should be 3:1 or flatter to reduce sloughing and gradual filling in of the zone by sediments. The depth of these zones should be 1.0 to 1.5 m deeper than the operating depths in the emergent zones to prohibit intrusion by emergent vegetation. Top width of internal deep-zones will depend upon the wetland scale, but a width of at least 5 m is recommended.

Internal deep-zones can be left open or vegetated with submerged aquatic plants, floating aquatic plants, or vegetation that roots in the bank and then spreads over the water surface to enhance habitat values and to provide more surface area within the water column for the attachment of bacteria and algal organisms (Fig. 10.10). The open-water surface of these deep-zones also provides an area for gas exchange with the atmosphere and allows access to wind action for mixing. Depending upon vegetation selection, the plants themselves will help to re-aerate wetland waters during daylight hours due to their photosynthetic activity. Conversely at night the vegetation will switch respiration pathways and consume oxygen resulting in a sag in dissolved oxygen (D.O.) in the pre-dawn hours. In either configuration, internal deep-zones can mitigate the aforementioned short-circuiting of flows in the emergent zones by allowing a place for waters to remix (Fox et al., 2000). By alternating emergent zones and deep-zones, the water is more evenly spread across the width of the wetland basin and the efficiency of the system should be increased.

Internal deep-zones will also provide substantial wildlife habitat value to a treatment wetland. Open-water areas and the submerged aquatic plants grown there are attractive to many species of waterfowl and can also provide forage area for swallows and other insectivorous birds. These deep-zones also provide refuge for aquatic organisms such as fish, turtles, and macroinvertebrates when the emergent areas are dried for maintenance reasons or during dry parts of the year.

10.7.3   Wetland Vegetation

Emergent aquatic macrophytes are rooted in the bottom muds and typically grow in saturated soils to water depths of approximately 1 m. Some species can tolerate complete saturation on a year-round basis, but others may require a disturbance, for example, drying, burning, or some means of re-aerating soils. Treatment wetland genera include but are certainly not limited to species such as cattails, bulrush, and common reed. In the past, monocultures have been extensively used but the current trend is towards the use of a diverse assemblage to meet water quality goals while providing greater habitat value. A diverse assemblage of macrophytes will also be more sustainable in the event of plant disease or insect infestation.

Emergent plants can influence the water quality by shading the water surface, thus dampening temperature and wind-induced turbulence. They also provide dense stands which aid settling and permit interception and impaction of particulate matter. Live and dead emergent shoots exchange gases between the bottom sediments, water column, and atmosphere. Oxygen can be transported through stem structures to the roots and rhizomes where some may "leak" out. This sets up an oxidizing zone immediately adjacent to an anoxic or anaerobic zone, which are very important for nutrient transformations, degradation of organic compounds, and if sufficient sulfur is present in the system, heavy-metal removal as metal-sulfide complexes. The decaying emergent vegetation provides a "litter-zone" where labile carbon is formed and subsequently used by bacteria for metabolic requirements and things like denitrification.

Transitional plant species are those plants that favor littoral areas and are capable of following the capillary fringe up the bank. These plants require periods when they are not completely inundated. They can provide cover for wildlife access and egress and also serve as forage materials. Several semi-aquatic grasses including knot-grass (*Paspalum* sp.), have been used, as have sedges, spike-rushes, and riparian trees and shrubs.

Submerged aquatic plants can persist in water depths deeper than 1 m. Submerged aquatic plants are established such that they "fill" the water column of internal deep zones. They will be typically rooted in the bottom and thrive if sufficient light penetrates the water column. Submerged plants provide surface in the water column for the attachment of microbes that do the brunt of the nutrient removal/transformations and can help in the assimilation of BOD produced from the natural decay of plant biomass. Further, submerged aquatic plants can impart significant amounts of dissolved oxygen to the water column during daylight hours. In addition, submerged plants provide an ideal "nursery" for zooplankton growth, mediate water temperatures and provide refuge and forage for wildlife. Several genomes have been successfully utilized in constructed wetlands including: *Cerratophylum*, *Potamogeton*, and *Myriophylum* species. A good algal cover or even duckweed covers can preclude light transmission and make it difficult to sustain these plants over time unless nutrients are managed in emergent vegetation zones and retention times in open-water areas are minimized.

Spreading/floating plants are species that root in moist substrates (shoreline or banks of islands) and the spread out over the water surface. These plants can provide numerous benefits to wildlife and other wetland vegetation. For example, *Hydrocotyle* (pennywort) has been used in several operating treatment wetlands to: (1) protect emergent macrophytes from predation during the spring re-growth periods. In essence emergent species are grown through the pennywort cover. Muskrats consume the pennywort, thereby allowing the bulrush to mature; (2) shade the water surface and providing nursery habitat for zooplankton, fish, and amphibians; and (3) provide forage for waterfowl (Fig. 10.11).

10.7.4 Islands

Islands placed within a treatment wetland can serve multiple roles. In many instances islands are placed within the system to provide habitat. When sufficiently isolated from the shore, they can provide refuge for loafing, courting, and nesting activities. Islands can also serve

FIGURE 10.11   Floating aquatic plants such as Pennywort provide water quality and habitat benefits in constructed treatment wetlands.

as a platform for terrestrial or riparian plantings that in turn can provide a vertical component to the habitat via canopy development. The canopy provides habitat for birds, insects, and reptiles and by shading the water surface can dampen water temperature fluctuations.

In systems subject to wind and wave action, the island shoreline may need to be stabilized. Stabilization can be accomplished using rock, grass, or even woody plants such as willow (*Salix goodingii*). If desirable, the islands can be configured into linear forms and situated to aid in internal flow control. If connected to the shoreline, islands can also be used to access the interior of the wetland for monitoring and maintenance activities.

10.7.5 Walkways and Access

Access to, around, and within the wetland is sometimes desirable for monitoring, O&M, and educational purposes. All weather access to the stormwater treatment wetlands should be provided. This can be accomplished by providing a stabilized surface capable of supporting construction-size equipment. Usually a width of 3 m or more will suffice. The surface course can be gravel, gravel stabilized with a binding agent, roller-compacted concrete, or even asphalt, but cost will likely dictate the choice of materials to be used. In areas where mosquitoes may become or are an issue, access around the complete perimeter of the wetland basin is recommended to allow truck-mounted broadcasting of control agents. Elevated walkways permitting access to interior densely vegetated sections of the wetland may also be desirable from the standpoint of monitoring larval mosquito populations and assessing efficiency of controls. For systems that public education as a goal, gravel, or otherwise stabilized walkways can be provided. Attention should be given to providing clear walkways to keep visitors safe from potential undesirable encounters with wildlife. If public use is a goal, disabled persons access requirements should be given full consideration during the design of site access features.

10.8 STORMWATER TREATMENT WETLAND O&M CONSIDERATIONS

As was previously stated, stormwater treatment wetlands should be designed with all-weather access for among other things routine operation and maintenance activities. Access should allow personnel to assess the structural integrity of containment structures and ensure inlets, outlets, and other water transfer structures are debris free. Access should also be provided to the forebay for periodic removal of accumulated sediments. If water-quality sampling and monitoring are required, ready access should also be supplied to the inlet and outlet areas.

In geographic regions where vector control is likely to be an issue, the wetland basin should be designed with deep-zones and stocked with fish. *Gambusia affinis* (common name, mosquito fish) is a good candidate as it is a very robust fish that can survive environmental extremes, reproduces quickly, and has an appetite for mosquito larvae. If fish cannot control mosquito populations, the operator has the option of drying the emergent areas of the facility or using one of several commercially available larvicides. If chemical mosquito control is required in the management of the stormwater treatment wetland, individuals educated and certified in aquatic pesticide application must be used. A larval and adult mosquito surveillance program should be instituted to determine adult populations, assess the potential for human health threats, and provide insight into the effectiveness of larval controls (e.g., fish, macroinvertebrates), and larvicide usage.

To maintain vegetation, fish, and aquatic organism populations during dry-weather periods, a supplemental water source can be provided. Candidate water sources could include surface water diversion, potable supplies, reclaimed wastewater effluents, and groundwater.

For supplemental water supplies that require pumping, consideration may be given to solar-powered facilities that would supply water to the system(s) during daylight hours.

10.9   TREATMENT WETLAND COST CONSIDERATIONS

The cost to design and construct a wetland for water treatment can be variable and often depends on a litany of factors. Initial costs will be incurred during the planning, engineering design, and resolution of legal matters. There is also the cost of the land and easements necessary to house the facilities. Costs will be incurred during site preparation activities for the treatment wetland, as well as, for pre-treatment structures and the stormwater runoff conveyance system. Heavy construction costs will be incurred during excavation and trenching activities and for placement of construction materials. Preparation for and the actual planting of the wetland will also need funding. Costs associated with terrestrial landscape features, walkways, interpretative areas, and other public use amenities if desirable should also be accounted.

After construction, O&M cost will be incurred. It is likely that personnel will have to be supplied for inspection and monitoring activities. If vector control is necessary, costs to monitor, treat, and even alter the wetland configuration will also be incurred. Disposal of sediments and debris should also be expected.

Because there are various ways of contracting and conducting work and providing services a good rule-of-thumb cost estimate for stormwater treatment wetland design, construction, and operation is difficult to provide (Bingham, 1994). A review of the North American Wetland Database does provide a range of costs for treatment wetlands systems implemented across the nation. Although this database includes wetland treatment systems that receive waste waters other than stormwater, some insight to the magnitude of wetland construction costs is provided that could be assumed to be similar for systems receiving stormwater runoff. Table 10.4 provides a cost summary in 1993 dollars for constructing FWS and SSF wetland, or for adapting natural wetlands for use as treatment systems. In the case of FWS wetlands, most of the construction cost is incurred because of wetland size and the associated earth-moving activities, while for SSF systems it is the gravel media. In natural wetlands, most of the cost is incurred because of land purchase and pipeline construction.

10.10   SUMMARY

Constructed and natural wetlands have been used for water quality treatment for over 25 years and there are operational stormwater wetlands with monitoring data that strongly sug-

TABLE 10.4   Summary of Treatment Wetland Capital Costs, 1993 U.S. Dollars

System configuration	Capital costs ($/ha)			
	Low	Median	High	Major contributors
Constructed FWS	$10,000	$44,600	$100,000	Earthwork & size
Constructed SSF	$80,000	$350,000	$800,000	Gravel media
Natural surface flow	$5,000	$18,000	$25,000	Land purchase & piping

Source: From NADB, 1993.

gests wetlands improve the quality of urban stormwater runoff. How to optimize pollutant removal in stormwater wetlands however, remains to be quantified. Treatment of conventional waste waters using wetland is still an art, and design guidance and criteria are still being refined. These criteria relate hydraulic and mass loading parameters to wetland size and pollutant removal efficiency. Since they were developed for waste streams that have well-defined and predictable hydraulic and mass loading rates, some question the applicability to stormwater runoff.

As stormwater treatment receives more regulatory scrutiny and numeric water quality standards are imposed, the relationships already developed or are in development for the conventional waste streams may be adapted and applied to the design of stormwater runoff treatment wetlands. This cause will be furthered if stormwater runoff is collected and controlled to the extent that the operator has a margin of control over how much, what quality, and how long runoff can be retained in a treatment wetland. If this becomes the case, volumetric and aereal-loading rates used to size conventional treatment wetlands might be adaptable for stormwater runoff wetlands.

Data necessary to quantify the performance of currently operating stormwater treatment wetlands are needed to develop new or refine pollutant removal estimates in constructed wetlands. Information is needed such as watershed size to treatment wetland area or volume, measured flowrates and hydraulic retention times, inlet and outlet sampling for pollutants. Efforts to catalog and document stormwater treatment wetland performance such as the *National Stormwater Best Management Practices* (*BMP*) *Database* sponsored by the American Society of Civil Engineers and the U.S. EPA and the *Stormwater Managers Resource Center,* operated by Center for Watershed Protection, Inc. under a grant from the U.S. EPA are two developing resources that will contribute to and likely speed the development of stormwater runoff treatment wetland technology. Designers are encouraged to utilize these resources and obtain the most current design and performance data by contacting them online at *http://www.bmpdatabase.org/* and *http://www.stormwatercenter.net/*."

REFERENCES

Anderson, J., Unpublished paper delivered at "A Short Course on the Role of Constructed Wetlands in Watershed Management." Environmental Resources Engineering Department, May 16–17, 2000, Humboldt State University, Arcata, CA, 2000.

Bingham, D. R., "Wetlands for Stormwater Treatment," in *Applied Wetlands Science and Technology,* D. M. Kent, ed., Lewis Publishers, New York, 1994,

Camp Dresser and McKee, Larry Walker Associates, Uribe and Associates, and Resources Planning Associates, "California Stormwater Best Management Practice Handbooks: Municipal," Prepared for the Storm Water Quality Task Force, 1993.

Driver, N. E., and G. D. Tasker, "Techniques for Estimation of Storm-Runoff Loads, Volumes, and Selected Constituent Concentrations in Urban Watersheds in the United States," United States Geological Survey Water-Supply Paper 2363, 1990.

Flood Control District of Maricopa County, "Drainage Design Manual for Maricopa County, Arizona," Vol. I, AZ, 1992.

Fox, P., Fernando, H. J. S., Rodriguez, R. V., Serra, T., and Arzabe, M. 2000. "Evaluation of Flow Fields in Wetlands Using Physical Models." Draft Final Report, Desalination Research and Development Program, U.S. Department of the Interior Bureau of Reclamation, Denver Technical Center. Environmental Resources Team Water Treatment Engineering and Research Group, Denver, CO.

Gearheart, R., Personal communication, 2000.

Grant, D. M., *Isco Open Channel Flow Measurement Handbook,* 3rd ed., Isco Inc., Lincoln, Nebraska, 1992.

Hoos, A. B., and J. K. Sisolak, "Procedures for Adjusting Regional Regression Models of Urban-Runoff Quality Using Local Data," United States Geological Survey Open-File Report 93-39, 1993.

Humboldt State University (HSU), "A Short Course on the Role of Constructed Wetlands in Watershed Management," Environmental Resources Engineering Department, May 16–17, 2000, Arcata, CA, 2000.

Kadlec, R. H., and R. L. Knight, *Treatment Wetlands,* Boca Raton, FL, Lewis-CRC Press, 1996a.

Kadlec, R. H., and R. L. Knight, "Short Course–Design and Implementation of Treatment Wetlands," sponsored by CH2M-Hill and the City of Show Low, AZ, Sept. 4–7, 1996, Show Low, AZ, 1996b.

Knight, R. L., "Wildlife Habitat and Public Use Benefits of Treatment Wetlands," *Water Science and Technology* 35(5): 35–43.

Lee, F. G., "The Right BMPs? Another Look at Water Quality," *Stormwater, The Journal for Surface Water Quality Professionals, http://www.forester.net/sw_0011_right_html,* 2000.

Linsley, R. K., and J. B. Franzini, *Water-Resources Engineering,* 2nd ed., McGraw-Hill Book Company, New York, 1972.

Lopes, T. J., K. D. Fossum, J. V. Phillips, and J. E. Monical, "Statistical Summary of Selected Physical, Chemical, and Microbial Characteristics, and Estimates of Constituent Loads in Urban Stormwater, Maricopa County, Arizona." United States Geological Survey Water-Resources Investigations Report 94-4240, 1995.

Maidment, D. R., *Handbook of Hydrology,* McGraw-Hill Inc., New York, 1992.

Metcalf and Eddy, *Wastewater Engineering—Treatment, Disposal, Reuse,* 3rd ed., G. Tchobanoglous and F. L. Burton, ed., Metcalf and Eddy, Inc., Irwin McGraw-Hill, 1991.

North American Treatment Wetland Database, electronic database created by R. Knight, R. Ruble, R. Kadlec, and S. Reed for the U.S. Environmental Protection Agency, copies available from Don Brown, EPA (513) 569-7630, 1993.

Payne, N. F., *Wildlife Habitat Management of Wetlands,* Krieger Publishing Company, Malabar, FL, 1998.

Reddy, K. R., and W. H. Smith, eds., *Aquatic Plants for Water Treatment and Resource Recovery,* Orlando, FL. Magnolia, 1987.

Reed, S. C., R. W. Crites, and E. J. Middlebrooks, *Natural Systems for Waste Management and Treatment,* 2nd ed., McGraw-Hill, New York, 1995.

Replogle, J. A., A. J. Clemmens, and C. A. Pugh, "Hydraulic Design of Flow Measuring Structures," Chap. 21 in Hydraulic Design Handbook, (ed. by L. W. Mays), McGraw-Hill, NY, 1999.

Shaver, E., and J. Maxted, "Construction of Wetlands for Stormwater Treatment," in Proceedings, *Symposium on Stormwater Runoff and Quality Management,* C. Y. Kuo, ed., Penn State University, University Park, PA, 1994.

Schueler, T. R., *Controlling Urban Runoff: A Practical Manual for Planning and Designing Best Urban Management Practices,* Metropolitan Council of Governments, Washington, DC, 1987.

Schueler, T. R., *Design of Stormwater Wetland Systems: Guidelines for Creating Diverse and Effective Stormwater Wetlands in the Mid-Atlantic Region,* Metropolitan Council of Governments, Washington, DC, 1992.

Seidel, K., "Macrophytes and Water Purification," in *Biological Control of Water Pollution,* J. Tourbier and R. W. Pierson, Jr., eds., Philadelphia: University of Pennsylvania Press, 1976.

Stowell, R., R. Ludwig, J. Colt, and G. Tchobanoglous, "Towards the Rational Design of Aquatic Treatment Systems," presented at the American Society of Civil Engineers Spring Convention, Portland, OR, April 14–18, 1980. 43 pp. 1980.

Strecker, E. W., J. M. Kersner, E. D. Driscoll, and R. R. Horner, *The Use of Wetlands For Controlling Stormwater Pollution,* EPA/600, The Terrene Institute, Washington, DC, 1992.

Tammi, C. E., "Onsite Identification and Delineation of Wetlands." In *Applied Wetlands Science and Technology,* D. M. Kent ed., Lewis Publishing, pp. 35–54, 1994.

Wass, R., *Tres Rios Demonstration Constructed Wetland Project—1996/1997 Operation & Water Quality Report,* prepared for the City of Phoenix, SROG, and Bureau of Reclamation, Phoenix, AZ, 1997.

Wilson, M., personal communication, Humboldt Water Resources, Engineering and Science, Arcata, CA, 2000.

U.S. Environmental Protection Agency, Environmental Technology Initiative, 1999. *Free Water Surface Wetlands for Municipal Wastewater Treatment A Technology Assessment,* Office of Water document EPA 832-S-99-002.

U.S. Environmental Protection Agency, *Subsurface Flow Constructed Wetlands for Wastewater Treatment: A Technology Assessment,* Office of Water, EPA Report 832-R-93-001, 1993.

CHAPTER 11

DESIGN OF DISTRIBUTED STORMWATER CONTROL AND RE-USE SYSTEMS

Leonard T. Wright
James P. Heaney
Department of Civil, Environmental and Architectural Engineering
University of Colorado, Boulder

This chapter describes design strategies for incorporating recent stormwater management innovations into an integrated, sustainable approach to urban drainage and the urban water budget. These innovations represent a departure from common practice and, as such, only a limited number of case studies are available to guide future designs. Nevertheless, forward-looking municipalities, drainage engineers, and landscape architects are designing and approving systems characterized by small-scale, distributed stormwater controls. Innovative stormwater systems are in place around the world, in watersheds with diverse climates and institutional arrangements, such as Australia, Japan, northern Europe, and over many regions of the United States, including the Chesapeake Bay region, Florida, Texas, and the Pacific Northwest.

The goal of these systems is to reduce the hydrologic impact of urban land use on water quality and quantity by distributing small-scale storage and infiltration systems throughout the watershed as an integral part of the urban hydrologic cycle and the landscape design. The most promising of these designs feature integration of the urban landscape with hydrologic function over a wide range of precipitation frequencies (Candaras et al., 1995; Heaney et al., 1999a; Roesner, 1999). These features complement traditional drainage structures and reduce the hydrologic impact of urban surfaces, as well as reduce the demand for downstream infrastructure.

A fundamental change in design philosophy has accompanied these innovations. Where previous thinking led to the conclusion that "all runoff is costly and unwanted," innovative designers have recognized that "runoff has value, at least in small quantities." By valuing small runoff events within the total urban water budget, savings may be achieved in other areas such as outdoor irrigation and other nonpotable uses. Innovators have also found aesthetic value in urban water, achieving landscape architectural benefits through the use of runoff in an urban setting.

The recommended strategy for implementing an innovative stormwater management plan is to follow four key steps:

1. Do a detailed evaluation of the urban water budget, including precipitation, infiltration, evapotranspiration, stormwater runoff, indoor and outdoor water use, and waste water.

2. Plan for the desired amount of storage and infiltration required for each land use, including the right-of-way as a separate use. Account for water budget items that may be amenable to water re-use across budget items.

3. Integrate quantity controls with quality controls to meet objectives for micro as well minor and major events.

4. Use stormwater controls to enhance the urban ecology and the urban aesthetic value.

The following section develops the need for sustainable distributed systems by reviewing hydrologic impacts associated with urban land development and how traditional drainage practice may fail to meet long-term societal goals. Subsequent sections address the urban water budget and the value of runoff, the value of planned land use decisions to meet hydrologic objectives, example micro-storm controls, and the integration of these innovative systems with traditional conveyance and storage facilities.

11.1 THE NEED FOR DISTRIBUTED STORMWATER CONTROLS IN THE URBAN ENVIRONMENT

Urban development is characterized by drastic changes to the earth's surface. From porous soils with vegetative cover, the land surface is transformed to small rectilinear sheets of nearly impenetrable pavement and roofing. The remaining porous surfaces are reduced to small islands of disturbed soils, generally covered with homogeneous turf. The resulting post-development urban land sheds runoff rapidly and in greater quantity. Drainage networks are an integral part of this hydrologic environment, installed to reduce local nuisance flooding on the developed land surface. Runoff has traditionally been treated as a nuisance with no value to the landowner, and a source of liability for the site designer who neglects to provide enough drainage to avoid flood damages and subsequent lawsuits.

Stormwater drainage systems in urban and suburban areas typically consist of a conveyance network and perhaps a downstream storage basin. Conveyance elements may include a curb and gutter system coupled with a sub-surface pipe network. The goal of these systems is to move runoff from developed areas as quickly and efficiently as possible to prevent local nuisance flooding, usually in the form of flooded basements and streets (Walesh, 1989). Historically, the objective of these systems was purely flood control, and the design was based around large, infrequent events (Walesh, 1989; ASCE, 1992). However, drainage networks also exacerbate distortions introduced to the long-term urban hydrologic cycle by reducing flow paths and increasing the velocity of runoff for all events, not just large, design-scale events.

After draining from upland urban areas, runoff may be stored at a central downstream location to reduce the peak rate of runoff prior to entering the receiving water (Prince Georges County, 2000). In doing so, traditional conveyance/storage drainage systems coupled with the impervious surfaces they drain tend to increase downstream flooding, prolong high storm flows, which may promote stream erosion, and deliver nonpoint source pollution to receiving waters during smaller events (Walesh, 1989; Heaney et al., 1999a; Roesner, 1999). These phenomena run counter to principles of sustainability, as recognized by many researchers in the field (Ellis, 1995; Roesner, 1999; Heaney et al., 2000a).

11.1.1 Sustainability Issues in Urban Hydrology

Sustainability (and *sustainable development, sustainable systems,* etc.) has been the focus of attention in many areas of research over the past decade, and the field of urban stormwater management is no exception (e.g., Rowney et al., eds., 1999; Butler and Maksimovic, eds., 1998; Sieker and Verworn, eds., 1996). The American Society of Civil Engineers (ASCE) Task Committee on Sustainability Criteria (1998) of the Water Resources Planning and Management Division proposed the following definition: "*Sustainable water resource systems are those designed and managed to fully contribute to the objectives of society, now and in the future, while maintaining their ecological, environmental, and hydrological integrity*" (p. 44). The ASCE Task Committee highlights the need for demand management to meet this goal. *Demand management* is an often overlooked alternative in addressing stormwater problems. Applied to stormwater management, managing demand means limiting the demand for services or decreasing the amount of runoff from urban lands over the long term that requires disposal.

Butler and Parkinson (1997) identify four objectives to improve the sustainability of modern urban drainage systems: (1) maintain public health and protection from flooding, (2) avoid polluting local as well as distant environs, (3) minimize resource use, and (4) be operable and adaptable in the long term. Beck et al. (1994) define the ideal sustainable drainage system as compensating for distortions introduced to the hydrologic cycle by the activities of the city.

A key to designing a sustainable drainage system is to recognize the value of all components of the water budget. The value of runoff changes with quantity: in small amounts it has value on-site; in increasing amounts the value becomes a cost, as damages accrue as a result of excess runoff. Traditional systems focus on large events and ignore the value of small runoff events. For this reason, small events are drained along with the large, and a valuable commodity is lost. In addition to discarding a valuable commodity, small runoff events also carry pollution to the receiving water and exacerbate other urban hydrologic impacts. The next section discusses the impacts of urbanization on the long-term hydrologic cycle.

11.1.2 Impacts of Urban Drainage Infrastructure on the Long-Term Hydrologic Cycle

It is well-documented that traditional drainage practices, coupled with the imperviousness of urban land, deleteriously affect the quantity, quality, and timing of runoff (Whipple, 1981; McCuen, 1989; Hollis, 1975, 1989; Wigmosta and Burges, 1990; Stephenson, 1996; Burges et al., 1998). Impervious surfaces generate larger volumes of runoff than the pervious surface they cover (Leopold, 1968). Over the long term, this has the effect of altering the recurrence relationship of runoff (Hollis, 1975). Small, frequent events that would produce little or no runoff under natural conditions may generate significant runoff under developed conditions (Hollis, 1975; Heaney et al., 2000d). The runoff lag time, characterized by the delay between rainfall and subsequent runoff, tends to decrease under developed conditions due to increased overland flow and channel velocities and decreased flow-path lengths. Runoff events carry *nonpoint source* (NPS) *pollution* to receiving waters that otherwise would receive neither runoff or NPS under undeveloped conditions. Furthermore, the receiving water may not have the benefit of large diluting flows that would occur for larger events.

Burges et al. (1998) and Wigmosta and Burges (1990) report on two small, neighboring experimental watersheds in the state of Washington. The first watershed remained forested and undeveloped during the study period, while the second was developed with suburban residential land use. Using an annual water budget approach, the percentage of annual runoff ranged from 12 to 3% for the undeveloped watershed and 44 to 48% in the suburban watershed. For peak runoff, the unit area runoff rate was nearly 10 times higher from the suburban catchment than from the forested catchment for an approximate 24-hour, 50-year rainfall event. The frequencies of a wide range of runoff events are significantly altered due to fundamental changes in the land surfaces affecting hydrologic processes. Both the long-term runoff volume and peak runoff rates were significantly affected (Burges et al., 1998). To mitigate these impacts, designs must take into account the full range of event frequencies. The next section briefly describes the root causes of these changes to the long-term urban water budget.

11.1.3 Causes of Urban Hydrologic Impacts

The most significant changes to the hydrologic function of the landscape result from those lands that are rendered impervious and are connected directly to a drainage system. The vast majority of the road network in an urban area falls into this category. In dense, urban areas, many roofs and parking lots are also directly connected to a drainage network (i.e., runoff is conveyed directly, and does not pass over or through any area that may reduce runoff due to infiltration or depression storage). *Directly connected impervious area* (DCIA) has the greatest hydrologic impact per unit area; however, due to its link with the road network it is often masked in land use evaluations of urban areas.

Heaney (2000) discusses the profound impact the automobile has on the urban landscape. A primary function of an urban drainage network is to provide drainage for the automobile; therefore, a large portion of DCIA impacts are attributable to transportation. Heaney (2000) discusses the impact of masking the road network within other land-use categories for urban hydrologic planning. For instance, a low-density residential area of 500 hectares may have only 25% DCIA. However, upon inspection, nearly all of this DCIA is either from the local road network, driveways, and parking areas. Integration of land-use planning and hydrologic design is called for at the planning phase to determine the transportation and drainage needs of a community simultaneously, along with the true costs of providing for those needs.

11.1.4 Urban Surfaces and Micro-storm Hydrology

Jones (1971), Walesh (1989), Geiger (1993), Urbonas and Roesner (1992), ASCE (1992), and Grigg (1996) all describe the dual nature of stormwater infrastructure as *minor* or *major drainage systems.* The minor system is typically designed to alleviate nuisance flooding of streets and developed areas. The major system (whether designed or not) conveys larger, infrequent runoff events and is typified by larger surface channels, floodways, etc. (Walesh, 1989). Various types of drainage and flood control works are given in Table 11.1 along with typical storm frequencies used for design.

TABLE 11.1   Typical Design Storm Frequencies

Land use	Design storm return period
Minor Drainage Systems	
Residential	2–5 years
High value commercial	2–10 years
Airports (terminals, roads, aprons)	2–10 years
High value downtown business areas	5–10 years
Major Drainage System Elements	50–200 years

Adapted from ASCE, 1992.

A common control strategy designed to reduce peak runoff from urban areas is detention storage. Storage ponds are typically designed to reduce the peak runoff from a moderately sized design event (usually 5- to 20-year return period) (Walesh, 1989). The goal of this type of detention is to maintain an estimated pre-development peak flow rate for the same design event (Chow et al., 1988; McCuen, 1989; ASCE, 1992). An idealized set of hydrographs is shown in Fig. 11.1 that shows a pre-development design hydrograph along with post-development runoff and post-development with a detention pond. This type of design does little to reduce runoff volume and may exacerbate stream erosion problems by sustaining relatively high flows for longer periods. While still an effective control measure for hydrograph peak shaving, detention does not appear to be the sole solution to mitigate urban impacts, but rather one component of a stormwater management plan.

Modifications and innovations to detention pond design arose out of concern for water quality (Whipple, 1981; Urbonas and Stahre, 1993). While storage ponds may be an efficient means of gravity pollutant removal, detention times needed for peak shaving were generally not long enough to provide adequate settling times. Retention is defined as extended detention, and many more recent designs maintain a permanent pool to increase pollutant removal capability.

The gradual move from flood protection to water quality improvement has led researchers to conclude that the full range of rainfall frequencies must be addressed by a sustainable urban stormwater management plan. Though designed for extreme conditions, drainage systems usually function under smaller events than those given in Table 11.1. For example, nearly 95% of runoff producing events will be smaller (less rainfall depth) than a 2-yr design storm as demonstrated by the rainfall event distribution for Denver, CO (Guo and Urbonas, 1996).

The vast majority of annual stormwater volume is derived from small storms. These small rainfall events are termed *micro-storms* by Candaras et al. (1995). They describe an innovative municipal program designed to mitigate water quality impacts from small storms, in conjunction with the major and minor drainage systems. Pitt (1999) shows examples of cumulative probability density functions of rainfall events and simulated runoff volumes using the Source Load and Management Model (SLAMM) from 25 National Oceanographic and Atmospheric Administration (NOAA) rain stations. This analysis shows that the majority of runoff and runoff pollutant loads are derived from small to moderate, but frequent, rains (Pitt, 1999).

The most promising stormwater management innovations recognize the relationship between runoff quantity and quality, and between flood control and receiving water health

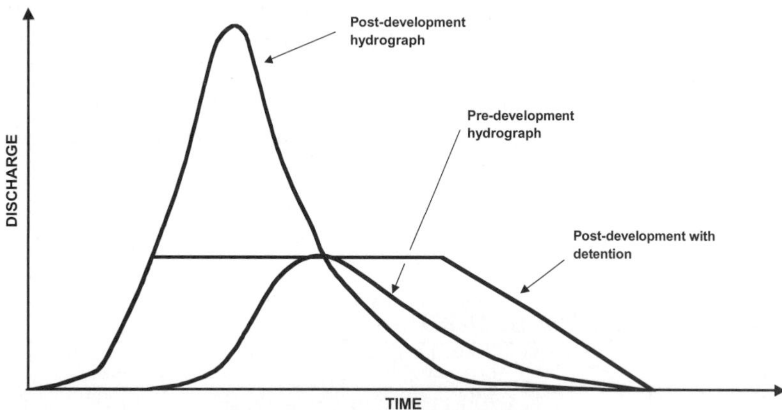

FIGURE 11.1 Idealized pre- and post-development hydrograph.

(Roesner, 1999). An integrated approach to stormwater management may also better achieve sustainability goals, as well as reducing the demand for downstream infrastructure (but rarely replacing it). It also provides a framework for research, to integrate land-use planning decisions that encompass transportation needs, the level of major, minor and micro drainage service, the potential to integrate stormwater management with outdoor water re-use and urban irrigation management, as well as provide a platform to showcase important and valuable aesthetic benefits of civic ecology (Wenk, 1999).

11.1.5 Distributed Control as a Functional Part of Land Development Design

Innovations in the field of stormwater management have been driven by the need to reduce the amount of downstream detention volume by increasing headwater storage and the need to improve downstream water quality (Prince Georges County, 2000). The former is a flood control goal, the second a water quality goal. The separation of these two goals has impeded the acceptance of *best management practices* (BMPs), as researchers have published opposing accounts of their effectiveness (Roesner, 1999). Along with the integration of quality and quantity control comes the need to value runoff and to decide, "How much is too much?"

Innovations in stormwater management have come about by designers attempting to integrate concepts of traditional conveyance and downstream storage with some level of upstream storage and infiltration. To achieve this integration over the long term without excessive maintenance problems, a careful balance is sought between aesthetic and hydrologic values, between cost and sizing constraints, and between upstream and downstream investments. Complete mitigation of hydrologic impacts of urban development through upstream controls is not likely to be cost-effective, and pure downstream control/conveyance does not meet performance goals of sustainability. Designs featuring an integration of small-scale, upstream controls with downstream conveyance and controls appear to have the greatest promise.

To mitigate the hydrologic impacts described in this section, it is critical to evaluate the urban water budget to identify sources and sinks of water, and when runoff changes from an asset to a liability (i.e., the transition from valued small events to troublesome and costly flooding). Designs that value and utilize a portion of the annual precipitation on-site provide many potentially important benefits. To design a more sustainable, integrated stormwater management plan, four key steps have been identified:

1. Estimate the long-term water budget, including indoor water use, irrigation needs, waste water production, precipitation, evapotranspiration, infiltration, and runoff.

2. Evaluate several (or many) land-use scenarios that will satisfy local and regional needs with an eye towards flood control and water quality objectives.

3. Identify feasible on-site controls for each land use.

4. Integrate on-site, distributed micro-storm control with minor and major drainage infrastructure.

The next section provides information on the various parts of the urban water budget and how they may be important for stormwater management.

11.2 THE URBAN WATER BUDGET

11.2.1 Introduction

Water supply, waste water, and stormwater systems are explored in this section, first individually and then looking at them in an integrative manner. Key areas of potential integration of these three functions are in re-use of waste water and stormwater to reduce the required net import of water for water supply. The traditional design conditions for water supply, waste water, and stormwater systems use rare design events such as the 100-yr storm for flood control or the 100-yr drought for water supply. The drawback of systems designed for rare events is that they may not function well during normal operations. For urban stormwater, the recommended approach is to design multi-purpose stormwater systems that are optimal for major floods, minor floods, and micro-storms. Another more recent change is the desire to manage the demand for these services by re-using waste water and stormwater. The sections presented below outline a methodology for doing these evaluations. The first recommended approach is to do a water budget for water supply, waste water, and stormwater.

McPherson (1973) argued that developing an urban water budget was an essential first step in using a systems approach. Stephenson (1996) compares water budgets for an undeveloped catchment with an urbanized catchment in Johannesburg, South Africa. The results show the expected increase in direct runoff and the need to import water for water supply. Herrmann and Klaus (1996) do general water and nutrient budgets for urban water systems including stormwater. Imbe et al. (1996) performed a water budget analysis to determine the impact of urbanization on the hydrological cycle of a new development near Tokyo, Japan. This development is trying to minimize hydrologic impacts by encouraging infiltration systems and storing rainwater. Mitchell et al. (1996) describe a water budget approach to integrated water management in Australia. Budgeting is done at the individual parcel, neighborhood, and wider catchment scale. On-site management options include providing rain and graywater storage. Clark et al. (1997) use a water budget approach to evaluate decentralized urban water infrastructure for Adelaide, Australia. Heaney et al. (2000a) do monthly water budgets for Denver and New York to evaluate the potential for recycling and re-using waste water and stormwater. Heaney et al. (2000b) used the Thornthwaite water budget method to estimate monthly water budgets for individual urban parcels in numerous U.S. cities. This method can be used to get a preliminary estimate of the storage required to achieve various levels of waste water and stormwater reuse. Heaney et al. (2000c) performed an hourly water budget for water supply, waste water, and stormwater for the City of Boulder and Boulder Creek Watershed. This analysis looked at critical periods from the point of view of the watershed as well as the urban area. An hourly time step was needed for this study because the temporal scale of the impacts in this mountainous area were quite rapid. Also, stream flow in Boulder Creek varies widely during a single day due to hydropower peaking.

11.2.2 Results of Selected Water Budget Case Studies

11.2.2.1 Melbourne, Australia.   Mitchell et al. (1996) used a daily water budget simulation model to evaluate the impact of on-site water management. They evaluated water use for two blocks in Melbourne, Australia. The attributes of each block are shown in Table 11.2.

The following retrofits were evaluated in these two areas:

- 13 kiloliter rain tank for storage of roof runoff for laundry, toilet, and garden water uses; spillage is directed to the storm drainage network.

- Graywater from bathrooms and laundry is used for gardening through a sub-surface irrigation system. Overflows go to the waste water sewer.

The simulated performance of the modified system is summarized in Table 11.3.

The reduction in demand for imported water was 41 and 49% for the two systems while off-site stormwater runoff was reduced by 56 and 49% for the two neighborhoods. These results indicate the potentially major impact of on-site water management on overall water use.

TABLE 11.2 Attributes of Two Neighborhoods in Melbourne, Australia

	Neighborhood	
Attribute	Essendon	Scoresby
Rainfall, mm/yr	591	887
Rain, days/yr	196	215
Evaporative demand, mm/yr	1054	1054
Soil Type	clay	silty clay
Area, sq m	750	750
Roof plan area, sq m	203	203
Paved area, sq m	113	113
Garden area, sq m	434	434
People/house	3	3
Type of garden	standard	standard

Source: Mitchell et al., 1996.

TABLE 11.3 Simulated Performance of Modified Urban Systems

	Neighborhood	
Attribute	Essend	Scoresby
Water demand, kl/yr	278	265
Reduced demand for imported water, %	41	49
Reduced off-site stormwater runoff, %	56	49
Reduced wastewater runoff, %	11	8
Usage from rainwater tank, kl/yr	84	107
Rain tank deficit/demand	0.48	0.3
Use of graywater, kl/yr	28	24
Graywater deficit/demand	0.65	0.65

Source: Mitchell et al., 1996.

TABLE 11.4 Assumed Common Attributes of Representative Neighborhoods in Denver, CO and New York, NY

Area, acres	Impervious area total	Impervious area directly connected	Total
Roof area, acres	15	5	
Driveway area, acres	10	5	
Local street area, acres	10	10	
Major street area, acres	5	5	
Lawn area, acres			60
Directly connected imperviousness, DCI, %			25
People			1,000

Source: Heaney et al., 2000a.

TABLE 11.5 Monthly Water Budget for Denver, CO. All Entries Are in Inches per Month

	Inputs (in./month)				Outputs (in./month)				
Month	Indoor water use	Outdoor water use	Precipitation	Total	DWF	I/I	Urban runoff	Recharge	Total
1	0.69	0	0.5	1.19	0.69	0.34	0.13	0.03	1.19
2	0.62	0	0.7	1.32	0.62	0.43	0.18	0.09	1.32
3	0.69	0.17	1.3	2.16	0.69	0.89	0.33	0.25	2.16
4	0.66	0.55	1.5	2.71	0.66	0.99	0.38	0.68	2.71
5	0.69	1.03	2	3.72	0.69	1.37	0.5	1.16	3.72
6	0.66	1.93	2.5	5.09	0.66	1.66	0.63	2.14	5.09
7	0.69	2.4	2.4	5.49	0.69	1.64	0.6	2.56	5.49
8	0.69	2	2.2	4.89	0.69	1.51	0.55	2.14	4.89
9	0.66	0.77	1.2	2.63	0.66	0.8	0.3	0.87	2.63
10	0.69	0.23	1	1.92	0.69	0.69	0.25	0.29	1.92
11	0.66	0	0.7	1.36	0.66	0.46	0.18	0.06	1.36
12	0.69	0	0.6	1.29	0.69	0.41	0.15	0.04	1.29
Total	8.09	9.08	16.6	33.77	8.09	11.19	4.18	10.31	33.77

11.2.2.2 Simulated Monthly Urban Water Budgets for Denver and New York. This section presents the results of monthly simulations of water budgets for cities with climates similar to New York and Denver (Heaney et al., 2000a). The results should not be construed to be accurate representations of actual conditions in these two cities. The purpose of presenting these case studies is to show the relative importance of the various terms in the water budget and to show the impact of climatic conditions. The common assumptions for the comparative studies of representative urban neighborhoods in Denver and New York are presented in Table 11.4.

Water use and waste water flows are typically expressed in terms of gallons per day. Stormwater runoff is usually expressed in inches averaged over the entire catchment. All flows were converted to inches averaged over the 100-acre catchment with 1,000 residents. The common assumed values, presented earlier in this analysis, are

1. Population   1,000

2. Area, acres   100

3. Indoor water use, gpcd   60

4. Runoff coefficient   0.25

5. Conversion factors: 7.48 gallons = 1 cu ft
43,560 sq ft = 1 acre

The summary results for Denver, CO are presented in Table 11.5 and Figs. 11.2 (inputs) and 11.3 (outputs). The results for Denver indicate a natural input from precipitation of 16.6 in per year and imported water of 17.17 in per year, slightly more than the natural input. The majority of the imported water is used for lawn watering. On the output side for Denver, infiltration/inflow (I/I) at 11.19 in is the largest source of the 19.28 in of water going to the waste water treatment plant (WWTP). Urban runoff contributes an additional 4.15 in of water leaving the system. Nearly 40% of the urban runoff falls on roofs and driveways. A significant portion of that water could be retained on-site and infiltrated and/or used for lawn watering. Urban runoff alone is insufficient to provide sufficient water for lawn watering. However, urban runoff and graywater do provide enough water to meet essentially all of the lawn watering needs.

Monthly water inputs-Denver climate

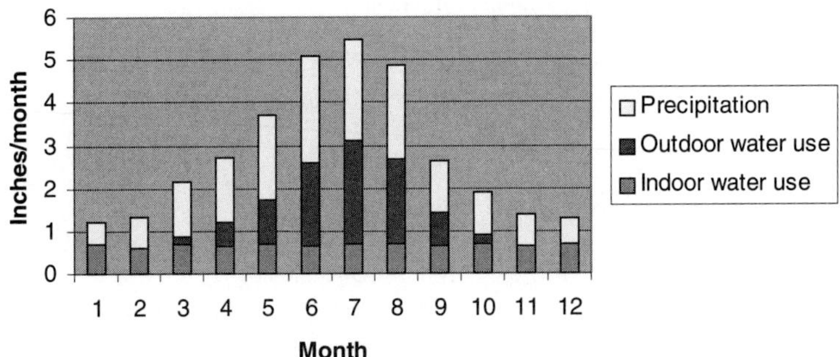

FIGURE 11.2   Monthly water budget input, Denver, CO.

Monthly outputs-Denver climate

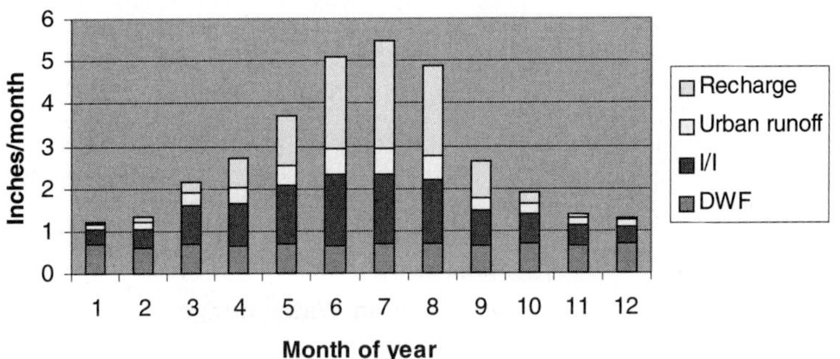

FIGURE 11.3   Monthly water budget output, Denver, CO.

TABLE 11.6   Monthly Water Budget for New York, NY. All Entries Are in Inches per Month

| | Inputs | | | | Outputs | | | | |
Month	Indoor water use	Outdoor water use	Precipitation	Total	DWF	I/I	Urban runoff	Recharge	Total
1	0.69	0	3	3.69	0.69	0.69	0.75	1.56	3.69
2	0.62	0	4	4.62	0.62	0.82	1	2.18	4.62
3	0.69	0	4	4.69	0.69	0.91	1	2.09	4.69
4	0.66	0	3	3.66	0.66	0.66	0.75	1.59	3.66
5	0.69	0.46	3	4.15	0.69	0.69	0.75	2.02	4.15
6	0.66	0.77	3.5	4.93	0.66	0.77	0.88	2.62	4.93
7	0.69	1.14	4	5.83	0.69	0.91	1	3.23	5.83
8	0.69	0.8	3.8	5.29	0.69	0.87	0.95	2.78	5.29
9	0.66	0.33	4.2	5.19	0.66	0.93	1.05	2.55	5.19
10	0.69	0	4	4.69	0.69	0.91	1	2.09	4.69
11	0.66	0	3.8	4.46	0.66	0.84	0.95	2.01	4.46
12	0.69	0	3	3.69	0.69	0.69	0.75	1.56	3.69
Total	8.09	3.5	43.3	54.89	8.09	9.69	10.83	26.28	54.89

Monthly water inputs-New York climate

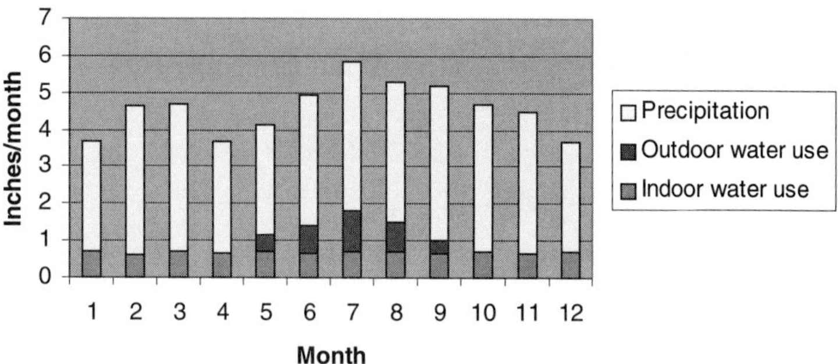

FIGURE 11.4   Monthly water budget input, New York, NY.

The results for New York, are presented in Table 11.6 and Figs. 11.4 (inputs) and 11.5 (outputs). The results for New York indicate a natural input from precipitation of 43.3 in per year and imported water of 11.59 in per year, slightly more than a quarter of the natural input. The majority of the imported water is used for indoor purposes. On the output side for New York, I/I at 9.69 in is the largest source of the 17.78 in of water going to the WWTP. Urban runoff contributes an additional 10.83 in of water leaving the system. Nearly 40% of the urban runoff falls on roofs and driveways. A significant portion of that water could be retained on-site and infiltrated and/or used for lawn watering. Urban runoff alone is sufficient to provide sufficient water for lawn watering.

11.2.3   Specialized Software for Doing Urban Water Budget Analyses

Mitchell (1998) developed software called Aquacycle as part of her doctoral research at Monash University in Australia. Aquacycle is a daily urban water balance model that has

Monthly outputs-New York climate

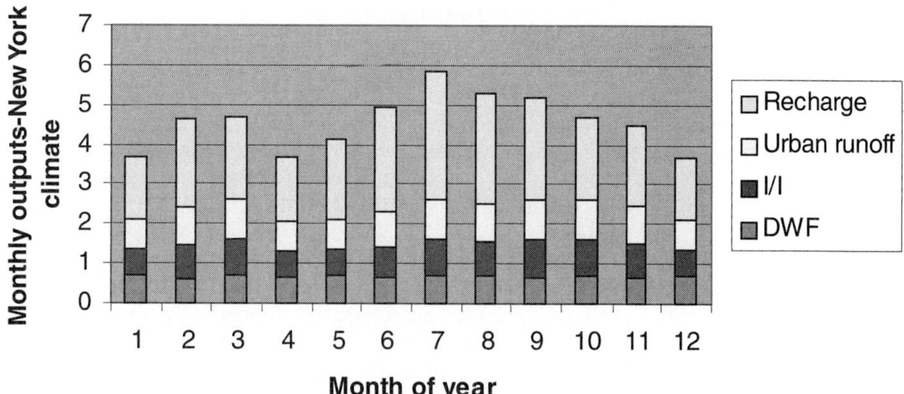

FIGURE 11.5   Monthly water budget output, New York, NY.

been developed to investigate the use of locally generated stormwater and waste water as a substitute for imported water. Water quality is not included. Aquacycle can be used at the single parcel up to the entire urban catchment scale. A catchment can be disaggregated into up to 50 sub-areas. Information about Aquacycle can be obtained from: http://www.catchment.crc.org.au/products/models/the_models/aquacycle/aquacycle.htm.

11.2.4 Conclusions on Urban Water Budgets

Urban water budgets are an essential first step in evaluating the integrated impacts of urban water systems. Opportunities for reusing treated waste water and stormwater depend on local climatic conditions. These budgets should be done at the micro (parcel) scale as well as for the overall system. Computationally, the water budgets can be done on a spreadsheet or using specially developed software such as Aquacycle.

11.3 LAND-USE EVALUATION

11.3.1 Introduction

One catalyst for changes in approaches to urban stormwater management is new attitudes towards overall urban water management. The purpose of this section is to review the literature on innovative urban developments in general, evaluate principles of sustainability, and present the urban stormwater management problem within this broader context. Each major functional component of urban land use including streets, driveways, roofs, etc., will be evaluated as to its potential for modifying its design to reduce impacts on urban water systems. A more detailed description of these options is contained in Heaney (2000).

11.3.2 Trends in Urbanization

11.3.2.1 Impact of the Automobile.   The growing influence of the automobile, beginning in the 1920s, forced cities to accommodate this new mode of transportation. Prior to World War II, U.S. cities developed around the concept of mixed neighborhoods as part of villages, towns, and cities. Beginning in the late 1940s, suburbia began to dominate urban America. From 1915 to 1994, the U.S population grew by a factor of 2.6 from 100 to 261 million people, while the number of automobiles grew by a factor of 80 from 2.5 million to nearly 200 million. At present, there are 0.76 vehicles per capita. Perhaps this is a saturation level based on the percentage of the population that is older than the minimum driving age. Annual vehicle miles traveled (VMT) per capita have continued to rise at a steady rate since 1945. Projections for the State of Colorado indicate that the 1995 VMT of 10,000 is expected to increase to 11,130 by the year 2020 (Yuhnke, 1997). Goldstein (1997) estimates that 25% or more of newly developed land is committed to roads, parking, driveways, and garages. In order to accommodate more cars and higher rates of utilization, the sizes and proportion of property devoted to vehicles has increased dramatically. One example is the shift from one- to two- and even three-car garages. Parking and other support services have similarly expanded. The net result is much higher amounts of impervious surfaces in urban areas on a per capita basis. Also, a large portion of these impervious surfaces are directly connected, thereby increasing the impacts of urban runoff, especially urban runoff quality.

11.3.2.2 Contemporary Neighborhoods and Urban Sprawl.   The National Commission on the Environment (1993) criticizes contemporary urban land-use patterns:

> Meanwhile, sprawling housing developments, shopping centers, highways, and myriad other developments have proceeded virtually unfettered by any sense of respect for the environment and humankind's relation to it. As a result, pollution from non-point sources continues to grow and is increasingly difficult to control; biological diversity is destroyed as habitats are fragmented and eliminated; sprawl development blighted the landscape and precludes cost-effective and environmentally beneficial means of providing transportation and other services.

Urban areas in the United States are using land four to eight times faster than the growth in population. The New York metropolitan area's population increase over the past 25 years has been only 5%, but the developed land has increased by 61%, replacing nearly 25% of the region's forests and farmlands (Peirce, 1994). Cities are spreading over the natural landscapes far faster than population increases or economic progress requires, while older urban districts with their valuable infrastructures are under used or abandoned (Barnett, 1995). In spite of an aggressive program to control urban sprawl and acquire greenways, Portland, OR has grown by nearly 25% since 1980 while expanding its urban area by only 1%. Without such management strategies, the Chicago area's population has grown only 4% in the past 20 years but expanded its urban land by 35%. Between 1960 and 1990, the population of the Baltimore metropolitan area increased by 33% but the amount of land in the region used for urban purposes grew fivefold-by 170% (Katz, 1997).

The impacts of sprawl in the Chesapeake Bay area include (Chesapeake Bay Foundation, 1996):

- Five to seven times the sediment and phosphorus as a forest
- Nearly twice as much sediment and nitrogen as compact development
- Each person uses four to five times as much land as 40 years ago
- Twice as much road building as compact development
- Three to four times as many automobile trips per day
- Much more air pollution as compact development
- Lower tax revenues than the cost of providing these services
- Induced re-location of people from central cities and inner suburbs

11.3.3 Possible New Approaches

The following principles are suggested for *sustainable infrastructure systems* for the 21st Century:

- Ideally, individual urban activities should minimize the external inputs to support their activities at the parcel level: For water supply, import only essential water for highly valued uses such as drinking water, cooking, showers, and baths. Re-use waste water and stormwater for less important uses such as lawn watering and toilet flushing. Minimize the demand for water by utilizing less water intensive technologies where possible. For transportation, minimize the generation of impervious areas, especially directly connected impervious areas, for providing traffic flow and parking especially in low use areas.
- Minimize the export of residuals from individual parcels and local neighborhoods: For waste water, export only highly concentrated wastes that need to be treated off-site. Re-use less contaminated water such as shower water for lawn watering. For stormwater, minimize off-site discharge by encouraging infiltration of less contaminated stormwater

and using cisterns or other collection devices to capture and re-use stormwater for lawn watering and toilet flushing.

- Structure the economic evaluation of infrastructure options to maximize the incentive to manage demand by using commodity use charges instead of fixed charges: For stormwater, assess charges based on the cost of providing stormwater quality control for smaller storms and flood control for larger storms. Charges should be based on imperviousness, with higher charges for directly connected imperviousness with regard to the nature of the use of the areas and their pollution potential. Some charges should be assessed for pervious areas. Credit should be given for on-site storage and infiltration.

- For the right-of-way, assess charges for transportation related imperviousness directly to a transportation utility in order to reflect true costs of service and encourage demand management in generating impervious area in general and directly connected impervious area in particular.

- Assess new development for the full cost of providing the water infrastructure that it demands, not only within the development, but also external support services.

11.3.4   Origins of Stormwater in Urban Areas

11.3.4.1   Introduction.   The purpose of this section is to evaluate the nature of the quantity of stormwater runoff in urban areas and to evaluate the relative importance of various sources. Stormwater falls onto pervious or impervious areas. Runoff occurs after the infiltration capacity has been exceeded. Impervious areas have a very small amount of initial storage capacity whereas pervious areas have much larger initial storage capacities depending on the soil type and antecedent conditions. A primary goal of sustainable water infrastructure systems is to maximize the management of the problem at the source, that is, the parcel or local level. Thus, it is important to understand the movement of water at this scale. An evaluation of the nature of the rainfall-runoff relationship at the parcel and neighborhood levels is presented in the next section. Then, detailed discussions of the nature of impervious and pervious areas are presented in the later sections.

11.3.4.2   Rainfall-runoff Relationships at the Neighborhood Scale.   An integrated urban stormwater management program should provide a sustainable solution to the problem of handling storms of all sizes from micro-storms to major floods. Early studies in Chicago showed that most of the annual volume of runoff is associated with smaller storms (APWA, 1968). Similar results were reported later by Heaney et al. (1977) and Roesner et al. (1991). Other early studies in Chicago by Harza Engineering and Bauer Engineering (1966) demonstrated that runoff is a nonlinear function of precipitation. Up to 2 inches of rainfall with corresponding runoff of about 0.6 in, the relationship is linear with contributions only from the impervious areas, approximately an equal mix of runoff from directly connected roofs and streets and alleys. For rainfalls greater than 2 inches, runoff from pervious areas begins and becomes the major source for rainfalls greater than 4 inches. Pitt and Voorhees (1994) show the nature of runoff for a residential area in Milwaukee. For this case study, all of the runoff came from streets, driveways, and roofs up to precipitation depths of 0.1 inches. In this range, about 80% of the runoff came from transportation related imperviousness. As the rainfall depths increase, the landscaped areas become more significant sources of total runoff. At the 1-in depth, landscaped areas contribute about 40% of the runoff. These relative contributions are site-specific but it is safe to conclude that the initial runoff is from the directly connected impervious areas. Impervious area (IA) is defined as land area that infiltrates less than 2% of precipitation that falls onto its surface directly or runs onto this surface. Directly connected impervious area (DCIA) is the IA that drains directly to the storm drainage system.

11.3.4.3   Nature of Imperviousness.   Imperviousness has been suggested as a good single indicator of the extent of urbanization for stormwater impacts (WEF-ASCE, 1998). For example, Schueler (1994) shows the dependence of the runoff coefficient on imperviousness.

This relationship is based on evaluation of more than 40 runoff monitoring sites as part of the Nationwide Urban Runoff Program (NURP) studies.

While a generally positive trend is evident, a large variability remains indicating that imperviousness alone is not an adequate predictor of runoff. Population density has been used to predict imperviousness (Heaney et al., 1977). A primary unresolved source of variability in these results is the use of different bases for defining the service area. Some of these studies used small areas on the scale of blocks while others used aggregate data for much larger areas that included other land uses such as schools, parks, and commercial areas. Some studies exclude right of way in calculating the service area. Thus, the results vary widely. A parcel-level evaluation of the nature of imperviousness is presented next.

11.3.4.4 Sources of Urban Runoff.   Each parcel consists of the development on the lot itself plus the adjacent development in the right-of-way (ROW) that provides infrastructure services for this parcel. ROW occupies about 25% of the developable land area in contemporary developments. When calculating development densities, it is important to define whether the denominator is the lot area only, the lot plus ROW area, or the lot plus ROW plus other land uses including open space. DCIA is the most important component as far as causing stormwater runoff quantity and quality problems. It is now standard practice to discharge roof runoff onto pervious areas, particularly in lower density developments with well-drained soils. Thus, rooftops are no longer a predominant source of DCIA; rather streets and driveways have grown in relative importance as the number of vehicles has increased.

11.3.4.5 How Imperviousness Varies for Different Types of Urban Developments.
Heaney (2000) evaluated total and directly connected imperviousness for a cross section of contemporary 1- and 2-story houses located on typical lots and ROWs. Using this database, relationships between total and directly connected imperviousness as a function of the dwelling unit density per gross acre were developed. The results are shown in Fig. 11.6. Gross

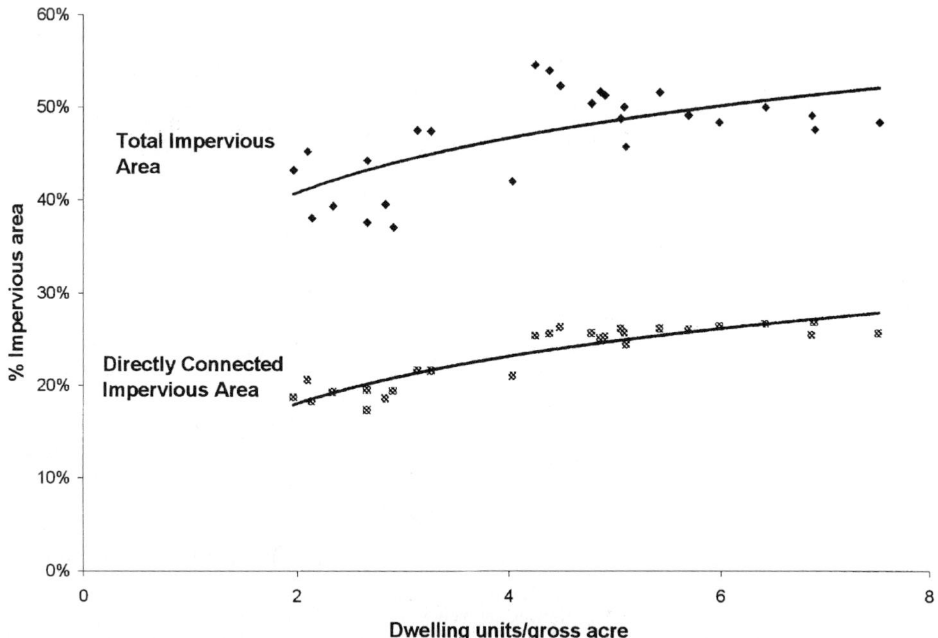

FIGURE 11.6   Effect of dwelling unit density on total and directly connected imperviousness.

acres is the lot plus the right-of-way. Typically densities in older portions of existing urban areas would be in the 8 unit/acre range. The associated total IA is about 50%. Newer lower density developments are in the 2 unit/acre range with total IAs of about 40%, only about 20% lower in spite of the fact that the dwelling units/acre have dropped on the average from 8 to 2 units/acre. Correspondingly, the amount of imperviousness generated, per dwelling unit, for low density developments is much higher than for traditional higher density developments. Similarly, the directly connected imperviousness is about 25% at 8 units/gross acre and decreases to about 20% at 2 units/gross acre. Thus, low-density developments have a much greater impact in terms of both total IA and DCIA. Most of the DCIA is due to the use of curb and gutter design for streets. Thus, a major goal for improving stormwater quality is to use curb and gutter systems only when necessary.

11.3.4.6 General Conclusions Regarding the Effect of Changing Land Use. The major trend over the 20th century has been towards decreased development densities. Densities greater than about 8 dwelling units per gross acre are difficult to achieve with automobiles since insufficient parking, by contemporary standards, is available. Therefore, the earlier impact of the automobile was to retrofit existing neighborhoods and foster growth in nearby suburbs that could accommodate automobiles as a major user of land. The development of expressways allowed people to move even farther out of the core urban areas. This movement resulted in even more dependence on automobiles and even lower development densities. Thus, the overall results of this analysis can be captured by showing the effect of density on infrastructure demand. The results are summarized below.

Higher densities significantly reduce the lengths of streets, water mains, and sanitary and storm sewers needed per dwelling unit as shown in Fig. 11.7 for the 5-acre block studied as part of traditional developments. The general equation for feet of street per dwelling unit for this 5-acre case is

$$L = \frac{198}{DUD} \tag{11.1}$$

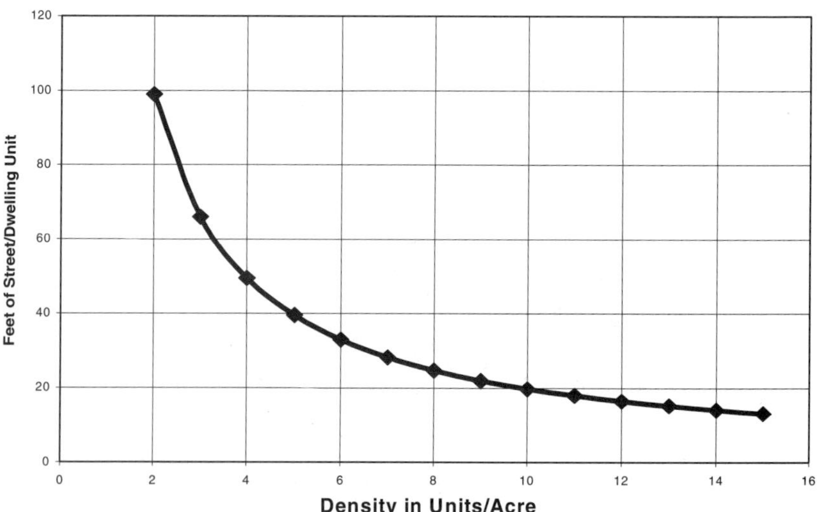

FIGURE 11.7 Relationship between street length and dwelling unit density for a 5-acre rectangular block of dimensions 660 ft by 330 ft.

where

L = feet of street per dwelling unit
DUD = dwelling units per gross acre

The length shown in Eq. (11.1) consists of one-half of the street frontage per dwelling unit plus a prorated share of the side street length. Urban sprawl is considered to be lot densities of 3 per acre or less. As indicated by Fig. 11.7, the street length per dwelling unit increases rapidly at lower densities reaching 100 ft per dwelling unit at 2 units per acre, 4 times the length at 8 units per acre. This length per dwelling unit is a critical parameter because the street, water main, sanitary sewer, and storm sewer lengths all increase in the same proportion.

The service area per household increases according to the same type of relationship as for infrastructure length, that is,

$$A = \frac{43560}{DUD} \tag{11.2}$$

where

A = square feet of area per dwelling unit
DUD = dwelling units per gross acre

The results are shown in Fig. 11.8. Lot area per dwelling unit is also a critical parameter in determining infrastructure costs. Larger lots generate an increased demand for lawn watering, the largest source of variability in urban water supply.

Another significance of lot area is that storm sewer peak design flows for small catchments are typically calculated using the Rational formula,

$$Q = CiA \tag{11.3}$$

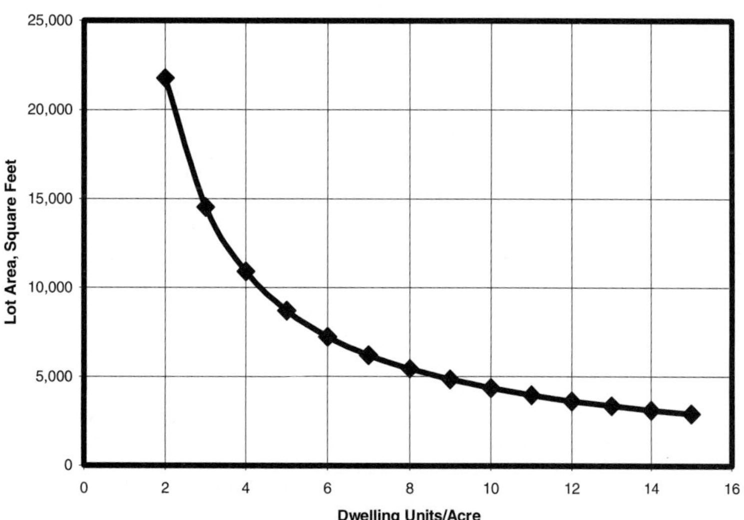

FIGURE 11.8 Relationship between dwelling unit density and area per lot.

where

Q = peak discharge rate
C = runoff coefficient that depends on the land use
i = rainfall intensity
A = drainage area

Q increases linearly with drainage area in Eq. (11.3). One offsetting factor is if the runoff coefficient decreases as A increases. The runoff coefficient is assumed to equal the imperviousness. The results shown in Fig. 11.6 indicate that total imperviousness decreases from about 50% at a DUD of 8 to about 40% at a DUD of two. The net effect, shown in Table 11.7, is more than a three-fold increase in CA as densities decrease from 10 to 2 *DU*/gross acre. This increase is moderated somewhat by the longer time of concentration for the lower density development.

The preceding results imply that serving contemporary lower density residential developments is significantly more expensive per dwelling unit than for higher density developments. Is this cost reflected in the charges for services rendered? If the new users paid *system development charges* (SDC) that covered the cost of the local improvements, then a significant part of this added cost is equitably assigned. Most of the charges for water supply are assessed based on water use. Per capita indoor water use is fairly constant. However, outdoor water use depends on the demand for irrigation water that ranges from insignificant in the northeastern United States to dominant in the arid southwestern United States. If irrigation is not a significant water use and SDCs were not assessed, then the lower density developments are being subsidized since they require more piping per unit of water delivered. If irrigation is significant, then the equity of the charges depends on the charge for outdoor water use. Waste water charges per household are either fixed or assessed based on indoor water use. This charging procedure is unfair to people living in higher density areas since they use less piping per family. Stormwater charges are a fixed amount per month or are based on impervious area. Only in the latter case are charges assessed in proportion to the contribution to the problem.

In summary, overall dwelling unit density is a good measure of the impact of residential development on infrastructure. Densities above about 8 dwelling units per acre are difficult to achieve in areas that are dependent on the automobile for transportation since there is insufficient space to accommodate the automobile with existing land-use zoning requirements. The quantity of stormwater runoff per person has grown dramatically during the past century. The following factors are the major causes of this growth:

1. The introduction of automobiles into cities: Automobiles are relatively inefficient people movers in cities with regard to the space and generation of pollutants. A vehicle weighing

TABLE 11.7   Effect of Dwelling Unit Density on CA in the Rational Formula

DUD Dwelling Unit Density (dwelling units/ gross acre)	A Lot Area per Dwelling (sq ft)	I Imperviousness (%)	CA In (sq ft)
2	21,780	40	8,712
10	4,356	60	2,616

2,500 to 4,000 pounds is used to carry a 150-pound person around the city. This vehicle is only used about 1 to 5% of the time. When not in use, it must be parked. Each off-street parking space uses 300–400 square ft of impervious area. In residential areas, transportation-related imperviousness accounts for over 65% of total imperviousness and nearly 65% of the DCIA. Within residential neighborhoods alone, about 1.25 to 2.0 square ft of impervious area is generated for transportation for every square foot of living area. Similar ratios exist for commercial areas.

2. The trend towards larger houses: House sizes have grown significantly in the past 40 years from about 1,000 square ft to over 2,000 square ft as families move to outlying areas.

3. The trend towards larger lots: Lot sizes have also grown significantly as families provide recreation and open space on each lot as opposed to using common areas. Lot sizes have also had to grow to accommodate larger garages and driveways.

4. The trend towards smaller families: Smaller family sizes and larger houses cause the need for support infrastructure per capita to increase accordingly.

5. The green trend of providing more open space as part of the development: This open space further reduces densities and increases sprawl. Properly designed, some or all of this open space could provide essential water infrastructure functions such as stormwater retention.

Given that demands for stormwater management have increased dramatically due to the trend towards lower density development, and the desire for open space, can any of these patterns be changed? The individual sources of imperviousness and their nature are discussed next.

11.3.5 Components of Urban Land Use and Stormwater Problems

The components of urban land use are examined in this section. For each component, its relative importance as a source of stormwater quantity and quality problems is discussed. The controllability of stormwater from each component is then analyzed. A popular way to classify urban land uses is to define various categories of residential land use (e.g., low density, commercial, industrial, and public land uses). Associated with each land use is an estimated imperviousness. A limitation of such general measures is that they don't provide a breakdown on the nature of the imperviousness. Another limitation is lack of specificity in how the area is defined as discussed above. A more functional way to partition urban areas is by the nature of the imperviousness and whether it is directly connected to the storm drainage system. The total land area can be divided into two major components: parcels and right-of-way, and functional units within these categories as listed below.

Parcels

The parcel or lot portion of the area can be divided into the following functional components:

- House
- Garage
- Part of driveway
- Yard
- Walkway to dwelling unit
- Pool
- Deck/shed

Right-of-Way

The ROW portion of the area is divided into the following components:

- One-half of street consisting of driving and parking lanes
- Curb and gutter, part of which is used as part of the parking lane
- Pervious area between curb and sidewalk
- Sidewalk
- Pervious area between sidewalk and property line
- One half of an alley in some neighborhoods
- Part of driveway
- Alley

Each of these components is described below.

11.3.5.1 Parcels. Lot sizes and associated dwelling unit densities were discussed previously with regard to estimating imperviousness. Lot size is seen to be a very good overall indicator of the amount of infrastructure needed to support residential development. Trends toward more automobiles and larger houses and a desire for privacy have resulted in much larger lot sizes. Demand for larger lot sizes might be reduced if the full costs of these larger lots were assessed on the property owners. In addition to promulgating regulations with regard to ROWs, cities often specify lot densities and minimum requirements (Schueler, 1995). These minimum setback and related requirements further reduce allowable densities. As with ROWs, it is advisable to revisit these requirements for larger lot sizes.

11.3.5.1.1 Dwelling Unit Footprint. Urban dwelling units vary greatly in size as illustrated by these typical units and size ranges:

Single room: 100–300 sq. ft

Studio apartment: 300–500 sq. ft

One-bedroom unit: 400–700 sq. ft

Two-bedroom unit: 600–1,200 sq. ft

Three-bedroom unit: 1,200–2,500 sq. ft

Four-bedroom unit: 1,800–4,000 sq. ft

Because of increasing affluence and more affordable housing, the median size of a dwelling unit per family has steadily increased since World War II. The footprint of the dwelling unit (DU) is the amount of land it occupies. For single story DUs, the sizes of the DU and the footprint are very similar. The footprint is slightly larger due to roof overhang. The footprint is much less than the DU area if multi-level construction is used.

Stormwater runoff from buildings depends upon the roof area and whether the roof downspouts are directly connected to the storm sewer system. At densities of eight or more units per gross acre, the roof area should probably be connected directly to the stormwater control system because insufficient pervious area exists on the property itself. Treatment of roof runoff consists of controlling sources of atmospheric deposition, changing to more benign roofing materials, periodic cleaning of gutters, and disconnecting downspouts. The primary demand management approach is to encourage smaller roof areas by constructing multi-level buildings.

11.3.5.1.2 Covered Porches and Patios. The footprint of the DU is increased if covered porches are included in the house. Covered porches are an icon of traditional neighborhood development. One reason that porches fell out of favor is traffic noise. Porches add imperviousness to the property and appear to be regaining popularity. However, porches are a minor source of imperviousness and much of this imperviousness is not directly connected. Thus, no detailed evaluation of porches is included.

Patios may be constructed of permeable or impermeable material. They typically drain to adjacent pervious areas. Also, patios are not a major source of pollutant loadings. Thus, no separate analysis of patios is included.

11.3.5.1.3 Garages and Carports. Garages have emerged as an important land use in urban areas during the 20th Century. Automobiles require about 200 square ft of garage space per car. As the number of automobiles has increased, so has the number of garage spaces. Two and three car garages are now the norm for new house construction. The primary runoff from garage areas is from the rooftop. Thus, the impact depends upon whether the roof downspouts are directly connected to the sewer system or discharge to adjacent imperviousness such as driveways. Treatment of roof runoff consists of controlling sources of atmospheric deposition, changing to more benign materials, and disconnecting downspouts. The primary demand management technique for garages and carports is to reduce the demand for the number of cars. In the United States, there are over 200 million cars for 250 million people. This corresponds to about one vehicle for every licensed driver in the United States. It is possible to have the number of cars per capita continue to increase as people have more than one car per capita.

11.3.5.1.4 Driveways. Driveways have become an important source of imperviousness in the 20th Century as new developments had to accommodate a growing number of automobiles. The Institute of Transportation Engineers (Southworth and Ben-Joseph, 1995) recommends minimum driveway widths of 10 ft for one car, with a 20-foot-wide curb cut (5-ft flare on each end). Driveways associated with garages are also an important land use. Four types of driveways need to be considered based on the location and orientation of the garage:

1. Attached, front facing garage

2. Attached, side facing garage

3. Attached, rear facing garage

4. Detached garage in rear of lot

Attached, front facing garage: If the garage faces the street and is attached to the house, then the driveway width is usually the width of the number of garage spaces, or about 9 to 10 ft of width per car. The length of the driveway depends on the house setback. Minimum driveway lengths are dictated by having sufficient length so that a car can pull into the driveway and not block the sidewalk. Thus, a minimum driveway length is the sum of the distance from the street to the sidewalk (0 to 15 ft) plus the width of the sidewalk (4 to 6 ft) if there is one plus the length of a car space or about 20 ft, or a total minimum driveway length of 20 to 41 ft. The extra house setback distance must be added to this minimum distance to get the total distance. For many houses, the paved area for the driveway exceeds the impervious area of the garage. Some, if not all, of the driveway drains to the street, thereby creating a significant source of directly connected impervious area.

Attached, side or rear facing garage: If the garage entrance faces the side of the house, then a narrower driveway from the street to the house can be used (e.g., 12 ft). However, this savings in width is offset by the need to provide a turning area so that the cars can maneuver to enter and exit the garage. This added turning area adds significant paved area.

Detached garage in rear of lot: If the garage is detached and located at or near the rear of the lot, then a longer driveway is needed to extend from the street to the rear of the house. The width of this driveway increases in front of the garage to allow cars to enter the various bays. Of course, if an alley exists, then the driveway distance is minimal.

As a low-intensity use, driveways are good candidates for porous and permeable pavements or simply paving only parallel strips for the wheels. Another effective control is to route driveway runoff onto adjacent pervious areas instead of directly to the street. This can be done by putting a crown on the driveway as is done for streets. An effective demand management to reduce the demand for driveways is to reduce the demand for automobiles. Another possibility is to better utilize on-street parking.

11.3.5.1.6   Pervious Area on Property.   The pervious area on the property is used primarily for lawns, gardens, and wooded areas. This land is used for aesthetic appeal, and recreation for people and pets. Under proposed innovations, this pervious area will be used more intensively to infiltrate stormwater from adjacent impervious areas as well as from precipitation directly onto its surface. At present, pervious areas do receive some of the runoff from impervious areas, primarily from roofs, patios, and some parts of the driveway. Thus, it is important to determine the infiltration capacity of these soils. The infiltration capacity depends on the soil type. Pervious areas can be graded to provide some on-site detention of stormwater, that could then be re-used for lawn watering or other purposes. Prince George's County (2000), MD has developed the idea of "functional landscapes" for on-site management of stormwater.

11.3.5.2   Right-of-Way

11.3.5.2.1   Streets and Highways.   Urban street patterns have changed during the 20th Century, with the automobile having a major influence on street design at all levels (Southworth and Ben-Joseph, 1995). They trace the major change in philosophy for street design to the 1930s when the federal government became involved in developing guidelines for sub-divisions as part of its program to insure home mortgages. The traditional pattern is the gridiron with typical block dimensions of 1/8 by 1/16 of a mile. In 1936, the Federal Housing Administration (FHA) rejected the grid pattern for residential neighborhoods, and has continued this policy of preferring other street layouts (Southworth and Ben-Joseph, 1995). Their primary reasons for rejecting the gridiron pattern are the following:

- It requires more paved area than necessary because all residential streets are built to the same specifications.
- It requires more expensive type of pavement since the traffic is dispersed throughout the neighborhood and thus the streets must be designed to a higher standard.
- This heavier traffic demand creates a hazard.
- The gridiron layout is monotonous and uninteresting.

The FHA recommended a hierarchical street pattern. For residential streets, they recommended curvilinear alignments, cul-de-sacs, and courts. Desirable design criteria promulgated by the FHA included the following (Southworth and Ben-Joseph, 1995):

- Layout should discourage through traffic.
- Minimum width of a residential street should be 50 ft with 24 ft of pavement, 8-ft planting/ utility strips and 4-ft walks.
- Cul-de-sacs are the most attractive street layout for family dwellings.
- Minimum setbacks for streets should be 15 ft.
- Front yard should avoid excessive planting, for a more pleasing and unified effect along the street.

These early FHA guidelines had a tremendous influence on residential development in the United States because of their financial leverage over developers and home buyers. The Institute of Transportation Engineers (ITE) has also had a major influence on residential street design. Their perspective is heavily influenced by traffic flow and parking considerations. They recommend (Southworth and Ben-Joseph, 1995):

- Right-of-way should be a minimum of 60 ft.
- Pavement width should be 32 to 34 ft.
- Cul-de-sacs should have a maximum length of 1,000 ft with a 50-ft radius at the end.
- Parking lanes should be 8 ft in width.

The influence of these street design standards on drainage and stormwater quality does not seem to have been a significant factor in the decision-making process.

The American Association of State Highway and Transportation Officials (AASHTO) has been responsible for developing the design standards for highways and streets. The primary reference is *A Policy on Geometric Design of Highways and Streets* (AASHTO, 1984). According to Khisty (1990), 10 to 13 ft lane widths predominate in the United States with 12 ft being the most common. The use of 11-ft lane widths is acceptable in urban areas due to higher right-of-way costs. Ten-ft lane widths are only acceptable on low-speed urban streets.

Southworth and Ben-Joseph (1995) bemoan the consequences of current standard street design practices stating:

> Attempts to reshape the form of the American city are often thwarted by the standards and procedures that have become embedded in planning and development. Particularly troublesome are standards for streets that virtually dictate a dispersed, disconnected community pattern providing automobile access at the expense of other modes. The rigid framework of current street standards has resulted in uniform, unresponsive suburban environments.

The current residential street design standards that are accepted virtually throughout the United States necessitate a large amount of impervious area per family that consists of wide streets, sidewalks, and driveways.

Southworth and Ben-Joseph (1997) provide a history of urban streets, a critique on current practices, and project the expected nature of streets in urban areas. They estimate that, worldwide, more than one-third of all developed urban land is devoted to roads, parking lots, and other automobile infrastructure. In urban United States, about one-half of the land is used for this purpose. In automobile-oriented cities such as Los Angeles, the percentage increases to two-thirds (Hanson, 1992; Renner, 1988). These estimates are compatible with the results presented in the previous section.

Transportation engineers tend to design streets to maximize convenience for the automobile subject to safety constraints. Recently, designers have attempted to re-cast the purpose of streets as multi-purpose components of the community with much more of a pedestrian orientation. Shared streets provide a multi-purpose use of residential streets. These streets have gained favor internationally but have not yet gained widespread acceptance in the United States. Key impediments in the United States include dependency on automobiles, and concerns of liability if existing street standards are changed. Portland, OR is one of the few cities in the United States that is rethinking its approach to residential streets with its *skinny streets program* (Southworth and Ben-Joseph, 1997). They have reduced street widths to 20 to 26 ft and have installed many traffic calming devices.

Streets have the potential to play a major role in stormwater management. Skokie, IL implemented an innovative approach to its streets by using them to intentionally convey and store stormwater in a controlled fashion so that combined sewers do not surcharge and back up into basements (Walesh and Carr, 1998). Stormwater control is achieved in this cost-effective system using on-street berms coupled with catch basin flow regulators and, where needed, sub-surface tanks.

11.3.5.2.2 Street Classification and Utilization. The Federal Highway Administration (FHWA) tabulates a variety of street-related statistics that can be obtained at *http:/ www.bts.gov/cgi-bin/stat/final_out.pl*. Results for urban areas in the United States are shown in Table 11.8. The major traffic carrying components of the urban highway system constitute only about 9% of the road mileage in urban areas. Local streets that carry little traffic constitute nearly 70% of the mileage. Parking is allowed on the less-used streets; thus, most of the parking is associated with local and collector streets. While the interstates, freeways, other expressways, and principal arterial streets constitute only 9.1% of the miles, they carry 58% of the traffic. At the other extreme, local streets, constituting 69.5% of the street length, carry only 13.8% of the traffic. Thus, in terms of managing imperviousness, the less-used local streets are the prime candidates for evaluating whether they could be reduced in size.

The results of Table 11.8 also suggest that the primary sources of traffic-related stormwater pollution are the intensively used street systems. This may suggest a control strategy of providing more treatment for these intensively used streets. This much smaller impervious area may be more amenable to control than trying to deal with the entire impervious area of the city.

11.3.5.2.3 Recommendations for Residential Streets. Southworth and Ben-Joseph (1997) recommend the following principles for future residential streets:

- Support varied uses of residential streets including children's play and adult recreation.
- Design and manage street space for the comfort and safety of residents.
- Provide a well-connected, interesting pedestrian network.
- Provide convenient access for people who live on the street, but discourage through traffic; allow traffic movement, but not facilitate it.
- Differentiate streets by function.
- Relate street design to the natural and historical setting.
- Conserve land by minimizing the amount of land devoted to streets.

Contemporary texts on highway engineering do not deal with urban runoff problems. Khisty (1990) cautions of the need to evaluate air pollution and noise impacts as part of highway design. He does not mention highway runoff as a problem. Wright and Paquette (1996) describe conventional highway drainage design but do not discuss stormwater quality problems or the detrimental off-site impacts from highway runoff. The FHWA has sponsored several studies to address the issue of stormwater problems associated with highways. Young et al. (1996) present a detailed overview of highway runoff quality problems. For a more current view from FHWA on whether they consider highway runoff to be a serious problem, see *http://www.tfhrc.gov/hnr20/runoff/runoff.html.*

11.3.5.2.4 Streets and Stormwater Runoff. Whether residential streets are laid out in a gridiron, curvilinear, or cul-de-sac format does not appear to have a major impact on the quantity of stormwater runoff per capita. The curvilinear and cul-de-sac layouts tend to have a larger impact per capita because of lower development densities. Schueler (1995) sum-

TABLE 11.8 Urban Street Mileage in the United States

Urban	Miles of road	% of urban
Interstate	13,307	1.6%
Other freeways/expressways	9,022	1.1%
Other principal arterial	53,044	6.4%
Minor arterial	89,013	10.8%
Collector	87,918	10.6%
Local	574,119	69.5%
Total Urban	826,423	100.0%

TABLE 11.9 Condensed Summary of National Design Standards for Residential Streets

Design criteria	AASHTO	ITS	Headwater streets
Residential street categories	1	3, depending on use density	4, depending on ADT
Minimum street width	26 ft	22–27 ft > 2 du 28–34 ft @ 2–6 du 36 ft < 6 du	16 ft (<100 ADT) 20 ft (100–500 ADT) 26 ft (500–3000 ADT) 32 ft (>6 du/ac)
Additional right of way	24 ft	24 ft	8 to 16 ft
Design speed, level terrain	30 mph	30 mph	15 to 25 mph
Curb and gutter	Generally required	Generally required	Not required on collectors
Cul-de-sac radii	30 ft	40 ft	30 ft
Turning radii in cul-de-sac	20 ft	25 ft	17 ft

Source: From Schueler, 1995.

marizes current national design standards for residential streets as shown in Table 11.9. Parking requires about 8 ft of space and traffic lanes require about 10 to 12 ft per lane. Thus, streets with two-way traffic and parking on both sides of the street would be 36 to 40 ft wide, if multi-purpose use is not incorporated into the design.

Average daily traffic (ADT) in vehicles per day is the common indicator of the utilization of streets for traffic. Schueler (1995) summarizes the expected traffic flow for various ADTs assuming 10 trips per dwelling unit per day and that the number of trips in the peak hour is 10% of the daily trips. The results are presented in Table 11.10 (Schueler, 1995). As Schueler points out, for ADTs of 25 or less, it is reasonable to share parking and traffic lanes. Unfortunately, many cities have adopted regulations that require wide residential streets even in areas with little or no traffic.

11.3.5.2.5 Parking. The ITE recommends (Southworth and Ben Joseph, 1995) that on-street parking lanes should be 8 ft in width and that driveway widths should be a minimum of 10 ft for one car, with a 20 ft-wide curb cut (5-ft flare on each end). According to Shoup

TABLE 11.10 Relationship Between Number of Dwelling Units, Traffic Generation, and Residential Congestion

No. of single family homes	Average daily trips	Peak trips per hour	Minutes between cars (average)	Minutes between cars (peak)
5	50	5	30	12
10	100	10	15	6
25	250	25	6	4
20	500	50	3	1.5
75	750	75	2	45 secs
100	1000	100	1.5	35 secs
150	1500	150	1	20 secs
300	3000	300	30 secs	10 secs

Source: From Schueler, 1995.

(1995), off-street parking space per vehicle ranges from 300 to 350 square ft per space. This square footage includes the space itself, the access aisles, and the entry, exit area. Shoup (1995) and Wilson (1995) summarize the origin of parking "requirements" in urban areas and the overall impact. Between 1975 and 1993, the average number of parking spaces required by cities per 1,000 square ft of office space increased from 3.6 to 3.8 spaces (Shoup 1995). According to Wilson (1995), zoning codes typically require between 3 and 5 spaces per 1,000 gross square ft of office building area, with 4 spaces being the most popular requirement. At 350 square ft per parking space, this corresponds to 1.05 to 1.75 square ft of parking per square foot of office space. Similar ratios have been obtained for residential areas.

The actual estimate of saturation demand for parking is 2.4 spaces per 1,000 square ft of office space for driver-paid parking to 3.1 spaces per 1,000 square ft for employer-paid parking (Shoup, 1995). According to Shoup (1995), over 91% of cities required more than this saturation demand. Wilson (1992) estimated an average requirement of 4.1 spaces per 1,000 square ft in southern California, with the average peak parking demand being only 56% of this capacity.

The primary justification for high parking requirements is to avoid spillover of parking from one parcel of land to others. However, if all facilities are designed for peak demand, often specified as the demand that only occurs 15 to 30 hrs per year, then, by definition, large amounts of excess capacity will exist in the system since these peaks are not coincident. According to the Urban Land Institute (1982), specifying a design hour of the 20th busiest hour of the year, leaves spaces vacant more than 99% of the time and leaves half the spaces vacant at least 40% of the time. Many residential streets carry relatively few vehicles each day. For example, streets serving less than 25 homes are so lightly traveled each day (and during peak hours) that shared parking and moving lanes make sense. The requirement for

TABLE 11.11   Parking Demand Ratios for Selected Land Uses and Activities

Land use	Parking space ratio used	Range
Single family homes	2 spaces/du	1.5–2.5
Townhouses	2.25 spaces/du	1.5–2.5
Professional office	1 space/200 sf gfa	150–330
Hotel/motel	1 space/guest room	0.8–1.25
Retail	1 space/250 sf gfa	200–300
Convenience store	1 space/300 sf gfa	100-500 + es
Shopping center	1 space/200 sf gfa	150–250
Movie theatre	1 space/4 seats	3.3–5
Gas station	2 spaces/pump (and 3 spaces)	
Industrial	1 space/1000 sf gfa	500–1200
Golf course	4 spaces/hole	3–6.5
Nursing home	1 space/3 beds	2–4+es
Day care center	1 space/8 children	4–10 + es
Restaurant	1 space/50 sf gla	0–200
Marina	0.5 space/slip	0.26–0.7 + es
Health club	1 space/100 gfa+es	100–150
Church	1 space/5 seats	4–6
High school	Many diverse ratios	
Medical/dental office	1 space/175 sf gfa	100–225

Source: From Schueler, 1995.
Notes: du = dwelling unit, sf = square feet, gla = gross leasable area, es = employee spaces, gfa = gross floor area.

parking is typically estimated from the *ITE* parking manual (ITE 1987). Sample parking requirements are shown in Table 11.11, from Schueler (1995). According to Arnold and Gibbons (1996), the City of Olympia, WA found not only parking oversupply with vacancy rates of 60 to 70%, but also developers building an average of 51% more spaces than required by the City of Olympia.

11.3.5.2.6 Effect of Parking Fees. According to Shoup (1995), motorists report free parking for 99% of all automobile trips. About 95% of automobile commuters say that they park free at work. A primary reason for such high use of cars to commute to work is that employers pay for parking. The average for 7 case studies of the impact of parking fees on driving behavior is that 72 cars are driven to work per 100 employees if the employer pays for parking while only 53 cars are driven to work per 100 employees if the employee pays for parking (Shoup, 1995). Recent state legislation in California requires employers to allow nonauto using employees to receive an equivalent cash payment to the amount of the subsidy for parking. Existing parking guidelines have evolved from observing practice around the United States. However, the database is observations on consumer behavior in lots where parking is provided free of charge. Thus, the existing standards are for the demand for parking if parking is free. According to Shoup (1995), virtually no research has been done to determine the optimal amount of parking since parking requirements are usually mandated by the local government agency. If a private developer was free to establish the amount of spaces to provide for his development, the developer would be expected to do a benefit-cost analysis and determine the number of spaces such that his net revenue was maximized.

11.3.5.2.7 Alleys. Alleys are regaining popularity as part of new urbanism designs. Alleys can be found in older neighborhoods. They provide access for garages and garbage pickup and other deliveries. Alleys eliminate the need for driveways and thereby permit narrower lot widths. Typical alley widths range from 12 to 16 ft. In addition to this pavement width, aprons to the garages on either side of the alley are needed. Boulder, CO specifies a 20-ft right-of-way width for alleys. The width of the alley is controlled by the required turning radius for vehicles entering and exiting from the garages and open parking areas. From a safety point of view, alleys greatly minimize the traffic and pedestrian safety hazards associated with vehicles entering and backing out of driveways onto the street. Runoff from alleys is directly connected to the storm sewer system. Runoff moves along the alley by overland flow until it reaches the street inlet. Treatment options would be the same as for other impervious areas with low traffic and parking rates. The demand for alleys can be eliminated by using driveways. The tradeoff on the amount of pavement used for alleys versus driveways depends on the lot geometry.

11.3.5.2.8 Sidewalks. Attractive sidewalks are an inducement to walking. According to Chellman (1997), about 10% of Americans walked to work in 1960. By 1990, the percentage walking to work had decreased to 4%. Sidewalks are an integral part of older cities. With lower density urban development, the need for sidewalks is less critical. If the housing density is very low, then people can walk in the street. Also, a single sidewalk can be used instead of having a sidewalk on either side of the street. Sidewalks can be located adjacent to the street or separated by a 6-ft or wider planting area. The ITE (Southworth and Ben Joseph, 1995) recommends sidewalks with a minimum width of 5 ft on both sides of the street. Sidewalks are typically constructed of reinforced concrete. The Urban Land Institute, ULI (1968), recommends sidewalks on both sides of the street if the density exceeds 6 houses per net acre. They recommend 5-ft wide sidewalks along collector streets and 4-ft sidewalks on minor streets. Chellman (1997) recommends sidewalk widths of 5 ft to provide sufficient room for pedestrians to pass without crowding.

Sidewalks typically drain to pervious areas allowing the runoff to infiltrate into the ground. The notable exception is when the sidewalks are located immediately adjacent to the streets; then the sidewalk runoff becomes directly connected since the drainage goes directly onto the streets. A traditional treatment is sweeping the sidewalk areas to keep them clean and to provide trash containers to discourage littering. Sidewalks can be eliminated if the street is safe for nonvehicular use.

11.3.5.2.9 Curb and Gutter and Swales. According to Khisty (1990), curbs are used for the following reasons:

- Drainage control
- Pavement-edge delineation
- Right-of-way reduction
- Aesthetics
- Delineation of pedestrian walkways
- Reduction of maintenance operations

Two primary types of curb and gutter are the *barrier curb* and the *rolling curb*. An alternative is to eliminate curb and gutter and allow street runoff to flow onto adjacent pervious areas. The curb and gutter are about 2 ft in width. The ITE (Southworth and Ben Joseph, 1995) recommends vertical curb with gutters. Rolled curbs are not recommended. However, the ULI (1968) recommends rolled curbs for most residential areas because they avoid curb cuts for driveways. Another option is to use curb and gutter but have periodic openings to adjacent pervious areas.

11.3.5.2.10 Planting Strip Between Street and Sidewalk. Many subdivision regulations require a planting strip to separate the sidewalk and the street. The ITE (Southworth and Ben Jospeh 1995) recommends planting strips on both sides of the street with a minimum width of 6 to 7 ft and with the planting strip draining towards the street. A 1990 revision of these standards decreased the minimum planting width to 5 ft. Boulder, CO specifies an 8-ft wide planting area. Planting strips with a width of 15 ft are popular in the western suburbs of Chicago. These planting strips provide a buffer between the street and sidewalk. They also provide a planting area within the right-of-way for trees. Early subdivision regulations promulgated by the federal government suggested two trees should be planted on each lot. Drainage from these planting areas is directed towards the street. No citations could be found regarding how these areas could function as part of the stormwater drainage system. They could be expected to attenuate noise and air pollution effects to a limited degree.

11.3.5.2.11 Overall Right-of-Way. Required right-of-way width dimensions for residential street design standards in Boulder, CO are (Boulder, 2000):

- Bikeway: 12 ft
- Alley: 16 ft
- Residential: 60 ft
- Residential collector: 60 ft

To this base are added medians, added travel lanes and speed changing lanes, and turn lanes. These ROW requirements are typical. The key control option is to take a hard look at existing ROW requirements, especially in residential areas, to see whether the requirements could be modified to reduce the generation of impervious area that is providing little or no added value and to encourage the more effective use of pervious areas within the ROW.

11.3.5.2.12 Will Americans Reduce Auto Use? Dittmar (1995) outlines a broader context for transportation planning that incorporates some of the above concepts for developing more sustainable transportation systems. In his conclusions, he discusses the feasibility of reversing the trend since World War II of increasing reliance on the automobile. Dittmar says:

> In discussions of the issues with transportation officials, their most frequent initial assertion is that Americans love cars and cherish driving, and that any reform effort is therefore somehow doomed. Running a close second are the assertions that Americans are voting with their gas pedals by choosing exurbia, and that building more roadways is simply giving folks what they want. I don't believe this is true. People are responding to a set of signals our society gives them

by building ring roads and beltways, subsidizing free parking and suburban development through utility infrastructure, and providing tax incentives that favor car use and suburban home ownership. These signals favor continued sprawl and reliance on cars. Changing these endemic signals by creating incentives to live in the city, eliminating tax biases toward cars, and enhancing livability can send the public new signals.

With regard to streets, parking, and other major sources of imperviousness, engineers have been the ones who have promulgated these regulations. Hopefully, they can also take the lead in modifying them to create more sustainable communities.

11.3.6 Summary and Conclusions

The results of this discussion on the nature of imperviousness in urban areas show that the quantity of urban stormwater generated per dwelling unit has increased dramatically during the 20th Century due to the trend towards more automobiles, which requires more streets and parking, and the trend towards larger houses, all combined on larger lots. Commercial and industrial areas likewise need much more parking per unit of office space than they did before automobiles. Interestingly, the square footage for residential and commercial areas is less than the support parking requirements. Modern practices dictate devoting more of the city landscape to parking than to human habitat and commercial activities. The net result of this major shift in urban land use is low-density sprawl development that generates over three times as much stormwater runoff per family than did pre-automobile land use patterns. Much of these requirements for more and wider streets and parking have been mandated in order to improve the transportation system. Ironically, unlike water infrastructure, these services are not charged directly to the users. Rather, they are paid for by the general public including nonusers.

11.4 SMALL-SCALE, DISTRIBUTED STORMWATER CONTROL

The breadth of innovations in distributed stormwater precludes a listing of the details of all published designs. Innovative designs for *micro-storm control* by enhanced infiltration systems and on-site detention have been reported in the domestic and international literature over the past 20 years. A broad array of techniques have been proposed, including porous and permeable pavement materials, infiltration pipes and channels, on-site detention devices, and functional landscapes. Known by various names throughout the world (low-impact development (LID) (Prince Georges County, 2000), on-site stormwater detention (UPRCT, 2000), BMPs (Roesner, 1999; Urbonas, 2000)), all encompass design methods for controls that are meant to store and/or infiltrate runoff from a small area, though some (like BMPs) are also used to describe larger scale controls as well. The theme is to distribute small systems throughout an urbanizing area, and integrate them into larger urban systems—the transportation network, the urban landscape, the urban hydrologic budget, etc. The design objectives integrate water quality and quantity, and the most promising innovations affect both in a positive manner without sacrificing landscape function.

A stormwater management plan that addresses micro-storms should provide the following:

1. Minimized DCIA

2. Increased flow paths and time of concentration

3. Increased infiltration, but not at the expense of nuisance flood damage

4. Increased visibility of the hydrologic cycle in urban environs for aesthetic benefits and an identifiable urban ecology

5. A maintainable and sustainable system without sacrificing large amounts of usable space

These principles provide an overall theme to the specific designs described in this section. While not all individual designs meet all criteria, collectively they should satisfy each of these principles.

The land-use decisions described in Sec. 11.3 may be made at two land-use levels: publicly owned ROW and privately owned developed parcels. This division is useful for examining the various micro-storm controls available in each sector. Aside from the land use-ownership division, controls may be described by process or function. By hydrologic function, distributed stormwater controls are divided into decreased DCIA through land-use modification, distributed physical storage, enhanced infiltration systems, and alternative conveyance systems. The next two sections present an overall summary of the state of the art of micro-storm management within the public ROW and the private parcel.

Prince Georges County, MD has been a leader in the United States developing distributed stormwater controls for over 10 years. Their design methods are given in two publications available on the Internet, "Low-Impact Development Design Strategies" and "Low Impact Development Hydrologic Analysis" (Prince Georges County, 2000). Many of the designs outlined in this section are from these volumes as well as several innovative designs from Europe, Australia, and Japan.

11.4.1 Distributed Control of Runoff from Publicly Owned Right-of-way

As demonstrated in Sec. 11.3, the publicly owned transportation corridors of urban areas are one of the leading sources of DCIA. The traditional storm drainage system design is largely driven by transportation performance specifications (e.g., allowable spread across a road). Therefore, it is instructive to review various ways designers have sought to minimize the hydrologic impacts of roadways.

11.4.1.1 Storage-based Control.   Carr et al. (1999) present an innovative use of residential streets to store stormwater in Skokie and Wilmette IL that exploits the idea of regulated surcharge. These cities had combined sewer problems, including overflows and basement flooding. A cheaper alternative to sewer separation was sought. Street storage has been used successfully in these two cities to reduce the rate at which stormwater enters the sewer. By regulating the inflow to the sewer and backing up runoff onto the street, basement sewer backups and combined sewer overflows have been reduced. Sub-surface storage may be needed where street flooding is not acceptable. The level of street flooding is controlled, however, and is designed to be shallow enough to allow for vehicular traffic to continue at slow speeds. Furthermore, the duration of inundation is designed to only last a short, publicly acceptable length of time. During dry weather, roadway modifications act as traffic speed mitigation devices as well.

11.4.1.2 Infiltration Controls.   Porous asphalt was originally designed as a surface for airport runways, where it was used to control dangerous surface ponding (Sorvig, 1993). This safety feature also makes it attractive for automobile service (Urbonas and Roesner, 1992). Porous pavements are distinguished from permeable pavement by Pratt (1997). Porous pavements allow infiltration through the entire surface of the material, and are made of porous material (Pratt, 1997). They are comprised of stone aggregate, with a binder of asphalt or Portland cement (Sorvig, 1993). Permeable pavements, on the other hand, are made up of impervious material but are constructed with permeable inlets within the surface. An example of this type of surface would be modular concrete paving blocks with grass and soil in the spaces between the concrete, allowing infiltration (Pratt, 1997).

Sorvig (1993) gives site-design criteria for porous pavement including infiltration rates and percent fines of the underlying soils. Furthermore, Sorvig (1993) and Ferguson (1998) describe the underlying infiltration bed needed for proper function. This is crucial for the performance under freeze-thaw cycles. Urbonas and Roesner (1992) discuss the potential for failure of these surfaces in cold climates. However, as reported by Sorvig (1993), failures have occurred under freeze-thaw cycles due to an inadequate underlying storage reservoir. When designed properly with an underlying infiltration reservoir, porous as well as permeable pavements have performed through freeze-thaw testing (Sorvig, 1993).

Diniz (1993) presents runoff results from a detailed study of various pavement types under simulated rainfall conditions using a sprinkler. Study findings indicate that porous and permeable pavements significantly increase the time of concentration of runoff. This has significant impacts on the design flow of the system. Water quality measurements indicate that loading reductions occur as a result of runoff volume reduction, not through reductions in pollutant concentrations. The long-term water quality impacts of porous and permeable pavements are not well-understood (Diniz, 1993). This is due to the complex adsorption/desorption processes of pollutants on the sub-surface media as well as the long-term buildup and release of particulates. Silverman and Stenstrom (1988) postulate that porous pavement and the underlying storage media have the potential to trap oil and grease and provide a favorable environment for biodegradation.

Costs of porous pavements may be approximately 10% higher than nonporous equivalents (Sorvig, 1993). However, a major factor that should be considered is the downstream system savings (Sorvig, 1993). For example, if a parking lot is constructed with porous pavement with a large storage/infiltration reservoir underlying it, significant savings may be realized through reduced downstream capacity requirements (Sorvig, 1993). Furthermore, if the porous pavement design negates the need for surface detention, tremendous savings may be realized by use of the land otherwise used for detention (Sorvig, 1993). This was accomplished by a design by Andropogon Associates of a porous parking lot for the Morris Arboretum in Philadelphia, PA (Sorvig, 1993; Ferguson, 1994).

11.4.1.3 *Groundwater Recharge Systems.*

Groundwater recharge systems may also be attractive for runoff disposal from highways with favorable groundwater conditions (Ferguson and Debo, 1990; Ferguson, 1994). An early example of a groundwater recharge system is in Nassau County, Long Island (Ferguson and Debo, 1990). This area began using groundwater recharge systems in the 1930s to avoid costly sub-surface storm sewers. Since then it was realized that these systems perform a valuable hydrologic function, replenishing the groundwater supply. Ferguson and Debo (1990) show that the average annual volume infiltrated in Long Island exceeds the average annual volume extracted for municipal use.

Another example of groundwater injection is described by Peterson and Hargis (1973). Four drilled wells on the island of Maui were used to discharge stormwater from a 400-acre site based on a 10 year design storm. These wells discharged stormwater deep into porous rock 280 ft deep, well beneath the freshwater/saltwater interface, to prevent stormwater from entering the fresh groundwater.

Pokrajac and Deletic (2000) report on a stormwater "soakaway" at the University of Aberdeen in the United Kingdom. Street runoff is directed towards a sub-surface chamber designed to drain in 72 hrs. Preliminary monitoring and modeling shows the chamber never completely empties due to surrounding soil moisture and hydraulic conductivity. Performance was satisfactory for small and medium-sized storms; however as is the case with all storage-based systems, performance was a strong function of the inter-event time and antecedent conditions. Not only does the storage remain partially full at the start of the second event, but the ability of the surrounding soil to infiltrate is greatly decreased due to higher initial water content.

11.4.1.4 *Conveyance.*

Grass-lined swales are a common means of stormwater conveyance and treatment where land costs are not high and the potential for groundwater contamination is low. They are not recommended for certain industrial and commercial areas where oil and grease may contaminate the groundwater (Pitt et al., 1996; Urbonas, 2000). Furthermore, they are generally not acceptable at slopes over 3 to 4%, though some may be used with slopes as high as 6% depending on soil conditions (Urbonas, 2000).

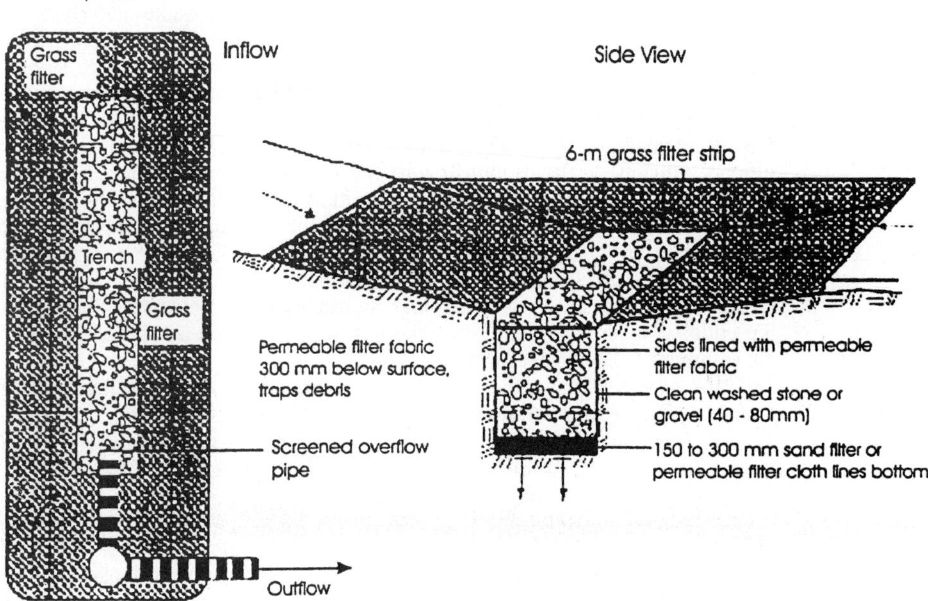

FIGURE 11.9   Median strip infiltration trench. (Prince Georges County 2000, adapted from MWCOG, 1987)

Ferguson (1998) shows how grass-lined swales may be designed in place of curb and gutter and with the ability to provide in-line storage. This could greatly increase the infiltration potential of the swale; however, sedimentation and clogging could result if the upstream sediment load was too high.

Standard gravel trenches may be an effective conveyance design for stormwater runoff. Diniz (1993) tested a gravel trench parking lot infiltration design along with the standard impervious, porous, and permeable pavement. Prince Georges County recommends a gravel filled trench surrounded by a grass filter strip, as shown in Fig. 11.9. This design may be incorporated in a highway median design as well as perimeters and islands in parking lots.

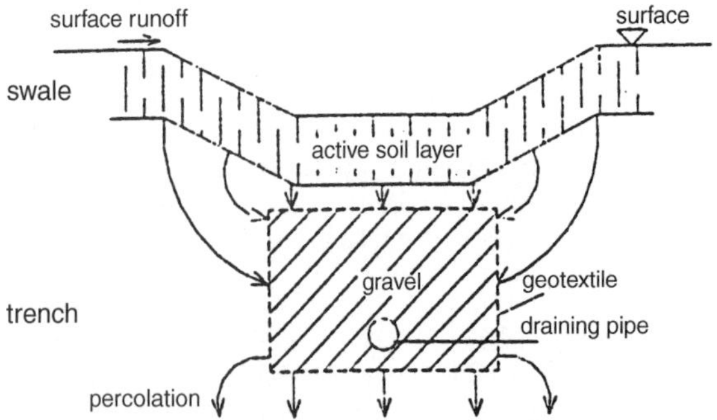

FIGURE 11.10   Cross section of the M-R element. (Seiker, 1998)

Urbonas and Stahre (1993) highlight the importance of the seasonal high groundwater table in infiltration trench performance. They provide details of a design that includes conveyance via a grass-lined swale for flows that exceed the capacity of the infiltration trench. Walesh (1989) shows a higher flow alternative to this design, where a sub-surface pipe is used to convey higher flows. This design could be made to retain runoff through the use of a downstream control to restrict flow. This would create an in-line storage effect as well, and if a perforated pipe were used, would create soil storage in the surrounding fill.

A more aggressive version of these designs was developed in Germany. This design exploits the in-line storage idea and downstream control as described above. Known as the M-R system (Mulden-Rigolen System), this design uses a gravel trench and pipe beneath a grass swale, as shown in Figs. 11.10 and 11.11 (Seiker, 1998).

These different versions of coupling filter media, conduits, and swales have the advantage of maintaining capacity for major flows even during periods of high groundwater (Seiker, 1998). For example, if the water table rose above the gravel layer in the M-R system depicted in Figs. 11.10 and 11.11, the swale would maintain emergency flow capacity. Similarly, if the gravel media or the pipe clogged and failed, the swale would maintain drainage function. The redundancy inherent in integrated designs is an important consideration for systems that must operate with little maintenance for long periods of time.

11.4.2 On-site Control for Privately Owned Urban Land Parcels

Privately owned urban parcels suggest a different approach to micro-storm control. Low- and medium-density areas may have enough land to utilize some forms of enhanced infiltration without sacrificing a significant portion of the utility of the land. As density increases, the value of the land tends to increase, and infiltration systems become less attractive. In purely urban lands with little or no area available for infiltration, the only options available are physical storage.

In low- and medium-density residential, commercial, or industrial areas where a portion of the area is "green" in some form, there is more flexibility for the designer to integrate land use with hydrologic function and value runoff as part of the urban water budget. The following sections suggest some designs for onsite stormwater management on privately owned urban lands.

11.4.2.1 Storage-based Control.   These systems are characterized by physical storage of runoff with little or no volumetric losses by themselves, though small volumes of runoff stored at the parcel level may be used for lawn and garden watering during dry weather, thus providing runoff volume reduction through delayed infiltration. In dense urban areas,

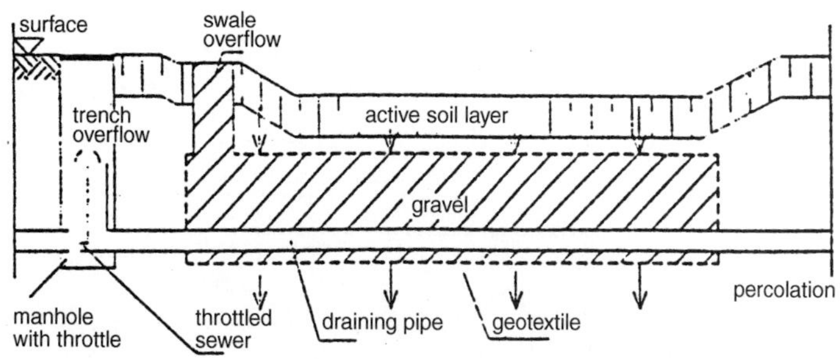

FIGURE 11.11   Longitudinal section of the M-R element. (Seiker, 1998)

infiltration may not be feasible due to the land required or the groundwater conditions, and physical storage may be the only option applicable. In this case, only the peak runoff rate would be reduced, thereby decreasing the demand for downstream centralized storage.

Prince Georges County design guidelines give several design strategies for physical storage at the residential lot level. The simplest storage is the rain barrel, used to collect roof runoff at the downspout exit. This water may then be used for lawn watering and gardening. On a larger scale, cisterns may be used to store runoff from roofs and driveways in underground storage tanks for later use.

From Australia, The Upper Parramatta River Catchment Trust published the "Onsite Stormwater Detention Handbook" (*http://www.uprct.nsw.gov.au*). This guidance manual was developed for the sole purpose of flood peak reduction, to alleviate the demand for downstream storage basins by investing in storage systems distributed throughout the watershed on individual private land. This guidance goes far beyond on-site micro-storm control, to recommend storage of runoff from 100-yr average recurrence interval events. This aggressive policy aims to maintain site runoff rates below 80 l/s/ha. Surface storage in landscaped areas as well as structural storage may satisfy design requirements for residential lots.

11.4.2.2   Infiltration Controls.   It has been recognized for some time that urban landscapes may be designed with hydrologic function in mind. Dunne and Leopold (1978) discuss the management aspects of small urban catchments: "The best solution to the urban runoff problem is to detain the stormwater in small volumes as near to the source as possible, and then to release it slowly to natural streams or to the groundwater system" (p. 414).

Prince Georges County, MD has used this idea to develop several manuals on *low-impact development* (LID). LID methods utilize a suite of BMPs designed systematically to reduce hydraulic and pollutant loads and retain pre-development hydrology (Prince Georges County, 2000). The hydrologic basis of LID as defined by Prince Georges County is the *SCS Curve Number (CN) hydrologic method.* By using appropriate BMPs, the CN of the developed land is designed close to the pre-development CN (Prince Georges County, 2000). LID systems may use many of the BMP designs presented in the following sections. Because of the system-view of site development presented, LID is probably the most aggressive BMP planning method used today in the United States.

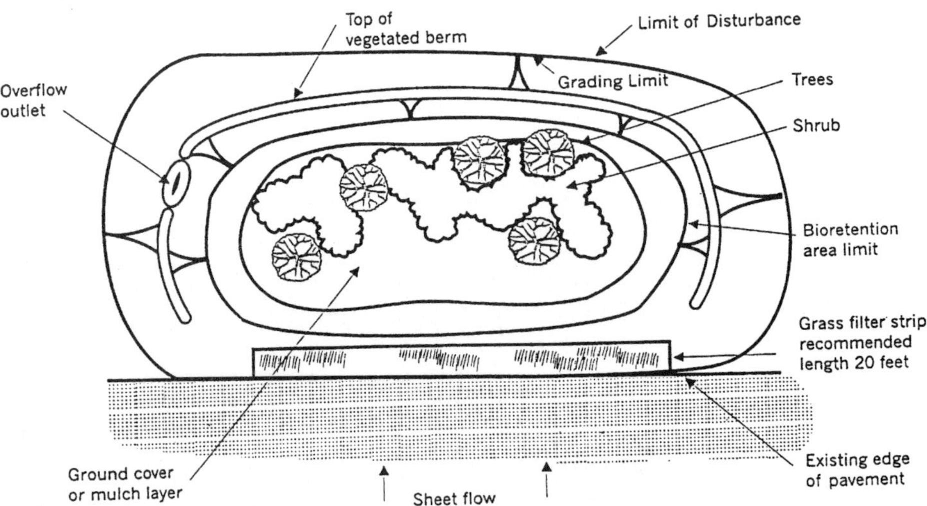

FIGURE 11.12   Plan view of a typical bioretention area. (Prince Georges County, 2000)

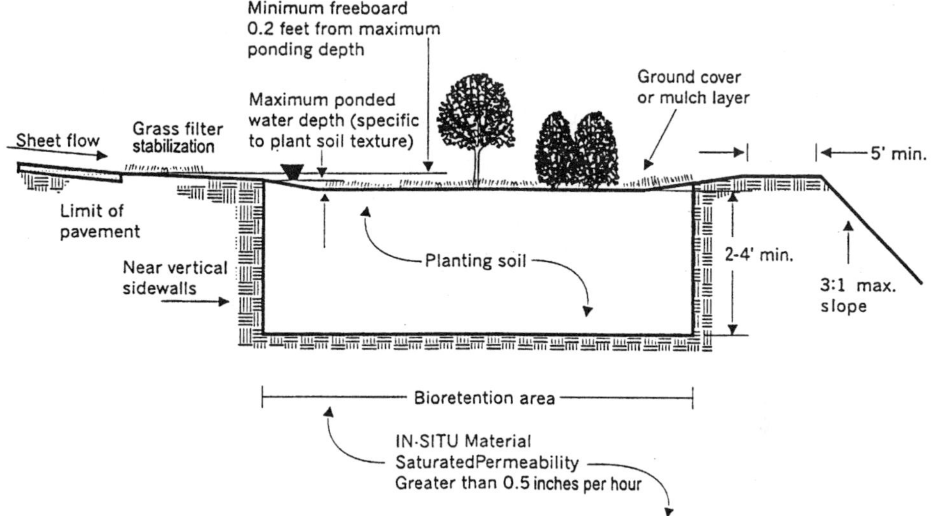

FIGURE 11.13 Example section view of a typical bioretention facility. (Prince Georges County, 2000)

Coffman (1993), Coffman et al. (1994), and Prince Georges County (2000) provide details of on-site stormwater infiltration techniques designed to retain water on permeable land. *Bioretention* is defined by the use of native plant species and soil conditioning to manage stormwater (Coffman et al., 1994). Bioretention areas typically receive water from impervious surfaces as shown in Fig. 11.12. These areas provide areas for on-site nutrient uptake as well as increasing the runoff travel time off the site (Coffman et al., 1994). Another example lot plan is shown in Fig. 11.13 with several small bioretention areas (Prince Georges County, 2000). Bioretention is a major part of the LID concept developed by Prince Georges County (2000).

Another design from Prince Georges County for enhanced infiltration is the dry-well. This design routes roof runoff to a sub-surface reservoir for subsequent infiltration to the surrounding soils. An example design is shown in Fig. 11.14. The success of this design is contingent on the groundwater level being below the reservoir. The performance of this type of design may vary with seasonal groundwater conditions as well. For example, if during spring high groundwater conditions inundate the bottom half of the storage reservoir, capacity is greatly reduced. If the groundwater level is lower during the summer, performance will improve. This information should be included in the urban water budget described in Sec. 11.2 if enhanced infiltration systems are to be used.

An important control for improved infiltration and decreased DCIA at the residential lot level is the use of permeable and/or porous media for driveway and parking areas. Due to the low traffic volume on these surfaces, porous and permeable pavement designs may be applicable and aesthetically pleasing. See Sec. 11.4.1 for more details on the types of surfaces.

11.4.2.3 Conveyance.   Generally there is limited formal conveyance needed at the residential lot level. Conveyance from the residential lot may be designed as a landscaped storage/infiltration bioretention zone with a designed overflow for larger flows. This combined hydrologic function is an important part of controlling micro-storms. One of the goals of micro-storm management is to increase flow paths and decrease conveyance velocities to promote infiltration. This runs counter to the objectives of traditional conveyance structures

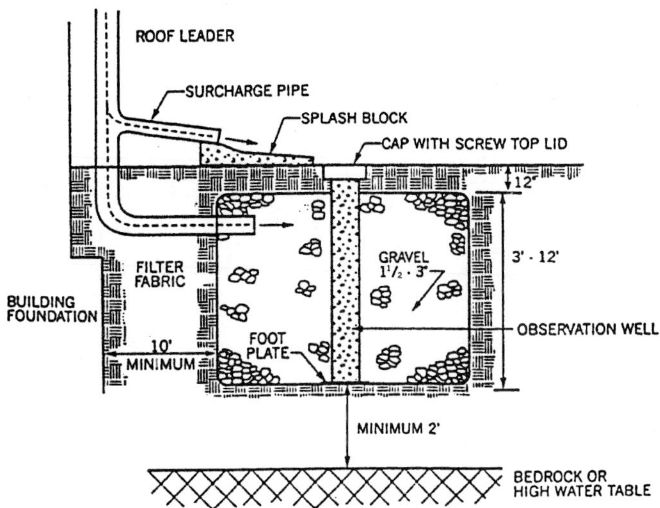

FIGURE 11.14   Example dry-well design. (Prince Georges County, 2000)

designed to move water quickly; therefore, conveyance at the parcel level is limited to minor and major event drainage.

More defined conveyance might be needed for larger commercial and industrial lands due to the size of drainage area. For these areas grass swales or combined conveyance designs described in Sec. 11.4.1 may be very suitable for conveying runoff offsite.

11.4.3 Summary of Distributed Stormwater Micro-storm Controls

The array of feasible alternative designs is growing larger every year, as this is an active research area. The designer may easily be overwhelmed when comparing the broad array of alternatives available compared to traditional drainage conventions. With the likelihood of many new ideas available to the designer in the near future, a summary comparison of micro-storm management versus conventional drainage is given in Table 11.12 to help the designer judge the applicability of future designs.

11.4.4 Discussion on Control Performance

The challenge facing stormwater designers is to integrate the principle of keeping runoff as close as possible to the source area (e.g., LID designs) while maintaining the functionality of the site provided by traditional site drainage. For certain site conditions, these may be conflicting goals. With careful planning and design, a controlled volume of runoff may be maintained close to the site while draining excess runoff that would inhibit site functionality in an unacceptable manner. This is a far more difficult design problem and the design is strongly determined by site conditions. Furthermore, the risk of failure due to poor maintenance of BMPs is troublesome (Urbonas, 2000). Designs should be carefully planned with either little or with no maintenance required or the maintenance responsibility clearly planned.

Roesner (1999) identifies five hindrances to an integrated approach to stormwater management that address the wide spectrum of event frequencies described in this chapter.

TABLE 11.12 Comparison of Conventional and Integrated Micro-storm Controls

Hydrologic attribute	Conventional drainage	Goals of integrated stormwater management
	Onsite	
Impervious cover	Encouraged to achieve effective drainage	Minimize to reduce impacts
Vegetative/natural cover	Reduced to improve efficient site drainage	Maximize to maintain predevelopment hydrology
Time of concentration	Shortened, reduced as a by-product of drainage efficiency	Maximize to approximate predevelopment conditions
Runoff volume	Large increases in runoff volume not controlled	Control to predevelopment conditions
Peak discharge	Controlled to a pre-development design storm	Control to predevelopment conditions for all storms
Runoff frequency	Greatly increased, especially for small frequent storms	Control to predevelopment conditions
Runoff duration	Increased for all storms, because volume is not controlled	Control to predevelopment conditions
Rainfall abstraction (interception, infiltration, depression storage)	Large reduction in all areas of rainfall abstraction	Maintained to predevelopment conditions
Groundwater recharge		
	Offsite	
Water quality	Reduction in pollutant loads for some controls, though limited performance on runoff rates less than design level	Improved pollutant loading reductions, full control of micro storms
Receiving waters	Channel erosion, sediment deposition, reduced base flow, impaired habitat	Maintain stream ecology to predevelopment conditions
Downstream flooding	Peak discharge controlled immediately downstream, though exacerbated flooding further downstream due to hydrograph superposition and cumulative impacts	Controlled to predevelopment conditions

Source: Modified from Prince Georges County, 2000.

1. Is the land required to accommodate enhanced infiltration, distributed storage, and surface conveyance systems acceptable and cost-effective?

2. Who is responsible for maintenance?

3. Who accepts liability for failure?

4. What is the fate of contaminated sediments?

5. Will the long-term performance of these systems compare well with conventional systems?

There is no question that some cost will be associated with the additional design criteria suggested by micro-storm management, and this cost is frequently in the form of land. Prince Georges County (2000), Wenk (1999), and Fujita (1999) have shown that a small-scale design can be an amenity to the land rather than a liability, thereby reducing the negative effect of the land cost. The reduced demand for downstream infrastructure, as well as downstream benefits in the form of improved water quality and reduced flooding potential should also be considered.

Maintenance may be a long-term problem and performance will suffer if not considered at the design stage. Institutional arrangements must be made for ROW maintenance, and controls on private land require a level of understanding and stewardship by the land owner. This must be planned at the design stage, not as an afterthought to the management plan.

Liability is one of the driving forces for providing drainage in the traditional manner. If water is conveyed quickly off-site, the land owner has little reason to come back to the designer for a failed site drainage plan. Unfortunately this does not consider the downstream costs associated with this philosophy. By collectively transferring the problem downstream, the problems of developed lands are enhanced without regard to liability for downstream costs. Progressive designers must face this challenge, and design redundant, integrated systems to prevent damages to nuisance flooding.

Another concern regarding infiltration-based BMPs is groundwater contamination (Pitt et al., 1996; Forester, 1996a, b; Martinelli and Alfakih, 1998). Forester (1996a, 1996b) reports on metal roofs contributing to the long-term buildup of metals in the soils where stormwater is disposed. Some biological pollutants (e.g., microbes, nutrients) may be treated in the vadose zone, and therefore do not reach the saturated zone (Ferguson and Debo, 1990). Pitt et al., "Groundwater Contamination from Stormwater Infiltration" (1996), provides a thorough treatment of the subject. Pitt et al. (1996) concludes that runoff from residential areas tends to be the least polluted, and is not likely to require treatment.

Under traditional drainage schemes, contaminated sediments are transferred directly to a receiving water. One of the advantages of micro-storm control is that many of the small events will not produce runoff. However, for systems serving high volumes of traffic and other contaminated areas, the contaminated sediment load may be a difficult problem that must be addressed in the maintenance plan at the design stage.

The long-term performance of these systems is still largely a research question. The recent volume "Urban Runoff Quality Management" (ASCE/WEF, 1998) discusses problems that may be associated with infiltration systems. In general these controls require very specific site requirements, including soil, vegetation, slope, etc. Absent these conditions, performance will likely suffer. Attempts have been made to develop simulation and planning tools to assist in BMP site development.

Mehler and Ostrowski (1998) argue that to fully exploit the advantages of local BMPs, they must be integrated within the larger, traditional regional drainage and flood control systems. However, this requires a large and speculative expenditure. To overcome this obstacle, Mehler and Ostrowski (1998) used long-term simulation to estimate the pollutant removal from various BMPs in a combined sewer system in Germany. Guther et al. (1997) used continuous simulation to test the effectiveness of stormwater management practices on maintaining baseflow in receiving waters. They developed an HSPF model of a future urban watershed under various conditions to estimate the effectiveness of various controls, including source controls. Based on this simulation, they concluded that for the City of Hamilton, ON, the preferred management strategy included infiltrating roof runoff and treating road and parking lot runoff using end-of-pipe methods (Guther et al., 1997). Tveit et al. (1996) used scenario simulation to estimate the reduction in needed future capacity extensions for the Munich-East Combined Sewer System (CSS) if infiltration systems were developed. Tveit et al. (1996) estimate a 20% decrease in runoff volume from East Munich if all residential roofs were disconnected by the year 2027.

One of the few post-audits done for the costs of onsite control is given by Liptan (1994). Two case studies are examined from the Portland, OR area and one from Davis, CA: a municipal parking lot, a light industrial site, and a residential community. This is one of the few studies that have compared the costs of on-site controls with a typical stormwater drainage design. The conventional wisdom is that on-site water quality controls will increase total project costs (Liptan, 1994). The three case studies prove otherwise. Due to the effects that each design had on the reduced quantity of stormwater to be conveyed off-site, significant project costs were saved. The key concept is that the water quality improvements are made by reducing the quantity of stormwater conveyed offsite. This is true of all BMPs based on infiltration.

The first case study is a municipal parking lot for the Oregon Museum of Science and Industry in Portland. This 20-acre site with 50% imperviousness receives 1.1 million visitors annually (Liptan, 1994). Approximately $78,000 was saved by using bioswales (a hybrid swale and biorention area) in place of manholes, piping, trenches, and catch basins (Liptan, 1994). The total project costs were not given. Liptan (1994) noted that maintenance requirements were the same as for other landscaped areas.

The second case study from Liptan (1994) is a light industrial area in north Portland, OR. A conventional design that was already constructed was used for a "what-if?" analysis. The design was modified to include on-site quantity/quality controls. The BMP-based design would have reduced the piping required by 80%, and reduced the number of catch basins needed from 8 to 2 (Liptan, 1994). It was estimated that approximately $10,000 would have been saved over the conventional design.

The third case study is the residential community of Village Homes in Davis, CA. Village Homes is a 70-acre development, of which 20 acres are impervious. It was built in 1981 as "the world's first solar housing development," and was designed to conserve resources and provide a greater sense of community amongst the residents. "According to the developer, the use of natural drainages, narrow tree-lined streets and cul-de-sacs created a stormwater system construction cost savings of $800 per unit, for a total of $192,000 for 240 units" (Liptan, 1994). These examples by Liptan (1994) are important in that they provide evidence of real savings in infrastructure requirements when integrated stormwater management is used. The Village Homes experience is also well-documented in Corbett and Corbett (2000).

A recent database of nationwide BMP performance has been published by the ASCE (1999). One of the difficulties in establishing design guidelines for BMPs is the lack of reliable performance data. This database demonstrates this dearth of information. The database contains 240 BMP sites from around the United States. To be included the site must have had adequate flow measurement and water quality sampling.

The focus of most BMP performance measurement has been water quality. Several sources recognize the potential benefits on-site control may have on the quantity of runoff; however, few have attempted to quantify these hydrologic impacts. While Liptan's (1994) work provided good cost data, it did not quantify flow reductions. The sole reference may be the Prince Georges County (2000) guidance manual; however, this is only used to estimate hydrologic performance for design purposes rather than report *in situ* measurements. This lack of follow-up work is very common throughout the field of water resources management. White (1998), in a paper for the World Water Council, states, "It is a sad fact that the number of truly incisive and comprehensive post-audits of completed water management efforts is very small." He goes on to state, "They [reports on water management projects] focus on direct inputs and outputs to projects rather than on inter-related outcomes of the type embedded in the notion of sustainable development. Only rarely do they indicate such aspects as the effects on quality of life or stability of environmental systems" (White, 1998).

11.5 INTEGRATION OF MICRO-COMPONENTS WITH MINOR AND MAJOR DRAINAGE INFRASTRUCTURE

As discussed in previous sections, the objectives of a stormwater management plan should be to

- Protect human health
- Protect property from excessive and frequent damages
- Protect and sustain hydrologic, aesthetic and functional resources

To meet these objectives, a stormwater management plan should be viewed within the context of the long-term urban water budget while meeting flood control and water quality objectives. After identifying urban water budget items, land use decisions are made, and feasible on-site controls are identified, the last step in developing a sustainable, integrated stormwater management plan is to coordinate upland, micro-storm controls with traditional minor and major infrastructure.

One of the themes of stormwater management integration is to properly account for all upstream controls. For example, an on-site control designed to infiltrate micro-storms will directly affect the downstream runoff hydrograph. Likewise, if there is a significantly smaller proportion of DCIA in an urban watershed than in a traditionally designed urban area, downstream minor and major infrastructure designs should reflect some savings (Wright, 1999). The integrated stormwater management plan may be seen as a series of cascading controls with each control impacting the sizing and design decisions of all subsequent downstream controls. When sizing downstream facilities, true upstream conditions should be used to calculate capacities.

Four primary questions need to be answered by simulating the integrated system:

1. How should the investments in controls be distributed amongst the micro, minor and major infrastructure?
2. How much collective storage and infiltration should the micro-storm controls provide?
3. What are the downstream sizing requirements for minor and major infrastructure given the hydrologic influence of upstream controls?
4. What are the impacts of the developed area on downstream receiving waters?

The design strategy will vary depending on the relation of the new infrastructure with the rest of the watershed. For a new upland development, much of the investment in infrastructure may be in on-site, micro-storm controls with less minor and major infrastructure than a traditional design would need. For a newly urbanized area with significant developed area upstream draining through the design area, more emphasis may be placed on minor and major controls. This section addresses areas of GIS and modeling required to account for the downstream effects of upstream stormwater controls.

11.5.1 Data Needed to Design an Integrated Stormwater Management Plan

As described in Sec. 11.2, design emphasis should be on the total urban water budget within the context of specific detailed land use decisions. To perform the *water budget accounting,* several data sets are needed:

1. Spatial information of the existing land, including topography, hydrography and soils
2. Time series meteorological information, including precipitation, temperature, wind, etc.
3. Surrounding land-use information, including land that drains onto the developing area and the land the developing area drains to, if not directly to a receiving water
4. Ultimate receiving water information, including a time series of flow and water quality parameters if available

To account for water budget items at the design stage, a detailed GIS is useful to evaluate different land design scenarios and water demands. A key factor in the land use analysis is the transportation needs of the community, due to the strong link between transportation surfaces, DCIA, NPS, and receiving water health. For example, some areas have designed narrow streets for a wide variety of aesthetic and community safety reasons. This may have a strong positive influence on the amount of DCIA in an urban watershed. These integrated factors should be included in the water budget analysis of land use scenario building.

Detailed meteorological data are needed for a water budget that includes evapotranspiration, outdoor water use, and precipitation. Once the primary layout of land uses is complete, detailed process modeling may be used to refine the landscape design and to begin to design specific micro-storm controls described in Sec. 11.4. Land-use scenarios including on-site micro-storm controls may then be used to develop input data sets for the design modeling required for minor and major storm drainage infrastructure.

Sample et al. (2000) used a GIS to develop an integrated stormwater design for a hypothetical watershed. GIS is an effective tool to isolate the various scales needed to address a multiple storm frequency design. For example, a low-density section of a developing area may be laid out in relation to surrounding land uses. A more detailed view of the area is developed to evaluate the attributes of the low-density area as a whole, for example, imperviousness, soils, slopes, etc. A suite of feasible controls may be used to develop the on-site characteristics, that is, biorentention, soil amendments, permeable driveway, etc. The GIS is then used to develop the collective hydrologic characteristics of the proposed land-use design.

11.5.2 Modeling and Decision Support Tools

The hydrologic and hydraulic modeling problems encountered for designing micro-storm controls are far more complex than a standard stormwater drainage plan based on large infrequent events. Not only are the controls distributed throughout the watershed, but also the performance of the controls is dependent on the antecedent conditions of the watershed. Specifically, the soil conditions prior to rainfall are extremely important for storage-based systems. Dry weather soil moisture accounting is important to estimate the amount of soil storage available in bioretention areas. A further complication is lawn watering. This may drastically affect the antecedent soil moisture conditions.

To circumvent these problems, the most common approach to simulate design conditions is to use a physically based empirical model such as the USDA NRCS (SCS) methods. The SCS method is a popular design tool because it is sensitive to site-specific conditions, and a great deal of data (generally based on soils, slope and land use) are available. Various surface conditions may be simulated and paired with cost information for further consideration. Methods presented in Prince Georges County (2000), Wright (1999), Wright et al. (2000), and Sample et al. (2000) make use of SCS-based ideas to design distributed micro-storm control systems.

The large amount of soils data collected for the SCS method makes it particularly attractive from a soil storage standpoint, especially for planning purposes when the only available site-specific data may be soils and slope. Heaney et al. (1999b) use the depth of initial abstraction (I_a) for undeveloped soils as the management goal of local stormwater management.

Due to its widespread use and available soils database, the Soil Conservation Service Curve Number method is well-suited for initial screening of micro-storm controls. For example, using changes in the catchment curve number resulting from innovative practices may be appropriate. However, CN methods should not be used to calculate runoff hydrographs from micro-storms, due to high variability of the CN with storm size. Tabulated CNs may be inappropriate for micro-storm estimation.

For more detailed process modeling, those methods that explicitly simulate storage seem to be the most applicable to modeling distributed micro-storm systems. The most important requirement when using these methods is obtaining reasonable estimates of storage parameters. The USEPA SWMM RUNOFF model uses a nonlinear storage element to simulate a runoff hydrograph. SWMM can also be used to simulate hydraulic routing in pipes, open channels, and storage ponds. Integration of the LID land surface modifications may then be combined with traditional drainage structures, resulting in a more complete process model.

Difficulties may arise when modeling distributed processes systems with SWMM. One important problem arises when attempting to simulate indirectly connected impervious surfaces. These impervious surfaces drain onto permeable ground. The U.S. EPA version of SWMM does not currently have the capability to model "run-on" directly as a physical process, and therefore the modeler must resort to modeling "tricks" such as altering the rain record or effective areas to approximate the resulting runoff response. Since run-on may be an important part of a distributed micro-storm control system, the simulation of run-on should be a research need for integrated simulation.

Wright and Heaney (2000) made the following recommendations regarding modeling for micro-storm management:

1. *Micro-storm management* is concerned primarily with estimating storage, not flow, processes that dominate current stormwater modeling. The standard problem of stormwater modeling is to estimate downstream hydrographs, typically from a single, large precipitation event. Thus, channel and reservoir routing are the key elements in traditional stormwater models that simulate large storm events.

2. The primary measure of the performance of the micro-management systems is the percent of the total annual or seasonal runoff that was controlled by the system. This measure is distinctly different than the traditional measure of performance in terms of attenuation of hydrograph peaks.

3. Given that a continuous water budget is essential, irrigation needs to be modeled as part of the simulation process since it is the corollary problem to minimizing urban stormwater runoff.

4. Stormwater management is not the primary purpose of landscape design and operation. Thus, it is vital to understand the overall landscaping process and to better integrate stormwater management into this process.

A recent example of extremely fine spatial resolution modeling was performed in Portland, OR on three sub-catchments in a combined sewer system (Hoffman, 2000). Using a GIS, detailed sub-parcel level spatial information was used to develop a hydrologic simulation model using SWMM, as demonstrated in Fig. 11.15. Historically the level of detail used in this study was unthinkable due to the data handling requirements and the computational expense needed to simulate a detailed, distributed system. With the combination of advanced database techniques, GIS, SWMM, and fast computing capability, this hindrance has been breached. While not a simple or cheap exercise, it does show that a simulation model of all pipes in a combined system, *including laterals,* may be done.

For integrated stormwater management, this study shows that it is technically feasible to model distributed parcel level processes using integrated databases and simulation models. While this level of detail is not required to design these systems, it does point to where simulation and scenario analysis is headed in the near future. The goal of estimating the performance of upstream designed conditions including distributed controls for sizing downstream infrastructure is now feasible, albeit at a considerable effort.

11.5.3 Summary of Integrated Stormwater Management Design Methods

The previous sections have presented a wide array of possible methods of mitigating the long term distortions to the hydrologic cycle that result from urbanization. The theme of the chapter is to integrate distributed, small-scale controls designs to control on small runoff events, with larger infrastructure designed to handle larger infrequent events. The ideas involved with designing minor and major infrastructure are covered elsewhere in this volume.

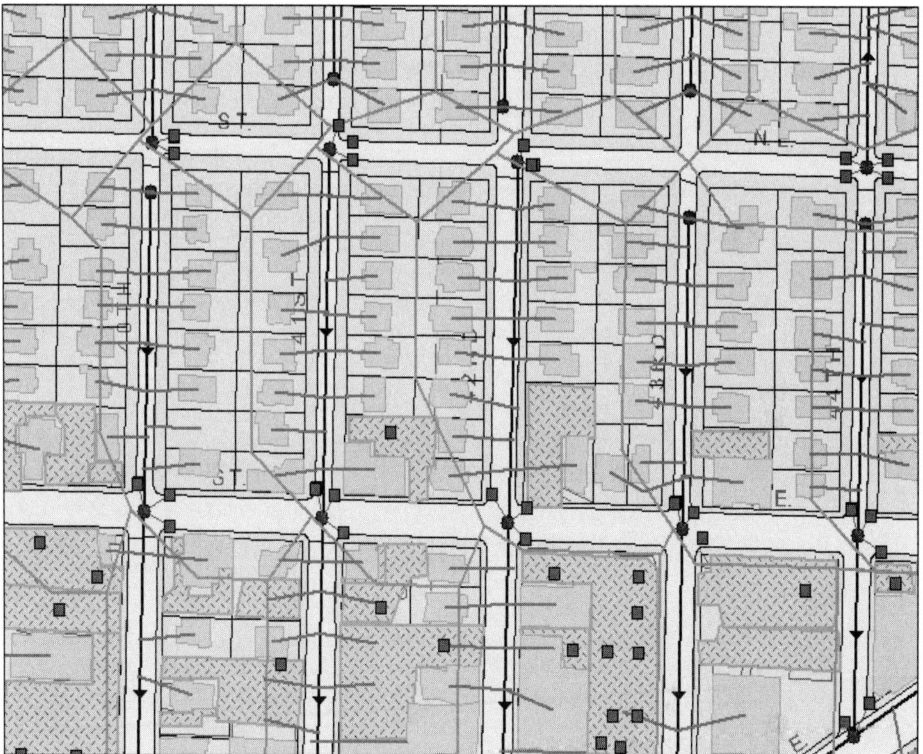

FIGURE 11.15 Example level of spatial discretization, Portland OR. (Hoffman 2000)

The ideas presented suggest the following strategy for designing a more complete, sustainable system.

1. First, identify the social needs of the planned development. Take into account the surrounding lands and cultural aspects of transportation, housing, and environmental valuation.

2. Identify land-use scenarios on the catchment scale using a GIS with underlying databases of topography, soils, and hydrography. For each land use scenario, identify key components of the urban water budget, including expected indoor and outdoor water use, precipitation, infiltration, runoff, etc. Pay particular attention to the expected transportation needs and identify components of the water budget amenable to micro-storm control, for example, DCIA.

3. Using the GIS as a database, perform a hydrologic analysis of various land-use scenarios. Include in this analysis land-use scenarios with micro-storm control. The hydrologic analysis may be based on the lumped effect of micro-storm controls using SCS-based methods, or with more detailed process modeling tool such as SWMM. SWMM may also be used as a lumped model, depending on the level of spatial discretization. The amount of feasible upstream control will depend greatly upon site and climate conditions. The value of micro-storms will also vary depending on the climate and expected outdoor water use.

4. An array of land uses and the hydrologic output of scenarios with upstream control will result. Use this information to design and size downstream minor and major infrastructure.

The integrated design should minimize DCIA, have longer times of concentration, include adequate site drainage and infiltrate runoff where feasible, as well as provide drainage and protection expected from the minor and major events.

5. The small scale designs should be completed using the wide array of techniques described in Sec. 11.4. Porous and permeable pavements, bioretention, and minimal DCIA should be planned at this stage. Fairly detailed site plans should also be feasible at this point.

6. Institutional arrangement should be made at the planning and design stage to satisfy expected maintenance requirements. Education for privately owned controls should also be expected if the systems are to remain in place and operational for an extended period of time.

7. The final design should show a spatially distributed system, reducing runoff from small storms and retaining runoff on-site where possible, while providing traditional drainage protection from major and minor flooding.

REFERENCES

American Association of State Highway Transportation Officials (AASHTO), *A Policy on Geometric Design of Highways and Streets,* Washington, DC, 1984.

American Public Works Association (APWA), *The Causes and Remedies of Water Pollution From Surface Drainage in Urban Areas-Research,* Project No. 120, Final Report to Federal Water Pollution Control Administration-WP-20-15, Washington, DC, 1968.

American Society of Civil Engineers (ASCE), *Design and Construction of Urban Stormwater Management Systems,* No. 77, WEF FD-20, New York, NY, 1992.

American Society of Civil Engineers (ASCE) Task Committee on Sustainability Criteria, *Sustainability Criteria for Water Resource Systems,* New York, NY, 1998.

American Society of Civil Engineers (ASCE) Urban Water Resources Research Council, and Wright Water Engineers, Inc., *National Stormwater Best Management Practices (BMP) Database, User's Guide,* New York, NY, 1999.

American Society of Civil Engineers (ASCE)/Water Environment Federation (WEF), *Urban Stormwater Quality: Planning, Management and BMP Design,* Manual of Practice, New York, NY, 1998.

Arnold, Jr., C. C., and C. J. Gibbons, "Impervious Surface Cover, The Emergence of a Key Environmental Indicator." *Journal of the American Planning Association,* Vol. 62, No. 2, pp. 242–258, 1996.

Barnette, C. H., "Revitalizing Brownfield Sites," *Iron Age New Steel,* 1:6 at 88, 1995.

Beck, M. B., J. Chen, A. J. Saul, and D. Butler. Urban Drainage in the 21st Century: Assessment of New Technology on the Bbasis of Global Material Flows, *Water, Science, & Technology,* 30(2):1–12, 1994.

Boulder, CO., Design and Construction Standards. Department of Building Services, City of Boulder, CO. *http://www.ci.boulder.co.us/buildingservices/dcs/index.htm.* November 16th, 2000.

Boulder, CO. *Design Criteria and Standard Specifications,* 11/9, 1982.

Burges, S. J., M. S. Wigmosta, and J. M. Meena, "Hydrological Effects of Land-Use Change in a Zero-order Catchment," *Journal of Hydrologic Engineering,* 3(2):86–97, 1998.

Butler, D., and C. Maksimovic, eds., *UDM '98 Fourth International Conference on Developments in Urban Drainage Modelling,* Vol. 2, Imperial College of Science, Technology & Medicine, London, UK, 1998.

Butler, D., and J. Parkinson, "Towards Sustainable Urban Drainage," *Water, Science, & Technology,* 35(9):53–63, 1997.

Carr, R. W., C. Esposito, and S. G. Walesh, "Street Surface Storage for Control of Combined Sewer Surcharge: Case Studies," *Proceedings of the 26th Annual ASCE Water Resources Planning & Management Conference,* Tempe, AZ, American Society of Civil Engineers, Reston VA, 1999.

Candaras, A. M., L. Carvalho, and M. K. Koo, "City of Etobicoke Exfiltration and Filtration Systems Pilot/Demonstration Project," in W. James, ed., *Modern Methods for Modeling the Management of Stormwater Impacts,* CHI, Guelph, ON, pp. 399–419, 1995.

Chellman, C. E., *Traditional Neighborhood Development Street Design Guidelines.* Institute of Transportation Engineers, Washington, DC, 1997.

Chesapeake Bay Foundation, *Growth, Sprawl and the Bay-Simple Facts About Growth and Land Use,* Chesapeake Bay Foundation, September 1996.

Chow, V. T., D. R. Maidement, and L. W. Mays, *Applied Hydrology,* McGraw-Hill Inc., New York, NY, 1988.

Clark, R., A. Perkins, and S. E. Wood, *Water Sustainability in Urban Areas—An Adelaide and Regions Case Study, Report One—An Exploration of the Concept,* Department of Environmental and Natural Resources, Adelaide, Australia, Draft, 1997.

Coffman, L. S., "Bioretention, an Innovative Urban Stormwater Treatment Option," *Proceedings of the 3rd Biennial Stormwater Research Conference,* Tampa, FL, Southwest Florida Water Management District, Brooksville, FL, pp. 210–218, 1993.

Coffman, L., R. Green, M. Clar, and S. Bitter, "Development of Bioretention Practices for Stormwater Management," in W. James, ed., *Current Practices in Modelling the Management of Stormwater Impacts,* CHI, Guelph, ON, pp. 23–42, 1994.

Corbett, J., and M. Corbett, *Designing Sustainable Communities—Learning from Village Homes,* Island Press, Washington, DC, 2000.

Diniz, E. V., "Hydrologic and Water Quality Comparisons of Runoff from Porous and Conventional Pavements," in R. Field, M. L. O'Shea, and K. K. Chin, eds., *Integrated Stormwater Management,* Lewis Publishers, Boca Raton, FL, pp. 149–162, 1993.

Dittmar, H., "A Broader Context for Transportation Planning," *Journal of the American Planning Association,* 61(1):7–13, 1995.

Dunne, T., and L. B. Leopold, *Water in Environmental Planning,* W. H. Freeman and Company, New York, NY, 1978.

Ellis, J. B., "Integrated Approaches for Achieving Sustainable Development of Urban Storm Drainage," *Water, Science, & Technology,* 32(1):1–6, 1995.

Ferguson, B. K., *Stormwater Infiltration,* Lewis Publishers, Boca Raton, FL, 1994.

Ferguson, B. K., *Introduction to Stormwater,* John Wiley & Sons, Inc., New York, NY, 1998.

Ferguson, B. K., and T. Debo, *On-Site Stormwater Management, Applications for Landscape and Engineering,* 2nd ed., Van Nostrand Reinhold, New York, NY, 1990.

Forster, J., "Heavy Metal and Ion Pollution Patterns in Roof Runoff, in F. Sieker and H. R. Verworn, eds., *Proceedings of the 7th International Conference on Urban Storm Drainage,* Hannover, Germany, IAHR/IAWQ Joint Committee Urban Storm Drainage, pp. 241–246, 1996a.

Forster, J., "Patterns of Roof Runoff Contamination and Their Potential Implications on Practice and Regulation of Treatment and Local Infiltration," *Water Science and Technology,* 33(6), 1996b.

Fujita, S., "Restoration of Polluted Urban Watercourses in Tokyo for Community Use, in A. C. Rowney, P. Stahre, and L. Roesner, eds., *Sustaining Urban Water Resources in the 21st Century, Proceedings of an Engineering Foundation Conference,* September 1997, Malmo, Sweden, 1999.

Geiger, W. F., "Concepts for Flood Control in Highly Urbanized Areas," in R. Field, M. L. O'Shea, and K. K. Chin, eds., *Integrated Stormwater Management,* Lewis Publishers, Boca Raton, FL, 1993.

Goldstein, H., "A City of Dreams," *Colorado Commons,* Spring, p. 42–44, 1997.

Grigg, N. S., *Water Resources Management,* McGraw-Hill, New York, NY, 1996.

Guo, J., and B. Urbonas, "Maximized Detention Volume Determined by Runoff Capture Ratio," *Journal of Water Resources Planning and Management,* 122(1):33–39, 1996.

Guther, R. T., R. B. Scheckenberger, and W. R. Blackport, "Use of Continuous Simulation for Evaluation of Stormwater Management Practices to Maintain Base Flow and Control Erosion," in W. James, ed., *Advances in Modeling the Management of Stormwater Impacts,* CHI, Guelph, ON, pp. 77–100, 1997.

Hanson, M., "Automobile Subsidies And Land Use," *Journal of the American Planning Association,* 58:1, 1992.

Harza Engineering Company and Bauer Engineering Company, Incorporated, *Flood and Pollution Control: A Deep Tunnel Plan for the Chicago Metropolitan Area,* 1966.

Heaney, J., W. Huber, and S. Nix, *Storm Water Management Model: Level I-Preliminary Screening Procedures,* U.S. Environmental Protection Agency. EPA–600/2-76-275, Cincinnati, OH, 1977.

Heaney, J. P., L. Wright, D. Sample, R. Pitt, R. Field, and C.-Y. Fan, "Innovative Wet-Weather Flow Collection/Control/Treatment Systems for Newly Urbanizing Areas in the 21st Century," in C. A. Rowney, P. Stahre, and L. A. Roesner, eds., *Sustaining Urban Water Resources in the 21st Century, Proceedings of an Engineering Foundation Conference,* Malmo, Sweden, September 7–12, 1997, American Society of Civil Engineers, Reston, VA, 1999a.

Heaney, J. P., L. Wright, and D. Sample, *Innovative Methods for Optimization of Urban Stormwater Systems,* Final Report to the U.S. Environmental Protection Agency, Edison, NJ, 1999b.

Heaney, J. P., "Principles of Integrated Urban Water Management," chapter 2, in R. Field, J. P. Heaney, and R. Pitt, *Innovative Urban Wet-Weather Flow Management Systems,* Technomics Press, Lancaster, PA, 2000.

Heaney, J. P., L. Wright, and D. Sample, "Sustainable Urban Water Management," chapter 3, in R. Field, J. P. Heaney, and R. Pitt, eds., *Innovative Urban Wet-Weather Flow Management Systems,* Technomic Publishing Co., Lancaster, PA, 2000a.

Heaney, J. P., L. Wright, and D. Sample, "Stormwater Storage-Treatment-Reuse Systems," chapter 8, in R. Field, J. P. Heaney, and R. Pitt, eds., *Innovative Urban Wet-Weather Flow Management Systems,* Technomic Publishing Co., Lancaster, PA, 2000b.

Heaney, J. P., L. Wright, and D. Sample, "Urban Stormwater and Watershed Management," chapter 9, in R. Field, J. P. Heaney, and R. Pitt, eds., *Innovative Urban Wet-Weather Flow Management Systems,* Technomic Publishing Co., Lancaster, PA, 2000c.

Heaney, J. P., L. T. Wright, and N. Weinstein, "Modeling of Low Impact Development Stormwater Practices," *ASCE Conference on Water Resources Engineering and Water Resources Planning and Management,* Minneapolis, American Society of Civil Engineering, New York, NY, 2000d.

Herrmann, T., and U. Klaus, "Fluxes of Nutrients in Urban Drainage Systems: Assessment of Sources, Pathways and Treatment Technologies," *Proceedings of the 7th International Conference on Urban Storm Drainage,* Hannover, Germany, pp. 761–766, 1996.

Hoffman, J., *Detailed Documentation of Subcatchment Definitions and Process for Creating Runoff Deck,* Appendix 1, Holliday, Stark and Sullivan Basin Characteristic Report, City of Portland, OR, 2000.

Hollis, G. E., "The Effect of Urbanization on Floods of Different Recurrence Interval," *Water Resources Research* 11(3):431–435, 1975.

Hollis, G. E., "Rain, Roads, Roofs and Runoff: Hydrology in Cities," *Geography,* 73(1):9–18, 1989.

Imbe, M., T. Ohta, and N. Takano, "Quantitative Assessment of Improvement in Hydrological Water Cycle in Urbanized River Basin," *Proceedings of the 7th International Conference on Urban Storm Drainage,* Hannover, Germany, pp. 1085–1090, 1996.

Institute of Transportation Engineers, *Parking Generation,* 2nd ed., ITE, Washington, DC, 1987.

Jones, D. E., "Some Urban Water Resource Management Dimensions," in *Urban Water Resources Management,* Report on the Third Conference on Urban Water Resources Research, American Society of Civil Engineers, Reston, VA, pp. 14–31, 1971.

Katz, B., *Brookings National Issues Forum on Metropolitan Solutions to Urban and Regional Problems,* www.smartgrowth.org/library/katz.html, 1997.

Khisty, C. J., *Transportation Engineering,* Prentice-Hall, Inc., Englewood Cliffs, NJ, 1990.

Kunstler, J. H., *Home From Nowhere,* Simon and Schuster, New York, NY, 1996.

Leopold, L. B., *Hydrology for Urban Land Planning—A Guidebook on the Hydrologic Effects of Urban Land Use,* USGS Water Supply Paper 1591-C, 60 pp., 1968.

Liptan, T., "Stormwater Design: A Cost Comparison of Conventional and Water Quality-based Designs," in *Proceedings of the 4th Biennial Stormwater Research Conference,* Southwest Florida Water Management District, Brooksville, FL, pp. 222–231, 1994.

Martinelli, I., and E. Alfakih, "Inclusive Modelling Methodology for Urban Stormwater Infiltration and Associated Pollutant Transfer, in D. Butler and C. Maksimovic, eds., *UDM '98 Fourth International Conference on Developments in Urban Drainage Modelling,* Vol. 2, Imperial College of Science, Technology & Medicine, London, UK, 1998.

McCuen, R. H., *Hydrologic Analysis and Design,* Prentice-Hall, Englewood Cliffs Inc., NJ, 1989.

McPherson, M. B., "Need for Metropolitan Water Balance Inventories," *Journal of the Hydraulics Division,* American Society of Civil Engineers, 99, HY10, pp. 1837–1848, 1973.

Mehler, R., and M. W. Ostrowski, "Comparison of the Efficiency of Best Stormwater Management Practices in Urban Drainage Systems," in D. Butler and C. Maksimovic, eds., *UDM '98 Fourth Inter-*

national Conference on Developments in Urban Drainage Modelling, Vol. 2, Imperial College of Science, Technology & Medicine, London, UK, pp. 561–568, 1998.

Mitchell, V. G., *Development of an Urban Water Balance Model to Assess the Re-use Potential of Stormwater and Wastewater,* Ph.D. dissertation, Department of Civil Engineering, Monash University, 273 pp., 1998.

Mitchell, V. G., R. G. Mein, and T. A. McMahon, "Evaluating the Resource Potential of Stormwater and Wastewater: An Australian Perspective," *Proceedings of the 7th International Conference on Urban Storm Drainage,* Hannover, Germany, pp. 1293–1298, 1996.

National Commission on the Environment, *Choosing a Sustainable Future,* Island Press, Washington DC, 180 p. 1993.

Peirce, N., *The Dawn Of Civic Environmentalism,* Tampa Tribune, January 10, 1994.

Peterson, F. L., and D. R. Hargis, "Subsurface Disposal of Storm Runoff," *Journal of the Water Pollution Control Federation,* 45(8):1663–1670, 1973.

Pitt, R. E., "Small Storm Hydrology and Why It Is Important for the Design of Stormwater Control Practices," in W. James, ed., *New Applications in Modeling Urban Water Systems,* Computational Hydraulics, Inc., Guelph, ON, pp. 61–91, 1999.

Pitt, R. E., S. Clark, K. Parmer, and R. Field, *Groundwater Contamination from Stormwater Infiltration,* Ann Arbor Press, Inc., Chelsea, MI, 1996.

Pitt, R. E., and J. Voorhees, *Source Loading and Management Model (SLAMM).* Department of Civil and Environmental Engineering, University of Alabama at Birmingham, 1994.

Pokrajac, D., and A. Deletic, "Hydraulic Behavior of a Stormwater Soakaway," *Hydroinform, International Conference on Urban Drainage via Internet (ICUDI), http://www.hydroinform.com/ICUDI/ pages/papers/deletic.htm,* 2000.

Pratt, C. J., "Design Guidelines or Porous/Permeable Pavements," in A. C. Rowney, P. Stahre, and L. Roesner, eds., *Sustaining Urban Water Resources in the 21st Century, Proceedings of an Engineering Foundation Conference,* Malmo, Sweden, 1997.

Prince Georges County, *Low-Impact Development Design Manual,* Department of Environmental Resources, Prince Georges County, MD, 2000.

Renner, M., *Rethinking the Role of the Automobile,* World Watch Institute, Washington, DC, 1988.

Roesner, L. A., "Urban Runoff Pollution—Summary Thoughts—The State of Practice Today and for the 21st Century," *Water Science & Technology,* 39(12):353–360, 1999.

Roesner, L. A., et al., "Hydrology of Urban Runoff Quality Management," *Proceedings 18th National Conf Water Resources Planning and Management,* Symposium Urban Water Resources, American Society of Civil Engineers, New York, NY, 1991.

Rowney, C., P. Stahre, and L. Roesner, eds., *Sustaining Urban Water Resources in the 21st Century, Proceedings of an Engineering Foundation Conference,* September 1997, Malmo, Sweden, American Society of Civil Engineers, Reston, VA, 1999.

Sample, D., J. Heaney, L. Wright, and R. Koustas, "Geographical Information Systems, Decision Support Systems, and Urban Stormwater Management," accepted for publication, *Journal of Water Resources Planning and Management,* 2000.

Schueler, T. R., *Site Planning for Urban Stream Protection,* Watershed Protection Techniques, 1(3):137–140, 1994.

Schueler, T. R., *Site Planning for Urban Stream Protection,* Metro. Washington Council of Governments, Washington, DC, 232 pp., 1995.

Seiker, F., "On-site Stormwater Management as an Alternative to Conventional Sewer Systems: A New Concept Spreading in Germany," *Water, Science, & Technology,* 38(10):65–71, 1998.

Shoup, D. C., "An Opportunity To Reduce Minimum Parking Requirements," *Journal American Planning Association,* 61(1):14–28, 1995.

Sieker, F., and H. R. Verworn, eds., *Proceedings of the Seventh Annual Conference on Urban Storm Drainage,* Hannover, Germany, Vols. I, II, and III, IAHR/IAWQ Joint Committee on Urban Storm Drainage, 1996.

Silverman, G. S., and M. L. Stenstrom, "Source Contol of Oil and Grease in an Urban Area," in L. A. Roesner, B. Urbonas, and M. B. Sonnen, eds., *Design of Urban Runoff Quality Controls, Proceedings of an Engineering Foundation Conference on Current Practices and Design Criteria for Urban Quality Control,* Potosi, MO, American Society of Civil Engineers, Reston, VA, pp. 403–420, 1988.

Sorvig, K., "Porous Paving," *Landscape Architecture,* 83(2):66–69, 1993.

Southworth, M., and E. Ben-Joseph, "Street Standards and the Shaping of Suburbia," *American Planning Association Journal,* 61(1):65–81, 1995.

Southworth, M., and E. Ben-Joseph, *Streets and the Shaping of Towns and Cities,* McGraw-Hill, New York, NY, 1997.

Stephenson, D., "Evaluation of Effects of Urbanization on Storm Runoff," *Proceedings of the 7th International Conference on Urban Storm Drainage,* Hannover, Germany, pp. 31–36, 1996.

Tveit, O. A., H. Zhu, D. S. Wirth, K. Gaffgen, and W. Schilling, "Storm Water Infiltration in Large Urban Areas as an Alternative to Capacity Extensions," in F. Sieker and H. R. Verworn, eds., *Proceedings of the Seventh Annual Conference on Urban Storm Drainage,* Hannover, Germany, Vol. I., IAHR/IAWQ Joint Committee on Urban Storm Drainage, pp. 383–394, 1996.

Upper Parramatta River Catchment Trust, *On-site Stormwater Detention Handbook,* Parramatta, New South Wales, Australia, *http://www.uprct.nsw.gov.au,* 2000.

Urban Land Institute, *The Community Builders Handbook,* Urban Land Institute, Washington, DC, 526 pp., 1968.

Urban Land Institute, *Parking Requirements for Shopping Centers,* Washington, DC, Urban Land Institute, 1982.

Urbonas, B., "BMP Planning and Management," chapter 7 in R. Field, J. P. Heaney, and R. Pitt, *Innovative Urban Wet-Weather Flow Management Systems,* Technomics Press, Lancaster, PA, 2000.

Urbonas, B. R., and L. A. Roesner, "Hydrologic Design for Urban Drainage and Flood Control," in D. R. Maidment, ed. *Handbook of Hydrology,* McGraw-Hill Inc., New York, NY, 1992.

Urbonas, B., and P. Stahre, *Stormwater, Best Management Practices and Detention for Water Quality, Drainage and CSO Management,* PTR Prentice-Hall, Englewood Cliffs, NJ, 1993.

Walesh, S., *Urban Surface Water Management,* John Wiley & Sons, Inc., New York, NY, 1989.

Walesh, S. G., and R. Carr, *Controlling Stormwater Close to the Source: An Implementation Case Study,* American Public Works Congress, Las Vegas, NV, September 1998.

Water Environment Federation—American Society of Civil Engineers, *Urban Runoff Quality Management,* WEF Manual of Practice No. 23, American Society of Civil Engineers Manual and Report on Engineering Practice No. 87, Alexandria, VA and Reston, VA, 259 pp., 1998.

Wenk, W., "Stormwater as Civic and Ecological Urban Framework," in C. Rowney, P. Stahre, and L. Roesner, eds., *Sustaining Urban Water Resources in the 21st Century, Proceedings of an Engineering Foundation Conference,* September 1997, Malmo, Sweden, American Society of Civil Engineers, Reston, VA, 1999.

Whipple, W., "Dual Purpose Detention Basins in Stormwater Management," *Water Resources Bulletin,* 17(4):642–646, 1981.

White, G. F., "Reflections on the 50-year International Search for Integrated Water Management," *Water Policy,* 1, 21–27, 1998.

Wigmosta, M. S., and S. J. Burges, "Proposed Model for Evaluating Urban Hydrologic Change, *Journal of Water Resources Planning and Management,* 116(6):742–763, 1990.

Wilson, R., *Suburban Parking Economics and Policy: Case Studies of Office Worksites in Southern California,* Report No. FTA-CA-11-0036092-1, Washington, DC, U.S. Department of Transportation, 1992.

Wilson, R. W., "Suburban Parking Requirements: A Tacit Policy for Automobile Use and Sprawl," *Journal of the American Planning Association,* 61(1):29–42, 1995.

Wright, L. T., *Design of Integrated Stormwater Management Systems,* Report to the Association of State Floodplain Managers, Madison, WI, 1999.

Wright, L. T., J. P. Heaney, and D. J. Sample, "Integrating Functional Landscapes with Stormwater Management Systems," *Proceedings of the 23rd Annual Conference of the Association of State Floodplain Managers,* Portland, OR, Association of State Floodplain Managers, Madison, WI, 1999.

Wright, L. T., and J. P. Heaney, *Suitability of Existing Stormwater Models for Microscale Simulations,* Report to the Low Impact Development Center, Rockville, MD, 2000.

Wright, L. T., J. P. Heaney, and N. Weinstein, "Modeling of Low Impact Development Stormwater Practices," *American Society of Civil Engineers Conference on Water Resources Engineering and Water Resources Planning and Management,* Minneapolis, MN, American Society of Civil Engineers, New York, NY, 2000.

Wright, P. H., and R. J. Paquette, *Highway Engineering,* J. Wiley & Sons, New York, NY, 1996.

Young, G. K., S. Stein, P. Cole, T. Kammer, F. Graziano, and G. Bank, *Evaluation and Management of Highway Runoff Quality,* Publication No. FHWA-PD-96-032, Federal Highway Administration, Washington, DC, p. 480, 1966.

Yuhnke, B., "Take a Deep Breath," *Colorado Commons,* Spring, p. 20–21, 1997.

CHAPTER 12
DESIGN OF PUMPS AND PUMP FACILITIES

Kevin Lansey
Dept. of Civil Engineering and Engineering Mechanics
University of Arizona
Tucson, Arizona

Walid El-Shorbagy
Dept. of Civil Engineering
University of United Arab Emirates
Al Ain, United Arab Emirates

Most often stormwater removal is a drainage process using gravity as the driving force to move water through the system. However, in some instances, to reduce hazards caused by excess water, terrain dictates that stormwater must be lifted. For example, a levee system acts to maintain water in a channel but the levees also act as retention structures for lateral inflow to the channel. Depending upon the channel water levels, water collected behind the levee may be pumped to the river.

This chapter provides background on the hydraulics of pipeline and pumping systems, pump types, hydraulics, and selection and pump station layout and design. The second part of the chapter considers design and configuration of the pumping station. An example design analysis is presented including wet-well sizing and pump operations.

Several references are available to supplement this material. The Hydraulic Institute Standards (1994) and the American Petroleum Institute (1995) provide excellent manuals for pump design and selection. Similarly, Sanks et al. (1998) and the Water and Environment Federation (1993) have published guidance on complete pumping stations including hydraulics, layout, and practical considerations.

12.1 HYDRAULIC AND ENERGY CONSIDERATIONS IN STORMWATER PUMPING SYSTEMS

The layout of a stormwater pump-force main system is typically a pump house with wet-well connected to a single pipeline (Fig. 12.1). The pump supplies additional required energy to overcome elevation changes and energy losses in the piping and valve system for the pump and pipeline.

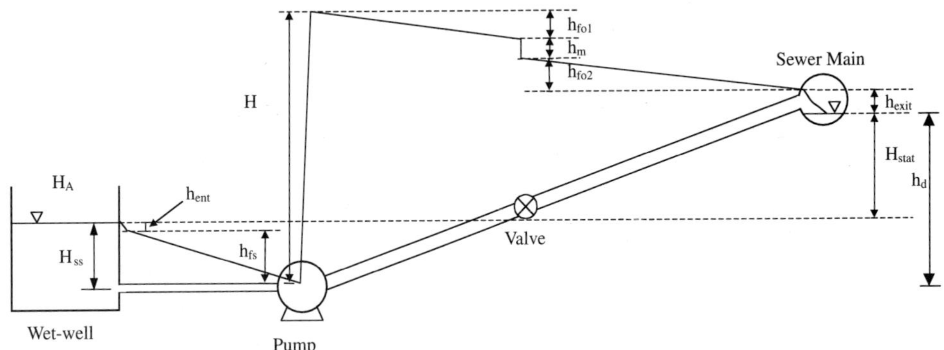

FIGURE 12.1   Typical layout of wet-well, pump and discharge pipeline. Also shown is the energy grade line with entrance, friction, minor, and exit losses and pump head gain.

To provide background on the overall system and its components, an energy balance can be performed between the inlet and outlet of the system. All energy terms are described as equivalent heads with units of energy per unit weight of water and dimensions of length. The energy relationship is shown graphically using the *energy grade line* (EGL) and its corresponding *hydraulic grade line* (HGL). The difference between the EGL and HGL is the *velocity head* (energy per unit weight of water due to its velocity) given by $V^2/2g$ where V is the water velocity and g is gravitational acceleration.

The energy available in the inlet tank is the *static suction head, h_{ss}*, which is measured above a datum defined as the center of the pump impeller. The pressure head at the datum in the source tank (wet-well) is p_{ss}/γ corresponding to depth of water of h_{ss}. This pressure is given in gage pressure using a pressure datum of atmospheric pressure at the surface of the tank. Absolute pressure is equal to gage pressure plus atmospheric pressure, H_A. A critical design element is ensuring that the minimum pressure remains above the vapor pressure of water at the water's temperature.

The water's energy drops as it enters the pipe inlet due to turbulent losses at the entrance (h_{ent}). In addition, it drops linearly in each pipe segment due to friction losses (h_f). Energy losses known as minor losses occur at all appurtenances such as valves, bends, and the exit (h_m). The hydraulic relationships for different components will be presented in the next section. The energy added by the pump, H, is known as the total dynamic head (*TDH*). Water leaves the pipeline with some discharge energy, h_d.

Conservation of energy can be written between the suction and outlet tanks assuming the water velocities in the reservoir are approximately zero. Starting with the upstream suction level:

$$h_{ss} - h_{ent} - h_{fs} - h_m + H - h_{fd} - h_{exit} = h_d \qquad (12.1)$$

where h_{fs} and h_{fd} = friction losses on the suction and discharge sides of the pump, respectively

h_m = total of the minor losses excluding entrance and exit losses, h_{ent} and h_{exit}

Rewriting Eq. (12.1) with H as the only term on the left hand side gives

$$H = (h_d - h_{ss}) + [h_{ent} + h_{fs} + h_m + h_{fd} + h_{exit}] = H_{stat} + H_{loss}$$

where H_{stat} = static head difference between the suction and discharge tanks = $(h_d - h_{ss})$

H_{loss} = sum of the energy loss terms in square brackets

Thus, as noted above, a pump is selected such that the pump energy supplied equals the sum of the static head difference (lift) and the energy losses.

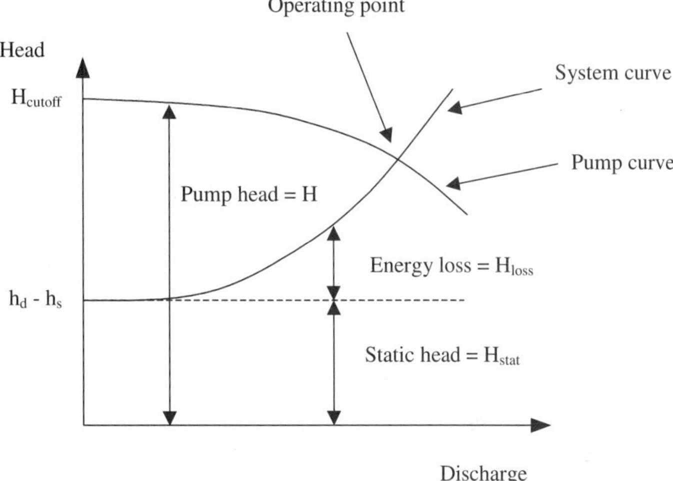

FIGURE 12.2   Pump and system curves.

As will be presented in the next section, H and H_{loss} are related to the flowrates in the system. At higher flowrates, the pump's ability to add energy decreases while the energy loss in the pipeline system increases. Graphically, this relationship is represented by the pump characteristic curve and the system curve (Fig. 12.2).

To maintain conservation of energy, the pump will operate at the intersection point between the system and pump curves. Changing the pump or its motor will shift the pump curve and the operating point. Similarly, modifying the pipe diameter, valve configurations, or the outlet elevation will change the system curve. The engineer must address three major issues to design this system.

1. Select a pump/pipeline combination that can supply the required discharge to the pipeline outlet in the most efficient manner.

2. Design the inlet structures to ensure that the liquid pressure always exceeds the water's vapor pressure, thus avoiding cavitation.

3. Design the wet-well and determine pump operations such that the pump operates near maximum efficiency given the variable inflow rates.

12.2   SELECTION OF A PUMP/PIPELINE COMBINATION

The primary decisions in designing a stormwater pumping system are the pipe diameter and pump/motor size. Component selection poses hydraulic and cost tradeoffs. The cost relationship is shown in Fig. 12.3. A small pipe diameter (with appropriate valves and fittings) will have relatively large energy losses, that is, the system curve is steep but at a relatively low cost. To provide this energy a high lift pump/motor combination is necessary that may be expensive. Increasing the pipe size and its cost will reduce the energy loss allowing a smaller pump to be used. The upper curve in Fig. 12.3 is the total system costs, that is, the sum of pipe and pump costs. The curve's minimum point is the optimal combination.

Details of computing the system and modified pump curves and the resulting system operating point are described in this section. A brief summary of hydraulic considerations is first presented as background.

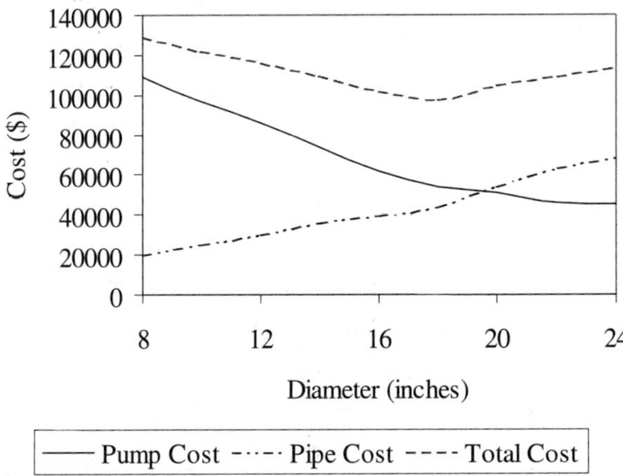

FIGURE 12.3 Comparison of pump and pipe costs to compute total system cost.

12.2.1 System Curve

The *system curve* is the summation of the energy requirements as a function of flowrate. The energy requirements defined in the system curve are the static lift and the energy losses that increase with flowrate. The static lift is controlled by topography and the system objectives but will vary during an event as the water level in the wet-well changes. H_{loss} accounts for the friction and losses due to minor system components. Relationships for estimating those losses are described in this section. Examining the exponents on the flowrates one expects the head loss (and the system curve) to be approximately quadratic with respect to the flowrate.

12.2.1.1 Friction Losses in Pipes.
Energy losses in pipes are related to the flowrate, pipe diameter and length, and the pipe's internal roughness. Two relationships are typically used for estimating friction losses in pipes: the Darcy-Weisbach and Hazen-Williams equations. The *Darcy-Weisbach equation* is a general, theoretically based relationship that is applicable to laminar and turbulent flow conditions. The *Hazen-Williams equation* is an empirical equation that is appropriate for fully turbulent flow only.

The Darcy-Weisbach equation is

$$h_f = f \frac{L}{D} \frac{V^2}{2g} \tag{12.2}$$

where L = the pipe length
  D = the pipe diameter $[L]$
  V = the flow velocity $[L/T]$
  g = gravitational acceleration $[L/T^2]$
  f = the Darcy-Weisbach friction factor [ ] that is a function of the Reynolds' number and the relative pipe roughness

The *Reynolds' number* equals VD/ν where ν is the kinematic viscosity of the fluid; ν of water at 20°C is 10^{-6} m²/s (10^{-5} ft²/s). The dimensionless relative pipe roughness $[L]$ is absolute roughness, ε $[L]$, divided by the pipe diameter or ε/D. Values of ε are given in Table 12.1 for different pipe materials.

TABLE 12.1 Typical Pipe Friction Coefficients[a]

| | | Relative roughness, ε[b] | |
Material	Hazen-Williams C	Millimeters	Inches
New pipe or lining			
Smooth glass or plastic[c]	150	0.019	0.00075
Centrifugally spun cement-mortar lining[d]	145	0.038	0.0015
Cement-mortar lining troweled in place	140	0.076	0.003
Commerical steel or wrought iron	140	0.076	0.003
Galvanized iron	135	0.13	0.005
Ductile or cast iron, uncoated	130	0.19	0.0075
Asbestos-cement, coated	145	0.038	0.0015
Asbestos-cement, uncoated	140	0.076	0.003
Centrifugally cast concrete pressure pipe	135	0.13	0.005
Ten-State Standards (1978)			
Cement mortar or plastic lining	120	0.41	0.016
Unlined steel or ductile iron	100	1.5	0.060
Old pipe or lining (in moderate service (>20 yrs) nonaggressive water)[e]			
Smooth glass or plastic[f]	135	0.13	0.005
Centrifugally spun cement-mortar lining	130	0.19	0.0075
Cement mortar troweled in place	125	0.28	0.011
Asbestos cement, coated	130	0.19	0.0075
Asbestos cement, uncoated	125	0.28	0.011
Ductile iron or carbon steel, uncoated	100	1.5	0.060
Centrifugally cast concrete pressure pipe	130	0.19	0.0075
Wood stave	110	0.89	0.035
Riveted steel	80	5.6	0.22
Concrete, formed	80	5.6	0.22
Wrought iron	100	1.5	0.060
Galvanized iron	90	0.30	0.012

[a] For critical problems, consult other sources.

[b] Values calculated from C coefficients for 300-mm pipe with a velocity 1.1 to 1.2 m/s and a temperature of 20°C.

[c] PVC, polyethylene, polypropylene, polybutylene, reinforced thermosetting resin pipe, and polyvinyl chloride.

[d] Average value of pipes 150 to 900 mm diameter

[e] For conservative design, reduce old pipe C values by 0.02%/mm for pipes less that 450 mm. Note that the Hazen-Williams equation predicts low headlosses for small pipes and/or low velocities.

[f] Conservative values for water pipe 150 to 500 mm.

Source: From Sanks et al. (1998).

The friction factor can be determined using the Moody diagram (Fig. 12.4). The Moody diagram has three regions corresponding to laminar, mixed, and fully turbulent flow. Given the flowrate and pipe size and type, the Reynolds' number is computed and the Moody diagram is entered at that point on the x-axis. A vertical line is drawn until the curve corresponding to the pipe's relative roughness is intersected. The f at the intersection point is determined by drawing a horizontal line to the y-axis and is substituted in Eq. (12.2) to compute the energy losses. Alternatively, the *Swanee-Jain equation* can be solved to directly determine the friction factor:

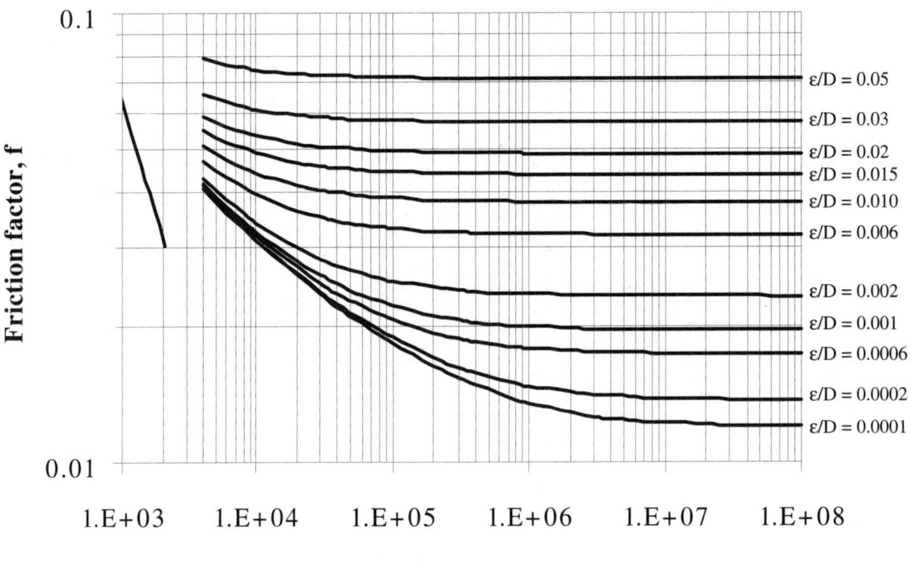

FIGURE 12.4 Moody diagram.

$$f = \frac{0.25}{\left[\log\left(\dfrac{\varepsilon}{3.7D} + \dfrac{5.74}{\mathrm{Re}^{0.9}}\right)\right]^2} \tag{12.3}$$

If flow is fully turbulent, as is the case in many systems, the Hazen-Williams equation is appropriate to compute friction losses. This empirical equation is written as

$$h_f = \frac{K_u L Q^{1.85}}{D^{4.87} C^{1.85}} \tag{12.4}$$

where K_u = a unit constant (4.73 for English units and 10.66 for SI)
$\quad h_f$ = the head loss in m (ft)
$\quad L$ = the pipe length in m (ft)
$\quad Q$ = the flowrate in m^3/s (cfs)
$\quad D$ = the pipe diameter in m (ft)

C, the Hazen-Williams roughness coefficient, is a single value for a given pipe, which is consistent with the turbulent flow assumption (Table 12.1). C and ε vary as a pipe ages and the pipe roughness increases (Table 12.1).

12.2.1.2 Minor Losses.   Head losses also occur due to flow disruption and turbulence in bends, inlets, outlets, and other appurtenances such as valves, contractions, and expansions. These so-called *minor losses* are usually estimated by

$$h_m = K\frac{V^2}{2g} = K\frac{Q^2}{2gA^2} \tag{12.5}$$

h_{ent} and h_{exit} are written in the same form. Values for K are given in Tables 12.2 and 12.3.

TABLE 12.2   Minor Loss Coefficients

Type of minor loss	K coefficient
Pipe fitting	
90° elbow, regular	0.21–0.30
90° elbow, long radius	0.14–0.23
45° elbow, regular	0.2
Return bend, regular	0.4
Return bend, long radius	0.3
AWWA tee, flow through side outlet	0.5–1.8
AWWA tee, flow through run	0.1–0.6
AWWA tee, flow split side inlet to run	0.5–1.8
Valves (fully open except as noted)	
Gate valves (4 to 12 in) fully open	0.07–0.14
¼ closed	0.47–0.55
½ closed	2.2–2.6
¾ closed	12–16
Angle	1.8–2.9
Ball	0.04
Gate	
Double disc	0.1–0.2
Resilient seat	0.3
Check valves	
Swing gate	0.6–2.5
Double door (8 in or smaller)	2.5
Double door (10 to 16 in)	1.2
Foot (hinged disc)	1–1.4
Foot (poppet)	5–14
Slanting disc	1.1

Sources: From Sanks et al. (1998) and Velon and Johnson (1993).

Example 12.1

A 3000-m long, 300-mm diameter cast-iron pipeline is connected at the outlet of a pump station. Flowrates are expected to vary between 0.03 and 0.20 m^3/s. A fully open gate valve is included in the line that exits to a full collection main. The difference in energy between the wet-well and the collection main is 40 m. Compute the system curve using the Darcy-Weisbach equation and a kinematic viscosity for water (ν) of 1×10^{-6} m^2/s.

Solution

The total energy requirements for the system are the static head plus the energy losses.

$$H_{system} = H_{stat} + H_{loss}$$

H_{stat} = 40 m and the losses are the sum of pipe friction, and the turbulent losses at the gate valve and exit or

$$H_{loss} = h_f + h_{valve} + h_{exit}$$

The pipe friction losses are computed using the Darcy-Weisbach equation after the friction factor f is determined. For a flowrate of 0.09 m^3/s, the flow velocity is $Q/A = Q/(\pi D^2/4) = 0.09/(\pi(0.3)^2/4) = 1.27$ m/s. The Reynolds' number equal to Re = VD/ν =

TABLE 12.3   Energy Loss Coefficients for Flanged Pipe Fittings

Fitting	K	Fitting	K
Bell mouthed entrance	0.05	Rounded entrance	0.25
Sharp edged entrance	0.5	Projecting entrance	0.8
Conical increaser $h = K(1 - (D_1/D_2)^2)(V_2^2/2g)$ $K = 3.5(\tan \theta)^{1.22}$		Sudden expansion $h = (V_1^2 - V_2^2)/2g$	
Conical reducer $H = KV_2^2/2g$ $K = 0.03 +/- 0.01$		Sudden contraction $h = \tfrac{1}{2}[1 - (D_1/D_2)^2]V_2^2/2g$	

$1.27(0.3)/10^{-6} = 382000$. The relative roughness for the pipe is equal to $\varepsilon/D = (0.19/1000)/0.3 = 0.0006$. Entering the Moody diagram with Re $= 382000$ and $\varepsilon/D = 0.0006$ gives a friction factor of about 0.019. Substituting in the Swanee-Jain equation (Eq. (12.3)) gives $f = 0.0188$.

The minor loss coefficients from Table 12.2 are 0.1 and 1.0 for the open-gate valve and exit loss, respectively.

The total losses are then

$$H_{loss} = f\frac{LV}{D2g} + K_v\frac{V^2}{2g} + K_{exit}\frac{V^2}{2g} = 0.0188\frac{1500*1.27^2}{0.3(2*9.81)} + 0.1\frac{1.27^2}{2*9.81} + 1.0\frac{1.27^2}{2*9.81}$$

$$H_{loss}(Q = 0.09 \text{ m}^3/\text{s}) = 7.86 \text{ m}$$

The system head requirement at 0.09 m^3/s is then

$$H_{system} = H_{stat} + H_{loss} = 40 + 7.86 = 47.86 \text{ m}$$

The system head requirements for the full range of flow is shown in Table 12.4. Note

TABLE 12.4   System Curve for Darcy-Weisbach Friction Loss and Minor Losses (Example 12.1)

Flow (m³/s)	Reynolds' number, Re	Velocity (m/s)	f	Friction loss (m)	Gate valve loss (m)	Exit loss (m)	Total loss (m)	System head (m)
0.001	4244	0.01	0.0406	0.00	0.00	0.00	0.00	40.00
0.03	127300	0.42	0.0204	0.94	0.00	0.01	0.95	40.95
0.06	254600	0.85	0.0193	3.54	0.00	0.04	3.58	43.58
0.09	382000	1.27	0.0188	7.77	0.01	0.08	7.86	47.86
0.12	509300	1.70	0.0185	13.62	0.01	0.15	13.78	53.78
0.15	636600	2.12	0.0184	21.10	0.02	0.23	21.35	61.35
0.20	848800	2.83	0.0182	37.17	0.04	0.41	37.62	77.62

that the friction factor must be recomputed for each flowrate since energy loss varies with flow rate.

Example 12.2

A 2000-ft long, 16-in pipeline with Hazen-Williams roughness coefficient of 120 is connected at the outlet of a pump station. Flowrates are expected to vary between 2 and 12 ft³/s. An open swing check valve and a long radius 90-degree elbow is included in the line that exits to a full collection main. The difference in energy between the wet-well and the collection main is 75 ft. Determine the system curve for the range of expected flows.

Solution

As in the previous example the system requirement equals the static lift plus the energy losses. The energy losses are

$$H_{loss} = K_u \frac{LQ^{1.85}}{C^{1.85}D^{4.87}} + K_v \frac{V^2}{2g} + K_{bend} \frac{V^2}{2g} + K_{exit} \frac{V^2}{2g}$$

$$= \frac{4.73*2000*Q^{1.85}}{120^{1.85}(16/12)^{4.87}} + 0.8 \frac{(Q/A)^2}{2*9.81} + 0.2 \frac{(Q/A)^2}{2*9.81} + 1.0 \frac{(Q/A)^2}{2*9.81}$$

Using the Hazen-Williams (H-W) equation does not require re-computing the friction factor for each flow condition. In this example, the Reynolds' numbers are above 10^6, which would

TABLE 12.5   System Curve Using Hazen-Williams Friction Loss and Minor Losses (Example 12.2)

Flow (cfs)	Velocity (ft/s)	Friction loss (ft)	Elbow loss (ft)	Butterfly valve loss (ft)	Exit loss (ft)	Total loss (ft)	System head (ft)
0.01	0.01	0.00	0.00	0.00	0.00	0.00	75.00
2.00	1.43	1.20	0.01	0.03	0.03	1.26	76.26
4.00	2.86	4.31	0.03	0.10	0.13	4.57	79.57
6.00	4.30	9.13	0.06	0.23	0.29	9.70	84.70
8.00	5.73	15.55	0.10	0.41	0.51	16.57	91.57
10.00	7.16	23.49	0.16	0.64	0.80	25.09	100.09
12.00	8.59	32.92	0.23	0.92	1.15	35.21	110.21

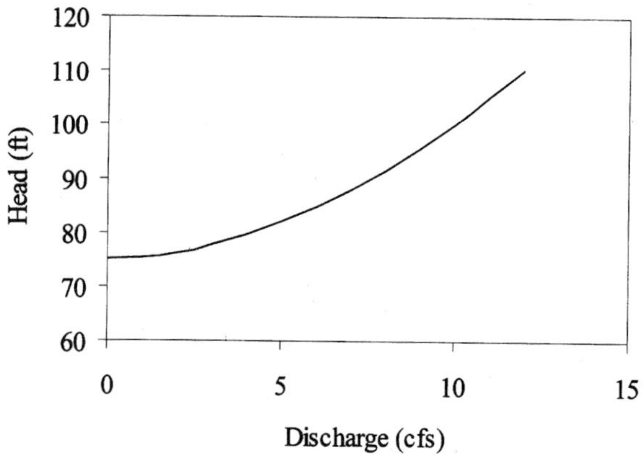

FIGURE 12.5   System curve for example 12.2.

suggest likely the flow is fully turbulent as assumed for the H-W equation. Note the flow area is 1.40 ft². For a flowrate of 6 cfs, the required system head is

$$H_{system} = H_{stat} + H_{loss} = 75 + 9.13 + 0.23 + 0.06 + 0.29 = 84.70 \text{ ft}$$

The system curve for the range of flows is given in Table 12.5 and plotted in Fig. 12.5.

12.3 PUMP TYPES AND DEFINITIONS

Centrifugal pumps can operate over a wide range of flow and head conditions. The U.S. usage of the term centrifugal pumps is used here to describe radial, axial, and mixed-flow pumps. In radial pumps, fluid enters through the center of the impeller and is displaced radially (perpendicular to the pump shaft) (Fig. 12.6). Axial pumps displace fluid axially through a propeller-type impeller with flow moving along the impeller shaft (Fig. 12.7) while

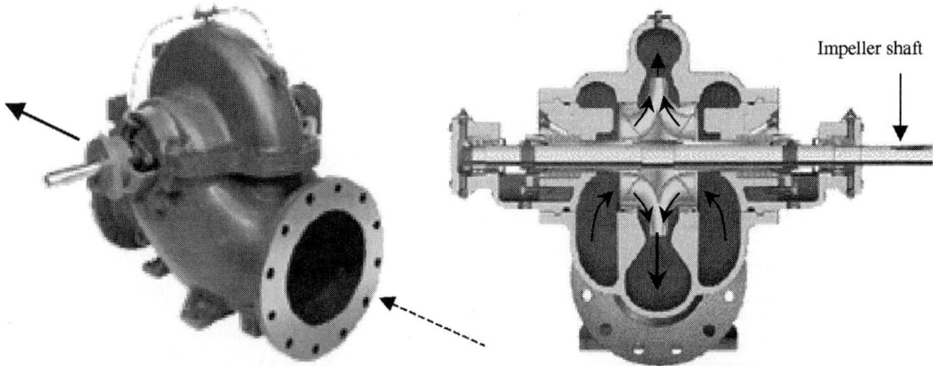

FIGURE 12.6   Single stage, double suction radial flow pump. Flow enters from front suction flange (dashed arrow) and leaves through back discharge flange (solid arrow on exterior view). Internally (see cutaway section), incoming flow laterally enters impeller from the outer portion of the casing. The impeller accelerates the water outward. Flow is collected in the central portion of the casing and moves to the outlet. (From Goulds Pumps)

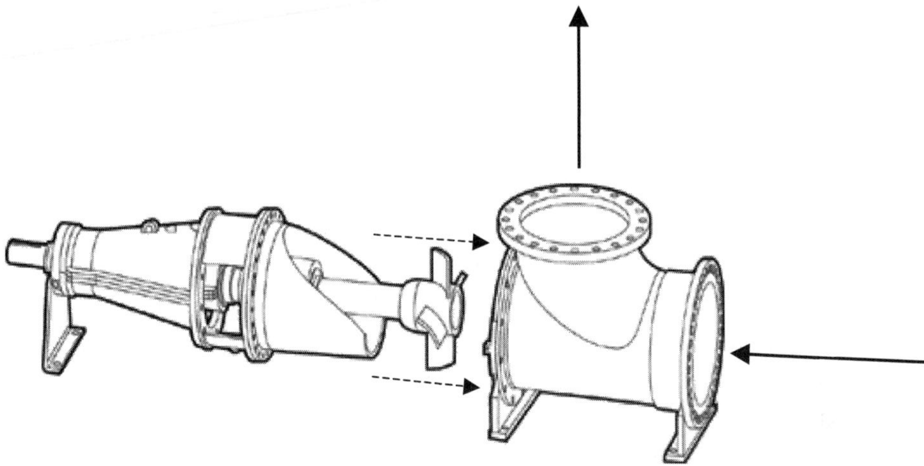

FIGURE 12.7 Axial pump. Pump/motor assembly is inserted and mounted in elbow (dashed arrows). When assembled, flow enters from right and energy is imparted by impeller to lift to outlet (solid arrows). (From Goulds Pumps)

mixed-flow pumps displace fluid both axially and radially. Multi-stage pumps contain more than one impeller (stage). The stages act in series with the outlet of one stage being the inlet to the next stage.

The choice of pump is dependent upon the system demands. Axial-flow and mixed-flow pumps are used in applications requiring large discharge rates with a low head gain. Radial-flow pumps are more efficient for large head, low-flow conditions, and are the most commonly used for stormwater pumping.

12.4 PUMP CHARACTERISTICS

12.4.1 Pump Curves

A pump is selected based on its ability to impart the desired energy to the flow to overcome the change in elevation and energy losses in the pipeline and piping components. In addition, the head must be efficiently supplied to maintain low energy costs. The energy supplied decreases with flowrate as seen in the *pump head-characteristic* (or *performance*) *curve* shown in Fig. 12.8*a*.

As supplied by the pump manufacturer, the pump head-characteristic or simply pump curve relates the TDH with the pump discharge. The TDH is the total head against which the pump operates. Other references points on the pump curve are the shutoff head (H_{cutoff}) and the normal discharge or rated capacity. The shutoff head is the head output corresponding to zero discharge while the rated capacity is the pumped head and discharge that is supplied with maximum efficiency (discharge at the best efficient point). Also shown in Fig. 12.8*a* is a typical efficiency curve that peaks at the best efficient point (bep). Efficiency is related to how well input energy is transferred to the pumped water.

The lowest curve in Fig. 12.8*a* defines the *net positive suction head* (NPSH$_R$) that is required at the eye of the impeller to avoid cavitation. NPSH$_R$ is discussed further later in this chapter. Finally, input energy requirements are defined with the *brake horsepower* (*bhp*)

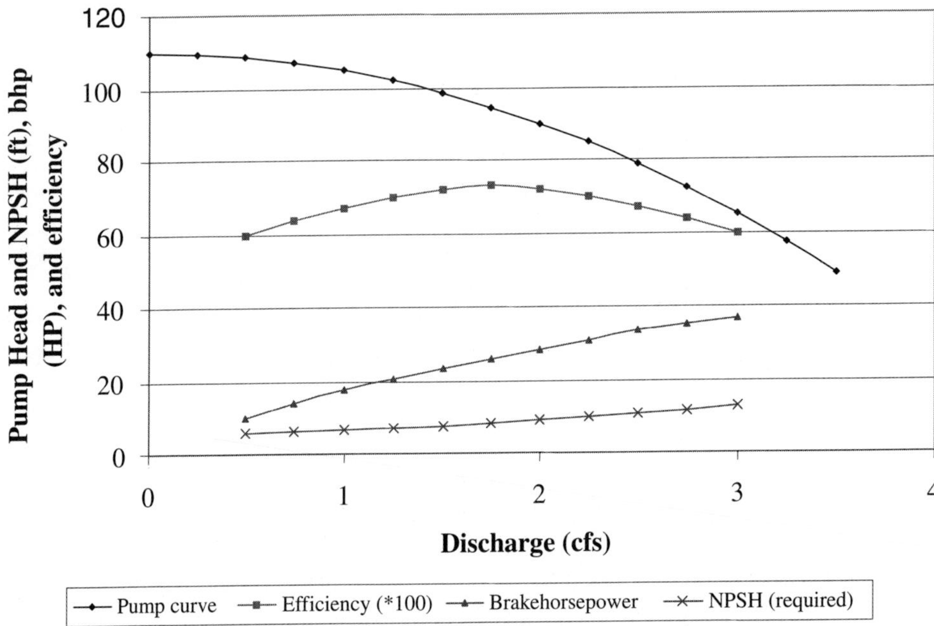

FIGURE 12.8a   Typical pump performance curve. Also shown are efficiency, brake horsepower and required net positive suction head curves versus flowrate.

curve. Brake horsepower is the power required by the pump to provide the corresponding discharge and TDH. The required power input can be computed by

$$\text{bhp} = \frac{\gamma QH}{550\eta} \qquad \text{for English units} \qquad (12.6)$$

$$\text{bhp} = \frac{\gamma QH}{\eta} \qquad \text{for SI units} \qquad (12.7)$$

where Q = pump discharge in m³/s (cfs)
H = total dynamic head in m (ft)
η = pump efficiency
γ = specific weight in kN/m³ (lb/ft³)

Figure 12.8*b* shows performance curves for a set of homologous (geometrically similar) pumps produced by a manufacturer. The same information is supplied in a slightly different compact form for the same impeller design with different diameters. The dashed vertical lines are the bhp, the concave upward curves overlain on the performance curves are efficiencies, and the lower curves provide the NPSH_R.

12.4.2   Performance Curves for Combinations of Pumps

Pumps may be installed in series or parallel to provide higher outputs. Pumps in *parallel* that draw water from a common source and discharge to a common downstream pipe add the same amount of energy to the flow through each individual pump. To determine the *total*

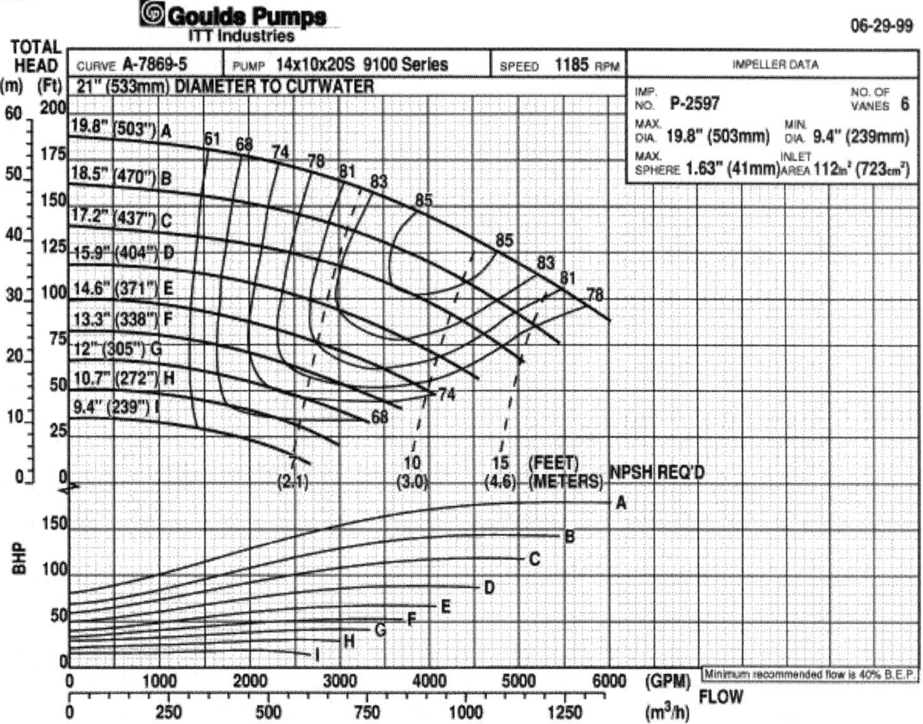

FIGURE 12.8b Typical manufacturer pump curves showing pump performance for set of homologous pumps with varying impeller diameter (9.4–19.8″). Solid efficiency curves are overlain on pump performance curves. Dashed lines represent NPSH$_R$. Lower curves with different scale present brake horsepower requirements for each impeller diameter. (From Goulds Pumps)

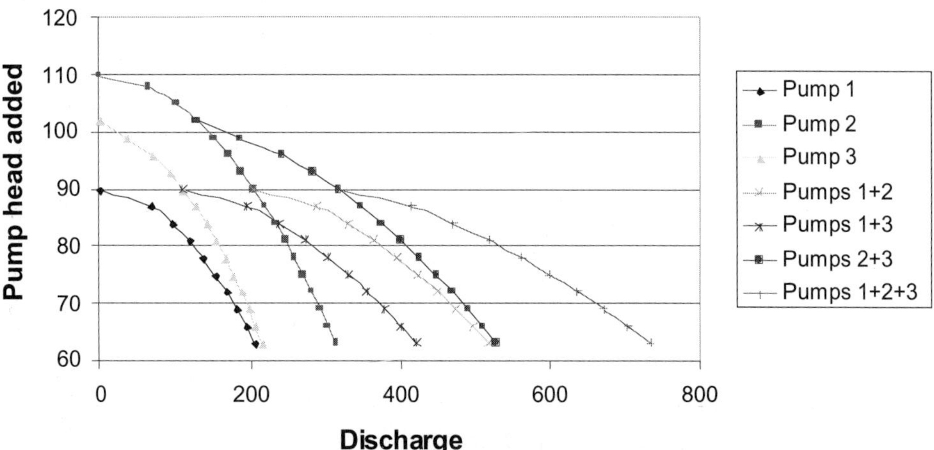

FIGURE 12.9 Performance curves for three pumps in parallel based on manufacturer pump curves. Corrections for pump station energy are made after pump curves are combined.

TABLE 12.6 Discharge for Pump Station with One or More Pumps Operating in Parallel (Example 12.3)

Head (ft)	Pump 1 (cfs)	Pump 2 (cfs)	Pump 3 (cfs)	Pumps 1 and 2 (cfs)	Pumps 1 and 3 (cfs)	Pumps 2 and 3 (cfs)	Pumps 1, 2, and 3 (cfs)
110	—	0	—	0	—	0	0
108	—	65	—	65	—	65	65
105	—	102	—	102	—	102	102
102	—	129	0	129	0	129	129
99	—	151	35	151	35	186	186
96	—	171	71	171	71	242	242
93	—	188	94	188	94	282	282
90	0	204	112	204	112	316	316
87	68	219	127	287	195	346	415
84	97	233	141	330	238	374	471
81	119	246	154	365	273	400	519
78	137	258	166	395	303	424	561
75	153	270	177	423	330	447	600
72	168	281	187	449	355	468	636
69	181	292	197	473	378	489	670
66	194	303	206	497	400	509	703
63	205	313	215	518	420	528	733

pump curve the flows that may be different for each pump are added for the same head as shown in Fig. 12.9 and Table 12.6 (see example 12.3).

Pumps in *series* or *staged pumps* have same discharge passing through each pump (stage). Each pump (stage) adds some amount of energy so the total energy added is the sum of the individual pumps (stages). Development of a pump characteristic curve for three pumps in series is seen in Fig. 12.10 and Table 12.7 (see example 12.4).

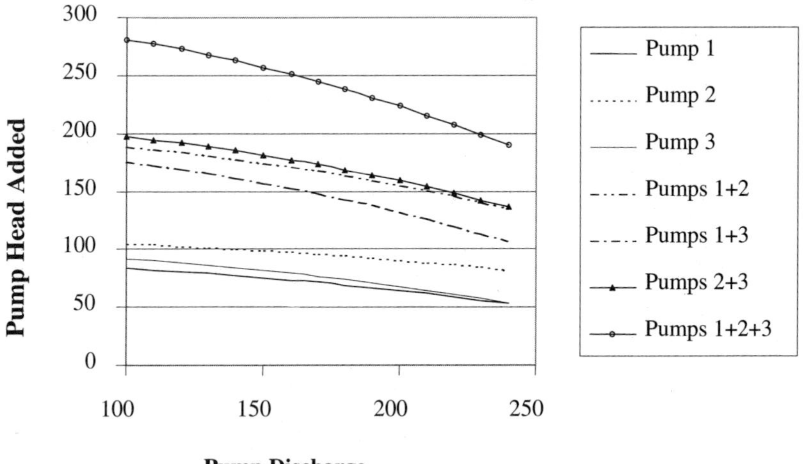

FIGURE 12.10 Addition of pump performance curves for three pumps operating in series. Corrections for energy losses is made to resulting combination pump curves.

TABLE 12.7 Pump Head Added for Different Combinations of Pumps Operating in Series (Example 12.4.2.2)

Discharge	Pump 1	Pump 2	Pump 3	Pumps 1 and 2	Pumps 1 and 3	Pumps 2 and 3	Pumps 1, 2, and 3
100	84	105	92	189	176	197	281
110	82	104	90	186	172	194	276
120	81	103	88	184	169	191	272
130	79	102	86	181	165	188	267
140	77	101	84	178	161	185	262
150	76	99	82	175	158	181	257
160	74	98	80	172	154	178	252
170	72	96	77	168	149	173	245
180	69	94	74	163	143	168	237
190	67	93	71	160	138	164	231
200	64	91	68	155	132	159	223
210	62	89	65	151	127	154	216
220	59	87	61	146	120	148	207
230	56	85	58	141	114	143	199
240	53	82	54	135	107	136	189

Example 12.3 (Pump Curves for Parallel Pumps)
Three pumps with different capacities are operating in parallel with pump curves given in the columns 1 to 4 of Table 12.6. Determine the pump curves for each pump combination.

Solution
Table 12.6 and Fig. 12.9 present tabular and graphical representations of the pump curves. Since the pumps are operating in parallel each pump must supply the same added head. The flows at that head can be added to determine the total discharge for the pump station with those pumps operating. For example, if all three pumps are operating and the added head is 75 ft, the discharge will be

$$Q_{1+2+3}(H = 75) = Q_1(H = 75) + Q_2(H = 75) + Q_3(H = 75) = 153 + 270 + 177 = 600$$

Note that pumps will only contribute to flow if the head that they are operating against is less than their cutoff head. For example for pumps 1 and 3 operating at against a head of 96′ is:

$$Q_{1+3}(H = 96) = Q_1(H = 96) + Q_3(H = 96) = 0 + 71 = 71$$

Example 12.4 (Pump Curves for Pumps in Series)
Three pumps with different capacities (columns 1 to 4 in Table 12.7) are operating in series. Determine the pump curves for each pump combination.

Solution
Table 12.7 and Fig. 12.10 presents tabular and graphical representations of the pump curves. Since the pumps are operating in series, each pump boosts the same flow and the heads are

added. For example, if all three pumps are operating and the discharge is 170 gpm, the total head added will be

$$H_{A+B+C}(Q = 170) = H_A(Q = 170) + H_B(Q = 170) + H_C(Q = 170)$$

$$= 72 + 96 + 77 = 245$$

12.4.3 Pump Similarity

Selecting pumps and motors to match a system curve can require sorting through numerous sets of manufacturer pump specifications. The selection process can be simplified if an engineer applies the similarity laws for pumps. If pumps have impellers and casings that are geometrically similar, their pump curves will vary in a predictable manner. Changing the motor speed driving a particular pump will also have a predictable effect. The relationships describing those effects are known as the pump similarity or affinity laws.

Leading to the affinity laws, three coefficients can be developed from dimensional analysis. Each coefficient is constant for a set of geometrically similar pumps. The *head coefficient,* C_H, relates the head added to the impeller diameter, D, and the pump speed, n, by

$$C_H = \frac{gH}{D^2 n^2} \tag{12.8}$$

Dimensionless *discharge* (C_Q) and *power* (C_p) *coefficients* relate flow and input power requirements to D and n as follows:

$$C_Q = \frac{Q}{nD^3} \tag{12.9}$$

$$C_p = \frac{P}{\rho n^3 D^5} \tag{12.10}$$

The units for the coefficients are $Q = [\text{m}^3/\text{sec (ft}^3/\text{min)}]$, $H = [\text{m (ft)}]$, $n = [\text{radians}/\text{second (rpm)}]$, and $P = [\text{kw (hp)}]$. Thus, changes in pump output can be predicted for varying impeller sizes and pump speeds. These relationships can be used to manipulated to form the *similarity laws:*

$$\frac{Q_2}{Q_1} = \frac{n_2}{n_1} \left(\frac{D_2}{D_1}\right)^3 \tag{12.11}$$

$$\frac{H_2}{H_1} = \left(\frac{n_2}{n_1}\right)^2 \left(\frac{D_2}{D_1}\right)^2 \tag{12.12}$$

$$\frac{P_2}{P_1} = \left(\frac{n_2}{n_1}\right)^3 \left(\frac{D_2}{D_1}\right)^5 \tag{12.13}$$

where the subscripts correspond to the operating points on pumps 1 and 2. These relationships describe how a pump curve will change as the pump motor or impeller size is altered. Typically, the efficiency for two corresponding or homologous points on the two pump curves is assumed to be the same.

Example 12.5
At peak efficiency a pump with a 12 in impeller is operating at 2000 rpm and produces a discharge of 750 gpm (1.67 cfs) at 100 ft of head. Compute the discharge, head and power coefficients for this pump. Note unit conversions of rpm to rps to remain consistent with Q in cfs.

$$C_Q = \frac{Q}{nD^3} = \frac{1.67}{(2000/60)(12/12)^3} = 0.05$$

$$C_H = \frac{gH}{n^2D^2} = \frac{32.2*100}{(2000/60)^2(12/12)^2} = 2.90$$

$$C_P = \frac{P}{\rho n^3 D^5} = \frac{\gamma QH}{\rho n^3 D^5} = \frac{(62.4*16.7*100)}{1.94*(2000/60)^3(12/12)^5} = 0.145$$

If a 2500-rpm motor was used with the same pump, what output could be expected at peak efficiency?

$$\frac{n_1}{n_2} = \frac{2000}{2500} = \frac{Q_1}{Q_2} = \frac{750}{Q_2} \Rightarrow Q_2 = 940 \text{ gpm}$$

$$\left(\frac{n_1}{n_2}\right)^2 = \left(\frac{2000}{2500}\right)^2 = \frac{H_1}{H_2} = \frac{100}{H_2} \Rightarrow H_2 = 156 \text{ ft}$$

$$\left(\frac{n_1}{n_2}\right)^3 = \left(\frac{2000}{2500}\right)^3 = \frac{P_1}{P_2} = \frac{189}{P_2} \Rightarrow P_2 = 369 \text{ HP}$$

If a larger geometrically similar pump with an 18″ diameter impeller was selected instead of the 12″ pump, what would its expected output be at peak efficiency?

$$\left(\frac{D_1}{D_2}\right)^3 = \left(\frac{12}{18}\right)^3 = \frac{Q_1}{Q_2} = \frac{750}{Q_2} \Rightarrow Q_2 = 2531 \text{ gpm}$$

$$\left(\frac{D_1}{D_2}\right)^2 = \left(\frac{12}{18}\right)^2 = \frac{H_1}{H_2} = \frac{100}{H_2} \Rightarrow H_2 = 225 \text{ ft}$$

$$\left(\frac{D_1}{D_2}\right)^5 = \left(\frac{12}{18}\right)^5 = \frac{P_1}{P_2} = \frac{189}{P_2} \Rightarrow P_2 = 1435 \text{ HP}$$

12.4.4 Specific Speed

Combining the discharge and head coefficients result in another parameter known as *specific speed*, N_s. The specific speed provides guidance when selecting the type of centrifugal pump to use in a particular application. It is computed by

$$N'_s = \frac{C_Q^{1/2}}{C_H^{3/4}} = \frac{nQ^{1/2}}{gH^{3/4}} \tag{12.14}$$

for a given pump speed, H and Q (the pump's *rated head* and *discharge*). N'_s, as written, is

TABLE 12.8   Specific Speed Ranges for Different Impeller Types

Type of impeller	Specific speed range (U.S. units)
Radial flow	500–4200
Mixed-flow	4200–9000
Axial-flow	9000–15000

Source: After Bosserman, 1999.

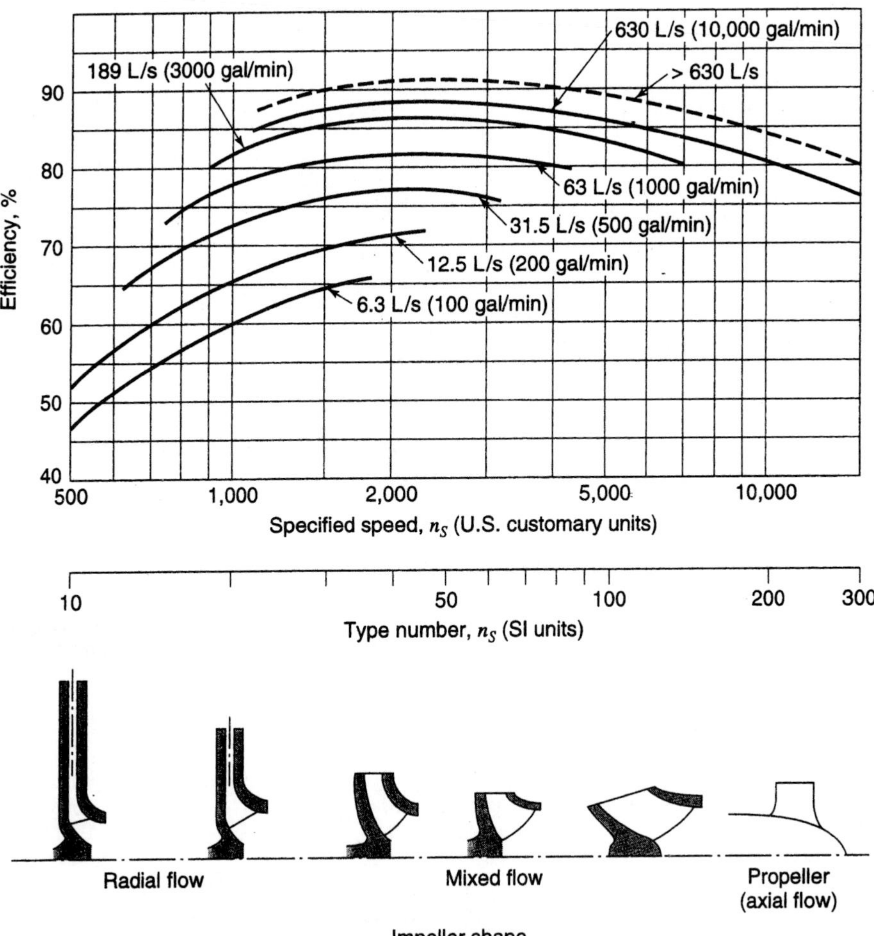

FIGURE 12.11   Pump efficiency as related to specific speed and discharge (from Tchobanoglous (1998)).

dimensionless with n in rpm, Q in cfs, and H in ft. Practicing engineers typically do not include the gravitational acceleration and use Q in gal/min and n in revolutions/min or

$$N_s = \frac{nQ^{1/2}}{H^{3/4}} = \left[\frac{(r/min)(gal/min)^{1/2}}{ft^{3/4}}\right] = 17182 N_s' \qquad (12.15)$$

The conversion of N_s to SI units with H in meters and Q in m³/s is 51.64. The optimum expected pump performance can be predicted using N_s for different types of pumps. The general ranges for the different impeller designs are shown in Table 12.8. Figure 12.11 shows the expected efficiency for different specific speed pumps to provide guidance on expected operation cost and pump selection.

Example 12.6
At peak efficiency a pump with a 12-in impeller is operating at 2000 rpm and produces a discharge of 750 gpm at 100 ft of head. What is the specific speed for this pump? What type of pump would you expect it to be?

$$N_s = \frac{nQ^{0.5}}{H^{0.75}} \; \frac{2000*750^{0.5}}{100^{0.75}} = 1732$$

$$N_S = 17182 N_S' \Rightarrow N_S' = 0.10$$

Based on the specific speed of 1732, the pump is likely a radial flow pump.

Example 12.7
Determine the type of pump appropriate to supply a flow of 0.025 m³/s against a head of 40 m. The pump motor tentatively selected has a speed of 1170 rpm. What is the likely efficiency for the pump?

$$N_s = \frac{nQ^{0.5}}{H^{0.75}} \; \frac{1170*0.025^{0.5}}{40^{0.75}} = 11.6$$

The specific speed of 11.6 in SI and 600 in U.S. customary units corresponds to a radial flow pump with efficiency around 62%.

12.4.5 Modified Pump Curve

The pump manufacturer's pump characteristic curve describes the H-Q relationship for the pump itself. Valves and other fittings are installed around the pump to protect it and connect it to the system. The manufacturer's pump curve must be modified to correct for energy losses that occur in those components for the specific system.

Example 12.8
The manufacturer's characteristic curve for a stormwater pump can be approximated by

$$H = 100 - 500Q^2$$

The discharge line is 4 m of a 250-mm pipe ($C = 130$). It contains control, swing gate and isolating gate valves, and a 90-degree bend to the discharge line. Flow enters the suction line through a sharp edged entrance. The line contains a partially open gate valve, a 90° bond and a reducer.
 Compute the modified pump curve for a range of flows.

Component	K	Component	K
Suction line		*Discharge line*	
Sharp-edged entrance	0.5	Control valve	0.25
Open gate valve	0.30	Swing gate check valve	1.5
90° bend	0.25	Gate valve	0.10
Reducer	0.03	90° bend	0.25
Total suction line K	1.08	Total discharge line K	2.1

The head loss through the pump can be computed by

$$H_L = h_{fs} + K_{ent}\frac{V_s^2}{2g} + K_{gate}\frac{V_s^2}{2g} + K_{bend}\frac{V_s^2}{2g} + K_{red}\frac{V_s^2}{2g}$$

$$+ h_{fo} + K_{cont}\frac{V_o^2}{2g} + K_{check}\frac{V_o^2}{2g} + K_{gate}\frac{V_o^2}{2g} + K_{bend}\frac{V_o^2}{2g}$$

where h_{fs} and h_{fo} = the frictional losses in the pipe on the suction and outlet side of the pump, respectively.

Similarly, V_s and V_o are the suction and outlet velocities. We assume that the pipe velocity is acceptable for use in the reducer loss relationship.

Pipe friction losses are estimated by the Hazen-Williams equation or

$$h_{fs} = \frac{10.66LQ^{1.85}}{D^{4.87}C^{1.85}} = \frac{10.66(3)Q^{1.85}}{(0.25)^{4.87}(130)^{1.85}} = 3.36Q^{1.85}$$

By a similar substitution, $h_{fo} = 4.48Q^{1.85}$. By continuity, $V_s = V_o = Q/A_s = Q/(\pi D_s^2/4) = 20.37Q$. So the suction velocity head is

$$\frac{V^2}{2g} = \frac{(20.37Q)^2}{2g} = 21.15Q^2$$

Substituting these terms and the minor loss coefficients in the head loss relationship gives

$$H_L = h_{fs} + (K_{ent} + K_{gate} + K_{bend} + K_{red})\frac{V_s^2}{2g} + h_{fo}$$

$$+ (K_{cont} + K_{check} + K_{gate} + K_{bend})\frac{V_o^2}{2g}$$

$$= 3.36Q^{1.85} + (0.5 + 0.30 + 0.25 + 0.03)(21.15\ Q^2)$$

$$+ 4.48Q^{1.85} + (0.25 + 1.5 + 0.10 + 0.25)(21.15Q^2)$$

$$= (3.36Q^{1.85} + 22.84Q^2)_{suction\ line} + (4.48Q^{1.85} + 44.42Q^2)_{outlet\ line}$$

$$= 7.84Q^{1.85} + 67.26Q^2$$

Based on this final equation, the energy losses through the pump housing can be computed for a range of discharge rates. The modified pump curve is then computed by subtracting these losses from the manufacturer's performance curve. The result is shown in Table 12.9 and in Fig. 12.12. Note that changing the pump and using the same valve and bend config-

TABLE 12.9 Manufacturer's and Corrected Pump Curves (Example 12.4.5.1)

Pump discharge (m³/s)	Manufacturer's head added (m)	Head losses (m)	Corrected head added (m)
0	100	0	100
0.03	99.55	0.07	99.48
0.06	98.20	0.29	97.91
0.09	95.95	0.64	95.31
0.12	92.80	1.12	91.68
0.15	88.75	1.75	87.00

uration will result in the same head loss corrections. Also, corrections for series or parallel pumps must be made prior to adding the pump curves.

12.4.6 Net Positive Suction Head

The principle behind a radial pump is that work is performed on the pumped fluid by increasing its velocity as the water moves out from the center of the impeller. As the fluid velocity is increased, the pressure drops as can be seen in the conservation of energy equation. At the edge of the impeller the velocity decreases and the pressure increases above the inlet pressure due to the work performed by the pump (Fig. 12.13).

Considering the fluid moving along the impeller vane, here the pressure is low and the design engineer must ensure that the pressure does not drop below vapor pressure or cavitation will occur. *Cavitation* is the formation of gas bubbles in the fluid. These bubbles will continue to move along the vane until the increased pressure causes them to collapse. The

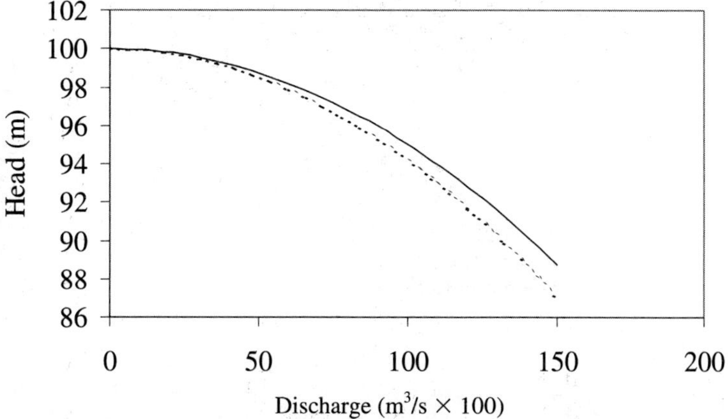

FIGURE 12.12 Manufacturer and corrected pump curves for example 12.8.

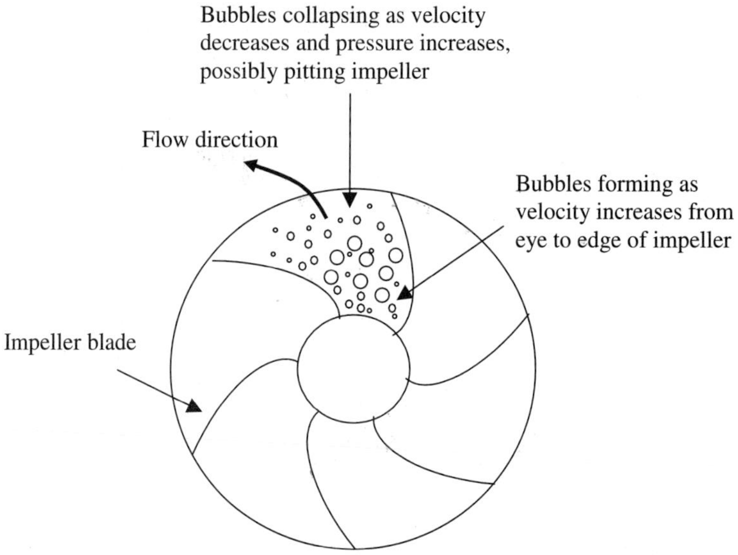

FIGURE 12.13   Pressure and velocity variations across an impeller in a radial flow pump.

bursting bubbles release energy that can pit the impeller blade. In addition to physical pump damage, the process also reduces pump efficiency.

From experiments, pump manufacturer's determine and provide the *required net positive suction head* (NPSH_R). NPSH_R is the minimum absolute pressure head required at the eye of the impeller to ensure that cavitation will not occur. The *available net positive suction head* (NPSH_A) is computed for the pump/valve/fitting configuration at the given flowrate by subtracting the friction and minor losses from the source head for a selected flowrate or

$$\text{NPSH}_\text{A} = H_A + h_{ss} - h_{fs} - h_{ms} - h_{ent} - H_v \tag{12.16}$$

where
$\quad H_A$ = available atmospheric pressure
$\quad h_{ss}$ = static suction head in the wet well above the impeller
$\quad h_{ent}$ = turbulent losses at the pipe inlet
h_{fs} and h_{ms} = friction and minor losses between the wet-well and the eye of the impeller
$\quad H_v$ = the vapor pressure for water at the given temperature

The computed NPSH_A is compared then to NPSH_R. If NPSH_A is greater than NPSH_R, cavitation should not occur at the given flowrate. The available NPSH must exceed the required NPSH for all operating flowrates. Given the uncertainty in estimating the available NPSH, Sanks et al. (1998) recommends that NPSH_A be at least 5 ft or 1.35 times NPSH_R as a safety factor. The Hydraulics Institute (HI, 1994) suggests that 2 to 20 times the NPSH_R may be necessary to completely suppress cavitation.

Example 12.9
Determine the available net positive suction head for flows ranging from 0 to 0.15 m³/s for the system in Example 12.8 assuming the atmospheric pressure is 101000 Pa, the static suction head is 2 m and the vapor pressure is 2450 Pa (0.25 m of water) (water temperature is 20°C).

TABLE 12.10 Available Net Positive Suction
Head for Different Flows (Example 12.9)

Pump discharge (m³/s)	Net positive suction head-available (m)
0	12.05
0.03	12.02
0.06	11.95
0.09	11.83
0.12	11.66
0.15	11.44

Solution

To determine NPSH_A, the energy available at the eye of the impeller is computed so losses only on the suction side of the pump are considered. Flow enters the suction pipe through a sharp-edged entrance and passes through a 3-m, 250-mm diameter pipe containing a gate valve, bend, and reducer. NPSH_A can be computed from Eq. (12.16) by substituting the appropriate losses as

$$\text{NPSH}_A = H_A + h_{ss} + h_{fs} - K_{ent}\frac{V_s^2}{2g} - K_{gate}\frac{V_s^2}{2g} - K_{bend}\frac{V_s^2}{2g} - K_{red}\frac{V_s^2}{2g} - H_V$$

As in the previous example, lumping the minor losses as one term written in terms of Q for the suction line only, this equation becomes

$$\text{NPSH}_A = H_A + h_{ss} - 3.36Q^{1.85} - 22.84Q^2 - H_V$$

$$= \left(\frac{101000}{9810}\right) + 2 - 3.36Q^{1.85} - 22.84Q^2 - 0.25$$

$$= 12.05 - 3.36Q^{1.85} - 22.84Q^2$$

H_A is the pressure in pascals converted to the equivalent pressure head in meters by dividing by the specific weight of water (assumed to be 9810 N/m³). Table 12.10 gives values of NPSH_A for the range of flows. As expected NPSH_A decreases with higher flows as the energy loss increases.

12.4.7 Matching Pump and System Curves

The operating point of a pipeline/pump combination is the intersection point of the system and modified pump curves (Fig. 12.2). Optimally, this point will correspond to the most efficient flow point for the pump. Alternative impeller sizes and pump motor speeds shift the pump curves and may provide better matches.

Several other points should be considered when selecting a pump. First, the assumptions in developing both the system and pump curves are uncertain. The system curve assumes an inlet pressure and an outlet condition. These assumptions may be in error. Second, pipe friction losses are based on an assumed friction factor. Pipe roughness increases over time depending upon the pipe type and water chemistry. Each of these factors will alter the system curve and the resulting operating point (Figs. 12.14 and 12.16).

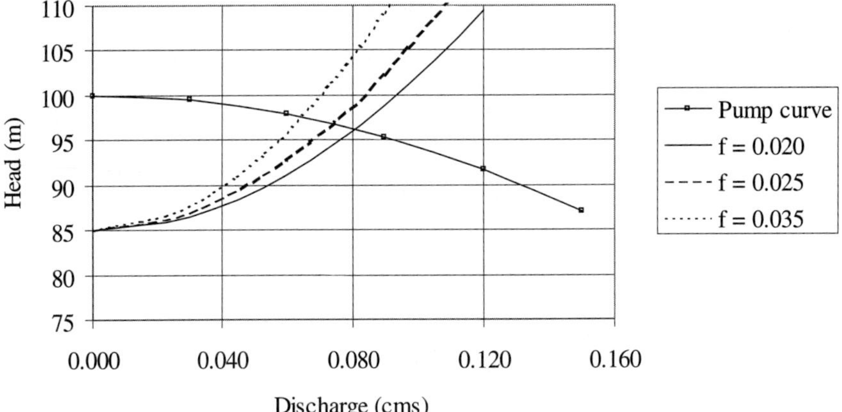

FIGURE 12.14   Variation in operating point with changes in the system curve resulting from different pipe roughnesses (example 12.11).

In addition, the pump curve supplied by the manufacturer may be for a polished pump that is on the upper end of the potential for the pump. As the pump impeller wears over time the pump curve will shift downward. Aging will have more effect at higher flowrates. Finally, pump impeller size also changes the pump output and resulting operation point (Fig. 12.15).

To account for these changes, the design operating point should be selected to the right of the most efficient point so that as pump and pipeline aging occurs the pump efficiency will improve. To reduce the impact of pump aging, very flat pump curves are to be avoided since a shift in the pump or system curve will have a significant impact on pump output. Also, steep pump curves around the operating point are likely not acceptable since errors in

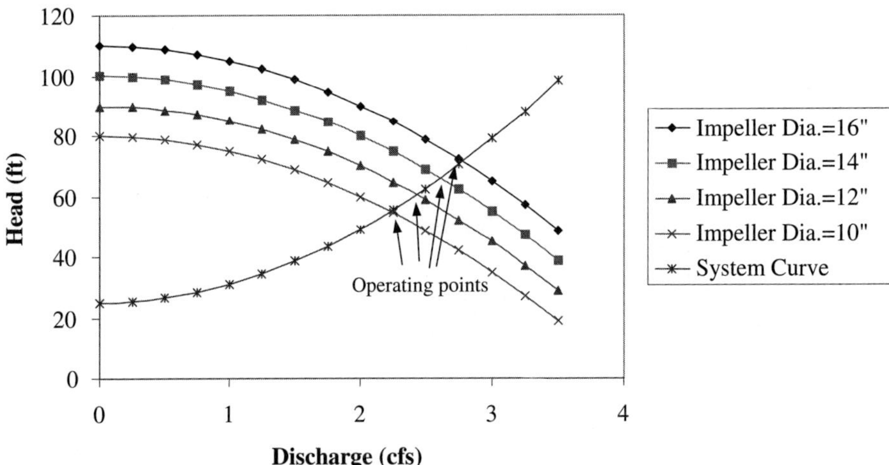

FIGURE 12.15   Pump performance curves and operating points for a single system curve and various impeller sizes.

assumptions developing the system curve will have a dramatic effect on the operating point and the head supplied by the pump.

Maintaining operation near the *best efficient point* (bep) will allow a pump to function for years with little maintenance. If a pump operates well above the bep, cavitation becomes a concern as the $NPSH_R$ increases rapidly with higher flows. Operating near the cutoff head may result in recirculation within the pump causing vibration and excessive wear on the impeller. As the operating point moves away from the bep, pump thrust and radial loads increase which increases the wear on the pump bearings and shaft. For all these reasons, pump operation should be maintained between 60 to 70 and 125 to 130% of the bep flowrate.

Example 12.10

The pump considered in Example 12.8 is used to supply a 2500-m, 300-mm pipeline system with a friction factor equal of 0.02. Assume the static head lift required is 85 m. What is the operating point for the pump?

Solution

The corrected pump curve defines the net head supplied by the pump across the pump station accounting for the friction and turbulence losses in the pump station piping and in the pump itself. The relationship is

$$H = H_{pump} - H_1 = 100 - 500Q^2 - (67.26Q^2 + 7.84Q^{1.85})$$

$$= 100 - 567.26Q^2 - 7.84Q^{1.85}$$

The energy required for the system is the static head gain (85 m) plus the friction losses. The friction losses in the system considering turbulent flow for the full range of flows ($f = 0.020$) are given by the Darcy-Weisbach equation and the system curve is then

$$H_{system} = 85 + 1700Q^2 = 85 + f\frac{LQ^2}{DA^22g}$$

The operating point is the intersection of these two curves or $H = H_{system}$ as shown in Fig. 12.14. Solving this equation for Q gives, $Q = 0.081$ m³/s. This discharge is the operating discharge for this pump/pipeline combination. The pump head added will be 96.2 m.

Example 12.11

As the pipe ages, the friction factor increases. What is the new operating point if f increases to 0.025 or 0.035?

Solution

As the f value increases, the system curve becomes higher and moves to the right while and the pump curve remains constant. By substituting the new f values, the system curve equations becomes:

$$H = 85 + 2125Q^2 \qquad \text{for } f = 0.025$$

$$H = 85 + 2975Q^2 \qquad \text{for } f = 0.035$$

Solving for the discharge for the new system curves gives, $Q(f = 0.025) = 0.074$ m³/s and $Q(f = 0.035) = 0.065$ m³/s (Fig. 12.14). Note that the discharge is lower as the pipe roughness increases. This result suggests that a pump should be selected to operate to the right of the best efficient point so that as the system ages the efficiency will increase.

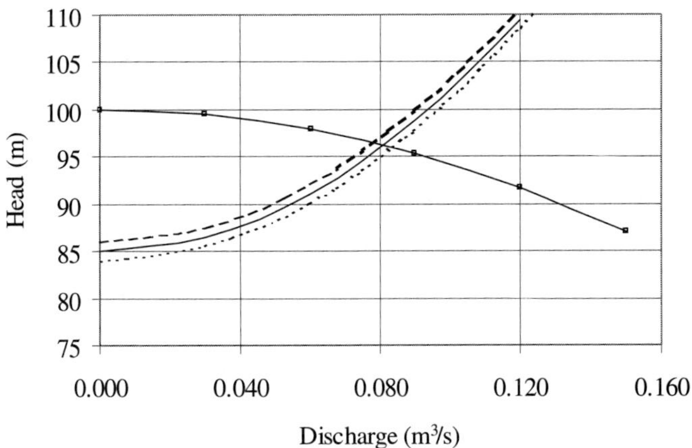

FIGURE 12.16   Shift of operating point due to change in wet-well water surface elevation (h_{ss}) (Example 12.12).

Example 12.12
The pump in Problem 12.10 draws water from the wet-well. During an event the water level will vary which affects the static lift required. Assuming the range for the wet-well is 2 m (84–86 m). Determine the change in flowrate for the range of static lifts.

Solution
As the static lift required changes, its corresponding term in system curve equation changes. The system equations for the wet-well level of 84 and 86 m for an f value of 0.020 are

$$H_{system} = 84 + 1700Q^2$$

$$H_{system} = 86 + 1700Q^2$$

To determine operating flowrate, these equations are equated with the pump curve equation or

$$H = 100 - 567.26Q^2 - 7.84Q^{1.85} = H_{system} = 84 + 1700Q^2$$

Solving for the flowrates gives $Q(h_{stat} = 84 \text{ m}) = 0.084 \text{ m}^3/\text{s}$ and $Q(h_{stat} = 86 \text{ m}) = 0.078$ m³/s (Fig. 12.16). The proportional difference in flowrate is not very great (3%). However, if the pump curve was flatter the difference in discharge would be more significant suggesting the operating in a flat region of the pump curve is not preferable.

12.4.8 Pump Startup and Shutdown Considerations

12.4.8.1 Startup.   During pump startup the flow velocity in the pipeline is low and the equilibrium point between the system curve and the pump has not been reached. Since the velocity is low the energy losses are small while the pump is able to supply a large head at low flowrates. This condition is not critical for radial-flow pumps because the brake horsepower requirement increases with flowrate. As such, the maximum brake horsepower occurs at the design point.

Motor overload, however, can occur in axial flow pumps that may be used in stormwater applications. The brake horsepower in axial pumps decreases with flowrate; thus, the highest demand is when the flowrate is small. Thus, the design engineer should check the efficient range and the low flow range when selecting a pump driver.

12.4.8.2 Shutdown.   Normal pump shutdown can pose the same difficulty as startup for axial flow pumps and the check noted above should be completed. Pipeline and pump casing design are a more critical concern during rapid pump shutdown, say by power failure or rapid valve closure. Rapid shutdown causes surges of high and low pressures known as a pressure transient. High pressures may rupture the pump casing or pipeline while low pressures may cause pipe collapse. Metal pipes have more strength to resist the surge compared to plastic piping. Joints between pipe segments must also be considered. In all, this issue should not be overlooked. Negative pressures and column separation can also occur during pump failures.

Analysis of the magnitude of the pressure wave and its possible impact on discharge line design and pump controls are complex and beyond the scope of this chapter.

12.4.9 Step-by-Step Hydraulic Design

To select a pump the following steps should be completed:

1. Develop system head curves using range of minor loss and friction coefficients and static inlet heads.

2. Select candidate pump/motor configurations and develop corrected pump curves to account for energy losses in the pump house.

3. Plot pump and system curves to identify operating point for most likely operating conditions for new pump and an average static head. Select a pump that operates at a flowrate above the bep. Avoid pumps with steep or flat performance curves in the operating range.

4. Check that the available NPSH exceeds the required NPSH with an appropriate safety factor for all conditions and operating points.

5. If a single pump cannot supply the desired flow-head range (approx. 70 to 125% of bep flowrate), complete checks for all pump combinations.

6. For axial pumps, verify that pump startup and shutdown will not overload pump motor, that is, bhp required does not exceed available.

12.5 PUMP STATION CONFIGURATION

As shown in Fig. 12.17, stormwater pump stations can be configured as a *dry-pit* or *wet-pit design.* In dry-pit stations, two separate chambers are used. The first chamber, called a storage box or wet-well, receives the stormwater and delivers it to the pumps positioned in the floor of the second chamber (dry-well) through a connecting suction line. The dry-pit station is more common in wastewater pumping stations since the dry-well protects personnel against the health hazards related to sewage during routine and emergency repairs.

In wet-pit stations, both the pump and sump are located in one chamber. The pump is connected to an overhead motor by a vertical drive shaft located in the center of the riser pipe that delivers the stormwater to the discharge line. Submersible pumps have special motors designed to operate in liquid. They are efficiently used in wet-pit stations due to their low requirements of maintenance and horsepower since the long drive shaft required in the

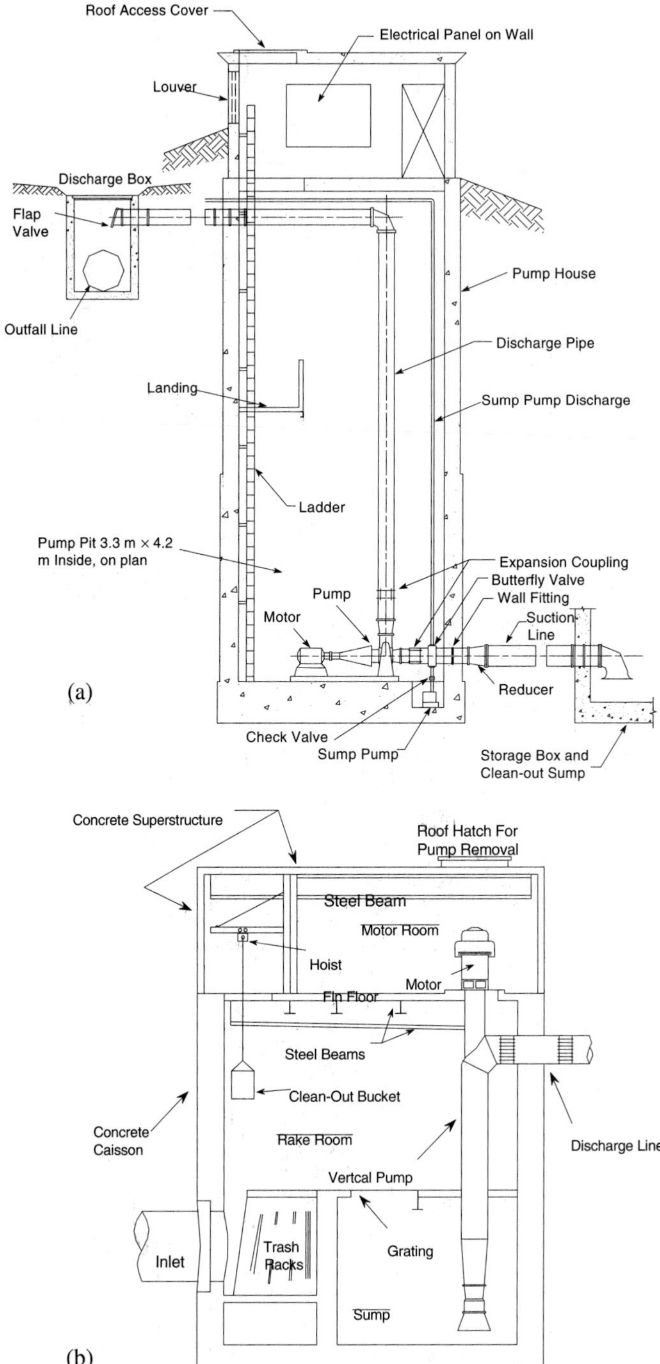

FIGURE 12.17 Layouts of (*a*) dry-pit station and (*b*) wet-pit station (after Brown et al. 1996).

regular pumps is eliminated. In the past, submersible units were produced in small sizes and were used in dry wells to evacuate seepage water. However, recent developments allow production of large submersible pumps and permit them to be used in broader applications.

12.6 PUMP STATION OPERATION

Pump operations are the driving factor of pump station design. Operations are particularly important with multiple pumps. In this case, the operation is characterized by pump starting-stopping cycles that depend on the pump performance and the defined starting-stopping control elevations. For a pump configuration and operating policy, a mass curve routing diagram is developed given an inflow hydrograph that represents the design storm. This diagram is a plot of the inflow mass curve and the pump discharge curve and describes the pumping units' operations during the storm. It also provides important information about the required wet-pit storage to accommodate the storm inflow for the selected number and ca pacities of the pumping units.

To provide a better understanding of pump operations, consider a wet-pit pump station housing three pumps. Figure 12.18 shows a sketch of the pump station connected to the suction and discharge lines. Each pump has a rated capacity of 0.15 m^3/s (5.3 ft^3/s).

The storage volume within the pumping system consists of storage in the upstream storm sewer line and storage in the wet-well. The pump pit is a 7.0 m (23 ft) diameter cylindrical wet well that is empty at the beginning of the storm (elevation 0.0 m). The influent storm sewer is a 250-m (820-ft) long, 1420-mm (56-in) diameter pipe laid on a downward slope of 0.3%. A stage-storage relationship (Table 12.11) relating the storage volume available with elevations was developed using geometrical relations like the ungula of a cone and cylinder formulas (see Appendix). The defined starting-stopping control elevations for the three pumps are also shown in Table 12.11.

The *mass inflow curve* for the design inflow hydrograph is plotted as the cumulative inflow volumes (in m^3) versus time (Fig. 12.19). A tabular summary is given in Table 12.12. The incremental inflow volume during a time increment (column 5) is the product of average flow during that time increment (column 3) and the time increment (column 4). The cumulative inflow volume at any time (column 6) is the summation of incremental inflows to the wet-well from the beginning of the storm until that time and is plotted in Fig. 12.19.

The pump discharge curve in Fig. 12.19 also shows the cumulative volume discharged from the wet-well over time. These volumes are calculated by adding the current incremental

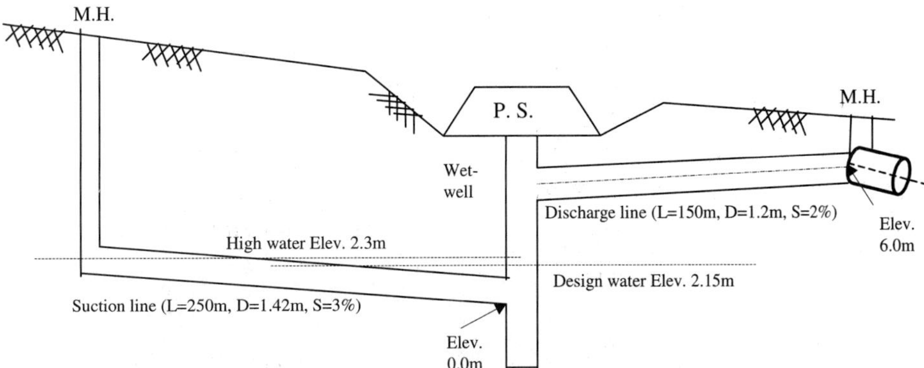

FIGURE 12.18 Schematic of pump station with supply and discharge lines. Note P.S. = pump station and M.H. = manhole.

TABLE 12.11 Stage-Storage Relationship and the Pump Control Elevations

Elevation (m)	Pipe storage (m³)	Wet-well storage (m³)	Total storage (m³)	Pump operating conditions		
				Pump 1	Pump 2	Pump 3
0.00	0.0	0.0	0.0	Off		
0.15	4.2	5.8	10.0			
0.30	20.5	11.6	32.1		Off	
0.45	47.1	17.3	64.4	On		
0.60	73.2	23.1	96.3			Off
0.75	90.5	28.9	119.4		On	
0.90	115.2	34.7	149.9			
1.05	140.1	40.4	180.5			
1.20	164.6	46.2	210.8			On
1.35	187.9	52.0	239.9			
1.50	201.6	57.8	259.4			
1.65	219.4	63.5	282.9			
1.80	235.7	69.3	305.0			
1.95	243.3	75.1	318.4			
2.10	245.6	80.9	326.5			
2.25	245.6	86.6	332.2			

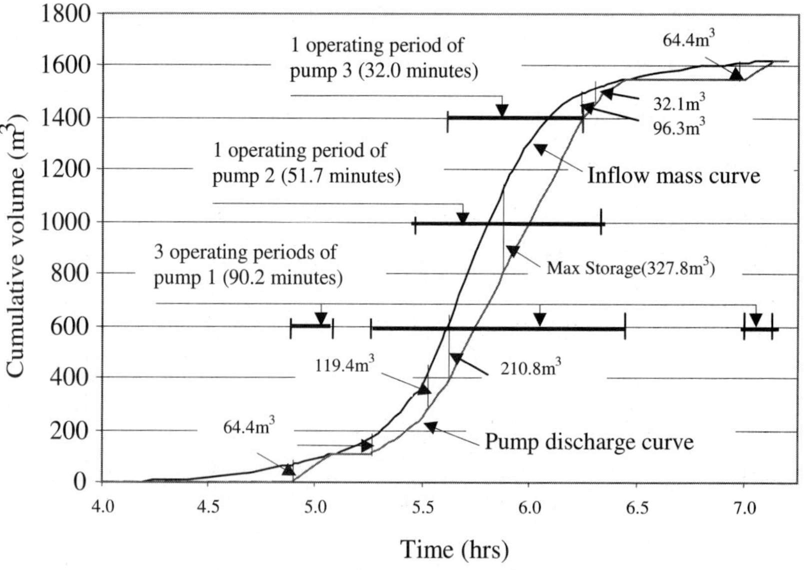

FIGURE 12.19 Mass inflow/outflow curves for design storm.

TABLE 12.12 Mass Curve Routing and Storage Calculations for Wet-well Design Example

(1) Time	(2) Inflow	(3) Average inflow rate	(4) Time step	(5) Incremental inflow vol.	(6) Cumul. inflow volume	(7) Cumul. pumping volume	(8) Storage volume	(9) Operating pump conditions {0 = off and 1 = on}		
(hrs)	(m³/s)	(m³/s)	(sec.)	(m³)	(m³)	(m³)	(m³)	Pump 1	Pump 2	Pump 3
4.00	0.000	—	—	—	0.0	0.0	0.0	0	0	0
4.10	0.004	0.002	360	0.7	0.7	0.0	0.7	0	0	0
4.20	0.008	0.006	360	2.2	2.9	0.0	2.9	0	0	0
4.30	0.011	0.010	360	3.4	6.3	0.0	6.3	0	0	0
4.40	0.014	0.013	360	4.5	10.8	0.0	10.8	0	0	0
4.50	0.020	0.017	360	6.1	16.9	0.0	16.9	0	0	0
4.60	0.027	0.024	360	8.5	25.4	0.0	25.4	0	0	0
4.70	0.035	0.031	360	11.2	36.5	0.0	36.5	0	0	0
4.80	0.042	0.039	360	13.9	50.4	0.0	50.4	0	0	0
4.885	0.049	0.046	306	14.0	64.4	0.0	64.4	1	0	0
4.90	0.050	0.050	54	2.7	67.0	8.1*	58.9	1	0	0
5.00	0.057	0.054	360	19.3	86.3	62.1	24.2	1	0	0
5.078	0.071	0.064	281	18.0	104.3	104.2	0.0	1	0	0
5.10	0.075	0.073	79	5.8	110.0	104.2	5.8	0	0	0
5.20	0.110	0.093	360	33.3	143.3	104.2	39.1	0	0	0
5.257	0.136	0.123	205	25.2	168.6	104.2	64.4	1	0	0
5.30	0.160	0.148	155	22.9	191.5	127.4	64.1	1	0	0
5.40	0.250	0.205	360	73.8	265.3	181.4	83.9	1	0	0
5.47	0.332	0.291	252	73.3	338.6	219.2	119.4	1	1	0
5.50	0.400	0.366	108	39.5	378.2	251.6**	126.5	1	1	0
5.60	0.580	0.490	360	176.4	554.6	359.6	194.9	1	1	0
5.615	0.609	0.595	54	32.1	586.7	375.8	210.8	1	1	1
5.70	0.620	0.615	306	188.0	774.7	513.5***	261.2	1	1	1
5.80	0.550	0.585	360	210.6	985.3	675.5	309.8	1	1	1
5.90	0.450	0.500	360	180.0	1165.3	837.5	327.8	1	1	1
5.92	0.420	0.435	72	31.3	1196.6	869.9	326.7	1	1	1
6.00	0.360	0.390	288	112.3	1308.9	999.5	309.4	1	1	1
6.10	0.220	0.290	360	104.4	1413.3	1161.5	251.8	1	1	1
6.20	0.130	0.175	360	63.0	1476.3	1323.5	152.8	1	1	1
6.247	0.102	0.116	169	19.6	1496.0	1399.7	96.3	1	1	1
6.30	0.088	0.095	191	18.1	1514.1	1456.9	57.2	1	1	0
6.332	0.076	0.082	115	9.5	1523.5	1491.5	32.1	1	1	0
6.398	0.065	0.071	236	16.6	1540.2	1526.9	13.3	1	0	0
6.439	0.055	0.060	148	8.9	1549.0	1549.0	0.0	1	0	0
6.50	0.049	0.052	221	11.5	1560.5	1549.0	11.5	0	0	0
6.605	0.040	0.045	377	16.8	1577.3	1549.0	28.3	0	0	0
6.70	0.032	0.036	343	12.3	1589.7	1549.0	40.7	0	0	0
6.80	0.024	0.028	360	10.1	1599.7	1549.0	50.7	0	0	0
6.90	0.019	0.022	360	7.7	1607.5	1549.0	58.5	0	0	0
7.00	0.014	0.016	360	5.9	1613.3	1549.0	64.4	1	0	0
7.10	0.010	0.012	360	4.3	1617.6	1603.0	14.6	1	0	0
7.129	0.010	0.010	104	1.0	1618.6	1618.7	0.0	1	0	0
7.20	0.009	0.010	360	3.4	1621.0	1618.7	2.4	0	0	0

*8.1 = 0.0 + 0.15*54; **251.6 = 219.2 + (0.15 + 0.15)*108; ***513.5 = 375.8 + (0.15 + 0.15 + 0.15)*306

pumped volume to the cumulative pumped volume at the previous time step (column 7 of Table 12.12). The incremental pumped volume is calculated by multiplying the time step by total pumping rate for all operating pumps.

The operating pumps and their operation periods are governed by the assigned starting-stopping control elevations. Each pump either operates or stops according to the wet-well storage. The storage is computed by subtracting the cumulative pumped volume (column 7) from the cumulative inflow volume (column 6). Therefore, the number of operating pumps is determined by the storage and the total pumping rate is computed (column 8). For example, pump 1, the lead operating pump, was turned on once the storage reached 64.4 m³ (2275 ft³) at time 4.885 hrs. After that time, the storage receded until the wet-well was empty and the pump was shut down. Pump 1 was then restarted at time 5.257 hrs when the storage again reached 64.4 m³. Here, however, the water level continued to rise eventually reaching 119.4 m³ (4217 ft³). This storage triggered the on-setting for pump 2 at time 5.47 hrs. Due to the continued high inflows, the two operating pumps are still not capable of reducing the storage that increases to 210.8 m³ (7445 ft³) at time 5.615 hrs. This storage volume activates the third pump and the storage begins to decrease. Pump 3 is turned off at time 6.247 hrs when the wet well storage was 96.3 m³ (3401 ft³). Since the inflow hydrograph is receding, the storage then drops rapidly until pump 2 stops at time 6.332 hrs. Pump 1 finally stops at time 6.439 hrs. Near the end of the storm at time 7.0 hrs, pump 1 operates to remove the small amount of accumulated inflow and stops at time 7.129 hrs.

The maximum storage during the design storm was 327.8 m³ (11,578 ft³) at time 5.9 hrs. This storage corresponds to an elevation of about 2.15 m that is defined as the design water elevation (Table 12.11). Table 12.12 shows that the lead pump was operated three times during the storm. Two short cycles occurred at the beginning and end of the storm. The elapsed time between starts, known as the cycle time, is dependent upon the pumping rate, the inflow, and the starting-stopping control elevations. The cycle time decreases with the volume present between the start and stop control elevations, known as the usable volume. Short cycle times may result in pump overheating and wear. The minimum permissible cycle time is usually obtained from the pump manufacturer based on size of the motor. Table 12.13 lists permissible cycle time estimates to be used during early design stages as a function of motor size.

The *cycle time* is comprised of the time required to empty the usable volume ($V_u/(Q - I)$) and the time to fill that volume (V_u/I). It has been found that the minimum permissible cycle time that avoids pump wear is when the inflow (I) is one-half of the pump output (Q). A relationship for the usable volume can be found by:

$$t_{min} = V_u/(Q - 0.5\ Q) + V_u/0.5\ Q = 4\ V_u/Q \tag{12.17}$$

or

$$V_u = 0.25\ Q\ t_{min} \tag{12.18}$$

TABLE 12.13 Estimates of Permissible Cycle Time

Motor size		Cycling time (t_{min})
HP	KW	minutes
0–1.5	0–11	5
20–30	15–22	6.5
35–60	26–45	8
65–100	49–75	10
150–200	112–149	13

where t_{min} = minimum permissible cycle time in seconds
V_u = usable volume in m^3 (ft^3)
Q = pump output in m^3/s (ft^3/s)

First estimates for control elevations can made based on this usable volume.

For illustrative purposes in the example problem, each pump was assumed to operate at its design discharge (0.15 m^3/s (5.3 ft^3/s)). In reality, the actual pump output varies according to the total dynamic head associated with the storage level. An analysis considering varying pump flows with input head (h_{ss}) is recommended for the final design. Each pump's permissible cycle time should be calculated based on the pump's average operating flow.

The pump control elevations can be determined using the minimum cycle time from Table 12.13 and the usable volume calculated from Eq. (12.18) for the lead pump. The stop elevation of the lead pump is usually taken at the floor of the storage facility so that it can completely evacuate the wet-well. The starting elevation is then set above the stop elevation by the estimated operating elevation range, V_u. For other pumps, a typical elevation difference between pump starts is in the range of 0.3 to 1.0 m (1 to 3.3 ft) for stations with a small storage and 0.08 to 0.15 m (0.3 to 0.5 ft) for stations with large storage. Knowing this distance and the full operating range, the control elevations for other pumps can be determined.

12.6.1 Variable Speed Pumps

Variable speed pumps are designed to discharge a flowrate equal to the incoming flow. The maximum speed is usually maintained at the high wet-well water level while the minimum speed is run at the low water level. They, therefore, require less storage and excavation and eliminate frequent pump starts and stops. However, constant speed pumps (intermittent pumps) are preferred in storm drainage systems.

12.7 DESIGN OF PUMP SUMPS

Pump sump design has profound effects on pump performance. A poor designed sump and piping system can significantly reduce pump efficiency. Although a systematic rational procedure is not readily available, the proposed sump design features should be reviewed for potential problems and modified accordingly. Model studies are often employed to ensure good sump design and performance for pumps larger than 1.3 m^3/s (46 ft^3/s). Some useful design criteria and guidelines for more typical and small units are summarized below.

12.7.1 Pump Intake

Each pump should have a separate suction line and be kept straight and free of valves and fittings that are directly connected to the pump. Suction intake velocities should be less than 1.5 m/s (5 ft/s) and the entrance velocity at the mouth of the inlet should be less than 1.1 m/s (3.6 ft/s). As noted in the hydraulics section, the net positive suction head available (NPSH$_A$) must exceed NPSH$_R$. The pump manufacturer's guaranteed NPSH$_R$ should be exceeded by the available NPSH$_A$ by at least about 1.5 m (5 ft).

A down-turned suction bell inlet (Fig. 12.20) is very common and recommended (Hydraulic Institute's Standards[2] and Langteau[3]) to avoid vortexing. The suction bell should be positioned at a clearance distance above the floor of $D/4$ to $D/2$ where D is the diameter of the flare. Vortices can be also inhibited by adding large fillets in the corners of the wet well to eliminate the possibility of flow discontinuities in these zones. Many experts insist

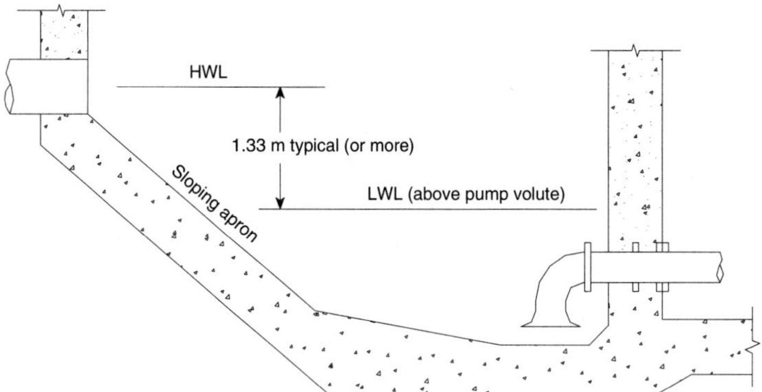

FIGURE 12.20 Sump with down-turned suction bell and sloping apron. (From Sanks et al., 1998)

on using vortex suppressors when using horizontal inlets (Sanks et al. (1998)) (Fig. 12.21). If the required piped inlets are larger than 0.5-m (20-in) draft tube inlets are recommended.

12.7.2 Sump Size and Geometry

To determine the required storage capacity, the high-water elevation and the sump dimensions must be determined. The design high-water level is usually set at least 0.3 to 0.6 m (1 to 2 ft) below ground surface. An initial storage capacity is estimated based on the wet-well dimensions and the influent sewer line. The final capacity is based on a mass curve routing described in Sec. 12.6.

Sump dimensions for wet-well stations are either obtained from the manufacturer literature or from general guidelines (e.g., Hydraulic Institute, 1994 (Fig. 12.22)). These guides

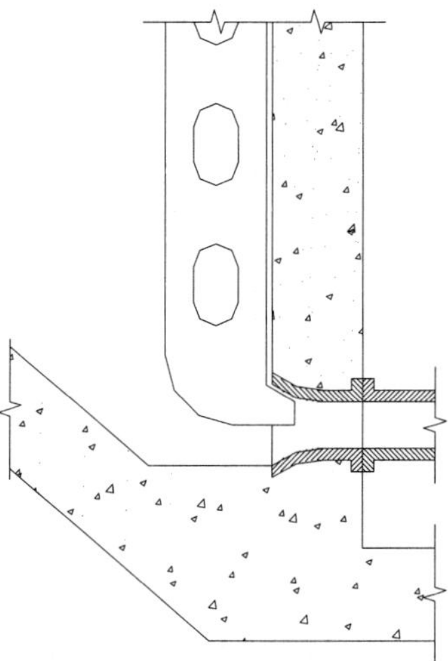

FIGURE 12.21 Plate vortex suppressor on a horizontal intake (after Sanks et al., 1998).

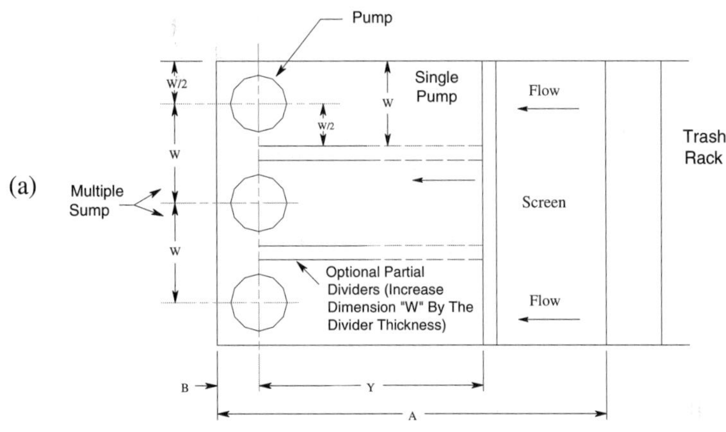

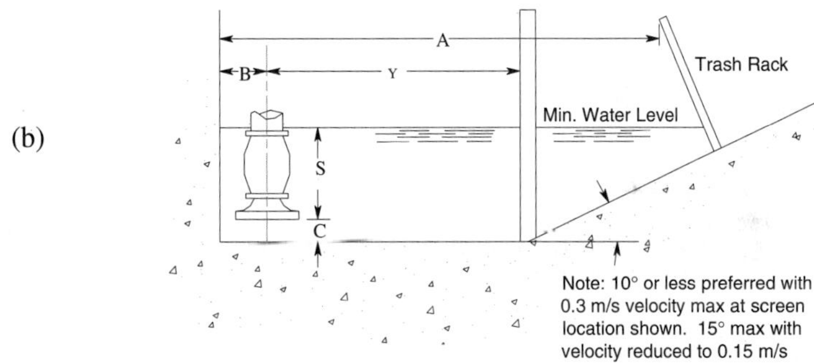

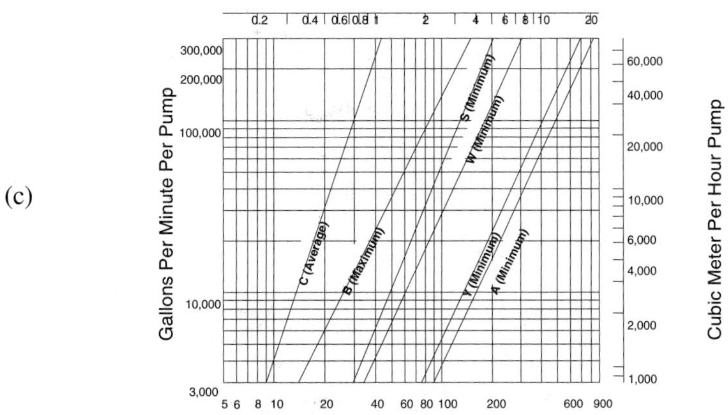

FIGURE 12.22 Sump layout and dimensions of wet-pit type pump stations: (*a*) plan view, (*b*) Elevation view, (*c*) Dimensions (after Brown et al., 1996).

identify the minimum clearances from pump to pump, pump to back wall, and pump to sidewall based on the discharge rate per pump.

The free fall of incoming flow into the wet-well produces entrained air bubbles that can be drawn into the pump greatly reducing the pump efficiencies (especially those with low specific speed). One way to avoid the free fall and associated air bubbles is to receive the incoming flow on a sloping apron as shown in Fig. 12.22.

12.7.3 Debris and Sediment Protection

If large debris are anticipated, coarse screens should be installed ahead of the wet-well. Simple inclined screens with bar spacing of 3.8 cm (1.5 in) are usually adequate for storm-water pumping stations. They should be manually raked at regular maintenance periods and following storm events. In some cases, screens are provided with overflow bypass arrange-

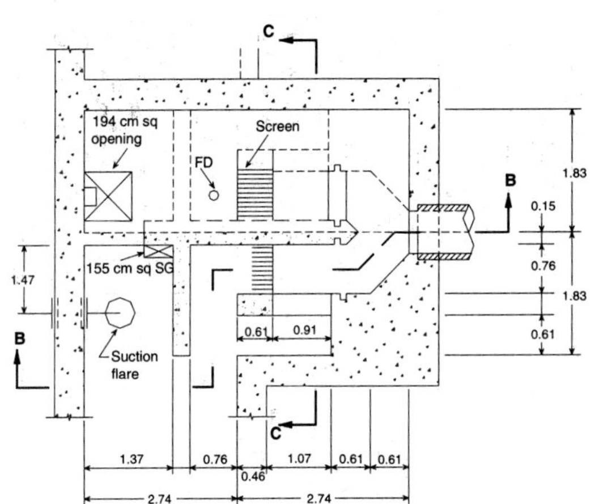

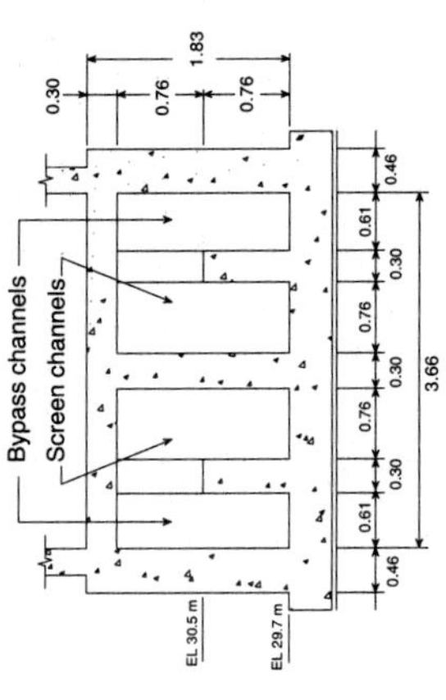

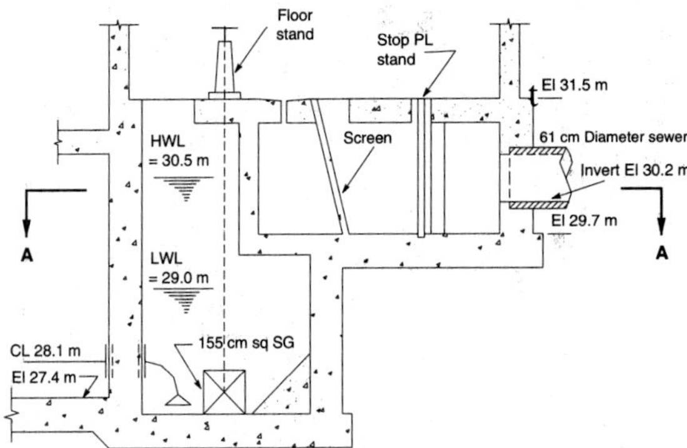

FIGURE 12.23 Trash rack in inlet facility (after Sanks et al., 1998).

ments to protect against clogging and subsequent surcharging of the upstream collection system. For example, Fig. 12.23 shows an inlet structure with two screens; one for each half of the wet-well accompanied by an adjacent overflow bypass chamber. A grit chamber may be needed to collect the solids if substantial sediment loads are expected.

12.8   DESIGN OF DISCHARGE LINES

The *discharge line* is an integral part of the pumping station design. In general, the designer should keep the discharge piping as simple as possible. Long pipeline lengths should be avoided to reduce hazards associated with flow reversal. For short outfall lengths, it is always economical and effective to have an individual discharge line for each pump. If long discharge lines are necessary, they should be combined into one force main using a manifold. A cost analysis should be conducted to determine the appropriate lengths of discharge lines that require manifolds.

Discharge lines should be sized to maintain a minimum line velocity of 0.6 m/s (2 ft/s) and a maximum line velocity of 2.5 m/s (8.2 ft/s). The size should also consider the manufactured pump outlet size. The diameter should be selected larger than the outlet size to reduce the line losses. Therefore, the preliminary design should propose some pumps before sizing the discharge line then the line size can be modified when pump selection is refined.

Another consideration of the discharge line selection is surge and acceleration head requirements. In systems characterized by high proportions of static head and friction head loss relative to the total head, the acceleration head requirement is of concern and should be reviewed. The time required to accelerate the liquid in the discharge pipe from zero velocity to design velocity, described as the pause time, is derived from Newton's second law:

$$t = \frac{LV}{g(H_{cutoff} - h_{ss})} \tag{12.19}$$

where t = pause time in seconds
  H_{cutoff} = pump shutoff head, m (ft)
  h_{ss} = static head available in the wet-well, m (ft)
  L = discharge line length, m (ft)
  V = flow velocity, m/s (ft/s)
  g = gravity acceleration, m/s² (ft/s²)

If the pause time exceeds 10 seconds, problems related to high radial thrust forces, when the pumps operate close to their zero flow, are expected. In this case, the design should be modified in coordination with the pump manufacturer. Finally, the design should ensure that surge pressures do not adversely affect the system operation. Due to the extended nature of the surge topic, design engineers should consult other detailed references addressing the surge causes and surge attenuating devices (Thorley 1992, Streeter and Wiley 1993).

12.9   OTHER PRACTICAL CONSIDERATIONS

12.9.1   Station Siting

The site of a stormwater pumping station is initially selected after studying the sewer layout and determining potential low points. A cost analysis of pumping versus increasing the gravity collection sewer depth should be performed to screen potential locations. The final selection should be made after considering many factors including availability of property, proximity to residences, architectural appearance, noise control, character of the architecture, ornamental plantings, access to and availability of parking, avoidance of existing utilities and obstruction, and land cost.

12.9.2 Valves and Fittings

Valves and fittings are needed in storm pumping stations for several purposes. *Check valves* are installed in horizontal positions to prevent flow from returning to force mains, manifolds, or pumps and motors. *Swing* and *ball check valves* are most often used for preventing flow reversal. However, rapid closure can cause water hammer problems. These can be avoided by using valves that close very quickly to immediately eliminate flow reversal (preferred) or very slowly to gradually stop the water column. Powered actuators are usually required for large valves.

Isolating valves are shut-off devices used to stop flow and allow for the removal of pumps, motors, or line portions for repair and maintenance. They are either fully open or fully closed. The common types are *gate* and *cone valves*. *Air/vacuum* and *combination valves* are usually placed at high points in the discharge line to release accumulated air or to remove vacuum by admitting air under down-surge operating conditions. *Release valves* are particularly important in large lines and should be carefully sized to avoid potential water hammer effects.

Failure to select the proper type of valve and/or in installing it in the appropriate location may result in poor and troublesome operation. For example, all valves should be avoided in wet wells. Valves that readily clog, such as butterfly valves, should not be used. Using swing checks or gate valves in vertical piping can lead to clogging. Also, using throttled valves for flow controls wastes energy and should be replaced by other methods such as spillage, bypassing, or use of variable speed pumps. Throttling, that produces low velocities, results in poor control and should be replaced by a reducer, a small valve, and an expander.

12.9.3 Pipe Connections and Material

Pipes in stormwater pumping stations are usually joined using sleeve-type couplings to allow for axial expansion and misalignment. The joints should be harnessed with restraining bolts and provided with restraints at bends. Common fittings in the suctions of the wet-well include 90-degree long radius flange and flare fittings. An eccentric reducer can be used to avoid high spots of air trapping in the pump intake line. All pump connections and joints are flanged. The discharge connections include a concentric increaser, a spring-loaded, lever-operated swing check, a sleeve-type coupling, a manually operated plug or gate valve, and a connection to the discharge header. These connections control the clearance between the pump and the wall of the adjacent wet-well.

The most common materials used in stormwater station piping are ductile iron and steel. Concrete pressure pipes are sometimes used in large discharge lines outside of the pumping station. Piping and fittings on the suction line are often ductile iron while discharge fittings are most often steel or ductile iron. Steel pipes must be protected against corrosion both externally and internally. The selection of a pipe or fitting material must be based on pipe strength, corrosion resistance, regulations, and cost.

12.9.4 Other Considerations

Numerous other concerns beyond hydraulics must be considered during pump station design. The major issues are noted here. Sanks et al. (1998) provides a detailed discussion of these and other concerns. Pumps and motors must be removed from the pump station for periodic maintenance. Although expensive, removable roof hatches in the pump station are sometimes provided for that purpose. Monorails are often used in large stations and mobile cranes are used to lift small equipment. In general, the maintenance program requires monitoring and testing of all equipment so that an efficient and problem-free operation is guaranteed.

Personnel safety during maintenance and routine checks should also be considered during pumping station design. Two major issues are ventilation and hazardous spills. Ventilation of wet-wells is achieved via exhaust fan systems that draw air from the bottom of the well. Hazardous spills are a potential problem in highway conditions, particularly gasoline spills. Other safety considerations include providing adequate headroom, removing obstructions in walkways, avoiding ladders and circular stairs, and installing guards around moving machinery.

Aesthetic considerations of the building structure are wall materials and construction, character of the building, and roof materials and design. Structural considerations include soil bearing, earth pressure against the walls, local groundwater characteristics, concrete quality against loads and water leaks, and framing for the roof and trolley beam.

12.10 *SUMMARY OF PUMP STATION DESIGN*

Pump station design is not a simple and straightforward process due to the interdependence between pumping rate and wet-well capacity. The design engineer should have a good deal of experience with pump station designs, be aware of different related considerations, and, above all, remain in consultation with the pump manufacturer. Given the design hydrograph, the inflow sewer dimensions, and downstream discharge conditions, an iterative design process can be posed consisting of the following tasks.

1. Initial estimate: Estimate the number of pumping units and their expected pumping capacities.
2. Design of major components: Size the pumps, pump sump, storage capacity, and discharge lines. Select pump control elevations for the pumps.
3. Check performance: Complete a mass curve routing as described in Sec. 12.6. Evaluate the performance of the station design for the design storm and/or other extreme storm events.
4. Evaluate the performance: The results of step 3 will suggest one of following: (a) refine the design of major components (step 2), (b) modify the number of pump units and/or their capacities (step 1) and redesign the major components (step 2), or (c) accept the design (stop).

To determine the required storage volume, the high water elevation and the sump dimensions should be determined first. The design high water level is usually set at least 0.3 to 0.6 m (1 to 2 ft) below the ground surface. An initial storage capacity is estimated based on the wet-well dimensions and the influent sewer line.

Once a hydraulically acceptable design is determined, a cost analysis should be completed to estimate the capital and operation costs. The capital cost usually represents the major portion of the total cost since stormwater pump stations only run for short periods of time. Therefore, it is typical to minimize the capital cost by seeking a small number of pumps with low capacities while increasing pump sump storage capacity to offset the pumping capacity. A minimum of two pumps is recommended if low storm inflows are expected and the likelihood of higher inflows is low. As a precaution against pump failure, the pumping rate of each pump should exceed 66% of the total design flow. If three pumps are used, the pumping rate of each pump is set at 50% of the design flow. In these cases, identical pumps are recommended. This approach provides flexibility to change the pump operation sequence after each storm, or even after each cycle in a storm and reduces pump wear compared to a non-alternating system.

A final design should provide an economic balance between the desired pumping and storage capacities, accommodate the design and extreme storm scenarios, and account for relevant practical considerations such as structural, architectural, electrical, heating and ventilation, mechanical, as well as maintenance and safety considerations.

APPENDIX

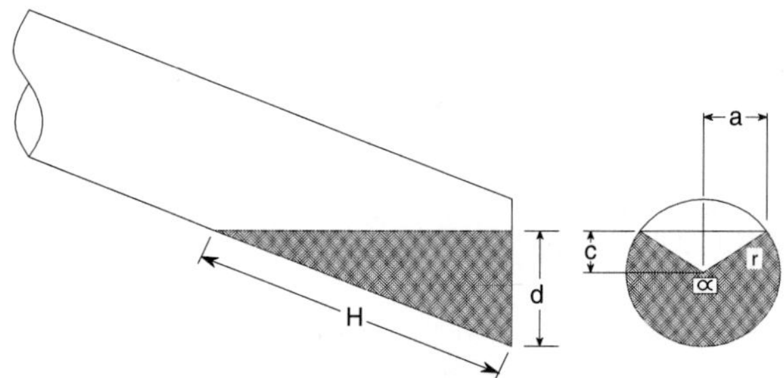

The storage volume in a circular pipe is computed by:

$$V = H \frac{2/3a^3 \pm cB}{r \pm c}$$

where V = volume of ungula, m² (ft²)
  r = pipe radius, m (ft),
  H = wetted pipe length, m (ft),
  B = wetted cross-sectional area m² (ft²)
  $c = d - r$
  d = flow depth in pipe, m (ft)
  $\alpha = 2 \sin^{-1}\left(\dfrac{a}{r}\right)$

and

$$a = \sqrt{(2r - d)d}$$

The wetted cross-sectional area, B, is computed as follows:

$$\text{If } d \leq r, \text{ then } \quad B = A_s$$
$$\text{If } d > r; \text{ then } \quad B = A - A_s$$

A_s = the segment area. m² (ft²) = $(\alpha - \sin \alpha)\dfrac{r^2}{2}$

REFERENCES

American Petroleum Institute, *Centrifugal Pumps for Petroleum and Heavy Duty, Chemical and Gas Industry Services,* API Standard 610, 8th ed., American Petroleum Institute, Washington, DC, 1995.

Bosserman, B., "Pump System Hydraulic Design," chapter 5, in *Water Distribution Systems Handbook,* L. W. Mays, ed., McGraw-Hill, NY, 1999.

Brown, S., S. Stein, and J. Warner, "Urban Drainage Design Manual, Hydraulic Engineering Circular No. 22, Report No. FHWA-SA-96-078," U.S. Department of Transportation, Federal Highway Administration, Office of Technology Applications, Washington, DC, 1996.

Hydraulic Institute, *American National Standards for Centrifugal Pumps for Nomenclature, Definitions, Applications and Operation,* ANSI/HI 1.1-1.5-1994, Hydraulics Institute, Parsippany, NJ, 1994.

Langteau, R. R., "Wet Well Design," in *Conference Proceedings, Pumping Station Design for the Practicing Engineer,* Vol. II, pp. 662–669, R. L. Sanks, C. W. Reh, and A. Amirtharajah, eds., Department of Civil Engineering and Engineering Mechanics, Montana State University, 1981.

Sanks, R. L. et al., *Pump Station Design,* Butterworth Publishers, London, UK, 2nd ed., 1998.

Streeter, V., and E. Wylie, *Fluid Transients,* Prentice-Hall, Englewood Cliffs, NJ, 1993.

Thorley, A., *Fluid Transients in Pipeline Systems,* D&L George, Ltd., Barnet Herts, UK, 1992.

Velon, J., and T. Johnson, "Water Distribution and Treatment," Davis' *Handbook of Applied Hydraulics,* 4th ed., Zipparo and Hasan, eds., McGraw-Hill, NY, 1996.

Water Environment Federation, "Design of Wastewater and Stormwater Pumping Stations," Manual of Practice FD-4, Technical Practice Committee, T. Krause, ed., 1993.

CHAPTER 13
SUBSURFACE STORMWATER MANAGEMENT SYSTEMS

Brian C. Roberts
National Corrugated Steel Pipe Association
Washington, DC

Subsurface stormwater management systems are increasing in popularity as a means of achieving the urban drainage objectives. This widespread use has resulted from a combination of high land values and limited above ground space in urban environments. Liability issues associated with surface systems have also lead to increased use of underground facilities. Subsurface facilities may also be incorporated into existing developments where flooding problems due to sewer surcharging are occurring. These "retrofit" systems can also be designed to provide water quality enhancement. Each proposed development should be carefully examined in order to determine which method of storm water detention or combination of methods could be best applied. Subsurface systems may be categorized under three classifications: 1) *detention,* 2) *infiltration,* 3) *filtration, or any combination of these three. Infiltration and filtration systems are designed as water quality measures.*

13.1 SUBSURFACE DETENTION

Corrugated steel pipe (CSP) detention structures have been used predominantly to control the quantitative aspects of stormwater management. Large pipes are installed underground, often beneath parking lots, where above ground use can be maintained. The quantitative design follows basic hydraulic procedures as described in detail in other chapters of this book. An inflow hydrograph is developed, and based on an allowable release rate, the required storage volume is estimated. Various pipe lengths and diameters are chosen to meet this storage volume. Once the release structure has been sized based on the allowable release rate, the inflow hydrograph is routed through the structure to obtain the outflow hydrograph.

These detention systems can be further classified as in-line or off-line systems. In-line storage incorporates the storage facility directly into the sewer system. Off-line detention collects storm water runoff before it enters the minor system and then discharges it into either a sewer or open water course at a controlled rate. By making use of the major system and connecting all tributary catch basins to a detention tank, approximately 80% of storm runoff may be prevented from directly entering conventional sewer systems. In areas where roof drains are discharged to the surface, close to 100% of the storm runoff may be controlled.

"

FIGURE 13.1   108″ diameter detention system.

13.2 SUBSURFACE WATER QUALITY SYSTEMS

Until recently, water quality enhancement using CSP was done primarily through infiltration and groundwater recharge practices. Perforated pipe and stone backfill provided a means of replacing a portion of the runoff that would otherwise have left the site as an increase in runoff volume. In using infiltration practices, it is extremely critical to have the proper soils and construction techniques to ensure success.

13.2.1 Sand Filter Systems

Sand filter technology originates from water treatment plants and has been adapted to treat runoff from urban areas. *Sand filters* are most often used as a "first flush" treatment system for drainage areas less than 5 hectares. Runoff is filtered through a sand bed and collected through a series of underdrains then discharged to the receiving system. These systems can be extremely efficient at removing pollutants but are susceptible to clogging if sediment is present in high concentrations.

While a variety of types of sand filters are used in stormwater management, the *DC* (District of Columbia) *sand filter* has been specifically adapted to CSP (Figure 13.3).

13.2.2 Pretreatment for Sand Filters

For impervious sites such as parking lots and pavements, sediment volumes may be so low that the sediment chamber requires little consideration. For higher sediment loads, pretreatment may be required to prevent clogging of the filter media. While the underground sand filter contains a sedimentation chamber that can be sized to promote settling, other pretreatment measures can be designed into the site. These include:

FIGURE 13.2 60″ perforated CSP infiltration system.

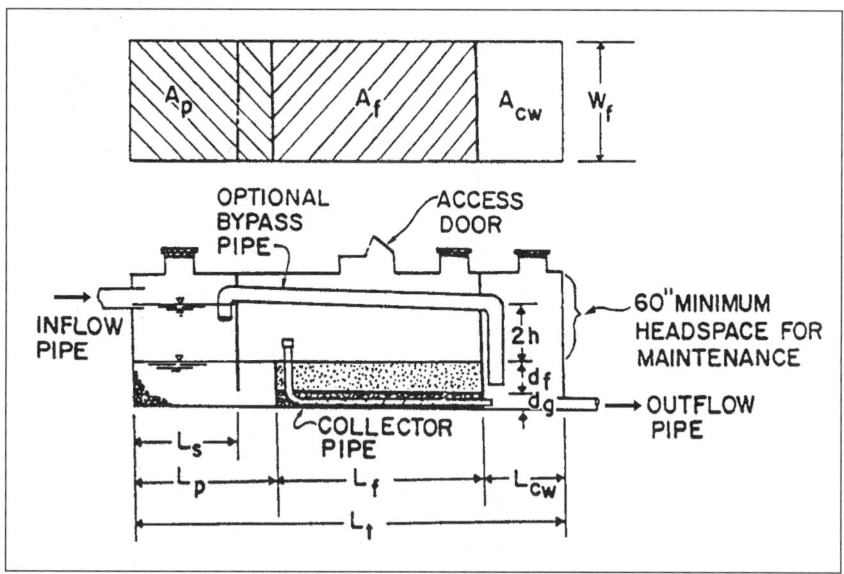

FIGURE 13.3 DC sand filter.

FIGURE 13.4   120″ CSP sand filter.

Sedimentation basins

Vegetative filter strips

Grass swales

Catch basins with sumps

Water quality structures (oil/grit separators)

Some jurisdictions suggest sizing the sedimentation chamber to provide approximately 20% of the WQV storage. Clayton and Schuler (1996) recommend that the following equation be used to size the sedimentation chamber surface area when a desired trap efficiency is known:

$$A_S = (Q_0/w) \cdot \text{Ln}(1 - E)$$

where A_s = sedimentation basin surface area, m²
      E = target trap efficiency (70% to 90% with 90% recommended)
      w = particle settling velocity (for silt use 0.00012 m/s; use 0.001 m/s if site is more than 75% impervious)
      Q_0 = rate of outflow from basin (WQV/detention time)

13.2.3 Design of Sand Filters

Most water quality design practices are sized based on the *water quality volume* (WQV) which is the volume of runoff to be treated. This approach applies to both infiltration and filtration systems. One common method used to determine the WQV is illustrated below (Schueler, 1987). First, the *volumetric runoff coefficient, R_v,* is first computed as:

$$R_V = 0.05 + 0.009(I)$$

where I = site percent impervious.

Then,

$$WQV = P \cdot R_V$$

where P = rainfall depth, in mm
  WQV = Water Quality Volume, in watershed mm.

Selection of the rainfall depth, P, is often made using the 90% rule. That is, selecting a rainfall depth so that 90% of average daily rain events are captured. In humid regions of North America, this often corresponds to between 13 and 25 mm. Rainfall data can be studied for a particular region to determine the appropriate rainfall depth.

A critical step in the design of a sand filter system is sizing the sand filter bed area. The following equation can be used based on Darcy's Law:

$$A_f = WQV \cdot (d_f)/[k * (h_f + d_f) \cdot t_f]$$

where A_f = surface area of the sand filter bed, m²
  WQV = water quality treatment volume, m³
  d_f = sand filter bed depth, m; 0.45 to 0.6 m
  k = coefficient of permeability for filter media (sand), m/day
  h_f = average height of water above the sand bed, m;
  $h_f = 3/4\ h_{max}$
  t_f = time required for the WQV to filter through the sand bed, typically 40 hours

k values for sand are available in text books but usually do not account for clogging associated with accumulated sediments. The City of Austin, Texas recommends a value of 1.05 m/day. Monitoring of these Austin facilities demonstrated values ranging from 0.15 to 0.81 m/day, with an average of 0.45 m/day (Clayton and Schueler, 1996).

The following design steps can be used a guide in sizng the system:

1. Compute the Water Quality Treatment Volume, WQV.
2. Compute the size of the sand filter bed surface area, A_f.
3. Compute the size of the sedimentation basin surface area, A_s.
4. Compute the required minimum storage within the practice ($V_{min} = 3/4 \cdot WQV$).
5. Compute the water volume within the filter bed, $V_f = A_f \cdot d_f \cdot n$.
6. Compute the minimum wet pool volume in the settling basin, $V_w = A_s \cdot 0.9m$.
7. Compute the temporary storage volume required within both chambers, $V_{temp} = V_{min} - (V_f + V_w)$.
8. Compute the total surface area of both chambers, $A_f + A_s$.
9. Compute the additional temporary storage height, Haddtl $= V_{temp}/(A_f + A_s)$. Check to make sure haddtl $> 2 \cdot h_f$, if not, decrease h_f and recompute.

13.2.4 Pollutant Removal

The primary pollutant removal mechanisms for sand filters are mechanical filtration, adsorption, and microbial decomposition on the filter material. Larger sediments are removed in the pretreatment process. Table 13.1 provides expected pollutant removal efficiencies based on monitoring studies (Bell, 1995. Clayton and Schueler, 1996).

TABLE 13.1 Estimated Pollutant Removal Rates for Sand Filters

Pollutant	Removal rate %
Sediment	85
Total phosphorus	40
Total nitrogen	35
Trace metals	50–90
BOD	70–80
Oil and grease	70–85

13.2.5 Inspection and Maintenance

The performance of the sand filter system should be monitored regularly during the first year of operation. After that, the systems are typically inspected on an annual basis unless the performance dictates otherwise. The sedimentation chamber of the sand filter system should be cleaned out when the sediment depth exceeds 300 mm. Trash and debris should be removed on an annual or semi-annual basis. Sediment should be removed from the sand filter bed when accumulation exceeds approximately 25 to 50 mm or if the draw down time significantly exceeds the design period. In most cases, sediment or sand removed from sand filters is disposed of in landfills. Liquid waste is often disposed of at wastewater treatment facilities. It should be noted that these systems are considered confined spaces and must follow Occupational Safety and Health Administration (OSHA) requirements.

13.3 STRUCTURAL DESIGN

After the stormwater management system has been hydraulically designed to meet the required storage volume, its structural design must be considered. Specifically, the *corrugation profile* and the *steel thickness* must be determined, so that the final installation will have strength and stiffness to adequately resist the live and dead loads present. The tables subsequently presented in this chapter simplify this process. The following discussion of loads and design considerations provides a background for the tables.

13.3.1 Depth of Cover

Tables for the selection of the wall thickness, depending upon the pipe diameter or span and the depth of cover requirements, are presented as Tables 13.2 and 13.3. The tables are for circular pipes of a particular corrugation profile. The tables include the effect of live loads that do not exceed a H-20 or H-25 or live load, as indicated.

13.4 LOADS

Underground conduits are subject to two principal kinds of loads:

1. *Dead loads*—developed by the embankment or trench backfill, as well as stationary (permanent) superimposed uniform or concentrated surface loads; and
2. *Live loads*—moving (temporary) loads, including an allowance for impact.

TABLE 13.2 Depth of Cover for CSP Sewers—H20 or H25 Live Load
68mm × 13mm (2⅔ × ½ in.) Corrugation

Diameter of pipe[3]		Specified thickness										Minimum cover	
		1.63 mm / .064 (in.)		2.01 mm / .079 (in.)		2.77 mm / .109 (in.)		3.51 mm / .138 (in.)		4.27 mm / .168 (in.)			
		Maximum cover											
mm	(in.)	m	(ft)	m	(ft)	m	(ft)	m	(ft)	m	(ft)	mm	(in.)
300	12	75	246	94	308							300	12
450	18	50	164	63	207							300	12
600	24	37	121	47	154	66	216					300	12
750	30	30	98	37	121	53	174					300	12
900	36	25	82	31	102	44	144	56	184			300	12
1050	42	21	69	27	89	37	121	48	157	59	194	300	12
1200	48	18	59	23	75	33	108	42	138	52	171	300	12
1350	54	16	52	21	69	29	95	37	121	46	151	300	12
1500	60	—		18	59	26	85	34	112	41	134	300	12

Notes:
1. Backfill around pipe must be compacted to a specified AASHTO T-99 density of 90%.
2. Use reasonable care in handling and installation.
3. Diameters for this corrugation are available up to 2400 mm (96 in.).

13.4.1 Dead Load

The *dead load* is considered to be the soil prism over the pipe. The unit pressure of this prism acting on the horizontal plane at the top of the pipe is given by:

$$DL = w\,H$$

where w = unit weight of soil, kN/m^3, (lb/ft^3)
H = height of fill over pipe, m, (ft)
DL = dead load pressure, kPa, (lb/ft^2)

13.4.2 Live Load

Live loads are greatest when the height of cover over the top of the pipe is small and decrease as the fill height increases. Standard highway loads are referred to as AASHTO H-2O and H-25 live loads. Table 13.4 gives the live load pressure, LL, on the pipe for these loads at various heights of cover.

13.5 STRUCTURAL DESIGN STEPS

The structural design of underground detention systems consists of the following steps:
1. Select the backfill density to be required or expected.
2. Calculate the design pressure.

TABLE 13.3 Depth of Cover for CSP Sewers—125mm × 25mm and 75mm × 25mm (5 × 1 and 3 × 1 in.) Corrugation H20, H25 Live Load

Diameter of pipe		Specified thickness										Minimum cover	
		mm	(in.)	mm	(in.)	mm	(in.)	mm	(in.)	mm	(in.)		
		1.63	.064	2.01	.079	2.77	.109	3.51	.138	4.27	.168		
mm	(in.)	m	(ft)	m	(ft)	m	(ft)	m	(ft)	m	(ft)	mm	(in.)
1350	54	17	56	21	69	29	95	38	125	47	154	300	12
1500	60	15	49	19	62	26	85	34	112	42	138	300	12
1650	66	14	46	17	56	24	79	31	102	38	125	300	12
1800	72	12	39	16	52	22	72	28	92	35	115	300	12
1950	78	11	36	14	46	20	64	26	85	32	105	300	12
2100	84	10	33	13	43	19	62	24	79	30	98	300	12
2250	90	10	33	12	39	17	56	23	75	28	92	300	12
2400	96	9	30	12	39	16	52	21	69	26	85	450	18
2550	102	9	30	11	36	15	49	20	66	25	82	450	18
2700	108	8	26	10	33	14	46	19	62	23	75	450	18
2850	114	8	26	10	33	13	43	17	56	21	69	450	18
3000	120			9	30	12	39	20	52	20	66	450	18
3150	126			9	30	11	36	18	49	18	59	450	18
3300	132			8	26	10	33	17	46	17	56	450	18
3450	138			8	26	10	33	16	43	16	52	450	18
3600	144					9	30	14	36	14	46	450	18

Notes:
1. Backfill around pipe must be compacted to a specified AASHTO T-99 density of 90%.
2. Use reasonable care in handling and installation.

3. Compute the compression in the pipe wall.

4. Calculate the allowable compressive stress.

5. Determine the wall thickness required.

6. Check minimum handling stiffness.

7. Check seam strength requirements.

13.5.1 Backfill Density

Select a percent compaction of pipe backfill for design. The value chosen should reflect the importance and size of the structure, and the quality of backfill that reasonably can be expected. The recommended value for routine use is 90%. However, for more important structures in higher fill situations, select higher quality backfill and specify the same for construction. This can increase the allowable fill height and/or save on metal thickness, although stormwater detention systems are usually under low cover.

TABLE 13.4 Highway and Railway Live
Loads (LL)

Depth of cover, (m)	Highway loading[1]	
	Load, kPa	
	H-20	H-25
0.30	86.2	107.8
0.61	38.3	47.9
0.91	28.7	35.9
1.22	19.2	24.0
1.52	12.0	15.0
1.83	9.6	12.0
2.13	8.4	10.5
2.44	4.8	6.0

Notes:
1. Neglect live load when less than 5 kPa; use dead load only.

13.5.2 Design Pressure

When the height of cover is equal to or greater than the span or diameter of the structure, the total load (total load is the sum of the dead and live loads) can be reduced by a factor of K which is a function of backfill density and which accounts for soil arching over the pipe.

- 85% Standard Proctor Density, $K = 0.86$
- 90% Standard Proctor Density, $K = 0.75$
- 95% Standard Proctor Density, $K = 0.65$

A conservative design uses the K value for a Standard Proctor Density of 85% when 90% density is specified for the backfill.

If the height of cover is less than one pipe diameter, which is often the case with underground detention systems, the total load is assumed to act on the pipe.

$$Pv = K (DL + LL), \text{ when } H \geq S$$

$$Pv = (DL + LL), \text{ when } H < S$$

where Pv = design pressure, kPa, (lb/ft^2)
  K = load factor
  DL = dead load pressure, kPa, (lb/ft^2)
  LL = live load pressure, kPa, (lb/ft^2)
  H = height of cover, m, (ft)
  S = diameter or span, m, (ft)

13.5.3 Ring Compression

With the inherent flexibility of corrugated steel pipe, the vertical load on the top of the pipe causes a slight downward deflection of the top of the pipe and a resulting outward deflection of the sides of the pipe. The sides of the pipe push against the compacted side fill material to mobilize passive earth pressure. Thus, the pipe is loaded by radial pressure. For round pipes, the pressure around the periphery tends to be approximately equal, particularly at deep fill heights.

The radial pressure develops a compressive thrust in the pipe wall, and the pipe must have sufficient structural strength to resist this load. Accordingly, the stress in the pipe wall may be determined and compared to recognized allowable values to prevent yielding and buckling. Such allowable values have been derived from destructive tests done in extensive research programs, against which a safety factor of 2.0 is applied.

The compressive thrust in the conduit wall, or "ring compression," is equal to the radial pressure acting on the wall multiplied by the pipe radius. The ring compression is an axial force acting tangentially to the conduit wall. For conventional structures in which the top arc approaches a semicircle, it is convenient to substitute half the span for the wall radius. The formula for ring compression is:

$$C = \frac{PvS}{2}$$

where C = ring compression, kN/m (lb/ft)
Pv = design pressure, kPa (lb/in^2)
S = diameter or span, m (ft)

13.5.4 Allowable Wall Stress

The *ultimate compressive stress*, f_b, for corrugated steel structures with backfill compacted to 85% Standard Proctor Density and a minimum yield strength of 33,000 psi, is expressed by equations. The first is the specified minimum yield strength of the steel which represents wall crushing or yielding. The second represents the interaction of yielding and ring buckling, and the third represents ring buckling. These equations are, respectively:

$$f_b = f_y = 230 \text{ MPa } (33{,}000 \text{ psi}), \text{ when } D/r \leq 294$$

$$f_b = 40{,}000 - .081(D/r)^2,$$

when $294 < D/r \leq 500$

$$f_b = (4.93 \times 10^9)/(D/r)^2, \text{ when } D/r > 500$$

where f_b = ultimate compressive stress, MPa, (lb/in^2)
f_y = specified minimum yield strength, MPa, (lb/in^2)
D = diameter or span, mm, (in)
r = radius of gyration, mm (in) for an assumed
corrugation profile and wall thickness

The radius of gyration is presented with other corrugation section properties in Table 13.5 for various corrugations and wall thicknesses. A factor of safety of 2 is applied to the ultimate compressive stress to obtain the design stress.

$$f_c = \frac{f_b}{2}$$

where f_c = design stress, MPa (lb/in^2)
f_b = ultimate compressive stress, MPa (lb/in^2)

TABLE 13.5 Moment of Inertia (I) and Cross-Sectional Area (A) of Corrugated Steel for Underground Conduits

Corrugation profile (mm)	Specified thickness[1], mm									
	1.32	1.63	2.01	2.77 / 2.82	3.51 / 3.56	4.27 / 4.32	4.79	5.54	6.32	7.11
Moment of inertia, I, mm⁴/mm										
38×6.5	5.62	7.19	9.28	14.06	19.79	26.75				
51×13	25.11	31.80	40.27	58.01	79.99	98.14				
68×13	24.58	31.00	39.20	56.13	74.28	93.82				
75×25	112.9	141.8	178.3	253.3	330.6	411.0				
125×25		145.0	181.8	256.5	332.9	411.2				
152×51				990.1	1281	1576	1770	2080	2395	2718
$19 \times 19 \times 190^2$		46.23	60.65	90.74	121.81					
$19 \times 25 \times 292^2$		75.05	99.63	151.7						
Cross-sectional wall area, mm²/mm										
38×6.5	1.287	1.611	2.011	2.817	3.624	4.430				
51×13	1.380	1.725	2.157	3.023	3.890	4.760				
68×13	1.310	1.640	2.049	2.870	3.692	4.515				
75×25	1.505	1.884	2.356	3.302	4.250	5.203				
125×25		1.681	2.100	2.942	3.785	4.627				
152×51				3.294	4.240	5.184	5.798	6.771	7.743	8.719
$19 \times 19 \times 190^2$		1.077	1.507	2.506	3.634					
$19 \times 25 \times 292^2$		0.792	1.109	1.869						

Notes:
1. Where two thicknesses are shown, top is corrugated steel pipe and bottom is structural plate.
2. Ribbed pipe. Properties are effective values.

13.5.5 Wall Thickness

The required wall area is computed from the calculated ring compression in the pipe wall and the allowable stress.

$$A = \frac{C}{f_c}$$

where A = area of pipe wall, mm²/mm (in²/ft)
C = ring compression, kN/m (lb/ft)
f_c = design stress, MPa (lb/in²)

Values of A are given in Table 13.5 for various corrugations and wall thicknesses.

13.5.6 Handling Stiffness

Minimum pipe stiffness requirements, for practical handling and installation without special considerations or attention, have been established through experience and formulation. The resultant flexibility factor, *FF*, limits the size of each combination of corrugation and metal thickness.

$$FF = \frac{D^2}{EI}$$

where FF = flexibility factor, mm/N (in/lb)
$\quad\quad E$ = modulus of elasticity, MPa (lb/in^2)
$\quad\quad\quad$ = 200×10^2 MPa (30×106 lb/in^2)
$\quad\quad D$ = diameter or span, mm (in)
$\quad\quad I$ = moment of inertia of wall, mm^4/mm (in^4/in)

Recommended maximum values of FF, for various corrugations, are:

68×13 mm ($2\frac{2}{3} \times \frac{1}{2}$ in.) corrugation, FF = 0.245 mm/N (0.043 in./lb)

125×25 mm (5 x 1 in.) corrugation, FF = 0.245 mm/N (0.043 in./lb)

75×25 mm (3×1 in.) corrugation, FF = 0.245 mm/N (0.043 in./lb)

152×51 mm (6×2 in.) corrugation, FF = 0.114 mm/N (0.020 in./lb)

Higher values can be used with special care or where experience allows. A trench installation condition, as in sewer design, and arches on footings are two examples where flexibility limits may be exceeded. Guidance can be found in ASTM A796 "Standard Practice for Structural Design of Corrugated Steel Pipe, Pipe-Arches and Arches for Storm and Sanitary Sewers and Other Buried Applications" and in the "Handbook of Steel Drainage and Highway Construction Products" available from the Corrugated Steel Pipe Institute.

13.5.7 Longitudinal Seam Strength

Most pipe seams develop the full yield strength of the pipe wall. However, there are exceptions in standard pipe manufacture. The maximum ring compression should not exceed the ultimate longitudinal seam strength divided by a factor of safety of 2.

13.6 *FITTINGS REINFORCEMENT*

A common practice in storm drain projects is to connect branch lines to the main line. Similarly, underground detention systems are often fabricated with several branches connected to a larger header or manifold. When a section of the main or larger pipe is cut away to make the connection, the ring compression capacity of the pipe is reduced. It is then necessary to evaluate the need for reinforcement of the main or larger pipe in order to re-establish adequate ring compression capacity. The evaluation and design of fittings reinforcement should follow the procedures in the ASTM Standard A998/A998M-98, "Standard Practice for Structural Design of Reinforcements for Fittings in Factory-Made Corrugated Steel Pipe for Sewers and Other Applications." These procedures are also described in the National Corrugated Steel Pipe Association's Design Data Sheet 18, "Pipe Reinforcement at Fittings and Intersections."

These procedures were based on a finite element analysis to determine whether reinforcement is required for an installation. The next step involved establishing reinforcement methods and the actual calculations of the reinforcement requirements for each method. This was done for main pipe diameters from 1200 mm (48 in.) to 3600 mm (144 in.) and branch pipe diameters from 600 mm (24 in.) up to the diameter of the main pipe. Main pipe with wall thickness from 1.63 mm (0.064 in. or 16ga) to 4.27 mm (0.168 in. or 8ga), and depths of cover of 3050, 7000 and 9150 mm (10, 20 and 30 ft.) were considered. The results of this research showed that the need for reinforcement increases with increasing branch diameter, increasing depth of cover and decreasing wall thickness. In Phase 2 of this effort, three methods were developed for providing the required reinforcement. The three methods are: (1) increasing the wall thickness of the main pipe, (2) mounting a saddle plate and branch stub assembly on the main pipe for connecting the branch, and (3) adding reinforcement around the direct connection of the branch pipe to the main pipe.

13.6.1 General Design Assumptions

It is assumed that the main and branch pipes have been designed to the appropriate specifications and that the wall thickness specified is appropriate for the corrugation profiles, backfill material and installation methods specified. Backfill and live load pressures on the pipes must be thoroughly evaluated. The possibility of higher construction loads on the pipe, when the cover is shallow must also be considered in selecting the pipe wall thickness and corrugation. Structural design of the pipes should be in accordance with ASTM A796, AASHTO Section 12 or the procedures in this chapter. It is further assumed that the branch pipe is adequately connected to the main pipe by direct welding or with a saddle branch that has been properly attached to the main pipe.

REFERENCES

American Society of Testing and Materials. A998/A998M-98, "Standard Practice for Structural Design of Reinforcements for Fittings in Factory-Made Corrugated Steel Pipe for Sewers and Other Applications," 1999.

American Iron and Steel Institute. *Modern Sewer Design,* Fourth Edition, Washington, DC, 1999.

Bell, W. *Applying BMP Technologies in Urban and Ultra-urban Settings,* Alexandria, VA.

Bell, W., Stokes, G., and Nguyen. Assessment of the Pollutant Removal Efficiencies of Delaware Sand Filter BMP's. Final Report. Department of Transportation and Environmental Services. City of Alexandria, VA, 1995.

Clayton, R. and T. Schueler. *Design of Stormwater Filtering Systems,* The Center for Watershed Protection. Silverspring, MD, (301) 598-1890, 1996.

National Corrugated Steel Pipe Association, Design Data Sheet 18, "Pipe Reinforcement at Fittings and Intersections," Washington, DC, 1999.

Roberts, B. C. *Design of Underground Detention Systems for Stormwater Management,* Corrugated Steel Pipe Institute, Canada, 1998.

Schueler, T. *Controlling Urban Runoff: A Practical Manual for Planning and Designing Urban BMP's.* Department of Environmnetal Programs. Metropolitan Washington Council of Governments. Washington, DC, 1987.

CHAPTER 14
SEDIMENT MOVEMENT IN DRAINAGE SYSTEMS

Jacques W. Delleur
School of Civil Engineering
Purdue University
West Lafayette, Indiana

14.1 INTRODUCTION

Sediment deposits in sewers create negative effects on the hydraulic performance of the system and on the environment. Problems that arise include blockage, surcharge, early overflows, large pollutant discharges and costly removal. The erosion of sediments in sewers can release pollutants in concentration that exceed the levels found in the various contributing sources of the sediments and pollutants (Ashley et al., 1992). Solid particles constitute the support on which other pollutants are adsorbed, among them heavy metals and hydrocarbons. Intermittent discharges from combined sewer overflows often involve sediments and associated pollutants and are recognized as a major pollution source in urban receiving waters.

There are several recent reviews of literature: Bertrand-Krajewski et al. (1993), Ashley et al. (2000) and Delleur (2001).

14.2 SEWER SEDIMENTS

The supply of sediments to the drainage system occurs primarily during storm events. The runoff washes off the particles accumulated on the tributary surfaces. Other sediment supply mechanisms include wind action and vehicle motion. Sanitary solids are also found in combined drainage systems and of course in sanitary sewers.

14.2.1 Surface Deposits

The *sediment load* from the upstream watersheds can be estimated using the *Universal Soil Loss Equation,* USLE (Wischmeier and Smith, 1965, 1978) or the more recent revised version, RUSLE (Renard et al., 1993). The RUSLE and USLE are multiplicative relationships of the form

$$A = R\,K\,L\,S\,C\,P$$

where A = the average soil loss per unit of area
  R = the rainfall/runoff factor
  K = the soil erodibility factors
  L = the slope length factor
  S = the slope steepness factor
  C = the cover and management factor
  P = the conservation practice factor

A summary of the procedure can be found in Haan et al. (1994) and a computer based version of RUSLE is available from the U.S. Department of Agriculture. Other sources of sediment include construction zones and street and highway de-icing operations.

The first major study of street surface accumulation of dust and dirt passing through a quarter inch (0.635 cm) mesh screen was performed by the American Public Works Association (1969) on 18 catchments in Chicago. Most of the surface dust and dirt materials are the product of local erosion, and wear of the street surface while some materials are the result of vehicle emission and wear. The accumulations are summarized in Table 14.1. These values assume no initial loading.

TABLE 14.1   Dust and Dirt Accumulation in Chicago (APWA 1969)

Land use	Pounds/dry day per 100 ft-curb	gm/dry day per m-curb
Single family residential	0.7	10
Multi-family residential	2.3	34
Commercial	3.3	49
Industrial	4.6	68
Undeveloped or park	1.5	22

These data imply a linear build-up. Sartor and Boyd (1972) indicated that the build-up since cleaning was non-linear and of an inverse exponential form over a period of up to 10 days. Table 14.2 lists the equations used in the Stormwater Management Model (SWMM) for the accumulation of dust and dirt, DD (lb), (Huber and Dickinson, 1988). In this Table, t is time, usually in days, and $DDFACT$, $DDPOW$ and $DDLIM$ are, respectively, a coefficient, an exponent and the build up limit.

Later it was recognized that street loadings are the result of deposition and removal rates plus a permanent storage that is not removed naturally or by cleaning equipment. Table 14.3

TABLE 14.2   Dust and Dirt Build-up Equations (Huber and Dickinson, 1988)

Type	Equation
Power-Linear	$DD = DDFACT \cdot t^{DDPOW}$ $DD \leq DDLIM$
Exponential	$DD = DDLIM\,(1 - e^{-DDPOW \cdot t})$
Michaelis-Menton	$DD = DDLIM \cdot t/(DDFACT + t)$

TABLE 14.3 Street Dirt Loadings (grams/curb-meter) and Deposition Rates (grams/curb-meter-day) (Pitt et al., 1999)

Location	Initial loading	Daily deposition rate	Maximum observed loading	Days to observed maximum loading	Reference
Smooth and Intermediate Textures Streets					
Reno/Sparks, NV—good condition	80	1	85	5	Pitt and Sutherland 1982
Reno/Sparks, NV—good with smooth gutters (windy)	250	7	400	30	Pitt and Sutherland 1982
San Jose, CA—good condition	35	4	>140	>50	Pitt 1979
US Nationwide—residential streets—good condition	110	6	140	5	Sartor and Boyd 1972 corrected
US Nationwide—commercial streets—good condition	85	4	140	5	Sartor and Boyd 1972 corrected
Reno/Sparks, NV—moderate to poor condition	200	2	200	5	Pitt and Sutherland 1982
Reno/Sparks, NV—new residential area (construction)	710	17	910	15	Pitt and Sutherland 1982
Reno/Sparks, NV—poor condition, with lipped gutters	370	15	630	35	Pitt and Sutherland 1982
San Jose, CA—fair to poor condition	80	4	230	70	Pitt 1979
Castro Valley, CA—moderate condition	85	10	290	70	Pitt and Shawley 1982
Ottawa, Ontario—moderate condition	40	20	na	na	Pitt 1983
Toronto, Ontario—moderate condition, residential	40	32	100	>10	Pitt and McLean 1986
Toronto, Ontario—moderate condition, industrial	60	40	351	>10	Pitt and McLean 1986
Bellevue, WA—dry period, moderate condition	140	6	>230	20	Pitt 1984
Bellevue, WA—heavy traffic	60	1	110	30	Pitt 1984
Bellevue, WA—other residential sites	70	3	140	30	Pitt 1984
average	150	9	>270	>25	
range	35–710	1–40	85–910	5–70	
Rough and Very Rough Textured Streets					
San Jose, CA—oil and screen overlay	510	6	>710	>50	Pitt 1979
Ottawa, Ontario—very rough	310	20	na	na	Pitt 1983
Reno/Sparks, NV	630	10	860	35	Pitt and Sutherland 1982
Reno/Sparks, NV—windy	540	34	>1400	>40	Pitt and Sutherland 1982
San Jose Ca—poor condition	220	6	430	30	Pitt 1970
Ottawa, Ontario—rough	200	20	na	na	Pitt 1983
US nationwide—industrial streets (poor condition)	190	10	370	10	Sartor and Boyd 1972 corrected
average	370	15	>750	>30	
range	190–630	6–34	370 > 1400	10 > 50	

summarizes accumulation rate measurements in North America corrected, where necessary, for the permanent storage (Pitt et al., 1999).

14.2.2 Washoff of Street Dirt

The washoff of street dirt is usually estimated using an equation based on the experiments of Sartor and Boyd (1972). This equation states that the rate of particle removal of a given size is proportional to the street dirt loading and the constant rain intensity:

$$\frac{dN}{dt} = -k\,r\,N \tag{14.1}$$

where dN/dt = the change in street dirt loading per unit time
$\quad\quad\quad k$ = the proportionality constant (1/in)
$\quad\quad\quad r$ = the rain intensity (in/hr)
$\quad\quad\quad N$ = the street loading (lb/curb-mile)

Equation (14.1) can be integrated as

$$N = N_0 e^{-krt} = N_0^{-kR} \tag{14.2}$$

where N = the residual street dirt load after the rain
$\quad\quad\quad N_0$ = the initial street dirt load
$\quad\quad\quad t$ = the rain duration
$\quad\quad\quad R$ = the total rain depth (in)

The *street washoff* is thus $N_o - N$. Table 14.4 gives values of the suspended solids washoff coefficient k and the availability factor (available loading/total loading) obtained by Pitt (1987) in Toronto, Ontario. Washoff is thus more efficient for the higher rain energy and the smoother pavement.

TABLE 14.4 Suspended Solids Washoff Coefficients (Pitt et al., 1999)

Rain intensity category	Street dirt loading category	Street texture category	k (calculated)[1]	Standard error for k	Available load/total initial load
high	clean	rough	0.832	0.064	0.11
low	clean	rough	0.344	0.038	0.061
high	dirty	rough	0.077	0.008	0.032
low	dirty	rough	0.619	0.052	0.028
high	clean	smooth	1.007	0.321	0.26
low	clean	smooth	0.302	0.024	0.047
high	dirty	smooth	0.167	0.015	0.13
low	(actually clean)	smooth	0.335	0.031	0.11

[1] in in^{-1} · k corresponds to *RCOEF* in SWMM, where it is taken as 4.6 in^{-1} or 0.18 mm^{-1} (Huber and Dickinson, 1988).

14.2.3 Chemical Quality of Street Dirt

The chemical quality of street dirt materials is generally expressed as a particulate *potency factor* formulated as mg of pollutant/kg of dry particulate residue. Table 14.5 summarizes the information compiled by Pitt et al., (1999). In this table, TKN designates total Kjeldahl nitrogen and COD designates chemical oxygen demand.

TABLE 14.5 Mean Street Dirt Chemical Quality in mg Constituent/kg solids (Pitt et al., 1999)

Constituent	Residential	Commercial	Industrial
P	620[4]		670 [4]
	540[6]	400[6]	
	1100[5]	1500[5]	
	710[1]	910[1]	
	810[3]		
TKN	1030[4]		560 [4]
	3000[6]	1100[6]	
	290[5]	340[5]	
	2630[3]	4300[2]	
	3000[2]		
COD	100 000 [4]		65 000[4]
	150 000[6]	110 000[6]	
	180 000[5]	250 000[5]	
	280 000[1]	340 000[1]	
	180 000[3]	210 000[2]	
	170 000[2]		
Cu	162[4]		360[4]
	110[6]	130[6]	
	420[2]	220[2]	
Pb	1010[4]		900[4]
	1800[6]	3500[6]	
	530[5]	2600[5]	
	1200[1]	2400[1]	
	1650[3]	7500[2]	
	3500[2]		
Zn	460[4]		500[4]
	260[5]	750[5]	
	325[3]	1200[2]	
	680[2]		
Cd	<3[5]	5[5]	
	4[2]	5[2]	
Cr	42[4]		70[4]
	31[5]	65[5]	
	170[2]	180[2]	

References; location; particle size:
[1] Bannerman et al., 1983 (Milwaukee, WI) $<31\,\mu$m
[2] Pitt, 1979 (San Jose, CA) $<45\,\mu$m
[3] Pitt, 1985 (Bellevue, WA) $<63\,\mu$m
[4] Pitt and McLean, 1986 (Toronto, Ontario) $<125\,\mu$m
[5] Pitt and Sutherland, 1982 (Reno/Sparks, NV) $<63\,\mu$m
[6] Terstriep et al., 1982 (Champaign/Urbana, IL) $>63\,\mu$m

14.2.4 Sanitary Solids

Solids of domestic origin are present in sanitary and combined sewers. In the U.S.A., the concentration of suspended solids in domestic sewage varies from 100 to 350 mg/l (Metcalf and Eddy, Inc., 1991).

14.2.5 Sediment Cohesion

Sediments in sanitary and combined sewers generally exhibit cohesive-like strength when subject to shear. This is due to the organic nature of the particles and the presence of biological slimes and greases. Table 14.6 shows a classification of sewer sediments used in the U.K. and the associated characteristics (Butler et al., 1996a)

14.2.6 Stormwater Inlets and Catch Basins

Stormwater inlets guide the surface runoff from the streets towards the underground storm or combined sewers. The inlets are placed at the curb or in the gutter. Sometimes combination of curb and gutter inlets are used. Catch basins (called *gully pots* in the U.K.) are grit chambers designed to retain the heavy debris from being carried into the sewer. Their primary purpose is to retain large solids in order to reduce the problems of sediment deposition downstream.They require periodic cleaning and maintenance so that they do not become a source of sediment rather than a sink.

Butler and Memon (1999) have developed a dynamic model that simulates the processes occurring in catch basins. These are: dilution, dispersion, sedimentation, sediment bed build-up and erosion, washout of suspended and dissolved pollutants from the basin liquid and reaeration of the basin liquid.

14.3 MECHANICS OF SEDIMENT MOVEMENT IN SEWERS

The knowledge on sediment transport engineering up to 1975 was compiled in the ASCE Manual 54 under the leadership of Vito Vanoni (1975) and still is an excellent source of basic information on sediment transport. The ASCE Committee on sedimentation engineering is currently preparing a revised edition. A summary of transport processes in urban drainage systems can be found in Delleur (1998).

TABLE 14.6 U.K. Sewer Sediment Characteristics (Butler et al., 1996a)

Type	Normal Transport Mode	Concentration (mg/l)			Median Particle Size, d_{50} (μm)			Specific Gravity		
		Low	Med	High	Low	Med	High	Low	Med	High
Sanitary solids	Suspension	100	350	500	10	40	60	1.01	1.4	1.6
Stormwater solids	Suspension	50	350	1000	20	60	100	1.1	2.0	2.5
Grit	Bedload	10	50	200	300	750	1000	2.3	2.6	2.7

Once the sediment particles have been eroded and are entrained by the flow, the particles can be transported in suspension or as bedload along the invert of the sewer. The mechanics of sediment movement in sewers differs from that in natural streams. This is because the shape of the cross section, the significant variations in hydraulic conditions and the limited supply of sediment influence the flow. Sediment coarser than 20 to 30 μ are non-cohesive. Finer sediments and the presence of organic materials can result in varying degrees of co-hesiveness.

14.3.1 Boundary Shear and Friction Coefficient

Many of the formulas concerned with entrainment of particles involve the *boundary shear stress* τ_o. In steady flow it is related to the mean velocity V by

$$\tau_o = \frac{f}{8}\, \rho\, V^2 \tag{14.3}$$

where ρ = the fluid density
$\quad\; f$ = the Darcy-Weisbach friction coefficient for the bed

For pipes flowing full, the friction coefficient, in turn, can be determined from the *Colebrook (1939) formula*

$$\frac{1}{\sqrt{f}} = -2\,\log\left(\frac{k/D}{3.7} + \frac{2.51}{R\sqrt{f}}\right) \tag{14.4}$$

where k = the boundary roughness
$\quad\; D$ = the diameter
$\quad\; R$ = the Reynolds number

This equation can be rewritten as

$$\frac{1}{\sqrt{f}} = -2\,\log\left(\frac{k}{3.7\,D} + \frac{2.51\,\nu}{D[2_g DS]^{0.5}}\right) \tag{14.4a}$$

where S = the slope of the hydraulic grade line
$\quad\; \nu$ = the fluid kinematic viscosity

When the viscous effects are negligible this equation reverts to the *rough pipe flow formula*

$$\frac{1}{\sqrt{f}} = -2\,\log\left(\frac{k}{3.7\,D}\right) \tag{14.5}$$

For the case of a duned or rippled bed, Butler et al. (1996a) suggest the use of a formula obtained by May (1993) to estimate the bed roughness k_b:

$$k_b = 2.4\, D^{0.61} d_{50}^{0.39} \tag{14.6}$$

where d_{50} is the median particle size. All three variables in this equation are in mm. The composite friction coefficient is f_c for a pipe with a sediment bed of width W_b, and friction coefficient f_b is

$$f_c = \frac{f\,P + f_b W_b}{P + W_b} \tag{14.7}$$

where f and P are the friction factor and the perimeter of the pipe exposed to the flow, respectively.

It should be pointed out that the actual shear stress distribution can deviate from the average value given by Eq. (14.3) for steady flows. Berlamont et al. (2000) have studied the effects of non-uniformity of the shear stress and flow unsteadiness on the modeling of the sediment transport. They have found that the maximum shear stress is a function of the bed roughness and the water depth and can be as high as 1.3 times the average value.

14.3.2 Non-Cohesive Sediment Movement

Sediment transport in sewer pipes can occur in the following modes (May 2000):

1. In suspension

2. In *flume traction* with particles moving individually in contact with the invert of the pipe

3. As separated slow moving dunes

4. As a layer of particles moving over a continuous deposit of stationary material

14.3.2.1 Suspended Sediments.   A particle remains in suspension when the vertical velocity component of the turbulent eddies exceeds the particle fall velocity. The *suspended load vertical distribution* can be estimated from the *Rouse-Vanoni relationship* (Shen and Julien, 1993)

$$\frac{D_y}{C_a} = \left(\frac{D - y}{y}\frac{a}{D - a}\right)^z \tag{14.8}$$

where C_a and C_y = the concentrations at distances a and y from the bed, respectively
$\quad\quad\quad\quad D$ = the depth of the stream
$\quad\quad\quad\quad z = w/(\kappa U_*)$ is a suspension number given as the ratio of the fall velocity w of the particle and shear velocity $U*$ multiplied by the von Karman constant κ (0.4 for clear water)

The *shear velocity* is defined as $U* = [\tau_o/\rho]^{1/2} = [gRS]^{1/2}$

where τ_o = the average shear stress around the periphery of the sewer
$\quad\quad\quad R$ = the hydraulic radius
$\quad\quad\quad S$ = the slope of the energy grade line

If $z \rightarrow 0$, the vertical suspended sediment concentration given by Eq. (14.8) is uniform. If $z \rightarrow 5$, the suspended load is not significant and bedload is the dominant transport form. Ackers et al. (1996) use

$$\frac{U*}{w} > 0.75 \tag{14.9}$$

as a criterion to indicate when the transport in suspension begins to be important. With a value of $\kappa = 0.4$, this corresponds to $z < 3.3$. Figure 14.1 shows the relation between the fall velocity in water, w, of sand grain with a shape factor SF for water temperatures varying from 0 to 40°C (notice the different velocity scales for the different values of the shape factor). The *shape factor* is defined as $SF = c/[ab]^{0.5}$ where a, b, c are the length of the longest, intermediate and shortest mutually perpendicular axes of the particle. Pisano (1996) gives settling curves for different types of sewer solids and discusses the several methods for measuring them. Ackers et al. (1996) quote an equation cited by Sakhuja (1987) as

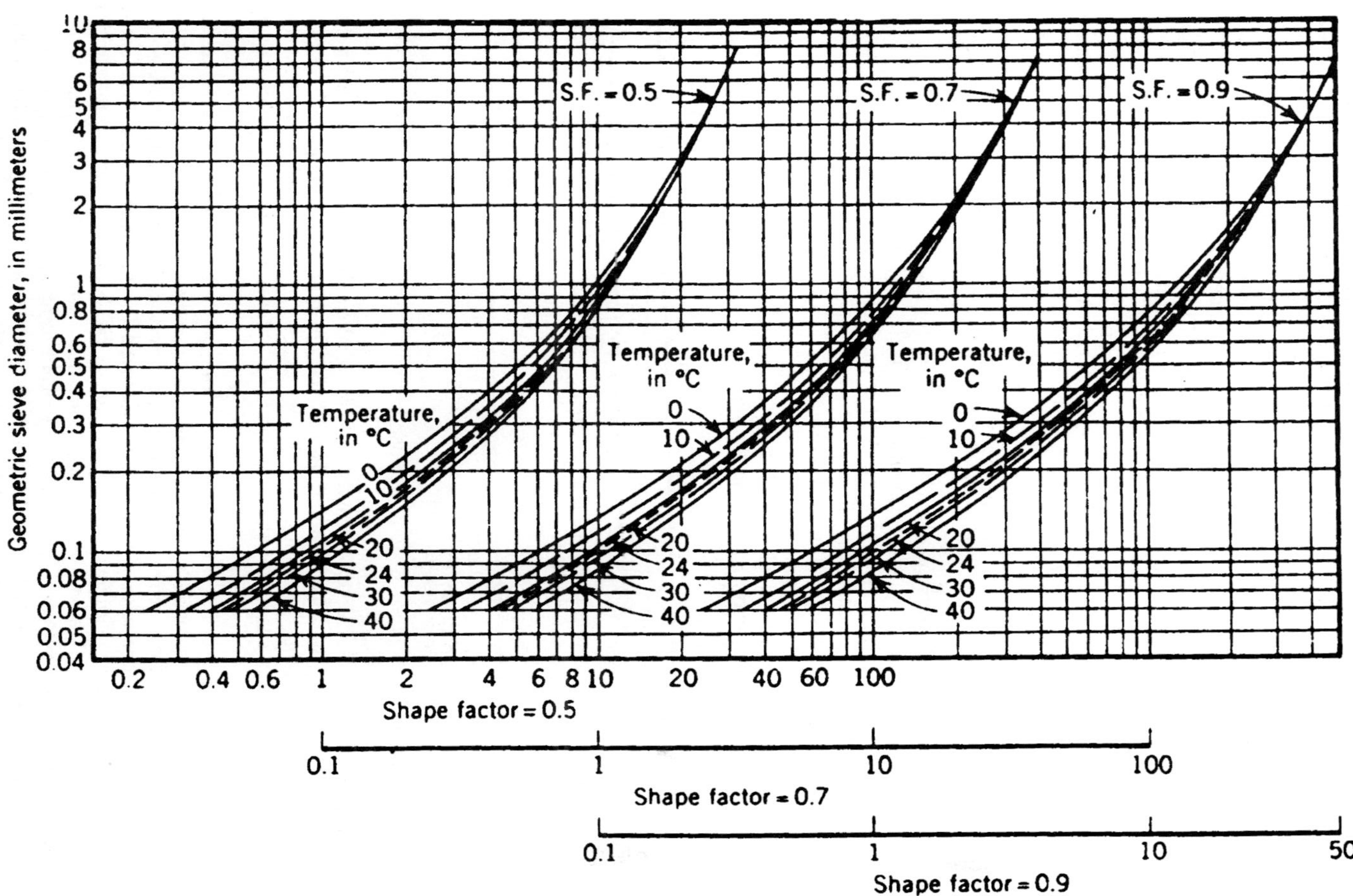

FIGURE 14.1 Relation between fall velocity, particle size and shape factor. (Vanoni, 1975)

$$w = \frac{[9\nu^2 + 10^9 \, d^2 \, g(s - 1)(0.03869 + 0.0248d) - 3\nu]^{0.5}}{10^{-3}(0.11607 + 0.074405d)}$$

where the fall velocity w is in m/s, the particle size d is in mm, the kinematic viscosity of the water ν is in m²/s and s is the specific gravity. The suspended load can be calculated as the depth-integration of the product of the local concentration and the flow velocity.

The British CIRIA study (Ackers et al. 1996) recommended the Macke (1982) formula for the suspended sediment transport with low concentrations at the limit of deposition and the Ackers-White (Ackers, 1991) formula for suspended sediment transport over a deposited bed.

The *Macke formula* predicts the sediment volumetric concentration C_v (volumetric discharge of sediments/volumetric discharge of fluid) as

$$C_v = \frac{f_c^3 V_L^3}{30.4(s - 1) \, w^{1.5} \, A} \tag{14.10}$$

where f_c = the Darcy-Weisbach friction coefficient for sediment carrying flow in the pipe
V_L = the limiting velocity without deposition
s = the specific gravity of the sediment particles
w = the fall velocity of sediments
A = the cross sectional area of the flow

Note that the equation is in SI units. It is valid for $\tau_c > 1.07$ N/m², where τ_c is the shear stress for the sediment carrying flow in the pipe.

The *Ackers-White formula* is

$$C_v = J\left[W_e \frac{R}{A}\right]^\alpha \left[\frac{d}{R}\right]^\beta f_c^\gamma \left\{\frac{V}{[g(s - 1)R]^{1/2}} - K f_c^\delta \left[\frac{d}{R}\right]^\varepsilon\right\}^m \tag{14.11}$$

where W_e = the effective width of the sediment bed (see Appendix)
R = the hydraulic radius
A = the cross sectional area of flow
d = the particle size
f_c = the composite Darcy-Weisbach friction coefficient
V = the mean flow velocity

The coefficients J, K, m, α, β, γ, δ, ε depend upon the dimensionless grain size D_{gr} and are given in the Appendix.

The Macke and the Ackers formulas can be used to calculate the full pipe velocity for given particle size and concentration. The Macke formula is recommended to ensure suspended sediment transport without deposition, except for very large sewers, where some amount of deposition can be tolerated, in which case the Ackers-White formula is suggested.

14.3.2.2　Bedload.　In the presence of a flowing fluid, the hydrodynamic forces that tend to dislodge a sediment particle are the drag force and the lift force. The forces resisting the motion are the particle submerged weight, the interlocking of particles and possibly the cohesive forces. If the hydrodynamic forces exceed the resisting forces, then *entrainment* occurs. Movement of particles then takes place at the boundary between the fluid and the sediment. Because of the velocity fluctuations due to the turbulence of the flow, not all particles of a given size are dislodged at the same time. The flow condition below which there is a negligible amount of sediment movement is the *threshold of movement* and is usually expressed in terms of a critical bed shear stress or a critical erosion velocity. Once the particles have been entrained, the lighter material will move in suspension as *suspended load* and the heavier ones will be rolling, sliding and saltating as *bedload.*

The hydraulic criterion for the beginning of motion in uniform cohesionless sediment in plane beds is

$$\tau^* = \frac{\tau_o}{g(\rho_s - \rho)d_e} = f\left(\frac{U^*d_e}{v}\right) = \tau_c^* \qquad (14.12)$$

The left hand side of Eq. (14.12) is known as the Shields parameter, τ^*. It is a dimensionless ratio between the bed shear stress τ_o, and $g(\rho_s - \rho)d_e$, where ρ_s and ρ are the densities of the sediments and of the water, respectively, and d_e is the equivalent grain size. The right hand side is a function of a Reynolds number in which $U^* = [\tau_o/\rho]^{1/2} = [gRS]^{1/2}$ is the shear velocity and v is the kinematic viscosity of the water, R is the hydraulic radius and S is the slope of the stream. The graphical representation of Eq. (14.12) is known as the *Shields curve* and it is shown in Fig. 14.2. The dimensionless shear stress obtained from the Shields curve is designated by τ_c^*. Equilibrium exists when $\tau^* = \tau_c^*$. Motion takes place when the dimensionless shear stress τ^* in the left hand side is larger than the critical value τ_c^* from the Shields curve in the right hand side of Eq. (14.12). Thus for $\tau^* > \tau_c^*$, there is sediment movement. A more conservative criterion for the critical dimensionless shear stress is $\tau_c^* = 0.03$ for no motion on a plane bed. For ripples and dune beds, one may take $\tau_c^* = 0.06$ (Shen and Julien, 1993).

The classical bedload transport formulas, such as the *Einstein equation,* the *Meyer-Peter* and *Muller equation* are of the form

$$q_s = F[(\tau^* - \tau_c^*)^\beta] \qquad (14.13)$$

where q_s = the sediment discharge per unit width
$F[\cdot]$ = a function
β = an exponent that is equal to 1.5 for the Einstein and the Meyer-Peter and Muller formulas

These formulas were derived for two-dimensional steady stream flow and generally are not applicable to flow in sewers. For mostly granular material (2 to 10 mm) and steep slopes (>2%) Lin and Le Guennec (1996) showed that the Shields criterion (Eq. (14.12)) and Meyer-Peter equation adequately predicted the incipience of the motion and the bed-load. But, in general, the incipient motion and the rate of transport of non-cohesive sediments in circular conduits are different from those in two-dimensional open channel flow. They are greatly influenced by the shape of the cross section and by the typical unsteadiness of sewer flows. Experiments made at Newcastle upon Tyne, in the U.K., by Kleijwegt et al. (1990) show that the Shields critical shear stress criterion for incipient motion (which is used in models such as SWMM) is not valid. The critical shear stress was found to be about 70% of that predicted by Shield's curve. The circular cross section causes various distributions of the shear stress over the width of the bed. One, two or three local maxima can exist (Kleijwegt, 1992). In subsequent experiments with non-cohesive sediments at Delft, in the Netherlands, Kleijwegt (1993) observed four different types of bed forms: continuous flat bed; continuous bed with ripples; continuous bed with ripples and isolated bed forms. Dunes and ripples migrate downstream. Antidunes, which are bed forms traveling upstream in two dimensional channels, were not observed in circular cross section. The flow resistance differs for each type of bed form because the form drag has to be added for ripples and dunes.

For unsteady flows, Berlamont et al. (2000) showed that the variation of flow rate, mean flow velocity and water depth are not synchronous. They found that better sediment transport predictions can be made with the classical equations by including the correct shear stress distribution at each time step and the phase lag between flow rate, velocity and water depth. These new developments, however, have not yet been incorporated in design procedures.

For the case of *limit of deposition* (*no bed*), which corresponds to the transition from transport in flume traction to transport on the surface layer of deposited bed, May (1993)

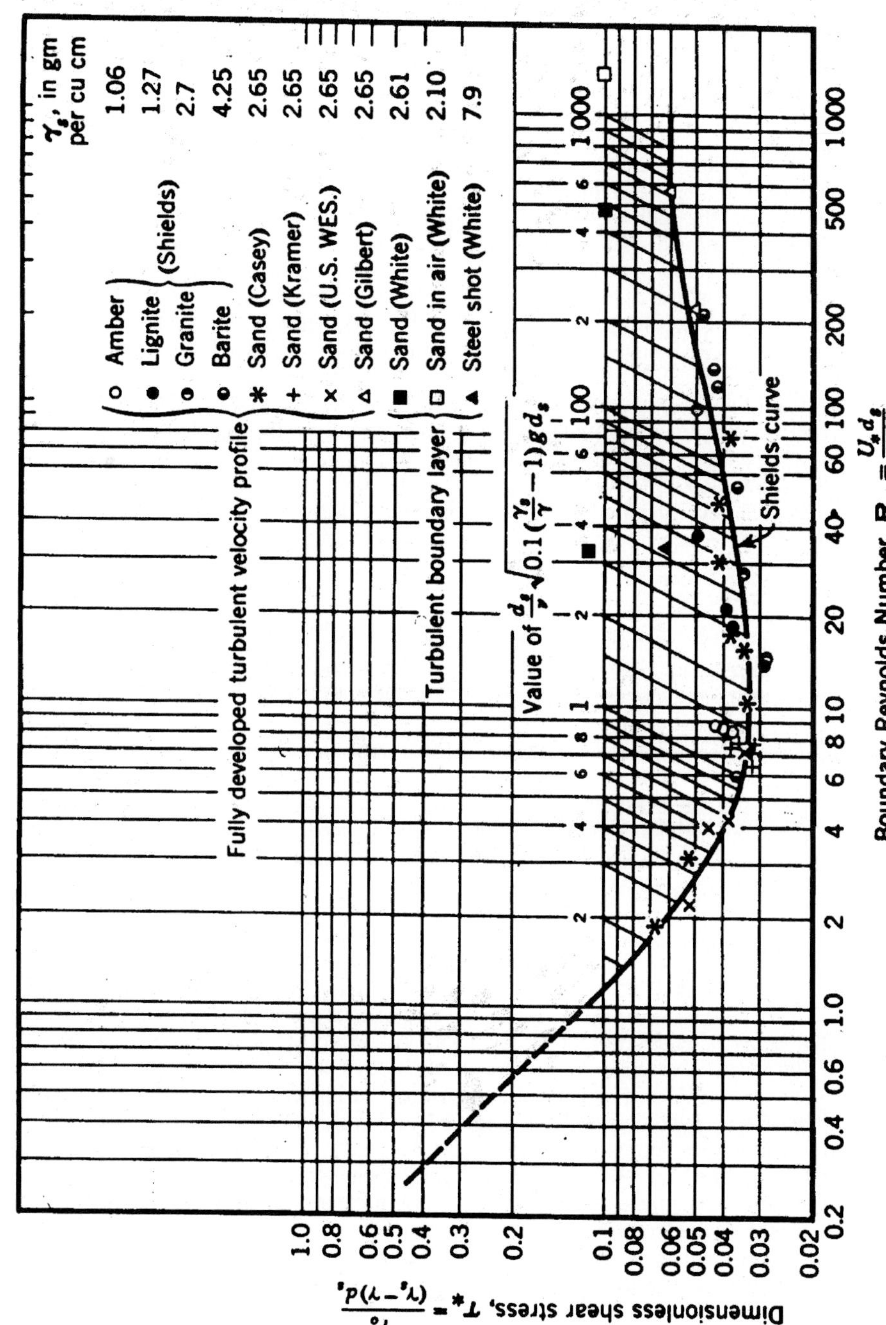

FIGURE 14.2 Shields diagram. (Vanoni, 1975)

developed a new equation within the framework of the CIRIA study (Ackers et al., 1996) as:

$$C_v = 3.03 \times 10^{-2} \left[\frac{D^2}{A}\right]\left[\frac{d}{D}\right]^{0.6}\left[1 - \frac{V_t}{V}\right]^4 \left[\frac{V^2}{g(s-1)D}\right]^{1.5} \tag{14.14a}$$

where V_t is the mean flow velocity at the threshold of movement and it is given by

$$V_t = 0.125[g(s-1)d]^{0.5}\left[\frac{y}{d}\right]^{0.47} \tag{14.14b}$$

where y is the flow depth. With the help of this formula, it can be shown that the required velocity increases rapidly with pipe size, therefore, it might be more economical to tolerate some bed deposition for large pipes. May (2000) has recently given an extended version of Eq. (14.14a) for the case of inverted sewer siphons as

$$C_v = (3.03 \times 10^{-2} - 0.0169 \sin \phi)\left[\frac{D^2}{A}\right]\left[\frac{d}{D}\right]^{0.6}\left[1 - \frac{\sigma V_t}{V}\right]^4 \left[\frac{V^2}{g(s-1)D \cos \phi}\right]^{1.5}$$

$$\tag{14.14c}$$

where ϕ is the angle of upward inclination, and σ is given by

$$\sigma = \left[\frac{\sin \phi + \mu \cos\phi}{\mu}\right]^{0.5} \tag{14.14d}$$

where μ the effective friction coefficient between the sediment particles and the pipe wall.

For the case of a *deposited bed,* the CIRIA team (Ackers et al., 1996) selected an equation developed by May (1995):

$$C_v = \eta \left[\frac{W_b}{D}\right]\left[\frac{D^2}{A}\right]\left[\frac{\theta f_g V^2}{8g(s-1)D}\right] \tag{14.15}$$

where η = a transport parameter for continuous bed (see Appendix)
$\quad W_b$ = the sediment bed width
$\quad f_g$ = the Darcy-Weisbach friction coefficient for grain shear stress (see Appendix)
$\quad \theta$ = a transition coefficient for particle Reynolds number (see Appendix)

Results obtained with this equation confirm that allowing a small amount of deposition greatly reduces the minimum flow velocity as compared to the limit of deposition case (Butler et al., 1996a).

14.3.2.3 Near Bed Solids.   Arthur and Ashley (1997, 1998) formulated the terminology *near bed solids* to describe the most concentrated layer in combined sewer where material in transport exhibits strong biochemical characteristics. These near bed solids can have pronounced pollutant impacts if discharged in water courses. Based on field data from Dundee, U.K., they developed the following relationship to predict the rate of near bed solids in a combined sewer system:

$$C_v = -105.73 + 2.55 \times 10^{-3}\left(\frac{I_t\, TSSS)}{D_t}\right) + 0.2023\left(\frac{y_o}{y_{max}}\right)$$

$$+ 47.808\left(\frac{\tau_o}{\tau_b}\right) + 120.45\left(\frac{\rho_d}{\rho_w}\right) \tag{14.16}$$

where C_v = the volumetric sediment concentration
I_t = the rainfall intensity
$TSSS$ = the time since the start of the last storm
D_t = the total depth of rainfall
y_o = the average flow depth
y_{max} = the maximum flow depth for an average dry weather flow day
τ_o = the average shear
τ_b = the shear at the bed
ρ_d = near bed dry solids density
ρ_w = the sewage density

A consistent set of units must be used so that each term in parenthesis is dimensionless. This relationship has not been tested on interceptors other than those in Dundee, U.K., but it is likely to indicate the variables which control the movement of near bed solids. The importance of the near bed solids transport lies in the fact that it contains a significant portion of the dry weather flow COD and BOD concentration even though it is a small portion of the total amount of solids transported. This makes it possible to make an estimation of the concentration of the first foul flush.

14.3.3 Cohesive Sediment Movement

If the fraction of clay in a sediment sample is $\geq 10\%$, the sample is considered to possess cohesive properties. Sediment particles in the silt and clay sizes can interact with each other to form floc. The size and specific weight of cohesive sediment flocs are a function of the flow field (Bui, 2000). Sediment particles $< 20\ \mu\text{m}$ are subject to the van der Waals attractive forces and the double layer repulsive forces. When brought together, the particles tend to position themselves at the minimum of the net interaction (Parker 1994). These electrochemical forces acting between the particles generate the cohesion that may exceed the hydrodynamic forces. The principal parameters governing the cohesion are: particle size, type of clay, organic matter content, water content of sediment and the salinity, and temperature and pH of the free and intersticial water (Hamm and Migniot, 1994). Mehta (1994) recognized three modes of erosion or resuspension of mud in estuarine situations: surface erosion, mass erosion of the bed and entrainment of high concentration fluid mud suspension. For consolidated cohesive beds, the *erosion rate, E,* is estimated in terms of the bed shear stress, τ_b, the bed shear strength, τ_s, assumed uniform with depth and a rate constant, M, as

$$E = M\,\frac{\tau_b - \tau_s}{\tau_s} \tag{14.17}$$

Mehta (1994) gives a graph showing the variation of M and τ_s with the bulk density. M can vary from 10^{-4} to 10^{-2} g cm^{-2} min^{-1} while τ_s can vary from 0 for organic rich floc to 10 N m^{-2} for hard soil.

14.3.3.1 Cohesive Sediment Movement in Sewers.   Most of the basic knowledge about the transport of sediments in drainage systems is concerned with non-cohesive sediments. But this is clearly not the case in most sewers or combined sewers where the sediment deposits usually are *cohesive.* The principal type of cohesive sediment is made of organic, mobile fine grained material. Other types include inorganic, coarse granular material cemented with fats and tars, organic wall slimes and deposits found in tanks (Crabtree, 1988). Thus the cohesion in sewers arises from agglutination due to tars, greases, chemical cementation and biological processes rather that the elctrostatic forces found in estuaries. The resistance to erosion in sewers with "cohesive-like" sediments is much greater than that anticipated from the application of non-cohesive sediment transport models. This increase is thought to be primarily due to the organic nature of the sediment (Skipworth et al. 1999).

Since 1988, a comprehensive program of research investigating the movement of sediments in sewers including their cohesive-like properties has been underway in Dundee, U.K. Continuous measurements of sediment depth were made. On the basis of these data and on the cohesive properties of sediments, Ashley et al. (1993) proposed a model for the prediction of the erosion from the bed and of the mass of sediments released into the flow.

Other studies were made in Germany, Belgium and France. The results of the detailed studies in Hildesheim, Germany were reported by Ristenpart et al. (1995). Torfs (1994) performed experiments at two Belgian universities with mixtures of cohesionless (fine sand) and cohesive (kaolinite, montmorillonite) materials in a rectangular flume and in a circular conduit. The critical shear stresses were not significantly different in the two channels. However, the erosion rates in the circular flumes were an order of magnitude larger than those in the rectangular flume. The mean bed shear stress is therefore higher in the circular cross section due to the presence of secondary flows. The position of the maximum shear stress is not necessarily located at the center. Local maxima may occur in the corner of the bed resulting in deeper erosion along the borders.

14.3.3.2 *Incipient Motion in Circular Sewers.*

Cohesion in sediments has a marked effect on the incipience of motion. The critical shear stress is increased by several orders of magnitude depending upon the concentration of the cohesive additive. When the critical shear stress is exceeded, a sudden collapse of the bed structure takes place. Clusters of various sizes are eroded and are carried downstream like noncohesive particles (Nalluri and Alvarez, 1992).

An extension of the Shields critical shear stress criterion (Eq. (14.12)) has been proposed by Mehta and Dyer (1994) and applied to circular sewers by Torfs (1995) as

$$\frac{\tau_o}{g(\rho_s - \rho)d_e} = \tau_c^* + \frac{\tau_s}{g(\rho_s - \rho)d_e} \tag{14.18}$$

where τ_c^* is the Shields parameter obtained from Shields diagram, and τ_s is the shear strength of the material, which can be obtained experimentally. Wotherspoon and Ashley (1992) provided a rheological methodology for the measurement of the yield strength of combined sewer deposits.

Nalluri and Alvarez (1992) tested cohesive sediments with various concentrations of clay gel simulating inorganic and coarse material cemented with fats. They obtained the following *Shields parameter:*

$$\frac{\tau_b}{(\rho_s - \rho)gd_{50}} = 0.964 C_v^{0.457} \left(\frac{d_{50}}{R_b}\right)^{-0.765} f_{sb}^{0.41} \tag{14.19}$$

where $\tau_b = \rho g R_b S_o$ is the bed shear stress
R_b = the bed hydraulic radius
S_o = the bed slope
f_{sb} = the bed friction factor with transport

(See French, 1985, p. 178–181 for the calculation of the bed shear stress by the side wall elimination technique). For a weak cohesive sewer sediment, they found that the maximum critical shear stress is approximately 2.5 N/m² and for freshly deposited sewer sediment, it can be as high as 6.7 N/m². Ackers et al. (1996) suggest a design value of $\tau_b = 2$ N/m².

Skipworth et al. (1996, 1999) studied the transport rate of cohesive sediment subject to variable flow rate such as a hydrograph. They found that the final flow rate and the rate of initial temporal change of the flow rate as well as the steady state flow variables influence the rate of transport of cohesive-like materials. They found that Eq. (14.17) is appropriate to describe the erosion and transport of an organic cohesive-like sediment bed in a sewer pipe, and suggest a relationship for the parameter M in terms of the flow rate in the pipe and the pipe slope.

14.4 *DESIGN CRITERIA*

The established design criterion for the design of self-cleansing sewers is the *minimum velocity V_m*. The typical U.S. and U.K. criteria for the self-cleansing velocity are listed below as well as the CIRIA design criteria which take into account much of the recent research results.

14.4.1 U.S. Design Criteria

The ASCE/WCPF manual (1969) recommends the use of the *Camp formula*

$$V_m = \left[\frac{8\,Kg(s-1)d}{f} \right]^{0.5} = \frac{\zeta}{n} R^{1/6}[K(s-1)d]^{0.5} \tag{14.20a}$$

where g = the acceleration of gravity
$\quad s$ = the specific gravity of the sediment
$\quad d$ = its particle size
$\quad f$ = the Darcy Weisbach friction coefficient for pipe flow

A value of K of 0.8 is recommended to ensure self-cleansing, but a value of 0.04 corresponds to the Shields criterion for fine sands. On the right hand side of Eq. (14.20a), ζ has a value of 1 for metric units or 1.486 for customary English units, n is Manning's roughness coefficient and R is the hydraulic radius. It is seen that the self-cleansing velocity depends primarily on the particle size and the specific gravity of the sediment. The specific gravity of sand is 2.65, while that of organic material is about 1.01. The minimum pipe slope, S_m, to achieve the minimum velocity is thus

$$S_m = \frac{f}{4R} \frac{V_m^2}{2\,g} \tag{14.20b}$$

With the help of Eq. (14.3), expression (14.20a) for the minimum velocity can be transformed into an expression for the minimum boundary shear stress τ_m as

$$\tau_m = \rho g K(s-1)d \tag{14.20c}$$

The ASCE/WPCF further states that in the design of sanitary sewers, an effort should be made to obtain adequate scouring velocities at the average or at least at the maximum flow. For storm sewers, scouring velocities may not be possible at minimum flow, but storm sewers should be self-cleansing during moderate storms. Finally, combined sewers may be subject to deposition during dry weather flow and are depending on wet weather for flushing. Figure 14.3 shows hydraulic elements for partly full circular sewers.

Fair et al. (1966) extended Eq. (14.20a) for partly full flows and obtained

$$\frac{S}{S_f} = \frac{R_f}{R}, \quad \frac{V_s}{V_f} = \frac{n_f}{n}\left(\frac{R}{R_f}\right)^{1/6}, \quad \frac{Q_s}{Q_f} = \frac{n_f}{n}\frac{A}{A_f}\left(\frac{R}{R_f}\right)^{1/6} \tag{14.21}$$

where the subscript f indicates full flow condition and the subscript s indicates self-cleansing condition. These equations are plotted in Fig. 14.4. The figure shows that to maintain self-cleansing, the slope must be doubled when the depth is 0.2 of the full depth and must be quadrupled when the depth is 0.1 of the full depth.

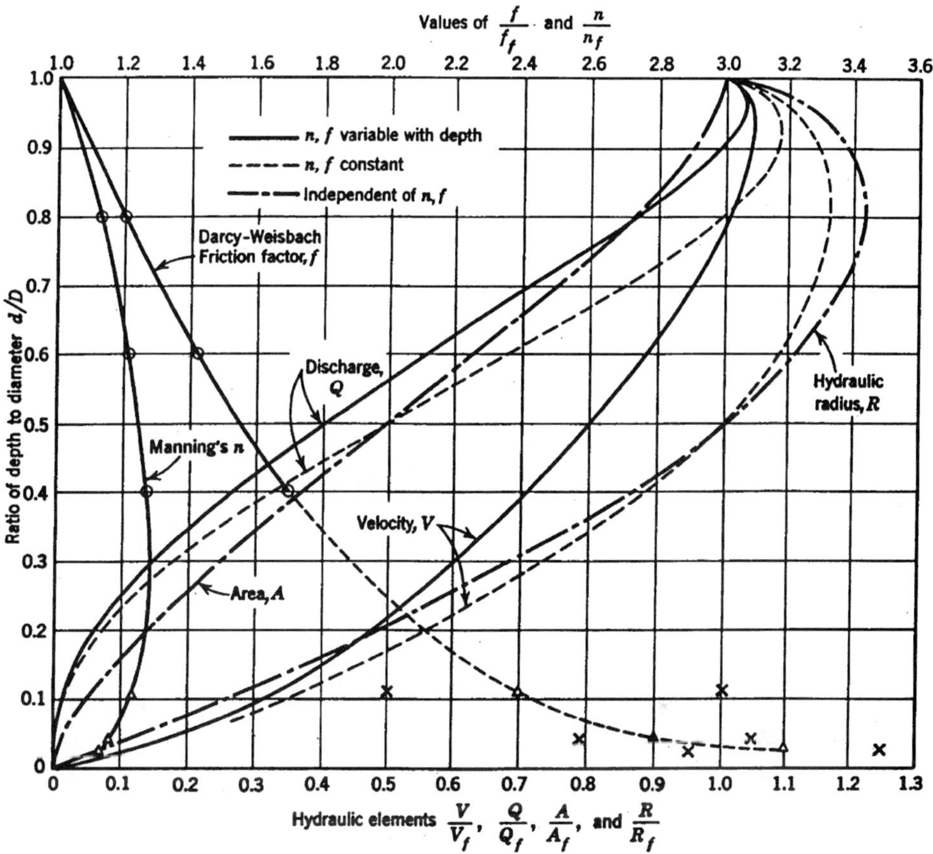

FIGURE 14.3 Hydraulic elements for circular sewers. (ASCE WPCF, 1969)

14.4.2 U.K. Design Criteria

14.4.2.1 Existing Criteria. The publication "Sewers for Adoption" (WAA, 1989) recommends a value of V_m of 0.75 m/s for sanitary sewers and 1.00 m/s for combined sewers. The British Standard (BS 8005, Part 1, 1987) recommends the equivalent of 1.00 m/s for sanitary and combined sewers and of 0.75 m/s for surface water. These are full pipe velocities.

14.4.2.2 Criteria Recommended By CIRIA. The CIRIA report (Ackers et al., 1996) uses three main design criteria. It assumes that the design discharge corresponds to full pipe condition and that the sewer is freely discharging, that is, that there are no backwater effects.

Criterion 1: On suspended sediment transport
For suspended sediment transport without deposition use the Macke Eq. (14.10) to obtain the diameter/minimum velocity relation. If limited deposition is permitted, use the Ackers-White Eq. (14.11).

Criterion 2: On bedload sediment transport

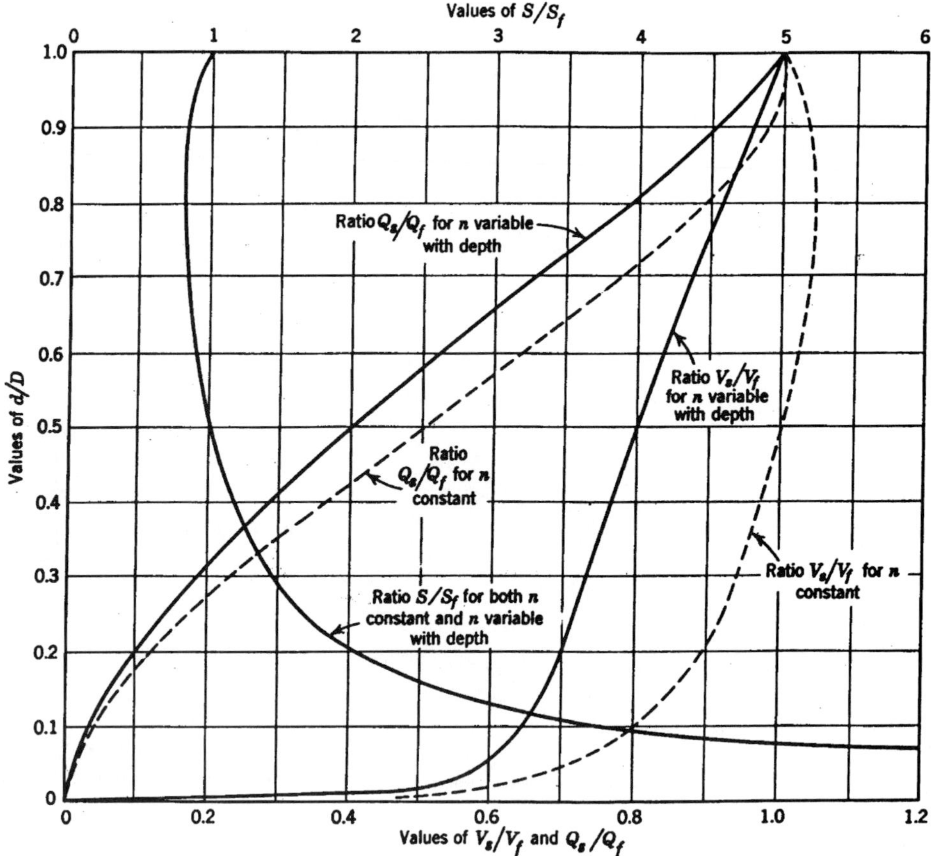

FIGURE 14.4   Hydraulic elements of circular sewers that possess equal self-cleansing properties at all depths. (ASCE/WPCF 1969)

For bedload transport at limit of deposition use the May Eq. (14.14a,b) to obtain the diameter/minimum velocity relation. See Eq. (14.9) for limit above which particles begin to move in suspension. If there is both suspended sediment and bedload movement, use the Ackers-White formula (14.11). This criterion is not applicable to sanitary sewers. For large circular sewers, a maximum bed deposition of 2% of the diameter is allowed. For egg-shaped sewers, a maximum of 5% of the diameter of a circular pipe with the same radius as the invert channel is allowed. An economic evaluation may change slightly the allowable maximum depth.

Criterion 3: On cohesive sediment erosion
Use grain shear stress 2 N/m^2 and obtain diameter/minimum velocity relationship.

Figure 14.5 shows the minimum required velocities for different sediment transport criteria (Arthur et al., 1999). It is seen that the erosion criterion dominates the design of small pipes, whereas the transport of material in suspended and bedload dominates the design of larger pipes.

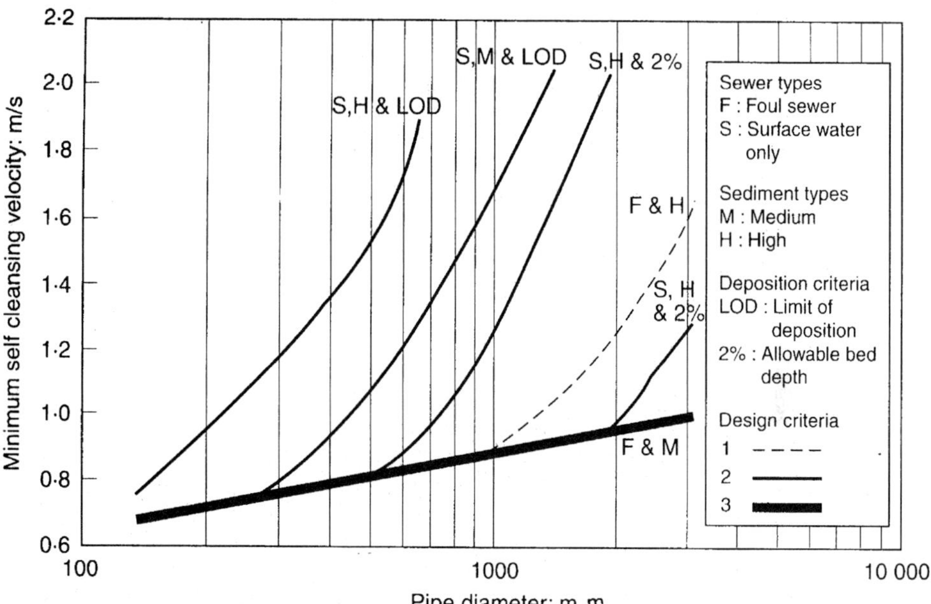

FIGURE 14.5   Minimum required velocities for different sediment transport criteria. (Arthur et al. 1999)

It is recommended that those interested in pursuing this design methodology consult the original CIRIA report (Ackers et al., 1996) and/or its summary (Butler et al., 1996a, b).

14.5 MODELS

As sediment movement is a non-linear function of the hydrodynamic parameters, the task of developing accurate simulation models is indeed a laborious one. Nevertheless, a number of recent attempts have been made in Europe to incorporate some of the recent research results on pollutant and sediment transport. Some of these simulation models are: MOSQITO and HYDROWORKS developed in the U.K., MOUSE TRAP in Denmark, FLUOPOL and HYPOCRAS developed in France.

The model HYPOCRAS simulates combined sewer systems. It models separately the transport of fine particles ($d_{50} = 40\ \mu m$) which are transported in suspension and the coarse particles ($d_{50} = 400\mu m$) which are transported as suspension or as bedload depending upon the hydraulic conditions. (Bertrand-Krajewski, 1994).

MOUSE TRAP is a component of the MOUSE sewer system package. Its purpose is the modeling of the transport of pollutants. It consists of several modules: ST for sediment transport, SRQ for surface runoff quality, AD for advection-dispersion and WQ for water quality (Crabtree et al., 1994). According to Mark et al., (1995), ST includes a continuity equation for the bed non-cohesive sediment transport, a source/sink description of cohesive sediments, a formulation for non-uniform sediment in which some fraction can be modeled as non-cohesive and cohesive simultaneously, and a fixed bed model which gives a first indication of the non-cohesive sediment transport capacity. MOUSETRAP has the ability to predict storm event pollutographs by modeling pollutants attached to different fractions of

TABLE 14.7 Comparison of Capabilities of Some Models (adapted from Ashley et al., 1999)

Model	Hydraulic solution	Sediment buildup	Multiple sediment fractions	Bed load transport	Biochemical processes
SWMM/Extran	Full	No	No	No	No
SWMM/Transport	Simplified	Somewhat	No	No	No
FLUOPOL	Simplified	Somewhat	Yes	No	No
HYDRO-WORKS	Full	Somewhat	Yes (2)	As total load	No
MOUSE-TRAP	Full	Yes	Yes	Yes	Yes

different types of sediment and pollutants associated with dissolved phase liquid transport, Crabtree et al. (1995). The prominent capabilities of the principal models are summarized in the following Table 14.7 (Ashley et al., 1999).

14.6 BEST MANAGEMENT PRACTICES TO CONTROL SEWER SEDIMENTS

Some of the best management practices cited by WEF/ASCE (1998) are:

1. Source controls including construction sites
2. Street cleaning
3. Catch basin cleaning
4. Vegetation control
5. Storm drain flushing
6. Roadway and bridge maintenance
7. Detention and infiltration device maintenance
8. Storm channel and creek maintenance
9. Prevention, detection and removal of illicit connections

U.S EPA projects for controlling sediments in sewers and associated pollution have been reviewed by Fan et al. (2000, 2001). They suggest an integrated approach including solids source management, inline control and treatment facilities.

14.6.1 Flushing of Combined Sewers

Flushing of combined sewers has been studied by Pisano et al. (1979) for the purpose of controlling combined sewer overflow (CSO) pollution. Sewer flushing has been found to be effective in reducing dry weather grit/inorganic deposition material in small combined sewer laterals with removal of 75% for single manhole to manhole segments. For serial segments of 200 m, removal of 55 to 65% of solids can be expected. Passive automatic flushing systems and grit control structures in use in the U.S., France and Germany have been described by Pisano et al. (2000) and Fan et al. (2001).

14.7 APPENDIX

14.7.1 Coefficients in Ackers Formula (14.11) (Ackers et al., 1996)

1. The effective width of the sediment bed, W_e is related to the width of the sediment bed W_b by

$$W_e = \left[0.2 + 3.33 \left(\frac{y_s}{D} - 0.01 \right) \right] W_b, \text{ when } 1\% \le y_s/D \le 10\%,$$

otherwise $W_e = 0.5\ W_b$

where y_s is the mean depth of the sediment depth over the invert and D is the internal pipe diameter.

2. The dimensionless grain size is

$$D_{gr} = \left[g\, \frac{s-1}{\nu^2} \right]^{1/3} d$$

where ν is the kinematic viscosity and d is the sediment grain size.

3. The coefficients are

$$m = 1.67 + \frac{.83}{D_{gr}}$$

$$\log H = 2.79 \log D_{gr} - 0.98(\log D_{gr})^2 - 3.46$$

$$A_{gr} = 0.14 + \frac{0.23}{D_{gr}^{0.5}}$$

$$J = \frac{8^{n(1-m)/2} H}{113^{m(1-n)}\, A_{gr}^m}$$

$$\alpha = 1 - n$$

$$\beta = (10 - 4\,m - mn)/10$$

$$\gamma = n(m - 1)/2$$

$$K = 11.3^{(1-n)} g^{n/2} A_{gr}$$

$$\delta = -n/2$$

$$\varepsilon = (4 + n)/10$$

for $D_{gr} > 60$ then $m = 1.78$, $n = 0$, $A_{gr} = 0.17$, $H = 0.025$

14.7.2 Coefficients η and θ in May's Equation (14.15) (Ackers et al., 1996)

1. The grain mobility factors is

$$F_g = \left[\frac{f_g V^2}{8g(s - 1)d_{50}} \right]^{0.5}$$

where f_g is calculated from eqn. (14.4) with $k = 1.23\ d_{50}$.

2. The particle Reynolds number is

$$R_{*_C} = \left[\frac{f_c}{8}\right]^{0.5} \left[\frac{Vd_{50}}{v}\right]$$

3. The particle transition factor is

$$\theta = \frac{\exp\left[\dfrac{R_{*_C}}{12.5}\right] - 1}{\exp\left[\dfrac{R_{*_C}}{12.5}\right] + 1}$$

4. The effective sediment mobility is

$$F_s = F_g \sqrt{\theta}$$

5. The parameter η depends on F_s as follows:

1. $F_s \leq 0.1$	1. $\eta = 0$
2. $0.1 < F_s \leq 0.225$	2. $\eta = 1.2\,(F_s - 0.1)$
3. $0.225 < F_s \leq 0.275$	3. $\eta = 0.15 + 9.0\,(F_s - 0.225)$
4. $0.275 < F_s \leq 0.4$	4. $\eta = 0.6 + 3.2\,(F_s - 0.275)$
5. $0.4 < F_s \leq 0.7$	5. $\eta = 1.0 - (F_s - 0.4)$
6. $0.7 < F_s \leq 0.8$	6. $\eta = 0.7$

REFERENCES

1. Ackers, P., "Sediment Aspects of Drainage and Outfall Design," *Proceedings of the International Symposium on Environmental Hydraulics,* Hong Kong, A. A. Balkema, Rotterdam, 1991.

2. Ackers, J. C., D. Butler, and R. W. P. May, *Design of Sewers to Control Sediment Problems,* Construction Industry Research and Information Association (CIRIA), 6 Storey's Gate, Westminister, London, SWIP 3AU, U.K. website: www.ciria.org.uk, 1996.

3. American Public Works Association, *Water Pollution Aspects of Urban Runoff,* Contract No. WA66-23, U.S. Dept. of the Interior, Federal Water Pollution Control Administration, Washington, DC 1969.

4. American Society of Civil Engineers/Water Pollution Control Federation, *Design and Construction of Sanitary and Storm Sewers,* ASCE Manuals and Reports on Engineering Practice, No. 37, 1969.

5. Arthur, S., and R. M. Ashley, Near Bed Solids Transport Rate Prediction in a Combined Sewer Network, *Water Science and Technology,* 36(8–9): 129–134, 1997.

6. Arthur, S., and R. M. Ashley, "The Influence of Near Bed Solids Transport on First Foul Flush in Combined Sewers, *Water Science and Technology,* 37(1): 131–138, 1998.

7. Arthur, S., S. Tait, and C. Nalluri, "Sediment Transport in Sewers—A Step Towards the Design of Sewers to Control Sediment Problems," *Proceedings, Institution of Civil Engineers, Water, Maritime & Energy,* 136: 9–19, 1999.

8. Ashley, R. M., D. J. J. Wotherspoon, and B. P. Coghlan, "The Erosion and Movement of Sediments and Associated Pollutants in Combined Sewers," *Water Science and Technology,* 25(8):101–114, 1992.

9. Ashley, R. J., D. J. J. Wotherspoon, and B. P. Coghlan, "Cohesive Sediment Erosion in Combined Sewers," *Proceedings,* Sixth International Conference on Urban Storm Drainage, Niagara Falls, Ontario, Canada, vol. 1, 644–651, 1993.

10. Ashley, R. M., T. Hitved-Jacobsen, and J.-L. Bertrand-Krajewski, "*Quo vadis* Sewer Processes Modelling?,"*Water Science and Technology,* 39(9):9–22, 1999.

11. Ashley, R., B. Crabtree, and A. Fraser, "Recent European Research into the Behaviour of Sewer Sediments and Associated Pollutants and Processes," *Proceedings ASCE 2000 Joint Conference on Water Resources Engineering and Water Resources Planning and Management,* CD-ROM, 2000.

12. Bannerman, R., K. Baun, M. Bohn, P. E. Hughes, and D. A. Graczyk, *Evaluation of Urban Nonpoint Source Pollution Management in Milwaukee County, Wisconsin,* PB 84-114164, U.S. Environmental Protection Agency, Chicago, IL., 1983.

13. Berlamont, J. E., K. Trouw, and G. Luyckx, "Shear Stress Distribution in Partially Filled Pipes and its Effect on the Modeling of Sediment Transport in Storm Drains," *Proceedings ASCE 2000 Joint Conference on Water Resources Engineering and Water Resources Planning and Management,* CD-ROM, 2000.

14. Bertrand-Krajewski, J. L., P. Briat, and O. Scrivener, "Sewer Sediment Production and Transport Modelling: A Literature Review," *Journal of Hydraulic Research* 31(4):435–460, 1993.

15. Bertrand-Krajewski, J.-L., "Solid Transport Modelling in Combined Sewer Systems: The HYPO-CRAS Model," *La Houille Blanche,* 4:243–255 (in French), 1994.

16. British Standards Institution, *Sewerage—Guide to New Sewerage Construction,* BS 8005: Part 1, 1987.

17. Bui, T. D., "Cohesive Sediment Transport in Natural Stream: State of Knowledge," *Proceedings ASCE 2000 Joint Conference on Water Resources Engineering and Water Resources Planning and Management,* CD-ROM, 2000.

18. Butler, D., R. W. P. May, and J. C. Ackers, "Sediment Transport in Sewers, Part 1: Background," Proceedings, Institution of Civil Engineers, Water, Maritime & Energy, 118:103–112, 1996a.

19. Butler, D., R. W. P. May, and J. C. Ackers, "Sediment Transport in Sewers, Part 2: Design," Proceedings, Institution of Civil Engineers, Water, Maritime & Energy, 118:113–120, 1996b.

20. Butler, D., and F. A. Memon, "Dynamic Modelling of Roadside Gully Pots During Wet Weather," *Water Research,* 33(15): 3364–3372, 1999.

21. Colebrook, C. F., Turbulent Flow in Pipes, with Particular Reference to the Transition Region between the Smooth and Rough Pipe Laws," *Journal of Inst. Civil Engineers,* London, 1939.

22. Crabtree, R. W., *A Classification of Combined Sewer Sediments Types and Characteristics,* WRC Report No. ER 324E, 1988.

23. Crabtree, R., Garsdal, R. Gent, O. Mark, and J. Dórge, "MOUSETRAP, A Deterministic Sewer Flow Quality Model," *Water Science and Technology,* 30(1):107–115, 1994.

24. Crabtree, R. W., R. Ashley, and R. Gent, "MOUSETRAP: Modelling of Real Sewer Sediment Characteristics and Attached Pollutants," *Water Science and Technology,* 31(7):43–50, 1995.

25. Delleur, J. W., "Water Quality Modelling in Sewer Networks," in *Hydroinformatics Tools for Planning, Design, Operation and Rehabilitation of Sewer Systems,* J. Marsalek, C. Maksimovic, E. Zeman, and R. Price (eds.), Kluwer Academic Publishers, NATO ASI Series, 286–324, 1998.

26. Delleur, J. W., "New Results and Research Needs on Sediment Movement in Urban Drainage," *Journal of Water Resources Planning and Management,* ASCE, 2001.

27. Fair, G. M., J. C. Geyer, and D. A. Okun, "Water and Wastewater Engineering," Vol. 1. *Water Supply and Wastewater Removal,* John Wiley & Sons, 1996.

28. Fan, C. Y., R. Field, F. Lai, and D. Sullivan, "Sewer-Sediment Control: Overview of an EPA Wet-Weather Flow (WWF) Research Program," *Proceedings of the ASCE 2000 Conference on Water Resources Engineering and Water Resources Planning and Management,* CD-ROM, 2000.

29. Fan, C. Y., R. Field, W. C. Pisano, J. Barsanti, J. J. Joyce, and H. Sorenson, "Sewer and Tank Flushing for Corrosion and Pollution Control," *Journal of Water Resources and Management,* ASCE, 2001.

30. French, R. H., *Open Channel Hydraulics,* New York, NY, McGraw-Hill, 1985.

31. Haan, C. T., B. J. Barfield, and J. C. Hayes, *Design Hydrology and Sedimentology for Small Catchments,* Academic Press, San Diego, CA, 1994.

32. Hamm, L., and C. Migniot, "Elements of Cohesive Sediment Deposition, Consolidation and Erosion," in Abbott and Price (ed.) *Coastal, Estuarial and Harbour Engineer's Reference Book,* E. & F. N. Spon, Chapman & Hall, London, 93–106, 1994.

33. Huber, W. C., and R. E. Dickinson, *Storm Water Management Model, User's Manual,* Environmental Research Laboratory, Office of Research and Development, US EPA, Athens, GA, 1988.

34. Kleijwegt, R. A., "Sewer Sediment Models and Basic Knowledge," *Water Science and Technology,* 25(8):123–130, 1992.

35. Kleijwegt, R. A., "On the Prediction of Sediment Transport in Sewers with Deposits," *Water Science and Technology,* 27(5–6):69–80, 1993.

36. Kleijwegt, R. A., R. G. Veldkamp, and C. Nalluri, "Sediment in Sewers: Initiation of Transport," *Water Science and Technology,* 22(10/11):239–246, 1990.

37. Lin, H., and B. Le Guennec, "Sediment Transport Modelling in Combined Sewers" *Water Science and Technology,* 33(9):61–67, 1996.

38. Macke, E., "About Sedimentation at Low Concentrations in Partly Filled Pipes," *Mitteilungen, Leichtweiss—Institut für Wasserbau der Tecnischen Universität Braunschweig,* Heft 76, 1–151, 1982.

39. Mark, O., C. Applegreen, and T. Larsen, "Principles and Approaches for Numerical Modelling of Sediment Transport in Sewers," *Water Science and Technology,* 31(7):107–115, 1995.

40. May, R. W. P., *Sediment Transport in Pipes and Sewers with Deposited Beds,* HR Wallingford, Report SR320, 1993.

41. May, R. W. P. *Sediment Transport Equations for Self-cleansing Sewers,* HR Wallingford, Report SR 436, 1995.

42. May, R. W. P., Minimum Self-Cleansing Velocities for Inverted Sewer Siphons," *Proceedings ASCE 2000 Joint Conference on Water Resources Engineering and Water Resources Planning and Management,* CD-ROM, 2000.

43. Metcalf and Eddy, Inc., *Wastewater Engineering: Treatment, Disposal and Reuse,* 3rd ed., McGraw-Hill, New York, NY, 1991.

44. Mehta, A. J., "Hydraulic Behaviour of Fine Sediment," in M. B. Abbott and W. A. Price (ed.) *Coastal, Estuarial and Harbour Engineer's Reference Book,* E. & F. N. Spon, Chapman & Hall, London, 577–584, 1994.

45. Mehta, A. J. and K. R. Dyer, "Cohesive Sediment Transport in Estuaries and Coastal Waters," *The Sea, Ocean Engineering Science,* vol. 9, Part B, *Interscience,* 815–839, John Wiley & Sons, 1990.

46. Nalluri, C., and E. M. Alvarez, "The Influence of Cohesion on Sediment Behavior," *Water Science and Technology,* 25(8):151–164, 1992.

47. Parker, W. R., "Cohesive Sediments—Scientific Background, in Abbott, M. B. and W. A. Price (ed.) *Coastal, Estuarial and Harbour Engineer's Reference Book,* E. & F. N. Spon, Chapman & Hall, London, 571–576, 1994.

48. Pisano, W. C., "Summary: United States "Sewer Solids" Settling Characterization Methods, Results, Uses and Perspectives," *Water Science and Technology,* 33(9):109–115, 1996.

49. Pisano, W. C., G. L. Aronson, C. S. Queiros, F. C. Blanc, and O'Shaughnessy, *Dry-Weather Deposition and Flushing for Combined Sewer Overflow Pollution Control;* EPA-600/2-79-133, 1979.

50. Pisano, W. C., S. L. White, and O. Riordan, "Implementation of Sewer and Drainage Flushing Systems in Cambridge, Mass.", *Proceedings ASCE 2000 Joint Conference on Water Resources Engineering and Water Resources Planning and Management,* CD-ROM, 2000.

51. Pitt, R., *Demonstration of Nonpoint Pollution Abatement Through Improved Street Cleaning Practices,* EPA-600/2-79-161, 1979.

52. Pitt, R., *Characterization and Controlling Urban Runoff Through Street and Sewerage Cleaning,* EPA/600/S2-85/038. PB85-186500, 1985.

53. Pitt, R., *Small Storm Urban Flow and Particulate Washoff Contribution to Outfall Discharges,* Ph.D. thesis, University of Wisconsin, Madison, WI, 1987.

54. Pitt, R., and J. McLean, *Toronto Area Watershed Management Strategy Study: Humber River Pilot Watershed Project,* Ontario Ministry of the Environment, Toronto, Ontario, 1986.

55. Pitt, R., and G. Shawley, *A Demonstration of Non-Point Source Pollution on Castro Valley Creek,* Alameda County Flood Control and Water Conservation District (Hayward, CA) for the Nationwide Urban Runoff Program, U.S. Environmental Protection Agency, Water Planning Division, Washington, DC, 1982.

56. Pitt, R., and R. Sutherland, *Washoe County Urban Stormwater Management Program; vol. 2, Street Particulate Data Collection and Analyses,* Washoe Council of Governments, Reno, Nevada, 1982.

56a. Pitt, R., *Urban Bacteria Sources and Control in the Lower Rideau River Watershed,* Ottawa, Ontario, Ontario Ministry of the Environment, ISBN 0-7743-8487-5, 1983.

56b. Pitt, R., *Characterizing and Controlling Runoff through Street and Sewerage Cleaning,* USEPA Report EPA/600/52-85/038, 1985.

57. Pitt, R., B. Robertson, P. Barron, A. Ayyoubi, and S. Clark, *Stormwater Treatment at Critical Areas: the Multi-chambered Treatment Train,* EPA/600R-99/017, 1999.

58. Renard, K. G., G. R. Foster, G. A. Weesies, D. K. McCool, and D. C. Yoder, *Predicting Soil Erosion by Water—A Guide to Conservation Planning with the Revised Soil Loss Equation RUSLE,* Publication ARS, U.S. Department of Agriculture, Washington, DC, 1993.

59. Ristenpart, E., R. M. Ashley, and M. Uhl, "Organic Near-bed Fluid and Particle Transport in Combined Sewers," *Water Science and Technology,* 31(7):61–68, 1995.

60. Sakhuja, V. S., *A Compilation of Methods for Predicting Friction and Sediment Transport in Alluvial Channels,* HR Wallingford, Technical Note OD/TN 27, 1987.

61. Sartor, J. D. and G. B. Boyd, *Water Pollution Aspects of Street Surface Contaminants,* 1972, USEPA, Report No. EPA-R2-72-081.

62. Shen, W. H., and P. Y. Julien, "Erosion and Sediment Transport," in *Handbook of Hydrology,* D. R. Maidment (ed.), 12.1–12.61, McGraw-Hill, New York, NY, 1993.

63. Skipworth, P. J., S. J. Tait, and A. J. Saul, "Laboratory Investigation into Cohesive Sediment Transport in Pipes," *Water Science and Technology,* 33(9):187–193, 1996.

64. Skipworth, P. J., S. J. Tait, and A. J. Saul, "Erosion of Sediment Beds in Sewers: Model Development," *Journal of Environmental Engineering,* ASCE 125(6):566–573, 1999.

65. Terstriep, M. L., G. M. Bender, and D. C. Noel, *Final Report—NURP Project, Champaign, Illinois: Evaluation of the Effectiveness of Municipal Street Sweeping in the Control of Urban Storm Runoff Pollution,* State Water Survey Division, Illinois Dept. of Energy and Natural Resources, Champaign-Urbana, Illinois, 1982.

66. Torfs, H., "Shape and Scale Effects on Secondary Currents and Shear Stress Distribution in Laboratory Flumes," *Proceedings, 2nd International Conference on Hydraulic Modelling, Development and Application of Physical and Mathematical Models,* Stratford-upon-Avon, U.K., June 1994, 287–295, 1994.

67. Torfs, H., *Erosion of Mud/sand Mixtures,* Ph.D. thesis, Katholieke Universiteit Leuven, Belgium, 1995.

68. Vanoni, V., *Sedimentation Engineering,* ASCE Manuals and Reports on Engineering Practice, No. 54, 1975.

69. Water Authorities Association, *Sewers for Adoption,* Water Research Centre, Medmenham, UK, 1989.

70. Water Environment Federation/American Society of Civil Engineers, *Urban Runoff Quality Management,* WEF Manual of Practice No. 23, ASCE Manual and Report on Engineering Practice No. 87.

71. Wischmeier, W. H., and D. D. Smith, *Predicting Rainfall Erosion Losses from Cropland East of the Rocky Mountains—A Guide for Selection of Practices for Soil and Water Conservation,* Agricultural Handbook No. 282, U.S. Dept. of Agriculture, Washington, DC, 1965.

72. Wischmeier, W. H., and D. D. Smith, *Predicting Rainfall Erosion Losses—A Guide to Conservation Planning,* Agricultural Handbook No. 537, U.S. Dept. of Agriculture, Washington, DC, 1978.

73. Wotherspoon, D. J. J., and R. M. Ashley, "Rheological Measurement of the Yield Strength of Combined Sewer Sediment Deposits," *Water Science and Technology,* 25(8):165–169, 1992.

CHAPTER 15
HYDRAULIC DESIGN OF CULVERTS AND HIGHWAY STRUCTURES

I. Kaan Tuncok
Daniel, Mann, Johnson, Mendenhall (DMJM)
Phoenix, Arizona

Larry W. Mays
Department of Civil and Environmental Engineering
Arizona State University
Tempe, Arizona

15.1 INTRODUCTION

The aim in highway drainage is to prevent on-site water standing on the surface and convey the off-site storm runoff from one side of the roadway to the other. To accomplish the off-site drainage either a culvert or a bridge can be used. *Culverts* are closed conduits in which the top of the structure does not form part of the roadway. Bridges are mainly provided for large streams and the road is practically a part of the span or drainage structure. National Bridge Inspection Standards (NBIS) define *bridges* as those structures that have at least 20 ft of length along the roadway centerline. The main operational differences between culverts and bridges may be described in terms of:

- Economics
- Hydraulics
- Structural aspects
- Maintenance attention requirements

Economically, the initial and operating costs of culverts are considerably less than that of bridges. Thus, the total investment of public funds for culverts constitutes a substantial share of the highway budget in relation to the investment for bridges.

The hydraulic properties for both the culverts and bridges can be computed by using the conservation of mass, energy, and momentum principles defined in Chap. 3. In the case of bridges the magnitude of energy losses must be carefully computed. One part of this loss is due to the contraction and expansion that occur in reaches immediately upstream and downstream from the bridge. The other part is due to the losses at the structure itself and is calculated with either the momentum or energy principles (U.S. Army Corps of Engineers,

1990). To compute the structural losses, the flow must be identified as low flow, pressure flow, or weir flow and equations based on the momentum principle, orifice flow, and a standard weir equation can be used, respectively, to separate the associated losses. The hydraulics of culverts is complicated and the concept of inlet control and outlet control is commonly used to simplify the analysis. Inlet and outlet control culverts have a barrel capacity larger and smaller than the capacity of the culvert entrance, respectively. This chapter will focus on the hydraulic design of culverts and stream stability at highway structures.

Structurally, culverts are designed with a heavy dependence on good and proper backfill around the perimeter of the culvert conduit. Often culverts must accommodate significant dead loads (i.e., embankment loads) in addition to the live loads (i.e., vehicles and pedestrians). The earth load transmitted to a culvert is largely dependent on the type of installation, and the four common types are:

- Trench, used in the construction of sewers, drains and water mains
- Positive projecting embankment, used in relatively flat streambed or drainage path
- Negative projecting embankment, used in relatively narrow and deep streambed or drainage path
- Induced trench, used in the construction of culverts placed under high embankment

For the details of these different installation types the reader should refer to the American Concrete Pipe Association (1992). Bridges barely have significant embankment loads and therefore the live load on bridges is the predominant structural consideration.

Maintenance requirements for culverts are considerable. Constant efforts must be made to ensure:

- Clear and open conduits
- Protection or repair against corrosion and abrasion
- Repair and protection against local and general scour
- Structural distress repair
- Maintenance of traffic safety devices pertinent to the culvert

Bridges require some of the same items of maintenance attention, but not to the extent required by culverts.

15.2 DESIGN PARAMETERS

To better understand the design of the culverts, the basic design parameters must be carefully studied. The headwater, outlet velocity, and tailwater are factors of significant importance.

15.2.1 Headwater and Tailwater

Headwater is the vertical distance from the culvert invert at the entrance, to the energy line (depth 1 velocity head) of the headwater. Because of the low velocities in most entrance pools and the difficulty in determining the velocity head for all flows, the water surface and the energy line at the entrance are assumed to be coincident (American Iron and Steel Institute, 1983). Therefore, *headwater* is the depth of water at the upstream face of the culvert. The main reason for the accumulation of water is to build up the energy required to pass the water through the culvert opening and to overcome the friction, entrance, and exit losses along the culvert. By this process, the potential energy accumulated in water is transformed into kinetic energy through the culvert. Figure 15.1 shows a typical inlet control flow condition with the headwater depicted on it.

The designer should not ignore the design limitations on the *maximum* or *allowable headwater (AHW)*. AHW is the level to which the culvert headwater may rise before causing an unwanted inundation or damage under the circumstances of the design flood (Reagan, 1993). The level of AHW may be based on the fill heights, elevation level of property, or features which would be affected by ponded water due to culvert headwater levels. In cases with a heavy load of debris in the runoff, provision for its passage (or retention) must be made. The designer should avoid the use of very high allowable headwater that may result in unacceptable turbulence and objectionable velocities of flow into and through the culvert. The criteria in establishing the AHW can be summarized as:

1. AHW should not give damage to upstream properties.
2. AHW should be below the traffic lanes of interest or lower than the shoulder.
3. AHW should be lower than the low point in the road grade.
4. AHW should not be equal to the elevation where flow diverts around the culvert.

The ponding affect at the upstream face of the culvert may cause an attenuation in the peak discharges, therefore resulting in smaller culvert size requirements for the design.

The *tailwater* is the depth of water downstream of the culvert measured from the outlet invert. It can be a very significant factor in culvert hydraulic operation. The depth of the tailwater can affect the outlet velocity, and depending on the character of the culvert flow it can also affect the operating headwater on the culvert.

15.2.2 Outlet Velocity

For design purposes the outlet velocity should be similar in magnitude to the velocity in the channel to provide appropriate protection at the downstream end (American Concrete Pipe Association, 1992). But generally due to the constriction inside the culvert opening, the outlet velocity is higher than the channel velocity. High channel velocities can be mitigated in several different ways such as the following:

1. Stabilization of the channel
2. Construction of energy dissipation structures, such as hydraulic jump basin, drop structure, stilling basin, riprap, and sill. For a detailed description of the individual methods, the designer should refer to Corry et., al. (1983).
3. Changing the size or the roughness of the culvert

A more detailed discussion of outlet velocity will be given in Sec. 15.3.3. Another important factor is the minimum velocity that must be maintained within the culvert for an

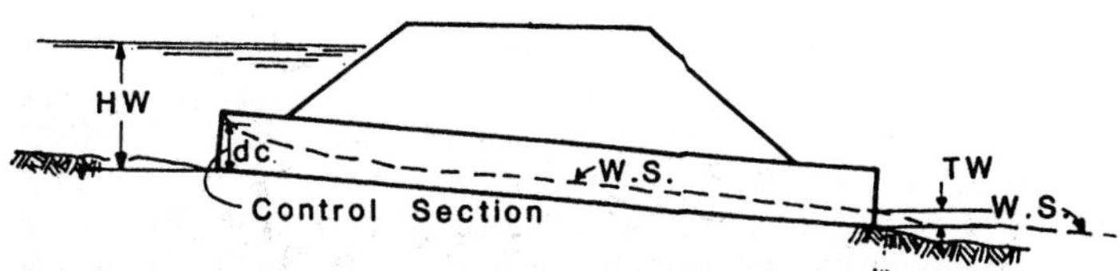

FIGURE 15.1 Typical inlet control flow condition. d_c = critical depth; HW = headwater; TW = tailwater; WS = water surface. (From Normann et al., 1985)

efficient operation. The ideal minimum velocity should be adjusted such that the sediment particles transported through the culvert should not be allowed to settle.

The culvert barrel should result in a tractive force, τ, greater than the critical τ of the transported stream bed material at low flow rates. In case of unknown stream bed material size a culvert velocity of 2.5 ft/s should be maintained within the culvert. If clogging is probable, installation of a sediment trap or sizing the culvert to facilitate the cleaning should be considered.

15.3 *CHARACTERISTICS of FLOW*

A *culvert* is a type of structure that can transmit water as full or partly full. Full flow is not common for culverts unless governed by a high downstream or upstream water surface elevation. Full flow can be described by the fundamentals of pipe flow. Partly full flow culverts follow the rules of open-channel flow and need to be classified as either subcritical or supercritical flow to accomplish the design procedure. The reader should refer to Chap. 3 for the details of open-channel flow and the related terminology that will be used in this chapter. An exact theoretical analysis of culvert flow is difficult due to the following factors:

1. The culvert may have both gradually and rapidly varied flow zones and non-uniform flow conditions must be considered to analyze the flow.
2. The location of hydraulic jump, if it occurs, must be identified.
3. The results of hydraulic model studies must be applied.
4. The change in the flow type as a function of discharge and hydraulic characteristics of the culvert must be studied.

In the culvert design procedures of this chapter, the concept of control section, as related to the relationship between the flow rate and the upstream water surface elevation will be used. A culvert can operate either under inlet or outlet control depending on whether the barrel or the inlet has a greater hydraulic capacity. Under inlet control, the cross sectional area of the culvert barrel, the inlet geometry, and the amount of headwater at the entrance are of primary importance. Outlet control involves the additional consideration of the tailwater elevation in the outlet channel, the slope, the roughness and the length of the culvert barrel.

When a culvert operates under inlet control, the barrel will flow partly full and depending on the headwater, the flow may be divided either by using the weir or orifice equations. For outlet control, the culvert barrel is intended to flow full for design conditions. A more detailed description of both the inlet and outlet control culverts will be described in the following sections.

15.3.1 Inlet Control

A culvert operates under *inlet control* when the barrel hydraulic capacity is higher than that of the inlet. A typical flow condition is critical depth near the inlet and supercritical flow in the culvert barrel. Depending on the tailwater, a hydraulic jump may occur downstream of the inlet. The barrel geometry and roughness have no direct influence on the hydraulic characteristics of the culvert.

Due to the severe constriction of the flow at the culvert entrance, the inlet configuration have a significant effect on the hydraulic performance. To increase the capacity of the culvert the designer can use beveled inlet edges, side-tapered inlet, or slope-tapered inlet. The details of different inlet conditions can be found in Sec. 15.7.4.

Typically both the inlet and outlet control culverts can be studied in four different types and can be summarized as:

1. Inlet and outlet unsubmerged
2. Inlet unsubmerged, outlet submerged
3. Inlet submerged, outlet unsubmerged
4. Inlet and outlet submerged

Culverts operating under steep slope regimes usually operate with inlet control of the headwater, and as a result supercritical flow is a very common flow condition. In cases where the specific energy within the culvert barrel is in the vicinity of minimum specific energy, the flow depth is sensitive to the changes (i.e., roughness) in the culvert. In application, if the normal depth and critical depth are within 5% of each other, hydraulic computations should be performed separately to identify the worst case for the design. In the case of inlet control, these four cases are depicted in Fig. 15.2. The flow will be governed by weir flow when the entrance is unsubmerged, and by orifice flow in the case of a submerged entrance. Laboratory investigations showed that for an entrance, either inlet or outlet control, to be considered as unsubmerged, the headwater must be less than a critical value H_{cr} and can be

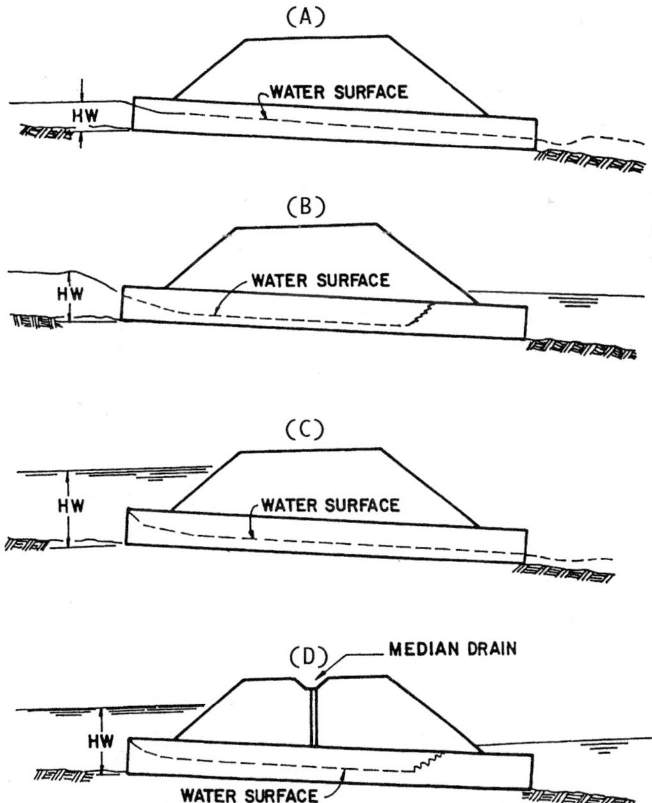

FIGURE 15.2 Types of inlet control: (**A**) outlet unsubmerged, (**B**) outlet submerged, inlet unsubmerged, (**C**) inlet submerged (**D**) outlet submerged. (From Normann et al., 1985)

approximated as (1.2–1.5) times the height of the barrel. Another important factor is the identification of the culvert as hydraulically short or long (Chow, 1959), which depends on the length, slope, size, entrance geometry, headwater, entrance, and outlet conditions.

The most typical case of inlet submerged and outlet unsubmerged condition is shown in Fig. 15.2c. As soon as the flow enters the culvert, critical depth is observed and the free-flow conditions at the downstream end results in supercritical normal depth. Supercritical flow is maintained along the culvert barrel. A very similar condition with both inlet and outlet unsubmerged is shown in Fig. 15.2a. The difference from the previous condition is the low headwater at the culvert entrance, but again the flow passes through critical depth at the culvert entrance. In the case of inlet unsubmerged and outlet submerged (Fig. 15.2b), even though the high downstream water surface depth cannot enforce outlet control conditions, a hydraulic jump is observed within the culvert. Similar to the previous cases the flow passes through critical depth at the culvert entrance, but the partly full flow is maintained only upstream of the hydraulic jump (ARMCO, 1950). An uncommon condition shown in Fig. 15.2d depicts submerged inlet and outlet with partly full flow. Similar to the previous case, a hydraulic jump will be observed within the culvert.

15.3.2 Outlet Control

The culvert will operate under outlet control conditions if the culvert barrel has a smaller hydraulic capacity than the inlet does. The flow regime is always subcritical; as a result, the control of the flow is either at the downstream end of the culvert or further downstream of the culvert outlet, depending on the depth of the tailwater. (Portland Cement Association, 1964). Typical flow conditions include a full or partially full culvert barrel for all or part of its length.

Similar to the inlet control culverts, four types of control are available for the outlet control culverts (Fig. 15.3). In the first flow condition in which both the inlet and the outlet are unsubmerged, the flow is subcritical and the culvert barrel is partly full over its entire length. The unsubmerged flow condition results in critical depth at the culvert outlet. For the second flow condition, in which the inlet is unsubmerged and outlet is submerged, headwater is an important factor. Typically, the water surface will drop at the culvert's entrance and experience a contraction through the culvert's opening. This is depicted in Fig. 15.3B and is a result of shallow headwater. The unusual case shown in Fig. 15.3C can be established with high submergence at the culvert inlet. Although there is no tailwater, the submergence at the inlet maintains the pressure flow along the culvert barrel. The condition where the inlet is submerged and outlet is unsubmerged, (Fig. 15.3D) is a result of tailwater that does not submerge the culvert depth but is greater than the critical depth at the outlet. Finally, the common condition where both the inlet and the outlet are submerged (Fig. 15.3A) results in pressure flow through the culvert opening.

In all of these flow types, in addition to the factors that affect the inlet control culverts, the barrel characteristics such as roughness, area, length, and slope also play an important role.

Hydraulics.   To evaluate the outlet control hydraulics the condition of full flow in the culvert barrel will be used. The energy equation, Eq. (15.1), must incorporate the losses due to entrance (h_e), friction (h_f), exit (h_{ex}), bend (h_b), junctions (h_j), and grates (h_g) and can be written as

$$H_L = h_e + h_f + h_{ex} + h_b + h_j + h_g \tag{15.1}$$

where H_L = total energy loss (ft).

The velocity head, (h_v), can be expressed as

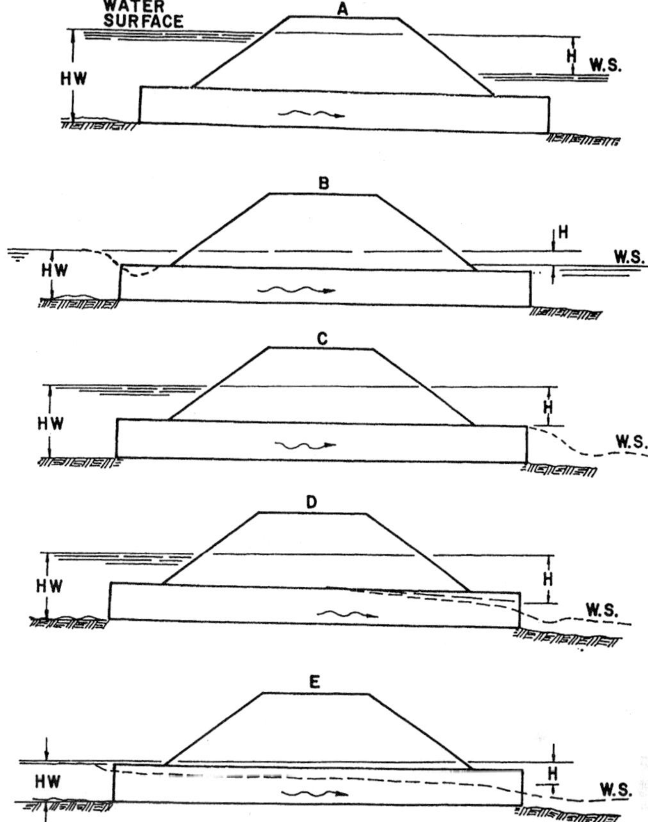

FIGURE 15.3   Types of outlet control. (From Normann et al., 1985)

$$h_v = \frac{(V^2 - V_a^2)}{2g} \tag{15.2}$$

where V = velocity of flow in the culvert barrel (ft/s), and can be computed as

$$V = \frac{Q_b}{A} \tag{15.3}$$

where Q_b = discharge per barrel (ft³/s)
   A = cross-sectional area of flow with the barrel full (ft²)
   V_a = approach velocity of flow (ft/s)
   g = acceleration due to gravity (32.2 ft/s²)

For design purposes, the approach velocity usually is ignored and the velocity head can be expressed as

$$h_v = \frac{V^2}{2g} \tag{15.4}$$

The entrance loss can be computed in terms of velocity head as

$$h_e = k_e \left(\frac{V^2}{2g}\right)$$
(15.5)

where k_e = entrance loss coefficient (Table 15.1).

Similar to the entrance loss, the exit loss can be computed as

TABLE 15.1 Entrance Loss Coefficients–Outlet Control, Full or Partly Full

$$h_e = k_e \left[\frac{V^2}{2g}\right]$$

Type of structure and design of entrance	Coefficient k_e
Pipe, concrete	
Mitered to conform to fill slope	0.7
End-section conforming to fill slope*	0.5
Projecting from fill, sq. cut end	0.5
Headwall or headwall and wingwalls	
Square-edge	0.5
Rounded (radius = 1/12D)	0.2
Socket end of pipe (groove-end)	0.2
Projecting from fill, socket end (groove-end)	0.2
Beveled edges, 33.7° or 45° bevels	0.2
Side- or slope-tapered inlet	0.2
Pipe, or pipe-arch, corrugated metal	
Projecting from fill (no metal)	0.9
Mitered to conform to fill slope, paved or unpaved slope	0.7
Headwall or headwall and wingwalls aquare-edge	0.5
End-section conforming to fill slope*	0.5
Beveled edges, 33.7° or 45° bevels	0.2
Side-or slope-tapered inlet	0.2
Box, reinforced concrete	
Wingwalls parallel (extension of sides)	
Square-edges at crown	0.7
Wingwalls at 10°–25° or 30°–75° to barrel	
Square-edged at crown	0.5
Headwall parallel to embankment (no wingwalls)	
Square-edged on three edges	0.5
Rounded on three edges to radius of 1/12 barrel dimension, or beveled edges on 3 sides	0.2
Wingwalls at 30°–75° to barrel	
Crown edge rounded to radius of 1/12 barrel dimension, or beveled top edges	0.2
Side- or slope-tapered inlet	0.2

Source: From Normann et al., (1995).

*Note: "End section conforming to fill slope," made of either metal or concrete, are the sections commonly available from manufacturers. From limited hydraulic tests they are equivalent in operation to a headwall in both *inlet* and outlet control. Some end sections, incorporating a *closed* taper in their design have a superior hydraulic performance. These latter sections can be designed using the information given for the beveled inlet.

$$h_{ex} = \frac{V^2}{2g} \tag{15.6}$$

The friction loss can be expressed as

$$h_f = S_f\, L \tag{15.7}$$

where S_f = friction slope, and can be computed by a manipulation of Manning's equation (Brater, et al 1996) and
L = length of the culvert barrel (ft).

$$S_f = \left[\frac{Qn}{1.486\, A\, R^{0.67}}\right]^2$$

in American customary units where, n = Manning's roughness coefficient, A = cross-sectional area (ft^2), R = hydraulic radius (ft), A and R are based on full-flow conditions.

When the expression for friction slope is inserted into Eq. (15.7), the expression for friction loss can be written as

$$h_f = \left(\frac{29n^2\, L}{R^{1.33}}\right)\frac{V^2}{2g} \tag{15.8}$$

The bend losses are also a function of velocity head and can be computed as

$$h_b = K_b\left(\frac{V^2}{2g}\right) \tag{15.9}$$

where K_b = bend loss coefficient (Table 15.2).

Junction losses, h_j, are significant whenever more than two culverts flow into a single culvert downstream, as shown in Fig. 15.4. In such cases the energy and momentum principles can be used to calculate h_j.

The energy equation for the junction can be written as

$$h_j = y + h_{vu} - h_{vd} \tag{15.10}$$

where h_j = junction loss in the main culvert (ft)
h_{vu} = velocity head in the upstream culvert (ft)
h_{vd} = velocity head in the downstream culvert (ft)
y = change in hydraulic grade line through the junction (ft)

TABLE 15.2   Loss Coefficients for Bends

Radius of bend equivalent diameter (/)	Angle 90	of 45	Bend,* 22.5
1	0.50	0.37	0.25
2	0.30	0.22	0.15
4	0.25	0.19	0.12
6	0.15	0.11	0.08
8	0.15	0.11	0.08

Source: From Normann, et al. (1985)

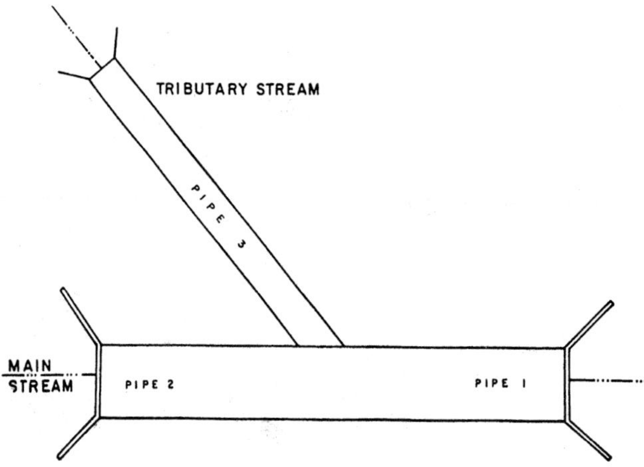

FIGURE 15.4 Culvert junction. (From Normann et al., 1985)

The momentum principle can be used to compute y as

$$y = \frac{Q_2 V_2 - Q_1 V_1 - Q_3 V_3 \cos\theta}{0.5(A_1 + A_2)g} \tag{15.11}$$

where A_1 and A_2 = cross-sectional areas for culverts 1 and 2, respectively
V_1 and V_2 = velocity for culverts 1 and 2
θ = angle between culverts 1 and 3

If only the entrance, friction, and exit losses are considered, and the respective equations are inserted into Eq. (15.1), the total headloss can be expressed in American customary units as

$$H = \left(1 + k_e + \frac{29n^2 L}{R^{1.33}}\right)\frac{V^2}{2g} = h_{ex} + h_e + h_f \tag{15.12}$$

By using the schematic in Fig. 15.5 that depicts the hydraulic and energy grade lines, the energy equation with the entrance, exit, and friction loss terms for a full flow, can be written between the upstream and downstream ends of a culvert system as

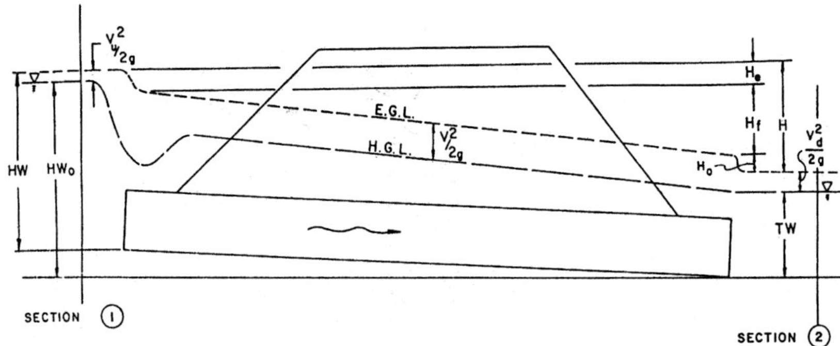

FIGURE 15.5 Full–flow energy and hydraulic grade lines. (From Normann et al., 1985)

$$HW_o + \frac{V_u^2}{2g} = TW + \frac{V_d^2}{2g} + H \tag{15.13}$$

where HW_o = headwater depth above the outlet invert
TW = tailwater depth above the outlet invert, and the subscripts u and d denote the upstream and downstream conditions, respectively.

Because of the smaller values of upstream and downstream velocities, compared to the culvert velocity, they can be neglected. Then the energy equation can be expressed as

$$HW = TW + H - S_oL \tag{15.14}$$

where S_oL = drop in elevation from the inlet invert to the outlet invert

A practical way to compute the headloss H is to use the outlet control nomographs (Figs. 15.6 and 15.7). Outlet control nomographs presented for full flow can also be used for the partly full flow for the computation of H. Details on the use of outlet control nomographs will be given in Sec. 15.4.2.2.

Equations (15.1–15.14) were developed for full barrel flow in which TW is greater than or equal to the culvert diameter, D. However in the case of partly full flow, backwater

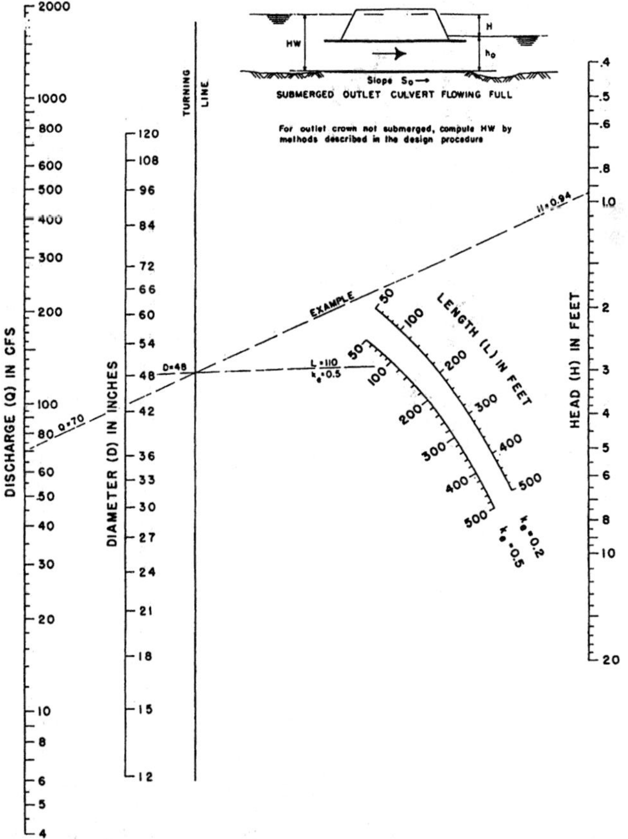

FIGURE 15.6   Head for concrete pipe culverts (n = 0.012). (From Normann et al., 1985)

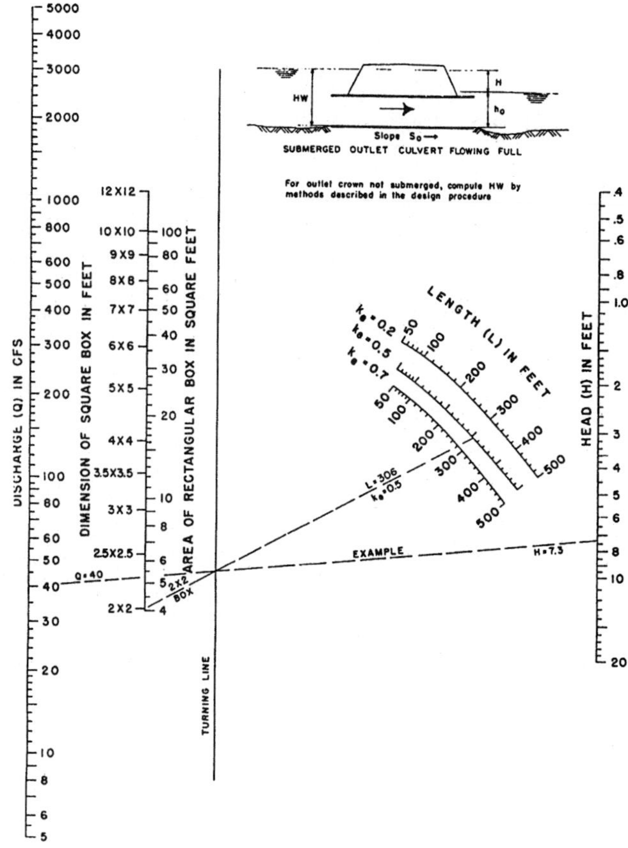

FIGURE 15.7 Head for concrete box culverts flowing full ($n = 0.012$).
(From Normann, et al., 1985)

calculations may be required starting at the downstream end of the culvert and proceeding upstream. During the backwater computations if the hydraulic grade line cuts the top of the barrel, full flow will be observed upstream of this intersection point. FHWA has also developed an approximate method to overcome the tedious backwater computations for partly full flow. They found out that the hydraulic grade line at the culvert outlet is at a point between the critical depth and the culvert diameter, and can be computed as $(d_c + D)/2$. It was oncluded that the observed TW should only be used if it is higher in magnitude than $(d_c + D)/2$. In such a case the following equation should be used:

$$HW = h_o + H - S_oL \qquad (15.15)$$

where $h_o = \max[TW, (d_c + D)/2]$.

The use of these two equations reasonable results for HW depths greater than $0.75D$.

15.3.3 Outlet Velocity

Outlet velocity is an important factor to define the type of outlet protection. Because of the decreased area of flow in the culvert, the velocity increases through the culvert barrel and

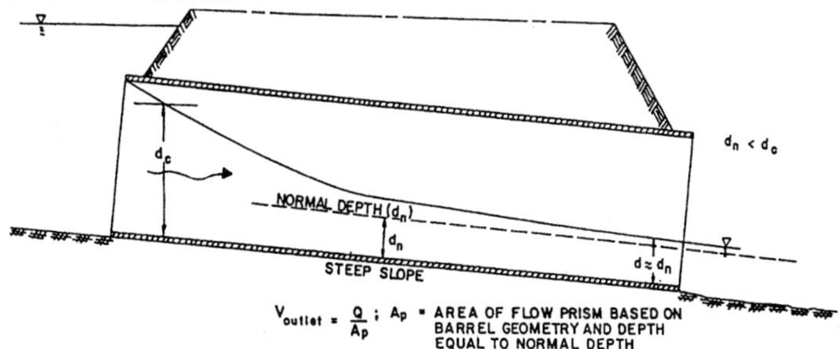

FIGURE 15.8   Inlet control outlet velocity. (From Normann et al., 1985)

results in higher velocities than that of the natural stream. Because of the high velocities, outlets must be protected by using riprap or an energy dissipator. To accurately compute the outlet velocity, the type of control must be defined.

For inlet control culverts, either an exact or approximate method can be used. In the exact method, water surface profile computations start at the upstream and proceed downstream and the velocity is computed by using the cross-sectional area at the exit. For the approximate method, normal depth is assumed to occur at the culvert outlet as depicted in Fig. 15.8. This method is more commonly used and results in more conservative estimates of the outlet velocity. Normal depth can be computed by using Manning's equation.

For outlet control culverts, the outlet velocity depends on the outlet geometry and the magnitude of tailwater depth with respect to either the critical depth or the barrel diameter. The following procedure can be used to define the tailwater depth and to compute the outlet velocity with the given geometry.

If	$TW < d_c$,	Depth at culvert outlet = d_c.
If	$d_c < TW < D$	Depth at culvert outlet = TW.
If	$TW > D$	Depth at culvert outlet = D.

The schematic in Fig. 15.9 also summarizes these three flow conditions.

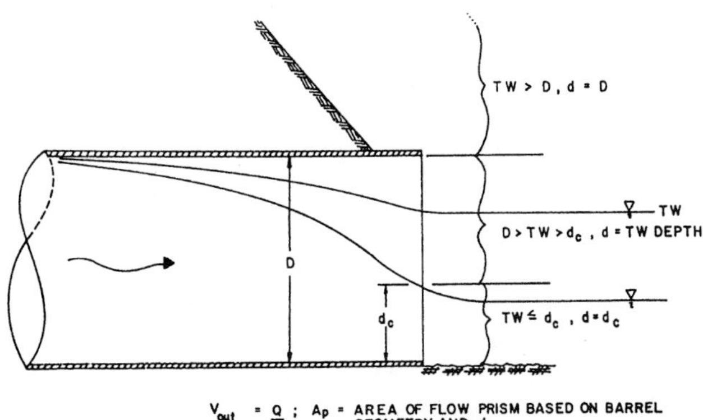

FIGURE 15.9   Outlet control outlet velocity. (From Normann et al., 1985)

15.3.4 Roadway Overtopping

Roadway overtopping starts soon after the headwater elevation reaches the top of roadway elevation. To compute the magnitude of the overtopping flow, the following broad-crested equation can be used:

$$Q_r = C_d L (HW_r)^{1.5} \tag{15.16}$$

where Q_r = overtopping flow rate (ft^3/s)
C_d = overtopping discharge coefficient = $k_t C_r$, the discharge coefficient C_r, and the submergence factor k_t, are presented in Fig. 15.10
L = length of roadway crest, (ft)
HW_r = the upstream depth, measured above the roadway crest, (ft)

The main reason for this type of equation is the similarity of this flow to the broad-crested weir flow. One of the most important factors in the computation of Q_r is a good estimate of the length of roadway crest. The length can be established in two different ways. The first method, although time consuming, results in more accurate estimates of the length, especially if the elevation of the roadway crest varies. It is based on an approximation of

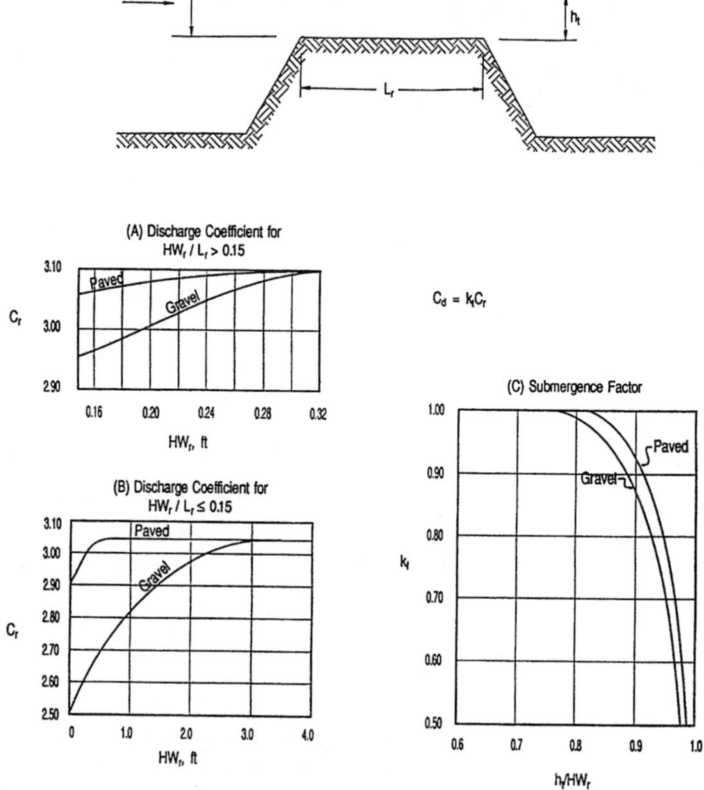

FIGURE 15.10   Discharge coefficient and submergence factor for roadway overtopping. (From Flood Control District of Maricopa County, 1996)

the vertical curve as a series of horizontal segments and can be applied to cases where the crest is defined by a roadway sag vertical curve (for definition see Sec. 15.7.3) and is depicted in Fig. 15.11. Then the flow over each segment is calculated for a given headwater and finally the flows for each segment are added together to determine the total flow. In the second method, which is more practical, the length can be represented by a single horizontal line at a constant roadway elevation. For correct estimates of Q_r, this horizontal line must span the roadway profile accurately. The depth of flow used in this method is the average depth of upstream pool above the roadway.

Given the headwater, length and the coefficient, Q_r can be computed by using Eq. (15.16). The difficult task is to determine the magnitude of total flow which equals roadway overflow plus the culvert flow and can be determined by using a trial and error procedure. Performance curves, as defined in Sec. 15.5, can also be used to compute the total flow.

15.4 *METHOD OF CULVERT DESIGN*

The culvert design procedures for both the inlet and outlet control culverts must consider some important factors (AASHTO, 1990) namely,

- Establishment of hydrology
- Design of downstream channel
- Assumption of a trial configuration
- Computation of inlet control headwater
- Computation of outlet control headwater at inlet
- Evaluation of the controlling headwater
- Computation of discharge over the roadway and then the total discharge
- Computation of outlet velocity and normal depth
- Comparison of headwater and velocity to limiting values

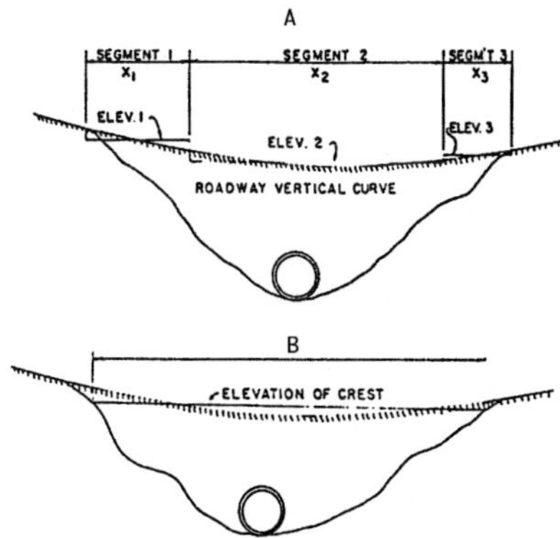

FIGURE 15.11 Weir crest length determinations for roadway overtopping. (**A**) Method 1—subdivision in to to segments. (**B**) Method 2—use of a single segment. (From Normann et al., 1985)

- Adjustment of configuration (if necessary)
- Recomputation of hydraulic characteristics (if necessary)

15.4.1 Inlet Control

The computation of headwater for inlet control culverts is based on either the design equations or the nomographs.

15.4.1.1 Design Equations.   The design equations to be used depend on the condition of the inlet control. A culvert performs as an orifice when the inlet is submerged and as a weir when it is unsubmerged. The *submerged (weir) equation* can written as

$$\left[\frac{HW_i}{D}\right] = c\left[\frac{Q}{AD^{0.5}}\right]^2 + Y + Z \quad \text{for} \quad \left[\frac{Q}{AD^{0.5}}\right] \geq 4.0 \tag{15.17}$$

where HW_i = headwater depth above the inlet control section invert (ft)
 D = interior height of the culvert barrel (ft)
 Q = discharge (ft^3/s)
 A = full cross-sectional area of the culvert barrel (ft^2)
 c, Y = constants from Table 15.3
 Z = term for culvert barrel slope (ft/ft). For mitered inlets, $Z = 0.7S$. For all other conditions, $Z = -0.5S$.

The *unsubmerged (weir) condition* can be presented in two forms. The first one is based on the specific head at critical depth, and can be written as

$$\left[\frac{HW_i}{D}\right] = \left[\frac{H_c}{D}\right] + K\left[\frac{Q}{AD^{0.5}}\right]^M + Z \quad \text{for} \quad \left[\frac{Q}{AD^{0.5}}\right] \leq 3.5 \tag{15.18}$$

where H_c = specific head at critical depth ($H_c = d_c + V_c^2/2g$) (ft)
 K, M = constants from Table 15.3.

The second form is similar to a weir equation and has a simpler form:

$$\left[\frac{HW_i}{D}\right] = \left[\frac{Q}{AD^{0.5}}\right]^M \quad \text{for} \quad \left[\frac{Q}{AD^{0.5}}\right] \leq 3.5 \tag{15.19}$$

The latter form is the only documented form for some of the design inlet control nomographs in Normann, et al., (1985).

Nomographs.   As an alternative procedure, the *inlet control nomographs* presented in Figs. 15.12 and 15.13 can be used as described in the following:

1. Identify the culvert size and flow rate to be used. It is important to note that for box culverts, the flow rate per foot of barrel width is used.

2. Connect the culvert size and discharge and extend a straight line to cut the *HW/D* axis. In case *HW/D* is required in another scale, extend the value at the original intersection horizontally to cut the appropriate scale.

3. Once the *HW/D* ratio is computed either by the design equations or the nomograph, compute the inlet control headwater depth, *HW$_i$*, by multiplying the barrel diameter by the ratio *HW/D*.

TABLE 15.3 Constants for Inlet Control Design Equations

Shape and material	Inlet edge description	Unsubmerged equation No.	K	M	Submerged	
					c	Y
Circular concrete	Square edge w/headwall	15.18	0.0098	2.000	0.0398	0.67
	Groove end w/headwall		0.0078	2.000	0.0292	0.74
	Groove end projecting		0.0045	2.000	0.0317	0.69
Circular CMP	Headwall	15.18	0.0078	2.000	0.0379	0.69
	Mitered to slope		0.0210	1.330	0.0463	0.75
	Projecting		0.0340	1.500	0.0553	0.54
Circular ring	Beveled ring, 45° bevels	15.18	0.0018	2.500	0.0300	0.74
	Beveled ring 33.7° bevels		0.0018	2.500	0.0243	0.83
Rectangular box	30°–75° wingwall flares	15.18	0.0260	1.000	0.0385	0.81
	90° and 15° winwall flares		0.0610	0.750	0.0400	0.80
	0° wingwall flares		0.0610	0.750	0.0423	0.82
Rectangular box	45° wingwall flare	15.19	0.5100	0.667	0.0309	0.80
	18°–33.7° wingwall flare		0.4860	0.667	0.0249	0.83
Rectangular box	90° headwall w/3/4 in chamfers	15.19	0.5150	0.667	0.0375	0.79
	90° headwall w/45° bevels		0.7950	0.667	0.0314	0.82
	90° headwall w/33.7° bevels		0.4860	0.667	0.0252	0.87
Rectangular box	3/4 in chamfers, 45° skewed headwall	15.19	0.5220	0.667	0.0402	0.73
	3/4 in chamfers, 30° skewed headwall		0.5330	0.667	0.0425	0.71
	3/4 in chamfers, 15° skewed headwall		0.5450	0.667	0.0451	0.68
	45° bevels, 10– 45° skewed wall		0.4980	0.667	0.0327	0.75
Rectangular box, 3/4 in. chamfers	45° non offset wingwall flares	15.19	0.4970	0.667	0.0339	0.80
	18.4° non offset wingwall flares		0.4930	0.667	0.0361	0.81
	18.4° non offset wingwall flares, 30° skewed barrel		0.4930	0.667	0.0386	0.71
Rectangular box, top bevels	45° wingwall flares—offset	15.19	0.4970	0.667	0.0302	0.84
	33.7° wingwall flares—offset		0.4950	0.667	0.0252	0.88
	18.4° wingwall flares—offset		0.4930	0.667	0.0227	0.89
Corrugated metal boxes	90° headwall	15.18	0.0083	2.000	0.0379	0.69
	Thick wall projecting		0.0145	1.750	0.0419	0.64
	Thin wall projecting		0.0340	1.500	0.0496	0.57

Source: From Normann et al. (1985).

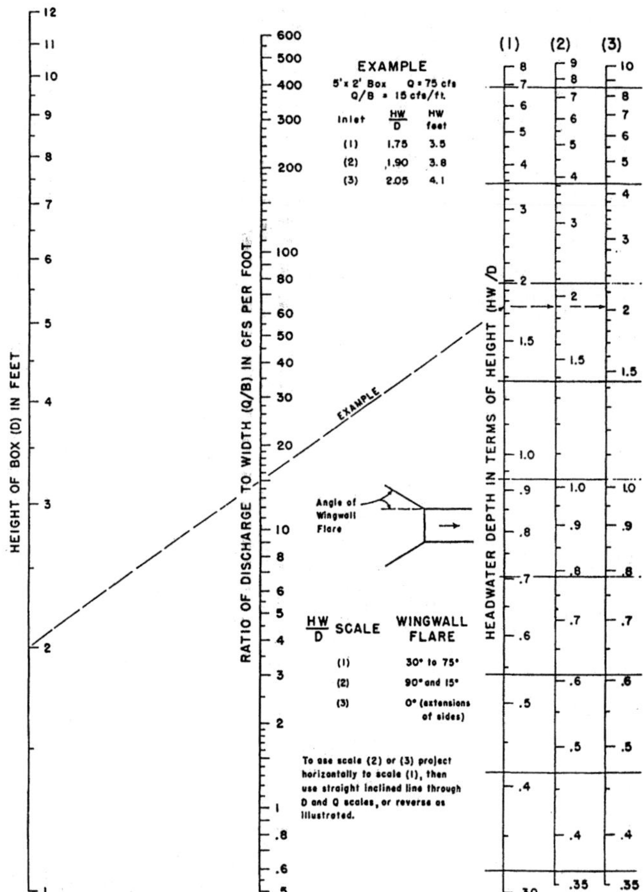

FIGURE 15.12 Headwater depth for box culverts with inlet control. (From Normann et al., 1985)

15.4.2 Outlet Control

Similar to the inlet control culverts, the headwater can be determined either by the design equations or the outlet control nomographs.

15.4.2.1 Design Equations.   Unlike the inlet control culverts, the headwater required for an outlet culvert cannot be computed by using a single equation, but Eq. (15.12) can be used to compute the energy losses. This concept is illustrated in the Example in Sec. 15.4.2.2.

15.4.2.2 Nomographs.   The nomographs can be used as described in the following:

1. Identify the culvert size, D, and length, L.

2. Use the appropriate scale for k_e to connect D and L with a straight line and identify the intersection of this line with the turning line (i.e., point X).

3. Draw a new line that connects the flow rate Q and headloss H and goes through point X.

4. The value read from this nomograph for H, includes the entrance, friction, and exit losses.

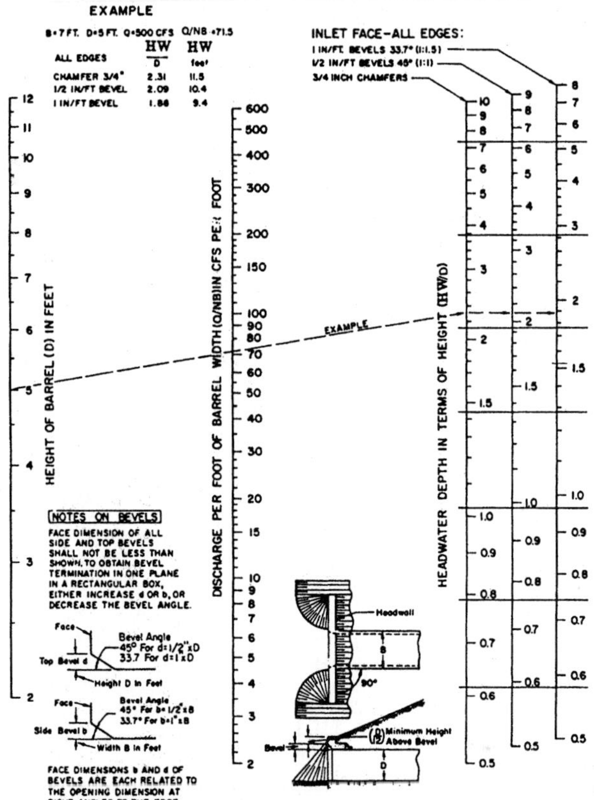

FIGURE 15.13   Headwater depth for inlet control rectangular box culverts, 90° headwall chamfered or beveled inlet edges. (From Normann et al., 1985)

Some of the important steps for design of outlet control culverts can be summarized as

1. Compute the tailwater depth.
2. Compute the critical depth, and remember the following restrictions:
   - d_c cannot exceed the diameter of the culvert, D.
   - If $d_c < 0.9D$, use Fig. 15.14
   - If d_c 0.9D, (use the guidelines in Chapter 3 for a more accurate estimate).
3. Compute $(d_c + D)/2$.
4. Compute h_o.

$$h_o = \max[TW, (d_c + D)/2]\tag{15.20}$$

5. Determine the entrance loss coefficient, k_e.
6. Determine H either by using the design equations or the nomograph.

For Manning's n value different from that of the outlet nomograph, a modified length L_1 is used as the length scale, given by

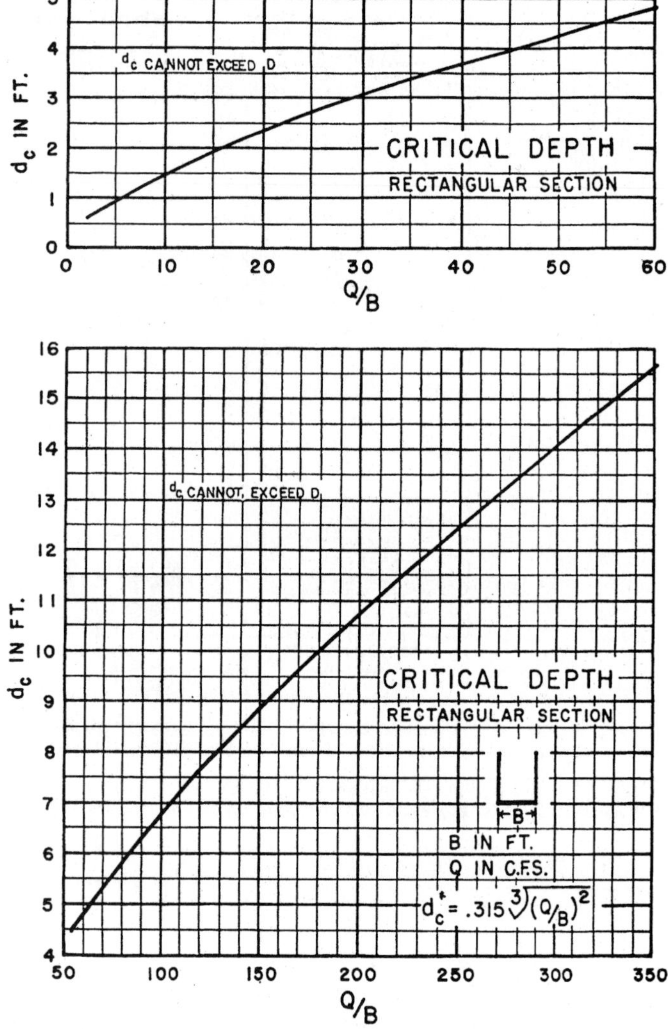

FIGURE 15.14   Critical depth rectangular section. (From Normann et al., 1985)

$$L_1 = L \left(\frac{n_1}{n}\right)^2 \tag{15.21}$$

where L_1 = adjusted culvert length (ft)
    L = actual culvert length (ft)
    n_1 = desired Manning n value
    n = Manning n value from the chart

8. Compute the outlet control headwater by using the following equation

$$HW_{out} = H + h_o - S_oL \qquad (15.22)$$

Once the inlet control headwater, HW_i and the outlet control headwater, HW_{out} are computed, the controlling headwater is determined by comparing HW_i and HW_{out*},

if $HW_{out} > HW_i$, the culvert is inlet control;
if $HW_{out} > HW_i$, the culvert is outlet control.

Then the discharge over the roadway is computed as given in Sec. 15.3.4. After computing the total discharge, Q_t, the outlet velocity, V_o, and the normal depth, d_n, the results are evaluated so that the barrels have adequate cover, the headwalls and wingwalls fit to the site conditions, the allowable headwater is not exceeded, the allowable overtopping flood frequency is not exceeded.

Example 15.1
Design a reinforced concrete box culvert for a roadway crossing to pass a 50-year discharge of 400 ft^3/s. The site conditions are as follows:

Shoulder elevation = 155 ft

Streambed elevation at culvert face = 140 ft

Natural stream slope = 1.5%

Tailwater depth = 3.0 ft

Approximate culvert length = 200 ft

Downstream channel is a 10 × 10 ft rectangular channel

The inlet is not depressed.

Solution
The design will be performed by using the procedures given in Sec. 15.4
Step 1. The 50-year design discharge is given as 400 ft^3/s.

Step 2. The geometry for the downstream channel is given (10 ft × 10 ft rectangular).

Step 3. Select a 7 ft × 5 ft reinforced concrete box culvert with 45° beveled edges in a headwall.

Step 4. Determine inlet control headwater HW_i

1. Either by using the inlet control nomograph (Fig. 15.13)
   a) $D = 5$ ft
   b) $Q/B = 400/7 = 57$ ft^3/s/ft
   c) $HW/D =$ (1) 1.80 for 3/4 in chamfer
              (2) 1.64 for 45° bevel
   d) $HW_i = (HW/D) D = 1.64 × 5 = 8.2$ ft
      Neglect the approach velocity.
2. Or the design equations (Eqs. 15.17, 15.18, or 15.19).
   To identify whether the inlet is submerged or unsubmerged,

$$\text{Compute} \left[\frac{Q}{AD^{0.5}} \right] = \left[\frac{400}{(35)\,(5)0.5} \right] = 5.11 > 4.0$$

Therefore the inlet is submerged; use Eq. 15.17

$$\left[\frac{HW_i}{D}\right] = c\left[\frac{Q}{AD^{0.5}}\right]^2 + Y - 0.5S$$

$$\left[\frac{HW_i}{5.0}\right] = 0.0314\left[\frac{400}{(35)\,(5)^{0.5}}\right]^2 + 0.82 - 0.5(0.015)$$

$$HW_i = 8.20 \text{ ft}$$

Step 5. Determine the outlet control headwater depth at inlet.

1. The tailwater depth is specified as 3.0 ft which is obtained either from a backwater computation or normal depth calculation.
2. Compute critical depth, d_c.
   a. Either by using the following equation (Chaudry, 1993)

$$d_c = \sqrt[3]{\frac{q^2}{g}}$$

where q (ft^3/s/ft), unit discharge = total discharge/culvert width, and g = gravitational acceleration, 32.2 ft/s^2; then

$$d_c = \sqrt[3]{\frac{(400/7)^2}{32.2}}$$

$$= 4.7 \text{ ft}$$

   b. Or by using Fig. 15.14, d_c = 4.7 ft
3. $(d_c + D)/2 = (4.7 + 5.0)/2 = 4.85$ ft
4. h_o = max $[TW, (d_c + D)/2]$ = max $[3.0, 4.85]$ = 4.85 ft
5. k_e = 0.2 from Table 15.1
6. Determine H
   a. Either by using Eq. (15.12)

$$H = \left(1 + k_e + \frac{29n^2L}{R^{1.33}}\right)\frac{V^2}{2g}$$

where $A = 7 \times 5 = 35$ ft^2
$V = 400/35 = 11.4$ ft/s
$R = A/P = 30/(7 + 7 + 5 + 5) = 1.25$ ft, then

$$H = \left(1 + 0.2 + \frac{29(0.012)^2(200)}{(1.25)^{1.33}}\right)\frac{(11.4)^2}{2(32.2)} = 3.7 \text{ ft}$$

   b. Or by using Fig. 15.7

where k_e scale = 0.2, culvert length
L = 200 ft (n = 0.012, same as on chart)
Area = 35 ft^2, H = 3.7 ft.

7. Compute the outlet control headwater by using Eq. (15.22)

$$HW_o = H - h_o - S_oL = 3.7 + 5.0 - (0.015)\,(200) = 5.7 \text{ ft}$$

Step 6. Determine the controlling headwater, HW_c:

$HW_i = 8.2$ ft $> HW_{out}$ 5.7 ft. Therefore $HW_c = 8.2$ ft and the culvert is in inlet control.

Step 7. Compute the discharge over the roadway, Q_r

1. Calculate the depth over the roadway, HW_r:

$HW_r = HW_c - h_c$ where h_c (*shoulder height*) $= 155 - 140) = 15$ ft, then
$HW_r = 8.2 - 15 = -6.8$ ft

2. As $HW_r < 0.$ $Q_r = 0$

Step 8. Compute the total discharge.

$$Q_t = Q_d - Q_r = 400 - 0 = 400 \text{ ft}^3/\text{s}$$

Step 9. Compute the outlet velocity, V_o and the normal depth, d_n.

Normal depth for the culvert can be computed using Manning's equation

$$Q = 1.49 \frac{A}{n} R^{0.3} S^{1/2}$$

$$400 = 1.49 \frac{(7d_n)}{0.012} \left(\frac{(7d_n)}{(7 + 2d_n)} \right)^{2/3} (0.015)^{1/2}$$

By trial and error, $d_n = 2.8$ ft.

Normal depth in the culvert is less than the critical depth, therefore, the type of flow in the culvert is supercritical, a typical case for inlet control culverts.

Compute the velocity at the culvert outlet, $V_o = 400/(7 \ 2.8) = 20.4$ ft/s

Compute the velocity at the downstream channel, $V = 400/(7) \ (3) = 13.3$ ft/s

Since $V = 1.5 \ V_o$, an armoring riprap as defined in Sec. 15.8.1 can be appropriate for erosion protection at culvert outlet.

Step 10. Review the results: the culvert barrel has 5 ft of cover, which is adequate and the allowable headwater (10 ft) is greater than $HW_i = 8.2$ ft, which is appropriate for design.

15.5 PERFORMANCE CURVES

A *performance curve* is a plot of flow rate versus headwater depth or elevation for a culvert. Because a culvert has several possible control sections (inlet, outlet, throat), a given installation will have a performance curve for each control section and one for roadway overtopping, as shown in Fig. 15.15. In addition to these individual performance curves, an overall culvert performance curve can be constructed by the controlling portions of the individual performance curves for inlet, outlet, and overtopping.

Inlet Control.   The curves for the inlet control culverts can be plotted either by using the design equations or the inlet control nomographs described in Sec. 15.4.1.

Outlet Control.   For the outlet control culverts there are a number of ways to plot the performance curves; the use of Eqs. (15.1–15.14), outlet control nomographs, or backwater calculations. In the first two cases, the flow rates that will be used for the design of the culvert are chosen, then the corresponding total losses are computed and are added to the elevation of the hydraulic grade line at the culvert outlet to obtain the headwater. In the backwater method the headwater elevations are computed for the corresponding flow

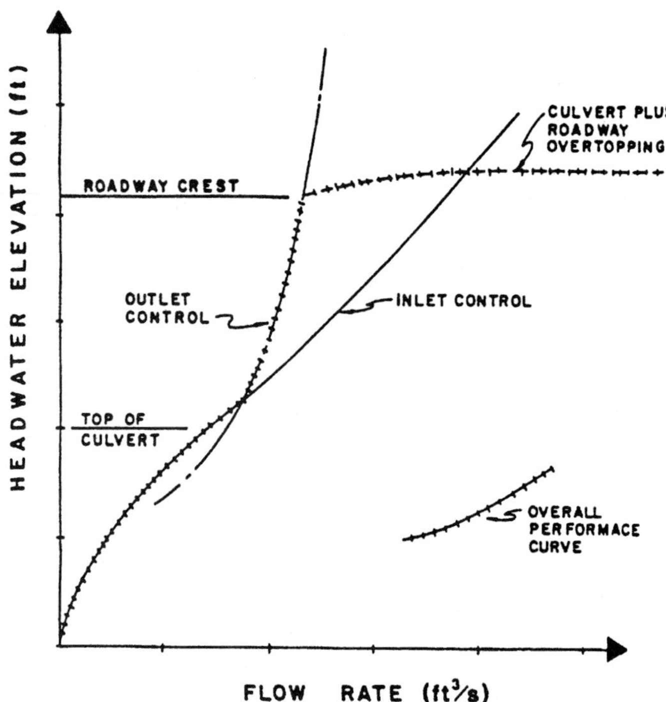

FIGURE 15.15   Culvert performance curve with roadway overtopping. (From Normann et al., 1985)

rates by adding the inlet losses to the energy equation. The energy grade line at the culvert inlet should reflect this adjustment.

Roadway Overtopping.   An overall performance curve is a useful tool to separate the culvert flow from the roadway overtopping flow and to compute their respective magnitudes. In order to develop a typical overall performance curve as shown in Fig. 15.15, the step-by-step procedure as defined in the following can be used:

1. A set of discharge values that will be used in the design of the culvert must be selected. Then the corresponding inlet and outlet headwater values must be computed.

2. The inlet and outlet control performance curves must be combined to develop a single performance curve for the culvert.

3. To compute the flow rates of the overtopping flow; calculate the depth of upstream water surface above the roadway for each one of the flow rates selected in Step 1. and use Eq. (15.16) and the depth calculated in Step 3.

4. Then the overall performance curve will be generated by adding the overtopping and culvert flow at the corresponding backwater elevations.

Example 15.2

Develop a performance curve for two 48-in corrugated metal pipe culverts (Manning's n = 0.024) with metal end sections. The culvert is 150 ft long and on a 0.08 % slope. The roadway is a 30-ft-wide paved crossing that can be approximated as a broad-crested weir 130 ft long. The roadway centerline elevation is 109 ft. The culvert invert elevations are 100 ft. at the inlet, and 99.98 ft at the outlet. Tailwater discharges and depths are listed below:

Q (ft^3/s)	80	120	160	200	240	280	320	360
TW (ft)	102.4	103.0	103.3	103.5	103.7	104.1	104.3	104.5

Step 1. Compute inlet and outlet control headwater elevations as tabulated below.
Step 2. Develop the rating curve for the overtopping flow by using Eq. (15.16)

$$Q_r = C_d L (HW_r)^{1.5}$$

HW_r	C_r	k_t	L	Q_r (Overtopping flow)	Q_c (Culvert flow)	Q_t (Total flow)
0.30	3.00	1	130	64	235	299
0.50	3.05	1	130	184	240	424

Step 3. Draw the performance curve for the inlet control, outlet control, and overtopping flow, shown in Fig. 15.17, by using the headwater calculations and Step 2.

Total flow	Flow per barrel Q/N (1)	Inlet control HW/D (2)	HW_i	ELh_i (3)	T_w	d_c (4)	$(d_c + D)/2$	h_o (5)	k_e (6)	H (7)	ELh_o (8)	ELh (9)
80	40	0.68	2.72	102.72	2.40	1.84	2.92	2.92	0.50	0.65	103.55	103.55
120	60	0.88	3.52	103.52	3.00	2.31	3.16	3.16	0.50	1.50	104.64	104.64
160	80	1.08	4.32	104.32	3.30	2.73	3.37	3.37	0.50	2.60	105.95	105.95
200	100	1.30	5.20	105.20	3.50	3.11	3.56	3.56	0.50	4.20	107.74	107.74
240	120	1.58	6.32	106.32	3.70	3.47	3.74	3.74	0.50	5.80	109.52	109.52

(1) Use Q/NB for box culverts
(2) Use Figure 15.16, Scale 1
(3) $ELh_i = HW_i + 100$
(4) $d_c = 0.325 \, (Q/ND)^{0.66} + 0.083$
(5) Use Equation 15.1
(6) Use Table 15.1
(7) Use Figure 15.17
(8) $ELH_o = 99.98 + H + h_o$
(9) $Elh = \max(ELh_i, ELh_o)$

15.6 MATERIALS AND CULVERT GEOMETRY

Culverts are required to serve under various conditions including wide temperature variations, heavy abrasion, erosion and sedimentation within the culvert barrel. Because of these factors there are a variety of culvert materials used to maintain the long life and durability of

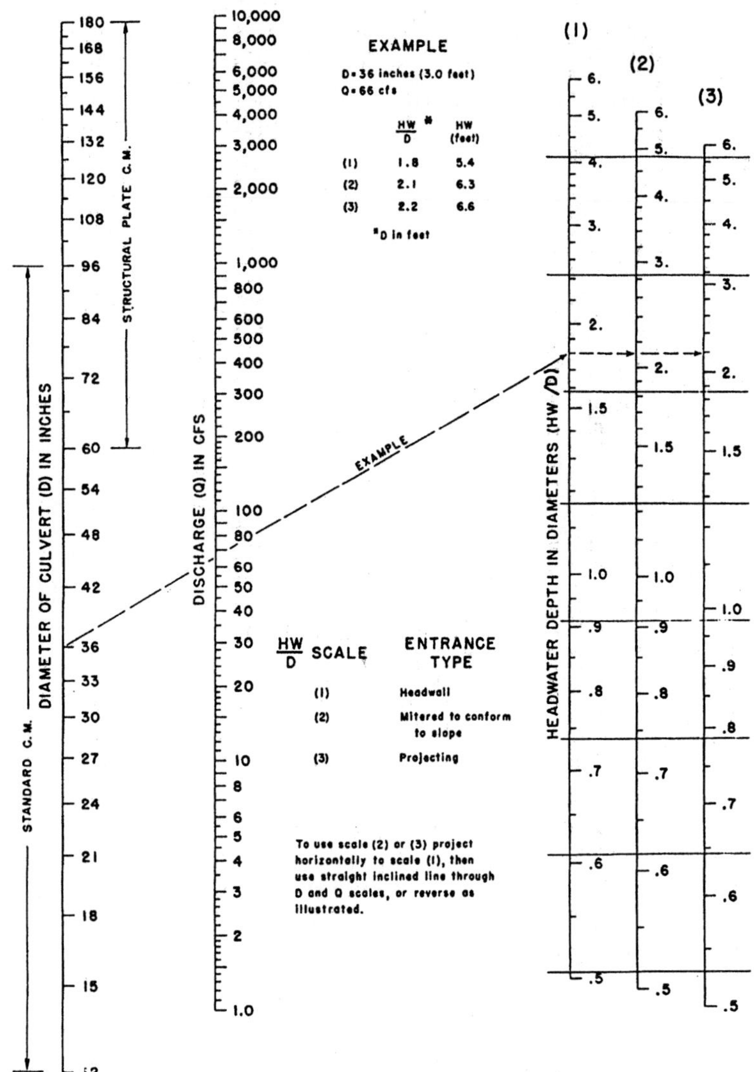

FIGURE 15.16 Headwater depth for CM pipe culverts with inlet control. (From Normann et al., 1985)

culverts. The three main types used in application are the precast concrete, monolithic or cast-in-place concrete, and corrugated aluminum and steel. Some other options are listed in Table 15.4. The culvert must be able to carry loads without large deflection and efficiently convey the water through the culvert opening. Therefore, the selection of culvert material should consider the desired hydraulic efficiency, structural, and operational characteristics.

One of the factors in the physical strength of the culverts is the type of the material used. The two main applications are with rigid and flexible culverts. Rigid culverts are usually of reinforced concrete pipe and their primary strength is built into the pipe itself. Special designs are possible that permit the pipe to support any magnitude of overload. Their durability

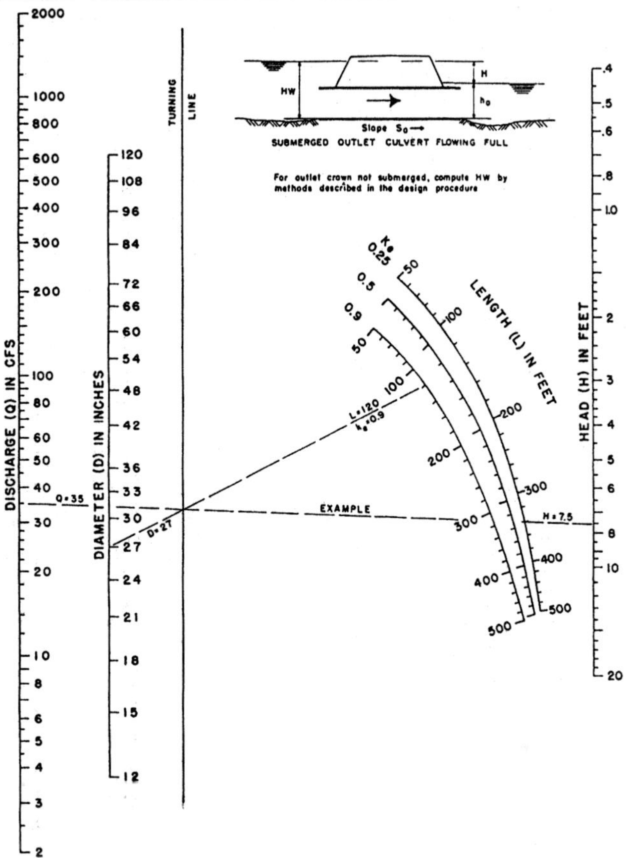

FIGURE 15.17   Head for standard CM pipe culverts flowing full ($n = 0.024$). (From Normann et al., 1985)

TABLE 15.4   Culvert Materials

Shape	Material	Protective coating
Circular	Steel	Galvanizing
Rectangular	Aluminium	Aluminized
Arch	Al/Fe Alloy	Bituminous
Pipe-arch	Concrete	Polymer
Elliptical	Plastic	Concrete

Source: From Reagan (1993)

against freezing, thawing and erosion are major advantages, combined with good hydraulic efficiency. In design with smaller culvert size requirements and lighter loads, nonreinforced concrete pipe can be used. Flexible culverts, generally corrugated metal pipe or plastic pipe, provide their strength by the interaction of the pipe and the surrounding backfill material. Corrosion is a concern with the corrugated metal pipes.

In many applications the operational characteristics of a culvert are as important as the hydraulic and structural elements. The culvert designer must select the material with the consideration of hydraulics, structures, maintenance, and geotechnical characteristics of the culvert site. The proper culvert design should optimize the hydraulic and structural characteristics of the culvert while minimizing the total cost. The culvert material which directly affects the service life, is a key parameter in the total cost of the culvert.

Another important factor for the efficient design is the shape of the culvert conduit. The shape has an impact on the hydraulics and structural characteristics. The type of the maintenance and accessibility for deterioration and deformation repairs are also affected by the shape. The most common culvert conduit shapes are circular, rectangular, pipe-arch, and arch.

15.7 *LOCATION AND ALIGNMENT OF CULVERTS*

The location and alignment has an important affect on the economics of culvert design. As a general criteria, culvert location should not alter the natural drainage, and be placed at the bottom of the ravine. It is almost impossible to find a unique rule that could be applied to

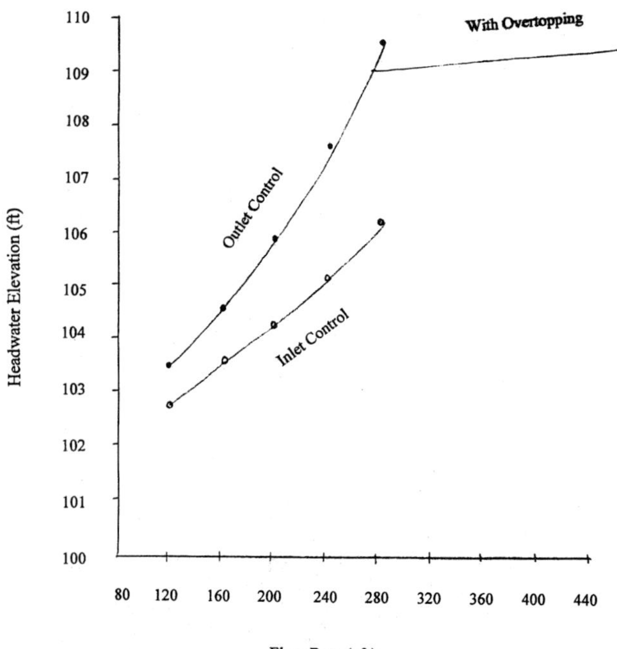

FIGURE 15.18   Example performance curve.

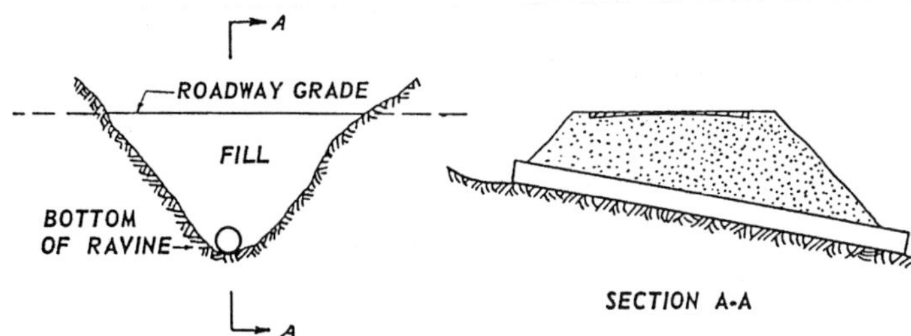

FIGURE 15.19   Bottom location culverts. (From Hendrickson, 1957)

every culvert. The final location and alignment of the culverts depend on the designer's own experience in hydrology, hydraulics and structural aspects of culvert design. This will help the designer to achieve maximum economy, utility, and safety (Hendrickson, 1957).

There are different options to locate the culvert; namely, bottom location placement, and top location placement.

15.7.1 Bottom Location Placement

The invert of the culvert with bottom location placement follows the line and grade of the natural stream channel as closely as possible (Fig. 15.19). Although this type of installation may result in some curvature and possible changes in grade to obtain satisfactory bedding, it has a good hydraulic performance. However, there are several disadvantages to this type of location. In applications with steep natural gradients, the velocity at the culvert outlet can be high and may require erosion and scour protection. In cases where the bottom of the stream channel lies in a ravine, the culverts can be under high fills resulting in long culverts and heavy loads. Cost comparison with other locations must include consideration of headwalls, spillways, energy dissipation structures, channel changes, and extra maintenance.

In cases of steep channels, it is possible to reduce the length and required strength of culvert by modifying the natural slope and alignment. This can be done by keeping the culvert inlet at the bottom of the channel, but shifting the culvert outlet horizontally and vertically and placing it in the downstream face of the embankment as shown in Fig. 15.20. Although this configuration can result in a more economical solution, an appropriate design must be provided to return the flow at the culvert outlet to the natural drainage channel. It may be necessary to protect the downstream slope against erosion, depending on the type of the material in the fill. In some instances, a form of spillway may be required, particularly if the return channel is steep. In other cases it may be feasible to project the culvert beyond the toe of the embankment and allow the water to drop into a pool.

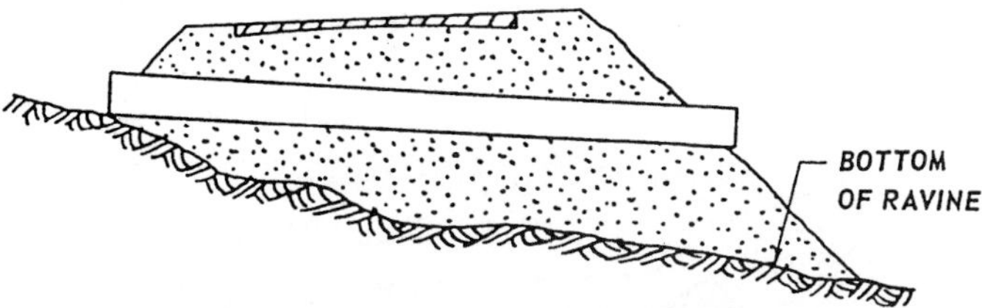

FIGURE 15.20   Modified bottom location culverts. (From Hendrickson, 1957)

15.7.2 Top Location Placement

In some applications the culvert can be placed near the top of the embankment fill just beneath the roadway grade. The main advantage over the bottom installation is its shorter length and considerably reduced load that the culvert has to carry. This type of installation forces the embankment to serve as a dam and results in a rise in the level of the upstream pool until water flows through the culvert. Therefore the embankment material must ensure stability during flood events. One other concern is the possible accumulation of sediment and debris due to stagnant ponding on the upstream side when the water level drops below the level of the culvert invert. Similar to the second solution of the bottom location culverts, spillways or auxiliary channels might be required to carry water from the outlet without undercutting or eroding the embankment.

15.7.3 Siphons

The site conditions might require the modification in the horizontal and vertical orientation, and under certain conditions a culvert may become a siphon. A *siphon* is defined as a tube by which liquid can be transferred by means of atmospheric pressure from a higher to a lower level over an intermediate elevation. This is the case for a culvert laid on a broken grade line as shown in Fig. 15.21A. Due to the change in grade, the hydraulic grade line drops below the crown of the culvert when the inlet and outlet are submerged. The negative pressure or vacuum in the center section of the culvert results in a siphon.

In certain cases the designer may have to use the inverted siphon alternative, which is also named as sag culvert. The inverted siphon is a consequence of a hydraulic grade line above the crown of the culvert, thus causing the culvert to flow under pressure. These culverts can be constructed under low fills where it is necessary to excavate down to get the required cross-sectional area for hydraulic capacity (Fig. 15.21B), or they can result from the gradual accumulation of sediment and debris at the upstream and downstream ends. In irrigated areas, inverted siphons are used to carry the water from open ditches and canals below highways, railroads, and other obstructions. This type of culvert must be periodically maintained as a protection against sediment and debris accumulation. Also they are not suggested for intermittent streams, because stagnant water can lay in them for a long period of time and can result significant health hazards.

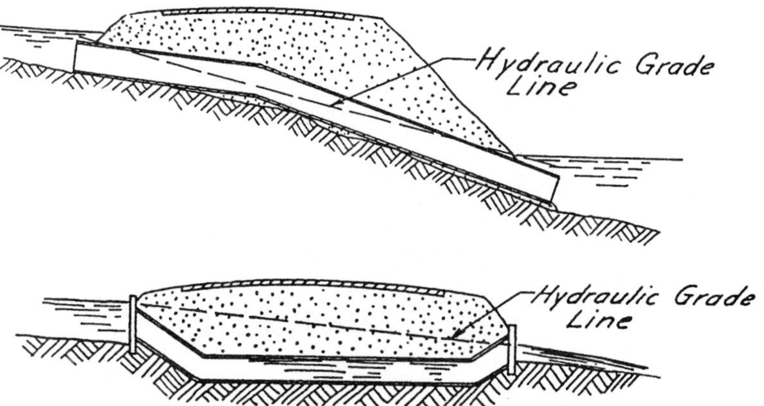

FIGURE 15.21   **A.** Culvert siphons: culvert siphon, **B.** inverted siphon culvert. (From Hendrickson, 1957)

15.7.4 End Treatments

End treatments are applied to culverts for many purposes, including; retention of roadway embankment, improvement of hydraulic efficiency, protection of highway embankment from discharge momentum, protection against piping, debris control, traffic safety.

Headwall end treatments are very common structures standardized with construction details and are economical and practical. In some instances, hydraulic efficiency can be improved by minor details such as flared wingwalls on the headwall. Prefabricated end sections are very popular due to their convenience of installation and economy. Their application is limited to smaller culverts.

The flowline length of a mitered end culvert (Fig. 15.22) corresponds to the distance between the upstream and downstream roadway toe-of-slope locations. The top length is measured between the points of interception with the roadway side slopes and the barrel top. Mitered end treatments are economical and relatively simple in construction.

Projecting end treatments are similar to mitered end treatments except that the upper portion of the culvert barrel protrudes from the side slopes of the roadway embankment as shown in Fig. 15.22. Projecting end treatments are convenient and economical. However, they have serious problems regarding, hydraulics (greater entrance losses), structural (inadequate perimeter containment), operational (potential for collision with maintenance operations).

Another alternative is the use of different improved inlet conditions. Since a culvert usually represents a severe constriction to the stream flow, the physical characteristics of the

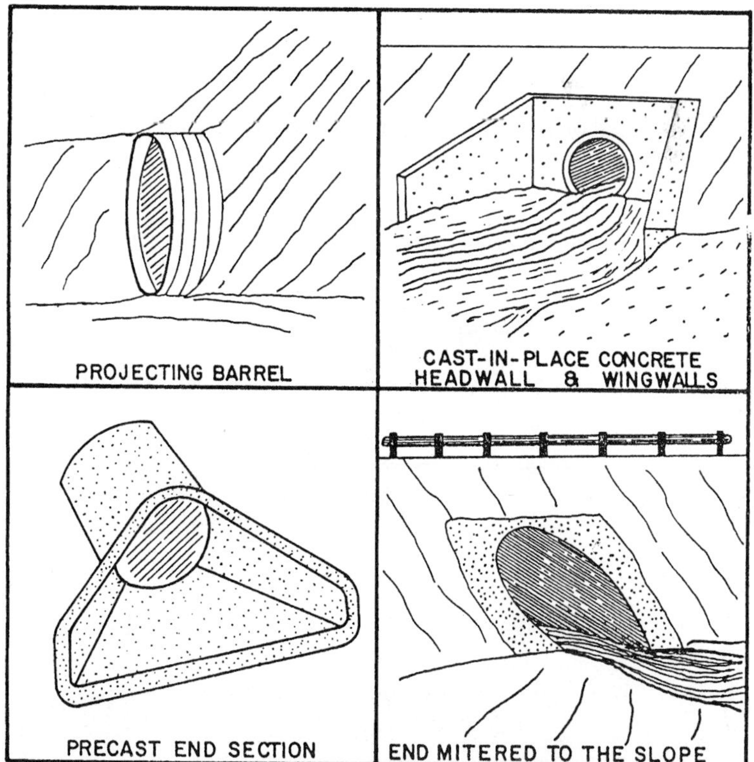

FIGURE 15.22 Four standard inlet types. (From Norman et al., 1985)

entrance to the culvert have a significant effect on the operational capacity. This effect is reflected by the headloss coefficient, k_e, associated with the entrance type used. With reference to the table of entrance loss coefficients in Table 15.1, note that some k_e are much lower than others depending on the type of entrance. Often, improvement in the hydraulic capacity of the culvert can be realized by simple and inexpensive expedients such as, beveling or rounding of the inside perimeter of the entrance to the conduit and ensuring installation of reinforced concrete pipe with the grooved end upstream.

Beveled edges are commonly used with box culverts and headwall structures. As given in Normann, et al., (1985), design charts are available for two bevel angles, 45° and 33.7° (Fig. 15.23). Although the 33.7° bevels result in a better inlet performance than the 45° bevel, it requires additional structural modifications. As a result, the 45° bevel is the preferred alternative. Groove ends provide similar hydraulic performance as the bevels. Figs. 15.24 and 15.25 demonstrate bevels with other inlet improvements. Side-tapered inlets provide enlarged culvert entrance with a transition to the original barrel dimensions. The test results of FHWA (Harrison et al., 1972) resulted in a section geometry as shown in Fig. 15.24. The throat section as identified in this figure is used to control the capacity and maximize the efficiency of side-tapered inlets. Slope-tapered inlets have steeper slopes at the entrance as compared to the culvert barrel. The steeper slope is provided to increase the head on the throat section and create additional fall. Depending on the magnitude of the available fall, the inlet capacities can be significantly higher than the conventional culvert with square edges.

Culvert inlet performance also can be improved cost effectively, by the incorporation of special entrance geometries. Effectively, such improvements are used to reduce the headlosses due to flow contractions and to increase capacity of culverts operating under steep slope regimes.

Expanded entrances have a "funneling" effect on the flow. These flared entrances allow more discharge into the conduit, thus significantly improving the efficiency of the conduit. Their main advantage is in the reduction of constriction loss coefficients. While there may be a minor loss reduction in outlet control culvert operations, they are mostly effective in situations of steep slope regime. Culverts operating in a supercritical flow regime often

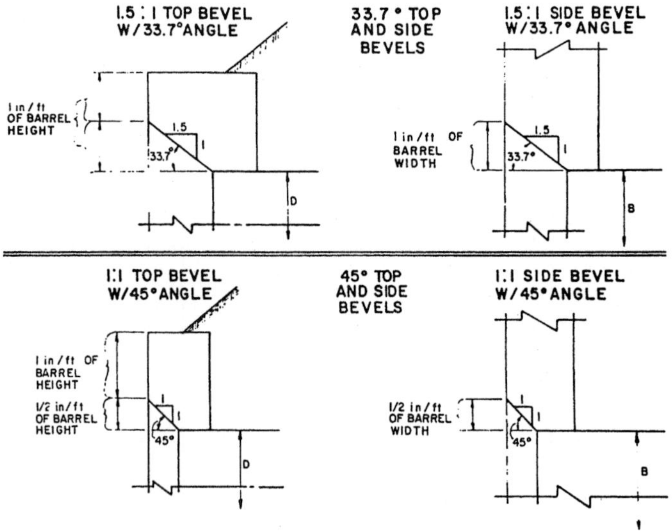

FIGURE 15.23   Beveled edges. (From Normann et al., 1985)

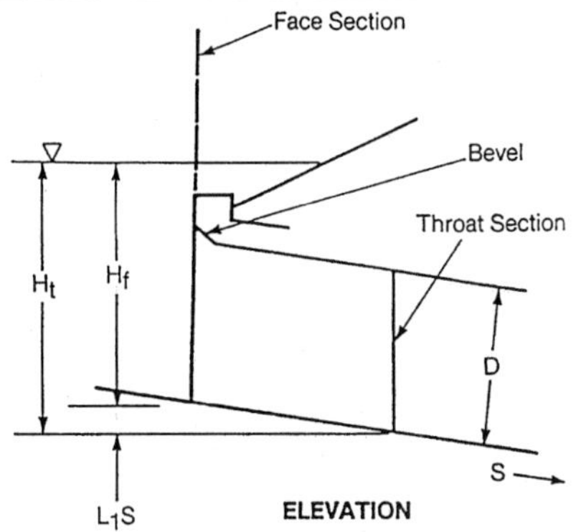

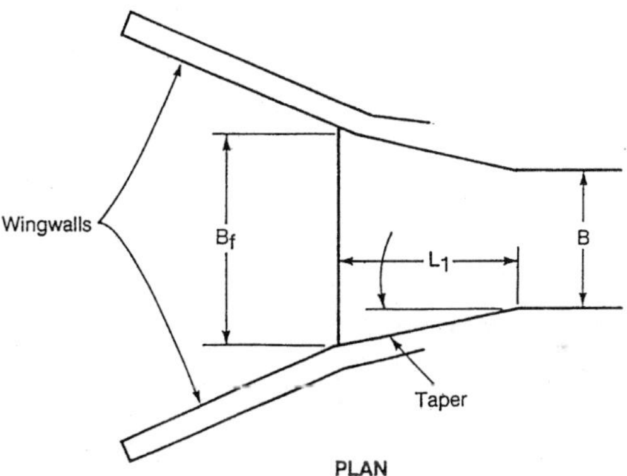

FIGURE 15.24 Typical side-tapered inlet detail. (From State of Florida Department of Transportation, 1987)

exhibit excessive outlet velocities. The usual effect of an improved entrance on such a configuration results in more efficient use of the culvert barrel with the further result of a lower outlet velocity.

Transitions of flow from the relatively wide channel approach to the constricted culvert opening may be useful in some instances. Both expanded entrances and transitions have the potential of clogging due to debris. Larger debris is allowed to enter the culvert partially, and subsequently becomes lodged in the smaller opening provided by the main conduit.

Improved inlets are useful in some situations but the following considerations should be made in design:

- The improved inlet often is a relatively expensive item, sometimes costing more than the rest of the culvert.
- Improved inlets are effective only in supercritical flow situations.
- Debris, erosion, and other maintenance problems can represent significant operational costs when improved inlets are used.
- The design charts and methods for improved inlet design are applicable to one or two barrels only of reinforced concrete box culvert configurations.

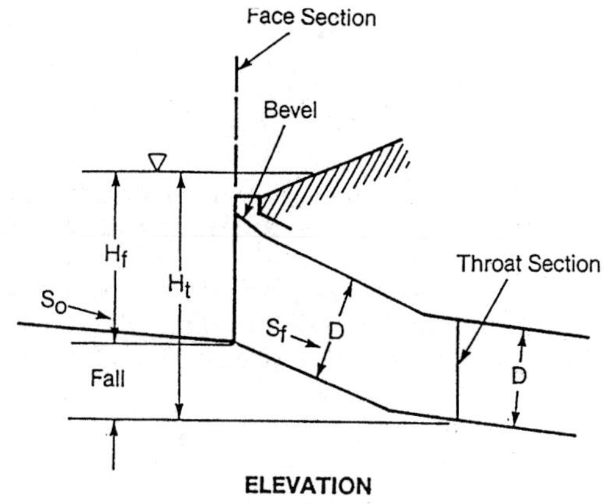

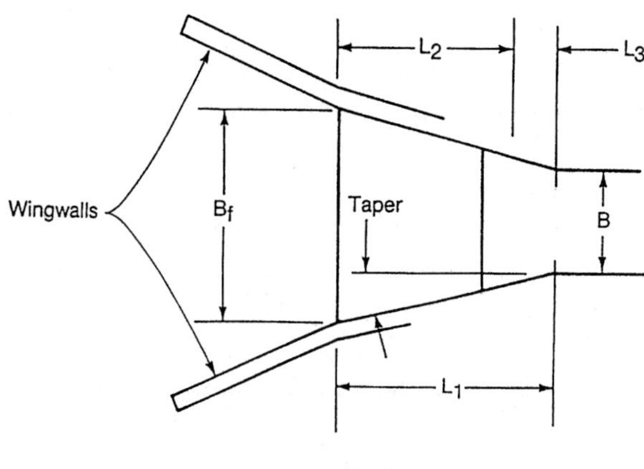

FIGURE 15.25 Typical slope-tapered inlet detail. (From State of Florida Department of Transportation, 1987)

- The design charts require a culvert face which is perpendicular to the stream flow. This can be a construction and maintenance problem when the culvert is skewed to the roadway centerline.

15.8 SPECIAL CONSIDERATIONS

15.8.1 Erosion

The high outlet velocities observed at the culvert outlets may result in excessive scour of the channel in the vicinity of the outlet. The variety in the soil type of natural channels and varying flow characteristics at the culvert outlet enforces the use of different methods to control or protect the channel against potential damaging effects. A method to predict and analyze scour conditions is given by Reagan (1993). Some of the commonly used techniques to provide protection against scour are: minor structural elements, velocity protection devices, velocity control device.

Minor Structural Elements.   In the case of low outlet flow velocities (outlet velocity less than 1.3 times the average natural stream velocity) cutoff walls are commonly used. They are provided in streams with adequate downstream protection or where the outlet hydraulics and soil conditions do not result in structural instability. Cutoff walls should be constructed such that its depth equals the depth of scour, and its width covers at least one-third the scour width. In cases where the cutoff wall depth exceeds 6.0 ft, either a different form of protection must be provided or the structural stability of the walls must be maintained (Flood Control District of Maricopa County, 1996).

Velocity Protection Device.   For culvert outlet velocities greater than 1.3 times the natural stream velocity, but less than 2.5 times, armoring riprap is used as a protection against the damaging potential. Different types of armoring material can be used including concrete riprap, rock riprap, vegetation and synthetic sodding methods.

The details on the size and placement of riprap for various types of design conditions can be found in Corry et al. (1983), Maccaferri Gabions, Inc. (1997), and Simons and Senturk (1977).

Velocity Control Device.   Energy dissipators are appurtenances that may be used at the downstream end of a culvert for the purpose of reducing the damaging effects of outlet velocities greater than 2.5 times the natural stream velocity. The control is accomplished by the use of an energy dissipator. The turbulent outlet flow is contained within this dissipator, resulting in a more tranquil flow downstream of it. Due to the high cost of an energy dissipator structure, it must be properly designed and constructed to control the excessive outlet velocity efficiently. For more detailed information on the design of energy dissipators the reader is referred to Corry et al. (1983) and Simons and Senturk. (1977).

A consequence of high outlet velocities is the local scour in the vicinity of the culvert exit. A detailed discussion of the local scour at culvert outlets is presented by Breusers and Raudkivi (1991). Also some of the research on scour at culvert outlets by Ruff (1982) can be summarized as follows:

- The dimension of the scour hole can not be predicted as a unique function of outlet velocity or exit Froude number, because the variable tailwater has a significant effect on the scouring process.

- The scour depth is dependent on the dimensionless ratio (outlet velocity/outlet critical shear velocity), whereas scour length and width are correlated to the exit Froude number.

- The effect of gradation on scour depth is significant at low exit velocities only, therefore it is appropriate to use the median grain diameter, d_{50}, in computations.

15.8.2 Sedimentation

Sedimentation is a concern in the design of culverts and may result in excessive loss in efficiency at the inlet, outlet, or within the culvert barrel. The two major factors that influence the sedimentation characteristics of a culvert are the barrel roughness and the slope. Sediment depositing within the culvert will change it's roughness (i.e., Manning's coefficient, n) and will adversely affect the flow conditions within the culvert. In terms of the stability of the culvert system, it is desirable to align the culvert at the natural stream slope. In cases with milder culvert gradients, low-flow velocities within the culvert will force the sediment particles to deposit. If the culvert is set at a steeper slope than that of the natural channel, the increased velocities within the culvert may result in excessive erosion along the culvert.

To maintain the stability and erosion/sedimentation characteristics of a culvert, the sediment characteristics of the channel system and the specific site conditions must be carefully studied. A channel system with a degrading pattern may not result in a sedimentation problem at the outlet but will have the potential to provide excessive sediment to the culvert inlet. For aggrading systems, sediment accumulation in the channel will reach to the culvert outlet and will gradually move upstream within the culvert. The site-specific conditions may result in the use of multibarrel installation or culverts with depression. Multibarrel culverts are used with wide and shallow channel systems or low fills and when used in a curved alignment, sedimentation is a natural consequence. In such cases the recommended installation is straight culverts aligned with the channel upstream. Culverts with depression may experience sediment due to a milder culvert slope than that of the upstream channel. Storm events with increased culvert velocities result in a self-cleansing effect. Details of the erosion and sedimentation characteristics of the channel and the culvert system can be found in Raudkivi (1990, 1993), Richardson et al. (1990), Simons and Senturk (1977), Simons, Li & Associates (1982), Vanoni (1977), and Yang (1996).

15.8.3 Control of Debris

To control the debris that flows through culverts, certain aspects of the site such as stream velocity, natural channel slope and alignment, existing and future vegetation and land use patterns, and frequency of flood events must be carefully analyzed.

Debris can either accumulate at the culvert entrance or within the culvert barrel, resulting in upstream flooding and roadway overtopping. For small volumes of debris accumulation, regular maintenance and inlet conditions will affect the hydraulic performance of culverts. Smooth inlets with single barrel configuration, aligned with the natural stream will enhance the hydraulic efficiency of culverts.

Typically three debris control methods are used for sites with higher volumes of debris accumulation:

1. Debris is intercepted in the vicinity of the culvert inlet and can be accomplished by the use of debris racks, floating drift booms, and debris basins.

2. Debris is deflected away from the entrance and collected in a basin and generaly debris risers and cribs are used.

3. Debris is transported through the barrel by using an oversized culvert. The guidance for considerations and design of debris control structures can be found in FHWA HEC-No. 9 (Reihsen and Harrison., 1971).

15.9 STREAM STABILITY AT HIGHWAY STRUCTURES

15.9.1 Basic Engineering Analysis

The factors that affect stream stability and potentially bridge stability at highway stream crossings can be classified as geomorphic factors (Fig. 15.26) and hydraulic factors (Fig. 15.27). The FHWA Hydraulic Engineering Circular (HEC) No. 20 (Lagasse et al., 1995) presents detailed descriptions of the various factors and how they affect stream stability. Much of the material presented in this section is adapted from the FHWA HEC No. 20 which recommends a three-level procedure for analyzing stream stability:

Level 1. Steps in qualitative and other geomorphic analysis, (1) stream characteristics, (2) land use changes, (3) overall stability, (4) lateral stability, (5) vertical stability, and (6) stream response

Level 2. Steps in basic engineering analysis (see Fig. 15.28).

Level 3. Mathematical and physical model studies.

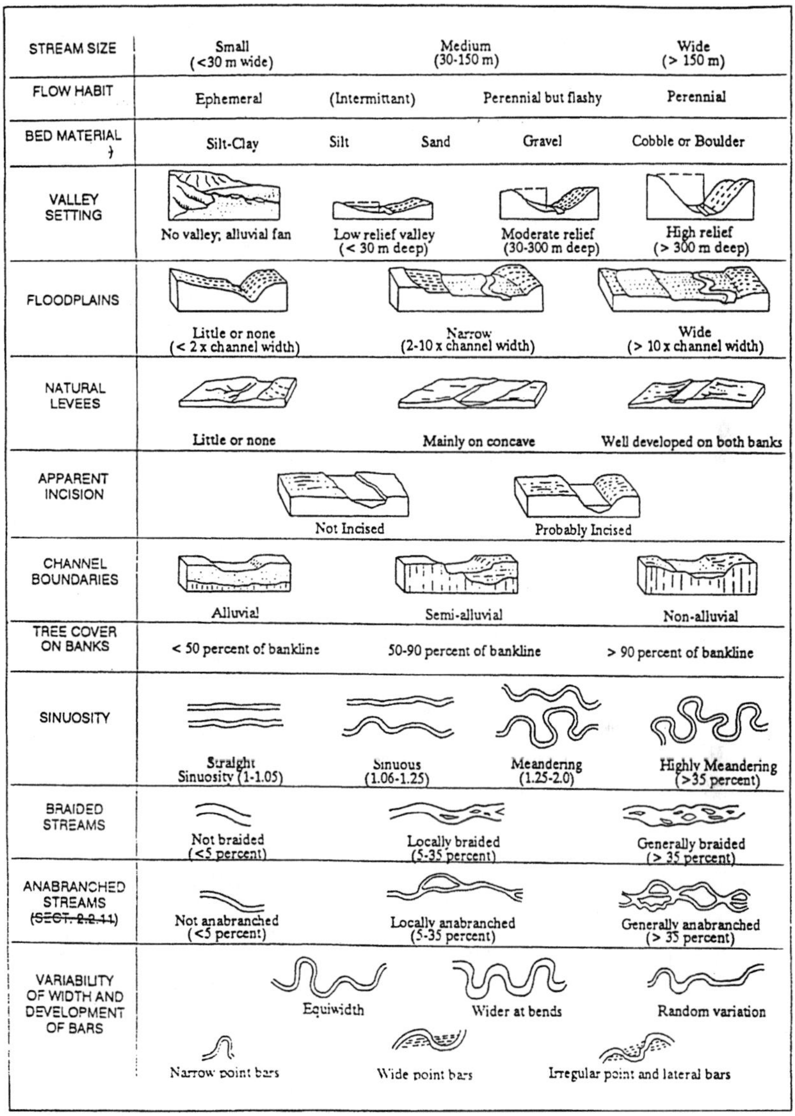

STREAM SIZE	Small (<30 m wide)		Medium (30-150 m)		Wide (> 150 m)
FLOW HABIT	Ephemeral	(Intermittant)		Perennial but flashy	Perennial
BED MATERIAL	Silt-Clay	Silt	Sand	Gravel	Cobble or Boulder
VALLEY SETTING	No valley; alluvial fan	Low relief valley (< 30 m deep)		Moderate relief (30-300 m deep)	High relief (> 300 m deep)
FLOODPLAINS	Little or none (< 2 x channel width)		Narrow (2-10 x channel width)		Wide (> 10 x channel width)
NATURAL LEVEES	Little or none		Mainly on concave		Well developed on both banks
APPARENT INCISION		Not Incised		Probably Incised	
CHANNEL BOUNDARIES	Alluvial		Semi-alluvial		Non-alluvial
TREE COVER ON BANKS	< 50 percent of bankline		50-90 percent of bankline		> 90 percent of bankline
SINUOSITY	Straight Sinuosity (1-1.05)		Sinuous (1.06-1.25)	Meandering (1.25-2.0)	Highly Meandering (>35 percent)
BRAIDED STREAMS	Not braided (<5 percent)		Locally braided (5-35 percent)		Generally braided (> 35 percent)
ANABRANCHED STREAMS (SECT. 2.2.11)	Not anabranched (<5 percent)		Locally anabranched (5-35 percent)		Generally anabranched (> 35 percent)
VARIABILITY OF WIDTH AND DEVELOPMENT OF BARS	Equiwidth		Wider at bends		Random variation
	Narrow point bars		Wide point bars		Irregular point and lateral bars

FIGURE 15.26 Geomorphic factors that affect stream stability. (Fom Brice and Blodgett, 1978 as presentel in Laggasse, et al., 1995)

The remaining discussion will be on the steps in the basic engineering analysis (Fig. 15.28). Evaluation of flood history (Step 1) is an integral step in characterizing watershed response and morphologic change, particularly in arid regions. It is important to study flood records and corresponding stream responses using aerial photography or other physical information. Development of the rainfall-runoff relation is important to understand watershed conditions and historical changes in the watershed.

To evaluate the hydraulic conditions (Step 2), it is common practice to compute water surface profiles and hydraulic conditions using a computer code. For the analysis and design

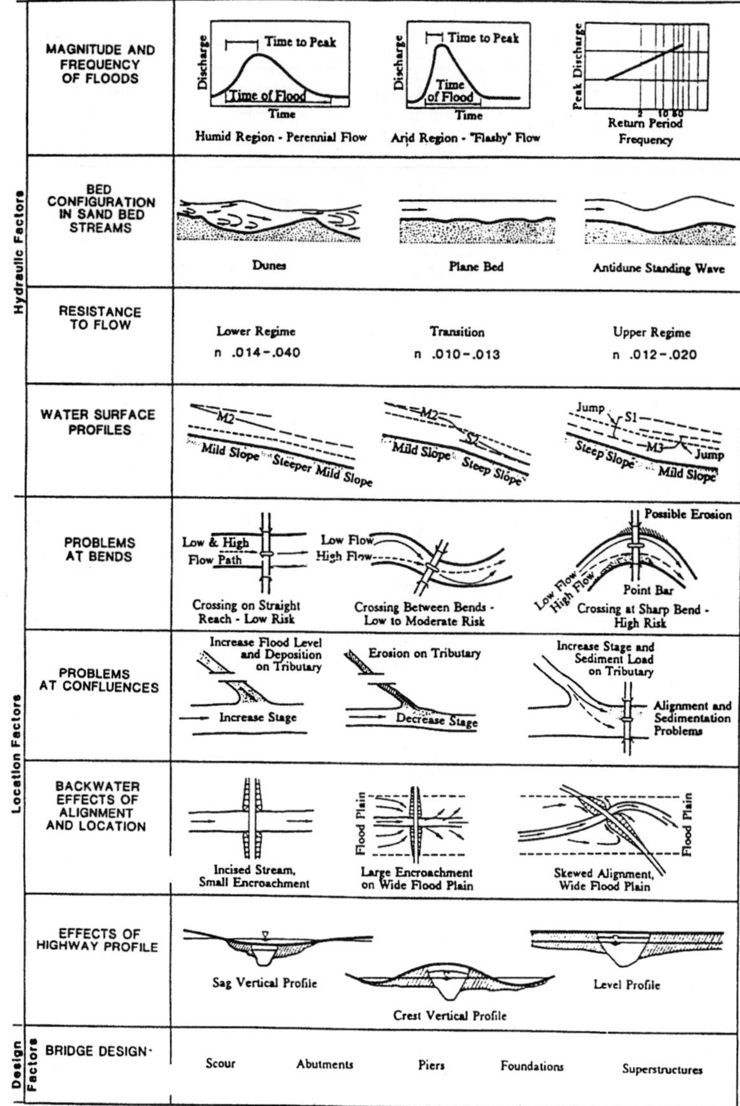

FIGURE 15.27   Hydraulic and location factors that affect stream stability. (From Laggasse et al., 1995)

of bridge crossings, the Federal Highway Administration's WSPRO (see Shearman, 1987) computer program is recommended because the computational procedure for evaluating bridge hydraulics is superior to other models such as the U.S. Army Corps of Engineers' HEC-2 (1991) and The U.S. Army Corps of Engineers' (1995) HEC–RAS. (River Analysis System). The input structure for WSPRO was specifically developed to facilitate bridge design.

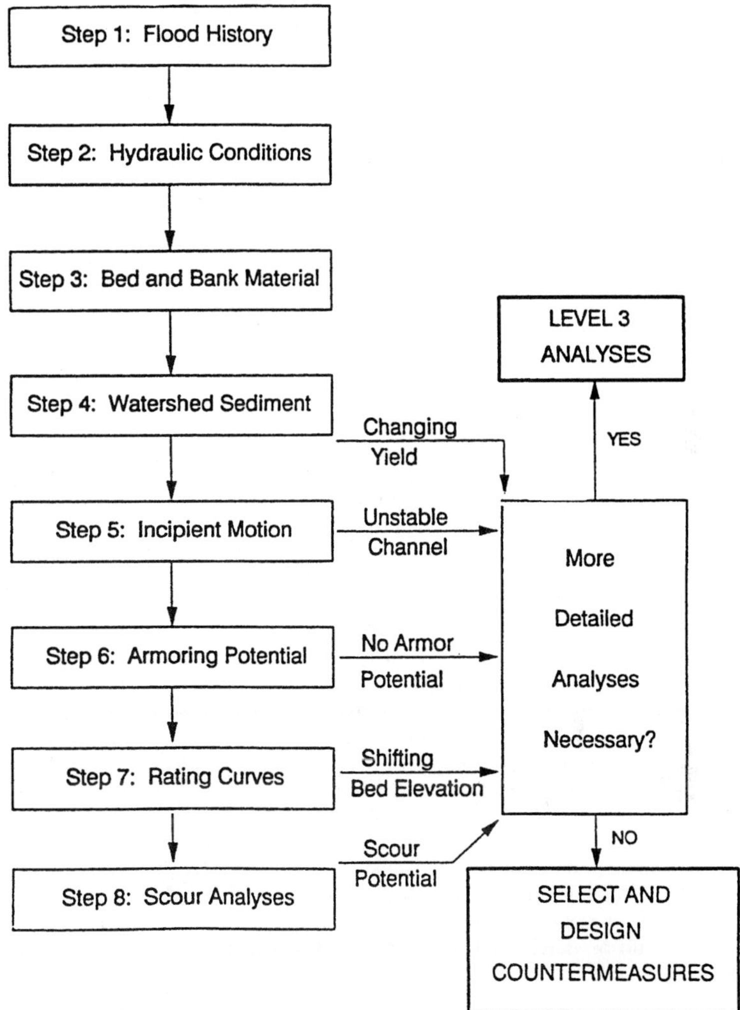

FIGURE 15.28 Flowchart for level 2: Basic Engineering Analyses. (From Lagasse et al., 1995)

Step 3 involves performing an analysis of the bed and bank material to determine particle size gradation and other properties such as shape, fall velocity. Step 4 involves evaluating the watershed sediment yield using a method such as the *Universal Soil Loss Equation* (USLE). Step 5 involves performing an incipient motion analysis to evaluate the relative channel stability. For most river conditions the following equation derived from the Shields diagram (see Chap. 6) can be used (Lagasse et al., 1995).

$$D_c = \frac{\tau}{0.047(\gamma_s - \gamma)} \tag{15.23}$$

where D_c = the diameter of the sediment particle at incipient motion condition in (m)
τ = the boundary shear stress in N/m^2 (pa)
γ_s and γ = the specific weights of the sediment and water, respectively, in N/m^3, and 0.047 (for sand-bed channels) is the dimensionless coefficient referred to as the Shields parameter.

Step 6 is to evaluate armoring potential. *Armoring* is the natural process where by an erosion-resistant layer of relatively large particles is formed on a streambed due to the removal of finer particles by the stream flow. An *armoring layer* which is sufficient to protect the bed against moderate discharges can be disrupted during high discharges and then may be restored as flows diminish. By determining the percentage of bed material equal to or larger than the armor size, D_c, computed using Eq. (15.23) for incipient motion, the depth of scour y_s necessary to establish an armor layer can be determined using (Pemberton and Lara, 1984) the following equation as

$$y_s = y_a \left[\frac{1}{P_c} - 1 \right] \tag{15.24}$$

where y_a = the thickness of the armor layer
P_c = the decimal fraction of material coarser than the armoring size
y_a = ranges from one to three times the armor size, P_c, depending on the value of D_c (Lagasse et al., 1995).

Step 7 is the evaluation of stage-discharge rating curve shifts. Scour and deposition is the most common cause of rating curve shifts. Rating curves that continually shift indicates channel instability. Step 8 is to evaluate the scour condition at bridge crossings. Section 15.10 provides a detailed description of procedures based on the FHWA-HEC No. 18. (Richardson and Davis, 1995). Calculation of the three components of scour (local scour, contraction scour, and regional aggradation/degradation) quantifies the potential instability at a bridge crossing.

Once the above steps have been completed, if a more detailed analysis is needed, then the level 3 analysis is performed in which mathematical and/or physical models are used to evaluate and assess stream stability. If a more detailed analysis is not needed then the counter measures are selected and designed. Historically the need for a level 3 type analysis has been for high-risk locations and for extraordinarily complex problems. However, because of the importance of stream stability to the safety and integrity of bridges, the level 3 analysis should be performed routinely according to Lagasse et al. (1995).

15.9.2   Countermeasures (Flow Control Structure) for Stream Instability

Countermeasures are measures incorporated into a highway-stream crossing system to monitor, control, inhibit, change, delay, or minimize stream and bridge stability problems or action plans for monitoring structures during and/or after floods events (Lagasse et al., 1995). Selection of a countermeasure for a bank erosion problem is dependent on the erosion mechanism, stream characteristics, construction and maintenance requirements, potential for vandalism and costs. Of particular interest are *flow control* structures which are defined as structures, either within or outside a channel, that acts as a countermeasure by controlling the direction, velocity, or depth of flowing water. (Richardson et al., 1990). These types of structures are also referred to as *river training works*. Various types of flow control structures are shown in Fig. 15.29.

15.9.2.1   Spurs.   *Spurs* are structures or embankments that are projected into streams at the same angle in order to deflect flowing water away from critical zones to prevent erosion of the banks, and to establish a more desirable channel alignment or width (Richardson et al., 1990). Spurs have several functions:

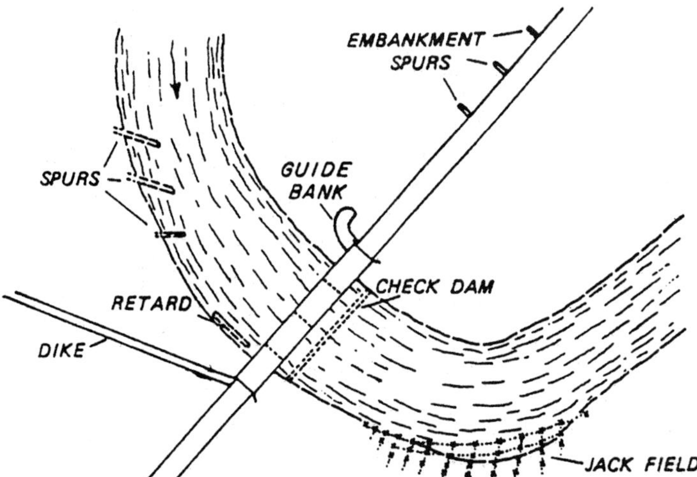

FIGURE 15.29 Placement of flow control structures relative to channel banks, crossing, and floodplain. Spurs, retards, dikes, and jack fields may be either upstream or downstream from the bridge. (From Brice and Blodgett, 1978)

1. To protect highway embankments that form approaches to bridge crossings

2. To channelize wide, poorly defined streams into well-defined channels

3. To establish and maintain new alignments through deposition of sediments

4. To halt meander migration at banks

Spurs are classified as retarder spurs, retarder/deflector spurs, and deflector spurs based on their *permeability,* which is defined as the percentage of the spur surface area facing the streamflow. Accordingly, *retarder spurs* are permeable and function by retarding flow velocities near stream banks. *Retarder/deflector spurs* are permeable and function by retarding flow velocities at stream banks and direct flow away from the banks. *Deflector spurs* are impermeable, and deflect flow currents away from banks.

Guide banks (*spur dikes*) are placed at or near the ends of approach embankments to guide the streamflow through a bridge opening. They are used to prevent erosion of the approach embankments by cutting off the flow adjacent to the embankment, and by guiding flow through the opening. Guide banks reduce separation of flow at the upstream abutment face and reduce abutment scoar as a result of less turbulent flow at the abutment face. Figure 15.30 illustrates a typical guide bank plan view. Figure 15.31 can be used to determine guide bank length, L_s, for designs greater than 15 m and less than 75 m. Lagasse et al. (1995) recommend that 15 m is the minimum length of guide banks and if the figure indicates a length larger than 75 m the design should be set at 75 m. FHWA practice has shown that a standardized length of 46 m has performed well (Lagasse et al., 1995). Other guidelines for design include the following (refer to Lagasse et al. (1995) for more detailed design concerns):

- A minimum freeboard of 0.6 m above the design water surface elevation should be used.

- Generally top widths of 3–4 m are used.

- Upstream end of the guide bank should be round nosed.

- Side slopes should be 1V:2H or smaller.

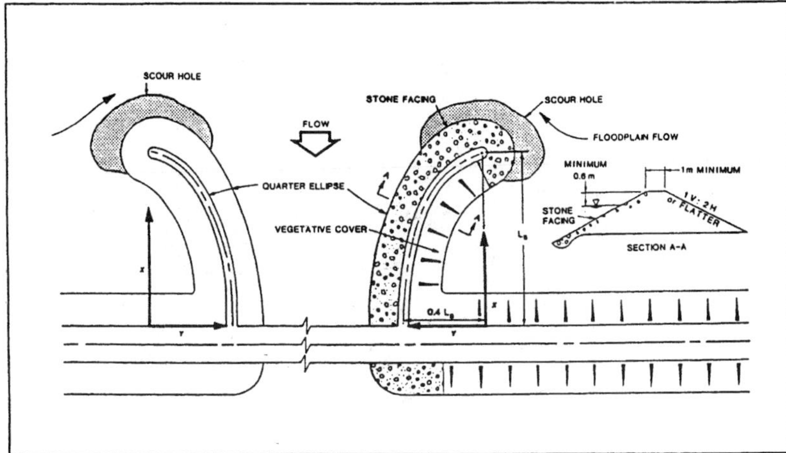

FIGURE 15.30 Typical guide bank. Guide banks should start at and be set parallel to the abutment and extend upstream from the bridge opening. The distance between the guide banks at the bridge opening should be equal to the distance between bridge abutments. Best results are obtained by using the guide banks with a plan form shape in the form of a quarter of an ellipse, with the ratio of the major axis (length L_s) to the minor axis (offset) of 2.5:1. This allows for a gradual constriction of the flow. If the length of the guide bank measured perpendicularly from the approach embankment to the upstream nose of the guide bank is denoted as L_s, the amount of expansion of each guide bank (offset), measured from the abutment parallel to the approach roadway, should be $0.4L_s$. (Fom Lagasse et al. 1995 as modified from Bradley 1978)

- Rock riprap should be placed on the stream side face and around the end of the guide bank.
- A gravel, sand, or fabric filter may be required to protect the underlying embankment material.

To use Fig. 15.31 the following parameters are needed: Q_f is the lateral or floodplain discharge of either floodplain intercepted by the embankment in m^3/s; $Q_{30\,m}$ is the discharge in 30 m of stream adjacent to the abutment in m^3/s; Q is the total discharge of the stream in m^3/s; A_{n2} is the cross-sectional flow area at the bridge opening at normal stage in m^2; $V_{n2} = Q/A_{n2}$ is the average velocity through the bridge opening in m/s, and Q_f/Q_{30m} is the guide bank discharge ratio.

To design the guide bank for the situation shown in Fig. 15.32 the following steps are followed:

1. Determine hydraulic design parameters (depths and velocities) using computer program WSPRO, HEC–2 or HEC–RAS.
2. Determine Q_f in the left overbank and right overbank.
3. Determine $Q_{30\,m}$ and $Q_f/Q_{30\,m}$ for the left and right overbank.
4. Determine length of the guide bank, L_s.
5. Select crest width, crest elevation, and riprap design.

15.9.2.2 Check Dams (Channel Drop Structures). *Check dams or channel drop structures* are countermeasures placed downstream of highway crossings to arrest head cutting and to maintain stable streambed elevation in the vicinity of the bridge. They are typically constructed of rock riprap, concrete, sheet piles, gabions, or treated timber piles. A typical

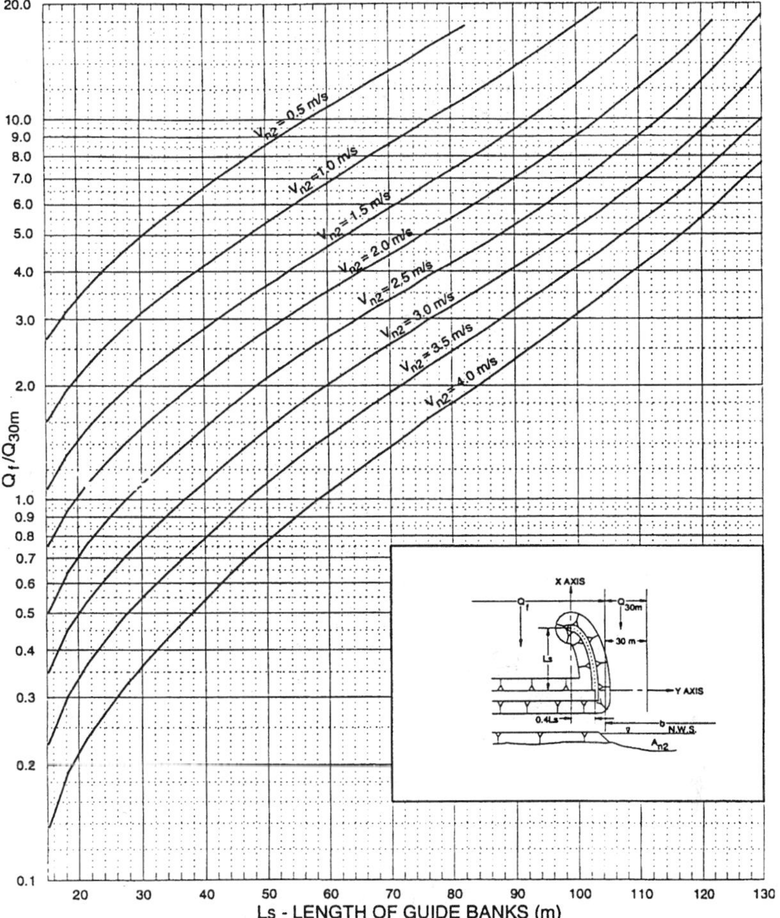

FIGURE 15.31   Nomograph to determine guide bank length. (From Lagasse et al., 1995)

vertical drop structure with a free overfall is shown in Fig. 15.33. Pemberton and Lara (1984) recommended the following equation to estimate the depth of scour downstream of a vertical drop

$$d_s = KH_t^{0.225} q^{0.54} - d_m \tag{15.25}$$

where d_s = the local scour depth for a free overfall, measured from the streambed downstream of the drop in m
    q = the discharge per unit width in m³/s/m
    H_t = the total drop in head, measured from the upstream to the downstream energy grade line in m
    d_m = the tailwater depth in m
    K = 1.90

Equation (15.25) is independent of grain size of the bed material and will acknowledge that the bed will scour regardless of the type of material. Based on the energy equation

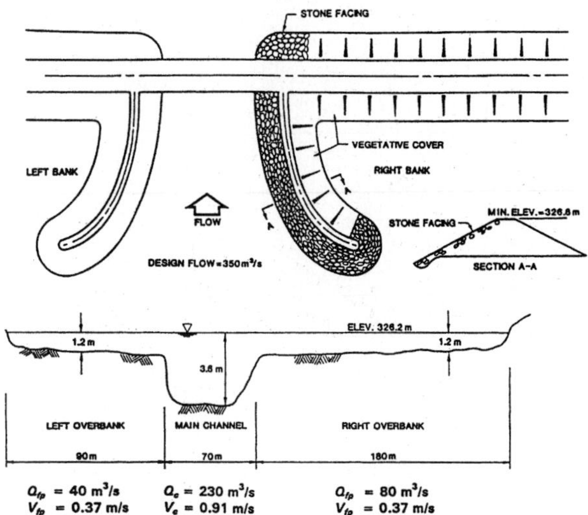

FIGURE 15.32 Example guide bank design. (From Lagasse et al., 1995)

$$H_t = \left(Y_u + \frac{V_u^2}{2g} + Z_u \right) - \left(Y_d + \frac{V_d^2}{2g} + Z_d \right) \tag{15.26}$$

where Y = the flow depth in m

V = the velocity in m/s

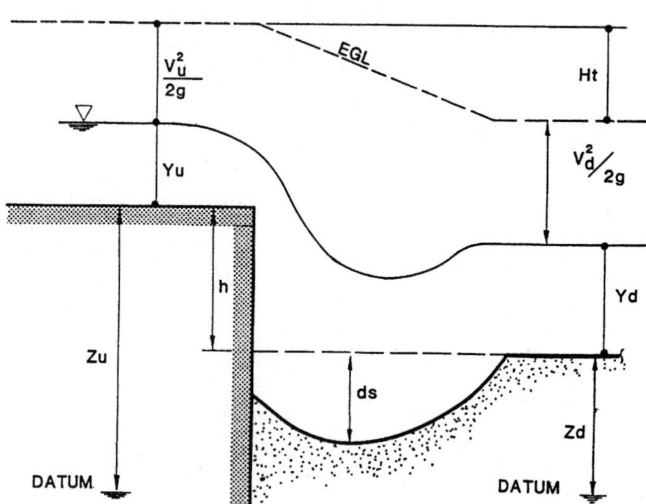

FIGURE 15.33 Schematic of a vertical drop caused by a check dam. (From Laggasse et al., 1995)

Z = the bed elevation in m
g = the acceleration due to gravity (9.81 m/s^2)

Subscripts u and d refer to upstream and downstream of the channel drop as shown in Fig. 15.33.

Example 15.3   (adapted from Lagas, et al.,1995). For the following hydraulic parameters, determine the depth of local scour. Design discharge = 167 m^3/s, channel width = 32 m; Y_u = 3.22 m; d_m = 2.9 m; Y_d = 2.9 m; unit discharge (q) = 5.22 m^3/s/m; V_u 1.62 m/s; V_d = 1.80 m/s; drop height (h) = 1.4 m. The results are presented in Fig. 15.33B.

Solution   Compute H_t using Eq. (15.26)

$$H_t = \left(3.22 + \frac{1.62^2}{2(9.81)} + 1.4\right) - \left(2.9 + \frac{1.80^2}{2(9.81)} + 0\right)$$

$$= 1.69 \text{ m}$$

Compute d_s using Eq. (15.25)

$$d_s = 1.90(1.69)^{0.225}(5.22)^{0.54} - 2.9 = 2.3 \text{ m}$$

The unsupported height of the structure is $h + d_s$ = 3.7 m.

15.10 BRIDGE SCOUR

This section presents methods and equations for determining total scour at a bridge, that is, long-term aggradation or degradation, contraction scour, and local scour. *Bridge scour* is the erosion or removal of streambed or bank material from bridge foundations due to flowing water, usually considered as long-term bed degradation, contraction, and local scour. Much of the material in this section has been adapted from the FHWA, Hydraulic Engineering Circular No. 18 by (Richardson and Davis 1995).

15.10.1 Design Approach

Before applying the various methods for estimating scour, it is necessary to (1) determine the fixed-bed channel hydraulics, (2) determine the long-term impact of degradation or aggradation on the bed profile, (3) if degradation occurs, adjust the fixed-bed hydraulics to reflect this change, and (4) compute the bridge hydraulics are 1995. The seven steps recommended in "Evaluating Scour at Bridges" (Richardson and Davis, 1995) are.

1. Determine the scour analysis variables.
2. Analyze long-term bed elevation change.
3. Evaluate the scour analysis method.
4. Compute the magnitude of contraction scour.
5. Compute the magnitude of local scour at piers.
6. Compute the magnitude of local scour at abutments.
7. Plot and evaluate the total scour depths.

Many of the hydraulic variables used in estimating scour can be obtained from computer models for water profile computation such as WSPRO developed by the U.S. Department of Transportation (1990). The following definitions are from the FHWA Hydraulic Engineering Circular No. 18 by Richardson and Davis (1995).

Aggradation and degradation are long-term streambed elevation changes due to natural or human-induced causes which can affect the reach of the river on which the bridge is located. Aggradation involves the deposition of material eroded from the channel or watershed upstream of the bridge; whereas, degradation involves the lowering or scouring of the streambed due to a deficit in sediment supply from upstream.

Contraction scour in a natural channel or at a bridge crossing, involves the removal of material from the bed and banks across all or most of the channel width. This component of scour can result from a contraction of the flow area, an increase in discharge at the bridge, or both. It can also result from a change in downstream control of the water surface elevation. The scour is the result of increased velocities and shear stress on the channel bed. Contraction of the flow by bridge approach embankments encroaching onto the floodplain and/or into the main channel is the most common cause of contraction scour. Contraction scour can be either clear-water or live-bed.

Live-bed contraction scour occurs when there is transport of bed material in the approach reach, whereas clear-water contraction scour occurs when there is no bed material transport in the approach reach or the bed material being transported in the upstream reach is so fine that it washes through the contracted section. Live-bed contraction scour typically occurs during the rising stage of a runoff event, while refilling of the scour hole occurs during the falling stage. Also, clear-water scour at low or moderate flows can change to live-bed scour at high flows. This cyclic nature creates difficulties in measuring contraction scour after a flood event.

Local scour involves removal of material from around piers, abutments, spurs, and embankments. It is caused by an acceleration of flow and resulting vortices induced by the flow obstructions. Local scour can also be either clear-water or live-bed scour. Live-bed local scour is cyclic in nature; that is, the scour hole that develops during the rising stage refills during the falling stage.

Lateral stream migration of the main channel of a stream within a floodplain may increase pier scour, erode abutments or the approach roadway, or change the total scour by changing the flow angle of attack at piers. Factors that affect lateral stream movement also affect the stability of a bridge. These factors are the geomorphology of the stream, location of the crossing on the stream, flood characteristics, and the characteristics of the bed and bank materials.

15.10.2 Contraction Scour

15.10.2.1 *Live-bed Contraction Scour.* *Live-bed contraction scour* occurs at a bridge when there is transport of bed material in the upstream reach into the bridge cross section, so that the area of the contracted section increases until, the transport of sediment out of the contracted section equals the sediment transported into the section (Richardson and Davis, 1995). The width of the contracted section is constrained and depth increases until the limiting conditions are reached. Laursen (1960) derived the following live-bed contraction scour equation based on a simplified transport function, transport of sediment in a long contraction, and other simplifying assumptions.

$$\frac{y_2}{y_1} = \left(\frac{Q_2}{Q_1}\right)^{6/7} \left(\frac{W_1}{W_2}\right)^{k_1} \left(\frac{n_2}{n_1}\right)^{k_2} \tag{15.27}$$

The average scour depth is $y_s = y_2 - y_0$,

where y_1 = average flow depth in the upstream main channel (m)

$\quad\quad y_2$ = average flow depth in the contracted section (m)

$\quad\quad y_0$ = existing flow depth in the contracted section before scour (m)

$\quad\quad Q_1$ = flow in the upstream channel transporting sediment (m³/s)

$\quad\quad Q_2$ = flow in the contracted channel (m³/s), which is often equal to the total discharge unless the total flood flow is reduced by relief bridges, water overtopping the approach roadway, or in the setback area

$\quad\quad W_1$ = bottom width of the upstream main channel (m)

$\quad\quad W_2$ = bottom width of main channel in the contracted section (m)

$\quad\quad n_2$ = Manning's n for contracted section

$\quad\quad n_1$ = Manning's n for upstream main channel

k_1 and k_2 = exponents depending on the mode of bed material transport (Table 15.5)

$\quad\quad V_*$ = $(gyS_1)^{1/2}$ shear velocity in the upstream section (m/s)

$\quad\quad \omega$ = median fall velocity of the bed material based on the D_{50}, (m/s) (see Fig. (15.34)

$\quad\quad g$ = acceleration of gravity (9.81 m/s²)

$\quad\quad S_1$ = slope of energy grade line of main channel (m/m)

$\quad\quad D_{50}$ = median diameter of the bed material (m)

The location of the upstream section needs to be located with engineering judgment. If WSPRO is used to obtain the values of the quantities, (y_1, Q_1, W_1, and n_1) the upstream channel section is located a distance equal to one bridge opening from the upstream face of the bridge.

A modified version of Laursen's (1960) equation for live-bed scour at a long contraction is recommended to predict the depth of scour in a *contracted section* (Richardson and Davis, 1995). The original equation given as Eq. (15.27) is modified by eliminating the ratio of Manning's n for the contracted and upstream sections. The equation assumes that bed material is being transported in the upstream section, and is given as

$$\frac{y_2}{y_1} = \left(\frac{Q_2}{Q_1}\right)^{6/7} \left(\frac{W_1}{W_2}\right)^{k_1} \tag{15.28}$$

and the average scour depth (m) is $y_s = y_2 - y_o$ and k_1 is obtained from Table 15.5(b).

The Manning's n ratio in Eq. (15.27) can be significant for the condition of a dune bed in the main channel and a corresponding plane bed, washed-out dunes or antidunes in the

TABLE 15.5   Exponents for Live-Bed Contraction Scour Equation

(a) For Long Section

V/w	k_1	k_2	Mode of bed material transport
<0.50	0.59	0.066	Mostly contact bed material discharge
0.50–2.0	0.64	0.21	Some suspended bed material discharge
>2.0	0.69	0.37	Mostly suspended bed material discharge

(b) For Contracted Section

V/w	k_1	Mode of bed material transport
<0.50	0.59	Mostly contact bed material discharge
0.50–2.0	0.64	Some suspended bed material discharge
>2.0	0.69	Mostly suspended bed material discharge

Source: From Laursen (1960).

contracted channel (Richardson and Davis, 1995). However, Eq. (15.27) does not adequately account for the increase in transport which occurs as a result of the bed planing out (which decreases resistance to flow, increases the velocity and the transport of bed material at the bridge). Laursen's Eq. (15.27) results in a decrease in scour for this case, whereas in reality, there would be an increase in scour depth. In addition, at flood flows, a plain bedform will usually exist upstream and through the bridge waterway, and the values of Manning's n will be equal. Consequently, the *n* value ratio is not recommended by Richardson and Davis (1995) as presented in the recommended Eq. (15.28). Scour depths with live-bed contraction scour may be limited by coarse sediments in the bed material armoring the bed. Where coarse sediments are present, Richardson and Davis (1995) recommend that scour depths be calculated for live-bed scour conditions using the clear-water scour equation (given in Sec. 15.10.2.2) in addition to the live-bed equation, and that the smaller calculated scour depth be used.

15.10.2.2 Clear-water Contraction Scour. *Clear-water contraction scour* occurs in a long contraction when (1) there is no bed material transport from the upstream reach into the downstream reach or (2) the material being transported in the upstream reach is transported through the downstream reach mostly in suspension and at less than capacity of the flow (Richardson and Davis, 1995). The area of the contracted section increases until, the velocity of the flow (V) or the shear stress (τ_o) on the bed is equal to the critical velocity (V_c) or the critical shear stress (τ_c) of a certain particle size (D) in the bed material. Widths (W) of contracted section usually are constrained and the depth (y) increases until the limiting conditions are reached.

Following the development given by Laursen (1963), equations for determining the clear-water contraction scour in a *long contraction* were developed in SI units. For equilibrium in the contracted reach, $\tau_o = \tau_c$ where τ_o is the average bed shear stress, contracted section, Pa (N/m^2) and τ_c is the critical bed shear stress at incipient motion, Pa (N/m^2). The average bed shear stress based on the hydraulic radius ($R = y$) and Manning's equation to determine the slope (S_f) is expressed as

$$\tau_o = \gamma y S_f = \frac{\rho g n^2 V^2}{y^{1/3}} \tag{15.29}$$

For noncohesive bed materials and fully developed clear-water contraction scour, the critical shear stress can be determined using Shields relation (Laursen, 1963)

$$\tau_c = K_s(\rho_s - \rho)gD \tag{15.30}$$

The bed in a long contraction scours until $\tau_o = \tau_c$ so that using Eqs. (15.29) and (15.30) implies

$$\frac{\rho g n^2 V^2}{y^{1/3}} = K_s(\rho_s - \rho)gl) \tag{15.31}$$

so that the depth (y) in the contracted section is

$$y = \left[\frac{n^2 V^2}{K_s(S_s - 1)D} \right]^3 \tag{15.32}$$

and in terms of discharge (Q), the depth is

$$y = \left[\frac{n^2 Q^2}{K_s(S_s - 1)DW^2} \right]^{3/7} \tag{15.33}$$

where y = average flow depth in the contracted section after contraction scour (m)
S_f = slope of the energy grade line (m/m)
V = average velocity in the contracted section (m/s)
D = diameter of smallest nontransportable particle in the bed material (m)
Q = discharge (m³/s)
W = bottom width of contracted section (m)
g = acceleration of gravity (9.81 m/s²)
n = Manning's roughness coefficient
K_s = Shield's coefficient S_s = specific gravity (2.65 for quartz)
γ = unit weight of water (9800 N/m³)
ρ = density of water (1000 kg/m³)
ρ_s = density of sediment (quartz, 2647 kg/m³)

Equations (15.32) and (15.33) are the basic equations for the clear-water scour depth (y) in a long contraction. Laursen used a value of 4 (in American customary units) for $K_s(\rho_s - \rho)g$; D_{50} for the size (D) of the smallest nonmoving particle in the bed material and Strickler's (1923) approximation for Manning's n ($n = 0.034 D_{50}^{1/6}$). Laursen's assumption that $\tau_c = 4D_{50}$ with $S_s = 2.65$ is equivalent to assuming a Shield's parameter $K_s = 0.039$. Shield's coefficient (K_s) to initiate motion ranges from 0.01 to 0.25 and is a function of particle size, Froude number, and size distribution. Some typical values for K_s for $F_r < 0.8$ and as a function of bed material size are (1) $K_s = 0.047$ for sand (0.065 mm $< D_{50} \leq 2$ mm); (2) $K_s = 0.03$ for medium coarse bed material (2 mm $< D_{50} \leq 40$ mm) and (3) $K_s = 0.02$ for coarse-bed material ($D_{50} > 40$ mm).

In SI units, Strickler's equation for n as given by Laursen is $n = 0.041 D_{50}^{1/6}$, where D_{50} is in meters. Richardson et al. (1990) recommend the use of the effective mean bed material size (D_m) in place of the D_{50} size for the beginning of motion ($D_{50} = 1.25 D_m$). Changing D_{50} to D_m in the Strickler's equation gives $n = 0.040 D_m^{1/6}$. Substituting $K_s = 0.039$ into Eq. (15.32) and (15.33) gives the following equation for y:

$$y = \left[\frac{V^2}{40D_m^{2/3}} \right]^3 \tag{15.34}$$

$$y = \left[\frac{Q^2}{40D_m^{2/3}W^2} \right]^{3/7} \tag{15.35}$$

$y_s = y - y_0$ = (average scour depth, m)

where Q = discharge through contraction (m³/s)
D_m = diameter of the bed material ($D_{50}/1.25$) in the contracted section (m)
W = bottom width in contraction (m)
y_0 = existing depth in the contracted section before scour (m)

The above clear-water contraction scour equations (15.32–15.35) assume homogeneous bed materials. For clear-water scour in stratified materials, using the layer with the finest D_{50} would result in the most conservative estimate of contraction scour. The clear-water contraction scour equations can be used sequentially for stratified bed materials.

The distribution of the contraction scour in the cross section can not be determined from the above scour equations. Assuming a uniform contraction scour depth across the opening may not be in error (e.g., short bridges, relief bridges, bridges with simple cross sections, and on straight reaches). For wide bridges, bridges on bends, bridges with large overbank flow, or crossings with a large variation in bed material size distribution, the contraction scour depths will not be uniformly distributed across the bridge opening (Richardson and Davis, 1995). In these cases, Eq. (15.32) and (15.34) can be used if the distribution of the velocity and/or the bed material is known. WSPRO uses the concept of stream tubes to

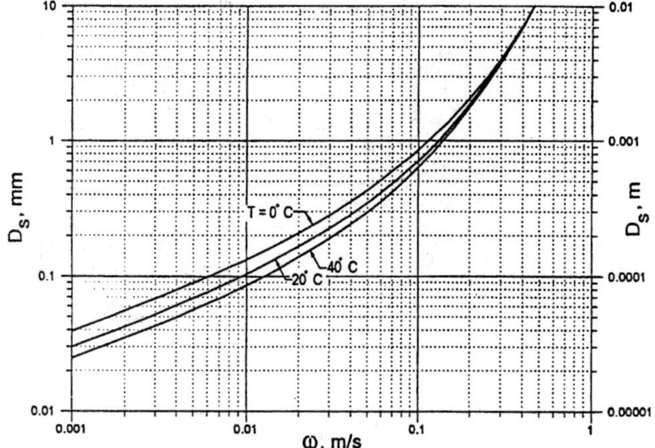

FIGURE 15.34 Fall velocity of sand-sized particles. (From Richardson and Davis, 1995)

determine the discharge and velocity distribution in cross sections (Fig. 15.35). Eqs. (15.32) and (15.34) can be used to estimate the distribution of the contraction scour depths with the WSPRO produced discharge and velocity distribution. Then Equations (15.33) and (15.35) can be used to determine the average contraction scour depth in the section.

The velocity and depth given in Eq. (15.32) are associated with initiation of motion of the indicated particle size (D). Eq. (15.32) can be rearranged to give the critical velocity (V_c) for beginning of motion of bed material of size D results in

$$V_c = \frac{K_s^{1/2}(S_s - 1)^{1/2}D^{1/2}y^{1/6}}{n} \tag{15.36}$$

Equation 15.16 can be simplified using $K_s = 0.039$, $S_s = 2.65$, and $n = 0.041\ D^{1/6}$, to obtain

$$V_c = 6.19y^{1/6}D^{1/3} \tag{15.37}$$

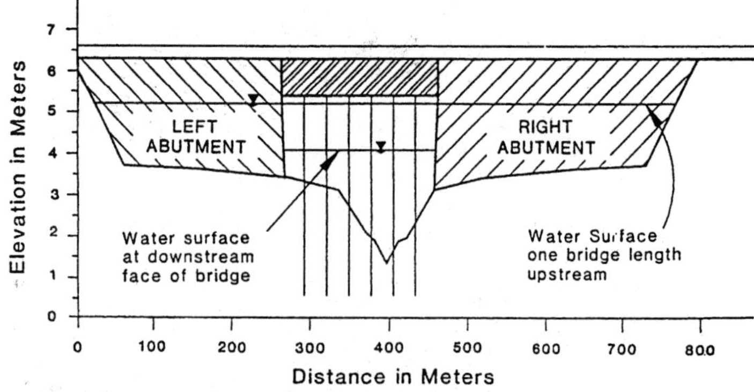

FIGURE 15.35 Cross section of proposed bridge. (From Richardson and Davis, 1995)

where V_c = critical velocity above which bed material of size D and smaller will be transported (m/s)

$\quad K_s$ = Shields parameter

$\quad S_s$ = specific gravity of the bed material

$\quad D$ = size of bed material (m)

$\quad y$ = depth of flow (m)

$\quad n$ = Manning's roughness coefficient

Example 15.4
This example problem and the succeeding examples in Sec. 15.10 are based on the examples taken from Arneson et al. (1991) and also used by Richardson and Davis (1995) in FHWA Hydraulic Engineering Circular No. 18. WSPRO was used to obtain the hydraulic variables listed in Tables 15.6–15.10.

A 198.12-m long bridge (Fig. 15.36) is to be constructed over a channel with spill-through abutments (slope of $1V{:}2H$). The left abutment is set approximately 60.5 m back from the channel bank. The right abutment is set at the channel bank. The bridge deck is set at elevation 6.71 m and has a girder depth of 1.22 m. Six round-nose piers are evenly spaced in the bridge opening. The piers are 1.52 m thick, 12.19 m long, and are aligned with the flow. The 100-year design discharge is 849.51 m³/s. The 500-year flow of 1444.16 m³/s was estimated by multiplying the Q_{100} by 1.7 since no hydrologic records were available to predict the 500-year flow. D_{50} is 0.002 m.

Determine whether the flow condition in the main channel is a live-bed or a clear-water condition.

Solution. Determine whether the flow condition in the main channel is live-bed or clear-water by comparing the critical velocity for sediment movement and the average channel velocity at the approach section. The discharge in the main channel at the approach section, Q_1, is computed as

TABLE 15.6 Hydraulic Variables from WSPRO for Estimation of Live-Bed Contraction Scour

		Remarks
Q (m³/s)	849.51	Total discharge
K_1 (approach)	19,000	Conveyance of main channel of approach
K_{total} (approach)	39,150	Total conveyance of approach section
W_1 or TOPW (approach) (m)	121.9	Top width of flow (TOPW). Assumed to represent active live bed width of approach
A_c (approach) (m²)	320	Area of main channel approach section
WETP (approach) (m)	122.0	Wetted perimeter of main channel approach section
K_c (bridge)	11,330	Conveyance through bridge
K_{total} (bridge)	12,540	Total conveyance through bridge
A_c (bridge) (m²)	236	Area of the main channel, bridge section
W_c (bridge) (m)	122	Channel width at the bridge. Difference between subarea break-points defining banks at bridge
W_2 (bridge) (m)	115.9	Channel width at the bridge. less four channel pier widths (6.08 m)
S_f (m/m)	0.002	Average unconstrictected energy slope (SF)

Source: From Richardson et al. (1995).

TABLE 15.7 Hydraulic Variables from WSPRO for Estimation of Clear-Water Contraction Scour on Left Overbank

		Remarks
Q (m³/s)	849.51	Total discharge, (see Table 16.10.2).
Q_{chan} (bridge)	767.54	Flow in main channel at bridge. Determined in live-bed computation of Step 5A.
Q_2 (bridge)	81.97	Flow in left overbank through bridge. Determined by subtracting Q_{chan} (listed above) from total discharge through bridge.
D_m (bridge overbank) (m)	0.0025	Grain size of left overbank area. $D_m = 1.25\ D_{50}$.
$W_{setback}$ (bridge) (m)	68.8	Top width of left overbank area. (SA #1) at bridge
$W_{contracted}$)bridge) (m)	65.8	Setback width less two pier widths (3.04 m).
A_{left} (bridge) (m²)	57	Area of left overbank at the bridge

Source: From Richardson et al. (1995).

$$Q_1 = Q\left(\frac{K_1}{K_{total}}\right) = 849.51\left(\frac{m^3}{s}\right)\left[\frac{19{,}000}{39{,}150}\right] = 412.28\ \text{m}^3/s$$

where K_1 and K_{total} are the conveyance of the main channel of approach and the total conveyance of the approach section, respectively, from Table 15.6. The average velocity in the main channel of the approach section is

$$V_1 = Q_1/A_c = 412.28/320 = 1.29\ \text{m/s}$$

where A_c is the area of the main channel approach section from Table 15.6. The average depth of flow y_1 in the approach section is

$$y_1 = A_1/W_1 = 320/121.9 = 2.63\ \text{m,}$$

where W_1 is the top width of flow.
     The critical velocity for D_{50} size sediment is computed using Eq. (15.37) thus:

$$V_c = 6.19 y_1^{1/6} D_{50}^{1/3}$$

$$= 6.19(2.63\ \text{m})^{1/6}(0.002\ \text{m})^{1/3}$$

$$= 0.92\ \text{m/s}$$

Comparing V_1 and V_c, $V_1 > V_c$ indicating that the flow condition is live-bed

TABLE 15.8 Hydraulic Variables from WSPRO for Estimation of Pier Scour (Conveyance Tube Number 12)

		Remarks
V_1 (m)	3.73	Velocity in conveyance tube #12
V_1 (m)	2.84	Mean depth of tube #12

TABLE 15.9 Hydraulic Variables from WSPRO for Estimation of Abutment Scour Using Froehlich's Equation for Left Abutment

		Remarks
Q (m³/s)	849.51	Total discharge (see Table 16.10.2).
q_{tube} (m³/s)	42.48	Discharge per equal conveyance tube, defined as total discharge divided by 20.
#Tubes	3.5	Number of approach section conveyance tubes which are obstructed by left abutment. Determined by superimposing abutment geometry onto the approach section.
Q_e (m³/s)	148.68	Flow in left overbank obstructed by left abutment. Determined by multiplying # tubes and q_{tube}.
A_e (left abut) (m²)	264.65	Area of approach section conveyance tubes number 1, 2, 3, and half of tube number 4.
L' (m)	232.80	Length of abutment projected into flow, determined by adding top widths of approach section conveyance tubes number 1, 2, and 3, and half of tube number 4.

Source: From Richardson et al. (1995).

Example 15.5
Compute the live-bed contraction scour for the main channel for the above example problem.

Solution
Equation 15.28 is used to compute y_2/y_1, and the scour depth is computed using $y_s = y_2 - y_0$ so that

$$y_s = \left(\frac{y_2}{y_1}\right) y_1 - y_o$$

$$= \left(\frac{Q_2}{Q_1}\right)^{6/7} \left(\frac{W_1}{W_2}\right)^{k_1} y_1 - y_o$$

The first step is to compute all the parameters in this equation to evaluate y_s.

Q_1 was evaluated as 412.28 m³/s in the above Example, and Q_2, the discharge in the main channel at the bridge, is

TABLE 15.10 Hydraulic Variables from WSPRO for Estimation of Abutment Scour Using HIRE equation for left abutment

		Remarks
V_{tube} (m/s) (bridge section)	1.29	Mean velocity of conveyance tube #1, adjacent to left abutment
y_1 (m) (bridge section)	0.83	Average depth of conveyance tube #1

Source: From Richardson et al. (1995).

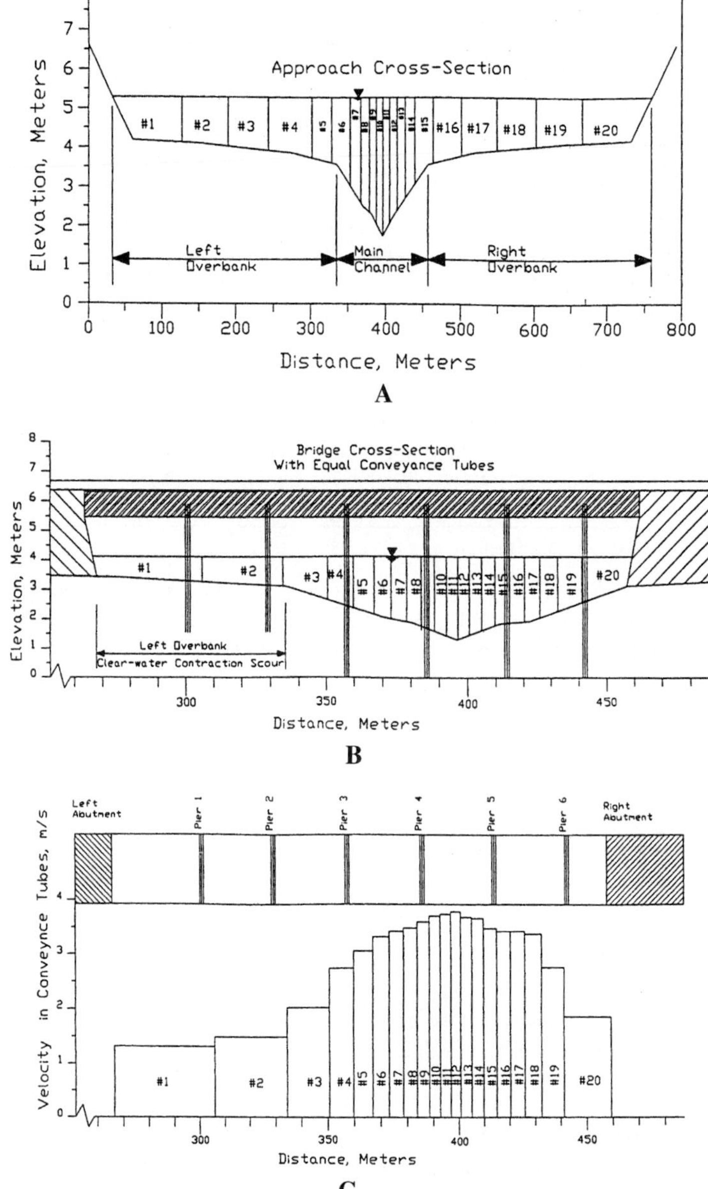

FIGURE 15.36 **A.** Equal conveyance tubes of approach section. **B.** Equal conveyance tubes of bridge section **C.** Velocity distribution at bridge crossing. (From Richardson and Davis, 1995)

$$Q_2 = Q\left(\frac{K_2}{K_{total}}\right) = 849.51\left(\frac{m^3}{s}\right)\left[\frac{11{,}330}{12{,}540}\right] = 767.54 \text{ m}^3/\text{s}$$

The channel widths are $W_1 = 121.9$ m and $W_2 = 115.9$ m in Table 15.6, $y_1 = 2.63$ m was computed in Example 15.10.1. The bridge channel flow depth is the area divided by the top width $y_0 = 236$ m^2/122 m $= 1.93$ m.

The only remaining unknown is K_1 which can be obtained from Table 15.5.B for V_*/ω. To compute V_*/ω, first compute the shear velocity V_* using $V_* = \sqrt{\tau_0/\rho}$ where $\tau_0 = \gamma RS = 9810$ (N/m^3)(2.62 m) (0.002 m/m) $= 51.4$ N/m$^2 = 51.4$ Pa. The shear velocity is then $V_* = \sqrt{51.4/1000} = 0.227$ m/s. The bed material is sand with $D_{50} = 0.002$ m (2 mm). Fall velocity $\omega = 0.21$ m/s from Fig. 15.34. Then $V_*/\omega = 0.227/0.21 = 1.08$. From Table 15.5B $K_1 = 0.64$ which indicates that the mode of bed transport is a mixture of suspended and contact bed material discharge.

The live-bed contraction scour of the channel is computed using Table 15.5A as

$$y_s = (767.54/412.28)^{6/7}(121.9/115.9)^{0.64}(2.63) - 1.93 = 2.7 \text{ m}$$

This scour is large and could be minimized through a larger bridge opening, by putting relief bridges in the overbank, or possibly in some cases providing for highway approach overtopping (Richardson and Davis, 1995).

Example 15.6
Clear-water contraction scour will occur in the overbank area between the left abutment and the left bank of the bridge opening Example 15.4. Compute the clear-water contraction scour in the left bank based on the discharge and depth of flow passing under the bridge and use the hydraulic variables from the example in Table 15.7.

Solution
Using Eq. (15.35),

$$y = \left[\frac{Q^2}{40D_m^{2/3} W_{contacted}^2}\right]^{3/7}$$

$$= \left[\frac{0.025(81.97 \text{ m}^3/\text{s})^2}{(0.0025 \text{ m})^{2/3} (65.8 \text{ m})^2}\right]^{3/7}$$

$$= 1.38 \text{ m}$$

Average flow depth (y_0) in the left overbank bridge section is $y_0 = a/W_{setback} = (57.0$ m$^2)/(68.8$ m$) = 0.83$ m. The clear-water contraction scour in the left overbank of the bridge opening is $y_s = y_2 - y_0 = 1.38 - 0.83$ m $= 0.55$ m.

15.10.3 Local Scour at Piers

Local scour at piers or abutments is caused by the formation of horseshoe vortices at their base (Fig. 15.37). The *horseshoe vortex* results from the increased head water on the upstream surface of the obstruction and subsequent acceleration of the flow around the nose of the pier or abutment. Vortex action removes bed material from around the base of the obstruction, where the transport rate of sediment away from the base region is greater than the transport rate into the region, and consequently, a scour hole develops. (Richardson and Davis, 1995).

There are also vertical vortices called *wake vortices* downstream of the pier (Fig. 15.37). which also remove material from the pier base region. The intensity of wake vortices diminishes rapidly as the distance downstream of the pier increases, so that there is often deposition of material, immediately downstream of a long pier. (Richardson and Davis, 1995)

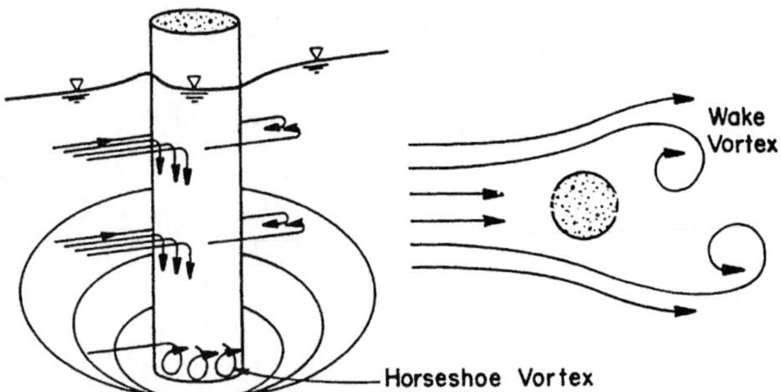

FIGURE 15.37 Schematic representation of scour at a cylindrical pier. (From Richardson and Davis, 1995)

Factors that affect the magnitude of local scour depth at piers and abutments are (1) velocity of the approach flow, (2) depth of flow, (3) width of the pier, (4) discharge intercepted by the abutment and returned to the main channel at the abutment, (5) length of the pier if skewed to flow, (6) size and gradation of bed material, (7) angle of attack of the approach flow to a pier or abutment, (8) shape of a pier or abutment, (9) bed configuration, and (10) ice formation or jams and debris (Richardson and Davis, 1995).

There have been many scour equations reported in the literature for live-bed scour in cohesionless sand-bed streams. The equation presented here is recommended for both live-bed and clear-water pier scour, given as (Richardson et al., 1990)

$$\frac{y_s}{y_1} = 2.0K_1\,K_2\,K_3\,K_4\left[\frac{a}{y_1}\right]^{0.65}\mathrm{FR}_1^{0.43} \tag{15.38}$$

In terms of y_s/a, where a is the pier width, Eq. (15.38) is

$$\frac{y_s}{a} = 2.0K_1\,K_2\,K_3\,K_4\left[\frac{y_1}{a}\right]^{0.35}\mathrm{FR}_1^{0.43} \tag{15.39}$$

where y_s = scour depth (m)
y_1 = flow depth directly upstream of the pier (m)
K_1 = correction factor for pier nose shape (Fig. 15.38 and Table 15.11)
K_2 = correction factor for angle of attack of flow from Table 15.12

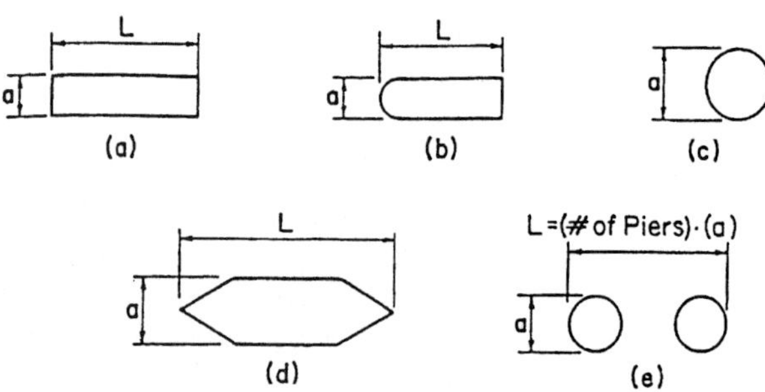

FIGURE 15.38 Common pier shapes **(A)** square nose, **(B)** round nose, **(C)** cylinder, **(D)** sharp nose, **(E)** group of cylinder (see Multiple Columns). (From Richardson and Davis, 1995)

TABLE 15.11 Correction Factor, K_1, for Pier Nose Shape*

Shape of pier nose	K_1
(a) Square nose	1.1
(b) Round nose	1.0
(c) Circular cylinder	1.0
(d) Group of cylinders	1.0
(e) Sharp nose	0.9

Source: Richardson et al. (1995).

The correction factor K_1 for pier nose should be determined using Table 15.11 for angles of attack up to 5. For greater angles. K_2 dominates and K_1 should be considered as 1.0. If L/a is larger than 12, use the values for $L/a = 12$ as a maximum in Table 15.12

K_3 = correction factor for bed condition from Table 15.13
K_4 = correction factor for armoring by bed material size from the list inmediately below
a = pier width (m)
L = length of pier (m)
Fr_1 = Froude number directly upstream of the pier = $V_1/(gy_1)^{1/2}$
V_1 = mean velocity of flow directly upstream of the pier (m/s)
g = acceleration of gravity (9.81 m/s^2)

The limits for bed material size and K_4 value are (Richardson and Davis, 1995).

TABLE 15.12 Correction Factor, K_2, for Angle of Attack, θ, of the Flow*

Angle	$L/a = 4$	$L/a = 8$	$L/a = 12$
0	1.0	1.0	1.0
15	1.5	2.0	2.5
30	2.0	2.75	3.5
45	2.3	3.3	4.3
90	2.5	3.9	5.0

Source: From Richardson and Davis. (1995).

*The value of the correction factor K_2 should be applied only when the field conditions are such that the entire length of the pier is subjected to the angle of attack of the flow. Use of this factor directly from the table will result in a significant overprediction of scour if (1) a portion of the pier is shielded from the direct impingement of the flow by an abutment or another pier, or (2) an abutment or another pier redirects the flow ia a direction parallel to the pier. For such cases, judgment must be exercised to reduce the value of the K_2 factor by selecting the effective length of the pier actually subjected to the angle of attack of the flow. (Richardson and Davis, 1995).

TABLE 15.13 Increase in Equilibrium Pier Scour Depths, K_3, for Bed Condition*

Bed dondition	Dune height m	K_3
Clear–water scour	NA	1.1
Plane bed and antidune flow	NA	1.1
Small dunes	$3 > H \geq 0.6$	1.1
Medium dunes	$9 > H \geq 3$	1.2–1.1
Large dunes	$H \geq 9$	1.3

Source: From Richardson and Davis (1995).
Abbreviations: NA, not aplicable.
*The correction factor K_3 results from the fact that for plane-bed conditions, which is typical of most bridge sites for the flood frequencies employed in scour design, the maximum scour may be 10% greater than computed with Eq. (15.38). In the unusual situation where a dune bed configuration with large dunes exists at a site during flood flow, the maximum pier scour may be 30% greater than the predicted equation value. This may occur on very large rivers, such as the Mississippi. For smaller streams that have a median dune configuration at flood flow, the dunes will be smaller and the maximum scour may be only 10 to 20% larger than equilibrium scour. For antidune bed configuration the maximum scour depth may be 10% greater than the computed equilibrium pier scour depth. (Richardson and Davis, 1995)

Minimum Bed Material Size	Minimum K_4 Value	$V_R > 1.0$
$D_{50} > 0.06$	0.7	1

Example 15.7
Compute the magnitude of the local scour at the pier. Anticipating that any pier under the bridge could be subject to the maximum flow depth and velocities, only one computation of pier scour of the side pier is needed. This assumption is acceptable because the thalweg is prone to shift and there is a possibility of lateral channel migration.

Solution
This solution requires the parameters in Eq. (15.38)

$$\frac{y_s}{y_1} = 2.0 K_1\, K_2\, K_3\, K_4 \left[\frac{a}{y_1}\right]^{0.65} \mathrm{Fr}_1^{0.43}$$

The Froude number, Fr_1, for the pier scour computation is based on the hydraulic characteristics of conveyance tube number 12. The velocity $V_1 = 3.73$ m/s, and $y_1 = 2.89$ m are from Table 15.8, so Fr_1 is:

$$\mathrm{Fr}_1 = \frac{V_1}{(gy_1)^{0.5}} = \frac{3.73 \text{ m/s}}{[(9.81 \text{ m/s}^2)(2.84 \text{ m})]^{0.5}} = 0.71$$

For a round-nose pier, aligned with the flow and sand-bed material (see Tables 15.11 and 15.12), $K_1 = K_2 = K_4 = 1.0$.

For plane-bed condition (Table 15.13), $K_3 = 1.1$.
Using Eq. (15.39)

$$\frac{y_s}{2.84} = 2.0(1)\ (1)\ (1.1)\ (1) \left[\frac{1.52\ \text{m}}{2.84\ \text{m}}\right]^{0.65} (0.71)^{0.43}$$

and

$$y_s = 3.3\ \text{m}$$

The maximum local pier scour depth will be 3.3 m.

15.10.4 Live-Bed Scour at Abutments

When abutments obstruct flow, local scour occurs as a result of a horizontal vortex that starts at the upstream end of the abutment and runs along the toe of the abutment, similar to the horizontal vortex that forms at piers. A vertical wake vortex forms at the downstream end of an abutment similar to that which forms downstream of a pier or downstream of any flow separation. Figure 15.39 is a definition sketch for abutment scour. There are three general abutment shapes as indicated in Fig. 15.40.

The potential for lateral channel migration, long-term degradation, and contraction scour should be considered in determining abutment foundation depths near the main channel. The FHWA, HEC-18 recommends that abutment scour equations be used to develop insight as to the scour potential of an abutment. Then, the abutment may be designed to resist the computed scour or as an alternative, riprap and guide banks can be used to protect the abutment from scour and erosion (Richardson and Davis, 1995). Two approaches, Froelich's live-bed scour equation and a U.S. Army Corps of Engineers' equation, are discussed in this section. To check the potential depth of scour for the design of the foundation and placement of rock riprap or guide banks, Froelich's live-bed scour equation can be used. Froelich (1989) analyzed 170 live-bed scour measurements in laboratory flumes and developed the following regression equation:

$$\frac{y_s}{y_a} = 2.27\ K_1\ K_2 \left[\frac{L'}{y_a}\right]^{0.43} Fr_1^{0.61} + 1 \tag{15.40}$$

where K_1 = coefficient for abutment shape (see the list immediately following)
$\quad\ \ K_2$ = coefficient for angle of embankment to flow
$\quad\quad\ \ = (\theta/90)^{0.13}$ (see Fig. 15.41 for definition of θ)
$\quad\quad\ \ = \theta < 90°$ if embankment points downstream

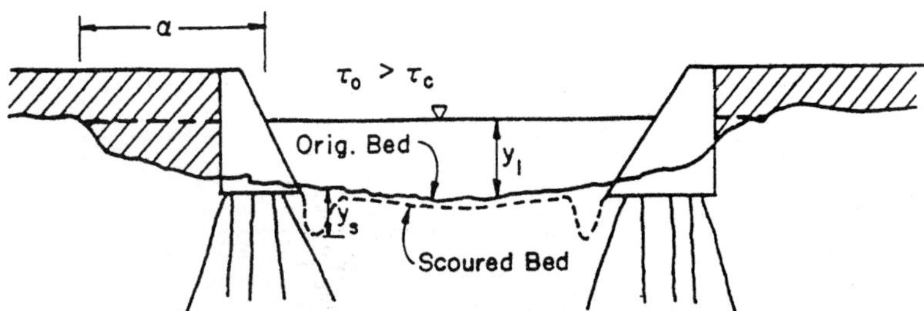

FIGURE 15.39   Definition sketch for abutment scour. (From Richardson et al., 1990)

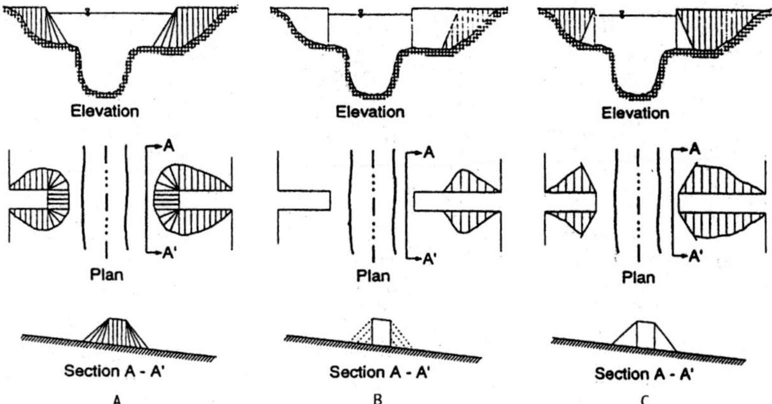

FIGURE 15.40 Abutment shapes (**A**) spill through, (**B**) vertical wall, (**C**) vertical wall with flared wingwalls. (From Richardson and Davis, 1995)

$$= \theta > 90° \text{ if embankment points upstream}$$

L' = length of abutment (embankment) projected normal to flow (m)

A_e = flow area of the approach cross section obstructed by the embankment (m²)

Fr_1 = Froude number of approach flow upstream of the abutment

$\quad = V_e/(gy_a)^{1/2}$

$V_e = Q_e/A_e$, m/s

Q_e = flow obstructed by the abutment and approach embankment (m³/s)

y_a = average depth of flow on the floodplain (m)

y_s = scour depth (m)

The abutment shape coefficients are (Richardson and Davis, 1995):

Description	K_1
Vertical-wall abutment	1.00
Vertical-wall abutment with wing walls	0.82
Spill-through abutment	0.55

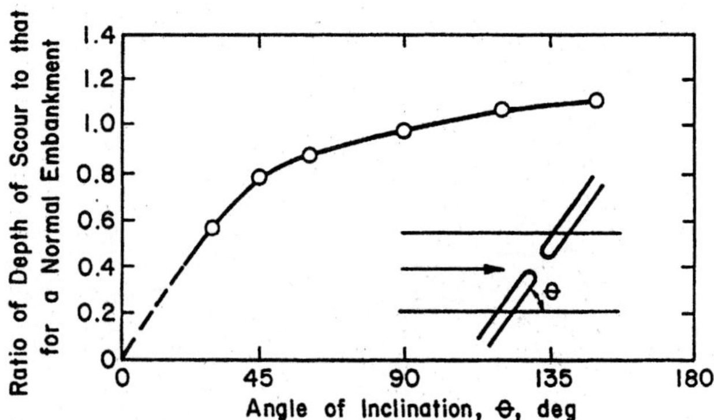

FIGURE 15.41 Adjustment of abutment scour estimate for skew. (From Richardson et al., 1990 as presented in Richardson and Davis, 1995)

Equation (15.40) is not consistent because as L' tends to 0, y_s also tends to 0. The 1 was added to the equation so as to encompas 98 % of the data. Field data of scour at the end of spurs in the Mississippi River (obtained by the U.S. Army Corps of Engineers) can also be used for estimating abutment scour. This field situation closely resembles the laboratory experiments for abutment scour in that the discharge intercepted by the spurs was a function of the spur length. Equation (15.41), referred to as the Highways in the River Environment (HIRE) equation (Richardson et al., 1990), is applicable when the ratio of projected abutment length (L') to the flow depth (y_1) is greater than 25.

This equation can be used to estimate scour depth (y_s) at an abutment where conditions are similar to the field conditions from which the equation was derived (Richardson and Davis, 1995):

$$\frac{y_s}{y_1} = 4 Fr_1^{0.33} \frac{K_1}{0.55} \tag{15.41}$$

where y_1 = depth of flow at the abutment on the overbank or in the main channel (m)
$\quad Fr_1$ = Froude number based on the velocity and depth adjacent to and upstream of the abutment
$\quad K_1$ = abutment shape coefficient (see list detailing abutment shape coefficients, above)

To correct Eq. (15.41) for abutments skewed to the stream, use Figure 15.41.

For clear-water scour at an abutment, use either Eq (15.40) or (15.41) because clear-water scour equations potentially decrease scour at abutments due to the presence of coarser material, which is unsubstantiated by field data (Richardson and Davis, 1995). Froelich's equation will generally result in deeper scour predictions than experienced in the field, according to Richardson and Davis (1995). These scour depths could occur if the abutments protruded into the main channel flow, or when a uniform velocity field is cut off by the abutment in a manner that most of the returning overbank flow is forced to return to the main channel at the abutment end. All of the abutment scour computations (left and right abutments) assumed that the abutments were set perpendicular to the flow. If the abutments were angled to the flow, a correction utilizing K_2 would be applied to Froelich's equation or, using Fig. 15.41 would be applied to Eq. (15.41). However, the adjustment for skewed abutments is minor when compared to the magnitude of the computed scour depths.

Example 15.8
Compute the magnitude of the local scour at the left abutment for the previous example in this section. Use Froelich's live-bed scour equation for the computation.

Solution
First determine all the parameters for Froelich's equation

$$\frac{y_s}{y_a} = 2.27 \, K_1 \, K_2 \left[\frac{L'}{y_a} \right]^{0.43} Fr_1^{0.61} + 1$$

For spill-through abutments, $K_1 = 0.55$, and for abutments perpendicular to the flow, $K_2 = 1.0$. Abutment scour can be estimated using Froelich's equation with data derived from the WSPRO output (Table 15.9). y_a at the abutment is assumed to be the average flow depth in the overbank area, computed as the cross-sectional area of the left overbank cut off by the left abutment divided by the distance the left abutment protrudes into the overbank flow

$$y_a = \frac{A_e}{L'} = \frac{264.65 \text{ m}^2}{232.80 \text{ m}} = 1.14 \text{ m}$$

The average velocity of the flow in the left overbank which is cut off by the left abutment is computed as the discharge cutoff by the abutment divided by the area of the left overbank cut off by the left abutment.

$$V_e = \frac{Q_e}{a_e} = \frac{148.68 \text{ m}^3/\text{s}}{264.65 \text{ m}^2} = 0.56 \text{ m/s}$$

The Froude number of the overbank flow is

$$Fr = \frac{V_e}{(gy_a)^{1/2}} = \frac{0.56 \text{ m/s}}{[(9.81 \text{ m/s}^2)(1.14 \text{ m})]^{0.5}} = 0.17$$

$$\frac{y_s}{1.14} = 2.27 \, (0.55) \, (1.0) \left[\frac{232.8}{1.14}\right]^{0.43} (0.17)^{0.61} + 1$$

$$\frac{y_s}{1.14} = 5.17$$

$$y_s = 5.9 \text{ m}$$

The abutment scour at the left abutment based on Froelich's equation is 5.9 m.

Example 15.9
Compute the magnitude of the local scour at the left abutment for the first example in this section using the HIRE Equation (Eq. 15.41).

Solution
The HIRE Equation (Eq. 15.41) is

$$\frac{y_s}{y_1} = 4Fr_1^{0.33} \frac{K_1}{0.55}$$

which is based on the velocity and depth of flow passing through the bridge opening adjacent to the abutment listed in Table 15.10. The Froude number is

$$Fr_1 = \frac{V_{\text{tube}}}{(gy_1)^{0.5}} = \frac{1.29 \text{ m/s}}{[(9.81 \text{ m/s}^2)(0.83 \text{ m})]^{0.5}} = 3.07$$

and from the list abutment shape coefficients, above, $K_1 = 0.55$, using Equation (15.41):

$$\frac{y_s}{0.83 \text{ m}} = 4Fr_1^{0.33} \left(\frac{0.55}{0.55}\right) = 4(0.45)^{0.33} = 3.07$$

$$y_s = 2.6 \text{ m}$$

The depth of scour at the left abutment, as computed using the HIRE equation, is 2.6 m.

Example 15.10
Use the HIRE equation (15.41) and the data below to compute the magnitude of the local scour at the right abutment using $V_{\text{tube}} = 2.19 \text{ m/s}$ and $y_1 = 1.22 \text{ m}$.

Solution
The HIRE equation is also applicable to the right abutment since L/y_1 is greater than 25. The HIRE equation is based on the velocity and depth of the flow passing through the bridge opening adjacent to the end of the right abutment listed above. The Froude number is

$$Fr_1 = \frac{2.19 \text{ m/s}}{[(9.81 \text{ m/s}^2)(1.22 \text{ m})]^{0.5}} = 0.63$$

and $K_1 = 0.55$ (see list detailing abutment shape coefficients), using the HIRE equation, is

$$\frac{y_s}{1.22 \text{ m}} = 4Fr_1^{0.33}\left(\frac{0.55}{0.55}\right) = 4(0.63)^{0.33} = 3.43$$

$$y_s = 4.2 \text{ m}$$

The depth of scour at the right abutment is 4.2 m.

Example 15.11
For the previous examples in this section 15.10 the final step is to plot the results of the scour computation.

Solution
Figure 15.42 is a plot of the total scour on the bridge cross section. Only the computation for pier scour with piers aligned with the flow was used, and only the abutment scour computation for Eq. (15.41) was used. The top width of the local scour holes suggested is 2.0 y_s (Richardson and Davis, 1995).

15.11 COMPUTER MODELS FOR CULVERTS AND SEDIMENTATION

Numerous computer programs are available to aid in the design and analysis of highway culverts and sedimentation issues. Familiarity with culvert hydraulics and sediment transport is necessary to provide a solid basis for designers to take advantage of the speed, accuracy, and increased capabilities of hydraulic design computer programs.

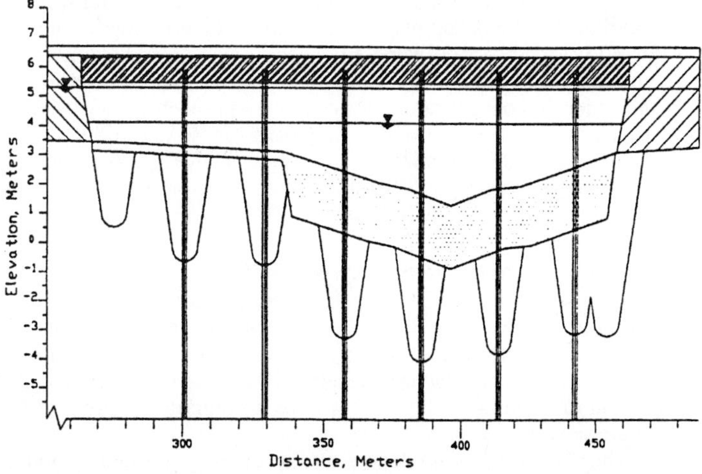

FIGURE 15.42   Plot of total scour for example problem. (From Richardson and Davis, 1995)

15.11.1   Computer Models for Culverts

15.11.1.1   Integrated Drainage Design Computer System.   The computer programs for the hydraulic design of culverts available from the FHWA are more fully described in the HDS No. 5 (Normann et al., 1985). More detailed information on the FHWA publications can be obtained from http://www.fhwa.dot.gov. The recommended personal computer system is the HYDRAIN-Integrated Drainage Design Computer System. There are currently two culvert design and analysis programs within the HYDRAIN system. They are the Culvert Design System (CDS) and the HY8-Culvert Analysis program. HYDRAIN Version 6.0 was developed by GKY and Associates, Inc. (1996), under contract with the FHWA. HYDRAIN software and user support is available through, McTrans at the University of Florida (*mctrans@ce.uf.edu*), and PCTrans at the University of Kansas (*pctrans@kuhub.cc.ukans.edu*).

15.11.1.2   Culvert Design System (CDS)   CDS is a program that can be used for the hydraulic design or the analysis of an existing or proposed culvert. The model can accommodate a variety of hydrograph relationships, culvert shapes, materials, and inlet types (Brater et al., 1996). Any of the six culvert types defined in Section 15.6 can be used. The model starts its design process by selecting a culvert size and number of barrels that are compatible with geometric, environmental and construction constraints. The culvert size is based on the design headwater, headwater-to-diameter ratio, inundation, outlet velocity or cover limitations.

15.11.1.3   HY8.   HY8 is a program that uses FHWA procedures as defined in HDS No. 5 for the hydraulic design of culverts, and incorporates factors such as hydrological inputs, storage and routing considerations, and energy dissipation devices (Norman et al.,) 1985. The program can accommodate various culvert shapes including circular, rectangular, elliptical, arch, and user-defined geometry. Improved inlets can be specified and various design conditions can be analyzed including inlet and outlet control for full and partially full culverts, flow over the roadway embankment, and balance of flows through multiple parallel culverts. The program features include the development of performance curves, generation of rating curves for uniform flow, velocity, and maximum shear for the downstream channel.

15.11.1.4   CULVERT2 (English units), CULVERT3 (metric units).   CULVERT2 and 3 are provided by Caltrans [California Department of Transportation (DOT)], have corrosion criteria for culverts and present alternative culvert materials and material thickness acceptable for 50 years of service using site-specific test data. The minimum resistivity test data (ohm-cm) and pH of the site soils and/or water are required for all analysis. Water-soluble sulfate and chloride concentrations (ppm) are required for aggressive sites when the minimum resistivity is less than 100 ohm-cm. Culvert materials addressed include corrugated steel pipe, (CSP), corrugated aluminized steel pipe, (CASP); corrugated aluminum pipe, (CAP); and reinforced concrete pipe, (RCP). Documentation and software can be ordered from http://www-mctrans.ce.ufl.edu/info-cen.

15.11.1.5   Culvert Analysis Program (CAP).   The Culvert Analysis Program (CAP) is a program (Fulford, 1995) used to compute discharges through a culvert and develop stage-discharge relationships for a culvert from the measurements of upstream and downstream water surface elevations. Types of culverts that can be studied include rectangular, circular, pipe arch, and nonstandard shaped culverts. CAP will not analyze culverts that vary in cross section or material, culverts that have a nonuniform slope or break, or culverts that have a severe adverse slope.

   The computation procedure that CAP uses is based upon the U.S. Geological Survey (USGS) method described by Bodaine (1968) entitled *Measurement of Peak Discharge of*

Culverts by Indirect Methods found in the USGS document *Techniques of Water-Resources Investigations, Book 3, Chapter A3.* The method is based upon USGS field investigations and laboratory investigations conduced by thde USGS, Bureau of Public Roads, and various universities.

CAP Version 97-01 is written in FORTRAN 77. The source program has been compiled for IBM personal computers, Macintosh personal computers, and mainframe computers, such as UNIX. To execute CAP, preparation of a input file utilizing a text editor that produces American Standard Code for Information Interchange (ASCII) files is required. A computer with 386 MHz and 2 Megabytes ramdon access memory (RAM) is also required.

15.11.2 Computer Models for Sedimentation

15.11.2.1 HEC-6. HEC-6 (U.S. Army Corps of Engineers, 1993) is a one-dimensional, moveable-boundary, steady-state, open-channel flow model used to compute degradation (scour) and aggradation (deposition) in river systems by simulating the interaction between hydraulics of the flow and the rate of sediment transport (Morris et al., 1996). A river system consisting of a main stem, tributaries, and local inflow/outflow points can be simulated. The program processes a water discharge hydrograph as a sequence of steady flows of variable durations. Using continuity of sediment, changes are calculated with respect to time and distance along the study reach for the total sediment load, volume and gradation of sediment that is scoured or deposited, armoring of the bed surface, and the cross-sectional elevations. HEC-6 is designed to analyze long-term scour and/or deposition. The bed-material transport algorithms assume that equilibrium conditions are reached within each time step. HEC-6 raises or lowers cross-sectional elevations to reflect deposition and scour. The horizontal locations of the channel banks are considered fixed and the floodplains on each side of the channel are considered as having fixed ground elevations. The hydraulic parameters needed to calculate sediment transport potential are obtained by using the standard step method. The hydraulic parameters are calculated at each cross-section for each successive discharge. Manning's equation and n values for overbank and channel areas may be specified by discharge or elevation. Sediment transport rates are calculated for each flow in the hydrograph for each grain size (upper limit for grain size is 2048 mm). The transport potential is calculated for each grain size in the bed as though that size comprised 100 % of the bed material. Transport potential is then multiplied by the fraction of each size class present in the bed at that time to yield the transport capacity for that size class. For deposition and erosion of clay and silt sizes up to 0.0625 mm, Krone's (1962) method is used for deposition and Ariathurai and Krone's (1976) adaption of Parthenaides's (1965) method is used for scour. The model and the user's manual can be downloaded from http://www.hec.usace.army.mil/software.

15.11.2.2 Generalized Stream Tube Model Alluvial River Simulation. (GSTARS 2.0). Most of the sediment and water routing models, such as the HEC-6 model were developed for solving one-dimensional alluvial river problems. Although there are truly two-dimensional and three-dimensional models for alluvial river simulation, they are too computationally intensive for engineering applications. The Generalized Stream Tube model for Alluvial River Simulation (GSTARS) was developed by Molinas and Yang (1986) to simulate the flow conditions in a semi-two-dimensional manner and the change of channel geometry in a semi-three-dimensional manner. GSTARS, Version 2.0, is an enhanced personal computer version of the original GSTARS program. The stream tubes are imaginary channels that are bounded by streamlines and convey the same discharge. They are used to compute the lateral variation of hydraulic and sediment parameters within the cross-section. The sediment transport results for stream tubes may vary and one stream tube may be aggrading and the other degrading. Stream boundaries are updated for each cross section and time step to result in equal hydraulic conveyance. The model can also solve for an unknown channel width or compute the changes in the channel width based on minimum stream power theory. The GSTARS 2.0 program and the user's manual can be requested from http://www.usbr.gov/srhg/gstars/2.0.

15.11.2.3 Surface Water Modeling System. Surfacewater Modeling System (SMS) is a powerful, and comprehensive two-dimensional surfacewater modeling package. The software models the water surface elevation, flow velocity, contaminant transport and dispersion, and sediment transport and deposition for complex two-dimensional horizontal flow problems. SMS provides complete support for the U.S. Army Corps of Engineers RMA two-dimensional hydrodynamic and contaminant transport, SED-2D two-dimensional sediment transport and deposition, HIVEL-2D two dimensional hydrodynamic supercritical and sub-critical flow, and U.S Federal Highway Administration FESWMS two-dimensional hydro-dynamic and bridge scour finite element models. Single- and multiple-opening bridge and culvert roadway crossings, channel networks, sinuous rivers, harbors, bays, estuaries, wet-lands, irregular floodplains, split flows, and other complex simulations can be modeled. The analysis results can be output or displayed graphically using a variety of plots, including vector plots, contour plots, color-shaded contour plots, and time-history plots. Time-history plots can be requested at any location to illustrate fluctuations in water surface elevation, velocity, discharge, contaminant concentration, and bed elevation. Along the flow boundaries and at each node in the finite element mesh, the water surface elevation, flow velocity, pollution contaminant concentration, and bed scour and deposition are computed. Flow sep-arations and eddy currents are accurately modeled. The software can find solutions for a single instance in time (steady-state solution), or during a series of time-steps (transient solution). Transient solutions can be used to model flow fluctuations caused by inflow hy-drographs, tidal cycles, and storm surges. For subcritical-supercritical mixed flow regimes, hydraulic jumps are automatically located. The software and the user's manual can be re-quested from http://www.bossintl.com.hk/html/ products.html.

15.11.2.4 Bridge Stream Tute Model for Alluvial River Simulation (BRI-STARS). Bridge Stream Tube model for Alluvial River Simulation (BRI-STARS, Version 3.3) is a program developed for the National Cooperative Highway Research Program Project 15-11, *Computer-Aided Analysis of Highway Encroachments on Mobile Boundary Systems.* It is a semi-two-dimensional model capable of computing alluvial scour/deposition through subcritical, supercritical, and a combination of both flow conditions involving hydraulic jump. This model, unlike conventional water and sediment routing computer models, is capable of simulating channel widening/narrowing phenomenon as well as local scour due to highway encroachments. It couples a fixed-width stream tube computer model, which simulates the scour/deposition process taking place in the vertical direction across the chan-nel, with a total stream power minimization algorithm. The decision-making algorithm, using rate of energy dissipation or total stream power minimization, determines whether the sim-ulated sediment erosion satisfying the sediment continuity equation should take place in the lateral or vertical direction. It is this second component that allows the lateral changes in channel geometries. Finally, the bridge component allows computation of the hydraulic flow variables and the resulting scour due to highway encroachments. The model also contains a rule-based expert system program for classifying streams by size, bed and bank material stability, platform geometry, and other hydraulic and morphological features. Documentation and software can be ordered from http://www-mctrans.ce.ufl.edu/info-cen.

15.11.2.5 HY9. HY9 is based on the FHWA *Publication Interim Procedures for Evalu-ating Scour at Bridges.* It computes contraction scour, pier scour, and abutment scour using the equations presented in this chapter. *Scour at Bridges* (HY-9), Version 4.0 and the accom-panying reports HEC-18, *Evaluating Scour at Bridges* and HEC-20, *Stream Stability at Highway Structures,* are available from http://www-mctrans.ce.ufl.edu/info-cen.

REFERENCES

American Association of State Highway and Transportation Officials (*AASHTO*) *Drainage Manual.* AASHTO Task Force on Hydraulics and Hydrology, 1990.

American Iron and Steel Institute, *Handbook of Steel Drainage and Highway Construction Products,* W. P. Reyman Associates, New York, 1983.

American Concrete Pipe Association, *Concrete Pipe Design Manual,* Vienna, VA, 1992.

Ariathurai R., and R. B. Krone, "Finite Element Model for Cohesive Sediment Transport," *Journal of the Hydraulics Division, American Society of Civil Enqineers,* 323–338, March 1976.

ARMCO Drainage and Metal Products, *Handbook of Culvert and Drainage Practice,* R. R. Donnelley & Sons, Chicago, IL, 1950.

Arneson, L., Shearman, J. O., and Jones, J. S., "Evaluating Scour at Bridges Using WSPRO," unpublished paper presented at the 71st Annual Transportation Research Board meeting, Washington, DC, January, 1991.

Bodaine, G. L., "Measurement of Peak Discharge at Culverts by Indirect Methods," *Techniques of Water—Resources Investigations,* U.S. Geological Survey, 1968. A3.

Brater E. F., H. W. King, J. E. Lindell, and C. Y. Wei, *Handbook of Hydraulics,* McGraw Hill, New York, 1996.

Brice, J. C., and J. C. Blodgett, *Countermeasures for Hydraulic Problems at Bridges,* Vol. 1, Analysis and Assessment, FHWA/RD-78-162, Federal Highway Administration, Washington, DC, 1978.

Bradley, J. N. Hydraulics of bride Waterways, Hydraulic Design Series No. 1 U.S. Department of Transportation, FHWA, 1978.

Breusers, H. N. C., and A. J. Raudkivi, *Scouring,* A. A. Balkema, Rotterdam, the Netherlands, 1991.

Brice, J. C., and J. C. Blodgett, *Countermeasures for Hydraulic Problems at Bridges,* Vol. 2, Case Histories for Sites I-283, FHWA/RD-78-163, Federal Highway Administration, Washington, DC, 1978b.

Chaudry, M. H., *Open-Channel Flow,* Prentice-Hall, NJ, 1993.

Chow, V. T., *Open-Channel Hydraulics,* McGraw-Hill, New York, 1959.

Corry, M. L., P. L. Thompson, F. J. Watts, J. S. Jones, and D. L. Richards, *Hydraulic Design of Energy Dissipators for Culverts and Channels,* Hydraulic Engineer Circular No. 14, Hydraulics Branch, Bridge Division, Office of Engineering, Federal Highway Administration, Washington, DC, 1983.

Flood Control District of Maricopa County, *Drainage Design Manual for Maricopa County, Vol. II, Hydraulics,* Phoenix, AZ, 1996.

Froelich, D. C., *Abutment Scour Prediction,* Presentation, Transportation Research Board, Washington, DC, 1989.

Fulford, J. M., *User's Guide to the Culvert Analysis Program, U.S.* Geological Survey, Open file report 95–137, 1995.

GKY and Associates, Inc., *Culvert Analysis Program HY8 Version 6.0,* Springfield, VA, 1996.

Harrison, L. J., J. L. Morris, J. M. Normann, and F. L. Johnson, *Hydraulic Design of Improved Inlets for Culverts,* Hydraulic Engineering Circular No. 13, Federal Highway Administration, U.S. Department of Transportation, Washington, DC, 1972.

Hendrickson, J. G., *Hydraulics of Culverts,* American Concrete Pipe Association, Chicago, IL, 1957.

Krone, R. B., "Flume Studies of the Transport of Sediment in Estuarial Shoaling Processes," Hydraulic Engineering Laboratory, University of California, Berkeley, CA, 1962.

Lagasse, P. F., Schall, F., Johnson, F., Richardson, E. V. and Chang, F., *Stream Stability at Highway Structure,* Department of Transportation, Federal Highway Administration, Hydraulic Engineering Circular No. 20, Washington, DC, 1995.

Laursen, E. M., "Scour at Bridge Crossings," *Journal of the Hydraulics Division, American Society of Civil Enqineers,* 86(1142) 1960.

Maccaferri Gabions Inc., "Gabion and Reno Mattress Short Course," West Sacramento, CA, 1997.

Molinas, A., and Yang, C. T., "Computer Program User's Manual for GSTARS," U.S. Bureau of Reclamation, Denver, Col, 1986.

Morris G. L., and J. Fan, *Reservoir Sedimentation Handbook,* McGraw-Hill, New York, 1997.

Normann, J. M., R. J. Houghtalen, and W. J. Johnston, *Hydraulic Design of Highway Culverts,* HDS No. 5, Federal Highway Administration (FHWA), U. S. Department of Transportation, Norfolk, VA, 1985.

Parthenaides, E., "Erosion and Deposition of Cohesive Soils," *Journal of the Hydraulics Division, American Society of Civil Enqineers,* 755–771, 1965.

Pemberton, E. L., and J. M. Lara, *Computing Degradation and Local Scour,* Technical Guideline for Bureau of Reclamation, Engineering Research Center, Denver, C, 1984.

Portland Cement Association, *Handbook of Concrete Culvert Pipe Hydraulics,* Skokie, IL, 1964.

Raudkivi, A. J., *Loose Boundary Hydraulics,* Pergamon Press, New York, 1990.

Raudkivi, A. J., *Sedimentation: Exclusion and Removal of Sediment from Diverted Water,* A. A. Balkema, Rotterdam, the Netherlands, 1993.

Reagan, D., "Highway Drainage Design," C-TAP Course, 1993.

Reihsen, G., and L. J. Harrison, *Debris Control Structures,* Hydraulic Engineering Circular No. 9, Federal Highway Administration, U.S. Department of Transportation, Washington, DC, 1971.

Richardson, E. V., and S. R. Davis, *Evaluating Scour at Bridges,* 3rd ed., Hydraulic Engineering Circular No. 18, Publication No. FHWA-IP-90-017, Federal Highway Administration, U.S. Department of Transportation, Washington, DC, 1995.

Richardson, E. V., Simons, D. B., and P. Y. Julien, *Highways in the River Environment,* FHWA-HI90-016, Federal Highway Administration, U.S. Department of Transportation, Washington, DC, 1990.

Ruff, J. F., *Scour at Culvert Outlets in Mixed Bed Materials,* Federal Highway Administration/RD-82/011, 1982.

Shearman, J. O. Bridge Waterways Analysis Model for Mainframe and Micro computers, WSPRO/HY-7, Federal Higway Administration, U.S. Department of Transportation, Washington, D.C., 1987.

Simons, D. B., and F. Senturk, *Sediment Transport Technology,* Water Resources Publications, Fort Collins, Co, 1977.

Simons, Li & Associates, *Engineering Analysis of Fluvial Systems,* Fort Collins, Co, 1982.

State of Florida Department of Transportation, *Drainage Manual, Vol. 3, Theory,* Drainage Design Office, FL, 1987.

Strickler, A., "Beiträge zur Frage der Geschwindigkeitsformel und der Rauhigkeitszahlen für ströme, Kanäle und geschlossene Leitungen," Mitteilungen des Eidgenössischen Amtes für Wasserwirtschaft 16, Bern, Switzerland, 1923 (Translated as "Contributions to the Question of a Velocity Formula and Roughness Data for Streams, Channel, and Closed Pipelines," by T. Roesgan and W. R. Brownie, Translation T-10, W. M. Keck Laboratory of Hydraulics and Water Resources, California Institute of Technology, Pasadena, CA, January 1981).

U.S. Army Corps of Engineers CA, *HEC-2, Water Surface Profiles,* Hydraulic Engineering Center, Davis, CA, 1991.

U.S. Army Corps of Engineers, HEC-6, *Scour and Deposition in River and Reservoirs,* Hydrologic Engineering Center, Davis, CA, 1993.

U.S. Army Corps of Engineers, River Analysis System, HEC–RAS, User's Manual Version 1.0, Hydrologic Engineering Center, Davis, CA, 1995

U.S. Department of Transportation, *User's Manual for WSPRO–A Computer Model for Water Surface Profile Computation,* Report No. FHWA-IP-89-027, Federal Highway Administration, Washington, DC, 1990.

Vanoni, V. A., ed., *Sedimentation Engineering,* ASCE Manual and Reports on Engineering Practice, 54, American Society of Civil Engineers, New York, 1975.

Yang, C. T., *Sediment Transport: Theory of Practice,* McGraw-Hill, New York, 1996.

CHAPTER 16
HYDRAULIC DESIGN OF FLOOD CONTROL CHANNELS

George K. Cotton
Simons & Associates, Inc.
Fort Collins, Colorado

16.1 INTRODUCTION

Flood protection is essential to the economic and environmental integrity of most civil engineering projects. The reliability of transportation systems, the livability of urban and suburban developments, the long-term stability of landfills, the operation of mining excavations, the reclamation of disturbed lands, and the management of forest and agricultural lands require careful planning for flood hazards. In nearly all these cases, design of a flood-control system will include a variety of conveyance channels referred to as flood-control channels.

To perform reliably, flood-control channels must behave in a stable, predictable manner. This ensures that a known flow capacity will be available for a planned flood event. In most cases, the design goal is a noneroding channel boundary, although, in certain cases, a dynamic channel is sought. Since most soils erode under a concentrated flow, channel linings are needed either temporarily or permanently to achieve channel stability.

Channel linings can be classified in two broad categories: rigid or flexible. *Rigid lining* include channel pavements of concrete or asphaltic concrete and a variety of precast interlocking blocks and articulated mats. *Flexible linings* include such materials as loose stone (riprap), vegetation, manufactured mats of light-weight materials, fabrics, or combinations of these materials. The selection of a particular lining is a function of the design context, involving issues related to the consequences of flooding, the availability of land, and environmental needs.

A rigid lining is capable of high conveyance and high-velocity flow. Flood-control channels with rigid linings are often used to reduce the amount of land required for a surface drainage system. When land is costly or unavailable because of restrictions, use of rigid channel linings is preferred.

Flexible channel linings are distinguished from rigid linings because they can respond to a change in channel shape. They can sustain some adjustment of the channel's shape and still maintain their integrity. This would not be the case for a rigid lining, where local damage to the lining may lead to a general unraveling. Damage to a rigid lining can result from secondary forces, such as frost heave, piping, or slumping, although uplift and shear forces are often causes of failure.

Flexible linings are used as temporary channel linings for control of erosion during construction or reclamation of disturbed areas. Also, when environmental requirements are part of the design, flexible lining materials have several advantages. Flexible linings are inexpensive, permit infiltration and exfiltration, and allow growth of vegetation. Hydraulically, flow conditions in the channel lined with flexible materials generally can be made to conform to conditions found in a natural channel. This provides better habitat opportunities for local flora and fauna. By permitting the growth of vegetation in the channel, flexible linings can provide a buffering effect for runoff contaminant's and sediment.

We are fortunate that a large number of innovative and traditional products are currently available for channel linings. In most cases, these products have received extensive testing, including laboratory flume studies, large-scale prototype modeling, and documentation of field case histories. From this effort has come a better understanding of complicated flow conditions associated with each type of lining and a better understanding of product performance and of methods needed for design. Construction experience has resulted in better specifications for channel lining materials and their installation.

The presentation in this chapter covering flexible lining materials is based on work in preparing *Design of Roadside Channels with Flexible Linings* (Hydraulic Engineering Circular No. 15) for the Federal Highway Administration. Chen and Cotton (1988), Since 1988, when that manual was published, the commercial market for channel-lining products has expanded. Because product testing and performance monitoring also have increased, this chapter also serves to update the reader on the status of current practice.

The so-called *tractive force* (or *boundary shear stress*) that acts on the channel's perimeter describes the basic mechanics of channel stability. Design and field methods based on tractive force are well suited for the evaluation of the stability of smaller channels where the grade of the channel dominates. In these cases, the calculation of shear stress is simple, the determination of channel slope is the most difficult estimation. For field observations on a channel of known grade, depth alone needs to be measured to estimate the maximum shear stress. Field estimates of flow velocity require some type of current meter. In design, the performance criteria are simple to recall for a specific type of lining because it is represented by a single permissible shear stress value. This permissible shear stress value is applicable over a wide range of channel slopes and shapes. Permissible velocity criteria, on the other hand, are a function of channel slope, lining roughness, and channel shape.

16.2 DESIGN CONCEPTS

The design methods are based on the concept of maximum permissible tractive force, coupled with the hydraulic resistance of the particular lining material. The method includes two parts: computation of the flow conditions for a given design discharge and determination of the degree of erosion protection required. The flow conditions are a function of the channel geometry, design discharge, channel roughness, and channel slope. The erosion protection required can be determined by computing the shear stress on the channel at the design discharge and by comparing the calculated shear stress to the permissible value for the type of channel lining used.

16.2.1 Flood-Control Channel Design

The methods given in this chapter are for small (less than 1 m^3/s) to medium-sized flood-control channels (less than 10 m^3/s). The design of larger flood-control channels is increasingly complex in terms of the physical environment, the nature of the flood risk, and the number of issues and constraints that a design team must address. Small channels can often be designed on the basis of simplified hydrologic analysis (i.e., the rational method), without detailed geotechnical data, and by assuming uniform flow. A medium-sized flood-control channel typically requires hydrologic routing, information on soil properties, and a hydraulic analysis based on nonuniform flow.

The level of effort for design increases steadily as the size of the channel increases. It is expected that it may require approximately four times the hours for technical staff to accomplish a medium-sized flood channel compared with a small channel. Since costs will increase by a factor of about 10, there is a definite economy of scale in the design of larger channels. However, the larger a channel becomes, the more tributary and appurtenant features will be needed for a complete design. Therefore, the design of a medium-sized channel often includes the design of several small channels and other features.

16.2.2 Open-Channel Flow

16.2.2.1 Types of Flow.   Open-channel flow can be classified according to three general conditions: (1) uniform or nonuniform flow, (2) steady or unsteady flow, and (3) subcritical or supercritical flow. In uniform flow, the depth and discharge remain constant along the channel. In steady flow, no change in discharge occurs over time. Most natural flows are unsteady and are described by a runoff hydrography. One can assume, in most cases, that the flow will vary gradually and can be described as steady, uniform flow for short periods. Subcritical flow is distinguished from supercritical flow by a dimensionless number called the *Froude number* (Fr), which is defined as the ratio of inertial forces to gravitational forces in the flow. *Subcritical flow* (*Fr* < 1.0) is characterized as tranquil and has deep, slower velocity flow. *Supercritical flow* (Fr > 1.0) is characterized as rapid and has shallow, high-velocity flow.

For design purposes, uniform flow conditions also are considered to be steady. The channel grade S_o, water-surface grade S_w, and the energy grade S_f are assumed to be equal. This allows the development of a flow equation based on a friction loss formula (such as Manning's equation) and channel grade.

Computation of uniform flow is suitable for small channels and a useful approximation for medium-sized channels. In medium channels, steady uniform flow is rare, and a more complete analysis using gradually varied flow methods is required. In solving gradually varied flow, the designer should rely on a current computer program, the accuracy of which is well documented.

Design of steep channels with supercritical flow presents a number of special concerns. Waves can form in a channel bed with a flexible lining (beginning near a Fr of 0.8) that ultimately approach the depth of flow. In extremely steep channels, the flow may splash and surge in a violent manner and additional freeboard is required. Ultimately, at Fr's near 2.0, the flow becomes unstable. Channel designs in this range should be avoided.

16.2.2.2 Resistance to Flow and Boundary Shear Stress.   Flow resistance is the result of the drag of moving water against the channel boundary. For practical purposes, the flow in most channels will be fully turbulent and the velocity, V, will be proportional to the square root of the shear stress, τ, on the channel boundary. This gives the following simple formula for estimating channel velocity:

$$V = C_f \sqrt{\tau/\rho} \tag{16.1}$$

where C_f = the conveyance factor and ρ = the density of water.

For uniform flow, the average shear stress on the channel boundary acts opposite to the weight component of the flow, as shown in Fig. 16.1. For channel grades of less than 10°, the sine of the grade can approximated by the tangent and, hence, the slope of the bed.

This leads to the equation for *mean boundary shear:*

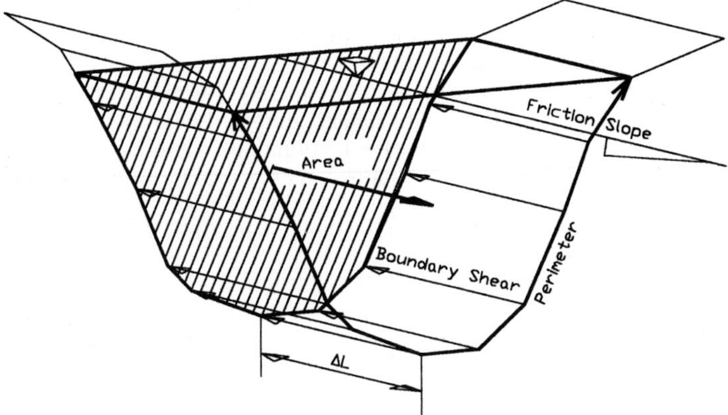

FIGURE 16.1   Definition sketch for boundary shear stress.

$$\tau \, \Delta L P = \gamma \, A \, \Delta L \, S_f$$

$$\tau = \gamma \, R \, S_f \tag{16.2}$$

where ΔL = the incremental reach length (Fig. 16.1)
P = the wetted perimeter
γ = the specific weight of water
A = the cross sectional area of flow
S_f = the friction slope (energy grade)

The conveyance factor in Eq. 16.1 is not constant but is a function of both boundary and flow conditions. If this were not the case, there would be little difference between a velocity-based approach and a tractive shear-based approach to channel design. In Manning's formulation, the *conveyance factor* varies as the one-sixth power of the hydraulic radius, R, giving

$$C_f = \frac{1}{n} R^{1/6} \tag{16.3}$$

Note: Because C_f is dimensionless, Manning's coefficient n has the dimension of length to the one-sixth power.

Combining Eqs. 16.1, 16.2, and 16.3 gives the standard form of the Manning's resistance equation:

$$V = \frac{1}{n} R^{2/3} S_f^{1/2} \tag{16.4}$$

Not all channel linings behave according to the Manning's formulation for the conveyance factor. Most rigid channel linings have a Manning's n that is approximately constant. For shallow flows, the n value increases in rigid channels, but this effect is often neglected even for small flood-control channels. However, for flexible types of channel linings, the conveyance factor is difficult to describe using Eq. 16.3.

A channel lined with a good stand of vegetation cannot be described by a single n value. The flow resistance is complicated by the mechanics of the vegetation since the stem of the plant will bend because of the drag force, changing the height of the plant relative to the flow depth. The Soil Conservation Service (SCS) (USDA, 1954), through the work of Ree and Palmer (1949), developed standard classifications for *vegetative flow retardance.* Grasses are classified into five broad categories of flow retardance: Class A identifies grasses with the highest flow retardance and Class E identifies the grasses with the lowest flow retardance. In general, high flow-retardance species of grass form a dense cover, are tall, and have stiff stems. Sparse cover or short, flexible grasses have lower flow retardance.

Temple (1980; Temple et al., 1987) and Kouwen, (1988; Kouwen and Li, 1980; Kouwen and Unny, 1969) have developed modern approaches to the flow resistance of vegetation. Temples approach is based on a mathematical fit to the SCS retardance charts and a generalized retardance index that is a function of two properties of gress: stem length and cover density. Temple compiled standard values of these properties for selected species for reference stem densities representing "good" cover conditions. Adjusting these reference properties for local conditions and expected growth allows the designer to determine an appropriate index for specific design conditions.

Kouwen analyzed the biomechanics of vegetation and developed conveyance factors that are a function of the density, height, and stiffness of the grass. He showed that the resulting method gives results that are consistent with the SCS retardance curves. Since Kouwen's method is physically based, it is a useful tool for extending retardance curves to channel conditions that were not studied by the SCS, including stiff vegetation and mild channel grades.

16.2.3 Flood-Control Channel Components

A basic flood-control channel consists of a minimum of five components. There are specific design issues associated with the local flow conditions at each component. More complex flood-control systems also may include additional components for flood storage (detention basins) and grade control (low dams).

Channel Inlet.   At the point where water enters a flood-control channel, local shear stresses can develop as a result of local acceleration of the flow. For small flood-control channels, the inlet may consist of small berms or dikes that collect runoff and direct this diffuse flow to the channel. A medium-sized flood-control channel will typically be the result of the combination of several tributaries or the extension of an existing channel. Collection and control of incoming flow is essential to the proper operation of a larger channel. In most cases, a flood-control channel will discharge as a tributary to a larger channel.

Reach.   The reach component is the length of channel, with only minor variation in the properties of grade, discharge, cross section, and lining material. A reach may have a straight or curving alignment. In larger flood-control channels, a reach also may include a bridge or culvert.

Flow around a bend in an open channel induces centrifugal forces created by the change in the direction of flow. This results in a superelevation of the water surface, with its surface being higher at the outside of the bend than at the inside of the bend. Flow around a channel bend also imposes higher shear stress on the channel bottom and banks. The increased shear stress requires additional design considerations within and downstream of the bend.

Confluence.   A *confluence* is the site where two or more flows merge without a significant grade difference between each channel. The process of merging flow involve turbulent mixing and a local increase in energy loss and boundary shear stress. Flow separation, wave formation, and bank impingement can occur in a confluence that creates much higher local boundary shear stresses and a potential for erosion. Control of the velocity and direction of tributary inflows often is required to prevent damage to the channel within a confluence. Where a large tributary joins the main channel, performance of the confluence can be improved if the direction of the two flows is as nearly parallel as possible.

Side Inlet.   A *side inlet* allows flow to enter the channel over the bank. Example of the need for a side inlet to a channel are flow from a field or a street. Often, the flow enters the channel from a chute constructed on the channel bank. When the bank channel is at a mild grade, vegetation may be a sufficient lining for the chute. For steeper bank slopes, however, riprap or gabion linings are common.

Channel Crossing.   A channel crossing is required where a private or public road passes over the channel. Structures used for this purpose are stream fords, culverts, and bridges. Stream fords are best suited for low-traffic-volume roads crossing channels with an intermittent type of flow. Fords can be hazardous if flow depths of more than 0.30 to 0.45 m. are expected. Culverts of concrete or metal pipe provide an economical crossing for small channels. Since the shape of the culvert is typically circular, elliptical, or arched, the culvert will obstruct a portion of the channel section. Bridges of concrete, steel, or timber are used when the channel must remain largely unobstructed.

Transition.   A *transition* is a gradual expansion or contraction between two channel sections. Transitions can occur between one reach and another or between a reach and the channel inlet or outlet.

Channel Outlet.   At the point where water exits a flood control channel, local shear stresses can develop as aresult of deceleration of the flow. At the outlet, the flow changes to match the local velocity and depth of the receiving channel.

16.2.4 Stable Channels

16.2.4.1 Stable Channel Modes.   A stable channel can be either static or dynamic, but over time the net effect of scour and deposition must be zero. In a *static stable channel,* scour and deposition occur within the limits of a channel boundary that effectively resists the erosive force of the flow. Although the transport of sediment can be significant in a static channel, the effective change in channel section at any given time is small. This implies equilibrium between the incoming supply of sediment and the transport of sediment within a channel reach. When the supply of sediment is small, the principles of rigid boundary hydraulics can be applied to evaluate the channel's capacity and the lining's stability. When the supply of sediment is large, then the determination of the channel's capacity must consider the transport of both water and sediment. The transport of sediment affects the flow resistance and therefore the shear forces and the stability of the channel lining.

In a *dynamic stable channel,* some change in the channel bed and banks are expected. The channel is considered to be stable if the morphology (i.e., shape) of the channel remains unchanged over time. Important measures of channel shape include the channel width, depth, slope, pattern (straight, meandering, or braided), hydraulic variables (flow velocity and discharge), and sediment variables (particle size, suspended concentration of sediment). A dynamic channel can be regarded as stable as long as local changes in channel shape do not have adverse consequences. For most small- or moderate-capacity flood-control systems, design of a dynamic channel is often too complex or it presents too many uncertainties. Consequently, development of static stable channels is usually preferable to using dynamic approaches to the design.

Several empirical methods are commonly applied to determine whether a channel is stable. These methods are defined as permissible velocity methods. Permissible velocity approaches were first developed in the early 1920s (Lacey, 1920, Lindley, 1919) and are related to the successful development of regime theory for large irrigation canals beginning in 1885 with Kennedy's formula. Regime theory continues to be a basic tool for evaluating large rivers and irrigation canals (Blench, 1969; Simons and Albertson, 1963); however, its empirical nature limits its application in the case of flood control channels.

In the 1950s, under the direction of Lane at the U.S. Bureau of Reclamation, research was conducted that refined the permissible tractive force method (Glover and Florey, 1951; Lane, 1955). This work and subsequent research clarified the actual physical processes occurring for flow in stable channels. This method extended the range of stable channel design beyond the empirical limits of regime theory. In most cases, a more realistic model of channel stability is based on permissible tractive force.

16.2.4.2 Tractive Force.   The flow resistance of water moving against a channel boundary results in shear force along the channel boundary referred to as the tractive force. As was discussed in Sec. 16.1.2, there is a close relationship between the tractive force and the flow velocity. In a uniform flow with a normal (semilog) velocity distribution, the tractive force is equal to the effective component of the gravitational force acting on the body of water parallel to the channel bottom (see derivation of Eq. 16.2).

However, shear stress is not uniformly distributed along the wetted perimeter in a channel, and Eq.16.2 describes only the average shear force on the channel. For design, it is necessary to determine the maximum shear along the channel perimeter so that the lining material is suitable for resisting the shear. At the same time, the stability of a soil particle decreases as the lateral slope of the channel increases. The relative combination of particle stability and boundary shear is needed to attain a stable channel. Simply stated, the principle of stable channel design using tractive force is to have the shear strength of the lining material τ_p, exceed the boundary shear force τ_b at every point in the channel's wetted perimeter:

$$\tau_p > \tau_b \tag{16.5}$$

The *permissible shear stress of the lining material* τ_p is the maximum shear stress that the lining can safely withstand. Using the critical shear stress τ_c for a soil particle as a reference value, the permissible shear stress can be defined as

$$\tau_p = C_a \, \tau_c, \tag{16.6}$$

where C_a is the critical shear-stress adjustment factor. The design shear stress is affected by a number of factors, including: the density of the soil particles, the submerged soil friction coefficient ($\mu = \tan \phi$, where ϕ is the angle of repose), the lateral (side slope) and down-slope components of soil particle weight, and the direction of flow.

The boundary shear stress τ_b varies within a channel section primarily as a function of depth and also in response to the lateral diffusion of momentum in the channel section. Using the mean shear stress τ_o as a reference value, the local boundary shear stress can be defined as

$$\tau_b = K_a \, \tau_o \tag{16.7}$$

where K_a is the boundary shear-stress adjustment factor. Channel alignment, the relative channel width and depth (aspect ratio), and the channel side slope affect the boundary shear stress.

Figure 16.2 shows a typical distribution in a parabolic channel. Boundary shear is greatest at the middle of the channel (maximum flow depth) and tends toward zero along the channel bank. The distribution shear of in a channel bend is shown in Fig. 16.3.

16.2.5 Design Parameters

Flood Frequency.   The most important question facing the designer of a flood-control channel is—What is the probability of failure? Because flood-control channels are built to protect valuable facilities, damage should occur only rarely. The probability of flood damage, although never zero, can be low enough to make the risk to the facility acceptable.

When flows are small and potential damages are small, such as in the case of a roadway drainage system, more frequent flooding is allowed. For larger flood-control channels, where flood damage is severe, a low frequency of flooding is desired. Although economic analysis can be used to determine the optimal range of flood frequency for a flood-control facility, only larger projects warrant such an effort. In most cases, standard flood frequencies are

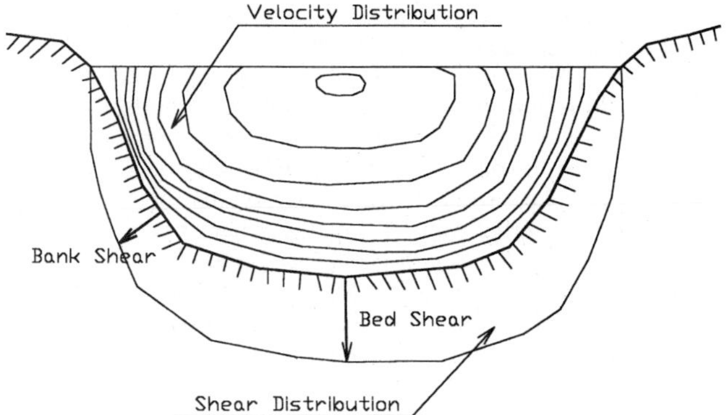

FIGURE 16.2   Boundary shear distribution in channel section.

used. Table 16.1 provides a brief summary of common flood frequencies and their use in the design of various types of facilities.

Duration of Flooding.   Small flood-control channels with intermittent flooding tend to have short times of flooding, typically on the order of several hours. Larger flood-control channels, or channels with sustained flows, require a design that considers the persistence of boundary stresses. Channels that sustain only short periods of boundary stress often can sustain higher stresses without damage.

Channel Profile.   Major project features, such as roads generally dictate the slope of smaller channels. If channel stability cannot be maintained for these conditions, it may be feasible to reduce the channel gradient slightly relative to other grading. This typically results in short, steep reaches that are constructed as rigid chutes or low drops.

Channel slope is among the major parameters affecting boundary shear stress. Thus at a given discharge, the shear stress is less for a mild (subcritical regime) gradient compared with a steep (supercritical regime) gradient. It is not uncommon for smaller flood-control channels to operate mainly in supercritical regime. However, extremely steep channels (above 10 percent) and larger channels require additional design considerations.

Channel Section.   The most common shapes of drainage channel are trapezoidal or triangular. These shapes are a close approximation of the most efficient hydraulic section, a semicircle. For smaller channels with flexible linings, these shapes tend to be constructed with slightly rounded corners, which gives the as-built section a slightly parabolic shape.

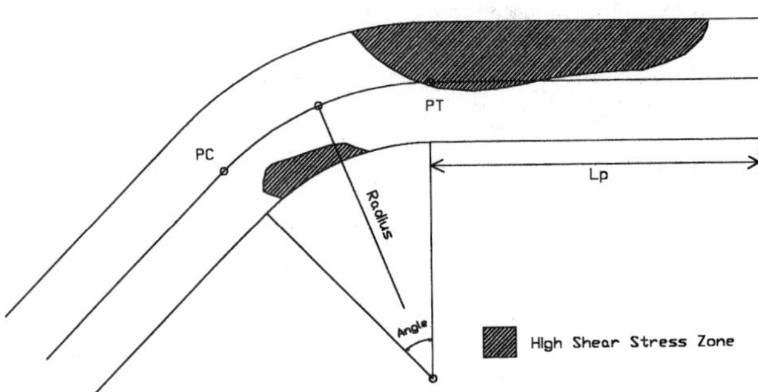

FIGURE 16.3   Boundary shear distribution in a bend.

TABLE 16.1 Commonly Used Flood Frequencies

Type of facility	Potential flood damage	Flood frequency (return period, years)
Collector road Roadside or median ditch	Delay costs Impeded traffic	2 to 5 years
Urban collector streets	Delay costs Impeded emergency access Detour costs	10 to 25 years
Rural flood control	Crop damage Road damage	25 to 50 years
Urban flood control	Property damage Infrastructure damage	100 years

The banks are the least stable portions of the channel section. Bank stability increases as the side slope flattens. For smaller channels, a 3:1 side slope is sufficient to prevent erosion in excess of the channel bed. For larger channels, or for small channels that require steep side slopes, the stability of the side slope should be checked.

In larger flood-control channels with persistent low flows, a low-flow channel often is included within the channel section. It is advisable to situate the low-flow channel near the center of the main channel to avoid creating increased shear near the channel bank. Low-flow channels are often designed to meander within the larger section. The meander reduces the gradient of the channel, allowing the enhancement of the channel section for habitat or appearance. The potential effect at meander bends on the main channel bank line should be considered.

Although a channel section can be widened to reduce boundary shear, channels with large width depth to ratios typically are not stable. Over time, a high stress area will develop within the section and cause erosion that will form a narrower, deeper channel, the channel eventually fails.

16.3 CHANNEL LININGS

Considerable research and development has produced a variety of rigid and flexible lining materials. Before the late 1960s, channel linings were constructed primarily of natural materials, such as rock riprap, stone masonry, concrete and vegetation. Material manufactured or fabricated into rolls offered several advantages for erosion control, particularly during construction and during the establishment of vegetation. Use of rolled erosion-control products for temporary erosion control assured improved long-term performance of vegetative linings.

16.3.1 Lining Types

Because of the large number of channel stabilization materials currently available, it is useful to classify materials according to their performance characteristics. Lining types are classified as either rigid or flexible. Flexible linings are grouped further as either permanent or temporary.

Rigid linings
   Cast-in-place concrete
   Cast-in-place asphaltic concrete
   Stone masonry
   Soil cement
   Fabric formed concrete
   Grouted riprap

Flexible linings (Long-term nondegradable)
   Riprap
   Wire-enclosed stone
   Vegetation
   Gravel
   Synthetic mat

Flexible linings (temporary nondegradable)
   Straw with net
   Curled wood mat
   Jute net
   Woven paper net
   Fiberglass roving

16.3.2 Performance Data

16.3.2.1 *Rigid Linings.* Rigid linings are useful in applications where high shear-stress or nonuniform flow conditions exist, such as at transitions in channel shape or at an energy dissipation structure. In areas where loss of water or seepage from the channel is undesirable, a rigid lining can provide an impermeable barrier. Because rigid linings are nonerodible, the designer can use any channel shape that provides adequate conveyance. Rigid linings may be the best option if right-of-way limitations restrict the channel size.

Despite the nonerodible nature of rigid linings, they are highly susceptible to failure from structural instability. For example, cast-in-place or masonry linings often break up and deteriorate if the foundation is poor. Once a rigid lining deteriorates, it is highly susceptible to erosion because it forms large, flat, broken slabs that are easily moved by flow.

The major causes of structural instability and failure of rigid linings are freeze-thaw, soil swelling, and excessive soil pore-water pressure. Freeze-thaw and swelling soils exert upward forces against the lining, and the cyclic nature of these conditions eventually causes failure. Excessive soil-pore pressure may occur when the flow levels in the channel drop quickly but the soil behind the lining remains saturated. This can result in instability of the bank's slope caused by the high water-table gradients within the channel bank.

Construction of rigid linings requires specialized equipment using relatively costly materials. As a result, the cost of rigid channel linings is high. Prefabricated linings can be a less expensive alternative if shipping distances are not excessive.

There often are significant environmental issues associated with rigid channel linings. In environmentally sensitive areas, replacement of concrete linings with articulated block mats is a partial solution, because openings between the blocks allows vegetation to take hold.

16.3.2.2 *Flexible Linings.* Riprap and gabion are suitable linings for hydraulic conditions similar to those requiring rigid linings. Because flexible linings are permeable, they may require protection of the underlying soil to prevent washout. For example, filter cloth is often used with riprap to inhibit soil piping.

Vegetative and temporary linings are suited to hydraulic conditions where uniform flow exists and shear stresses are moderate. Most grass linings cannot survive sustained flow conditions or long periods of submergence. Grass-lined channels with sustained low flow and intermittent high flows often are designed with a composite lining of a riprap or concrete low-flow section.

The primary use of temporary linings is to provide protection from erosion until vegetation is established. In most cases, the lining will deteriorate over the period of one growing season, which means that successful revegetation is essential to the overall channel stabilization effort. Temporary channel linings can be used without vegetation to control erosion on construction sites temporarily.

16.3.3 Information About Flexible Linings

The erosion control industry has grown in response to continued infrastructure development and increased awareness of water-quality problems. Traditional methods of erosion control are well documented, and a variety of specifications exist for these types of linings. In 1997, the Erosion Control Technology Council (ECTC) established standard terminology and testing methods for the variety of manufactured products used for erosion control. The ECTC terminology for rolled erosion control products (RECPs) is used here for clarity.

16.3.3.1 *Long-term, Nondegradable Flexible Linings.* *Vegetation.* Vegetative linings consist of planted or sodded grasses placed in and along the drainage (Fig. 16.4). If planted, grasses are seeded and fertilized according to the requirements for the particular variety or mixture and to soil conditions. Sod is laid parallel to the direction of flow and can be secured with staples or stakes.

Rock Riprap. Rock riprap is dumped or hand placed on prepared ground with a filter blanket or a prepared bedding material interface (Fig. 16.5). The stone layer is placed to form a well-graded mass with a minimum of voids. Stones should be hard, durable, preferably angular in shape, and free from overburden, shale, and organic material. Resistance to disintegration from channel erosion should be determined from service records or from specified field and laboratory tests.

Wire-enclosed Stone. Wire-enclosed stone consists of a wire basket or tube filled with stone (Fig. 16.6). The wire basket is made of steel wire woven in a uniform pattern and reinforced on corners and edges with heavier wire. Common forms of wire-enclosed stone include boxlike baskets, thin mattresses, and tubes. The containers are filled with stone, connected together, and anchored to the channel side slope. Stones are graded fairly uni-

FIGURE 16.4 Vegetative channel lining.

FIGURE 16.5   Rock channel lining.

formly, with the smallest size larger than the wire mesh opening. The stones should be hard, durable, and free from overburden, shale, and organic material. Wire-enclosed stone typically is used when rock riprap is either not available or not large enough to be stable

   Gravel.   Gravel consists of coarse gravel or crushed stone placed on filter fabric or pre-pared bedding material to form a well-graded lining with a minimum of voids (Fig. 16.7). The stones should be hard, durable, and free from overburden, shale, and organic material.

FIGURE 16.6   Wire-enclosed stone channel lining.

FIGURE 16.7   Gravel channel lining.

Turf Reinforcement Mat.   Turf reinforcement mat is composed of ultraviolet-stabilized, nondegradable, synthetic fibers, netting, or filaments processed into a three-dimensional reinforcement matrix ranging in thickness from 6 to 20 mm (Fig. 16.8). The mat provides sufficient thickness, strength, and void space to permit soil filling and development of vegetation within the matrix. The mat is laid parallel to the direction of flow is stapled or staked to the channel surface, and is anchored into cutoff trenches at regular intervals along the channel. The mat is top-dressed with fine soil at a depth equal to the mat's thickness and is seeded and fertilized. Turf reinforcement occurs as root growth penetrates and entangles with the mat.

16.3.3.2   *Temporary Degradable Flexible Linings.*   Woven paper net.

Woven paper net is erosion-control net that consists of knotted plastic netting interwoven with paper strips (Fig. 16.9). The net is applied evenly on the channel slopes, with the fabric running parallel to the channel's direction of flow. The net is stapled to the ground and placed into cutoff trenches at regular intervals along the channel. Woven paper net is usually installed immediately after seeding operations.

Jute Net.   Jute net is an erosion control mat that consists of jute yarn approximately 6 mm in diameter that is woven into a net with openings approximately 10 to 20 mm (Fig. 16.10). The jute net is loosely laid in the channel parallel to the direction of flow. The net is secured with staples and is placed int cutoff trenches at regular intervals along the channel. Jute net is usually installed immediately after seeding operations.

Curled Wood Mat.   Curled wood mat is an erosion control blanket that consists of wood fibers, 80 percent of which are 150 mm or longer, with a consistent thickness and an even distribution of fiber over the entire mat (Fig. 16.11). The topside of the mat is covered with biodegradable plastic mesh. The mat is placed in the channel parallel to the direction of flow and is secured with staples and cutoff trenches.

Straw with Net.   Straw with net can be manufactured as an erosion control blanket or can be constructed from straw mulch secured with erosion control net. The constructed version consists of plastic mesh with 20-mm^2 square openings overlaying straw mulch (Fig. 16.12). Straw is spread uniformly over the channel surface at a rate of approximately 4.5

FIGURE 16.8   Turf-reinforcement mat channel lining.

metric tons/hectare and can be crimped into the soil. Plastic mesh is placed over the straw and stapled to the soil. The manufactured version consists of straw, either sewn to a layer of erosion control net or sandwiched between to two layers of erosion control net. Straw weight is approximately 2.5 metric tons/hectare. The mat is placed parallel to the direction of flow and is secured with staples and cutoff trenches.

Fiberglass Roving.   Fiberglass roving consists of continuous fibers drawn from molten glass that is coated and lightly bound together into roving (Fig. 16.13). The roving is ejected by compressed air, forming a random mat of continuos glass fibers. The material is spread uniformly over the channel and anchored with asphaltic materials.

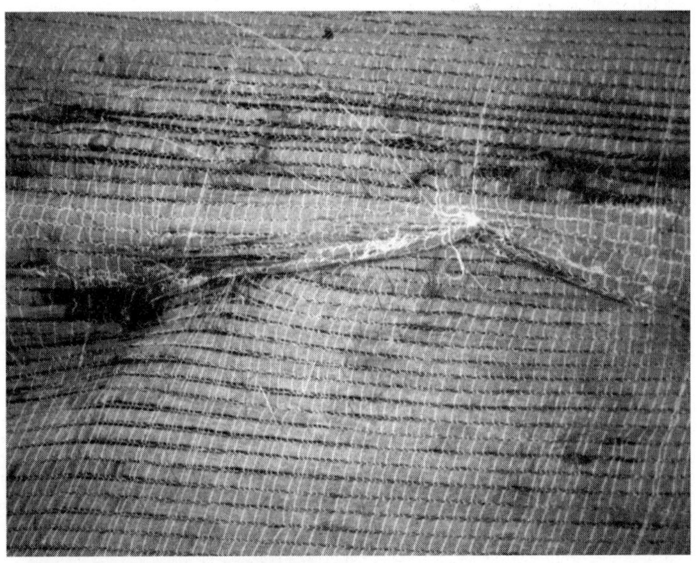

FIGURE 16.9 Woven-paper net channel lining.

16.4 MILD-GRADIENT CHANNEL DESIGN (SYMMETRIC SECTION)

This section outlines a method of stable channel design for channel gradients of less than 10 percent. Most of the lining types presented will perform acceptably up to grades of about 4 percent, whereas fewer are effective as the grade approaches 10 percent. On steeper grades, stone lining (riprap and wire-tied rock) is the most common lining type.

FIGURE 16.10   Jute net channel lining.

16.4.1 Resistance to Flow

16.4.1.1 *Rigid and Flexible Lining Materials.* Channel roughness is affected by the relative height of the roughness compared to the flow depth. As a result, channel roughness increases for shallow flow depths and decreases as flow-depth increases. For flow depths between 0.15 and 0.60 m (a typical range for small drainage channels), roughness is higher compared to deeper flows. Recommended values of Manning's roughness coefficients are summarized in Table 16.2. A more detailed table is presented in Chapter 3, (Appendix 3.B) These values are means of the range; the maximum and minimum of the range are about ±10% of the mean value.

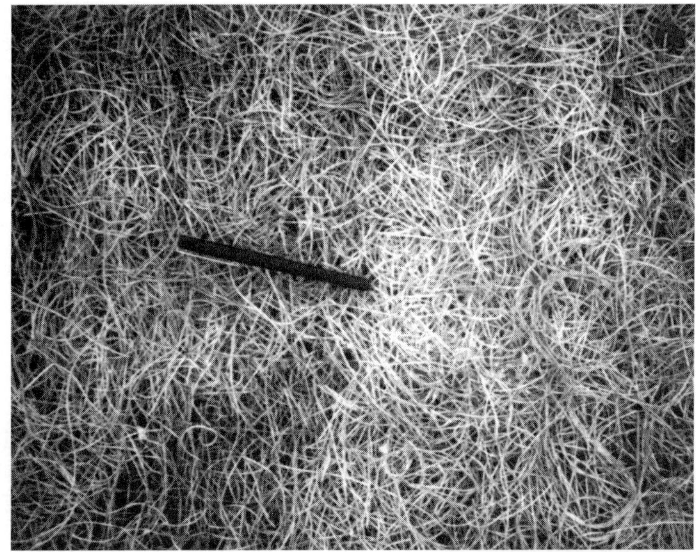

FIGURE 16.11 Curled-wood mat channel lining.

16.4.1.2 *Vegetative Linings*

For vegetative linings, the channel roughness varies significantly, depending on the amount of submergence of the vegetation. Because vegetation is flexible, the amount of submergence will increase as the drag force bends the plant stems toward the channel bed. For shallow flow, plants are erect and may protrude from the flow. As flow depth and velocity increase, plants bend and are submerged by the flow. At higher velocities, the plants are bent flat to the channel bed. Kouwen's method (1990) provides a practical means of determining Manning's roughness value for a wide range of flow conditions and vegetation types.

FIGURE 16.12   Straw with net channel lining.

The mechanical properties of individual plants (height and stiffness) and the density of plant cover affect flow resistance. Kouwen and Li (1980) found that the *effective roughness height k_s* of vegetation varies as a function of shear stress exerted on the lining and density-stiffness properties:

$$k_s = 0.14\, h \left[\frac{\left[\dfrac{\text{MEI}}{\tau_o} \right]^{0.25}}{h} \right]^{1.59} \tag{16.8}$$

Values of the vegetation height h and the *density stiffness parameter* MEI that are equivalent to the five standard retardance classifications (Class A representing vegetation types with the highest flow resistance and Class E, with the lowest flow resistance) are listed below:

Retardance Class	Average Height, h, mm	Stiffness, MEI, $N\ m^2$
A	900	300.0
B	600	20.0
C	200	0.5
D	100	0.05
E	40	0.005

With grasses, there is a good correlation between the length of vegetation and the density-stiffness parameter MEI. Empirical relationships (Kouwen, 1988) are useful in the field estimation and determination of seasonal effects (note that h is in meters):

FIGURE 16.13   Fiberglass roving channel lining.

$$\text{Green grass: } MEI = 319\ h^{3.30} \tag{16.9}$$

$$\text{Dormant grass: } MEI = 24.5\ h^{2.26} \tag{16.10}$$

The Manning's roughness coefficient for vegetation is a function of the relative roughness (ratio of flow depth to roughness height). The variation of Manning's roughness with relative roughness follows the well-known semilog relationship:

TABLE 16.2 Mannings Roughness Coefficients–Mild Gradients (<10%)

Lining category	Lining type	k_s mm	Depth ranges		
			0.0–0.15 m	0.15–0.60 m	>0.60 m
Rigid	Concrete	– – –	0.015	0.013	0.013
	Grouted riprap	– – –	0.040	0.030	0.030
	Stone masonry	– – –	0.042	0.032	0.030
	Soil cement	– – –	0.025	0.022	0.020
	Asphalt	– – –	0.018	0.016	0.016
Unlined	Bare soil (compacted)	– – –	0.023	0.020	0.020
	Rock cut	– – –	0.045	0.035	0.025
Flexible Linings	Woven paper net	1.2	0.016	0.015	0.015
Temporary	Jute net	11.6	0.028	0.022	0.019
(degradable)	Straw with net	36.6	0.065	0.033	0.025
	Curled wood mat	33.5	0.066	0.035	0.028
	Fiberglass roving	10.7	0.028	0.021	0.019
Long-term	Synthetic mat (unvegetated)	19.8	0.036	0.025	0.021
Gravel	25 mm D_{50}	25.0	0.044	0.033	0.030
	50 mm D_{50}	50.0	0.066	0.041	0.034
Large stone	150 mm D_{50}	150.	0.104	0.069	0.035
	300 mm D_{50}	300.	– – –	0.078	0.040

$$n = \frac{R^{1/6}}{\sqrt{g}\,(a + b\,\log\,(R/k_s))} \tag{16.11}$$

The a and b coefficients are based on a classification of the three types of flow conditions with vegetation: erect, submerged (bent), and flattened. The initial shear stress that bends the vegetation from an erect position is referred to as the *vegetative critical shear stress*.

$$\tau_{cv} = \text{minimum of } (0.78 + 354\,MEI^2 + 40100\,MEI^4,\ 53\,MEI^{0.212}) \tag{16.12}$$

As a practical matter, the first term of Eq. (16.12) controls for vegetative stiffness of less than 0.16 N-m² (D retardance). Stiffer vegetation (A, B, and C retardances) has critical shear stresses governed by the second term of Eq. (16.12). The minimum vegetative critical shear stress is about 0.80 N/m², which occurs for stiffness values lower than approximately 0.010 N–m² (E retardance).

The values of a and b as a function of square root of the ratio of the mean boundary shear stress to the vegetative critical shear stress are listed in Table 16.3. For design, it is acceptable to assume that the grass would be bent completely flat. As long as the underlying soil and plant roots are not eroded, damage to the plant stems is not a critical factor in the lining's performance.

To simplify the estimation of flow resistance for vegetated linings, HEC-15 (Chen and Cotton, 1988) uses simplified formulas for Class A through E vegetation, as shown below:

TABLE 16.3   Values of "a" and "b"

Classification	Criteria	"a"	Parameter "b"
Erect	$\sqrt{\dfrac{\tau_o}{\tau_{cv}}} \leq 1.0$	0.42	5.23
Submerged (Bent)	$1.0 < \sqrt{\dfrac{\tau_o}{\tau_{cv}}} \leq 2.5$		Linearly interpolate between erect and prone parameter values
Flat	$2.5 < \sqrt{\dfrac{\tau_o}{\tau_{cv}}}$	0.82	9.90

Source: (Kowen and Li, 1980)

$$n_A = \frac{R^{1/6}}{15.8 + 19.97\,\log(R^{1.4}S^{0.4})} \qquad n_B = \frac{R^{1/6}}{23.0 + 19.97\,\log(R^{1.4}S^{0.4})}$$

$$n_C = \frac{R^{1/6}}{30.2 + 19.97\,\log(R^{1.4}S^{0.4})} \qquad n_D = \frac{R^{1/6}}{34.6 + 19.97\,\log(R^{1.4}S^{0.4})}$$

$$n_E = \frac{R^{1/6}}{37.7 + 19.97\,\log(R^{1.4}S^{0.4})}$$

16.4.1.3 *Flexible Linings*

Flexible linings have more variation in resistance compared with rigid linings (see Table 16.2). This is particularly true of linings with a large relative roughness, such as mats. HEC-15 (Chen and Cotton, 1988) gives the coefficients in Table 16.4 for estimation of Manning's n using Eq. (16.11).

Example 16.1   Flow Resistance
The following example illustrate the variation in Manning's n for increasing shear stress on a grass-lined channel.determine Manning's n for a range of hydraulic radius 0.1 to 0.6.

TABLE 16.4   Flexible Lining Coefficients for the Resistance Equation (k_s in meters)

	Woven paper net	Jute net	Straw with net	Curl wood mat	Fiberglass roving	Synthetic mat (unveg.)
a	0.73	0.74	0.72	0.65	0.73	0.96
b	8.00	8.04	7.83	7.10	8.00	8.13
k_s	0.0012	0.0116	0.0366	0.0335	0.0107	0.0198

Inputs: S_o = 0.015 m/m
 R = 0.1 to 0.6
 h = 200
Condition: Green grass

Outputs: a. Estimate stem stiffness using Eq. (16.9):

$$MEI = 1.57 \quad Nm^2.$$

b. Compute relative roughness height using Eq. (16.8):

R	τ_0 N/m^2	k_s/h	k_s m
0.1	14.7	0.74	0.149
0.2	29.4	0.57	0.113
0.3	44.1	0.48	0.096
0.4	58.8	0.43	0.086
0.5	73.5	0.39	0.079
0.6	88.2	0.37	0.073

c. Compute vegetative critical shear stress using Eq. (16.12):

$$\tau_{cv} = 58.4 \quad N/m^2$$

d. Compute semilog coefficients (Table 16.3):

τ_0/τ_{cv}	a	b
0.252	0.42	5.23
0.504	0.42	5.23
0.756	0.42	5.23
1.008	0.42	5.25
1.260	0.49	6.04
1.511	0.56	6.82

e. Compute Manning's n Eq. (16.11)

R	R/k_s	n
0.1	1.00	0.518
0.2	1.77	0.142
0.3	3.12	0.087
0.4	4.66	0.070
0.5	6.37	0.053
0.6	8.21	0.043

16.4.2 Tractive Force Design

To achieve a stable channel lining, an anchoring force must counter the hydraulic force that acts to detach the lining. The unit force acting to detach the lining is referred to as the *tractive force,* or *boundary shear stress* τ_b. The unit force acting to prevent movement of the lining is the component of the anchoring force perpendicular to the slope and is referred to as the *permissible shear stress* τ_p. To provide a stable lining, the permissible shear sress must exceed the boundary shear stress:

$$\tau_p > \tau_b \tag{16.5}$$

16.4.3 Permissible Shear Stress

The *critical shear stress* τ_c is the force required to initiate movement of the lining material or the underlying soil. In most cases, before movement of the lining, the underlying soil is relatively protected from erosion. However, for vegetative linings and other types of flexible lining, soil movement may occur before movement of the lining. Once the lining material is lost, the underlying soil is exposed to the full erosive force of the flow. The consequence of lining failure on highly erodible soils is great because the erosion rate after failure is high compared with the rate for soils of low erodibility.

Values for critical shear stress for soils and lining materials are based on research conducted in laboratory flumes and in the field. The values presented here are judged to be conservative and appropriate for design use. The critical shear stress for noncohesive soils is a function of the mean diameter of the particle gradation, as shown in Fig. 16.14. For larger stones, the critical shear stress (assuming a stone density of 2.65 and a Shields parameter of 0.047) can be determined by the following equation:

$$\tau_c = 0.76 \, D_{50} \tag{16.13}$$

For cohesive material, the variation in critical shear stress depends on the concentration of the clay particles within the soil. The plasticity index of cohesive soil provides a simple guide to the permissible shear stress (Fig. 16.15). Table 16.5 presents the Erosion Control Technology Council (ECTC, 1997) guidelines for maximum shear stress values for various rolled erosion control products.

Since the above guidelines present a large range and are not product specific, it is important for product information to be carefully reviewed. In general, the upper end of the ECTC range is based on linings with good vegetative cover, short-duration flow, and excellent installation. The lower end of the ECTC range is typical of linings with little or no vegetation.

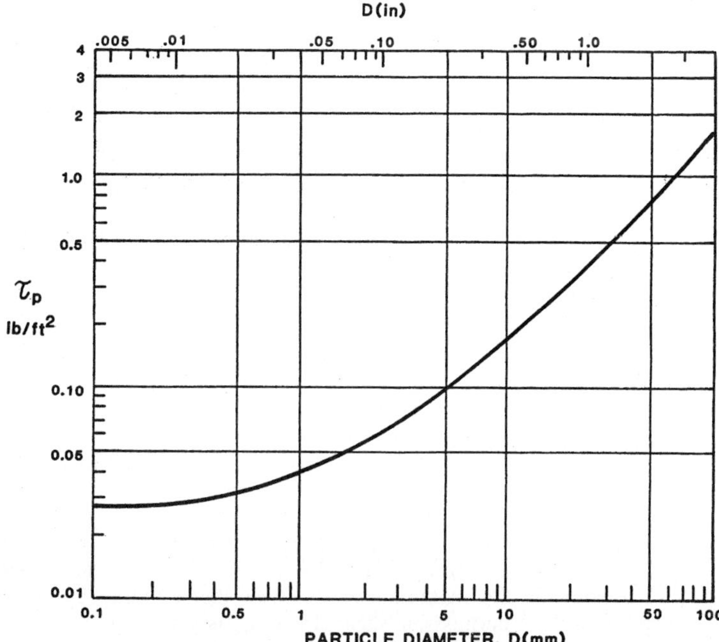

FIGURE 16.14 Permissible shear stress for non-cohesive soils. (From Lane, 1995)

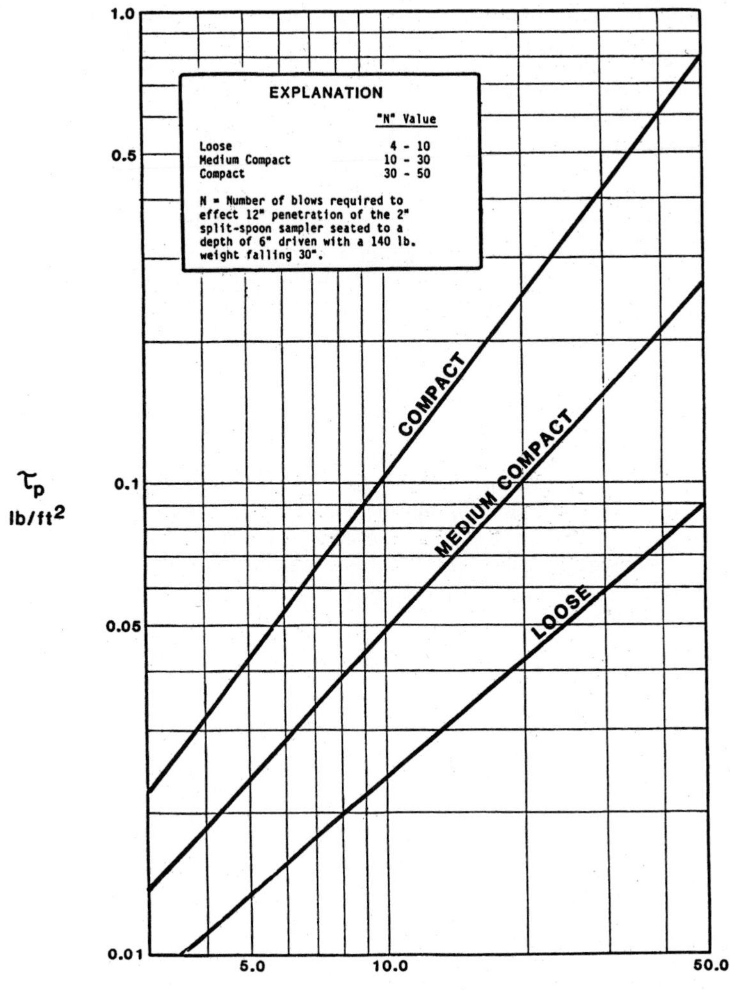

FIGURE 16.15   Permissible shear stress for cohesive soils. (From Smerdon and Beaseley, 1959)

In arid climates or semiarid climates, it often is difficult to achieve good vegetative cover in a channel. A lining with a good vegetative cover density in excess of 70 percent is typically required to resist shear stresses in excess of approximately 100 N/m^2.

The critical shear stress is the maximum shear force that the lining system can sustain. For thin lining systems, such as rolled erosion control products, the maximum shear force is typically controlled by the erodibility of the underlying soil. Often the shear strength of a manufactured product will exceed the erodibility of the underlying soil.

For noncohesive soils, several factors reduce the critical shear stress. The *permissible shear stress* is the maximum shear stress that a lining can safely withstand given local conditions. The permissible shear stress, τ_p, is given by:

TABLE 16.5 Permissible Shear Stresses for Lining Materials

Lining category	Lining type	Permissible shear (N/m^2)
Temporary degradable	Erosion control net	4.5 to 9.5
	Erosion control mat	20 to 140
	Erosion control blanket	
	Single net	70 to 95
	Double net	95 to 140
Long-term nondegradable	Turf reinforcement mat	
	Unvegetated	140 to 280
	Vegetated	240 to 380

Source: ECTC (1997)

$$\tau_p = C_a\, \tau_c \tag{16.6}$$

Factors that are known to affect the permissible shear stress include the density of the soil particles, the angle of repose, the channel side slope, the flow angle, and the required factor of safety. On channel side slopes, the soil particle has a reduced weight component with which to anchor the particle and resist detachment. The angle of repose and the required safety factor also modify the allowable shear stress. *The critical shear adjustment factor* is defined as follows (see appendix for complete derivation):

$$C_a = \left(\frac{1}{SF}\right) Cl\ C_2 Cw\ (1 - Cr\ CB\ SF) \tag{16.14}$$

where $Cz = \cos(Az)$
$\quad Cw = (\rho_s - 1)/1.65$
$\quad Cr = \tan(Az)/\tan(Ar)$
$\quad Cl = 2([1 + \sin(Al + Ab)]$
$\quad Cb = \cos(Ab)$
$\quad Az = \tan^{-1}(1/z)$
$\quad Ab = \tan^{-1}\{\cos[Al]/[E + \sin(Al)]\}$
$\quad E = 2\ (\tau_c/\tau_b)\ Cz\ Cr$

where Az = angle of the channel side slope, Ar = angle of repose of the soil, Ab = angle of particle movement, Al = oblique flow angle, SF = design safety factor, ρ_s = soil particle density, τ_c = critical shear, and τ_b = local boundary shear.

The factor Cz accounts for the effect of the channel side slope on particle stability. As is summarized below, the factor Cz decreases, as the side slope becomes steeper. The value Z is defined as the ratio of the side slope's horizontal to vertical distance.

Z	0.5	1.0	1.5	2.0	3.0	4.0	5.0	horizontal
Cz	0.45	0.71	0.83	0.89	0.95	0.97	0.98	1.00

The factor Cw is an adjustment for the effect of stone density on particle stability. The density of a sediment particle is a function of the mineral composition. The specific gravity of sediments ranges from 2.3 for coal to 7.5 for galena. Alluvial sediments consist mainly of common minerals of quartz and feldspar with a specific gravity between 2.6 and 2.8. The default stone density for computation of critical shear stress is 2.65. The variation of factor Cw with specific gravity, ρ_s.

ρ_s	2.3	2.5	2.65	2.8	3.0	4.0
Cw	0.79	0.91	1.00	1.09	1.21	1.82

The factor Cr accounts for the effect of the size and shape of soil particles on a slope. The flatter the channel side slope, or the larger and more angular the particle size, the more stable the lining. The size and angularity of the stone effect the slope at which particles will stabilize (the angle of repose). Angle of repose for noncohesive sediments and stones ranges from a low of 30° to a maximum of about 42.5° for large crushed rock (see Fig. 16.14). Table 16.6 provides values of Cr for a range of side slopes and angle of repose conditions.

The factors Cl and Cb account for the hydraulic forces resulting from flow past soil layer particles. Because the direction of flow at a particle need not be parallel to the stream bank, the angle at which forces resolve must be calculated before Cl and Cb can be determined. The angle Al is the oblique flow angle. Oblique flow typically occurs in contraction and expansion reaches, such as at bridge or culvert openings.

The angle Ab is the direction at which the combined forces of lift and drag on the particle and the down-slope weight of the particle push a particle. The angle depends on the angle of the side slope, the particle's size and shape, and the ratio of the local boundary shear stress to the critical shear stress. For a low boundary shear stress, the direction of particle movement tends to follow the bank slope, whereas at higher shear, the direction of the particle movement tends to be more in the direction, of flow.

Tables 16.7 and 16.8 summarize the values of Cl and Cb coefficients for a range of shear and bank slope conditions without oblique flow ($Al = 0°$). In general, values of Cl and Cb decrease with decreasing side slope and increasing shear stress. Cl approaches a value of 1.0, and Cb approaches zero.

For cohesive soils, particle stability is affected by the same factors that effect non cohesive soils, but chemical forces tend to hold the soil together as a larger mass. For well-compacted homogeneous soils forming small channels, the critical shear adjustment factor is equal to

TABLE 16.6   Values of Factor Cr

	Very rounded		Crushed
Side slope, Z	Small $Ar = 36°$	Large $Ar = 39°$	Large $Ar = 42°$
0.5	2.75	2.47	2.18
1.0	1.38	1.23	1.09
1.5	0.92	0.82	0.73
2.0	0.69	0.62	0.55
3.0	0.46	0.41	0.36
4.0	0.34	0.31	0.27
5.0	0.28	0.25	0.22
Horizontal	0.00	0.00	0.00

TABLE 16.7 Values of Factor Cl (assumes $Ar = 39°$)

Side Slope, Z	τ_o/τ_c		
	1/2	2/3	l
0.5	1.640	1.554	1.418
1.0	1.568	1.474	1.336
1.5	1.487	1.389	1.256
2.0	1.416	1.318	1.197
3.0	1.299	1.212	1.118
4.0	1.220	1.148	1.077
5.0	1.167	1.108	1.054
horizontal	1.000	1.000	1.000

the inverse of the safety factor ($1/SF$). For larger channels, the various components of Eq. (16.11) should be considered in detail.

16.4.4 Boundary Shear Stress

The *mean boundary shear stress* is the tractive force acting parallel to the energy gradient for the flow per unit of wetted area:

$$\tau_o = \gamma R S_f \tag{16.2}$$

The *average boundary shear stress* also can be calculated from the Manning's Eq. (16.4):

$$\tau_o = \left\{\frac{n}{R^{1/6}}\right\}^2 \gamma V^2 \tag{16.15}$$

The *local boundary shear stress* varies within a river reach as a consequence of the nonuniform distribution of velocity in the cross section or because of other channel features. For design, it is important to assess the maximum shear stress that can occur at specific

TABLE 16.8 Values of Factor Cb (assumes $Ar = 39°$)

Side slope, Z		τ_o/τ_c	
	1/2	2/3	l
0.5	0.976	0.958	0.912
1.0	0.961	0.934	0.868
1.5	0.939	0.898	0.806
2.0	0.911	0.856	0.741
3.0	0.842	0.760	0.615
4.0	0.769	0.670	0.515
5.0	0.700	0.592	0.440
horizontal	0.000	0.000	0.000

locations in the reach. The local boundary shear stress is determined by multiplying the mean boundary shear stress by an associated adjustment factor that accounts for local hydraulic conditions. The maximum boundary shear stress is then calculated using Eq. (16.7).

For a simple channel section on a mild slope, there are four basic types of local boundary shear-stress adjustments, which are summarized as follows:

K_a	Description
K_{bs}	Channel bed in a straight, symmetric reach
K_{ss}	Channel bank in a straight, symmetric reach
K_f	Flexible lining
K_r	Channel bed or bank in a bendway, symmetric reach

Channel-bed Adjustment Factor in a Straight, Symmetric Reach. In a straight reach of channel, the shear stress on the channel bed varies across the channel section. The maximum boundary shear stress is a function of the bank slope and the cross-sectional aspect ratio B/R (the ratio of the channel bottom width to the hydraulic radius). Figure 16.17 provides the boundary shear-stress adjustment factor for the channel bed for a trapezoidal section. (*Note:* For aspect ratios greater than 5, the average flow depth can approximate the hydraulic radius.)

Channel-bank Adjustment Factor in a Straight, Symmetric Reach. The maximum boundary shear stress on the bank of a channel also is a function of the bank slope and the cross-sectional aspect ratio B/R. Figure 16.16 provides the boundary shear-stress adjustment factor for the channel bank for a trapezoidal section.

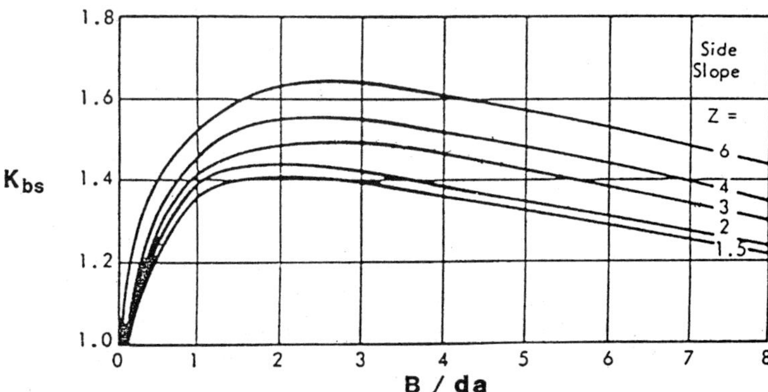

$$K_{bs} = \frac{\dfrac{B}{da} + 2\sqrt{Z^2 + 1}}{\dfrac{B}{da} + Z}$$

$$\text{for} \quad \frac{B}{da} > 8$$

FIGURE 16.16 Boundary shear stress adjustment factor for channel bed. (From Anderson, et al. 1970)

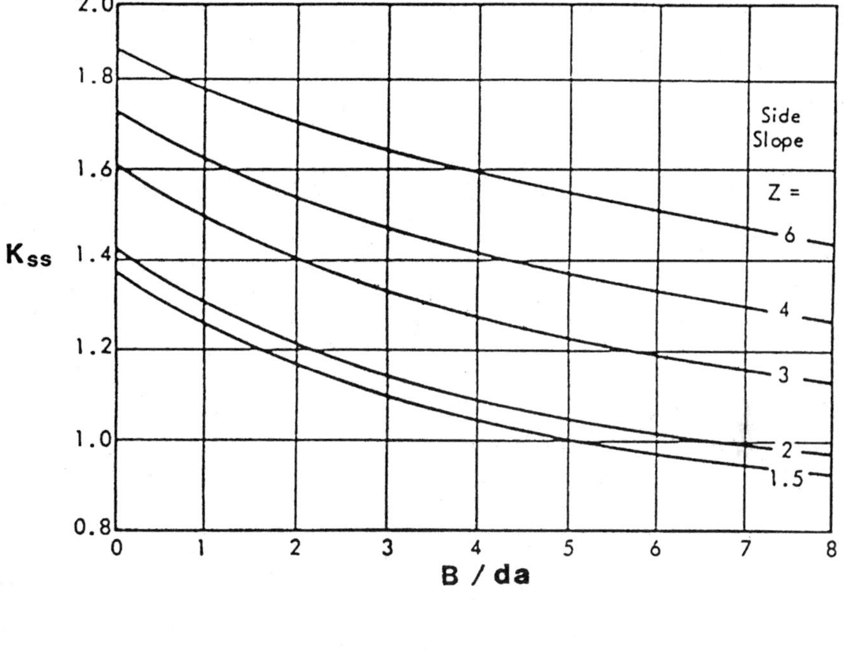

$$K_{ss} = C \left[\frac{\dfrac{B}{da} + 2\sqrt{Z^2 + 1}}{\dfrac{B}{da} + Z} \right]$$

$$\text{for} \quad \frac{B}{da} > 8$$

Where

Z	C
1.5	0.76
2	0.785
3	0.85
4	0.935
6	0.97

FIGURE 16.17   Boundary shear stress adjustment factor for channel side slope. (From Anderson, et al., 1970)

Flexible Lining Adjustment Factor.   Flexible linings tend to deflect shear away from the channel surface. This effect creates a condition where part of the shear is absorbed by the movement of the lining material, resulting in a lower shear stress on the soil surface. This is the primary mechanism by which-flexible lining reduces erosion in a channel. The lining should be anchored well to the soil to provide the shear reduction either by mechanical means in the case of synthetic materials or in the case of well-established roots in the case of vegetation.

Temple (1980) developed a method to estimate the shear stress acting on the soil for a vegetative channel lining. The adjustment factor is

$$K_f = (1 - C_F) \left(\frac{n_s}{n_1} \right) \tag{16.16}$$

Temple (1980) estimated C_F values for a range of vegetative densities and grass covers. C_F values for untested covers can be estimated by recognizing that density and uniformity of cover near the soil surface dominate the cover factor. Thus, the sod-forming grasses (top of Table 16.9) exhibit higher C_F values than do the bunch grasses and annuals (near the bottom of Table 16.9). For large-stemmed plants, such as forbs and shrubs, the effective stem density should be increased. In the case of the legumes, the effective stem density is about five times the actual stem density. For turf-reinforcing mats, increasing the stem density by 4/3 is recommened.

Bend Adjustment Factor. A channel bend causes an increase in local boundary shear on the bank of a channel from the beginning of the bendway, through the bend for a significant distance downstream of the bend. The boundary shear-stress adjustment factor for the channel bank in the bend is a function of the radius of the curvature of the bend and the width of flow:

$$K_r = \frac{3.16}{\sqrt{\dfrac{Rc}{T}}} R_o/T < 10.0 \tag{16.17}$$

and
$$K_r = 1.0 \qquad R_o/T > 10.0.$$

The extent of the downstream effect of a bend depends on the channel roughness in the bend, and on the average depth of flow, as described by the hydraulic radius.

$$L_P = 0.604 \left(\frac{R^{1/6}}{n_b} \right) R \tag{16.18}$$

Example 16.2 Channel Lining Evaluation
This example illustrates the hydraulic analysis of a grass-lined channel. A rating is performed in order to illustrate the effect of increasing depths of flow on channel performance.

TABLE 16.9 Properties of Grass Channel Linings (good, uniform stands of each cover) (after Temple, 1987)

Cover Factor (C_F)	Grass Cover	Growth form*	Stem Density M (stems/m²)[f]
0.90	Bermudagrass	S	5380
	Centipedegrass	S	5380
0.87	Buffalograss	S	4300
	Kentucky bluegrass	S	3770
	Blue grama	S	3770
0.75	Grass mixture	B-S	2150
0.50	Weeping lovegrass	B	3770
	Yellow bluestem	F	2690
	Alfalfa†	F	5380
	Lespedeza sericea†	B	3230
	Common lespedeza	B	1620
	Sudangrass	B	540

[f] Multiply the stem densities given by 1/3, 2/3, 1, 4/3, and 5/3 for poor, fair, good, very good, and excellent covers, respectively.
† Legumes, large-stemmed.
*B = Bunchgrass; S = sod-former; F = forb.

Inputs: Channel attributes

$S_o = 0.020$ m/m   Grade

  Straight   Alignment
Compact, Cohesive Soil PI $= 20$ Soil type
$B = 1.5$ m   Bottom width
$z = 3:1$ Side slope
Lining attributes
   Green grass   Condition
$h = 0.250$ m   Grass height
   Cover factor $= 0.75$

Outputs: Range of rating
$d = 0.1$ to 0.9 m

a. Estimate the stem stiffness Eq. (16.9):

$$\text{MEI} = 3.29 \quad \text{Nm}^2$$

b. Compute geometry rating for the channel

d	Area m^2	Perimeter m	Hyd. Rad. R, m	Mean Depth da, m
0.1	0.18	2.13	0.08	0.09
0.2	0.42	2.76	0.15	0.16
0.3	0.72	3.40	0.21	0.22
0.4	1.08	4.03	0.27	0.28

d	Area m^2	Perimeter m	Hyd. Rad. R, m	Mean Depth d_a, m
0.5	1.50	4.66	0.32	0.33
0.6	1.98	5.29	0.37	0.39
0.7	2.52	5.93	0.43	0.44
0.8	3.12	6.56	0.48	0.50
0.9	3.78	7.19	0.53	0.55

c. Compute relative roughness height using Eq. (16.8):

R	τ_0	k_s/h	ks
0.08	16.5	0.67	0.17
0.15	29.8	0.53	0.13
0.21	41.5	0.46	0.12
0.27	52.5	0.42	0.11
0.32	63.1	0.39	0.10
0.37	73.3	0.37	0.09
0.43	83.3	0.35	0.09
0.48	93.2	0.34	0.08
0.53	103	0.32	0.08

d. Compute vegetative critical shear stress using Eq. (16.12)

$$\tau_{cv} = 68.2 \text{ N/m}^2$$

e. Compute semi-log coefficients using Table 16.3:

d	τ_0/τ_{cv}	a	b
0.1	0.243	0.42	5.23
0.2	0.436	0.42	5.23
0.3	0.609	0.42	5.23
0.4	0.770	0.42	5.23
0.5	0.924	0.42	5.23
0.6	1.074	0.44	5.46
0.7	1.222	0.48	5.92
0.8	1.367	0.52	6.37
0.9	1.510	0.56	6.82

f. Compute Manning's n using Eq. (16.11)

d	R/k_s	n
0.10	1.00	0.504
0.20	1.15	0.317
0.30	1.83	0.138
0.40	2.54	0.101
0.50	3.28	0.085
0.60	4.05	0.072
0.70	4.84	0.061

d	R/k_s	n
0.80	5.67	0.053
0.90	6.51	0.047

g. Compute channel velocity (Eq. 16.4) and discharge:

d	$R^{2/3}$	V, m/s	Q, m³/s
0.1	0.215	0.060	0.01
0.2	0.342	0.15	0.06
0.3	0.448	0.46	0.33
0.4	0.543	0.76	0.82
0.5	0.630	1.05	1.58
0.6	0.711	1.39	2.76
0.7	0.788	1.83	4.60
0.8	0.862	2.30	7.16
0.9	0.932	2.80	10.60

h. Maximum boundary shear (Figs. 16.16 and 16.17):

d	B/da	k_{bs}	k_{ss}	τ_{bed} N/m²	τ_{side} N/m²
0.1	17.5	1.16	0.99	19.2	16.3
0.2	9.6	1.26	1.07	37.6	32.0
0.3	6.9	1.34	1.16	55.5	48.2
0.4	5.4	1.40	1.21	73.5	63.6
0.5	4.5	1.45	1.25	91.4	78.8
0.6	3.9	1.47	1.27	108	93.1
0.7	3.4	1.49	1.30	124	108
0.8	3.0	1.49	1.33	139	124
0.9	2.7	1.49	1.35	153	139

i. Shear stress below lining (Eq. 16.16):

Depth m	n_s	K_f	τ_{bed} N/m^2	τ_{side} N/m^2	
0.1	0.023	0.011	0.22	0.19	
0.2	0.020	0.016	0.59	0.50	
0.3	0.020	0.036	2.0	1.8	
0.4	0.020	0.049	3.6	3.1	
0.5	0.020	0.059	5.4	4.6	
0.6	0.020	0.069	7.5	6.4	
0.7	0.020	0.082	10.2	8.9	
0.8	0.020	0.094	13.1	11.7	exceeds permissible shear
0.9	0.020	0.106	16.3	14.8	exceeds permissible shear

j. Permissible shear stress (Fig. 16.15):
11.0 N/m^2
Note: The permissible shear is first exceeded on the channel bed at a depth of approximately 0.73 meters.

16.5 STEEP-GRADIENT CHANNEL DESIGN

The maximum permissible shear stress is less for channels on steep slopes. There are a couple of reasons for this effect on the channel lining. One reason is that as velocity increases and flow depth decreases the exchange of momentum between portions of the channel becomes more efficient. In low-gradient channels, the maximum shear stress is concentrated over a small portion of the channel boundary. For a steep slope, the channel boundary away from the zone of maximum shear stress receives increased shear stress that approach the maximum. A second reason is the creation of localized shear zones near irregularities in the lining. Steep slopes typically require linings with larger particle sizes, that make construction of a uniform lining difficult. Use of large stones to construct a relatively small channel section should be avoided. Design flow depths should exceed the largest stone size by several times. Protrusions of large stones from the lining are generally a problem. If large stones are desired, then the stability of these stones should be assessed individually.

16.5.1 Resistance to Flow in Steep-Gradient Channels

Most of the flow resistance in channels with large-scale roughness is derived from the form drag of the roughness elements and the distortion of the flow as it passes around roughness elements. Consequently, a flow-resistance equation for these conditions must account for skin friction and form drag. Because of the shallow depths of flow and the large size of the roughness elements, the flow resistance will vary with relative roughness area and roughness geometry, the Fr (the ratio of inertial forces to gravitational forces), and the Reynolds number (Re) (the ratio of inertial forces to viscous forces).

Bathhurst et al., (1979) quantified these relationships experimentally. This work showed that for Re's in the range of 4×10^4 to 2×10^5, the effect on resistance remains constant. When roughness elements protrude through the free surface, resistance increases significantly because of the effects of the Fr (standing waves, hydraulic jumps, and free-surface drag). For the channel as a whole, free-surface drag decreases as the Fr and the relative submergence increases. Once the elements are submerged, ther effect of Fr related to free-surface drag is small, but those related to standing waves are important.

The Manning's form of the *Bathurst equation* is

$$ n = \frac{R^{1/6}}{\sqrt{g}\, f(Fr)\, f(REG)\, f(CG)} \tag{16.19}$$

The function $f(Fr)$ accounts for the free surface drag of the elements:

$$ f(Fr) = \left[\frac{0.28}{b} Fr \right]^{c_1}, \quad \text{where } c_1 = \log\left(\frac{0.755}{b}\right) \quad \begin{array}{l} \text{for } b < 0.755 \\ b \geq 0.755. \end{array} \tag{16.20}$$
$$ = 1.0 $$

The parameter b describes the relationship between the effective roughness concentration and the relative submergence of the roughness bed:

$$ b = 1.14 \left(\frac{D_{50}}{T}\right)^{0.453} \left(\frac{R}{D_{50}}\right)^{0.814} $$

The function $f(REG)$ accounts for the roughness geometry:

$$ f(REG) = c_2 \left(\frac{T}{D_{50}}\right)^{0.492} \tag{16.21}$$

where $c_2 = 13.434\, b^{1.025(T/D_{50})^{0.118}}$

The function $f(CG)$ accounts for the relative roughness area:

$$ F(CG) = \left(\frac{T}{R}\right)^{-b} \tag{16.22}$$

16.5.2 Permissible Shear Stress in Steep Gradient-Channels

Flow in steep channels is generally a shallow high-velocity flow. The main force acting on a lining composed of large particles, such as stone, is the drag force. The lift force will be absent until there is a sufficient depth of flow over the particle. The effect is to increase the shear parameter to approximately 0.094, or roughly twice the value for fully submerged flow. The *critical shear stress for shallow flow* is determined by the following equation:

$$ \tau_c = 1.5\, D_{50} \quad \text{for } R/D_{50} < 2.0 \tag{16.23}$$

16.5.3 Boundary Shear Stress in Steep Gradient Channels

Correction for Slope. For a steep channel ($>10\%$), the approximation of the energy gradient by the tangent of the grade is not accurate. The mean boundary shear-stress adjustment for steep slopes is:

$$ K_\theta = \cos(\theta) \tag{16.24}$$

Correction for High Froude Number. For Fr's greater than 0.6, the distribution of shear stress in the channel is affected by increased flow turbulence (Davidian and Cahal, 1963). An increase in the Fr will cause

- The average wall shear to approach the maximum wall shear, thereby decreasing the ratio of the maximum wall shear to the average wall shear toward unity
- The average total cross section shear to approach the average bottom shear, thereby decreasing the ratio of the average floor shear to the average total cross-sectional shear toward unity

- The average wall shear to approach the average bottom shear, thereby increasing the ratio of the average wall shear to the average bottom shear toward unity

The result is a more uniform distribution of shear on the boundary. The channel banks are adjusted for the high Fr by using a secondary factor that adjusts the bed shear stress. The resulting channel-bank shear-adjustment factor is then compared to the bank adjustment factor that is determined without consideration of the Froude effect. The higher of the two factors is then used. The boundary stress adjustment factor for the channel bank is

$$K_{sf} = \text{maximum of } (Fsf{*}K_{bs}, K_{ss}) \tag{16.25}$$

$$\text{where } Fsf = 0.23 \log(Fr) + 0.91 \qquad \text{for } Fr < 2.5$$
$$= 1.0 \qquad\qquad\qquad\qquad \text{for } Fr \geq 2.5.$$

Example 16.3 Steep Channel Lining Evaluation
This example illustrates hydraulic analysis of a steep riprap channel. A rating is performed in order to illustrate the effect of increasing depths of flow on channel performance.

Inputs:
    Channel attributes
    So = 0.200 m/m   Grade
    Sine = 0.196
    Straight Alignment
    B = 1.2 m   Bottom width
    z = 3:1   Side slope

    Lining attributes
    Angular riprap   Type
    D_{50} = 0.450 m   Size

Outputs:
    Range of rating
    d = 0.1 to 0.9 m

a. Compute the geometry rating for the channel:

d	Area m²	Perimeter m	Top width m	Hyd.Rad. R, m
0.1	0.15	1.83	1.80	0.082
0.2	0.36	2.46	2.40	0.15
0.3	0.63	3.10	3.00	.20
0.4	0.96	3.73	3.60	0.26
0.5	1.35	4.36	4.20	0.31
0.6	1.80	4.99	4.80	0.36

b. Bathhurst resistance equation coefficients:

d	b	c_1	$f(Fr)$	c_2	$f(REG)$	$f(CG)$
0.1	0.152	0.696	$1.653v^{c_1}$	1.382	2.733	0.625
0.2	0.214	0.548	$1.051v^{c_1}$	1.955	4.455	0.550
0.3	0.253	0.475	$0.891v^{c_1}$	2.305	5.862	0.506
0.4	0.282	0.428	$0.818v^{c_1}$	2.559	7.119	0.475
0.5	0.306	0.393	$0.777v^{c_1}$	2.763	8.290	0.451
0.6	0.326	0.365	$0.752v^{c_1}$	2.936	9.409	0.430

c. Manning n function—$1/n = a\,V^b$ and hydraulics:

d	a	b	n	V m/s	Q m³/s	τ_0 N/m²	Fr
0.1	13.4	0.696	0.075	1.5	0.22	157	1.62
0.2	11.1	0.548	0.090	2.0	0.72	281	1.66
0.3	10.8	0.475	0.093	2.6	1.64	391	1.85
0.4	10.9	0.428	0.092	3.2	3.07	495	2.02
0.5	11.1	0.393	0.090	3.8	5.09	595	2.17
0.6	11.3	0.365	0.088	4.3	7.8	693	2.30

d. Maximum boundary shear using Figs. 16.16 and 16.17 and Eq. (16.25):

d	B/da	K_{bs}	K_{ss}	Fsf	τ_{bed} N/m²	τ_{bank} N/m²	
0.1	14.7	1.19	1.01	0.958	187	179	OK
0.2	8.2	1.30	1.10	0.961	364	350	OK
0.3	5.9	1.34	1.13	0.971	524	509	
0.4	4.7	1.40	1.18	0.980	693	679	
0.5	3.9	1.44	1.22	0.987	856	846	
0.6	3.3	1.47	1.25	0.993	1018	1011	

e. Permissible shear stress (Eqs. 16.13, 16.14, and 16.21);

D_{50} 450 mm	τ_c 675 N/m²	SF 1.3	Cz 0.95	Cw 1.00	Cr 0.36	τ_{pbed} 519 N/m²

d	E	Ab	Cl	Cb	Ca	τ_{p_bank} N/m²	
0.1	2.58	0.370	1.47	0.93	0.61	408	OK
0.2	1.32	0.648	1.25	0.80	0.57	386	OK
0.3	0.91	0.834	1.15	0.67	0.58	389	
0.4	0.68	0.973	1.09	0.56	0.59	398	
0.5	0.55	1.071	1.07	0.48	0.60	408	
0.6	0.46	1.143	1.05	0.42	0.62	416	

Note: The riprap lining is adequate to a depth of slightly over 0.2 m. The channel bank is weaker than the bed and is the limiting condition.

16.6　COMPOSITE-SECTION CHANNEL DESIGN

Because the highest shear stresses occur on the bed of the channel or on the channel bank near the bed, it is often desirable to use a combination of linings in a channel section. A composite lining consists of two or more lining materials with a higher strength lining material used selectively in high shear areas. Other requirements may dictate the design of a composite lining. Many times, low-flow channels are included within the main channel or channel banks are vegetated or allowed to remain in a natural condition as part of a bioengineering approach.

16.6.1 Resistance to Flow

Computation of flow conditions in a composite channel requires the use of an equivalent Manning's n-value for the entire perimeter of the channel. For determination of equivalent roughness, the channel perimeter is divided into similar parts. For channels consisting of a bed of one lining type and banks of a second lining type, the Manning's n-value for the entire section is

$$n = K_c n_s \tag{16.26}$$

$$K_c = \left[\left(\frac{P_s}{P} \right) + \left(1 - \frac{P_s}{P} \right) \left(\frac{n_r}{n_s} \right)^{1.5} \right]^{0.667} \tag{16.27}$$

16.6.2 Boundary Shear

The maximum boundary shear acting on the bed and the bank of the channel is determined using Eq. 16.7, and the permissible shear stress for each lining is determined using Eq. 16.6.

16.6.3 Special Considerations

When two lining materials have significantly different roughnesses and are adjacent to each other in the flow, erosion may occur near the boundary of the two linings and erosion of the weaker lining material may damage the lining as a whole. In the case of a composite channel lining with bank vegetation, this problem can occur in the early stages of establishing the vegetation. Thus a temporary lining should be used adjacent to the channel bed to provide temporary erosion protection until the lining vegetation is well established.

Low-flow channels are needed when a flow of long duration is expected, such as the outlet from a detention pond or from areas with sustained groundwater inflow to the channel. Establishing vegetative linings under these conditions can be difficult, particularly if the soil is highly erodible without the lining material. In erodible soils, gullies form in the channel invert that weakens the lining during higher flows. One solution is to provide a nonvegetative channel lining, such as concrete or riprap, for the channel invert. The dimensions of the low-flow channel are determined in the basis of the low flow only, with the remainder of the channel covered in vegetation.

16.7 CHANNELS WITH SEDIMENT TRANSPORT

All flood-control channels will transport some sediment and will require routine maintenance as necessary to prevent excessive accumulation. However, when the supply of sediment to a flood-control channel increases, the channel must be designed with a sufficient capacity to transport the load of both the water and the sediment. The purpose of this section is to present general concepts for approaching sediment transport analysis for flood-control channels.

16.7.1 Sediment Supply

For a flood-control channel to operate effectively, it must be able to transport the supply of sediment that it receives from the upper watershed. Ideally, the transport capacity of the channel will equal the supply over the duration of a flood event. Watershed sediments can be derived from upstream erosion of the stream network (erosion of stream banks and bed) or from erosion of catchment areas (typically a combination of soil detachment by the impact of rain drops and small-scale rilling of the soil surface). For undisturbed watersheds, the supply of sediment can be estimated based on the sediment transport capacity of the existing stream network, assuming that the entire system is near equilibrium. The existing drainage network often is a good guide to the type of channel sections and grades required to achieve adequate sediment transport in a flood-control channel.

The development of land can greatly alter the supply of watershed sediment. Construction can result in a large temporary increase in the potential supply of sediment. Long-term change in land use, such as agricultural development, urbanization, and deforestation, can greatly increase or decrease the yield of watershed sediment. This can result in a complex adjustment within the stream network that may require many years for equilibrium to be reestablished.

Estimation of sediment yield for disturbed watersheds should consider potential sources of sediment from both catchment erosion and erosion within the stream network. Both are typically altered in the course of development. The amount of erosion can be affected by installation of erosion control systems. Methods of analyzing the supply of catchment-level sediment have been developed and tested primarily for agricultural lands. These methods have been adapted successfully to a variety of other land alterations. The widely used *Universal Soil Loss Equation,* USLE (Wischmeier and Smith, 1978) has recently been updated as the *Revised Universal Soil Loss Equation,* RUSLE (Renard et. al., 1992).

Methods of watershed stream-network sediment supply depend on an evaluation of stream transport capacity relative to catchment sediment supply. A qualitative analysis can be conducted for channels in non-cohesive sediments using *Lane's relationship* (Lane, 1955).

$$Q \, S \, \alpha \, G_s d_s \tag{16.28}$$

The Lane relationship is used by assuming that two of the variables are either constant or has a known behavior. For example, an increase in bed load $+G_s$, with no change in sediment size d_s or stream discharge Q, will result in an increase in stream slope $+S$. This is interpreted as a potential for sediment deposition in the existing stream channel. In another case, an increase in stream discharge $+Q$, with no change in sediment supply G_s, or d_s, will result in a decrease in stream slope $-S$. This is interpreted as a potential for scour within the existing stream channel.

16.7.2 Sediment Transport

Sediment is transported in a stream channel as a combination of *bed load* (sediment that is in frequent contact with the bed of the channel) and *suspended bed* load (sediment from the bed that is mixed with the water flow by turbulence). The combined transport is referred to as the *total sediment load.* Incidental to the bed load is wash load, which consists of fine particles that stay in suspension and are derived from the erosion of watershed soil.

Sediment transport begins when the critical shear stress of the bed sediment is exceeded. As the mean shear stress and flow velocity increase, the total rate of sediment-transport increases. Sediment is continuously exchanged vertically in the water column and, to a lesser extent, laterally. Areas of the channel where critical shear stress has been exceeded are referred to as active channel areas. The sediment discharge for the channel section is

$$G_s = g_s \, W_b' \tag{16.29}$$

Numerous formulas for estimating sediment transport exist (Yang, 1996). These formulas need to be evaluated carefully before use; because many are designed for specific sizes of sediment or a specific range of hydraulic conditions. Total transport often is computed by computing portions of the transport process in separate steps (bed load first, then suspended load). Many formulas require careful calibration to obtain a valid result. Similarly, it is possible to find formulas developed for use in similar conditions that differ significantly in estimates sediment transport. Therefore, once a method of analysis is selected, it is important to maintain consistency within that analysis.

It is extremely difficult to generalize sediment-transport formulas. However, it is important to have an understanding of the relation between the dominant hydraulic variables (velocity and flow depth) and the total transport of bed material. From Eq. (16.28), we can state the dominant variables for sediment transport are stated qualitatively as

$$g_s \propto f\left(\frac{qS}{d_s}\right) = f\left(\frac{V\tau_o}{d_s}\right) \tag{16.30}$$

The product of velocity and shear stress is the stream power per unit of bed area. The functional form of Eq. (16.30) can be expressed as a basic power function:

$$g_s = a_1(V\tau_o - V_c\tau_c)b_1 \tag{16.31}$$

For most channels, the exponent b_1 typically ranges from 1.5 to 2.0. Since boundary shear stress is proportional to the square of the velocity, sediment transport is roughly proportional to velocity to the 4.5 to 6.0 power. In other words, small increases or decreases in velocity can greatly increase or decrease capacity for the sediment transport.

It is often useful to develop sediment-rating functions for selected channel reaches using a sediment transport formula, such as Eq. (16.31) or actual measurements at gaging stations.

$$G_s = a_2 Q^{b_2} \tag{16.32}$$

Comparison of successive rating relationships can be used to give a preliminary estimate of aggradation and degradation trends in the channel. The drawback to this type of analysis is that, in reality, the rating will shift with changes in the channel bed and adjustment of hydraulic conditions. However, this approach, although essentially qualitative, can be used as a refinement of the Lane relationship Eq. (16.29).

16.7.3 Aggradation-Degradation

Sediment is transported from the watershed through stream channels to a place of deposition. If we observe sediment movement at a cross section in a stream we can note two conditions (Einstein, 1964): (1) every sediment particle must have been eroded somewhere in the watershed above the cross section and (2) the sediment particle must be transported by the flow to the cross section. If the supply of sediment from the watershed exceeds the sediment transport at a cross-section, then sediment will accumulate: i.e., aggrade. If the transport at the cross section exceeds the supply of sediment, then sediment will be removed from the channel: i.e., *degrade*.

Degradation will generally occur within the active area of the channel bed, whereas aggradation will be distributed fairly uniformly over the entire channel bed. The process of aggradation-degradation involves a mass balance of sediment within a channel reach:

$$G_{\text{sup}} - G_{\text{sxs}} - \left(\frac{W_b'\Delta L\,\Delta z)\gamma_s}{\Delta t}\right) = 0 \tag{16.33}$$

Accurate solution of Eq. (16.32) involves a repetitive solution of stream hydraulics and sediment transport for a sequence of many time intervals. This is accomplished through computer programs (Fan, 1988).

16.7.4 Resistance to Flow

Among the most important aspects of flood-control channels that transport sediment is the effect of that transport on flow resistance. The transport of sediment creates features in the channel bed referred to as *bed forms*. As shear increases on the channel, bed forms increase in size and have an increasing effect on flow resistance in the channel. The general progression of bed forms is from lower regime (ripples to dunes) through a transition (washed out dunes to plane bed) to upper regime (antidunes). Allen (1978) determined the height of lower regime and transition bed forms relative to flow depth experimentally as follows:

$$\frac{H}{d} = 0.080 + 2.24 \left(\frac{\theta}{3}\right) - 18.13 \left(\frac{\theta}{3}\right)^2 + 70.90 \left(\frac{\theta}{3}\right)^3 - 88.3 \left(\frac{\theta}{3}\right)^4 \quad \theta \le 1.5$$

$$\frac{H}{d} = 0 \qquad\qquad\qquad \theta \ge 1.5 \tag{16.34}$$

$$\theta = \frac{\tau_o}{\gamma_s' D_{50}} \tag{16.35}$$

(*Note:* $\theta < 1.1$ lower regime, $1.1 < \theta < 1.5$ transition, and $\theta > 1.5$ upper regime.)

For sand, Karim (1990) developed a simple equation for estimating the Manning's roughness coefficient based on Allen's bed-form height Eq. (16.34):

$$n = 0.037 \, D_{50}^{0.126} \left(\frac{f}{f_o}\right)^{0.465} \qquad (D_{50} \text{ in m}) \tag{16.36}$$

where

$$\frac{f}{f_o} = 1.20 + 8.92 \frac{H}{d}$$

For upper regime flow, a typical range of Manning's resistance coefficients is 0.015 to 0.030. Flow resistance increases with increasing Fr as antidunes become unstable and form breaking waves and eventually form chutes and pools.

REFERENCES

Allen, J. R. L., "Computational methods for dune. Calculations using Stein's rule for dune height," Sedimentary Geol., 20(3), 165–216, 1978)

Anderson, A. G., A. S. Paintal, and J. T. Davenport, "Tentative Design Procedure for Riprap–Lined Channel," Highway Research Board, National Cooperative Highway Research Program, Report 108, Washington, DC., 1970.

Apperly, L. W., *Effect of Turbulence on Sediment Entrainment,* doctoral dissertation, University of Auckland, New Zealand, 1968.

Austin, D. N., and L. E. Ward, ECTC Provides Guidelines for Rolled Erosion-Control Products, Geotechnical Fabrics Report, Industrial Fabrics Association International, Jan/Feb 1996.

Bathurst, J. C., R. M. Li, and D. B. Simons, *Hydraulics of Mountain Rivers,* Colorado State University Experiment Station, CER78–79JCB–RML–DBS55, May 1979.

Benedict, B. A., and B. A. Christensen, "Hydrodynamic Lift on a Stream Bed," Chap. 5, *Sedimentation,* H W. Shen, ed. and Pub., 1972.

Benedict, B. A.,"*Hydrodynamic Lift in Sediment Transport,*" Doctoral Dissertation, University of Florida, Gainesville, Florida, 1968.

Blench, T., *Mobile-Bed Fluviology: A Regime Treatment of Canals and Rivers,* University of Alberta Press, Edmonton, Canada, 168, 1969.

Blodgett, J. C., and C. E. McConaughy, *Rock Riprap Design for Protection of Stream Channels near Highway Structures,* Vols. 1 and 2, Water-Resources Investigations Report No. 86–4128, U.S. Geological Survey, 1986.

Chen, Y. H., and G. K. Cotton, "Design of Roadside Channels with Flexible Linings," Hydraulic Engineering Circular No. 15 (HEC-15), FHWA, Publication No. FHWA–IP–87–7, U.S. Department of Transportation, McLean, VA., 1988.

Davidian, J., and D. I. Cahal, "Distribution of Shear in Rectangular Channels," Article 113, U.S. Geological Survey Professional Paper No. 475–C, pp. C206–C208, Washington, DC, 1963.

Chepil, W. S. *"The Use of Evenly Spaced Hemispheres to Evaluate Aerodynamic Forces on a Soil Surface"* Transactions American Geophysical Union, 39(3), 397–404, 1958.

Einstein, H. A., "Sedimentation, Part II. River Sedimentation," *Handbook of Applied Hydrology,* V. T. Chow, ed., Sec 17, McGraw-Hill, New York, 1964.

El–Samni, E. A., *"Hydrodynamic Forces Acting on the Surface Particles of a Stream Bed,"* doctoral dissertation, University of California, Berkley, California, 1949.

Erosion Control Technology Council (ECTC), *Technical Guidance Manual: Terminology and Index Testing Procedures for Rolled Erosion Control Products,* ECTC Testing and Standards Committee, January 1997.

Fan, S. S., ed., "Twelve Selected Computer Stream Sedimentation Models Developed in the United States," Interagency Sedimentation Work Group, Washington, DC, 1988.

Glover, R. E., and Q. L. Florey, Stable Channel Profiles, Vol. 235, U.S. Bureau of Reclamation, Denver, CO., 1951.

Karim, M. F., "Menu of Coupled Velocity and Sediment Discharge Relations for Rivers," *Journal of Hydraulic Engineering, American Society of Civil Engineers.* 116: 987–996, 1990.

Kennedy, R. G., "The Prevention of Silting in Irrigation Canals," *Proceedings of the Institute of Civil Engineers,* 1885.

Kouwen, N., "Field Estimation of the Biomechanical Properties of Grass," *Journal of Hydraulic Research,* 26: 559–568, 1988.

Kouwen, N , and R. M. Li, "Biomechanics of Vegetated Channel Linings," *Journal of the Hydraulics Division, American Society of Civil Enginneers,* 106, 1085–1103, 1980.

Kouwen, N., and T. E. Unny, "Flexible Roughness in Open Channels," *Journal of the Hydraulics Division, American Society of Civil Enginneers,* 99: 713–728, 1969.

Lacey, G., "Stable Channels in Alluvium," *Proceedings of the Institute of Civil Engineers,* 229:259–292, 1920.

Lane, E. W., "Design of Stable Channels," *Transactions of the American Society Of Civil Engineers,* 120:1234–1260, 1955.

Lane, E. W., "The Importance of Fluvial Morphology in Hydraulic Engineering," *Proceedings of the American Society of Civil Engineers,* 81:1–17, 1920.

Lane, E. W., and E. J. Carlson, "Some factors affecting the stability of canals constructed in coarse granular material," *Proceedings of the Minnesota International Hydraulics Convention,* September 1953.

Lindley, E. S., "Regime channels," *Proceeding of the Punjab Engineers Congress,* Vol. 7, Punjab, India, 1919.

Ree, W. O., and V. J. Palmer, *Flow of Water in Channels Protected by Vegetative Linings,* U.S. Soil Conservation Service Bulletin No. 967, 1949.

Renard, K. G., G. R. Foster, G. A. Weesies, D. K. McCool, and D. C. Yoder, "Predicting Soil Erosion by Water: A Guide to Conservations Planning with the Revised Universal Soil Loss Equation (RUSLE)," *USDA–ARS Agricultural Handbook,* No. 537, 1992.

Simons, D. B., and M. Albertson, "Uniform Water Conveyance Channels in Alluvial Material," *Transactions of the American Society of Civil Engineers,* 128(1): 65–167, 1963.

Simons, D. B., Y. H. Chen, and L. J. Swenson, "Hydraulic Test to Develop Design Criteria for the Use of Reno Mattress," Maccafferri Steel Wire Products, March 1984.

Simons, Li & Associates, Inc., "Designing Stable Channels with Armorflex Articulated Concrete Block Revetment Systems," Nicolon Corporation, February 1990.

Smerdon, E. T. and R. P. Beaseley, "The Tractive Force Theory Applied to stability of Open Channels in Cohesive soils," Agricultural Experiment Station, Research Bulletin No. 715, University of Missouri, Columbia, Missouri, October 1959.

Temple, D. M., "Tractive force design of vegetated channels," *Transactions of the American Society for Agricultural Engineers,* 23: 884–890, 1980.

Temple, D. M., K. M. Robinson, R. M. Ahring, and G.S. Davis, "Stability design of grass-lined channels," Agricultural Handbook No. 667, Agricultural Research Service, U.S. Department of Agriculture, Washington, D.C., 1987.

U.S. Dept. of Agriculture, (USDA), *Handbook of Channel Design for Soil and Water Conservation,* SCS–TP–61, U.S. Department of Agriculture, Washington, D.C., Soil Conservation Service, as revised 1954.

Yang, C. T., *Sediment Transport, Theory and Practice,* McGraw-Hill, New York, 1996.

Wischmeier, W. H. and D. D. Smith, "Predicting Rainfall Erosion Losses: A Guide to Conservation Planning," *USDA Agricultural Handbook* 537, 1978.

DESIGN FOR EROSION AND SEDIMENT CONTROL

Carlos Carriaga
Stantec Consulting Inc.
Phoenix, Arizona

I. Kaan Tuncok
Daniel, Mann, Johnson, & Mendenhall (DMJM)
Phoenix, Arizona

17.1 INTRODUCTION

When land is disturbed for any reason, soil erosion becomes a problem. According to Goldman et al. (1986) such disturbance would increase the soil erosion rate from 2 to 40,000 times. This magnitude of increase would translate into millions of tons of soil loss that would end up in our rivers, lakes, and reservoirs.

Most problems associated with erosion and sedimentation are avoidable. Numerous books have been written just to address the subject of erosion and sediment control, offering simple and low-cost control measures. Also, many attempts have been made to develop models that are accessible to both planners and designers to predict potential scenarios allowing better erosion and sediment control practices to be implemented. This chapter in this book presents various materials that could be used by any technical person to understand the problem associated with soil erosion and sedimentation and their causes. A section has been included to identify some of the best management practices to prevent and minimize the occurrence of erosion and sedimentation. Most important, as an aid to the engineer or designer, a section on predicting and estimating soil loss was covered. A catalog of empirical relations and models addressing soil erosion prediction is presented for various users from the casual to the analytical. The presentation was intended to highlight the existing prediction equations and models for soil erosion evaluation.

Soil erosion is the process by which soil particles are dislodged by the action of rainfall, wind, gravity, or runoff. The dislodging and removal rate of soil particles depend on a number of contributing factors such as

- Ground cover
- Climate
- Topography
- Soil characteristics

These four factors and their impact on soil erosion are briefly explained in Sec. 17.5 below.

Sedimentation, on the other hand, is the natural process of deposition of the eroded and transported soil particles. Examples are sedimentation basins and sediment traps, which are used for collecting transported and eroded sediments through settling action in detention facilities or by trapping or catching them by gravity means.

17.1.1 Definition of Terms

Annual grass refers to grass that grows for one year only.

Berms are small embankments.

Channel erosion is the dislodgment of soil particles from the streambed and the streambank.

Check dams are small temporary dams or embankment built across a swale or drainage ditch to reduce the channel flow velocity and to effectively reduce channel bed erosion.

Chemical stabilization is the application or spraying of materials (such as vinyl, asphalt, or rubber) on the ground surface to hold the soil in place and to provide protection against erosion from stormwater runoff and wind.

Chutes are open-channel structures that carry concentrated runoff down steep slopes without causing erosion.

Cohesive sediments are sediments whose resistance to initial movement or erosion is affected mostly by cohesive bonds between particles.

Contour refers to the imaginary line on the surface of the earth connecting points of the same elevation or a line drawn on a map connecting points of the same elevation.

Contour farming is the conduct of farming operations on sloping, cultivated land in such a way that plowing, land preparation, planting, and cultivation are done on the contour.

Contour furrowing is the plowing of furrows on the contour at intervals varying with the slope and ground cover.

Contour strip-cropping is the growing of crops in a systematic arrangement of strips or bands on the contour.

Debris refers to the loose materials arising from the disintegration of rocks and vegetative material; also, leftover vegetative material from clearing operations on construction projects.

Debris basin is a reservoir designed specifically to trap sediment and debris (Vanoni, 1977).

Dike is a ridge of compacted earth materials.

Effluent refers to the discharge or outflow of water from a construction work area.

Entrainment is the carrying away of bed material produced by erosive action of moving water.

Fall velocity is the falling or settling rate of a particle in a given medium.

Gabions are rock-filled, wire-mesh cages that are used to shield the ground surface from erosion and scour. It serves as the interface medium between the flowing water and the soil.

Grading is the process of modifying or changing the natural topography to control the movement of water onto or from the land surface.

Gradient terraces is the construction of earth embankments or a series of ridges and channels across a slope at suitable spacing and with accepted grades. These are installed to reduce erosion damage and sediment yield by shortening slope lengths and by intercepting surface runoff and directing it to a stable outlet at a nonerrosive flow velocity.

Grassed waterways are natural waterways or depressions that are re-shaped or graded and on which suitable vegetation is established. They provide for disposal of excess surface water from terraces, diversions, contoured fields, or natural concentrations without damage by erosion or flooding.

Gully erosion is the removal of soil by concentrations of flowing water sufficient to cause the formation of channels that can not be smoothed completely by normal cultivation methods. *Gully erosion* is an advanced stage of rill erosion.

Hydroseeding or hydraulic seeding and mulching is the process of spraying a slurry of seed, fertilizer, wood fiber mulch, and water in a one-step process. The most important consideration in hydroseeding is the ability of the fiber to adhere to the soil and hold the seed in place against the action of raindrops and wind.

Isoerodent maps refer to developed maps that show lines of constant erosion rainfall factors, R.

Levees are embankments with heights of 15 ft and less.

Organic matters are plant and animal litter in various stages of decomposition.

Permanent seeding is the practice of planting grass or trees on a land surface to hold soil particles together, and to increase the stability of soil.

Mulching is the use of cover materials to stabilize the soil surface for seeding and/or prevention of erosion. Mulches include organic materials, straw, wood chips, bark or other wood fibers, decomposed granite, gravels, a variety of netting or mats of organic, or non-organic materials, and chemical soil stabilization.

Pipe slope drains are closed-conduit structures that carry concentrated runoff down slopes without causing erosion.

Rate of erosion is the rate at which soil is eroded from a given area. This is usually expressed in volume or weight units per unit area per time. When the areas involved are small, the rate of erosion, soil loss, and sediment yield are equivalent, i.e., the sediment delivery ratio is unity.

Rill erosion is the removal of soil by water from small but well-defined channels or streamlets when there is a concentration of overland flow. *Rill erosion* is an advanced stage of sheet erosion.

Riprap refers to stone- and rock-sized materials used to protect the ground surface from erosion and scour.

Scour is the removal of bed material through the action of flowing water.

Sediment delivery ratio is a measure of the diminution of eroded sediments, by deposition, as they move from the point of erosion to any designated downstream location. This is also expressed as a percentage of the on-site eroded material that reaches a given measuring point.

Sediment yield is the measure of sediment production from a watershed exiting the watershed at some point in the drainage network. It refers to the total sediment outflow from a watershed or drainage basin, measurable at a cross section of reference and in a specified period of time.

Settling velocity is the falling or settling rate of a particle in a given medium.

Shear stress is the frictional force per unit of bed area exerted on the bed by the flowing water. An important factor in the movement of bed material.

Sheet erosion is the uniform removal of soil in thin layers from sloping land, resulting from overland flow occurring in thin layers.

Shrubs are woody plants less than 15 ft tall usually with several trunks rising from a common base.

Sod stabilization is the practice of covering the ground surface with vegetation aimed at providing areas where stormwater can infiltrate the ground.

Soil erosion is the process by which soil particles are dislodged by the action of rainfall, wind, gravity, or runoff.

Soil loss is the quantity of soil actually removed by erosion from land areas.

Soil permeability is the ability of the soil to allow air and water to move through the soil.

Soil structure is the arrangement of soil particles into aggregates.

Soil texture refers to the sizes and proportions of the particles making up a particular soil.

Specific weight of a sediment deposit is the ratio of its dry weight to its associated volume.

Swale is a ditch or depression that is cut into the soil.

Temporary seeding is the practice of growing vegetative cover on disturbed site areas for a short period of time.

Top soil is the surface layer of the soil profile where most of the nutrients and water required for plant growth and development are stored.

Trackwalking is the process of leaving track or tread imprints on the soil surface by the use of heavy equipment such as bulldozer or tractor with track wheels.

Trap efficiency is the proportion of sediment inflow to a stream reach (or reservoir) that is retained within that reach (or reservoir). Computed as inflowing sediment volume minus outflowing sediment volume divided by inflowing sediment volume.

Watershed is a topographically defined area drained by a river/stream or system of connecting rivers/streams such that all outflow is discharged through a single outlet. It is also called a *drainage area*.

17.1.2 Purposes and Objectives

The purpose of an erosion and sediment control program is to allow land development and construction activities to take place while also accomplishing the threefold objectives identified by the American Association of State Highway Transportation officials (AASHTO) (1997):

- Limit off-site impacts of the erosion and sedimentation
- Facilitate development and construction activities at minimum cost
- Comply with federal, state, and local regulations and standards

Off-site impacts associated with erosion and sedimentation are numerous. Without the administration of an effective erosion and sediment control program, problems such as those enumerated in Sec. 17.5 will likely occur. Limiting those problems' occurrence requires assessment and inventory of possible and potential off-site impacts. The designer of the erosion and sediment control measures should attempt to make some assessment of the type and the magnitude of off-site effects to be expected. This involves the determination of whether the impacts would be favorable, unfavorable, or neutral, so that the design or measure could be tempered accordingly. The assessment may require some prediction of estimates of the quantity of materials eroded from the watershed. Such information will allow for an evaluation of what, if any, control measures are required and adequate. The measure or measures adopted may involve some design efforts to effectively handle the conditions anticipated. Several acceptable procedures for predicting soil loss quantities are presented in Sec. 17.4 of this chapter.

The second objective deals with facilitating the land development and construction activities while affordable erosion control measures are applied. For erosion control measures to be economical, they should be simple to construct, use building materials that are locally available, afford as little or minimum interruption to normal activities as practicable, and effective.

Paramount to the preparation of an erosion and sediment control plan is an evaluation of each site for possible practices to implement and assessment of their potential consequences. It is necessary to analyze the probable impacts associated with the implementation of the control measures. From this analysis, an effective plan for erosion and sediment control could be established.

The third objective is compliance with federal, state, and local regulations or standards regarding control of erosion and sedimentation. At the federal level, several executive orders have been issued and various regulations have been established to address erosion and sediment control requirements especially on federally supported activities. There are also federal control requirements imposed by various agencies like those from the U.S. Army Corps of Engineers, Fish and Wildlife Service and Environmental Protection Agency. These are related to the administration and submittal of necessary permitting requirements such as sections 404 and section 402 of the *Federal Water Pollution Control Act* (FWPCA), and sections 9 and 10 of the *River and Harbor Act.*

17.1.3 Problems Associated with Erosion and Sedimentation

Some of the most significant problems associated with erosion and sedimentation are identified as follows:

1. Eroded soils in the form of sediment particles that find their way to reservoirs, streams, lakes, ponds and other water bodies, become contaminants and pollutants. Eroded soil contains nitrogen, phosphorus, and other nutrients that when carried into water bodies would trigger algal bloom that reduce water clarity, deplete oxygen, lead to fish kills, and create odors (Goldman et al., 1986). Also, turbidity from sediment reduces in-stream photosynthesis leading to reduced food supply and habitat for aquatic lives.

2. Eroded soils that settle in reservoirs, streams, lakes, and other water bodies significantly reduce their design capacities, paving a way for possible structural failures, overflows, and potential floods. Uncontrolled sedimentation may require frequent maintenance and dredging operations to restore affected water bodies to their best operating conditions.

3. Eroded soils restrict and reduce effective flow areas for drainage ways, plug culverts, damage adjacent properties, and affect the ecosystems of streams.

4. Erosion along streambanks and adjacent areas destroys streamside vegetation that provides food to aquatic and wildlife habitats.

5. Uncontrolled or excessive sedimentation in streams blankets the bottom fauna, "paves" stream bottoms, and destroys spawning areas for fish.

6. Erosion severely diminishes the ability of the soil to support plant growth. Restoration of vegetation is often very costly. Erosion removes the smaller and less dense constituents of topsoil comprising clay, fine silt particles and organic material, which hold nutrients required by plants. The remaining subsoil is often hard, rocky, infertile, and droughty, making the re-establishment of vegetation difficult.

17.1.4 Factors Affecting Erosion

Four principal factors affect soil erosion: (1) soil characteristics, (2) topography, (3) climate, and (4) soil cover. These factors form the foundation for the development and use of the

Universal Soil Loss Equation (USLE), which comprises five factors. The USLE is a widely used empirical method developed by the U.S. Department of Agriculture (USDA) Agricultural Research Service to evaluate the interaction of the erosion factors and to estimate the soil loss from a land area per year. The erosion control practice (for the P) factor is the fifth factor in the USLE. It describes the roughness condition or the resistance of the ground surface to erosion. This P factor, although evaluated separately in the USLE, is a component of both the topography and the soil cover.

Figure 17.1 summarizes the major active and passive factors that impact erosion and transport of sediments from a land surface (1964). These active and passive factors encompass the four principal erosion factors identified earlier. The following sections describe the influence of these four factors on the amount of predicted soil loss evaluated from a given watershed:

17.1.4.1 Soil Characteristics. The four important soil characteristics that determine the degree of soil erodibility are (1) soil texture, (2) organic matter content, (3) soil structure, and (4) permeability.

1. *Soil texture* refers to the sizes and proportions of the particles comprising the soil sample. Sizes of individual grains or particles determine the soil type or soil class. Clay, silt, and

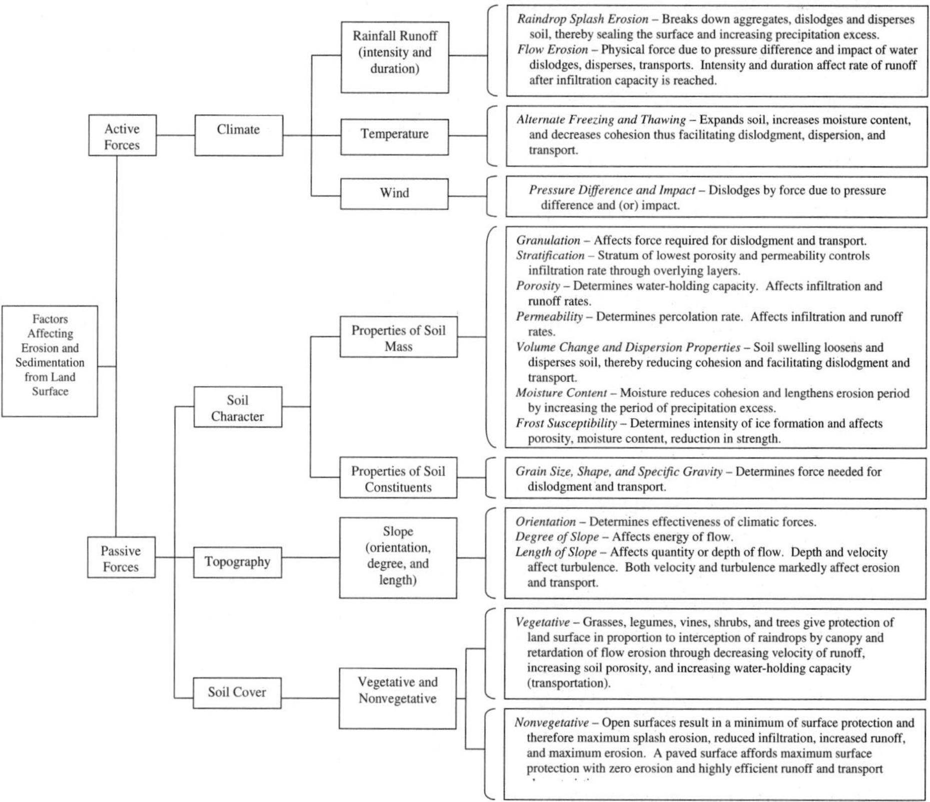

FIGURE 17.1 Factors affecting erosion and sediment transport from a land surface. (From Brooks et al., 1997)

sand are the three major classes of soil particles. Their proportion and composition in a given soil mass describe the soil texture involved. For example, soils high in silt and clay content are identified as fine-textured or heavy soils, while soils high in sand are referred to as coarse-textured or heavy soils, while soils high in sand are referred to as coarse-textured materials. In terms of erodibility, clay possesses binding characteristics that bond soil particles together making the soil resistant to erosion. Once the fine particles are dislodged or dispersed, however, the particles travel great distances before they settle. The sand particles, on the other hand, do not have binding characteristics to keep particles together. Their erosion resistance is very low compared to clay particles. Generally, soils that are high in silt and fine sand particles and low in clay and organic matters are the most erodible (Goldman et al., 1986). In contrast, well-drained sandy and rocky soils are the least erodible.

2. *Organic matters* are the materials comprising all plant and animal litter that are in various stages of decomposition. Organic matter generally improves the condition of the soil. Soil structure greatly benefits from organic matter and likewise for other soil characteristics such as soil permeability, water-holding capacity, and soil fertility. In general, organic matter reduces surface runoff and lowers the erosion potential of the ground surface.

3. *Soil structure* refers to the arrangement of soil particles into aggregates. The most desirable soil structure is granular because it increases the soil's ability to absorb water. The ability of loose, granular soils to absorb and retain water inhibits surface runoff and encourages vegetation growth. A compacted soil, on the other hand, tends to encourage surface runoff, thus increasing the erosion hazard.

4. *Soil permeability* is the ability of the soil to transmit water and air through the soil column. Soil structure, texture, and organic matter all affect the permeability of the soil. High soil permeability indicates high absorption capacity, which indicates less surface runoff potential. Low surface runoff also means low erosion hazard.

It is important that erodible soil types need to be identified so that diligent erosion control efforts could be administered. Equally important is the knowledge and understanding of the four soil characteristics that determine the degree of erodibility of the soil being investigated.

17.1.4.2 Topography. Three topographic factors form the most critical information that determine the degree of erodibility of any watershed. These three factors are (1) orientation of slope, (2) degree of slope, and (3) length of slope.

1. The *orientation of slope* determines the effectiveness of climatic forces to influence soil erosion.

2. The *degree of slope* affects the flow velocity and the energy of flow. The steeper the degree of slope, the greater the erosive energy is available to dislodge and disperse and transport the soil particles.

3. The *length of slope* affects both the quantity and the depth of flow. For long, continuous slopes, runoff builds up momentum as flow travels downstream.

17.1.4.3 Climate. Climate impacts the erosion potential of any land surface. Principal climatic factors affecting soil erosion are rainfall, runoff, temperature, and wind.

Rainfall causes the breakdown of aggregates and the dislodgment and dispersal of soil particles, thereby sealing the surface and increasing the amount of excess precipitation.

Runoff, which is translated from the excess precipitation, exerts force against the soil interface is gets in contact with causing further dislodgment and dispersal of particles and their eventual transport downstream.

Rainfall intensity and droplet size are directly proportional to the errosive power of rain. For example, 1.60-mm raindrops from moderate rain have over 28 times as much kinetic energy per unit time as drizzle with about 0.96-mm raindrops (Lull, 1959). Also, a high rainfall intensity of short duration can produce far more erosion than a low intensity rainfall with longer duration. Also, storms of larger raindrops are more erosive than storms of smaller droplets.

Rainfall intensity and duration and droplet size are determinants to the erosivity index of a given land area. Distribution pattern of rainfall intensity in the United States illustrates that the southeastern states have the most intense rainfalls (USGS, 1970). This indicates that the southeastern region has the greatest erosivity index of all the regions in the United States.

Rainfall intensity and duration impacts the rate of runoff after the soil has reached its infiltration capacity. Dislodgment and dispersal of soil particles are the direct result of flow conditions from runoff exceeding either the permissible velocity or the critical tractive stress associated with the bank and bed materials (Chow, 1959).

Temperature affects the degree of dislodgment, dispersion, and transport of particles. Alternate freezing and thawing from extreme weather conditions cause the expansion of soil, the increase of moisture content, and the eventual decrease of cohesive strength. Such alternate freezing and thawing obviously facilitate the erosion of soil particles. Also, temperature plays a significant role in the settling action and transport of soil particles. At lower temperatures, settling and transport rates of soil particles are slower.

Wind velocity and eddies cause the dislodgment of dry soil particles from land surface. Wind makes larger particles move and jump along the ground. The jumping particles exert energy to the soil surface, dislodging other particles that are hit from the *saltation* action. Like water, the erosive power of wind increases exponentially with velocity, but, unlike water, the erosive power of wind is not influenced by gravity (Brooks et al., 1997).

17.1.4.4 Soil Cover.   *Soil cover* generally refers to the surface covering of the land by vegetation or by nonvegetative materials such as mulches, jute netting, wood chips, crushed rocks, etc.

Vegetative cover, no doubt, is the most effective form of erosion control. This involves the application and use of trees, shrubs, vines, legumes, and grasses that protect the land surface. Vegetation accomplishes remarkable inhibiting conditions against soil erosion as follows: (1) serves as a natural shield against impact of raindrops, (2) slows down the velocity of surface runoff, (3) improves soil aeration and absorption of water, and (4) holds and stabilizes the soil particles.

Nonvegetative cover involves the application of some surface treatments on the ground to protect the soil from erosive actions associated with rainfall, runoff, wind, and temperature. Such surface covering materials minimizes the splash erosion from raindrops, retards flow erosion by decreasing the runoff velocity, shields the surface from wind erosion, and prevents the freezing and thawing of the soil particles during extreme weather conditions.

17.2 GENERAL STABILIZATION GUIDELINES

Stabilization is a key factor in reducing the erosion potential of rainfall or stormwater discharge over an unprotected area. This is accomplished by forcing the water particles to infiltrate into soil and subsequently reduce the sediment carrying capacity of surface runoff. The key factor is to provide appropriate cover and maintenance to the existing cover soil. The following paragraphs summarize some general guidelines that can be used to select the stabilization practices.

If an impervious area is used in a project as a vegetative or nonvegetative erosion control measure, the designer must consider both the advantages and disadvantages. Even though impervious areas will stabilize the cover soil, they may increase surface runoff in the adjacent areas and result in destabilization. Therefore, a good practice is to minimize impervious areas.

During construction activities, the disturbed areas should also be minimized. This will reduce the potential of eroding the soil and providing control measures to remove the soil particles from the surface runoff. If the construction activities take place over a large area, staging the project and matching the existing land contours will also help minimize the disturbed area. Once the construction activities have been completed, the disturbed areas must be permanently stabilized. In areas that will not be re-disturbed for a long period, temporary seeding and mulching can be used as stabilization measures.

In the following sections, vegetative practices, such as temporary seeding, sod stabilization, permanent seeding and planting, buffer zones, and preservation of natural vegetation, and nonvegetative practices, such as mulching, geotextiles, chemical stabilization, and stream bank stabilization will be described in detail.

17.2.1 Vegetative Stabilization Practices

Preserving existing vegetation or re-vegetating disturbed soil soon after construction is the most effective way to control erosion. This practice can provide a six-fold reduction in discharge of suspended-sediment levels (Metropolitan Washington Council of Governments, 1990). Vegetation helps reduce erosion potential through a series of protection measures including rainfall and flow protection, soil retention, and infiltration.

- Rainfall protection by covering the soil surface from the direct erosive impacts of raindrops
- Flow protection, by decreasing the velocity of surface flow and reducing the erosion potential
- Soil retention, by physically holding the soil in place with plant roots; infiltration by improving the soil's water storage capacity by providing additional roots and plant residues, so that more water can infiltrate the ground
- Transpiration, by conducting moisture into plants, where it is eliminated by evaporation

Typical vegetative cover practices include grass, trees, vines, or shrubs. Grasses are very commonly used for re-vegetation because they grow quickly and provide erosion protection in a short time frame. If shrubs and trees are used, the designer should give the priority to maintaining the existing ones. Since shrubs and trees establish their root systems more slowly, keeping existing ones is a more practical and economical measure.

Vegetative and other site stabilization practices can be provided in the form of either temporary or permanent controls (Dodson, 1999). Temporary controls provide protection only for a short time frame or until permanent erosion controls are placed. Permanent vegetative practices are commonly used following the completion of activities that disturb the soil.

17.2.1.1 Temporary Seeding. *Temporary seeding* is the practice of growing vegetative cover on disturbed site areas for a short time frame.

Temporary seeding is generally accomplished by fast-growing grasses that have root systems holding down the soil. As a result, the soil is less vulnerable to stormwater runoff. In most climates temporary seeding typically is appropriate for areas exposed by grading or clearing for more than 7 to 14 days. Temporary seeding practices have been found to be up to 95% effective in reducing the erosion (USDA Natural Resources Conservation Service, 1985).

Rye grass and sterile wheat are commonly used for temporary seeding. Vetiver, a grass native to India, has also received recognition. Sterile varieties of vetiver are available (National Academy Press, 1992).

Temporary seeding is typically used in areas such as de-nuded areas, soil stockpiles, dikes, dams, sides of sediment basins, and temporary road banks. Its application is granted if these areas have been disturbed by construction or are likely to be redisturbed, but not for at least 21 days. In arid and semiarid regions where the climate prevents fast plant growth, particularly during the dry seasons, temporary seeding may not be an efficient practice.

The key factor in seeding is placing the seeds evenly distributed and in contact with the soil. The results are improved when seeds are covered with a shallow layer of soil. Seed should be covered to a depth of ¼ to ½ in (0.6 to 1.3 cm), and the soil cover should not be greater than 1 in (2.5 cm). The three techniques described in the following paragraphs can be used for seeding.

1. *Hand broadcast seeding:* The best results are achieved on small, gently sloping or flat areas that can be effectively seeded by hand. In most applications labor is 1 to 3 hr/acre (2.5–7.5 hr/hectare).

2. *Hydraulic seeding and mulching* (*hydroseeding*): This process is more applicable to slopes 2:1 or steeper. It is basically spraying a slurry of seed, fertilizer, wood fiber mulch, and water in a one-step process. A basic application is shown in Fig. 17.2. The most important factor in the application of hydroseeding is the ability of the fiber to adhere to the soil and hold the seed in place during rainfall and wind.

3. *Seed drilling:* Seed drills are used on flat or gently sloping areas. Seeded areas should be covered with mulch to provide protection from the weather. On slopes of 2:1 or steeper, in adverse soil conditions, during excessively hot or dry weather, seeding should be followed by the spreading of the mulch. A key aspect is to keep seeded areas adequately moist. In cases where seasonal rainfall will not be sufficient, mulching, matting, and

FIGURE 17.2   Hydraulic seeding. (From IECA, 1998)

controlled watering should be applied. Both temporary and permanent seeding are one of the lest expensive stabilization measures. Average cost for both of these measures is around $1/yd^2($0.84/m^2)$, excluding the costs for mulch.

17.2.1.2 *Permanent Seeding and Planting.*   Permanent seeding of grass and planting trees has two primary functions:

- Holding soil particles together
- Increasing the stability of impacted and adjacent areas

These practices also result in a reduced surface runoff and ultimately reduces the sediment load carried to downstream areas. This reduction is mainly accomplished by reducing flow velocities and increasing the infiltration potential of surface runoff. Vegetation provides additional benefits by filtering sediments, helping the soil absorb water, improving wildlife habitats, and enhancing the aesthetics of a site. Basic applications are shown in Fig. 17.3.

FIGURE 17.3   Temporary seeding applications. (From IECA, 1998)

17.2.1.3 Effective Use of Topsoil.   The surface layer of the soil profile is referred to as *topsoil*. Most of the nutrients and water required for plant growth and development is stored within this layer.

Although topsoil has significant benefits to plants, the following disadvantages have to be considered as well:

- Topsoiling can delay seeding or sodding operations, which will eventually increase the exposure time of de-nuded areas.
- Stripping, stockpiling, and re-applying topsoil or importing topsoil may not always be cost-effective.

In certain applications, such as shallow soils, soil containing potentially toxic material, and soils of critically low pH levels, topsoiling is a required process to establish vegetation.

The following items should be considered in the application of topsoiling:

1. The designer should be able to provide adequate volumes of topsoil. Typically, compacted topsoil depths of 50 to 100 mm are used. Table 17.1 can be used as a guidelines to determine topsoil volume requirements. This table should be used as a first step to establish the topsoil volume, and should be finalized by using the site-specific conditions.
2. The application of topsoil must allow for adequate spread and bonding prior to seeding, sodding, or planting.
3. In cases where there is a subsoil of contrasting texture, topsoil should not be applied. A particularly undesirable condition is clayey topsoil over sandy subsoil, which might allow the water creep along the junction between the soil layers, and cause the topsoil to slough.
4. Topsoil should have an organic matter content of at least 1.5% by weight.
5. The pH range should be 5.0 to 7.5. If pH is less than 6.0, a good practice is to add lime in accordance with soil test results or in accordance with the recommendations of the vegetative established practice being used.
6. Soluble salts should not exceed 500 parts per million.

17.2.1.4 Sod Stabilization.   *Sod stabilization* covers the surface with vegetation and provides areas where stormwater can infiltrate the ground. If installed and maintained properly, sodding can be 99% effective in reducing erosion (USDA Natural Resources Conservation Service, 1985). Since sod stabilization has a higher cost compared to other vegetative controls, its application is limited to exposed soils where a quick vegetative cover is desired and to sites that can be maintained with ground equipment. Sodding can be used in buffer

TABLE 17.1   Required Volumes of Topsoil

Depth		Topsoil volume	
mm	in	m³/ha	yd³/ac
25	1	253	134
51	2	506	268
76	3	761	403
102	4	1014	537
127	5	1270	672
152	6	1523	806

Source: From Dodson (1999).

zones, stream banks, dikes, swales, slopes, outlets, level spreaders, and filter strips. The average cost of sod is about $4/yd^2 ($3.35/m^2). Based on the climate, sod may require intensive watering and fertilizing.

The designer should not apply sod on slopes greater than 3% if the slopes are to be mowed. If the site conditions require the application on these slopes, sod should be laid with staggered joints. In areas such as steep slopes or next to running waterways jute or other netting can be placed over the sod for extra protection against lifting (Dodson, 1999).

17.2.1.5   Buffer Zones.   *Buffer zones* are preserved or planted strips of vegetation at the top and bottom of a slope, outlining property boundaries, or adjacent to receiving waters such as streams or wetlands (Dodson, 1999). Buffer zones can provide several advantages including: reduced runoff flows, reduced erosion potential, and increased sediment deposition.

Application of a buffer zone is commonly seen on floodplains, next to wet lands, along stream banks, and on steep, unstable slopes. They can be of significant importance for construction projects such as underground utilities or pipelines. In applications where buffer zones are established by planting, the costs may be estimated at about $1/yd^2 ($0.84/m^2), excluding the costs for mulching.

17.2.2   Nonvegetative Stabilization Practices

Typical applications of nonvegetative practices include mulching, geotextiles, degradable soil stabilization blankets, nondegradable soil stabilization mats, chemical stabilization, and stream bank stabilization. Similar to vegetative covers the basic processes that help reduce erosion potential are rainfall and flow protection, soil retention, and infiltration.

17.2.2.1   Mulching.   *Mulching* is a commonly used temporary soil stabilization/erosion control measure. Typically materials such as grass, hay, wood chips, wood fibers, straws, fabric, or gravel are placed on the soil surface. Mulching can also help reduce stormwater runoff velocities. In certain applications mulching is used with seeding or planting to enhance plant growth. This combination holds the seeds, fertilizers, and topsoil in place, which helps to retain moisture and insulate the soil against extreme temperatures. Table 17.2 summarizes typical mulch materials, quality standards, and application rates.

Mulching becomes more effective when secured by a binder or netting, or by tracking the mulch to the ground. Mulch costs can be estimated at about $1.25/yd^2 ($1.05/m^2).

TABLE 17.2   Application Rates and Quality Standards for Various Mulch Materials

Mulch material	Quality standards	Application rates
Straw	Air-dried; free from undesirable seed and coarse material	2–3 in thick: 2–3 bales per 1000 ft^2 or 2–3 tons/acre
Wood-fiber cellulose	No growth-inhibiting factors	Approx. 25–30 lb/1000 ft^2 or 1000–1500 lb/acre
Compost	No visible water or dust during handling. Must be purchased from supplier with solid waste handling permit	2-in thick minimum; approx. 100 tons/acre (approx. 800 lb/yd)
Chipped site vegetation	Average size shall be several inches	2-in minimum thickness

Source: From Dodson 1999.

Basic considerations prior to the application of mulching are site characteristics, product availability, and cost. Straw and wood fibers are commonly used at construction sites.

Straw mulch is effective at absorbing raindrop impact and moderating the climate on the soil surface because of its length and bulk. Even though straw is an excellent mulch material as a highly effective and inexpensive erosion control measure, it has certain disadvantages, including weed growth, and a higher possibility of fire hazard compared to wood fiber mulch.

Wood fiber mulch is a relatively inexpensive material that has an easy application with a hydroseeder. Even though wood fiber mulch is not as effective as straw, it is preferable to straw on cut slopes 2:1 or steeper, and where weed growth or fire hazard is a major problem.

17.2.2.2 Geotextiles.   *Geotextiles* are porous fabrics used for erosion control and other construction purposes. Typically, synthetic materials such as polypropylene, polyester, polyethylene, nylon, polyvinyl chloride, and glass are used in the manufacturing process (Dodson, 1999).

Geotextiles are typically used as mulch netting and matting. They can also be used as separators as in the case of a geotextile placed between riprap and soil to maintain the base of the riprap and to prevent the soil from being eroded. A basic application is shown in Fig. 17.4.

17.2.2.3 Degradable Soil Stabilization Blankets.   *Degradable soil stabilization blankets* are used to help establish vegetation on slopes 3:1 or steeper. These blankets may include a photo-degradable plastic netting that covers and is intertwined with a natural organic or artificial mulch. The mulching material may consist of wood fibers, wood excelsior, straw, artificial fibers, or a combination of these. The mulching material and fibers must interlock to form a dense layer, which not only resists raindrop impact, but also will allow vegetation to penetrate the blanket (Dodson, 1999).

FIGURE 17.4   A typical geotextile application. (From IECA, 1998)

If properly applied, degradable soil stabilization blankets can provide the following benefits:

- Protect the seed and soil from raindrop impact and subsequent displacement
- Enhance moisture retention of seedbed area
- Provide stronger and faster germination of grasses and legumes
- Prevent sloughing of topsoil added to steeper slopes

Soil stabilization blankets are typically anchored to the ground with staples. Typical applications require staples made of No. 11 gauge wire or heavier. The typical staple length is 150 mm (6 in). A larger staple with a length of up to 300 mm (12 in) should be used on loose, sandy, or unstable soils.

17.2.2.4 Nondegradable Soil Stabilization Mats.

17.2.2.4 Nondegradable Soil Stabilization Mats.   A soil stabilization mat may consist of a nondegradable nylon or polyethylene, which can be filled with soil prior to planting. This configuration provides an efficient environment for root development, and ultimately enhances the energy dissipation process. In addition to the benefits of degradable soil stabilization blankets, soil stabilization mats cause soil to settle out of stormwater and fill the mat with fine solid, which becomes the growth medium of roots.

Turf reinforcement mats (TRMs), which reinforce roots of grasses to provide additional resistance to erosion, are applied more commonly as environmentally and economically attractive alternatives to concrete and rock. Figures 17.5 and 17.6 show typical applications.

Since soil stabilization mats are nondegradable, they can be used in permanent conveyance channels. This practice will help the channel carry higher velocities of flow. However, a maximum flow velocity of 3 m/second (10 ft/second) should be used in the conveyance system. At higher velocities structural lining might be used to provide a better protection.

FIGURE 17.5   TRM being rolled out and installed in an urban channel. (From IECA, 1998)

FIGURE 17.6   A channel that has been stabilized with TRM. (From IECA, 1998)

17.2.2.5 Chemical Stabilization.   *Chemical stabilization* can be used as a temporary erosion control practice. Typical materials used in application are vinyl, asphalt, or rubber. These materials are sprayed onto the surface of the soil to hold the soil in place and to provide protection against erosion from stormwater runoff and wind. In areas and climates that limit the use of temporary seeding, chemical stabilization can be used as an alternative. Chemical stabilization's major advantage is its immediate, effective, and inexpensive impact on the surface of the soil.

17.2.2.6 Streambank Stabilization Using Riprap.   Streambank erosion and bank failures are one of the primary causes of erosion problems.

There are various streambank stabilization methods, including riprap, gabions, slope paving and bulkheads, log cribbing, grid pavers, and asphalt paving. These measures help prevent streambank erosion due to high velocities and quantities of stormwater runoff. The two commonly used methods are riprap and gabions.

Riprap.   Riprap is comprised of large angular stones placed along the bank lines. Its main function is to protect soil from erosion. Alternative uses include inlet and outlet protection at culverts, channel liners, stream bank protection, and protection of shorelines subject to wave action. Riprap is generally placed over a filter blanket. Figure 17.7 shows a riprap application in sediment basin. Riprap costs may be estimated at about $45/yd^2 ($38/m^2).

A. Design Guidelines for Riprap

    1. *Hydraulics.* Some of the significant hydraulic parameters that should be considered in riprap design include flow velocities, discharge, flow depth and channel bed and side slopes. Tractive force and flow velocity criteria are the two basic methods used to compute riprap size. There are also several empirical equations adopted by public agencies. The basic methods and empirical equations are described in detail in several

FIGURE 17.7   Riprap application. (From IECA, 1998)

references (U.S. Department of Transportation (DOT), 1989, Dickinson, 1968; ASCE, 1975).

2. *Design criteria.* In general, riprap is used on channel slopes of less than 10%. For steeper slopes, the use of riprap may not be economical. Riprap selection must consider some basic factors such as design runoff, stone size and location, riprap gradation, thickness of riprap lining, filter requirements, and side slopes.

   a. Design runoff. The design discharge should be the basis for riprap size. However, as recommended by National Academy of Science of Highway Research Board (1970), a minimum design return interval of 25 yrs should be used.

   b. Stone size and location. As indicated above there are a variety of procedures for riprap design. The methodology that will be presented in this section is based on the recommendation of U.S. DOT (1989), which uses a tractive force method.

   c. Rock gradation. In order to provide adequate resistance to erosion, rocks should be well-graded throughout the riprap layer thickness.

   Tables 17.3 and 17.4 below present riprap gradation limits and classes, respectively. The gradation classes provided in Table 17.4 is based on AASHTO speci-

TABLE 17.3   Rock Riprap Gradation Limits

Stone size range (ft)	Stone weight (lbs)	Percent of gradation smaller than
$1.5\,D_{50} - 1.7\,D_{50}$	$3.0\,W_{50} - 5.0\,W_{50}$	100
$1.2\,D_{50} - 1.4\,D_{50}$	$2.0\,W_{50} - 2.75\,W_{50}$	85
$1.0\,D_{50} - 1.15\,D_{50}$	$1.0\,W_{50} - 1.5\,W_{50}$	50
$0.4\,D_{50} - 0.6\,D_{50}$	$0.1\,W_{50} - 0.2\,W_{50}$	15

Source: From U.S. DOT (1989).

TABLE 17.4 Riprap Gradation Classes

Riprap class	Rock size[1] (ft)	Rock size[2] (lb)	Percent of riprap smaller than
Facing	1.30	200	100
	0.95	75	50
	0.40	5	10
Light	1.80	500	100
	1.30	200	50
	0.40	5	10
1/4 ton	2.25	1000	100
	1.80	500	50
	0.95	75	10
1/2 ton	2.85	2000	100
	2.25	1000	50
	1.80	500	5
1 ton	3.60	4000	100
	2.85	2000	50
	2.25	1000	5
2 ton	4.50	8000	100
	3.60	4000	50
	2.85	2000	5

[1] Assuming a specific gravity of 2.65.
[2] Based on AASHTO gradations.
Source: From U.S. DOT (1989).

fications (1987). In Table 17.3, D_{50} is the diameter of a particle by which 50% by weight of the material is smaller. W_{50} is the weight of the D_{50} particle size.

At small quarries, the site conditions may not allow the use of a four-level gradation presented in Table 17.3. In such cases the 85% specification can be eliminated, because a uniform gradation between D_{50} and D_{100} will result in an appropriate D_{85} (U.S. DOT, 1989). In application, the designer must ensure that each load of riprap is reasonable graded from the smallest to the maximum size specified. Stones smaller than the specified 5 or 10% size should not be permitted in an amount exceeding 20% by weight of each load.

d. Layer thickness (LT). A key aspect in erosion resistance is to contain all stones within the riprap layer thickness. Oversize stones should be removed and replaced with proper size stones. This will eliminate the possibility of exposing filter and bedding materials, and creating excessive local turbulence that removes smaller stones.

The following criteria can be used to establish the riprap layer thickness:

- The higher of D_{100} or 1.5 (D_{50}) stones should be used as the layer thickness.
- $LT \geq 12$ in (30 cm)
- If the riprap is placed under water, the layer thickness should be increased by 50% to account for uncertainties associated with this type of placement.
- If debris or ice attack to riprap revetment is possible, an increase in thickness of 6 to 12 in (15 to 30 cm), accompanied by an appropriate increase in stone sizes, should be provided.

e. Filter design. A filter is a transitional layer of gravel, small stones, or fabric placed between the underlying soil and the structure (Dodson, 1999).

The filter is mainly used for the following reasons:

- To prevent the migration of the fine soil particles through voids in the structure
- To reduce erosion potential on noncohesive soils subject to significant subsurface drainage
- To distribute the weight of the armor units to provide more uniform settlement
- To permit relief of hydrostatic pressures within the soils

The effectiveness of a filter is significantly impacted by its opening. If the filter openings are too large, excessive flow piping through the filter can cause erosion and failure of the bank material below the filter. On the other hand, small filter openings may result in a buildup of hydrostatic pressures behind the filter. These excessive pressures can cause a slip plane to form along the filter resulting in massive translational slide failure.

The following design guidelines for granular and fabric filters are critical to the stability of riprap installations on channel banks.

Granular Filters

The need for a filter blanket is a function of particle-size ratios between the riprap and the underlying soil that comprise the channel bank (FCDMC, 1993). The inequalities that must be satisfied are as follows:

$$\frac{(D_{15})_{\text{filter}}}{(D_{85})_{\text{base}}} < 5 < \frac{(D_{15})_{\text{filter}}}{(D_{15})_{\text{base}}} < 40 \tag{17.1}$$

$$\frac{(D_{50})_{\text{filter}}}{(D_{50})_{\text{base}}} < 40 \tag{17.2}$$

In these relationships, filter refers to the overlying material and base refers to the underlying material.

The inequalities in Eq. (17.1) are provided for the following reasons:

- $\dfrac{(D_{15})_{\text{filter}}}{(D_{85})_{\text{base}}} < 5$, to prevent piping through the filter,
- $5 < \dfrac{(D_{15})_{\text{filter}}}{(D_{15})_{\text{base}}} < 40$, to provide adequate permeability for structural bedding layers
- $\dfrac{(D_{15})_{\text{filter}}}{(D_{15})_{\text{base}}} < 40$, to provide a uniformity criteria

The following conditions should also be considered to increase the efficiency of filter design.

- If the requirements in Eqs. (17.1) and (17.2) are not met with a single layer of filter material, additional layers of filter must be used.
- The filter requirements provided in Eqs. (17.1) and (17.2) apply to the following layers:
  i. bank material and the filter blanket
  ii. successive layers of filter material if more than one layer is used
  iii. filter basket and the riprap cover
- The grain size curves for the various layers are as significant as the filter requirements. These curves should be approximately parallel to minimize the infiltration of the fine material from the fine layer to the coarse layer. In general, the portion of filter material passing the No. 200 sieve should be limited to a maximum of 5%. The standard gradation guidelines are provided in Table 17.5.

TABLE 17.5 Gradation for Gravel
Bedding

Standard sieve size	Type I[1]	Type II[1]
3 inches	—	90–100
1½ inches	—	—
3/4 inches	—	20–90
3/8 inches	100	—
#4 (4.75 mm)	95–100	0–20
#16 (1.18 mm)	45–80	—
#50 (0.30 mm)	10–30	—
#100 (0.15 mm)	2–10	—
#200 (0.075 mm)	0–2	0–3

[1] Percent passing by weight.
Source: From FCDMC (1996).

The criteria provided in Eqs. (17.1) and (17.2) were used to develop the type-I and type-II bedding specifications. The designer can use the following guidelines to choose between the two filter types.

- Type-I bedding is designed to be the lower layer in a two-layer filter for protecting fine grained soils.
- If the channel is excavated in coarse sand and gravel, only the Type-II filter is required. Otherwise, two bedding layers (Type I topped by Type II) are required.

Table 17.6 can be used for the required bedding thickness.

Fabric Filters

Synthetic fabric filters can be used as an alternative to granular filters. The fabric filter has the following advantages:

TABLE 17.6 Thickness Requirements for Gravel
Bedding

Riprap classification	Minimum bedding thickness, (inches)		
	Fine grain native soils		Coarse grain native soils
	Type I	Type II	Type III
6″, 8″	4	4	6
12″	4	4	6
18″	4	6	8
24″	4	6	8
30″	4	8	10
36″	4	8	10

Source: From FCDMC (1996).

- Installation—generally quick and labor-efficient
- More economical than granular filters
- Consistent and more reliable material quality
- Good inherent tensile strength

The disadvantages of fabric filter include:

- Fabric filters can be difficult to lay under water.
- There are uncertainties to the life of the fabric in a soil environment over the lifetime of a project.
- The hydraulic responses of a fabric filter system may be dependent on the bacterial activity within the soil or upon the filter.
- In the case of noncohesive bank material, experimental studies showed that when the channel banks were subject to wave action, the bank material had a tendency to migrate downslope beneath fabric filters. This behavior was not observed with granular filters.
- Fabric filters may induce translational or modified slump failures when used under rock riprap installed on steep slopes.

Even though fabric filters are provided for both drainage and filtration, specification of proper piping and clogging resistance, and construction strength requirements are required to achieve these functions.

The design criteria for fabric filters are a function of the permeability of the fabric and the effective opening size. The permeability of the fabric must exceed the permeability of the underlying soil, and the *apparent opening size* (AOS) must be small enough to retain the soil.

The criteria for apparent opening size are as follows:

- For soil with less than 50% of the particles, by weight, passing a No. 200 sieve, the AOS should be less than 0.6 mm (a No. 30 sieve).
- For soil with more than 50% of the particles, by weight, passing a No. 200 sieve, the AOS should be less than 0.3 mm (a No. 50 sieve).

   f. Side slopes. A maximum design side slope of 2:1 is recommended.

B. Design Relationships

This section presents a riprap design relationship based on tractive force theory. The design relationship is based on the assumption of uniform, gradually varied flow.

$$D_{50} = \frac{0.001V^3}{d^{0.5}K_c^{1.5}} \qquad (17.3)$$

where D_{50} = median riprap particle size
   C_F = correction factor (described in the following paragraphs)
   V = average velocity in the main channel (ft/s)
   d = average flow depth in the main flow channel (ft)
   K_c = correction factor for the bank slope

K_c is defined as

$$K_c = \sqrt{1 - \left(\frac{\sin^2\phi}{\sin^2\phi}\right)} \qquad (17.4)$$

where ϕ = bank angle with the horizontal
   θ = riprap material's angle of repose

Equation (17.3) is based on a rock riprap specific gravity of 2.65, and a stability factor of 1.2. Equations (17.5) and (17.6) present correction factors for other specific gravities and stability factors.

$$C_{sg} = \frac{2.12}{(G_s - 1)^{1.5}}$$ (17.5)

where G_s = specific gravity of the rock riprap

$$C_{sf} = \left(\frac{SF}{1.2}\right)^{1.5}$$ (17.6)

where SF = stability factor to be applied

By using these two correction factors, a single correction factor C_F is computed as

$$C_F = C_{sg}\, C_{sf}$$ (17.7)

This correction factor is then multiplied by the riprap size computed from Eq. (17.3) to arrive at a stable riprap size.

The stability of riprap can be maintained if SF is less than 1. SF can be computed by using the following equation:

$$SF = \frac{F_t}{\tau_c}$$ (17.8)

where F_t = average tractive force exerted by the flow field
τ_c = riprap materials' critical shear stress

Table 17.7, which presents guidelines for the selection of an appropriate value for the stability factor, may be a more practical tool.

1. *Application.* As indicated previously, the application of Eq. (17.3) is limited to uniform or gradually varying flow conditions. In order to identify the rock size at bridge piers

TABLE 17.7   Guidelines for the Selection of Stability Factors

Condition	Stability factor range	Design value[1]
Uniform flow	1.0–1.2	1.2
(a) Straight or mildly curving reach (curve radius/channel width > 30)		
(b) Impact from wave action and floating debris is minimal		
(c) Little or no uncertainty in design parameters		
Gradually varying flow	1.3–1.6	
(a) Moderate bend curvature (10 < curve radius/channel width <30)		
(b) Impact from wave action and floating debris is moderate		
Approaching rapidly varying flow	1.6–2.0	1.7
(a) Sharp bend curvature (curve radius/channel width <10)		
(b) Significant impact potential from floating debris and/or ice		
(c) Significant wind and/or waves (1–2 ft (0.3–0.61 m))		
(d) Significant uncertainty in design parameters		

[1] Recommended design value.
Source: From U.S. DOT (1989).

and abutments the following recommendations are provided.

Bridge Piers
The following relationships can be used at bridge piers.
a. D_{50} size can be computed by using the following equation:

$$D_{50} = \left(\frac{1}{2}\right)\left(\frac{1.384\ V_b^2}{(G_s - 1)2g}\right) \tag{17.9}$$

where V_b = velocity of flow at the base of the pier (ft/s)
    g = 32.2 ft/s²
b. To calculate V_b, first the velocity of flow just upstream of the pier must be determined. This value must be multiplied by a factor of 1.5 to 2.0 to compute the velocity of flow at the base of the pier. U.S. DOT's preliminary research indicated that a factor of about 1.5 might be used as a reasonable design value.

- The mat width should extend horizontally at least 2 times the pier width measured from the pier face.
- The mat should be placed below a depth equivalent to the expected scour. The thickness should be 3 stone diameters or more.

Abutments
If Eq. (17.3) is used for riprap design at abutments, a velocity in the vicinity of the abutments should be used instead of the average section velocity. A stability factor of 1.6 to 2.0 should also be used for turbulently mixing flow at bridge abutments.

2. *Wave erosion.* Waves generated by wind or boat traffic might cause bank erosion on inland waterways. The most widely used measure of riprap's resistance to wave is the one developed by Hudson (1959). The Hudson relationship can be written as

$$W_{50} = \frac{\gamma_s\ H^3}{2.2(G_s - 1)^3\ cot\,\theta} \tag{17.10}$$

where H = wave height.

Assuming G_s = 2.65 and γ_s = 165 lb/ft³, Eq. (17.10) can be reduced to

$$W_{50} = \frac{16.7\ H^3}{cot\,\theta} \tag{17.11}$$

In terms of equivalent diameter Eq. (17.11) can be reduced to

$$D_{50} = \frac{0.75\ H}{cot^{1/3}\theta} \tag{17.12}$$

Equations (17.11) and (17.12) can be used for preliminary or final design when H is less than 5 ft, and there is no major overtopping of the embankment.

C. Rock Riprap Design Procedure
Rock riprap design procedure outlined in the following steps is comprised of two primary steps: preliminary data analysis and rock sizing.

1. Preliminary Data Analysis
   Step 1. Compile all necessary field data including channel cross-section surveys, soil data, aerial photographs, history of problems at site, etc.
   Step 2. Determine design discharge.
   Step 3. Develop design cross section(s).

2. Rock Sizing
   Step 4. Compute design water surface.
   Step 5. Determine design average velocity and depth.
   Step 6. Compute the bank angle correction factor K_c, Eq. (17.4).
   Step 7. Determine riprap size required to resist particle erosion, Eq. (17.3).
   a. Use Eq. (17.6) to compute D_{50}.
   b. Evaluate correction factor for rock riprap specific gravity and stability factor, Eq. (17.7).
   c. If designing riprap for piers or abutments, use the guidelines provided in Sec. 17.2.2.6B above.
   d. Compute corrected rock riprap size.
   Step 8. If surface waves are to be evaluated:
   a. Determine wave height
   b. Use Eq. (17.10) to determine rock size required to resist wave action.
   Step 9. Design riprap size, gradation and layer thickness.

17.2.2.7 Streambank Stabilization Using Gabions. *Gabions* are rock-filled wire-mesh cages that are used to create a new streambank. They can be used at locations where the available rock size is too small for riprap, or where steeper side slopes are required. A typical application in a grass-lined channel is shown in Fig. 17.8.

The durability of wire-enclosed rock is generally limited by the service life of the galvanized binding wire which, under normal conditions, is considered to be about 15 yrs. The designer should not ignore factors such as water carrying silt, sand, or gravel that can reduce the service life of the wire, or water that rolls or moves cobbles and large stones, which breaks the wire.

FIGURE 17.8   Grass-lined channel with gabion basket. (From IECA, 1998)

The following criteria should be met when gabions are designed:

- The gabions should not be exposed to abrasion from larger debris and/or bedload transported by the flow. The abrasion can damage the wire mesh and lead to failure.
- Rock used to fill the gabion should be angular, block-shaped rock. Minimum rock sizes are generally 3 to 4 in or 1/3 the basket depth. The maximum stone size should not exceed 2/3 the basket depth or 12 in, whichever is smaller.

17.2.2.8 Soil-retaining Measures.   *Soil-retaining measures* can be in the form of structural or vegetative stabilization practices, which holds the soil firmly to its original place or confines as much as possible within the site boundary. Typical applications include reinforced soil retaining systems, windbreaks, and streambank protection by shrubs or reeds.

Reinforced soil-retaining measures are structural measures used to hold the soil in place. Skeleton sheeting, which is the least expensive soil-bracing system, is commonly used in cohesive soils. An alternative to skeleton sheeting is continuous sheeting, which involves use of a material that covers the face of the slope in a continuous manner. Struts and boards are placed along the slope, which provide continuous support to the slope face. Steel, concrete or wood can be used. Another alternative is the permanent retaining walls, which may be used to provide support to the slope after the construction is completed.

17.3   BEST MANAGEMENT PRACTICES

Stormwater management involves the effective handling of the quantity and/or the quality of runoff water being discharged into a land or water area. Effective management requires that possible pollution conditions of stormwater are addressed adequately, as these significantly impact large bodies of water downstream. Also, erosion and sedimentation assert detrimental impact on existing drainage and floodway structures as the deposited silt and soil particles render these structures incapable to operate as originally designed. Management practices are then suggested to reduce or to possibly eliminate these detrimental impacts resulting from uncontrolled erosion and sedimentation from land areas upstream.

There are 21 management practices listed below, addressing both the agricultural and nonagricultural aspects to erosion and sediment control. Each management practice is presented and described in the following sections for specific use and application. Design considerations are also given as guidelines for implementation. The 21 most common management practices for erosion and sediment control covered in this book are

1. Temporary diversion dike
2. Check dams
3. Cut-to-fill slope transition
4. Straw bale barriers
5. Erosion protection at structures
6. Rock outlet protection
7. Storm drain inlet protection
8. Stabilized construction entrance
9. Erosion control matting
10. Pipe slope drain
11. Silt fence

12. Protection of trees and natural areas in construction areas
13. Sandbag berm
14. Construction road stabilization
15. Temporary drainage swale
16. Temporary sediment basin
17. Temporary sediment trap
18. Gravel filter berm
19. Surface roughening
20. Trees, shrubs, vines, and ground covers
21. Mulching

17.3.1 Temporary Diversion Dike

A. *Description.* A temporary diversion dike is comprised of either a berm or a ridge of compacted soil material, which is primarily used to direct runoff to a desired location as illustrated in Fig. 17.9.

B. *Purpose.* To direct runoff to a stabilized outlet to reduce or minimize erosion potential. It also serves to divert clear water and sheet flows away from unprotected slopes and disturbed sites.

C. *Applications.* One most common use of a temporary diversion dike is along the top of erodible fill slopes prior to the construction or installation of embankment curbs for roads. Also, earthen dikes are provided around construction sites and upstream of disturbed

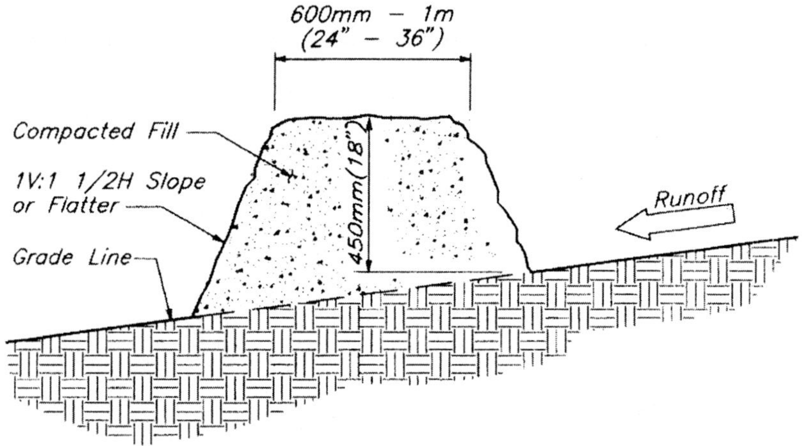

Size of dike is based on individual site.

Dimensions shown are for general information only.

FIGURE 17.9   Temporary diversion dike. (From FCDMC, 1993)

areas to maintain the historical drainage pattern at the site. These diversion dikes should remain in place until the disturbed areas are stabilized.

D. *Limitations.* Earthen dikes are limited for drainage areas not exceeding 4 hectares and for slopes not more than 5%. Permanent dikes should be used for larger drainage areas. Each dike can become a barrier to construction equipment. They should be stabilized immediately, which adds to costs and maintenance concerns.

E. *Design considerations.* Design criteria developed for temporary dikes are presented by Goldman et al. (1986) and Arizona Department of Transportation (ADOT) (1998).

   1. The minimum height for a dike is 18 in (46 cm) of compacted material.
   2. Side slopes should have a 2:1 or flatter slope and should be stabilized by seeding with grass if the dike is to serve for more than 30 days.
   3. Top width should be at least 2 ft (0.6 m). For construction traffic crossings, however, the top width may be wider.
   4. Dikes should be adequately compacted with stabilized outlets to prevent or reduce erosion.

17.3.2 Check Dams

A. *Description.* A *check dam* is a small temporary dam that is constructed in an unlined channel or swale as illustrated in Fig. 17.10.

B. *Purpose.* To reduce the velocity of concentrated stormwater flows which reduce erosion and allow for the settling of sediments in small channels and temporary swales.

C. *Applications.* One most common use of the check dam is along newly constructed channels or swales where flow velocities are regulated to an acceptable range while grass or channel vegetation has not sufficiently matured. Also, on channels with slopes greater than 4%, check dams should remain in place to continue to provide erosion control in the channel.

D. *Limitations.* Check dams should only be used in channels or swales that are designed for drainage areas not exceeding 3 hectares. Check dams are not appropriate for live streams and are not useful for lined or vegetated channels. Check dams become cost-prohibitive on steeper slopes because of greater number of dams required.

E. *Design considerations.* ADOT (1998) has provided the following design specifications for check dams:

   1. The maximum spacing between check dams in succession should be the horizontal distance from the toe upstream dam to the same elevation at the top of the downstream dam.
   2. Rock check dams should be constructed of approximately sized rock of D_{50} equal to 25 to 75 mm (1 to 3 in) minimum.

17.3.3 Cut-to-Fill Slope Transition

A. *Description.* The stabilization at the cut-to-fill slope transition with the use of riprap materials as illustrated in Figure 17.11.

B. *Purpose.* To reduce runoff erosion and sedimentation at cut-to-fill slope transitions where potential of erosion exists

C. *Application.* One most common use is where large amount of concentrated surface runoff must be conveyed from a cut ditch to the toe of a fill slope. Riprap materials are used in the cut-to-fill slope transition to reduce or minimize runoff erosion.

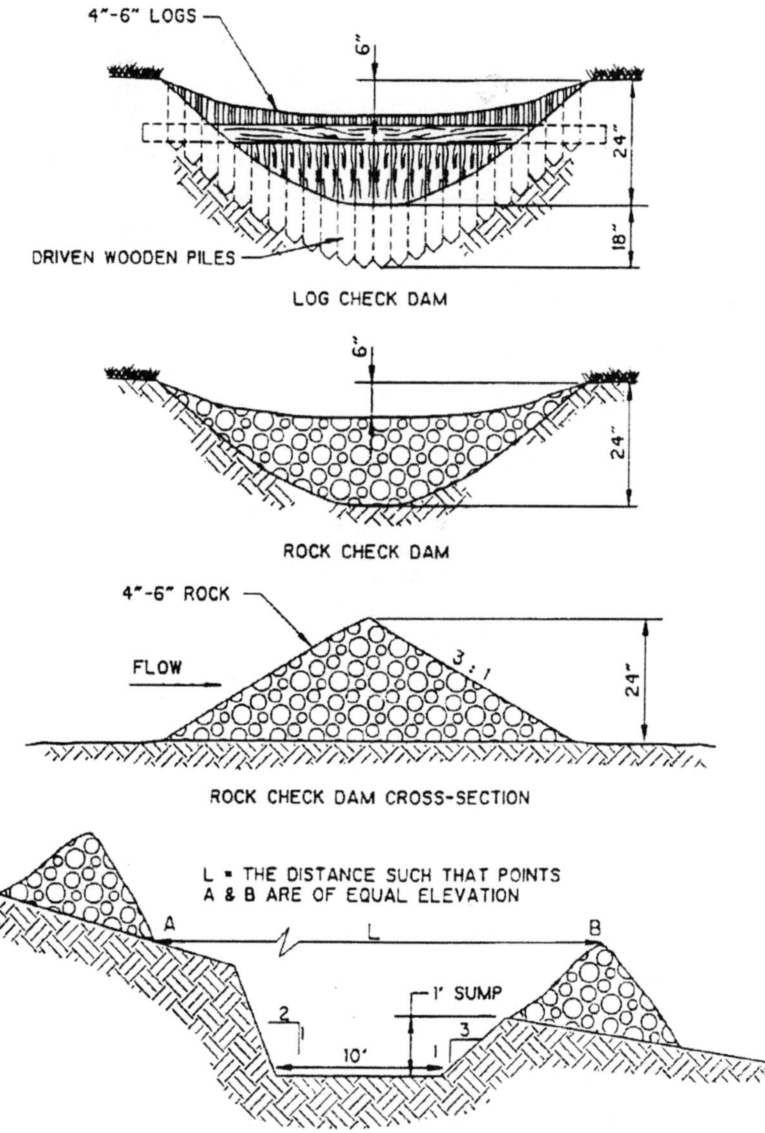

FIGURE 17.10 Check dam. (From FCDMC, 1993)

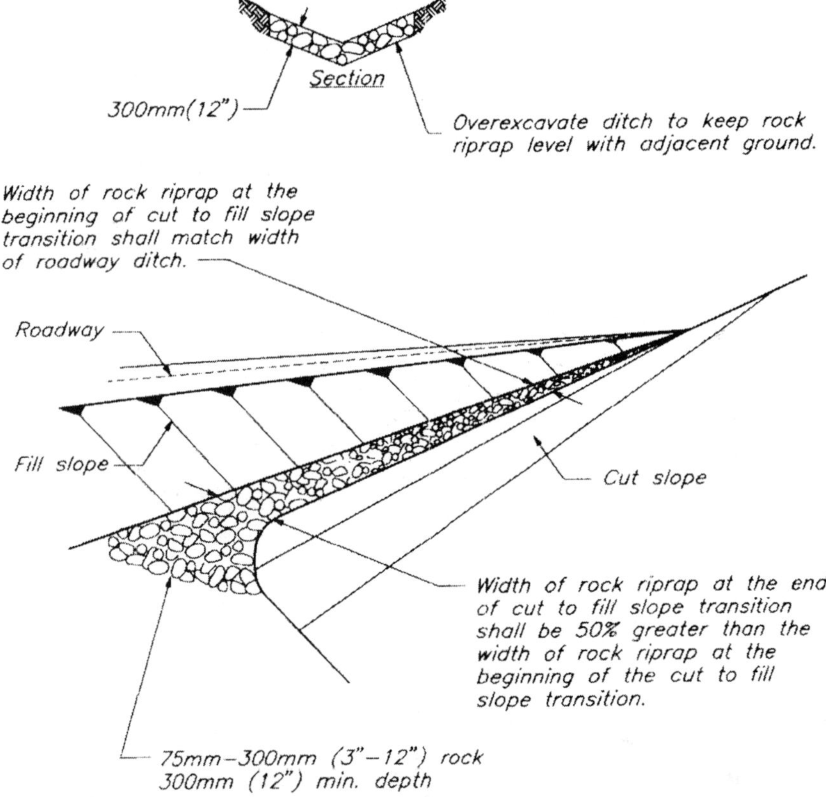

FIGURE 17.11   Cut-to-fill slope transition. (From ADOT, 1998)

D. *Limitations.* Availability of sufficient rock materials for the riprap. Since cut-to-fill transitions are relatively steep, it follows that flow regime may be in the supercritical range that larger rock sizes are necessary for stabilized condition.

E. *Design considerations.* Design specifications for cut-to-fill transition are provided as follows:

1. Rock riprap transitions should be made up of rocks with D_{50} from 75 mm to 300 mm (3 to 12 in) minimum (ADOT, 1998). Appropriate size of riprap, however, must be made based on the design flow velocity.
2. The width of rock riprap coverage at the beginning of the cut-to-fill transition should match the width of the roadside ditch.
3. The width of rock riprap coverage at the end of cut-to-fill transition should be 1.5 times the width of rock riprap coverage at the beginning of the cut-to-fill transition.

17.3.4 Straw Bale Barriers

A. *Description.* A temporary barrier placed in the direction of flow to intercept sediment-laden runoff from small drainage areas of disturbed soil as illustrated in Fig. 17.12.

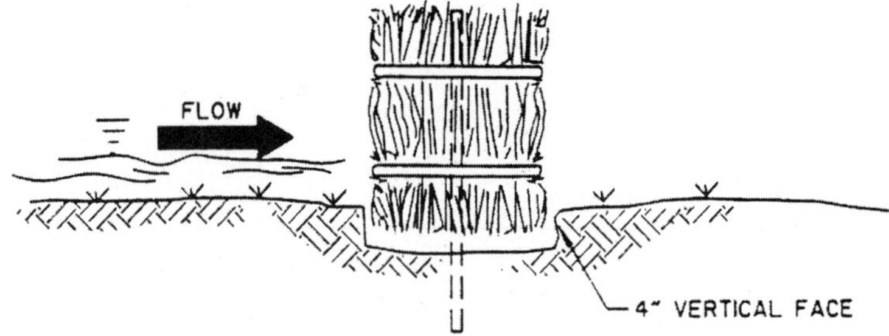

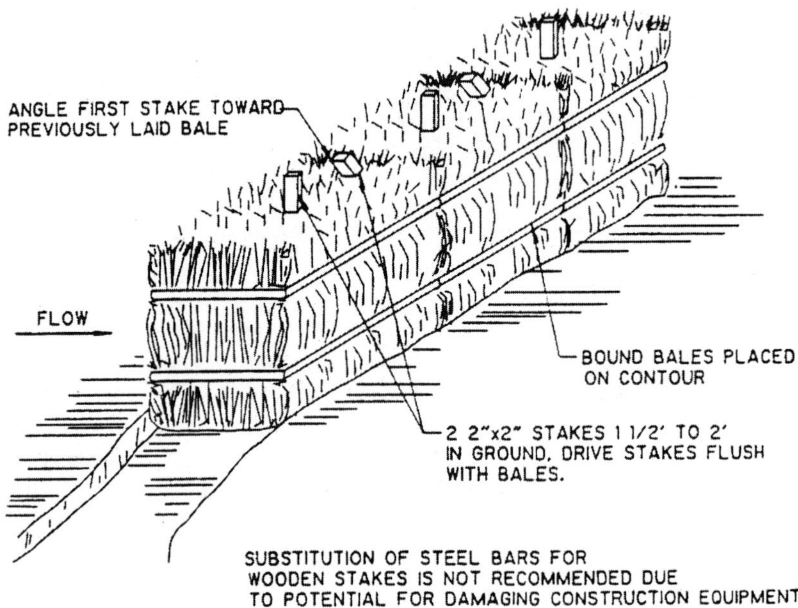

FIGURE 17.12 Straw bale barrier. (From FCDMC, 1993)

B. *Purpose.* To reduce runoff velocity and allow the disposition of the transported sediment load behind the barrier.

C. *Application.* Straw bale barriers are provided in the direction of flow for the detention and interception of sediments that may impact the design conditions of outlet or conveyance structures. They are not recommended for live streams.

D. *Limitations.* Straw bale dikes should not be used for an extended period of time because they lose their effectiveness rapidly due to decomposition.

E. *Design considerations.* ADOT (1998) has provided the following design information:

1. Use straw bale barriers for drainage areas not exceeding 1 acre (0.4 hectare).
2. Straw bale barriers should be used for slopes not exceeding 2%.

17.3.5 Erosion Protection at Structures

A. *Description.* An assemblage of rocks or large stones placed along soil interface of concrete and metal structures as illustrated in Figures 17.13a and b.

B. *Purpose.* To reduce erosion that occurs particularly in drainage outlets when concentrated rainwater flows undercut a structure.

C. *Application.* For drainage structure outlets where the concentration and velocity of stormwater cause erosive actions.

D. *Limitations.* Rock sizes should be sized properly based on the design discharge. Since rock protection requires stabilization, grouted or wire-tied riprap, called gabion mattress, can absolutely minimize the maintenance requirements.

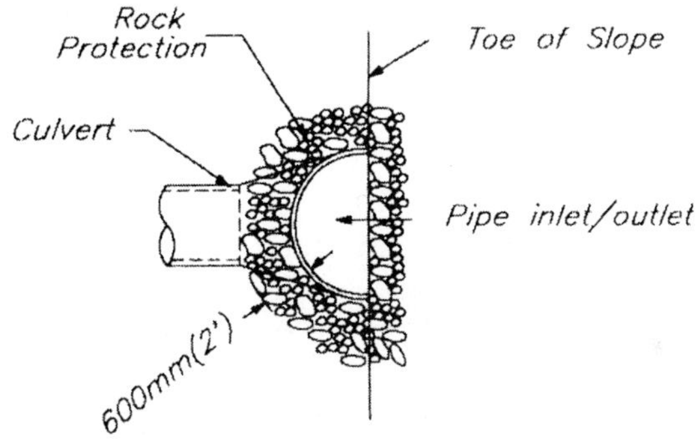

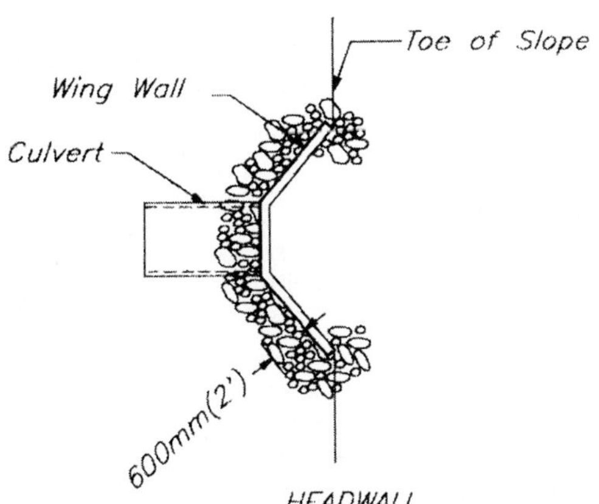

FIGURE 17.13a Erosion protection at structures. (From ADOT, 1998)

FIGURE 17.13*b*   Erosion protection at structures. (From FCDMC, 1998)

E. *Design considerations.* Recommendations on the use of riprap for erosion protection are listed as follows:

1. Use angular rocks as riprap material.
2. Adequate sizing of riprap is necessary based on the design flow velocity.

17.3.6   Rock Outlet Protection

A. *Description.* Rock riprap, grouted riprap, or concrete rubble assembled or placed at the base of outlet ends of culverts, conduits, or channels as illustrated in Fig. 17.14.

B. *Purpose.* To dissipate the energy and to reduce the velocity by increasing the surface roughness at the outlet, thus eliminating or reducing erosion potential in the downstream channel.

C. *Application.* Outlet protection is necessary when hydraulic conditions at the structure outlets are sufficient to cause erosion and bed movement.

D. *Limitations.* Rock outlet protection may require periodic check ups especially after storm events when potential dislodgment of rocks and stones trigger further erosion. Grouted riprap or wire-tied rock assemblage like gabion mattresses can minimize maintenance. Availability of proper rock sizes for riprap may be prohibitive.

E. *Design considerations.* Use the guideline presented in Sec. 17.2.2.6 for riprap sizing.

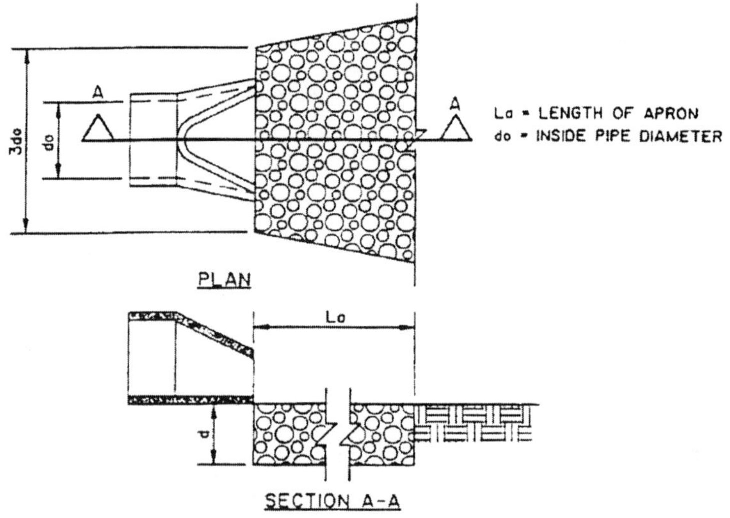

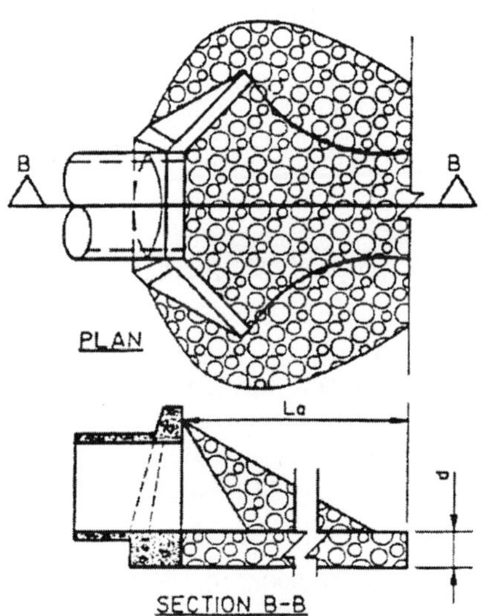

PIPE SIZE	AVERAGE ROCK DIA.	La
12″	6″	12′
15″	10″	18′
18″	12″	21′
21″	15″	25′
24″	15″	30′

FIGURE 17.14   Rock outlet protection. (From FCDMC, 1993)

Example 17.1
Design example
Problem
A 1250-ft channel reach is to be realigned to make room for the widening of an existing highway. Realignment of the channel reach will necessitate straightening the channel and

reducing its length from 1250 ft to 1000 ft. The channel is to be sized to carry 500 ft³/ second within its banks. Additional site conditions are as follows:

- Flow conditions can be assumed to be uniform or gradually varying,
- The straightened reach is at a uniform slope of 0.0014.
- The natural soils are graded from medium sands to coarse with the following characteristics:
  $D_{85} = 2.2$ in
  $D_{50} = 1.1$ in
  $D_{15} = 0.09$ in
- The permeability, $k = 0.035$ cm/second
- Available rock riprap has a specific gravity of 2.65

Design a stable trapezoidal riprap lined channel for this site.

Solution

Step 1. Compile field data
Data other than the given information would include site history, geometric constraints, roadway crossing profiles, site topography, etc.

Step 2. Design discharge, $Q = 500$ ft³/second

Step 3. Design cross section
Use a trapezoidal section with 21.86 ft bottom width and 2:1 side slopes.

Step 4. Design water surface
Compute design water depth by using the information in the problem statement and Manning's equation:

$$Q_D = \left(\frac{1.49}{n}\right) A_f R_f^{2/3} S_f^{1/2}$$

Since both the cross-sectional area, A_f, and hydraulic radius, R_f, are functions of flow depth, a trial-and-error process was used to compute the flow depth. The resulting depth is equal to 3.77 ft.

Step 5. Determine the design parameters
Given the cross-sectional area, $A_f = 110.8$ ft², the design parameters are
Average Design Velocity,

$$V = Q_D/A_f = 500/110.8 = 4.51 \text{ ft/second}$$

Average Design Depth,

$$d = 3.77 \text{ ft}$$

Step 6. Bank angle correction factor

Given $\phi = 26.5$ and $\theta = 33.5$, use Eq. (17.4) to compute K_c,

$$K_c = \sqrt{1 - \left(\frac{\sin^2 \theta}{\sin^2 \phi}\right)} = \sqrt{1 - \left(\frac{\sin^2 26.5}{\sin^2 33.5}\right)} = 0.587$$

Step 7. Determine riprap size.

    a. Use Eq. (17.3) to compute D_{50}

$$D_{50} = \frac{0.001\ V^3}{d^{0.5}K_c^{1.5}} = \frac{0.00(4.51)^3}{3.77^{0.5}\ 0.587^{1.5}} = 0.11 \text{ ft} = 1.32 \text{ in}$$

    b. Riprap specific gravity = 2.65 (given)
        Use Table 17.7 to determine the stability factor (SF)
        SF = 1.2 (uniform flow, little or no uncertainty in design)
    c. No piers or abutments to evaluate for this example; therefore, no need to adjust the safety factor
    d. Corrected riprap size

$$D_{50} = 0.11 \text{ ft}$$

Step 8. Surface waves
       Surface waves determined not to be a problem at this site

Step 9. Select design riprap size, gradation, and layer thickness

$$D_{50} \text{ size: } D_{50} = 0.11 \text{ ft}$$

Gradation: Use the guidelines in Sec. 17.2.2.6.

Layer thickness (LT): Layer thickness is the greater of $1.5D_{50}$ or D_{100}

$$1.5(D_{50}) = 1.5(0.11) = 0.17 \text{ ft}$$

$$D_{100} = 1.3 \text{ ft}$$

Use a layer thickness of 2 ft in design.

17.3.7 Storm Drain Inlet Protection

    A. *Description.* Sediment filtering set-up assembled around a storm drain, drop inlet, or curb inlet as illustrated in Figs. 17.15a and b.

    B. *Purpose.* To prevent and control sediment materials from entering and clogging storm drain systems prior to attaining permanent stabilization for a disturbed site

    C. *Application.* Protection for storm drain inlets to filter sediment out at the crucial time when the disturbed drainage site has not reached its stabilized and controlled condition

    D. *Limitations.* The effectiveness of storm drain inlet protection is limited to small drainage areas of 5 acres (2 hectares) or less. For larger drainage areas, sediment catchment areas may be appropriate. Ponding may result at the inlet with possible short-term flooding when drainage areas are over the size limit.

    E. *Design consideration.* ADOT (1998) has listed the following recommendations for storm drain inlet protection:

        1. Sediment sumps of 1 to 2 feet (0.3 to 0.6 m) in depth with side slopes at a maximum of 2:1 should be included in inlet sediment traps.
        2. If filter fabrics are used, they should be kept in place firmly. Filter fabric is limited to storm drain inlets for small drainage of 5 acres (2 hectares) or less.
        3. For larger drainage areas of more than 5 acres (2 hectares), smaller sediment basins are recommended.

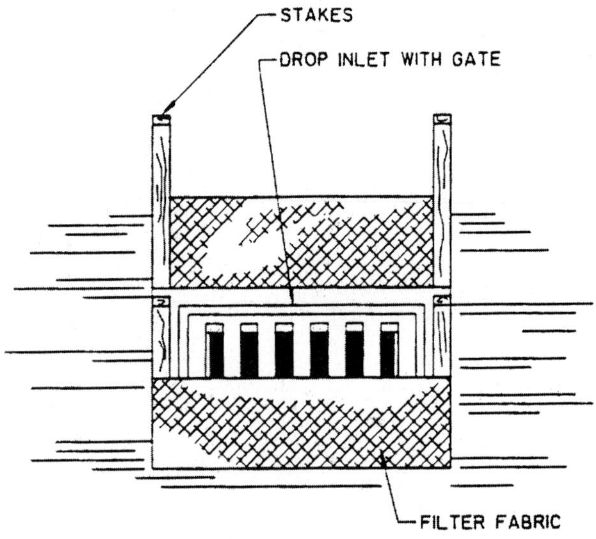

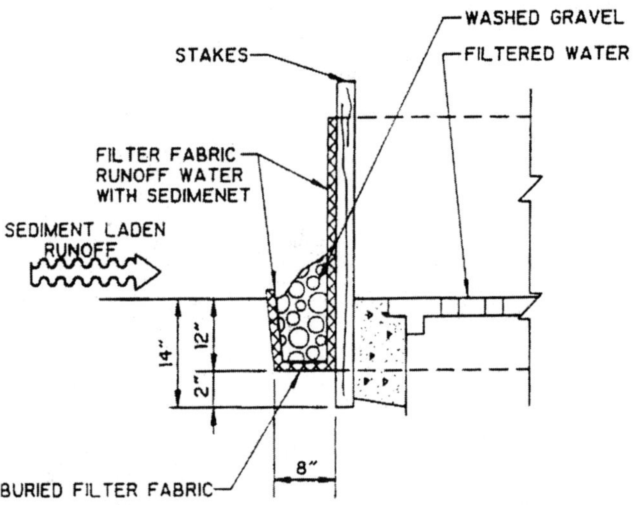

FIGURE 17.15a Storm drain inlet protection. (From FCDMC, 1993)

17.3.8 Stabilized Construction Entrance

A. *Description.* A well-stabilized construction entrance placed where traffic enters or leaves the construction site to or from a public right-of-way, street, sidewalk, or parking lot as illustrated in Fig. 17.16.

B. *Purpose.* To reduce and possibly eliminate tracking of sediment onto public right-of-way, streets, sidewalks, or parking lots. A well-stabilized construction entrance reduces sedi-

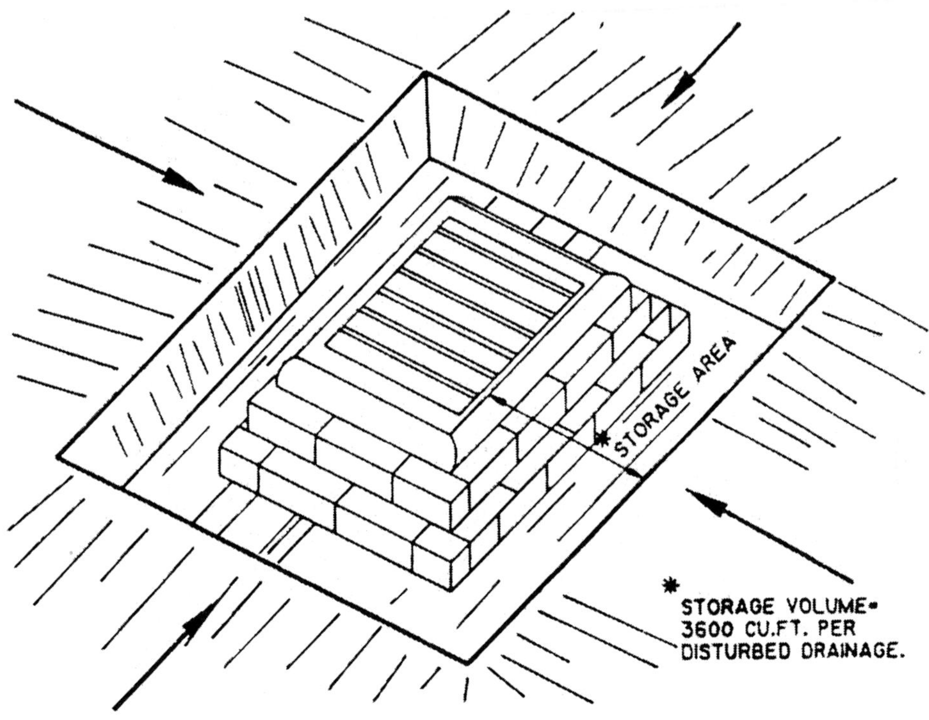

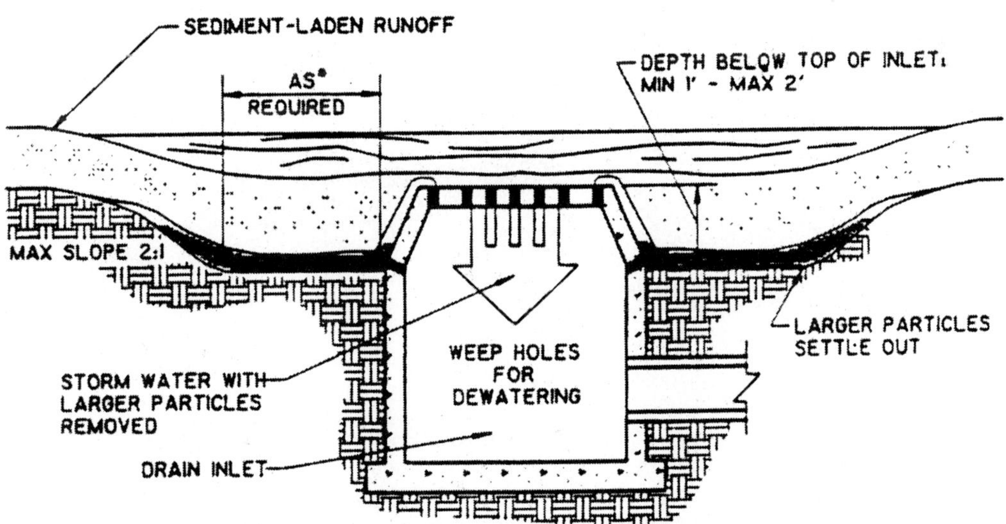

FIGURE 17.15*b* Storm drain inlet protection. (From FCDMC, 1993)

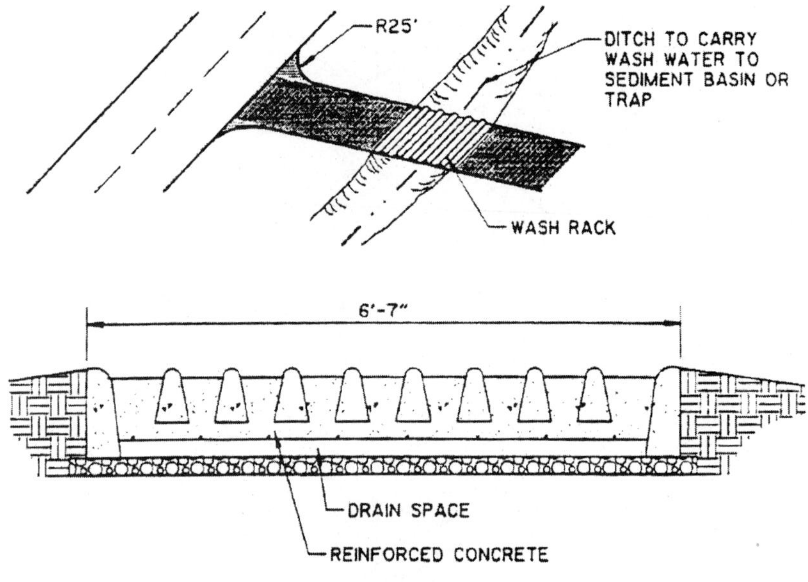

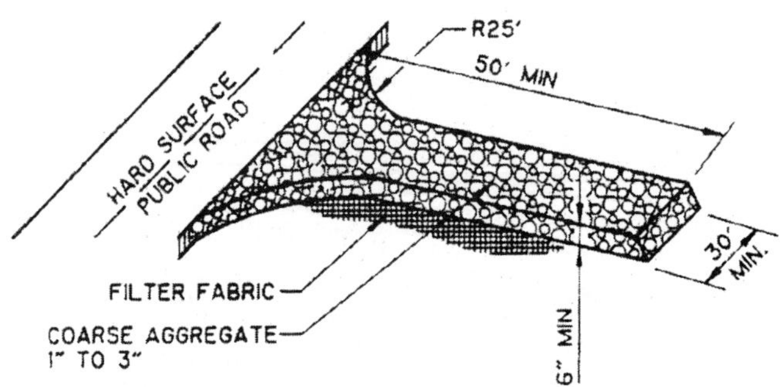

FIGURE 17.16 Stabilized construction entrance. (From FCDMC, 1993)

ment trackouts and other pollutants onto paved roads, thus preventing entry and deposition of sediments into the local storm drain system.

C. *Application.* A stabilized construction entrance should be used at all points of traffic ingress and egress. NPDES permit requirements demands that appropriate measures be implemented or guidelines are in place to prevent sediment trackouts onto paved roadways.

D. *Limitations.* Cost is a limiting factor to adopt and implement stabilized construction entrances for vehicles.

E. *Design considerations.* ADOT (1998) has provided the following design information:

1. Aggregates for stabilized construction entrance apron shall comprise 1 to 3 in (0.25 to 0.75 m) well-graded gravel or crushed rock.
2. Recommended apron dimensions are 30 ft × 50 ft (20 m × 15 m) and 6 in (150 mm) deep.
3. Entrance should be properly graded to prevent runoff from leaving the construction site.
4. When wash areas are provided, washing shall be done on areas stabilized with crushed stones which drain into properly constructed sediment traps or basins.

17.3.9 Erosion Control Matting

A. *Description.* A natural mat comprised of prepared vegetative seedbeds or a synthetic mat installed on steep slopes to reduce soil surface erosion as illustrated in Figs. 17.17a and b.

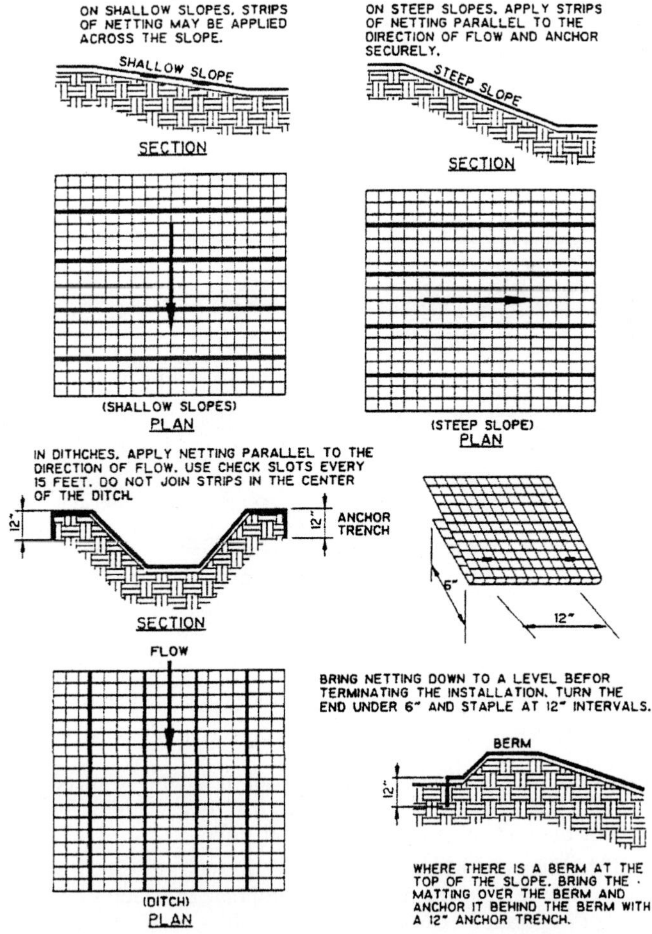

FIGURE 17.17a Erosion control matting. (From FCDMC, 1993)

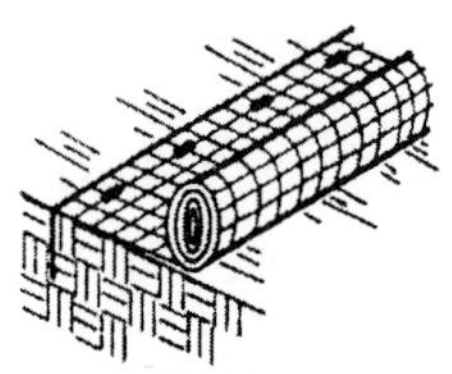

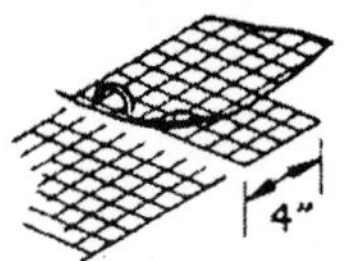

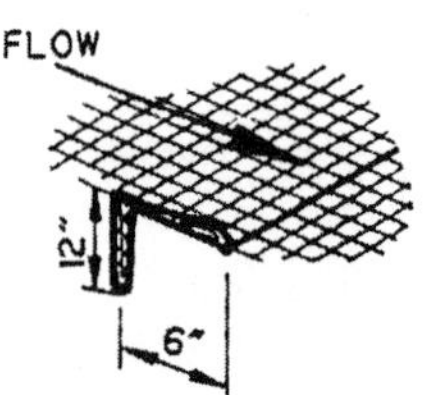

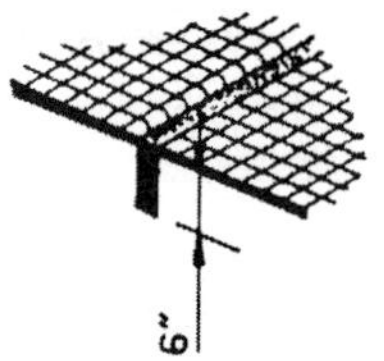

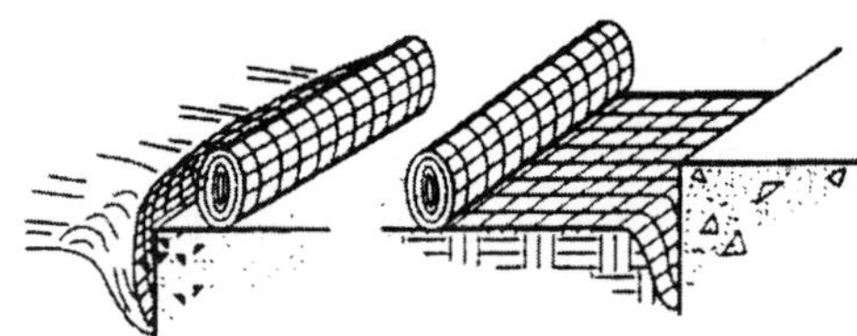

FIGURE 17.17b Erosion control matting. (From FCDMC, 1993)

B. *Purpose.* To reduce rainfall impact, hold soil in place, and absorb and hold moisture at the soil surface. Mulch may be used together with the matting during the establishment of the protective cover on critical slopes.

C. *Application.* Matting is best in channels where flowrates are beyond the permissible velocities allowed. It is very appropriate for steep slopes and in areas where establishment of vegetation is likely to be slow. It is recommended on streambanks where flow will likely wash out new plant growth. Typically, matting is used in areas of concentrated channel flows and steep slopes.

D. *Limitations.* Cost is a limiting factor for matting, despite its excellent potential to control soil surface erosion.

E. *Design considerations.*

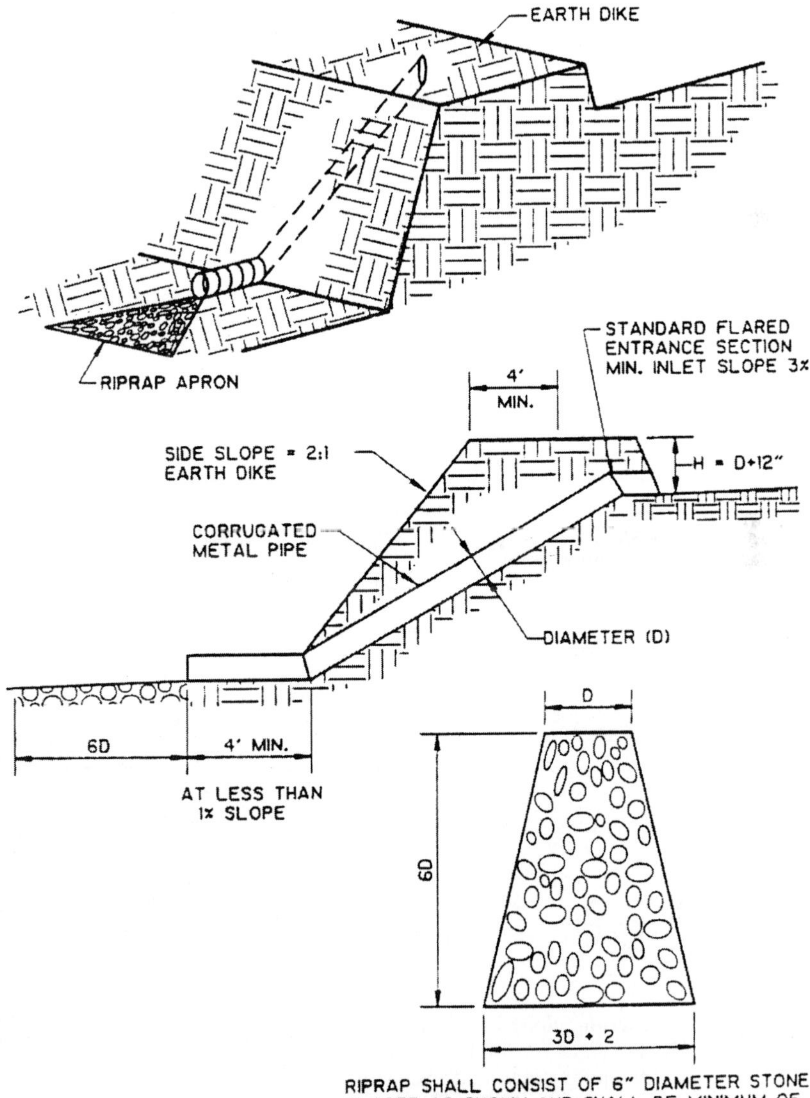

FIGURE 17.18a   Pipe slope drain (rigid). (From FCDMC, 1993)

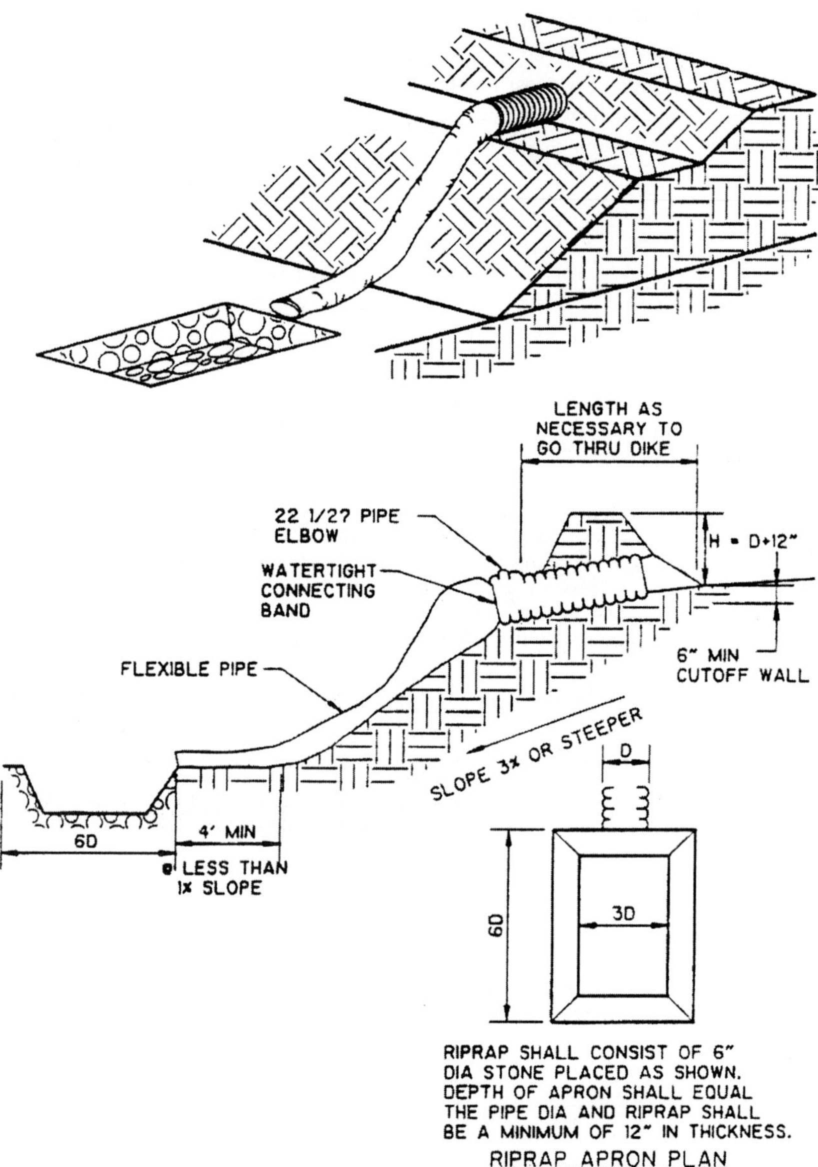

FIGURE 17.18*b* Pipe slope drain (flexible). (From FCDMC, 1993)

1. Matting alone is not acceptable for flow velocities greater than 6 fps (1.8 m/second).
2. The mats recommended by the Flood Control District of Maricopa County (FCDMC) are jute, straw, and excelsior. It is best to use and install these named mats per the manufacturer's specifications.
3. Generally in the installation of mats, there are guidelines to follow for optimum performance. These installation guidelines are provided by ADOT (1998) from site preparation to final inspection check.

17.3.10 Pipe Slope Drain

A. *Description.* A temporary drain pipe placed along the slope as illustrated in Figs. 17.18a and b.

B. *Purpose.* To convey concentrated runoff from the top of the slope to the bottom of slope.

C. *Application.* Typically, pipe slope drains are used in conjunction with top of slope diversion berms, swales, or dikes to drain concentrated flows down the slope. It could be used as an emergency spillway for a sediment basin.

D. *Limitations.* Drainage areas for effective use of pipe slope drains are 2 hectares (5 acres). Larger drainage areas require paved spillways, rock-lined chutes or channels. Multiple pipe slope drains could be used for sites serving larger drainage areas.

E. *Design considerations.* Design specifications for pipe slope drain provided by ADOT (1998) follow:

    1. Temporary pipe drains should be designed to convey 10-year, 24-hr peak flow.
    2. Pipe slope drains should not be smaller than 610 mm (24 in).
    3. The drainage area size served by the pipe slope drain should not exceed 2 hectares (5 acres).
    4. The inlet should consist of a standard flared end section for pipe size of 610 mm (24 in) and larger with a minimum 150-mm (6-in) metal toe plate to prevent runoff from undercutting the pipe inlet. The flared inlet section should be securely connected to the slope drain with watertight connecting bands.
    5. The slope of the entrance should be at least 3%.
    6. The soil around and under the pipe and entrance section should be thoroughly compacted.
    7. Slope drain pipe sections should be securely fastened together and anchored into the soil.
    8. Slope drain may require interceptor dikes to direct runoff to the pipe. The height of the dike should be at least 0.3 m (1 ft) higher than the top of the inlet pipe.
    9. The outlet section must be stabilized with riprap apron to prevent undercutting.
    10. In cases where pipe slope drains are conveying sediment laden water, all flows should be directed into a sediment trapping device.

17.3.11 Silt Fence

A. *Description.* A temporary sediment barrier comprising a filter fabric attached to supporting posts and entrenched or secured into the soil as illustrated in Fig. 17.19.

B. *Purpose.* To intercept and detain small amounts of sediment from disturbed areas during construction operation or prior to attaining a stable surface condition. The silt fence serves to collect and detain silts or sediments from leaving the disturbed area, thus preventing the materials to impact downstream system.

C. *Application.* Typically, silt fences are placed upstream of a point of discharge where silts or sediments have the potential to impact the system. They are appropriate to be installed below disturbed areas where runoff may occur in the form of sheet and rill erosion and wherever the runoff has the potential to impact downstream resources.

D. *Limitations.* Problem associated with undercutting, overtopping, or collapsing is prevalent for improperly installed fences.

E. *Design considerations.* Silt fence must be designed based on the following guidelines provided by Goldman et al. (1986):

    1. Maximum drainage area served by silt fence should be 1 acre (0.4 hectare) or less.
    2. Silt fence is effective for concentrated flows not exceeding 1.0 cfs (0.028 m^3/second).

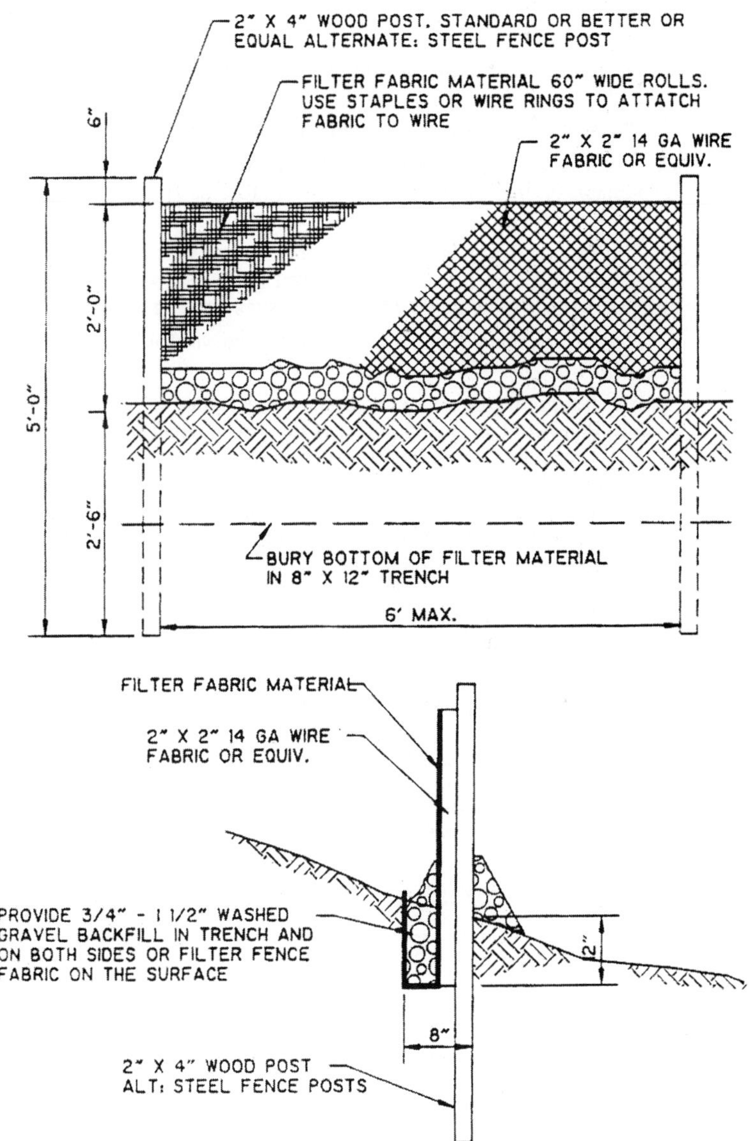

FIGURE 17.19   Silt fence. (From FCDMC, 1993)

3. Maximum allowable flow path length to the fence is 100 ft (30 m).
4. Maximum slope steepness is 2:1.

17.3.12   Protection of Trees and Natural Areas in Construction Areas

A. *Description.* Trees and natural areas are marked to be salvaged or protected from mechanical injury while construction activity in the site is taking place as illustrated in Fig. 17.20.

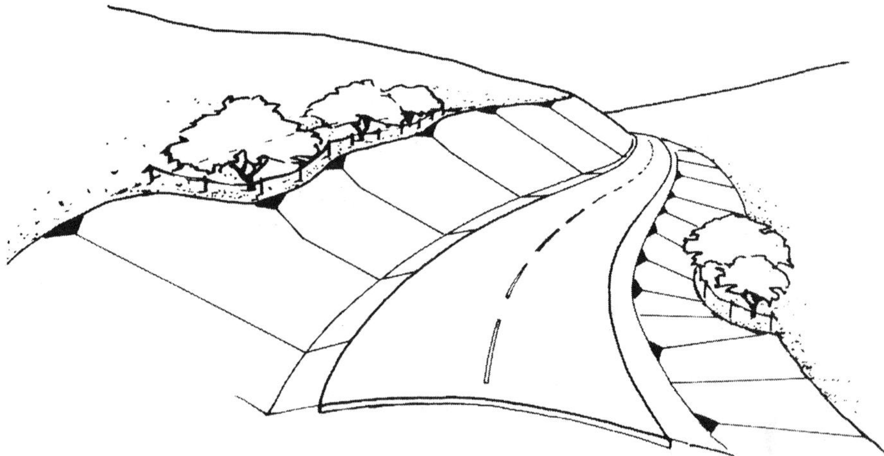

FIGURE 17.20   Protection of trees/natural areas in construction areas. (From ADOT, 1998)

B. *Purpose.* To provide protective measures that will ensure the survival of desirable trees and natural areas

C. *Application.* For trees and natural areas that will not interfere with construction activities; their protection is necessary to maintain natural environment, health, and habitat value.

D. *Limitations.* Salvaging trees and natural areas is often difficult and expensive for sites with very diverse topography and terrain. Saving existing trees is sometimes very impractical to implement specially if sites are to be graded satisfactorily.

E. *Design considerations.* ADOT (1998) has provided the following guidelines:

    1. Trees can be protected from construction grading by building rock-lined tree wells. The well size depends on the size of the tree to be protected as well as the surrounding grades.

    2. Mature trees have large root system that helps hold soil in place and thus reduce erosion. In many cases, a tree cannot be saved if a significant portion of its root system must be removed during construction.

17.3.13 Sandbag Berm

A. *Description.* A temporary berm made up of stacked sandbags installed across a channel or right-of-way in a developing or disturbed area as illustrated in Fig. 17.21.

B. *Purpose.* Stacked sandbags functioning as a berm would serve to intercept sediment-laden flows from disturbed areas.

C. *Application.* Sandbag berms could be used to create temporary sediment traps, retention basins and in place of straw bales or silt fences to filter sediments. Sandbags stacked up could also serve as check dams across stream channels to slow down flow velocity.

D. *Limitations.* Use is restricted to construction of low berms with 18 in maximum.

E. *Design considerations.* The FCDMC (1993) has provided the following criteria as design guide for sandbag berms:

    1. Sandbag berms are effective for drainage areas less than 10 acres.

    2. Berm height should not exceed 24 in. Berm width should be 48 in minimum at the base and 18 in minimum at the top.

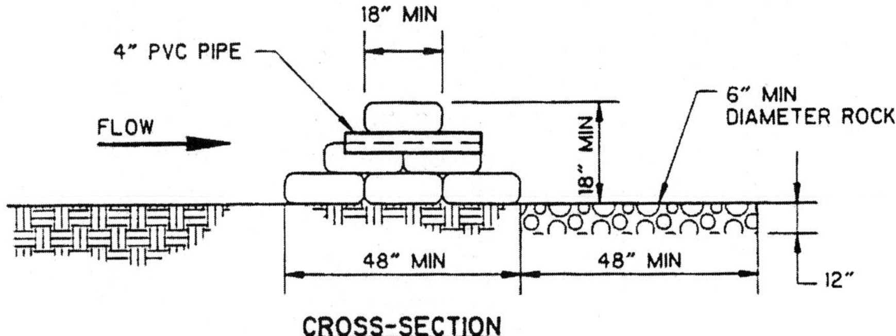

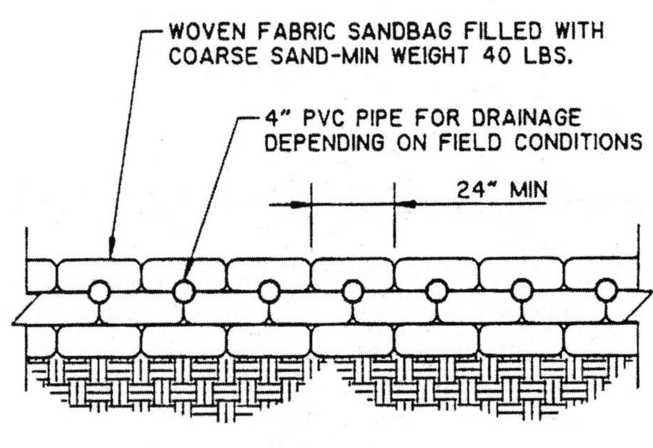

FIGURE 17.21   Sandbag berm. (From FCDMC, 1993)

3. Recommended sandbag sizes are: length 24 to 36 in; width 16 to 18 in, and thickness 6 to 8 in. Weight should be 90 to 125 pounds.

4. Coarse sand and gravel are used as materials.

5. Sandbag material should be made up of one of the following: polypropylene, polyethylene, or polyamide woven fabric. Minimum unit weight is four ounces per square yard. Mullen burst strength exceeding 300 psi and ultraviolet stability exceeding 70%.

6. When silt accumulation has reached 6 in deep, remove and properly dispose of accumulated sediments.

17.3.14 Construction Road Stabilization

A. *Description.* Temporary stabilization of access roads, subdivision roads, parking areas, and other on-site vehicle transportation routes with gravel or by chemical stabilization immediately after grading.

B. *Purpose.* To reduce erosion of temporary road beds due to construction traffic.

C. *Application.* For phased construction projects where roadways are graded but are not paved immediately. Road stabilization is also appropriate for detour roadways where roads are often unpaved. Stabilization is also recommended for roadway constructions during wet weather conditions.

D. *Limitations.* Roads graded for construction vehicles are especially susceptible to erosion. The exposed soil surface is continually disturbed by traffic resulting in erosion, dust problems, and transport runoff waters along their surfaces. During wet weather, the roads may generate significant quantities of sediment that may be transported off-site in surface runoff or on the wheels of construction vehicles.

E. *Design considerations.* The FCDMC (1993) has provided the following recommendations for road stabilization:

1. A 6-in course of 2- to 4-in crushed rock, gravel base, or crushed surfacing base course should be applied immediately after surface grading or after the completion of utility installation within the right-of-way. A 4-in course of aggregate base course might be used in lieu of the crushed rock.

2. Chemical stabilization may also be applied upon compacted native sub-grade. Application of chemical controls should be made per the manufacturer's directions.

3. Temporary roads should follow closely the contour of the natural terrain and slopes used should not exceed 15%.

4. Drainage swales should be provided on each side of the roadway for normal crown road sections. For super-elevated road sections, drainage swales are placed on the downstream side.

17.3.15 Temporary Drainage Swale

A. *Description.* A temporary drainage way with a lining of grass, stone, asphalt, concrete, or other appropriate materials as illustrated in Figure 17.22.

B. *Purpose.* Swales are used as perimeter control or as slope protection to convey runoff effectively without causing erosion. This is achieved by intercepting the runoff from above unprotected slopes or at the perimeter and directing the flow to a stabilized outlet or to a sediment trapping device.

C. *Application.* For disturbed areas or exposed slope surfaces needing to divert flows to minimize erosion. Swales are also used to direct on-site sediment laden water to a sediment trapping device for settlement.

D. *Limitations.* Adequate design capacity is essential for swales to effectively function as intended. Swales could be costly as lining may be required.

E. *Design considerations.* The FCDMC (1993) recommended the following specifications for swale design:

1. For swale stability, a lining material comprised of properly sized rocks is recommended as provided in the table below.

Treatment	Channel grade	Rock size, D_{50}[1]	Rock size, D_{50}[2]
1	0.5–1.0%	4″	4″
2	1.1–2.0%	6″	6″
3	2.1–3.0%	8″	6″–12″ riprap
4	3.1–5.0%	8″–12″ riprap	Engineered

Note:
[1]For drainage areas less than 5 acres.
[2]For drainage areas from 5 to 10 acres.

2. Side slopes should be 3:1 or flatter.

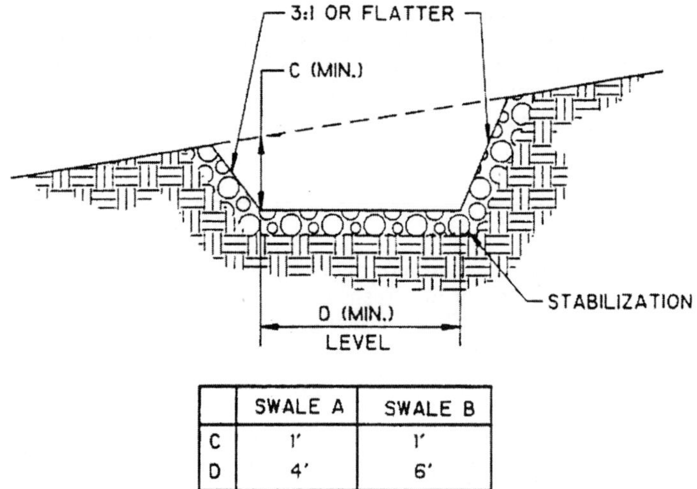

	SWALE A	SWALE B
C	1'	1'
D	4'	6'

CROSS SECTION

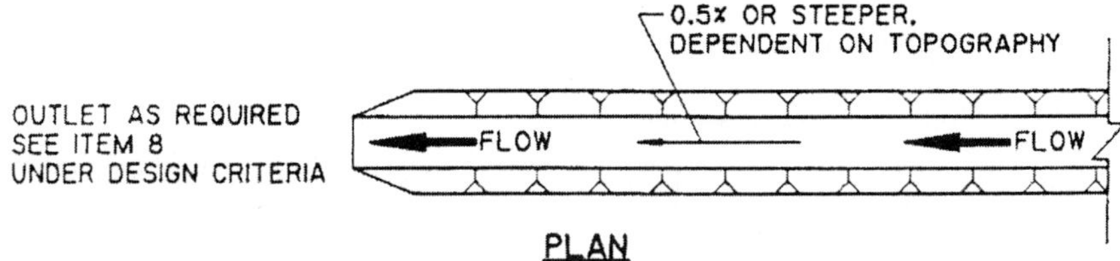

PLAN

FIGURE 17.22   Temporary drainage swale. (From FCDMC, 1993)

3. For drainage areas less than 5 acres, a minimum bottom width of 4.0 ft is recommended. For larger drainage areas of 5 to 10 acres, a minimum bottom width of 6 ft should be used.

4. For larger channel systems, either the method of permissible velocity or the method of tractive force can be used. A summary of these methodologies will be provided in the following paragraphs.

Maximum Permissible Velocity
As defined in Chow (1959), the maximum permissible velocity is the greatest mean velocity that will not cause erosion of the channel body.

In the design of stable channels Table 17.8 can be used to establish the maximum permissible velocities for a variety of soil materials. This table also provides equivalent unit tractive force values converted by the U.S. Bureau of Reclamation, and Manning's roughness (*n*) values. The permissible velocities presented in Table 17.8 were established by Fortier and Scobey (1925) for well-seasoned, straight channels of small slopes with flow depths less than 3 ft. A more comprehesive range of values for noncohesive and cohesive soils are provided in Figs. 17.23 and 17.24, respectively.

For sinous channels, the velocities provided in Table 17.8, and Figs. 17.23 and

TABLE 17.8 Maximum Permissible Velocities Recommended by Fortier and Scobey and Unit Tractive Force Values Converted by the U.S. Bureau of Reclamation[1]

Material	Manning's n value[2]	Clear water V_p (ft/s)	Clear water τ_o (lb/ft²)	Water transporting colloidal silts V_p (ft/s)	Water transporting colloidal silts τ_o (lb/ft²)
Fine sand, colloidal	0.020	1.50	0.027	2.50	0.075
Sandy loam, noncolloidal	0.020	1.75	0.037	2.50	0.075
Silt loam, noncolloidal	0.020	2.00	0.048	3.00	0.11
Alluvial silts, noncolloidal	0.020	2.00	0.048	3.50	0.15
Ordinary firm loam	0.020	2.50	0.075	3.50	0.15
Volcanic ash	0.020	2.50	0.075	3.50	0.15
Stiff clay, very colloidal	0.025	3.75	0.26	5.00	0.46
Alluvial silts, noncolloidal	0.025	3.75	0.26	5.00	0.46
Shales and hardpans	0.025	6.00	0.67	6.00	0.67
Fine gravel	0.020	2.50	0.075	5.00	0.32
Graded loam to cobbles when noncolloidal	0.030	3.75	0.38	5.00	0.66
Graded silts to cobbles when colloidal	0.030	4.00	0.43	5.50	0.80
Coarse gravel, noncolloidal	0.025	4.00	0.30	6.00	0.67
Cobbles and shingles	0.035	5.00	0.91	5.50	1.10

[1] The Fortier and Scobey values were recommended for use in 1926 by the Special Committee on Irrigation Research of the American Society of Civil Engineers.

[2] See Chow (1959) for a more comprehensive range of roughness values.

Source: From Chow (1959).

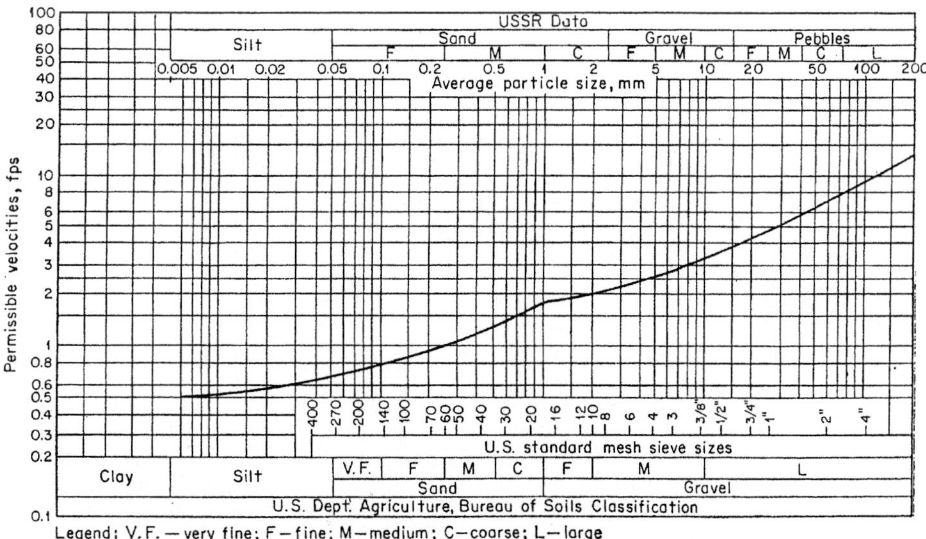

FIGURE 17.23   U.S. and U.S.S.R. data on permissible velocities for noncohesive velocities. (From Chow, 1959)

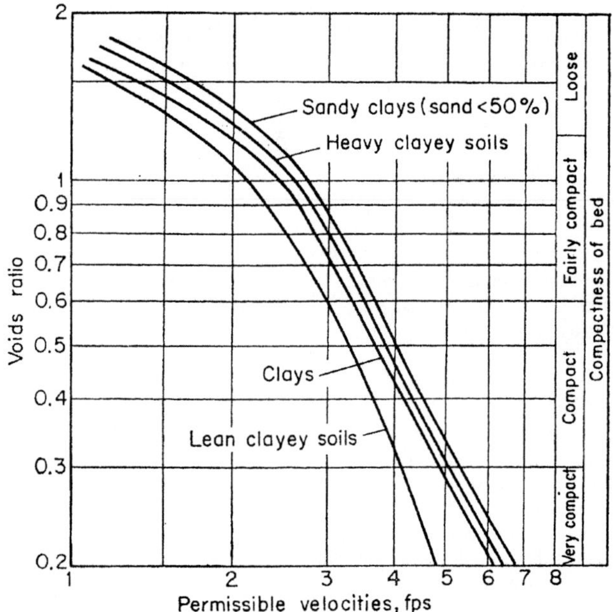

FIGURE 17.24 Curves showing U.S.S.R. data on permissible velocities for cohesive soils. (From Chow, 1959)

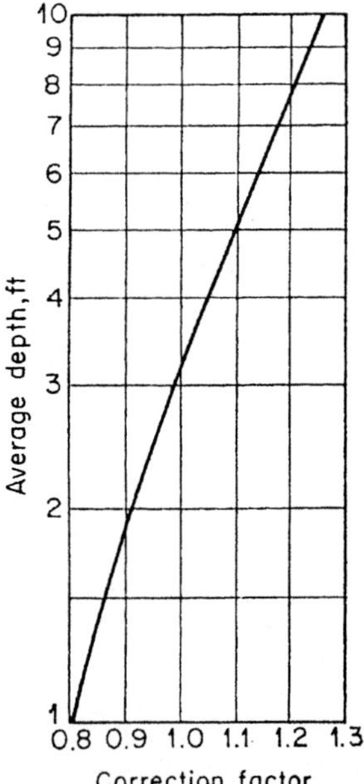

FIGURE 17.25 Curves showing U.S.S.R. corrections of permissible velocity for depth for both cohesive and noncohesive materials. (From Chow, 1959)

17.24 should be reduced by using the correction factors shown in Fig. 17.25. Lane (1955), in a later study suggested a reduction of 5% for slightly sinous channels, and 22% for very sinous channels. These corrections should be used as a preliminary guideline and the final design values should be based on specific site conditions.

Method of Permissible Velocity
The following step-by-step procedure can be used to design channels by using the permissible velocity approach:

Step 1. Determine the roughness coefficient n (Table 17.8), side slope z, and the maximum permissible velocity V_p (Table 17.8 and Figs. 17.23 to 17.25). For natural channels Table 17.9 can be used to estimate the side slope z.

Step 2. Compute the hydraulic radius, R_f, by using Manning's equation,

$$V_p = \left(\frac{1.49}{n}\right)R_f^{2/3} S_f^{1/2} \qquad \text{(English units)} \qquad (17.13)$$

$$V_p = \left(\frac{1.0}{n}\right)R_f^{2/3} S_f^{1/2} \qquad \text{(metric units)} \qquad (17.14)$$

where V_p = maximum permissible velocity (ft/s or m/second)
n = roughness factor
R_f = hydraulic radius (ft or m)
S_f = average channel bottom slope (ft/ft or m/m)

Step 3. Compute the water area, A_f, corresponding to the design discharge and maximum permissible velocity by using

$$A_f = Q_D/V_p \qquad (17.15)$$

where Q_D = Design discharge (ft³/second or m³/second)
V_p = Maximum permissible velocity (ft/second or m/second)

Step 4. Compute the wetted perimeter, $P_f = A_f/R_f$.

Step 5. Using the expression for A_f and P_f computed in steps 3 and 4, solve simultaneously for b and d.

Tractive force. As defined by Chow (1959) tractive force acts along the wetted perimeter of the channel and is equal to the effective component of the gravity force. The magnitude

TABLE 17.9 Suitable Side Slopes for Channels Built in Natural Channels

Material	Side slope
Rock	Nearly vertical
Muck and peat soils	¼:1
Stiff clay or earth with concrete lining	½:1 to 1:1
Earth with stone lining, or earth for large channels	1:1
Firm clay or earth for small ditches	1½:1
Loose sandy earth	2:1
Sandy loam or porous clay	3:1

Source: From Chow (1959).

of this force can be computed by using the following equation:

$$\tau = \gamma A L_c S \tag{17.16}$$

where τ = tractive force (lb/ft² or N/m²)
γ = unit weight of water (lb/ft³ or N/m³)
L_c = length of the channel reach (ft or m)

The average value of the tractive force per unit wetted area (unit tractive force), τ_o, can be computed as

$$\tau_o = \gamma R_f S_f \tag{17.17}$$

In wide open-channels, the hydraulic radius (R_f) is equal to flow depth (d) and Eq. (17.17) can be expressed as

$$\tau_o = \gamma d S_f \tag{17.18}$$

where d = flow depth (ft or m)

Except for wide open channels, the unit tractive force is not uniformly distributed along the wetted perimeter. Olsen and Florey's (1952) study for distribution of tractive force in a trapezoidal channel ($z = 1.5$, bottom width/flow depth ratio, $b/d = 4$) is shown in Fig. 17.26. For other channel configurations, Fig. 17.27 must be used to compute the unit tractive force values. The values provided in Fig. 17.27 are independent of the channel size.

An important design parameter is the ratio of the unit tractive force that causes impending motion on a level surface, τ_L, to the unit tractive force that causes impending motion on a sloping surface, τ_s, and can be expressed as

$$K_c = \frac{\tau_s}{\tau_L} = \cos\phi \sqrt{1 - \frac{\tan^2\phi}{\tan^2\theta}} \tag{17.19}$$

where ϕ = angle of the side slope
θ = angle of repose of the material

Simplifying Eq. (17.19),

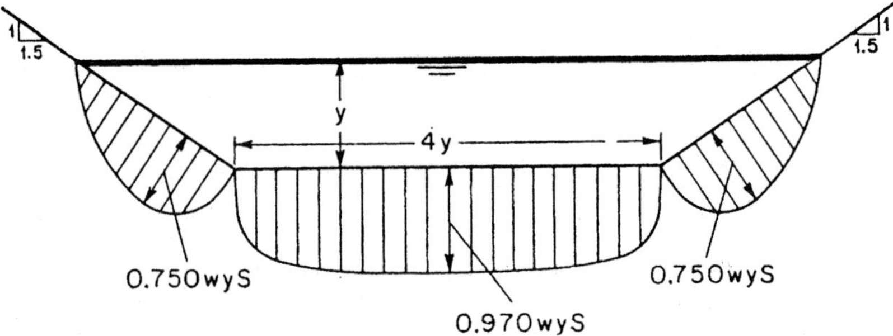

FIGURE 17.26   Distribution of tractive force in a trapezoidal channel section. (From Chow, 1959)

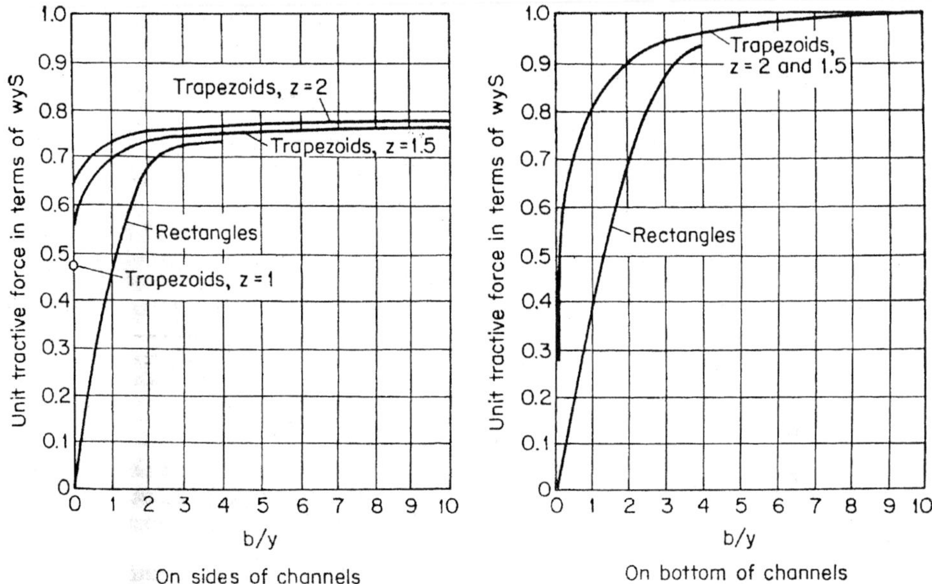

FIGURE 17.27   Maximum unit tractive forces in terms of $\gamma y s$. (From Chow, 1959)

$$K_c = \sqrt{1 - \frac{\sin^2\phi}{\sin^2\theta}} \tag{17.20}$$

As stated by Chow (1959), for cohesive and fine noncohesive materials, the cohesive forces, even with comparatively clear water, become so great in proportion to the gravity force component causing the particle to roll down that the gravity force can safely be neglected. Therefore, the angle of repose should only be considered for coarse noncohesive materials and for cohesive materials the following equation should be used to estimate K_c.

$$K_c = \sqrt{1 - \sin^2\phi} \tag{17.21}$$

A U.S. Bureau of Reclamation (USBR) study showed that angle of repose is a function of the size and angularity of the material. The results of this study are presented in Fig. 17.28. The particle size shown in this figure is the D_{75} particle size, the diameter of a particle by which 75% (by weight) of the material is smaller.

Permissible Tractive Force
The permissible tractive force is the maximum tractive force that will not cause severe erosion of the material forming the channel bed on a level surface (Chow, 1959). The following equations can be used to compute the permissible tractive force for coarse and fine noncohesive, and cohesive soils.

Coarse noncohesive soil. A preliminary guideline provided by USBR is:

$$\tau_p = 0.4 \, D_{75} \tag{17.22}$$

where τ_p = permissible tractive force (lb/ft²)

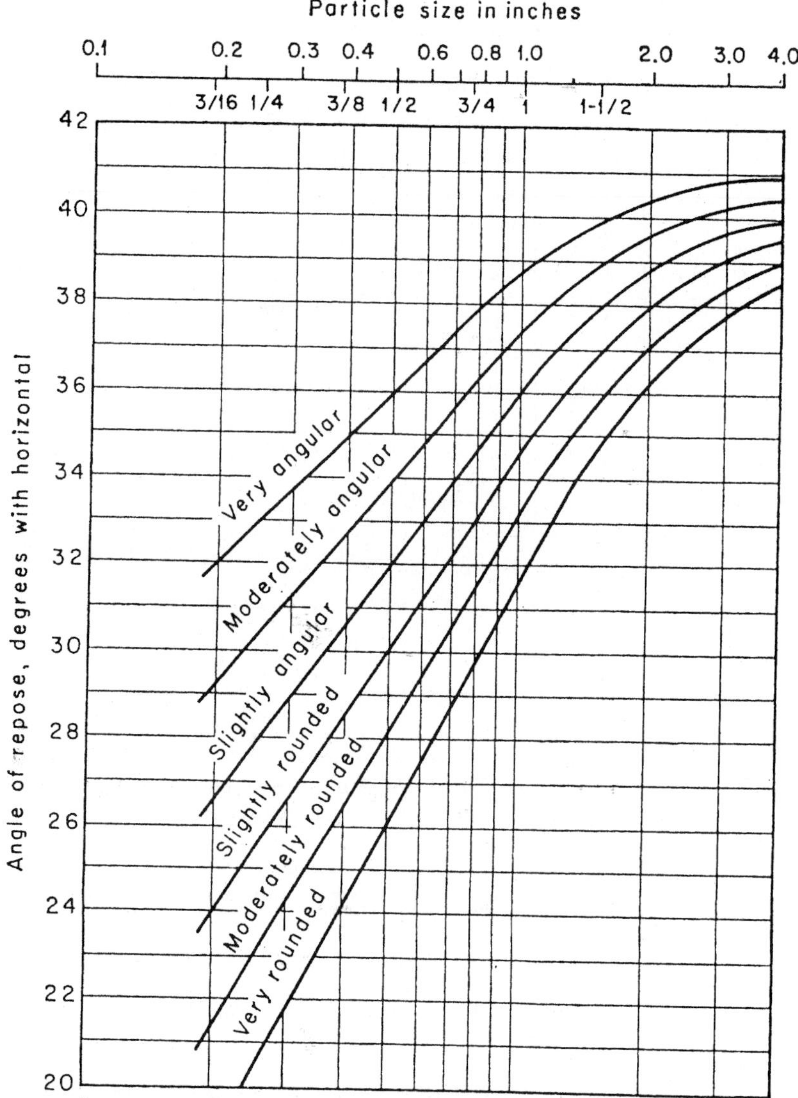

FIGURE 17.28   Angles of repose of noncohesive material. (From Chow, 1959)

This recommendation is shown by the straight line in Fig. 17.29. The average particle diameter that will be used for coarse noncohesive soils is D_{75}.

Fine noncohesive soil. The three design curves associated with fine noncohesive materials are also presented in Fig. 17.29. The average particle diameter that will be used for fine noncohesive soils is D_{50}, particle size by which 50% (by weight) of the material is smaller. These design curves represent for a variety of canal sediment content including:

- Canals with high content of fine sediment in water
- Canals with low content of fine sediment in water
- Canals with clear water

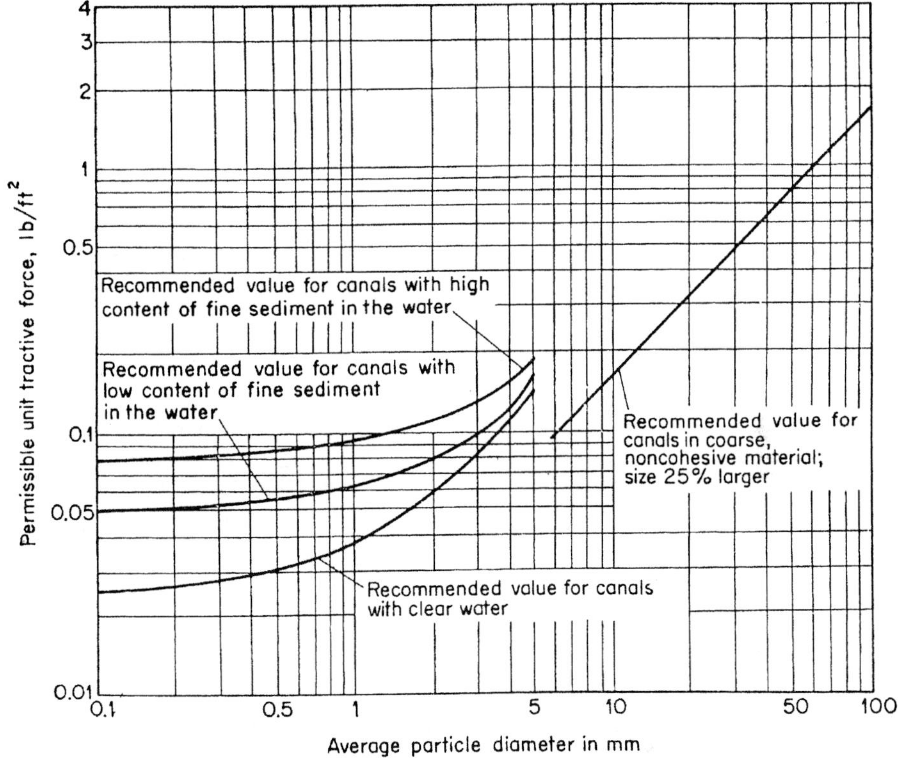

FIGURE 17.29   Recommended permissible unit tractive forces for canals in noncohesive material. (From Chow, 1959)

Cohesive soil. For cohesive soils the unit tractive forces given in Table 17.8 and Fig. 17.30 are recommended for design.

The permissible tractive forces recommended in the previous paragraphs are applicable to straight channels. For sinuous channels, the values should be adjusted as suggested by Lane (1955). 10-, 25-, and 40% reductions can be applied to slightly sinuous, moderately sinuous and very sinuous canals, respectively.

Method of Permissible Tractive Force

The following step-by-step procedure can be used to design channels by using the permissible tractive force approach:

Step 1.   Make initial assumptions of channel dimension.
- Assume a width-to-depth ratio, b/d, and side slope, z.
- Use Fig. 17.27 to establish the maximum unit tractive force.
  Note: Compare the maximum unit tractive force on the sides, τ_{mos}, with the maximum unit tractive force on channel bottom, τ_{mom}. The controlling value for design is the smaller of the two unit tractive forces.

Step 2.   Compute the tractive force ratio, K_c.
  For noncohesive materials: Use the following procedure for the determination of K_c.
  a. Based on field observations and/or soil experiments identify the typical angularity of soil particles with D_{75} size.

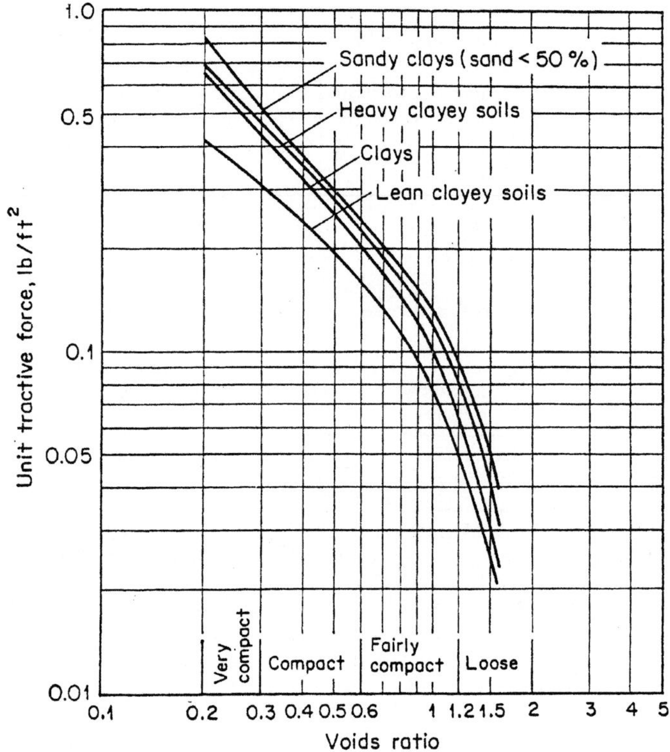

FIGURE 17.30 Permissible unit tractive forces for canals in cohesive material as converted from the U.S.S.R. data on permissible velocities. (From Chow, 1959)

    b. Given the particle size, D_{75}(in), and angularity, determine the associated angle of repose from Fig. 17.28.
    c. Use Eq. (17.20) to compute K_c.
    For cohesive materials: Use Eq. (17.21) to compute K_c.

Step 3. Determine the permissible unit tractive force on channel bottom, τ_{mp}
For noncohesive soils, use either Eq. (17.22) or Fig. 17.29.
For cohesive soils, use Table 17.8 and Fig. 17.30.

Step 4. Compute the permissible unit tractive force on channel sides, τ_{ms}.
Since Eq. (17.19) represents the ratio of unit tractive forces on the sides and bottom of the channel, it can be used to compute the ratio of permissible unit tractive forces on the sides and bottom of the channel as

$$K_c = \frac{\tau_{ms}}{\tau_{mp}}$$

Step 5. For a state of impending motion of the particles, the maximum unit tractive force should be equal to the permissible unit tractive force. Use this condition to compute depth of flow, d.

Step 6. Use the b/d ratio in step 1 and result of step 5 to compute b.

Step 7. Use the results of steps 5 and 6 to compute cross-sectional area, A_f, and hydraulic radius, R_f.

Step 8. Given the roughness coefficient, n, channel bottom slope, S_f, cross-sectional area, A_f, and hydraulic radius, R_f, compute the design discharge, Q_D, by using Manning's equation.

Step 9. Compare the design discharge, Q_D (step 8) with the design capacity, Q_C, of the channel given in step 1.

    a. If $| (Q_D - Q_C)/Q_C | < 5\%$, continue with step 10.

    b. If $| (Q_D - Q_C)/Q_C | \geq 5\%$, change the *b*/*d* ratio assumed in step 1 and repeat steps 5–9 until the requirement in step 9a is satisfied.

Step 10. Check the maximum unit tractive force. The maximum unit tractive force must be less than the permissible tractive force. If the controlling value identified in step 1 is the channel bottom,

- Use Fig. 17.27 and channel dimensions from step 9a to compute the maximum unit tractive force on the channel bottom, τ_{mom}.
- Compare τ_{mom} to the permissible tractive force on the channel bottom, τ_{mp}, computed in step 3.

If $\tau_{mom} < \tau_{mp}$, design dimensions computed in step 9a are adequate. The designer should finalize the channel dimensions by incorporating any other design requirements enforced by local government agencies.

If $\tau_{mom} \geq \tau_{mp}$, change the channel dimensions to lower the maximum shear tractive force on the channel bottom.

If the controlling value identified in step 1 is for the channel sides,

- Use Fig. 17.27 and design channel dimensions from step 9a to compute the maximum unit tractive force on the sides, τ_{mos}.
- Compare τ_{mos} to the permissible tractive force on the channel sides, τ_{ms}, computed in step 4.

If $\tau_{mos} < \tau_{ms}$, design dimensions computed in step 9a are adequate. The designer should finalize the channel dimensions by incorporating any other design requirements enforced by local government agencies.

If $\tau_{mos} \geq \tau_{ms}$, change the channel dimensions to lower the maximum shear tractive force on the channel bottom.

F. *Design examples.* The following examples apply the two design methods presented above.

Example 17.2
Design Problem Using Permissible Velocity Method
Problem
Compute the bottom width and depth of flow of a trapezoidal channel laid on a slope of 0.0014, carrying a design discharge of 500 cfs. The soil material contains noncolloidal coarse gravels and pebbles.

Solution

Step 1. By using the given conditions the following parameters can be estimated:

    where the soil material is noncolloidal coarse gravels and pebbles
    Average particle size = 25 mm
$$n = 0.025 \text{ (Table 17.8)}$$
$$z = 2 \text{ (Table 17.9)}$$
$$V_p = 4.5 \text{ ft/s. (Fig. 17.23)}$$

Step 2. Compute R_f by using Manning's equation;

$$V_p = \left(\frac{1.49}{n}\right) R_f^{2/3} S_f^{1/2}$$

$$4.5 = \left(\frac{1.49}{0.025}\right) R_f^{2/3}\, 0.0014^{1/2} \Rightarrow R_f = 2.86 \text{ ft}$$

Step 3. Compute the water area, $A_f = Q_D/V = 500/4.5 = 88.8 \text{ ft}^2$

Step 4. Compute the wetted perimeter, $P_f = A_f/R_f = 88.8/2.86 = 31.1 \text{ ft.}$

Step 5. For a trapezoidal channel with side slope $z = 2{:}1$,

$$A_f = (b + 2d)\, d = 88.8 \text{ ft}^2$$

$$P_f = b + 2\sqrt{5}(d) = 31.1 \text{ ft}$$

Solving the above two equations simultaneously for b and d gives, $b = 11.5$ ft and $d = 4.38$ ft.

The channel depth should be finalized by considering the freeboard and any other design requirements enforced by local government agencies.

Example 17.3
Design Problem Using Tractive Force Method
Problem
Design a trapezoidal channel laid on a slope of 0.0014 and carrying a discharge of 500 cfs (Q_c). The channel will be excavated on noncolloidal coarse gravels and pebbles, $D_{75} = 1.25$ in and Manning's $n = 0.025$.

Solution

Step 1.   Initial assumptions.
  Side slopes (2:1), $z = 2$
  Base-to-depth ratio, $b/d = 5$

  As shown in Fig. 17.27, the maximum unit tractive force on the sides is usually less than that on the bottom. Therefore, the design should be based on the maximum unit tractive force on the sides of the channel.
    From Fig. 17.27, the maximum unit tractive force on the sloping sides is 0.775 $\gamma\, d\, S_f$, and can be calculated by using the given data as

$$0.775 \times 62.4 \times 0.0016\, d = 0.077d \text{ lb/ft}^2$$

Step 2.

  a. Assume a very rounded material 1.25 in diameter
  b. From Fig. 17.28, Angle of repose, $\theta = 33.5°$
  c. Use Eq. (17.20) to compute the tractive force ratio, K_c

$$K_c = \sqrt{1 - \frac{\sin^2\phi}{\sin^2\theta}} = \sqrt{1 - \frac{\sin^2 26.5}{\sin^2 33.5}} = 0.587$$

Step 3.   For a size of 1.25 in, the permissible tractive force on the channel bottom can be computed either from Eq. (17.22) or from Fig. 17.29.

$$\tau_{mp} = 0.4(1.25) = 0.5 \text{ lb/ft}^2 \text{ (same as Fig. 17.29)}$$

Step 4. The permissible tractive force on the sides can be computed by using Eq. (17.19) as:

$$K_c = \frac{\tau_{ms}}{\tau_{mp}} \Rightarrow \tau_{ms} = 0.587 \times 0.5 = 0.294 \text{ lb/ft}^2$$

Step 5. For a state of impending motion of the particles on the sides, (maximum unit tractive force on the sloping sides) = (permissible unit tractive force on the sides)

$$0.077 \, d = 0.294$$

$$d = 3.82 \text{ ft.}$$

Step 6. The bottom width, $b = 3.82 \times 5 = 19.09$ ft.

Step 7. For this trapezoidal section: $A_f = 102.1$ ft^2 and $R_f = 2.82$ ft.

Step 8. Use Manning's equation to compute design discharge, Q_D,

$$Q_D = \left(\frac{1.49}{n}\right) A_f \, R_f^{2/3} \, S_f^{2/3}$$

$$Q_D = \left(\frac{1.49}{0.025}\right)(102.1)(2.82)^{2/3} \, (0.0014)^{1/2}$$

$$Q_D = 455 \text{ ft}^3/\text{s}$$

Step 9. Compare Q_D with Q_C

$$|\,(Q_d - Q_c)/Q_c\,| = |\,(455 - 500)/500\,| = 9\% > 5\%, \text{ therefore revise the } b/d \text{ ratio}$$

Since $Q_d < Q_c$, a higher b/d ratio will be used.
If $Q_d \geq Q_c$, a smaller b/d ratio would be used.

Further computations will show that, for $z = 2$ and $b/d = 5.8$, the section dimensions are $d = 3.77$ ft and $b = 21.86$ ft

Step 10. With $z = 2$ and $b/d = 5.8$. Use Fig. 17.27 to compute the maximum unit tractive force on the channel bottom

$$\tau_{mom} = 0.98 \, \gamma \, y \, S = 0.98 \times 62.4 \times 3.77 \times 0.0014 = 0.327 \text{ lb/ft}^2$$

The permissible tractive force on the channel bottom was computed in step 3,

$$\tau_{mp} = 0.5 \text{ lb/ft}^2$$

Since $\tau_{mom} < \tau_{mp}$, channel dimensions specified in step 9 are acceptable.

17.3.16 Temporary Sediment Basin

A. *Description.* A temporary basin with a controlled stormwater outlet or drainage structure formed by constructing an embankment of compacted soil across a drainageway or other suitable locations as illustrated in Figure 17.31.

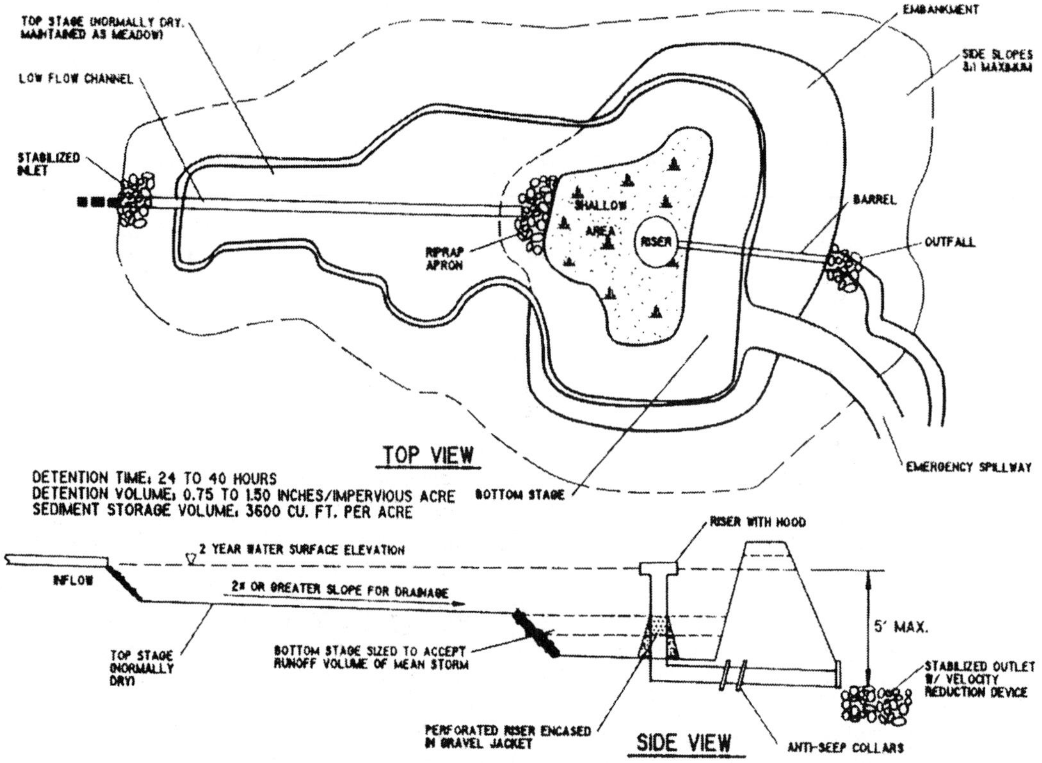

FIGURE 17.31 Temporary sediment basin. (From FCDMC, 1993)

B. *Purpose.* Sediment basin serves to collect and store sediment from sites cleared and/or graded during construction or for extended periods of time before reestablishment of permanent vegetation and/or construction of structures. Intended as a temporary measure with less than 1 yr of design life, sediment basins are maintained until the site is permanently protected against erosion or until a permanent detention basin is constructed.

C. *Application.* Sediment basins are appropriate for nearly all types of construction projects where erosion from exposed land areas becomes a factor to be addressed. Typically located at the stormwater outlet, basins include temporary dikes (berms) and/or channel to divert runoff into the basin inlet. Development projects often need stormwater detention basin which easily becomes a sedimentation pond that impounds sediments.

D. *Limitations.* Sediment basins and ponds are limited in terms of location where possible failure would not result in the loss of life, damages to buildings and homes, or cause interruptions to public roads and utilities. Depending on the size of the basin, fencing may be required for public safety considerations. Sediment basins should not hold stormwater more than 36 hrs as this becomes a public health issue. Since basins are at best only 70 to 80% effective in trapping sediments, they should be used in conjunction with other erosion control practices.

E. *Design considerations.* The FCDMC (1993) has provided the following recommendations for sediment basins:

1. Drainage areas of 10 acres or more are served by sediment basins.

2. The total sediment pond volume is 3600 cubic ft/acre of upstream disturbed soil. The pond volume includes a 1-ft deep sediment storage, and 2 to 4 ft of a settling volume.
3. Baffles to spread the flow throughout the basin should be included in the design. Stabilized inlet sections into the basin are necessary to prevent erosion.
4. Discharge mechanism for basins includes a securely anchored riser pipe as principal outlet along with an emergency overflow spillway. Stabilized outlet should be provided for the principal and emergency spillways to reduce or prevent erosion.
5. Basin side slopes are typically built with a 3:1 or flatter slope.
6. Typically for sediment basins, emergency spillway crests are about 1 ft above the top of the riser pipe.
7. A 6:1 aspect ratio between the basin length and width of the pond is desirable.
8. Other detailed design considerations include

Basin surface area. Based on a U.S. Environmental Protection Agency publication on erosion control for surface mining (1976) sediment basin surface area is

$$A_s = \frac{1.2Q}{V_s}$$

where A_s = the appropriate surface area for trapping particles of a certain size
V_s = settling velocity for that particle size

Basin efficiency. Basin efficiency is primarily a function of the particle size distribution of the inflowing sediment. Theoretically, the particles larger in size and density than the design particle size will be retained in the basin. During the de-watering process some particles of smaller sizes will also settle out. Since this additional settling accounts for a maximum of 10% of the total sediment load, and can be offset by flow events larger than the design storm, it will not be considered in the design.

Therefore, ideal basin efficiency corresponds to the percent of soil equal to or larger than the design particle size. For example, if a sediment basin on a site is designed to capture the 0.02 mm particle and 64% of the particles on this site are greater than or equal to 0.02 mm, the maximum efficiency of the basin is 64%. The only practical way to increase this efficiency is to increase the surface area of the basin (Goldman, 1986).

Design particle size. Table 17.10 can be used as a guideline in selecting the design particle size. As shown in this table, for soils with high content of clay or fine silt, increasing the size of a basin does not result in a proportional increase in basin efficiency. The key factor is to balance the sediment capturing capacity of a basin with the cost-effectiveness of a certain basin size.

Settling depth. In cases where the basin is shallow, water flowing through the basin

TABLE 17.10 Surface Area Requirements of Sediment Traps and Basins

Particle size mm	Soil type	Settling velocity ft/s	Surface area requirements ft² per ft³/s discharge
0.5	Coarse sand	0.19	6.3
0.2	Medium sand	0.067	17.9
0.1	Fine sand	0.023	52.2
0.05	Coarse silt	0.0062	193.6
0.02	Medium silt	0.00096	1,250.0
0.01	Fine silt	0.00024	5,000.0
0.005	Clay	0.00006	20,000.0

may resuspend settled particles and decrease efficiency of capture. Therefore, the ratio of the basin length, L_B, to the basin depth, D_B is an important design factor. A relation presented by Goldman (1986) can be expressed as

$$\frac{L_B}{D_B} \leq 3.1 \, \frac{R^{1/6}}{n}$$

where n = Manning's roughness coefficient (typically 0.016 to 0.020)
$\quad\ \ R_f$ = hydraulic radius, ft

A conservative basin design criterion would be to keep the ratio of length to settling zone depth less than 200. A basic design recommendation is to maintain the settling zone of a sediment basin free of accumulated sediment for a depth of at least 2 ft (0.61 m).

Sediment storage depth. The storage zone must be large enough to contain sediment deposits without decreasing the settling volume. The sediment yield to the basin can be estimated by using the universal soil loss equation presented in Sec. 17.4.1. The depth required for storage can be determined by using the following equation:

$$\text{Required Basin Depth} = \frac{\text{Estimated Annual Sediment Yield}}{\text{Surface Area of the Basin}}$$

This is the depth that will provide storage for one year's accumulated sediment. In cases where the basin will be cleaned more frequently than once per year, the storage depth may be proportionately shallower.

De-watering. A sediment basin should have a means for de-watering it after a storm. De-watering is required to avoid accumulation of water in basins that can result in health and safety hazards, and to remove the accumulated sediment easier

---

Example 17.4
Design example
Problem
A construction site, on a hillside, has a drainage area of 15 acres (6.1 hectares). The area will be smoothly graded without seeding or mulching. House pads will be terraced up the slope with the following final gradients and slope lengths:

- 5 acres (2 hectares): cut and fill embankments, 4:1 slopes, 50 ft (15.2 m) long

- 10 acres (4.1 hectares): 20:1 slopes, 200 ft (61 m) long

Diversions will carry runoff away from the steep slopes and convey all site runoff to a single sediment basin. The particles of 80% of the soil, by weight, are equal to or larger than 0.02 mm. Average annual rainfall is 30 in (762 mm), and the 10-year, 6-hr rainfall is 3.0 in (76.2 mm). The rainfall erosion index (R factor) is 30, and the K factor is 0.35.
Find

- Required surface area to capture the 0.02 mm and larger particles
- Storage volume required to retain 1 year's soil loss
- Basin depth required for once per year cleaning

Solution
Required Surface Area

Step 1. Calculate average runoff by using the 10-yr, 6-hr rainfall and the rational method.

$$Q_{\text{avg}} = C_r \, i \, A_D$$

TABLE 17.11 Runoff Coefficients for Use with the Rational Method

	Minimum design values				Maximum design values			
Stormwater frequency (years)	2–10	25	50	100	2–10	25	50	100
Streets and roads								
Asphaltic	0.70	0.77	0.84	0.88	0.76	0.84	0.91	0.95
Concrete	0.75	0.83	0.90	0.94	0.80	0.88	0.96	0.95
Brick	0.70	0.77	0.84	0.88	0.72	0.79	0.86	0.90
Gravel roadways and shoulders	0.60	0.66	0.72	0.75	0.70	0.77	0.84	0.88
Industrial area								
Light areas	0.60	0.66	0.72	0.75	0.70	0.77	0.84	0.88
Heavy areas	0.70	0.77	0.84	0.88	0.80	0.88	0.96	0.95
Business areas								
Downtown	0.75	0.83	0.90	0.94	0.85	0.94	0.95	0.95
Neighborhood	0.55	0.61	0.66	0.69	0.65	0.72	0.78	0.81
Lawns								
Sandy soil, flat, 2%	0.05	0.06	0.06	0.06	0.10	0.11	0.12	0.13
Sandy soil, average, 2–7%	0.10	0.11	0.12	0.13	0.15	0.17	0.18	0.19
Sandy soil, steep, 7%	0.15	0.17	0.18	0.19	0.20	0.22	0.24	0.25
Heavy soil, flat, 2%	0.13	0.14	0.16	0.16	0.17	0.19	0.20	0.21
Heavy soil, average, 2–7%	0.18	0.20	0.22	0.23	0.22	0.24	0.26	0.28
Heavy soil, steep, 7%	0.25	0.28	0.30	0.31	0.40	0.44	0.48	0.50
Residential areas								
Suburban	0.30	0.33	0.36	0.38	0.40	0.44	0.48	0.50
Single family	0.45	0.50	0.54	0.56	0.55	0.61	0.66	0.69
Multi-unit	0.50	0.55	0.60	0.63	0.60	0.66	0.72	0.75
Apartments	0.60	0.66	0.72	0.75	0.70	0.77	0.84	0.88
Parks/cemetries	0.10	0.11	0.12	0.13	0.25	0.28	0.30	0.31
Playgrounds	0.40	0.44	0.48	0.50	0.50	0.55	0.60	0.63
Agricultural land, 0–30%								
Bare packed soil								
Smooth	0.30	0.33	0.36	0.38	0.48	0.53	0.58	0.60
Rough	0.20	0.22	0.24	0.25	0.40	0.44	0.48	0.50
Cultivated rows								
Heavy soil, no crop	0.30	0.33	0.36	0.38	0.48	0.53	0.58	0.60
Heavy soil with crop	0.20	0.22	0.24	0.25	0.40	0.44	0.48	0.50
Sandy soil, no crop	0.20	0.22	0.24	0.25	0.32	0.35	0.38	0.40
Sandy soil with crop	0.10	0.11	0.12	0.13	0.20	0.22	0.24	0.25
Pasture								
Heavy soil	0.15	0.17	0.18	0.19	0.36	0.40	0.43	0.45
Sandy soil	0.05	0.06	0.06	0.06	0.20	0.22	0.24	0.25
Woodlands	0.05	0.06	0.06	0.06	0.20	0.22	0.24	0.25
Bare ground	0.20	0.22	0.24	0.25	0.30	0.33	0.36	0.38
Undeveloped desert	0.30	0.33	0.36	0.38	0.40	0.44	0.48	0.50
Mountain terrain (slopes > 10%)	0.60	0.66	0.72	0.75	0.80	0.88	0.96	0.95

Source: From FCDMC (1993).

where C_r = runoff coefficient, From Table 17.11, choose a value of 0.4 for bare, smooth earth and moderate slopes.

i = The average intensity is 3 in (76.2 mm) in 6 hr, or 0.50 in (12.7 mm) in 1 hr.

A_D = the drainage area is 15 acres (6.7 hectares).

$$Q_{avg} = C_r\, i\, A_D$$
$$= 0.4 \times 0.5 \times 15$$
$$= 3.0 \text{ ft}^3/\text{s} \ (0.1 \text{ m}^3/\text{second})$$

Step 2. Sieve analysis indicates that 80 % by weight of the soil is equal to or larger than 0.02 mm. The settling velocity of the 0.02 mm particle, V_s, is 0.00096 ft/second (0.00029 m/second)

Step 3. The minimum required surface area, A_s, to capture the 0.02 mm and larger particles is

$$A_s = \frac{1.2\, Q_{avg}}{V_s} = \frac{1.2(3.0)}{0.00096} = 3750 \text{ ft}^2/\text{acre}$$

This formula means that there should be 3750 ft^2 of sediment trap surface area (when the trap is full of water) for each acre of drainage area to the trap.

Storage Volume Required to Retain 1 Year's Soil Loss

Step 4. The storage depth is estimated by using the USLE. Apply the USLE to each of the two length-slope categories on the site. The terms of the USLE equation are described in detail in Sec. 17.4.1.

$$A = R \times K \times LS \times C \times P$$

R, K, C, and P will be the same for the two slope types; LS varies.

R factor. A rainfall index of 30 is given.
K factor. A K factor of 0.35 is given.
C factor. Since no vegetation or mulch is planned, C is 1.0
P factor. For smooth, compacted soil, P is 1.3.

LS factor (see Sec. 17.4.1.3)

For the 5 acres segment, 4:1, 50 ft: LS is 3.73
For the 10 acres segment, 20:1, 200 ft: LS is 0.76

Estimate Soil Loss for 1 Year
On 5 acres (2 hectares)

$$A = 30 \times 0.35 \times 3.73 \times 1.0 \times 1.3 = 50.91 \text{ tons/acre} \ (125.8 \ t/\text{hectares})$$

Assuming 1 ton of soil = 1 yd^3 ($1t$ = 0.843 m^3),
Soil loss = (50.91 yd^3/acre) (5 acres) = 255 yd^3 (195 m^3)

On 10 acres (4 hectares)

$$A = 30 \times 0.35 \times 0.76 \times 1.0 \times 1.3 = 10.37 \text{ tons/acre} \ (25.6 \ t/\text{hectares})$$

Soil loss = (104 yd^3/acre) (10 acres) = 104 yd^3 (79.5 m^3)

Therefore,

$$\text{Total soil loss} = 255 + 104 = 359 \text{ yd}^3 \ (274.5 \text{ m}^3)$$

To compute volume retained in basin use the following equation:

$$\text{Volume retained in basin} = (\text{soil loss}) \ (\text{trap efficiency})$$

As stated in the problem, 80% of the soil particles is larger than or equal to 0.02 mm, trap efficiency for this soil type is, ideally, about 80%.
Assuming ideal basin performance:

$$\text{Volume retained in basin} = 359 \ (0.8) \ (27 \text{ ft}^3/\text{yd}^3) = 7754 \text{ ft}^3 \ (220 \text{ m}^3)$$

Basin Depth Required for Once per Year Cleaning

$$\text{Basin depth} = (\text{Volume Retained in Basin})/(\text{Minimum Required Surface Area})$$

$$= 7754 \text{ ft}^3 / 3750 \text{ ft}^2 = 2.1 \text{ ft} \ (0.6 \text{ m})$$

17.3.17 Temporary Sediment Trap

A. *Description.* A small temporary ponding area or depression, with a gravel outlet, formed by an excavation into the ground and/or by constructing an earthen embankment as illustrated in Figure 17.32.

B. *Purpose.* To trap and store sediments from sites cleared and/or graded during construction. Sediment traps are intended as temporary on-site erosion control measure with a design life of approximately 6 months. These traps are maintained until the area is permanently protected from erosion by vegetation or by other means.

C. *Application.* Sediment traps are typically built along drainage ways and/or at points of discharge from disturbed areas mainly to trap sediments.

D. *Limitations.* Sediment traps are only effective for limited areas. They serve to trap only bed load materials. Wash loads will pass through untreated requiring the need for other erosion control measures to work together with sediment traps.

E. *Design considerations.* The FCDMC (1993) has provided the following guidelines for sediment traps:

1. Sediment traps are for drainage areas not larger than 10 acres.
2. Formed by either an excavation into the ground or by the construction of a compacted embankment or a berm, sediment traps require an 18-in deep sump for sediment storage.
3. The effectiveness of sediment traps is associated with the trap size. In Maricopa County, the recommended sediment trap size is 3600 cubic ft per acre of disturbed upstream drainage area. This trap volume roughly equates to pond the precipitation of a 1-in rain event.

Example 17.5
Design example
Given:
The site conditions of the sediment basin design example. A sediment trap will be constructed to capture sediment eroded from the 5-acre segment of the site.

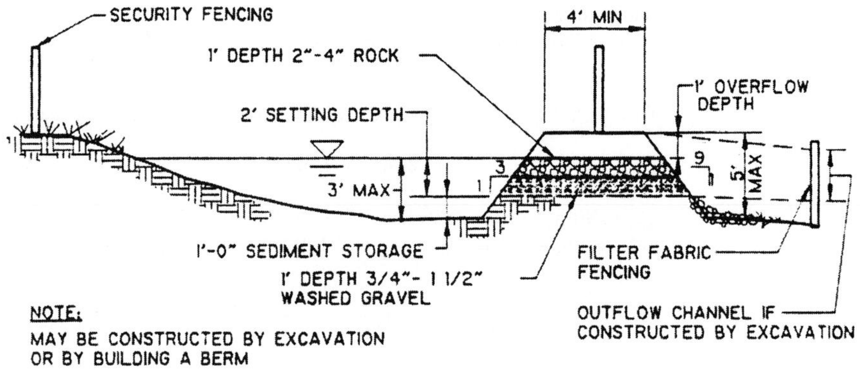

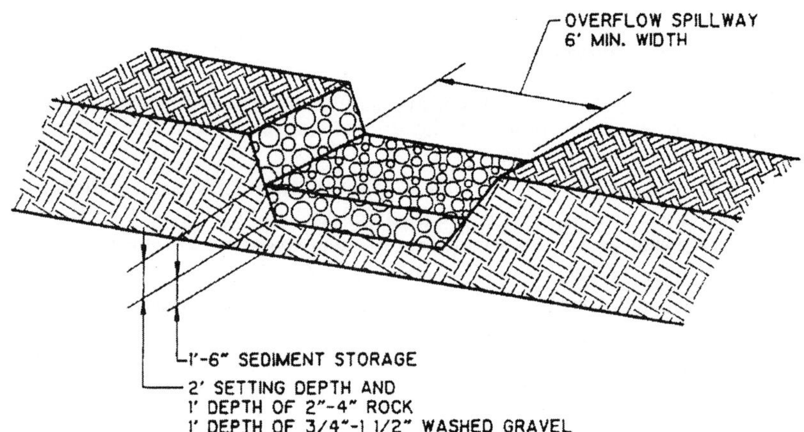

FIGURE 17.32   Temporary sediment trap. (From FCDMC, 1993)

Find:

- Required frequency of cleaning the sediment trap
- Alternative practice to reduce soil loss

Solution
Required frequency of cleaning. The trap will be designed to capture particles 0.02 mm. As shown in the sediment basin example, the soil loss for the 5-acre segment of the site is 255 yd^3 (195 m^3).

Step 1. Sediment Capture

$$\text{Volume of soil captured in the trap} = (\text{soil loss})(\text{trap efficiency})$$

Trap efficiency is the percent by weight of soil particles larger than or equal to the design particle size. As stated in the sediment basin example, trap efficiency is about 80%.

$$\text{Volume of soil captured in the trap} = (255 \text{ yd}^3)\,(0.8) = 204 \text{ yd}^3 \ (156 \text{ m}^3)$$

Step 2. Cleaning Frequency

To determine a standard trap size per acre of drainage area, the following procedure will be used. The rainfall data is the same as the sediment basin example.

$$\text{where } Q_{avg} = C_r\, i\, A_D$$
$$= 0.4 \times 0.5 \times 1$$
$$= 0.20 \text{ ft}^3/\text{s} \ (0.0057 \text{ m}^3/\text{second})$$

The settling velocity of the 0.02 mm particle V_s is 0.00096 ft/second (0.00029 m/second).

The minimum required area A_s is

$$A_s = \frac{1.2Q_{avg}}{V_s} = \frac{1.2(0.20)}{0.00096} = 250 \text{ ft}^2/\text{acre}$$

Assume 1 ft of settling depth and 1 ft of storage for the sediment trap. Therefore,

$$\text{the available storage} = (250 \text{ ft}^2/\text{acre})\,(1 \text{ ft})\,(5 \text{ acres}) = 1250 \text{ ft}^3 \ (35.4 \text{ m}^3)$$

$$\text{Required cleaning frequency} = \frac{\text{Volume of soil captured in the trap}}{\text{Available storage}}$$

$$\text{Required cleaning frequency} = \frac{(204 \text{ yd}^3)(27 \text{ ft}^3/\text{yd}^3)}{1250 \text{ ft}^3} = 4.4$$

Thus, sediment would have to be cleaned out of the trap at least 4 times in a normal year.

Reduction in soil loss. There are several ways to reduce soil loss. Straw mulch is very effective at reducing erosion. If 1.5 tons/acre (3.7 t/hectare) of straw mulch were applied and tacked into the soil, the cropping management factor, C, would decrease from 1.0 to 0.2. Thus the soil loss would be reduced by 80%.
By using $C = 0.2$,

$$A = 30 \times 0.35 \times 3.73 \times 0.2 \times 1.3 = 10.18 \text{ tons/acre} \ (25.2 \text{ t/hectare})$$

Multiplying by site acreage and trap efficiency and comparing with the storage volume gives us

$$(10.18 \text{ tons/acre})\,(5 \text{ acres})\,(0.80)\,(27 \text{ ft}^3/\text{yd}^3)/1250 = 0.9$$

With 1.5 tons/acre (3.7 t/hectare) of straw mulch punched into the soil, soil loss is extremely small.

17.3.18 Gravel Filter Berm

A. *Description.* A temporary berm constructed of open graded rock installed at the toe of a slope, or the perimeter of a developing or disturbed area as illustrated in Figure 17.33.

B. *Purpose.* To effectively intercept and detain sediment-laden flows from unprotected areas; trap the sediments and release the water as sheet flows

C. *Application.* Gravel filter berms are installed to effectively filter or trap sediments from sediment-laden flows. Detention and collection of sediments can be made near the toe of slopes or at construction site perimeters.

D. *Limitations.* Removal or clean up of temporary gravel berms may be difficult.

E. *Design considerations.* The FCDMC (1993) has provided the following criteria as design guide for gravel filter berms:

  1. For areas involving no traffic, the following considerations are provided:
    a. Maximum flow through rate per square foot of berm is 60 gpm;
    b. Berm dimensions are: height 18 in (min.), base width 24 in (min.), side slopes of 2:1 or flatter;
    c. Built on a level contour
    d. Rock sizes: ¾ in to 3 in open graded for sheet flow and 3 in to 5 in open graded for concentrated flow.
    e. Woven wire sheating (poultry netting) is recommended in areas of concentrated flow.

  2. For areas involving traffic, the following design criteria are provided:
    a. Berm height is 12 in (max.)
    b. Provide multiple berms in series with the following frequencies: every 300 ft on slopes less than 5%; every 200 ft on slopes 5 to 10%; and every 100 ft on slopes greater than 10%.

  3. Remove sediments when depth reaches about 1/3 of the berm height or 1 ft, whichever occurs first.

17.3.19 Surface Roughening

A. *Description.* Provision of a roughened soil surface with horizontal depressions created by operating a tiller or other suitable mechanical equipment on the contour or by leaving slopes in a roughened state by not fine-grading them as illustrated in Figure 17.34.

B. *Purpose.* To reduce runoff velocity and to establish a vegetative cover on the surface. Surface roughening is designed to increase infiltration and to provide trapping for sediments.

C. *Application.* For nonvegetated surfaces with erodible slopes steeper than 3:1 to be stabilized with vegetation. Surface roughening of such surfaces is achieved by stair-step grading, grooving, furrowing, or tracking. For bare soil, surface roughening provides erosion protection while vegetative cover is being established.

D. *Limitations.* Surface roughening is of limited effectiveness to erosion control specially for events stronger than moderate storms.

E. *Design considerations.* The FCDMC (1993) has provided the following recommendations for surface roughening:

  1. Graded areas with slopes greater than 3:1 should be roughened before seeding. Surface roughening could be accomplished either by track walking or driving a crawler tractor up and down the slope leaving a pattern of cleat imprints which are parallel to slope contours.

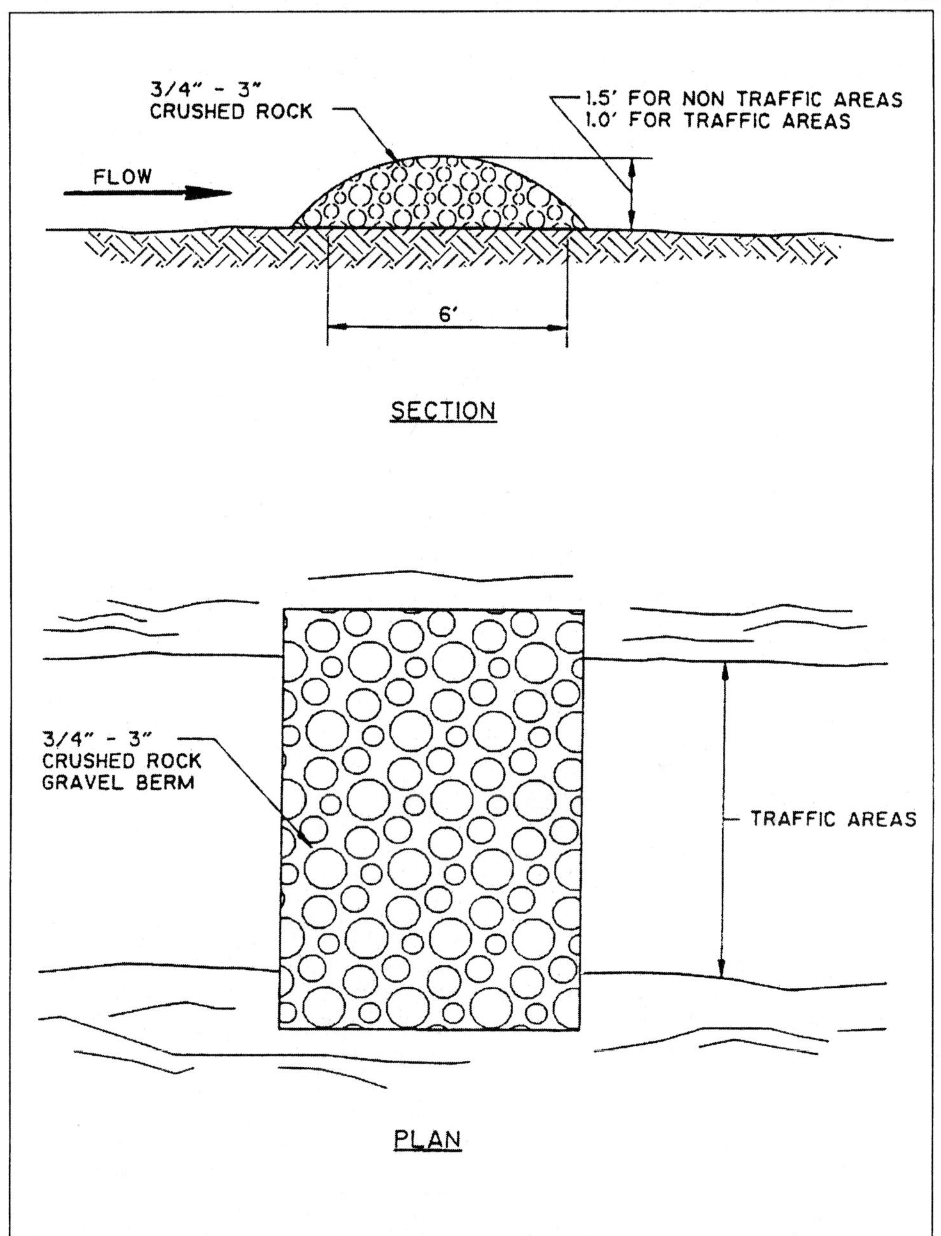

FIGURE 17.33   Gravel filter berm. (From FCDMC, 1993)

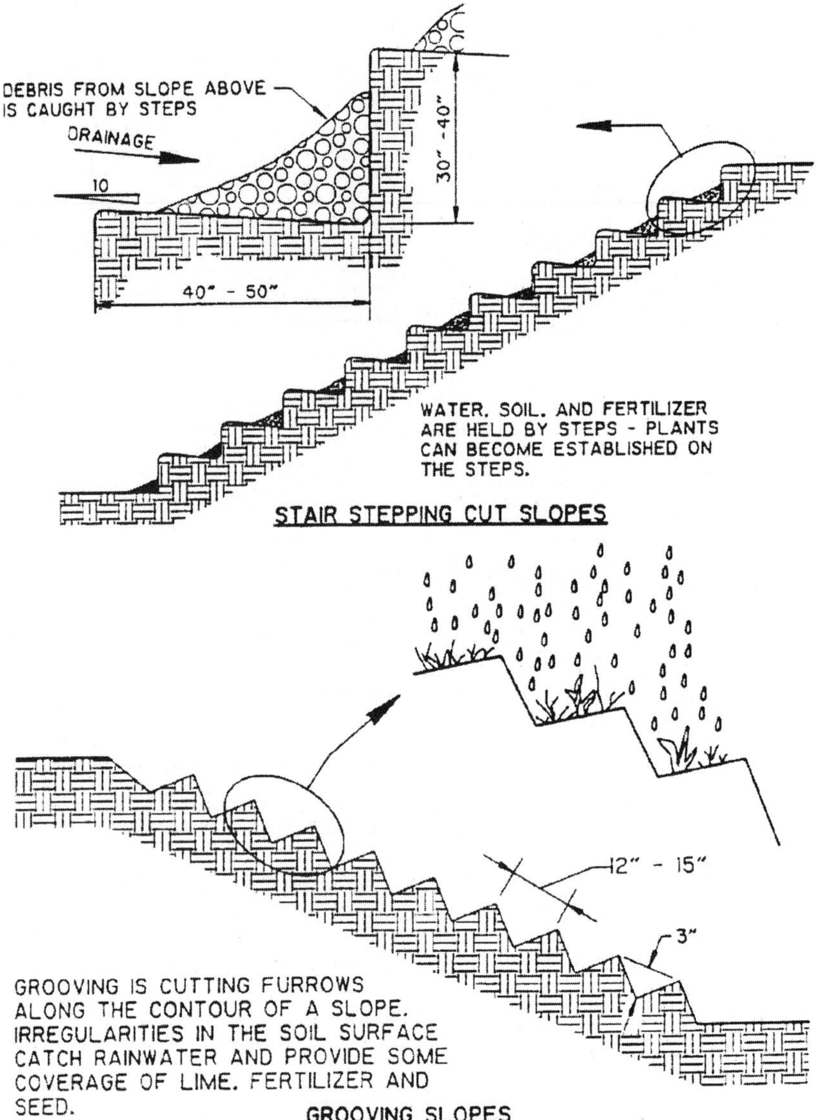

FIGURE 17.34 Surface roughening. (From FCDMC, 1993)

2. For slopes greater than 2:1, surface roughening is achieved by stair-stepping with benches. Stair-stepping aids vegetation to become established while it traps eroded soil from upstream.

17.3.20 Trees, Shrubs, Vines, and Ground Covers

A. *Description.* Trees, shrubs, vines, and ground covers can provide excellent, low-maintenance, and long-term erosion protection as illustrated in Figure 17.35.

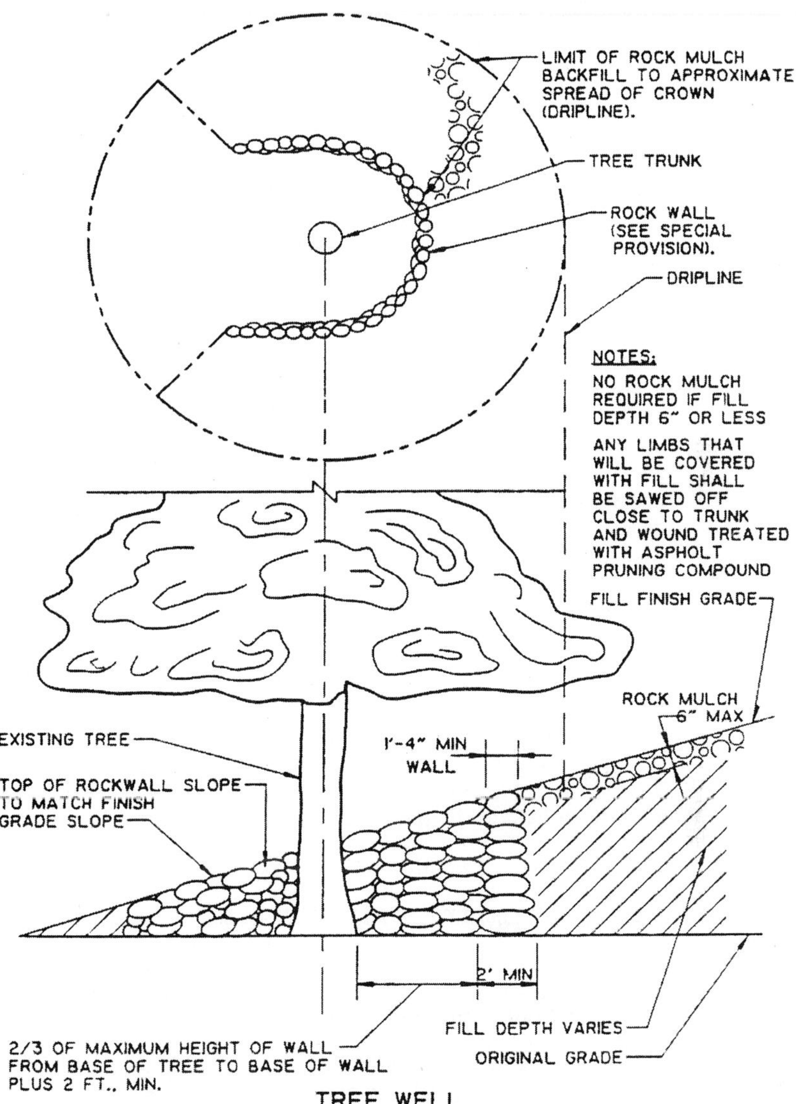

FIGURE 17.35 Trees, shrubs, vines, and ground covers. (From FCDMC, 1993)

B. *Purpose.* To stabilize the soil and help prevent erosion, decrease stormwater runoff, reduce or moderate temperatures, provide buffers and screens, filter pollutants from the air, supply oxygen, provide habitat for wildlife, and provide site aesthetics.

C. *Application.* Planting of trees, shrubs, and vines are appropriate on steep or rocky slopes where moving is not feasible, on sites where ornamentals are desirable for landscaping purposes, and on bare lands where woody plants are desirable for soil conservation or to establish wildlife habitats.

D. *Limitations.* Construction activities are likely to injure or kill trees unless adequate measures are considered. The most obvious cause of problem is direct contact by equipment, but damages are most often caused by root stress from filling, excavating, or compacting too close to the plants.

E. *Design considerations.* FCDMC (1993) has provided useful guidelines in planting and transplanting trees and shrubs, and outlined some viable considerations in the use of vines and ground covers to control soil erosion. There are vast species of plants that are used for erosion control purposes but climate, water, and soil factors are probably the three serious factors for species selection.

17.3.21 Mulching

A. *Description.* Mulching involves organic materials, straw, wood chips, bark or other wood fibers, decomposed granite, gravels, a variety of netting or mats of organic or nonorganic materials, and chemical soil stabilization principally to stabilize the soil surface for seeding and for soil erosion prevention.

B. *Purpose.* To prevent erosion by protecting the soil surface from raindrop impact and to reduce overland flow velocities. Mulching is also used to foster vegetation growth by increasing available moisture and providing insulation against extreme weather conditions.

C. *Application.* Areas that have been seeded should be mulched immediately following seeding. Mulching is also for areas protected for seeding in a future date when current time is not suitable for seeding operation. It is also used for temporary soil stabilization when inadequate erosion control is considered.

D. *Limitations.* The choice of materials for mulching is based on the type of soil to be protected, site conditions, landscape requirements, and economics.

E. *Design considerations.* Compliance to various requirements stipulated for different mulches is essential and recommendations suggested should be observed very closely. This is to ensure that mulches used are safe and will not cause undue harm to the environment and to the public.

17.4 PREDICTION AND ESTIMATION METHODS OF SEDIMENT YIELD

17.4.1 Universal Soil Loss Equation (USLE)

Developed at the Runoff and Soil Loss Data Center of the Agricultural Research Service (ARS), U.S. Department of Agriculture (USDA) in Purdue University in the 1950s, the Universal Soil Loss Equation (or USLE) was originally intended for soil conservation work in cropland areas. The USLE was a modification of earlier equations, which were found to be too localized for general use. In recent years, USLE has been adopted and interpreted for other erosion loss problems as presented in Table 17.12. The USLE is the product of six important factors, which is expressed as follows:

$$A = R \times K \times L \times S \times C \times P \tag{17.23}$$

where A = calculated soil loss in tons/acre/yr
R = rainfall factor or erosion index, in 100 ft-tons/acre $\times$ in/hr
K = soil-erodibility factor, in tons/acre per unit of R
L = slope-length factor, dimensionless
S = slope-gradient factor, dimensionless

TABLE 17.12 Application and Use of USLE Under Variety of Conditions

Application conditions	Geographical location	Reference	Comments
(1)	(2)	(3)	(4)
Forest Vegetation		Dissmeyer and Foster (1980)	Manual in the use of USLE for forest lands
Rangelands	Arizona	Osborn et al. (1977)	Estimation of soil loss
	California	Singer et al. (1977)	Estimation of soil loss
	West region (USA)	Renard and Foster (1985)	Estimation of soil loss
Surface Mining (Coal)	Alabama	Shown et al. (1982)	Evaluation of soil loss at various stages of mining and reclamation
	Pennsylvania	Khanabilvardi et al. (1983)	Evaluation of soil loss in interrill areas on reclaimed and mined areas
	West region (USA)	USDA Soil Conservation Service (1977)	Manual in the use of USLE for surface-mined lands
	Wyoming	Frickel et al. (1981)	Evaluation of soil loss at various stages of mining and reclamation
Watershed	Arizona	Fogel et al. (1977)	Evaluation of soil loss using the modified USLE (MUSLE) for stochastic sediment yield measurements
	Arizona, Colorado, Idaho	Jackson et al. (1986)	Evaluation of soil loss using MUSLE for rangeland drainage basins
	Maryland	Stephens et al. (1977)	Evaluation of soil loss for erosion and sediment inventory in basins
	Mississippi	Murphree et al. (1976)	Evaluation of soil loss for comparison with sediment yield data
	Tennessee	Dyer (1977)	Evaluation of soil loss for comprehensive resource planning of river basin

Source: From Brooks et al. (1997).

C = cropping management factor or vegetative cover, dimensionless
P = erosion-control practice factor, dimensionless

In metric units,

$$A(\text{in } t/\text{hectares}) = R(\text{in } J/\text{hectares}) \times K \text{ (in } t/J) \times L \times S \times C \times P \qquad (17.24)$$

Unit conversion for R, K, and A from English to metric or vice versa could be made using the multiplying factors provided in Table 17.13.

The USLE above identifies the six important factors that have direct impact on the soil loss. They are (1) rainfall or erosion index, (2) soil-erodibility, (3) slope-length, (4) slope-gradient, (5) cropping management or vegetative cover and (6) erosion-control practice. Individually, these factors and their significance to the USLE are presented in the following sections.

TABLE 17.13   Conversion Table

To convert	From	To	Multiply
(1)	(2)	(3)	(4)
R	100 ft-tons/acre $\times$ in/hr	10^7 J/ha $\times$ mm/hr	1.70
K	tons/acre per unit of R	t/ha per unit of R	1.32
A	tons/acre	t/ha	2.24

17.4.1.1   The Rainfall or Erosion Index Factor (R).   The rainfall factor or erosion index (R) is a direct measure of the erosive energy and intensity of rainfall estimated in a normal year. This factor is evaluated from the product of the total energy, E, and the maximum 30-minute rainfall intensity, I_{30}, of storms. Values of the rainfall factor (R) for various regions of the continental United States could be evaluated from the following sources:

1. Erosion index distribution curves for the north central region of Florida by Griffin (1975).

2. The isoerodent map for the continental United States by Wischmeier and Smith (1965, 1972, and 1978). Specifically, the map is used for determining *R-values* for areas east of the Rocky Mountains (approximately 104° west longitude).

3. West of the 104° west meridian, the predominant topographic irregularity in the region makes the use of the generalized isoerodent map (Wischmeier and Smith, 1965, 1972, and 1978) impractical. Estimation of *R-values* for the western states could be made from the regression equations developed for three storm types according to the U.S. Bureau of Plant Industry (1951).

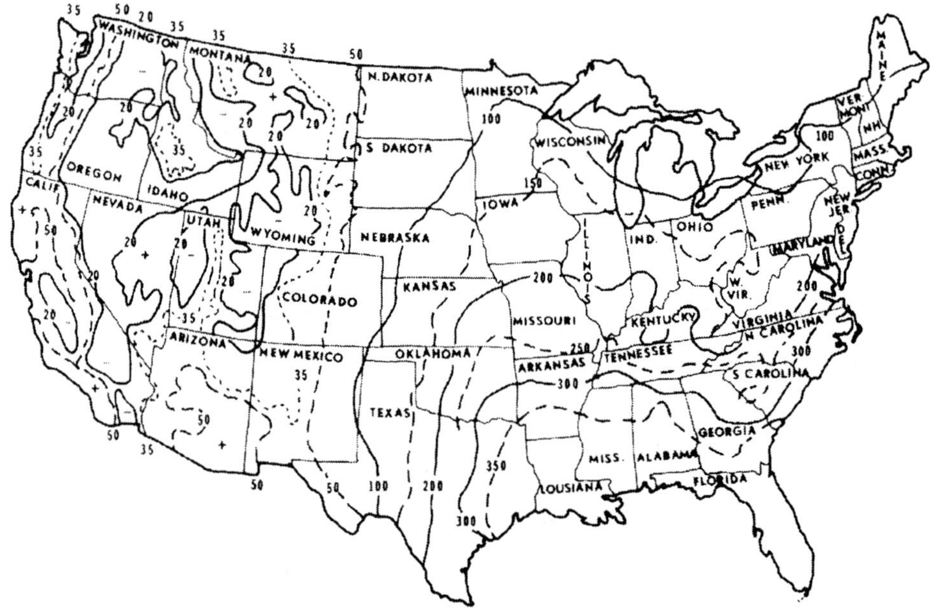

FIGURE 17.36   Isoerodent map for the continental United States. (From Wanielista, 1978)

4. Suggested values of R are provided through maps and charts developed throughout the United States from accumulated data collected over the years by Stewart (1975).

5. Guidelines on the evaluation of R values for individual storms with numerous years of data for local applications are provided by Israelson (1980).

The isoerodent map presented by Wischmeier and Smith in 1972 is illustrated in Fig. 17.36. Because of irregular terrain and topography in the west United States (particularly west of the 104° meridian), R values are evaluated from local rainfall data. This leads to the importance of using three storm types in the western states as shown in Fig. 17.37. These three storm types are used for the numerical evaluation of the R values for the western states. These storm types, whose distributions are shown in Fig. 17.38, are described below.

Type IA: Maritime climate typical to the coastal areas of northern California, Oregon, Washington, and the western slopes of the Sierra Nevada, characterized by a low broad peak in the rainfall distribution

Type I: Maritime climate typical to the storms that occur in southern and central California, and characterized by a more well-defined peak than type IA.

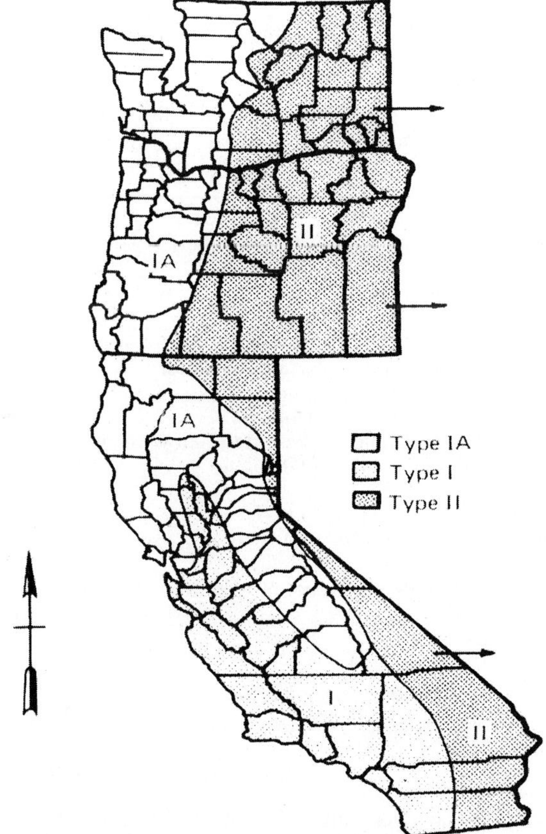

FIGURE 17.37 Distribution of storm types in the western United States. (From Goldman et al., 1986)

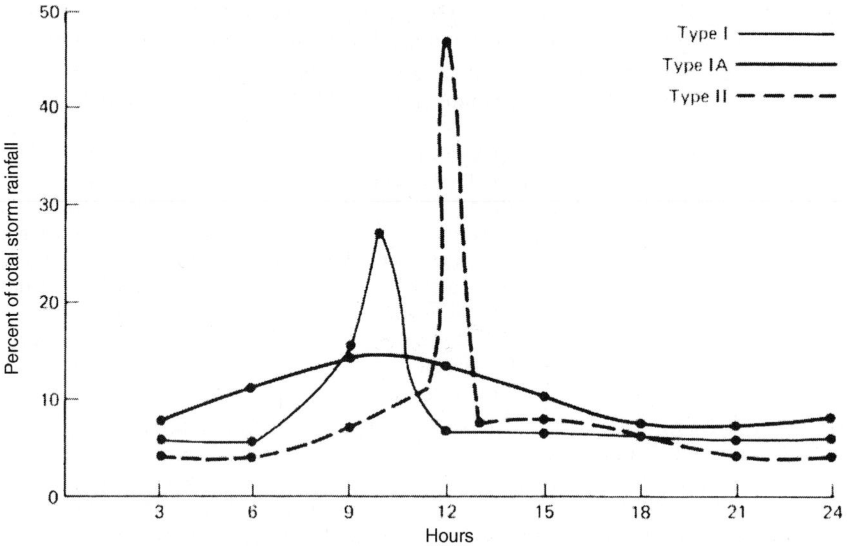

FIGURE 17.38   Time distribution of rainfall within storm types. (From Goldman et al., 1986)

Type II: Inland climate typical to the storms in the eastern parts of Washington, Oregon, California (east of Sierra Nevada), Idaho, Montana, Nevada, Utah, Wyoming, Arizona, and New Mexico

Evaluation of R values from the three storm types are made using the general expression provided by Eq. (17.25) (Brooks et al., 1975).

$$R = ap^{2.2} \qquad (17.25)$$

where a = coefficient specific to storm type as listed in Table 17.4
$\quad\ R$ = erosion index or rainfall factor
$\quad\ p$ = 2-year, 6-hour rainfall depth in inches (mm)

An illustrative example is presented below to demonstrate how Eq. (17.25) and Table 17.4 are used in the evaluation of R value.

Example 17.6   Determination of R
Problem
Determine the average annual R value for North California if the 2-year, 6-hr rainfall is 1.2 in.

Solution
Type-IA storm characterizes the type of rainfalls in North California. Using Eq. (17.25) and the coefficients in Table 17.14, we have

$$R = 27\,(p,\text{ in inches})^{2.2}\quad [0.00828\,(p,\text{ in mm})^{2.2}] \qquad (17.26)$$

Substituting the given value of p into the equation, the value of R is determined to be equal to 40.3 or 40. (A value range from 35 to 50 for the R factor could be verified in Fig. 17.36).

TABLE 17.14 Coefficients in the Evaluation of R for Three Storm Types in the Western United States

	a coefficients	
Storm type	English unit	Metric unit
Type IA	27	0.00828
Type I	16.55	0.0134
Type II	10.2	0.0219

Source: From Goldman et al. (1986).

17.4.1.2 The Soil Erodibility Factor (K).

The soil erodibility factor (K) is a measure of susceptibility to detachment and transport for soil particles by the action of rainfall and surface runoff. Various factors are identified to influence the erodibility of soil particles. They are soil texture, organic matter content, structure, and permeability.

General values of the soil-erodibility factor, K, for different soil textures and organic matter content are shown in Table 17.15. As shown, the soil erodibility factor, K, generally decreases with the increase in organic matter content.

The two methods used in determining the soil-erodibility factor (K) are presented as follows:

TABLE 17.15 Soil-erodibility Factor, K, for Various Soil Types.

	Organic matter content		
Texture class	<0.5%	2%	4%
Sand	0.05	0.03	0.02
Fine sand	0.16	0.14	0.10
Very fine sand	0.42	0.36	0.28
Loamy sand	0.12	0.10	0.08
Loamy fine sand	0.24	0.20	0.16
Loamy very fine sand	0.44	0.38	0.30
Sandy loam	0.27	0.24	0.19
Fine sandy loam	0.35	0.30	0.24
Very fine sandy loam	0.47	0.41	0.33
Loam	0.38	0.34	0.29
Silt loam	0.48	0.42	0.33
Silt	0.60	0.52	0.42
Sandy clay loam	0.27	0.25	0.21
Clay loam	0.28	0.25	0.21
Silty clay loam	0.37	0.32	0.26
Sandy clay	0.15	0.13	0.26
Silty clay	0.25	0.23	0.19
Clay	—	0.13–0.29	—

Source: From Wanielista (1978).

TABLE 17.16 USDA Particle Size Class

Particle name	Size, mm
Clay	<0.002
Silt	0.002–0.05
Very fine sand	0.05–0.10
Sand	0.10–2.0
Gravel	>2.0

Nomograph Method. This method requires size distribution analysis of soil particles to evaluate the percentages of sand, silt, and clay. Size ranges for soil classes were specified by the USDA and are listed in Table 17.16. The nomograph developed by Erickson (1977) based on the original nomograph provided by Wischmeier and Smith (1965) is reproduced in Fig. 17.39. The triangular nomograph was developed based on the following soil conditions:

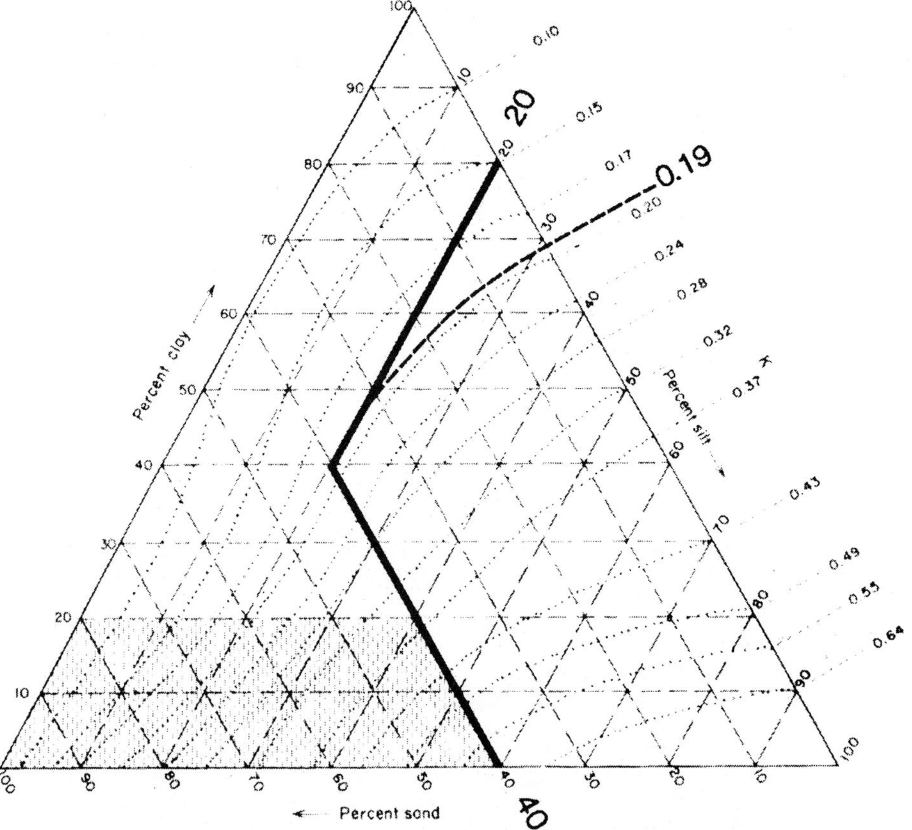

FIGURE 17.39 Triangular nomograph for estimating *K* value. (From Goldman et al., 1986)

- 2% organic matter content
- Undisturbed soil structure
- 0 to 15% rock content (i.e., grain size greater than 2 mm)

The procedure in the use of the nomograph shown in Fig. 17.39 is provided as follows:

Step 1. Enter the triangle with the percentage information of two particle sizes (e.g., sand and silt, or silt and clay, or clay and sand).

Step 2. Determine the K value that corresponds to the intersection of the two particle size information.

An illustrative example showing how the nomograph method is used to determine the soil erodibility factor, K, is presented in Example 17.7.

Example 17.7   Determination of K
Problem
Determine the K value of the soil collected in Scottsdale, AZ with the following particle size distribution.

Component	Size, mm	Fraction, %
Sand	2.0–0.1	30
Very fine sand	0.1–0.05	10
Silt	0.05–0.002	20
Clay	< 0.002	40

Solution
Entering the nomograph with silt and clay percents of 20% and 40%, the corresponding K value is 0.19.

Adjustments to the soil erodibility factor (K) should be made for the following conditions:

1. For soils having more than 15% of very fine sand

2. For soil conditions that are different from the conditions used to construct the nomograph involving organic matter content, rock content, soil structure, and permeability

Goldman et al. (1986) outlined the adjustments necessary to correct soil erodibility factor (K) for the conditions listed above. The guidelines in the use of corrections to adjust the soil erodibility factor, K, are listed in Table 17.17.

An illustrative example showing how the nomograph method and the adjustment procedure outlined in Table 17.17 is presented in Example 17.8.

Example 17.8   Adjustment in the K value
Problem
Determine the adjusted value of K considering the soil data information provided below. Compare this value with the K evaluated without the adjustment.

TABLE 17.17   Adjustments for the Erodibility Factor (K)

1. For soils with high very fine sand content of greater than 15%.
    (a) For soil textures coarser than loam (represented by the shaded area in the nomograph), subtract 5% from the percentage of very fine sand and add the difference to the silt content. The 5% remaining very fine sand should be added to the % total sand.
    (b) For soil textures finer than loam (represented by the areas outside of the shaded area in the nomograph), subtract 10% from the % very fine sand and add the difference to the silt content. The 10% remaining very fine sand should be added to the % total sand.
    (c) Determine the K value from the nomograph using the corrected sand and silt contents.
2. For soil conditions outside the ideal conditions used to develop the nomograph as follows:
    (a) If organic matter content is different from 2%, add the correction values provided in the following table.

K value	Correction values when organic matter is				
	0%	1%	2%	3%	4%
Greater than 0.40	+0.14	+0.07	0.00	−0.07	−0.14
0.20–0.40	+0.10	+0.05	0.00	−0.05	−0.10
Less than 0.20	+0.06	+0.03	0.00	−0.03	−0.06

(b) If rock content greater than 15%, the following table is used to adjust the K values. Rock content is the amount of soil particles by volume greater than 2.0 mm.

K value	Correction values when rock content is about		
	15–35%	35–60%	60–75%
0.10	+0.06	+0.05	+0.02
0.15	+0.10	+0.05	+0.02
0.17	+0.10	+0.05	+0.02
0.20	+0.10	+0.05	+0.02
0.24	+0.15	+0.10	+0.05
0.28	+0.15	+0.10	+0.05
0.32	+0.17	+0.10	+0.05
0.37	+0.20	+0.10	+0.05
0.43	+0.24	+0.15	+0.10
0.49	+0.28	+0.15	+0.10
0.55	+0.32	+0.17	+0.10
0.64	+0.37	+0.20	+0.15

(c) If soil structures are disturbed, the following corrections are added to the K values.

Soil structure	Corrections
Very fine granular	−0.09
Fine granular	−0.06
Moderate or coarse granular	−0.03

(d) K corrections for permeability are provided as follows:

Soil characteristics	Corrections
Compact soil or pH greater than 9.0	+0.03
Many medium or coarse pores	−0.03

Source: From Goldman et al. (1986).

Particle	Size, mm	Fraction, %
Sand	2.0–0.1	15
Very fine sand	0.1–0.05	20
Silt	0.05–0.002	25
Clay	< 0.002	40

Solution

Entering the nomograph in Fig. 17.39 with silt and sand percents of 25% and 35%, respectively, the intersection lies outside the shaded area classifying the soil texture as loam. Since the very fine sand content is greater than 15%, adjustment is made using 1(b) from Table 17.17. Following the guideline provided, the revised data to be used for the nomograph are as follows:

Particle	Size, mm	Fraction, %
Sand	2.0–0.05	15 + 10 = 25
Silt	0.05–0.002	25 + 10 = 35
Clay	< 0.002	40

Entering the nomograph with the updated values of 25% sand and 35% silt, K is 0.29. Without adjustments, the K value would have been 0.23.

Soil Survey Method.   The use of this method is advantageous when the grain size distribution of the soil particle is not available and that the soil surface is minimally disturbed. County soil surveys available through the Soil Conservation Service now contain K values for various soil series listed with other soil properties (USDA, 1977b). The use of K values from the SCS publications should be considered only when minimal information on the soil is available.

17.4.1.3 *The Slope Length and Slope Gradient (LS) Factor.*

The slope length (l) and gradient (s) are major factors in determining the susceptibility of any watershed to erosion, with steeper slopes being more susceptible to soil loss. Combined together, these two factors are referred to as the soil loss ratio, or the topographic factor, symbolized by LS. In relatively flat areas, runoff and flow velocity are slow, making the transport and movement of dislodged particles very limited and insignificant. For steep slopes, soil movement and transport increase dramatically. As an example, the topographic factor (LS) evaluated for a 100-ft slope length will triple (2.98 times) in value if the slope ratio is increased from 10:1 to 5:1. This means that doubling the slope gradient from 10% to 20% would triple the estimated soil loss. The impact of slope length, on the other hand, is not as great as the impact of the slope gradient—indicating LS to be less sensitive to the change in the slope length. On a 5:1 slope, for example, the topographical factor (LS) is only doubled at each time the slope length is quadrupled.

A very common formula in estimating the topographic factor (LS) for any watershed is suggested by Wischmeier and Smith (1965). Verification of the LS values provided by the equation is corrected for steep slopes by Israelson (1980) using a rainfall simulator. The equation is presented as follows:

$$LS = \left(\frac{65.41s^2}{s^2 + 10,000} + \frac{4.56s}{\sqrt{s^2 + 10,000}} + 0.065 \right) \left(\frac{l}{72.5} \right)^m \qquad (17.27)$$

where l = field slope length, ft ($m \times 0.3048$)
$\quad\quad s$ = slope gradient in percent
$\quad\quad m$ = exponent dependent on the slope gradient
$\quad\quad\quad$ = 0.2 for $s \leq 1.0\%$
$\quad\quad\quad$ = 0.3 for $1.0\% < s \leq 3.5\%$
$\quad\quad\quad$ = 0.4 for $3.5\% < s \leq 5.0\%$
$\quad\quad\quad$ = 0.5 for $s > 5.0\%$

The topographic factor (LS) can be calculated from the above equation. Other references present the values of the topographic factor (LS) in both the tabular as well as in graphic forms (Goldman et al., 1986; Wanielista, 1978). Additional discussion on the slope length (l) is necessary to define the extent covered by the parameter. It should be understood that the slope length (l) is the distance traveled by overland flow from the point of origin to either of the following terminal points:

1. To the point where the slope decreases so that sediment deposition is expected to occur

2. To the point where runoff exits into a well-defined channel

Whichever point is limiting becomes the basis for selecting the slope length. In many cases, of course, there is a variety of slope gradients and lengths. Under such conditions, the area must be broken up for separate calculations or some average value can be used.

Foster and Wischmeier (1974) proposed an approach for determining the topographic factor (LS) when dealing with irregular slopes. The resulting modifications can cause variations from 30% to 40% in the predicted soil loss. True (1974) had used the following simplified topographic factor equation for all slope gradients and slope lengths:

$$LS = (l/75)^{0.6} \, (s/9)^{1.4} \qquad (17.28)$$

where l = the slope length in ft
$\quad\quad s$ = the slope in percent

17.4.1.4   The Cropping Management (C) Factor.   The cropping management factor, C, also called the cover factor, is the ratio of the soil loss from land under certain cover and cropping management conditions to the soil loss from tilled and bare soil. Recommended erosion control practices such as sodding, seeding of grasses, mulching, and matting are often considered to reduce the amount of soil exposed to the action of raindrop and overland flow. Suggested values for the cropping management factor (C) have been compiled from the works of Wanielista (1978) and Goldman et al. (1986) and are listed in Table 17.18. The list of values shows that the C factor can vary through several orders of magnitude, from a very low value of 0.01 for sodding and permanent seeding to a high value of 1.0 for fallow (or tilled and bare) ground.

For general land use categories, typical cover factor values are listed in Table 17.19 taken from the 1973 Council of Environmental Quality (True, 1974).

17.4.1.5   The Erosion Control Practice (P) Factor.   The erosion control practice factor, P, is a measure of the soil loss for any given surface condition relative to the soil loss with uphill and downhill plowing. This factor accounts for various mechanical erosion control practices widely adopted such as contouring, terracing, and strip-cropping. Practices that effectively reduce flow velocity and overland runoff and alter the direction of runoff to flow downslope would reduce the P factor. The erosion control practice factor (P) is well understood for agricultural uses and applications as this describes the extent of plowing and tillage

TABLE 17.18 Cropping Management Factor (C) Values for Various Types of Cover

Type of cover	C	Reduction, %
None (fallow ground)	1.00	0
Native Vegetation (undisturbed)	0.01	99
Temporary Seedings (90% stand)—after 60 days	0.10	90
90% cover, annual grasses, no mulch		
Wood fiber mulch, 0.75 ton/acre, with seed	0.50	50
Ryegrass (perennial type)	0.05	95
Ryegrass (annulus)	0.10	90
Millet or Sudan Grass	0.05	95
Small Grain	0.05	95
Permanent Seedings (90% stand)		
First 60 days	0.40	60
60 to 365 days	0.05	95
After 365 days	0.01	99
Sod (laid immediately)	0.01	99
Mulch (rate of application in tons/ac)		
Hay at 0.5 ton/ac	0.25	75
Hay at 1.0 ton/ac	0.13	87
Hay at 1.5 tons/ac	0.07	93
Hay at 2.0 tons/ac	0.02	98
Small Grain Straw at 2.00 tons/ac	0.02	98
Straw at 1.5 tons/ac, tacked down (for slopes up to 2:1)	0.20	80
Straw at 4.0 tons/ac, tacked down (for slopes up to 2:1)	0.05	95
Wood Chips at 6.0 tons/ac	0.06	94
Wood Cellulose at 1.75 tons/ac	0.10	90
Fiberglass at 0.5 ton/ac	0.05	95
Asphalt Emulsion (1250 gal/ac)	0.02	98
Excelsior mat, jute (for slopes up to 2:1)	0.30	70

Unit conversion: 1 ton/ac = 2.267 t/ha.
Sources: From Wanielista, 1978 and Goldman et al. (1986).

practices considered. This factor as applied in construction, however, reflects the various practices considered to roughen the soil surface by mechanical means like treading, grading, raking, or disking.

Recommended values of the erosion control practice factor applicable in agricultural applications are listed in Table 17.20. For general land use, the 1973 Council of Environmental

TABLE 17.19 Cropping Management Factor (C) for General Land Use

General land use	C
Crop Land	0.08
Pasture Land	0.01
Forest Land	0.005
Urban Land	0.01
Other	1.00

Source: Wanielista (1978).

TABLE 17.20   Erosion Control Practice Factors for Agricultural Applications

Land slope (%)	Strip cropping/ ridge planting	Terracing	Contour strip cropping		
			R-W	R-R-M-M	R-S
1.1–2.0	0.30	0.12	0.52	0.30	0.60
2.1–7.0	0.25	0.10	0.44	0.25	0.50
7.1–12.0	0.30	0.12	0.52	0.30	0.60
12.1–18.0	0.40	0.16	0.70	0.40	0.80
18.1–24.0	0.45	0.18	0.90	0.45	0.90

R = row crop; W = fall-seeded grain; S = spring-seeded grain; and M = meadow.
Source: From Wischmeier and Smith (1965).

Quality report recommended various values for P as listed in Table 17.21. For construction site applications, Goldman et al. (1986) recommended appropriate values for P as listed in Table 17.22. The justifications offered on the listed values are:

- A compacted and smoothed surface by grading equipment is highly susceptible to sheet runoff so that a P value of 1.3 is appropriate.

- Trackwalking along the contour creates tread marks that are oriented up and down the slope. These tread marks become depressions resembling up- and down-furrows that worsen runoff conditions. This creates tread marks that are oriented up and down the slope. A P value of 1.2 is assigned.

- Trackwalking up and down the slope reduces the P value to 0.9. The created tread marks, which are parallel to the contours, serve as slope benches, which reduce flow velocity and effectively trap dislodged soil particles.

- Punched straw on the soil surface roughens the surface condition that creates a similar trackwalking effect. A P value of 0.9 is assigned.

17.4.1.6   Limitations of the USLE.   A number of limitations in the use of the USLE have been identified, which serve as warning signs to be noted down for users and designers alike. Understanding of the limitations is very crucial to avoiding possible pitfalls surrounding the use of the equation. These limitations are listed and discussed by Goldman et al. (1986) as follows:

TABLE 17.21   Recommended Erosion Control Factor for General Land Use

General land use	P
Crop land	0.5
Pasture land	1.0
Forest land	1.0
Urban land	1.0
Other	1.3

Source: From Wanielista (1978).

TABLE 17.22 Erosion Control Practice
Factors for Construction Sites

Surface condition	P
Compacted and smooth	1.3
Trackwalked along contour[1]	1.2
Trackwalked up and down slope[2]	0.9
Punched straw	0.9
Rough, irregular cut	0.9
Loose to 12-in (30-cm) depth	0.8

[1] Tread marks oriented up and down slope.
[2] Tread marks oriented parallel to contours.
Source: From Goldman et al. (1986).

1. *The USLE predicts the average soil loss.* The data used in the estimation of soil loss should be thoroughly evaluated, particularly those factors that significantly impact the soil loss estimate. The factors that vary significantly from year to year must be examined carefully before attaching values to them. Any anticipated deviation from the design conditions and practices considered must be taken into account before the equation is applied.

2. *The USLE considers only sheet and rill erosion and not gully erosion.* Extreme soil mass erosion such as gully erosion is not among the ideal conditions which USLE was designed to model. Only sheet and rill erosions are the two scenarios considered by the soil loss equation. Adjustment of the soil loss estimate is necessary if field conditions show potential of a more extreme case of erosion.

3. *The USLE does not calculate sediment deposition.* The equation evaluates the potential soil loss from a given land area or watershed but deposition within the basin is not taken into account. The soil loss estimate from the USLE should appear to be greater than the actual or observed soil loss at the basin outlet because sediment deposition within the watershed or land area was not factored out. It may be interesting to note that soil loss estimate for elongated watersheds are overestimated than are evenly size watersheds. The longer the runoff travels, the smaller the percentage of eroded soil materials being transported to the basin outlet due to deposition.

17.4.1.7 Step-by-Step Procedure to Use USLE. Estimation of soil loss using USLE is outlined by Goldman et al. (1986) as provided below:

Step 1. Determine the R factor.

Step 2. Determine the K value from the nomograph in Fig. (17.39) based on the particle size distribution analysis of the soil sample. If more than one soil sample is involved, determine the K values for each of the soil samples.

Step 3. Divide the area into sub-areas of uniform slope gradient and length (LS). Each sub-area must be assigned an LS value.

Step 4. Choose appropriate values C to represent the seasonal average of the effect of mulch and vegetation.

Step 5. Use recommended values of P based on the erosion control practice being considered.

Step 6. Evaluate the product of the five factors to obtain the soil loss per unit area.

Step 7. Multiply the soil loss per unit area by the total basin area to obtain the total sediment volume.

An example is presented below (see Example 17.9) to apply and use the above procedure. This example is adopted from the work of Goldman et al. (1986).

Example 17.9   Illustrative Example Using the Procedure
Problem
A 4-acre (1.6-hectare) site on a 400-ft long (122-m), 20%, smooth-graded slope. No mulching or seeding is planned, but the slope will be trackwalked by driving a tractor up and down the slope. A sediment trap will be built at the lower edge of the slope. R is 34. Determine the soil loss estimate from the site in one year.
A representative sample of the site soil contains:

Particle and size	Fraction, %
Sand, greater than or equal to 0.10 mm	51.0
Very fine sand, 0.05–0.10 mm	18.6
Silt, 0.002–0.05 mm	20.9
Clay, less than 0.002 mm	9.5

Solution
Following the procedure outline above, we have:

Step 1. R is given in the problem. $R = 34$.

Step 2. Using the given soil data would classify the soil texture to be sandy loam. Since the very fine sand percentage is more than 15% and the texture is coarser than loam, adjustment 1(a) is used. The adjustment is made by subtracting 5% from the very fine sand percentage and adding it to the 51% sand fraction. Add the remaining very fine sand of 13.6% to the silt percent fraction. These adjustments could be summarized in the following table:

Particle	Original (%)	Change (%)	Adjusted (%)
Sand	51.0	+5.0	56.0
Very fine sand	18.6	−18.6	0.0
Silt	20.9	+13.6	34.5

Using the nomograph in Fig. 17.39, find 56% on the "percent sand" side of the triangle and 35% on the "percent silt" side. Two line drawn parallel to the dashed lines intersect near the base of the triangle. The point of intersection is below the 0.32 curve (dotted line), so we find a K value of 0.34.

Step 3. For the given data of $L = 400$-ft (122-m) long and $S = 20\%$ slope, the LS value using equation (17.2.7) gives an LS value of 8.16.

Step 4. For no mulch or no ground cover, the recommended value of C factor in Table 17.21 is 1.0.

Step 5. The trackwalking employed creates tread marks parallel to the slope contours which gives P factor to take a value of 0.9 (Table 17.22).

Step 6. Using the soil loss equation with the five factors determined,

$$A = R \times K \times LS \times C \times P$$

$$= 34 \times 0.34 \times 8.16 \times 1.0 \times 0.9$$

$$= 84.9 \text{ tons/ac/year} \quad [\text{or } 190.5 \text{ t/ha/year}]$$

Step 7. The given acreage of 4.0 acres is multiplied with the value obtain in step 6. The soil loss in tons/ac is

$$A = 84.9 \, (4.0)$$

$$= 340 \text{ tons/year} \quad [\text{or } 305 \text{ t/year}]$$

If mulching is recommended for the site as a practice to control erosion, significant change on the soil loss estimate could be observed. Assuming a straw mulch is applied at the rate of 1.5 tons/acre (3.4 t/hectares) and tacked down on the soil surface, the new C value becomes 0.20. The change in the soil loss estimate is determined as follows:

$$A = R \times K \times LS \times C \times P$$

$$= 34 \times 0.34 \times 8.16 \times 0.2 \times 0.9$$

$$= 17 \text{ tons/acres/yr} \quad [\text{or } 38 \text{ t/hectares/yr}]$$

The reduction of the soil loss estimate from *no mulch* to *with mulch* is about 80% (i.e., $(84.9 - 17)/(84.9)) \times 100$). This indicates that mulching practice is an effective way to reduce erosion.

17.4.2   Modified Versions of the USLE

Brooks et al. (1997) presented some modified versions of the USLE. One modification is developed for application in forest and rangeland environments. The other modification involves the use of the runoff volume and peak flow to replace the rainfall factor R.

17.4.2.1   *Modified Soil Loss Equation* (*MSLE*).

This modification replaces the cropping management (C) factor and the erosion control practice (P) factor from the USLE with a vegetation management (VM) factor. The VM factor is defined as the ratio of soil loss from land managed under specified conditions of vegetative cover to the soil loss from the fallow condition on which the K factor was based. The MSLE involving the VM factor is expressed as

$$A = R \times K \times LS \times VM \tag{17.29}$$

where VM = the vegetation management factor

Both vegetative cover and soil surface conditions of natural ecosystems, regardless of the degree of surface disturbance, are accounted for in the VM factor. Three sub-factors were identified to impact or effect change on the VM values involving the following vegetative cover and soil surface conditions:

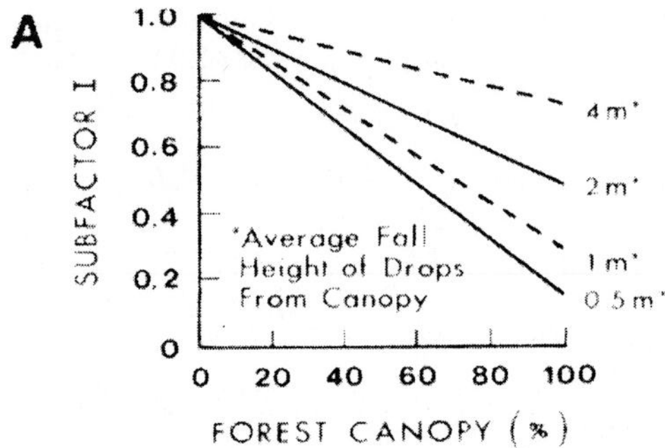

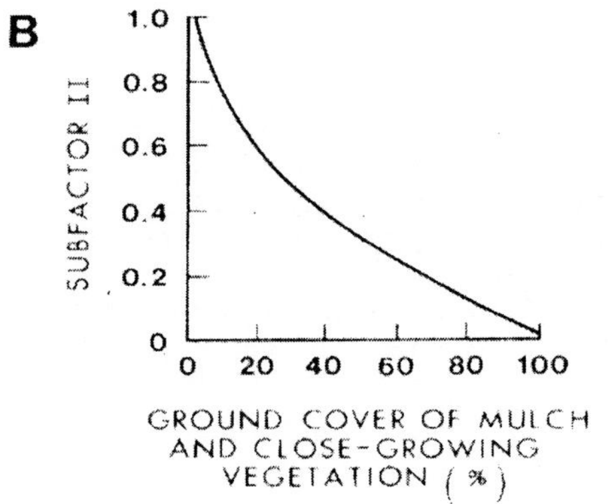

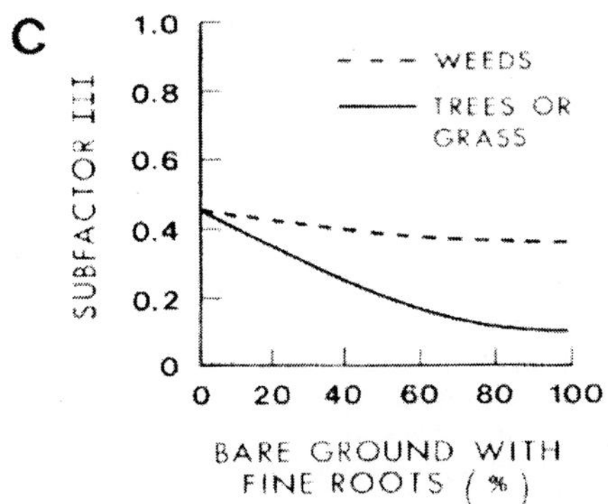

FIGURE 17.40 Relationships of (A) forest canopy cover, (B) ground cover, and (C) fine roots in the topsoil used to determine the VM factor. (From Brooks et al., 1997, and Wischmeir, 1975)

1. Forest canopy cover

2. Ground cover (i.e., low-growing vegetative cover, mulch and litter)

3. Bare ground with fine roots

Relationships for each of the three sub-factors were established by Wischmeier (1975) and are shown in Fig. 17.40. Figs. 17.40 (A), (B) and (C), respectively, show the developed relationships between the forest canopy cover and sub-factor I, ground cover and sub-factor II, and bare ground with fine roots and sub-factor III. The product of the values evaluated from the three sub-factors provide the estimate for the *VM* factor to be used for the equation defined above.

In many rangeland conditions where forest canopy cover is not present, the relationship developed by Clyde et al. (1976) shown in Fig. 17.41 is used to determine the value of *VM* factor in place of Fig. 17.40.

Appropriate values for the *VM* factor for permanent pasture, rangeland, idle land, and grazed woodland recommended by USDA Soil Conservation Service (1977) are listed in Table 17.23.

17.4.2.2 Modified Universal Soil Loss Equation (MUSLE). Another modification of the USLE was suggested by Williams (1975) to replace the *R* factor with a runoff energy factor. The modification assumes that the estimated soil loss is dependent upon the total discharge volume and peak discharge resulting from a given storm that fell on the watershed. This being the case, the runoff energy factor is largely dependent upon the storm duration, the rainfall depth and intensity. This modified USLE (or MUSLE) involving the runoff energy factor is generally expressed as follows:

$$Y_s = [95q_pQ]^w \times K \times LS \times C \times P \tag{17.30}$$

where Y_s = sediment yield in tons
$\quad Q$ = volume of storm runoff in acre-ft

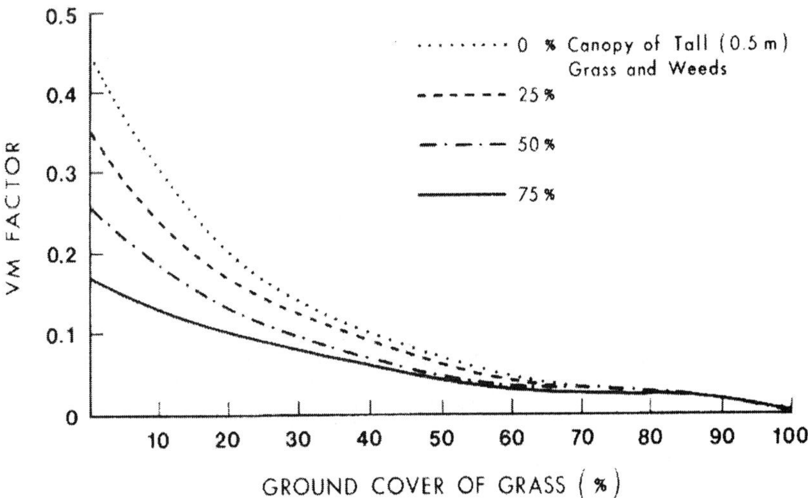

FIGURE 17.41 Relationships between ground cover conditions and the *VM* factor for the Modified Soil Loss Equation (MSLE). (From Brooks et al., 1997 and adapted from Clyde et al. 1976)

TABLE 17.23   *C* or *VM* Factors for Permanent Pasture, Rangeland, Idle Land, and Grazed Woodland

| | | | Cover that contacts the surface (% ground surface) | | | | | |
| Type and height of raised canopy[1] | Canopy cover[2] | Type[3] | 0 | 20 | 40 | 60 | 80 | 90–100 |
(1)	(2)	(3)	(4)	(5)	(6)	(7)	(8)	(9)
No appreciable		G	0.45	0.20	0.10	0.042	0.0013	0.003
canopy		W	0.45	0.24	0.15	0.090	0.043	0.011
Canopy of tall	25	G	0.36	0.17	0.09	0.038	0.012	0.003
weeds or short		W	0.36	0.20	0.13	0.082	0.041	0.011
brush (0.5 m fall	50	G	0.26	0.13	0.07	0.035	0.012	0.003
height)		W	0.26	0.16	0.11	0.075	0.039	0.011
	75	G	0.17	0.10	0.06	0.031	0.011	0.003
		W	0.17	0.12	0.09	0.067	0.038	0.011
Appreciable brush	25	G	0.40	0.18	0.09	0.040	0.013	0.003
or bushes (2 m		W	0.40	0.22	0.14	0.085	0.042	0.011
fall height)	50	G	0.34	0.16	0.085	0.038	0.012	0.003
		W	0.34	0.19	0.13	0.081	0.041	0.011
	75	G	0.28	0.14	0.08	0.036	0.012	0.003
		W	0.28	0.17	0.12	0.077	0.040	0.011
Trees but no	25	G	0.42	0.19	0.10	0.041	0.013	0.003
appreciable low		W	0.42	0.23	0.14	0.087	0.042	0.011
brush (4 m fall	50	G	0.39	0.18	0.09	0.040	0.013	0.003
height)		W	0.39	0.21	0.14	0.085	0.042	0.011
	75	G	0.36	0.17	0.09	0.039	0.012	0.003
		W	0.36	0.20	0.013	0.083	0.041	0.011

Note: All values assume (1) random distribution of mulch or vegetation and (2) mulch of appreciable depth where it exists. Idle land refers to land with undisturbed profiles for at least a period of 3 consecutive years. Also to be used for burned forest land and forest land that has been harvested less than 3 years ago.

[1] Average fall height of water drops from canopy to soil surface.

[2] Portion of total area surface that would be hidden from view by canopy in a vertical projection (a bird's eye-view).

[3] G = cover or surface is grass, grasslike plants decaying compacted duff, or litter at least 2 in. deep; W = cover at surface is mostly broadleaf herbaceous plants (as weeks with little lateral-root network near the surface) and/or undecayed residue.

Source: From USDA Soil Conservation Service (1977).

$$q_p = \text{peak discharge in ft}^3/\text{s}$$
$$w = \text{power coefficient } (w = 0.56 \text{ for watershed application and } w = 1.0 \text{ for farming applications})$$

The above equation estimates the sediment yield at the outlet of a watershed directly in weight unit on per storm event basis. The peak flow and total discharge volume are estimated by some convenient methods if runoff data are not available. Satisfactory results have been obtained with the equation above when tested for a wide range of watershed sizes and slopes. However, it tends to overestimate sediment yield from small storms and to underestimate sediment yield from large storms.

17.4.3   The Revised Universal Soil Loss Equation (RUSLE)

The USLE is a state-of-the-art tool designed for predicting the average soil loss in agricultural fields. Through the years since its release in the 1950s, its broad use and extensive applications according to Brooks et al. (1997) revealed significant weaknesses. The apparent need to address the weaknesses found from the equation has led the USDA Agricultural Research Service to initiate the development of the Revised Universal Soil Loss Equation (or RUSLE) in the 1980s (Yoder and Lown, 1995). Although the development of the RUSLE was largely based on the original formulation of the USLE, attempts were made to accommodate a much-expanded data set. Some necessary corrections were also incorporated into the RUSLE to improve the initial formulations considered in the standard USLE (Brooks et al., 1997).

The new and revised version of the USLE is computer-based. The tables, figures, and keyboard entries that were very common in the past have been replaced by better user interfaces. Renard et al. (1994) has outlined the many improvements adopted by the RUSLE over the original USLE. Despite the significant strides that the RUSLE program took to improve the original USLE the revised and greatly improved model version still relies heavily on the USLE data and its documentation.

Key improvements made in the RUSLE program over the original USLE according to Brooks et al. (1997) and Mitasova et al. (1996) include

- Availability of more data from different locations for different crops and cropping patterns and for forest and rangeland erosion calculations have been incorporated
- Addressing errors in the USLE analysis on soil loss prediction and resolving gaps in the data inputs
- Increased flexibility of and better user interface built for the RUSLE program for greater use and application for a variety of ecosystems and management alternatives
- Use of monthly factors and the incorporation of the influence of profile convexity/concavity using segmentation of irregular slopes, and the use of much-improved empirical equations for the computation of LS factor

Application of the RUSLE program is very extensive. Use has been documented for cropland, rangeland, disturbed forest lands, landfills, construction sites, reclaimed lands, parks, military training lands, mining sites, land disposal sites for waste, and other land uses where mineral soil material is exposed to the erosive forces of raindrop impact and overland flow. Also the RUSLE program is being implemented in geographic information systems (GIS) technology that offers better handling and analysis of database information from land areas interfaces with powerful mapping tools (Mitasova et al., 1998).

Like any model, some precautions should be noted as to limitations when using the RUSLE program. Established to predict annual soil erosion by water, the soil loss predictions are averages for many storms and years. The estimated soil loss represents the average over an entire area, and not just a single point in the area.

As a planning and a soil conservation tool, the RUSLE program offers better modeling capabilities than the standard USLE largely because of its improved accuracy, extensive database, and flexibility of use. The most current version of the RUSLE program can be downloaded from the official RUSLE web site at *www.sedlab.olemiss.edu/rusle*. The RUSLE program and the supporting database set are made available in a self-extracting zipped executable file format.

17.4.4 Water Erosion Prediction Project Model

Process-based models designed to evaluate cause-and-effect relationships in systems have made their significant mark in soil erosion modeling. One remarkable example of such a process-based, computer-driven modeling program is the Water Erosion Prediction Project of the WEPP model, which is being developed and currently updated by the Agricultural Research Service of the U.S. Department of Agriculture (Savabi et al., 1995). The model is applicable to hillslope erosion processes (sheet and rill erosion), as well as simulation of the hydrologic and erosion processes on small watershed. It includes options for single storm, continuous simulation, single crop, crop rotation, irrigation, contour farming, and strip-cropping.

Figure 17.42 depicts a small watershed on which the WEPP erosion model could be applied. As shown, individual hillslopes (1 to 5), or the entire watershed (composed of 5 hillslopes, 2 channel segments, and 3 impoundments) could be simulated.

17.4.4.1 Model Components. The WEPP model includes components for weather generation, frozen soils, snow accumulation and melt, irrigation, infiltration, overland flow hydraulics, water balance, plant growth, residue decomposition, soil disturbance by tillage, consolidation, and erosion and deposition. Included as part of the WEPP erosion prediction system are user interface programs, input file building programs, a climate database, a soil database, a crop parameter database, and a tillage implement database. These programs and databases make the WEPP model a very powerful tool for users involved in natural resource conservation and environmental assessment.

17.4.4.2 Program Design and Features. The WEPP model is designed to run on a daily time-step. It allows for the incorporation of temporal changes on various data inputs, such as soil erodibility, management practices, plant height, canopy cover, and soil cover. These temporal changes in data inputs provide increased flexibility in the prediction of soil erosion on agricultural and rangeland watersheds. The WEPP model designed to model every known situation involving soil erosion resulting from rainfall, snowmelt, irrigation, and ephemeral gullies.

The WEPP model is intended to evaluate sediment delivery from watersheds, sediment transport off hillslopes, and deposition and dislodgment of soil particles in small channels and their impoundment within a watershed.

Figure 17.43 shows the major calculation blocks and decision sequences involved in the WEPP model.

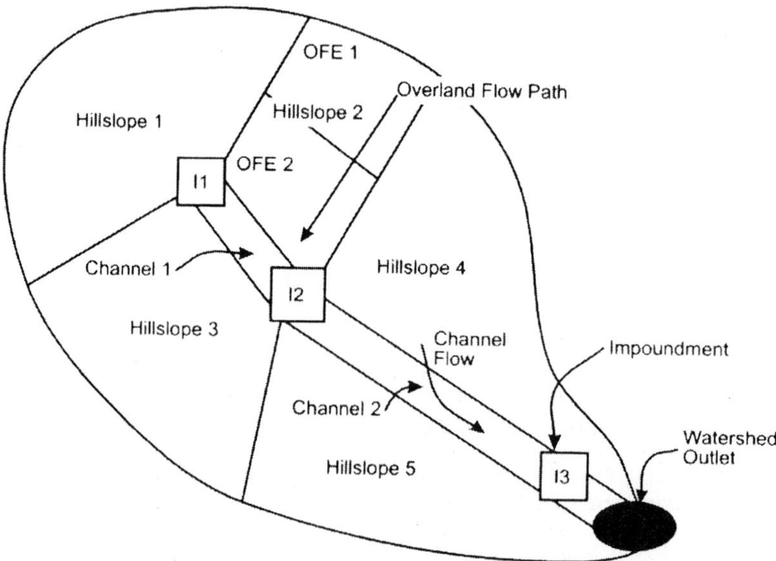

FIGURE 17.42 Schematic of a small watershed, which the WEPP model, could be applied to. (From NSERL, 1995)

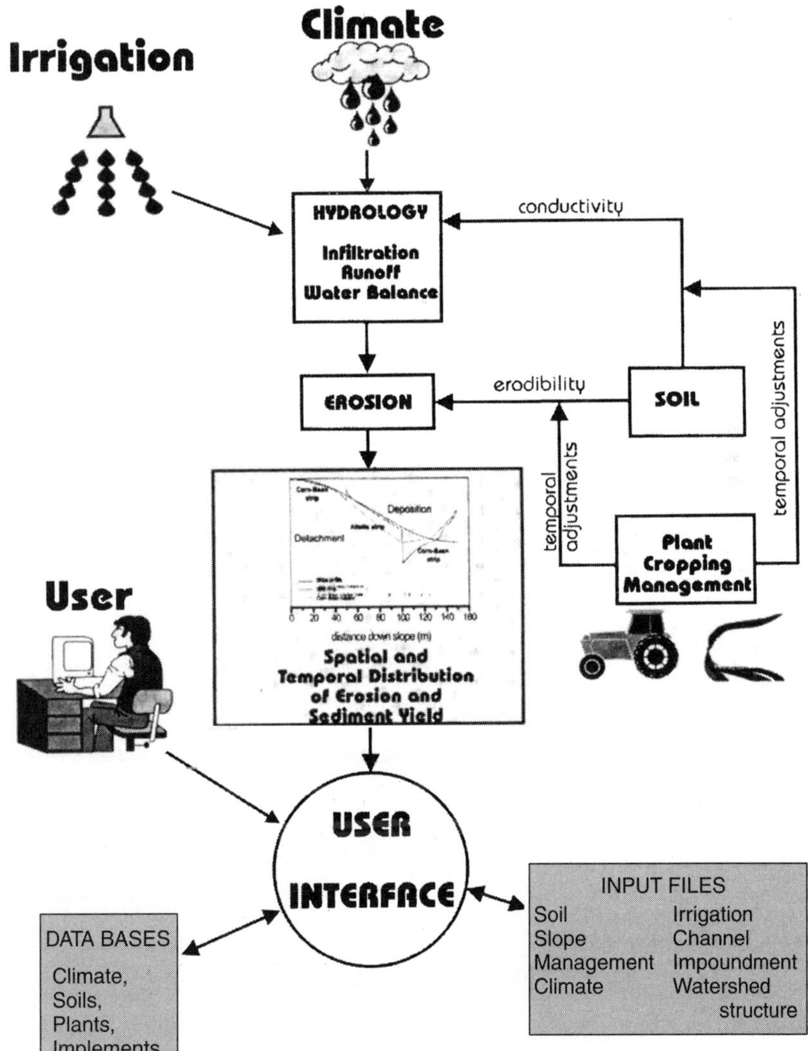

FIGURE 17.43 Flow chart for the WEPP erosion prediction model system. (From NSERL, 1995)

17.4.4.3 Model Updates and Current Versions. The WEPP model has been in development for many years now and beta versions of its Windows interface have already been released for their testing and evaluation. Once the program is fully tested and thoroughly evaluated, the model could become a powerful tool for soil conservation planning. One important model feature to explore is its capability to do an impact-and-assessment study of various management practices on surface runoff, erosion and sedimentation, and crop or forage production (Brooks et al., 1997).

Current versions of the WEPP model can be downloaded from the National Soil Erosion Research Laboratory (NSERL) web site at *http://topsoil.nserl.purdue.edu/nserlweb/*

weppmain/wepp.html. Figure 17.44 shows the Windows interface of the current version of the WEPP model.

17.4.4.4   Model Limitations.   The erosion predictions from the WEPP model are meant to be applicable to "field-sized" areas or conservation treatment units. The maximum size "field" is about a section (640 acres), although an area as large as 2000 acres is needed for some rangeland applications. Application of the model should not be made for areas having permanent channels such as classical gullies and perennial streams, since the processes occurring in these types of channels are not simulated in WEPP. Although the WEPP model could be used to run in a single storm mode, the results have limited value as the model is intended to predict long-term average annual soil loss.

17.4.5   Other Empirical Relations for Estimating Sediment Yield

The following empirical relations are the common equations that are used in the estimation of sediment yield. The first five relations are often used to estimate the wash load component of the sediment yield while the last relation, Zeller and Fullerton equation (1983), is used to estimate the bed load component. These two sediment load components, when combined together, comprise the total sediment yield.

17.4.5.1   Dendy/Bolton Method.   The soil loss prediction study conducted by Dendy and Bolton (1976) evaluated the effect of drainage area and mean annual runoff on sediment yield in the contiguous United States. The sediment yield data were represented by the sediment deposition rates obtained from 800 ponds, debris basins, and reservoirs. Drainage

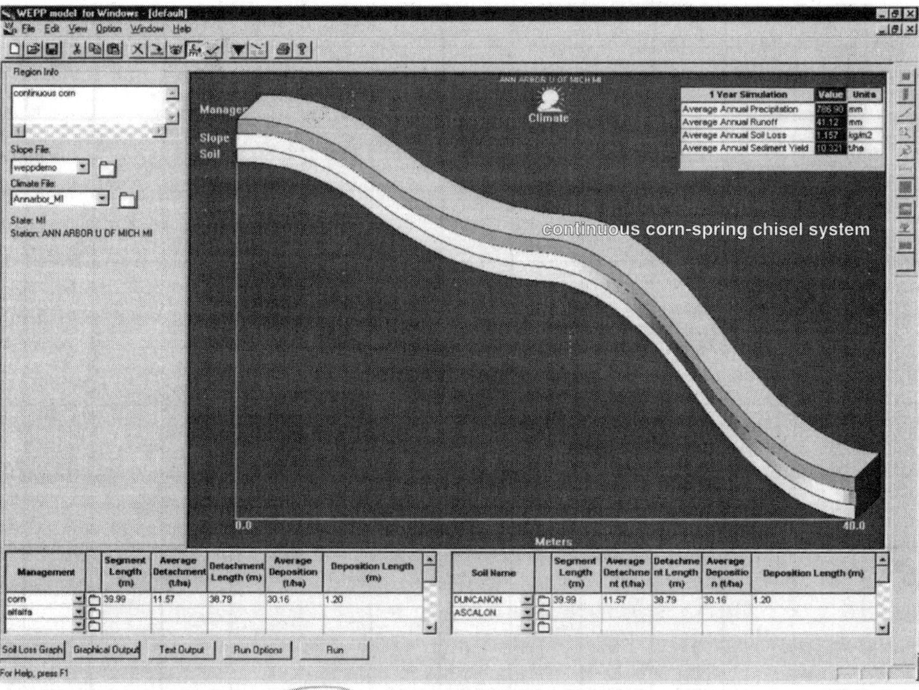

Management	Segment Length (m)	Average Detachment (t/ha)	Detachment Length (m)	Average Deposition (t/ha)	Deposition Length (m)
corn	39.99	11.57	38.79	30.16	1.20
alfalfa					

Soil Name	Segment Length (m)	Average Detachment (t/ha)	Detachment Length (m)	Average Deposition (t/ha)	Deposition Length (m)
DUNCANON	39.99	11.57	38.79	30.16	1.20
ASCALON					

FIGURE 17.44   The Windows interface for the current version of the WEPP model.

areas studied were at least one square mile where smaller drainage basins were excluded because of highly variable sediment yields. Soils, local terrain, vegetation, land use, and agricultural practices greatly influence erosion in small drainage areas and variability in sediment yield is normally much greater than in large basins, even in the same geographic area. The reservoirs selected in the study are distributed throughout the United States. The accuracy of the sedimentation surveys varied, ranging from reconnaissance-type measurements of sediment deposits to detailed surveys consisting of closely spaced cross sections or contours. Assumption was made that the reservoir sediment deposition rates equaled watershed sediment yield, that is, trap efficiency equaled unity.

Sediment Yield vs. Drainage Area.   The linear relationship between sediment yield and drainage area obtained by Dendy and Bolton (1976) is expressed as follows:

$$\frac{S_Y}{S_R} = \left[\frac{B}{B_R}\right]^{-0.16}$$

(17.31)

where S_Y = sediment yield in tons/mi^2/yr
$\quad\ S_R$ = reference sediment yield value equal to 1645 tons/mi^2/yr
$\quad\ B$ = drainage area, mi^2
$\quad\ B_R$ = reference drainage area, mi^2

Sediment Yield versus Runoff.   Two relationships were determined based on the threshold runoff value of 2 in. For the 505 runoff data evaluated, a general behavior was noted. Sediment yield increased sharply to about 1860 tons/mi^2/yr as runoff increased from 0 to 2 in. As runoff is increased from 2 to 50 in, sediment yield decreased exponentially. The two relationships derived are
For runoff less than 2 in (i.e., $Q_m < 2$ in):

$$\frac{S_Y}{S_R} = 1.07 \left[\frac{Q_m}{Q_R}\right]^{0.46}$$

(17.32)

For runoff that is equal or greater than 2 in and less than 50 in (i.e., $2.0 \le Q_m < 50$ in):

$$\frac{S_Y}{S_R} = 1.19 e^{(-0.11 Q/Q_R)}$$

(17.33)

where S_Y = sediment yield in tons/mi^2/yr
$\quad\ S_R$ = reference sediment yield value equal to 1645 tons/mi^2/yr
$\quad\ Q_m$ = mean annual runoff, in
$\quad\ Q_R$ = reference mean annual runoff equal to 2 in

Sediment Yield versus Runoff and Drainage Area.   Dendy and Bolton (1976) introduced a factor based on the correction necessary to adjust the calculated sediment yield to resemble or match the observed sediment yield data. The evaluated adjustment factor is then introduced to Eqs. (17.32) and (17.33). The sediment yield equations with the adjustment factors considered are:
For runoff less than 2 in (i.e., $Q_m < 2$ in):

$$\frac{S_Y}{S_R} = 1.07 \left[\frac{Q_m}{Q_R}\right]^{0.46} x\left(1.43 - 0.26 \log\left[\frac{B}{B_R}\right]\right)$$

(17.34)

For runoff that is equal or greater than 2 in (i.e., $Q_m \ge 2$ in):

$$\frac{S_Y}{S_R} = 1.19 e^{(-0.11 Q_m/Q_R)} x\left(1.43 - 0.26 \log\left[\frac{B}{B_R}\right]\right)$$

(17.35)

where S_Y = sediment yield in tons/mi^2/yr
S_R = reference sediment yield value equal to 1645 tons/mi^2/yr
Q_m = mean annual runoff, in
Q_R = reference mean annual runoff equal to 2 in
B = drainage area, mi^2
B_R = reference drainage area, mi^2

Limitations.   The developed equations relating mean sediment yield to mean annual runoff and drainage area size explain only about 75% of the variation in average sediment yields (i.e., regression correlation coefficient = 0.75). Since the data used were average values of grouped data, caution must be exercised in the use of these equations to predict sediment yield for specific locations due to the wide variability caused by local factors not considered in the equation development. These equations should be used to estimate sediment yields on a regional basis and not for individual drainage basins.

17.4.5.2 *Flaxman Method.*

The predictive sediment yield equation by Flaxman (1972) is established from multiple regression analysis involving various independent variables such as climate, watershed slope, coarse soil particles, and soil aggregation. The data used for the regression analysis were obtained from 11 western states that include Arizona, California, Colorado, Idaho, Montana, Nevada, New Mexico, Oregon, Utah, Washington, and Wyoming. Watershed sizes vary from a few acres to more than 50 square miles. The empirical relation evaluated by Flaxman (1972) to predict the sediment yield from a given watershed is expressed as follows:

$$\log(Y + 100) = 6.21301 - 2.19113 \log(X_1 + 100) + 0.06034 \log(X_2$$

$$+ 100) - 0.01644 \log(X_3 + 100) + 0.04250 \log(X_4 + 100) \qquad (17.36)$$

where Y = average annual sediment yield in ac-ft/sq.mi./yr
= antilog [log(Y + 100)] − 100
X_1 = ratio of average annual precipitation (in) to average annual temperature (°F)
X_2 = average watershed slope in percent
X_3 = percent of soil particles greater than 1.0 mm
X_4 = soil aggregation index (X_4 is zero if more than 25% of the soil sample is coarser than 1 mm)

The independent variables used in Eq. (17.36) are described in the following sections.
Climate.   Variable X_1 or the P/T ratio equals average annual precipitation (in inches) divided by average annual temperature (in °F). Variable X_1 is a measure of the natural response of vegetation to climate. On this assumption, the higher the value of the P/T ratio, the greater the vegetation protection, providing no disturbance was made such as clearing for cultivation, urban development, or excessive grazing. In this relation, a field stripped of vegetation for purposes of development is presumed to have a zero value of X_1, even though the climatic indicator may be relatively large. Details of the modifications for variable X_1 or the P/T ratio are discussed by the author (Flaxman, 1972).
Watershed Slope.   Variable X_2 is the weighted average slope of the watershed expressed in percent. These values were evaluated chiefly from topographic maps of the U.S. Geological Survey. Each watershed is subdivided into five contour sub-areas. Each sub-area is planimetered and contour lengths are evaluated. The five contour areas divided by the average contour lengths provide the average width of each sub-area. The contour interval divided by the average width gives the degree of steepness from which the slope in percent is determined. By multiplying the sub-areas with their respective slopes, summing the results for the entire watershed, and dividing by the total area, the average weighted watershed slope is obtained.

Coarse Soil Particles.   Variable X_3 is the percent of soil particles coarser than 1 mm in the top 2-in surface layer of the soil profile. The variable is a measure of the resistance of coarse materials to entrainment and transport and the influence of bed armoring to the erosion process and to the sediment yield. Since the presence of rock fragments, gravel, or cobbles on and near the surface is common in many parts of the western United States, particularly in the arid and semiarid areas, consideration of this variable into the predictive sediment yield equation is important.

Soil Aggregation.   Variable X_4 indicates the aggregation or dispersion characteristics of clay-size particles 2 μ (2.0×10^{-6} m) or finer in size obtained from the 2-in surface soil sample. Field data indicates that particles that tend to aggregate resist erosion while particles that disperse are easily erodible. The soil pH is used to generally classify which soils tend to aggregate and disperse. Soils with pH values greater than 7 exhibit the following general characteristics because of low precipitation and generally sparse vegetative cover: (1) low organic matter content, (2) high base saturation, and (3) high exchangeable sodium percentage. Soils with high pH are poorly aggregated. Conversely, soils with pH values lower than 7 are associated with higher precipitation and plentiful vegetative cover. Theses soils are generally well-aggregated. There are exceptions that have been identified especially for soils that behave differently from the above general characteristics. The reader is directed to the original publication for those exceptions. Also, it was noted that the aggregation or dispersion of soil samples are influenced by the amount of clay (i.e., < 2.0 μ in size) present in the sample. These soil properties and characteristics were used to assess the values of X_4 as follows:

1. When soil pH > 7.0 (alkaline), a positive value is assigned for the percent of soil finer than 2 μ. These are dispersive soils.

2. When soil pH ≤ 7.0 (neutral and acidic), a negative value is assigned for the percent of soil finer than 2 μ. These are aggregated soils.

3. When percent of soil particles greater than 1.0 mm exceeds 25 percent, X_4 is assigned a zero value. This was done on the theory that the coarser sizes then dominate the erosion characteristics no matter what the dispersion or aggregation characteristics of the clay particles.

An example is presented in the next section to use Flaxman equation to estimate the sediment yield.

Limitations.   The Flaxman method is an empirical relation determined from data collected from 11 western states. Application of the equation is appropriate for the climate, topography, soils and land condition of the arid and semiarid areas typical of the western United States. The equation assumes that the amount of erosion from snowmelt runoff will not result in an amount of erosion equivalent to the amount produced by rainfall runoff. It should be emphasized that the watershed characteristics considered in the equation exclude the effect of substantial gully and stream channel erosion. Sediments from these sources need to be added to the amounts evaluated from the equation.

Example 17.10   Using Flaxman Equation
Estimate the sediment yield for a watershed, given the following data:

1. *P/T* ratio, $X_1 = 0.46$ in/°F
2. Watershed slope, $X_2 = 16.5\%$
3. Percent of soil particles coarser than 1 mm, $X_3 = 8\%$
4. Soil aggregation index, $X_4 = +15$

Solution
Using Eq. (17.36), each component is evaluated as follows:

Constant/Variables	Original (%)	Values
Constant	6.21301	+6.21301
X_1	$-2.19113 \log(X_1 + 100)$	-4.38662
X_2	$+0.06034 \log(X_2 + 100)$	$+0.12486$
X_3	$-0.01644 \log(X_3 + 100)$	-0.03343
X_4	$+0.04250 \log(X_4 + 100)$	$+0.08758$
Difference		$+2.00522$

Using the values to evaluate the sediment yield, we have

$$Y = \text{antilog } [\log(Y + 100)] - 100$$

$$Y = \text{antilog } [\log(2.0052 + 100)] - 100$$

$$Y = 1.09 \text{ ac-ft/sq. mile or } [5.19 \times 10^{-4} \text{ m}^3/\text{m}^2]$$

17.4.5.3   *Bureau of Reclamation Sediment Surveys.*   Reservoir survey data are an excellent source for determining sediment yield rates for any part of the United States (USDA, 1978). Corrections and adjustments in the sediment yield rate are often necessary to account for variation in drainage area or watershed characteristics. One of the most important variations is the drainage basin size. Some investigators have established and found that the sediment yield rate varies with the 0.8 power of the drainage area size (Chow, 1964). This is equivalent sediment yield rate varying with -0.2 power of the drainage area.

Sediment Yield versus Drainage Area.   Compiling and analyzing the sediment yield data from survey of 28 reservoirs in the semiarid climate of the southwestern United States, a relationship was established. The data were plotted on a log-log paper as shown in Fig. 17.45. The derived relationship between drainage basin size and the sediment yield rate is expressed as:

$$Y = 1.84B^{-0.24} \text{ (English units)} \tag{17.37}$$

$$Y = 1098B^{-0.24} \text{ (metric units)} \tag{17.38}$$

where Y = annual sediment yield in ac-ft/mi^2/yr (m^3/km^2/yr)
$\quad B$ = drainage area in square miles (sq. km)

Limitations.   Application of the equation is appropriate for the sediment yield estimation of arid to semiarid areas in the southwest region of the United States. If sediment yield measurements are available from the area, it is necessary to plot those data points in Fig. 17.45 to verify if those measurements reflect the general trend of the plot. Confidence in the use of the equation could be established from such an important step.

Example 17.11   Illustrative Example
Problem
Determine the sediment yield rate at Cave Creek Dam in Phoenix, Arizona as a basis for planning purposes. The watershed draining into the flood control structure is about 121 square miles. Other known sediment yield estimates from Arizona are provided below:

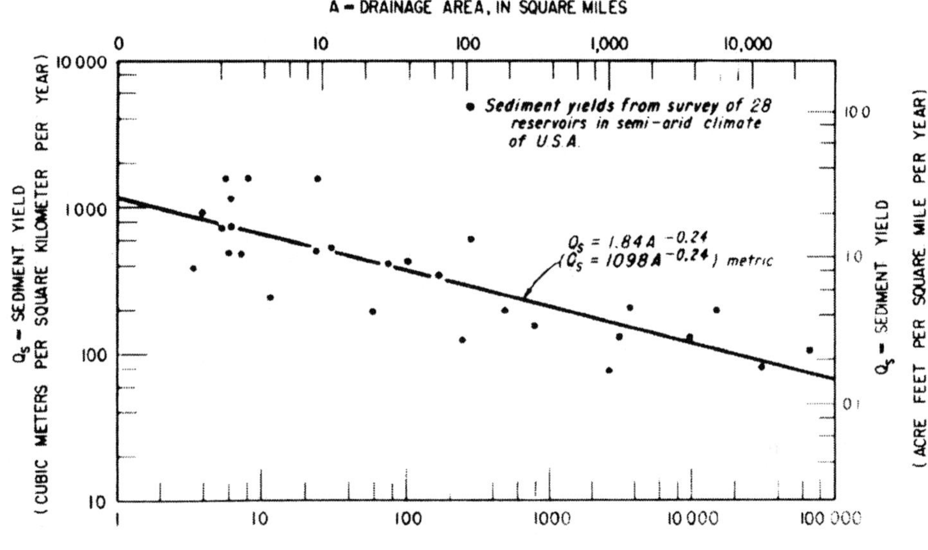

FIGURE 17.45   Average annual sediment yield rate versus drainage area. (From US Bureau of Reclamation, 1987)

No	Location	Drainage area (sq. miles)	Sediment yield (ac-ft/mi²/ year)
1	Davis Tank, AZ	0.21	0.96
2	Black Hills Tank, AZ	1.56	0.58

Solution

The estimation of sediment yield using the USBR equation (1987) is done in two steps.

Step 1. Using the sediment yield data provided, verify if the data points follow the data trend or are within the data envelope plotted in Fig. 17.45. Both of the data points are within the envelope of the historical data used to establish the curve.

Step 2. Use the predictive equation if there is a good correlation between the data and the fitted curve. Having established the good correlation of the data collected with the fitted curve, the equation is used as follows:

$$Y = 1.84 B^{-0.24}$$

$$Y = 1.84 \, (121)^{-0.24}$$

$$Y = 0.582 \text{ ac-ft/sq.mi/year.}$$

The above estimated sediment yield means that about 70 acre ft volume of the reservoir storage is taken by sediment per year.

17.4.5.4   The Pacific Southwest Inter-Agency Committee (PSIAC) Method.   The Pacific Southwest Inter-Agency Committee (PSIAC) developed this method to deal with specific conditions prevalent to the southwest region of the United States. This approach is intended as an aid to the estimation of sediment yield for the variety of conditions encountered in the area. The method considers five classes of average annual yields which are the result of the soil loss data evaluated from the region.

Sediment Yield Classification and Rating.   The five classifications recommended by PSIAC for average annual sediment yields, in the southwest area are listed in Table 17.24. Estimates of sediment yields should fall within the range of values listed.

The Rating Chart for Sediment Yield Factors.   Nine factors were identified by the Committee to have impact on the amount of the sediment yield. These are:

1. Surface geology

2. Soils

3. Climate

4. Runoff

5. Topography

6. Ground cover

7. Land use

8. Upland erosion

9. Channel erosion and sediment transport

For each of the nine factors, three rating levels—high, moderate, and low—are considered. Each level is assigned a numerical rating from which watershed characteristic data are correlated. These numerical ratings are provided in Table 17.25 which is known as the rating chart for the nine factors identified by PSIAC. The chart also shows the three rating levels.

Each factor, depending on the rating, is assigned a numerical value representing the conditions and characteristics inherent to the watershed being studied. Once each of the factors is rated, the nine ratings are summed. The summation of the ratings from the nine factors determines the total rating for the watershed from which the annual sediment yield is estimated. The sum should fall between 1 and 130 and into one of the five classifications as listed in Table 17.24.

It is recommended that a feedback process should be considered when evaluating the total rating. The sum of the values from factors A to G must be checked with the sum of factors H and I. In most instances, high values in the former should correspond to high

TABLE 17.24 Sediment Yield Classification and Rating

Classification	Rating	Sediment yield (ac-ft/sq. mile)
1	>100	>3.0 ac-ft/sq. mile
2	75–100	1.0–3.0
3	50–75	0.5–1.0
4	25–50	0.2–0.5
5	0–25	<0.2

Source: From Pacific Southwest Inter-Agency Committee (1968).

TABLE 17.25   Rating Chart for Sediment Yield Factors

| Factors (1) | Sediment Yield Levels | | |
	High (2)	Moderate (3)	Low (4)
(A) **Surface** **Geology**	(10) a. Marine shales and related mud-stones and silt stones	(5) a. Rocks of medium hardness. b. Moderately weathered. c. Moderately fractures.	(0) a. Massive, hard formations.
(B) **Soils**	(10) a. Fine-textured; easily dispersed; saline-alkaline; high shrink-swell characteristics. b. Single grain silts and fine sands.	(5) a. Medium textured soil. b. Occasional rock fragments. c. Caliche layers.	(0) b. High percentage of rock fragments. c. Aggregated clays. d. High inorganic matter.
(C) **Climate**	(10) a. Storms of several days' duration with short periods of intense rainfall. b. Frequent intense convection storms. c. Freeze-thaw occurrence.	(5) a. Storms of moderate duration and intensity. b. Infrequent convective storms.	(0) a. Humid climate with rainfall of low intensity. b. Precipitation in form of snow. c. Arid climate, low intensity storms. d. Arid climate; rare convective storms.
(D) **Runoff**	(10) a. High peak flows per unit area. b. Large volume of flow per unit area.	(5) a. Moderate peak flows. b. Moderate volume of flow per unit area.	(0) a. Low peak flows per unit area. b. Low volume of runoff per unit area. c. Rare runoff events.
(E) **Topography**	(20) a. Steep upland slope (in excess of 30%); High relief; little or no floodplain development.	(10) a. Moderate upland slopes (less than 20%) b. Moderate fan or floodplain development.	(0) a. Gentle upland slopes (less than 5%) b. Extensive alluvial plains.
(F) **Ground** **Cover**	(10) Ground cover does not exceed 20% a. Vegetation space; little or no litter b. No rock in surface soil.	(0) Cover not exceeding 40% a. Noticeable litter. b. If trees present understory not well developed.	(−10) a. Area completely protected by vegetation, rock fragments, and litter. Little opportunity for rainfall to reach erodible material.
(G) **Land Use**	(10) a. More than 50% cultivated. b. Almost all of area intensively grazed. c. All of area recently burned.	(0) a. Less than 25% cultivated b. 50% or less recently logged. c. Less than 50% intensively grazed. d. Ordinary road and other constructor.	(−10) a. No cultivation. b. No recent logging. c. Low intensity grazing.
(H) **Upland** **Erosion**	(25) a. More than 50% of the area characterized by rill and gully or landslide erosion.	(10) a. About 25% of the area characterized by rill and gully or landslide erosion. b. Wind erosion with deposition in stream channels.	(0) a. No apparent signs of erosion.
(I) **Channel** **Erosion** **and** **Sediment** **Transport**	(25) a. Eroding banks continuously or at frequent intervals with large depths and long flow duration. b. Active headcuts and degradation in tributary channels.	(10) a. Moderate flow depths, medium flow duration with occasionally eroding banks or bed.	(0) a. Wide shallow channels with flat gradients, short flow duration. b. Channels in massive rock, large boulders or well vegetated c. Artificially controlled channels.

Source: From Pacific Southwest Inter-Agency Committee (1968).

values in the latter. If they do not, either special erosion conditions exist and that factors A through G should be re-evaluated. Although numerical values are attached to the high, moderate, and low sediment yield levels in Table 17.25, interpolation between levels may be made.

An example is presented below to demonstrate how the rating chart in Table 17.25 is used to estimate the sediment yield.

Limitations.   PSIAC method is used for the estimation of sediment yield for conditions that are typical of the characteristics in the Pacific Southwest region. Estimates of soil loss are intended for broad planning purposes only, rather than for specific projects where intensive investigation of sediment yield would be required. Use of the methodology is appropriate for areas not smaller than 10 square miles.

Example 17.12   Illustrative Example Using PSIAC Method
Problem
A watershed of 20 square miles in northeastern Arizona has the field characteristics and sediment yield levels in the data below. Estimate the average sediment yield of the watershed using the PSIAC method.

ID	Factors	Field conditions
A	Surface geology	Rocks of medium hardness.
B	Soils	Easily dispersed, high shrink-swell characteristics
C	Climate	Storms of moderate duration and intensity, freeze-thaw occurrence.
D	Runoff	High peak flows; low volumes
E	Topography	Moderate slopes
F	Ground cover	Sparse, little or no litter.
G	Land use	More than 50% cultivated.
H	Upland erosion	More than 50% rill and gully erosion
I	Channel erosion and sediment transport	Moderately flow depths, medium flow duration.

Solution
Using Table 17.25, the total rating is determined as follows:

ID	Factors	Rating	Field conditions
A	Surface geology	10	Rocks of medium hardness.
B	Soils	10	Easily dispersed, high shrink-swell characteristics
C	Climate	7	Storms of moderate duration and intensity, freeze-thaw occurrence.
D	Runoff	5	High peak flows; low volumes
E	Topography	10	Moderate slopes
F	Ground cover	10	Sparse, little or no litter.
G	Land use	10	More than 50% cultivated.
H	Upland erosion	25	More than 50% rill and gully erosion
I	Channel erosion and sediment transport	10	Moderately flow depths, medium flow duration.
	TOTAL	97	Classification 2

The total rating of 97 would indicate that the sediment yield is under Classification No. 2 from Table 17.24. This classification translates to a sediment yield of 2.76 acre-ft/square mile.

17.4.5.5 *Zeller and Fullerton Equation.*

An alternative approach to the determination of sediment yield is through the use of watershed hydrographs and a sediment transport rate equation. The hydrograph provides an incremental flow rate distribution, which can be utilized with Manning's equation to calculate parameters needed to produce a unit sediment transport rate of q_s in cfs/ft. A sediment transport function applicable to the arid regions is that relation developed by Zeller and Fullerton (1983). The equation relates grain particle size (weight), the effect of saltation, and kinetic energy (velocity) to approach the problem from a somewhat more physical approach than other methods. The equation is expressed as

$$q_s = \frac{0.0064 n^{1.77} \, V^{4.32} \, G^{0.45}}{Y_b^{0.30} \, D_{50}^{0.61}} \tag{17.39}$$

where q_s = unit sediment transport rate in cfs/ft
$\quad\ n$ = Manning's roughness value
$\quad\ V$ = velocity in ft/second
$\quad\ G$ = gradation coefficient
$\quad Y_b$ = hydraulic depth in ft
$\quad D_{50}$ = median diameter of bed material in mm

A q_s is calculated for each incremental flow rate. That result is then multiplied by the natural channel bottom width and the time increment between flow rates to produce a volume of sediment yielded by the basin.

Step-By-Step Procedure. The steps in the use of the sediment transport equation to estimate the sediment yield from the watershed is outlined as follows:

Step 1. Generate the event hydrograph expected from the watershed. Discretize the hydrograph.

Step 2. Determine the typical channel section in the watershed, and the representative gradient of the stream. Using the discretized hydrograph, and the representative roughness data, evaluate the flow hydraulics using Manning's equation. For each discharge event, take note of the flow velocity, hydraulic depth, and water surface top width.

Step 3. Evaluate the soil parameters such as the mean grain size, D_{50}, and the gradation coefficient, G.

Step 4. For each discharge event, use the Zeller and Fullerton equation to evaluate corresponding unit sediment transport rate, q_s. Multiply q_s with the top width to determine the total sediment transport rate.

Step 5. Evaluate the sum of the evaluated transport rates and convert the result to the desired soil loss units.

Limitations. Table 17.26 lists the range of parameters considered in the development of the sediment transport function presented in Equation (17.39). Before the function is used, it is necessary that the watershed data being applied should be verified and compared with the range of data listed.

Further, it should be noted that the methodology presented using Zeller and Fullerton equation evaluates the per event soil loss and not the typical annual sediment yield expected from other equations.

TABLE 17.26 Range of Parameters for Zeller and Fullerton Equation

No	Parameter	Value range
1	Froude Number, F_N	Unlimited
2	Velocity, V	3–30 fps
3	Manning's n	0.018–0.035
4	Bed Slope, S_o	0.001–0.040 ft/ft
5	Unit Discharge, q	10–200 cfs/ft
6	Particle Size, D_{50}	0.5–10 mm
7	Depth, d	1–20 ft
8	Gradation Coefficient, G	2–5

Source: From Arizona Department of Water Resources (1985).

APPENDIX

LIST OF SYMBOLS

a = coefficient specific to storm type used in the evaluation of the rainfall factor, R

A = calculated soil loss in tons/acre/yr (t/ha/yr)

A_f = flow area, sq. ft

A_s = appropriate surface area for trapping particles of a certain size, sq. ft/acre

A_D = drainage area, acres

b = bottom width of the channel, ft

B = drainage area, mi^2 (acre)

B_R = reference drainage area, mi^2

C = cropping management factor or vegetative cover, dimensionless

C_F = correction factor

C_r = runoff coefficient for the rational method

C_{sf} = correction factor for the stability factor

C_{sg} = correction factor for the specific gravity

d = average flow depth in the flow channel, ft

D_{15} = grain size in which 15 percent of the materials are finer, mm (in)

D_{35} = grain size in which 35 percent of the materials are finer, mm (in)

D_{50} = the median diameter of bed material, mm (in)

D_{75} = grain size in which 75 percent of the materials are finer, mm (in)

D_{85} = grain size in which 85 percent of the materials are finer, mm (in)

D_{100} = maximum grain size, mm (inch)

D_B = basin depth, ft

F_t = average tractive force exerted by the flow field, psf

g = gravitational constant, 32.2 ft/s^2

G = gradation coefficient, dimensionless

G_s = specific gravity of soil particles

H = wave height, ft

i = average rainfall intensity, in/hr

k = permeability of soil, ft/s (cm/s)

K = soil-erodibility factor, tons/acre per unit of R

K_c = bank angle correction factor

l = field slope length, ft (m $\times$ 0.3048)

L = slope-length factor, dimensionless

L_c = length of the channel, ft
L_B = basin length, ft
LS = the slope length and slope gradient factor
m = the exponent dependent on the slope gradient
n = Manning's roughness value, dimensionless
p = 2-yr, 6-hr rainfall depth, inches (mm) used for the evaluation of the rainfall
P = erosion-control practice factor, dimensionless
P_f = wetted perimeter of the flow, ft
q_p = peak discharge, ft^3/s
q_s = unit sediment transport rate, cfs/ft
Q = volume of storm runoff in acre-ft
Q_{avg} = estimated discharge from rational method, cfs
Q_C = channel design discharge capacity, cfs
Q_D = design flow, cfs
Q_m = mean annual runoff, inches
Q_R = reference mean annual runoff equal to 2 in
R = rainfall factor or erosion index, in 100 ft-tons/acre $\times$ in/hr
R_f = hydraulic radius of the flow, ft
s = slope gradient, %
S = slope-gradient factor, dimensionless
S_f = energy slope, ft/ft
SF = stability factor
S_Y = computed sediment yield, tons/mi^2/yr
S_R = reference sediment yield value equal to 1645 tons/mi^2/yr
VM = vegetation management factor
V = flow velocity in ft/second
V_b = velocity of flow at the base of the pier, ft/second
V_p = permissible flow velocity, ft/second
V_s = settling velocity for a given particle size, ft/second
w = power coefficient for the runoff energy factor used in the MUSLE
W_{50} = weight of soil materials associated with D_{50} size, lbs
X_1 = ratio of average annual precipitation (in) to average annual temperature (°F)
X_2 = average watershed slope, percent
X_3 = % of soil particles greater than 1.0 mm
X_4 = soil aggregation index (X_4 is zero if more than 25% of the soil sample is coarser than 1 mm)
Y = average annual sediment yield, ac-ft/sq. miles/yr
Y_b = hydraulic depth, ft
Y_s = sediment yield, tons
z = side slope of the channel bank, z unit horizontal to 1 unit vertical
ϕ = angle the bank makes with the horizontal, degrees
θ = angle of repose of the riprap material, degrees
τ = tractive stress in the channel, psf
τ_c = critical shear stress for the riprap materials, psf
τ_L = unit tractive force on a level surface, psf
τ_{mos} = maximum unit tractive force on the channel side, psf
τ_{mp} = permissible unit tractive force on the channel bottom, psf
τ_{ms} = permissible unit tractive force on the channel sides, psf
τ_{mom} = maximum unit tractive force on channel bottom, psf
τ_o = unit tractive force, psf
τ_p = permissible tractive stress, psf
τ_s = unit tractive force on a sloping surface, psf
γ = specific weight of water, pcf
γ_s = specific weight of soil materials, pcf

REFERENCES

American Association of State Highway and Transportation Officials (AASHTO), *Drainage Handbook,* Washington, DC, 1987.

American Association of State Highway and Transportation Officials (AASHTO), *Highway Drainage Guidelines* (metric ed.), Vol. III–AASHTO Guidelines for Erosion and Sediment Control in Highway Construction, prepared by AASHTO, Washington, DC, 1999.

Arizona Department of Transportation, *ADOT Erosion and Pollution Control Manual,* for Highway Design and Construction, prepared by WLB Group, Inc. for Arizona Department of Transportation, Phoenix, AZ, pp. 100, June 1995.

Arizona Department of Water Resources, *Design Manual for Engineering Analysis of Fluvial Systems,* prepared by Simons, Li & Associates, Inc., March 1985.

American Society of Civil Engineers (ASCE), "Sedimentation Engineering," Manuals and Reports on Engineering Practice No. 54, New York, 1975.

American Society of Civil Engineers (ASCE), "Design and Construction of Urban Stormwater Management Systems," Manuals and Reports on Engineering Practice No. 77, New York, 1992.

Brooks, K. N., P. F. Ffolliott, H. M. Gregersen, and L. F. DeBano, *Hydrology and the Management of Watersheds,* 2nd ed., Iowa State University Press, Ames, IA, 1997.

Chow, V. T., *Open Channel Hydraulics,* McGraw-Hill, New York, 1959.

Debo, T. N., and A. J. Reese, *Municipal Stormwater Management,* Lewis Publishers, Boca Raton, FL, 1995.

Dendy, F. E., and G. C. Bolton, "Sediment Yield-Runoff-Drainage Area Relationships in the United States," *Journal of Soil and Water Conservation,* 31(6), November-December 1976.

Dickinson, W. T., "Design of Rock Riprap for Streambank Stabilization," *Canadian Agricultural Engineering,* 10(2), 1968.

Dissmeyer, G. E., and G. R. Foster, *A Guide for Predicting Sheet and Rill Erosion on Forest Lands,* USDA Technical Publications SA-TP 11, Atlanta, GA, 1980.

Dodson, R. D., *Stormwater Pollution Control,* McGraw-Hill Inc., New York, 1999.

Dyer, E. B., "Use of the Universal Soil Loss Equation: A River Basin Experience," In *Soil Erosion: Prediction and Control,* G. R. Foster, ed., pp. 292–297, Soil Conservation Society of America, Ankeny, IA, 1977.

Erickson, A. J., *Aids for Estimating Soil Erodibility—"K" Value Class and Soil Loss Tolerance,* U.S. Department of Agriculture, Soil Conservation Service, Salt Lake City, UT, 1977.

Flaxman, E. M., Predicting Sediment Yield in Western United States," *Journal of the Hydraulics Division,* American Society of Civil Engineers, Vol. 98, No. HY12, Dec. 1972, pp. 2073–2085, 1972.

Flood Control District of Maricopa County, *Drainage Design Manual for Maricopa County, Arizona, Volume III—Erosion Control,* prepared by Camp Dresser & McKee Inc., Phoenix, AZ, January 1993.

Flood Control District of Maricopa County, *Drainage Design Manual for Maricopa County, Volume II—Hydraulics,* Phoenix, Arizona, 1996.

Fogel, M. M., L. H. Hekman, and L. Duckstein, "A Stochastic Sediment Yield Model Using the Modified Universal Soil Loss Equation, *In Soil Erosion: Prediction and Control,* G. R. Foster, ed., pp. 226–233, Soil Conservation Society of America, Ankeny, IA, 1977.

Fortier, S., and Scobey, F. C. "Permissible Canal Velocities," *Transactions,* American Society of Civil Engineers, Vol. 89, pp. 940–956, 1926.

Frickel, D. G., L. M. Shown, R. F. Hatley, and R. F. Miller, *Methodology for Hydrologic Evaluation of a Potential Surface Mine: The Red Rim Site, Carbon and Sweetwater Counties, Wyoming,* USGS Survey—Water Resources Investigation, Open File Report 81–75, 1981.

Goldman, S. J., K. Jackson, and T. Bursztynsky, *Erosion and Sediment Control Handbook,* McGraw-Hill Book Company, New York, NY, 1986.

Griffin, J. D., "Predicting Soil Loss on Construction and Other Similar Non-Vegetative Areas," *Proceedings of the Storm-Water Management Workshop,* Gainesville, FL, February 1975.

International Erosion Control Association (IECA), Photo gallery from IECA web site: *http:// www.ieca.org,* 1998.

Israelson, E., *Erosion Control During Highway Construction—Manual on Principles and Practices,* Transportation Research Board, National Cooperative Highway Research Program Report No. 221, TRB, Washington, DC, 1980.

Jackson, W. L., K. Gebbardt, and B. P. Van Haveren, "Use of the Modified Universal Soil Loss Equation for Average Annual Sediment Yield Estimates on Small Rangeland Drainage Basins." in *Drainage Basin Sediment Delivery,* T. F. Hadley, ed., pp. 413–422. Int. Assoc. Hydrol. Sci. Publ. No. 159. IAHS, Washington, DC, 1986.

Khanbilvardi, R. M., A. S. Rogowski, and A. C. Miller, *Predicting Erosion and Deposition on a Strip-mined and Reclaimed Area,* Water Resources Bulletin 19:585–593.

Hudson, R. Y., "Laboratory Investigations of Rubble-Mound Breakwaters," *Proceedings of the Waterways and Harbors Division,* American Society of Civil Engineers, Vol. 85, No. WW3, 1959.

Lane, E. W., "Design of Stable Channels," *Transactions,* American Society of Civil Engineers, Vol. 120, pp. 1234–1260, 1955.

Lull, H. W., *Soil Compaction of Forest and Rangelands,* U.S. Department of Agriculture Miscellaneous Publication 768, Washington DC, 1959.

Metropolitan Washington Council of Governments, *Performance of Current Sediment Control Measures at Maryland Construction Sites,* Washington, DC, 1998.

Mills, T. R., and M. L. Clar, *Erosion and Sediment Control, Surface Mining in the Eastern U.S.,* U.S. Environmental Protection Agency, Washington, DC, 1976.

Mitasova, H., J. Hofierka, M. Zlocha, and R. L. Iverson, "Modeling Topographic Potential for Erosion and Deposition Using GIS," *International Journal of Geographical Information Science,* 10(5), 629–641, 1996.

Mitasova, H., L. Mitas, W. M. Brown, and D. Johnson, "Multidimensional soil erosion/deposition modeling and visualization using GIS," Final Report for USA CERL, University of Illinois, Urbana-Champaign, IL, 1998.

Murphree, C. E., C. K. Mutchler, and L. L. McDowell, "Sediment Yields from a Mississippi Delta Watershed," In *Proceedings of the Third Interagency Sedimentation Conference.* 1–99 to 1–109, PB-245-100, Water Resources Council, Washington, DC, 1976.

National Academy of Science of Highway Research Board, "Tentative Design Procedure for Riprap-Lined Channels," National Cooperative Highway Research Program Report No. 108, Washington, DC, 1970.

National Academy Press, *Vetiver: A Thin Green Line Against Erosion,* Washington, DC, 1992.

National Soil Erosion Research Laboratory (NSERL), Report No. 10, (http:/topsoil.nserl.purdue.edu/nserlweb/).

Olsen, O. J., and Q. L. Florey, "Sedimentation Studies in Open Channels," U.S. Bureau of Reclamation, Laboratory Report No. Sp-34, 1952.

Osborn, H. B., J. R. Simanton, and K. G. Renard, "Use of the Universal Soil Loss Equation in the Semiarid Southwest," in *Soil Erosion: Prediction and Control,* G. R. Foster, ed., Soil Conservation Society of America, Ankeny, IA, pp. 19–41, 1977.

Pacific Southwest Inter-Agency Committee, Report of the Water Management Subcommittee on Factors Affecting Sediment Yield in the Pacific Southwest Area and Selection and Evaluation of Measures for Reduction of Erosion and Sediment Yield, Oct. 1968.

Renard, K. G., and G. R. Foster, Managing Rangeland Soil Resources: The Universal Soil Loss Equation, *Rangelands* 7:118–122, 1985.

Richards, D. L., and L. M., Middleton, "Best Management Practices for Erosion and Sediment Control," Department of Transportation, Federal Highway Administration, Region 15, Dec. 1978.

Savabi, M. R., D. C. Flanagan, B. Hegel, and B. A. Engel, Application of WEPP and GIS-GRASS to a Small Watershed in Indiana," *Journal of Soil Water Conservation* 50:477–483, 1995.

Shown, L. M., D. G. Frickel, R. F. Miller, and F. A. Branson, *Methodology for Hydrologic Evaluation of a Potential Surface Mine: Loblolly Branch Basin, Tuscaloosa County, Alabama,* USGS Survey, Water Resources Investigation, Open File Report 82-50, 1982.

Singer, M. J., G. L. Huntington and H. R. Sketchley, "Erosion Prediction on California Rangelands *in Soil Erosion: Prediction and Control,* G. R. Foster, ed., Soil Conservation Society of America, Ankeny, IA, pp. 143–151, 1977.

Smoot, J. L., and R. D. Smith, *Soil Erosion Prevention and Sediment Control,* The University of Tennessee, Knoxville, TN, December 1999.

Stephens, H. V., H. E. Scholl, and J. W. Gaffney, "Use of the Universal Soil Loss Equation in Wide-area Soil Loss Surveys in Maryland," in *Soil Erosion: Prediction and Control,* G. R. Foster, ed., pp. 277–282, Soil Conservation Society of America, Ankeny, IA, 1977.

Stewart, B. A. et al., *Control of Water Pollution from Cropland: A Manual for Guideline Development,* Vol. I, Agricultural Research Service and Environmental Protection Agency, Washington, DC, November 1975.

Toy, T. J., and W. R. Osterkamp, "The Applicability of RUSLE to Geomorphic Studies," *Journal of Soil Water Conservancy* 50:498–503, 1995.

True, H. A., *Erosion, Sedimentation and Rural Runoff: A Gross Assessment Process,* Environmental Protection Agency, Surveillance and Analysis Division, Athens, GA, July 1974.

Urbonas, B. and P. Stahre, *Stormwater: Best Management Practices and Detention,* PTR Prentice-Hall, Englewood Cliffs, NJ, 1993.

U.S. Army Corps of Engineers, *Handbook for the Preparation of Stormwater Pollution Prevention Plans for Construction Activities,* EP 1110-1-16, Washington, DC, 1997.

U.S. Bureau of Reclamation, *Design of Small Dams,* 3rd ed., 1987.

USDA Natural Resources Conservation Service, *Guides for Erosion and Sediment Control in California,* Davis, CA, 1985.

USDA Soil Conservation Service, *Preliminary Guidance for Estimating Erosion on Area Disturbed by Surface Mining Activities in the Interior Western United States,* Interim Final Report, EPA-908/4-77-005, 1977a.

USDA Soil Conservation Service, *Soil Erosion: Prediction and Control, Proceedings of a National Conference on Soil Erosion,* May 24–26, 1976, Purdue University, IN, published by the Soil Conservation Service, USDA, Ankeny, IA, 1977b.

U.S. Department of Transportation, "Design of Riprap Revetments," Hydraulic Engineering Circular No. 11, Washington, DC, 1989.

U.S. Geologic Survey, *The National Atlas of the United States of America,* U.S. Department of the Interior, GPO, Washington, DC, 1970.

Vanoni, V. A., ed., *Sedimentation Engineering,* prepared by the ASCE Task Committee for the Preparation of the Manual on Sedimentation of the Sedimentation Committee of the Hydraulics Division, ASCE, New York, NY, 1977.

Wanielista, M. P., *Stormwater Management,* Ann Arbor Science Publishers, Inc., MI, 1978.

Wischmeier, W. H., and D. D. Smith, *Predicting Rainfall Erosion Losses—A Guide to Conservation Planning,* Agriculture Handbook No. 537, U.S. Department of Agriculture, Science and Education Administration, Washington, DC, 1978.

Wischmeier, W. H., and D. D. Smith, *Predicting Rainfall Erosion Losses from Cropland East of the Rocky Mountains,* Agriculture Handbook No. 282, ARS_USDA, May 1965 (reprinted 1972), 1978.

Wischmeier, W. H., and D. D. Smith, *Predicting Rainfall Erosion Losses from Cropland East of the Rocky Mountains,* Agriculture Handbook No. 282, U.S. Department of Agriculture, Washington, DC, 1965.

Wischmeier, W. H. "Estimating the Soil Loss Equation's Cover and Management Factor for Undisturbed Areas," in *Present and Prospective Technology for Predicting Sediment Yields and Sources,* USDA-ARS-S-40, 118–124, 1975.

Yoder, D., and J. Lown, "The Future of RUSLE: Inside the New Revised Universal Soil Loss Equation," *Journal of Soil Water Conservation* 50:484–489, 1995.

Zeller, M. E., and W. T. Fullerton, "A Theoretically Derived Sediment Transport Equation for Sand-bed Channels in Arid Regions," *Proceedings of the D. B. Simons Symposium on Erosion and Sedimentation,* R. M. Li and P. F. Lagasse, eds., 1983.

CHAPTER 18

STORM AND COMBINED SEWER OVERFLOW: FLOW REGULATION AND CONTROL

Larry W. Mays
Department of Civil and Environmental Engineering
Arizona State University
Tempe, Arizona

18.1 COMBINED SEWER OVERFLOW (CSO)

Sewer systems that convey both sanitary sewerage and stormwater through a single pipe are referred to as *combined sewer systems* (CSSs). During dry weather a CSS is able to convey all flows to the waste water treatment plant. However, during periods of heavy rainfall or *wet weather flow* (WWF), the CSS capacity may be exceeded, resulting in a combination of untreated sewage and stormwater flowing in the same pipe(s), and may be discharged directly into nearby receiving water. These overflows are referred to as *combined sewer overflows* (CSOs). Manuals on CSO control have been published by the Water Environment Federation (WEF) (1989) and the U.S. Environmental Protection Agency (USEPA) (1993).

Retention storage facilities capture and dispose stormwater runoff through infiltration, percolation, and evaporation, whereas *detention storage* is temporary storage for stormwater runoff and CSO. In CSOs, temporarily stored flows are subsequently returned to the sewerage system at a reduced rate of flow, or the flow is discharged to a receiving water with or without treatment. The common elements of a typical CSS that are generally found in older cities are illustrated in Fig. 18.1.

The solution to WWF problems most often include a combination of *best management practices* (BMPs). The backbone of integrated WWF management plans should be storage and/or sedimentation facilities (Perdek et al., 1998). Storage/sedimentation is the most commonly and perhaps the most effectively practiced method of urban CSO and stormwater runoff control in terms of the number of operating installations and length of service (Perdek et al., 1998). Downstream storage/sedimentation facilities vary functionally from total containment, which experience only a few overflows per year, to flow-through treatment systems. Removal of settled solids from an in-line storage facility is a problem that requires an auxiliary flushing system. Figure 18.2 shows an innovative self-cleaning storage/sedimentation basin used in Zurich, Switzerland.

The flow from storage/sedimentation facilities can be controlled by specially designed flow control devices. These devices can provide more effective flow control than the conventional static devices and they include the following (Perdek et al., 1998):

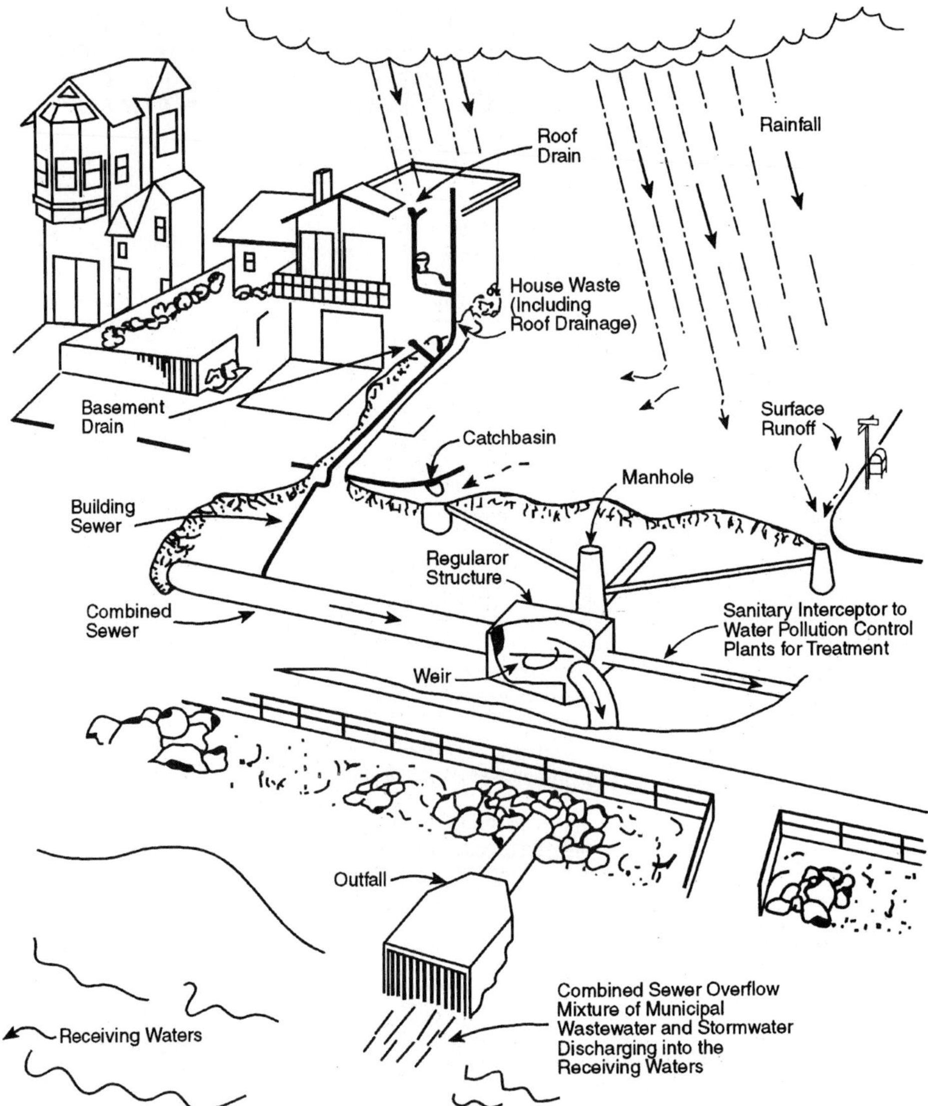

FIGURE 18.1 Common elements of a combined sewer system. (From USEPA, 1998)

- Steinscrew flow regulator
- Hydrobrake
- Wirbeldrossel or turbulent throttle
- Central outlet pipe surrounded by a pressure chamber filled with air

These devices are described in more detail in Sec. 18.5.

The USEPA has a *CSO Control Policy,* under which discharge permits that are issued to communities with CSSs are expected to include nine minimum controls (USEPA, 1995). These measures can reduce the magnitude, frequency, and impacts of CSOs without signif-

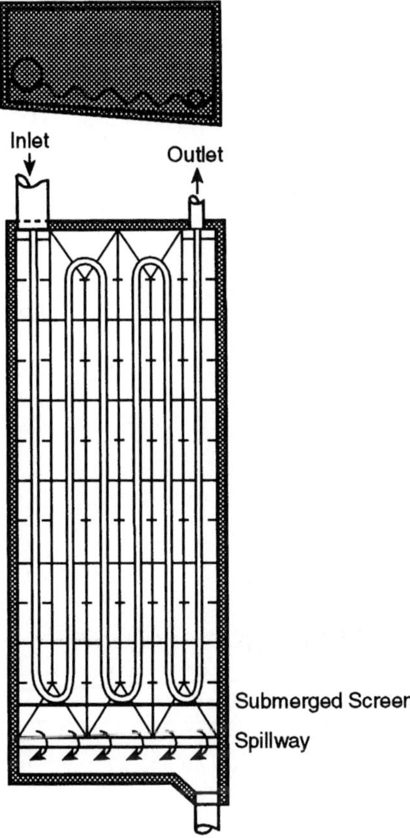

FIGURE 18.2   Self-cleaning storage/sedimentation basin used in Zurich, Switzerland; a continuous dry-weather channel, which is an extension of the tank's combined sewer inlet, is formed by a number of parallel grooves connected at their end points similar to the configuration shown. Any solids that have settled in the basin during its storage operation are resuspended by the channelized high-velocity flow during the drawdown following a storm event. (From USEPA, 1998)

icant construction or expense. Maximization of storage is one of the nine minimum CSO controls. In order to maximize in-line storage in the collection system, control measures downstream of the excess capacity typically are used. These include the following (USEPA, 1999):

- Collection system inspection and removal of obstructions
- Tide and control gate maintenance, repair, and replacement
- Regulator installation and adjustment
- Reduction/retardation of inflows and infiltration
- Upgrade and adjustment of pumps

TABLE 18.1 Cost Comparison for Control Measures

Control measure	City	Unit	Capital cost	Annual O & M
Regulator installation (inflatable dams)	Washington D.C.	Per dam	$1,219,000[1]	N/A
Pump installation	San Francisco, CA	gpd	$0.10–$0.11 gpd[2]	N/A
Install raised weirs	Boston, MA area	Per weir	$3,250 brick & mortar[3] $12,050 formed concrete[3] $18,100 stop logs[3]	N/A
Real time monitoring system	Cleveland, OH area	Per unit	$264,000 auto regulators[4] $12,900 remote level monitor[3]	Automated regulator $11,975[3] Rain gauge $2,830[3] Flow and level monitor $3,270[3]
	Philadelphia, PA	System	$1,906,000[5]	N/A
Collection system inspection	Norwalk, CT		None	Reduction in O & M for microscreening devices; increase in O & M for collection system
Tide gate replacement	Boston, MA area	Per gate	$24,100[3]	N/A

ENR = 5750
[1] Average cost per dam of 12 dams installed at 8 sites; complete cost includes all monitoring equipment.
[2] Used to convey flows from CSO treatment facilities to WWTP as conditions allow.
[3] Typical cost; costs vary by site.
[4] Average bid cost of automated regulators (usually inflatable dams) in Southerly WWTP district; costs vary by site.
[5] System = 1 central computer, 8 computer-controlled regulators, level-measuring devices at 53 regulators, 22 rain gages, flow metering at 18 stations, several other level devices.
Sources: From AMSA (1994), Hudson (1996), Metcalf & Eddy (1993), Northeast Ohio Regional Sewer District (1997), and Treffeisen (1997).

- Raising existing weirs and installation of new weirs
- System of real-time monitoring/network

A comparison of costs for control measures implemented by various communities was reported by the USEPA (1999) as shown in Table 18.1. Techniques to maximize in-line storage are the most cost-effective in trunk sewers and interceptors with mild slopes and large storage capacities. Some in-line storage measures, such as regulation of a main control valve, do not require capital improvements.

18.2 TYPES OF FLOW REGULATORS

Flow regulation by *flow regulating structures* in storm drainage systems and CSO systems are very important components of these systems. Flow regulating structures are used for several major functions including (1) the control of releases from detention storage facilities; (2) the prevention of unacceptable storm sewer surging; (3) the control of flow between

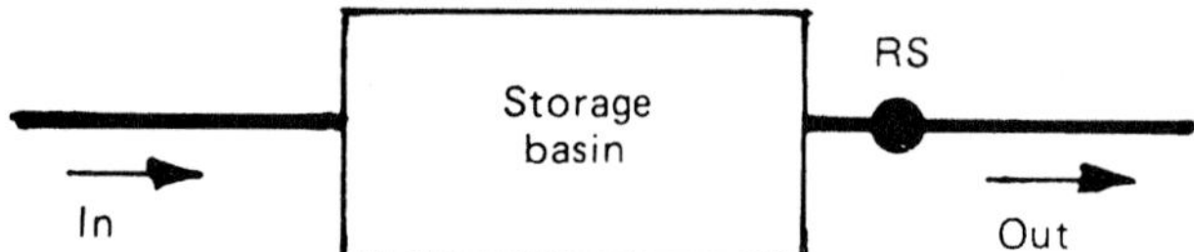

FIGURE 18.3 Location of a flow regulating structure in an in-line connected storage. (From Stahre and Urbonas, 1990)

combined sewers and companion facilities such as interceptor(s); and (4) the diversion of flow to CSO treatment facilities. Flow regulators can range from simple fixed orifices and weirs to remotely controlled, automated mechanically operated gates or valves.

18.2.1 Control of Releases from Storage Facilities

Flow regulating structures for detention facilities are located according to how the storage basin is connected to the stormwater conveyance system. For *in-line storage* facilities the flow regulation is located at the downstream end of the storage facility as shown in Fig. 18.3. The flow regulator is typically a flow restrictor located at or near the bottom of the storage basin. In some cases pumping is also used to regulate the flow. For *off-line storage* facilities (which are parallel to the conveyance system), flow regulation is accomplished through limiting the flow that bypasses the storage basin (excess flow is diverted to the storage facility) as shown in Fig. 18.4. Figure 18.5 illustrates an underground detention tank with a regulated outflow.

Flow regulators for outlet control devices can be classified as fixed and movable flow regulators (Stahre and Urbonas, 1990). *Fixed flow regulators* are permanently attached to the basin structure and have a constant cross section. *Movable flow regulators* have an outlet

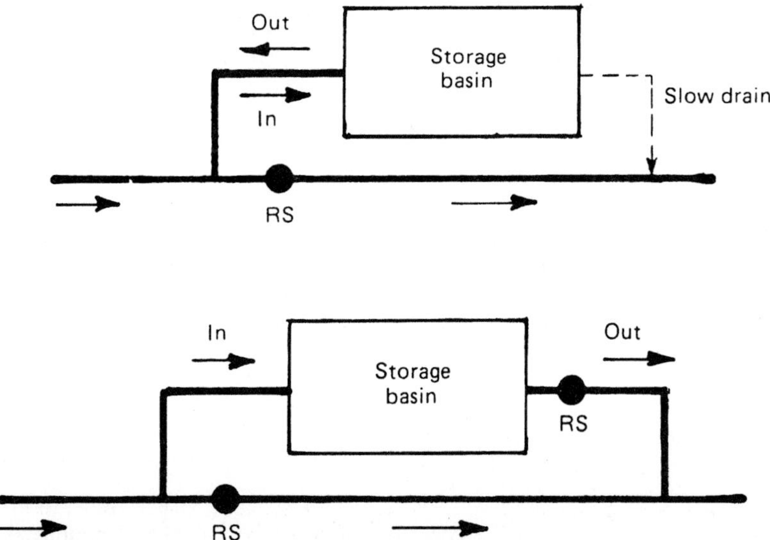

FIGURE 18.4 Location of a flow regulating structure in an off-line connected storage system. (From Stahre and Urbonas, 1990)

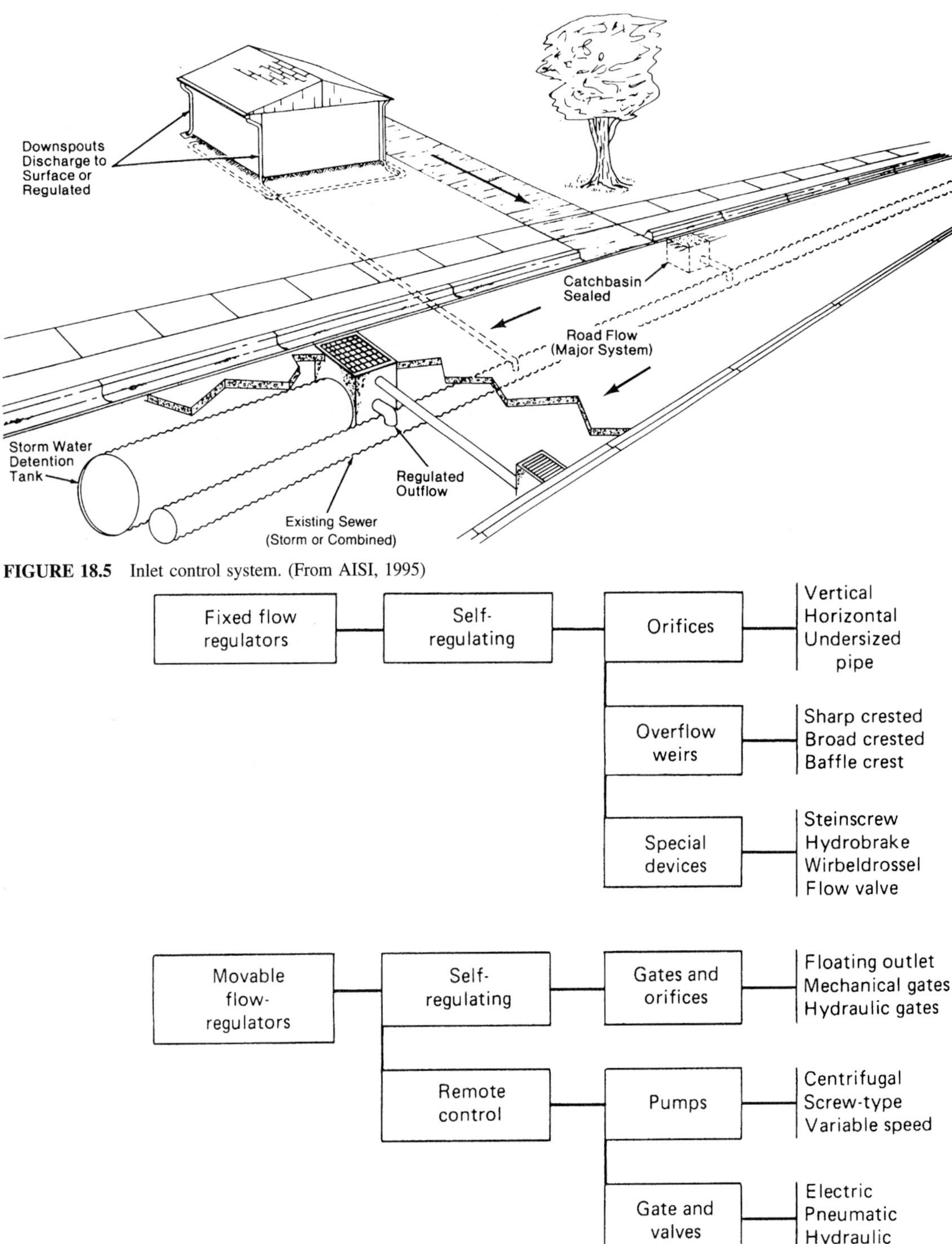

FIGURE 18.5   Inlet control system. (From AISI, 1995)

FIGURE 18.6   Various types of outflow control devices. (From Stahre and Urbonas, 1990)

that varies in size or elevation. Movable flow regulators can be self-regulating or they can be used for remote control. Figure 18.6 classifies some of the various types of regulating structures for outflow control.

Vortex valves are used to discharge high flows through a spiral action valve, while maintaining low flows. These valves have been used to control discharges from storage facilities to divert flows to CSO treatment facilities and to replace mechanical regulators that have failed (USEPA, 1993). The flows through vortex valves remain constant so that it essentially serves to dam high flows. Trash and debris pass through the vortex so that maintenance costs are reduced over traditional valves.

18.2.2 Prevention of Storm Sewer Surging

Flow regulators at the inlets to storm sewers are used to prevent unacceptable storm sewer surging. When the capacity of a storm sewer is exceeded, water may be temporarily ponded on the road surface. When this is not feasible, water may be stored temporarily in off-line detention basins or in an underground detention tank. Figure 18.7 illustrates the placement of a regulator between an underground storage tank and a sewer so that only the predetermined rate of flow, which the sewer can handle without excessive surcharging, would be released.

18.2.3 Control Flow between Collection Sewers and Interceptors

Flow regulators are frequently used to control the flow between collection sewers and interceptors. During dry weather conditions regulators allow waste water flows to be conveyed to a downstream treatment facility. During WWF, regulators (also referred to as *diversion structures*) are used to divert flows away from an interceptor to a CSO outlet or to a specially

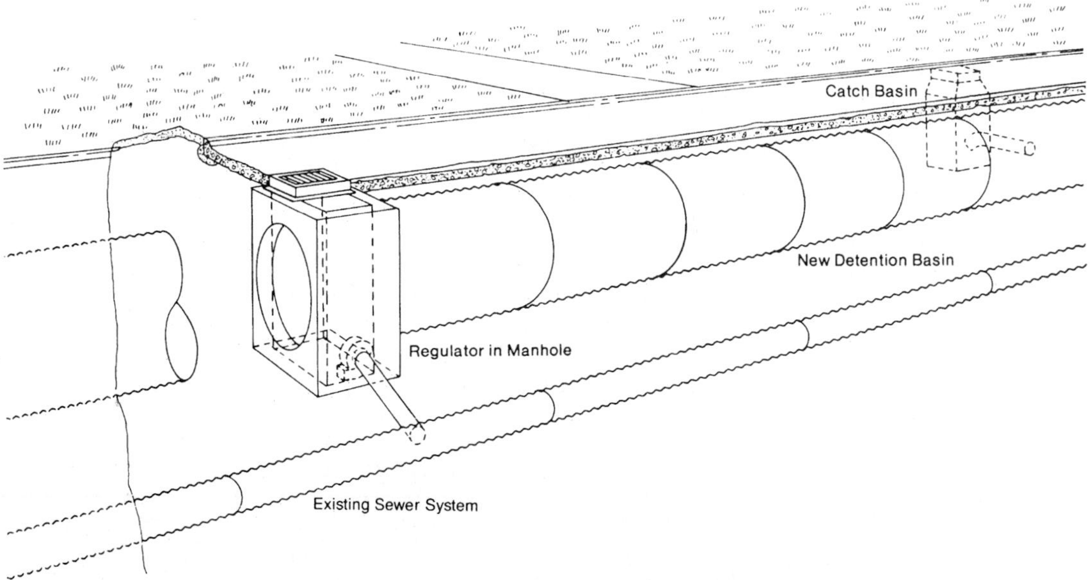

FIGURE 18.7 Typical installation of regulator for underground storage. (From AISI, 1995)

TABLE 18.2 Description of Some Typical Flow Regulators

Regulator	Description
Side weir	Typically consists of a weir parallel to the wastewater flow located in the side of the sewer pipe (Fig. 18.8*a*). The weir should be high enough to prevent any discharge of dry-weather flows, but low and long enough to discharge the required excess flow during wet weather.
Transverse weir	A weir or small dam placed directly across the sewer, perpendicular to the line of flow, is used to direct dry-weather flow to the interceptor sewer (Fig. 18.8*b*). Increase of flow during wet weather results in flow overtopping the weir and discharging to the overflow outlet.
Orifice	These diversion structures allow flow from the combined sewer to pass through a circular or rectangular orifice and enter the interceptor. The orifice is sized to allow the dry-weather flow, and possibly some of the wet weather flow, to pass. The orifices can be oriented in a variety of ways, including horizontally at the invert of the sewer ("drop inlet"), and vertically on the side of the sewer (often used in conjunction with a transverse weir as in Fig. 18.8*b*).
Leaping weir	A leaping weir is formed by an opening in the invert of a sewer of such dimensions as to permit dry-weather waste water flow to fall through the opening and pass to the interceptor (Fig. 18.8*c*). During storms, the increased velocity and depth of flow cause most of the flow to leap the opening and enter the overflow outlet. The steel weir plate is normally designed so that it can be adjusted for various flow conditions.
High outlet regulator	A commonly used orifice-type regulator in which the invert of the overflow pipe is typically above the crown of the combined sewer (Fig. 18.8*d*).
Relief siphon	The relief siphon (Fig. 18.8*e*) affords a means of regulating the maximum water-surface elevation in a sewer with smaller variations in high-water level than can be obtained with other devices. A siphon works automatically and does not require any auxiliary mechanisms. The siphon inlet is typically set as far below the top water level as possible to minimize the carryover of floating scum and debris to the overflow outlet.
Mechanical regulator	The mechanical regulator (also known as an automatic regulator or reverse taintor gate) responds to the water level in the combined sewer or interceptor sewer (Fig. 18.9*a*). In either case, the float travel and the corresponding gate travel may be adjusted to regulate closely the flow to the interceptor.
Tipping-plate regulator	In these devices, the plate is pivoted off-center, and its motion is controlled by the difference of water levels above and below the gate (Fig. 18.9*b*). Multiple gates can be used to increase the capacity of an installation.
Hydro-brake regulator	The patented configuration of the hydro-brake (Fig. 18.9*c*) imparts a more-or-less centrifugal motion to the entering fluid. This action, which commences when a predetermined liquid head has been reached, effectively reduces the rate of discharge. This device has been used extensively on combined sewers to limit the flow to the interceptor, thus maximizing storage of flows in the combined sewer.

Source: From Metcalf & Eddy, Inc. (1991).

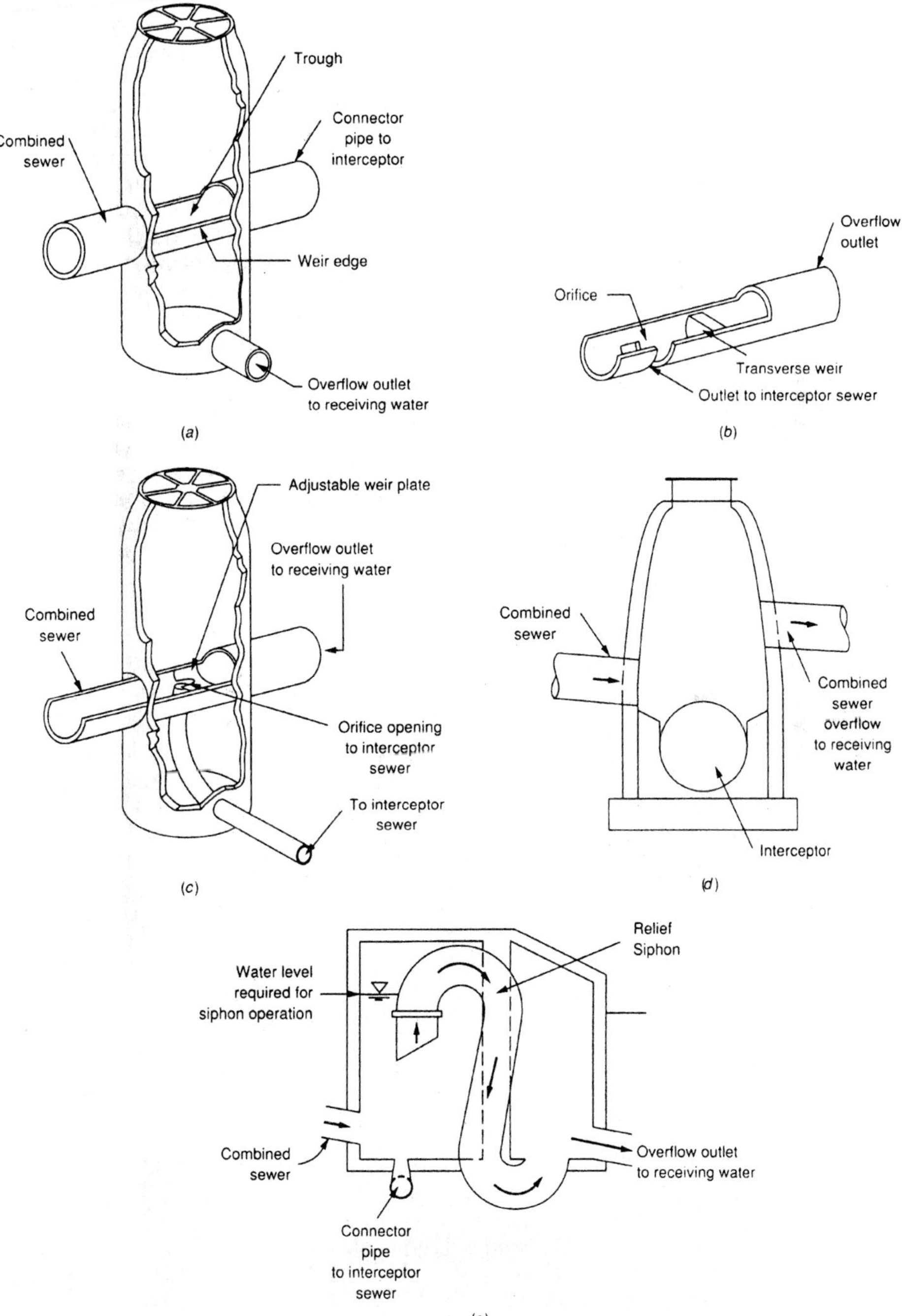

FIGURE 18.8 Typical diversion regulators: (*a*) side weir, (*b*) transverse weir with orifice, (*c*) leaping weir, (*d*) high outlet regulator, and (*e*) relief siphon. (From Metcalf and Eddy, Inc., 1991)

designed CSO treatment facility or storage facility. The diverted flows are a pre-determined flow or a design discharge. Table 18.2 describes some of the typical regulators used for CSOs, which are further illustrated in Figs. 18.8 and 18.9.

Inflatable dams, shown in Fig. 18.10, are control structures consisting of a rubberized fabric device that are inflated and deflated. They can be used to maximize storage at designated points in a CSS. These dams are usually activated by automatic sensors that measure flow levels at specified locations in the CSS. Generally, very little maintenance is required for inflatable dams; however, the air or water supply should be inspected regularly (USEPA, 1999).

18.2.4 Diverting Flows to CSO Treatment Facilities

Similar types of devices can be used to divert flows to CSO treatment facilities. (See Fig. 18.11.) The regulator consists of a weir constructed across the combined sewer with a vertical orifice just upstream of the weir. The weir diverts the dry weather flow through the vertical orifice to the interceptor. During wet weather flows, discharges exceeding the capacity of the orifice flow over the weir and through the outfall to the receiving water or to the CSO storage/treatment. The size of the (circular or rectangular) orifice is designed to intercept peak dry weather flow; however, the flow diverted to the interceptor during wet weather is higher. This regulator, which is generally used for flow less than 2 cfs is not recommended because of its tendency to clog with debris and because of the variation of discharge with respect to head (ASCE, 1992).

18.3 ORIFICES

An *orifice* is an opening (usually circular) in the wall of a tank or in a plate normal to the axis of a pipe. The plate is either at the end or in some intermediate location. The thickness of the wall or plate is very small compared to the size of the opening. A *standard orifice* has a sharp edge as shown in Fig. 18.12a or an absolutely square shoulder as shown in Fig. 18.12b. The orifices shown in Figs. 18.12a and 18.12b are not standard, as the flow is affected by the thickness of the plate and by the roughness of the surface. Flow through the rounded orifice is also affected by the radius of the curvature of the orifice.

In Fig. 18.12, C_c is the *coefficient of contraction,* which is the ratio of the area of the jet, A, to the area of the opening, A_o; therefore, $A = C_c A_c$. C_v is the *coefficient of velocity,* which is the ratio of the actual average velocity, V, to the ideal (without friction) velocity, V_i; therefore, $V = C_v V_i$. C_d is the *coefficient of discharge,* which is the ratio of the actual discharge, Q, to the ideal (no friction and no contraction) discharge, Q_i; therefore, $Q = C_d Q_i$. The coefficient of discharge is then $C_d = C_c C_v$. The equation to determine the discharge through an orifice is

$$Q = C_d A_o [2g(\Delta H)^{1/2}] \tag{18.1}$$

ΔH is the total difference in energy head between the upstream section to the minimum section of the jet, which is section A in Fig. 18.12. The total energy head is $H = z + p/\gamma + V^2/2g$. If the flow is from a tank, the velocity of approach is negligible and may be neglected. If the discharge is into the atmosphere (free jet), the downstream pressure head is zero, whereas if the jet is submerged, the downstream pressure head is equal to the depth of submergence. Table 18.3 lists coefficients of discharge for circular and square orifices with full contraction. Table 18.4 lists coefficients of discharge for a 20-cm-wide rectangular orifice with partially suppressed contractions.

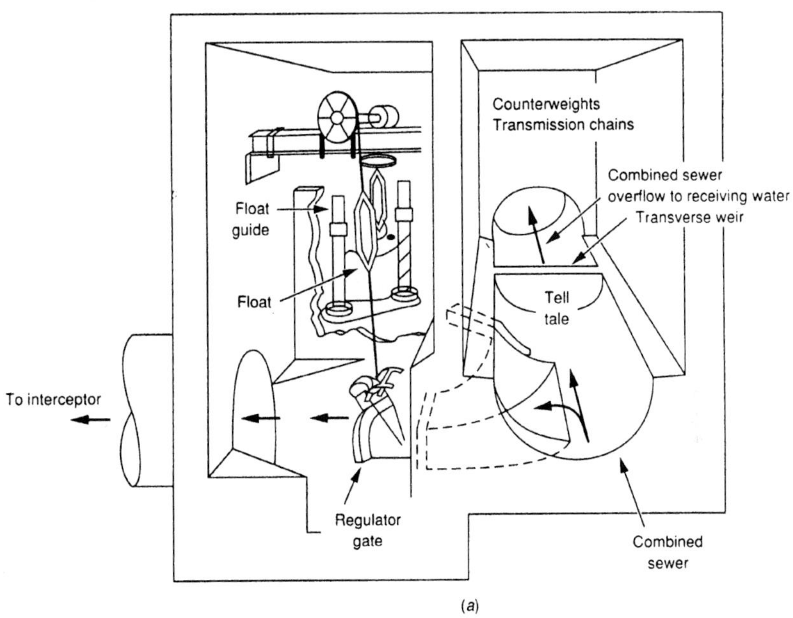

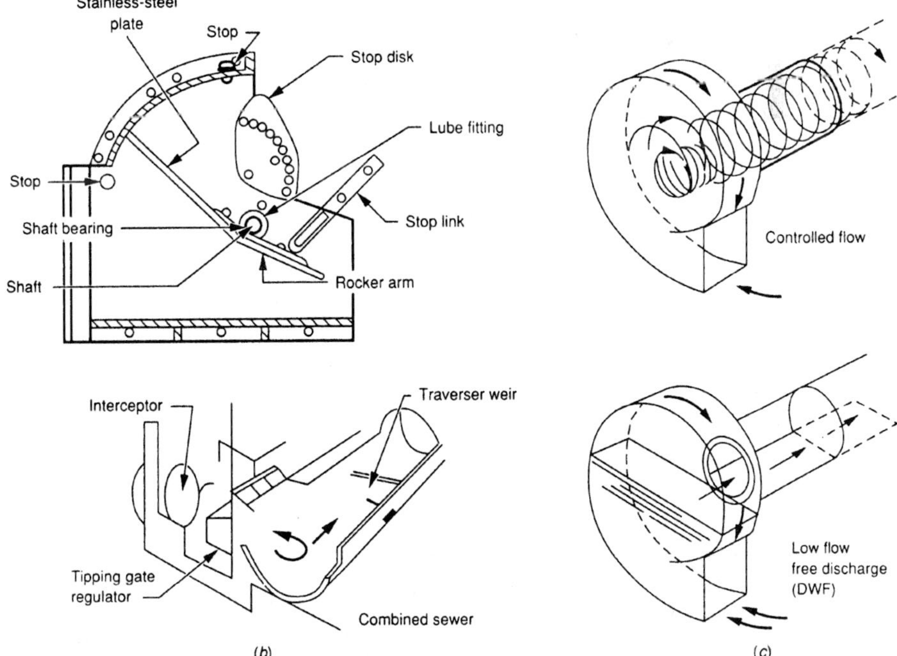

FIGURE 18.9 Typical CSO regulators: (*a*) mechanized regulator, (*b*) tipping plate regulator, and (*c*) hydro-brake regulator. (From Metcalf and Eddy, Inc., 1991)

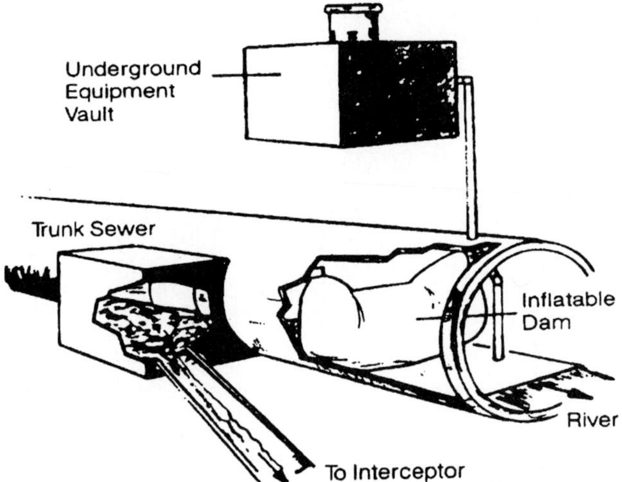

FIGURE 18.10 Typical inflatable dam. (From WEF, 1989)

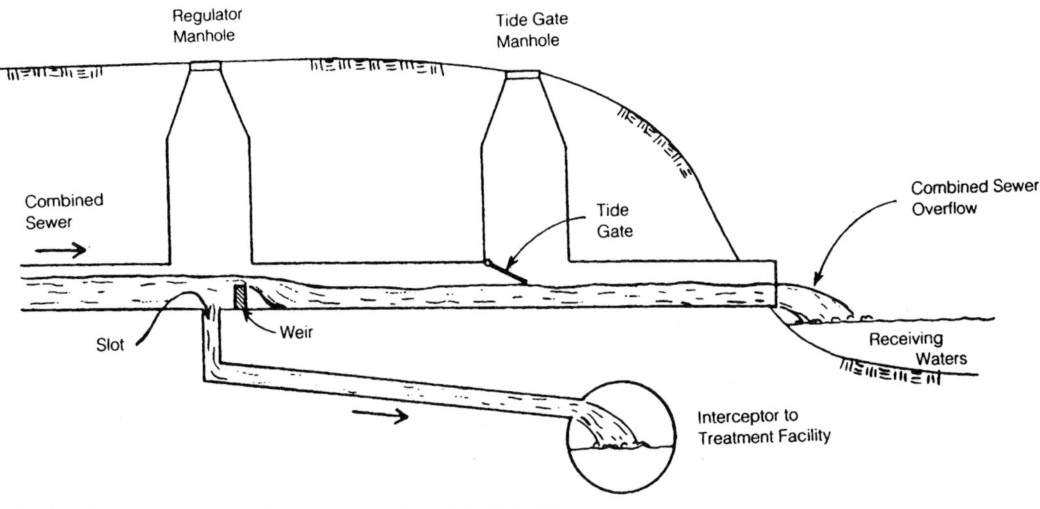

FIGURE 18.11 Fixed orifice flow regulator. (From ASCE, 1992)

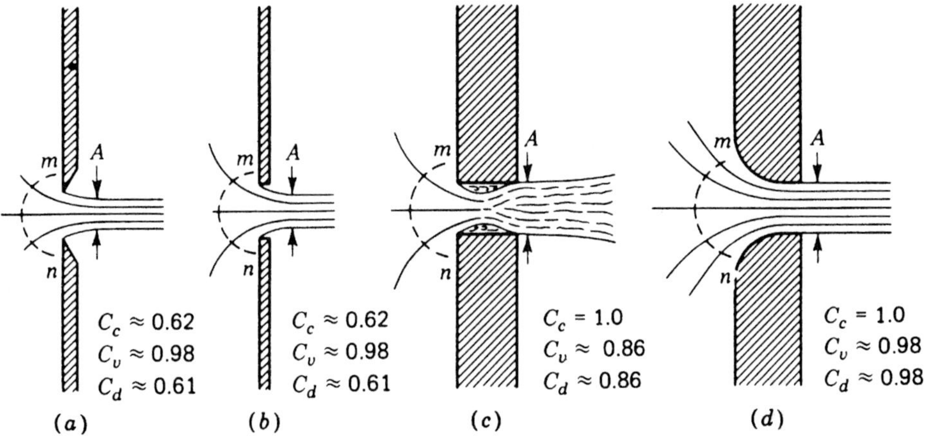

FIGURE 18.12 Types of orifice. (From Brater et al., 1996)

TABLE 18.3 Smith's Coefficients of Discharge for Circular and Square Orifices with Full Contraction

Head, m	Diameter of circular orifices, cm						
	0.5	1.0	2.0	5.0	10.0	15.0	30.0
0.1		0.642	0.627	0.606			
0.2	0.657	0.637	0.618	0.604	0.598	0.595	
0.3	0.648	0.630	0.613	0.603	0.598	0.596	0.591
0.4	0.643	0.627	0.611	0.601	0.598	0.597	0.592
0.5	0.638	0.623	0.608	0.601	0.599	0.597	0.594
1	0.629	0.616	0.604	0.600	0.599	0.598	0.597
2	0.619	0.610	0.602	0.599	0.598	0.597	0.596
3	0.612	0.606	0.599	0.597	0.597	0.596	0.595
4	0.609	0.604	0.599	0.597	0.596	0.596	0.595
5	0.606	0.603	0.598	0.596	0.596	0.596	0.594
10	0.599	0.598	0.596	0.595	0.595	0.595	0.594
20	0.595	0.594	0.593	0.593	0.593	0.593	0.593
30	0.593	0.592	0.592	0.592	0.592	0.592	0.592

Head, m	Side of square orifices, cm						
	0.5	1.0	2.0	5.0	10.0	15.0	30.0
0.1		0.649	0.631	0.606			
0.2	0.661	0.642	0.623	0.609	0.603	0.601	
0.3	0.651	0.635	0.619	0.608	0.604	0.602	0.599
0.4	0.647	0.631	0.617	0.606	0.604	0.603	0.600
0.5	0.642	0.627	0.614	0.606	0.604	0.603	0.602
1	0.634	0.621	0.609	0.605	0.605	0.604	0.603
2	0.624	0.615	0.607	0.604	0.604	0.603	0.602
3	0.617	0.611	0.605	0.603	0.603	0.602	0.601
4	0.614	0.609	0.604	0.603	0.602	0.602	0.601
5	0.611	0.608	0.603	0.602	0.602	0.602	0.600
10	0.605	0.603	0.602	0.601	0.601	0.600	0.600
20	0.601	0.600	0.600	0.599	0.599	0.599	0.599
30	0.599	0.598	0.598	0.598	0.598	0.598	0.598

Source: From Brater et al. (1996).

TABLE 18.4 Coefficients of Discharge for 20-cm-Wide Rectangular Orifices with Partially Suppressed Contraction

Description of contraction	Height, cm	Head, m 0.5	Head, m 1.0	Head, m 1.5
Complete contraction	20	0.600	0.604	0.603
	10	0.616	0.614	0.611
	5	0.630	0.626	0.620
	3	0.631	0.627	0.623
	1	0.646	0.632	0.620
Suppressed at bottom only	20	0.621	0.624	0.625
	10	0.648	0.645	0.634
	5	0.670	0.668	0.666
	3	0.679	0.677	0.677
	1	0.708	0.704	0.696
Suppressed on both sides only	20	0.631	0.628	0.628
	10	0.635	0.630	0.630
	5	0.639	0.634	0.635
	3	0.650	0.642	0.639
	1	0.677	0.665	0.655
Suppressed at bottom and partly on one side	20	0.634	0.636	0.637
	10	0.657	0.656	0.654
	5	0.675	0.673	0.672
	3	0.682	0.683	0.681
	1	0.707	0.704	0.695
Suppressed at bottom and partly on two sides	20	0.674	0.665	0.663
	10	0.678	0.675	0.672
	5	0.685	0.679	0.673
	3	0.692	0.687	0.683
	1	0.707	0.704	0.698
Suppressed at bottom and two sides	20	0.686	0.676	0.672
Complete suppression	20	—	0.950	

Source: From Brater et al. (1996).

18.3.1 Gates

Gates are openings in hydraulic structures for allowing discharge from the structure and also include a means of regulating the outflow. Gates have hydraulic properties similar to orifices with varying degrees of contraction. Discharge through gates may be either free or submerged with the following equation,

$$Q = Ca(2gh)^{1/2} \qquad (18.2)$$

where h = head on the center of the gate opening (orifice) for gates with free discharge, a = area of opening; and for submerged condition h is the difference in elevation of the water surfaces. Table 18.5 presents coefficients of discharges for wooden models of gates for the arrangements shown in Fig. 18.13.

TABLE 18.5 Coefficients of Discharge for Various Models, Fig. 18.13

Model	Depth of opening, cm	Depth of water above top of orifice, m										
		0.02	0.03	0.10	0.15	0.20	0.30	0.60	1.00	1.50	2.00	3.00
A	40			0.598	0.604	0.610	0.616	0.618	0.610	0.608	0.594	0.592
	20			0.633	0.638	0.640	0.641	0.640	0.638	0.637	0.636	0.634
	5			0.691	0.688	0.684	0.683	0.678	0.674	0.672	0.670	0.668
	3			0.709	0.700	0.695	0.692	0.688	0.681	0.677	0.675	0.672
B	40			0.644	0.650	0.654	0.656	0.649	0.634	0.620	0.615	0.611
	20			0.665	0.670	0.674	0.675	0.676	0.674	0.673	0.671	0.669
	5			0.665	0.681	0.688	0.693	0.695	0.694	0.692	0.691	0.689
	3			0.694	0.700	0.705	0.708	0.710	0.704	0.699	0.695	0.693
C	40			0.649	0.654	0.658	0.660	0.652	0.635	0.622	0.616	0.612
	20			0.668	0.673	0.676	0.678	0.679	0.677	0.674	0.672	0.670
	5			0.667	0.682	0.690	0.695	0.697	0.696	0.693	0.692	0.690
	3			0.696	0.702	0.707	0.710	0.712	0.705	0.699	0.695	0.693
D	20	0.487	0.495	0.543	0.562	0.577	0.588	0.601	0.601	0.601	0.601	0.601
	5	0.495	0.550	0.621	0.630	0.631	0.630	0.625	0.623	0.619	0.612	0.606
E	20	0.487	0.495	0.534	0.554	0.573	0.580	0.595	0.599	0.602	0.602	0.601
	5	0.495	0.544	0.602	0.612	0.618	0.623	0.627	0.628	0.627	0.622	0.617
F	20	0.530	0.535	0.571	0.584	0.595	0.600	0.608	0.610	0.610	0.609	0.608
	5	0.590	0.600	0.630	0.640	0.645	0.649	0.652	0.651	0.650	0.650	0.649

Source: From Brater et al. (1996).

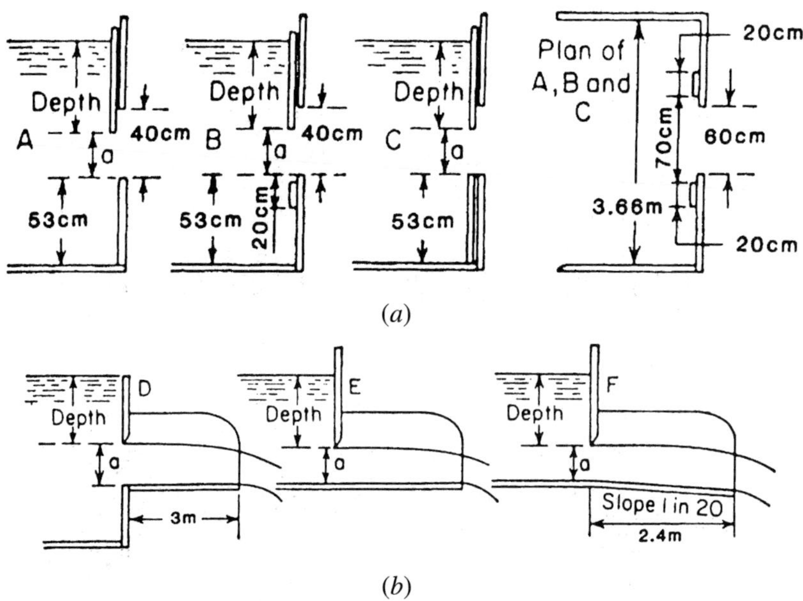

FIGURE 18.13   Gates. (*a*) Gates discharging freely into air. (*b*) Gates with prolonged bottoms and sides. (From Chow, 1959)

18.4 WEIRS

Weirs are devices (or overflow structures) that are placed normal to the direction of flow. Weirs essentially back up water so that in flowing over the weir, the water goes through critical depth. Weirs have been used for the measurement of water flow in open channels for many years. Weirs can generally be classified as *sharp-crested weirs* and *broad-crested weirs,* and are discussed in detail in Brater et al. (1996).

A *sharp-crested weir* is basically a thin plate, mounted perpendicular to the flow, with the top of the plate having a beveled, sharp edge, which makes the nappe spring clear from the plate (see Figs. 18.14 and 18.15).

18.4.1 Suppressed Rectangular Weirs

Sharp-crested weirs are as wide as the channel and the width of the nappe is the same length as the crest. Referring to Fig. 18.14, consider an elemental area, $dA = Bdh$, and assume that the velocity is $\sqrt{2gh}$; then, the elemental flow is

$$dQ = Bdh\sqrt{2gh} = B\sqrt{2g}h^{1/2}\,dh \tag{18.3}$$

The discharge is expressed by integrating Eq. (18.3) over the area above the top of the weir crest

$$Q = \int_0^H dQ = \sqrt{2g}B \int_0^H h^{1/2}\,dh = \frac{2}{3}\sqrt{2g}BH^{3/2} \tag{18.4}$$

Friction effects have been neglected in the derivation of Eq. (18.4). The drawdown effect shown in Fig. 18.14 and the crest contraction indicate that the streamlines are not parallel

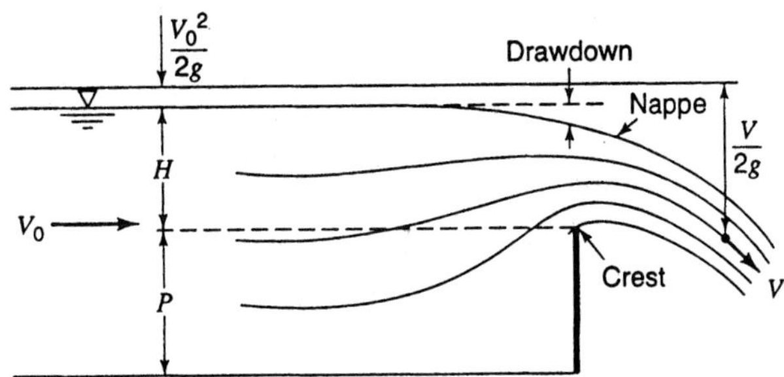

FIGURE 18.14   Flow over sharp-crested weir.

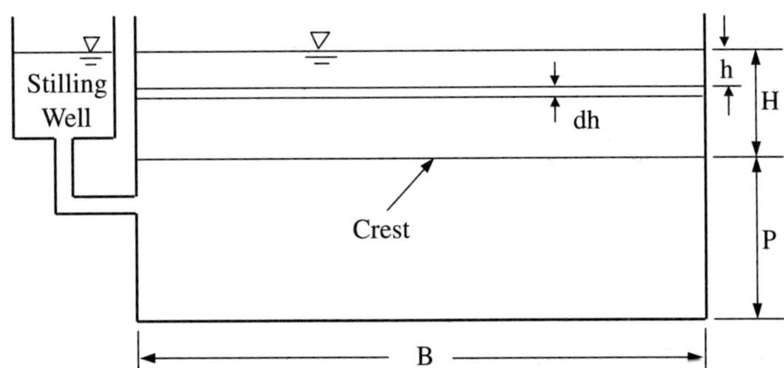

FIGURE 18.15   Rectangular sharp-crested weir without end contraction.

or normal to the area in the plane. To account for these effects, a coefficient of discharge, C_d, is used so that

$$Q = C_d \frac{2}{3} \sqrt{2g} B H^{3/2} \tag{18.5}$$

where C_d = approximately 0.62.

This is the basic equation for suppressed rectangular weir, which can be expressed more generally as

$$Q = C_w B H^{3/2} \tag{18.6}$$

where C_w = weir coefficient, $C_w = C_d 2/3\sqrt{2g}$. For U.S. customary units $C_w \approx 3.33$ and for SI units $C_w \approx 1.84$.

If the velocity of approach, V_a, where H is measured is appreciable then the integration limits are

$$Q = \sqrt{2g}B \int_{V_a^2/2g}^{H+V_a^2/2g} h^{1/2}\, dh = C_w B \left[\left(H + \frac{V_a^2}{2g} \right)^{3/2} - \left(\frac{V_a^2}{2g} \right)^{3/2} \right] \tag{18.7}$$

or when $(V_a^2/2g)^{3/2} \approx 0$, Eq. (18.7) can be simplified to

$$Q = C_w B \left(H + \frac{V_a^2}{2g} \right)^{3/2} \tag{18.8b}$$

18.4.2 Contracted Rectangular Weirs

A *contracted horizontal weir* is another sharp-crested weir with a crest that is shorter than the width of the channel and one or two beveled end sections so that water contracts both horizontally and vertically. This forces the nappe width to be less than B. The effective crest length

$$B' = B - 0.1nH \tag{18.9}$$

where $n = 1$, if the weir is placed against one side wall of the channel so that the contraction on one side is suppressed and $n = 2$, if the weir is positioned so that it is not placed against a side wall.

18.4.3 Triangular Weirs

Triangular or *V-notch weirs* are sharp-crested weirs that are used for relatively small flows, but also have the advantage that they can also function for reasonably large flows as well. Referring to Fig. 18.16, the rate of discharge through an elemental area, dA, is

$$dQ = C_d \sqrt{2gh} \, dA \tag{18.10}$$

where $dA = 2xdh$
    $x = (H - h) \tan \theta/2$

so $dA = 2(H - h) \tan(\theta/2) \, dh$, then

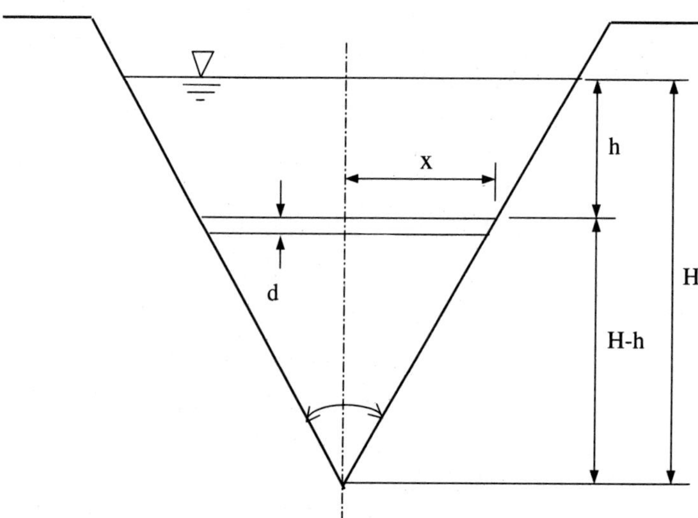

FIGURE 18.16   Triangular sharp-crested weir.

$$dQ = C_d\sqrt{2gh}\left[2(H - h)\tan\left(\frac{\theta}{2}\right)dh\right]$$ (18.11)

and

$$Q = C_d 2\sqrt{2g}\,\tan\left(\frac{\theta}{2}\right)\int_0^H (H - h)h^{1/2}\,dh$$

$$= C_d\left(\frac{8}{15}\right)\sqrt{2g}\,\tan\left(\frac{\theta}{2}\right)H^{5/2}$$

$$= C_w H^{5/2}$$ (18.12)

The value of C_w for $\theta = 90°$ (the most common) is $C_w = 2.50$ for U.S. customary units and $C_w = 1.38$ for SI units. Figure 18.17 gives coefficients of V-notched sharp-crested weirs for other angles.

18.4.4 Submerged Sharp-Crested Weirs

The flow profile over a submerged sharp-crested weir is illustrated in Fig. 18.18. Discharge, Q, over the weir is related to both the head, H_1, on the upstream side of the weir and the head, H_2, on the downstream side of the weir, and, to a lesser extent, the height, P, of the weir crest above the floor of the channel. For an unsubmerged condition, the discharge Q_1 is

$$Q_1 = C(H_1)^n$$ (18.13)

The discharge, Q, for a submerged weir can be determined using Fig. 18.18, which shows plots of Q/Q_1 versus H_2/H_1 for 90° V-notch weirs, rectangular weirs, and all sharp-crested weirs. The n is the exponent in the free discharge equation (for unsubmerged conditions). Villemonte (1947) developed a single equation, given as

$$Q/Q_1 = [1 - (H_2/H_1)^n]^{0.385}$$ (18.14)

where n = exponent in the free discharge equation. He showed that this equation applied to rectangular, triangular, parabolic, cusped, and proportional weirs.

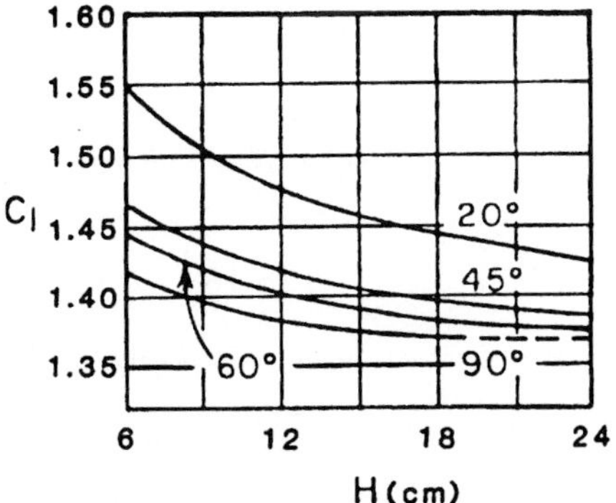

FIGURE 18.17 Coefficients, V-notch sharp-crested weirs. (From Chow, 1959)

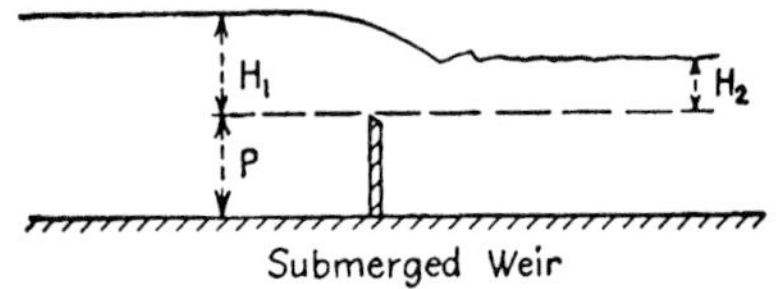

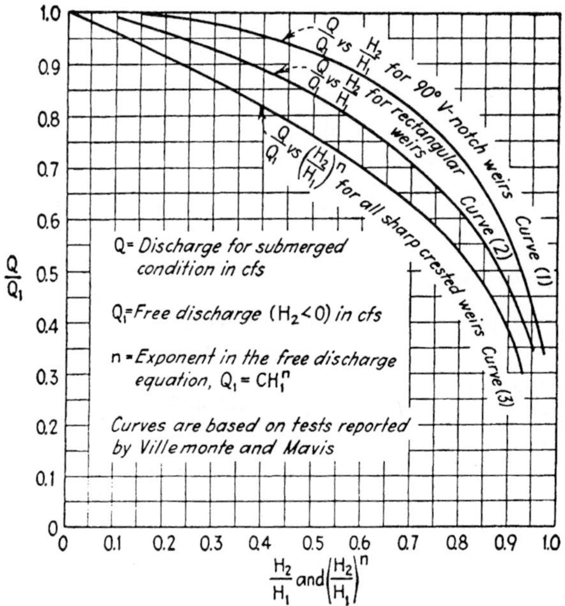

FIGURE 18.18   Discharge for submerged weirs. (From Chow, 1959)

18.4.5   Broad-crested Weirs

Broad-crested weirs (refer to Fig. 18.19) are essentially *critical-depth weirs* in that if the weirs are high enough, critical depth occurs on the crest of the weir. For critical-flow conditions $y_c = (q^2/g)^{1/3}$ and $E = 3/2y_c$ for rectangular channels.

$$Q = B \cdot q = B\sqrt{gy_c^3} = B \sqrt{g \left(\frac{2}{3}E\right)^3}$$

$$= B \left(\frac{2}{3}\right)^{3/2} \sqrt{g}E^{3/2}$$

or assuming the approach velocity is negligible

$$Q = B \left(\frac{2}{3}\right)^{3/2} \sqrt{g}H^{3/2}$$

$$Q = C_w BH^{3/2} \tag{18.15}$$

Figure 18.19 illustrates a broad-crested weir installation in a concrete-lined canal. Values of C_w are listed in Table 18.6.

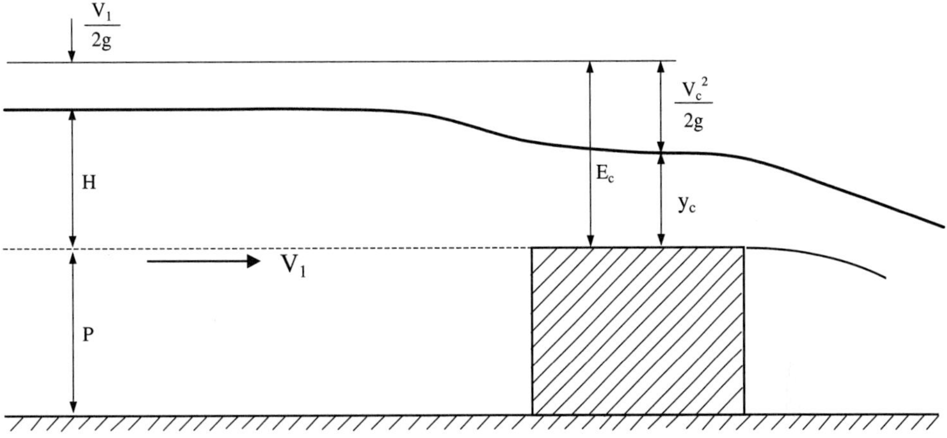

FIGURE 18.19   Broad-crested weir.

18.4.6   Side Weirs

Flow through *side weirs,* also known as *lateral weirs,* is a case of *spatially varied flow* with decreasing discharge as illustrated in Fig. 18.20. de Marchi (1934) developed a spatially varied flow equation for side weirs, based upon a horizontal frictionless channel, and assumed that the lateral outflow was unaffected by the channel flow. Ackers (1957) and Frazer (1957) presented theoretical developments similar to that of de Marchi (1934). Allen (1957) reported on experimental results on side weirs in circular channels and developed a semi-empirical formula for such structures. According to Frazer (1957), five types of flow profiles can occur on side weirs. For a spatially varied flow with decreasing discharge, the specific energy can be considered constant along a channel (Chow, 1959). Smith (1973) extended de Marchi's analysis to include energy losses due to friction.

TABLE 18.6   Values of C in Formula $Q = CLH^{3/2}$ for Broad-Crested Weirs

Measured head, m	Breadth of crest of weir, m										
	0.15	0.20	0.30	0.45	0.60	0.75	0.90	1.20	1.50	3.00	4.50
0.10	1.61	1.55	1.50	1.46	1.44	1.44	1.43	1.40	1.38	1.41	1.49
0.20	1.70	1.60	1.52	1.46	1.44	1.44	1.48	1.49	1.49	1.49	1.49
0.30	1.83	1.73	1.65	1.52	1.47	1.46	1.46	1.48	1.48	1.48	1.45
0.40	1.83	1.80	1.77	1.61	1.53	1.48	1.46	1.46	1.46	1.48	1.46
0.50	1.83	1.82	1.81	1.70	1.60	1.52	1.48	1.47	1.46	1.46	1.45
0.60	1.83	1.83	1.82	1.67	1.57	1.52	1.50	1.48	1.46	1.46	1.45
0.80	1.83	1.83	1.83	1.81	1.70	1.60	1.55	1.50	1.48	1.46	1.45
0.90	1.83	1.83	1.83	1.83	1.77	1.69	1.61	1.51	1.47	1.46	1.45
1.0	1.83	1.83	1.83	1.83	1.83	1.76	1.64	1.52	1.48	1.46	1.45
1.2	1.83	1.83	1.83	1.83	1.83	1.83	1.70	1.54	1.49	1.46	1.45
1.4	1.83	1.83	1.83	1.83	1.83	1.83	1.83	1.59	1.51	1.46	1.45
1.5	1.83	1.83	1.83	1.83	1.83	1.83	1.83	1.70	1.54	1.46	1.45
1.7	1.83	1.83	1.83	1.83	1.83	1.83	1.83	1.83	1.59	1.46	1.45

Source: From Brater et al. (1996).

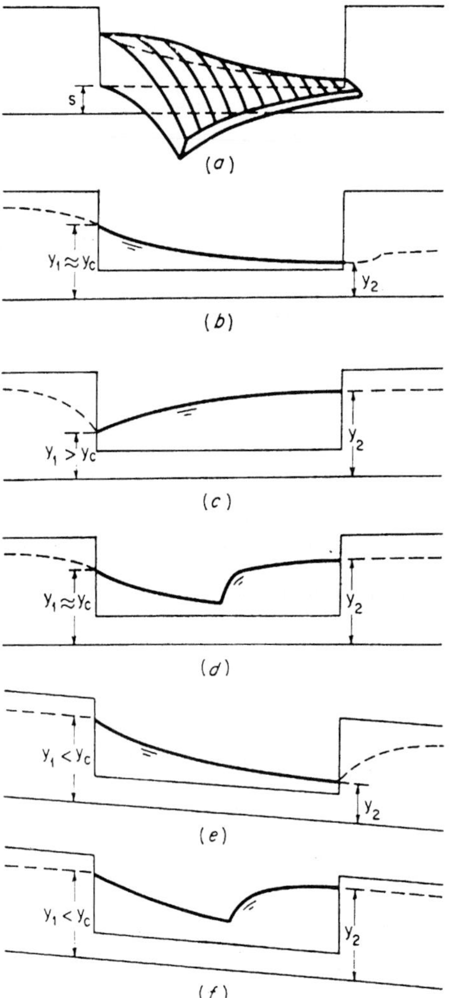

FIGURE 18.20   Various flow profiles along a side weir. (From Chow, 1959)

Type a.   Critical conditions at or near the entrance with supercritical flow in the weir section, the depth of flow decreasing along the weir.

Type b.   Depth of flow greater than critical at the entrance with subcritical flow in the weir section, the depth of flow increasing along the weir section.

Type c.   Type *a* flow at the beginning of the weir section with a hydraulic jump occurring in the weir section, and type *b* flow after the jump at a lower specific-energy level owing to jump losses.

Type d.   Depth of flow less than critical at the entrance with supercritical flow in the weir section, the depth of flow decreasing along the weir section.

Type e.   Type *d* flow at the entrance section with a hydraulic jump occurring in the weir section and type *b* flow after the jump at a lower specific-energy level owing to jump losses.

Assuming that the specific energy along a side weir is constant (friction slope S_f = channel bottom slope S_o) and assuming the channel is prismatic and horizontal, then the change in depth, y, with respect to the change in distance, x, along the side weir is expressed as (Chow, 1959)

$$dy/dx = [Qy(-dQ/dx)]/[gb^2y^3 - Q^2] \qquad (18.16)$$

where b = width of the channel.

The discharge over a side weir, Q_w, per unit length of the weir can be expressed as

$$dQ_w/dx = -dQ/dx = C_{sw}(2g)^{1/2}(y - s)^{3/2} \qquad (18.17)$$

where C_{sw} = discharge coefficient and s = the height of the weir sill above the bottom of the channel

The sill of the weir is parallel to the bottom of the channel as shown in Fig. 18.20. At any section, the discharge Q is expressed as

$$Q = by[2g(E - y)]^{1/2} \tag{18.18}$$

where b = width of the channel and E = specific energy $E = y + Q^2/(2gb^2y^2)$.

Numerous studies have been reported in the literature (Cheong, 1991; Kumar and Pathak, 1987; Nadesamoorthy and Thomson, 1972; Ranga Raju et al., 1979; Subramanya and Awasthy, 1972; Yu-Tech, 1972) on determining discharge coefficients for side weirs. Each of these previous studies invariably developed discharge coefficients as a function of the upstream Froude number. Nadesamoorthy and Thompson (1972) claimed that their equation applies to both subcritical and supercritical flow. Subramanya and Awasthy (1972) recognized that through dimensional analysis the discharge coefficient was also a function of the ratio of sill height to the upstream flow depth. Hager (1987) used the assumption of constant energy, coupled with a new technique for determining the weir coefficient as a function of the local Froude number. Other studies include El-Khashab and Smith (1976), Ramamurthy and Carballada (1980), Uyumaz and Smith (1991), and Robinson and McGhee (1993).

Singh et al. (1994) experimentally investigated the effect of the upstream Froude number and the sill height on the discharge coefficient of rectangular side weirs under sub-critical flow conditions in a prismatic rectangular main channel. They found the discharge coefficient to be a function of the upstream Froude number, F_1, and the ratio of the sill height to upstream depth, s/y_1, given as

$$C_{sw} = 0.33 - 0.18F_1 + 0.49s/y_1 \tag{18.19}$$

This equation resulted from a first order polynomial fit of data. The mean of the C_{sw} estimated from the above equation is 0.6 with a standard deviation of 0.07.

Swamee et al. (1994) also reported an equation for the discharge coefficient of side weirs for sub-critical flow over rectangular side weirs. They found the discharge coefficient to be a function of the depth, y, and weir height, s, as

$$C_e = 0.0447[(A^{6.67} + (B)^{6.67}]^{-0.15} \tag{18.20}$$

where

$$A = 44.7s/(49s + y) \tag{18.21}$$

and

$$B = (y - s)/y \tag{18.22}$$

The weir coefficient C_{sw} is

$$C_{sw} = 2/3(C_e) \tag{18.23}$$

The differential equation for spatially varied flow with decreasing discharge is

$$dy/dx = [S_o - S_f - (Q/gA^2)\, dQ_w/dx]/[1 - Q^2T/(gA^3) \tag{18.24}$$

where A = cross-sectional area of flow and T = top width of flow, with dQ_w/dx computed using Eq. (18.17). This equation can be used to determine the flow profile over a side weir.

18.5 *SPECIAL FLOW REGULATORS*

18.5.1 Steinscrew Flow Regulators

Steinscrew flow regulators are static devices that were developed in Sweden in the 1970s by Stein Bendixsen (Janson et al., 1976) to regulate waste water discharge rates in combined sewer pipes. Flow equalization is achieved by retaining excess flow in the unutilized volume of the main sewers by installing these regulators as damming devices in the pipes at certain points in the network. This regulator consists of a stationary, anchored, screw-shaped plate (twisted into a 270° spiral) that is attached inside the pipe (see Fig. 18.21). In the part of the plate that fits against the bottom of the pipe, a bottom opening releases a specified base dry-weather flow.

These devices are designed to be in pipes so that the pipe upstream of the regulator is the detention facility. Consequently, it is recommended that these be installed in larger pipes such as 36-in and larger (Stahre and Urbonas, 1990). Discharge characteristics of the Steinscrew are presented in Fig. 18.22. The regulator is self-regulating, has no moving parts, and does not require power.

These flow regulators have the following characteristics (Jansen et al., 1976):

1. A certain sewage flow (basic flow) is always discharged through a basic opening in the regulator. The size of the basic opening is calculated to obtain the most favorable use of the upstream storage volume and to maintain the pipe self-cleaning at dry-weather flow.

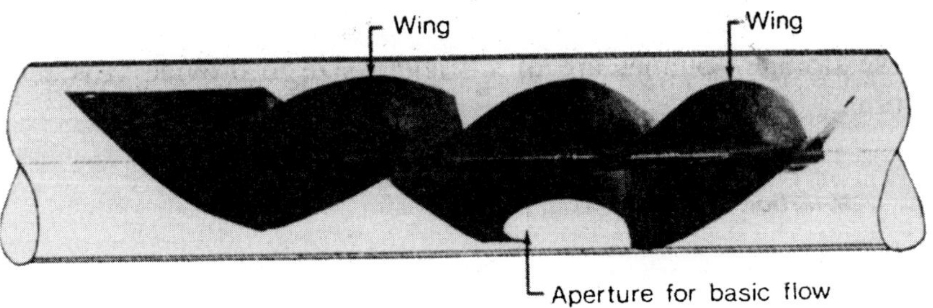

(a) Flow Regulator

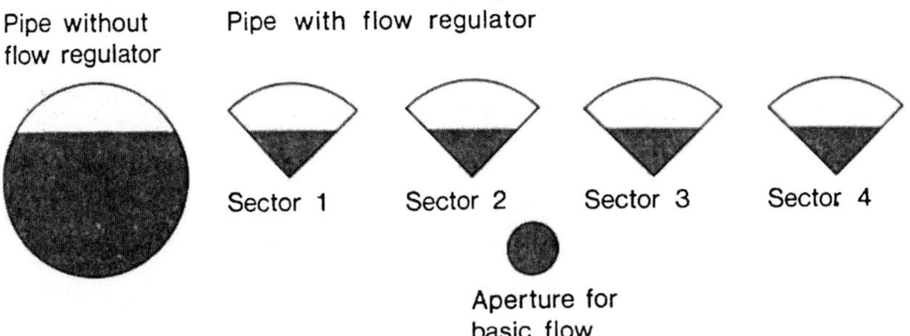

(b) Discharge Areas for Given Water Depth

FIGURE 18.21   Steinscrew. (From Janson et al., 1976)

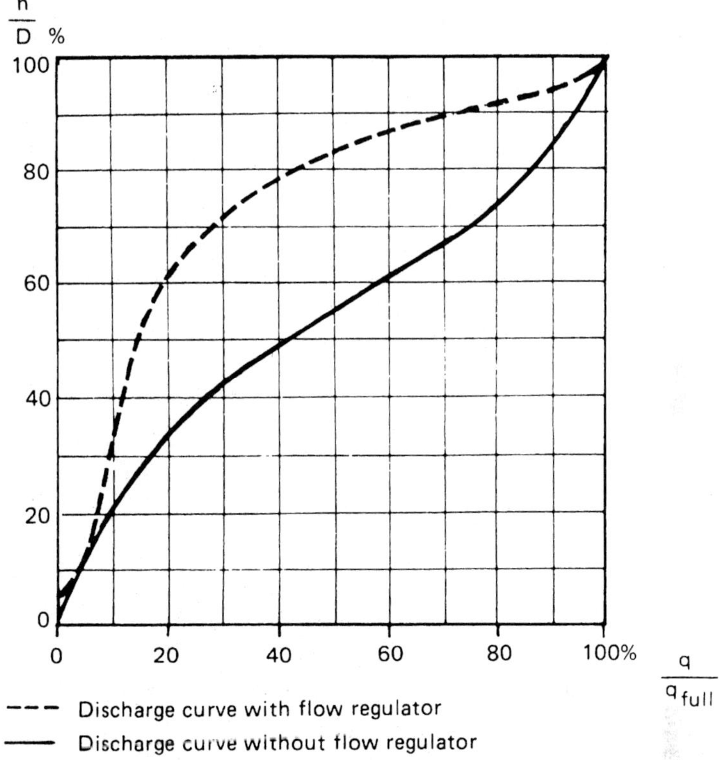

FIGURE 18.22 Discharge characteristics of the Steinscrew. (After Janson, Bendixen, and Harlaut, 1976)

2. For upstream water levels up to the center of the pipe, only the basic flow is discharged through the regulator, while the inflow excess is stored in the sewer upstream from the regulator.

3. When the water level rises above the pipe center, discharge starts over four triangular outlets, and the total discharge through the flow regulator is practically the same as the pipe capacity without a flow regulator.

4. At full flow in the sewer the entire inflowing sewage is discharged over the four outlets, and the total discharge through the flow regulator is practically the same as the pipe capacity without a flow regulator.

5. At decreasing flow the storage upstream from the regulator is slowly emptied through the triangular outlets and the basic opening.

18.5.2 Hydro-brakes

The *hydro-brake,* a self-regulating device, was developed in Denmark in the mid-1960s and is used to control outflows from storage structures. This device is typically used as a flow control outlet on CSS to limit the flow to interceptors in order to maximize the storage of flows in the CSS. Hydrostatic pressure (water level) controls the rate of flow through this device, which commences at a pre-determined head. This device consists of an eccentric cylinder housing with an inlet opening located on the side. Figure 18.23 illustrates three

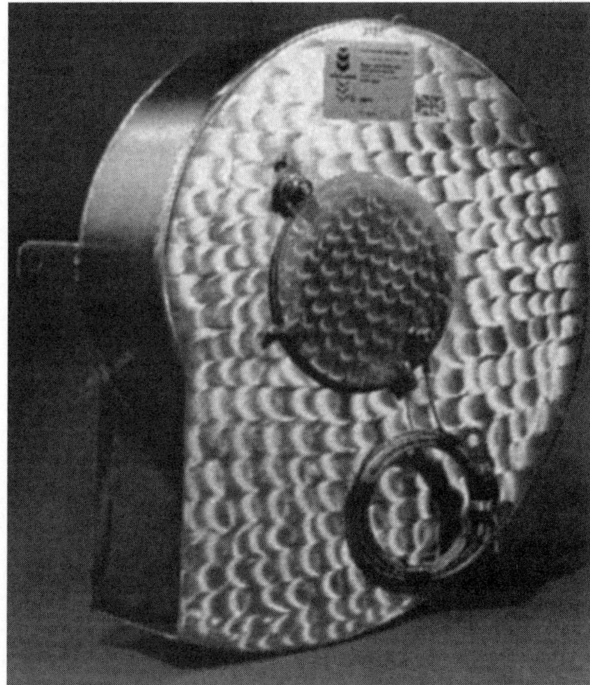

FIGURE 18.23 Hydro-brake configuration. (*a*) Type SH Hydro-Brake® flow control. (From HydroInternational, PLC)

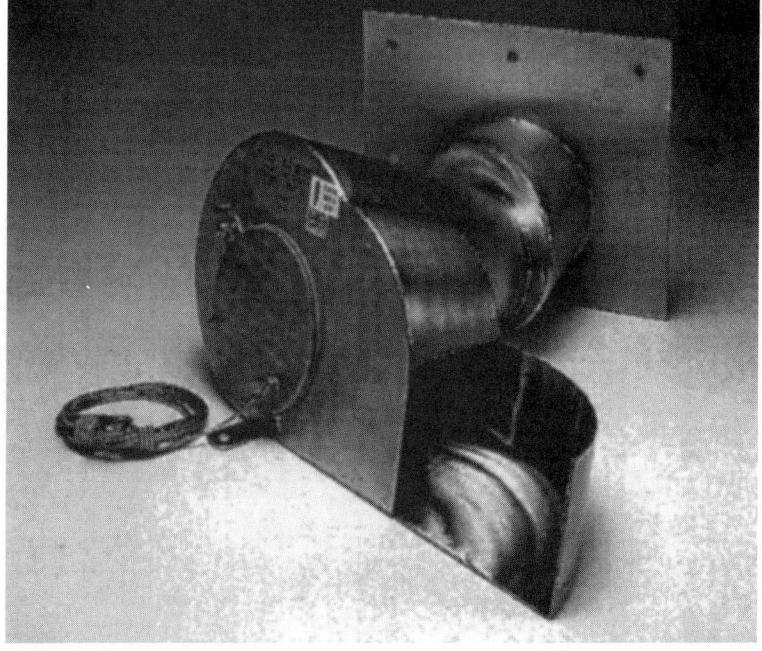

FIGURE 18.23 (*Continued*) (*b*) Type CH Hydro-Brake® flow control.

FIGURE 18.23   (*Continued*)   (*c*) Type CH Hydro-Brake® flow control with an adjustable intake gate and pivoting by-pass door.

types of configurations of hydro-brakes. Sewer flows enter the hydro-brake tangentially to the outlet pipe, which is installed normal to the housing cylinder. The flow patterns in hydro-brakes are illustrated in Fig. 18.24.

The head-discharge characteristics for two sizes of the standard hydro-brake manufactured by Vortechnics are presented in Fig. 18.25. A typical installation of a hydro-brake in a catch basin outlet is shown in Fig. 18.26. The head-discharge characteristics of the cone style of hydro-brake for two sizes are presented in Fig. 18.27. Figure 18.28 illustrates a typical in-

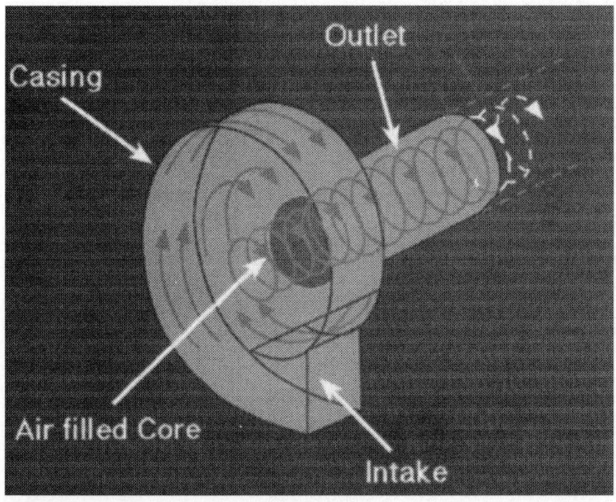

Type SH Vortex Flow Pattern

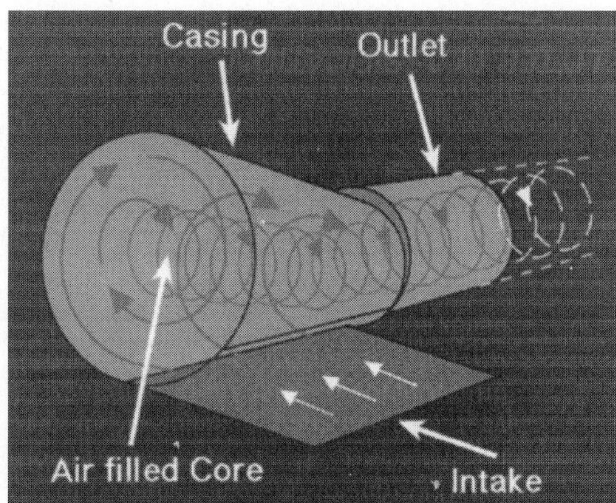

Type C Vortex Flow Pattern

FIGURE 18.24   Flow patterns. (From HydroInternational, PLC)

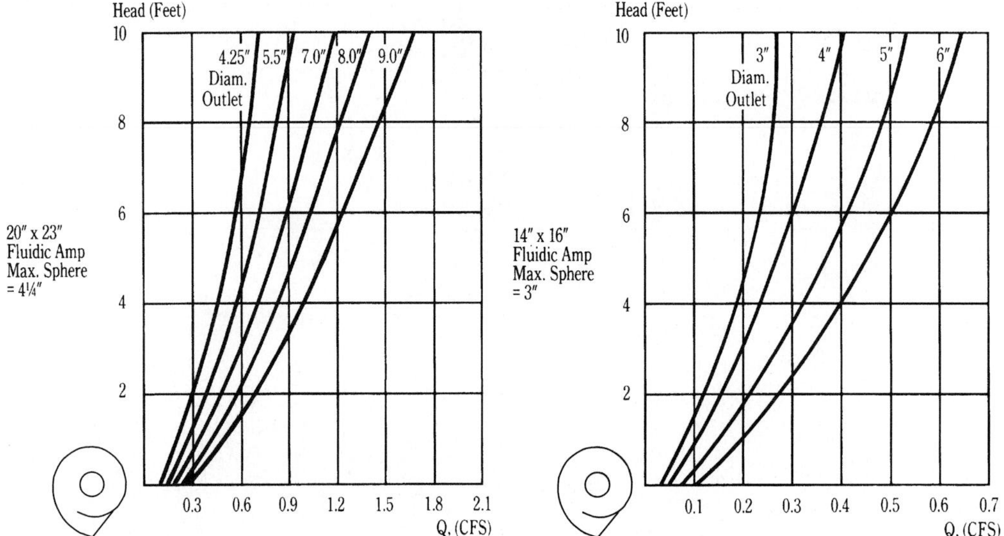

FIGURE 18.25   Head-discharge relationships for standard hydro-brakes; developed for installation in a catch basin outlet pipe. (From Vortechnics)

stallation of the cone style hydro-brake in a manhole without a sump. Figures 18.29–18.31 from HydroInternational PLC shows applications of hydro-brakes and Figure 18.32 shows three inch catchbasin control units.

Hydrobrakes have the following two advantages over an orifice (Stahre and Urbonas, 1990): (1) larger openings can be used for the same discharge rate, which can significantly reduce the risk of clogging the regulator, and (2) water depth causes a much smaller variation in the discharge resulting in a more constant release. Discharges depend on the design type and the size of the outlet, so that the manufacturer must supply the head-discharge characteristics for their particular regulator.

18.5.3   Wirbeldrossel

The *wirbeldrossel* or *turbulent throttle* has a similar operating principle as the hydrobrake and is used to control discharge from storage facilities. This device was developed at the University of Stuttgart in Germany in the mid-1970s. A wirbeldrossel has symmetrical cylinder housing with an inlet pipe connecting tangentially to the cylinder. Also shown in the figure is an air supply pipe. A circular cylinder in the bottom surface of the cylinder functions as the outlet, which can be adjusted using manufactured rings of varying sizes. Wirbeldrossels are installed so that the outlet fits on a vertical pipe. These devices are located in a separate chamber downstream of the catch basin, so that they can be inspected and maintained when the storage basin is filled with water.

A similar flow regulator is the *wirbelvalve,* which is actually a modified version of the wirbeldrossel. Wirbelvalves have a slanted rotation chamber and a cone-shaped bottom. The advantage of the wirbelvalve is that it requires less vertical height between the basin outlet and the downstream sewer (Stahre and Urbonas, 1990).

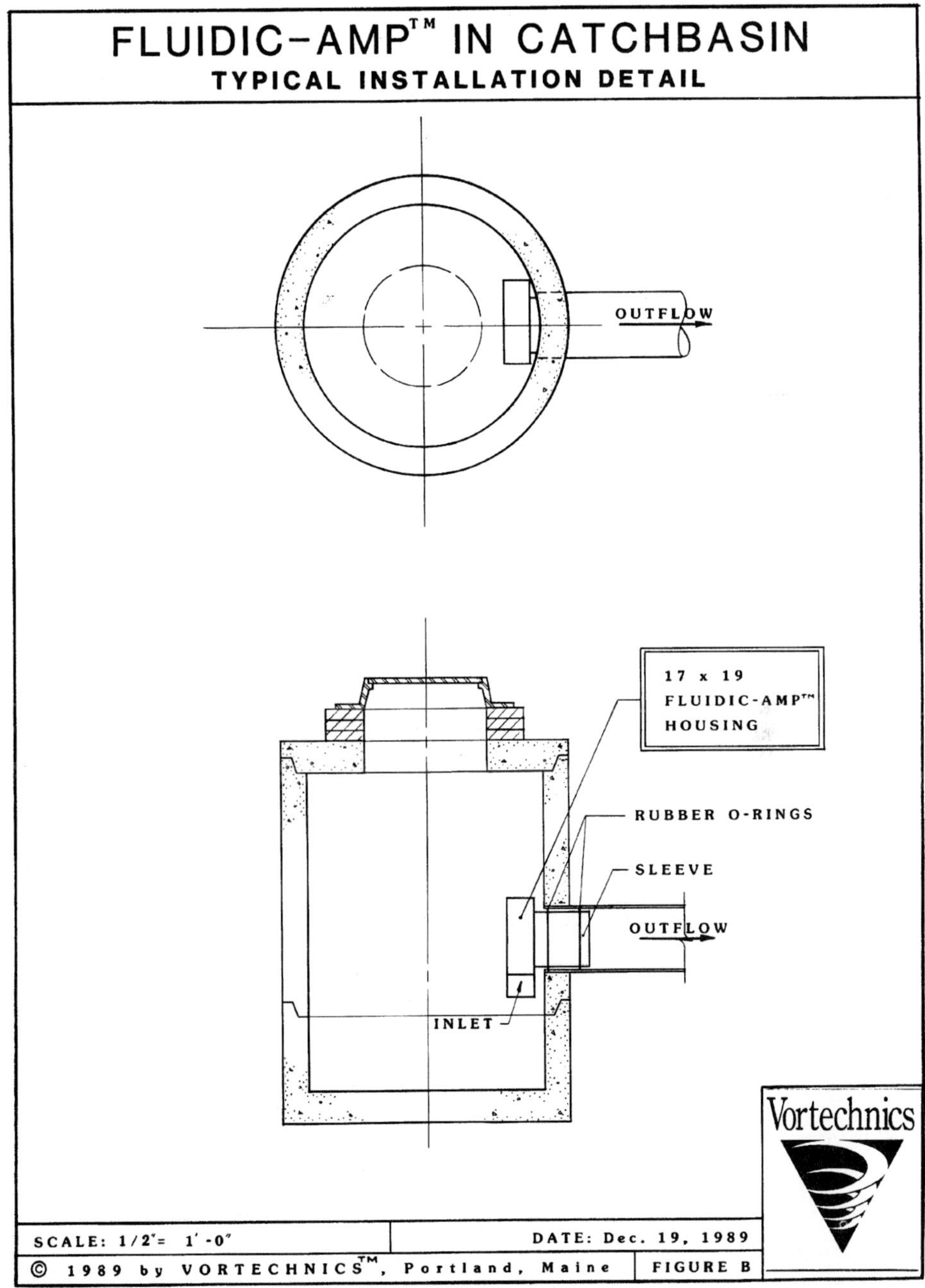

FIGURE 18.26 Typical installation of a standard hydro-brake in a catch basin outlet. (From Vortechnics)

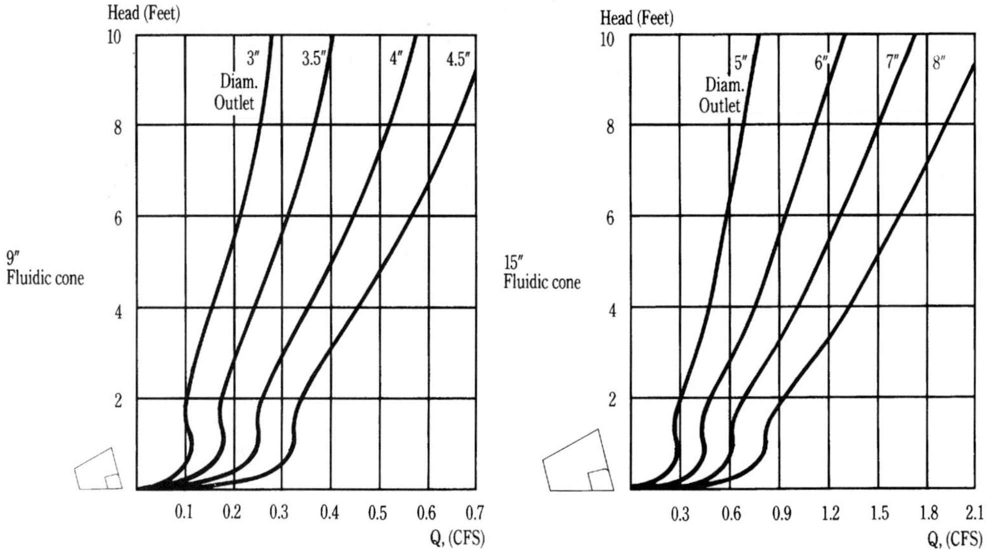

FIGURE 18.27   Head-discharge relationships for hydro-brakes (fluid-cone style); developed for installation in structure with no slump; typically fitted to outlets from manholes in storm drains and sanitary sewers. (From Vortechnics)

18.5.4   Central Outlet Pipe Surrounded by a Pressure Chamber Filled with Air

This flow regulator valve was developed in Sweden in the late 1970s. The device is a central outlet pipe surrounded by a pressure chamber filled with air in which water pressure on the upper portion of the device displaces the fabric at the outlet. Displacement of the fabric controls the discharge volume.

18.6   CSO OUTLETS

The outlets of CSO systems typically discharge into a receiving water body. Also, outlets are frequently located where a combined sewer intersects an interceptor sewer. If the CSO outlet discharges into a receiving water body that has high-water elevations greater than the elevation of the CSO outlet, a backwater or tide gate is needed to prevent the receiving water from entering the sewer. These control devices include flap gates and elastomeric gates. Also, these gates allow greater detention volume in collection systems by minimizing backwater and tidal inflows (USEPA, 1999).

18.6.1   Flap Gates

Flap gates or *backwater gates* are commonly used to resist backwater and tidal influences. As shown in Fig. 18.33, these gates consist of a flap hung against an inclined seat. Hinges may be placed at the top for single-leaf gates or they may consist of two leaves with side hinges. Even though these gates have a simple construction and operation these gates are subject to fouling and sticking and require appreciable hydraulic heads to operate against their own weight (WEF, 1989). Some of the commonly encountered problems with typical

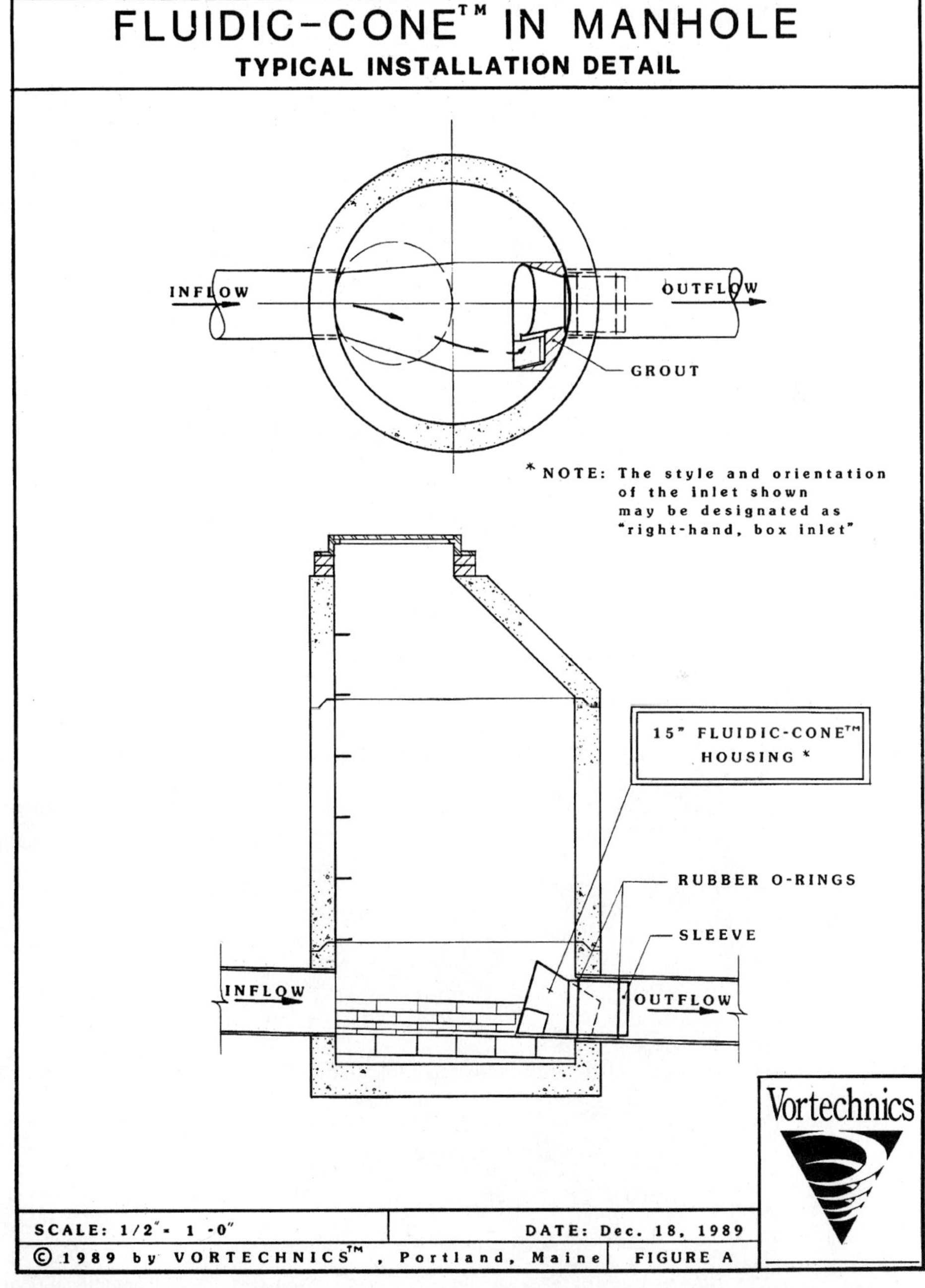

FIGURE 18.28 Typical installation of a cone-style hydro-brake in manhole without sump. (From Vortechnics)

FIGURE 18.29 Plate mounted Type "SH" Vortex Valve. Design discharge of 0.42 cfs (12 l/s) under a 5-foot head, City of Portland, Maine. The "SH" unit was installed in a 4-ft. diameter manhole. Intake to the unit is on the bottom left. This unit is plate mounted as the outlet pipe is larger than the casing of the control unit. Normally, this type of unit is fixed using 3 lugs and anchor bolts or a push-fit spigot/sleeve arrangement. (From HydroInternational, PLC)

FIGURE 18.30 6-Inch Type "C" Conical Reg-U-Flow Vortex Valve. Installed in a 4-ft diameter catchbasin. This unit was supplied with a sleeve pipe which was inserted into the outlet pipe and sealed by 2 neoprene rubber "O" rings. The consultant on this project wished to maintain a silt trap so the control unit was mounted 3-feet above the base of the chamber with an additional support member to anchor the base of the cone. (From HydroInternational, PLC)

FIGURE 18.31 Type "C" Reg-U-Flo Vortex Valve. A 22-inch Type "C" Reg-U-Flo Vortex Valve discharging 15 cfs (425 l/s), installed at the Eastern Mall Office Park project in South Portland, Maine. This photograph shows the valve being lowered into place in the precast concrete control chamber. This stormwater management project was designed by H.I.L. Technology, Inc. By storing the 2-year storm within the pipes and manholes and the 25-year storm on the parking lots, the developer was able to avoid the construction of a detention pond which would have required the purchase of additional land. (From HydroInternational, PLC)

FIGURE 18.32 3-Inch Catchbasin Control Units. The valve assembly comprises 2 parts: one which is pushed into the outlet pipe; and the valve itself which seats on the slide mount. The valve may be removed from its seating to facilitate maintenance. Many hundreds of controls such as these have been supplied. Stormwater attenuation techniques utilizing the vortex valve in this way have been shown to be extremely cost effective. (From HydroInternational, PLC)

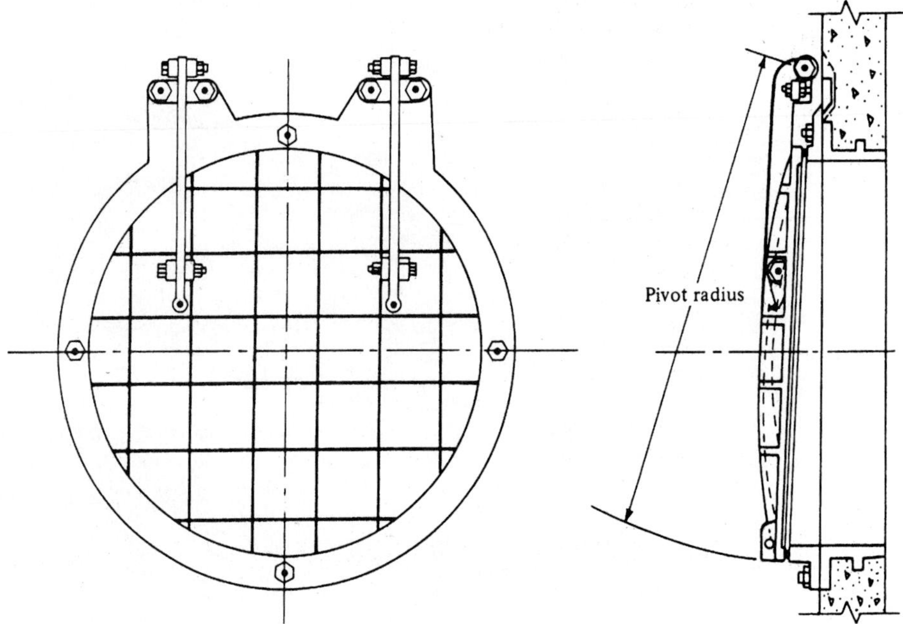

FIGURE 18.33   A single-leaf tide gate. (From Armco)

flap gates include warpage, corrosion, and the tendency to stick in one position (USEPA, 1993).

18.6.2   Elastomeric Check Valves

Elastomeric check valves or *duckbill gates* are positive displacement devices that force CSO flows to open an otherwise collapsed rubber-based valve. A typical elastomeric check valve is illustrated in Fig. 18.34. These valves can overcome some of the problems encountered

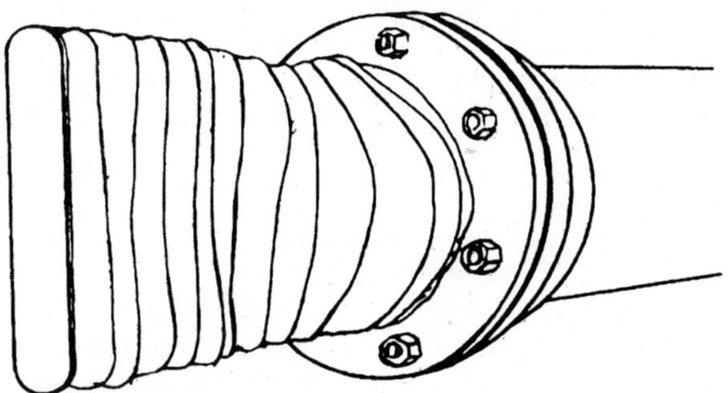

FIGURE 18.34   Typical elastomeric check valve. (From WEF, 1989)

with flap gates, requiring less maintenance. Also elastomeric check valves are designed to open with smaller head requirements and to close over large debris.

REFERENCES

Ackers, P., "A Theoretical Consideration of Side Weirs as Storm Water Overflows," *Proceedings, Institution of Civil Engineers,* London, England, Vol. 6, pp. 205–269, 1957.

Allen, J. W., "The Discharge of Water over Side Weirs in Circular Pipes," *Proceedings, Institution of Civil Engineers,* London, England, Vol. 6, pp. 270–287, 1957.

American Iron and Steel Institute (AISI), *Modern Sewer Design,* 3rd ed., Washington, DC, 1995.

American Society of Civil Engineers (ASCE) and Water Environment Federation (WEF), *Design and Construction of Urban Stormwater Management Systems,* Reston, VA, 1992.

Association of Metropolitan Sewerage Agencies (AMSA), *Approaches to Combined Sewer Overflow Program Development,* Washington, DC, 1994.

Brater, E. F., H. W. King, J. E. Lindel, and C. Y. Wei, *Handbook of Hydraulics,* 7th ed., McGraw-Hill, Inc., New York, 1996.

Cheong, H., "Discharge Coefficient of Lateral Diversion from Trapezoidal Channel," *Journal of Irrigation and Drainage Engineering,* ASCE, 117(4):461–475, 1991.

Chow, V. T., *Open-Channel Hydraulics,* McGraw-Hill, Inc., New York, 1959.

Collinge, V. K., "The Discharge Capacity of Side Weirs," *Proceedings of the Institution of Civil Engineers,* London, England, Vol. 6, 1957.

Daugherty, R. L., and J. B. Franzini, *Fluid Mechanics with Engineering Applications,* McGraw-Hill Book Company, New York, 1977.

De Marchi, G., "Essay on the Performance of Lateral Weirs," *L. Energia Elettrica,* Milan, Italy, 11, pp. 849–860, 1934.

El-Khashab, A., and K. V. H. Smith, "Experimental Investigation of Flow over Side Weirs," *Journal of the Hydraulics Division,* ASCE, Vol. 102, No. HY9, September 1976.

Frazer, W., "The Behavior of Side Weirs in Prismatic Rectangular Channels," *Proceedings, Institution of Civil Engineers,* London, England, Vol. 6, pp. 270–287, Feb., 1957.

Hager, W. H., "Lateral Outflow over Side Weirs," *Journal of Hydraulic Engineering,* ASCE, 113(4): 491–504, April 1987.

Hudson, D. M., "Protecting the Waters of Lake Erie through Real-time Control of CSO's in the Cleveland Area," *Proceedings, Urban Wet Weather Pollution Control Sewer Overflow and Stormwater Runoff,* Water Environment Federation, Quebec City, Quebec, 1996.

Janson, L. E., S. Bendixen, and A. Harlaut, "Equalization of Flow Variations in a Combined Sewer," *Journal of the Environmental Engineering Division,* ASCE, Dec. 1976.

Kumar, C. P., and S. L. Pathak, "Triangular Side Weirs," *Journal of Irrigation and Drainage Engineering,* ASCE, 113(1):98–106, 1987.

Lager, J. A., and W. G. Smith, "Urban Stormwater Management and Technology: An Assessment," EPA-670/2-74-040 (NTIS PB 240 687), U.S. Environmental Protection Agency, Cincinnati, Ohio, 1974.

Massachusetts Water Resources Authority, personal communication of the USEPA and David Parker of Parsons Engineering Science, Inc., 1997.

Metcalf and Eddy, Inc., *Waste Water Engineering: Treatment, Disposal, and Reuse,* McGraw-Hill, Inc., New York, 1991.

Metcalf and Eddy, Inc., *System Optimization Plans for CSO Control,* prepared for the Massachusetts Water Resources Authority, Wakefield, MA, 1993.

Nadesamoorthy, T., and A. Thomson, discussion of "Spatially Varied Flow over Side Weirs," by K. Subramanya and S. C. Awasthy, *Journal of Hydraulic Engineering,* ASCE, 98(12):2234–2235, 1972.

Northeast Ohio Regional Sewer District, Daniel Hudson, Northeast Ohio Regional Sewer District, personal communication with Parsons Engineering Science, Inc., regarding construction cost for auto regulations, 1997.

Perdek, J. M., R. Field, and S-L. Liao, *Storage/Sedimentation Facilities for Control of Storm and Combined Sewer Overflows: Design Manual, Environmental Research Brief,* EPA/600/s-98/004, National Risk Management Research Laboratory, Cincinnati, OH, May 1998.

Ranga Raju, K. G., B. Prasad, and S. K. Gupta, "Side Weirs in Rectangular Channels," *Journal of Hydraulic Engineering,* ASCE, 105(5):547–554, 1979.

Robinson, D. I., and T. J. McGhee, "Computer Modeling of Side-flow Weirs," *Journal of Irrigation and Drainage Engineering,* ASCE, 119(6):989–1005, Nov./Dec. 1993.

Singh, R., D. Manivannan, and T. Satyanarayana, "Discharge Coefficient of Rectangular Side Weirs," *Journal of Irrigation and Drainage Engineering,* ASCE, 120(4):814–819, July/Aug. 1994.

Stahre, P., and B. Urbonas, *Storm-water Detention for Drainage, Water Quality, and CSO Management,* Prentice-Hall, Englewood Cliffs, NJ, 1990.

Stallard, W. M., W. G. Smith, R. W. Crites, and J. A. Lager, *Storage/Sedimentation Facilities for Control of Storm and Combined Sewer Overflows: Design Manual,* EPA/600/R-98/006 PB98-132228, U.S. Environmental Protection Agency, Cincinnati, OH, 1998

Subramanya, K., and S. C. Awasthy, "Spatially Varied Flow over Side Weirs," *Journal of Hydraulic Engineering,* ASCE, 98(1):1–10, 1972.

Swamee, P. K., S. K. Pathak, M. Mohan, S. K. Agrawal, and M. S. Ali, "Subcritical Flow over Rectangular Side Weir," *Journal of Irrigation and Drainage Engineering,* ASCE, 120(1), Jan./Feb. 1994.

Treffeisen, F., City of Norwalk, CT, personal communication with Parsons Engineering Science, Inc., 1998.

U.S. Environmental Protection Agency (USEPA), *Manual: Combined Sewer Overflow Control,* EPA 625/R-93-007, Washington, DC, 1993.

U.S. Environmental Protection Agency (USEPA), *Combined Sewer Overflows: Guidance for Nine Minimum Controls,* EPA 832/B-95-003, Washington, DC, 1995.

U.S. Environmental Protection Agency (USEPA), *Combined Sewer Overflow Technology Fact Sheet, Maximization of In-line Storage,* EPA 832-F-99-036, Office of Water, Washington, DC, Sept. 1999.

Uyumaz, A., and R. H. Smith, "Design Procedure for Flow over Side Weirs," *Journal of Irrigation and Drainage Engineering,* ASCE, 117(1):79–90, 1991.

Villemonte, J. R., "Submerged-weir Discharge Studies," *Engineering News Record,* p. 866, Dec. 25, 1947.

Water Environment Federation (WEF), *Combined Sewer Pollution Control Abatement, Manual of Practice FD-17,* Alexandria, VA, 1989.

Yu-Tech, L., Discussion of "Spatially Varied Flow over Side Weirs," by K. Subramanya and S. C. Awasthy, *Journal of Hydraulic Engineering,* ASCE, 98(11):2046–2048, 1972.

THE REMOVAL OF URBAN LITTER FROM STORMWATER DRAINAGE SYSTEMS

Neil Armitage
Department of Civil Engineering
University of Cape Town
Private Bag Rondebosch, 7701, South Africa

19.1 INTRODUCTION

In years to come, archaeologists sifting through the remains of late 20th century civilization might well come to identify this period of history as one of waste, "the throw-away society." This is most clearly demonstrated by the large quantities of urban litter (alternatively called trash, debris, flotsam, jetsam, floatables, gross pollutants, rubbish, or solid waste) that is so often seen strewn about in public places.

The litter typically consists of manufactured materials, such as bottles, cans, plastic and paper wrappings, newspapers, shopping bags, cigarette packets, and hypodermic needles, but can include such items as used car parts, rubble from construction sites, and old mattresses. It accumulates in the vicinity of shopping centres, parking lots, fast food outlets, railway and bus stations, roads, schools, public parks, garbage cans, landfill sites, and recycling depots, where it remains until either someone removes it or it is transported by the wind and/or stormwater runoff into the drainage system.

Once in the drainage system, the litter potentially is able to travel via the stormwater conduits, streams, rivers, lakes, and estuaries until it eventually reaches the open sea. Along the way, however, items are frequently entangled among the vegetation along the banks of the streams, rivers, or lakes, or are strewn along the beaches. Some of this debris is picked up, often at great expense. Most of it is probably buried in the river, lake, or beach sediments (Hall, 1996).

The existence of such litter in the waterways and on the beaches has a number of impacts:

- Litter is aesthetically unattractive.

- There is a potential health hazard to humans associated with, for example, the putrefying contents of bottles and tins, or pathogenic organisms attached to discarded hypodermic needles.

- Aquatic fauna are at risk of becoming entangled in, or suffocating from, litter ingested in the course of their search for food.

- Pathogenic organisms or toxins, for example, heavy metals, may be taken into the food chain poisoning aquatic life and possibly later impacting on humans.
- Significant costs are incurred by local authorities in conducting clean-up operations.

It is a worldwide problem. For example, in 1991 South Africa produced some 40 million tons of solid waste annually—mostly of domestic origin (President's Council Report, 1991); 780,000 tons of this was believed to have entered the drainage system—195,000 tons reaching the sea (CSIR, 1991). Local governments in Texas spend upward of US$14 million per year to clean their beaches (Baur and Iudicello, 1990). An estimated 230,000 m^3 or 1.8 billion items of litter (approximately 60,000 tons of wet material) annually enter the waterways of greater Melbourne in Australia (Allison, 1997). This is unacceptable. Something needs to be done to

- Reduce the quantities of litter being deposited in the catchments
- Prevent the deposited litter from entering the drainage system wherever possible
- Remove the balance from the drains where necessary

19.2 LITTER CLASSIFICATION

For the purposes of this discussion, *urban litter* will be defined as visible solid waste emanating from the urban environment with an average dimension of greater than about 10 mm. Researchers, for example, Allison and Chiew (1995), Island Care New Zealand Trust (1996), Armitage et al. (1998), or Armitage and Rooseboom (2000a) have identified many different types of litter. A simplified classification system is proposed below:

Plastics, for example, shopping bags, wrapping, containers, bottles, crates, straws, polystyrene blocks, straps, ropes, nets, tape cassettes, syringes, and eating utensils

Paper, for example, wrappers, newspapers, advertising flyers, ATM receipts, bus tickets, food and drink containers, and cardboard

Metals, for example, foil, cans, bottle tops, and license plates

Glass, for example, bottles and various broken pieces

Vegetation, for example, branches, leaves, rotten fruit and vegetables

Animals, for example, dead dogs, cats, and sundry skeletons

Construction material, for example, shutters, planks, timber props, broken bricks, and lumps of concrete

Miscellaneous, for example, old clothing, shoes, rags, sponges, balls, pens and pencils, balloons, oil filters, cigarette butts, and old tires

19.3 FACTORS AFFECTING THE QUANTITY OF LITTER IN THE ENVIRONMENT

The most common sources of urban litter in the environment are probably the following (Hall, 1996):

- Anti-social behaviour of individuals in dropping litter on footpaths, throwing it from vehicles, and dumping household wastes
- Excessive packaging

- Failure of street sweeping services to rid pavements and public areas of litter
- Inadequate disposal facilities, including a breakdown in garbage collection or the provision of inappropriate garbage cans. Open garbage cans and garbage trucks may provide an opportunity for litter to be blown into the public domain.
- Failure of authorities to enforce effective penalties as a deterrent to offenders

It is obvious that litter is a problem associated with human habitation.

The rate of catchment litter production is highly variable—dependent on a large number of independent factors including:

- Type of development, for example, commercial, industrial, or residential
- Density of development
- Income level of the community. Poor people in some regions of the world don't have access to many consumer products and hence are not in a position to waste them or their containers. In other regions, the population can afford some consumer products, but the tax base is not large enough to support adequate garbage removal.
- Type of industry—some industries tend to produce more pollutants than others.
- Rainfall patterns: for example, does the rain fall in one season only or year-round? Litter will build up in the catchment until it is either picked up by garbage removal, or is swept into the drains by a downpour. Long, dry spells give greater opportunity to the local authority to pick up the litter, but also tend to result in heavy concentrations of accumulated rubbish being brought down the channels with the first rains of the season—the so-called "first flush."
- Type of vegetation in the catchment. In Australia, for example, leaves form the major proportion of "litter" collected in traps. Some species of trees cause more problems than others—for example, London Plane trees have relatively large leaves that are slow to decompose and are mostly shed over a very short period in autumn.
- Efficiency and effectiveness of garbage removal by the local authority, including both the removal of garbage from private properties as well as the cleaning of the streets and public garbage cans. It is also important that sweepers do not, for example, sweep or flush the street litter into the stormwater drains.
- Level of environmental concerns in the community, leading to, for example, the reduction in the use of certain products, and the recycling of others.
- Extent of legislation prohibiting or reducing waste, with which is associated the effectiveness of the policing of the legislation, and the level of the fines.

Anecdotal data gathered from around the world suggests that some of the more highly regulated countries of the developed world (e.g., Scandinavian countries, Germany, and Singapore) have littering largely under control. In other developed countries (e.g., the United Kingdom, Australia, Canada, and the United States) visible littering seems to be mainly limited to lower valued land. With middle-income countries (e.g., South Africa, India, Pakistan, and Brazil), however, the situation is often dramatically different. These countries have large manufacturing industries but often lack the resources to control the associated littering. Very poor countries (e.g., most of Africa) often don't have a particularly large litter problem, as their populations have not developed a consumer lifestyle. What little they have is either biodegradable or is extensively recycled.

The variability in the nature of the litter associated with different land uses has been identified by a number of researchers, for example, Allison and Chiew (1995). They showed that for a fully urbanized catchment at Coburg, which is situated about 10 km north of Melbourne's CBD, "garden debris" made up 85% of the litter collected from a residential site, but only 36% from a light industrial site. "Paper" and "plastics" made up 64% of the litter from the light industrial site, but only 13% from the residential site (Fig. 19.1). Similar profiles have been obtained for Auckland (Cornelius et al., 1994; Island Care New Zealand Trust, 1996).

Often, a single shop or factory, for example, a fast-food outlet, a bank, or a plastics-recycling factory is responsible for a large percentage of the litter collected in the drains. The amount of litter can be substantially reduced once the situation has been brought to the attention of the offending company (Island Care New Zealand Trust, 1996; Allison, 1996).

There is an infinite variety in the types and quantities of litter washed off a catchment. In fact, *each catchment has a unique litter "footprint" that is indicative of the state of the catchment at the time of measurement.* This is why there is such an enormous variation in the published data on measured litter loadings. What complicates the matter further is that there is no uniformity in the reporting of catchment litter data. The mass of a sample will vary with its moisture content, and frequently samples taken from drains are not dried before they are weighed. The density of the sample can increase by as much as five times from being loose in the drain to being compacted in the back of the garbage removal vehicle. Sometimes, leaves or silt or construction debris are taken as being part of the litter and sometimes not, and the process of removing the leaves, silt and/or construction debris will also change the density of the residual sample. Sometimes litter loadings are expressed in terms of the curb length. At other times they are expressed in terms of the land area.

The quantity of litter finding its way into the drainage system is also extremely sensitive to the efficiency of the garbage removal and street-cleaning and sweeping services in the

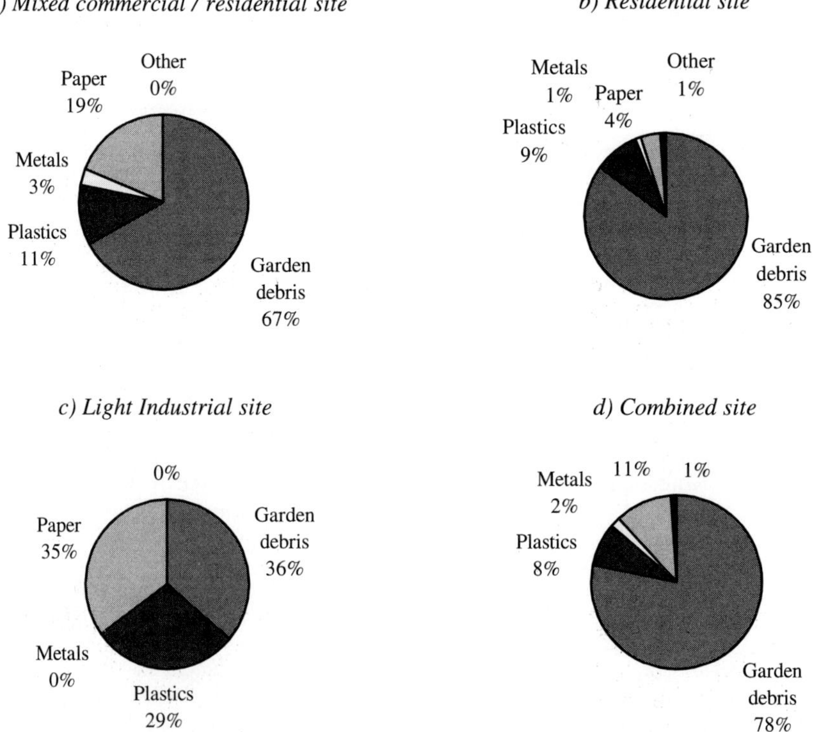

FIGURE 19.1 Composition of collected gross pollutants by dry mass from different catchments in Coburg. (After Allison and Chiew, 1995)

area, and these are frequently not reported. In many parts of the developing world formal domestic garbage removal is either non-existent or inadequate. Domestic litter is dumped onto the street or into the nearest open space. A stormwater canal is a particularly attractive candidate as the water washes away the evidence of illegal dumping. Even where there is an adequate domestic service, frequent street cleaning can reduce the quantity of street litter reaching the drainage system by a large percentage. On the other hand, if street cleaning is infrequent, the removal efficiency will then depend strongly on the rainfall patterns. If there are periods of major rainfall between cleaning events, a lot of litter is likely to end up in the drains. If not, the litter will tend to accumulate in the street until it is removed.

In consequence of the foregoing, litter data should ideally be collected from the site of interest prior to the design of any litter removal system. Failing that, reference may be made to data gathered in South Africa (Armitage et al., 1998, Armitage and Rooseboom, 2000a), Australia (Allison, 1997), and the United States (HydroQual, 1995), and summarized in Tables 19.1 and 19.2. The litter loadings in Tables 19.1 and 19.2 are expressed in terms of kg/hectare·yr or m³/hectare·yr. Although expressing the litter loading in terms of curb length is likely to be more meaningful in the more developed countries, it is not necessarily so in the less developed countries that have by far the bigger litter problem. Large areas of informal settlements surround many of the cities in the less developed countries. Many of these settlements do not even have a formal road system. The residents settle wherever they can find a piece of land big enough for the construction of a simple shelter, and movement through the settlement is mainly by means of narrow pedestrian ways. Under these circumstances, measuring the gross area of land becomes a much quicker and simpler proposition than trying to determine an equivalent curb length. Data on curb density in North America (Pitt, 2000) is presented in Table 19.3.

19.4   REDUCING LITTER LOAD IN THE STORMWATER DRAINAGE SYSTEM

Much can and should be done to reduce the quantity of litter that finds its way into the stormwater drainage system.

19.4.1   Reducing Total Litter Load

The most desirable solution to the problem of litter in the drainage system is to reduce the total litter load through some form of catchment litter management. Some of the various options that are available to local authorities are listed below (Senior, 1992; Melbourne Water, 1993; Hall, 1996; Island Care New Zealand Trust, 1996; Allison, 1997; Armitage et al., 1998, Armitage and Rooseboom, 2000a):

- Better placement of garbage cans
- Placing litter traps inside strategically located catch basins. Use the evidence provided by litter trapped in the catch basins to identify the polluters who may then be pressured into changing their ways.
- Organize volunteer clean-up days for cleaning the banks of urban streams and lakes. This also helps to raise public awareness of the problem.
- Organize a public education campaign to highlight the source of litter in urban waterways, its pathway, and environmental hazards. During 1990 a number of small informal public awareness surveys were conducted in offices and schools in Melbourne, Australia. It was readily apparent that a majority of children and adults in that city either did not appreciate that there are separate stormwater and sewerage systems, or did not understand that catch

TABLE 19.1   Typical Litter Wash-off Rates in the Developing World (South Africa)

Catchment description	Annual load (kg/ha.yr)	Density (kg/m³)	Annual load (m³/ha.yr)
South Africa:			
Springs (summer rainfall with a mean average precipitation of 750 mm): Central Business District, 85% (254 ha) commercial/industrial, 15% (45 ha) residential			
● Average catchment load before street sweeping	470		
● Average catchment load before street sweeping assuming no contribution from the residential area	550		
● Average load in the stream after daily street sweeping in the commercial and industrial areas with an overall removal efficiency of approximately 83%	82	95	0.86
● Average load in the stream after street sweeping assuming no contribution from the residential area	95	95	1.0
Measured dry litter densities: On the street		35	
On the banks of the stream		95	
In the trapping structure		95	
In the back of the refuse vehicle		150	
The average litter load of the first major storm of the rainy season is 3.6 times the average of all storms.			
Johannesburg (summer rainfall with a mean annual precipitation of 713 mm): 800 ha of the central districts with a mix of residential, commercial, industrial and informal trading areas. Limited street cleaning.	48	95 (assumed)	0.5 (measured)
The average litter load of the first major storm of the rainy season is 1.5–2.0 times the average of all storms.			
Cape Town (winter rainfall with an overall mean annual precipitation of 518 mm—but with large local variation):			
● Ocean View sub-economic residential area for poor people including both stand-alone dwellings and "hostels" (3-story, high-density apartment blocks), no street cleaning—excluding vegetation	104		
● Vegetation load for Ocean View	58		
● Summer Greens medium density, medium income residential area, no street cleaning—excluding vegetation	5.8		
● Vegetation load for Summer Greens	1.9		
● Welgemoed low density, high income residential area, no street cleaning – excluding vegetation	0.2		
● Vegetation load for Welgemoed	27		
● Montague Gardens light industrial park, no street cleaning—excluding vegetation	49		
● Vegetation load for Montague Gardens	5.0		
● Cape Town Central Business District including office blocks, hotels, line shops, informal traders and a bus terminus, extensive street cleaning (up to 3 times daily) with a removal efficiency of approximately 99%—excluding vegetation	49		
● Vegetation load for Cape Town Central Business District	29		

TABLE 19.2   Typical Litter Wash-off Rates in the Developed World

Catchment description	Annual load (kg/ha.yr)	Density (kg/m³)	Annual load (m³/ha.yr)
Australia:			
Melbourne (winter rainfall with a mean annual precipitation of 730 mm): 150 ha of the inner-city suburb of Coburg, 25% commercial, 65% residential, 5% light industrial, 5% park-lands. Daily street sweeping along the major commercial streets, fortnightly sweeping in the residential areas.			
● Average inclusive of leaf matter (dry)	30		
● Average inclusive of leaf matter (wet—from the trapping device)	100	250	0.4
● Average excluding leaves (dry)	6		
● Commercial areas excluding leaves (dry)	17		
● Residential areas excluding leaves (dry)	4.0		
● Light industrial areas excluding leaves (dry)	7.4		
United States			
New York City (year round rainfall with a mean annual precipitation of 1 092 mm): Based on a measured quantity of 11g litter per 100 ft of curb length per day measured along 90 blockfaces throughout the city. The assumed curb density is 370 m/ha. The frequency of street sweeping is unknown.	50	94	

basins in streets and surface grates in private properties connect to the drainage and stream systems. Even after an extensive radio and poster campaign, a more comprehensive market survey undertaken in 1991 revealed that at least a third of the population in Melbourne were still ignorant of the drainage systems role and its connection to waterways. Subsequent to this, a television advertising campaign was prepared, while kits were put together to educate school children (Senior, 1992).

● Encourage the formation of public interest/action groups to brainstorm new ideas for litter control and to act as environmental watch dogs.

TABLE 19.3   Typical Curb Densities in North America

Land use	Curb density (m/ha)	Average street width (m)
Apartments	300	14
Duplexes	230	11
High density residential	250	10
Low density residential	210	9
Older medium density residential	170	10
Typical medium density residential	220	12
New medium density residential	260	11
Office park	260	13
Shopping areas	60	13
Strip commercial	280	13
Light industrial	340	11
New York City Centre	370	12

Source: After Pitt, 2000.

- Force businesses to become responsible for the proper reduction and disposal of litter generated on their premises.
- Study the behavior of litter in the stormwater drainage system through the tracking of tagged litter items. Information from this study could be used to devise better ways of controlling litter in waterways as well as raising public awareness of litter pathways (Melbourne Water, 1993).
- Encourage commerce and industry to move to more environmentally friendly packaging. In 1991, the Board of Works in Melbourne staged a small exhibit as part of the Plastic Institute's Annual Conference in Melbourne. The display featured a number of polystyrene and plastic items—both unused and recovered from river litter traps. Also prominent was an enlarged photograph of the material trapped behind a litter boom, which illustrated many recognizable items. This was provocatively captioned: "Do you really want your product advertised in this way?" (Senior, 1992).
- Prevent businesses from imposing unwanted packaging or advertising on unwilling consumers.
- Set up proper solid-waste collection services in those urban areas that do not yet have such a service.
- Ensure that there is no loss of litter once it has been collected, for example, from inadequate disposal facilities or open collection vehicles.
- Force shops to institute a deposit on all containers.
- Place an "environment tax" on plastic shopping bags and packaging. Encourage the move back to large re-usable bags provided by the customer.
- Employ the jobless to collect garbage from more remote areas.
- Institute and enforce effective penalties to act as a deterrent to offenders.
- Encourage the formation of interest groups that will adopt areas/reaches of streams, etc., and help keep them free of litter.

19.4.2 Street Sweeping

There is little published data on the effectiveness of street sweeping in the removal of litter. In view of the relatively large size of most litter, it is relatively easy to pick up—whether by machine or by hand. The most difficult aspect is the problem of removing litter from between or underneath parked vehicles. Most important, however, is the frequency of sweeping which has a large impact on the efficiency of litter removal. Street sweeping carried out two to three times daily in the commercial business district of Cape Town (South Africa) appears to remove as much as 99% of the total litter load from the streets. On the other hand, sweeping selected streets in Springs (South Africa) once a day removes only about 83%. Where the cleaning frequency is less than the frequency of the runoff producing storms (with, say, more than 5 mm of rainfall), the litter removal efficiency is likely to drop below 50%. In the absence of data, the maximum expected efficiency of litter removal is indicated by Eqs. (19.1) and (19.2) or Fig. 19.2:

$$\eta_{removal} = F_{sw}/2F_s \text{ (for } F_{sw} < F_s) \tag{19.1}$$

$$\eta_{removal} = F_s/2F_{sw} \text{ (for } F_{sw} \leq F_s) \tag{19.2}$$

where $\eta_{removal}$ = estimated efficiency of street cleaning (fraction)
F_{sw} = average number of days between street sweeping (d)
F_s = average number of days between storms (d)

Where there is a significant difference in the sweeping and/or rainfall frequencies between

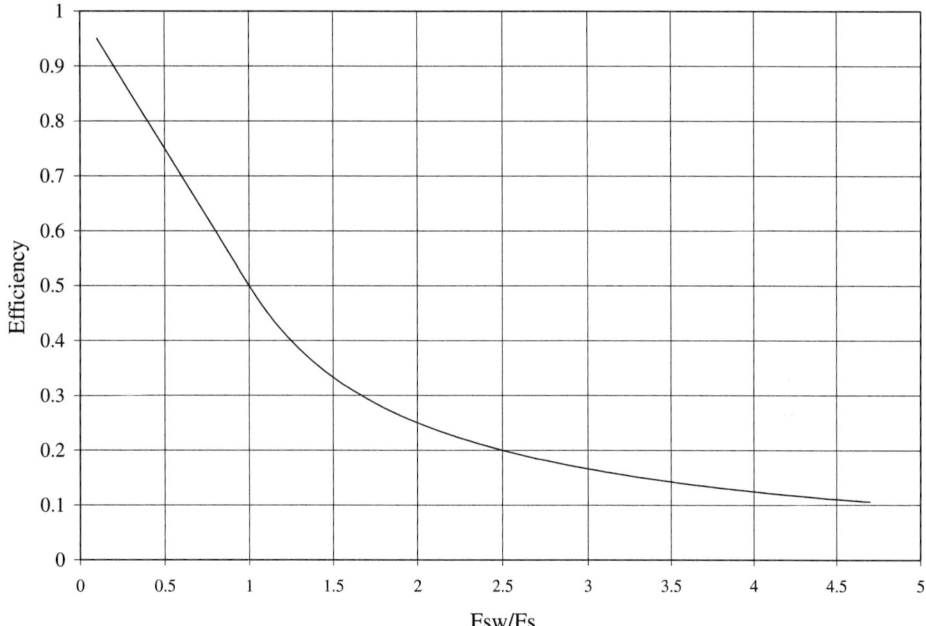

FIGURE 19.2   Plot of the maximum expected litter removal efficiency against the average number of days between street sweeping divided by the average number of days between significant storms.

different seasons, each season must be evaluated separately and an average taken over the full year.

It is important that street sweeping is carried out in an acceptable manner. A survey carried out by the Board of Works, Melbourne in 1990 revealed that, at that time, 67% of 54 councils in the metropolitan area used street flushing to some extent. Of these about half regularly and extensively used flushing equipment or street hydrants to clean shopping centres and similar litter accumulation areas (Senior, 1992). Under these circumstances, street sweeping could increase the quantities of litter reaching the drainage system.

In theory the sweeping frequency can be reduced during long, dry periods without a significant increase in the litter reaching the stormwater systems. In reality, though, other mechanisms, including wind and even poor street-sweeping techniques, tend to move a certain amount of litter into the drains. The first major runoff-producing rainfall then washes the accumulated litter away as a highly concentrated litter "plug." This is the so-called first flush. Data collected from South African catchments (Armitage et al., 1998) suggests that the first flush can contain up to four times the normal quantity of litter. This must be allowed for in the sizing of the storage capacity of any litter traps.

19.4.3   Use of Grids over Catch Basin Entrances

The most obvious method of preventing litter from getting into the drainage system is to ensure that some form of grid covers as many entrances as possible. This is the norm in some of the more developed parts of the world, for example, in Europe or in the northwest United States. In less developed countries, however, this is not always a satisfactory solution. High litter loads together with high rainfall intensities and unreliable maintenance programs frequently lead to blockages and the associated risk of flooding. The question of who is liable for damages in the event of flooding associated with such an eventuality is unclear, but the local authority is likely to be a focus of attention. For this reason, many local authorities allow some form of unrestricted overflow even when grids are provided. Where unrestricted overflows exist, litter will certainly be found in the drains.

Paradoxically, grids over the catch basin entrances may be a suitable solution in the very high-density, low-income informal urban settlements surrounding many of the major cities in the developing world. This is for the simple reason that if the residents can see the grids blocking, and if there is a risk that their own homes will be affected by the consequent flooding, they are likely to take the appropriate action to keep them clear. If the litter trap is hidden, or if local drainage is unaffected by moderate litter loads, it is unlikely that the residents will help keep the drains clear of litter, leaving it to the local authority to take full responsibility for maintenance. This has been observed in Khayelitsha near Cape Town, South Africa (Compion, 1998). There is, of course, a risk of serious flooding if a major storm occurs at night.

An alternative approach is to place grids over the entrances to high-lying drains, while placing litter traps into lower-lying drains. In this situation, the additional flood risk may be limited as stormwater can bypass blocked grids to enter the drains at another point. Traps (with a suitable high flow by-pass) may then be placed on the reduced number of open entrances.

Another strategy is illustrated by Santa Monica (California) where drain entrances are closed with wooden boards during the dry season (Allison, 2000).

19.5 ASPECTS TO CONSIDER IN LITTER TRAP DESIGN

Since it is difficult to prevent all litter from reaching the drainage system, litter traps might have to be placed in the waterways to remove the balance.

In those regions of the world where stormwater and wastewater is reticulated together in "combined sewers," this is most readily achieved at the waste water treatment works by means of mechanically raked screens and other forms of mechanical solids removal devices. Details of these devices are available from many other sources and are outside the scope of this discussion. During periods of high rainfall, however, the wastewater treatment works cannot handle the hydraulic load and the surplus flow is usually discharged to the most convenient nearby receiving water via a "combined sewer overflow" or CSO. The situation is then similar to that which pertains in most of the world where the stormwater is reticulated separately from the wastewater and is discharged directly into the nearest receiving water. Unless it is trapped along the way, litter is free to move with the stormwater flow through the drainage system and ultimately to the sea.

The ideal trap has, *inter alia,* the following features (Armitage et al., 1998):

- It is reliable.
- It is economical to construct and operate.
- It does not have any moving parts.
- It does not require an external power source.
- It requires minimal water head (i.e., it can be used in association with flat gradients).
- It does not increase flood levels in the vicinity of the structure.
- It has high removal efficiency.

Nothing remotely resembling this exists. All known structures represent some sort of compromise. It is thus the engineer's task to choose the most appropriate structure to fit the circumstances. Ideally this should fit into a total litter removal strategy that takes into account catchment litter management.

One of the biggest problems facing the designer of a litter trap is that litter can be just about anything—any size, any shape, any density, and any hardness. Furthermore, the characteristics of each item often change as it moves through the drainage system. Plastic bags deform and tear, bottles break, and aluminum cans fill with water and/or sediment.

For convenience, litter moving down an open channel is often considered as belonging to one of three factions:

- *Bed-load* (i.e., rolling or sliding along the floor of the channel)
- *Suspended material* (i.e., drifting somewhere in-between the floor of the channel and the surface of the stream)
- *Flotsam* (i.e., material floating on the surface)

This division is however somewhat arbitrary, as it can be shown (e.g., Rooseboom, 1992) that solids carried in a water matrix can adopt just about any position depending on the applied unit stream power. In addition, most types of litter change their density continuously as they are transported through the drainage system.

Three slightly more useful characteristics are size, density, and settling velocity.

- The *size* of a particular type of litter, for example, aluminum cans, plastic bottles, and plastic shopping bags is reasonably constant except, of course, when they are broken. It is a useful characteristic for design purposes because it determines the likelihood of the object being trapped by a screen.
- Although the *density* of a particular item of litter often changes as it moves down the drainage system, it generally tends asymptotically to some reasonably constant value that will determine whether, in slow moving water, it will either float to the surface or sink to the bottom.
- The *settling velocity* is a function of size, density, and shape factor (which is a measure of how streamlined a particular object is), electrical charge, viscosity and concentration (e.g., Raudkivi, 1990). The higher the settling velocity, the more likely the object under consideration is to be found near the bottom of the water column; the more negative it is, the more likely it is to be found near the surface.

The high degree of variability in litter characteristics makes it extremely difficult for the designer to design a structure that will cater for every eventuality. Many structures work extremely well in low flows, but not in high, or work well with certain types of litter, but not others. Some pose major maintenance problems. Many, if not most, litter trap designs currently available have major design limitations.

The vertical or near-vertical screen illustrates a typical example of a poor design (Fig. 19.3). Since there is little flow in the channel for much of the time, the litter load is trapped over a small area at the bottom of the screen to form a sharp-crested weir. Additional items of litter carried by the flow tend to collect on the screen just above the crest of the informal weir. The screen thus blocks from the bottom upward. Meantime, the water level in the channel upstream of the structure increases with the increasing height of the blockage. In times of flood, the litter quickly blinds out any remaining screen area and the water is forced to go over or around the structure.

Research carried out into some 50 different types of trapping devices for the Water Research Commission of South Africa (Armitage et al., 1998; Armitage & Rooseboom, 1999; Armitage and Rooseboom, 2000b) showed that litter is difficult to trap without the aid of a screen. Screenless traps, which usually make use of weirs or baffle walls to hold back the litter, only work in situations where the flow velocities are very low (typically the Froude number needs to be below 0.07). This usually requires the presence of a large tank or pond. Screens have a tendency to block, but they can be made to be effectively self-cleaning if the bars are declined in the direction of flow. Under these circumstances the litter is generally

FIGURE 19.3 Typical example of blockage on a vertical screen.

subjected to a thin sheet of high velocity flow which tends to keep it moving down the screen and away. In the event of temporary blockage, hydrostatic pressure helps to start the litter moving again. The bar design is also important. Bars should offer as little resistance as possible to litter sliding along their surfaces, and litter that does penetrate the openings must fall free of the bars.

19.6 A REVIEW OF SOME TYPICAL LITTER TRAP DESIGNS

Many designs of litter traps have entered the market in recent years. In the evaluation of litter traps undertaken for the Water Research Commission of South Africa in 1998 (Armitage et al., 1998; Armitage and Rooseboom, 1999; Armitage and Rooseboom, 2000c), options included in-line screens, self-cleaning screens, booms, baffles, detention/retention ponds, and vortex devices. They were evaluated in terms of the following criteria:

- Size of catchment that could be serviced by the device (which is related to the runoff and the litter loads)
- Typical cleaning frequency
- Hydraulic head requirement for operation
- Efficiency (expressed as a percentage of litter removed from the flow)
- Capital and operating costs
- Any other features that might make the structure attractive or unattractive to the potential user

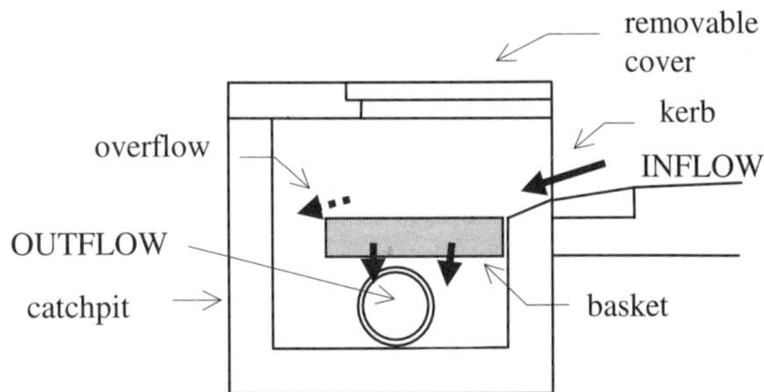

FIGURE 19.4 Cross-section through a typical SECT.

In the end, seven patented devices were identified as showing the most promise for use in typical urban drainage situations. A few other designs were suitable for specialised installations. While new products are constantly making their way into the marketplace, these seven devices are worth describing in greater detail as they exhibit features typical of many of the better designs.

19.6.1 Side-entry Catch Basin Trap (SECT)

A perforated tray is mounted on metal supports next to and underneath a catchpit opening (see Figs. 19.4 and 19.5). Stormwater either flows through the perforations, leaving the litter

FIGURE 19.5 View of a typical side-entry catchpit trap.

behind, or, if the perforations are blocked and/or the tray full, the stormwater flows over the back wall of the tray. To remove the litter, the basket is either manually cleaned, or it is vacuum educted ("sucked" clean) and washed with water under high pressure (Melbourne Water, 1995). Newer designs include more elaborate bypass systems, or attempt to reduce blockage of the screen by shaping it into an inverted V-section (Allison, 2000).

19.6.2 The North Sydney Litter Control Device (LCD)

A drop is provided in a pit between the invert of the inflow pipe and the invert of the outlet. This drop is in the order of a meter which caters for removable baskets and a small gap below them. Above the removable litter baskets is an inclined trash rack with vertical bars. This trash rack is inclined towards the litter baskets to prevent the inflow from scouring out previously deposited litter. It is hinged so that it can be pushed back to enable easy removal of the litter baskets. See Figs. 19.6 and 19.7 (Brownlee, 1995).

19.6.3 The HumeGuard In-line Litter Separator (ILLS)

A carefully shaped boom situated in the separator pit deflects the flow into the holding pit. Once in the holding pit, the flow is forced down under a suspended baffle wall and up over a weir before being returned to the separator pit downstream of the boom. The relatively large plan area of the holding pit ensures that the average vertical flow velocities are low enough to prevent carry-through of those objects, such as plastic bags, that have a negligible settling (or rising) velocity. See Figs. 19.8 and 19.9.

In the event of particularly high flows through the stormwater conduit, the increased water levels on both sides of the boom causes it to float out of the way. This ensures that upstream flood levels are not affected by the structure, and the litter already trapped in the holding pit is not washed out. The boom is restrained by rods, which are attached to its upper surface and the walls of the chamber above the pipe inlet in such a way that the boom is free to rotate about a hinge at the wall. (Swinburne University of Technology, 1996).

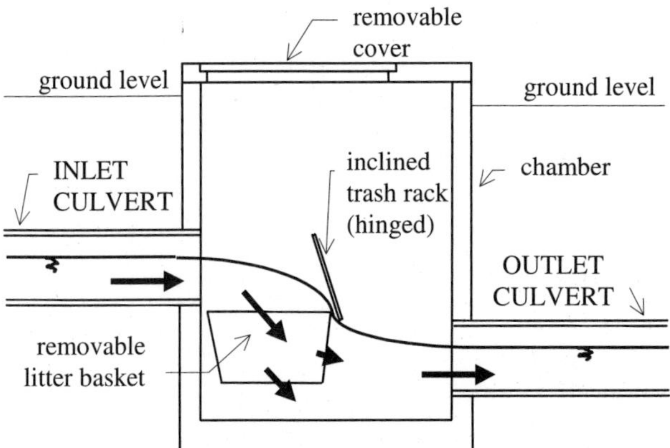

FIGURE 19.6   Section through a typical North Sydney Litter Control Device (LCD). (After Brownlee, 1995 and Hocking, 1996)

FIGURE 19.7   View inside a North Sydney Litter Control Device (LCD).

19.6.4 The Continuous Deflective Separation (CDS) Device

The flow in a stormwater conduit is deflected into a circular pollutant separation and containment chamber. Gross pollutants are separated within the upper separation portion of the inner chamber with the aid of a perforated plate screen, which allows the filtered water to pass into a volute return chamber and back to the outlet conduit. The water and associated pollutant contained within the inner chamber are kept in continuous motion by the vortex action generated by the incoming flow. This has the effect of keeping the gross pollutant in the containment chamber from blocking the perforated plate screen. The heavier pollutants

a) Plan

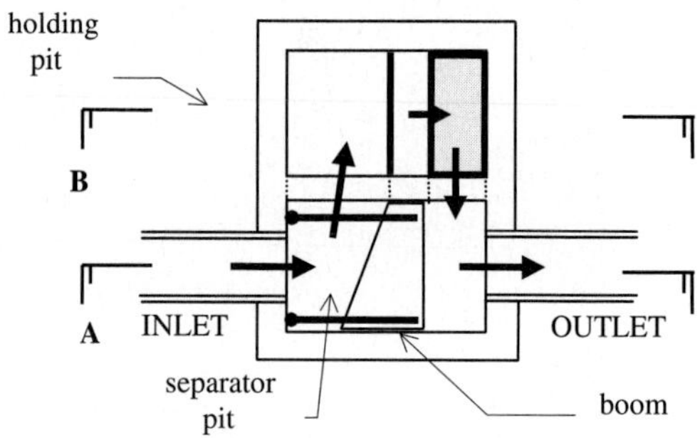

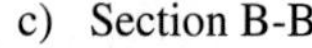

b) Section A-A   c) Section B-B

FIGURE 19.8   Plan of and cross-sections through the In-line Litter Separator (ILLS).

ultimately settle into the lower solids collection sump, whilst the flotsam floats on the surface of the containment chamber (Wong and Wooton, 1995). See Figs. 19.10 and 19.11.

19.6.5 The Baramy® Gross Pollutant Trap (BGPT)

Flow from a conduit is directed over a screen declined at an angle of about 20°. The water flows through the screen and either goes under the collection shelf (direct flow version), or around it (low profile version). The litter is separated from the water by the screen and is deposited on the collection shelf ready for removal by a skid-steer loader (Bobcat or similar), which gains access down a concrete ramp (Baramy). See Figs. 19.12 and 19.13.

19.6.6 The Stormwater Cleaning Systems (SCS) Structure

The SCS structure is similar to the BGPT, except that the screen is mounted on the crest of a weir and is declined at approximately 45° below the horizontal. There are two alternative layouts: with the weir directly in the path of flow for small flows emanating from, say, a pipe, and with the weir lying parallel to the initial flow direction for the larger flows in canals. A settling basin can be provided upstream of the weir to trap the bed-load separately if required (Armitage et al., 1998). See Figs. 19.14 and 19.15.

FIGURE 19.9   View of the In-line Separator (ILLS). Flow from left to right.

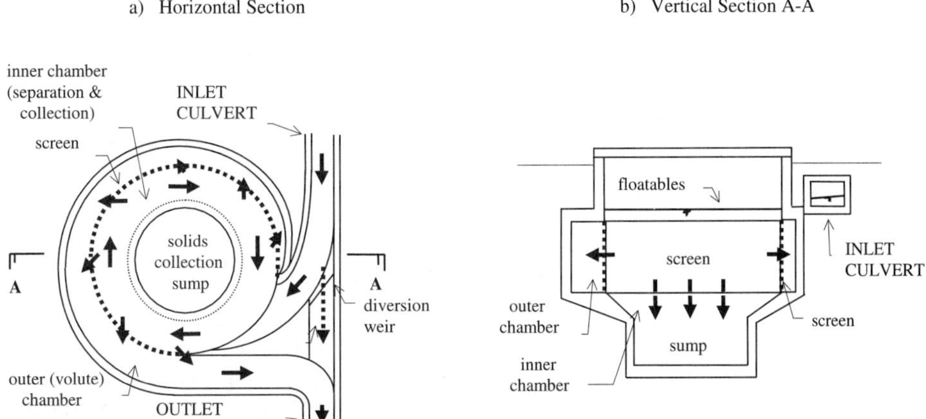

FIGURE 19.10   Horizontal and vertical sections through the CDS device.

FIGURE 19.11   View of the inside of a CDS device. The unit is being drained for cleaning. The inlet culvert and the screen are clearly visible.

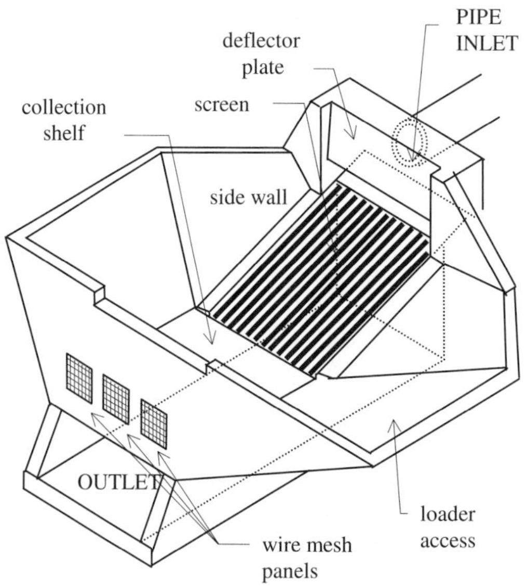

FIGURE   19.12 Baramy® Gross Pollutant Trap (BGPT)—typical direct flow version.

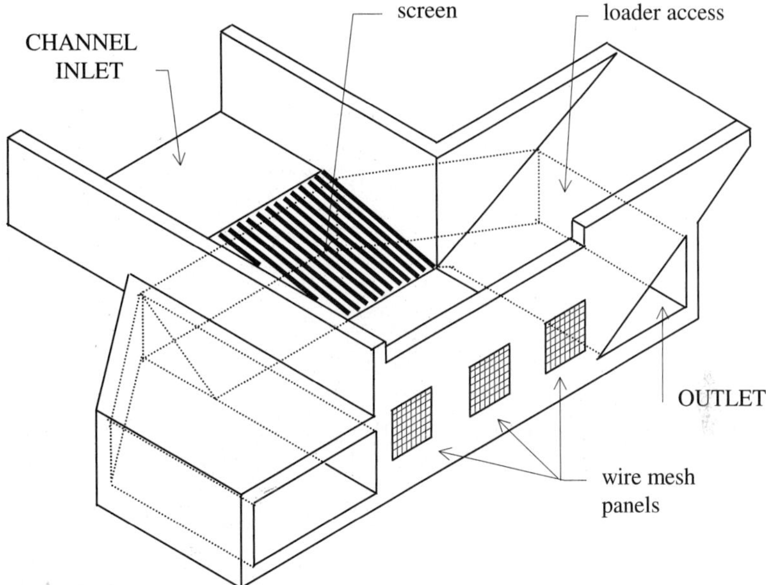

FIGURE 19.13 Baramy® Gross Pollutant Trap—typical low profile version.

19.6.7 The Urban Water Environmental Management (UWEM) Concept

A hydraulically controlled sluice gate is used to create the necessary head required to force the stormwater through a series of screens, under a suspended baffle wall, and over a weir. In the event of a major flood coming down the channel, the sluice gate automatically lifts to pass the peak and prevent upstream flood levels from being raised higher than they would have been, had there been no structure at all (Armitage et al., 1998). See Figs. 19.16 and 19.17.

The device is readily adapted to the removal of pollutants other than litter, for example, silt or sewage. It can also be designed to handle very large flows. Its chief advantage,

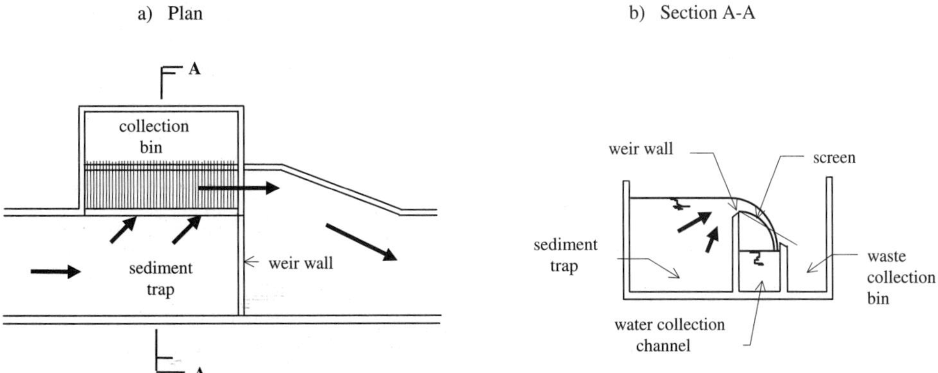

FIGURE 19.14 Plan of and long-section through the SCS structure.

FIGURE 19.15   View of an SCS structure being cleaned and repaired. The flow would normally be from right to centre and down. The collection bin is on the left.

however, is that it can be applied in areas with flat gradients, such as along the coast, since the head that is required to operate the trap is generated by the hydraulically operated sluice gate.

19.6.8 Other Potential Trapping Structures

The principal attributes of the seven typical litter traps described above are listed in Table 19.4. The investigation also indicated a potential application for some other designs for use in specialized situations. These included

a) Plan (not to scale)

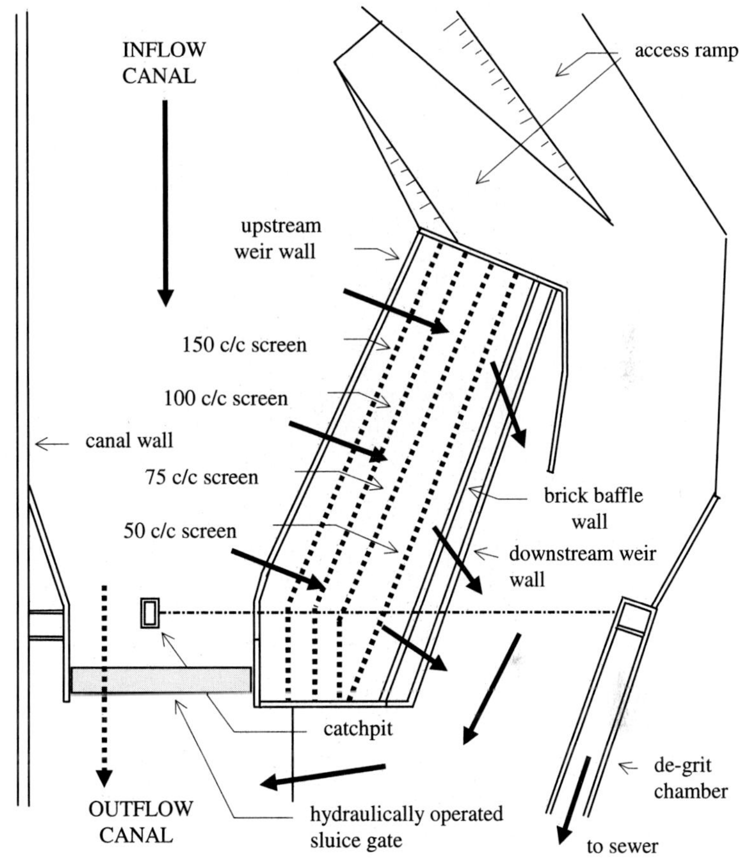

b) Section through the screens (not to scale)

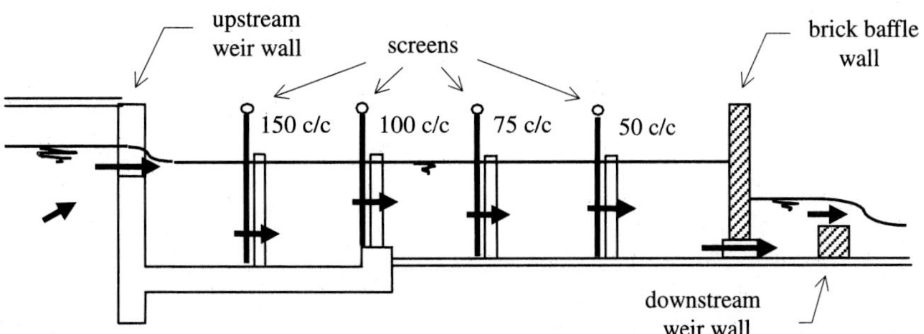

FIGURE 19.16 Plan of and section through the screens at the UWEM Pollution Control Works on the Robinson Canal, Johannesburg.

FIGURE 19.17   View of the Robinson Canal screens after cleaning. The hydraulically operated sluice gate is clearly visible on the top right-hand side of the picture.

- Booms (for the removal of surface material only)
- Fences or nets across slow-flowing channels (see Fig. 19.18)
- Ponds/wet lands

In every case, flow velocities must be kept low at all times—typically with the Froude number less than about 0.07. This equates to a maximum velocity that varies from about 0.15 m/second for a 500-mm deep channel to about 0.3 m/second for a 2-m deep channel. This implies large flow areas—from 10 to 20 times that of a typical concrete culvert. Cleaning and maintenance are also often major problems. See Armitage et al. (1998) for further information.

Another possibility is release nets. A strong net is attached to the outlet of a pipe. Under normal operating conditions, the net, with accumulated litter, is periodically replaced with an empty one. If however the upstream water levels rise above a certain predetermined level, the net is detached from the pipe outlet and swung to one side on the end of a tether to allow the flood flow go through. The litter is retained in the net for later removal (Allison, 2000).

19.7   SELECTING A LITTER TRAPPING SYSTEM

19.7.1   The Importance of Trap Location

Once due consideration has been given to reducing the litter load in the drainage system (Sec. 19.4), litter traps will probably be required to remove the balance of the material. The

TABLE 19.4 Summary of the Attributes of the Seven Typical Litter Traps

Device	Typical catchment area (ha)	Typical cleaning frequency	Head requirement	Maximum efficiency (%)	Comments on performance
SECT	0,1–1	Monthly or after every major storm	Low (effectively)	59–76 (50–100% coverage respectively)	Need to be able to target the catchpits with the highest loads. The efficiency of the unit is strongly affected by the number of un-trapped catchpits and the cleaning frequency.
LCD	20–150	Monthly or after every major storm	High	25	Inefficient in high flows but collects most material at low to medium flows. Likely to be a relatively expensive option. Relatively easy to clean.
ILLS	5–25	Monthly or after every major storm	Low	25	Little data available. Likely to be a relatively expensive option. Moving parts may cause problems.
CDS	10–200	4 times a year	Low	99	Very efficient trapping device, but very expensive to install and tedious to clean.
BGPT	10–500	4 times a year	High	95	Little prototype data available, but shows considerable promise. Compact. Easy to clean.
SCS	>1	Monthly or after every major storm	High	95	Works well providing the head is available. Easy to clean.
UWEM	>400	After every major storm	Low (effectively)	90	The concept of generating head in-situ via a hydraulically actuated sluice shows considerable promise for use with other structures e.g. Baramy®, SCS.

FIGURE 19.18   View of litter trapped behind a fence strung across a river.

choice of trapping structure is site-specific and therefore the location of the traps is an important decision. Clearly, the closer to the source a trap is located, the smaller the flow and therefore the smaller the structure required. On the other hand, many more of these structures will be required to cover the entire catchment. The construction and maintenance of large numbers of smaller traps will almost certainly be greater than the construction and maintenance of one or two larger traps situated at the mouth of the main canal or stream draining the entire catchment.

Trapping points and the typical associated structures may be loosely categorized as follows:

- Entry (SECT)
- In-pipe (CDS, ILLS, LCD)
- End-of-pipe (LCD, CDS, SCS, BGPT)
- Canal/stream (BGPT, SCS, UWEM, fences or nets across streams, ponds, and wet lands)

No trap is 100% effective. Furthermore, it is often more cost-effective to aim for a trap efficiency of, say, 70% and look to trap the balance at another point in the system. Many traps are only designed to handle peak flowrates in the region of 1:1 month recurrence interval (RI) (i.e., the structure is bypassed 12 times a yr on average). The surplus flow, with its associated litter, is bypassed. A design flow of 1:2 years (which is the capacity of many conduits) is probably the largest flowrate that would ever be considered for a litter trap. Consideration should therefore be given to providing at least two lines of traps (e.g., side-entry catch basins at key locations together with a number of in-pipe or end-of-pipe traps downstream).

Another important issue is access for cleaning and maintenance—particularly for the larger structures. *Ease of maintenance is crucial.* Trapping efficiency will rapidly fall to zero if the traps are not properly cleaned and maintained. In some instances, the cost of providing adequate access may be more than the structure itself.

19.7.2 Suitability of Particular Traps

Once suitable trapping points have been identified, the main criteria determining the suitability of a particular trap in that location are flowrate, allowable head loss, size, efficiency, reliability, ease of maintenance, and cost-effectiveness. The first three items are site constraints, while the balance depend on the structure under consideration.

Considering only the site constraints, the available structures may be roughly divided into

- "Low flow" or "high flow"
- "Low head" or "high head"
- "Small," "medium," or "large"

The division between "low" and "high" flow may be taken to be roughly 1 m^3/second. The division between "low" and "high" head may be taken to be roughly 0.5 m. Structures may be described as "small" if they are contained wholly within the channel, "medium" if they are only slightly larger than the channel, and "large" if they require considerable extra space or if the channel must be widened.

The typical litter trap designs may be loosely categorized as follows:

- Low-flow, low-head structures ("small"—SECT, "medium"—ILLS, "large"—CDS)
- Low-flow, high-head structures ("medium"—LCD, "large"—BGPT, SCS)
- High-flow, low-head structures ("small"—fences, nets, booms or baffles, "large"— UWEM, CDS)
- High-flow, high-head structures ("medium"—BGPT, "large"—SCS)

19.7.3 Recommended Selection Procedure

Once the designer has some idea of the potential trapping point and associated structures, the recommended selection procedure is as follows:

1. Identify each catchment with its associated drainage system/waterways. It may be necessary to divide the catchments into sub-catchments, depending on the number, type, and location of structures envisaged.

2. Identify and measure the area of each land use (A_i) within each catchment (the main categories being commercial, industrial, and residential).

3. Estimate the total litter load (T) in each catchment area. In the unlikely event that there are existing litter traps of known efficiency already operating in the catchment(s), information gleaned from these traps would be used to estimate the total litter load(s). Otherwise Tables 19.1 and 19.2 could be used to give a first estimate of likely litter loads. A typical litter density of 95 kg/m^3 can be used to convert masses to volumes if necessary. Then estimate the street cleaning service factor (f_{sci}), the vegetation load (V_i) and the basic litter load (B_i) for each land use in each catchment or sub-catchment, and apply Eq. (19.3):

$$T = \Sigma f_{sci}.(V_i + B_i)A_i \qquad (19.3)$$

where T = total litter load in the waterways (m^3/yr)
f_{sci} = street-cleaning factor for each land use
  (This factor relates the anticipated efficiency of street cleaning, including private refuse collection if applicable, to the efficiencies that pertained during data collection. If there is no difference, $f_{sci} = 1$. Otherwise,

$$f_{sci} = (1 - \eta_{design})/(1 - \eta_{data}) \qquad (19.4)$$

where η_{design} = anticipated efficiency of street cleaning (fraction)
η_{data} = efficiency of street cleaning during data collection (fraction)

For example, an improvement in refuse removal from, say, 15% of the total load to 75% of the total load will indicate $f_{sci} = 0.294$).

V_i = vegetation load for each land use
  (varies from 0.0 m^3/hectare/yr for poorly vegetated areas to, say 0.6 m^3/hectare/yr (58 kg/ha.yr) for densely vegetated areas)
B_i = basic litter load for each land use (from local data or Tables 19.1 and 19.2)
A_i = area of each land use (hectare)

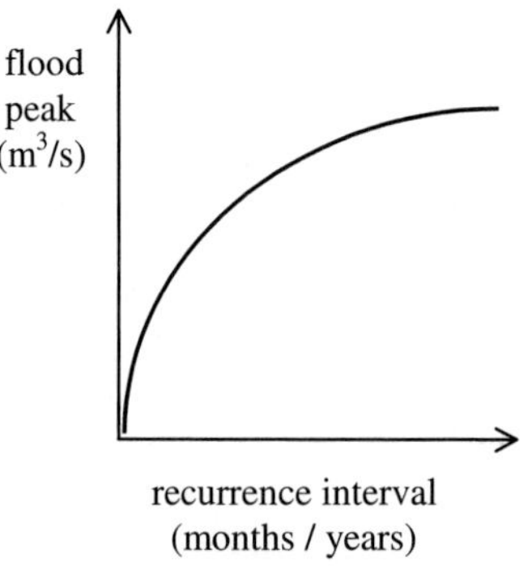

FIGURE   19.19   Schematic   flood peak/frequency curve.

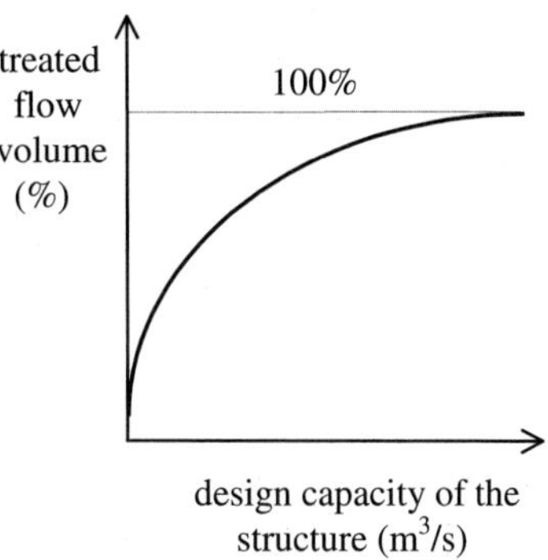

FIGURE 19.20 Schematic curve of treated flow volume/design capacity of structure.

4. For each potential trap site, carry out a hydrological assessment to estimate the relationship between the flood peaks and their frequency, and the relationship between the treated flow volume and the design capacity of the structure.

The flood peak/frequency curve (Fig. 19.19) is a plot of the flood peak in m^3/second versus the RI, while the treated volume/capacity curve (Fig. 19.20) expresses the percentage of the total flow volume intercepted by a structure versus its design capacity. The calculation of the flow volume is shown schematically in Fig. 19.21. Its significance lies in the fact that trapping structures are seldom designed to handle the maximum expected flood peak. Usually, they are designed to handle a much lower flow—typically with an RI in the order of one or two months. Under these circumstances, the total volume bypassing the structure will generally be a relatively small percentage of the total. If the

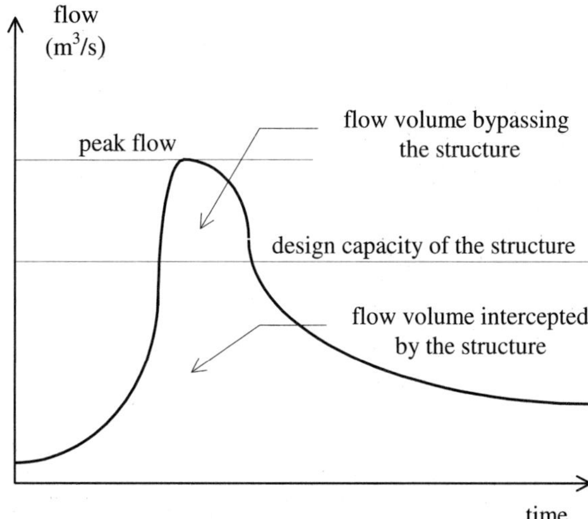

FIGURE 19.21 Typical flood hydrograph indicating the relative volumes intercepted by, and bypassing the structure.

usually conservative assumption is made that the concentration of litter is constant, then the overall trapping efficiency of the structure at any design capacity can be calculated from a knowledge of the proportion of flow passing through the structure. Considerable cost savings can often be made at the expense of a minimal drop in efficiency by selecting a smaller structure with a slightly higher bypass ratio.

The assumption of a constant litter concentration is made in the interests of simplicity. Initially there is insufficient flow to mobilize the litter and any runoff is essentially clear of large particles. Once, however, there is sufficient flow to move the litter, its concentration in the runoff increases sharply to a maximum that is related to the type and quantity of litter available and the carrying capacity of the flow. This maximum cannot be sustained. Even if the runoff rate remains constant, the majority of the litter is soon flushed out of the system and the concentration starts to drop back towards zero. Whilst this pattern is repeated for nearly every storm, the concentration values depend on the condition of the catchment at the time.

The hydrological assessment would typically be carried out with the assistance of one of the numerous urban hydrology computer packages. Care must be taken to ensure that the capacities of any conduits are taken into account.

5. Consideration is now given to the candidate trapping structures. Once a preliminary selection has been made, the patent holders/suppliers should be contacted for more up-to-date information on design and cost.

6. To calculate trap storage volumes and cleaning frequencies, it is suggested that the total litter load is assumed to be split between the significant downpours (with more than, say, 5 mm of rainfall) with the greater weighting given to those storms following long, dry periods. Storm litter loads may then be estimated from Eq. (19.5):

$$S = f_s . T / \Sigma f_{si} \tag{19.5}$$

where S = storm load in the waterways (m³/storm)
$\quad f_s$ = storm factor
$\qquad$ (Suggested factors are
$\qquad$ 1 for storms occurring less than 1 week after a previous downpour
$\qquad$ 2 for storms occurring after a dry period of about 1 month
$\qquad$ 3 for storms occurring after a dry period of about 2 months
$\qquad$ 4 for storms occurring after a dry period of more than about 3 months)
$\quad T$ = total litter load in the waterways (m³/yr)
Σf_{si} = the sum of all the storm factors for all of the storms in the year (since this information is generally not available, a suggested alternative is to count the average number of significant storms in a year and multiply by 1.1)

7. The cost-effectiveness of the candidate structures may now be determined by means of an economic analysis. There are many ways of carrying out this economic analysis, but the simplest is described below:

a. For each particular structure, consider several design capacities with RIs varying between, say, 1:1 month (the structure is bypassed 12 times a year) to 1:2 years (which is the capacity of many pipe conduits). For each design capacity, obtain an estimate of the overall efficiency of the trap by multiplying the published trap efficiency by the percentage of flow volume treated by the structure, as previously determined in step 4 above, using Eq. (19.6):

$$\eta_o = \eta_s . \eta_f \tag{19.6}$$

where η_o = overall efficiency of the installation (fraction)
$\quad \eta_s$ = published efficiency of the structure (fraction)
$\quad \eta_f$ = treated flow volume expressed as a fraction of the total flow

b. The required storage capacity can be calculated by multiplying the proposed average cleaning frequency in days by the average estimated storm load, S_{av}, (determined with the aid of Eq. (19.5) above utilizing a typical storm factor f_s for the area) and by the overall efficiency of the installation. This is then divided by the average storm frequency (in days) during the wet season determined from municipal records. The calculation is shown in Eq. (19.7):

$$V_t = F_c.\eta_o.S_{av}/F_s \qquad (19.7)$$

where V_t = required trap storage (m³)
F_c = average number of days between trap clearouts (d)
η_o = overall efficiency of the installation
S_{av} = average estimated storm load (m³)
F_s = average number of days between significant storms (d)

The storage capacity must also be more than the maximum expected storm load, S_{max}, which is determined from Eq. (19.5) utilizing the maximum expected value of f_s.

c. For each particular type and size of structure, decide on the re-payment period, and estimate the capital cost and the real interest rate (a reasonable approximation is to subtract the historic average inflation rate from the historic average nominal interest rate). The capital recovery amount may then be determined from Eq. (19.8):

$$A = P.i(1 + i)^n/((1 + i)^n - 1) \qquad (19.8)$$

where A = capital recovery amount ($/yr)
P = capital cost of the structure ($)
i = annual interest rate (expressed as a fraction)
n = repayment period (yrs)

If the payments are subject to inflation, the initial payments will be higher than the later payments in real terms, but this does not change the overall picture.

d. The total volume of litter that the trap is likely to intercept each year at each particular design capacity is obtained by multiplying the total litter load estimated in step 3 by the overall efficiency of the installation using Eq. (19.9):

$$L = T.\eta_o \qquad (19.9)$$

where L = load trapped by the structure (m³/yr)
T = total litter load (m³/yr)
η_o = overall efficiency of the installation (fraction)

e. The total annual cost of the structure is obtained by adding the annual capital recovery amount to the annual cost of clearing and maintaining the structure using Eq. (19.10):

$$C_t = A + C_c \qquad (19.10)$$

where C_t = total annual cost of the structure ($/yr)
A = capital recovery amount ($/yr)
C_c = annual cost of clearing and maintaining the structure ($/yr)

f. The unit cost of litter removal for any particular structure and design capacity is obtained by dividing the total annual cost of the structure by the estimated annual load that will be trapped by the structure as expressed in Eq. (19.11):

$$C = C_t/L \qquad (19.11)$$

where C = unit cost of litter removal ($\$/m^3$)
C_t = total annual cost of the structure ($\$/yr$)
L = load trapped by the structure ($m^3/hectare/yr$)

Unit costs in terms of $\$/kg$ or $\$/hectare$ may be obtained by dividing the unit cost of litter removal by the litter density (typically 95 kg/m^3), or by dividing the total annual cost of the structure by the catchment area respectively.

8. In theory, the trapping system may now be optimized to give the lowest overall unit cost of removal. In reality, a balance must be struck between the desire to achieve the lowest overall unit cost of removal and the overall objective of removing as much litter from the aquatic system as is reasonably possible—in other words, achieving the maximum efficiency. This is a political decision that requires input from all the role players concerned with the removal of litter from the environment, including engineers, hydrologists, aquatic scientists, environmental interest groups, taxpayers, and local government. It may also turn out that costs can be reduced and/or efficiencies raised by putting more money into catchment litter reduction measures such as street sweeping. One further caution: data on trapping structures are site-specific and highly variable. Costs and efficiencies may vary considerably from site to site.

The litter removal process is summarized in Fig. 19.22. The trap selection procedure is summarized in Fig. 19.23.

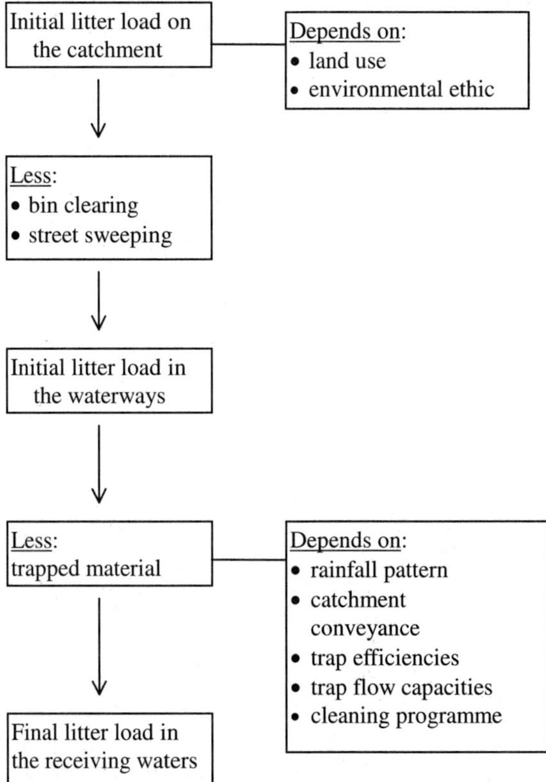

FIGURE 19.22   The litter removal process.

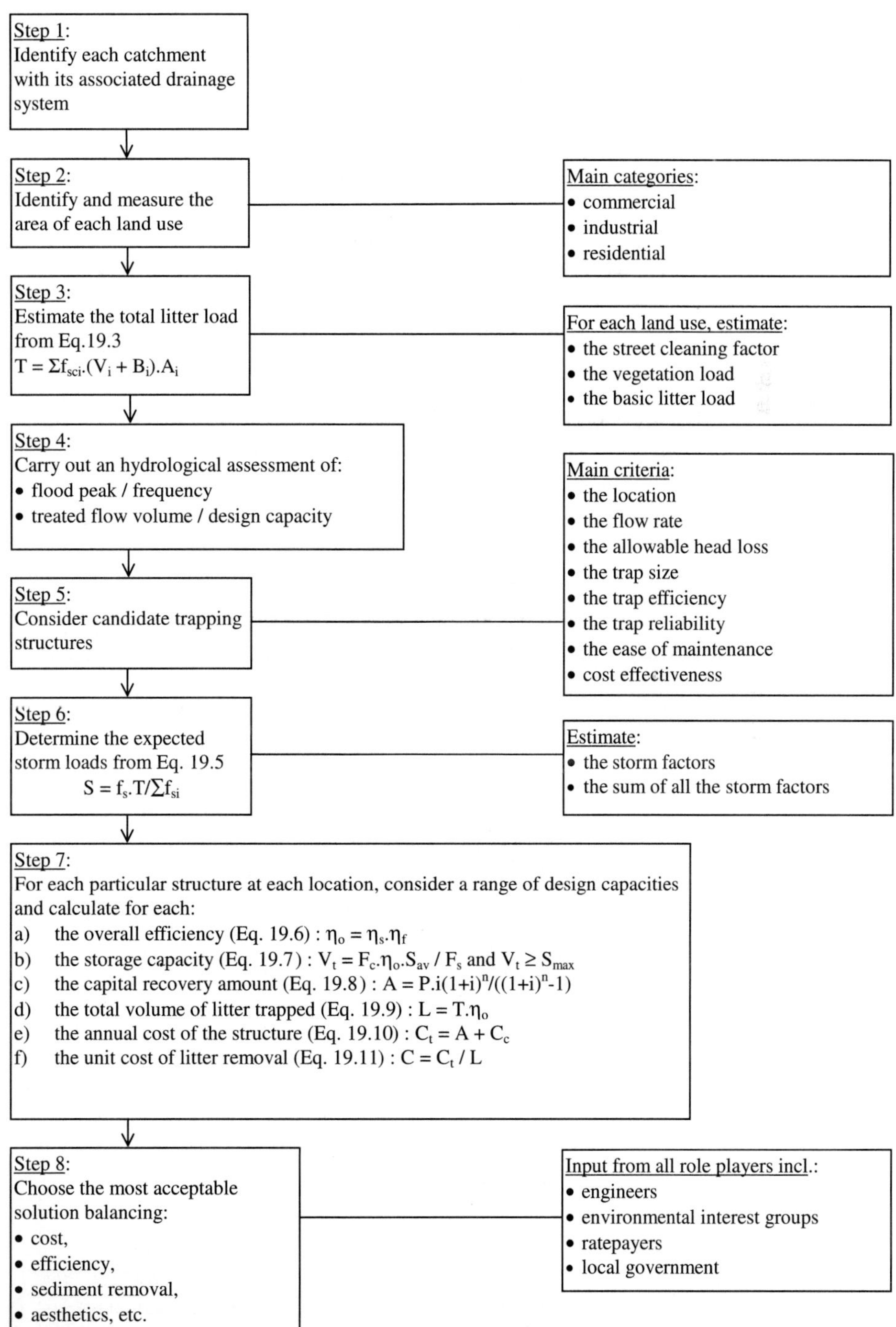

FIGURE 19.23   Summary of the trap selection procedure.

19.7.4 Preliminary Assessment of Some Litter Traps

The trap selection procedure was applied to a hypothetical catchment making the following assumptions (Armitage et al., 1998):

- The catchment comprises the central business district of a medium-sized town in South Africa (50% commercial, 30% industrial, 20% residential).
- The catchment area is 100 hectares (1 km^2).
- It is situated in a summer rainfall area with a mean annual precipitation = 850 mm.
- The topography and layout permits the installation of any of the seven typical litter traps previously described.
- The underground drainage system is designed for 1:2 year recurrence interval (RI) storms.
- There are 400 catch basins (a density of 4/hectare).
- There is regular street cleaning (η_{design} = 0.82).
- There is no vegetation load.
- The runoff coefficient is 0.7 (70% of the storm rainfall is transported by the drainage system during the storm).
- The time of concentration of the rainfall (the time theoretically taken for a rain drop falling on the most remote point of the catchment to reach the trap) = 30 min.
- There are 50 significant rainfall events (more than 5 mm rainfall) a year concentrated in the summer rainfall season.
- Recurrence intervals of 1:1 month and 1:2 years were considered. The associated critical rainfall intensities were assumed to be 21 and 51 mm/hr respectively.
- The economic analysis is to be carried out assuming a repayment period of 20 years and a real interest rate (after taking inflation into account) of 6%.

From this information, the total litter load for typical South African conditions was estimated to be 84.2 m^3/yr. The 1:1 month peak flow was estimated to be approximately 4 m^3/second, while the 1:2 yr peak flow was estimated to be 10 m^3/second.

The outcome of this analysis is summarized in Table 19.5. The unit costs/ton of litter removed were obtained by dividing the volumes by a litter density of 95 kg/m^3.

TABLE 19.5 The Seven Most Promising Litter Traps Ranked in Order of Unit Cost

#	Device	Traps	η (%)	$/m^3
1	BGPT (1:1 month RI)	1	86	20
2	SCS (1:1 month RI)	1	86	26
3	UWEM (1:1 month RI)	1	81	32
4	BGPT (1:2 year RI)	3	95	39
5	SCS (1:2 year RI)	1	95	44
6	UWEM (1:2 year RI)	1	90	46
7	SECTs (50% coverage)	200	59	276
8	SECTs (100% coverage)	400	78	284
9	CDS (1:1 month RI)	3	89	308
10	ILLS	20	25	314
11	LCD	—	25	551
12	CDS (1:2 year RI)	6	99	553

With the assumed catchment data and the cost data available at the time of the analysis, it appears that

- The BGPT, SCS, and UWEM devices have a lower unit cost than the remaining four structures.
- The CDS unit offers a very high removal efficiency, but at a high unit cost. Unit costs may, however, be lowered if high bypass ratios are used.
- SECTs offer the advantage of being a potential catchment management tool as they show where the bulk of the litter is being generated.
- The ILLS and LCD structures appear on the surface to be costly. On the other hand, they have the advantage that they are small and can be installed under streets in confined spaces. The ILLS has the additional advantage that it requires very little head.

In addition, though not considered in the evaluation above, fences, nets, weirs, booms, or baffles may be the most cost-effective structures of all provided a suitable slow-flowing stream or pond is available. A major problem with these devices is cleaning and maintenance. Ideally, it should be possible for the channel to be periodically drained for cleaning and maintenance purposes. Another avenue to explore is a mix of technologies. For example, the hydraulically actuated sluice gate that is used in the UWEM approach could be used to generate the required head to run a BGPT or a SCS structure.

Table 19.5 must not be read as valid for every situation. The results of the analysis will be strongly influenced by the site location and conditions. Furthermore, cost is not the only consideration. Lack of head may rule out the BGPT and SCS devices. Lack of space may rule out the UWEM approach. The desire for a catchment management tool may favour the choice of SECTs. A requirement for exceptionally high-removal efficiency may prompt the installation of a CDS unit. A small catchment may be best served by an ILLS or LCD. *The final choice of trapping structure will be specific to each site and situation.*

Most sites in South Africa, for example, would probably be best served by SECTs installed in key catch basins around the central business district, and a BGPT, SCS, or UWEM unit installed on the main outlet conduit to the catchment, with head provided by a hydraulically actuated sluice gate if required. The situation may very well be different in other countries.

19.8 CONCLUSIONS

- Litter in the waterways is a major environmental problem that will cost a lot of money to address.
- It is the direct result of human behavior.
- The amount and types of litter to be found at any one place are extremely variable as they depend on a large number of independent factors.
- Some form of catchment litter management might help to reduce the quantities of litter.
- Litter wash-off rates are extremely sensitive to the frequency of street sweeping. Major reductions in the quantity of litter reaching the drainage system can often be achieved with an increased frequency of sweeping.
- Many existing litter traps are poorly designed.
- It is difficult to achieve high removal efficiencies with screenless traps unless the flow velocities (Froude number) are kept low.
- Screens can be made to be self-cleaning if they are declined in the direction of the flow.
- There is a rational procedure for the selection of the optimum trapping system.

- A comparison of seven patented litter traps for use in South African conditions indicate that three are likely to be the optimal choice in the majority of urban drainage situations in South Africa and possibly the rest of the developing world. Two utilize declined self-cleaning screens (the Baramy® Gross Pollutant Trap and the Stormwater Cleaning Systems Structure). The other utilizes suspended screens in tandem with a hydraulically actuated sluice gate (the Urban Water Environmental Management concept). Other trap designs may be appropriate elsewhere.

DISCLAIMER

Most of the more successful structures have been patented and are available only from approved suppliers. Mention of a trade name does not indicate that the author necessarily supports the product in question. The structures are described here in an attempt to indicate some of the best available technologies. There may, of course, be other structures which might remove litter from drainage systems more efficiently and effectively than these.

ACKNOWLEDGEMENTS

The information contained in this paper emanates from a project funded by the Water Research Commission of South Africa (WRC) entitled *The Removal of Floating and Suspended Materials from Streams* (WRC Project No. 691) and published as WRC Report No. TT 95 /98: *"The Removal of Urban Litter from Stormwater Conduits and Streams."* The report is also available on the web at *http://www.wrc.org.za/reports/tt95_98.htm*.

REFERENCES

Allison R. A., and F. H. S. Chiew, "Monitoring of Stormwater Pollution from Various Land Uses in an Urban Catchment," *Proceedings of the 2nd International Symposium on Urban Stormwater Management,* Melbourne, Australia, pp. 551–516, 1995.

Allison, R. A., Personal communication with Rob Allison, Ph.D. student, Department of Civil and Environmental Engineering, University of Melbourne, Australia, 1996.

Allison, R. A., "Effective Gross Pollutant Removal from Urban Waterways," Ph.D. thesis, University of Melbourne, Australia, 1997.

Allison, R. A., Personal communication with Rob Allison, Melbourne Water, Australia, 2000.

Armitage, N. P., A. Rooseboom, C. Nel, and P. Townshend, "The Removal of Urban Litter from Stormwater Conduits and Streams," Water Research Commission (South Africa) Report No. TT 95/98, Pretoria, 1998.

Armitage, N. P., and A. Rooseboom, "The Removal of Litter from Stormwater Conduits and Streams in South Africa: Selecting a Trapping System," *Proceedings of the 3rd International Symposium on Stormwater Management and 6th International Conference on Hydraulics in Civil Engineering, Hydra Storm '98,* Adelaide, Australia, 1998.

Armitage, N. P., and A. Rooseboom, "The Removal of Litter from Stormwater Conduits in the Developing World," *Water Science Technology,* 39(9):277–284, 1999.

Armitage, N. P., and A. Rooseboom, "The Removal of Urban Litter from Stormwater Conduits and Streams: Paper 1—The Quantities Involved and Catchment Litter Management Options," *Water SA,* 26(2):181–187, 2000a.

Armitage, N. P., and A. Rooseboom, "The Removal of Urban Litter from Stormwater Conduits and Streams: Paper 2—Model Studies of Potential Trapping Structures," *Water SA,* 26(2):189–194, 2000b.

Armitage, N. P., and A. Rooseboom, "The Removal of Urban Litter from Stormwater Conduits and Streams: Paper 3—Selecting the Most Suitable Trap," *Water SA,* 26(2):195–204, 2000c.

Baramy, promotional literature, Baramy Engineering Pty. Ltd., Katoomba, New South Wales, Australia, undated.

Baur, D. C., and S. Iudicello, "Stemming the Tide of Marine Debris Pollution: Putting Domestic and International Control Authorities to Work," *Ecology Law Quarterly* 17:71–142, 1990.

Brownlee, R. P., "Evaluation of Effectiveness and Efficiency of North Sydney Council Litter Control Device Program," *Proceedings of the 2nd International Symposium on Urban Stormwater Management,* Melbourne, Australia, 1995.

Compion, J. K., personal communication with Cobus Compion, Liebenberg & Stander Western Cape (Pty.) Ltd., Consulting Engineers, Cape Town, South Africa, 1998.

Cornelius M., T. Clayton, G. Lewis, G. Arnold, and J. Craig, *Litter Associated with Stormwater Discharge in Auckland City, New Zealand,* Island Care New Zealand Trust, Auckland, New Zealand, 1994.

CSIR, "The Situation of Waste Management and Pollution Control in South Africa," Report CPE 1/91 to the Department of Environment Affairs by the CSIR (Council for Scientific and Industrial Research) Programme for the Environment, Pretoria, South Africa, 1991.

Hall, M., "Litter Traps in the Stormwater Drainage System," unpublished M. Eng. paper, Swinburne University of Technology, Melbourne, Australia, 1996.

Hocking, J., "Evaluation of Effectiveness and Efficiency of Smoothey Park Litter Control Device," *Proceedings of the 4th Annual Conference on Soil and Water Management for Urban Development,* Australia, 1996.

HydroQual Inc., "Floatables Pilot Program Final Report: Evaluation of Non-Structural Methods to Control Combined and Storm Sewer Floatable Materials," City-Wide Floatables Study, Contract II, prepared for New York City, Department of Environmental Protection, Bureau of Environmental Engineering, Division of Water Quality Improvement, NYDP 2000.

Island Care New Zealand Trust, "Reducing the Incidence of Stormwater Debris and Street Litter in the Marine Environment—A Co-operative Community Approach," Auckland, New Zealand, 1996.

Melbourne Water and Melbourne Parks & Waterways, "Backyard to Bay Program," Melbourne, Australia, 1993.

Melbourne Water Waterways and Drainage Group, "A Guide to Current Technology for Removing Litter and Sediments from Drains and Waterways," Melbourne Water, Australia, 1995.

Pitt, R. Personal communication with Robert Pitt, Professor of Civil and Environmental Engineering, University of Alabama at Birmingham, 2000.

President's Council Report, "Report of the Committee of the President's Council concerning a National Environment Management System," State Press, Cape Town, South Africa, 1991.

Raudkivi, A. J., "Loose Boundary Hydraulics, 3rd edition," Pergamon Press, Oxford, England, 1990.

Rooseboom, A., "Sediment Transport in Rivers and Reservoirs—A Southern African perspective," Water Research Commission (South Africa) Report No. 297/1/92, Pretoria, South Africa, 1992.

Senior J. C., "Litter Control in Urban Waterways," *Proceedings of the International Symposium on Urban Stormwater Management,* pp. 234–239, Sydney, Australia.

Swinburne University of Technology, "In-line Litter Separator Research and Development," Project 1996/8, Waste Management Council of Victoria, Australia, 1996.

Wong, T. H. F., and R. M. Wootton, "An Innovative Gross Pollutant Trap for Stormwater Treatment," *Proceedings of the 2nd International Symposium on Urban Stormwater Management,* Melbourne, Australia, 1995.

CHAPTER 20

FLOW TRANSITIONS AND ENERGY DISSIPATORS FOR CULVERTS AND CHANNELS

Larry W. Mays
Department of Civil and Environmental Engineering
Arizona State University
Tempe, Arizona

20.1 FLOW TRANSITIONS FOR CULVERTS

Flow transitions are changes in the cross section of an open channel over short distances. They are designed to have a minimum amount of flow disturbance. Figure 20.1 illustrates the various types of transitions; the two most common ones are the abrupt (headwall) and the straight line (wingwall).

Highway culverts typically are designed to operate with an upstream headwater pool that dissipates the of the channel approach velocity. This type of situation does not require an approach flow transition. *Outlet transitions (expansions)* should be considered in the design of all culverts, channel protection, and energy dissipators. *Transition design* can be categorized as:

- Culverts with outlet control (subcritical flow)
- Culverts with inlet control (supercritical flow)

20.1.1 Culverts with Outlet Control

Culverts with outlet control can have abrupt expansions or gradual transitions. In an *abrupt expansion,* the water surface plunges or drops rapidly and the flow spreads out. The potential energy stored as depth is converted to kinetic energy with a higher velocity of flow. The transition (apron) end velocity can be determined using the experimental results of Watts (1968) (Figs. 20.2 and 20.3). Figure 20.2 relates the average depth brink depth ratio (y_A/y_0) for a rectangular outlet to the Froude number. Figure 20.3 is a similar relationship for pipe culverts. These curves in Figs. 20.2 and 20.3 were developed for Fr's ranging from 1.0 to 2.5, the applicable range for most abrupt outlet transitions. Usually, a low tailwater is encountered at culvert outlets and the flow is supercritical on the outlet apron.

The design procedure for *abrupt expansions* is as follows (Corry et al., 1975):

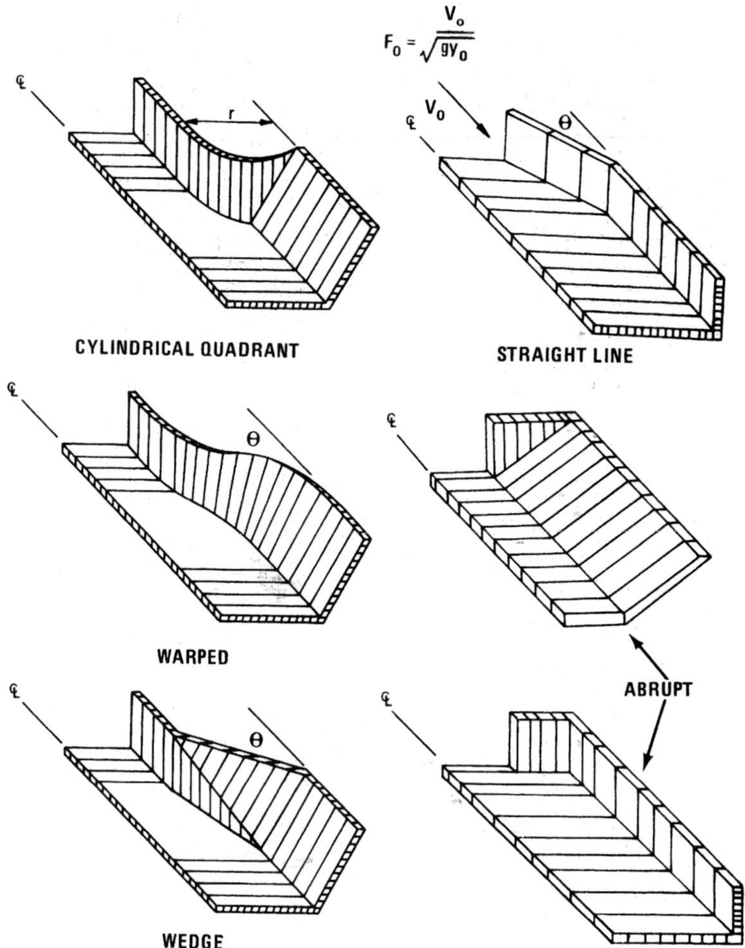

FIGURE 20.1   Transition types. (From Corry et al., 1975)

1. Determine V_0 and y_0 at the culvert outlet using Fig. 20.4 for box culverts or Fig. 20.5 for circular culverts.

2. Compute Froude number, $\text{Fr} = V_0/\sqrt{gy_0}$.

3. Determine optimum flare angle (θ) using $\tan\theta = Fr/3$ (Blaisdell and Donnelly, 1949). If the selected wingwall flare (θ_w) is greater than θ, consider reducing θ_w to θ.

4. Use Fig. 20.2 (for box culverts) or Fig. 20.3 (for pipe culverts) to determine the downstream average depth (y_A) knowing Fr and the desired distance L (expressed in multiples of D, diameter).

5. Determine the average velocity (V_A) for box culverts using

$$\frac{V_A}{V_0} = 1.65 - 0.3Fr$$

and for pipe culverts, (for $L \leq 3D$) using

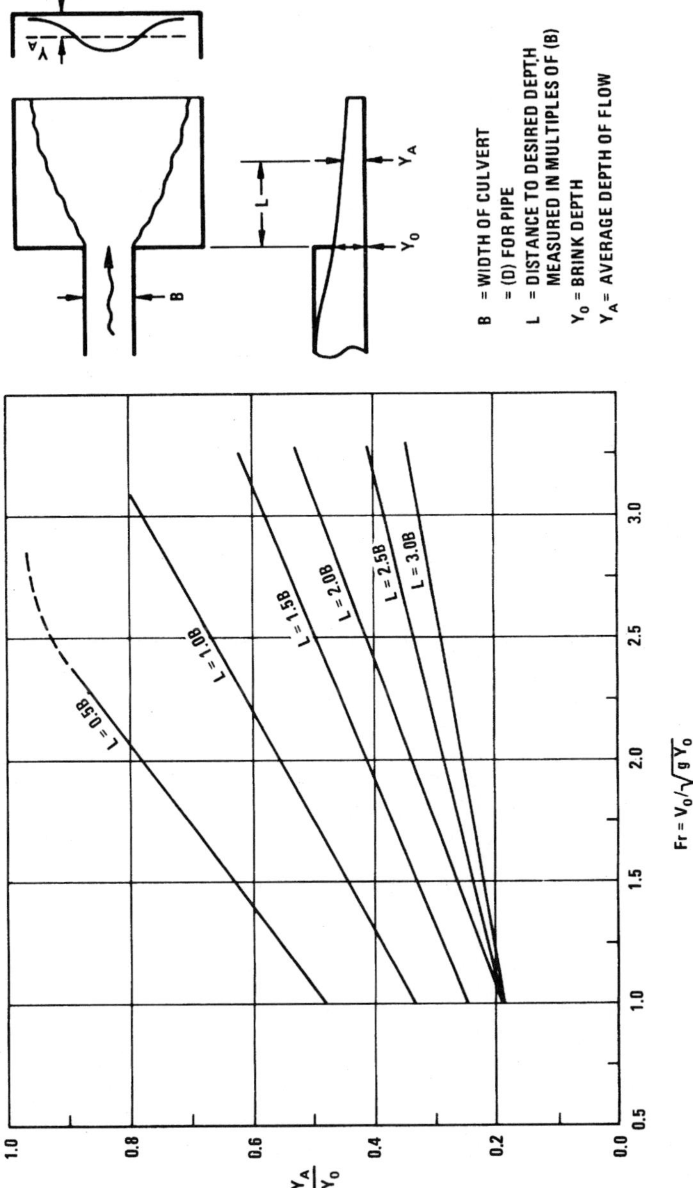

FIGURE 20.2 Average depth for abrupt expansion below rectangular culvert outlet. (From Corry et al., 1975)

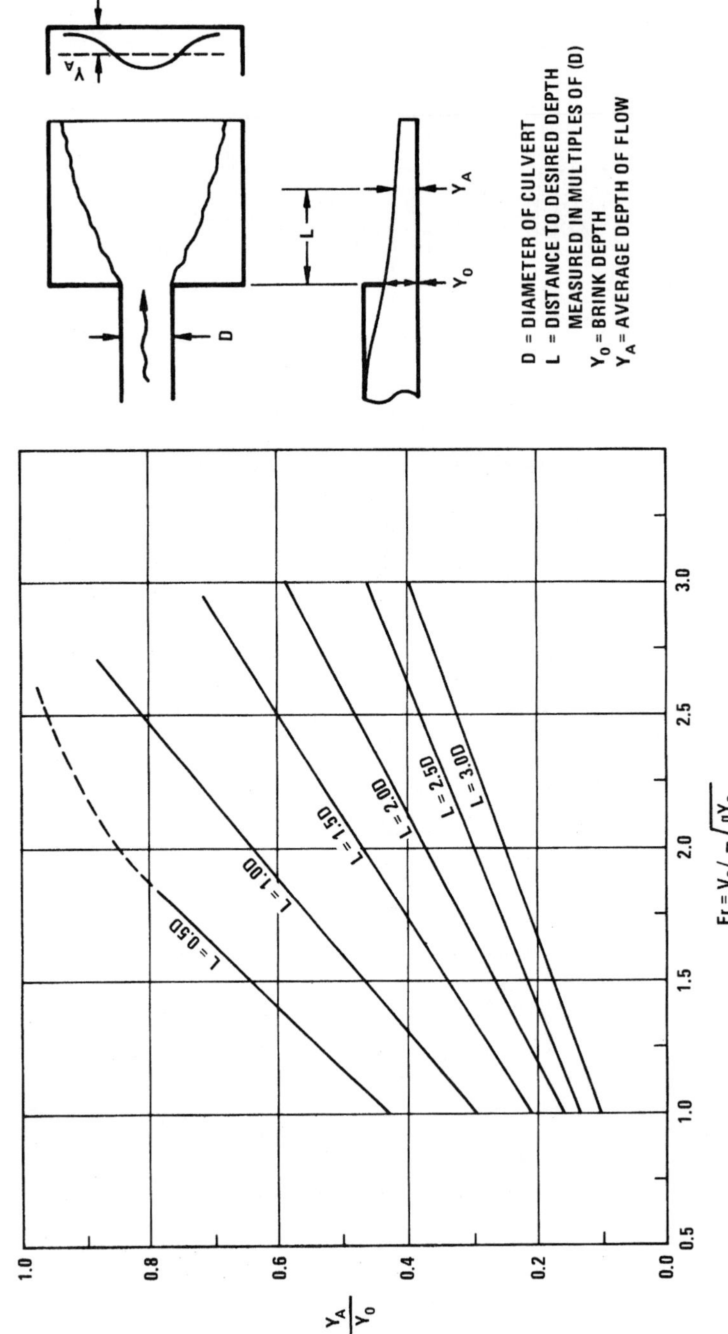

FIGURE 20.3 Average depth for abrupt expansion below circular culvert outlet. (From Corry et al., 1975)

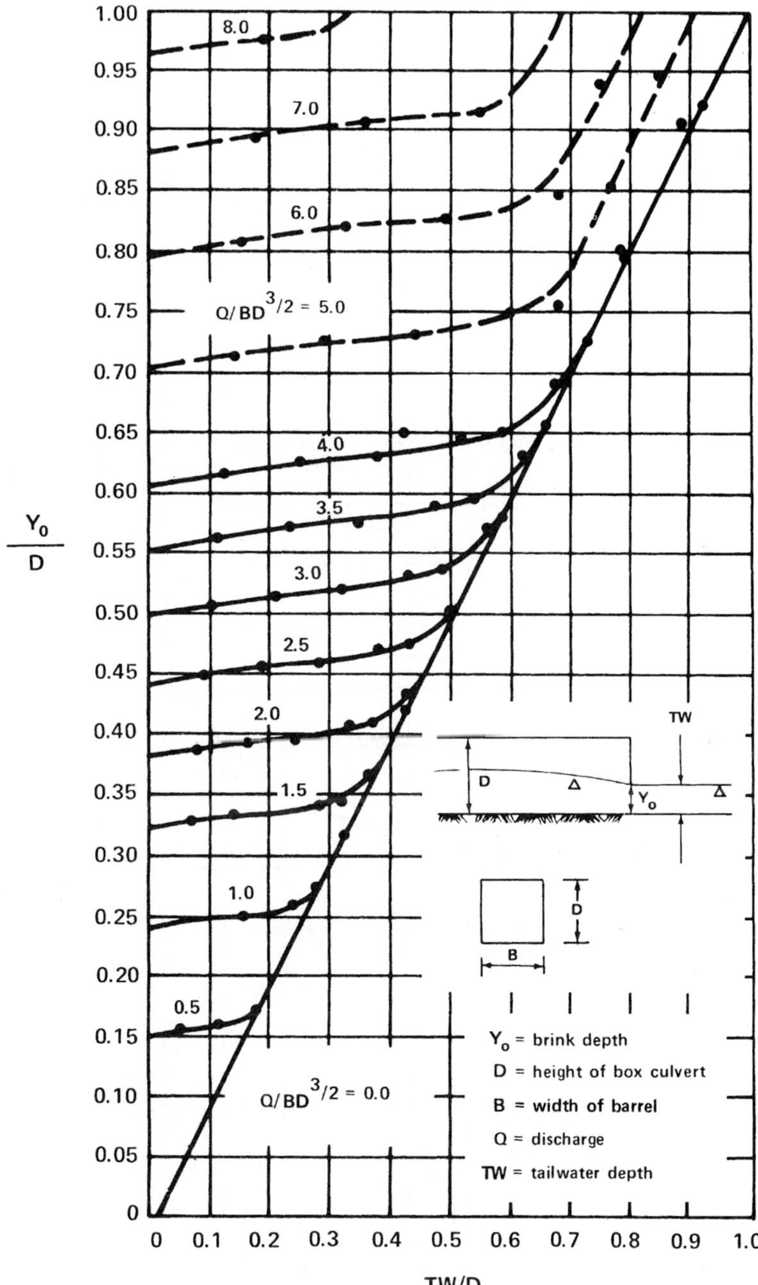

FIGURE 20.4 Dimensionless rating curves for the outlets of rectangulars culverts on horizontal and mild slopes. (From Simons et al; 1970)

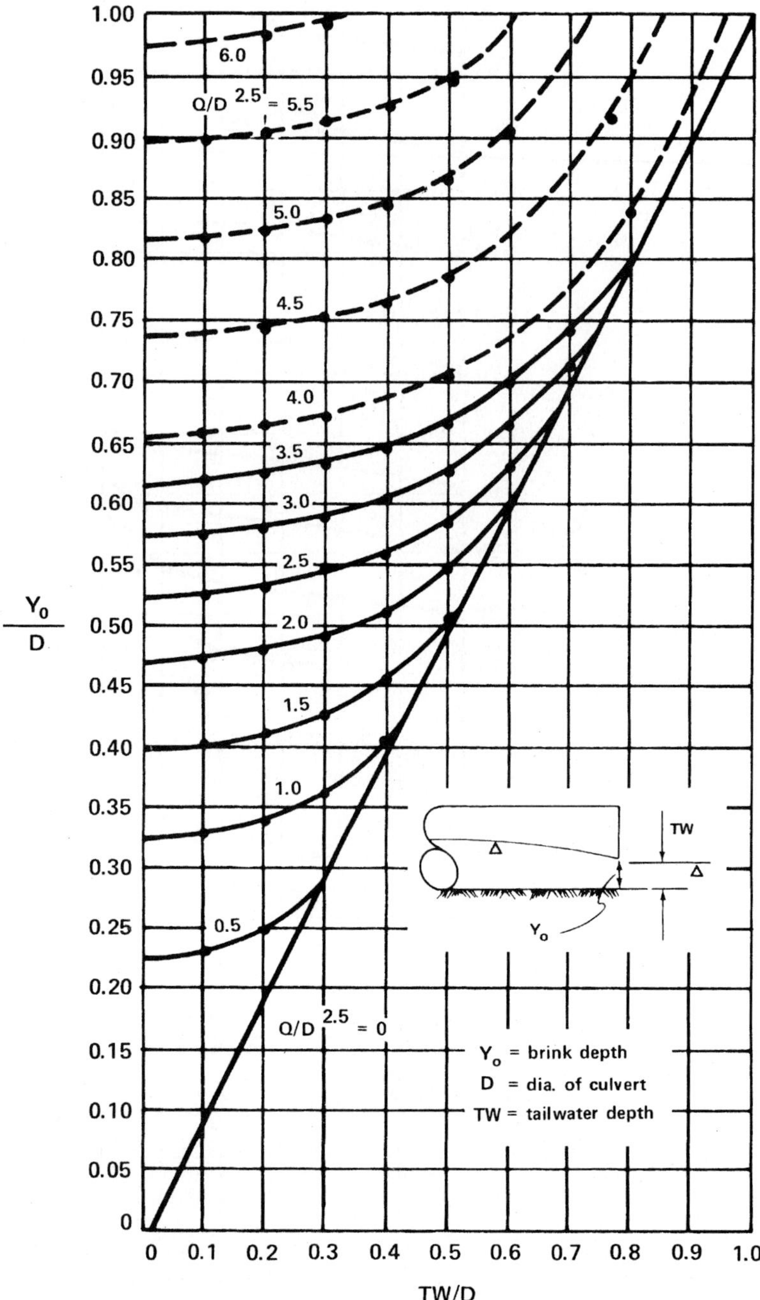

FIGURE 20.5 Dimensionless rating curve for the outlets of circular culverts on horizontal and mild slopes. (From Simons, et al., 1970)

$$\frac{V_A}{V} = 1.65 - 0.45 \left(\frac{Q}{\sqrt{gD^5}}\right)$$

6. Compute downstream width $W_2 = W_o + 2L \tan \theta$, where $\tan \theta = Fr/3$. If $\theta_w > \theta$, then use θ_w to compute W_2.

7. If θ is used to compute W_2, then compute the downstream depth y_2 using W_2 and V_A. Because the flow prism is laterally confined, y_2 will be larger than y_A.
If θ_w is used, $y_2 = y_A$, and the average flow width is $W_A = Q/V_A y_A$. If $W_A < W_2$, use W_2 to compute $y_2 = Q/V_A W_2$.

Example 20.1 (adapted from Corry et al., 1975)
Determine the width of an abrupt expansion and the hydraulic condition (y and V) at the end of an abrupt expansion for a 5 ft by 5 ft, 200-ft-long reinforced concrete culvert on a 0.2 percent slope ($S_0 = 0.002$ ft/ft). The discharge is $Q = 270$ cfs and a wingwall flare ($\theta_W = 45°$) with a $10 =$ ft apron is considered.

Given

$$d_c = 5 \text{ ft and } \frac{TW}{D} \approx 0$$

Solution

Step 1. $\dfrac{Q}{BD^{3/2}} = \dfrac{270}{5(5)^{3/2}} = 4.83; \ d_c = \sqrt[3]{\left(\dfrac{Q}{B}\right)^2 \dfrac{1}{g}} = \sqrt[3]{\left(\dfrac{270}{5}\right)^2 \dfrac{1}{32.2}} = 4.5 \text{ ft and } \dfrac{TW}{D} \approx 0.$

From Fig. 20.4, $y_0/D = 0.68$; therefore, y_0 and V_0 can be determined to be $y_0 = 0.68(5) = 3.40$ ft and $V_0 = 270/[3.4 \ (5)] = 15.88$ ft/s.

Step 2. The outlet Fr $= V_0/\sqrt{gy_0} = 15.88/\sqrt{32.2 \ (3.4)} = 1.52$

Step 3. The optimum flare angle is determined using $\tan \theta = \frac{1}{3}$ Fr $= \frac{1}{3}(1.52) = 0.507$, so $\theta = 26.9°$.

Step 4. With the apron length/width $= 10/5 = 2$ and $L = 2.0B$, use Fig. 20.2 to compute the average depth where $y_A/y_0 = 0.26$ and $y_A = 0.26 \ (3.4) = 0.88$ ft.

Step 5. Compute the average velocity, $V_A/V_0 = 1.65 - 0.3$ Fr $= 1.65 - .3 \ (1.52) = 1.19$, so $= V_A = 1.19 \ V_0 = 1.19 \ (15.88) = 18.90$ ft/s

Step 6. Determine the downstream width $W_2 = W_o + 2L \tan \theta_W$ using $\theta = \theta_W = 45°$ because $\theta_W > \theta$ and $W_0 = 5$ ft, then $W_2 = 5 + 2 \ (10) \ (\tan 45°) = 25$ ft.

Step 7. θ_W is used above, so $y_2 = y_A = 0.88$ ft and $W_2 = Q/V_A y_A) = 16.1 < 25$ ft.

When subcritical flow is maintained throughout a culvert, gradual transitions can be used. Referring to Fig. 20.6, upstream of section 1 where some backwater exists because of the culvert, flow is transitional from a channel into the culvert and out. According to the Federal Highway Administration (FHWA, 1978), a flare angle of 17.5° (4.5 to 1) or flatter provides a gradually varied transition that can be analyzed using the energy equation.

Referring again to Fig. 20.6, the headloss in the contraction H_{LC} is

$$H_{LC} = C_C \left(\frac{V_2^2}{2g} - \frac{V_1^2}{2g}\right) \tag{20.1}$$

and the headloss in the expansion is

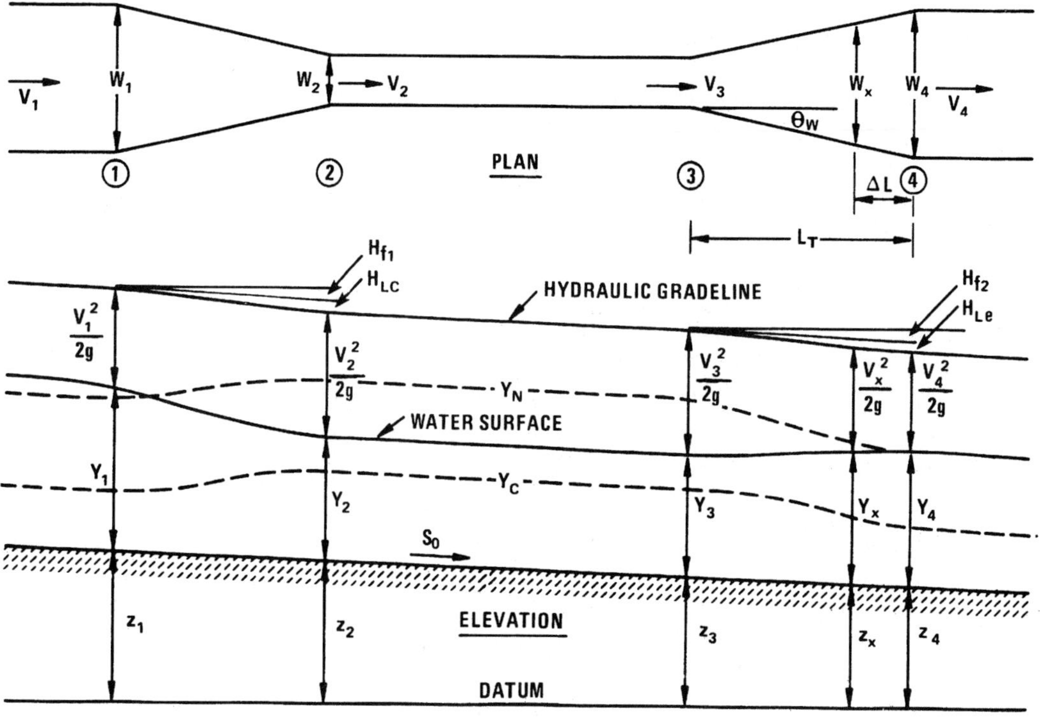

FIGURE 20.6   Definition sketch. (From Corry et al., 1975)

$$H_{LE} = C_e \left(\frac{V_3^2}{2g} - \frac{V_4^2}{2g} \right) \tag{20.2}$$

where transition loss coefficients are listed in Table 20.1 for the transition types (USACE, 1970).

The design procedure for *gradual transitions* is as follows (Corry et al 1975):

1. Use Manning's equation to compute $y_4 = y_n$ and V_4, knowing Q, S_o, n and the outlet geometry.

TABLE 20.1   Transition Loss Coefficients

Transition type	Contaraction C_r	Expansion C_e
Warped	1.10	0.20
Cylindrical quadrant	0.15	0.25
Wedge	0.30	0.50
Straight line	0.30	0.50
Square end	0.30	0.75

Source: Corry, et al (1975).

2. Compute the critical depth using $y_c = k_1(Q/W)^{2/3}$. For box culvert, $k_1 = 0.315$ where Q is in cfs and W is the box culvert width in ft. In SI units $k_1 = 0.467$ with Q in m³/s and W in m. Compare y_n and y_c to ensure subcritical flow.

3. For the chosen transition type, use Table 20.1 to obtain the transition loss coefficients C_c and C_e.

4. Choose $y_3 = 1.1\,y_c$ to have a culvert with a flow depth conservatively above, y_c.

5. Compute the culvert width using the following energy equation, ignoring the headloss caused by friction:

$$Z_4 + y_4 + \frac{V_4^2}{2g} + C_e\left(\frac{V_3^2}{2g} - \frac{V_4^2}{2g}\right) = Z_3 + y_3 + \frac{V_3^2}{2g}$$

$V_3 = Q/W_3 y_3$ where $y_3 = 1.1y_c = 1.1k_1(Q/W_3)^2$. For box culverts and $V_4 = Q/W_4 y_4$. Neglecting $Z_3 - Z_4$ for a short reach and a mild bottom slope,

$$y_4 + \left[\left(\frac{Q}{W_4 y_4}\right)^2 \frac{1}{2g}\right](1 - C_e) = y_3 + \left[\left(\frac{Q}{W_3 y_3}\right)^2 \frac{1}{2g}\right](1 - C_e)$$

$$y_4 + \left[\left(\frac{Q}{W_4 y_4}\right)^2 \frac{1}{2g}\right](1 - C_e) = 1.1k_1\left(\frac{Q}{W_3}\right)^{2/3} + \left[\left(\frac{Q}{W_3[1.1k_1(Q/W_3)^{2/3}]}\right)^2 \frac{1}{2g}\right](1 - C_e)$$

$$y_4 + \left[\left(\frac{Q}{W_4 y_4}\right)^2 \frac{1}{2g}\right](1 - C_e) = 1.1k_1 Q^{2/3}\left(\frac{1}{W_3}\right)^{2/3} + \frac{Q^{2/3}}{1.21k_1^2}\left(\frac{1}{W_3}\right)^{2/3}\frac{1}{2g}(1 - C_e)$$

$$y_4 + \left[\left(\frac{Q}{W_4 y_4}\right)^2 \frac{1}{2g}\right](1 - C_e) = \left(1.1k_1 Q^{2/3} + \frac{Q^{2/3}}{1.21k_1^2}\frac{1}{2g}(1 - C_e)\right)\left(\frac{1}{W_3}\right)^{2/3}$$

$$[A] = [B]\left(\frac{1}{W_3}\right)^{2/3}$$

and

$$W_3 = ([B]/[A])^{3/2} \tag{20.3}$$

where

$$[A] = y_4 + \left[\left(\frac{Q}{W_4 y_4}\right)^2 \frac{1}{2g}\right](1 - C_e)$$

and

$$[B] = 1.1k_1 Q^{2/3} + \frac{Q^{2/3}}{1.21k_1^2}\frac{1}{2g}(1 - C_e)$$

6. Compute y_1 with $y_2 = y_3$ using the energy equation

$$Z_2 + y_2 + \frac{V_2^2}{2g} + C_c\left(\frac{V_2^2}{2g} - \frac{V_1^2}{2g}\right) = Z_1 + y_1 + \frac{V_1^2}{2g}. \tag{20.4}$$

7. If the amount of backwater $(y_1 - y_n)$ exceeds a preferred or required freeboard, then select a larger culvert width and calculate y_3 using Eq. (20.3) given in Step 5 (return to Step 5).

8. If the culvert width is acceptable, use the flow conditions to compute the transition length using a 4.5:1 flare (USACE, 1970):

$$L_T = 4.5 \frac{(W_4 - W_3)}{2} \tag{20.5}$$

or

$$L_T = 4.5 \frac{(W_1 - W_2)}{2} \tag{20.6}$$

9. Compute the water surface profile through the structure using a standard step backwater analysis. This will include an evaluation of friction losses.

Example 20.2
Determine the dimensions for a culvert and gradual transition needed for a 3 m-wide rectangular flood control channel at a slope of 0.001 m/m. The culvert length is 30.5 m, and is to convey 8.5 m³/s. Use $n = 0.02$.

Solution
Step 1. Assuming normal depth at Section 4 (refer to Fig. 20.6), then use Manning's equation to compute $y_4 = y_n = 1.99$ m and $V_n = 1.42$ m/s.

Step 2. The critical depth for section 3 is $y_c = 0.467 \left(\dfrac{Q}{W}\right)^{2/3} = 0.467 \left(\dfrac{8.5}{3}\right)^{2/3} = 0.935$ m.

Step 3. Use a straight line transition with $C_e = 0.5$ and $C_c = 0.3$.

Step 4. $y_3 = 1.1 y_c = 1.1 (0.935) = 1.03$ m.

Step 5. Compute W_3 using Eq. (20.3).

$$W_3 = ([B]/[A])^{3/2}$$

$$[B] = 1.1 k_1 Q^{2/3} + \frac{Q^{2/3}}{1.21 k_1^2} \frac{1}{2g} (1 - C_e)$$

$$= 1.1(0.467)8.5^{2/3} + \frac{8.5^{2/3}}{1.21(0.467)^2} \frac{1}{2(9.81)} (1 - 0.05)$$

$$= 2.542$$

$$[A] = y_4 + \left(\frac{Q}{W_4 y_4}\right)^2 \frac{1}{2g} (1 - C_e) = 1.99 + \left(\frac{8.5}{3(1.99)}\right)^2 \frac{1}{2(9.81)} (1 - 0.5)$$

$$= 2.042$$

$$W_3 = (2.54, 42/2.042)^{3/2} = 1.39 \text{ m. Take } W_3 = 1.4 \text{ m.}$$

Step 6. Compute y_1 with $y_2 = y_3$ using Eq. (20.4):

$$Z_2 - Z_1 + y_2 + (1 + C_e) \left(\frac{Q}{W_2 y_2}\right)^2 \frac{1}{2g} = y_1 + (1 + C_e) \left(\frac{Q}{W_1 y_1}\right)^2 \frac{1}{2g}$$

$$1.03 + (1 + 0.3) \left(\frac{8.5}{1.4(1.03)}\right)^2 \frac{1}{2(9.81)} = y_1 + (1 + 0.3) \left(\frac{8.5}{3y_1}\right)^2 \frac{1}{2(9.81)}$$

$$3.33 = y_1 + \frac{0.53}{y_1^{\,2}}.$$

Then $y_1 = 3.28$ m, which is 1.29 m above y_n.

Step 7. Because y_1 is 1.29 m above the normal depth and because only 0.6 m free board is available, we go back to Step 5 and compute y_3. Assuming that the culvert width is 1.45 m: Repeat step 5,

$$y_4 + \left(\frac{Q}{W_4 y_4}\right)^2 \frac{1}{2g} (1 - C_e) = y_3 + \left(\frac{Q}{W_3 y_3}\right)^2 \frac{1}{2g} (1 - C_e)$$

$$1.99 + \left(\frac{8.5}{3(1.99)}\right)^2 \frac{1}{2(9.81)} (1 - 0.5) = y_3 + \left(\frac{8.5}{1.45 y_3}\right)^2 \frac{1}{2(9.81)} (1 - 0.5)$$

$$2.042 = y_3 + \frac{0.876}{y_3^{\,2}}$$

$$y_3 = 1.76 \text{ m}$$

Then, returning to Step 6, y_1 using Eq. (20.4):

$$1.76 + (1 + 0.3) \left(\frac{8.5}{1.45(1.76)}\right)^2 \frac{1}{2(9.81)} = y_1 + (1 + 0.3) \left(\frac{8.5}{3y_1}\right)^2 \frac{1}{2(9.81)}$$

$$2.50 = y_1 + \frac{0.53}{y_1^2}$$

$$y_1 = 2.41 \text{ m}.$$

y_1 is 0.42 m above the normal depth. Because a free board of 0.6 m is available; the design is OK.

Step 8. Determine transition length using Eq. (20.5) for a 4.5:1 flare.

$$L_T = 4.5 \frac{(W_1 - W_2)}{2} = 4.5 \frac{(3 - 1.45)}{2} = 3.49 \text{ m, or } 3.5 \text{ m.}$$

Step 9. Compute the water surface profile.

20.1.2 Culverts with Inlet Control

Culverts with inlet control require the transitioning of supercritical flow. (see Fig. 20.7) Because supercritical flow is difficult to control without creating a hydraulic jump or other surface irregularities, the full flow area should be maintained. Because changing the flow smoothly requires a long structure, model studies should be performed to determine the transition geometry if a hydraulic jump is not desired. When a hydraulic jump is acceptable,

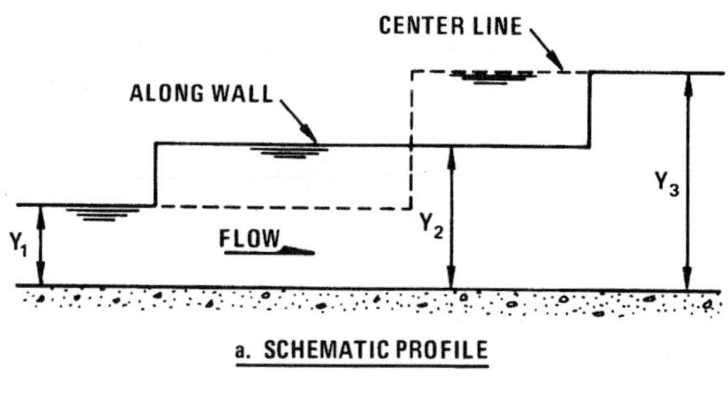

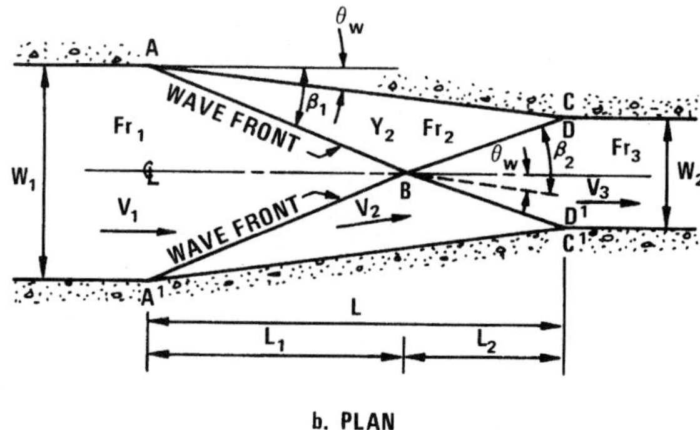

FIGURE 20.7   Supercritical inlet transition for rectangular channel. (From USACE, 1970)

Figs. 20.8 and 20.9 can be used (USACE, 1970). Such a design requires a rectangular channel and a long transition.

The design procedure for supercritical flow contractions is as follows (Corry et al, 1975):

1. Flow conditions for the approach channel should be computed assuming normal flow (y_n, V_n, Fr) using Manning's equation or other design aids.

2. Approach sections that are not rectangular should have a transition to the rectangular section with a bottom width of W_1. This bottom width should be approximately equal to the average of the water surface top width (T) and the trapezoidal section base width (B_w): i.e., $W = (T + B_w)/2$. Compute the normal flow condition for this rectangular section using Manning's equation.

3. Assume a trial culvert width W_2. Refer to Figure 20.7 for a definition of parameters.

4. Compute the contraction length needed to reduce W_1 to W_2. This is accomplished by varying the contraction using wall angle (θ_w) until $L = (W_1 - W_2)/2 \tan \theta_w$ is equal to $L_1 + L_2$ where $L_1 = W_1/2 \tan \beta_1$ and $L_2 = W_2/2 \tan(\beta_2 - \theta_w)$. To minimize surface disturbances, L should be equal to $L_1 + L_2$ (Corry et al, 1975).
   Choose θ_w.

   a. Compute $L = (W_1 - W_2)/2 \tan \theta_w$
   b. Find β_1, y_2/y_1, and Fr_1 from

$$\tan \theta_w = \frac{\tan \beta_1(\sqrt{1 + 8Fr_1^2 \sin^2 \beta_1} - 3)}{2 \tan^2 \beta_1 + \sqrt{1 + 8Fr_1^2 \sin^2 \beta_1} - 1} \tag{20.7}$$

and

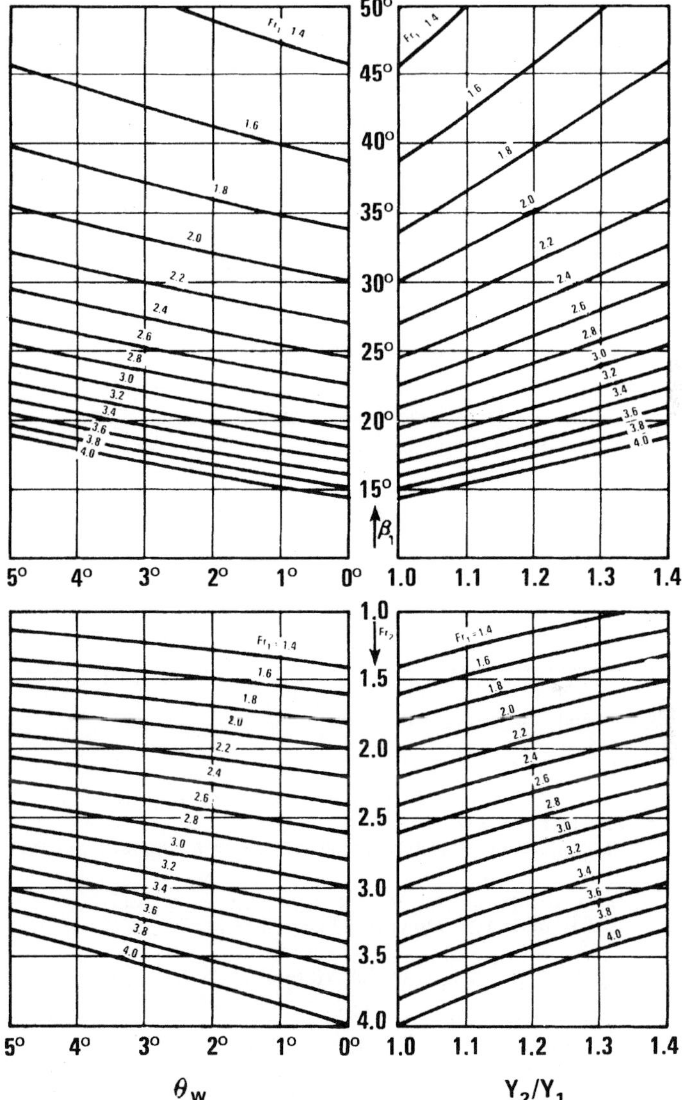

FIGURE 20.8 Supercritical inlet transition design curves for rectangular channels. (USACE, 1970)

$$\frac{y_2}{y_1} = \frac{1}{2}\left[\sqrt{1 + 8Fr_1^2 \sin^2 \beta_1} - 1\right] \tag{20.8}$$

and

$$Fr_2^2 = \left(\frac{y_1}{y_2}\right)\left[Fr_1^2 - \left(\frac{y_1}{2y_2}\right)\left(\frac{y_2}{y_1} - 1\right)\left(\frac{y_2}{y_1} + 1\right)^2\right] \tag{20.9}$$

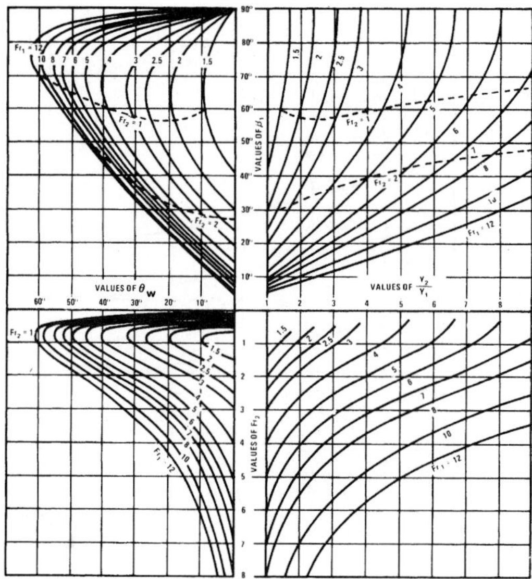

FIGURE 20.9   Supercritical inlet transition design curves for rectangular channels. (From Ippen, 1951)

Alternatively, Figs. 20.8 or 20.9 can be used to approximately solve Eqs. (20.7), (20.8), and (20.9).

c. Calculate $L_1 = W_1/2 \tan \beta$.

d. Compute β_2, y_3/y_2, and Fr_2 using the same procedure as in Step 4b (that is, using β_2 for β_1, Fr_2 for Fr_1, y_3/y_2 for y_2/y_1 and the same θ_w).

e. Calculate $L_2 = W_2/2 \tan (\beta_2 - \theta_w)$.

f. Compare L with $L_1 + L_2$ if $L > L_1 + L_2$, then increase θ_w and repeat Steps 4a–f until $L = L_1 + L_2$.

g. Compute y_3 from $y_3 = y_1 \left(\dfrac{y_2}{y_1}\right)\left(\dfrac{y_3}{y_2}\right)$.

5. Compare depth y_3 with the width W_2. Culvert should be of a standard dimension. If not return to Step 3 using another W_2 and repeat the design process until a better combination of y_3 and W_2 are found.

Example 20.3 (adapted from Corry et al., 1975)
Design the transition contraction and select the culvert size for a discharge of 300 cfs in a trapezoidal channel (6 ft bottom width, 2:1 side slope, $S_0 = 0.02$ ft/ft and $n = 0.012$)

Solution
Step 1. The normal depth, velocity, and Fr are $y_n = 1.67$ ft, $V_n = 19.2$ ft/s, and

$$Fr = \frac{V}{\sqrt{gA/T}} = \frac{19.2}{\sqrt{32.2(15.6)/12.7}} = 3.05\text{m}$$

Step 2. Because $(T + B_w)/2 = (12.7 + 6)/2 = 9.4$ ft, use $W_1 = 10$ ft, a rectangular channel. Use Manning's equation to compute $y_n = 1.54$ ft and $V_n = 19.5$ ft/s, then compute the Fr:

$$Fr = \frac{V}{\sqrt{gy}} = \frac{19.5}{\sqrt{32.2(1.54)}} = 2.77.$$

Step 3. Assume a trial culvert width $W_2 = 5$ ft.

Step 4. Try $\theta_w = 14°$ for Fr $= 2.8$:

$\quad$ L $= (10 - 5)/2 \tan 14° = 10$ ft

$\quad \beta_1 = 35°$, $y_2/y_1 = 1.8$, Fr$_2 = 1.8$

$\quad$ L$_1 = 10/2 \tan 35° = 7.1$ ft.

$\quad \beta_2 = 55°$, $y_3/y_2 = 1.6$, Fr$_3 = 1.1$

$\quad$ L$_2 = 5/2 \tan(55° - 14°) = 2.9$ ft.

$\quad$ L$_1$ + L$_2 = 10$ ft $= L$, O.K.

$\quad y_3 = 1.54(1.6)(1.8) = 4.4$ ft.

$\quad$ use $\theta_w = 14°$, $y_3 = 4.4$ ft, $V_3 = 13.6$ ft/s, Fr$_3 = 1.1$

Step 5. Because $y_3 = 4.4$ ft. and $W_3 = 5$ ft, a 5 × 5 box culvert will be satisfactory.

20.2 ENERGY DISSIPATION FOR CULVERTS AND CHANNELS

20.2.1 Hydraulic Jump Basins

For supercritical flow expansions, the procedure outlined in Sec. 20.1 is applicable if the exit Fr is less than 3, if the location where the flow conditions are desired within three culvert diameters of the outlet and if S_o is less than 10 percent (Corry, et al 1975).For expansions outside these limits, a hydraulic jump basin (Fig. 20.10) should be used. This type of basin allows the flow to expand, drop or both, resulting in depth decreases, velocity increases, and an Fr increase in. Higher Fr's result in more efficient jumps and shorter basins.

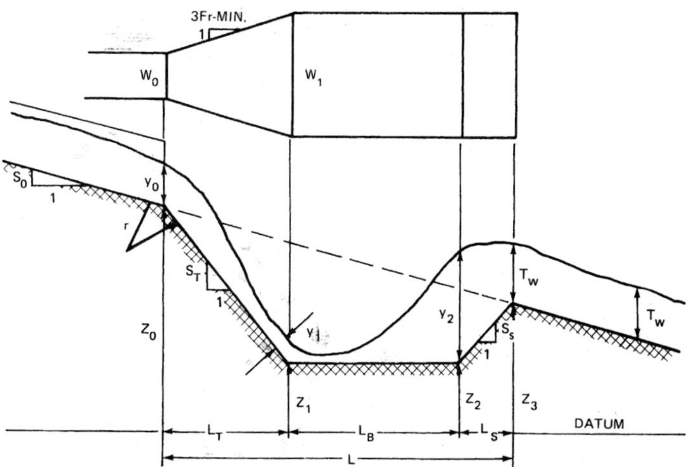

FIGURE 20.10 Definition sketch basin transition. (From Corry et al., 1975)

The design procedure for supercritical flow expansions with hydraulic jump basin is as follows (Corry, et al; 1975):

1. Compute the culvert brink depth y_o using Figs. 20.4 or 20.5.
2. Compute the tailwater depth T_w in the downstream channel assuming normal flow (using Manning's equation) or perform backwater analysis.
3. To determine the basin elevation, first select Z_1 and then use the following steps.
   a. Select basin width W_B (W_1 in Fig. 20.10) and basin slopes S_s and S_T (Fig. 20.10). Slope of S_s or $S_T = 0.5(2:1)$ or 0.33 $(3:1)$ are satisfactory (Corry et al, 1975).
   b. Check W_B using

$$W_B < W_0 + \left[\frac{2L_T\sqrt{S_T^2 + 1}}{3Fr_0}\right] \tag{20.10}$$

   where $L_T = (Z_0 - Z_1)/S_T$ and the right-hand side is the limit that flares naturally in the slope distance L.
   c. Compute y_1 using the following equation derived from the energy equation from the culvert outlet brink to the basin (Sec. 1 in Fig. 20.10). Use $V_1 = Q/y_1W_B$, to determine y_1 and then V_1:

$$Q = y_1 W_B[2g(Z_0 - Z_1 + y_0 - y_1) + V_0^2]^{1/2} \tag{20.11}$$

   d. Compute the Fr

$$Fr_1 = \frac{V_1}{\sqrt{gy_1}}$$

   e. Compute y_2 using Eq. (20.12) for the hydraulic jump basin:

$$y_2 = \frac{y_1}{2}[\sqrt{1 + 8Fr_1^2} - 1] \tag{20.12}$$

   f. Compute Z_3 from geometry using

$$Z_3 = \frac{\left[Z_0 - \left(L_T + L_B - \dfrac{Z_2}{S_s}\right)S_0\right]}{\left(\dfrac{S_o}{S_s} + 1\right)} \tag{20.13}$$

   where

$$L_T = \frac{(Z_0 - Z_1)}{S_T}$$

   and from Fig. 20.11, determine L, which is defined as L_B

$$L_B = f(y_1, Fr_1).$$

   g. Check value of Z_1 by computing $y_2 + Z_2$ and $Z_3 + T_w$. If $y_2 + Z_2 > Z_3 + T_w$, select another Z_1 and repeat steps 4a to 4g until a balance is reached.
4. Compute L_S and L:

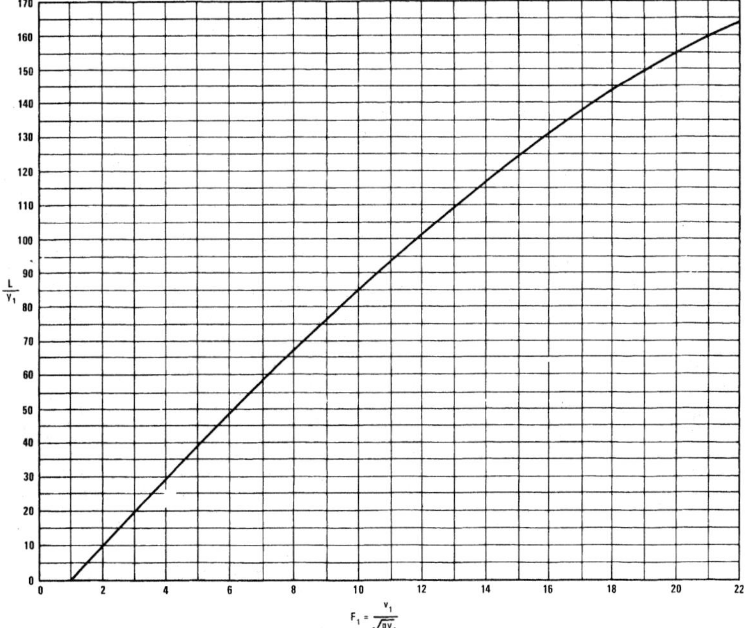

FIGURE 20.11   Length of jump in terms of y_1, rectangular channel. (From Corry et al., 1975)

$$L_S = \frac{(Z_3 - Z_2)}{S_s}$$

$$L = L_T + L_B + L_S = \frac{(Z_0 - Z_3)}{S_o}$$

Example 20.4
A supercritical flow expansion is to be designed for a reinforced concrete box culvert measuring 3 by 2 m. Determine the dimensions for the hydraulic jump basin using a design discharge of $Q = 11.8$ m³/s. The slope is 6.5 percent, the invert outlet elevation is 30.5 m, the downstream channel has a bottom width of 3 m and side slopes of 2:1, and Manning's $n = 0.03$. The brink depth is supercritical, $y_0 = 0.457$ m and $V_0 = 8.47$ m/s. $S_s = 0.5$ and $S_T = 0.5$. (Refer to Figure 20.12)

Solution
Step 1. The brink depth and velocity are given: $Fr_0 = 4.0$.

Step 2. The tailwater depth is computed assuming normal depth. Manning's equation is solved to obtain $y_n = 0.57$ m tailwater depth and $V_n = 4.85$ m/s. (Refer to Fig. 20.12)

Step 3. Assuming a basin elevation of $Z_1 = 25.9$ m:

a. Select $W_B = 3$ m and $S_s = S_T = 0.5$
b. Check W_B using Eq. 20.10, where $W_0 = 3$ m,

$$W_B = 3 < 3 + [2L_T\sqrt{S_T^2 + 1}]/(3Fr_0). \text{ OK.}$$

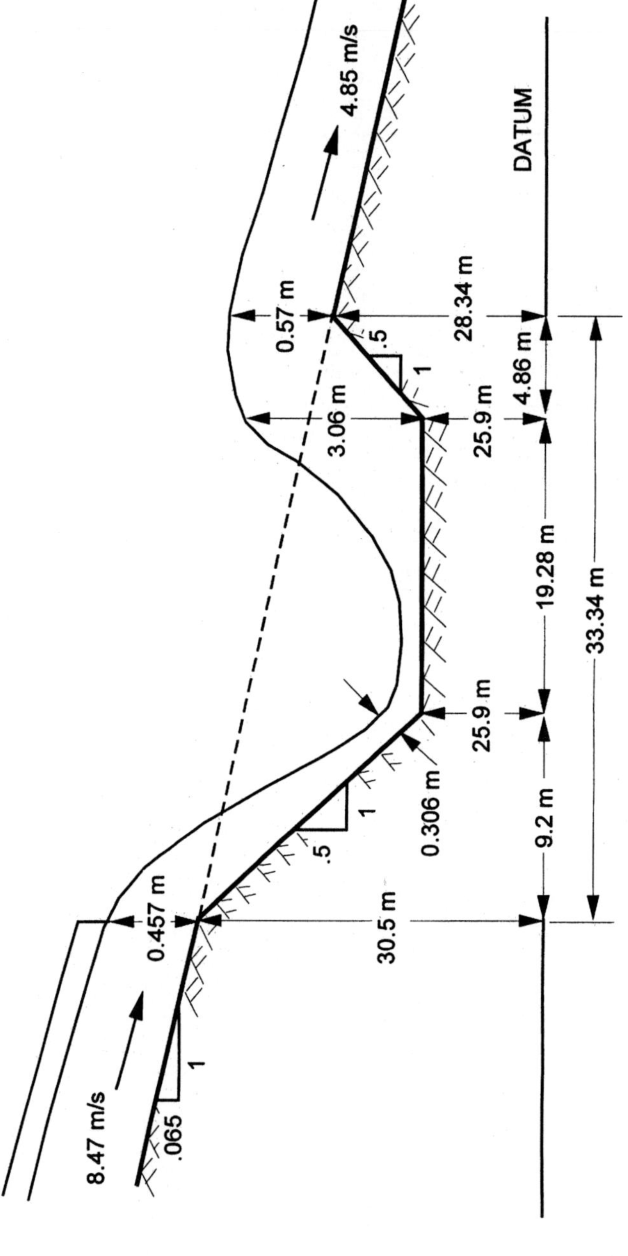

FIGURE 20.12 Example problem 18.2.1 hydraulic jump basin. (Net to scale.)

c. Compute y_1 using Eq. (20.11)

$$11.8 = y_1(3)[2(9.81)(30.5 - 25.9 + 0.457 - y_1) + (8.47)^2]^{1/2}$$

Solving $y = 0.306$ m and $V_1 = Q/A = 11.8/[3(0.306)] = 12.85$ m/s.

d. The Froude number is Fr $= 12.85/\sqrt{9.81(0.306)} = 7.42$

e. Compute y_2 using Eq. (20.12). $y_2 = 0.306[\sqrt{1 + 8(7.42)^2} - 1]/2 = 3.06$ m.

f. Compute Z_3 using Eq. (20.15). First compute $L_T = (Z_0 - Z_1)/2 = (30.5 - 25.9)/0.5 = 9.2$ m. From Fig. 20.11, $L_B/y_1 = 63$, so $L_B = 63y_1 = 63(0.306) = 19.28$ m, then

$$Z_3 = \frac{[30.5 - (9.2 + 19.28 - 25.9/0.5)0.065}{(0.065/0.5 + 1)} = 28.34 \text{ m}$$

g. Check value of Z_1: $y_2 + Z_2 = 3.06 + 25.9 + 28.96$ and $Z_3 + T_W = 28.34 + 0.58 = 28.92$, $y_2 + Z_2 > Z_3 + T_W$; therefore, $Z_1 = 25.9$ is OK.

Step 4. Compute L_s and then L. $L_s = (Z_3 - Z_2)/S_s = (28.33 - 25.9)/0.5 = 4.86$ m. Then

$$L = L_T + L_B + L_s = 9.2 + 19.28 + 4.86 = 33.34 \text{ m. Refer to Fig. 20.12}$$

20.2.2 Forced Hydraulic Jump Basins

20.2.2.1 Saint Anthony Falls Stilling Basin. The Saint Anthony Falls, (SAF) stilling basin is a generalized design based upon model studies conducted by the U.S. Soil Conservation Service at the St. Anthony Falls Hydraulic Laboratory. University of Minnesota. Figure 20.13 illustrates the SAF stilling basin design which is recommended for small structures such as spillway outlet works, and for canals where the Fr ranges from 1.7 to 17 (at the dissipator entrance). Through the use of chute blocks, baffle or floor blocks, and an end sill, the basin length is about 80 percent of the free hydraulic jump length.

The design procedure for SAF basins is as follows (Corry et al 1975):

1. Choose basin configuration and flare dimension, Z. (Refer to Fig. 20.13)

2. Use the design procedure presented in Sec. 20.2.1 for supercritical expansions into hydraulic jump basins to determine basin width (W_B), elevation (Z_1), length (L_B), total length (L), incoming depth (y_1), incoming Fr (Fr_1), and jump height (y_2). Steps 3e and 3f in Sec. 20.2.1 are modified; for Step 3e, determine y_2 using the sequent depth y_j:

$$y_j = \frac{y_1}{2} [\sqrt{1 + 8Fr_1^2} - 1] \tag{20.14}$$

$$y_2 = \left(1.1 - \frac{Fr_1^{\,2}}{120}\right) y_j \qquad Fr_1 = 1.7 \text{ to } 5.5 \tag{20.15}$$

$$y_2 = 0.85y_j \qquad Fr_1 = 5.5 \text{ to } 11 \tag{20.16}$$

$$y_2 = \left(1.0 - \frac{Fr_1^2}{800}\right) y_j \qquad Fr_1 = 11 \text{ to } 17 \tag{20.17}$$

For Step 3f, compute L_B:

$$L_B = \frac{4.5y_j}{Fr_1^{o76}} \tag{20.18}$$

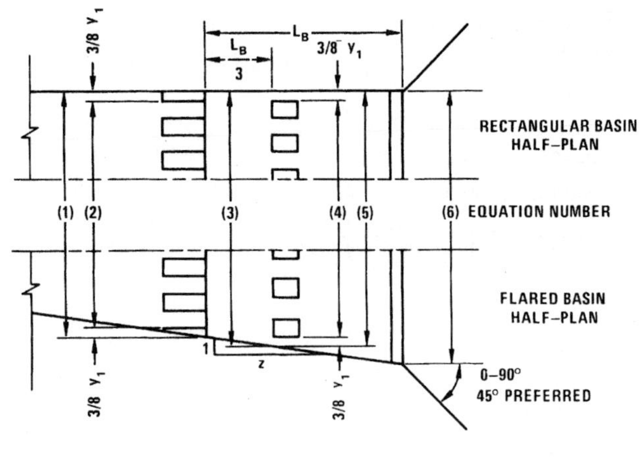

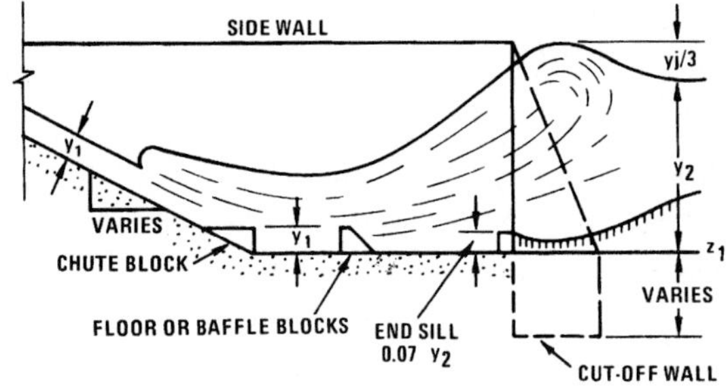

(1) W_B = BASIN WIDTH UPSTREAM

(2) n BLOCKS AT 3/4 y_1 ±

(3) $0.40\,W_{B2} \leq$ AGGREGATE BLOCK WIDTH $\leq 0.55\,W_{B2}$

(4) $\dot{n}$ BLOCKS AT 3/4 y_1 $\dfrac{W_{B2}}{W_B}$ ±

(5) $W_{B2} = W_B + 2L_B/3z$

(6) $W_{B3} = W_B + 2L_B/z$

FIGURE 20.13   St. Anthony Falls stilling basin. (From Blaisdell, 1959)

3. Determine the dimensions of the chute block:

Height: $y_1 = y_1$

Width: $W_2 = 0.75y_1$ and $W_1 =$ spacing

Number: $N_c \approx \dfrac{W_b}{2W_1}$ (rounded)

Adjusted: $W_1 = W_2 = \dfrac{W_B}{2nN_c}$ (N_c includes the 1/2 block at each wall)

4. Determine the dimensions of the baffle block:

Height: $h_3 = y_1$

Width: $W_3 =$ spacing and $W_4 = 0.75y_1$

Basin width at baffle blocks: $W_{B2} = W_B + 2L_B/3Z$

Number of blocks: $N_B = W_{B2}/2W_3$ rounded

Adjusted $W_3 = W_4 = W_{B2}/2N_B$

Check total block width to insure that at least 40 to 55 percent of W_{B2} is occupied by blocks.

Distance from chute blocks to baffle blocks $= L_B/3$

5. End sill height: $h_4 = 0.07y_j$.

6. Side wall height: $y_2 + y_j/3$.

Example 20.5
Determine the dimensions of an SAF basin for the supercritical flow expansion described in Example 20.3.

Solution
Step 1. Select a rectangular basin with no flare.

Step 2. Steps 1 through 3(a–f) (in Example 20.3) for a supercritical flow expansion into a hydraulic jump basin.

Given $V_0 = 8.47$ m/s, $y_0 = 0.457$ m, and $Fr_0 = 4.0$.

The tailwater depth $T_W = y_n = 0.57$ m and $V_n = 4.85$ m/s.

Assume that $Z_1 = Z_0 = 30.5$ m. (Refer to Fig. 20.14)

a. Compute y_j using Eq. 20.14 with $y_1 = y_0 = 0.457$ m and $Fr_1 = Fr_0 = 4.0$. This assumes that $Z_1 = Z_0$: i.e., the basin floor is the same elevation as the culvert outlet.

$$y_j = \frac{y_1}{2} [\sqrt{1 + 8Fr_1^2} - 1$$

$$= \frac{9.457}{2} [\sqrt{1 + 8(4)^2} - 1]$$

$$= 2.37 \text{ m}.$$

b. Next, use Eq. (20.15) for $Fr_1 = 1.7$ to 5.5 to compute y_2:

$$y_2 = \left(1.1 - \frac{Fr_1^2}{120}\right) y_1$$

$$= \left(1.1 - \frac{(4)^2}{120}\right) (2.37)$$

$$= 2.29 \text{ m}$$

Because $y_2 > T_W = 0.57$, we can lower the elevation of the basin. Use $Z_1 = 27.9$ m with $W_B = 3$ m and $S_T = S_s = 0.5$. W_B is OK and no flare is used.

c. Compute y_1 using the energy Eq. (20.11). The basin has been lowered so now y_1 is not y_0, the brink depth.

$$Q = y_1 W_B[2g(Z_0 - Z_1 + y_0 - y_1) + V_0^2]^{1/2}$$

$$11.8 = y_1(3)[2(9.81)(30.5 - 27.9 - 0.457 - y_1) + (8.47)^2]^{1/2}$$

Solving $y_1 = 0.348$ m. $V_1 = 11.8/(3 \times 3 \times 0.348) = 11.3$ m/s

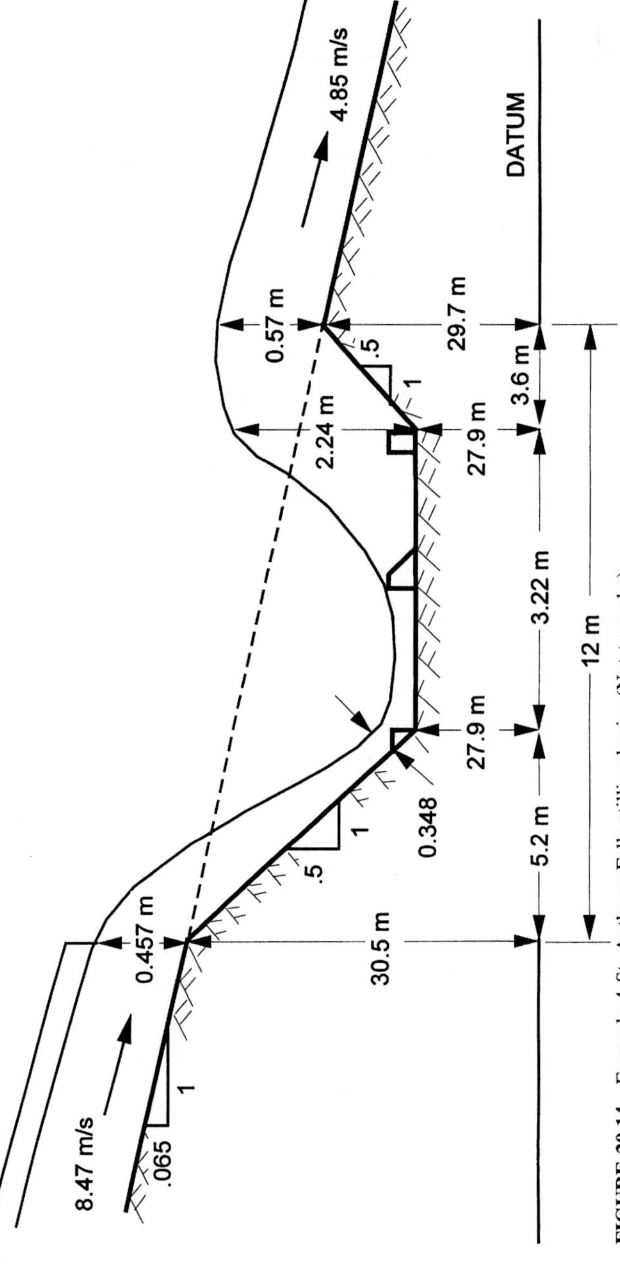

FIGURE 20.14 Example 4 St. Anthony Falls stilling basin. (Not to scale.)

d. The $Fr_1 = V_1/\sqrt{gy_1} = 11.3/\sqrt{9.81(0.348)} = 6.1$

e. Compute $y_j = \dfrac{y_1}{2}[\sqrt{1 + 8Fr_1^2} - 1] = \dfrac{0.348}{2}[\sqrt{1 + 8(6.1)^2} - 1] = 2.83$ m, and

$$y_2 = \left(1.1 - \frac{Fr_1^2}{120}\right) y_i = \left(1.1 - \frac{(6.1)^2}{120}\right) 2.83 = 2.24\text{m}.$$

f. Compute L_B using Eq. (20.18):

$$L_B = 4.5y_j/Fr_1^{0.76} = 4.5(2.83)/6.1^{0.76} = 3.22 \text{ m}.$$

Compute L_T using $L_T = (Z_0 - Z_1)/S_T = (30.5 - 27.9)/0.5 = 5.2$ m.
Compute Z_3 using Eq. (20.13)

$$Z_3 = \frac{Z_0 - (L_T + L_B - Z_2/S_s)S_0}{(S_0/S_s + 1)}$$

$$= \frac{30.5 - (5.2 + 3.22 - 27.9/0.5)0.065}{(0.065/0.5 + 1)}$$

$$= 29.7$$

g. Check the assumption of $Z_1 = 27.9$ m: $y_2 + Z_2 = 2.24 + 27.9 = 30.14$ and $T_W + Z_3 = 0.57 + 29.7 = 30.2$ m: $y_2 + Z_2 \approx T_W + Z_3$. Because they are so close, then Z_1 will be OK.

h. $L_T = 5.2$ m, $L_B = 3.22$ m, $L_s = (Z_3 - Z_2)/S_s = (29.7 - 27.9)/0.5 = 3.6$ m, and

$$L = L_T + L_B + L_s = 5.2 + 3.22 + 3.6 = 12 \text{ m}.$$

Step 3 Chute blocks: $h_1 \approx y_1 = 0.35$, $W_1 \approx 0.75y_1 = 0.26$ m $= W_2$. $N_c = W_B/2W_1 = 3/(2 \times 0.26) = 5.77$ so use 6 blocks. Adjusted $W_1 = W_B/(2N_c) = 3/(2 \times 6) = 0.25$ m. This provides 5 blocks, 6 spaces, and a half block at each wall.

Step 4. Baffle blocks: $h_1 \approx y_1 = 0.35$ m, $W_3 \approx 0.75y_1 = 0.26$ m $= W_4$. Basin width, $W_{B2} = W_B + 2L_B/(3Z) = 3 + 0 = 3$ m (no flare), $N_B = W_{B2}/(2W_1) = 3/(2 \times 0.26) = 5.77$ so use 6 blocks. Adjusted $W_3 = W_4 = 3/(2 \times 6) = 0.25$ m. Total block width $= 6(0.25) = 1.5$ m. Check percentage: $0.4 < 1.5/3.0 < 0.55$, OK. This provides six blocks, five spaces, and a half-space at each wall. Distance from chute block $= L_B/3 = 3.22/3 = 1.07$ m.

Step 5. End sills: $h_4 = 0.07y_j = 0.07(2.83) = 0.2$ m.

Step 6. Side walls: height $= y_2 + y_j/3 = 2.24 + 2.83/3 = 3.2$ m. Refer to Fig. 20.14 with the dimension for the chute block shown.

20.2.2.2 Type II, III, and IV Basins.   The U.S. Bureau of Reclamation's Type II, III, and IV basins are illustrated in Chapter 18, in Figs. 18.3, 18.4 and 18.7 respectively.

Type II Basin Design.   Use the design procedure presented in Sec. 20.2.1 for supercritical expansion into a hydraulic jump basin to determine W_B, Z_1, L_B, L, y_1, Fr_1 and y_2.

For Step 3e in that section, use $C = 1.1$ to find $y_2 = C_1 y_1[\sqrt{1 + 8Fr_1^2} - 1]/2$. For Step 3f, use Fig. 20.15 to determine L_B.

Determine the dimensions for the chute blocks and dentated sill height using the relations in Fig. 18.3.

Type III Basin Design.   Use the design procedure in Sec. 20.2.1 to determine basin dimension. For Step 3e use $C = 1.0$ to determine $y_2 = C_1 y_1[\sqrt{1 + 8Fr_1^2} - 1]/2$. For Step 3f, use Fig. 20.15 to determine L_B. Use dimensions in Fig. 20.16 to determine chute block dimensions and spacing.

Use Fig. 18.4 to determine dimensions for baffle blocks and the end sill height.

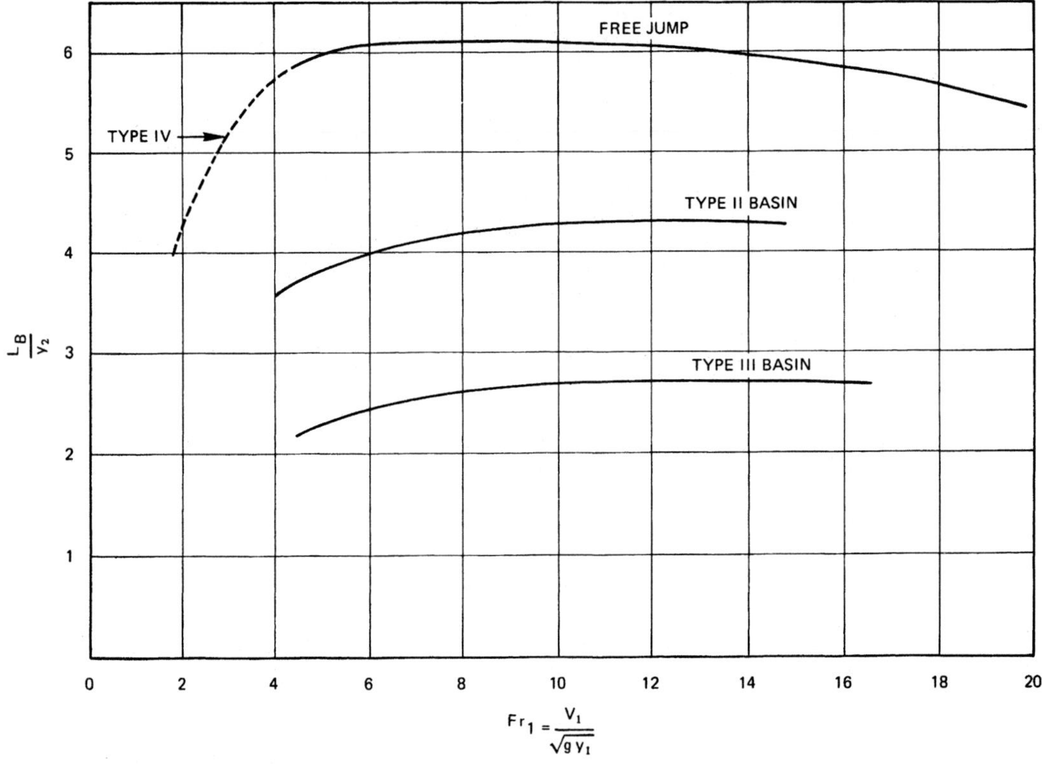

$$Fr_1 = \frac{V_1}{\sqrt{g\,y_1}}$$

FIGURE 20.15   U.S. Bureau of Reclamation Type II basin. (Corry et al, 1975)

Type IV Basin Design.   Use the same design procedure presented in Sec. 20.2.1 for supercritical expansions into a hydraulic jump basin to determine the dimensions the basin. For Step 3e in the section, use $C_1 = 1.0$ to determine $y_2 = C_1 y_1 [\sqrt{1 + 8Fr_i^2} - 1]/2$. For Step 3f, use Fig. 20.15 to determine L_B.

Use dimensions in Fig. 18.7 to determine chute block dimensions and spacing.

Use Fig. 18.7 to determine the end sill height.

Example 20.6
Determine the dimensions of U.S. Bureau of Reclamation's Type II basin for the supercritical flow expansion described in Example 20.3.

Solution
Step 1. Use the design procedure for a supercritical flow expansion into a hydraulic jump basin. (Refer to Examples 20.3 and 20.4).

Given $V_0 = 8.47$ m/s, $y_0 = 0.457$ m, and $Fr_0 = 4.0$. (Figure 20.17)

The tailwater depth is $T_w = y_n = 0.57$ m, and $V_n = 4.85$ m/s.

Assume $Z_1 = Z_0 = 30.5$ m, compute y_2

$$y_2 = C_1 y_1 [\sqrt{1 + 8Fr_0^2} - 1]/2$$

$$= 1.1(0.457)[\sqrt{1 + 8(4)^2} - 1]/2$$

$$= 2.6 \text{ m}$$

$y_2 > T_w (2.6 > 0.57)$. Therefore we need to lower elevation Z_1 of the basin floor.

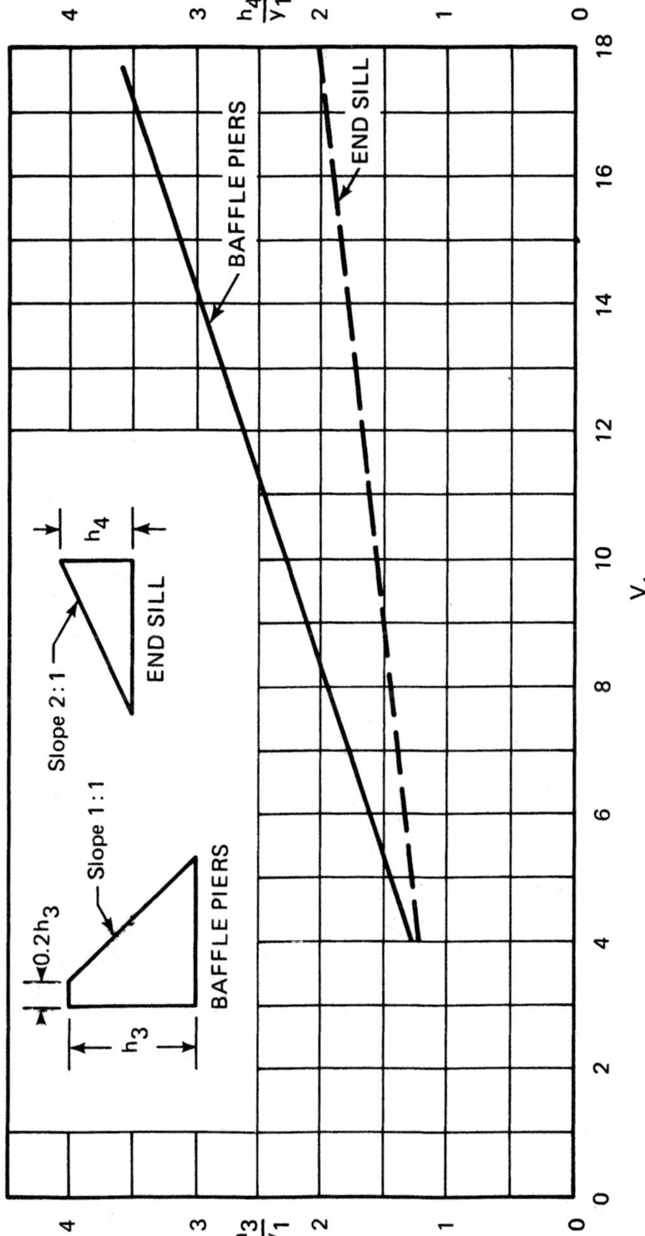

$$Fr_1 = \frac{V_1}{\sqrt{g\,y_1}}$$

FIGURE 20.16 Height of baffle piers and end sill (Type III basin). (From U.S. Bureau of Reclamation, 1987)

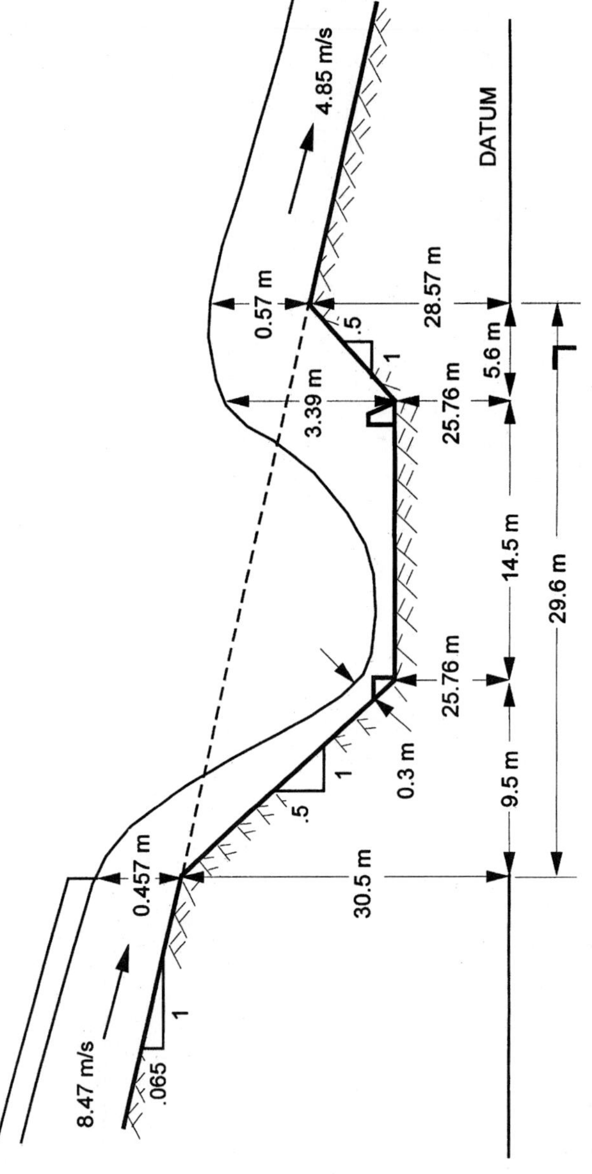

FIGURE 20.17 Example 5 U.S. Barean of Reclamation type II basin.

a. Use a basin floor elevation of $Z_1 = Z_2 = 25.76$ m, with $W_B = 3$ m, $S_T = S_s = 0.5$.

b. W_B is OK, no flare.

c. Compute y_1 using energy Eq. (20.11)

$$Q = y_1 W_B [2g(Z_0 - Z_1 + y_0 - y_1) + V_0^2]^{1/2}$$

$$11.8 = y_1(3)[2(9.81)(30.5 - 25.76 + 0.457 - y_1) + 8.47^2]^{1/2}$$

Solving $y_1 = 0.3$ m and $V_1 = 13.1$ m/s.

d. The $Fr_1 = 13.1/\sqrt{9.81(0.3)} = 7.6$

e. Compute $y_2 = 1.1(0.3)[\sqrt{1 + 8(7.6)^2} - 1]/2 = 3.39$ m.

f. Using Fig. 20.15, $L_B/y_2 = 4.3$, so $L_B = 4.3(3.39) = 14.6$ m:

$$L_T = (Z_0 - Z_1)/S_T = (30.5 - 25.76)/0.5 = 9.5 \text{ m.}$$

Using Eq. (20.13), compute Z_3

$$Z_1 = \frac{[30.5 - (9.5 + 14.6 - 25.76/0.5)0.0651}{(0.065/0.5 + 1)} = 28.57 \text{ m.}$$

g. Check Z_1, $y_2 + Z_2 = 3.39 + 25.76 = 29.15$ and $Z_3 + T_W = 28.57 + 0.57 = 29.14$, OK

h. Compute $L_s = (Z_3 - Z_2)/S_s = (28.57 - 25.76)/0.5 = 5.6$ m and

$$L = L_T - L_B + L_s = 9.5 + 14.6 + 5.6 = 29.6 \text{ m.}$$

Step 2. Chute blocks: $h_1 = W_1 = W_2 = y_1 = 0.3$ m, $N_C = 3/2(0.3) = 5$, OK. Side-wall spacing $= y_1/2 = 0.15$ m.

Step 3. Dentated sill: $h_2 = 0.2y_2 = 0.2(3.39) \approx 0.7$ m.

$$W_3 = W_4 = 0.15y_2 = 0.15(3.39) = 0.51 \text{ m.}$$

$$N_s = W_B/W_3 = 3/0.51 \text{ m} = 5.88.$$

Use 5, which provides three blocks and two spaces, each of which is 0.6 m wide. Refer to Fig. 20.17 for the dimensions of the basin.

20.2.3 Impact-Type Energy Dissipation (USBR Type VI Basin)

Figure 20.18 illustrates the U.S, Bureau of Reclamation's Type VI impact-type energy dissipator, which can be used with culverts. This basin is contained in boxlike structures requiring no tail water for operation. The structures can be used for open channels as well as culverts, and the basin can be used at sites where the entrance velocity to the basin does not exceed 50 ft/s, and the discharge is less than 400 cfs. This dissipator should not be used if the buildup of debris or ice can cause substantial clogging.

The design procedure is as follows (Corry et al, 1975).

1. Compute the flow area at the end of the culvert using the maximum design discharge and velocity. Compute the equivalent depth of flow entering the dissipator from the culvert as

$$y_e = \left(\frac{A}{2}\right)^{1/2} \tag{20.19}$$

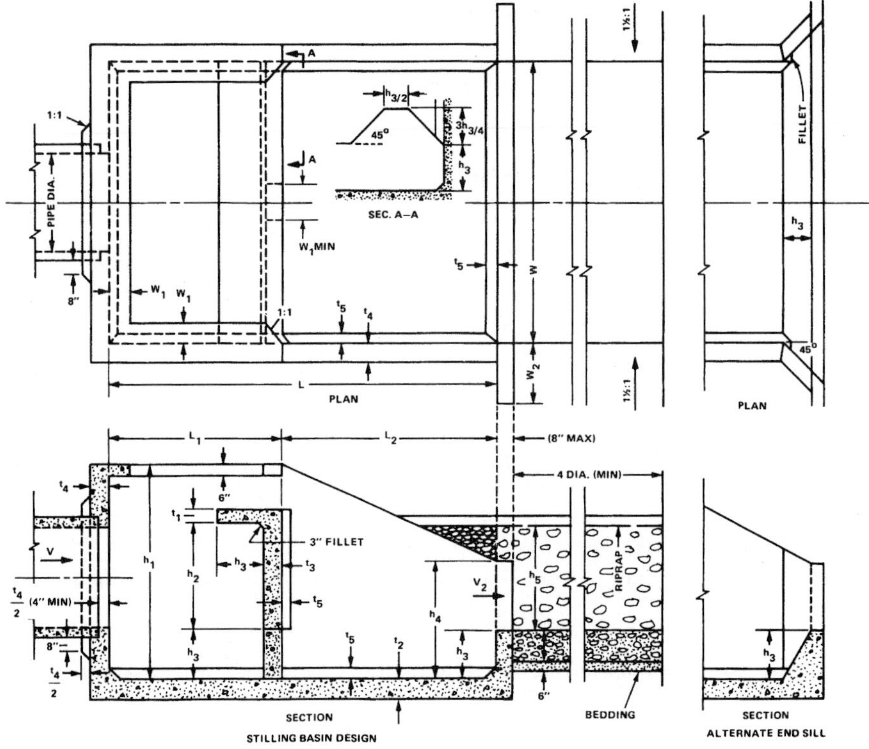

FIGURE 20.18   Baffle-wall energy dissipator of a U.S. Bureau of Reclamation type IV basin. (From Corry et al., 1975)

where A is the cross-sectional area of flow in the culvert. This converts the cross-sectional area of flow of a pipe into an equivalent rectangular cross section with a width twice the depth of flow. The culvert preceding the dissipator can be open, closed, or have any cross section. This approach ignores the size and shape of the culvert entirely except for the determination of flow entering the dissipator.

2. Compute Fr and the energy at the end of the culvert H_o:

$$H_0 = y_e + \frac{V_0^2}{2g}.$$

Then use Figure 20.19 to determine the basin width. Enter Figure with Fr to determine H_o/W then $W = H_o/(H_o/W)$.

3. Use Table 20.2 to determine the dimensions of the dissipator structure.

20.2.4 Drop Structures

20.2.4.1 Straight-Drop Spillway.   The straight-drop spillway shown in Figs 20.20 and 20.21 is generally used for subcritical flow in the upstream as well as the downstream channel. To describe the flow geometry, the following *drop number* is used:

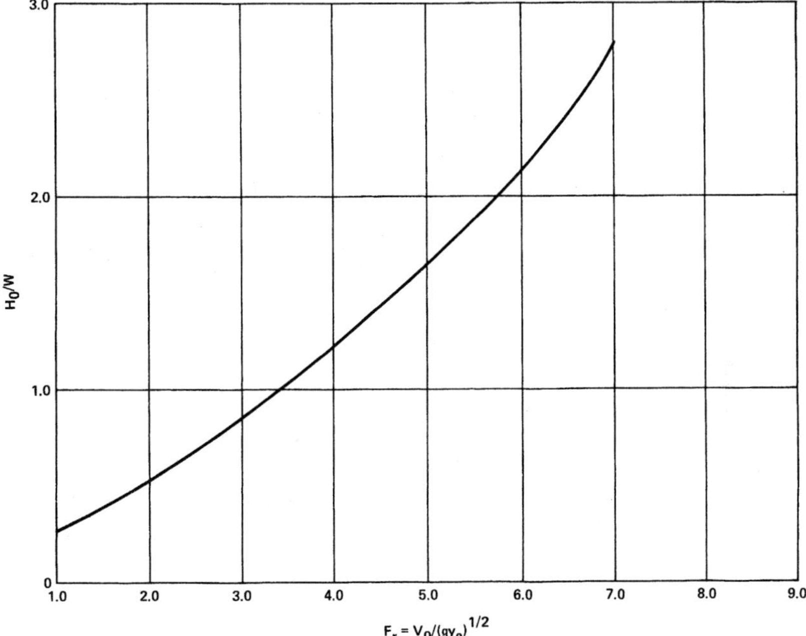

FIGURE 20.19   Design curve for a baffle-wall dissipator. (Corry et al., 1975)

TABLE 20.2   Baffle Wall Dissipator: Dimensions of The Basin in Feet and Inches

						Dimensions of basin in feet and inches								
W	h_1	L	h_2	h_3	L_1	L_2	h_4	W_1	W_2	t_3	t_2	t_1	t_4	t_5
4–0	3–1	5–5	1–6	0–8	2–4	3–1	1–8	0–4	1–1	0–6	0–6	0–6	0–6	0–3
5–0	3–10	6–8	1–11	0–10	2–11	3–10	2–1	0–5	1–5	0–6	0–6	0–6	0–6	0–3
6–0	4–7	8–0	2–3	1–0	3–5	4–7	2–6	0–6	1–8	0–6	0–6	0–6	0–6	0–3
7–0	5–5	9–5	2–7	1–2	4–0	5–5	2–11	0–6	1–11	0–6	0–6	0–6	0–6	0–3
8–0	6–2	10–8	3–0	1–4	4–7	6–2	3–4	0–7	2–2	0–7	0–7	0–6	0–6	0–3
9–0	6–11	12–0	3–5	1–6	5–2	6–11	3–9	0–8	2–6	0–8	0-7	0–7	0–7	0–3
10–0	7–8	13–5	3–9	1–8	5–9	7–8	4–2	0–9	2–9	0–9	0–8	0–8	0–8	0–3
11–0	8–5	14–7	4–2	1–10	6–4	8–5	4–7	0–10	3–0	0–9	0–9	0–8	0–8	0–4
12–0	9–2	16–0	4–6	2–0	6–10	9–2	5–0	0–11	3–0	0–10	0–10	0–8	0–9	0–4
13–0	10–0	17–4	4–11	2–2	7–5	10–0	5–5	1–0	3–0	0–10	0–11	0–8	0–10	0–4
14–0	10–9	18–8	5–3	2–4	8–0	10–9	5–10	1–1	3–0	0–11	1–0	0–8	0–11	0–5
15–0	11–6	20–0	5–7	2–6	8–6	11–6	6–3	1–2	3–0	1–0	1–0	0–8	1–0	0–5
16–0	12–3	21–4	6–0	2–8	9–1	12–3	6–8	1–3	3–0	1–0	1–0	0–9	1–0	0–6
17–0	13–0	22–6	6–4	2–10	9–8	13–0	7–1	1–4	3–0	1–0	1–1	0–9	1–0	0–6
18–0	13–9	23–11	6–8	3–0	10–3	13–9	7–6	1–4	3–0	1–1	1–1	0–9	1–1	0–7
19–0	14–7	25–4	7–1	3–2	10–10	14–7	7–11	1–5	3–0	1–1	1–2	0–10	1–1	0–7
20–0	15–4	26–7	7–6	3–4	11–5	15–4	8–4	1–6	3–0	1–2	1–2	0–10	1–2	0–8

Source: Corry, et al; (1975)

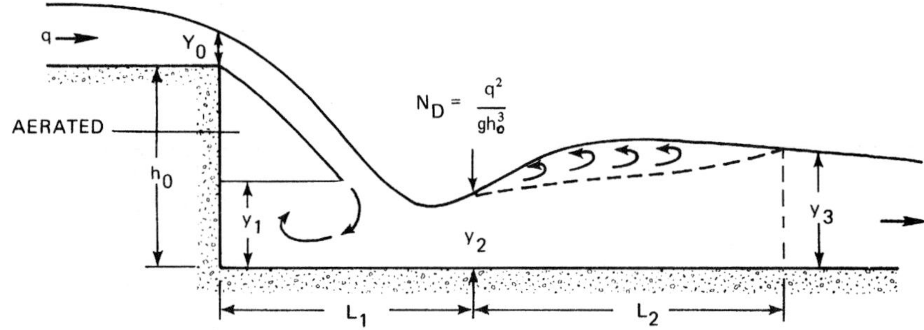

FIGURE 20.20   Flow geometry of a straight-drop spillway. (From Corry et al., 1975)

$$N_D = \frac{q^2}{gh_0^3} \tag{20.20}$$

where q is the discharge per unit width of the crest overfall, g is the acceleration caused by gravity, and h_o is the height of the drop. The dimensions L_1, y_1, y_2, and y_3 in Fig. 20.22 are determined using the following:

$$\frac{L_1}{h_0} = 4.30 N_D^{0.27} \tag{20.21}$$

$$\frac{y_1}{h_0} = 1.0 N_D^{0.22} \tag{20.22}$$

$$\frac{y_2}{h_0} = 0.54 N_D^{0.425} \tag{20.23}$$

$$\frac{y_3}{h_0} = 1.66 N_D^{0.27} \tag{20.24}$$

L_1 is the length of the jump, L_1 can be determined using Fig. 20.22. The sequent depth

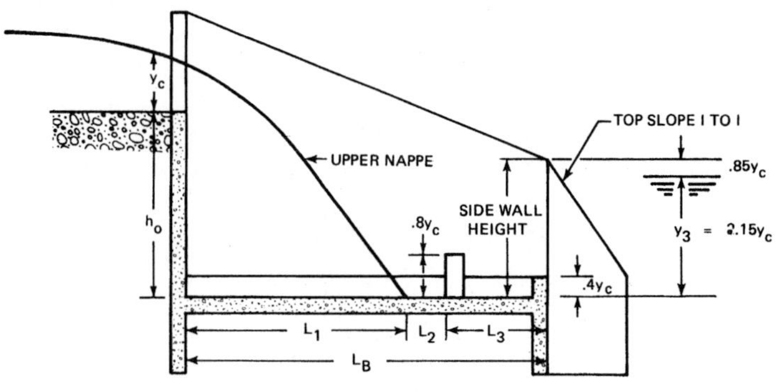

SECTION AT CENTER LINE

FIGURE 20.21   Straight-drop spillway stilling basin. (From Rand, 1955)

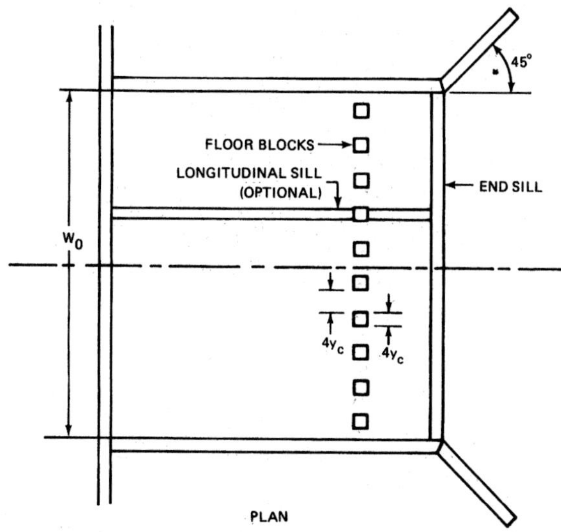

FIGURE 20.21 (*Continued*)

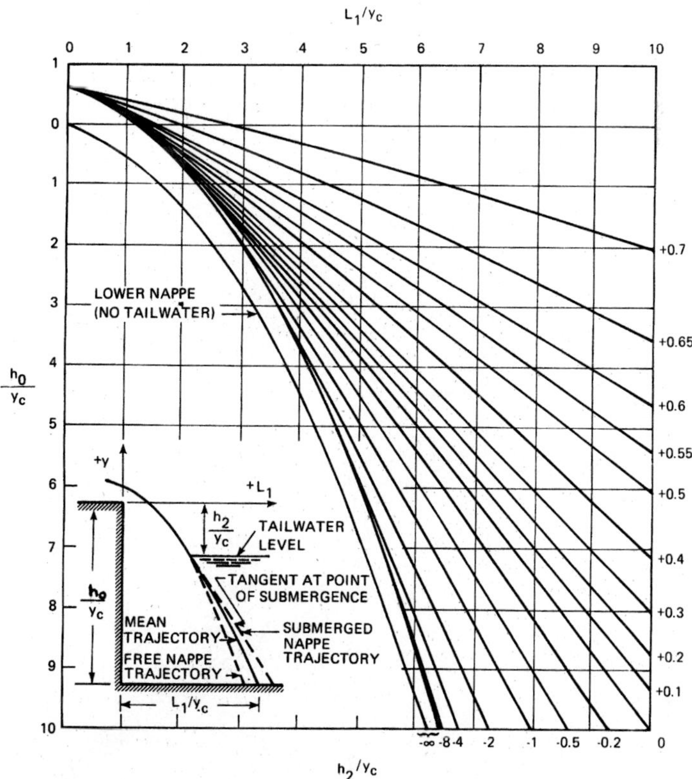

FIGURE 20.22 Design chart for determination of L_1. (From Corry et al., 1975)

and the tailwater depth T_W must be compared to determine whether $T_W < y_3$, or $T_W = y_3$, or $T_W > y_3$. If $T_W < y_3$, the hydraulic jump moves downstream. In this case, it is necessary to construct the apron at the bed level and an end sill or baffles or to construct the apron below the downstream bed level and an end sill. If $T_W > y_3$, the hydraulic jump may become submerged. If $T_W = y_3$, the hydraulic jump begins at depth y_2; there is no supercritical flow on the apron, and L_1 is a minimum.

20.2.4.2 Grated Energy Dissipators.   Energy dissipators with grates (Fig. 20.23) also can be used in conjunction with drop structures. The U.S. Bureau of Reclamation (1987) developed the following design recommendations for grates for incoming subcritical flow:

1. Select slot width with a full slot width at each wall.
2. Compute beam length L_G using

$$L_G = \frac{Q}{C(W)(N)\sqrt{2gy_0}} \tag{20.25}$$

where C is a coefficient equal to 0.245, W is the width of the slots in feet, and N is the number of slots or spaces between beams. Then compute the beam width $= 1.5W$. The quantity $(W)(N)$ can be adjusted until an acceptable beam length, L_G is determined.

3. For self-cleaning, the grate can be tilted approximately 3° in the downstream direction.

20.2.4.3 Straight Drop Structures.   The straight drop structure shown in Fig. 20.21 consists of a horizontal apron with blocks and sills to dissipate energy. This structure is for drops of less than 15 ft (4.57 m) and for sufficient tailwater. The design parameters include the length of the basin, the position and size of the floor blocks, the position and height of the end sill, the position of the wingwalls, and the geometry of the approach channel. This structure was developed by the Agricultural Research Service at the Saint Anthony Falls Hydraulic Laboratory.

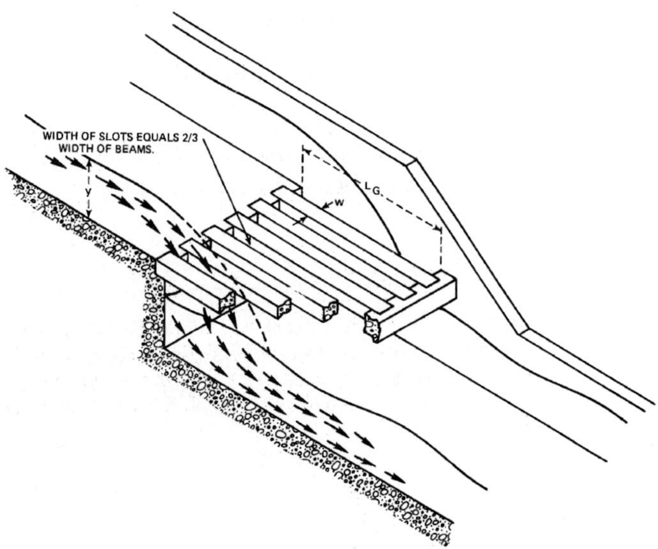

FIGURE 20.23   Energy dissipator with grate. (Corry et al, 1975)

The design procedure for a straight-drop structure is as follows (Corry et al 1975):

1. Compute the minimum length of stilling basin L_B

$$L_B = L_1 + L_2 + L_3 = L_1 + 2.55y_c \tag{20.26}$$

where distances are illustrated in Fig. 20.21. The distance from the headwall to the point where the surface of the upper nappe strikes the still basin floor L_1 is

$$L_1 = \frac{(L_f + L_s)}{2} \tag{20.27}$$

where

$$L_f = \left[-0.406 + \sqrt{3.195 - 4.368 \left(\frac{h_0}{y_c} \right)} \right] y_c \tag{20.28}$$

$$L_s = \frac{\left[0.691 + 0.228 \left(\frac{L_t}{y_c} \right)^2 - \left(\frac{h_0}{y_c} \right) \right] y_c}{\left[0.185 + 0.456 \left(\frac{L_t}{y_c} \right) \right]} \tag{20.29}$$

and

$$L_t = \left[-0.406 + \sqrt{3.195 - 4.368 \left(\frac{h_2}{y_c} \right)} \right] y_c \tag{20.30}$$

where $h_2 = h_0 - y_3$ (Fig. 20.21). Alternatively, L_1 can be determined using Fig. 20.22. The distance from the point where the surface of the upper nappe strikes the stilling basin floor to the upstream face of the floor blocks L_2 is determined using $L_2 = 0.8\ y_c$. The distance between the upstream face of the floor blocks and the end of the stilling basin, L_3, is determined using $L_3 \geqslant 1.75y_c$.

2. Floor blocks are proportioned as follows:
  a) height $= 0.8y_c$;
  b) width and spacing should be $0.4y_c$ with a variation of $+0.15y_c$ permitted, and
  c) blocks should be square in plan, and
  d) blocks should occupy between 50 to 60 percent of the stilling basin width.

3. Compute the end sill height as $0.4y_c$.

4. If longitudinal sills (for structural purposes only) are used, they should be constructed through the floor block, not between the floor blocks.

5. Compute sidewall height above the tailwater level as $0.85y_c$.

6. Wingwalls are constructed at an angle of $45°$ with the outlet center line with a top slope of 1:1.

7. Compute the minimum height of the tailwater surface above the floor of the stilling basin y_3 using $y_3 = 2.15y_c$.

8. Modification, to the approach channel are as follows: The crest of the spillway should be at the same elevation as the approach channel, the bottom width should be equal to the spillway notch length W_o at the headwall, and protection with riprap or paving should be provided for a distance upstream of the headwall of $3y_c$.

9. Using the recommendations in Step 8, no special provision for aeration is needed.

Example 20.7
Determine the dimensions of a straight-drop spillway stilling basin for a discharge of 7.08 m^3/s. The downstream trapezoidal channel has a 3:1 side slope with a 3.05 m bottom width $S_0 = 0.002$ m/m, $n = 0.03$, and a normal depth of 1.024 m. The drop $h_0 = 1.83$ m.

Solution
Step 1. Determine the minimum basin length L_B. The critical depth y_c is determined as $y_c = 0.655$ m. Then $h_0/y_c = 1.83/0.655 = 2.79$; $h_2 = h_0 - 2.15y_c = 1.83 - 2.15(0.655) = 0.422$ m; $h_2/y_c = 0.422/0.655 = 0.644$.
   Using Fig. 20.22, $L_1/y_c = 8.2$ for $h_0/y_c = 2.79$ and $h_2/y_c = 0.644$. Then $L_1 = 8.2(0.655) = 5.37$ m; $L_2 = 0.8y_c = 0.8(0.655) = 0.52$ m; and $L_3 = 1.75y_c = 1.75(0.655) = 1.15$ m; thus, $L_B = L_1 + L_2 + L_3 = 5.37 + 0.52 + 1.15 = 7.04$ m.

Step 2. Proportions the floor blocks are height = $0.8y_c = 0.8(0.655) = 0.524$ m, width = $0.4y_c = 0.262$ m, and spacing = $0.4y_c = 0.262$ m.

Step 3. Calculate the end sill height = $0.4y_c = 0.262$ m.

Step 4. Use the longitudinal sill passing through the floor blocks.

Step 5. Calculate the side-wall height above the tailwater = $0.85_{yc} = 0.85(0.655) = 0.557$ m.

Step 6. Local wingwalls are at a 45° angle with the outlet center line.

Step 7. Calculate the minimum height of tailwater above the floor of the stilling basin: $y_3 = 2.15y_c - 2.15(0.655) = 1.41$ m. The basin must be placed $1.43 - 1.024 = 0.406$ m below the downstream bed level.

20.2.4.4 Box-Inlet Drop Structure.

The box-inlet drop structure shown in Fig. 20.24 consists of two different sections that are effective in controlling flow: the crest of the box inlet and the opening in the headwall. This structure is based on experiments by the U.S. Soil Conservation Service at the Saint Anthony Falls Hydraulic Laboratory (Blaisdell and Donnelly, 1956). The design procedure of box-inlet drop structures is as follows: (FHWA, 1978):

1. Select h_o

2. Select L_1, W_2, and L_c where L_c is the length of the box-inlet crest, $L_c = W_2 + 2L_1$, where W_2 is the width of the box inlet, and L_1 is the length of the box inlet.

3. Compute the head y_o for the crest using the discharge equation for a rectangular weir:

$$y_o = \left(\frac{Q}{3.42L_c}\right)^{2/3} \tag{20.31}$$

4. Compute (h_o/W_2) and determine the coefficient of discharge C_2 from Fig. 20.25.

5. Compute (L_1/h_o) and determine the relative head correction, C_H, from Fig. 20.26.

6. Compute y_0 for the headwall opening using

$$y_0 = \left[\frac{Q}{C_2W_2\sqrt{2g}}\right]^{2/3} - C_H \tag{20.32}$$

which is based upon the rectangular weir equation, $Q_2 = C_2W_2\sqrt{2g}(y_0 + C_H)^{3/2}$.

7. Compare y_o from Step 3 for the crest and y_o from Step 6 for the headwall opening. The larger value of y_o controls. If the crest controls, adjust y_o from Step 3 using the following procedure:

   a. Compute y_o/W_2 and determine the correction for the head, C_1, using Fig. 20.27.

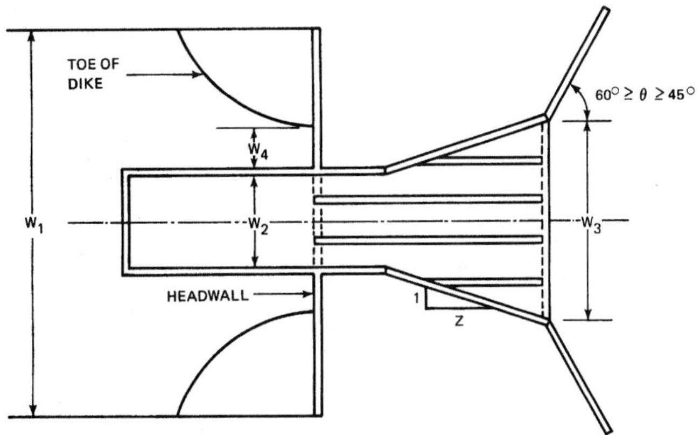

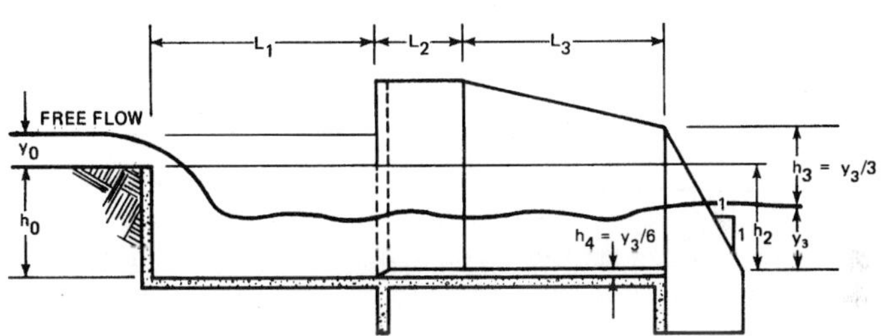

FIGURE 20.24   Box-inlet drop structure. (From Corry et al., 1975)

HYDRAULIC DESIGN OF THE BOX - INLET DROP SPILLWAY

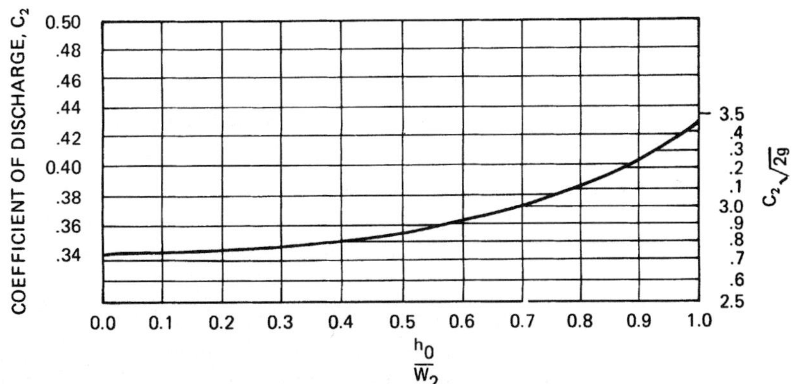

FIGURE 20.25   Coefficient of discharge, with control at headwall opening. (From Corry et al., 1975)

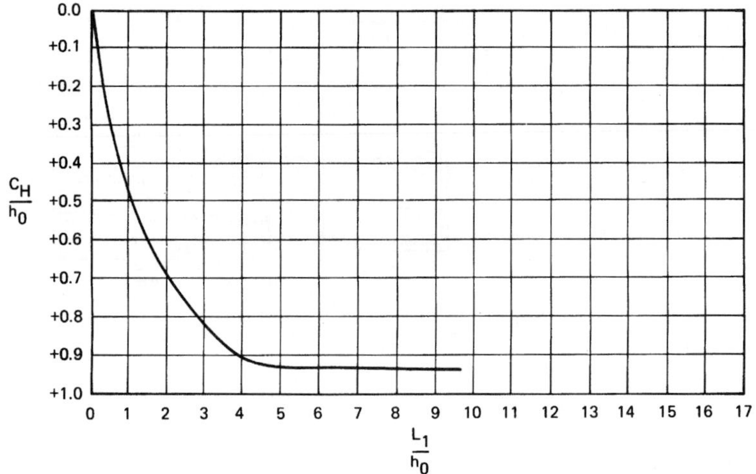

FIGURE 20.26 Relative head correction for $h_0/w_2 \geq \frac{1}{4}$ with control at headwall opening. (From Corry et al., 1975)

b. Compute L_1/W_2 and determine the correction for the box inlet shape, C_S, using Fig. 20.28

c. Compute W_1/L_c and determine the correction for the approach channel width C_A, using Fig. 20.29

d. Compute W_4/W_2 and determine the correction for dike effect (proximity of dike to box inlet crest) C_E, using Table 20.3

e. Determine the adjusted y_o:

$$y_o = \left(\frac{Q}{3.43C_1C_SC_AC_EL_C}\right)^{2/3} \tag{20.33}$$

8. Compute the critical depth in the straight section y_c using

$$y_c = \left[\left(\frac{Q}{W_2}\right)^2 \left(\frac{1}{g}\right)\right]^{1/3} \tag{20.34}$$

9. Compute the critical depth at the exit of the stilling basin y_{c3} using

$$y_{c3} = \left[\left(\frac{Q}{W_3}\right)^2 \left(\frac{1}{g}\right)\right]^{1/3} \tag{20.35}$$

10. Compute the minimum length of the straight section L_2 using

$$L_2 = y_c \left(\frac{0.2}{\left(\dfrac{L_1}{W_2}\right)} + 1\right) \tag{20.36}$$

for values of $L_1/W_2 \geq 0.25$.

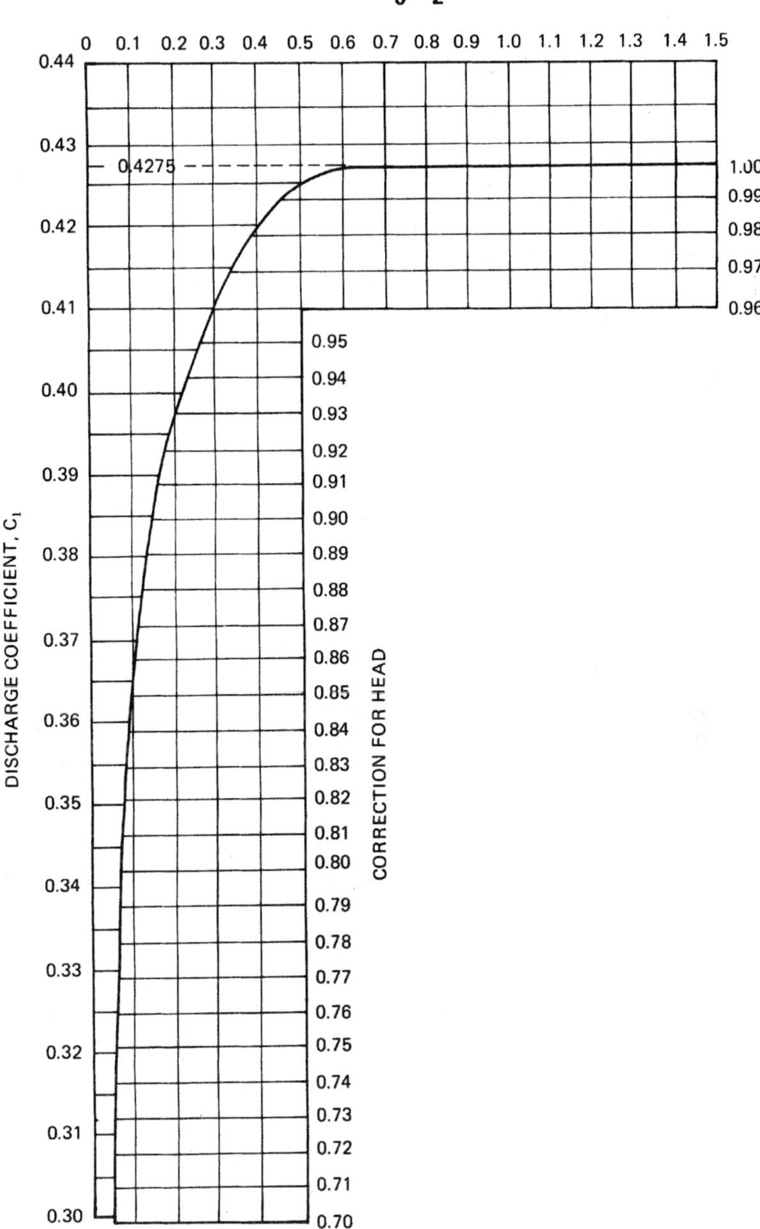

FIGURE 20.27 Discharge coefficient and correction for head, with control at box-inlet crest. (From Corry et al., 1975)

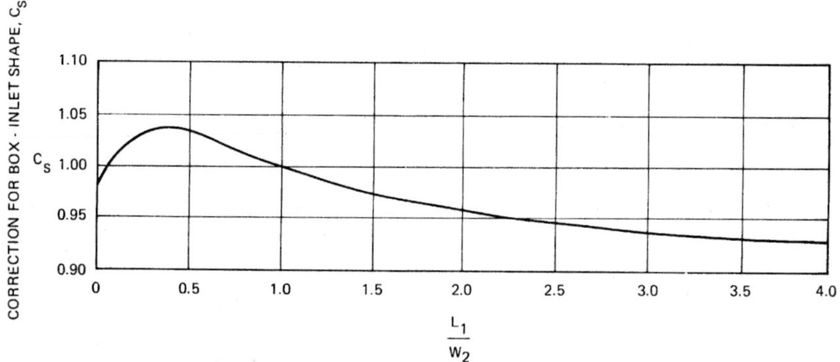

FIGURE 20.28   Correction for box-inlet shape, with control at the box-inlet crest ($W_1/L_c \geq 3$). From Corry et al., 1975)

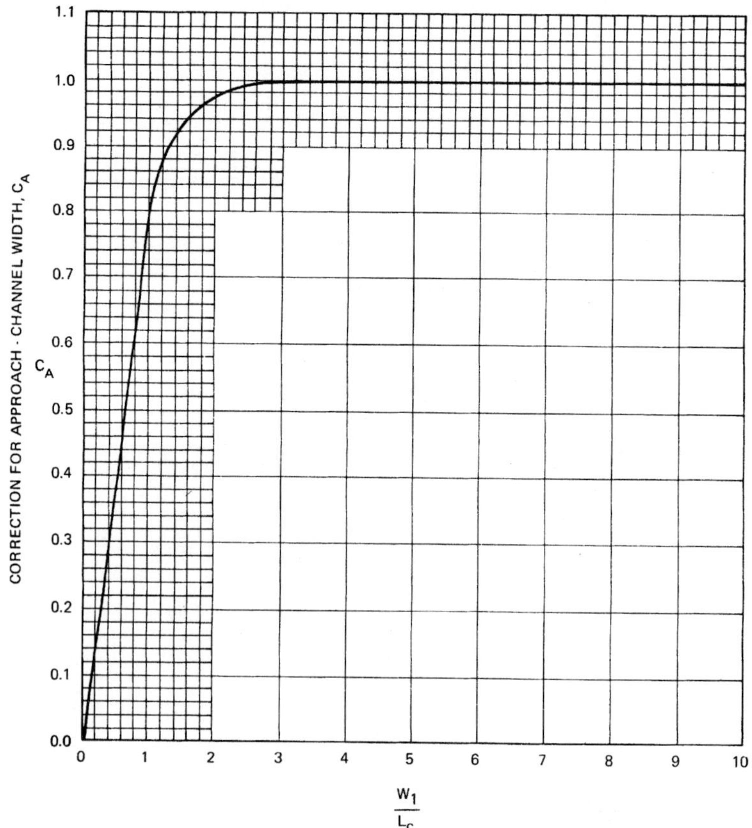

FIGURE 20.29   Correction for approach-channel width, with control at box-inlet crest. (From Corry et al., 1975)

TABLE 20.3 Correction for Dike Effect C_E: Control at the Box–Inlet Crest

L_1/W_2	(Control at Box–Inlet Crest) W_4/W_2						
	0.0	0.1	0.2	0.3	0.4	0.5	0.6
0.5	0.90	0.96	1.00	1.02	1.04	1.05	1.05
1.0	.80	.88	.93	.96	.98	1.00	1.01
1.5	.76	.83	.88	.92	.94	.96	.97
2.0	.76	.83	.88	.92	.94	.96	.97

Source: Corry et al (1975)

11. Compute the minimum length of the stilling basin using

$$L_3 = \left(\frac{L_c}{\dfrac{2L_1}{W_2}} \right) \tag{20.37}$$

and

$$L_3 = \frac{Z(W_3 - W_2)}{2} \tag{20.38}$$

and choose the larger value of L_3.

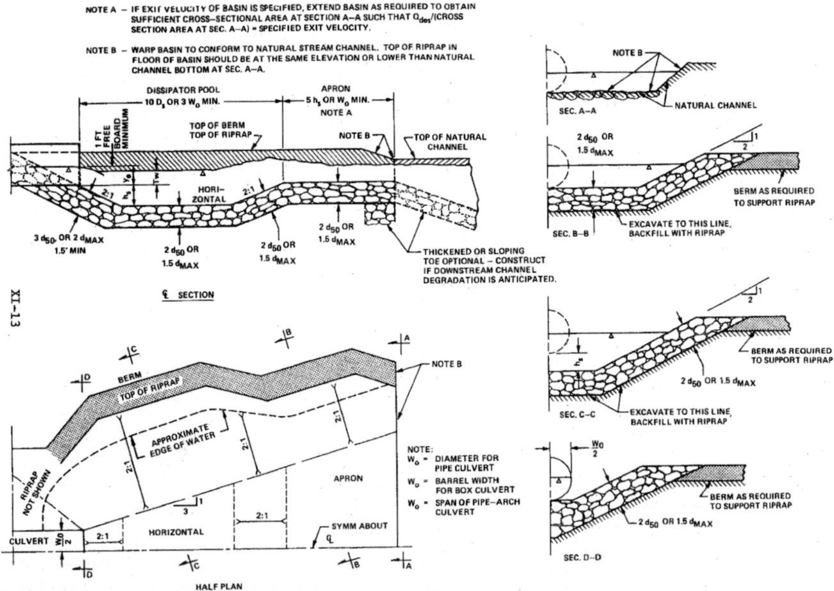

FIGURE 20.30 Details of riprapped culvert energy basin. (Corry et al., 1975)

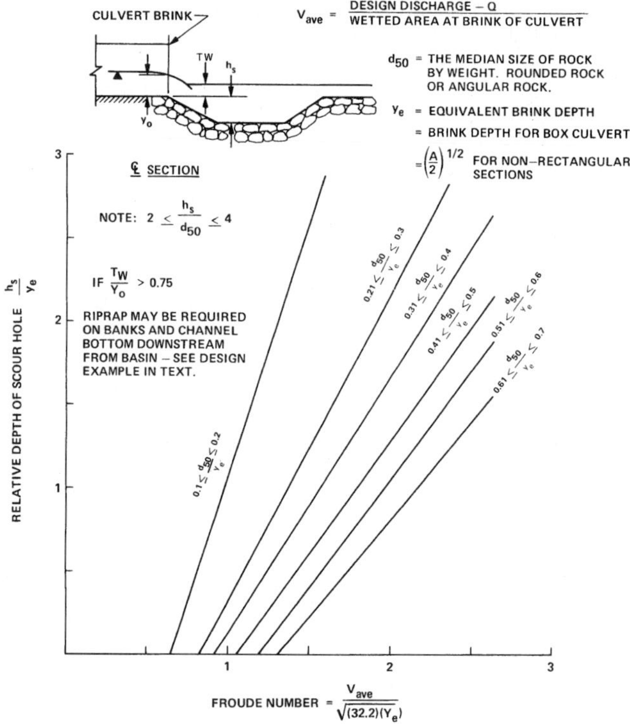

FIGURE 20.31   Relative depth of scour hole versus Froude number at brink of culvert with relative size of riprap as a third variable. (Corry et al., 1975)

12. Compute the minimum tailwater depth over the basin floor using

$$y_3 = 1.6y_{c3} \qquad \text{if} \qquad W_3 < 11.5y_{c3} \tag{20.39}$$

or

$$y_3 = y_{c3} + 0.052W_3 \qquad \text{if} \qquad W_3 > 11.5y_{c3} \tag{20.40}$$

13. Compute the height of the end sill h_4 using $h_4 = y_3/6$.

14. Determine the number of longitudinal sills:
If $W_3 < 2.5W_2$, use two sills;
if $W_3 > 2.5W_2$, use four sills.
When two sills are used they should be located at a distance W_5 on each side of the centerline. When four sills are used, the two additional sills should be located parallel to the outlet centerline and midway between the center sills and the sidewalls at the stilling basin exit.

15. Compute the minimum height of the sidewalls above the water surface at the exit of the stilling basin h_3 using $h_3 = y_3/3$. Sidewalls should extend above the tailwater surface under all conditions.

16. Wingwalls should be triangular in elevation and have a top slope of 45° with the horizontal. The top slopes can be as flat as 30°. Wingwalls should flare in plan at an angle

of 60° with the outlet centerline. The flare-wall angle can be as small as 45°. Wingwalls parallel to the outlet centerline should not be used.

20.2.5 Riprap Basins

The riprap basin recommended in Corry et al (1975) for culverts is shown in Fig. 20.30, which is a preshaped basin lined with riprap. The surface of the riprapped floor of the energy dissipating pool is at an elevation hs below the culvert invert, where h_s is the approximate depth of scour that would occur in a thick pad of riprap if subjected to the design discharge (Corry et al 1975). The length of the pool is the larger of $10h_s$ or W_o, and the overall length of the basin is the larger of $15h_s$ or $4W_o$. The ratio h_s/d_{50} should be less than 4 [$(h_s/d_{50}) < 4$]. Figure 20.31 provides design detaills for riprapped culvert energy basins.

REFERENCES

Blaisdell, F. W., *Flow Through Diverging Open Channel Transitions at Supercritical Velocities,* SCS Report No. SCS–TR–76, U.S. Department of Agriculture, April 1949.

Blaisdell, F. W., and C. A. Donnelly, "The Box Inlet Drop Spillway and Its Outlet", *Transactions, of the American Society of Civil Engineers,* 121:955–986, 1956.

Blaisdell, F. W., *The SAF Stilling Basin,* U.S. Goverment Printing Office, 1959.

Corry, M. L., P. L. Thompson, E. J. Watts, J. S., Jones, and D. L. Richards, *Hydraulic Design of Energy Dissipators for Culverts and Channels,* Enqineering Circular 145, Federal Highway Administration, U.S. Department of Transportation, Washington DC, 1975.

Federal Highway Administration, *Hydraulics of Bridge Waterways,* Hydraulic Design Series No. 1, Federal Highway Administration, U.S. Department of Transportation, Washington, DC, 1978.

Ippen, A. T., "Mechanics of Supercritical Flow", *Transactions of the American Society of Civil Engineers,* 11b; 268–295, 1951.

Rand, W., "Flow Geometry at Straight Drop Spillways," Paper No. 791, *Proceedings of the American Society of Civil Engineers,* Vol. 81, pp. 1–13, September 1955.

Simons, D. B., M. A. Stevens, and F. J. Watts, *Flood Protection at Culvert Outlets,* CER No. 69–70–DBS–MAS–FJW4, Colorado State University, Fort Collins, CO. 1970.

U.S. Army Corps of Engineers (USACE), *Hydraulic Design of Flood Control Channels,* Engineering and Design Manual EM1110–2–1601, pp. 20–26, July 1970.

U.S. Bureau of Reclamation, (USBR), *Design of Small Dams,* U.S. Government Printing Office, Denver, CO, 1987.

Watts, F. J., *Hydraulics of Rigid Boundary Basins,* doctoral dissertation, Colorado State University, Fort Collins, CO, 1968.

CHAPTER 21
COMPUTER MODELS FOR STORMWATER SYSTEMS DESIGN

Bing Zhao
Flood Control District of Maricopa County
Phoenix, Arizona

Computer models are very important to engineers because they can help engineers perform engineering tasks in a faster, cheaper and better way. There are numerous computer models related to stormwater systems design. In the United States, most computer models were developed by the federal government and local governments, which are public domain software and can be obtained without any fee. Most of them can be downloaded from INTERNET. Some private civil engineering software companies also developed propriety computer models. Many of these propriety computer models were developed by adding more graphical user interface features to the existing governmental computer models. These propriety computer models are not free.

This chapter will focus on the most commonly used public domain computer models for stormwater systems design. Some of the propriety computer models will be briefly mentioned if they are related to the public domain computer models. The sources for the public domain software to be discussed in this chapter are: U.S. Environmental Protection Agency (USEPA), U.S. Geological Survey (USGS), U.S. Hydrologic Engineering Center (HEC) of Army Corps of Engineers, Federal Highway Administration (FHA) of U.S. Department of Transportation, and Natural Resources Conservation Service (NRCS) of U.S. Department of Agriculture. The computer models are listed below in an alphabetic order:

1. BASINS (GIS ARCVIEW-based point and nonpoint sources modeling and analysis based on watershed management approach, USEPA)
2. FEQ (solving St. Venant equations in open channels and through control structures, USGS)
3. HEC-1 (rainfall-runoff modeling, HEC)
4. HEC-GeoRAS (GIS ARC/INFO and ARCVIEW-based tool for use with HEC-RAS for cross-section cutting and automatic floodplain delineation, HEC)
5. HEC-RAS (one-dimensional hydraulic calculations by solving energy and momentum equations, HEC)
6. HEC-HMS (rainfall-runoff modeling with graphical user interface, HEC)
7. HSPEXP (an Expert System for calibration of HSPF model, USGS)
8. HSPF (rainfall-runoff modeling and water quality modeling, USEPA and USGS)

9. HYDRAIN (a package of seven computer models for storm drain and sanitary sewer design, open channel water surface analysis, culvert design and analysis, channel lining design, USGS regression equations, FHA)
10. SWMM (urban runoff quantity and quality modeling, USEPA)
11. TR-20 (rainfall-runoff modeling, NRCS)
12. TR-55 (rainfall-runoff modeling for small watersheds, NRCS)
13. UNET (one-dimensional unsteady flow modeling, HEC)

21.1 BASINS

BASINS stands for Better Assessment Science Integrating Point and Nonpoint Sources. It integrates a geographic information system (GIS), national watershed data, and state-of-the-art environmental assessment and modeling tools into one convenient GIS ARCVIEW environment.

BASINS can be used to facilitate examination of environmental information, provide an integrated watershed and modeling framework, and support analysis of point and nonpoint source management alternatives. Several watershed and water quality models are integrated into BASINS within an ARCVIEW GIS environment. It supports the development of total maximum daily loads (TMDLs), which require a watershed-based approach that integrates both point and nonpoint sources. BASINS was originally released in September 1996 by U.S. Environmental Protection Agency (USEPA). The most current version is version 2.0. Version 3.0 is being developed by the USEPA.

BASINS' watershed and water quality analysis is based on five groups of components: (1) national databases, (2) assessment tools (TARGET, ASSESS, and Data Mining) for evaluating water quality and point source loadings at a variety of scales, (3) utilities, including local data import, land-use and DEM reclassification, watershed delineation, and management of water quality observation data, (4) watershed and water quality models including NPSM (HSPF), TOXIROUTE, and QUAL2E, and (5) post processing output tools for interpreting model results. By using GIS, a user can fully visualize, explore, and query to bring a watershed to life. The simulation models run in a Windows environment, using data input files generated in ARCVIEW.

The national databases include land use/land cover, urbanized areas, populated place locations, reach file, version 1 (RF1) and RF3 Alpha, soils (STATSGO), elevation (DEM), major roads, U.S. Geological Survey (USGS) hydrologic unit boundaries (accounting unit, cataloging unit), drinking water supply sites, dam sites, USEPA regional, state, and county boundaries, federal and Indian lands, eco-regions, water quality monitoring station summaries, water quality observation data, bacteria monitoring station summaries, weather station sites, USGS gaging stations, fish and wildlife advisories, national sediment inventory (NSI), shellfish classified areas, Clean Water Needs Survey, permit compliance system (PCS) sites and computed loadings, industrial facilities discharge (IFD) sites, toxic release inventory (TRI) sites, superfund national priority list (NPL) sites, Resource Conservation and Recovery Act (RCRA) sites, and mineral industry locations.

TARGET is a watershed targeting tool that allows environmental managers to make a broadly based evaluation of a watershed's water quality and/or point source loadings. It operates on a large area such as a region or a state. ASSESS operates on a single watershed (cataloging unit) or a limited set of watersheds and focuses on the status of specific water quality stations or discharge facilities and their proximity to water bodies. Data Mining dynamically links different data elements, using a combination of tables and maps. This unique linkage adds significant informational value to the raw data on water quality and loadings. Data Mining is a powerful tool that can assist in the integration and environmental interpretation of both geographic and historical information simultaneously.

The watershed and water quality models that are integrated into BASINS are NPSM, HSPF, QUAL2E, and TOXIROUTE. Nonpoint Source Model (NPSM) estimates land-use-specific nonpoint source loadings for selected pollutants at a watershed (cataloging unit or user-defined sub-watershed scale). The model uses landscape data such as watershed boundaries and land-use distribution to automatically prepare many of the input data it requires. NPSM combines a Windows-based interface with EPA's Hydrologic Simulation Program-FORTRAN (HSPF) model and is linked to ARCVIEW. QUAL2E is a one-dimensional, steady-state water quality and eutrophication model. It is integrated with ARCVIEW through a Windows-based interface. It allows fate and transport modeling for both point and nonpoint source loadings. TOXIROUTE is a screening-level stream routing model that performs simple dilution/decay calculations under mean or low-flow conditions for a stream system within a given watershed (cataloging unit).

21.1.1 Where to Get BASINS?

BASINS is public domain software. The software (version 2.01) including the data and the manuals can be downloaded without charge from *http://www.epa.gov/ost/basins.*

21.1.2 More about BASINS

BASINS 3.0 is being developed by the USEPA. The beta release of BASINS 3.0 is available at http://www.epa.gov/ost/basins. BASINS 3.0 will provide the user with a choice of watershed models: the Hydrological Simulation Program—Fortran (HSPF v.12) and the Soil and Water Assessment Tool (SWAT). SWAT simulates hydrology, pesticide and nutrient cycling, bacteria transport, erosion, and sediment transport . Both SWAT and HSPF are spatially distributed, lumped parameter models. The features of NPSM, the current interface to HSPF, are being coded into (GenScn) and expanded to provide the full functionality of HSPF (i.e., Special Actions, yield-based nutrient modeling, and air deposition). Generation and analysis of model simulation scenarios (GenScn) was developed by the U.S. Geological Survey (USGS) to create simulation scenarios, analyze the results of the scenarios, and compare scenarios. GenScn provides an interactive framework for analysis built around HSPF for simulating the hydrologic and associated water quality processes on pervious and impervious land surfaces and in streams and well-mixed impoundments. The GenScn graphical user interface (GUI) uses standard Windows 9x/NT components. This new version of GenScn, an USEPA/USGS product, provides for full HSPF model set-up, execution, post-processing, display, and analysis. GenScn will also support SWAT output time-series post processing. HSPF version 11 will be replaced by HSPF version 12, which includes a simplified snowmelt algorithm (i.e., degree-day approach), the ability to model land-to-land transfers, high water tables and surface ponding (wet lands), and the addition of new BMP and reporting modules. Another project is being done through a U.S. EPA Office of Administration and Resources Management (OARM) contract with Lockheed Martin Information Services (LMIS) to develop an interface between the BASINS and SWMM.

21.1.3 Training and Technical Support

The training information can be obtained from

Exposure Assessment Branch (4305)
Standards and Applied Science Division
Office of Science and Technology
U.S. Environmental Protection Agency
401 M Street, SW, Washington, DC 20460
web: *www.epa.gov/ost/basins*
Fax: 202-260-9830

EPA's Office of Science and Technology (OST) provides assistance and technical support to users of the BASINS system to facilitate its effective application. Technical support can be obtained as described below:

1. OST's Internet home page: BASINS users are encouraged to access OST's home page for information on new updates, answers to the most frequently asked questions, user tips, and additional documentation. As more real-world applications become available, references to case studies will also be posted. EPA OST's Internet home page address is *http://www.epa.gov/ost/basins*.

2. Telephone assistance: OST personnel are available to assist potential users with technical questions regarding system installation, applications, and data products. Enquiries on the BASINS system can be directed to

   Andrew Battin, tel. (202) 260-3061; e-mail *battin.andrew@epa.gov*

   Paul Cocca, tel. (202) 260-8614; e-mail *cocca.paul@epa.gov*

   William Tate, tel. (202) 260-7052; e-mail *tate.william@epa.gov*

21.2   FEQ

FEQ stands for Full EQuations computer model. The most current version of FEQ is version 8.92, 1999. FEQ simulates flow in a stream system by solving the full, dynamic equations of motion for one-dimensional unsteady flow (St. Venant equations) in open channels and through control structures. It was first developed in 1976 to model the Sanitary and Ship Canal, Chicago, IL, as part of the 208 Water Quality Management Studies conducted by the Northeastern Illinois Planning Commission with principal support from the USEPA. The USGS published documentation for version 8.10 in 1997. FEQ is based on the mass conservation and momentum conservation principles (St. Venant equations). An implicit finite-difference approximation method is used to calculate the flow and depth in the stream system.

A stream system that is simulated with FEQ is subdivided into three broad classes of flow paths: (1) stream reaches (branches), (2) parts of the stream system for which complete information on flow and depth are not required (dummy branches), and (3) level-pool reservoirs. These components are connected by special features or hydraulic control structures, such as junctions, bridges, culverts, dams, waterfalls, spillways, weirs, side weirs, pumps, and others. The hydraulic characteristics of channel cross sections and special features are stored in function tables calculated by the companion program FEQUTL. FEQUTL can read HEC-2 and WSPRO cross section input data and calculate cross section function tables for use in FEQ simulation. FEQ can interpolate hydraulic properties of cross sections between measured sections. FEQ can be applied in the simulation of a wide range of stream configurations (including loops), lateral-inflow conditions, and special features. Boundary conditions can be water-surface stage, discharge, or the stage-discharge relationship at a node. Wind stress terms are supported. The effects of lateral inflows can also be simulated in FEQ when given local runoff intensity data.

21.2.1 Where to Get FEQ?

FEQ is public domain software. The software (version 8.92, 1999) and the manuals can be downloaded without charge from *http://water.usgs.gov/software/*.

21.2.2 Training

Training course for FEQ modeling is offered in DuPage County. The contact information is

Department of Environmental Concerns
DuPage County Center
421 N. County Farm Road
Wheaton, IL 60187

21.3 HEC-1 (1998 VERSION)

HEC-1 is a public domain software package for simulating the precipitation-runoff process. HEC-1 was first developed by the U.S. Army Corps of Engineers (USACE) Hydrologic Engineering Center (HEC) in 1968 and has been updated over the past three decades. The current version, 1998, is anticipated to be the final release of HEC-1 (USACE, 1998). The functional differences between the 1998 version and 1991 version are not significant. The future hydrology model development efforts at HEC will be directed towards HEC-HMS which will supersede HEC-1.

The HEC-1 model is designed to simulate the surface runoff response of a river basin to precipitation by representing the basin as an interconnected system of hydrologic and hydraulic components (USACE, 1998). Each component models an aspect of the precipitation-runoff process within a portion of the basin, commonly referred to as a sub-basin. A component may represent a surface runoff entity, a stream channel, or a reservoir. Representation of a component requires a set of parameters that specify the particular characteristics of the component and mathematical relations that describe the physical processes. The result of the modeling process is the computation of stream flow hydrographs at desired locations in the river basin.

The precipitation hyetograph is the input for runoff computation. In HEC-1, historical storms precipitation data can be supplied to the program by either basin-average precipitation or weighted precipitation gages. The synthetic storms in HEC-1 include (1) standard project storm, (2) probable maximum precipitation, and (3) synthetic storms from depth-duration data. Snowfall and snowmelt can be modeled in HEC-1. The rainfall loss methods include (1) initial and uniform loss rate, (2) exponential loss rate, (3) SCS curve number, (4) Holtan loss rate, (5) Green and Ampt infiltration method, and (6) combined snowmelt and rain losses.

The unit hydrograph methods include (1) user-specified unit hydrograph, (2) Clark unit hydrograph, (3) Snyder unit hydrograph, and (4) SCS dimensionless unit hydrograph. Kinematic wave and Muskingum-Cunge routing can be used to model the distributed runoff. The channel/reservoir routing methods include Muskingum, Muskingum-Cunge, Modified Puls, Working R and D, Level-Pool Reservoir Routing, Average-Lag, Kinematic Wave, and Muskingum-Cunge. Channel infiltration can be modeled with channel routing methods. Flow diversion and pump simulations can also be performed.

In addition to above major features, HEC-1 allows users to calibrate parameters for unit hydrograph, rainfall loss, and routing by using the built-in optimization technique such that the differences between the observed discharge hydrograph and computed discharge hydrograph are minimized. Dam safety analysis, flood damage analysis, and multiPlan-MultiFlood analysis can also be performed in HEC-1.

21.3.1 Where to Get HEC-1?

HEC-1 (HEC, 1998) is public domain software and can be downloaded without charge from *http://www.hec.usace.army.mil.*

21.3.2 GIS and Graphics Resources for HEC-1

Watershed Modeling System (WMS), developed by Brigham Young Unversity, can be used to pre-process GIS data and digitial elevation data, delineate watersheds, compute the sub-basin parameters, and develop an HEC-1 input file. WMS can be purchased from *http://www.ems-i.com* or *http://www.bossintl.com* (about $1,800 per copy).

21.4 HEC-GeoRAS

HEC-GeoRAS is a public domain GIS codes, developed by using ESRI's GIS software for pre-processing/post-processing geospatial data for use with the Hydrologic Engineering Center's HEC-RAS. There are two kinds of HEC-GeoRAS: the first is based on ESRI's GIS ARCVIEW; the second is based on ESRI's GIS ARC/INFO.

21.4.1 ARCVIEW-based HEC-GeoRAS

The ARCVIEW-based HEC-GeoRAS version 3.0 is public domain software (ARCVIEW extension) for processing geospatial data for use with the Hydrologic Engineering Center's HEC-RAS (HEC, April 2000). It allows users with limited GIS ARCVIEW experience to create an HEC-RAS import file containing geometric data from an existing digital terrain model (DTM) and complementary data sets. The current version of HEC-GeoRAS (version 3.0) creates an HEC-RAS import file containing river, reach, and station identifiers; cross-sectional cut lines; cross-sectional surface lines; cross-sectional bank stations; downstream reach lengths for the left overbank, main channel, and right overbank; and cross-sectional roughness coefficients. Hydraulic structure data are not written to the import file. The flow depth and velocity results computed from HEC-RAS can be imported into HEC-GeoRAS for floodplain delineation.

GIS ARCVIEW 3.1 or higher and 3D Analyst 1.0 extension are required for the use of HEC-GeoRAS 3.0. The Spatial Analyst extension, though not required, significantly speeds up the post-processing (i.e., after HEC-RAS results are imported into HEC-GeoRAS). The full functionality of HEC-GeoRAS 3.0 requires HEC-RAS 3.0. However, HEC-RAS 2.2 can still be used except for importing roughness coefficients, exporting velocities, and filtering cross section data points. HEC-GeoRAS requires a digital terrain model (DTM) in the form of a TIN (triangulated irregular network). The DTM must be a continuous surface that includes the bottom of the river channel and the floodplain to be modeled.

The ARCVIEW-based HEC-GeoRAS has three major user environments: preRAS, postRAS, and GeoRAS Util. The preRAS menu option is for pre-processing geometric data which can then be exported to HEC-RAS for hydraulic computation. The postRAS menu option is for post-processing the results imported from HEC-RAS. The GeoRAS Util menu option provides utilities for editing themes and theme management. Herein, themes represent the geospatial data such as stream centerline theme, cross-section theme, and so on. Because of the inherent problem in one-dimensional flow modeling, the floodplain delineation in HEC-GeoRAS for relatively flat areas may not be accurate and requires manual modifications to the automatically delineated floodplains.

21.4.2   ARC/INFO-based HEC-GeoRAS

The ARC/INFO-based HEC-GeoRAS version 1.0 is a public domain package of ARC/INFO marcros for processing geospatial data for use with the Hydrologic Engineering Center's HEC-RAS (HEC, March 1999). It allows users with limited GIS ARC/INFO experience to create an HEC-RAS import file containing geometric data from an existing digital terrain model (DTM). The current version requires a DTM represented by a traiangulated irregular network (TIN). Water surface profile data exported from HEC-RAS simulations may also be viewed using HEC-GeoRAS. The ARC/INFO-based HEC-GeoRAS macros are written in the arc macro language (AML) and require the ARC/INFO program version 7.1 or higher and the ARC/INFO's TIN extension. ARC/INFO and its TIN extension are suported *http://www.esri.com*. The ARC/INFO-based HEC-GeoRAS requires a DTM in the form of a TIN. The DTM must be a continuous surface that includes the bottom of the river channel, and includes all of the floodplain to be modeled. HEC-GeoRAS does not build a TIN or provide tools for builting a TIN. Building a TIN can be performed in ARC/INFO.

The interface for the ARC/INFO-based HEC-GeoRAS is divided into three user environments: project management, pre-processing, and post-processing. The project manager allows the user to manage and access the data and pre- and post-processing windows. The pre-processing environment includes windows for creating a contour coverage from a DTM, creating and editing RAS coverages, and creating the HEC-RAS import file. The post-processing environment allows the user to read in a HEC-RAS export file, which contains the computed flow depths and delineates the floodplains automatically.

The current version creates an HEC-RAS import file containing river, reach, and station identifiers; cross section cut lines, cross section surface lines; cross section bank stations, and downstream reach lengths for the left overbank, main channel, and right overbank. Rough coefficients and hydraulic structure data are not written to the import file. Because of the inherent problem in one-dimensional flow modeling, the floodplain delineation in HEC-GeoRAS for relatively flat areas may not be accurate and requires manual modifications to the automatically delineated floodplains.

21.4.3   Where to Get HEC-GeoRAS?

HEC-GeoRAS (both ARC/INFO-based and ARCVIEW-based) are public domain software that can be downloaded without charge from from *http://www.wrc-hec.usace.army.mil/*.

21.5   HEC-RAS

HEC-RAS is a public domain software package for performing one-dimensional hydraulic calculations for a full network of natural and constructed channels. HEC-RAS was developed by the U.S. Army Corps of Engineers Hydrologic Engineering Center (HEC) as part of "next generation" software. The most current version is version 2.2 (HEC, 1998). It supersedes HEC's HEC-2. HEC-RAS is an integrated system of software, designed for interactive use in a multi-tasking, multi-user network environment. The system is comprised of a graphical user interface, separate hydraulic analysis components, data storage and management capabilities, graphics, and reporting facilities. Although the current version of HEC-RAS is a one-dimensional steady state hydraulics analysis tool, it will ultimately contain a one-dimensional unsteady state component and movable boundary sediment transport component. The one-dimensional unsteady state component is based on UNET program (USACE, 1997).

The major equation for HEC-RAS water surface profile computation is energy equation. The momentum equation is applied for the following situations: (1) the occurrence of a hydraulic jump, (2) low-flow hydraulics at bridges, and (3) stream junctions. HEC-RAS can perform multiple profile analysis, multiple plan analysis, mixed sub-critical and super-critical flow regime calculations with hydraulic jumps, stream junction modeling, and flow distribution calculations. The bridge modeling capabilities in HEC-RAS include modeling of low flow (Class A, B, and C), low flow and weir flow, pressure flow, pressure and weir flow, and highly submerged flows. The methods for computing losses through the bridge are energy equation, momentum equation, Yarnell equation, and FHWA WSPRO method (USFHWA, 1990). The culvert modeling in HEC-RAS is similar to the bridge modeling except that the U.S. Federal Highway Administration's (USFHWA, 1985) standard equations for culvert hydraulics are used to compute inlet control losses at the structure. Inline gated spillways, weirs, and drop structures can be modeled in HEC-RAS. Lateral gated spillways and weirs are not available in this version of the software, but will be included in a future version. Five floodplain encroachment methods are available in HEC-RAS for flood insurance study engineers and planners to evaluate the impact of floodplain encroachment on water surface profiles. The computation of scour at bridges is based on the methods outlined in FHWA's Hydraulic Engineering Circular No. 18 (HEC No. 18). HEC-RAS allows the user to model ice-covered channels at two levels. The first level is an ice cover with known geometry. The second level is a wide-river ice jam in which the ice jam thickness is determined at each section by balancing forces on it.

The limitations of current version of HEC-RAS (version 2.2) are (1) flow is steady; (2) flow is gradually varied except at hydraulic structures such as bridges, culverts, and weirs; (3) flow is one-dimensional; and (4) river channels' slopes are less than 1:10.

21.5.1 Where to Get HEC-RAS?

HEC-RAS version 2.2 is public domain software. The software including the manuals can be downloaded without charge from from *http://www.wrc-hec.usace.army.mil/*.

21.5.2 GIS and Graphical Resources About HEC-RAS

The preparation of cross section data in HEC-RAS is usually time-consuming. Recently HEC in cooperation with ESRI developed a GIS-based tool called HEC-GeoRAS (in both ARCINFO AML and ARCVIEW Avenue formats) to allow users to cut cross sections in a GIS environment and transfer cross section and stream-related data to HEC-RAS. After the HEC-RAS is run, the user can transfer the computed flow depths back to HEC-GeoRas for automatic floodplain delineation (HEC, April 2000). The ARCINFO-based HEC-GeoRAS (version 1.0) is an AML code. The ARCVIEW-based HEC-GeoRAS (version 3.0) is an extension for GIS ARCVIEW. It was developed through a cooperative research and development agreement between the Hydrologic Engineering Center (HEC) and ESRI. Both AR-CINFO-based and ARCVIEW-based HEC-GeoRAS are public domain software and can be downloaded without charge from *http://www.wrc-hec.usace.army.mil/*. Because of the inherent problem in one-dimensional flow modeling, the floodplain delineation in HEC-GeoRAS for relatively flat areas may not be accurate and requires manual modifications to the automatically delineated floodplains. A two-dimensional model should be used for flat areas where sheet flows exist. To run ARCVIEW-based HEC-GeoRAS, users must have ESRI's ARCVIEW and 3D Analyst extension (each costs about $2,000 per copy). To run

ARCINFO-based HEC-GeoRAS, users must have ESRI's ARC/INFO version 7.1 or higher and TIN module. For more information about ESRI's products, please contact ESRI (*http://www.esri.com*).

RMS and RiverCad developed by Boss International, Inc. are two CAD-based softwares that help cut cross sections for use with HEC-RAS. RMS requires AutoCad while RiverCad already has a built-in CAD function. RMS and RiverCad can be purchased from *http://www.bossintl.com*. AutoCad can be purchased from *http://www.autodesk.com*.

21.6 HEC-HMS

HEC-HMS version 2.0 is a public domain software package for hydrologic modeling of dendritic watershed systems and was developed by the U.S. Army Corps of Engineers Hydrologic Engineering Center (HEC) (USACE, 2000). It is HEC's "next generation" software for precipitation-runoff simulation that will supersede the HEC-1 Flood Hydrograph Package. Several important new capabilities that were not available in the 1998 version HEC-1 are (1) graphical user interface for model development and viewing modeling results; (2) continuous hydrograph simulaiton over long periods of time; (3) distributed runoff computation using a grid cell depiction of the watershed based on quasi-distributed linear transform (ModClark) of cell-based precipitation and infiltration; and (4) importing pre-processed hydrologic GIS data.

21.6.1 Major Features of HEC-HMS 2.0

In HEC-HMS 2.0, there are four different methods for analyzing historical precipitation: (1) user-specified hyetograph method, (2) user gage weights method, (3) inverse distance method, and (4) gridded precipitation method. There are three methods for producing synthetic precipitation: (1) frequency storm-based on Technical Paper 40 (U.S. National Weather Service, 1961) and NOAA Atlas 2 (National Weather Service, 1973); (2) standard project storm (USACE, 1952); and (3) the users-specified hyetograph method. The evapotranspiration method is based on monthly average values with an optional pan coefficient. Time series, paired, and gridded data are stored in the Data Storage Systems HEC-DSS (USACE, 1994).

HEC-HMS 2.0's rainfall loss methods include initial and constant, SCS curve number method, gridded SCS cruve number, and Green and Ampt method. The unit hydrograph methods include Clark unit hydrograph, Snyder unit hydrograph, SCS method, ModClark (modified Clark method), kinematic wave, and user-specified unit hydrograph method. The open-channel routing methods include lag method, Muskingum method, modified Puls method, and Muskingum-Cunger method.

Most parameters for methods used in subbasin and reach elements can be estimated automatically by using optimization techniques in HEC-HMS to fit the observed discharge data. Four objective functions are available for measuring the goodness-of-fit between the computed and observed discharge hydrographs. The objective functions are peak-weighted root mean square errors, sum of squared residuals, and sum of absolute residuals, and percent error in peak flow. The optimal parameters are found such that one of the objective funcions is minimized and the physical-based constraints for the parameters are satisfied.

21.6.2 Where to Get HEC-HMS?

HEC-HMS 2.0 is public domain software. The software (version 2.0), including the documentations, can be downloaded without charge from *http://www.hec.usace.army.mil*.

21.6.3   GIS Resources for HEC-HMS

Some GIS resources are available for pre-processing GIS data, which serve as data-pre-processor for HEC-HMS. HECPREPRO is a GIS preprocessor for HEC-HMS, and was developed by University of Texas at Austin (David Maidment at Civil and Environmental Engineering Department). HECPREPRO summarizes data from a GIS system so that they can be used as input for HMS. HECPREPRO takes a stream and sub-basin GIS layers as input data. The output data consists of an HMS Basin file. The system is written in ARC/INFO Arc Macro Language (AML) (version 3.1.ai) and ARCVIEW Avenue (version 4.0.av). HECPREPRO can be obtained from *http://www.ce.utexas.edu/prof/maidment/GISHYDRO/* without charge.

Watershed Modeling System (WMS), developed by Brigham Young Unversity, can be used to pre-process GIS data and digitial elevation data, delineate watersheds, compute the sub-basin parameters, and transfer data from WMS to HEC-HMS. WMS can be purchased from *http://www.ems-i.com* or *http://www.bossintl.com* (about \$1,800 per copy).

HEC-GeoHMS (version 1.0, July 2000) is a software package for use with the ArcView GIS. It produces a number of hydrologic inputs that are used directly in HEC-HMS. HEC-GeoHMS can be downloaded from http://www.hec.usace.army.mil.

21.7   HSPEXP

HSPEXP is an Expert System for calibration of HSPF model (Lumb, et al. 1994). It can be used to assist less-experienced modelers with calibration of HSPF and to facilitate the interaction between the modeler and the modeling process not provided by mathematical optimization. It allows the user to interactively edit the input file for the HSPF, simulate with HSPF, produce plots of HSPF output, compute error statistics for a simulation, and provide the user with expert advice on which parameters should be changed up or down to improve the calibration. Expert System or Artificial Intelligience was first applied to HSPF on a project at the Stanford Research Institute, Menlo Park, CA to estimate initial values for model parameters. HSPEXP is the result of a USGS project to create an expert system to assist with the calibration of a watershed model.

HSPEXP is based on over 35 rules involving over 80 conditions to recommend parameter adjustments. The rules are divided into four phases: annual volumes, low flows, storm flows, and seasonal flows. Rules in subsequent phases are not tested until all rules in the previous phase pass. The rules are based on the experience of experts in the use of HSPF in a wide range of climates and physiographic regions.

21.7.1   Where to Get HSPEXP?

HSPEXP is public domain software. The software (version 2.3, 1996) and the manuals can be downloaded without charge from *http://water.usgs.gov/software/*.

21.7.2   Training

The USGS National Training Center offers training for HSPEXP.

21.8   HSPF

HSPF stands for Hydrological Simulation Program—Fortran. HSPF was first developed in the early 1960s as the Stanford Watershed Model. Since the 1960s, many functions have been added to HSPF by EPA in cooperation with USGS (EPA, 1980, 1984). The current release is version 11 (EPA, 1997). An interactive version (see HSPEXP) based on Expert System technology was developed by the USGS in the 1990s.

HSPF simulates the rainfall-runoff process and associated water quality processes using continuous rainfall and other meteorologic records to compute streamflow hydrographs and pollutographs (EPA and USGS, 1996). HSPF can be used to assess the effects of land-use change, reservoir operations, point or nonpoint source treatment alternatives, flow diversions, etc. HSPF simulates interception soil moisture, surface runoff, interflow, base flow, snowpack depth and water content, snowmelt, evapotranspiration, groundwater re-charge, dissolved oxygen, biochemical oxygen demand (BOD), temperature, pesticides, conservatives, fecal coliforms, sediment detachment and transport, sediment routing by particle size, channel routing, reservoir routing, constituent routing, pH, ammonia, nitrite-nitrate, organic nitrogen, orthophosphate, organic phosphorus, phytoplankton, and zooplankton. HSPF can simulate one or many pervious or impervious unit areas discharging to one or many river reaches or reservoirs. Any period from a few minutes to hundreds of years may be simulated.

21.8.1 Where to Get HSPF?

HSPF is public domain software. The software (version 11, 1997) and the manuals can be downloaded without charge from *http://water.usgs.gov/software/*.

21.8.2 GIS Resources for HSPF

Private organizations have developed graphical user interface programs to assist the data input and output graphics/report. A graphical user interface was recently added to Watershed Modeling System (WMS), which is a graphical and GIS-based software sponsored by the U.S. Army Corps of Engineers Waterway Experiment Station and developed by Brigham Young University (*http://www.emrl.byu.edu*). WMS can be purchased from *http://www.ems-i.com* and *http://www.bossintl.com* (about $1,800 per copy).

21.8.3 More about HSPF

U.S. EPA's BASINS software (Better Assessment Science Integrating Point and Nonpoint Sources) has a Nonpoint Source Model (NPSM) which combines a Windows-based interface with EPA's Hydrologic Simulation Program-FORTRAN (HSPF) model and is linked to ARCVIEW. Based on U.S. EPA's web site, the features of NPSM, the current interface to HSPF, are being coded into (GenScn) which was developed by USGS to provide the full functionality of HSPF (i.e., Special Actions, yield-based nutrient modeling, and air deposition). This new version of GenScn, an USEPA/USGS product, provides for full HSPF model set-up, execution, post-processing, display, and analysis. HSPF version 11 will be replaced by HSPF version 12, which includes a simplified snowmelt algorithm (i.e., degree-day approach), the ability to model land-to-land transfers, high water tables and surface ponding (wet lands), and the addition of new BMP and reporting modules.

21.8.4 Training

The USGS National Training Center offers training of HSPF.

21.9 HYDRAIN

The HYDRAIN System version 6.1, developed by U.S. Department of Transportation Federal Highway Administration (1999), is a software package of seven hydrologic and hydaulic engineering/analysis programs, which include

HYDRA—Storm Drain and Sanitary Sewer Design and Analysis

WSPRO—Open Channel Water Surface Analysis

HYDRO—Design Event versus Return Period

HY8—FHWA Culvert Analysis and Design

HYCHL—Flexible & Rigid Channel Lining Design and Analysis

HYEQT—Flow Equation Program that allows the user to create and solve regression equations for solving peak flow (or any other formula of interest)

NFF—USGS National Flood Frequency Program that allows users to estimate flood-peak discharges and hydrographs for a given recurrence interval (return period).

21.9.1 HYDRA

HYDRA is one of the seven programs within HYDRAIN. HYDRA is a storm drain and sanitary sewer design and analysis computer program. The HYDRA program provides hydraulic design engineers a means of accurately, easily, and quickly designing and analyzing storm, sanitary, or combined collection systems. In the HYDRA design process, the program will select pipe size, slope, and invert elevations if given certain design criteria. Additionally, HYDRA will perform analyses on an existing system of pipes and/or ditches. When an existing system of pipes is overloaded, HYDRA will indicate suggested flow removal quantities, as well as an increased pipe size as an alternative remedy. Additionally, HYDRA can optionally consider the possibility of surcharged systems. The design procedure is not optimized, so alternatives should be examined. Chapter 23 provides more detail on HYDRA.

HYDRA operates in design mode and analysis mode for storm drain systems, sanitary (sewage) systems, and combined (storm and sanitary) sewer systems. The major features of HYDRA are

- *Cost estimation*—capabilities that allow for consideration of de-watering, traffic control, sheeting, shrinkage of backfill, costs of borrow, bedding costs, surface restoration, rock excavation, pipe zone costs, etc. HYDRA is also sufficiently flexible to allow cost criteria to be varied for any segment of pipe in a system, if desired. Ground profiles, either upstream or downstream from any specified point along the system, can also be accepted for consideration in cost estimation, if desired.

- *Models storm flow and offers choice of methods*—HYDRA is capable of "generating" storm flow based on the Rational method or modeling user-defined hydrographs in a hydrologic simulation, at the user's discretion. This may be particularly advantageous for engineers who wish to compare designs or analysis results based on different methods.

- Models' sanitary flow—HYDRA "generates" sanitary flow based on the traditional "peaking factor" concept.

- *Models' drainage systems of any size*—HYDRA has a data handling algorithm especially designed to accept a drainage system of any realistically conceivable design.

- *Infiltration/inflow analysis*—HYDRA is ideally suited for making these analyses.

- *Planning*—HYDRA can be used for determining the most practical alternative choices for unloading an existing overloaded storm drain or waste water system and for formulating master plans to allow for an orderly growth of these systems. The program's features and capabilities should have far-reaching implications for municipal agencies whose existing sewer systems are under stress from rapid population growth and/or changes in land-use patterns.

- *Easy data input structure and quick editing capacity*—All data needed to run HYDRA are in one user-supplied input file, simplifying data editing operations. Furthermore, if the program is run from within the HYDRAIN environment, the input file may be modified without leaving the HYDRAIN program by using the built-in editor. (The user documentation section describes the capabilities of this screen editor and provides instructions for its use.) Time required for data modification and job re-submission is thus minimized, which enables the user to spend more time on his or her own decision analysis.

21.9.2 WSPRO

The Water Surface Profile (WSPRO) program is one of the seven programs within HYDRAIN. It computes the water surface profile of one dimensional, gradually varied, and steady state flow for six major types of flow situations: (1) unconstricted flow, (2) single-opening bridge, (3) bridge opening(s) with guide banks, (4) single-opening, embankment overflow, (5) multiple alternatives for a single job, and 6) multiple openings. The major features for WSPRO are

- The water surface profile computations in the absence of bridges are generally consistent with the methods used in other models, such as the Corps of Engineers HEC-2.
- Any combination of sub-critical, critical, and supercritical flow profiles may be analyzed for one-dimensional, gradually varied steady flow.
- Discharge may be varied from cross section to cross section to account for tributary and lateral flow gains or losses.
- Up to 20 profiles for different discharges and/or initial water-surface elevations may be computed at one time.
- Initial water-surface elevations for each profile may be specified by the user or computed by the model.
- Variable Manning's roughness coefficients may be specified for any cross section to reflect roughness changes both horizontally and vertically in the cross section.
- Up to three different flow lengths for left, central, and right portions of a valley may be specified between any two valley cross sections.
- Users may select the friction-slope averaging technique to be used in the friction loss computations.
- Users may specify the coefficients used to compute energy losses associated with expansion or contraction of flow.
- The model can compute backwater for both free-surface and pressure flow situations at a bridge.
- The model can compute water-surface profiles through bridges for cases where road overflow occurs in conjunction with flow through the bridge opening.
- The effects of guide banks on the water-surface profile are estimated when guide banks data are entered.
- The model can analyze multiple waterway openings for a cross section, including culverts when used as one of the multiple openings.

WSPRO requires the creation of an input file, consisting of commands to describe the physical characteristics of a waterway. The program shell or WSPRO Input/Output program facilitates this activity. The commands are placed in a logical sequence, usually from a downstream section to upstream to facilitate a step-backwater computation methodology. The input file, established for a specific stream reach pattern by the user, is then executed using the WSPRO program, regardless of how the input file is created.

21.9.3 HYDRO

HYDRO is another program within HYDRAIN. It is a hydrology design and analysis computer program for single-event simulation based on the Federal Highway Administration (FHWA) Highway Engineering Circular 19, HYDROLOGY (HEC-19) and some areas of HEC-12. HYDRO's outputs can be used by other programs in the HYDRAIN as inputs. HYDRO can be used as a stand-alone application program.

HYDRO's capabilities are divided into three major hydrological components: (1) rainfall analysis, (2) Intensity-Duration-Frequency (IDF) curve generation, and (3) flow (runoff or stream flow) analysis. The rainfall analysis component computes rainfall intensities based on two databases. The first database, also the default database, contains the information from three sources: the National Weather Service (NWS) technical memorandum, HYDRO-35, the Rainfall Frequency Atlas of the United States, Technical Paper 40 (TP-40), and the National Oceanic and Atmospheric Administration (NOAA) Atlas 2 documents. The user is freed from needing the HYDRO-35, TP-40, and Atlas 2 documents. The second database was added to the HYDRO at the request from California Department of Transportation to address the problems in California and the western states. A triangular hyetograph can be generated based on Yen and Chow. The IDF curve can be generated based on either database for any location for the continental United States. The return periods are between 2 and 100 years, and the durations are beween 5 minutes and 24 hours.

HYDRO allows the user to use three flow methods: the Rational method, user-supplied regression equations, and the Log Pearson Type-III method. Each of these methods produce a single peak value. HYDRO also allows the user to combine the peak flow with the dimensionless hydrograph method so as to consider hydrographic or dynamic flow conditions. The first two flow determination methods, rational and regression, are techniques that are used on ungaged watersheds where runoff is the primary source of flow. The Log Pearson Type-III method requires gaged stream flow as the primary input.

21.9.4 HY8

HY8 is another of the seven programs within HYDRAIN. It can also be used as a stand-alone application program. HY8 allows the users to investigate the hydraulic performance of a culvert system. It automates the methods presented in "Hydraulic Design of Highway Culverts" (HDS-5), "Hydraulic Design of Energy Dissipators for Culverts and Channels" (HEC-14), "Hydrology" (HEC-19), and information published by pipe manufacturers pertaining to the culvert sizes and materials.

HY8 is composed of four different modules. These four modules are a culvert analysis and design module, a hydrograph generation module, a routing module, and an energy dissipation module.

21.9.4.1 Culvert Analysis and Design Module.   The culvert analysis and design module allows the user to review the hydraulic characteristics of user-supplied culvert data. This module also permits minimization of culvert size based on an allowable headwater elevation. The program will compute culvert hydraulics for circular, rectangular, elliptical, arch, and user-defined geometry. Improved inlets can also be specified. The user will have the options of entering either a regular or an irregular cross section for calculating tailwater or a fixed tailwater may be specified. A series of up to six parallel culvert systems, each having different inlets, inlet elevations, outlets, outlet elevations, lengths, and cross-sectional shape characteristics (e.g., diameter), can be analyzed. Each culvert system may consist of several barrels.

21.9.4.2 Hydrograph Generation Module.   The hydrograph geneation module generates a storm hydrograph that can be used singly or as input into culvert routing analyses. The hydrograph is generated using methods found in the FHWA HEC-19 document. Main input parameters in this module are watershed characteristics, such as drainage area, slope, curve number, watershed distribution, coefficient of abstraction, and base flow. From these parameters, a storm hyetograph and hydrograph are produced. An option allows a user-defined storm hyetograph to be entered, from which a hydrograph is produced. As a final alternative, the user can enter a hydrograph.

21.9.4.3 Routing Module.   The routing module uses the culvert data collected in the first module and the hydrograph generated in the second module to calculate storage and outflow hydrograph characteristics. The routing is performed using the storage indication (modified Puls) method. Four options are available to determine the upstream stage-storage relationships. These four options are: employing the prism method, entering a surface area-versus-elevation curve, entering a volume-versus-elevation curve, and providing stream cross section data.

21.9.4.4 Energy Dissipation Module.   The energy dissipation module permits the design and analysis of an energy dissipator at the culvert outlet. It follows the design procedures used in FHWA publication HEC-14. Similar to the routing module, this module needs the performance curve generated from a culvert analysis file to perform the energy dissipator design and analysis. If there is more than one culvert system in the culvert analysis file, the user has to specify which system is used for the design. The program will design a dissipator for only one culvert at a time. The user can select several options from within this module. These options are designing an external dissipator, designing an internal dissipator, estimating the scour hole geometry, and modifying hydraulic aspects of the culvert being analyzed.

21.9.5 HYCHL

HYCHL is another program in HYDRAIN. The HYCHL program can be used to design stable roadside channel and irregular channel riprap linings based on the Federal Highway Administration (FHWA) publications Hydraulic Engineering Circular 15 (HEC-15) and HEC-11. When provided design flow and channel conditions (i.e., slope, shape, and lining type), HYCHL can analyze drainage channels for stability. The HYCHL program determines stability through application of tractive force theory. The tractive force theory compares shear exerted on the lining, as a result of flow, with the permissible shear stress of the lining. HYCHL also calculates the maximum discharge that a particular channel can convey, given permissible shear stress and the corresponding allowable depth. HYCHL can analyze all linings with a known permissible shear for both stability and maximum discharge. The output generated by HYCHL includes flow depth, velocity, calculated shear stress, permissible shear stress, and maximum discharge.

Depending on channel function, material availability, costs, aesthetics, and desired service life, a designer may choose from a variety of lining types. HYCHL offers a variety of design and analysis options, including

- Rigid or flexible linings
- Temporary or permanent linings
- Single or composite linings

- Straight or bend channel sections
- Alternative regular channel shapes
- Constant or variable channel flow

HYCHL capabilities allow analysis of lining having both rigid and flexible composition. Rigid linings in HYCHL include concrete, grouted riprap, stone masonry, soil cement, and asphalt. Flexible linings in HYCHL include those that may be considered permanent and those considered temporary. Permanent flexible linings include vegetation, riprap, and gabions. Temporary linings include woven paper, jute mesh, fiberglass roving, straw with net, curled wood mat, synthetic mat, and bare soil (unlined). The lining can consist of a single material or a composite set of materials (i.e., a paved low-flow channel section with a vegetative upper channel section). Composite linings are typically a low-flow lining protecting the bottom of a channel where higher shear stresses are anticipated and a side-slope lining protecting the channel sides. Composite linings are used when lining side slopes with the same material applied to the bottom is undesirable for reasons of economics, aesthetics, or safety.

The designer of rigid, vegetative, and temporary linings may apply HYCHL to a variety of geometric configurations. HYCHL calculates the shear stresses on linings in straight channel sections as well as the higher stresses found in bend sections. Channel cross sections available in HYCHL for these lining types are trapezoidal, parabolic, triangular, and triangular with rounded bottom.

The performance of rigid, vegetative, gabion, and temporary linings can be evaluated using a design flow that is assumed to be constant for the entire channel length or a variable inflow. The variable inflow is characterized as a uniform lineal flow that results in an increasing discharge with channel length. Under such conditions, HYCHL provides the designer with an estimate of the length of channel for which a given lining may be suitable.

21.9.6 HYEQT

The Hydrain Equation Program (HYEQT) is another program in the HYDRAIN system. It allows the user to create and solve regression equations for solving peak flow (or any other formula of interest). It allows the users to supplement and modify the NFF's USGS peak flow regression equations.

The program groups a series of equations into an element designated a "region." A region may consist of a watershed, hydrologic district, or some other unique division in a State. Each region can contain up to 10 regression equations with a total of 15 independent variables. All regions are placed into a single ASCII database file. There is no practical limit to the number of regions that the program can access (the upper limit is a function of the available disk space).

The program uses two screen modes, edit and solve, which are used to create and solve the equations. The edit mode places the user into a screen with four sets of fields where users can create the equations based on dependent and independent variables. The solve mode screen shows the already-created equations. The equations cannot be modified within the solve mode. After entering all values in the fields next to the independent variable, the program automatically "plugs" them into the equations and displays the solutions.

21.9.7 NFF

The National Flood Frequency (NFF) program is a collection of all current statewide and metropolitanwide regression equations compiled by the United States Geological Survey (USGS) in cooperation with the Federal Highway Administration and the Federal Emergency

Management Agency to estimate flood-peak discharges and associated flood hydrographs for a given recurrence interval or exceedance probability for unregulated rural and urban watersheds (Jennings et al., 1993). These estimates are useful to engineers and hydrologists for planning and design applications. NFF includes the statewide regression equations for rural watersheds in each state, includes the applicable metropolitanwide or statewide regression equations for urban watersheds, and generates rural and urban frequency functions and hydrographs.

The rural regression equations in NFF are based on watershed and climatic characteristics that can be obtained from topographic maps or rainfall reports and atlases. The most frequently used watershed and climatic characteristics are drainage area, main-channel slope, and mean annual precipitation.

Seven-parameter urban equations are based on multiple regression analysis of urban flood frequency data from 199 urbanized basins. The variables for the urban equations are (1) urban peak discharges for the 2- and 500-yr return periods (cubic feet per second); (2) drainage area (square miles); (3) main channel slope (ft/mile) measured between points which are 10% and 85% of the main channel length upstream from the study site; (4) rainfall depth (in) for the 2-hr, 2-yr occurrence determined from the HYDRO program, from U.S. Weather Bureau (USWB) Technical Paper 40 (1961) (eastern U.S.), or from the National Oceanic and Atmospheric Administration (NOAA) Atlas 2 (Miller et al., 1973) (western U.S.); basin storage (the percentage of the drainage basin occupied by lakes, reservoirs, swamps, and wet lands; in-channel storage of a temporary nature, resulting from detention ponds or roadway embankments, should not be included in the computation), (5) basin development factor (an index of the prevalence of the urban drainage improvement), (6) percentage of the drainage basin occupied by impervious surfaces, such as houses, buildings, streets, and parking lots, and (7) peak discharges, in cfs, for an equivalent rural drainage basin in the same hydrologic area as the urban basin, for a recurrence interval of T years; equivalent rural peak discharges are computed from the rural equations for the appropriate state, in the NFF program, and are automatically transferred to the urban computations. The basin development factor is a very significant variable in the urban equations. The detailed discussion on how to determine the basin development factor can be found in Appendix D of HYDRAIN (USFHWA, 1999).

21.9.8   Where to Get HYDRAIN?

The HYDRAIN is public domain software. The software (version 6.1, 1999) and the manuals can be downloaded without charge from *http://www.fhwa.dot.gov/bridge/hydrain.htm*.

21.10   SWMM

The Stormwater Management Model (SWMM), developed by the U.S. Environmental Protection Agency (USEPA), is a comprehensive computer model for mathematical simulation of urban runoff quantity and quality in storm sewers, combined sewers (storm and sanitary), and natural drainage systems on the basis of rainfall (hyetograph) and other meteorological inputs and system characterization (catchment, conveyance, storage/treatment) (Huber and Dickinson, 1992). Both single-event and continuous-event (long-term) processes can be simulated by the SWMM. Rainfall, snowmelt, surface and sub-surface runoff, and flow routing can be simulated. The SWMM can be used for screening, planning, design, and operation purposes for stormwater management.

The SWMM consists of four computational blocks and six service blocks. The computational blocks are Runoff, Transport, Extran, and Storage/Treatment blocks. The services blocks are Executive, Rain, Temp, Graph, Statistics, and Combine blocks. A typical run of the SWMM involves one or two computational blocks with the Exective block. The Executive block acts as a "manager" of the SWMM. It controls the simulation sequence and time clock for all blocks. The Rain and Temp blocks handle the input and manipulation of precipitation, evaporation, and temperature time series. The Runoff block is a critical block in the SWMM. Based on the meteorological data received from Rain and Temp blocks, the Runoff block simulates the rainfall-runoff process by non-linear reservoir approach with an option of snowmelt simulation, and produces hydrographs and pollutographs at inlet locations.

Based on the results from the Runoff block, the TRANSPORT block routes the inlet hydrographs and pollutographs through the sewer system based on kinematic wave rouitng, first-order decay, Shield's criterion for scour and deposition within the sewer system for initiation of motion, and generation of dry-weather flow and quality. The Extended Transport (EXTRAN) block solves St. Venant equations (one-dimensional, unsteady state continuity and momentum partial differential equations) for dynamic wave simulation (Roesner et al., 1988). A maximum of 200 inlets can be simulated in Runoff, Transport, and Extran blocks. The Storage/Treatment block simulates the routing of flows and pollutants (maximum of 3 pollutants) through a dry- or wet-weather Storage/Treament tank containing up to 5 units or processes.

The Combine block adds the capability of modeling larger areas by combining the output of the SWMM runs from different blocks. The Statistics block performs simple statistical analyses on output time series. The Graph block outputs the graphical results on a line printer. SWMM produces a time history of flow, stage and constituent concentration at any point in the watershed for Runoff, Transport, Storage/Treatment Blocks.

21.10.1 Where to Get SWMM?

The SWMM is public domain software. EPA SWMM version 4.3 (1994) can be downloaded from the USEPA web site without charge at *http://www.epa.gov/ceampubl/softwdos.htm*. A beta version of SWMM 4.4 is available through the Oregon State University web site *http://www.ccee.orst.edu/swmm/*. The USEPA developed a Windows-based SWMM in 1994. However, it is no longer available from the USEPA.

21.10.2 GIS and Graphical Resources for SWMM

An ARCVIEW extension was developed by Camp, Dresser, and McKee to link Runoff and Extran blocks input/output routines with the GIS ARCVIEW. It can be downloaded from *http://www.esri.com/arcscripts* without charge. The file to be downloaded is arswmm.zip. A keyword of "avswmm" may be used to search on this website. Another project is being done through a USEPA Office of Administration and Resources Management (OARM) contract with Lockheed Martin Information Services (LMIS) to develop an interface between the SWMM and BASINS. BASINS is ARCVIEW-based software developed by the USEPA, and stands for Better Assessment Science Integrating Point and Nonpoint Sources. BASINS can be downloaded from the USEPA web site without charge. The interface between the SWMM and BASINS is expected to be completed by end of 2000.

Private organizations have developed propriety resources for SWMM to assist in data input and output graphics/report. PCSWMM is a menu-driven interface for running the model and providing hydro/pollutographs and animated hydraulic gradelines, plus sensitivity analysis and GIS options (about $400, *http://www.chi.on.ca*). MTVE is a post-processor for Extran (MTVE) and Runoff (MTVR), including the ability to show the dynamic movement of the hydraulic grade line and display Extran and Runoff block networks. (about $1,000, *http://web:people.mw.mediaone.net/10brooks/index.html*). XP-SWMM, Visual SWMM, or MIKE-SWMM is a complete graphical user interface and post-processor for SWMM (about $4,000, available at *http://www.xpsoftware.com*, *http://www.drainage.com*, *http://www.dhisoftware.com/mikeswmm/*).

21.11   *TR-20*

TR-20 is a large computer program for rainfall-runoff simulation, which is equivalent to HEC's HEC-1. It was developed by Natural Resources Conservation Service, formerly the Soil Conservation Service (SCS). The most current version is 2.04.

TR-20 develops flood hydrographs from runoff and routes the flow through stream channels and reservoirs (NRCS, 1992). The program uses procedures described in the SCS National Engineering Handbook, Section 4, Hydrology (SCS, 1985) except for the reach flood routing procedure. Peak discharges, their times of occurrence, water surface elevations, and duration of flows can be computed at any desired cross section or structure. Complete discharge hydrographs, as well as discharge hydrograph elevations, can be obtained if requested. The program provides for the analysis of up to nine different rainstorm distributions over a watershed under various combinations of land treatment, floodwater retarding structures, diversions, amd channel modifications. Such analysis can be performed on as many as 200 subwatersheds or reaches and 99 structures in any one continuous run.

The watershed should be divided into as many sub-watersheds as required to define hydrologic and alternative structural effects. Hydrologic effects are influenced by entrance of tributaries, watershed shape, valley slope changes, homogeneity of the runoff curve number, and existing or proposed water impoundment structures. Each sub-watershed is assumed to be hydrologically homogeneous and should not have an area greater than 25 square miles. Either actual or synthetic cumulative rainfall distributions may be used. The runoff volume (Q) in inches is computed using the SCS runoff equation outlined in the SCS National Engineering Handbook, Section 4, Hydrology (SCS, 1985). If the user requests, the CN will be adjusted for the low or high antecedent runoff condition (ARC). The unit hydrograph method is based on NEH-4 (SCS, 1985). The maximum of 400 coordinates applies to all hydrographs that are generated by the operations in the program. The composite flood hydrograph is routed through a reservoir using the Storage-Indication methods described in Chapter 17, NEH-4 (SCS, 1985). The composite flood hydrograph is routed through a valley reach using a Modified Attenuation-Kinematic (Modified Att-Kin) method as described in Hydrology Note 2.

21.11.1   Where to Get TR-20?

TR-55 is public domain software and can be downloaded without charge from *http://www.wcc.nrcs.usda.gov/water/quality/common/tr20/tr20.html*.

21.11.2   GIS and Graphical Resources for TR-20

Watershed Modeling System (WMS), developed by Brigham Young Unversity, can be used to pre-process GIS data and digitial elevation data, delineate watersheds, compute the sub-basin parameters, and develop a TR-20 input file. WMS can be purchased from *http://www.ems-i.com* or *http://www.bossintl.com* (about $1,800 per copy).

21.12   TR-55

TR-55 is a storm runoff computation software package for calculating storm runoff volume, peak discharge, hydrographs, and storage volumes for small urban and rural watersheds. It was developed by the Natural Resources Conservation Service (NRCS) of the U.S. Department of Agriculture. The current version is 2.0. TR-55 is consistent with the procedures documented in Technical Release 55 (NRCS, 1986).

TR-55 can be used to estimate the runoff volume based on the SCS runoff curve number method (SCS, 1985). The curve number is determined based on land-use cover type, land treatment, percent of impervious areas, soil hydrologic condition, and antecedent moisture content. The runoff volume is computed based on the computed curve number or composite curve number and the given total rainfall depth.

The peak discharge can also be computed in TR-55. There are two methods for computing peak discharge in TR-55: Graphical Peak Discharge and the Tabular Hydrograph. For both methods, the time of concentration needs to be first estimated. For sheet flow of less than 300 feet, Manning's kinematic solution is used (NRCS, 1986). For shallow concentrated flow (more than 300 feet wide), the average velocity must be first estimated based on a relationship between the average velocity and watercourse slope and type of channel. Dividing the watercourse length by the average velocity gives the time of concentration for shallow concentrated flow. For open-channel flow, Manning's equation is used to compute the average flow velocity, which is then used to compute the time of concentration. After the time of concentration is computed, the peak discharge can be estimated. For the Graphical Peak Discharge method, the required parameters are the time of concentration, drainage area, curve number, and unit peak discharge (csm/in) for different rainfall distribution (Type I, Type IA, type II, or Type III). Herein, csm/in stands for cfs/sq. mile/in of runoff. The unit peak discharge for different rainfall distribution can be found in NRCS (1986). For the Tabular Hydrograph method, the major required parameters are drainage areas for sub-basins, time of concentration for each sub-basin, travel time in the routing reaches, weighted curve number for each sub-basin, and total runoff. The relationships between the unit peak discharge (csm/in) and time of concentration for different rainfall types can be found in NRCS (1986). The Tabular Hydrograph method can also be used to compute the composite flood hydrograph.

21.12.1   Where to Get TR-55?

TR-55 is public domain software and can be downloaded without charge from *http:// www.wcc.nrcs.usda.gov/water/quality/common/tr55/tr55.html.*

21.12.2   GIS and Graphical Resources for TR-55

The Watershed Modeling System (WMS), developed by Brigham Young Unversity, can be used to pre-process GIS data and digitial elevation data, delineate watersheds, compute the sub-basin parameters, and develop a TR-55 input file. WMS can be purchased from *http:// www.ems-i.com* or *http://www.bossintl.com* (about $1,800 per copy).

21.13   UNET

UNET or HEC-UNET version 3.2 (HEC, August 1997) is public domain software. It simulates one-dimensional unsteady flow through a full network of open channels. UNET was developed by Dr. Robert L. Barkau and is maintained and distributed by the Hydrologic

Engineering Center (HEC). The mathematical equations in UNET are based on one-dimensional continuity and momentum unsteady state partial differential equations. The numerical solution method is based on linearized and implicit finite differences method.

The internal boundary conditions that can be dealt with in UNET include (1) levee failures and storage interactions, (2) gated spillways and weir overflow structures, (3) bridge and culvert hydraulics, and (4) pumped diversions. The UNET program has two procedures for modeling bridge crossings. The first procedure is the normal bridge procedure which is similar to the normal bridge procedure, used by the HEC-2 backwater program. The normal bridge procedure is preferred when the embankments are low and greatly submerged. The second procedure is the special bridge procedure, which models the bridge crossing as an interior boundary condition that substitutes a family of free and submerged rating curves for the unsteady flow equations. The UNET system models culverts using a set of free and submerged flow rating curves. UNET has the capability of calculating the free and submerged flow rating curves for a system of up to five parallel culverts and four overflow weirs. The culverts can be circular pipes, pipe arches, or box culverts. Corrugated metal and concrete materials are supported, as well as many different types of entrance conditions. Flows in closed conduits (circular and rectangular) are also modeled in UNET based on Preissmann Slot method. On-line and off-line storage areas can be modeled in UNET. An on-line storage area is located at the termination of a stream reach while an off-line storage area is connected to a channel reach by a lateral spillway. Levee modeling and levee failure simulation can be modeled in UNET based on a network of storage cells and parallel channel.

21.13.1 Where to Get UNET?

UNET is public domain software. The software (version 3.2, 1997) and the manuals can be downloaded without charge from *http://www.hec.usace.army.mil*.

21.14 *SUMMARY*

The major functions of thirteen most commonly used public domain computer models for stormwater systems design have been discussed in this chapter. The thirteen models are BASINS, FEQ, HEC-1, HEC-GeoRAS, HEC-RAS, HEC-HMS, HSPEXP, HSPF, HYDRAIN, SWMM, TR-20, TR-55, and UNET. The following is a list of the web sites from which the software and documentation can be obtained.

1. BASINS (*http://www.epa.gov/ost/basins*)
2. FEQ (*http://water.usgs.gov/software/*)
3. HEC-1 (*http://www.hec.usace.army.mil*)
4. HEC-GeoRAS (*http://www.hec.usace.army.mil*)
5. HEC-RAS (*http://www.hec.usace.army.mil*)
6. HEC-HMS (*http://www.hec.usace.army.mil*)
7. HSPEXP (*http://water.usgs.gov/software/*)
8. HSPF (*http://water.usgs.gov/software/*)
9. HYDRAIN (*http://www.fhwa.dot.gov/bridge/hydrain.htm*)
10. SWMM (*http://www.epa.gov/ceampubl/softwdos.htm*)
11. TR-20 (*http://www.wcc.nrcs.usda.gov/water/quality/common/tr20/tr20.html*)
12. TR-55 (*http://www.wcc.nrcs.usda.gov/water/quality/common/tr55/tr55.html*)
13. UNET (*http://www.hec.usace.army.mil*)

Due to rapid development in computer industry such as object-oriented programming, Geographical Information System (GIS), and INTERNET/INTRANET technologies, the computer models are constantly updated and improved, which gives model developers and users much more challenge than before. One of the positive aspects of this rapid development is that the computer models will be more powerful and easier-to-use. The revolution in information technology will definitely revolutionize civil engineering modeling and design. However, engineering judgement is still critical to the success in engineering analysis and design no matter how easy-to-use computer models may become. Data preparation and results interpretations require sound engineering judgement. Never assume the computer-generated results are always correct. Systematic review and testing are necessary for any new computer models for stormwater systems design.

REFERENCES

Bicknell, B. R., J. C. Imhoff, J. L. Kittle, Jr., A. S. Donigian, Jr., and R. C. Johanson, *Hydrological Simulation Program—Fortran: User's Manual for Version 11*, U.S. Environmental Protection Agency, National Exposure Research Laboratory, Athens, GA, EPA/600/R-97/080, 755 pp., 1997.

Donigian, A. S., Jr., J. C. Imhoff, Brian Bicknell, J. L. Kittle, Jr., *Application Guide for Hydrological Simulation Program—Fortran (HSPF)*: U.S. Environmental Protection Agency, Environmental Research Laboratory, Athens, GA, EPA-600/3-84-065, 177 pp., 1984.

Franz, D. D., and C. S. Melching, "Full Equations Utilities (FEQUTL) Model for the Approximation of Hydraulic Characteristics of Open Channels and Control Structures during Unsteady Flow," U.S. Geological Survey Water-Resources Investigations Report 97-4037, 205 pp., 1997.

Franz, D. D., and C. S. Melching, "Full Equations (FEQ) Model for the Solution of the Full, Dynamic Equations of Motion for One-dimensional Unsteady Flow in Open Channels and through Control Structures," U.S. Geological Survey Water-Resources Investigations Report 96-4240, 258 pp., 1997.

Huber, W. C., and R. E. Dickinson, "Stormwater Management Model (SWMM) User's Manual, Version 4.0," program Developed by Environmental Research Laboratory Office of Research and Development U.S. Environmental Protection Agency, 1992.

Johanson, R. C., J. D. Imhoff, and H. H. Davis, Jr., *User's Manual for Hydrological Simulation Program—Fortran (HSPF)*, Environmental Research Laboratory, EPA-600/9-80-015, Athens, GA, April 1980.

Lumb, A. M., R. B. McCammon, and J. L. Kittle, Jr., "User's Manual for an Expert System (HSPEXP) for Calibration of the Hydrologic Simulation Program—Fortran," U.S. Geological Survey Water-Resources Investigations Report 94-4168, 102 pp., 1994.

Roesner, L. A., J. A. Aldrich, and R. E. Dickinson, *Stormwater Management Model User's Manual, Version 4*, Addendum I, EXTRAN, Environmental Protection Agency, Athens, GA, 203 pp., 1988.

U.S. Army Corps of Engineers (USACE), Engineer Manual 1110-2-1411: Standard Project Flood Determinations. U.S. Army, Washington, DC, 1952.

U.S. Army Corps of Engineers (USACE), Hydrologic Engineering Center, *HEC-DSS User's Guide and Utility Manuals: User's Manual*, U.S. Army Corps of Engineers, Davis, CA, Oct. 1994.

U.S. Army Corps of Engineers (USACE), Hydrologic Engineering Center, *UNET version 3.2, One-dimensional Unsteady Flow through a Full Network of Open Channels, User's Manual*, U.S. Army Corps of Engineers, Davis, CA, August 1997.

U.S. Army Corps of Engineers (USACE), Hydrologic Engineering Center, *HEC-1 Flood Hydrograph Package: User's Manual*, U.S. Army Corps of Engineers, Davis, CA, June 1998.

U.S. Army Corps of Engineers (USACE), Hydrologic Engineering Center, *HEC-RAS, River Analysis System, Version 2.2, Hydraulic Reference Manual*, U.S. Army Corps of Engineers, Davis, CA, 1998.

U.S. Army Corps of Engineers (USACE), Hydrologic Engineering Center, *HEC-GeoRAS, An Application for Support of HEC-RAS using ARC/INFO, User's Manual*, U.S. Army Corps of Engineers, Davis, CA, Mar. 1999.

U.S. Army Corps of Engineers (USACE), Hydrologic Engineering Center, *Hydrologic Modeling System (HEC-HMS): Technical Reference Manual*, U.S. Army Corps of Engineers, Davis, CA, Mar. 2000.

U.S. Army Corps of Engineers (USACE), Hydrologic Engineering Center, *HEC-GeoRAS, An Extension for Support of HEC-RAS using ARCVIEW, User's Manual,* U.S. Army Corps of Engineers, Davis, CA, Apr. 2000.

U.S. Environmental Protection Agency, *Better Assessment Science Integrating Point and Nonpoint Sources, BASINS Version 2.0, User's Manual,* prepared by Tetra Tech for Exposure Assessment Branch Standards and Applied Science Division Office of Science and Technology Office of Water U.S. Environmental Protection Agency, 401 M Street, SW Washington, DC 20460, 1998.

U.S. Federal Highway Administration (USFHWA), *Hydraulic Design of Highway Culverts,* Hydraulic Design Series No. 5, U.S. Department of Transportation, Washington D.C., Sept. 1985.

U.S. Federal Highway Administration (USFHWA), *User's Manual for WSPRO—A Computer Model for Water Surface Profile Computations,* Publication No. FHWA-IP-89-027, Sept. 1990.

U.S. Federal Highway Administration (USFHWA), *User's Manual for HYDRAIN Integrated Drainage Design Computer System: Version 6.1,* Office of Engineering, Office of Technology Applications, 400 Seventh Street, SW, Washington, DC 20590, 1999.

U.S. Natural Resources Conservation Service (NRCS), *Urban Hydrology for Small Watersheds,* TR-55, U.S. Department of Agriculture, 1986.

U.S. Natural Resources Conservation Service (NRCS), *Computer Program for Project Formulation Hydrology,* TR-20, U.S. Department of Agriculture, 1992.

U.S. National Weather Service, Technical Paper 40: Rainfall Frequency Atlas for the United States for Durations from 30 Minutes to 24 Hours and Return Periods from 1 to 100 Years, U.S. Department of Commerce, Washington, DC, 1961.

U.S. National Weather Service, NOAA Atlas 2: Precipitation-Frequency Atlas of the Western United States. U.S. Department of Commerce, Silver Spring, MD, 1973.

U.S. Soil Conservation Service (SCS), *National Engineering Handbook, Section 4: Hydrology,* U.S. Department of Agriculture, 1985.

CHAPTER 22
RISK/RELIABILITY MODEL FOR DESIGN

Yeou-Koung Tung
Department of Civil Engineering
Hong Kong University of Science & Technology
Clearwater Bay, Kowloon, Hong Kong

Larry W. Mays
Department of Civil and Environmental Engineering
Arizona State University
Tempe, Arizona

Ben C. Yen
Department of Civil and Environmental Engineering
University of Illinois
Urbana, Illinois

22.1 INTRODUCTION

A stormwater collection system is often designed with the goal to effectively collect excess surface runoffs and efficiently remove them from areas posing threat to human safety or having adverse effects on human activities. Design of stormwater collection systems is a decision-making process normally involving multiple tasks, including data analysis, hydrological modeling, hydraulic investigations, structural design, and economic studies. In each task, uncertainties of various degrees exist that prohibit designers to put their complete confidence on the satisfactory performance of the system at all times.

When facing uncertainties, the designers' general attitude is that of conservatism, which leads to the use of safety factor by either inflating load to the system and/or discounting system capacity. Conventionally, the value of safety factor is subjectively determined on the basis of individual experience and judgement. Although the use of safety factor could lead to a safer design, the conventional approach is unable to quantitatively assess the reliability of the system which allows further consideration of the tradeoff in reliability and project costs.

The intent of this chapter is to provide an overview of some practical procedures applicable in dealing with risk and reliability analyses of stormwater collection systems. More detailed descriptions of the reliability assessment techniques can be found elsewhere (Tung, 1996, 1999).

22.1.1 Uncertainties in Stormwater Collection System Design

Uncertainty is primarily attributed to our lack of complete knowledge about the phenomena and processes involved in problem definition and resolution. Uncertainty due to inherent randomness of physical processes cannot be eliminated whereas those uncertainties associated with incomplete knowledge about processes, models, parameters, and data can be reduced through engaging research, data collection, and careful manufacturing.

Like any engineering design, uncertainties exist in the design of various types of stormwater collection systems, which may arise from hydraulic, hydrologic, structural, environmental, and socioeconomic aspects. Uncertainties in hydrologic aspect can further be classified into inherent, parameter, or model uncertainties. *Hydraulic uncertainty* refers to the uncertainty in the design of hydraulic structures and in the analysis of the performance of hydraulic structures. *Structural uncertainty* refers to failure from structural weaknesses and *economic uncertainty* can arise from uncertainties in various cost items, inflation, project life, and other intangible factors.

To categorize differently, uncertainties in any hydraulic engineering design could arise from various sources (Yen et al., 1986) including, but not limited to, nature processes, model, parameter, data, and manufacturing and operation of the system. *Model uncertainty* arises from a model that may not represent the real physical processes of flows due to simplifications and idealizations. As a result, peak discharges or hydrographs produced by a hydrologic or hydraulic model are not consistent with those observed in the field. Another example of model uncertainty can be found in the determination of proper probability distribution function in frequency analysis of hydrologic extremes on the basis on limited amount of data. *Parameter uncertainty* refers to our lack of confidence about the inputs and parameters of a model. *Data uncertainties* arise from errors in measurement, transmission, and recording, as well as possible inconsistency. *Operational* and *manufacturing uncertainties* refer to those changes in physical characteristics of a system, such as pipe settlement and sediment deposition, as well as manufacturing tolerance of producing components of hydraulic structures.

22.1.2 Measures of Uncertainty

The most complete and ideal way to describe the degree of uncertainty of a parameter, a function, a model, or a system in hydraulic engineering design is the probability density function (PDF) of the quantity subject to uncertainty. However, such a probability function cannot be derived or found in most practical problems. Alternative ways of expressing the uncertainty of a quantity include confidence intervals or statistical moments. In particular, the second-order moment, that is, the variance or standard deviation, is a measure of the dispersion of a random variable. The coefficient of variation, defined as the ratio of standard deviation to the mean, is also often used.

22.1.3 Implications of Uncertainty in Stormwater Collection Systems Design

The presence of various uncertainties (including inherent randomness of natural processes) is the main contributor to the potential structural or performance failure of hydraulic engineering systems. Knowledge of uncertainty features of hydraulic engineering systems is essential for assessing their reliability.

The layout, capacity, and operation of a stormwater collection system largely based on the system response under some anticipated design conditions. When inputs, parameters, or characteristics of the system are subject to uncertainty, the system's responses cannot be assessed with certainty. An engineer should consider various criteria including, but not limited to, the cost of the system, failure probability, and consequences of failure, such that a proper design can be made for the system. The task of uncertainty analysis is to determine the uncertainty features of the system performance as a function of uncertainties in model and involved random basic variables. *Uncertainty analysis* provides a formal and systematic framework to quantify the uncertainty associated with the system output. Furthermore, it offers the designer useful insights with regard to the contribution of each random basic variable to the overall uncertainty of the system outputs. Such knowledge is essential to identify the "important" parameters to which more attention should be given to better assess their values and, accordingly, to reduce the overall uncertainty of the system outputs.

22.1.4 Reliability of Stormwater Collection Systems

Hydraulic engineering systems are often placed in a natural environment subject to various external stresses or loads. The *resistance* or *strength* of a system is its ability to accomplish the intended mission satisfactorily without failure when subject to external loads. *Failure* will occur when the load exceeds the resistance of the system. For a stormwater collection system, from the viewpoint of hydraulic performance, *external loads* are surface runoff hydrographs produced by rainfall events, whereas the resistance is the flow carrying capacity of the system.

From the previous discussions, the capacity of a stormwater collection system and the imposed loads, more often than not, are random and are subject to uncertainty. Hence, the design and operation of any stormwater collection systems are always subject to uncertainties and the system could fail potentially.

The *reliability, p_s,* of a stormwater collection system is defined as the probability of nonfailure in which the resistance of the system exceeds the load, that is,

$$p_s = P_r[R > L] \tag{22.1a}$$

where $P_r[\cdot]$ = probability

The *failure probability, p_f,* is the complement of the reliability, which can be expressed as

$$p_f = P_r[R < L] = 1 - p_s \tag{22.1b}$$

In hydraulic engineering system design and analysis, loads generally arise from natural events, such as floods and storms, which occur randomly in time and in space. A common practice for determining the reliability of a hydraulic engineering system is to assess the return period or recurrence interval of the design event. Frequency analysis of hydrologic extremes, such as floods and rainstorms, using the annual maximum series is a typical example of this kind. The main disadvantage of using the return period method is that reliability is measured only in terms of time of occurrence of loads without considering the interactions with the system resistance (Melchers, 1987).

Two other types of safety measures that consider the relative magnitudes of resistance and load are frequently used in engineering practice. One is the *safety margin (SM)*, defined as the difference between the resistance and the load, that is,

$$SM = R - L \tag{22.2}$$

The other is the *safety factor (SF)*, a ratio of resistance to load, which is defined as

$$SF = R/L \tag{22.3}$$

Yen (1979) summarized several types of safety factors and discussed their applications in hydraulic engineering system design.

There are two basic probabilistic approaches to evaluate the reliability of a hydraulic engineering system. The most direct approach is a statistical analysis of data of past failure records for similar systems. The other approach is through the reliability analysis, which considers and combines the contribution of each factor potentially influencing the performance of the system. The former is a lumped-system approach requiring no knowledge about the behavior of the facility or structure, or its load and resistance.

Two major steps are involved in the second approach of the reliability analysis: (1) identifying and analyzing the uncertainties of each contributing factor, and (2) combining the uncertainties of the random factors to determine the overall reliability of the systems. The second step, in turn, may proceed in two different ways: (1) directly combining the uncertainties of all factors, or (2) separately dealing with the uncertainties of the factors belonging to different subsystems and evaluating the respective subsystem reliability before combining the subsystems' reliabilities to yield the overall system reliability. The first way applies to very simple structures whereas the second way is more suitable for complicated systems.

There exist many techniques applicable to uncertainty and reliability analyses of stormwater collection system design. The following sections describe and demonstrate the usage of some practical techniques through examples. For more detailed descriptions of the various other techniques, readers are referred to Tung (1996, 1999).

22.1.5   Uncertainty Analysis by First-Order Variance Estimation (FOVE) Method

The *FOVE method,* also called the *variance propagation method* (Berthouex, 1975), estimates the uncertainty features of a model output based on the statistical properties of random basic variables consisting of model inputs and parameters. The FOVE method approximates a model involving random variables by the Taylor series expansion.

Consider that a hydraulic or hydrologic model output W is related to N random basic variables $X_1, X_2, \ldots, X_N$ as

$$W = g(\mathbf{X}) = g(X_1, X_2, \ldots, X_N) \tag{22.4}$$

where $\mathbf{X} = (X_1, X_2, \ldots, X_N)^t$, an N-dimensional column vector of random basic variables, the superscript t represents the transpose of a matrix or vector.

The Taylor series expansion of the function $g(\mathbf{X})$ with respect to the mean values of random basic variables, $\mathbf{X} = \boldsymbol{\mu}$ can be expressed as

$$W = g(\boldsymbol{\mu}) + \sum_{i=1}^{N} \left(\frac{\partial g(\mathbf{X})}{\partial X_i} \right)_{\boldsymbol{\mu}} (X_i - \mu_i) + \varepsilon \tag{22.5}$$

where μ_i = the mean of the ith random basic variable X_i
ε = the higher-order terms

The first-order partial derivative terms are called the *sensitivity coefficients,* each representing the rate of change of model output W with respect to unit change of each variable at $\boldsymbol{\mu}$.

By truncating higher-order terms in Eq. (22.5), the first-order approximation of the mean and variance of W at $\mathbf{X} = \boldsymbol{\mu}$, respectively, are

$$\mu_W \approx g(\boldsymbol{\mu}) = \overline{w} \tag{22.6}$$

and

$$\sigma_W^2 \approx \sum_{i=1}^{N} \sum_{j=1}^{N} \left(\frac{\partial g(X)}{\partial X_i} \right)_{\boldsymbol{\mu}} \left(\frac{\partial g(X)}{\partial X_j} \right)_{\boldsymbol{\mu}} Cov(X_i, X_j) = \mathbf{s}^t \mathbf{C}(\mathbf{X}) \mathbf{s} \tag{22.7}$$

where $\mathbf{s} = \nabla_x W(\boldsymbol{\mu})$ = N-dimensional vector of sensitivity coefficients evaluated at $\boldsymbol{\mu}$ with $\nabla_x = (\partial/\partial X_1, \partial/\partial X_2, \ldots, \partial/\partial X_N)^t$ being the gradient operator
$\mathbf{C(X)}$ = the variance-covariance matrix of the random vector, $\mathbf{X}$

When all the random basic variables are statistically independent, the variance of model output W can be approximated as

$$\sigma_W^2 \approx \sum_{i=1}^{N} s_i^2 \sigma_i^2 = \mathbf{s}^t \mathbf{D} \mathbf{s} \tag{22.8}$$

where σ = the standard deviation
$\mathbf{D}$ = diag $(\sigma_1^2, \sigma_2^2, \ldots, \sigma_N^2)$, a diagonal matrix of variances of the random basic variables

From Eq. (22.8), the ratio of $s_i^2 \sigma_i^2 / \sigma_W^2$ indicates the proportion of overall uncertainty in the model output contributed by the uncertainty associated with the random basic variable X_i.

Example 22.1
Manning's equation is frequently used to determine the flow capacity of storm sewer by

$$Q_c = 0.463 \, \lambda_m \, n^{-1} \, d^{2.67} \, S^{0.5}$$

where Q_c = flowrate (in cfs)
λ_m = the model correction factor, representing model uncertainty
n = the roughness coefficient
d = the sewer diameter (in ft)
S = pipe slope (in ft/ft)

The uncertainty associated with the sewer slope due to installation error is 7% of its intended value 0.001. The model correction factor has a mean of 1.10 with a coefficient of variation of 0.11. Determine the uncertainty features of the sewer flow capacity using the FOVE method for a section of 5-ft sewer with a 1% error in diameter due to manufacturing tolerances. The roughness coefficient has the mean value 0.015 with a coefficient of variation 0.05. Assume that the correlation coefficient between the roughness coefficient n and sewer diameter d is -0.75. The sewer slope S is uncorrelated with the other two random variables.

Solution
The first-order Taylor series expansion of Manning's equation about $\lambda_m = \mu_{\lambda m} = 1.10$; $n = \mu_n = 0.015$, $d = \mu_d = 5.0$, and $S = \mu_S = 0.001$, according to Eq. (22.5), is

$$Q_c \approx 0.463 \, (1.10) \, (0.015)^{-1} \, (5)^{2.67} \, (0.001)^{0.5} + [\partial Q_c/\partial \lambda_m] \, (\lambda_m - 1.1)$$

$$+ [\partial Q_c/\partial n] \, (n - 0.015) + [\partial Q_c/\partial d] \, (d - 5.0)$$

$$+ [\partial Q_c/\partial S] \, (S - 0.0001)$$

$$= 78.49 + [0.463 \, (0.015)^{-1} \, (5.0)^{2.67} \, (0.001)^{0.5}] \, (\lambda_m - 1.1)$$

$$+ [0.463 \, (-1) \, (1.1) \, (0.015)^{-2} \, (5.0)^{2.67} \, (0.001)^{0.5}] \, (n - 0.015)$$

$$+ [0.463 \, (2.67) \, (1.1) \, (0.015)^{-1} \, (5.0)^{1.67} \, (0.001)^{0.5}] \, (d - 5.0)$$

$$+ [0.463 \, (0.5) \, (1.1) \, (0.015)^{-1} \, (5.0)^{2.67} \, (0.001)^{-0.5}] \, (S - 0.001)$$

$$= 78.49 + 71.74(\lambda_m - 1.1) - 5260.67(n - 0.015)$$

$$+ 42.14(d - 5.0) + 39455.03(S - 0.001)$$

Based on Eq. (22.6), the approximated mean value of the sewer flow capacity is

$$\mu_{Q_c} \approx 78.49 \text{ cfs}$$

According to Eq. (22.7) and knowing $Cov(n, S) = Cov(d, S) = 0$, the approximated variance of the sewer flow capacity is

$$\sigma^2_{Q_c} \approx (71.74)^2 \, \sigma^2_{\lambda m} + (-5260.67)^2 \, \sigma^2_n$$

$$+ (42.14)^2 \, \sigma^2_d + (39455.03)^2 \, \sigma^2_S + 2(5260.67)(42.14) \, Cov(n, d)$$

where $\sigma_{\lambda m} = (0.11)(1.1) = 0.121$
$\sigma_n = (0.05)(0.015) = 0.00075$
$\sigma_d = (0.01)(5.0) = 0.05$
$\sigma_S = (0.07)(0.001) = 0.00007$

Then, the variance of the sewer flow capacity can be computed as

$$\sigma^2_{Q_c} \approx (71.74 \times 0.121)^2 + (-5260.67 \times 0.00075)^2 + (42.14 \times 0.05)^2$$

$$+ (4100.99 \times 0.00007)^2 + 2(-0.75)(-5260.67 \times 0.00075)$$

$$(42.14 \times 0.05)$$

$$= 8.68^2 + 3.95^2 + 2.11^2 + 2.76^2 + 12.50 = 115.48 \text{ cfs}^2$$

The corresponding standard deviation of the sewer flow capacity is $\sqrt{115.48} = 10.75$ cfs, which is 13.6% of the estimated mean sewer flow capacity.

Without considering correlation between n and d, $\sigma_{Q_c}^2 = 115.48 - 12.50 = 102.98$ cfs^2, which underestimates the variance of the sewer flow capacity by approximately 10%. The percentages contribution of uncertainty of λ_m, n, d, and S to the overall uncertainty of the sewer flow capacity, under the uncorrelated condition, are 73.2%, 15.1%, 4.3%, and 7.4%, respectively. The most dominant random basic variable contributing to the total uncertainty of sewer flow capacity is the model correction factor λ_m, followed by Manning's roughness coefficient as a distant second. The combination of the two basic variables contributes close to 90% of the total variability associated with the sewer flow capacity. Uncertainty associated with the pipe diameter is the least significant among the four random basic variables in Manning's equation. It is interesting to note that the sewer flow capacity is most sensitive to the sewer slope, that is, $\partial Q_c / \partial S = 39455$. However, its contribution to $Var(Q_c)$ is only 7.4%. This is because the variance of S, σ_S^2, is significantly smaller than the variances of all other random basic variables.

22.2   RELIABILITY ANALYSIS METHODS

In stormwater collection system design and analysis problems, uncertainties in data and in theory, including design and analysis procedures, warrant a probabilistic treatment of the problems. The risk associated with the potential failure of a stormwater collection system is the result of the combined effects of two aspects: (1) inherent randomness of hydrological inflows, and (2) various uncertainties involved in the analysis, design, and construction, operational procedures that affect the capacity of the system. Hence, reliability evaluation of a stormwater collection system requires uncertainty analysis of system's input/output relation.

In stormwater collection system design, the resistance and load are frequently functions of several random variables, that is, $L = g(\mathbf{X}_L) = g(X_1, X_2, \ldots, X_m)$ and $R = h(\mathbf{X}_R) = h(X_{m+1}, X_{m+2}, \ldots, X_N)$, where $X_1, X_2, \ldots, X_N$ are random basic variables defining the load function, $g(\mathbf{X}_L)$, and the resistance function, $h(\mathbf{X}_R)$. Referring to Eq. (22.1a), the reliability can be written, in terms of random basic variables, as

$$p_S = P_r[g(\mathbf{X}_L) < h(\mathbf{X}_R)] \tag{22.9}$$

22.2.1 Performance Functions and Reliability Index

Equation (22.9) can alternatively be written, in terms of a *performance function, W(X)* = $W(\mathbf{X}_L, \mathbf{X}_R)$, as

$$p_s = P_r[W(\mathbf{X}_L, \mathbf{X}_R) > 0] = P_r[W(\mathbf{X}) > 0] \tag{22.10}$$

where $\mathbf{X}$ = the vector of random basic variables in the load and resistance functions.

In reliability analysis, the system state is divided into safe (satisfactory) set, defined by $W(\mathbf{X}) > 0$, and failure (unsatisfactory) set, defined by $W(\mathbf{X}) < 0$ (see Fig. 22.1). The boundary separating the safe set and failure set is the critical surface, defined by $W(\mathbf{X}) = 0$, which is also called the *limit-state function*. Since the performance function $W(\mathbf{X})$ defines the condition of the system, it is sometimes called the *system state function*.

The performance function $W(\mathbf{X})$ in Eq. (22.10) can be expressed in the form of safety margin as

$$W_1(\mathbf{X}) = R - L = h(\mathbf{X}_R) - g(\mathbf{X}_L) \tag{22.11a}$$

or of safety factor as

$$W_2(\mathbf{X}) = (R/L) - 1 = [h(\mathbf{X}_R)/g(\mathbf{X}_L)] - 1 \tag{22.11b}$$

$$W_3(\mathbf{X}) = \ln(R/L) = \ln[h(\mathbf{X}_R)] - \ln[g(\mathbf{X}_L)] \tag{22.11c}$$

Also, the *reliability index, β,* is a frequently used indicator in reliability analysis. The reliability index was first introduced by Cornell (1969) and was later formalized by Ang and Cornell (1974). It is defined as the ratio of the mean to the standard deviation of the performance function $W(\mathbf{X})$,

$$\beta = \frac{\mu_w}{\sigma_w} \tag{22.12}$$

where μ_w and σ_w = the mean and standard deviation of the performance function, respectively

From Eq. (22.10), assuming an appropriate PDF for the random performance function, $W(\mathbf{X})$, the reliability then can be computed as

$$p_S = \int_0^\infty f_W(u)\,du = 1 - F_W(0) = 1 - F_{w'}(-\beta) \tag{22.13}$$

where $f_W(\cdot)$ and $F_W(\cdot)$ = the probability density function (PDF) and cumulative distribution function (CDF) of the performance function W, respectively
W' = the standardized performance function defined as $(W - \mu_W)/\sigma_W$

For practically all probability distributions used in the reliability analysis, the value of the reliability, p_s, is a strictly increasing function of the reliability index, β. In practice, the normal distribution is commonly used for $W(\mathbf{X})$ in which case the reliability can be simply computed as

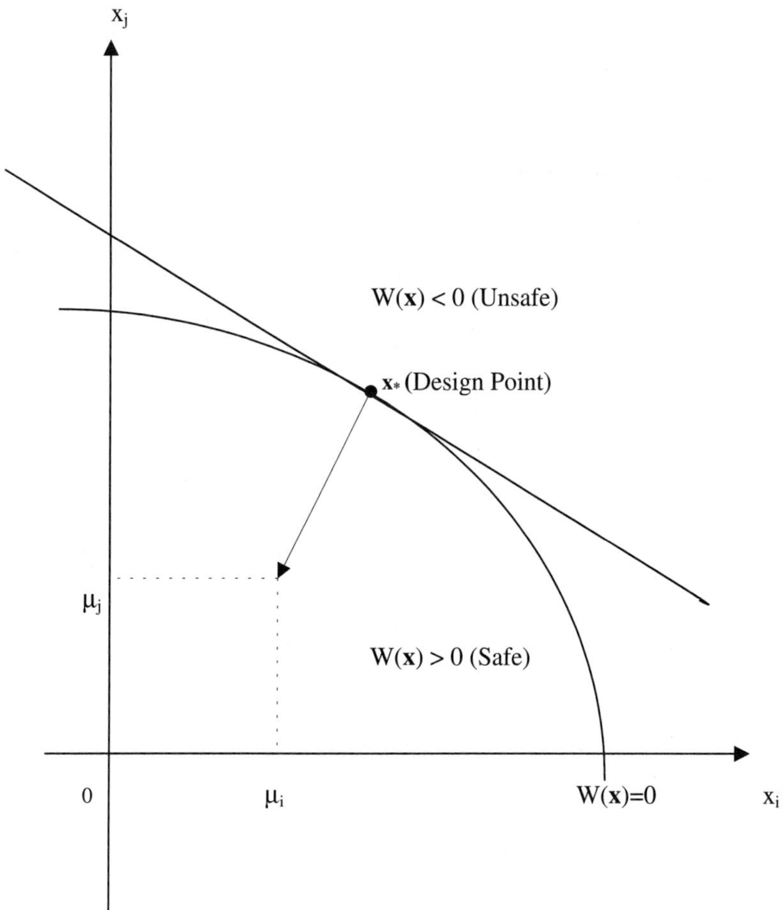

FIGURE 22.1   System state defined by the performance function $W(\mathbf{x})$.

$$p_s = 1 - \Phi(-\beta) = \Phi(\beta) \tag{22.14}$$

where $\Phi(\beta)$ = the standard normal CDF whose values can be found in many statistics textbooks

22.2.2   First-Order Second-Moment (FOSM) Reliability Methods

By the *FOSM reliability methods,* the performance function $W(\mathbf{X})$, defined on the basis of the load and resistance functions, $g(\mathbf{X}_L)$ and $h(\mathbf{X}_R)$, are expanded in Taylor series at a selected reference point at $\mathbf{X} = \mathbf{x}_o$ as

$$W(\mathbf{X}) = W(\mathbf{x}_o) + \sum_{i=1}^{N} \left(\frac{\partial(\mathbf{X})}{\partial X_i} \right)_{\mathbf{x}_o} (X_i - x_{oi}) + \varepsilon \tag{22.15}$$

Ignoring the higher-order terms in the series expansion, the mean and variance of the performance function $W(\mathbf{X})$ can be approximated, respectively, by

$$\mu_W \approx W(\mathbf{x}_o) + \sum_{i=1}^{N} \left(\frac{\partial W(X)}{\partial X_i} \right)_{\mathbf{x}_o} (\mu_i - x_{oi}) \tag{22.16}$$

$$\sigma_W^2 \approx \sum_{i=1}^{N} \left(\frac{\partial W(X)}{\partial X_i} \right)_{\mathbf{x}_0}^2 \sigma_i^2 \tag{22.17}$$

This simplification greatly enhances the practicality of the first-order methods because in many problems it is rather difficult, if not impossible, to find the PDF of the variables while it is relatively simple to estimate the first two statistical moments. In practice, two types of the FOSM reliability methods are available, namely, the *mean-value first-order second-moment (MFOSM) method* and the *advanced FOSM (AFOSM) method.*

22.2.2.1   MFOSM Reliability Method.   The MFOSM method for the reliability analysis selects the expansion point as the FOVE uncertainty method at the mean values of the random basic variables, that is $\mathbf{x}_0 = \boldsymbol{\mu}$. Once the mean and standard deviation of $W(\mathbf{X})$ are estimated, the reliability is computed according to Eq. (22.14) with the reliability index β_{MFOSM} computed as

$$\beta_{MFOSM} = \frac{W(\boldsymbol{\mu})}{\sqrt{\mathbf{s}'\mathbf{C}(\mathbf{X})\mathbf{s}}} \tag{22.18}$$

where $\boldsymbol{\mu}$ and $\mathbf{C}(\mathbf{X})$ = the vector of means and covariance matrix of the random basic variables $\mathbf{X}$, respectively

$\mathbf{s} = \nabla_x W(\boldsymbol{\mu})$ is the column vector of sensitivity coefficients with each element representing $\partial W/\partial X_i$ evaluated at $\mathbf{x}_0 = \boldsymbol{\mu}$

Example 22.2
Referring to the storm sewer in Example 22.1, suppose that the sewer is to be used for draining runoff produced by the 10-yr, 30-minute storm on a 10-acre basin in Urbana, IL. Determine the reliability that the sewer capacity could handle the peak runoff discharge from the storm. Assume that all random basic variables are statistically independent.

Solution
The load to the sewer is the peak runoff discharge resulting from the 10-yr, 30-minute storm on a 20-acre basin. The peak discharge can be estimated by the Rational formula, that is,

$$Q_L = \lambda_L \, CiA$$

where Q_L = the peak runoff discharge (in cfs)
λ_L = the model correction factor for the Rational formula
C = the runoff coefficient
i = the average rainfall intensity (in in/hr) for the specified return period (T) and storm duration (t_d)
A = the contributing basin area (in acres)

In Urbana, IL, the average rainfall intensity can be estimated by the following intensity-duration-frequency relation (Yen et al., 1976):

$$i\,(\text{in/hr}) = \frac{120 \, T^{0.175}}{27 + t_d}$$

where T = the return period (in years)
t_d = the rainfall duration (in minutes)

Based on the analysis of uncertainty for the Rational formula, the following information can be obtained:

$$\mu_{\lambda_L} = 1.00; \ \mu_C = 0.825; \ \mu_i = 3.15 \text{ in/hr}; \ \mu_A = 20 \text{ ac}$$

$$\Omega_{\lambda_L} = 0.15; \ \Omega_C = 0.071; \ \Omega_i = 0.177; \ \Omega_A = 0.05$$

where μ and Ω = the mean and coefficient of variation, respectively

Due to the multiplicative form of the Rational formula and statistical independence of the random basic variables, the uncertainty of peak runoff discharge, Q_L, in terms of its coefficient of variation, by the FOVE method can be expressed as

$$\Omega_{Q_L}^2 = \Omega_{\lambda_L}^2 + \Omega_C^2 + \Omega_i^2 + \Omega_A^2$$

The mean and coefficient of variation of peak runoff discharge corresponding to the 10-yr, 30-minute rainfall are, respectively,

$$\mu_{Q_L} \approx \mu_{\lambda_L} \mu_C \mu_i \mu_A = (1.00)(0.825)(3.15)(20.0) = 51.98 \text{ cfs}$$

$$\Omega_{Q_L}^2 = \Omega_{\lambda_L}^2 + \Omega_C^2 + \Omega_i^2 + \Omega_A^2 = (0.15)^2 + (0.071)^2 + (0.177)^2 + (0.05)^2 = (0.248)^2$$

Hence, the corresponding variance of runoff peak discharge is

$$\sigma_{Q_L}^2 = (\mu_{Q_L} \times \Omega_{Q_L})^2 = (51.98 \times 0.248)^2 = 12.90^2 = 166.31 \text{ cfs}^2$$

The percentages of contribution of each random basic variable in the Rational formula (that is $\Omega_j^2/\Omega_{Q_L}^2$, are 36.7%, 8.2%, 51.1%, and 4.0% for λ_L, C, i, and A, respectively.

From Example 22.1, the mean and standard deviation of the sewer capacity, under the statistical independence of the random basic variables, are 78.91 cfs and 10.15 cfs, respectively. Therefore, the mean and variance of the performance function $W = Q_C - Q_L$ are, respectively,

$$\mu_W = \mu_{Q_C} - \mu_{Q_L} = 78.49 - 51.98 = 26.51 \text{ cfs};$$

$$\sigma_W^2 = \sigma_{Q_C}^2 + \sigma_{Q_L}^2 = 12.90^2 + 10.15^2 = (16.41 \text{ cfs})^2 = 269.33 \text{ cfs}^2$$

The MFOSM reliability index is $\beta_{MFOSM} = 26.51/16.41 = 1.62$. Assuming a normal distribution for the performance function W, the reliability that the sewer capacity Q_C can accommodate the incoming peak discharge Q_L is

$$p_s = P_r[Q_C > Q_L] = P_r(W > 0) = \Phi(\beta_{MFOSM}) = \Phi(1.62) = 0.94738$$

The corresponding failure probability is $p_f = 1 - p_s = 0.05262$.

Yen and Ang (1971), Ang (1973), and Cheng et al. (1986) indicated that, if the calculated reliability or failure probability is in the extreme tail of a distribution, the shape of the tails of a distribution becomes very critical. In such cases, accurate assessment of the distribution of $W(\mathbf{X})$ should be used to evaluate the reliability or failure probability. The MFOSM method has been used widely in various hydraulic structural and facility designs such as storm sewers (Tang et al., 1975; Yen et al., 1976), culverts (Yen et al., 1980; Tung and Mays, 1980b), and levees (Tung and Mays, 1981a).

Although the MFOSM method is simple and straightforward to use, it possesses a few weaknesses:

1. Inappropriate choice of the expansion point: In reliability computation, one should be concerned with those points in the parameter space that fall on the failure surface or limit-state surface.

2. Inability to handle distributions with large skew coefficient: This is mainly due to the fact that MFOSM method incorporates only the first two moments of the random basic variables involved.

3. Generally poor estimation of the mean and variance of highly nonlinear functions: This is because the first-order approximation of a highly nonlinear function is not accurate.

4. Sensitivity of the computed failure probability to the formulation of the performance function W.

5. Inability to incorporate available information on probability distributions of the basic variables in load and resistance functions.

From the above arguments, the general rule of thumb is not to rely on the result of the MFOSM method if any of the following conditions exist: (1) high accuracy requirements for the estimated reliability or risk, (2) high nonlinearity of the performance function, or (3) many skewed random variables are involved in the performance function.

22.2.2.2 AFOSM Reliability Method.

The *AFOSM method* mitigates the deficiencies associated with the MFOSM method mentioned above, while keeping the simplicity of the first-order approximation. The difference between the AFOSM and MFOSM methods is that the expansion point in the Taylor series expansion by the AFOSM method is located on the failure surface defined by the limit-state equation, $W(\mathbf{x}) = 0$.

In cases where there are several random basic variables in a performance function, the number of possible combinations of such variables satisfying $W(\mathbf{x}) = 0$ is infinite. From the design viewpoint, one is more concerned with the combination of random basic variables that would yield the lowest reliability or highest failure probability. The point on the failure surface associated with the lowest reliability is the one having the shortest distance on the failure surface to the point defined by the means of the random basic variables (see Fig. 22.1). This point is called the *design point* (Hasofer and Lind, 1974) or the most probable failure point (Shinozuka, 1983).

In the uncorrelated standardized parameter space, that is, $\mathbf{X}' = \mathbf{D}^{-1}(\mathbf{X} - \boldsymbol{\mu})$, the design point is the one that has the shortest distance from the failure surface $W(\mathbf{x}') = 0$ to the origin $\mathbf{x}' = \mathbf{0}$. Such a point can be found by solving

$$\text{Minimize} \quad |\mathbf{x}'| = (\mathbf{x}'^t\,\mathbf{x}')^{1/2} \tag{22.19a}$$

$$\text{subject to} \quad W(\mathbf{x}') = 0 \tag{22.19b}$$

in which $|\mathbf{x}'|$ represents the length of the vector $\mathbf{x}'$. Utilizing the Lagrangian multiplier method, the design point can be determined as

$$\mathbf{x}'_* = -\frac{\nabla_{\mathbf{x}'}W(\mathbf{x}'_*)}{|\nabla_{\mathbf{x}'}W(\mathbf{x}'_*)|}\,|\mathbf{x}'_*| = -|\mathbf{x}'_*|\,\boldsymbol{\alpha}_* \tag{22.20}$$

where $\boldsymbol{\alpha}_* = \nabla_{\mathbf{x}'}W(\mathbf{x}'_*)/|\nabla_{\mathbf{x}'}W(\mathbf{x}'_*)|$ = a unit vector emanating from the design point $\mathbf{x}'_*$ and pointing towards the origin (see Fig. 22.2).

The elements of $\boldsymbol{\alpha}_*$ are called the directional derivatives representing the value of the cosine angle between the gradient vector $\nabla_{\mathbf{x}'}W(\mathbf{x}'_*)$ and axes of the standardized variables. Geometrically, Eq. (22.20) shows that the vector $\boldsymbol{\alpha}_*$ is perpendicular to the tangent hyperplane passing through the design point. Recall that $x_i = \mu_i + \sigma_i x'_i$, for $i = 1, 2, \ldots, N$. By the chain rule in calculus, the shortest distance, in terms of the original variables $\mathbf{X}$, can be expressed as

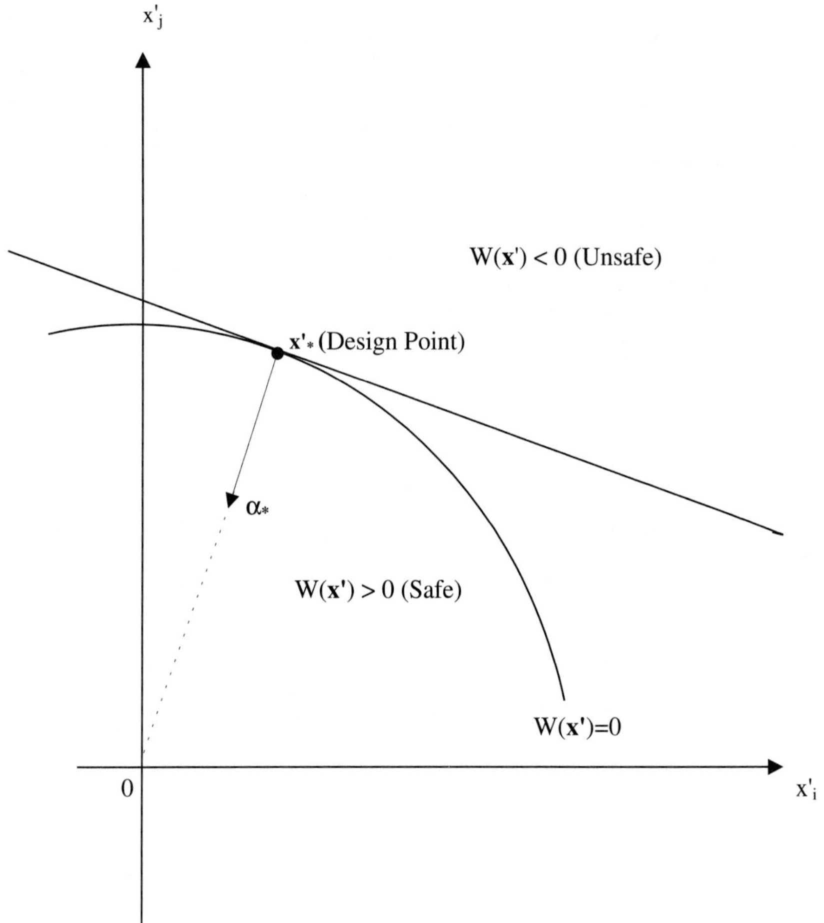

FIGURE 22.2   Design point in standardized basic variable space.

$$|\mathbf{x}'_*| = \frac{\displaystyle\sum_{i=1}^{N} \left(\frac{\partial W(\mathbf{x})}{\partial x_i}\right)_{\mathbf{x}^*} (\mu_i - x_{i*})}{\sqrt{\displaystyle\sum_{j=1}^{N} \left(\frac{\partial W(\mathbf{x})}{\partial x_i}\right)^2_{\mathbf{x}^*} \sigma_i^2}} \tag{22.21}$$

where $\mathbf{x}_* = (x_{1*}, x_{2*}, \ldots, x_{N*})'$ is the point in the original parameter $\mathbf{x}$-space, which can be easily determined from the design point $\mathbf{x}'_*$ in $\mathbf{x}'$-space as $\mathbf{x}_* = \boldsymbol{\mu} + \mathbf{D}^{1/2} \mathbf{x}'_*$

In fact, the shortest distance from the origin to the design point, $|\mathbf{x}'_*|$, is the reliability index based on the first-order Taylor series expansion of the performance function $W(\mathbf{X})$ with the expansion point at x_*.

Referring to Eq. (22.16), the first-order approximation of the performance function, $W(\mathbf{X})$, taking the expansion point $\mathbf{x}_o = \mathbf{x}_*$, is

$$W(\mathbf{X}) \approx \sum_{i=1}^{N} s_{i*} (X_i - x_{i*}) = \mathbf{s}_*^t(\mathbf{X} - \mathbf{x}_*) \tag{22.22}$$

where $\mathbf{s}_* = (s_{1*}, s_{2*}, \ldots, s_{N*})^t = $ a vector of sensitivity coefficients of the performance function $W(\mathbf{X})$ evaluated at the expansion point $\mathbf{x}_*$, that is,

$$s_{i*} = \left[\frac{\partial W(\mathbf{X})}{\partial X_i}\right]_{\mathbf{X}=\mathbf{x}_*} \tag{22.23}$$

The term $W(\mathbf{x}_*)$ is not on the right-hand-side of Eq. (22.22) because $W(\mathbf{x}_*) = 0$. Hence, at the expansion point $\mathbf{x}_*$, the expected value and the variance of the performance function $W(\mathbf{X})$ can be approximated as

$$\mu_W \approx \mathbf{s}_*^t(\boldsymbol{\mu} - \mathbf{x}_*) \tag{22.24a}$$

$$\sigma_W^2 \approx \mathbf{s}_*^t\mathbf{C}(\mathbf{X})\mathbf{s}_* \tag{22.24b}$$

where $\boldsymbol{\mu}$ and $\mathbf{C}(\mathbf{X}) = $ the mean vector and covariance matrix of the random basic variables, respectively

If the random variables are uncorrelated, Eq. (22.24b) reduces to

$$\sigma_W^2 = \sum_{i=1}^{N} s_{i*}^2\sigma_i^2 \tag{22.25}$$

where $\sigma_i = $ the standard deviation of the ith random basic variable

When the random variables are uncorrelated, the standard deviation of the performance function $W(\mathbf{X})$ can alternatively be expressed in terms of the directional derivatives as

$$\sigma_W = \sum_{i=1}^{N} \alpha_{i*}\, s_{i*}\, \sigma_i \tag{22.26}$$

where $\alpha_{i*} = $ the directional derivative for the ith random variable at the expansion point $\mathbf{x}_*$

$$\alpha_{i*} = \frac{s_{i*}\sigma_i}{\sqrt{\sum\limits_{j=1}^{N} s_{j*}^2\sigma_j^2}} \tag{22.27}$$

or in matrix form,

$$\boldsymbol{\alpha}_* = \frac{\mathbf{D}^{1/2}\nabla_\mathbf{x} W(\mathbf{x}_*)}{|\mathbf{D}^{1/2}\nabla_\mathbf{x} W(\mathbf{x}_*)|} \tag{22.28}$$

which is identical to Eq. (22.20). With the mean and standard deviation of the performance function $W(\mathbf{X})$ computed at $\mathbf{x}_*$, the AFOSM reliability index β_{AFOSM} can be determined as

$$\beta_{AFOSM} = \frac{\mu_W}{\sigma_W} = \frac{\sum\limits_{i=1}^{n} s_{i*}(\mu_i - x_{i*})}{\sum\limits_{i=1}^{n} \alpha_{i*}s_{i*}\sigma_i} \tag{22.29}$$

Equation (22.29) is identical to Eq. (22.21) indicating that the AFOSM reliability index β_{AFOSM} is identical to the shortest distance from the origin to the design point in the standardized parameter space. This reliability index β_{AFOSM} is called the *Hasofer-Lind reliability index.*

Once the value of β_{AFOSM} is computed, the reliability can be estimated by $p_s = \Phi(\beta_{AFOSM})$. Since $\beta_{AFOSM} = |\mathbf{x}'_*|$, the sensitivity of β_{AFOSM} with respect to the uncorrelated, standardized random variables is

$$\nabla_{\mathbf{x}'}\beta_{AFOSM} = \nabla_{\mathbf{x}'}|\mathbf{x}'_*| = \frac{\mathbf{x}'_*}{|\mathbf{x}'_*|} = -\alpha_* \tag{22.30}$$

Equation (22.30) shows that $-\alpha_{i*}$, is the rate of change in β_{AFOSM} due to a one standard deviation change in random variable X_i at $\mathbf{X} = \mathbf{x}_*$. Therefore, the relation between $\nabla_{\mathbf{x}'}\beta$ and $\nabla_{\mathbf{x}}\beta$ can be expressed as

$$\nabla_{\mathbf{x}}\beta_{AFOSM} = \mathbf{D}^{-1/2}\nabla_{\mathbf{x}'}\beta_{AFOSM} = -\mathbf{D}^{-1/2}\alpha_* \tag{22.31}$$

It also can be shown that the sensitivity of reliability or failure probability with respect to each random variable can be computed as

$$\left(\frac{\partial p_s}{\partial X_{i'}}\right)_{\mathbf{x}_*} = \alpha_{i*}\phi(\beta_{AFOSM}); \quad \left(\frac{\partial p_s}{\partial X_i}\right)_{\mathbf{x}'_*} = \frac{\alpha_{i*}\phi(\beta_{AFOSM})}{\sigma_i}; \quad i = 1,2,\ldots,N \tag{22.32}$$

where $\phi(\cdot)$ = the standard normal PDF or in matrix form as

$$\nabla_{\mathbf{x}'_*}p_s = \phi(\beta_{AFOSM})\alpha_*; \quad \nabla_{\mathbf{x}_*}p_s = \phi(\beta_{AFOSM})\nabla_{\mathbf{x}_*}\beta_{AFOSM} = -\phi(\beta_{AFOSM})\mathbf{D}^{-1/2}\alpha_* \tag{22.33}$$

These sensitivity coefficients reveal the relative importance of each random basic variable on reliability or failure probability.

22.2.2.3 *Algorithms of AFOSM for Independent Normal Parameters.*

In the case that the random basic variables X are independent normal random variables, standardization of $\mathbf{X}$ reduces to vector of independent standard normal random variables $\mathbf{Z}'$ with mean $\mathbf{0}$ and covariance matrix $\mathbf{I}$, with $\mathbf{I}$ being an $N \times N$ identity matrix. Hasofer and Lind (1974) proposed the following recursive equation for determining the design point $\mathbf{z}'_*$

$$\mathbf{z}_{(r+1)} = -[-\alpha^t_{(r)}\mathbf{z}'_{(r)}]\,\alpha_{(r)} - \frac{W(\mathbf{z}'_{(r)})}{|\nabla_{\mathbf{z}'}W(\mathbf{z}'_{(r)})|}\,\alpha_{(r)}, \quad r = 1,2,\ldots \tag{22.34}$$

where subscripts (r) and $(r + 1)$ = the iteration numbers; $-\alpha$ the unit gradient vector of the failure surface pointing to the failure region

It would be more convenient to rewrite the above recursive equation in the original x-space as

$$\mathbf{X}_{(r+1)} = \mu + \mathbf{D}\mathbf{s}_{(r)}\frac{[\mathbf{x}_{(r)} - \mu]^t\,\mathbf{s}_{(r)}\,W(\mathbf{x}_{(r)})}{\mathbf{s}^t_{(r)}\,\mathbf{D}\mathbf{s}_{(r)}}, \quad r = 1,2,3,\ldots \tag{22.35}$$

Based on Eq. (22.35), the Hasofer-Lind algorithm for the AFOSM reliability analysis for problems involving uncorrelated, normal random variables can be outlined as follows.

Step 1. Select an initial trial solution $\mathbf{x}_{(r)}$.

Step 2. Compute $W(\mathbf{x}_{(r)})$ and the corresponding sensitivity coefficient vector $\mathbf{s}_{(r)}$.

Step 3. Revise solution point $\mathbf{x}_{(r+1)}$ according to Eq. (22.35).

Step 4. Check if $\mathbf{x}_{(r)}$ and $\mathbf{x}_{(r+1)}$ are sufficiently close. If yes, compute the reliability index β_{AFOSM} and the corresponding reliability $p_s = \Phi(\beta_{AFOSM})$. Then, go to step 5; otherwise, update the solution point by letting $\mathbf{x}_{(r)} = \mathbf{x}_{(r+1)}$ and return to step 2.

Step 5: Compute the sensitivity of reliability index and reliability with respect to changes in random basic variables according to Eqs. (22.31) to (22.33).

Due to the nature of nonlinear optimization, the above algorithm does not necessarily converge to the true design point associated with the minimum reliability index. Therefore, Madsen et al. (1986) suggest that different initial trial points are used and the smallest reliability index is chosen to compute the reliability.

Sometimes, it is possible that a system might have several design points. Such condition could be due to the use of multiple performance functions or the performance function is highly irregular in shape. In the case that there are J such design points, the reliability of the system requires that the system perform satisfactorily at all design points. Assuming independence of the occurrence of individual design point, the reliability of the system is the survival of the system at all design points can be calculated as

$$p_s = [\Phi(\beta_{AFOSM})]^J \tag{22.36}$$

Example 22.3

Referring to the data given in Example 22.2, apply the Hasofer-Lind AFOSM method to evaluate the sewer reliability assuming all random basic variables listed below are statistically independent normal random variables.

Parameter	Mean	Coeff. of variation
λ_m	1.10	0.11
n (ft$^{1/6}$)	0.015	0.05
d (ft)	5.0	0.01
S (ft/ft)	0.001	0.07
λ_L	1.00	0.15
C	0.825	0.071
i (in/hr)	3.15	0.177
A (ac)	22.0	0.05

Solution

For this problem, the performance function can be defined as

$$W = 0.463\, \lambda_m\, n^{-1}\, d^{8/3}\, S^{1/2} - \lambda_L\, CiA$$

where vector of random basic variables is $\mathbf{X} = (\lambda_m, n, d, S, \lambda_L, C, i, A)^t$.

To start the iteration, the initial solution is taken to be the means of the random basic variables, namely, $\mathbf{x}_{(1)} = \boldsymbol{\mu} = (1.1, 0.015, 5.0, 0.001, 1.0, 0.825, 3.15, 22.0)^t$. The covariance matrix for the eight random basic variables is an 8×8 diagonal matrix,

$$\mathbf{D} = diag\ (0.121^2,\ 0.00075^2,\ 0.05^2,\ 0.00007^2,\ 0.15^2,\ 0.059^2,\ 0.558^2,\ 1^2)$$

At $\mathbf{x}_{(1)}$, the value of the performance function $W(\boldsymbol{\mu}) = 26.51$ (from Example 22.2), which is not equal to zero. This implies that the solution point $\mathbf{x}_{(1)}$ does not lie on the limit-state surface. By Eq. (22.35) a new solution, $\mathbf{x}_{(2)}$, can be obtained as

$$\mathbf{x}_{(2)} = (\lambda_m, n, d, S, \lambda_L, C, i, A)$$

$$= (0.933, 0.01547, 4.983, 0.000969, 1.101, 0.8437, 3.561, 22.45)$$

Then, one checks the difference between the two consecutive solution points as

$$\left| \mathbf{x}_{(1)} - \mathbf{x}_{(2)} \right| = 1.063$$

which is considered large and, therefore, the iteration continues searching for the design point. After a few iterations, the difference between the two consecutive solutions is sufficiently small and the value of the performance function is very close to zero. The solution of the Hasofer-Lind algorithm converges to the design point indicated by the second column of the table given below

Basic Var.	Design Pt, $\mathbf{x}_*$	$\partial\beta/\partial x_i$	$\partial p_s/\partial x_i$	$(\partial\beta/\partial x_i)(x_i/\beta)$	$(\partial p_s/\partial x_i)(x_i/p_s)$
λ_m	1.014	3.808	0.4653	0.9024	0.4718
n	0.01522	253.7	-31	-0.9024	-0.4718
D	4.992	2.063	0.252	2.4068	1.2580
S	0.0009852	1960	239.4	0.4513	0.2359
λ_L	1.119	-3.45	-0.4215	-0.9022	-0.4717
C	0.849	-4.548	-0.5557	-0.9024	-0.4718
i	3.655	-1.057	-0.1291	-0.9029	-0.4719
A	20.29	-0.1903	-0.02325	-0.9024	-0.4717

At the design point $\mathbf{x}_*$, the mean and standard deviation of the performance function W can be estimated, by Eqs. (22.24a) and (22.24b), respectively, as

$$\mu_{W*} = 28.08 \quad \text{and} \quad \sigma_{W*} = 18.25$$

The corresponding reliability index is $\beta_* = \mu_{W*}/\sigma_{W*} = 1.538$ with the reliability and failure probability as

$$p_s = \Phi(\beta_*) = 0.9380; \quad p_f = 1 - p_s = 0.06198$$

Finally, at the design point x_*, the sensitivity of reliability index and reliability with respect to each of the random basic variables can be computed by Eqs. (22.32) and (22.33). The results are shown in columns 3 and 4 of the above table.

From the table, the quantities $\partial\beta/\partial x_i$ and $\partial p_s/\partial x_i$ show the sensitivity of the reliability index and reliability for one unit change in the value of the random basic variable X_i. The sensitivities of β and p_s associated with Manning's roughness and those basic variables in the rational formula (load function) are negative indicate that an increase in those basic variables would result in a decrease in β and p_s. On the other hand, an increase in model correction factor λ_m in Manning's formula, slope, pipe size would increase β and p_s.

It is might be difficult to draw meaningful conclusion of the sensitivity of different basic variables based on the magnitude of $\partial\beta/\partial x$ and $\partial p_s/\partial x$ because they are unit-dependent. A commonly used sensitivity measure, called the *relative sensitivity*, is defined as

$$s_i\% = \frac{\partial y/y}{\partial x_i/x_i} = \left(\frac{\partial y}{\partial x_i}\right)\left(\frac{x_i}{y}\right), \quad i = 1, 2, ..., N \tag{22.37}$$

where $s_{i\%}$ = a dimensionless quantity measuring the percentage change in the dependent variable y due to 1% change in the variable x_i.

The last two columns of the table given above show the percentage change in β and p_s of the sewer capacity due to 1% change in various basic variables. Relatively speaking, the pipe diameter is the most important random basic variable affecting the reliability of the flow carrying capacity of the sewer.

22.2.2.4 Treatment of Correlated and Non-normal Random Variables. The AFOSM method described above is for a performance function defined by uncorrected and normal random basic variables. When some of the random basic variables in the performance function are correlated or non-normal, appropriate transformations of the variables are required. To transform correlated random basic variables to uncorrelated ones, several orthogonal transformations, such as the spectral (or eigenvalue-eigenvector) decomposition, can be applied (Tung, 1999).

When non-normal random variables are involved, it is advisable to transform them into equivalent normal variables. Rackwitz (1976) and Rackwitz and Fiessler (1978) proposed an approach that transforms a non-normal variable into an equivalent normal one by preserving its probability content at the design point $\mathbf{x}_*$ as

$$F_i(x_{i*}) = \Phi\left(\frac{x_{i*} - \mu_{i*N}}{\sigma_{i*N}}\right) = \Phi(z_{i*}), \quad i = 1,2,..., N \tag{22.38}$$

where
$$\begin{aligned}
F_i(x_{i*}) &= \text{the CDF of the random variable } X_i = x_{i*}\\
z_{i*} = \Phi^{-1}[F_i(x_{i*})] &= \text{the standard normal quantile}\\
\mu_{i*N} \text{ and } \sigma_{i*N} &= \text{the mean and standard deviation of the normal equivalent for } X_i \text{ at}\\
&\quad x_{i*} \text{ which can be obtained by}
\end{aligned}$$

$$\mu_{i*N} = x_{i*} - z_{i*}\sigma_{i*N} \text{ and } \sigma_{i*N} = \frac{\phi(z_{i*})}{f_i(x_{i*})} \tag{22.39}$$

where $f_i(\cdot)$ and $\phi(\cdot)$ = the PDFs of the random variable X_i and the standard normal variable Z_i, respectively.

The mean and standard deviation of the normal equivalent of the random variable X_i at any expansion point $\mathbf{x}_*$ must be updated as the search for the design point progress.

Note that the above normal transformation preserves only the marginal distributions of the random basic variables without considering their correlations. Therefore, it is suitable for problems involving independent, non-normal random variables. When random variables are non-normal and correlated, additional transformation on the correlation among the basic variables from their original space to normal space must be made (Der Kiureghian and Liu, 1985).

22.2.3 Monte Carlo Simulation Methods

Monte Carlo simulation is a general purpose method to estimate the statistical properties of a random variable that is related to a number of random variables which may or may not be correlated. In the Monte Carlo simulation, the values of basic variables are generated according to their distributional properties. The generated values of the basic variables are used to compute the value of performance function. After several simulated realizations of the performance function are generated, the reliability of the structure can be estimated by the ratio of the number of realizations with $W \geq 0$ to the total number of simulated realizations.

The major disadvantage of the Monte Carlo simulation is its computational intensiveness. The number of sample realizations required in simulation to accurately estimate the failure probability depends on the magnitude of the unknown risk itself. In general, as the failure probability value gets smaller, the required number of simulated realizations increases. Therefore, several efficient Monte Carlo simulations to accurately estimate the failure probability while reducing excessive computation time have been developed. They include *stratified sampling* and *Latin hypercubic sampling* (McKay et al., 1979), *importance sampling* (Schueller & Stix, 1986; Harbitz, 1983), and the *reduced space approach* (Karamchandani, 1987).

22.3 *RISK-BASED DESIGN OF STORMWATER COLLECTION SYSTEMS*

The conventional design approach has been to subjectively pre-select a "proper" design frequency or return period on the basis of individual's experience, perceived importance of the structure, and/or legal requirements. Once the design return period is determined, the corresponding design load to the stormwater collection system is used to determine the system configurations producing satisfactory hydraulic performance. The selection of the "proper" design frequency is a complex decision-making procedure, which involves the consideration of economic, social, legal, and other factors. However, the conventional design procedure does not account for these factors explicitly.

Reliability analysis methods can be applied to design a stormwater collection system with or without considering risk costs. *Risk costs* due to the failure of the system can be tangible or intangible. Tangible costs are those measurable by monetary units including damage to properties and structures, loss in business, cost of repair, etc. On the other hand, intangible costs are not measurable by monetary units such as psychological trauma, loss of lives, social unrest, and others.

Risk-based procedures integrate uncertainty, reliability, and economic aspects in the design practice. Therefore, the procedure explicitly considers tradeoffs among various factors such as failure probability, economics, and other performance measures in the design. When the risk-based design is embedded into an optimization framework, the combined procedure is called the *optimal risk-based design*. Applications of risk-based design to various hydraulic structures include storm sewers (Tang et al., 1975; Yen et al., 1976; Wenzel et al., 1979), levees (Tung and Mays, 1981b), dams (Pate-Cornell and Tagaras, 1986), floodplain encroachment (Corry et al., 1980), and highway drainage structures such as culverts (Young et al., 1974; Tung and Bao, 1990) and bridges (Tung and Mays, 1982).

22.3.1 Basic Concept

The basic concept of risk-based design is shown schematically in Fig. 22.3. The *risk function* considering various uncertainties can be obtained using an appropriate reliability assessment procedure. Alternatively, the risk function can consider the potential undesirable consequences associated with the failure of hydraulic structures.

Because risk costs associated with the performance failure of a hydraulic structure could vary from one year to another, a practical way to quantify these costs is to use expected values on an annual basis. For a stormwater collection system, the *total annual expected cost* (*TAEC*) is the sum of the annual installation cost, operation and maintenance costs, and *annual expected damage cost* which can be expressed as

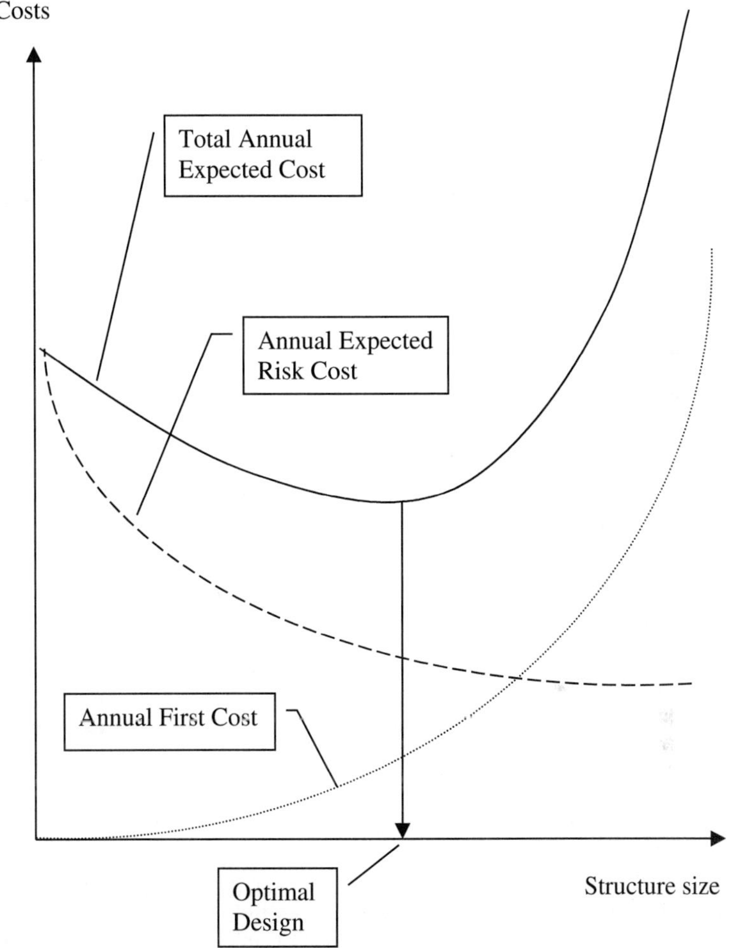

FIGURE 22.3 Schematic sketch of optimal risk-based design concept.

$$TAEC(\Theta) = FC(\Theta) \times CRF + E(D|\Theta) \tag{22.40}$$

where $FC(\ )$ = the first or total installation costs depending on the decision vector Θ that may include the size and configuration of the system

CRF = the capital recovery factor, which brings the present worth of the installation costs to an annual basis;

$E(D|\Theta)$ = the annual expected damage cost associated with the performance failure of the system

The *CRF* can be computed as

$$CRF = \frac{(1 + r)^k - 1}{r(1 + r)^k} \tag{22.41}$$

with k and r being the expected service life of the system and the interest rate, respectively.

In practice, the optimal risk-based design of a stormwater collection system determines the optimal system configuration, structural size, and operation such that the total annual expected cost is minimized. Referring to Fig. 22.3, as the design frequency or structural size increases, the annual first cost increases whereas the annual expected risk cost decreases. In risk-based design procedure, the return period is a decision variable instead of being a preselected design parameter value as in the conventional design procedure. The optimal risk-based design procedure attempts to determine the lowest point on the total annual expected cost curve. Mathematically, the *optimal risk-based design problem* can be stated as

$$\text{Minimize} \quad TAEC(\Theta) = FC(\Theta \times CRF + E(D|\Theta) \tag{22.42a}$$

$$\text{subject to} \quad g_j(\Theta) = 0, j = 1, 2, ..., m \tag{22.42b}$$

where $g_j(\Theta) = 0$ are constraints representing the design specifications that must be satisfied.

The solution to Eqs. (22.42a) and (22.42b) could be acquired through the use of appropriate optimization algorithms. The selection or development of the solution algorithm is largely problem specific, depending on the characteristics of the problem to be optimized.

The concept of risk-based design has been recognized as early as in 1936 by which time the U.S. Congress passed the *Flood Control Act* (U.S. Statutes 1570) advocating the consideration of failure consequences in the design procedure. The economic risks or the expected flood losses were not explicitly considered until the early 1960s. Pritchett's work (1964) was one of the early attempts to apply the risk-based hydraulic design concept to highway culverts.

22.3.2 Data Requirements in Risk-Based Design of Stormwater Collection Systems

Information needed for the risk-based design of a stormwater collection system can be categorized into the following types:

1. *Hydrologic/physiographical data* include rainfall frequency relationships and drainage basin features, such as land use and topography, for rainfall-runoff modeling. They produce the load and input conditions to the stormwater collection systems.

2. *Hydraulic data* include geometry and layout of various components in a stormwater collection system, and their hydraulic properties such as roughness coefficients. They are used for determining the flow carrying capacities (the resistance) of stormwater collection systems and to perform hydraulic analysis.

3. *Structural data* include material and installation of various system components that affect the capacity of the stormwater collection system.

4. *Tangible economic data* include (1) type, location, distribution, and value of properties in the area affected by the failure of the system, (2) costs of system components and construction, and (3) operation, maintenance, and repair costs.

In the design of stormwater collection systems, the installation cost often is dependent on the environmental conditions such as the geomorphic and geologic conditions, the soil type at the structure site, type and price of construction materials, hydraulic and flow conditions, recovery factor of the capital investment, and labor. In reality, these factors are subject to uncertainties in cost functions used in the analysis. The economic uncertainties have been incorporated in the risk-based design of flood control structures by the U.S. Army Corps of Engineers (1996).

22.3.3 Evaluations of Annual Expected Risk Cost

One major task in the risk-based design and analysis lies in the evaluation of annual expected economic losses. In the design of hydraulic systems, the common practice is that the economic risks are calculated considering only the randomness of hydrologic loads. When rainfall-runoff models are used to produce hydrologic loads to stormwater collection systems, the randomness of rainfall depth or intensity is generally represented by the intensity-duration-frequency (IDF) relationship. In reality, there are various types of uncertainties in a hydraulic structure design. Advances were made to incorporate various aspects of uncertainty in hydraulic structure designs.

In the risk-based design procedure, the evaluation of $E(D|\Theta)$ depends on the types of uncertainty to be considered in defining the load, resistance, and damage functions. The common practice considers only inherent hydrologic uncertainty in that the annual expected risk cost is computed as

$$E_1(D|\Theta, q_c^*) = \int_{q_c^*}^{\infty} D(q|q_c^*, \Theta)\, f(q)\, dq \tag{22.43}$$

where Θ = a vector containing design parameters of the system affecting the system capacity

q_c^* = the deterministic flow capacity of a system subject to random hydrologic loadings following a PDF, $f(q)$

$D(q|q_c^*, \Theta)$ = the damage function corresponding to the inflow magnitude q and hydraulic structure capacity q_c^*

In general, the integration by Eq. (22.43), in most practical applications, is difficult, if not impossible, due to the complexity of the damage function and the form of the PDF of hydrologic load. Hence, it is practical to compute Eq. (22.43) by a numerical approximation.

In reality, uncertainties also exist in hydraulic analysis in determining the flow carrying capacity of the hydraulic structure. Therefore, q_c is a quantity subject to uncertainty. To incorporate the uncertainty feature of q_c, the annual expected risk cost can be calculated as

$$E_2(D|\Theta) = \int_0^{\infty} E_1(D|q_c, \Theta) h(q_c|\Theta)\, d\,q_c \tag{22.44}$$

where $E_1(D|q_c, \Theta)$ = computable by Eq. (22.43) for a particular value of capacity q_c

$h(q_c|\Theta)$ = the PDF of random flow carrying capacity q_c, which can be assessed by uncertainty analysis of hydraulic computation

In addition to the inherent randomness of rainfall represented by the IDF relationship, there exists parameter uncertainty due to sampling error from the use of limited amount of hydrologic data. Mathematical expression to consider various types of uncertainty in evaluating $E(D)$ can be straightfordly extended from Eq. (22.43) or (22.44) (see Tung, 1999). Depending on the degree of uncertainty from various sources, the impact of their inclusion in $E(D)$ varies. Tung (1987) numerically shows that, without providing full account of uncertainties in the analysis, the resulting annual expected damage is significantly underestimated, even with a 75-yr long flood record.

Example 22.4
To illustrate the risk-based design of stormwater collection systems, a single storm sewer pipe section is considered. In this case, the pipe diameter with the least total annual expected cost is sought. Parameters and their uncertainties in Manning's formula for calculating flow carrying capacity identical to those in Example 22.1, except the mean pipe diameter, and are repeated below:

Parameter	Distribution	Mean	C.O.V.
λ_m	Lognormal	1.10	0.11
n (ft$^{1/6}$)	Lognormal	0.015	0.05
S (ft/ft)	Lognormal	0.001	0.07
D (ft)	Lognormal	——	0.01

The load to the sewer section is the peak runoff hydrograph computed by the Rational formula in which the rainfall intensity with a 30-minute duration is to be used. The natural randomness of rainfall intensity is described by the following rainfall IDF relationship:

$$i(\text{in/hr}) = \frac{120 \ T^{0.175}}{27 + t_d}$$

Parameters in the Rational formula and their uncertainty features are given below:

Parameter	Distribution	Mean	C.O.V.
λ_L	Lognormal	1.00	0.15
C	Lognormal	0.825	0.071
A (ac)	Lognormal	20.0	0.05
i (in/hr)	Lognormal	——	0.177

In risk-based design, one is required to evaluate the failure probability of the sewer section with different pipe sizes under random inflow peak discharge which, in turn, is dependent on the recurrence interval of the rainfall intensity. For simplicity, it is assumed here that only the mean of rainfall intensity is dependent on the return period whereas the coefficient of variation of rainfall intensity is a constant. Furthermore, all random basic variables are statistically independent.

For a pipe size under consideration, one needs to evaluate the failure probability of the sewer under peak discharges from rainfalls of different return periods. This can be done by using the MFOSM or AFOSM methods described previously. The performance function can be expressed as

$$W = \left[\ln(0.463) + \ln(\lambda_m) - \ln(n) + \frac{8}{3} \ln(D) + \frac{1}{2} \ln(S) \right]$$
$$- [\ln(\lambda_L) + \ln(C) + \ln(i) + \ln(A)]$$

Since all random basic variables are lognormally distributed, analytical solution for risk calculation can be performed. The mean and variance of the performance function W can be expressed, respectively, as

$$\mu_W = \left[\ln(0.453) + \mu_{\ln(\lambda_m)} - \mu_{\ln(n)} + \frac{8}{3} \mu_{\ln(D)} + \frac{1}{2} \mu_{\ln(S)} \right]$$
$$- [\mu_{\ln(\lambda_L)} + \mu_{\ln(C)} + \mu_{\ln(i)} + \mu_{\ln(A)}]$$

$$\sigma_W^2 = \left[\sigma_{\ln(\lambda_m)}^2 + \sigma_{\ln(n)}^2 + \frac{64}{9} \sigma_{\ln(D)}^2 + \frac{1}{4} \sigma_{\ln(S)}^2 \right] + [\sigma_{\ln(\lambda_L)}^2 + \sigma_{\ln(C)}^2 + \sigma_{\ln(i)}^2 + \sigma_{\ln(A)}^2]$$

For each terms in the above equations, the mean and variance of the random basic variables in the log-space can be obtained by the following two formulas:

$$\sigma^2_{\ln(X)} = \ln(1 + \Omega^2_x); \quad \mu_{\ln(X)} = \ln(\mu_x) - \frac{1}{2}\sigma^2_{\ln(X)}$$

For example, consider a 4-ft pipe under a 10-yr storm. The mean rainfall intensity, by the first-order approximation, is $\mu_i = 120(10)^{0.175}/(27 + 30) = 3.15$ in/hr. Form the above equations, the mean and variance of the performance function can be calculated as

$$\mu_W = 0.5476 \quad \text{and} \quad \sigma^2_W = 0.08014$$

Since W is a normal random variable, the reliability index and the corresponding failure probability for the 4-ft pipe under the 10-yr storm, respectively, are

$$\beta = 0.5476/(0.08014)^{0.5} = 1.935; \quad p_f = \Phi(-1.935) = 0.02652$$

Different plausible return periods can be considered to produce the following results for $D =$ 4-ft

	Return period (years)									
	1	2	5	10	25	50	100	200	500	1000
μ_W	0.9506	0.8293	0.6689	0.5476	0.3873	0.2660	0.1447	0.0234	−0.1370	−0.2583
β	3.358	2.929	2.363	1.935	1.368	0.940	0.511	0.083	−0.484	−0.912
p_f	3.92E-04	1.69E-03	9.06E 03	2.65E-02	8.56E-02	1.74E-01	3.05E-01	4.67E-01	6.86E-01	8.19E-01

The process can be repeated for different pipe sizes from which a series of failure probability-return period ($p_f - T$) curves such as Fig. 22.4 can be established. As expected, the failure probability increases with an increase in return period of rainstorm intensity and decreases with pipe size.

Based on the failure probability, conditioned on the rainfall of a specified return period, one can assess the corresponding economic loss. For illustration, assume that the damage associated with the performance failure of the sewer section is a constant, say, \$100/linear-ft of pipe (for simplicity). The conditional expected damage cost associated with the rainfall intensity (specified by the corresponding return period, T) and a pipe size (d) can be calculated by

$$E(D|T, d) = \$100 \times p_f(T, d)$$

To obtain the annual expected damage associated with a pipe size the above conditional expectation, $E(D|T, d)$, can be integrated over the probability domain as

$$E(D|d) = \int_0^1 E(D|F, d)\, dF \approx \sum_{k=1}^{K} \overline{E_k(D|T_k, T_{k+1}, d)} \times \left(\frac{1}{T_k} - \frac{1}{T_{k+1}}\right)$$

in which F is the cumulative probability which is related to the return period T by $F = 1 - 1/T$. The last term of the above equation is a numerical approximation of the integral. If one replaces $E(D|T, d)$ by $E(p_f|T, d)$ in the above equation, the annual expected failure probability for each pipe size, $E(p_f|d)$, can be obtained. After going through the annual expected damage evaluation for each pipe size, an annual expected damage (or failure probability) versus pipe diameter relation could be established as the following table,

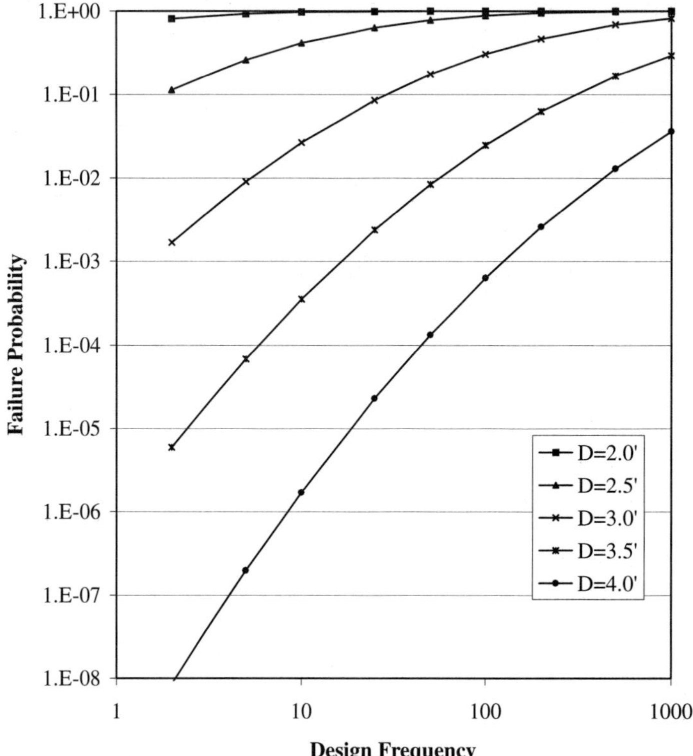

FIGURE 22.4   Failure probability—return period relationship for different pipe sizes.

Pipe Diam.	$E(p_f \vert d)$	$E(D \vert d)$	Pipe cost (\$/LF)	Total cost (\$/LF)
2.0′	4.54E-01	45.36	18.38	65.34
2.5′	1.58E-01	15.85	23.87	41.72
3.0′	1.86E-02	1.86	29.36	33.62
3.5′	1.27E-03	0.13	34.85	37.78
4.0′	6.34E-05	0.01	40.34	43.55

The pipe cost listed in column 4 is computed by the following simple empirical function (Yen et al., 1976), assuming a soil cover depth of 3 ft,

$$C_p = 10.98d + 0.8H - 5.98$$

where C_p = the unit pipe cost (in \$/linear ft.)
H = the sewer invert depth (in ft) below the ground surface

A plot of pipe cost and annual expected damage cost, $E(D \vert d)$ with respect to pipe size, as shown in Fig. 22.5, reveal the tradeoff between the installation cost and risk-cost. The total cost – pipe size curve in Fig. 22.5 (or from the above table) indicates that the optimal pipe size with the least annual total expected cost is 3 ft.

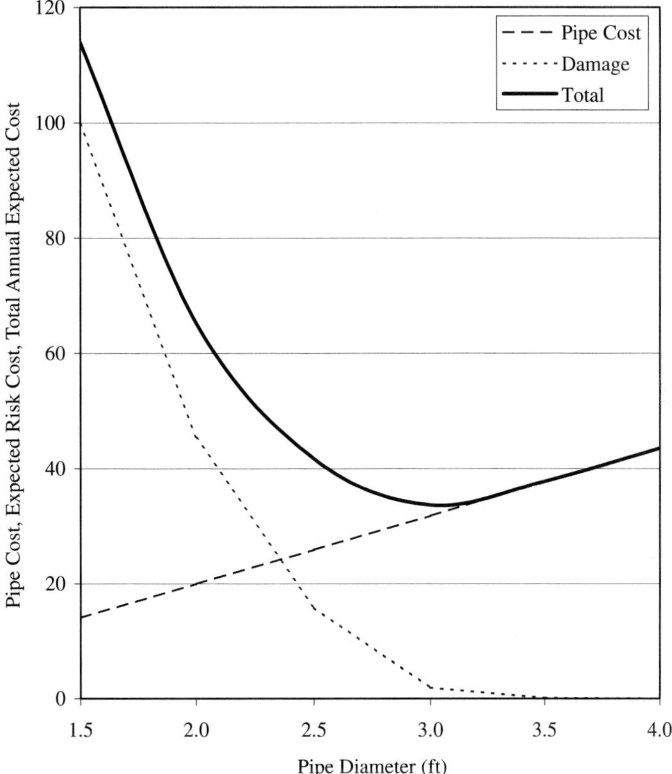

FIGURE 22.5   Trade-off between pipe cost and expected damage cost.

This simple example illustrates the procedures involved in the risk-based design of a pipe section in a storm sewer system. It has been incorporated in the design of storm sewer systems involving many pipes and manholes (Tang et al., 1975; Yen et al., 1976) in the framework of dynamic programming. A typical risk-based least-cost sewer system design model in ILSD-2 (Yen et al., 1984) listed in Table 6.3(b).

REFERENCES

Ang, A. H-S., "Structural Risk Analysis and Reliability-Based Design," *Journal of Structural Engineering Division*, American Society of Civil Engineers, 99(9):1891–1910, 1973.

Ang, A. H. S., and C. A. Cornell, "Reliability Bases of Structural Safety and Design," *Journal of Structural Engineering*, American Society of Civil Engineers, 100(9):1755–1769, 1974.

Berthouex, P. M. "Modeling Concepts Considering Process Performance, Variability, and Uncertainty," in *Mathematical Modeling for Water Pollution Control Processes*, 405–439, T. M. Keinath and M. P. Wanielista, eds. Ann Arbor Science, Ann Arbor, MI, 1975.

Cheng, S. T., B. C. Yen, and W. H. Tang, "Sensitivity of Risk Evaluation to Coefficient of Variation," *Stochastic and Risk Analysis in Hydraulic Engineering*, B. C. Yen, ed., pp. 266–273, Water Resources Publications, Littleton, CO, 1986.

Cornell, C. A. "A Probability-based Structural Code," *Journal of American Concrete Institute*, 66(12): 974–985, 1969.

Corry, M. L., J. S. Jones, and D. L. Thompson, "The Design of Encroachments of Floodplain Using Risk Analysis," *Hydraulic Engineering Circular*, No. 17, U.S. Department of Transportation Federal Highway Administration, Washington, DC, 1980.

Der Kiureghian, A., and P. L. Liu, "Structural Reliability Under Incomplete Probability Information," *Journal of Engineering Mechanics*, American Society of Civil Engineers, 112(1):85–104, 1985.

Harbitz, A., "Efficient and Accurate Probability of Failure Calculation by Use of the Importance Sampling Technique," *Proceedings, International Conference on Applications of Statistics and Probability in Soil and Structural Engineering,* University de Firenze, Florence, Italy, 1983.

Hasofer, A. M., and N. C. Lind, "Exact and Invariant Second-moment Code Format," *Journal of Engineering Mechanics Division*, American Society of Civil Engineers, 100(1):111–121, 1974.

Karamchandani, A., "Structural System Reliability Analysis Methods," Report to Amoco Production Company, Department of Civil Engineering, Stanford University, 1987.

Madsen, H. O., S. Krenk, and N. C. Lind, *Methods of Structrual Safety*, Prentice-Hall, Inc., Englewood Cliffs, NJ, 1986.

McKay, M. D., R. J. Beckman, and W. J. Conovre, "A Comparison of Three Methods for Selecting Values of Input Variables in the Analysis of Output from a Computer Code," *Technometrics*, 21, 1979.

Melchers, R. E., *Structural Reliability Analysis and Prediction*, Ellis Horwood Ltd., Chichester, England, 400 pp., 1987.

Pate-Cornell, M. E., and G. Tagaras, "Risk Cost for New Dams: Economic Analysis and Effects of Monitoring," *Water Resources Research*, American Geophysics Union, 22(1):5–14, 1986.

Pritchett, H. D., "Application of the Principles of Engineering Economy to the Selection of Highway Culverts," Stanford University, Report EEP-13, 1964.

Rackwitz, R., "Practical Probabilistic Approach to Design," *Bulletin* 112, Comité European du Beton, Paris, France, 1976.

Rackwitz, R., and B. Fiessler, "Structural Reliability under Combined Random Load Sequence," *Computers and Structures*, 9:489–494, 1978.

Schueller, G. I., and R. Stix, "A Critical Appraisal of Methods to Determine Failure Probabilities," Report, no. 4–86, Institute fur Mechanik, Universitat Innsbruck, Austria, 1986.

Shinozuka, M., "Basic Analysis of Structural Safety," *Journal of Structural Engineering Division*, American Society of Civil Engineers, 109(3):721–740, 1983.

Tang, W. H., L. W. Mays, and B. C. Yen, "Optimal Risk-based Design of Storm Sewer Networks," *Journal of Environmental Engineering,* American Society of Civil Engineers, 101(EE3):381–398, 1975.

Tung, Y. K., "Effects of Uncertainties on Optimal Risk-Based Design of Hydraulic Structures," *Journal of Water Resources Planning and Management*, American Society of Civil Engineers, 113(5):709–722, 1987.

Tung, Y. K., and L. W. Mays "Risk and Reliability Model for Levee Design," *Water Resources Research*, American Geophysics Union, 17(4):833–842, 1981a.

Tung, Y. K., and L. W. Mays, "Optimal Risk-based Design of Flood Levee Systems," *Water Resources Research*, American Geophysics Union, 17(4):843–852, 1986.

Tung, Y. K., and L. W. Mays, "Optimal Risk-based Hydraulic Design of Bridges," *Journal of Water Resources Planning and Management*, American Society of Civil Engineers, 108(2):191–202, 1982.

Tung, Y. K., "Chapter 7: Uncertainty and Reliability Analysis," in *Water Resources Handbook*, L. W. Mays, ed., McGraw-Hill Book Company, New York, NY, 1996.

Tung, Y. K. "Chapter 7: Risk/Reliability-based Hydraulic Engineering Design," in *Hydraulic Design Handbook*, L. W. Mays, ed., McGraw-Hill Book Company, New York, NY, 1999.

Tung, Y. K., and L. W. Mays, "Risk and Reliability Model for Levee Design," *Water Resources Research*, AGU, 17(4):833–842, 1981a.

Tung, Y. K., and L. W. Mays, "Optimal Risk-based Design of Flood Levee Systems," *Water Resources Research*, AGU, 17(4):843–852, 1981b.

Tung, Y. K., and L. W. Mays, "Optimal Risk-based Hydraulic Design of Bridges," *Journal of Water Resources Planning and Management*, American Society of Civil Engineers, 108(2):191–202, 1982.

Tung, Y. K., and Y. Bao, "On the Optimal Risk-based Designs of Highway Drainage Structures," *Journal of Stochastic Hydrology and Hydraulics*, 4(4):311–324, 1990.

US Army Corps of Engineers, "Risk-based Analysis for Flood Damage Reduction Studies," EM 1110-2-1619, Washington, DC, 1996.

Yen, B. C., "Safety Factor in Hydrologic and Hydraulic Engineering Design," *Reliability in Water Resources Management*, E. A. McBean, K. W. Hipel, and T. E. Unny, eds., pp. 389–407, Water Resources Publications, Littleton, CO, 1979.

Yen, B. C., and A. H.-S., Ang, "Risk Analysis in Design of Hydraulic Projects," *Stochastic Hydraulics, Proceedings of First International Symposium*, University of Pittsburgh, C. L. Chiu, ed., pp. 694–701, Pittsburgh, PA, 1971.

Yen, B. C., H. G. Wenzel, L. W. Mays, and W. H. Tang, "Advanced Methodologies for Design of Storm Sewer Systems," Research Report No. 112, Water Resources Center, University of Illinois at Urbana-Champaign, 1976.

Yen, B. C., S. T. Cheng, B.-H. Jun, M. L. Voorhees, H. G. Wenzel, Jr., and L. W. Mays, "Illinois Least-Cost Sewer System Design Model: ILSD-1 & 2 User's Guide," Water Resources Center Research Report, No. 188, University of Illinois at Urbana-Champaign, 1984.

Yen, B. C., S. T. Cheng, and C. S. Melching, "First-order Reliability Analysis," in *Stochastic and Risk Analysis in Hydraulic Engineering*, B. C. Yen, ed., pp. 1–36, Water Resources Publications, Littleton, CO, 1986.

Young, G. K., M. R. Childrey, and R. E. Trent, "Optimal Design of Highway Drainage Culverts," *Journal of Hydraulics Engineering*, American Society of Civil Engineers, 107(7):971–993, 1974.

CHAPTER 23
STORM DRAIN DESIGN AND ANALYSIS USING HYDRA

Stuart Stein
Ken Young
GKY & Associates, Inc.
Springfield, Virginia

23.1   INTRODUCTION

HYDRA is a storm drain and sanitary sewer analysis and design program. Originally designed and developed in 1975, the HYDRA program ran on mainframe computer systems. The objective of the HYDRA program was to provide hydraulic design engineers with a means of accurately, easily and quickly designing and analyzing storm, sanitary or combined collection systems. HYDRA achieved these objectives with a high degree of success, and for this reason, it was selected for incorporation into the HYDRAIN system which is supported by the Federal Highway Administration and 31 State Transportation Departments. Furthermore, if it is being used within HYDRAIN, there are those advantages associated with HYDRA's ability to interact with other related hydraulic design programs.

In the HYDRA design process, the program will select pipe size, slope and invert elevations if given certain design criteria. Additionally, HYDRA will perform analyses on an existing system of pipes (and/or ditches). When an existing system of pipes is overloaded, HYDRA will indicate suggested flow removal quantities as well as an increased pipe size as an alternative remedy. Additionally, HYDRA can optionally consider the possibility of surcharged systems. The design procedure is not optimized, so alternatives should be examined.

HYDRA requires the creation of an input file, consisting of commands to describe the drainage system. The commands are placed in a logical sequence, usually from higher to lower elevations. It is possible that several command sequences can produce the same result. The input file, established for a particular collection system by the user, is then executed using the HYDRA analysis program.

23.2   CAPABILITIES AND LIMITATIONS

HYDRA operates in two modes: **design** and **analysis.** There are three possible types of systems on which HYDRA can work:

- Storm drain systems
- Sanitary (sewage) systems
- Combined (storm and sanitary) sewer systems

These types are collectively referred to as storm drain systems or sewer systems, or simply, systems. As implied by the preceding text, HYDRA is made to perform the following tasks:

1. Analyze a drainage system design given user-supplied specifications.

2. "Free design" its own drainage system based on design criteria supplied by the user.

To meet these broad objectives, HYDRA was necessarily designed to be an extremely flexible and powerful program. The user is warned that care and responsibility should be exercised when using the program as a decision-making tool. *HYDRA is a design aid only and is not a substitute for sound engineering judgment.* This being mentioned, the following is a list of some of HYDRA's more useful features:

- Cost estimation—Capabilities that allow for consideration of de-watering, traffic control, sheeting, shrinkage of backfill, costs of borrow, bedding costs, surface restoration, rock excavation, pipe zone costs, etc. HYDRA is also sufficiently flexible to allow cost criteria to be varied for any segment of pipe in a system, if desired. Ground profiles, either upstream or downstream from any specified point along the system, can also be accepted for consideration in cost estimation, if desired.

- Models storm flow and offers choice of methods—HYDRA is capable of "generating" storm flow based on the Rational method or modeling user-defined hydrographs in a hydrologic simulation, at the user's discretion. This may be particularly advantageous for engineers who wish to compare designs or analysis results based on different methods.

- Models sanitary flow—HYDRA "generates" sanitary flow based on the traditional "peak ing factor" concept.

- Models drainage systems of any size and configuration—HYDRA has a data handling algorithm especially designed to accept a drainage system of any realistically conceivable design. System components may include pipes, junctions, gutters, inlets, channels, ponds, and pump stations.

- Infiltration/Inflow analysis—HYDRA is ideally suited for making these analyses.

- Planning—Use HYDRA for determining the most practical alternate choices for unloading an existing overloaded storm drain or wastewater system and for formulating master plans to allow for an orderly growth of these systems. The program's features and capabilities should have far-reaching implications for municipal agencies whose existing sewer systems are under stress from rapid population growth and/or changes in land use patterns.

- Easy data input structure and quick editing capacity—All data needed to run HYDRA is in one user-supplied input file, simplifying data editing operations. Furthermore, if the program is run from within the HYDRAIN environment, the input file may be modified without leaving the HYDRAIN program by using the built-in editor. (The User Documentation section describes the capabilities of this screen editor and provides instructions for its use.) Time required for data modification and job resubmission is thus minimized, which enables the user to spend more time on his or her own decision analysis.

23.3 HYDRA EXAMPLE: RATIONAL METHOD DESIGN WITH HYDRAULIC GRADELINE

23.3.1 Problem

The example shown in Fig. 23.1 illustrates the use of HYDRA for design of a storm drain using the Rational method. Table 23.1 is a glossary of commands through which HYDRA

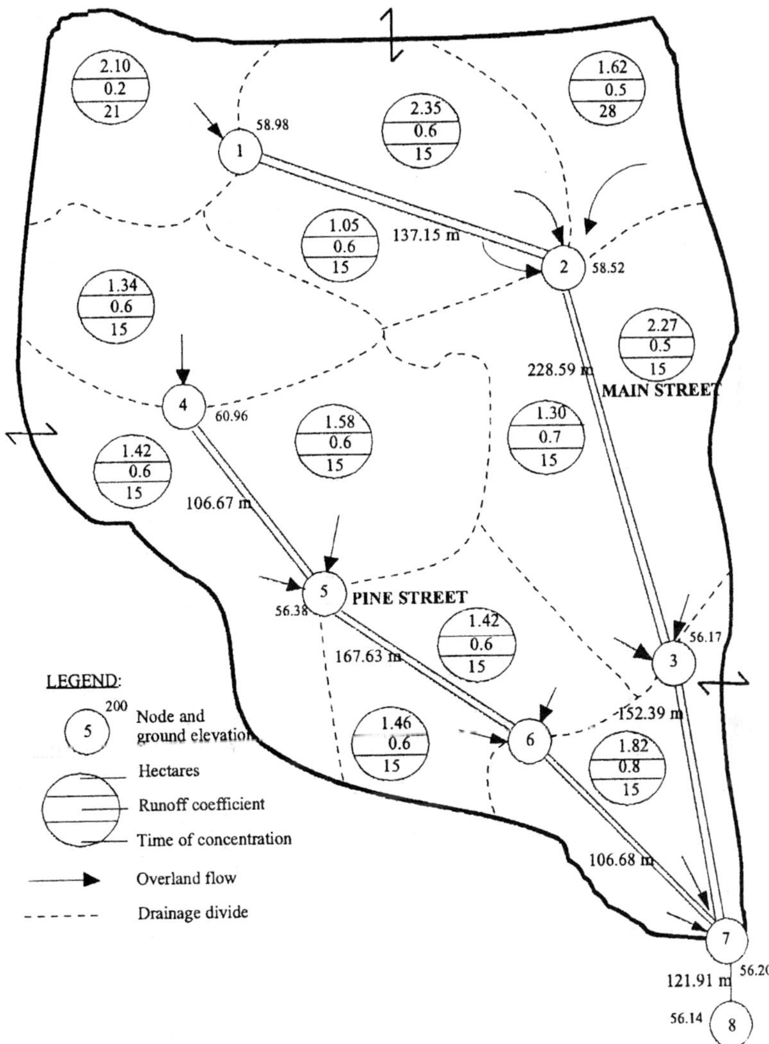

FIGURE 23.1   Rational method design.

operates. Table 23.2 is an example command string showing the command name, accompanying data field, and explanation of the command. As for any design problem, the cost commands (EXC, TSL, PCO, and CST) must be used if cost estimates are desired. Also, a PDA command must be provided to inform HYDRA what pipe characteristics are desired. This command can be used at least once. Four additional commands are introduced in this example: STO, RAI, HGL, and PNC. The STO command provides the physical characteristics of a drainage area required for the Rational method: size, runoff coefficient, and time of concentration. The RAI command supplies the Intensity-Duration-Frequency (IDF) curve for the return period and location the user desires to analyze. HYDRA uses the time of concentration for each individual area and then consults the IDF curve to find an intensity for the rational formula. As HYDRA proceeds through the system, it continually adjusts the time of concentration as more areas are being aggregated. The HGL and PNC commands

TABLE 23.1 Glossary of Commands

Command	Description
BEN—specifies Pipe **BEN**d data such as angle and radius.	
BOX—rectangular pipe that transport system flow from one point to another.	
CHA—allows definition of an open **CHA**nnel or ditch.	
CRI—determines whether inverts or crowns are to be matched (**CRI**teria).	
CST—sets geometry factors and unit prices (**C**o**ST**s in place).	
DIV—splits the system flow into two components (**DIV**erts flow).	
ECF—allows **E**xtra **C**osts per linear meter to be added to the pipe cost.	
EFF—describes inlet performance in ordered pairs of flow versus **EFF**iciency.	
ELP—horizontal or vertical **El**li**P**se that transports system flow.	
END—**END**s a command string.	
EXC—establishes trench **EXC**avation costs.	
FLO—adds or subtracts a constant **FLO**w to the system.	
GET—**GET**s a gutter hydrograph from storage.	
GPC—sets the **G**allons **P**er **C**apita per day for flow calculations.	
GUT—establishes **GUT**ter characteristics.	
HGL—signals that **H**ydraulic **G**rade**L**ine computations should be made.	
HOL—**HOL**ds system flow at the lower end of a lateral.	
IDY—**I**nitializes **D**epths for pressure flow.	
INF—inputs **INF**iltration flows by population or area.	
INL—sets parameters for a storm water **INL**et.	
IPU—establishes the number of **I**ndividuals **P**er sanitary **U**nit.	
IQV—**I**nitializes flows and **V**elocities.	
JOB—initiates **JOB** and enters **JOB** title.	
LOS—allows input of additional pipe **LOS**ses.	
LPC—calculates and **L**ists **P**ipe **C**osts in place.	
MAP—establishes factor for converting in^2 on a **MAP** to ac.	
NEW—clears some registers and loads **NEW** lateral name.	
PCF—establishes a **P**ipe **C**ost **F**actor.	
PCO—establishes **P**ipe **CO**sts per meter in place.	
PDA—establishes **P**ipe design **DA**ta.	
PEA—translates average daily flow into **PEA**k daily flow.	
PFS—specifies **P**ressure **F**low **S**imulation parameters.	
PFP—**P**rints **F**low in **P**ipes for pressure flow simulation.	
PHJ—**P**rints **H**ead of **J**unctions for pressure flow simulation.	
PIP—moves water from one point to another in a circular **PIP**e.	
PNC—specifies **P**ipe-**N**ode **C**onnections for hydraulic gradeline computation.	
PON—allows surface **PON**ding of flows **HYD, GUT, GET** commands.	
PSZ—sets **P**ipe **S**i**Z**es for design.	
PUM—lifts the hydraulic gradient a specified amount (**PUM**p).	
PUT—**PUT**s gutter flow into storage.	
RAI—sets the values on a **RAI**nfall intensity versus duration curve.	
REC—**REC**alls flow previously stored using the **HOL** or **DIV** commands.	
REM—allows a line for **REM**arks or comments.	
RES—allows the analysis of in-line storage (**RES**ervoir).	
SAF—applies **SAF**ety factors to calculated flows.	
SAN—enters **SAN**itary flow into the system.	
SDI—allows input of **S**tage-**DI**scharge curve.	
SST—allows input of **S**tage-**ST**orage curve.	
STE—sets the length of time increments (**STE**ps).	
STO—enters subbasin data for determining **STO**rm water design flow.	
SUN—enters the number of contributing **S**anitary **UN**its.	
SWI—sets **SWI**tch for determining method of storm/sanitary flow analysis.	
TRA—**TRA**nsfer system flow to surface flow.	
TSL—determines **T**rench side wall **SL**ope.	
TWE—allows for the input of a **T**ail**W**ater **E**levation at the system outfall.	
UHY—enables use of externally produced hydrographs (User **HY**drograph).	

Source: U.S. Federal Highway Administration (1999).

TABLE 23.2   Example of a Command String

HYDRA Command	Data	Comments
JOB	Command Example	The title of the HYDRA job.
PDA	0.014 300 2.5 1.5 0.5 0.005	Sets initial design criteria, such as Manning's **n**, minimum diameter, etc.
NEW	Main Street	Starts a new lateral, providing it with a unique name.
FLO	0.500	Places a 0.500 m^3/s FLOw into the system at this location.
PIP	100 33.500	Pipe parameters. (Transports the flow in a pipe 100-m long—ground elevation at both the upper and lower ends is 33.500 m.)
PIP	120 33.500 34.000	Continues to transport the flow into another pipe 120-m long. Notice that the ground elevation at the upstream end is 33.500 m and 34.000 m at the downstream end. HYDRA can model adverse ground elevations.
FLO	0.100	Adds another 0.100 m^3/s to the system at this point.
PDA	0.025 300 2.5 1.5 0.5 0.005	Change the pipe friction factor to 0.025 (corrugated metal pipe).
PIP	140 34.000 30.500	Transports the flow another 140 m.
END		Terminates the HYDRA run and prints the results.

Source: U.S. Federal Highway Administration (1999).

are related to the calculation of the hydraulic gradeline through the system. The HGL commands acts as a switch to "turn on" the computations. The PNC commands details the pipe-node connectivity through the system, as well as the other parameters required to calculate minor losses.

Data file: FOXHALL.IDF
```
FOXHALL CRESCENTS INTENSITY-DURATION-FREQUENCY
   8
         5     192.02
        10     160.02
        15     138.18
        30     100.33
        60      67.56
       100      48.77
       120      43.18
       360      19.81
```

Input file: HYDRA1.HDA
```
JOB Example One: Rational Method Design with Hydraulic Gradeline
SWI 2
PDA 0.013 300 1.22 0.914 0.61 0.001
RAI FOXHALL.IDF
EXC 1.52 0.94 7.62 1.48
```

```
TSL 0 0.25 3.05 0.15
PCO 300 15.03 900 81.50
CST 1.5 4.92 0 0 0 1.64 8.12 1 0 1.99 1.05 0.79 2.04 1.99 0
HGL 1
REM Begin Main Lateral
NEW MAIN STREET
REM Main Lateral: Junctions 1 to 2
STO 2.10 0.2 21
PIP 137.15 58.98 58.52
PNC 1 2 0.914 125 0 1
REM Main Lateral: Junctions 2 to 3
STO 1.62 0.5 28
STO 2.35 0.6 15
STO 1.05 0.6 15
PIP 228.59 58.52 56.17
PNC 2 3 1.52 180 0 0
REM Main Lateral: Junctions 3 to 7
STO 2.27 0.5 15
STO 1.30 0.7 15
PIP 152.39 56.17 56.20
PNC 3 7 2.13 180 0 3
HOL 1
REM Begin Second Lateral
NEW PINE STREET
REM Side Lateral: Junctions 4 to 5
STO 1.34 0.6 15
PIP 106.67 60.96 56.38
PNC 4 5 1.22 155 0 1
REM Side Lateral: Junctions 5 to 6
STO 1.42 0.6 15
STO 1.58 0.6 15
PIP 167.63 56.38 56.20
PNC 5 6 1.52 180 0 2
REM Side Lateral: Junctions 6 to 7
STO 1.46 0.6 15
STO 1.42 0.6 15
PIP 106.68 56.17 56.20
PNC 6 7 2.13 150 0 3
REC 1
STO 1.82 0.8 15
REM Outfall Link: Junctions 7 to 8
PIP 121.91 56.20 56.14
PNC 7 8 0 180 2
END
```

Discussion of Output

The gravity flow simulation portion of the results provides tables of designed pipe sizes, invert elevations, and costs. This pipe design scenario is then "sent" to the hydraulic gradeline component of the model. The hydraulic gradeline table indicates the pipe links that could possibly surcharge. However, since HGL is based on pipe full flow, the occurrence of super-critical flow in a surcharged pipe would be contradictory to the surcharged condition. In this case, it would be recommended to run the pressure flow simulation option.

Output file: HYDRA1.LST

```
****** HYDRA ******* (Version 6.1) *****              Date 07-16-1998
                                                      Page No 1


Example One: Rational Method Design with Hydraulic Gradeline

+++ Commands Read From File C:\HYDRA\HYDRA1.HDA
   JOB
   SWI 2
   PDA 0.013 300 1.22 0.914 0.61 0.001
   RAI FOXHALL.IDF
```

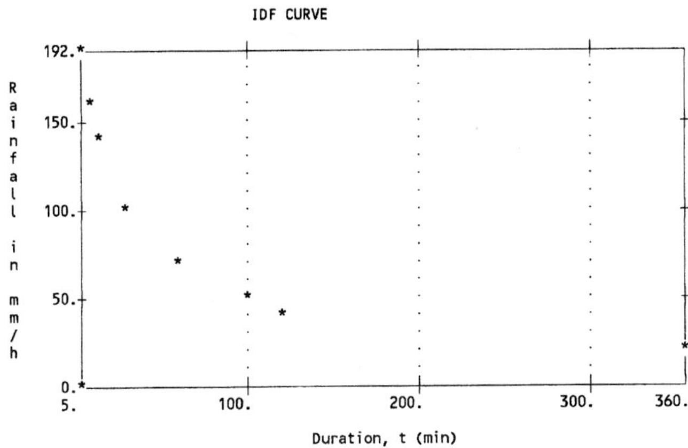


```
PLOT-DATA (Time, t(min) vs. Intensity, i(mm/h))
     t     i      t     i      t     i     t     i     t     i
   ────────────────────────────────────────────────────────────
     5.   192.    60.   68.    0.    0.    0.    0.    0.    0.
    10.   160.   100.   49.    0.    0.    0.    0.    0.    0.
    15.   138.   120.   43.    0.    0.    0.    0.    0.    0.
    30.   100.   360.   20.    0.    0.    0.    0.    0.    0.
   ────────────────────────────────────────────────────────────


EXC 1.52 0.94 7.62 1.48
TSL 0 0.25 3.05 0.15
PCO 300 15.03 900 81.50
CST 1.5 4.92 0 0 0 1.64 8.12 1 0 1.99 1.05 0.79 2.04 1.99 0
HGL 1
     ****** HYDRA ******* (Version 6.1) *****          Date 07-16-1998
                                                       Page No   2
   Example One: Rational Method Design with Hydraulic Gradeline

   REM Begin Main Lateral
   NEW MAIN STREET
```

```
    REM Main Lateral: Junctions 1 to 2
    STO 2.10 0.2 21
    PIP 137.15 58.98 58.52
+++ Tc = 21.0 min
+++ CA =   .4
+++ Top width of trench → UP 1.5 m
                        → DN 1.5 m
+++ Link # 1, Flow depth =    .324 m, Critical depth =    .264 m
    PNC 1 2 0.914 125 0 1
    REM Main Lateral: Junctions 2 to 3
    STO 1.62 0.5 28
    STO 2.35 0.6 15
    STO 1.05 0.6 15
    PIP 228.59 58.52 56.17
+++ Tc = 28.0 min
+++ CA =   3.3
*** WARNING: Pipe invert at D/S end dropped to meet minimum depth
    to invert criterion at D/S end
*** WARNING: Pipe invert at U/S dropped to meet cover criterion at U/S end
*** WARNING: Inverts at D/S & U/S dropped to meet cover criterion at D/S end
+++ Top width of trench → UP 2.0 m
                        → DN 1.9 m
+++ Link # 2, Flow depth = .547 m, Critical depth = .602 m
    PNC 2 3 1.52 180 0 0
    REM Main Lateral: Junctions 3 to 7
    STO 2.27 0.5 15
    STO 1.30 0.7 15
    PIP 152.39 56.17 56.20
+++ Tc = 29.4 min
+++ CA =   5.3
*** WARNING: Pipe invert at U/S dropped to meet cover criterion at U/S end
+++ Top width of trench → UP 2.8 m
                        → DN 2.8 m
+++ Link # 3, Flow depth =    .999 m, Critical depth = .645 m
    PNC 3 7 2.13 180 0 3
    HOL 1
    REM Begin Second Lateral
    NEW PINE STREET
    REM Side Lateral: Junctions 4 to 5
    STO 1.34 0.6 15
    PIP 106.67 60.96 56.38
+++ Tc = 15.0 min
+++ CA = .8
+++ Top width of trench → UP 1.4 m
                        → DN 1.4 m
+++ Link # 4, Flow depth =    .266 m, Critical depth =    .362 m
    PNC 4 5 1.22 155 0 1
****** HYDRA ******* (Version 6.1) *****
```


Example One: Rational Method Design with Hydraulic Gradeline

```
    REM Side Lateral: Junctions 5 to 6
    STO 1.42 0.6 15
    STO 1.58 0.6 15
    PIP 167.63 56.38 56.20
```

```
+++ Tc =    15.5 min
+++ CA =     2.6
*** WARNING: Pipe invert at U/S dropped to meet cover criterion at U/S end
*** WARNING: Inverts at D/S & U/S dropped to meet cover criterion at D/S end
+++ Top width of trench → UP 2.6 m
                        → DN 2.6 m
+++ Link # 5, Flow depth =   .816 m, Critical depth =     .537 m
   PNC 5 6 1.52 180 0 2
   REM Side Lateral: Junctions 6 to 7
   STO 1.46 0.6 15
   STO 1.42 0.6 15
   PIP 106.68 56.17 56.20
+++ Tc =    17.8 min
+++ CA =     4.3
*** WARNING: Pipe invert at U/S dropped to meet cover criterion at U/S end
+++ Top width of trench → UP 2.8 m
                        → DN 2.8 m
+++ Link # 6, Flow depth =   1.040 m, Critical depth =     .661 m
   PNC 6 7 2.13 150 0 3
   REC 1
   STO 1.82 0.8 15
   REM Outfall Link: Junctions 7 to 8
   PIP 121.91 56.20 56.14
+++ Tc =    31.3 min
+++ CA =    11.1
*** WARNING: Pipe invert at U/S dropped to meet cover criterion at U/S end
+++ Top width of trench → UP   3.4 m
                        → DN   3.5 m
+++ Link # 7, Flow depth −   1.260 m, Critical depth −     .852 m
   PNC 7 8 0 180 2
   END
END OF INPUT DATA.
   ****** HYDRA ******* (Version 6.1) *****          Date 07-16-1998
                                                     Page No    4

   Example One: Rational Method Design with Hydraulic Gradeline


   *** MAIN STREET                                      Pipe Design
```

Estimated Link	Length (m)	Diam (mm)	Invert Up/Dn (m)	Slope (m/m)	Depth Up/Dn (m)	Min. Cover (m)	Velocity Act/Full (m/s)	−Flow− Act/Full (m³/s)	Cost ($)
1	137	450	57.579	.003	1.402	.914	1.166	.142	5206.
			57.118		1.402		1.038	.165	
2	229	750	56.612	.009	1.908	.914	2.763	.949	17385.
			54.444		1.726		2.455	1.084	
3	152	1350	53.793	.001	2.377	.914	1.330	1.492	23687.
			53.641		2.559		1.179	1.688	

```
            Length =   518. m Total length =   518. m
            Cost   = 46279.  Total Cost   = 46279.
   *** PINE STREET                                      Pipe Design
```

Estimated Link	Length (m)	Diam (mm)	Invert Up/Dn (m)	Slope (m/m)	Depth Up/Dn (m)	Min. Cover (m)	Velocity Act/Full (m/s)	—Flow— Act/Full (m³/s)	Cost ($)
4	107	375	59.640	.043	1.320	.914	3.687	.306	3061.
			55.060		1.320		3.290	.363	
5	168	1200	54.154	.001	2.226	.914	1.209	.983	22588.
			53.986		2.214		1.090	1.233	
6	107	1350	53.793	.001	2.377	.914	1.338	1.565	16570.
			53.687		2.513		1.179	1.688	
7	122	1800	53.336	.001	2.864	.914	1.597	3.026	26560.
			53.214		2.926		1.429	3.636	

```
              Length =   5034. m Total length =   1021. m
              Cost  = 68780.  Total Cost  = 115058.
        ******  HYDRA  ******* (Version 6.1) *****         Date 07-16-1998
                                                              Page No 5
```

Example One: Rational Method Design with Hydraulic Gradeline

Hydraulic Gradeline Computations

Link #	Down-stream node #	Hydraulic gradeline elevation	Crown elev.	Possible surcharge	Ground elev.	Super-crit.?	Manhole depth	Loss coef
1	2	57.569	57.576	N	58.520	N	.957	.28
2	3	55.046	55.206	N	56.170	Y	1.090	.01
3	7	54.726	54.987	N	56.200	N	1.390	.08
	5	55.422	55.441	N	56.380	Y	.891	.01
5	6	54.925	55.180	N	56.170	N	1.132	.07
6	7	54.726	55.033	N	56.200	N	1.390	.09
7	8	54.474	55.017	N	56.140	N	1.260	.00

Link #	Terminal node #	Hydraulic gradeline elevation	Ground elevation	Loss coef.
1	1	58.007	58.980	1.50
4	4	60.946	60.960	1.50

NORMAL END OF HYDRA

REFERENCE

U.S. Federal Highway Administration (US FHWA), HYDRAIN—Integrated Drainage Computer System, Volume III. HYDRA—Storm Drains (www.fhwa.dot.gov/bridge/hydrain.htm), Office of Technology Applications, Washington, D.C., March 1999.

INDEX

Index note: The f or t following a page number refers to a figure or table, respectively on that page.

ABOUT THE EDITOR

Larry W. Mays is Professor of Civil and Environmental Engineering at Arizona State University and former chair of the department. He was formerly Director of the Center for Research in Water Resources at The University of Texas at Austin where he also held an Engineering Foundation Endowed Professorship. A registered professional engineer in seven states and a registered professional hydrologist, he has served as a consultant to many organizations. Professor Mays is author of *Water Resources Engineering* (John Wiley & Sons, Inc.) and *Optimal Control of Hydrosystems* (Marcel Dekker, Inc.), coauthor of *Applied Hydrology* and *Hydrosystems Engineering and Management* (both from McGraw-Hill), and editor-in-chief of *Water Resources Handbook, Hydraulic Design Handbook,* and the *Water Distribution Systems Handbook* (all from McGraw-Hill). He was also editor-in-chief of *Reliability Analysis of Water Distribution Systems* (ASCE) and co-editor of *Computer Modeling of Free Surface and Pressurized Flow* (Kluwer Academic Publishers). Among his honors include a distinguished alumnus award from the University of Illinois at Urbana-Champaign in 1999.